Student Solution

Single Variable Calculus

NINTH EDITION

James Stewart
McMaster University and University of Toronto

Daniel Clegg
Palomar College

Saleem Watson
California State University, Long Beach

© 2021 Cengage Learning, Inc.

Unless otherwise noted, all content is © Cengage

ALL RIGHTS RESERVED. No part of this work covered by the copyright herein may be reproduced or distributed in any form or by any means, except as permitted by U.S. Copyright law, without the prior written permission of the copyright holder.

For product information and technology assistance, contact us at
**Cengage Customer & Sales Support,
1-800-354-9706 or support.cengage.com.**

For permission to use material from this text or product, submit all requests online at
www.cengage.com/permissions.

ISBN: 978-0-357-04314-1

Cengage
200 Pier 4 Boulevard
Boston, MA 02210
USA

Cengage is a leading provider of customized learning solutions with employees residing in nearly 40 different countries and sales in more than 125 countries around the world. Find your local representative at: **www.cengage.com**.

Cengage products are represented in Canada by Nelson Education, Ltd.

To learn more about Cengage platforms and services, register or access your online learning solution, or purchase materials for your course, visit **www.cengage.com**.

Printed in Mexico
Print Number: 01   Print Year: 2020

CONTENTS

Diagnostic Tests 1

1 Functions and Limits 9

1.1 Four Ways to Represent a Function 9
1.2 Mathematical Models: A Catalog of Essential Functions 15
1.3 New Functions from Old Functions 19
1.4 The Tangent and Velocity Problems 26
1.5 The Limit of a Function 28
1.6 Calculating Limits Using the Limit Laws 31
1.7 The Precise Definition of a Limit 36
1.8 Continuity 40
Review 46

Principles of Problem Solving 51

2 Derivatives 55

2.1 Derivatives and Rates of Change 55
2.2 The Derivative as a Function 61
2.3 Differentiation Formulas 69
2.4 Derivatives of Trigonometric Functions 76
2.5 The Chain Rule 80
2.6 Implicit Differentiation 85
2.7 Rates of Change in the Natural and Social Sciences 92
2.8 Related Rates 97
2.9 Linear Approximations and Differentials 101
Review 105

Problems Plus 113

3 Applications of Differentiation 119

3.1 Maximum and Minimum Values 119
3.2 The Mean Value Theorem 124
3.3 What Derivatives Tell Us about the Shape of a Graph 127

3.4 Limits at Infinity; Horizontal Asymptotes 138
3.5 Summary of Curve Sketching 146
3.6 Graphing with Calculus *and* Technology 157
3.7 Optimization Problems 165
3.8 Newton's Method 177
3.9 Antiderivatives 182
    Review 187

Problems Plus 197

4 Integrals 203

4.1 The Area and Distance Problems 203
4.2 The Definite Integral 209
4.3 The Fundamental Theorem of Calculus 217
4.4 Indefinite Integrals and the Net Change Theorem 223
4.5 The Substitution Rule 226
    Review 230

Problems Plus 235

5 Applications of Integration 239

5.1 Areas Between Curves 239
5.2 Volumes 247
5.3 Volumes by Cylindrical Shells 257
5.4 Work 264
5.5 Average Value of a Function 267
    Review 268

Problems Plus 275

6 Inverse Functions: Exponential, Logarithmic, and Inverse Trigonometric Functions 279

6.1 Inverse Functions and Their Derivatives 279
6.2 Exponential Functions and Their Derivatives 282
6.3 Logarithmic Functions 289
6.4 Derivatives of Logarithmic Functions 293

6.2* The Natural Logarithmic Function 299
6.3* The Natural Exponential Function 305
6.4* General Logarithmic and Exponential Functions 312

- **6.5** Exponential Growth and Decay   316
- **6.6** Inverse Trigonometric Functions   319
- **6.7** Hyperbolic Functions   324
- **6.8** Indeterminate Forms and l'Hospital's Rule   329
  Review   337

Problems Plus   345

7 Techniques of Integration   349

- **7.1** Integration by Parts   349
- **7.2** Trigonometric Integrals   356
- **7.3** Trigonometric Substitution   360
- **7.4** Integration of Rational Functions by Partial Fractions   366
- **7.5** Strategy for Integration   376
- **7.6** Integration Using Tables and Technology   384
- **7.7** Approximate Integration   388
- **7.8** Improper Integrals   396
  Review   403

Problems Plus   413

8 Further Applications of Integration   417

- **8.1** Arc Length   417
- **8.2** Area of a Surface of Revolution   422
- **8.3** Applications to Physics and Engineering   427
- **8.4** Applications to Economics and Biology   434
- **8.5** Probability   436
  Review   439

Problems Plus   443

9 Differential Equations   447

- **9.1** Modeling with Differential Equations   447
- **9.2** Direction Fields and Euler's Method   449
- **9.3** Separable Equations   453
- **9.4** Models for Population Growth   459
- **9.5** Linear Equations   465
- **9.6** Predator-Prey Systems   468
  Review   471

Problems Plus   475

10 Parametric Equations and Polar Coordinates 479

- **10.1** Curves Defined by Parametric Equations 479
- **10.2** Calculus with Parametric Curves 487
- **10.3** Polar Coordinates 495
- **10.4** Calculus in Polar Coordinates 501
- **10.5** Conic Sections 510
- **10.6** Conic Sections in Polar Coordinates 516
  Review 519

Problems Plus 527

11 Sequences, Series, and Power Series 529

- **11.1** Sequences 529
- **11.2** Series 536
- **11.3** The Integral Test and Estimates of Sums 543
- **11.4** The Comparison Tests 547
- **11.5** Alternating Series and Absolute Convergence 551
- **11.6** The Ratio and Root Tests 555
- **11.7** Strategy for Testing Series 559
- **11.8** Power Series 562
- **11.9** Representations of Functions as Power Series 566
- **11.10** Taylor and Maclaurin Series 572
- **11.11** Applications of Taylor Polynomials 582
  Review 589

Problems Plus 595

Appendixes 601

- **A** Numbers, Inequalities, and Absolute Values 601
- **B** Coordinate Geometry and Lines 603
- **C** Graphs of Second-Degree Equations 606
- **D** Trigonometry 608
- **E** Sigma Notation 613

ABBREVIATIONS

CD	concave downward
CU	concave upward
D	the domain of f
FDT	First Derivative Test
HA	horizontal asymptote(s)
I	interval of convergence
IP	inflection point(s)
R	radius of convergence
VA	vertical asymptote(s)
$\stackrel{CAS}{=}$	indicates the use of a computer algebra system.
$\stackrel{PR}{\Rightarrow}$	indicates the use of the Product Rule.
$\stackrel{QR}{\Rightarrow}$	indicates the use of the Quotient Rule.
$\stackrel{CR}{\Rightarrow}$	indicates the use of the Chain Rule.
$\stackrel{H}{=}$	indicates the use of l'Hospital's Rule.
$\stackrel{j}{=}$	indicates the use of Formula j in the Table of Integrals on the back endpapers of the text.
$\stackrel{s}{=}$	indicates the use of the substitution $\{u = \sin x, du = \cos x \, dx\}$.
$\stackrel{c}{=}$	indicates the use of the substitution $\{u = \cos x, du = -\sin x \, dx\}$.
$\propto$	is proportional to

DIAGNOSTIC TESTS

Test A Algebra

1. (a) $(-3)^4 = (-3)(-3)(-3)(-3) = 81$

   (b) $-3^4 = -(3)(3)(3)(3) = -81$

   (c) $3^{-4} = \dfrac{1}{3^4} = \dfrac{1}{81}$

   (d) $\dfrac{5^{23}}{5^{21}} = 5^{23-21} = 5^2 = 25$

   (e) $\left(\dfrac{2}{3}\right)^{-2} = \left(\dfrac{3}{2}\right)^2 = \dfrac{9}{4}$

   (f) $16^{-3/4} = \dfrac{1}{16^{3/4}} = \dfrac{1}{\left(\sqrt[4]{16}\right)^3} = \dfrac{1}{2^3} = \dfrac{1}{8}$

2. (a) Note that $\sqrt{200} = \sqrt{100 \cdot 2} = 10\sqrt{2}$ and $\sqrt{32} = \sqrt{16 \cdot 2} = 4\sqrt{2}$. Thus $\sqrt{200} - \sqrt{32} = 10\sqrt{2} - 4\sqrt{2} = 6\sqrt{2}$.

   (b) $(3a^3b^3)(4ab^2)^2 = 3a^3b^3 \, 16a^2b^4 = 48a^5b^7$

   (c) $\left(\dfrac{3x^{3/2}y^3}{x^2y^{-1/2}}\right)^{-2} = \left(\dfrac{x^2y^{-1/2}}{3x^{3/2}y^3}\right)^2 = \dfrac{(x^2y^{-1/2})^2}{(3x^{3/2}y^3)^2} = \dfrac{x^4y^{-1}}{9x^3y^6} = \dfrac{x^4}{9x^3y^6y} = \dfrac{x}{9y^7}$

3. (a) $3(x+6) + 4(2x-5) = 3x + 18 + 8x - 20 = 11x - 2$

   (b) $(x+3)(4x-5) = 4x^2 - 5x + 12x - 15 = 4x^2 + 7x - 15$

   (c) $\left(\sqrt{a} + \sqrt{b}\right)\left(\sqrt{a} - \sqrt{b}\right) = \left(\sqrt{a}\right)^2 - \sqrt{a}\sqrt{b} + \sqrt{a}\sqrt{b} - \left(\sqrt{b}\right)^2 = a - b$

   Or: Use the formula for the difference of two squares to see that $\left(\sqrt{a}+\sqrt{b}\right)\left(\sqrt{a}-\sqrt{b}\right) = \left(\sqrt{a}\right)^2 - \left(\sqrt{b}\right)^2 = a - b$.

   (d) $(2x+3)^2 = (2x+3)(2x+3) = 4x^2 + 6x + 6x + 9 = 4x^2 + 12x + 9$.

   Note: A quicker way to expand this binomial is to use the formula $(a+b)^2 = a^2 + 2ab + b^2$ with $a = 2x$ and $b = 3$:
   $(2x+3)^2 = (2x)^2 + 2(2x)(3) + 3^2 = 4x^2 + 12x + 9$

   (e) See Reference Page 1 for the binomial formula $(a+b)^3 = a^3 + 3a^2b + 3ab^2 + b^3$. Using it, we get
   $(x+2)^3 = x^3 + 3x^2(2) + 3x(2^2) + 2^3 = x^3 + 6x^2 + 12x + 8$.

4. (a) Using the difference of two squares formula, $a^2 - b^2 = (a+b)(a-b)$, we have
   $4x^2 - 25 = (2x)^2 - 5^2 = (2x+5)(2x-5)$.

   (b) Factoring by trial and error, we get $2x^2 + 5x - 12 = (2x-3)(x+4)$.

   (c) Using factoring by grouping and the difference of two squares formula, we have
   $x^3 - 3x^2 - 4x + 12 = x^2(x-3) - 4(x-3) = (x^2 - 4)(x-3) = (x-2)(x+2)(x-3)$.

   (d) $x^4 + 27x = x(x^3 + 27) = x(x+3)(x^2 - 3x + 9)$

   This last expression was obtained using the sum of two cubes formula, $a^3 + b^3 = (a+b)(a^2 - ab + b^2)$ with $a = x$ and $b = 3$. [See Reference Page 1 in the textbook.]

   (e) The smallest exponent on x is $-\dfrac{1}{2}$, so we will factor out $x^{-1/2}$.
   $3x^{3/2} - 9x^{1/2} + 6x^{-1/2} = 3x^{-1/2}(x^2 - 3x + 2) = 3x^{-1/2}(x-1)(x-2)$

   (f) $x^3y - 4xy = xy(x^2 - 4) = xy(x-2)(x+2)$

2 ☐ **DIAGNOGSTIC TESTS**

5. (a) $\dfrac{x^2 + 3x + 2}{x^2 - x - 2} = \dfrac{(x+1)(x+2)}{(x+1)(x-2)} = \dfrac{x+2}{x-2}$

(b) $\dfrac{2x^2 - x - 1}{x^2 - 9} \cdot \dfrac{x+3}{2x+1} = \dfrac{(2x+1)(x-1)}{(x-3)(x+3)} \cdot \dfrac{x+3}{2x+1} = \dfrac{x-1}{x-3}$

(c) $\dfrac{x^2}{x^2 - 4} - \dfrac{x+1}{x+2} = \dfrac{x^2}{(x-2)(x+2)} - \dfrac{x+1}{x+2} = \dfrac{x^2}{(x-2)(x+2)} - \dfrac{x+1}{x+2} \cdot \dfrac{x-2}{x-2} = \dfrac{x^2 - (x+1)(x-2)}{(x-2)(x+2)}$

$= \dfrac{x^2 - (x^2 - x - 2)}{(x+2)(x-2)} = \dfrac{x+2}{(x+2)(x-2)} = \dfrac{1}{x-2}$

(d) $\dfrac{\dfrac{y}{x} - \dfrac{x}{y}}{\dfrac{1}{y} - \dfrac{1}{x}} = \dfrac{\dfrac{y}{x} - \dfrac{x}{y}}{\dfrac{1}{y} - \dfrac{1}{x}} \cdot \dfrac{xy}{xy} = \dfrac{y^2 - x^2}{x - y} = \dfrac{(y-x)(y+x)}{-(y-x)} = \dfrac{y+x}{-1} = -(x+y)$

6. (a) $\dfrac{\sqrt{10}}{\sqrt{5} - 2} = \dfrac{\sqrt{10}}{\sqrt{5} - 2} \cdot \dfrac{\sqrt{5} + 2}{\sqrt{5} + 2} = \dfrac{\sqrt{50} + 2\sqrt{10}}{(\sqrt{5})^2 - 2^2} = \dfrac{5\sqrt{2} + 2\sqrt{10}}{5 - 4} = 5\sqrt{2} + 2\sqrt{10}$

(b) $\dfrac{\sqrt{4+h} - 2}{h} = \dfrac{\sqrt{4+h} - 2}{h} \cdot \dfrac{\sqrt{4+h} + 2}{\sqrt{4+h} + 2} = \dfrac{4 + h - 4}{h(\sqrt{4+h} + 2)} = \dfrac{h}{h(\sqrt{4+h} + 2)} = \dfrac{1}{\sqrt{4+h} + 2}$

7. (a) $x^2 + x + 1 = \left(x^2 + x + \tfrac{1}{4}\right) + 1 - \tfrac{1}{4} = \left(x + \tfrac{1}{2}\right)^2 + \tfrac{3}{4}$

(b) $2x^2 - 12x + 11 = 2(x^2 - 6x) + 11 = 2(x^2 - 6x + 9 - 9) + 11 = 2(x^2 - 6x + 9) - 18 + 11 = 2(x - 3)^2 - 7$

8. (a) $x + 5 = 14 - \tfrac{1}{2}x \iff x + \tfrac{1}{2}x = 14 - 5 \iff \tfrac{3}{2}x = 9 \iff x = \tfrac{2}{3} \cdot 9 \iff x = 6$

(b) $\dfrac{2x}{x+1} = \dfrac{2x-1}{x} \implies 2x^2 = (2x-1)(x+1) \iff 2x^2 = 2x^2 + x - 1 \iff x = 1$

(c) $x^2 - x - 12 = 0 \iff (x+3)(x-4) = 0 \iff x + 3 = 0 \text{ or } x - 4 = 0 \iff x = -3 \text{ or } x = 4$

(d) By the quadratic formula, $2x^2 + 4x + 1 = 0 \iff$

$x = \dfrac{-4 \pm \sqrt{4^2 - 4(2)(1)}}{2(2)} = \dfrac{-4 \pm \sqrt{8}}{4} = \dfrac{-4 \pm 2\sqrt{2}}{4} = \dfrac{2(-2 \pm \sqrt{2})}{4} = \dfrac{-2 \pm \sqrt{2}}{2} = -1 \pm \tfrac{1}{2}\sqrt{2}$.

(e) $x^4 - 3x^2 + 2 = 0 \iff (x^2 - 1)(x^2 - 2) = 0 \iff x^2 - 1 = 0 \text{ or } x^2 - 2 = 0 \iff x^2 = 1 \text{ or } x^2 = 2 \iff$

$x = \pm 1 \text{ or } x = \pm\sqrt{2}$

(f) $3|x - 4| = 10 \iff |x - 4| = \tfrac{10}{3} \iff x - 4 = -\tfrac{10}{3} \text{ or } x - 4 = \tfrac{10}{3} \iff x = \tfrac{2}{3} \text{ or } x = \tfrac{22}{3}$

(g) Multiplying through $2x(4-x)^{-1/2} - 3\sqrt{4-x} = 0$ by $(4-x)^{1/2}$ gives $2x - 3(4 - x) = 0 \iff$

$2x - 12 + 3x = 0 \iff 5x - 12 = 0 \iff 5x = 12 \iff x = \tfrac{12}{5}$.

9. (a) $-4 < 5 - 3x \leq 17 \iff -9 < -3x \leq 12 \iff 3 > x \geq -4 \text{ or } -4 \leq x < 3$.

In interval notation, the answer is $[-4, 3)$.

(b) $x^2 < 2x + 8 \iff x^2 - 2x - 8 < 0 \iff (x+2)(x-4) < 0$. Now, $(x+2)(x-4)$ will change sign at the critical values $x = -2$ and $x = 4$. Thus the possible intervals of solution are $(-\infty, -2)$, $(-2, 4)$, and $(4, \infty)$. By choosing a single test value from each interval, we see that $(-2, 4)$ is the only interval that satisfies the inequality.

(c) The inequality $x(x-1)(x+2) > 0$ has critical values of $-2, 0$, and 1. The corresponding possible intervals of solution are $(-\infty, -2)$, $(-2, 0)$, $(0, 1)$ and $(1, \infty)$. By choosing a single test value from each interval, we see that both intervals $(-2, 0)$ and $(1, \infty)$ satisfy the inequality. Thus, the solution is the union of these two intervals: $(-2, 0) \cup (1, \infty)$.

(d) $|x - 4| < 3 \Leftrightarrow -3 < x - 4 < 3 \Leftrightarrow 1 < x < 7$. In interval notation, the answer is $(1, 7)$.

(e) $\dfrac{2x-3}{x+1} \leq 1 \Leftrightarrow \dfrac{2x-3}{x+1} - 1 \leq 0 \Leftrightarrow \dfrac{2x-3}{x+1} - \dfrac{x+1}{x+1} \leq 0 \Leftrightarrow \dfrac{2x-3-x-1}{x+1} \leq 0 \Leftrightarrow \dfrac{x-4}{x+1} \leq 0$.

Now, the expression $\dfrac{x-4}{x+1}$ may change signs at the critical values $x = -1$ and $x = 4$, so the possible intervals of solution are $(-\infty, -1)$, $(-1, 4]$, and $[4, \infty)$. By choosing a single test value from each interval, we see that $(-1, 4]$ is the only interval that satisfies the inequality.

10. (a) False. In order for the statement to be true, it must hold for all real numbers, so, to show that the statement is false, pick $p = 1$ and $q = 2$ and observe that $(1+2)^2 \neq 1^2 + 2^2$. In general, $(p+q)^2 = p^2 + 2pq + q^2$.

(b) True as long as a and b are nonnegative real numbers. To see this, think in terms of the laws of exponents:
$\sqrt{ab} = (ab)^{1/2} = a^{1/2}b^{1/2} = \sqrt{a}\sqrt{b}$.

(c) False. To see this, let $p = 1$ and $q = 2$, then $\sqrt{1^2 + 2^2} \neq 1 + 2$.

(d) False. To see this, let $T = 1$ and $C = 2$, then $\dfrac{1+1(2)}{2} \neq 1 + 1$.

(e) False. To see this, let $x = 2$ and $y = 3$, then $\dfrac{1}{2-3} \neq \dfrac{1}{2} - \dfrac{1}{3}$.

(f) True since $\dfrac{1/x}{a/x - b/x} \cdot \dfrac{x}{x} = \dfrac{1}{a-b}$, as long as $x \neq 0$ and $a - b \neq 0$.

Test B Analytic Geometry

1. (a) Using the point $(2, -5)$ and $m = -3$ in the point-slope equation of a line, $y - y_1 = m(x - x_1)$, we get
$y - (-5) = -3(x - 2) \Rightarrow y + 5 = -3x + 6 \Rightarrow y = -3x + 1$.

(b) A line parallel to the x-axis must be horizontal and thus have a slope of 0. Since the line passes through the point $(2, -5)$, the y-coordinate of every point on the line is -5, so the equation is $y = -5$.

(c) A line parallel to the y-axis is vertical with undefined slope. So the x-coordinate of every point on the line is 2 and so the equation is $x = 2$.

(d) Note that $2x - 4y = 3 \Rightarrow -4y = -2x + 3 \Rightarrow y = \frac{1}{2}x - \frac{3}{4}$. Thus the slope of the given line is $m = \frac{1}{2}$. Hence, the slope of the line we're looking for is also $\frac{1}{2}$ (since the line we're looking for is required to be parallel to the given line). So the equation of the line is $y - (-5) = \frac{1}{2}(x - 2) \Rightarrow y + 5 = \frac{1}{2}x - 1 \Rightarrow y = \frac{1}{2}x - 6$.

2. First we'll find the distance between the two given points in order to obtain the radius, r, of the circle:
$r = \sqrt{[3-(-1)]^2 + (-2-4)^2} = \sqrt{4^2 + (-6)^2} = \sqrt{52}$. Next use the standard equation of a circle, $(x-h)^2 + (y-k)^2 = r^2$, where (h, k) is the center, to get $(x+1)^2 + (y-4)^2 = 52$.

3. We must rewrite the equation in standard form in order to identify the center and radius. Note that
$x^2 + y^2 - 6x + 10y + 9 = 0 \Rightarrow x^2 - 6x + 9 + y^2 + 10y = 0$. For the left-hand side of the latter equation, we factor the first three terms and complete the square on the last two terms as follows: $x^2 - 6x + 9 + y^2 + 10y = 0 \Rightarrow (x-3)^2 + y^2 + 10y + 25 = 25 \Rightarrow (x-3)^2 + (y+5)^2 = 25$. Thus, the center of the circle is $(3, -5)$ and the radius is 5.

4. (a) $A(-7, 4)$ and $B(5, -12) \Rightarrow m_{AB} = \dfrac{-12 - 4}{5 - (-7)} = \dfrac{-16}{12} = -\dfrac{4}{3}$

(b) $y - 4 = -\frac{4}{3}[x - (-7)] \Rightarrow y - 4 = -\frac{4}{3}x - \frac{28}{3} \Rightarrow 3y - 12 = -4x - 28 \Rightarrow 4x + 3y + 16 = 0$. Putting $y = 0$, we get $4x + 16 = 0$, so the x-intercept is -4, and substituting 0 for x results in a y-intercept of $-\frac{16}{3}$.

(c) The midpoint is obtained by averaging the corresponding coordinates of both points: $\left(\frac{-7+5}{2}, \frac{4+(-12)}{2}\right) = (-1, -4)$.

(d) $d = \sqrt{[5 - (-7)]^2 + (-12 - 4)^2} = \sqrt{12^2 + (-16)^2} = \sqrt{144 + 256} = \sqrt{400} = 20$

(e) The perpendicular bisector is the line that intersects the line segment $\overline{AB}$ at a right angle through its midpoint. Thus the perpendicular bisector passes through $(-1, -4)$ and has slope $\frac{3}{4}$ [the slope is obtained by taking the negative reciprocal of the answer from part (a)]. So the perpendicular bisector is given by $y + 4 = \frac{3}{4}[x - (-1)]$ or $3x - 4y = 13$.

(f) The center of the required circle is the midpoint of $\overline{AB}$, and the radius is half the length of $\overline{AB}$, which is 10. Thus, the equation is $(x+1)^2 + (y+4)^2 = 100$.

5. (a) Graph the corresponding horizontal lines (given by the equations $y = -1$ and $y = 3$) as solid lines. The inequality $y \geq -1$ describes the points (x, y) that lie on or *above* the line $y = -1$. The inequality $y \leq 3$ describes the points (x, y) that lie on or *below* the line $y = 3$. So the pair of inequalities $-1 \leq y \leq 3$ describes the points that lie on or *between* the lines $y = -1$ and $y = 3$.

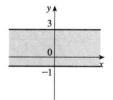

(b) Note that the given inequalities can be written as $-4 < x < 4$ and $-2 < y < 2$, respectively. So the region lies between the vertical lines $x = -4$ and $x = 4$ and between the horizontal lines $y = -2$ and $y = 2$. As shown in the graph, the region common to both graphs is a rectangle (minus its edges) centered at the origin.

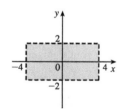

(c) We first graph $y = 1 - \frac{1}{2}x$ as a dotted line. Since $y < 1 - \frac{1}{2}x$, the points in the region lie *below* this line.

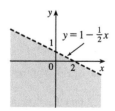

(d) We first graph the parabola $y = x^2 - 1$ using a solid curve. Since $y \geq x^2 - 1$, the points in the region lie on or *above* the parabola.

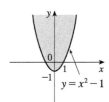

(e) We graph the circle $x^2 + y^2 = 4$ using a dotted curve. Since $\sqrt{x^2 + y^2} < 2$, the region consists of points whose distance from the origin is less than 2, that is, the points that lie *inside* the circle.

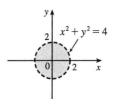

(f) The equation $9x^2 + 16y^2 = 144$ is an ellipse centered at $(0, 0)$. We put it in standard form by dividing by 144 and get $\dfrac{x^2}{16} + \dfrac{y^2}{9} = 1$. The x-intercepts are located at a distance of $\sqrt{16} = 4$ from the center while the y-intercepts are a distance of $\sqrt{9} = 3$ from the center (see the graph).

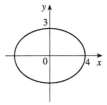

Test C Functions

1. (a) Locate -1 on the x-axis and then go down to the point on the graph with an x-coordinate of -1. The corresponding y-coordinate is the value of the function at $x = -1$, which is -2. So, $f(-1) = -2$.

   (b) Using the same technique as in part (a), we get $f(2) \approx 2.8$.

   (c) Locate 2 on the y-axis and then go left and right to find all points on the graph with a y-coordinate of 2. The corresponding x-coordinates are the x-values we are searching for. So $x = -3$ and $x = 1$.

   (d) Using the same technique as in part (c), we get $x \approx -2.5$ and $x \approx 0.3$.

   (e) The domain is all the x-values for which the graph exists, and the range is all the y-values for which the graph exists. Thus, the domain is $[-3, 3]$, and the range is $[-2, 3]$.

2. Note that $f(2 + h) = (2 + h)^3$ and $f(2) = 2^3 = 8$. So the difference quotient becomes
$$\frac{f(2+h) - f(2)}{h} = \frac{(2+h)^3 - 8}{h} = \frac{8 + 12h + 6h^2 + h^3 - 8}{h} = \frac{12h + 6h^2 + h^3}{h} = \frac{h(12 + 6h + h^2)}{h} = 12 + 6h + h^2.$$

3. (a) Set the denominator equal to 0 and solve to find restrictions on the domain: $x^2 + x - 2 = 0 \Rightarrow$ $(x - 1)(x + 2) = 0 \Rightarrow x = 1$ or $x = -2$. Thus, the domain is all real numbers except 1 or -2 or, in interval notation, $(-\infty, -2) \cup (-2, 1) \cup (1, \infty)$.

   (b) Note that the denominator is always greater than or equal to 1, and the numerator is defined for all real numbers. Thus, the domain is $(-\infty, \infty)$.

   (c) Note that the function h is the sum of two root functions. So h is defined on the intersection of the domains of these two root functions. The domain of a square root function is found by setting its radicand greater than or equal to 0. Now,

$4 - x \geq 0 \Rightarrow x \leq 4$ and $x^2 - 1 \geq 0 \Rightarrow (x-1)(x+1) \geq 0 \Rightarrow x \leq -1$ or $x \geq 1$. Thus, the domain of h is $(-\infty, -1] \cup [1, 4]$.

4. (a) Reflect the graph of f about the x-axis.

   (b) Stretch the graph of f vertically by a factor of 2, then shift 1 unit downward.

   (c) Shift the graph of f right 3 units, then up 2 units.

5. (a) Make a table and then connect the points with a smooth curve:

x	-2	-1	0	1	2
y	-8	-1	0	1	8

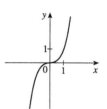

   (b) Shift the graph from part (a) left 1 unit.

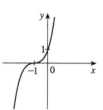

   (c) Shift the graph from part (a) right 2 units and up 3 units.

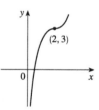

   (d) First plot $y = x^2$. Next, to get the graph of $f(x) = 4 - x^2$, reflect f about the x-axis and then shift it upward 4 units.

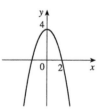

   (e) Make a table and then connect the points with a smooth curve:

x	0	1	4	9
y	0	1	2	3

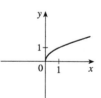

   (f) Stretch the graph from part (e) vertically by a factor of two.

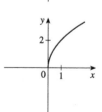

(g) First plot $y = 2^x$. Next, get the graph of $y = -2^x$ by reflecting the graph of $y = 2^x$ about the x-axis.

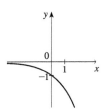

(h) Note that $y = 1 + x^{-1} = 1 + 1/x$. So first plot $y = 1/x$ and then shift it upward 1 unit.

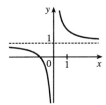

6. (a) $f(-2) = 1 - (-2)^2 = -3$ and $f(1) = 2(1) + 1 = 3$

(b) For $x \leq 0$ plot $f(x) = 1 - x^2$ and, on the same plane, for $x > 0$ plot the graph of $f(x) = 2x + 1$.

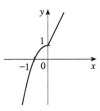

7. (a) $(f \circ g)(x) = f(g(x)) = f(2x - 3) = (2x - 3)^2 + 2(2x - 3) - 1 = 4x^2 - 12x + 9 + 4x - 6 - 1 = 4x^2 - 8x + 2$

(b) $(g \circ f)(x) = g(f(x)) = g(x^2 + 2x - 1) = 2(x^2 + 2x - 1) - 3 = 2x^2 + 4x - 2 - 3 = 2x^2 + 4x - 5$

(c) $(g \circ g \circ g)(x) = g(g(g(x))) = g(g(2x - 3)) = g(2(2x - 3) - 3) = g(4x - 9) = 2(4x - 9) - 3$
$= 8x - 18 - 3 = 8x - 21$

Test D Trigonometry

1. (a) $300° = 300° \left(\dfrac{\pi}{180°}\right) = \dfrac{300\pi}{180} = \dfrac{5\pi}{3}$ (b) $-18° = -18° \left(\dfrac{\pi}{180°}\right) = -\dfrac{18\pi}{180} = -\dfrac{\pi}{10}$

2. (a) $\dfrac{5\pi}{6} = \dfrac{5\pi}{6}\left(\dfrac{180}{\pi}\right)° = 150°$ (b) $2 = 2\left(\dfrac{180}{\pi}\right)° = \left(\dfrac{360}{\pi}\right)° \approx 114.6°$

3. We will use the arc length formula, $s = r\theta$, where s is arc length, r is the radius of the circle, and θ is the measure of the central angle in radians. First, note that $30° = 30°\left(\dfrac{\pi}{180°}\right) = \dfrac{\pi}{6}$. So $s = (12)\left(\dfrac{\pi}{6}\right) = 2\pi$ cm.

4. (a) $\tan(\pi/3) = \sqrt{3}$ [You can read the value from a right triangle with sides 1, 2, and $\sqrt{3}$.]

(b) Note that $7\pi/6$ can be thought of as an angle in the third quadrant with reference angle $\pi/6$. Thus, $\sin(7\pi/6) = -\frac{1}{2}$, since the sine function is negative in the third quadrant.

(c) Note that $5\pi/3$ can be thought of as an angle in the fourth quadrant with reference angle $\pi/3$. Thus,
$$\sec(5\pi/3) = \dfrac{1}{\cos(5\pi/3)} = \dfrac{1}{1/2} = 2,$$ since the cosine function is positive in the fourth quadrant.

8 ☐ **DIAGNOGSTIC TESTS**

5. $\sin\theta = a/24 \;\Rightarrow\; a = 24\sin\theta$ and $\cos\theta = b/24 \;\Rightarrow\; b = 24\cos\theta$

6. $\sin x = \frac{1}{3}$ and $\sin^2 x + \cos^2 x = 1 \;\Rightarrow\; \cos x = \sqrt{1 - \frac{1}{9}} = \frac{2\sqrt{2}}{3}$. Also, $\cos y = \frac{4}{5} \;\Rightarrow\; \sin y = \sqrt{1 - \frac{16}{25}} = \frac{3}{5}$.

So, using the sum identity for the sine, we have

$$\sin(x+y) = \sin x \cos y + \cos x \sin y = \frac{1}{3}\cdot\frac{4}{5} + \frac{2\sqrt{2}}{3}\cdot\frac{3}{5} = \frac{4+6\sqrt{2}}{15} = \frac{1}{15}\left(4+6\sqrt{2}\right)$$

7. (a) $\tan\theta\,\sin\theta + \cos\theta = \dfrac{\sin\theta}{\cos\theta}\sin\theta + \cos\theta = \dfrac{\sin^2\theta}{\cos\theta} + \dfrac{\cos^2\theta}{\cos\theta} = \dfrac{1}{\cos\theta} = \sec\theta$

(b) $\dfrac{2\tan x}{1+\tan^2 x} = \dfrac{2\sin x/(\cos x)}{\sec^2 x} = 2\dfrac{\sin x}{\cos x}\cos^2 x = 2\sin x\cos x = \sin 2x$

8. $\sin 2x = \sin x \;\Leftrightarrow\; 2\sin x\cos x = \sin x \;\Leftrightarrow\; 2\sin x\cos x - \sin x = 0 \;\Leftrightarrow\; \sin x\,(2\cos x - 1) = 0 \;\Leftrightarrow\;$
$\sin x = 0$ or $\cos x = \frac{1}{2} \;\Rightarrow\; x = 0, \frac{\pi}{3}, \pi, \frac{5\pi}{3}, 2\pi$.

9. We first graph $y = \sin 2x$ (by compressing the graph of $\sin x$ by a factor of 2) and then shift it upward 1 unit.

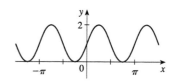

1 □ FUNCTIONS AND LIMITS

1.1 Four Ways to Represent a Function

1. The functions $f(x) = x + \sqrt{2-x}$ and $g(u) = u + \sqrt{2-u}$ give exactly the same output values for every input value, so f and g are equal.

3. (a) The point $(-2, 2)$ lies on the graph of g, so $g(-2) = 2$. Similarly, $g(0) = -2$, $g(2) = 1$, and $g(3) \approx 2.5$.

   (b) Only the point $(-4, 3)$ on the graph has a y-value of 3, so the only value of x for which $g(x) = 3$ is -4.

   (c) The function outputs $g(x)$ are never greater than 3, so $g(x) \leq 3$ for the entire domain of the function. Thus, $g(x) \leq 3$ for $-4 \leq x \leq 4$ (or, equivalently, on the interval $[-4, 4]$).

   (d) The domain consists of all x-values on the graph of g: $\{x \mid -4 \leq x \leq 4\} = [-4, 4]$. The range of g consists of all the y-values on the graph of g: $\{y \mid -2 \leq y \leq 3\} = [-2, 3]$.

   (e) For any $x_1 < x_2$ in the interval $[0, 2]$, we have $g(x_1) < g(x_2)$. [The graph rises from $(0, -2)$ to $(2, 1)$.] Thus, $g(x)$ is increasing on $[0, 2]$.

5. From Figure 1 in the text, the lowest point occurs at about $(t, a) = (12, -85)$. The highest point occurs at about $(17, 115)$. Thus, the range of the vertical ground acceleration is $-85 \leq a \leq 115$. Written in interval notation, the range is $[-85, 115]$.

7. We solve $3x - 5y = 7$ for y: $3x - 5y = 7 \Leftrightarrow -5y = -3x + 7 \Leftrightarrow y = \frac{3}{5}x - \frac{7}{5}$. Since the equation determines exactly one value of y for each value of x, the equation defines y as a function of x.

9. We solve $x^2 + (y-3)^2 = 5$ for y: $x^2 + (y-3)^2 = 5 \Leftrightarrow (y-3)^2 = 5 - x^2 \Leftrightarrow y - 3 = \pm\sqrt{5-x^2} \Leftrightarrow y = 3 \pm \sqrt{5-x^2}$. Some input values x correspond to more than one output y. (For instance, $x = 1$ corresponds to $y = 1$ and to $y = 5$.) Thus, the equation does *not* define y as a function of x.

11. We solve $(y+3)^3 + 1 = 2x$ for y: $(y+3)^3 + 1 = 2x \Leftrightarrow (y+3)^3 = 2x - 1 \Leftrightarrow y + 3 = \sqrt[3]{2x-1} \Leftrightarrow y = -3 + \sqrt[3]{2x-1}$. Since the equation determines exactly one value of y for each value of x, the equation defines y as a function of x.

13. The height 60 in ($x = 60$) corresponds to shoe sizes 7 and 8 ($y = 7$ and $y = 8$). Since an input value x corresponds to more than output value y, the table does *not* define y as a function of x.

15. No, the curve is not the graph of a function because a vertical line intersects the curve more than once. Hence, the curve fails the Vertical Line Test.

17. Yes, the curve is the graph of a function because it passes the Vertical Line Test. The domain is $[-3, 2]$ and the range is $[-3, -2) \cup [-1, 3]$.

19. (a) When $t = 1950$, $T \approx 13.8°C$, so the global average temperature in 1950 was about $13.8°C$.

(b) When $T = 14.2°C$, $t \approx 1990$.

(c) The global average temperature was smallest in 1910 (the year corresponding to the lowest point on the graph) and largest in 2000 (the year corresponding to the highest point on the graph).

(d) When $t = 1910$, $T \approx 13.5°C$, and when $t = 2000$, $T \approx 14.4°C$. Thus, the range of T is about $[13.5, 14.4]$.

21. The water will cool down almost to freezing as the ice melts. Then, when the ice has melted, the water will slowly warm up to room temperature.

23. (a) The power consumption at 6 AM is 500 MW, which is obtained by reading the value of power P when $t = 6$ from the graph. At 6 PM we read the value of P when $t = 18$, obtaining approximately 730 MW.

(b) The minimum power consumption is determined by finding the time for the lowest point on the graph, $t = 4$, or 4 AM. The maximum power consumption corresponds to the highest point on the graph, which occurs just before $t = 12$, or right before noon. These times are reasonable, considering the power consumption schedules of most individuals and businesses.

25. Of course, this graph depends strongly on the geographical location!

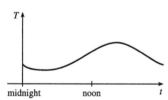

27. As the price increases, the amount sold decreases.

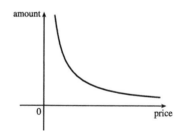

29.

Height of grass graph across Wed. to Wed.

31. (a) Temperature graph in °F from $t = 0$ to 14 hours, ranging approximately 65 to 85.

(b) 9:00 AM corresponds to $t = 9$. When $t = 9$, the temperature T is about $74°F$.

33. $f(x) = 3x^2 - x + 2$.

$f(2) = 3(2)^2 - 2 + 2 = 12 - 2 + 2 = 12$.

$f(-2) = 3(-2)^2 - (-2) + 2 = 12 + 2 + 2 = 16$.

$f(a) = 3a^2 - a + 2$.

$f(-a) = 3(-a)^2 - (-a) + 2 = 3a^2 + a + 2$.

$f(a+1) = 3(a+1)^2 - (a+1) + 2 = 3(a^2 + 2a + 1) - a - 1 + 2 = 3a^2 + 6a + 3 - a + 1 = 3a^2 + 5a + 4$.

$2f(a) = 2 \cdot f(a) = 2(3a^2 - a + 2) = 6a^2 - 2a + 4$.

$f(2a) = 3(2a)^2 - (2a) + 2 = 3(4a^2) - 2a + 2 = 12a^2 - 2a + 2$.

$f(a^2) = 3(a^2)^2 - (a^2) + 2 = 3(a^4) - a^2 + 2 = 3a^4 - a^2 + 2$.

$[f(a)]^2 = [3a^2 - a + 2]^2 = (3a^2 - a + 2)(3a^2 - a + 2)$
$= 9a^4 - 3a^3 + 6a^2 - 3a^3 + a^2 - 2a + 6a^2 - 2a + 4 = 9a^4 - 6a^3 + 13a^2 - 4a + 4$.

$f(a+h) = 3(a+h)^2 - (a+h) + 2 = 3(a^2 + 2ah + h^2) - a - h + 2 = 3a^2 + 6ah + 3h^2 - a - h + 2$.

35. $f(x) = 4 + 3x - x^2$, so $f(3+h) = 4 + 3(3+h) - (3+h)^2 = 4 + 9 + 3h - (9 + 6h + h^2) = 4 - 3h - h^2$,

and $\dfrac{f(3+h) - f(3)}{h} = \dfrac{(4 - 3h - h^2) - 4}{h} = \dfrac{h(-3 - h)}{h} = -3 - h$.

37. $f(x) = \dfrac{1}{x}$, so $\dfrac{f(x) - f(a)}{x - a} = \dfrac{\dfrac{1}{x} - \dfrac{1}{a}}{x - a} = \dfrac{\dfrac{a - x}{xa}}{x - a} = \dfrac{a - x}{xa(x - a)} = \dfrac{-1(x - a)}{xa(x - a)} = -\dfrac{1}{ax}$.

39. $f(x) = (x + 4)/(x^2 - 9)$ is defined for all x except when $0 = x^2 - 9 \Leftrightarrow 0 = (x+3)(x-3) \Leftrightarrow x = -3$ or 3, so the domain is $\{x \in \mathbb{R} \mid x \neq -3, 3\} = (-\infty, -3) \cup (-3, 3) \cup (3, \infty)$.

41. $f(t) = \sqrt[3]{2t - 1}$ is defined for all real numbers. In fact $\sqrt[3]{p(t)}$, where $p(t)$ is a polynomial, is defined for all real numbers. Thus, the domain is $\mathbb{R}$, or $(-\infty, \infty)$.

43. $h(x) = 1/\sqrt[4]{x^2 - 5x}$ is defined when $x^2 - 5x > 0 \Leftrightarrow x(x - 5) > 0$. Note that $x^2 - 5x \neq 0$ since that would result in division by zero. The expression $x(x - 5)$ is positive if $x < 0$ or $x > 5$. (See Appendix A for methods for solving inequalities.) Thus, the domain is $(-\infty, 0) \cup (5, \infty)$.

45. $F(p) = \sqrt{2 - \sqrt{p}}$ is defined when $p \geq 0$ and $2 - \sqrt{p} \geq 0$. Since $2 - \sqrt{p} \geq 0 \Leftrightarrow 2 \geq \sqrt{p} \Leftrightarrow \sqrt{p} \leq 2 \Leftrightarrow 0 \leq p \leq 4$, the domain is $[0, 4]$.

47. $h(x) = \sqrt{4 - x^2}$. Now $y = \sqrt{4 - x^2} \Rightarrow y^2 = 4 - x^2 \Leftrightarrow x^2 + y^2 = 4$, so the graph is the top half of a circle of radius 2 with center at the origin. The domain is $\{x \mid 4 - x^2 \geq 0\} = \{x \mid 4 \geq x^2\} = \{x \mid 2 \geq |x|\} = [-2, 2]$. From the graph, the range is $0 \leq y \leq 2$, or $[0, 2]$.

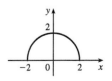

49. $f(x) = \begin{cases} x^2 + 2 & \text{if } x < 0 \\ x & \text{if } x \geq 0 \end{cases}$

$f(-3) = (-3)^2 + 2 = 11$, $f(0) = 0$, and $f(2) = 2$.

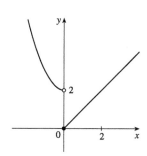

51. $f(x) = \begin{cases} x + 1 & \text{if } x \leq -1 \\ x^2 & \text{if } x > -1 \end{cases}$

$f(-3) = -3 + 1 = -2$, $f(0) = 0^2 = 0$, and $f(2) = 2^2 = 4$.

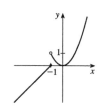

53. $|x| = \begin{cases} x & \text{if } x \geq 0 \\ -x & \text{if } x < 0 \end{cases}$

so $f(x) = x + |x| = \begin{cases} 2x & \text{if } x \geq 0 \\ 0 & \text{if } x < 0 \end{cases}$

Graph the line $y = 2x$ for $x \geq 0$ and graph $y = 0$ (the x-axis) for $x < 0$.

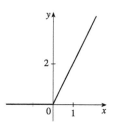

55. $g(t) = |1 - 3t| = \begin{cases} 1 - 3t & \text{if } 1 - 3t \geq 0 \\ -(1 - 3t) & \text{if } 1 - 3t < 0 \end{cases}$

$= \begin{cases} 1 - 3t & \text{if } t \leq \frac{1}{3} \\ 3t - 1 & \text{if } t > \frac{1}{3} \end{cases}$

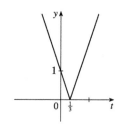

57. To graph $f(x) = \begin{cases} |x| & \text{if } |x| \leq 1 \\ 1 & \text{if } |x| > 1 \end{cases}$, graph $y = |x|$ [Figure 16]

for $-1 \leq x \leq 1$ and graph $y = 1$ for $x > 1$ and for $x < -1$.

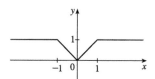

We could rewrite f as $f(x) = \begin{cases} 1 & \text{if } x < -1 \\ -x & \text{if } -1 \leq x < 0 \\ x & \text{if } 0 \leq x \leq 1 \\ 1 & \text{if } x > 1 \end{cases}$.

59. Recall that the slope m of a line between the two points (x_1, y_1) and (x_2, y_2) is $m = \dfrac{y_2 - y_1}{x_2 - x_1}$ and an equation of the line connecting those two points is $y - y_1 = m(x - x_1)$. The slope of the line segment joining the points $(1, -3)$ and $(5, 7)$ is $\dfrac{7 - (-3)}{5 - 1} = \dfrac{5}{2}$, so an equation is $y - (-3) = \frac{5}{2}(x - 1)$. The function is $f(x) = \frac{5}{2}x - \frac{11}{2}$, $1 \leq x \leq 5$.

61. We need to solve the given equation for y. $x+(y-1)^2=0 \Leftrightarrow (y-1)^2=-x \Leftrightarrow y-1=\pm\sqrt{-x} \Leftrightarrow y=1\pm\sqrt{-x}$. The expression with the positive radical represents the top half of the parabola, and the one with the negative radical represents the bottom half. Hence, we want $f(x)=1-\sqrt{-x}$. Note that the domain is $x\le 0$.

63. For $0\le x\le 3$, the graph is the line with slope -1 and y-intercept 3, that is, $y=-x+3$. For $3<x\le 5$, the graph is the line with slope 2 passing through $(3,0)$; that is, $y-0=2(x-3)$, or $y=2x-6$. So the function is

$$f(x)=\begin{cases}-x+3 & \text{if } 0\le x\le 3 \\ 2x-6 & \text{if } 3<x\le 5\end{cases}$$

65. Let the length and width of the rectangle be L and W. Then the perimeter is $2L+2W=20$ and the area is $A=LW$. Solving the first equation for W in terms of L gives $W=\dfrac{20-2L}{2}=10-L$. Thus, $A(L)=L(10-L)=10L-L^2$. Since lengths are positive, the domain of A is $0<L<10$. If we further restrict L to be larger than W, then $5<L<10$ would be the domain.

67. Let the length of a side of the equilateral triangle be x. Then by the Pythagorean Theorem, the height y of the triangle satisfies $y^2+\left(\tfrac{1}{2}x\right)^2=x^2$, so that $y^2=x^2-\tfrac{1}{4}x^2=\tfrac{3}{4}x^2$ and $y=\tfrac{\sqrt{3}}{2}x$. Using the formula for the area A of a triangle, $A=\tfrac{1}{2}(\text{base})(\text{height})$, we obtain $A(x)=\tfrac{1}{2}(x)\left(\tfrac{\sqrt{3}}{2}x\right)=\tfrac{\sqrt{3}}{4}x^2$, with domain $x>0$.

69. Let each side of the base of the box have length x, and let the height of the box be h. Since the volume is 2, we know that $2=hx^2$, so that $h=2/x^2$, and the surface area is $S=x^2+4xh$. Thus, $S(x)=x^2+4x(2/x^2)=x^2+(8/x)$, with domain $x>0$.

71. The height of the box is x and the length and width are $L=20-2x$, $W=12-2x$. Then $V=LWx$ and so
$V(x)=(20-2x)(12-2x)(x)=4(10-x)(6-x)(x)=4x(60-16x+x^2)=4x^3-64x^2+240x$.
The sides L, W, and x must be positive. Thus, $L>0 \Leftrightarrow 20-2x>0 \Leftrightarrow x<10$; $W>0 \Leftrightarrow 12-2x>0 \Leftrightarrow x<6$; and $x>0$. Combining these restrictions gives us the domain $0<x<6$.

73. We can summarize the amount of the fine with a piecewise defined function.

$$F(x)=\begin{cases}15(40-x) & \text{if } 0\le x<40 \\ 0 & \text{if } 40\le x\le 65 \\ 15(x-65) & \text{if } x>65\end{cases}$$

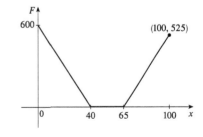

75. (a)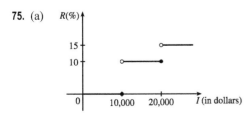

(b) On \$14,000, tax is assessed on \$4000, and $10\%(\$4000)=\$400$.

On \$26,000, tax is assessed on \$16,000, and
$10\%(\$10,000)+15\%(\$6000)=\$1000+\$900=\$1900$.

(c) As in part (b), there is $1000 tax assessed on $20,000 of income, so the graph of T is a line segment from $(10{,}000, 0)$ to $(20{,}000, 1000)$. The tax on $30,000 is $2500, so the graph of T for $x > 20{,}000$ is the ray with initial point $(20{,}000, 1000)$ that passes through $(30{,}000, 2500)$.

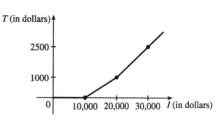

77. f is an odd function because its graph is symmetric about the origin. g is an even function because its graph is symmetric with respect to the y-axis.

79. (a) The graph of an even function is symmetric about the y-axis. We reflect the given portion of the graph of f about the y-axis in order to complete it.

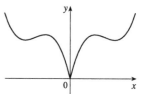

(b) For an odd function, $f(-x) = -f(x)$. The graph of an odd function is symmetric about the origin. We rotate the given portion of the graph of f through $180°$ about the origin in order to complete it.

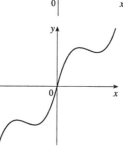

81. $f(x) = \dfrac{x}{x^2+1}$.

$f(-x) = \dfrac{-x}{(-x)^2+1} = \dfrac{-x}{x^2+1} = -\dfrac{x}{x^2+1} = -f(x)$.

Since $f(-x) = -f(x)$, f is an odd function.

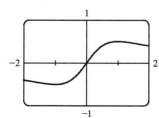

83. $f(x) = \dfrac{x}{x+1}$, so $f(-x) = \dfrac{-x}{-x+1} = \dfrac{x}{x-1}$.

Since this is neither $f(x)$ nor $-f(x)$, the function f is neither even nor odd.

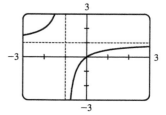

85. $f(x) = 1 + 3x^2 - x^4$.

$f(-x) = 1 + 3(-x)^2 - (-x)^4 = 1 + 3x^2 - x^4 = f(x)$.

Since $f(-x) = f(x)$, f is an even function.

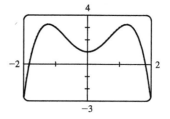

87. (i) If f and g are both even functions, then $f(-x) = f(x)$ and $g(-x) = g(x)$. Now

$(f+g)(-x) = f(-x) + g(-x) = f(x) + g(x) = (f+g)(x)$, so $f+g$ is an *even* function.

(ii) If f and g are both odd functions, then $f(-x) = -f(x)$ and $g(-x) = -g(x)$. Now
$(f+g)(-x) = f(-x) + g(-x) = -f(x) + [-g(x)] = -[f(x) + g(x)] = -(f+g)(x)$, so $f+g$ is an *odd* function.

(iii) If f is an even function and g is an odd function, then $(f+g)(-x) = f(-x) + g(-x) = f(x) + [-g(x)] = f(x) - g(x)$, which is not $(f+g)(x)$ nor $-(f+g)(x)$, so $f+g$ is *neither* even nor odd. (Exception: if f is the zero function, then $f+g$ will be *odd*. If g is the zero function, then $f+g$ will be *even*.)

1.2 Mathematical Models: A Catalog of Essential Functions

1. (a) $f(x) = x^3 + 3x^2$ is a polynomial function of degree 3. (This function is also an algebraic function.)

   (b) $g(t) = \cos^2 t - \sin t$ is a trigonometric function.

   (c) $r(t) = t^{\sqrt{3}}$ is a power function.

   (d) $v(t) = 8^t$ is an exponential function.

   (e) $y = \dfrac{\sqrt{x}}{x^2 + 1}$ is an algebraic function. It is the quotient of a root of a polynomial and a polynomial of degree 2.

   (f) $g(u) = \log_{10} u$ is a logarithmic function.

3. We notice from the figure that g and h are even functions (symmetric with respect to the y-axis) and that f is an odd function (symmetric with respect to the origin). So (b) $\left[y = x^5\right]$ must be f. Since g is flatter than h near the origin, we must have (c) $\left[y = x^8\right]$ matched with g and (a) $\left[y = x^2\right]$ matched with h.

5. The denominator cannot equal 0, so $1 - \sin x \neq 0 \Leftrightarrow \sin x \neq 1 \Leftrightarrow x \neq \frac{\pi}{2} + 2n\pi$. Thus, the domain of $f(x) = \dfrac{\cos x}{1 - \sin x}$ is $\{x \mid x \neq \frac{\pi}{2} + 2n\pi, n \text{ an integer}\}$.

7. (a) An equation for the family of linear functions with slope 2 is $y = f(x) = 2x + b$, where b is the y-intercept.

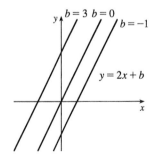

   (b) $f(2) = 1$ means that the point $(2, 1)$ is on the graph of f. We can use the point-slope form of a line to obtain an equation for the family of linear functions through the point $(2, 1)$. $y - 1 = m(x - 2)$, which is equivalent to $y = mx + (1 - 2m)$ in slope-intercept form.

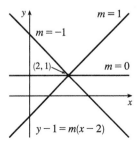

(c) To belong to both families, an equation must have slope $m = 2$, so the equation in part (b), $y = mx + (1 - 2m)$, becomes $y = 2x - 3$. It is the *only* function that belongs to both families.

9. All members of the family of linear functions $f(x) = c - x$ have graphs that are lines with slope -1. The y-intercept is c.

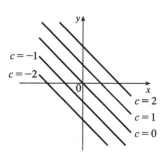

11. Because f is a quadratic function, we know it is of the form $f(x) = ax^2 + bx + c$. The y-intercept is 18, so $f(0) = 18 \Rightarrow c = 18$ and $f(x) = ax^2 + bx + 18$. Since the points $(3, 0)$ and $(4, 2)$ lie on the graph of f, we have

$$f(3) = 0 \Rightarrow 9a + 3b + 18 = 0 \Rightarrow 3a + b = -6 \quad \textbf{(1)}$$
$$f(4) = 2 \Rightarrow 16a + 4b + 18 = 2 \Rightarrow 4a + b = -4 \quad \textbf{(2)}$$

This is a system of two equations in the unknowns a and b, and subtracting **(1)** from **(2)** gives $a = 2$. From **(1)**, $3(2) + b = -6 \Rightarrow b = -12$, so a formula for f is $f(x) = 2x^2 - 12x + 18$.

13. Since $f(-1) = f(0) = f(2) = 0$, f has zeros of -1, 0, and 2, so an equation for f is $f(x) = a[x - (-1)](x - 0)(x - 2)$, or $f(x) = ax(x+1)(x-2)$. Because $f(1) = 6$, we'll substitute 1 for x and 6 for $f(x)$.
$6 = a(1)(2)(-1) \Rightarrow -2a = 6 \Rightarrow a = -3$, so an equation for f is $f(x) = -3x(x+1)(x-2)$.

15. (a) $D = 200$, so $c = 0.0417D(a+1) = 0.0417(200)(a+1) = 8.34a + 8.34$. The slope is 8.34, which represents the change in mg of the dosage for a child for each change of 1 year in age.

(b) For a newborn, $a = 0$, so $c = 8.34$ mg.

17. (a)

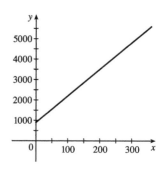

(b) The slope of $\frac{9}{5}$ means that F increases $\frac{9}{5}$ degrees for each increase of $1°C$. (Equivalently, F increases by 9 when C increases by 5 and F decreases by 9 when C decreases by 5.) The F-intercept of 32 is the Fahrenheit temperature corresponding to a Celsius temperature of 0.

19. (a) Let x denote the number of chairs produced in one day and y the associated cost. Using the points $(100, 2200)$ and $(300, 4800)$, we get the slope

$\frac{4800 - 2200}{300 - 100} = \frac{2600}{200} = 13$. So $y - 2200 = 13(x - 100) \Leftrightarrow y = 13x + 900$.

(b) The slope of the line in part (a) is 13 and it represents the cost (in dollars) of producing each additional chair.

(c) The y-intercept is 900 and it represents the fixed daily costs of operating the factory.

21. (a) We are given $\dfrac{\text{change in pressure}}{10 \text{ feet change in depth}} = \dfrac{4.34}{10} = 0.434$. Using P for pressure and d for depth with the point $(d, P) = (0, 15)$, we have the slope-intercept form of the line, $P = 0.434d + 15$.

(b) When $P = 100$, then $100 = 0.434d + 15 \Leftrightarrow 0.434d = 85 \Leftrightarrow d = \dfrac{85}{0.434} \approx 195.85$ feet. Thus, the pressure is 100 lb/in^2 at a depth of approximately 196 feet.

23. If x is the original distance from the source, then the illumination is $f(x) = kx^{-2} = k/x^2$. Moving halfway to the lamp gives an illumination of $f\left(\tfrac{1}{2}x\right) = k\left(\tfrac{1}{2}x\right)^{-2} = k(2/x)^2 = 4(k/x^2)$, so the light is four times as bright.

25. (a) $P = kAv^3$ so doubling the windspeed v gives $P = kA(2v)^3 = 8(kAv^3)$. Thus, the power output is increased by a factor of eight.

(b) The area swept out by the blades is given by $A = \pi l^2$, where l is the blade length, so the power output is $P = kAv^3 = k\pi l^2 v^3$. Doubling the blade length gives $P = k\pi(2l)^2 v^3 = 4(k\pi l^2 v^3)$. Thus, the power output is increased by a factor of four.

(c) From part (b) we have $P = k\pi l^2 v^3$, and $k = 0.214 \text{ kg/m}^3$, $l = 30$ m gives

$$P = 0.214 \,\dfrac{\text{kg}}{\text{m}^3} \cdot 900\pi \text{ m}^2 \cdot v^3 = 192.6\pi v^3 \,\dfrac{\text{kg}}{\text{m}}$$

For $v = 10$ m/s, we have

$$P = 192.6\pi \left(10\,\dfrac{\text{m}}{\text{s}}\right)^3 \dfrac{\text{kg}}{\text{m}} = 192{,}600\pi \,\dfrac{\text{m}^2 \cdot \text{kg}}{\text{s}^3} \approx 605{,}000 \text{ W}$$

Similarly, $v = 15$ m/s gives $P = 650{,}025\pi \approx 2{,}042{,}000$ W and $v = 25$ m/s gives $P = 3{,}009{,}375\pi \approx 9{,}454{,}000$ W.

27. (a) The data appear to be periodic and a sine or cosine function would make the best model. A model of the form $f(x) = a\cos(bx) + c$ seems appropriate.

(b) The data appear to be decreasing in a linear fashion. A model of the form $f(x) = mx + b$ seems appropriate.

Exercises 29–33: Some values are given to many decimal places. The results may depend on the technology used—rounding is left to the reader.

29. (a)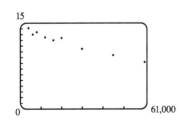

A linear model does seem appropriate.

(b) Using the points $(4000, 14.1)$ and $(60{,}000, 8.2)$, we obtain

$$y - 14.1 = \dfrac{8.2 - 14.1}{60{,}000 - 4000}(x - 4000) \text{ or, equivalently,}$$

$y \approx -0.000105357x + 14.521429$.

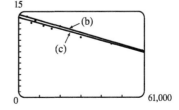

(c) Using a computing device, we obtain the regression line $y = -0.0000997855x + 13.950764$.

The following commands and screens illustrate how to find the regression line on a TI-84 Plus calculator.

Enter the data into list one (L1) and list two (L2). Press STAT 1 to enter the editor.

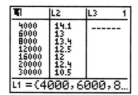 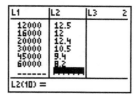

Find the regession line and store it in Y_1. Press 2nd QUIT STAT ▶ 4 VARS ▶ 1 1 ENTER.

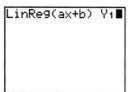 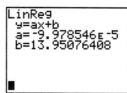 

Note from the last figure that the regression line has been stored in Y_1 and that Plot1 has been turned on (Plot1 is highlighted). You can turn on Plot1 from the Y= menu by placing the cursor on Plot1 and pressing ENTER or by pressing 2nd STAT PLOT 1 ENTER.

Now press ZOOM 9 to produce a graph of the data and the regression line. Note that choice 9 of the ZOOM menu automatically selects a window that displays all of the data.

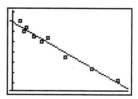

(d) When $x = 25{,}000$, $y \approx 11.456$; or about 11.5 per 100 population.

(e) When $x = 80{,}000$, $y \approx 5.968$; or about a 6% chance.

(f) When $x = 200{,}000$, y is negative, so the model does not apply.

31. (a)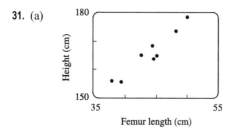

(b) Using a computing device, we obtain the regression line
$y = 1.88074x + 82.64974$.

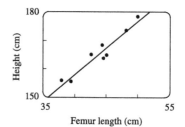

(c) When $x = 53$ cm, $y \approx 182.3$ cm.

33. (a) See the scatter plot in part (b). A linear model seems appropriate.

(b) Using a computing device, we obtain the regression line
$y = 1124.86x + 60{,}119.86$.

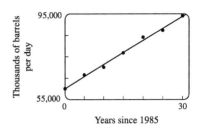

(c) For 2002, $x = 17$ and $y \approx 79{,}242$ thousands of barrels per day.

For 2017, $x = 32$ and $y \approx 96{,}115$ thousands of barrels per day.

35. (a) If $A = 60$, then $S = 0.7A^{0.3} \approx 2.39$, so you would expect to find 2 species of bats in that cave.

(b) $S = 4 \Rightarrow 4 = 0.7A^{0.3} \Rightarrow \frac{40}{7} = A^{3/10} \Rightarrow A = \left(\frac{40}{7}\right)^{10/3} \approx 333.6$, so we estimate the surface area of the cave to be 334 m^2.

37. We have $I = \dfrac{S}{4\pi r^2} = \left(\dfrac{S}{4\pi}\right)\left(\dfrac{1}{r^2}\right) = \dfrac{S/(4\pi)}{r^2}$. Thus, $I = \dfrac{k}{r^2}$ with $k = \dfrac{S}{4\pi}$.

1.3 New Functions from Old Functions

1. (a) If the graph of f is shifted 3 units upward, its equation becomes $y = f(x) + 3$.

(b) If the graph of f is shifted 3 units downward, its equation becomes $y = f(x) - 3$.

(c) If the graph of f is shifted 3 units to the right, its equation becomes $y = f(x - 3)$.

(d) If the graph of f is shifted 3 units to the left, its equation becomes $y = f(x + 3)$.

(e) If the graph of f is reflected about the x-axis, its equation becomes $y = -f(x)$.

(f) If the graph of f is reflected about the y-axis, its equation becomes $y = f(-x)$.

(g) If the graph of f is stretched vertically by a factor of 3, its equation becomes $y = 3f(x)$.

(h) If the graph of f is shrunk vertically by a factor of 3, its equation becomes $y = \frac{1}{3}f(x)$.

3. (a) *Graph 3:* The graph of f is shifted 4 units to the right and has equation $y = f(x - 4)$.

(b) *Graph 1:* The graph of f is shifted 3 units upward and has equation $y = f(x) + 3$.

(c) *Graph 4:* The graph of f is shrunk vertically by a factor of 3 and has equation $y = \frac{1}{3}f(x)$.

(d) *Graph 5:* The graph of f is shifted 4 units to the left and reflected about the x-axis. Its equation is $y = -f(x + 4)$.

(e) *Graph 2:* The graph of f is shifted 6 units to the left and stretched vertically by a factor of 2. Its equation is
$y = 2f(x + 6)$.

5. (a) To graph $y = f(2x)$ we shrink the graph of f horizontally by a factor of 2.

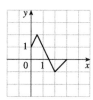

The point $(4, -1)$ on the graph of f corresponds to the point $\left(\frac{1}{2} \cdot 4, -1\right) = (2, -1)$.

(b) To graph $y = f\left(\frac{1}{2}x\right)$ we stretch the graph of f horizontally by a factor of 2.

The point $(4, -1)$ on the graph of f corresponds to the point $(2 \cdot 4, -1) = (8, -1)$.

(c) To graph $y = f(-x)$ we reflect the graph of f about the y-axis.

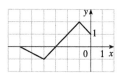

The point $(4, -1)$ on the graph of f corresponds to the point $(-1 \cdot 4, -1) = (-4, -1)$.

(d) To graph $y = -f(-x)$ we reflect the graph of f about the y-axis, then about the x-axis.

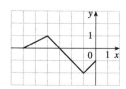

The point $(4, -1)$ on the graph of f corresponds to the point $(-1 \cdot 4, -1 \cdot -1) = (-4, 1)$.

7. The graph of $y = f(x) = \sqrt{3x - x^2}$ has been shifted 4 units to the left, reflected about the x-axis, and shifted downward 1 unit. Thus, a function describing the graph is

$$y = \underbrace{-1 \cdot}_{\substack{\text{reflect} \\ \text{about } x\text{-axis}}} \underbrace{f(x+4)}_{\substack{\text{shift} \\ \text{4 units left}}} \underbrace{-1}_{\substack{\text{shift} \\ \text{1 unit left}}}$$

This function can be written as

$$y = -f(x+4) - 1 = -\sqrt{3(x+4) - (x+4)^2} - 1$$
$$= -\sqrt{3x + 12 - (x^2 + 8x + 16)} - 1 = -\sqrt{-x^2 - 5x - 4} - 1$$

9. $y = 1 + x^2$. Start with the graph of $y = x^2$ and shift 1 unit upward

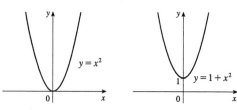

11. $y = |x + 2|$. Start with the graph of $y = |x|$ and shift 2 units to the left.

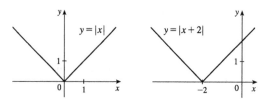

13. $y = \dfrac{1}{x} + 2$. Start with the graph of $y = \dfrac{1}{x}$ and shift 2 units upward.

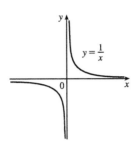
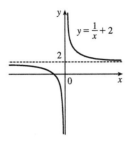

15. $y = \sin 4x$. Start with the graph of $y = \sin x$ and compress horizontally by a factor of 4. The period becomes $2\pi/4 = \pi/2$.

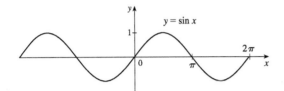
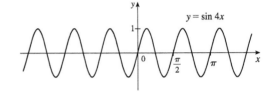

17. $y = 2 + \sqrt{x+1}$. Start with the graph of $y = \sqrt{x}$, shift 1 unit to the left, and then shift 2 units upward.

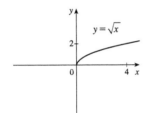
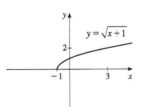
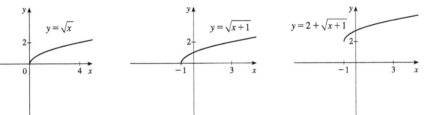

19. $y = x^2 - 2x + 5 = (x^2 - 2x + 1) + 4 = (x-1)^2 + 4$. Start with the graph of $y = x^2$, shift 1 unit to the right, and then shift 4 units upward.

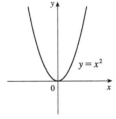
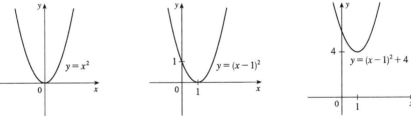
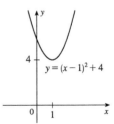

21. $y = 2 - |x|$. Start with the graph of $y = |x|$, reflect about the x-axis, and then shift 2 units upward.

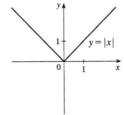
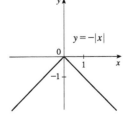
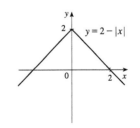

23. $y = 3\sin\frac{1}{2}x + 1$. Start with the graph of $y = \sin x$, stretch horizontally by a factor of 2, stretch vertically by a factor of 3, and then shift 1 unit upward.

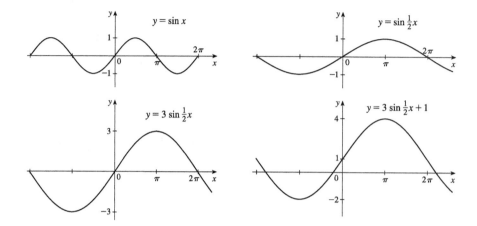

25. $y = |\cos \pi x|$. Start with the graph of $y = \cos x$, shrink horizontally by a factor of π, and reflect all the parts of the graph below the x-axis about the x-axis.

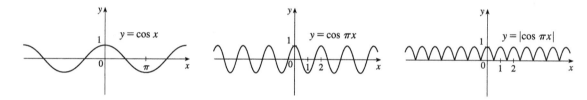

27. This is just like the solution to Example 4 except the amplitude of the curve (the 30°N curve in Figure 9 on June 21) is $14 - 12 = 2$. So the function is $L(t) = 12 + 2\sin\left[\frac{2\pi}{365}(t - 80)\right]$. March 31 is the 90th day of the year, so the model gives $L(90) \approx 12.34$ h. The daylight time (5:51 AM to 6:18 PM) is 12 hours and 27 minutes, or 12.45 h. The model value differs from the actual value by $\frac{12.45 - 12.34}{12.45} \approx 0.009$, less than 1%.

29. The water depth $D(t)$ can be modeled by a cosine function with amplitude $\frac{12-2}{2} = 5$ m, average magnitude $\frac{12+2}{2} = 7$ m, and period 12 hours. High tide occurred at time 6:45 AM ($t = 6.75$ h), so the curve begins a cycle at time $t = 6.75$ h (shift 6.75 units to the right). Thus, $D(t) = 5\cos\left[\frac{2\pi}{12}(t - 6.75)\right] + 7 = 5\cos\left[\frac{\pi}{6}(t - 6.75)\right] + 7$, where D is in meters and t is the number of hours after midnight.

31. (a) To obtain $y = f(|x|)$, the portion of the graph of $y = f(x)$ to the right of the y-axis is reflected about the y-axis.
(b) $y = \sin|x|$ (c) $y = \sqrt{|x|}$

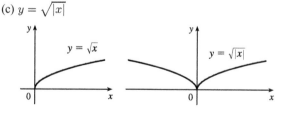

33. $f(x) = \sqrt{25 - x^2}$ is defined only when $25 - x^2 \geq 0 \Leftrightarrow x^2 \leq 25 \Leftrightarrow -5 \leq x \leq 5$, so the domain of f is $[-5, 5]$.

For $g(x) = \sqrt{x+1}$, we must have $x + 1 \geq 0 \Leftrightarrow x \geq -1$, so the domain of g is $[-1, \infty)$.

(a) $(f + g)(x) = \sqrt{25 - x^2} + \sqrt{x+1}$. The domain of $f + g$ is found by intersecting the domains of f and g: $[-1, 5]$.

(b) $(f - g)(x) = \sqrt{25 - x^2} - \sqrt{x+1}$. The domain of $f - g$ is found by intersecting the domains of f and g: $[-1, 5]$.

(c) $(fg)(x) = \sqrt{25 - x^2} \cdot \sqrt{x+1} = \sqrt{-x^3 - x^2 + 25x + 25}$. The domain of fg is found by intersecting the domains of f and g: $[-1, 5]$.

(d) $\left(\dfrac{f}{g}\right)(x) = \dfrac{\sqrt{25 - x^2}}{\sqrt{x+1}} = \sqrt{\dfrac{25 - x^2}{x+1}}$. Notice that we must have $x + 1 \neq 0$ in addition to any previous restrictions.

Thus, the domain of f/g is $(-1, 5]$.

35. $f(x) = x^3 + 5$ and $g(x) = \sqrt[3]{x}$. The domain of each function is $(-\infty, \infty)$.

(a) $(f \circ g)(x) = f(g(x)) = f\left(\sqrt[3]{x}\right) = \left(\sqrt[3]{x}\right)^3 + 5 = x + 5$. The domain is $(-\infty, \infty)$.

(b) $(g \circ f)(x) = g(f(x)) = g(x^3 + 5) = \sqrt[3]{x^3 + 5}$. The domain is $(-\infty, \infty)$.

(c) $(f \circ f)(x) = f(f(x)) = f(x^3 + 5) = (x^3 + 5)^3 + 5$. The domain is $(-\infty, \infty)$.

(d) $(g \circ g)(x) = g(g(x)) = g\left(\sqrt[3]{x}\right) = \sqrt[3]{\sqrt[3]{x}} = \sqrt[9]{x}$. The domain is $(-\infty, \infty)$.

37. $f(x) = \dfrac{1}{\sqrt{x}}$ and $g(x) = x + 1$. The domain of f is $(0, \infty)$. The domain of g is $(-\infty, \infty)$.

(a) $(f \circ g)(x) = f(g(x)) = f(x+1) = \dfrac{1}{\sqrt{x+1}}$. We must have $x + 1 > 0$, or $x > -1$, so the domain is $(-1, \infty)$.

(b) $(g \circ f)(x) = g(f(x)) = g\left(\dfrac{1}{\sqrt{x}}\right) = \dfrac{1}{\sqrt{x}} + 1$. We must have $x > 0$, so the domain is $(0, \infty)$.

(c) $(f \circ f)(x) = f(f(x)) = f\left(\dfrac{1}{\sqrt{x}}\right) = \dfrac{1}{\sqrt{1/\sqrt{x}}} = \dfrac{1}{1/\sqrt{\sqrt{x}}} = \sqrt{\sqrt{x}} = \sqrt[4]{x}$. We must have $x > 0$, so the domain

is $(0, \infty)$.

(d) $(g \circ g)(x) = g(g(x)) = g(x+1) = (x+1) + 1 = x + 2$. The domain is $(-\infty, \infty)$.

39. $f(x) = \dfrac{2}{x}$ and $g(x) = \sin x$. The domain of f is $(-\infty, 0) \cup (0, \infty)$. The domain of g is $(-\infty, \infty)$.

(a) $(f \circ g)(x) = f(g(x)) = f(\sin x) = \dfrac{2}{\sin x} = 2\csc x$. We must have $\sin x \neq 0$, so the domain is

$\{x \mid x \neq k\pi, k \text{ an integer}\}$.

(b) $(g \circ f)(x) = g(f(x)) = g\left(\dfrac{2}{x}\right) = \sin\left(\dfrac{2}{x}\right)$. We must have $x \neq 0$, so the domain is $(-\infty, 0) \cup (0, \infty)$.

(c) $(f \circ f)(x) = f(f(x)) = f\left(\dfrac{2}{x}\right) = \dfrac{2}{\dfrac{2}{x}} = x$. Since f requires $x \neq 0$, the domain is $(-\infty, 0) \cup (0, \infty)$.

(d) $(g \circ g)(x) = g(g(x)) = g(\sin x) = \sin(\sin x)$. The domain is $(-\infty, \infty)$.

41. $(f \circ g \circ h)(x) = f(g(h(x))) = f(g(x^2)) = f(\sin(x^2)) = 3\sin(x^2) - 2$

43. $(f \circ g \circ h)(x) = f(g(h(x))) = f(g(x^3 + 2)) = f[(x^3 + 2)^2] = f(x^6 + 4x^3 + 4)$
$= \sqrt{(x^6 + 4x^3 + 4) - 3} = \sqrt{x^6 + 4x^3 + 1}$

45. Let $g(x) = 2x + x^2$ and $f(x) = x^4$. Then $(f \circ g)(x) = f(g(x)) = f(2x + x^2) = (2x + x^2)^4 = F(x)$.

47. Let $g(x) = \sqrt[3]{x}$ and $f(x) = \dfrac{x}{1+x}$. Then $(f \circ g)(x) = f(g(x)) = f(\sqrt[3]{x}) = \dfrac{\sqrt[3]{x}}{1 + \sqrt[3]{x}} = F(x)$.

49. Let $g(t) = t^2$ and $f(t) = \sec t \tan t$. Then $(f \circ g)(t) = f(g(t)) = f(t^2) = \sec(t^2)\tan(t^2) = v(t)$.

51. Let $h(x) = \sqrt{x}$, $g(x) = x - 1$, and $f(x) = \sqrt{x}$. Then
$(f \circ g \circ h)(x) = f(g(h(x))) = f\left(g\left(\sqrt{x}\right)\right) = f\left(\sqrt{x} - 1\right) = \sqrt{\sqrt{x} - 1} = R(x)$.

53. Let $h(t) = \cos t$, $g(t) = \sin t$, and $f(t) = t^2$. Then
$(f \circ g \circ h)(t) = f(g(h(t))) = f(g(\cos t)) = f(\sin(\cos t)) = [\sin(\cos t)]^2 = \sin^2(\cos t) = S(t)$.

55. (a) $f(g(3)) = f(4) = 6$. (b) $g(f(2)) = g(1) = 5$.

(c) $(f \circ g)(5) = f(g(5)) = f(3) = 5$. (d) $(g \circ f)(5) = g(f(5)) = g(2) = 3$.

57. (a) $g(2) = 5$, because the point $(2, 5)$ is on the graph of g. Thus, $f(g(2)) = f(5) = 4$, because the point $(5, 4)$ is on the graph of f.

(b) $g(f(0)) = g(0) = 3$

(c) $(f \circ g)(0) = f(g(0)) = f(3) = 0$

(d) $(g \circ f)(6) = g(f(6)) = g(6)$. This value is not defined, because there is no point on the graph of g that has x-coordinate 6.

(e) $(g \circ g)(-2) = g(g(-2)) = g(1) = 4$

(f) $(f \circ f)(4) = f(f(4)) = f(2) = -2$

59. (a) Using the relationship *distance = rate · time* with the radius r as the distance, we have $r(t) = 60t$.

(b) $A = \pi r^2 \Rightarrow (A \circ r)(t) = A(r(t)) = \pi(60t)^2 = 3600\pi t^2$. This formula gives us the extent of the rippled area (in cm^2) at any time t.

61. (a) From the figure, we have a right triangle with legs 6 and d, and hypotenuse s. By the Pythagorean Theorem, $d^2 + 6^2 = s^2 \Rightarrow s = f(d) = \sqrt{d^2 + 36}$.

(b) Using $d = rt$, we get $d = (30 \text{ km/h})(t \text{ hours}) = 30t$ (in km). Thus, $d = g(t) = 30t$.

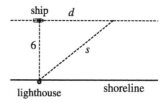

(c) $(f \circ g)(t) = f(g(t)) = f(30t) = \sqrt{(30t)^2 + 36} = \sqrt{900t^2 + 36}$. This function represents the distance between the lighthouse and the ship as a function of the time elapsed since noon.

63. (a)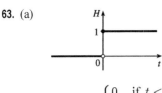

$$H(t) = \begin{cases} 0 & \text{if } t < 0 \\ 1 & \text{if } t \geq 0 \end{cases}$$

(b)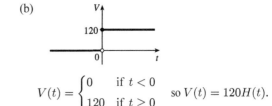

$$V(t) = \begin{cases} 0 & \text{if } t < 0 \\ 120 & \text{if } t \geq 0 \end{cases} \text{ so } V(t) = 120H(t).$$

(c)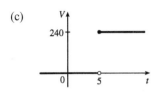

Starting with the formula in part (b), we replace 120 with 240 to reflect the different voltage. Also, because we are starting 5 units to the right of $t = 0$, we replace t with $t - 5$. Thus, the formula is $V(t) = 240H(t - 5)$.

65. If $f(x) = m_1 x + b_1$ and $g(x) = m_2 x + b_2$, then

$(f \circ g)(x) = f(g(x)) = f(m_2 x + b_2) = m_1(m_2 x + b_2) + b_1 = m_1 m_2 x + m_1 b_2 + b_1$.

So $f \circ g$ is a linear function with slope $m_1 m_2$.

67. (a) By examining the variable terms in g and h, we deduce that we must square g to get the terms $4x^2$ and $4x$ in h. If we let

$f(x) = x^2 + c$, then $(f \circ g)(x) = f(g(x)) = f(2x + 1) = (2x + 1)^2 + c = 4x^2 + 4x + (1 + c)$. Since

$h(x) = 4x^2 + 4x + 7$, we must have $1 + c = 7$. So $c = 6$ and $f(x) = x^2 + 6$.

(b) We need a function g so that $f(g(x)) = 3(g(x)) + 5 = h(x)$. But

$h(x) = 3x^2 + 3x + 2 = 3(x^2 + x) + 2 = 3(x^2 + x - 1) + 5$, so we see that $g(x) = x^2 + x - 1$.

69. We need to examine $h(-x)$.

$h(-x) = (f \circ g)(-x) = f(g(-x)) = f(g(x))$ [because g is even] $= h(x)$

Because $h(-x) = h(x)$, h is an even function.

71. (a) $E(x) = f(x) + f(-x) \Rightarrow E(-x) = f(-x) + f(-(-x)) = f(-x) + f(x) = E(x)$. Since $E(-x) = E(x)$, E is an even function.

(b) $O(x) = f(x) - f(-x) \Rightarrow O(-x) = f(-x) - f(-(-x)) = f(-x) - f(x) = -[f(x) - f(-x)] = -O(x)$.

Since $O(-x) = -O(x)$, O is an odd function.

(c) For any function f with domain $\mathbb{R}$, define functions E and O as in parts (a) and (b). Then $\frac{1}{2}E$ is even, $\frac{1}{2}O$ is odd, and we show that $f(x) = \frac{1}{2}E(x) + \frac{1}{2}O(x)$:

$$\frac{1}{2}E(x) + \frac{1}{2}O(x) = \frac{1}{2}[f(x) + f(-x)] + \frac{1}{2}[f(x) - f(-x)]$$
$$= \frac{1}{2}[f(x) + f(-x) + f(x) - f(-x)]$$
$$= \frac{1}{2}[2f(x)] = f(x)$$

as desired.

(d) $f(x) = 2^x + (x-3)^2$ has domain $\mathbb{R}$, so we know from part (c) that $f(x) = \frac{1}{2}E(x) + \frac{1}{2}O(x)$, where

$$E(x) = f(x) + f(-x) = 2^x + (x-3)^2 + 2^{-x} + (-x-3)^2$$
$$= 2^x + 2^{-x} + (x-3)^2 + (x+3)^2$$

and
$$O(x) = f(x) - f(-x) = 2^x + (x-3)^2 - [2^{-x} + (-x-3)^2]$$
$$= 2^x - 2^{-x} + (x-3)^2 - (x+3)^2$$

1.4 The Tangent and Velocity Problems

1. (a) Using $P(15, 250)$, we construct the following table:

t	Q	slope $= m_{PQ}$
5	$(5, 694)$	$\frac{694-250}{5-15} = -\frac{444}{10} = -44.4$
10	$(10, 444)$	$\frac{444-250}{10-15} = -\frac{194}{5} = -38.8$
20	$(20, 111)$	$\frac{111-250}{20-15} = -\frac{139}{5} = -27.8$
25	$(25, 28)$	$\frac{28-250}{25-15} = -\frac{222}{10} = -22.2$
30	$(30, 0)$	$\frac{0-250}{30-15} = -\frac{250}{15} = -16.\overline{6}$

(b) Using the values of t that correspond to the points closest to P ($t = 10$ and $t = 20$), we have

$$\frac{-38.8 + (-27.8)}{2} = -33.3$$

(c) From the graph, we can estimate the slope of the tangent line at P to be $\frac{-300}{9} = -33.\overline{3}$.

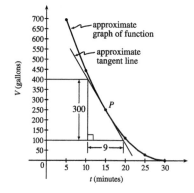

3. (a) $y = \dfrac{1}{1-x}$, $P(2, -1)$

	x	$Q(x, 1/(1-x))$	m_{PQ}
(i)	1.5	$(1.5, -2)$	2
(ii)	1.9	$(1.9, -1.111111)$	1.111111
(iii)	1.99	$(1.99, -1.010101)$	1.010101
(iv)	1.999	$(1.999, -1.001001)$	1.001001
(v)	2.5	$(2.5, -0.666667)$	0.666667
(vi)	2.1	$(2.1, -0.909091)$	0.909091
(vii)	2.01	$(2.01, -0.990099)$	0.990099
(viii)	2.001	$(2.001, -0.999001)$	0.999001

(b) The slope appears to be 1.

(c) Using $m = 1$, an equation of the tangent line to the curve at $P(2, -1)$ is $y - (-1) = 1(x - 2)$, or $y = x - 3$.

5. (a) $y = y(t) = 275 - 16t^2$. At $t = 4$, $y = 275 - 16(4)^2 = 19$. The average velocity between times 4 and $4 + h$ is

$$v_{\text{avg}} = \frac{y(4 + h) - y(4)}{(4 + h) - 4} = \frac{\left[275 - 16(4 + h)^2\right] - 19}{h} = \frac{-128h - 16h^2}{h} = -128 - 16h \quad \text{if } h \neq 0$$

(i) 0.1 seconds: $\quad h = 0.1$, $v_{\text{avg}} = -129.6$ ft/s

(ii) 0.05 seconds: $\quad h = 0.05$, $v_{\text{avg}} = -128.8$ ft/s

(iii) 0.01 seconds: $\quad h = 0.01$, $v_{\text{avg}} = -128.16$ ft/s

(b) The instantaneous velocity when $t = 4$ (h approaches 0) is -128 ft/s.

7. (a) (i) On the interval $[2, 4]$, $v_{\text{avg}} = \dfrac{s(4) - s(2)}{4 - 2} = \dfrac{79.2 - 20.6}{2} = 29.3$ ft/s.

(ii) On the interval $[3, 4]$, $v_{\text{avg}} = \dfrac{s(4) - s(3)}{4 - 3} = \dfrac{79.2 - 46.5}{1} = 32.7$ ft/s.

(iii) On the interval $[4, 5]$, $v_{\text{avg}} = \dfrac{s(5) - s(4)}{5 - 4} = \dfrac{124.8 - 79.2}{1} = 45.6$ ft/s.

(iv) On the interval $[4, 6]$, $v_{\text{avg}} = \dfrac{s(6) - s(4)}{6 - 4} = \dfrac{176.7 - 79.2}{2} = 48.75$ ft/s.

(b) Using the points $(2, 16)$ and $(5, 105)$ from the approximate tangent line, the instantaneous velocity at $t = 3$ is about

$$\frac{105 - 16}{5 - 2} = \frac{89}{3} \approx 29.7 \text{ ft/s.}$$

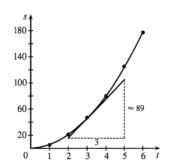

9. (a) For the curve $y = \sin(10\pi/x)$ and the point $P(1, 0)$:

x	Q	m_{PQ}
2	$(2, 0)$	0
1.5	$(1.5, 0.8660)$	1.7321
1.4	$(1.4, -0.4339)$	-1.0847
1.3	$(1.3, -0.8230)$	-2.7433
1.2	$(1.2, 0.8660)$	4.3301
1.1	$(1.1, -0.2817)$	-2.8173

x	Q	m_{PQ}
0.5	$(0.5, 0)$	0
0.6	$(0.6, 0.8660)$	-2.1651
0.7	$(0.7, 0.7818)$	-2.6061
0.8	$(0.8, 1)$	-5
0.9	$(0.9, -0.3420)$	3.4202

As x approaches 1, the slopes do not appear to be approaching any particular value.

(b)

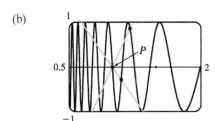

We see that problems with estimation are caused by the frequent oscillations of the graph. The tangent is so steep at P that we need to take x-values much closer to 1 in order to get accurate estimates of its slope.

(c) If we choose $x = 1.001$, then the point Q is $(1.001, -0.0314)$ and $m_{PQ} \approx -31.3794$. If $x = 0.999$, then Q is $(0.999, 0.0314)$ and $m_{PQ} = -31.4422$. The average of these slopes is -31.4108. So we estimate that the slope of the tangent line at P is about -31.4.

1.5 The Limit of a Function

1. As x approaches 2, $f(x)$ approaches 5. [Or, the values of $f(x)$ can be made as close to 5 as we like by taking x sufficiently close to 2 (but $x \neq 2$).] Yes, the graph could have a hole at $(2, 5)$ and be defined such that $f(2) = 3$.

3. (a) $\lim_{x \to -3} f(x) = \infty$ means that the values of $f(x)$ can be made arbitrarily large (as large as we please) by taking x sufficiently close to -3 (but not equal to -3).

   (b) $\lim_{x \to 4^+} f(x) = -\infty$ means that the values of $f(x)$ can be made arbitrarily large negative by taking x sufficiently close to 4 through values larger than 4.

5. (a) As x approaches 1, the values of $f(x)$ approach 2, so $\lim_{x \to 1} f(x) = 2$.

   (b) As x approaches 3 from the left, the values of $f(x)$ approach 1, so $\lim_{x \to 3^-} f(x) = 1$.

   (c) As x approaches 3 from the right, the values of $f(x)$ approach 4, so $\lim_{x \to 3^+} f(x) = 4$.

   (d) $\lim_{x \to 3} f(x)$ does not exist since the left-hand limit does not equal the right-hand limit.

   (e) When $x = 3$, $y = 3$, so $f(3) = 3$.

7. (a) $\lim_{x \to 4^-} g(x) \neq \lim_{x \to 4^+} g(x)$, so $\lim_{x \to 4} g(x)$ does not exist. However, there is a point on the graph representing $g(4)$. Thus, $a = 4$ satisfies the given description.

   (b) $\lim_{x \to 5^-} g(x) = \lim_{x \to 5^+} g(x)$, so $\lim_{x \to 5} g(x)$ exists. However, $g(5)$ is not defined. Thus, $a = 5$ satisfies the given description.

   (c) From part (a), $a = 4$ satisfies the given description. Also, $\lim_{x \to 2^-} g(x)$ and $\lim_{x \to 2^+} g(x)$ exist, but $\lim_{x \to 2^-} g(x) \neq \lim_{x \to 2^+} g(x)$. Thus, $\lim_{x \to 2} g(x)$ does not exist, and $a = 2$ also satisfies the given description.

   (d) $\lim_{x \to 4^+} g(x) = g(4)$, but $\lim_{x \to 4^-} g(x) \neq g(4)$. Thus, $a = 4$ satisfies the given description.

9. (a) $\lim_{x \to -7} f(x) = -\infty$ (b) $\lim_{x \to -3} f(x) = \infty$ (c) $\lim_{x \to 0} f(x) = \infty$

   (d) $\lim_{x \to 6^-} f(x) = -\infty$ (e) $\lim_{x \to 6^+} f(x) = \infty$

   (f) The equations of the vertical asymptotes are $x = -7$, $x = -3$, $x = 0$, and $x = 6$.

11. From the graph of f we see that $\lim_{x \to 1^-} f(x) = 0$, but $\lim_{x \to 1^+} f(x) = 1$, so $\lim_{x \to a} f(x)$ does not exist for $a = 1$. However, $\lim_{x \to a} f(x)$ exists for all other values of a. Thus, $\lim_{x \to a} f(x)$ exists for all a in $(-\infty, 1) \cup (1, \infty)$.

13. (a) From the graph, $\lim_{x \to 0^-} f(x) = -1$.

(b) From the graph, $\lim_{x \to 0^+} f(x) = 1$.

(c) Since $\lim_{x \to 0^-} f(x) \neq \lim_{x \to 0^+} f(x)$, $\lim_{x \to 0} f(x)$ does not exist.

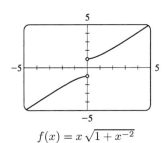

$f(x) = x\sqrt{1 + x^{-2}}$

15. $\lim_{x \to 1^-} f(x) = 3$, $\lim_{x \to 1^+} f(x) = 0$, $f(1) = 2$

17. $\lim_{x \to -1^-} f(x) = 0$, $\lim_{x \to -1^+} f(x) = 1$, $\lim_{x \to 2} f(x) = 3$, $f(-1) = 2$, $f(2) = 1$

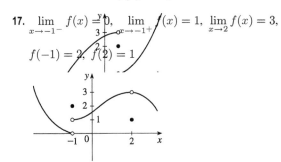

19. For $f(x) = \dfrac{x^2 - 3x}{x^2 - 9}$:

x	$f(x)$	x	$f(x)$
3.1	0.508 197	2.9	0.491 525
3.05	0.504 132	2.95	0.495 798
3.01	0.500 832	2.99	0.499 165
3.001	0.500 083	2.999	0.499 917
3.0001	0.500 008	2.9999	0.499 992

It appears that $\lim\limits_{x \to 3} \dfrac{x^2 - 3x}{x^2 - 9} = \dfrac{1}{2}$.

21. For $f(x) = \dfrac{\sin x}{x + \tan x}$:

x	$f(x)$
±1	0.329 033
±0.5	0.458 209
±0.2	0.493 331
±0.1	0.498 333
±0.05	0.499 583
±0.01	0.499 983

It appears that $\lim\limits_{x \to 0} \dfrac{\sin x}{x + \tan x} = 0.5 = \dfrac{1}{2}$.

23. For $f(\theta) = \dfrac{\sin 3\theta}{\tan 2\theta}$:

θ	$f(\theta)$
±0.1	1.457 847
±0.01	1.499 575
±0.001	1.499 996
±0.0001	1.500 000

It appears that $\lim\limits_{\theta \to 0} \dfrac{\sin 3\theta}{\tan 2\theta} = 1.5$.

The graph confirms that result.

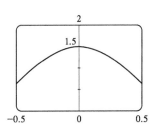

25. For $f(x) = x^x$:

x	$f(x)$
0.1	0.794 328
0.01	0.954 993
0.001	0.993 116
0.0001	0.999 079

It appears that $\lim\limits_{x \to 0^+} f(x) = 1$.

The graph confirms that result.

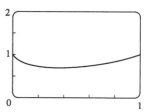

27. $\lim\limits_{x \to 5^+} \dfrac{x+1}{x-5} = \infty$ since the numerator is positive and the denominator approaches 0 from the positive side as $x \to 5^+$.

29. $\lim\limits_{x \to 2} \dfrac{x^2}{(x-2)^2} = \infty$ since the numerator is positive and the denominator approaches 0 through positive values as $x \to 2$.

31. $\lim\limits_{x \to -2^+} \dfrac{x-1}{x^2(x+2)} = -\infty$ since $(x+2) \to 0$ as $x \to -2^+$ and $\dfrac{x-1}{x^2(x+2)} < 0$ for $-2 < x < 0$.

33. $\lim\limits_{x \to (\pi/2)^+} \dfrac{1}{x} \sec x = -\infty$ since $\dfrac{1}{x}$ is positive and $\sec x \to -\infty$ as $x \to (\pi/2)^+$.

35. $\lim\limits_{x \to 1} \dfrac{x^2+2x}{x^2-2x+1} = \lim\limits_{x \to 1} \dfrac{x^2+2x}{(x-1)^2} = \infty$ since the numerator is positive and the denominator approaches 0 through positive values as $x \to 1$.

37. The denominator of $f(x) = \dfrac{x-1}{2x+4}$ is equal to 0 when $x = -2$ (and the numerator is not), so $x = -2$ is the vertical asymptote of the function.

39. (a) $f(x) = \dfrac{1}{x^3-1}$.

From these calculations, it seems that

$\lim\limits_{x \to 1^-} f(x) = -\infty$ and $\lim\limits_{x \to 1^+} f(x) = \infty$.

x	$f(x)$
0.5	−1.14
0.9	−3.69
0.99	−33.7
0.999	−333.7
0.9999	−3333.7
0.99999	−33,333.7

x	$f(x)$
1.5	0.42
1.1	3.02
1.01	33.0
1.001	333.0
1.0001	3333.0
1.00001	33,333.3

(b) If x is slightly smaller than 1, then $x^3 - 1$ will be a negative number close to 0, and the reciprocal of $x^3 - 1$, that is, $f(x)$, will be a negative number with large absolute value. So $\lim\limits_{x \to 1^-} f(x) = -\infty$.

If x is slightly larger than 1, then $x^3 - 1$ will be a small positive number, and its reciprocal, $f(x)$, will be a large positive number. So $\lim\limits_{x \to 1^+} f(x) = \infty$.

(c) It appears from the graph of f that

$\lim\limits_{x \to 1^-} f(x) = -\infty$ and $\lim\limits_{x \to 1^+} f(x) = \infty$.

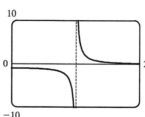

41. For $f(x) = x^2 - (2^x/1000)$:

(a)

x	$f(x)$
1	0.998 000
0.8	0.638 259
0.6	0.358 484
0.4	0.158 680
0.2	0.038 851
0.1	0.008 928
0.05	0.001 465

It appears that $\lim\limits_{x \to 0} f(x) = 0$.

(b)

x	$f(x)$
0.04	0.000 572
0.02	$-0.000\,614$
0.01	$-0.000\,907$
0.005	$-0.000\,978$
0.003	$-0.000\,993$
0.001	$-0.001\,000$

It appears that $\lim\limits_{x \to 0} f(x) = -0.001$.

43.

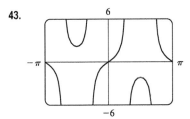

There appear to be vertical asymptotes of the curve $y = \tan(2 \sin x)$ at $x \approx \pm 0.90$ and $x \approx \pm 2.24$. To find the exact equations of these asymptotes, we note that the graph of the tangent function has vertical asymptotes at $x = \frac{\pi}{2} + \pi n$. Thus, we must have $2 \sin x = \frac{\pi}{2} + \pi n$, or equivalently, $\sin x = \frac{\pi}{4} + \frac{\pi}{2} n$. Since $-1 \le \sin x \le 1$, we must have $\sin x = \pm \frac{\pi}{4}$ and so $x = \pm \sin^{-1} \frac{\pi}{4}$ (corresponding to $x \approx \pm 0.90$). Just as $150°$ is the reference angle for $30°$, $\pi - \sin^{-1} \frac{\pi}{4}$ is the reference angle for $\sin^{-1} \frac{\pi}{4}$. So $x = \pm\left(\pi - \sin^{-1} \frac{\pi}{4}\right)$ are also equations of vertical asymptotes (corresponding to $x \approx \pm 2.24$).

45. $\lim\limits_{v \to c^-} m = \lim\limits_{v \to c^-} \dfrac{m_0}{\sqrt{1 - v^2/c^2}}$. As $v \to c^-$, $\sqrt{1 - v^2/c^2} \to 0^+$, and $m \to \infty$.

1.6 Calculating Limits Using the Limit Laws

1. (a) $\lim\limits_{x \to 2} [f(x) + 5g(x)] = \lim\limits_{x \to 2} f(x) + \lim\limits_{x \to 2} [5g(x)]$ [Limit Law 1]

$\qquad\qquad\qquad\qquad\quad = \lim\limits_{x \to 2} f(x) + 5 \lim\limits_{x \to 2} g(x)$ [Limit Law 3]

$\qquad\qquad\qquad\qquad\quad = 4 + 5(-2) = -6$

(b) $\lim\limits_{x \to 2} [g(x)]^3 = \left[\lim\limits_{x \to 2} g(x)\right]^3$ [Limit Law 6]

$\qquad\qquad\quad = (-2)^3 = -8$

(c) $\lim\limits_{x \to 2} \sqrt{f(x)} = \sqrt{\lim\limits_{x \to 2} f(x)}$ [Limit Law 7]

$\qquad\qquad\quad = \sqrt{4} = 2$

(d) $\lim\limits_{x \to 2} \dfrac{3f(x)}{g(x)} = \dfrac{\lim\limits_{x \to 2} [3f(x)]}{\lim\limits_{x \to 2} g(x)}$ [Limit Law 5]

$\qquad\qquad\quad = \dfrac{3 \lim\limits_{x \to 2} f(x)}{\lim\limits_{x \to 2} g(x)}$ [Limit Law 3]

$\qquad\qquad\quad = \dfrac{3(4)}{-2} = -6$

(e) Because the limit of the denominator is 0, we can't use Limit Law 5. The given limit, $\lim\limits_{x \to 2} \dfrac{g(x)}{h(x)}$, does not exist because the denominator approaches 0 while the numerator approaches a nonzero number.

(f) $\lim\limits_{x \to 2} \dfrac{g(x)\,h(x)}{f(x)} = \dfrac{\lim\limits_{x \to 2}[g(x)\,h(x)]}{\lim\limits_{x \to 2} f(x)}$ [Limit Law 5]

$= \dfrac{\lim\limits_{x \to 2} g(x) \cdot \lim\limits_{x \to 2} h(x)}{\lim\limits_{x \to 2} f(x)}$ [Limit Law 4]

$= \dfrac{-2 \cdot 0}{4} = 0$

3. $\lim\limits_{x \to 5}(4x^2 - 5x) = \lim\limits_{x \to 5}(4x^2) - \lim\limits_{x \to 5}(5x)$ [Limit Law 2]

$= 4\lim\limits_{x \to 5} x^2 - 5\lim\limits_{x \to 5} x$ [3]

$= 4(5^2) - 5(5)$ [10, 9]

$= 75$

5. $\lim\limits_{v \to 2}(v^2 + 2v)(2v^3 - 5) = \lim\limits_{v \to 2}(v^2 + 2v) \cdot \lim\limits_{v \to 2}(2v^3 - 5)$ [Limit Law 4]

$= \left(\lim\limits_{v \to 2} v^2 + \lim\limits_{v \to 2} 2v\right)\left(\lim\limits_{v \to 2} 2v^3 - \lim\limits_{v \to 2} 5\right)$ [1 and 2]

$= \left(\lim\limits_{v \to 2} v^2 + 2\lim\limits_{v \to 2} v\right)\left(2\lim\limits_{v \to 2} v^3 - \lim\limits_{v \to 2} 5\right)$ [3]

$= [2^2 + 2(2)][2(2)^3 - 5]$ [10, 9, and 8]

$= (8)(11) = 88$

7. $\lim\limits_{u \to -2} \sqrt{9 - u^3 + 2u^2} = \sqrt{\lim\limits_{u \to -2}(9 - u^3 + 2u^2)}$ [Limit Law 7]

$= \sqrt{\lim\limits_{u \to -2} 9 - \lim\limits_{u \to -2} u^3 + \lim\limits_{u \to -2} 2u^2}$ [2 and 1]

$= \sqrt{\lim\limits_{u \to -2} 9 - \lim\limits_{u \to -2} u^3 + 2\lim\limits_{u \to -2} u^2}$ [3]

$= \sqrt{9 - (-2)^3 + 2(-2)^2}$ [8 and 10]

$= \sqrt{25} = 5$

9. $\lim\limits_{t \to -1}\left(\dfrac{2t^5 - t^4}{5t^2 + 4}\right)^3 = \left(\lim\limits_{t \to -1} \dfrac{2t^5 - t^4}{5t^2 + 4}\right)^3$ [Limit Law 6]

$= \left(\dfrac{\lim\limits_{t \to -1}(2t^5 - t^4)}{\lim\limits_{t \to -1}(5t^2 + 4)}\right)^3$ [5]

$= \left(\dfrac{2\lim\limits_{t \to -1} t^5 - \lim\limits_{t \to -1} t^4}{5\lim\limits_{t \to -1} t^2 + \lim\limits_{t \to -1} 4}\right)^3$ [3, 2, and 1]

$= \left(\dfrac{2(-1)^5 - (-1)^4}{5(-1)^2 + 4}\right)^3$ [10 and 8]

$= \left(-\dfrac{3}{9}\right)^3 = -\dfrac{1}{27}$

11. $\lim\limits_{x \to -2} (3x - 7) = 3(-2) - 7 = -13$

13. $\lim\limits_{t \to 4} \dfrac{t^2 - 2t - 8}{t - 4} = \lim\limits_{t \to 4} \dfrac{(t-4)(t+2)}{t-4} = \lim\limits_{t \to 4}(t+2) = 4 + 2 = 6$

15. $\lim\limits_{x \to 2} \dfrac{x^2 + 5x + 4}{x - 2}$ does not exist since $x - 2 \to 0$, but $x^2 + 5x + 4 \to 18$ as $x \to 2$.

17. $\lim\limits_{x \to -2} \dfrac{x^2 - x - 6}{3x^2 + 5x - 2} = \lim\limits_{x \to -2} \dfrac{(x-3)(x+2)}{(3x-1)(x+2)} = \lim\limits_{x \to -2} \dfrac{x-3}{3x-1} = \dfrac{-2-3}{3(-2)-1} = \dfrac{-5}{-7} = \dfrac{5}{7}$

19. Factoring $t^3 - 27$ as the difference of two cubes, we have

$\lim\limits_{t \to 3} \dfrac{t^3 - 27}{t^2 - 9} = \lim\limits_{t \to 3} \dfrac{(t-3)(t^2 + 3t + 9)}{(t-3)(t+3)} = \lim\limits_{t \to 3} \dfrac{t^2 + 3t + 9}{t+3} = \dfrac{3^2 + 3(3) + 9}{3+3} = \dfrac{27}{6} = \dfrac{9}{2}.$

21. $\lim\limits_{h \to 0} \dfrac{(h-3)^2 - 9}{h} = \lim\limits_{h \to 0} \dfrac{h^2 - 6h + 9 - 9}{h} = \lim\limits_{h \to 0} \dfrac{h^2 - 6h}{h} = \lim\limits_{h \to 0} \dfrac{h(h-6)}{h} = \lim\limits_{h \to 0}(h-6) = 0 - 6 = -6$

23. $\lim\limits_{h \to 0} \dfrac{\sqrt{9+h} - 3}{h} = \lim\limits_{h \to 0} \dfrac{\sqrt{9+h} - 3}{h} \cdot \dfrac{\sqrt{9+h} + 3}{\sqrt{9+h} + 3} = \lim\limits_{h \to 0} \dfrac{\left(\sqrt{9+h}\right)^2 - 3^2}{h\left(\sqrt{9+h} + 3\right)} = \lim\limits_{h \to 0} \dfrac{(9+h) - 9}{h\left(\sqrt{9+h} + 3\right)}$

$= \lim\limits_{h \to 0} \dfrac{h}{h\left(\sqrt{9+h} + 3\right)} = \lim\limits_{h \to 0} \dfrac{1}{\sqrt{9+h} + 3} = \dfrac{1}{3+3} = \dfrac{1}{6}$

25. $\lim\limits_{x \to 3} \dfrac{\dfrac{1}{x} - \dfrac{1}{3}}{x - 3} = \lim\limits_{x \to 3} \dfrac{\dfrac{1}{x} - \dfrac{1}{3}}{x - 3} \cdot \dfrac{3x}{3x} = \lim\limits_{x \to 3} \dfrac{3 - x}{3x(x-3)} = \lim\limits_{x \to 3} \dfrac{-1}{3x} = -\dfrac{1}{9}$

27. $\lim\limits_{t \to 0} \dfrac{\sqrt{1+t} - \sqrt{1-t}}{t} = \lim\limits_{t \to 0} \dfrac{\sqrt{1+t} - \sqrt{1-t}}{t} \cdot \dfrac{\sqrt{1+t} + \sqrt{1-t}}{\sqrt{1+t} + \sqrt{1-t}} = \lim\limits_{t \to 0} \dfrac{\left(\sqrt{1+t}\right)^2 - \left(\sqrt{1-t}\right)^2}{t\left(\sqrt{1+t} + \sqrt{1-t}\right)}$

$= \lim\limits_{t \to 0} \dfrac{(1+t) - (1-t)}{t\left(\sqrt{1+t} + \sqrt{1-t}\right)} = \lim\limits_{t \to 0} \dfrac{2t}{t\left(\sqrt{1+t} + \sqrt{1-t}\right)} = \lim\limits_{t \to 0} \dfrac{2}{\sqrt{1+t} + \sqrt{1-t}}$

$= \dfrac{2}{\sqrt{1} + \sqrt{1}} = \dfrac{2}{2} = 1$

29. $\lim\limits_{x \to 16} \dfrac{4 - \sqrt{x}}{16x - x^2} = \lim\limits_{x \to 16} \dfrac{(4 - \sqrt{x})(4 + \sqrt{x})}{(16x - x^2)(4 + \sqrt{x})} = \lim\limits_{x \to 16} \dfrac{16 - x}{x(16 - x)(4 + \sqrt{x})}$

$= \lim\limits_{x \to 16} \dfrac{1}{x(4 + \sqrt{x})} = \dfrac{1}{16\left(4 + \sqrt{16}\right)} = \dfrac{1}{16(8)} = \dfrac{1}{128}$

31. $\lim\limits_{t \to 0} \left(\dfrac{1}{t\sqrt{1+t}} - \dfrac{1}{t} \right) = \lim\limits_{t \to 0} \dfrac{1 - \sqrt{1+t}}{t\sqrt{1+t}} = \lim\limits_{t \to 0} \dfrac{\left(1 - \sqrt{1+t}\right)\left(1 + \sqrt{1+t}\right)}{t\sqrt{t+1}\left(1 + \sqrt{1+t}\right)} = \lim\limits_{t \to 0} \dfrac{-t}{t\sqrt{1+t}\left(1 + \sqrt{1+t}\right)}$

$= \lim\limits_{t \to 0} \dfrac{-1}{\sqrt{1+t}\left(1 + \sqrt{1+t}\right)} = \dfrac{-1}{\sqrt{1+0}\left(1 + \sqrt{1+0}\right)} = -\dfrac{1}{2}$

33. $\lim\limits_{h \to 0} \dfrac{(x+h)^3 - x^3}{h} = \lim\limits_{h \to 0} \dfrac{(x^3 + 3x^2 h + 3xh^2 + h^3) - x^3}{h} = \lim\limits_{h \to 0} \dfrac{3x^2 h + 3xh^2 + h^3}{h}$

$= \lim\limits_{h \to 0} \dfrac{h(3x^2 + 3xh + h^2)}{h} = \lim\limits_{h \to 0}(3x^2 + 3xh + h^2) = 3x^2$

35. (a)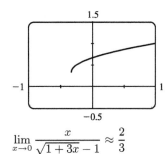

$$\lim_{x \to 0} \frac{x}{\sqrt{1+3x}-1} \approx \frac{2}{3}$$

(b)

x	$f(x)$
-0.001	$0.666\,166\,3$
$-0.000\,1$	$0.666\,616\,7$
$-0.000\,01$	$0.666\,661\,7$
$-0.000\,001$	$0.666\,666\,2$
$0.000\,001$	$0.666\,667\,2$
$0.000\,01$	$0.666\,671\,7$
$0.000\,1$	$0.666\,716\,7$
0.001	$0.667\,166\,3$

The limit appears to be $\frac{2}{3}$.

(c) $\lim_{x \to 0} \left(\frac{x}{\sqrt{1+3x}-1} \cdot \frac{\sqrt{1+3x}+1}{\sqrt{1+3x}+1} \right) = \lim_{x \to 0} \frac{x(\sqrt{1+3x}+1)}{(1+3x)-1} = \lim_{x \to 0} \frac{x(\sqrt{1+3x}+1)}{3x}$

$= \frac{1}{3} \lim_{x \to 0} (\sqrt{1+3x}+1)$ [Limit Law 3]

$= \frac{1}{3} \left[\sqrt{\lim_{x \to 0}(1+3x)} + \lim_{x \to 0} 1 \right]$ [1 and 7]

$= \frac{1}{3} \left(\sqrt{\lim_{x \to 0} 1 + 3 \lim_{x \to 0} x} + 1 \right)$ [1, 3, and 8]

$= \frac{1}{3} (\sqrt{1+3 \cdot 0} + 1)$ [8 and 9]

$= \frac{1}{3}(1+1) = \frac{2}{3}$

37. Let $f(x) = -x^2$, $g(x) = x^2 \cos 20\pi x$ and $h(x) = x^2$. Then

$-1 \le \cos 20\pi x \le 1 \;\Rightarrow\; -x^2 \le x^2 \cos 20\pi x \le x^2 \;\Rightarrow\; f(x) \le g(x) \le h(x)$.

So since $\lim_{x \to 0} f(x) = \lim_{x \to 0} h(x) = 0$, by the Squeeze Theorem we have

$\lim_{x \to 0} g(x) = 0$.

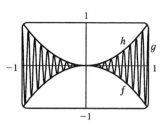

39. We have $\lim_{x \to 4} (4x - 9) = 4(4) - 9 = 7$ and $\lim_{x \to 4} (x^2 - 4x + 7) = 4^2 - 4(4) + 7 = 7$. Since $4x - 9 \le f(x) \le x^2 - 4x + 7$ for $x \ge 0$, $\lim_{x \to 4} f(x) = 7$ by the Squeeze Theorem.

41. $-1 \le \cos(2/x) \le 1 \;\Rightarrow\; -x^4 \le x^4 \cos(2/x) \le x^4$. Since $\lim_{x \to 0}(-x^4) = 0$ and $\lim_{x \to 0} x^4 = 0$, we have

$\lim_{x \to 0} [x^4 \cos(2/x)] = 0$ by the Squeeze Theorem.

43. $|x+4| = \begin{cases} x+4 & \text{if } x+4 \ge 0 \\ -(x+4) & \text{if } x+4 < 0 \end{cases} = \begin{cases} x+4 & \text{if } x \ge -4 \\ -(x+4) & \text{if } x < -4 \end{cases}$

Thus, $\lim_{x \to -4^+} (|x+4| - 2x) = \lim_{x \to -4^+} (x+4-2x) = \lim_{x \to -4^+} (-x+4) = 4+4 = 8$ and

$\lim_{x \to -4^-} (|x+4| - 2x) = \lim_{x \to -4^-} (-(x+4) - 2x) = \lim_{x \to -4^-} (-3x - 4) = 12 - 4 = 8$.

The left and right limits are equal, so $\lim_{x \to -4} (|x+4| - 2x) = 8$.

45. $|2x^3 - x^2| = |x^2(2x-1)| = |x^2| \cdot |2x-1| = x^2 |2x-1|$

$|2x - 1| = \begin{cases} 2x - 1 & \text{if } 2x - 1 \geq 0 \\ -(2x - 1) & \text{if } 2x - 1 < 0 \end{cases} = \begin{cases} 2x - 1 & \text{if } x \geq 0.5 \\ -(2x - 1) & \text{if } x < 0.5 \end{cases}$

So $|2x^3 - x^2| = x^2[-(2x-1)]$ for $x < 0.5$.

Thus, $\lim\limits_{x \to 0.5^-} \dfrac{2x-1}{|2x^3 - x^2|} = \lim\limits_{x \to 0.5^-} \dfrac{2x-1}{x^2[-(2x-1)]} = \lim\limits_{x \to 0.5^-} \dfrac{-1}{x^2} = \dfrac{-1}{(0.5)^2} = \dfrac{-1}{0.25} = -4.$

47. Since $|x| = -x$ for $x < 0$, we have $\lim\limits_{x \to 0^-} \left(\dfrac{1}{x} - \dfrac{1}{|x|} \right) = \lim\limits_{x \to 0^-} \left(\dfrac{1}{x} - \dfrac{1}{-x} \right) = \lim\limits_{x \to 0^-} \dfrac{2}{x}$, which does not exist since the denominator approaches 0 and the numerator does not.

49. (a)

(b) (i) Since $\text{sgn}\, x = 1$ for $x > 0$, $\lim\limits_{x \to 0^+} \text{sgn}\, x = \lim\limits_{x \to 0^+} 1 = 1.$

(ii) Since $\text{sgn}\, x = -1$ for $x < 0$, $\lim\limits_{x \to 0^-} \text{sgn}\, x = \lim\limits_{x \to 0^-} -1 = -1.$

(iii) Since $\lim\limits_{x \to 0^-} \text{sgn}\, x \neq \lim\limits_{x \to 0^+} \text{sgn}\, x$, $\lim\limits_{x \to 0} \text{sgn}\, x$ does not exist.

(iv) Since $|\text{sgn}\, x| = 1$ for $x \neq 0$, $\lim\limits_{x \to 0} |\text{sgn}\, x| = \lim\limits_{x \to 0} 1 = 1.$

51. (a) (i) $\lim\limits_{x \to 2^+} g(x) = \lim\limits_{x \to 2^+} \dfrac{x^2 + x - 6}{|x - 2|} = \lim\limits_{x \to 2^+} \dfrac{(x+3)(x-2)}{|x-2|}$

$= \lim\limits_{x \to 2^+} \dfrac{(x+3)(x-2)}{x-2}$ [since $x - 2 > 0$ if $x \to 2^+$]

$= \lim\limits_{x \to 2^+} (x + 3) = 5$

(ii) The solution is similar to the solution in part (i), but now $|x - 2| = 2 - x$ since $x - 2 < 0$ if $x \to 2^-$.

Thus, $\lim\limits_{x \to 2^-} g(x) = \lim\limits_{x \to 2^-} -(x + 3) = -5.$

(b) Since the right-hand and left-hand limits of g at $x = 2$ are not equal, $\lim\limits_{x \to 2} g(x)$ does not exist.

(c)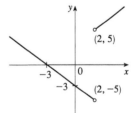

53. For the $\lim\limits_{t \to 2} B(t)$ to exist, the one-sided limits at $t = 2$ must be equal. $\lim\limits_{t \to 2^-} B(t) = \lim\limits_{t \to 2^-} (4 - \tfrac{1}{2}t) = 4 - 1 = 3$ and $\lim\limits_{t \to 2^+} B(t) = \lim\limits_{t \to 2^+} \sqrt{t + c} = \sqrt{2 + c}$. Now $3 = \sqrt{2 + c} \Rightarrow 9 = 2 + c \Leftrightarrow c = 7.$

55. (a) (i) $[\![x]\!] = -2$ for $-2 \leq x < -1$, so $\lim\limits_{x \to -2^+} [\![x]\!] = \lim\limits_{x \to -2^+} (-2) = -2$

(ii) $[\![x]\!] = -3$ for $-3 \leq x < -2$, so $\lim\limits_{x \to -2^-} [\![x]\!] = \lim\limits_{x \to -2^-} (-3) = -3.$

The right and left limits are different, so $\lim\limits_{x \to -2} [\![x]\!]$ does not exist.

(iii) $[\![x]\!] = -3$ for $-3 \leq x < -2$, so $\lim\limits_{x \to -2.4} [\![x]\!] = \lim\limits_{x \to -2.4} (-3) = -3.$

(b) (i) $[\![x]\!] = n-1$ for $n-1 \le x < n$, so $\lim\limits_{x \to n^-} [\![x]\!] = \lim\limits_{x \to n^-} (n-1) = n-1$.

(ii) $[\![x]\!] = n$ for $n \le x < n+1$, so $\lim\limits_{x \to n^+} [\![x]\!] = \lim\limits_{x \to n^+} n = n$.

(c) $\lim\limits_{x \to a} [\![x]\!]$ exists $\Leftrightarrow$ a is not an integer.

57. The graph of $f(x) = [\![x]\!] + [\![-x]\!]$ is the same as the graph of $g(x) = -1$ with holes at each integer, since $f(a) = 0$ for any integer a. Thus, $\lim\limits_{x \to 2^-} f(x) = -1$ and $\lim\limits_{x \to 2^+} f(x) = -1$, so $\lim\limits_{x \to 2} f(x) = -1$. However,

$f(2) = [\![2]\!] + [\![-2]\!] = 2 + (-2) = 0$, so $\lim\limits_{x \to 2} f(x) \ne f(2)$.

59. Since $p(x)$ is a polynomial, $p(x) = a_0 + a_1 x + a_2 x^2 + \cdots + a_n x^n$. Thus, by the Limit Laws,

$$\lim_{x \to a} p(x) = \lim_{x \to a} \left(a_0 + a_1 x + a_2 x^2 + \cdots + a_n x^n \right) = a_0 + a_1 \lim_{x \to a} x + a_2 \lim_{x \to a} x^2 + \cdots + a_n \lim_{x \to a} x^n$$

$$= a_0 + a_1 a + a_2 a^2 + \cdots + a_n a^n = p(a)$$

Thus, for any polynomial p, $\lim\limits_{x \to a} p(x) = p(a)$.

61. $\lim\limits_{x \to 1} [f(x) - 8] = \lim\limits_{x \to 1} \left[\dfrac{f(x) - 8}{x - 1} \cdot (x - 1) \right] = \lim\limits_{x \to 1} \dfrac{f(x) - 8}{x - 1} \cdot \lim\limits_{x \to 1} (x - 1) = 10 \cdot 0 = 0$.

Thus, $\lim\limits_{x \to 1} f(x) = \lim\limits_{x \to 1} \{[f(x) - 8] + 8\} = \lim\limits_{x \to 1} [f(x) - 8] + \lim\limits_{x \to 1} 8 = 0 + 8 = 8$.

Note: The value of $\lim\limits_{x \to 1} \dfrac{f(x) - 8}{x - 1}$ does not affect the answer since it's multiplied by 0. What's important is that

$\lim\limits_{x \to 1} \dfrac{f(x) - 8}{x - 1}$ exists.

63. Observe that $0 \le f(x) \le x^2$ for all x, and $\lim\limits_{x \to 0} 0 = 0 = \lim\limits_{x \to 0} x^2$. So, by the Squeeze Theorem, $\lim\limits_{x \to 0} f(x) = 0$.

65. Let $f(x) = H(x)$ and $g(x) = 1 - H(x)$, where H is the Heaviside function defined in Exercise 1.3.63.

Thus, either f or g is 0 for any value of x. Then $\lim\limits_{x \to 0} f(x)$ and $\lim\limits_{x \to 0} g(x)$ do not exist, but $\lim\limits_{x \to 0} [f(x)g(x)] = \lim\limits_{x \to 0} 0 = 0$.

67. Since the denominator approaches 0 as $x \to -2$, the limit will exist only if the numerator also approaches

0 as $x \to -2$. In order for this to happen, we need $\lim\limits_{x \to -2} \left(3x^2 + ax + a + 3 \right) = 0 \Leftrightarrow$

$3(-2)^2 + a(-2) + a + 3 = 0 \Leftrightarrow 12 - 2a + a + 3 = 0 \Leftrightarrow a = 15$. With $a = 15$, the limit becomes

$\lim\limits_{x \to -2} \dfrac{3x^2 + 15x + 18}{x^2 + x - 2} = \lim\limits_{x \to -2} \dfrac{3(x+2)(x+3)}{(x-1)(x+2)} = \lim\limits_{x \to -2} \dfrac{3(x+3)}{x-1} = \dfrac{3(-2+3)}{-2-1} = \dfrac{3}{-3} = -1$.

1.7 The Precise Definition of a Limit

1. If $|f(x) - 1| < 0.2$, then $-0.2 < f(x) - 1 < 0.2 \Rightarrow 0.8 < f(x) < 1.2$. From the graph, we see that the last inequality is true if $0.7 < x < 1.1$, so we can choose $\delta = \min\{1 - 0.7, 1.1 - 1\} = \min\{0.3, 0.1\} = 0.1$ (or any smaller positive number).

3. On the given graph, the leftmost question mark is the solution of $\sqrt{x} = 1.6$ and the rightmost, $\sqrt{x} = 2.4$. So the values are $1.6^2 = 2.56$ and $2.4^2 = 5.76$. On the left side, we need $|x - 4| < |2.56 - 4| = 1.44$. On the right side, we need $|x - 4| < |5.76 - 4| = 1.76$. To satisfy both conditions, we need the more restrictive condition to hold—namely, $|x - 4| < 1.44$. Thus, we can choose $\delta = 1.44$, or any smaller positive number.

5. From the graph, we find that $y = \sqrt{x^2 + 5} = 2.7\ [3 - 0.3]$ when $x \approx 1.513$, so $2 - \delta_1 \approx 1.513 \Rightarrow \delta_1 \approx 2 - 1.513 = 0.487$. Also, $y = \sqrt{x^2 + 5} = 3.3\ [3 + 0.3]$ when $x \approx 2.426$, so $2 + \delta_2 \approx 2.426 \Rightarrow \delta_2 \approx 2.426 - 2 = 0.426$. Thus, we choose $\delta = 0.426$ (or any smaller positive number) since this is the smaller of δ_1 and δ_2.

7. From the graph with $\varepsilon = 0.2$, we find that $y = x^3 - 3x + 4 = 5.8\ [6 - \varepsilon]$ when $x \approx 1.9774$, so $2 - \delta_1 \approx 1.9774 \Rightarrow \delta_1 \approx 0.0226$. Also, $y = x^3 - 3x + 4 = 6.2\ [6 + \varepsilon]$ when $x \approx 2.022$, so $2 + \delta_2 \approx 2.0219 \Rightarrow \delta_2 \approx 0.0219$. Thus, we choose $\delta = 0.0219$ (or any smaller positive number) since this is the smaller of δ_1 and δ_2.

For $\varepsilon = 0.1$, we get $\delta_1 \approx 0.0112$ and $\delta_2 \approx 0.0110$, so we choose $\delta = 0.011$ (or any smaller positive number).

9. (a) The graph of $y = \dfrac{x^2 + 4}{\sqrt{x - 4}}$ shows that $y = 100$ when $x \approx 4.04$ (more accurately, 4.04134). Thus, we choose $\delta = 0.04$ (or any smaller positive number).

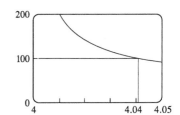

(b) From part (a), we see that as x gets closer to 4 from the right, y increases without bound. In symbols, $\lim\limits_{x \to 4^+} \dfrac{x^2 + 4}{\sqrt{x - 4}} = \infty$.

11. (a) $A = \pi r^2$ and $A = 1000\text{ cm}^2 \Rightarrow \pi r^2 = 1000 \Rightarrow r^2 = \dfrac{1000}{\pi} \Rightarrow r = \sqrt{\dfrac{1000}{\pi}}\ (r > 0) \approx 17.8412\text{ cm}$.

(b) $|A - 1000| \le 5 \Rightarrow -5 \le \pi r^2 - 1000 \le 5 \Rightarrow 1000 - 5 \le \pi r^2 \le 1000 + 5 \Rightarrow \sqrt{\dfrac{995}{\pi}} \le r \le \sqrt{\dfrac{1005}{\pi}} \Rightarrow 17.7966 \le r \le 17.8858$. $\sqrt{\dfrac{1000}{\pi}} - \sqrt{\dfrac{995}{\pi}} \approx 0.04466$ and $\sqrt{\dfrac{1005}{\pi}} - \sqrt{\dfrac{1000}{\pi}} \approx 0.04455$. So if the machinist gets the radius within 0.0445 cm of 17.8412, the area will be within 5 cm² of 1000.

(c) x is the radius, $f(x)$ is the area, a is the target radius given in part (a), L is the target area (1000 cm²), ε is the magnitude of the error tolerance in the area (5 cm²), and δ is the tolerance in the radius given in part (b).

13. (a) $|4x - 8| = 4\,|x - 2| < 0.1 \Leftrightarrow |x - 2| < \dfrac{0.1}{4}$, so $\delta = \dfrac{0.1}{4} = 0.025$.

(b) $|4x - 8| = 4\,|x - 2| < 0.01 \Leftrightarrow |x - 2| < \dfrac{0.01}{4}$, so $\delta = \dfrac{0.01}{4} = 0.0025$.

15. Given $\varepsilon > 0$, we need $\delta > 0$ such that if $0 < |x - 4| < \delta$, then

$|(\tfrac{1}{2}x - 1) - 1| < \varepsilon$. But $|(\tfrac{1}{2}x - 1) - 1| < \varepsilon \Leftrightarrow |\tfrac{1}{2}x - 2| < \varepsilon \Leftrightarrow$

$|\tfrac{1}{2}| |x - 4| < \varepsilon \Leftrightarrow |x - 4| < 2\varepsilon$. So if we choose $\delta = 2\varepsilon$, then

$0 < |x - 4| < \delta \Rightarrow |(\tfrac{1}{2}x - 1) - 1| < \varepsilon$. Thus, $\lim_{x \to 4} (\tfrac{1}{2}x - 1) = 1$

by the definition of a limit.

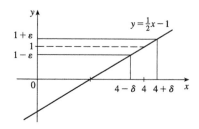

17. Given $\varepsilon > 0$, we need $\delta > 0$ such that if $0 < |x - (-2)| < \delta$, then

$|(-2x + 1) - 5| < \varepsilon$. But $|(-2x + 1) - 5| < \varepsilon \Leftrightarrow$

$|-2x - 4| < \varepsilon \Leftrightarrow |-2| |x - (-2)| < \varepsilon \Leftrightarrow |x - (-2)| < \tfrac{1}{2}\varepsilon$.

So if we choose $\delta = \tfrac{1}{2}\varepsilon$, then $0 < |x - (-2)| < \delta \Rightarrow$

$|(-2x + 1) - 5| < \varepsilon$. Thus, $\lim_{x \to -2} (-2x + 1) = 5$ by the definition of a

limit.

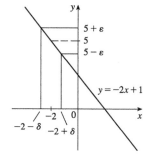

19. Given $\varepsilon > 0$, we need $\delta > 0$ such that if $0 < |x - 9| < \delta$, then $|(1 - \tfrac{1}{3}x) - (-2)| < \varepsilon$. But $|(1 - \tfrac{1}{3}x) - (-2)| < \varepsilon \Leftrightarrow$

$|3 - \tfrac{1}{3}x| < \varepsilon \Leftrightarrow |-\tfrac{1}{3}| |x - 9| < \varepsilon \Leftrightarrow |x - 9| < 3\varepsilon$. So if we choose $\delta = 3\varepsilon$, then $0 < |x - 9| < \delta \Rightarrow$

$|(1 - \tfrac{1}{3}x) - (-2)| < \varepsilon$. Thus, $\lim_{x \to 9} (1 - \tfrac{1}{3}x) = -2$ by the definition of a limit.

21. Given $\varepsilon > 0$, we need $\delta > 0$ such that if $0 < |x - 4| < \delta$, then $\left|\dfrac{x^2 - 2x - 8}{x - 4} - 6\right| < \varepsilon \Leftrightarrow$

$\left|\dfrac{(x - 4)(x + 2)}{x - 4} - 6\right| < \varepsilon \Leftrightarrow |x + 2 - 6| < \varepsilon \quad [x \neq 4] \Leftrightarrow |x - 4| < \varepsilon$. So choose $\delta = \varepsilon$. Then

$0 < |x - 4| < \delta \Rightarrow |x - 4| < \varepsilon \Rightarrow |x + 2 - 6| < \varepsilon \Rightarrow \left|\dfrac{(x - 4)(x + 2)}{x - 4} - 6\right| < \varepsilon \quad [x \neq 4] \Rightarrow$

$\left|\dfrac{x^2 - 2x - 8}{x - 4} - 6\right| < \varepsilon$. By the definition of a limit, $\lim_{x \to 4} \dfrac{x^2 - 2x - 8}{x - 4} = 6$.

23. Given $\varepsilon > 0$, we need $\delta > 0$ such that if $0 < |x - a| < \delta$, then $|x - a| < \varepsilon$. So $\delta = \varepsilon$ will work.

25. Given $\varepsilon > 0$, we need $\delta > 0$ such that if $0 < |x - 0| < \delta$, then $|x^2 - 0| < \varepsilon \Leftrightarrow x^2 < \varepsilon \Leftrightarrow |x| < \sqrt{\varepsilon}$. Take $\delta = \sqrt{\varepsilon}$.

Then $0 < |x - 0| < \delta \Rightarrow |x^2 - 0| < \varepsilon$. Thus, $\lim_{x \to 0} x^2 = 0$ by the definition of a limit.

27. Given $\varepsilon > 0$, we need $\delta > 0$ such that if $0 < |x - 0| < \delta$, then $||x| - 0| < \varepsilon$. But $||x|| = |x|$. So this is true if we pick $\delta = \varepsilon$.

Thus, $\lim_{x \to 0} |x| = 0$ by the definition of a limit.

29. Given $\varepsilon > 0$, we need $\delta > 0$ such that if $0 < |x - 2| < \delta$, then $|(x^2 - 4x + 5) - 1| < \varepsilon \Leftrightarrow |x^2 - 4x + 4| < \varepsilon \Leftrightarrow$

$|(x - 2)^2| < \varepsilon$. So take $\delta = \sqrt{\varepsilon}$. Then $0 < |x - 2| < \delta \Leftrightarrow |x - 2| < \sqrt{\varepsilon} \Leftrightarrow |(x - 2)^2| < \varepsilon$. Thus,

$\lim_{x \to 2} (x^2 - 4x + 5) = 1$ by the definition of a limit.

31. Given $\varepsilon > 0$, we need $\delta > 0$ such that if $0 < |x - (-2)| < \delta$, then $|(x^2 - 1) - 3| < \varepsilon$ or upon simplifying we need $|x^2 - 4| < \varepsilon$ whenever $0 < |x + 2| < \delta$. Notice that if $|x + 2| < 1$, then $-1 < x + 2 < 1 \Rightarrow -5 < x - 2 < -3 \Rightarrow |x - 2| < 5$. So take $\delta = \min\{\varepsilon/5, 1\}$. Then $0 < |x + 2| < \delta \Rightarrow |x - 2| < 5$ and $|x + 2| < \varepsilon/5$, so $|(x^2 - 1) - 3| = |(x + 2)(x - 2)| = |x + 2||x - 2| < (\varepsilon/5)(5) = \varepsilon$. Thus, by the definition of a limit, $\lim_{x \to -2}(x^2 - 1) = 3$.

33. Given $\varepsilon > 0$, we let $\delta = \min\left\{2, \frac{\varepsilon}{8}\right\}$. If $0 < |x - 3| < \delta$, then $|x - 3| < 2 \Rightarrow -2 < x - 3 < 2 \Rightarrow 4 < x + 3 < 8 \Rightarrow |x + 3| < 8$. Also $|x - 3| < \frac{\varepsilon}{8}$, so $|x^2 - 9| = |x + 3||x - 3| < 8 \cdot \frac{\varepsilon}{8} = \varepsilon$. Thus, $\lim_{x \to 3} x^2 = 9$.

35. (a) The points of intersection in the graph are $(x_1, 2.6)$ and $(x_2, 3.4)$ with $x_1 \approx 0.891$ and $x_2 \approx 1.093$. Thus, we can take δ to be the smaller of $1 - x_1$ and $x_2 - 1$. So $\delta = x_2 - 1 \approx 0.093$.

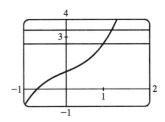

(b) Solving $x^3 + x + 1 = 3 + \varepsilon$ with a CAS gives us two nonreal complex solutions and one real solution, which is
$$x(\varepsilon) = \frac{\left(216 + 108\varepsilon + 12\sqrt{336 + 324\varepsilon + 81\varepsilon^2}\right)^{2/3} - 12}{6\left(216 + 108\varepsilon + 12\sqrt{336 + 324\varepsilon + 81\varepsilon^2}\right)^{1/3}}.$$ Thus, $\delta = x(\varepsilon) - 1$.

(c) If $\varepsilon = 0.4$, then $x(\varepsilon) \approx 1.093\,272\,342$ and $\delta = x(\varepsilon) - 1 \approx 0.093$, which agrees with our answer in part (a).

37. *1. Guessing a value for δ* Given $\varepsilon > 0$, we must find $\delta > 0$ such that $|\sqrt{x} - \sqrt{a}| < \varepsilon$ whenever $0 < |x - a| < \delta$. But $|\sqrt{x} - \sqrt{a}| = \dfrac{|x - a|}{\sqrt{x} + \sqrt{a}} < \varepsilon$ (from the hint). Now if we can find a positive constant C such that $\sqrt{x} + \sqrt{a} > C$ then $\dfrac{|x - a|}{\sqrt{x} + \sqrt{a}} < \dfrac{|x - a|}{C} < \varepsilon$, and we take $|x - a| < C\varepsilon$. We can find this number by restricting x to lie in some interval centered at a. If $|x - a| < \frac{1}{2}a$, then $-\frac{1}{2}a < x - a < \frac{1}{2}a \Rightarrow \frac{1}{2}a < x < \frac{3}{2}a \Rightarrow \sqrt{x} + \sqrt{a} > \sqrt{\frac{1}{2}a} + \sqrt{a}$, and so $C = \sqrt{\frac{1}{2}a} + \sqrt{a}$ is a suitable choice for the constant. So $|x - a| < \left(\sqrt{\frac{1}{2}a} + \sqrt{a}\right)\varepsilon$. This suggests that we let $\delta = \min\left\{\frac{1}{2}a, \left(\sqrt{\frac{1}{2}a} + \sqrt{a}\right)\varepsilon\right\}$.

2. Showing that δ works Given $\varepsilon > 0$, we let $\delta = \min\left\{\frac{1}{2}a, \left(\sqrt{\frac{1}{2}a} + \sqrt{a}\right)\varepsilon\right\}$. If $0 < |x - a| < \delta$, then $|x - a| < \frac{1}{2}a \Rightarrow \sqrt{x} + \sqrt{a} > \sqrt{\frac{1}{2}a} + \sqrt{a}$ (as in part 1). Also $|x - a| < \left(\sqrt{\frac{1}{2}a} + \sqrt{a}\right)\varepsilon$, so
$$|\sqrt{x} - \sqrt{a}| = \frac{|x - a|}{\sqrt{x} + \sqrt{a}} < \frac{\left(\sqrt{a/2} + \sqrt{a}\right)\varepsilon}{\left(\sqrt{a/2} + \sqrt{a}\right)} = \varepsilon.$$ Therefore, $\lim_{x \to a} \sqrt{x} = \sqrt{a}$ by the definition of a limit.

39. Suppose that $\lim_{x \to 0} f(x) = L$. Given $\varepsilon = \frac{1}{2}$, there exists $\delta > 0$ such that $0 < |x| < \delta \Rightarrow |f(x) - L| < \frac{1}{2}$. Take any rational number r with $0 < |r| < \delta$. Then $f(r) = 0$, so $|0 - L| < \frac{1}{2}$, so $L \le |L| < \frac{1}{2}$. Now take any irrational number s with $0 < |s| < \delta$. Then $f(s) = 1$, so $|1 - L| < \frac{1}{2}$. Hence, $1 - L < \frac{1}{2}$, so $L > \frac{1}{2}$. This contradicts $L < \frac{1}{2}$, so $\lim_{x \to 0} f(x)$ does not exist.

41. $\dfrac{1}{(x+3)^4} > 10{,}000 \Leftrightarrow (x+3)^4 < \dfrac{1}{10{,}000} \Leftrightarrow |x+3| < \dfrac{1}{\sqrt[4]{10{,}000}} \Leftrightarrow |x-(-3)| < \dfrac{1}{10}$

43. Let $N < 0$ be given. Then, for $x < -1$, we have $\dfrac{5}{(x+1)^3} < N \Leftrightarrow \dfrac{5}{N} < (x+1)^3 \Leftrightarrow \sqrt[3]{\dfrac{5}{N}} < x+1$.

Let $\delta = -\sqrt[3]{\dfrac{5}{N}}$. Then $-1-\delta < x < -1 \Rightarrow \sqrt[3]{\dfrac{5}{N}} < x+1 < 0 \Rightarrow \dfrac{5}{(x+1)^3} < N$, so $\lim\limits_{x \to -1^-} \dfrac{5}{(x+1)^3} = -\infty$.

1.8 Continuity

1. From Definition 1, $\lim\limits_{x \to 4} f(x) = f(4)$.

3. (a) f is discontinuous at -4 since $f(-4)$ is not defined and at -2, 2, and 4 since the limit does not exist (the left and right limits are not the same).

(b) f is continuous from the left at -2 since $\lim\limits_{x \to -2^-} f(x) = f(-2)$. f is continuous from the right at 2 and 4 since $\lim\limits_{x \to 2^+} f(x) = f(2)$ and $\lim\limits_{x \to 4^+} f(x) = f(4)$. The function is not continuous from either side at -4 since $f(-4)$ is undefined.

5. (a) From the graph we see that $\lim\limits_{x \to a} f(x)$ does not exist at $a = 1$ since the left and right limits are not the same.

(b) f is not continuous at $a = 1$ since $\lim\limits_{x \to 1} f(x)$ does not exist by part (a). Also, f is not continuous at $a = 3$ since $\lim\limits_{x \to 3} f(x) \neq f(3)$.

(c) From the graph we see that $\lim\limits_{x \to 3} f(x) = 3$, but $f(3) = 2$. Since the limit is not equal to $f(3)$, f is not continuous at $a = 3$.

7.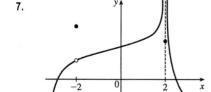

The graph of $y = f(x)$ must have a removable discontinuity (a hole) at $x = -2$ and an infinite discontinuity (a vertical asymptote) at $x = 2$.

9.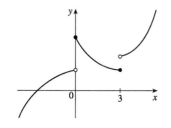

The graph of $y = f(x)$ must have discontinuities at $x = 0$ and $x = 3$. It must show that $\lim\limits_{x \to 0^+} f(x) = f(0)$ and $\lim\limits_{x \to 3^-} f(x) = f(3)$.

11. (a) The toll is $5 except between 7:00 AM and 10:00 AM and between 4:00 PM and 7:00 PM, when the toll is $7.

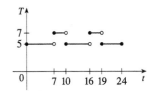

(b) The function T has jump discontinuities at $t = 7, 10, 16$, and 19. Their significance to someone who uses the road is that, because of the sudden jumps in the toll, they may want to avoid the higher rates between $t = 7$ and $t = 10$ and between $t = 16$ and $t = 19$ if feasible.

13. $\lim\limits_{x \to -1} f(x) = \lim\limits_{x \to -1} \left[3x^2 + (x+2)^5 \right] = \lim\limits_{x \to -1} 3x^2 + \lim\limits_{x \to -1} (x+2)^5 = 3 \lim\limits_{x \to -1} x^2 + \lim\limits_{x \to -1} (x+2)^5$

$= 3(-1)^2 + (-1+2)^5 = 4 = f(-1)$

By the definition of continuity, f is continuous at $a = -1$.

15. $\lim\limits_{v \to 1} p(v) = \lim\limits_{v \to 1} 2\sqrt{3v^2 + 1} = 2 \lim\limits_{v \to 1} \sqrt{3v^2 + 1} = 2\sqrt{\lim\limits_{v \to 1} (3v^2 + 1)} = 2\sqrt{3 \lim\limits_{v \to 1} v^2 + \lim\limits_{v \to 1} 1}$

$= 2\sqrt{3(1)^2 + 1} = 2\sqrt{4} = 4 = p(1)$

By the definition of continuity, p is continuous at $a = 1$.

17. For $a > 4$, we have

$\lim\limits_{x \to a} f(x) = \lim\limits_{x \to a} (x + \sqrt{x-4}) = \lim\limits_{x \to a} x + \lim\limits_{x \to a} \sqrt{x-4}$ [Limit Law 1]

$= a + \sqrt{\lim\limits_{x \to a} x - \lim\limits_{x \to a} 4}$ [8, 11, and 2]

$= a + \sqrt{a - 4}$ [8 and 7]

$= f(a)$

So f is continuous at $x = a$ for every a in $(4, \infty)$. Also, $\lim\limits_{x \to 4^+} f(x) = 4 = f(4)$, so f is continuous from the right at 4.

Thus, f is continuous on $[4, \infty)$.

19. $f(x) = \dfrac{1}{x+2}$ is discontinuous at $a = -2$ because $f(-2)$ is undefined.

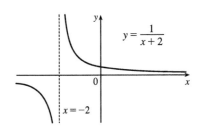

21. $f(x) = \begin{cases} 1 - x^2 & \text{if } x < 1 \\ 1/x & \text{if } x \geq 1 \end{cases}$

The left-hand limit of f at $a = 1$ is

$\lim\limits_{x \to 1^-} f(x) = \lim\limits_{x \to 1^-} (1 - x^2) = 0$. The right-hand limit of f at $a = 1$ is

$\lim\limits_{x \to 1^+} f(x) = \lim\limits_{x \to 1^+} (1/x) = 1$. Since these limits are not equal, $\lim\limits_{x \to 1} f(x)$

does not exist and f is discontinuous at 1.

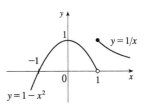

23. $f(x) = \begin{cases} \cos x & \text{if } x < 0 \\ 0 & \text{if } x = 0 \\ 1 - x^2 & \text{if } x > 0 \end{cases}$

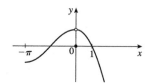

$\lim\limits_{x \to 0} f(x) = 1$, but $f(0) = 0 \neq 1$, so f is discontinuous at 0.

25. (a) $f(x) = \dfrac{x-3}{x^2 - 9} = \dfrac{x-3}{(x-3)(x+3)} = \dfrac{1}{x+3}$ for $x \neq 3$. $f(3)$ is undefined, so f is discontinuous at $x = 3$. Further,

$\lim\limits_{x \to 3} f(x) = \dfrac{1}{3+3} = \dfrac{1}{6}$. Since f is discontinuous at $x = 3$, but $\lim\limits_{x \to 3} f(x)$ exists, f has a removable discontinuity at $x = 3$.

(b) If f is redefined to be $\tfrac{1}{6}$ at $x = 3$, then f will be equivalent to the function $g(x) = \dfrac{1}{x+3}$, which is continuous at $x = 3$.

27. The domain of $f(x) = \dfrac{x^2}{\sqrt{x^4 + 2}}$ is $(-\infty, \infty)$ since the denominator is never 0. By Theorem 5(a), the polynomial x^2 is continuous everywhere. By Theorems 5(a), 7, and 9, $\sqrt{x^4 + 2}$ is continuous everywhere. Finally, by part 5 of Theorem 4, $f(x)$ is continuous everywhere.

29. $h(t) = \dfrac{\cos(t^2)}{1 - t^2} = \dfrac{\cos(t^2)}{(1+t)(1-t)}$. Now t^2 is continuous on $\mathbb{R}$ since it is a polynomial and cosine is continuous everywhere. Thus, $\cos(t^2)$ is continuous on $\mathbb{R}$ by Theorem 9. h is not continuous where its denominator is 0, so h is continuous on $(-\infty, -1) \cup (-1, 1) \cup (1, \infty)$ by part 5 of Theorem 4.

31. $L(v) = v\sqrt{9 - v^2}$ is defined when $9 - v^2 \geq 0 \Leftrightarrow v^2 \leq 9 \Leftrightarrow |v| \leq 3 \Leftrightarrow -3 \leq v \leq 3$. By Theorems 7 and 9, $\sqrt{9 - v^2}$ is continuous on its domain because it is a composite of a root function and a polynomial function. L is the product of a polynomial (continuous everywhere) and $\sqrt{9 - v^2}$, so it is continuous at every number in its domain, $[-3, 3]$, by part 4 of Theorem 4.

33. $M(x) = \sqrt{1 + \dfrac{1}{x}} = \sqrt{\dfrac{x+1}{x}}$ is defined when $\dfrac{x+1}{x} \geq 0 \Rightarrow x+1 \geq 0$ and $x > 0$ or $x+1 \leq 0$ and $x < 0 \Rightarrow x > 0$ or $x \leq -1$, so M has domain $(-\infty, -1] \cup (0, \infty)$. M is the composite of a root function and a rational function, so it is continuous at every number in its domain by Theorems 7 and 9.

35. Because x is continuous on $\mathbb{R}$ and $\sqrt{20 - x^2}$ is continuous on its domain, $-\sqrt{20} \leq x \leq \sqrt{20}$, the product $f(x) = x\sqrt{20 - x^2}$ is continuous on $-\sqrt{20} \leq x \leq \sqrt{20}$. The number 2 is in that domain, so f is continuous at 2, and $\lim\limits_{x \to 2} f(x) = f(2) = 2\sqrt{16} = 8$.

37. The function $f(x) = x^2 \tan x$ is continuous throughout its domain because it is the product of a polynomial and a trigonometric function. The domain of f is the set of all real numbers that are not odd multiples of $\tfrac{\pi}{2}$; that is, domain $f = \{x \mid x \neq n\pi/2,\ n\text{ an odd integer}\}$. Thus, $\frac{\pi}{4}$ is in the domain of f and

$$\lim_{x \to \pi/4} x^2 \tan x = f\left(\frac{\pi}{4}\right) = \left(\frac{\pi}{4}\right)^2 \tan\frac{\pi}{4} = \frac{\pi^2}{16} \cdot 1 = \frac{\pi^2}{16}$$

39. The function $f(x) = \dfrac{1}{\sqrt{1 - \sin x}}$ is discontinuous wherever

$1 - \sin x = 0 \;\Rightarrow\; \sin x = 1 \;\Rightarrow\; x = \frac{\pi}{2} + 2n\pi$, where n is any

integer. The graph shows the discontinuities for $n = -1,\ 0,$ and 1.

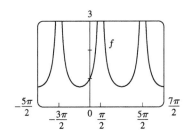

41. $f(x) = \begin{cases} 1 - x^2 & \text{if } x \leq 1 \\ \sqrt{x - 1} & \text{if } x > 1 \end{cases}$

By Theorem 5, since $f(x)$ equals the polynomial $1 - x^2$ on $(-\infty, 1]$, f is continuous on $(-\infty, 1]$. By Theorem 7, since $f(x)$ equals the root function $\sqrt{x - 1}$ on $(1, \infty)$, f is continuous on $(1, \infty)$. At $x = 1$, $\lim\limits_{x \to 1^-} f(x) = \lim\limits_{x \to 1^-} (1 - x^2) = 1 - 1^2 = 0$

and $\lim\limits_{x \to 1^+} f(x) = \lim\limits_{x \to 1^+} \sqrt{x - 1} = \sqrt{1 - 1} = 0$. Thus, $\lim\limits_{x \to 1} f(x)$ exists and equals 0. Also, $f(1) = 1 - 1^2 = 0$. Therefore, f

is continuous at $x = 1$. We conclude that f is continuous on $(-\infty, \infty)$.

43. $f(x) = \begin{cases} x^2 & \text{if } x < -1 \\ x & \text{if } -1 \leq x < 1 \\ 1/x & \text{if } x \geq 1 \end{cases}$

f is continuous on $(-\infty, -1)$, $(-1, 1)$, and $(1, \infty)$, where it is a polynomial,

a polynomial, and a rational function, respectively.

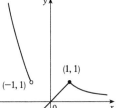

Now $\lim\limits_{x \to -1^-} f(x) = \lim\limits_{x \to -1^-} x^2 = 1$ and $\lim\limits_{x \to -1^+} f(x) = \lim\limits_{x \to -1^+} x = -1$,

so f is discontinuous at -1. Since $f(-1) = -1$, f is continuous from the right at -1. Also, $\lim\limits_{x \to 1^-} f(x) = \lim\limits_{x \to 1^-} x = 1$ and

$\lim\limits_{x \to 1^+} f(x) = \lim\limits_{x \to 1^+} \dfrac{1}{x} = 1 = f(1)$, so f is continuous at 1.

45. $f(x) = \begin{cases} x + 2 & \text{if } x < 0 \\ 2x^2 & \text{if } 0 \leq x \leq 1 \\ 2 - x & \text{if } x > 1 \end{cases}$

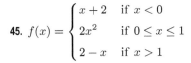

f is continuous on $(-\infty, 0)$, $(0, 1)$, and $(1, \infty)$ since on each of

these intervals it is a polynomial. Now $\lim\limits_{x \to 0^-} f(x) = \lim\limits_{x \to 0^-} (x + 2) = 2$ and

$\lim\limits_{x \to 0^+} f(x) = \lim\limits_{x \to 0^+} 2x^2 = 0$, so f is discontinuous at 0. Since $f(0) = 0$, f is continuous from the right at 0. Also

$\lim\limits_{x \to 1^-} f(x) = \lim\limits_{x \to 1^-} 2x^2 = 2$ and $\lim\limits_{x \to 1^+} f(x) = \lim\limits_{x \to 1^+} (2 - x) = 1$, so f is discontinuous at 1. Since $f(1) = 2$,

f is continuous from the left at 1.

47. $f(x) = \begin{cases} cx^2 + 2x & \text{if } x < 2 \\ x^3 - cx & \text{if } x \geq 2 \end{cases}$

f is continuous on $(-\infty, 2)$ and $(2, \infty)$. Now $\lim_{x \to 2^-} f(x) = \lim_{x \to 2^-} (cx^2 + 2x) = 4c + 4$ and

$\lim_{x \to 2^+} f(x) = \lim_{x \to 2^+} (x^3 - cx) = 8 - 2c$. So f is continuous $\Leftrightarrow$ $4c + 4 = 8 - 2c$ $\Leftrightarrow$ $6c = 4$ $\Leftrightarrow$ $c = \frac{2}{3}$. Thus, for f to be continuous on $(-\infty, \infty)$, $c = \frac{2}{3}$.

49. If f and g are continuous and $g(2) = 6$, then $\lim_{x \to 2}[3f(x) + f(x)g(x)] = 36$ $\Rightarrow$

$3 \lim_{x \to 2} f(x) + \lim_{x \to 2} f(x) \cdot \lim_{x \to 2} g(x) = 36$ $\Rightarrow$ $3f(2) + f(2) \cdot 6 = 36$ $\Rightarrow$ $9f(2) = 36$ $\Rightarrow$ $f(2) = 4$.

51. (a) $f(x) = \dfrac{x^4 - 1}{x - 1} = \dfrac{(x^2 + 1)(x^2 - 1)}{x - 1} = \dfrac{(x^2 + 1)(x + 1)(x - 1)}{x - 1} = (x^2 + 1)(x + 1)$ [or $x^3 + x^2 + x + 1$]

for $x \neq 1$. The discontinuity is removable and $g(x) = x^3 + x^2 + x + 1$ agrees with f for $x \neq 1$ and is continuous on $\mathbb{R}$.

(b) $f(x) = \dfrac{x^3 - x^2 - 2x}{x - 2} = \dfrac{x(x^2 - x - 2)}{x - 2} = \dfrac{x(x - 2)(x + 1)}{x - 2} = x(x + 1)$ [or $x^2 + x$] for $x \neq 2$. The discontinuity is removable and $g(x) = x^2 + x$ agrees with f for $x \neq 2$ and is continuous on $\mathbb{R}$.

(c) $\lim_{x \to \pi^-} f(x) = \lim_{x \to \pi^-} [\![\sin x]\!] = \lim_{x \to \pi^-} 0 = 0$ and $\lim_{x \to \pi^+} f(x) = \lim_{x \to \pi^+} [\![\sin x]\!] = \lim_{x \to \pi^+} (-1) = -1$, so $\lim_{x \to \pi} f(x)$ does not exist. The discontinuity at $x = \pi$ is a jump discontinuity.

53. $f(x) = x^2 + 10 \sin x$ is continuous on the interval $[31, 32]$, $f(31) \approx 957$, and $f(32) \approx 1030$. Since $957 < 1000 < 1030$, there is a number c in $(31, 32)$ such that $f(c) = 1000$ by the Intermediate Value Theorem. *Note:* There is also a number c in $(-32, -31)$ such that $f(c) = 1000$.

55. $f(x) = -x^3 + 4x + 1$ is continuous on the interval $[-1, 0]$, $f(-1) = -2$, and $f(0) = 1$. Since $-2 < 0 < 1$, there is a number c in $(-1, 0)$ such that $f(c) = 0$ by the Intermediate Value Theorem. Thus, there is a solution of the equation $-x^3 + 4x + 1 = 0$ in the interval $(-1, 0)$.

57. $f(x) = \cos x - x$ is continuous on the interval $[0, 1]$, $f(0) = 1$, and $f(1) = \cos 1 - 1 \approx -0.46$. Since $-0.46 < 0 < 1$, there is a number c in $(0, 1)$ such that $f(c) = 0$ by the Intermediate Value Theorem. Thus, there is a solution of the equation $\cos x - x = 0$, or $\cos x = x$, in the interval $(0, 1)$.

59. (a) $f(x) = \cos x - x^3$ is continuous on the interval $[0, 1]$, $f(0) = 1 > 0$, and $f(1) = \cos 1 - 1 \approx -0.46 < 0$. Since $1 > 0 > -0.46$, there is a number c in $(0, 1)$ such that $f(c) = 0$ by the Intermediate Value Theorem. Thus, there is a solution of the equation $\cos x - x^3 = 0$, or $\cos x = x^3$, in the interval $(0, 1)$.

(b) $f(0.86) \approx 0.016 > 0$ and $f(0.87) \approx -0.014 < 0$, so there is a solution between 0.86 and 0.87, that is, in the interval $(0.86, 0.87)$.

61. (a) Let $f(x) = x^5 - x^2 - 4$. Then $f(1) = 1^5 - 1^2 - 4 = -4 < 0$ and $f(2) = 2^5 - 2^2 - 4 = 24 > 0$. So by the Intermediate Value Theorem, there is a number c in $(1, 2)$ such that $f(c) = c^5 - c^2 - 4 = 0$.

(b) We can see from the graphs that, correct to three decimal places, the solution is $x \approx 1.434$.

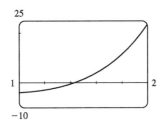
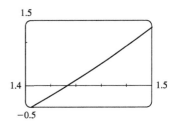

63. Let $f(x) = \sin x^3$. Then f is continuous on $[1, 2]$ since f is the composite of the sine function and the cubing function, both of which are continuous on $\mathbb{R}$. The zeros of the sine are at $n\pi$, so we note that $0 < 1 < \pi < \frac{3}{2}\pi < 2\pi < 8 < 3\pi$, and that the pertinent cube roots are related by $1 < \sqrt[3]{\frac{3}{2}\pi}$ [call this value A] < 2. [By observation, we might notice that $x = \sqrt[3]{\pi}$ and $x = \sqrt[3]{2\pi}$ are zeros of f.]

Now $f(1) = \sin 1 > 0$, $f(A) = \sin \frac{3}{2}\pi = -1 < 0$, and $f(2) = \sin 8 > 0$. Applying the Intermediate Value Theorem on $[1, A]$ and then on $[A, 2]$, we see there are numbers c and d in $(1, A)$ and $(A, 2)$ such that $f(c) = f(d) = 0$. Thus, f has at least two x-intercepts in $(1, 2)$.

65. ($\Rightarrow$) If f is continuous at a, then by Theorem 8 with $g(h) = a + h$, we have
$$\lim_{h \to 0} f(a + h) = f\left(\lim_{h \to 0} (a + h)\right) = f(a).$$

($\Leftarrow$) Let $\varepsilon > 0$. Since $\lim_{h \to 0} f(a + h) = f(a)$, there exists $\delta > 0$ such that $0 < |h| < \delta \Rightarrow$
$|f(a + h) - f(a)| < \varepsilon$. So if $0 < |x - a| < \delta$, then $|f(x) - f(a)| = |f(a + (x - a)) - f(a)| < \varepsilon$.
Thus, $\lim_{x \to a} f(x) = f(a)$ and so f is continuous at a.

67. As in the previous exercise, we must show that $\lim_{h \to 0} \cos(a + h) = \cos a$ to prove that the cosine function is continuous.
$$\lim_{h \to 0} \cos(a + h) = \lim_{h \to 0} (\cos a \cos h - \sin a \sin h) = \lim_{h \to 0} (\cos a \cos h) - \lim_{h \to 0} (\sin a \sin h)$$
$$= \left(\lim_{h \to 0} \cos a\right)\left(\lim_{h \to 0} \cos h\right) - \left(\lim_{h \to 0} \sin a\right)\left(\lim_{h \to 0} \sin h\right) = (\cos a)(1) - (\sin a)(0) = \cos a$$

69. *Proof of Law 6*: Let n be a positive integer. By Theorem 8 with $f(x) = x^n$, we have
$$\lim_{x \to a} [g(x)]^n = \lim_{x \to a} f(g(x)) = f\left(\lim_{x \to a} g(x)\right) = \left[\lim_{x \to a} g(x)\right]^n$$

Proof of Law 7: Let n be a positive integer. By Theorem 8 with $f(x) = \sqrt[n]{x}$, we have
$$\lim_{x \to a} \sqrt[n]{g(x)} = \lim_{x \to a} f(g(x)) = f\left(\lim_{x \to a} g(x)\right) = \sqrt[n]{\lim_{x \to a} g(x)}$$

71. $f(x) = \begin{cases} 0 & \text{if } x \text{ is rational} \\ 1 & \text{if } x \text{ is irrational} \end{cases}$ is continuous nowhere. For, given any number a and any $\delta > 0$, the interval $(a - \delta, a + \delta)$ contains both infinitely many rational and infinitely many irrational numbers. Since $f(a) = 0$ or 1, there are infinitely many numbers x with $0 < |x - a| < \delta$ and $|f(x) - f(a)| = 1$. Thus, $\lim_{x \to a} f(x) \neq f(a)$. [In fact, $\lim_{x \to a} f(x)$ does not even exist.]

73. $f(x) = x^4 \sin(1/x)$ is continuous on $(-\infty, 0) \cup (0, \infty)$ since it is the product of a polynomial and a composite of a trigonometric function and a rational function. Now since $-1 \leq \sin(1/x) \leq 1$, we have $-x^4 \leq x^4 \sin(1/x) \leq x^4$. Because $\lim_{x \to 0}(-x^4) = 0$ and $\lim_{x \to 0} x^4 = 0$, the Squeeze Theorem gives us $\lim_{x \to 0}(x^4 \sin(1/x)) = 0$, which equals $f(0)$. Thus, f is continuous at 0 and, hence, on $(-\infty, \infty)$.

75. Define $u(t)$ to be the monk's distance from the monastery, as a function of time t (in hours), on the first day, and define $d(t)$ to be his distance from the monastery, as a function of time, on the second day. Let D be the distance from the monastery to the top of the mountain. From the given information we know that $u(0) = 0$, $u(12) = D$, $d(0) = D$ and $d(12) = 0$. Now consider the function $u - d$, which is clearly continuous. We calculate that $(u - d)(0) = -D$ and $(u - d)(12) = D$. So by the Intermediate Value Theorem, there must be some time t_0 between 0 and 12 such that $(u - d)(t_0) = 0 \Leftrightarrow u(t_0) = d(t_0)$. So at time t_0 after 7:00 AM, the monk will be at the same place on both days.

1 Review

TRUE-FALSE QUIZ

1. False. Let $f(x) = x^2$, $s = -1$, and $t = 1$. Then $f(s + t) = (-1 + 1)^2 = 0^2 = 0$, but $f(s) + f(t) = (-1)^2 + 1^2 = 2 \neq 0 = f(s + t)$.

3. False. Let $f(x) = x^2$. Then $f(3x) = (3x)^2 = 9x^2$ and $3f(x) = 3x^2$. So $f(3x) \neq 3f(x)$.

5. True. See the Vertical Line Test.

7. False. Limit Law 2 applies only if the individual limits exist (these don't).

9. True. Limit Law 5 applies.

11. True. $\lim_{x \to 3} \dfrac{x^2 - 9}{x - 3} = \lim_{x \to 3} \dfrac{(x + 3)(x - 3)}{(x - 3)} = \lim_{x \to 3}(x + 3)$

13. False. Consider $\lim_{x \to 5} \dfrac{x(x - 5)}{x - 5}$ or $\lim_{x \to 5} \dfrac{\sin(x - 5)}{x - 5}$. The first limit exists and is equal to 5. By Example 1.5.3, we know that the latter limit exists (and it is equal to 1).

15. True. Suppose that $\lim_{x \to a}[f(x) + g(x)]$ exists. Now $\lim_{x \to a} f(x)$ exists and $\lim_{x \to a} g(x)$ does not exist, but $\lim_{x \to a} g(x) = \lim_{x \to a}\{[f(x) + g(x)] - f(x)\} = \lim_{x \to a}[f(x) + g(x)] - \lim_{x \to a} f(x)$ [by Limit Law 2], which exists, and we have a contradiction. Thus, $\lim_{x \to a}[f(x) + g(x)]$ does not exist.

17. True. A polynomial is continuous everywhere, so $\lim_{x \to b} p(x)$ exists and is equal to $p(b)$.

19. False. Consider $f(x) = \begin{cases} 1/(x-1) & \text{if } x \neq 1 \\ 2 & \text{if } x = 1 \end{cases}$

21. True. Use Theorem 1.8.8 with $a = 2$, $b = 5$, and $g(x) = 4x^2 - 11$. Note that $f(4) = 3$ is not needed.

23. True, by the definition of a limit with $\varepsilon = 1$.

25. True. $f(x) = x^{10} - 10x^2 + 5$ is continuous on the interval $[0, 2]$, $f(0) = 5$, $f(1) = -4$, and $f(2) = 989$. Since $-4 < 0 < 5$, there is a number c in $(0, 1)$ such that $f(c) = 0$ by the Intermediate Value Theorem. Thus, there is a solution of the equation $x^{10} - 10x^2 + 5 = 0$ in the interval $(0, 1)$. Similarly, there is a solution in $(1, 2)$.

27. False. See Exercise 1.8.76(c).

EXERCISES

1. (a) When $x = 2$, $y \approx 2.7$. Thus, $f(2) \approx 2.7$. (b) $f(x) = 3 \Rightarrow x \approx 2.3, 5.6$

(c) The domain of f is $-6 \leq x \leq 6$, or $[-6, 6]$. (d) The range of f is $-4 \leq y \leq 4$, or $[-4, 4]$.

(e) f is increasing on $[-4, 4]$, that is, on $-4 \leq x \leq 4$.

(f) f is odd because its graph is symmetric about the origin.

3. $f(x) = x^2 - 2x + 3$, so $f(a+h) = (a+h)^2 - 2(a+h) + 3 = a^2 + 2ah + h^2 - 2a - 2h + 3$, and

$$\frac{f(a+h) - f(a)}{h} = \frac{(a^2 + 2ah + h^2 - 2a - 2h + 3) - (a^2 - 2a + 3)}{h} = \frac{h(2a + h - 2)}{h} = 2a + h - 2.$$

5. $f(x) = 2/(3x - 1)$. Domain: $3x - 1 \neq 0 \Rightarrow 3x \neq 1 \Rightarrow x \neq \frac{1}{3}$. $D = \left(-\infty, \frac{1}{3}\right) \cup \left(\frac{1}{3}, \infty\right)$

Range: all reals except 0 ($y = 0$ is the horizontal asymptote for f.)

$R = (-\infty, 0) \cup (0, \infty)$

7. $y = 1 + \sin x$. Domain: $\mathbb{R}$.

Range: $-1 \leq \sin x \leq 1 \Rightarrow 0 \leq 1 + \sin x \leq 2 \Rightarrow 0 \leq y \leq 2$. $R = [0, 2]$

9. (a) To obtain the graph of $y = f(x) + 5$, we shift the graph of $y = f(x)$ 5 units upward.

(b) To obtain the graph of $y = f(x + 5)$, we shift the graph of $y = f(x)$ 5 units to the left.

(c) To obtain the graph of $y = 1 + 2f(x)$, we stretch the graph of $y = f(x)$ vertically by a factor of 2, and then shift the resulting graph 1 unit upward.

(d) To obtain the graph of $y = f(x - 2) - 2$, we shift the graph of $y = f(x)$ 2 units to the right (for the "-2" inside the parentheses), and then shift the resulting graph 2 units downward.

(e) To obtain the graph of $y = -f(x)$, we reflect the graph of $y = f(x)$ about the x-axis.

(f) To obtain the graph of $y = 3 - f(x)$, we reflect the graph of $y = f(x)$ about the x-axis, and then shift the resulting graph 3 units upward.

11. $f(x) = x^3 + 2$. Start with the graph of $y = x^3$ and shift 2 units upward.

13. $y = \sqrt{x+2}$. Start with the graph of $y = \sqrt{x}$ and shift 2 units to the left.

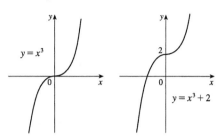

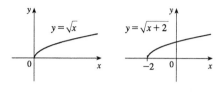

15. $g(x) = 1 + \cos 2x$. Start with the graph of $y = \cos x$, compress horizontally by a factor of 2, and then shift 1 unit upward.

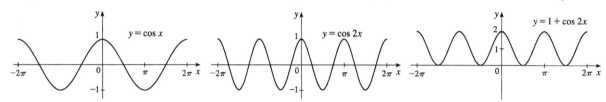

17. (a) $f(x) = 2x^5 - 3x^2 + 2 \Rightarrow f(-x) = 2(-x)^5 - 3(-x)^2 + 2 = -2x^5 - 3x^2 + 2$. Since $f(-x) \neq f(x)$ and $f(-x) \neq -f(x)$, f is neither even nor odd.

(b) $f(x) = x^3 - x^7 \Rightarrow f(-x) = (-x)^3 - (-x)^7 = -x^3 + x^7 = -(x^3 - x^7) = -f(x)$, so f is odd.

(c) $f(x) = 1 - \cos 2x \Rightarrow f(-x) = 1 - \cos[2(-x)] = 1 - \cos(-2x) = 1 - \cos 2x = f(x)$, so f is even.

(d) $f(x) = 1 + \sin x \Rightarrow f(-x) = 1 + \sin(-x) = 1 - \sin x$. Now $f(-x) \neq f(x)$ and $f(-x) \neq -f(x)$, so f is neither even nor odd.

(e) $f(x) = (x+1)^2 = x^2 + 2x + 1$. Now $f(-x) = (-x)^2 + 2(-x) + 1 = x^2 - 2x + 1$. Since $f(-x) \neq f(x)$ and $f(-x) \neq -f(x)$, f is neither even nor odd.

19. $f(x) = \sqrt{x}$, $D = [0, \infty)$; $g(x) = \sin x$, $D = \mathbb{R}$.

(a) $(f \circ g)(x) = f(g(x)) = f(\sin x) = \sqrt{\sin x}$. For $\sqrt{\sin x}$ to be defined, we must have $\sin x \geq 0 \Leftrightarrow x \in [0, \pi], [2\pi, 3\pi], [-2\pi, -\pi], [4\pi, 5\pi], [-4\pi, -3\pi], \ldots$, so $D = \{x \mid x \in [2n\pi, \pi + 2n\pi]$, where n is an integer$\}$.

(b) $(g \circ f)(x) = g(f(x)) = g(\sqrt{x}) = \sin \sqrt{x}$. x must be greater than or equal to 0 for $\sqrt{x}$ to be defined, so $D = [0, \infty)$.

(c) $(f \circ f)(x) = f(f(x)) = f(\sqrt{x}) = \sqrt{\sqrt{x}} = \sqrt[4]{x}$. $D = [0, \infty)$.

(d) $(g \circ g)(x) = g(g(x)) = g(\sin x) = \sin(\sin x)$. $D = \mathbb{R}$.

21.

More than one model appears to be plausible. Your choice of model depends on whether you think medical advances will keep increasing life expectancy, or if there is bound to be a natural leveling-off of life expectancy. A linear model, $y = 0.2441x - 413.3960$, gives us an estimate of 82.1 years for the year 2030.

23. (a) (i) $\lim_{x \to 2^+} f(x) = 3$ (ii) $\lim_{x \to -3^+} f(x) = 0$

(iii) $\lim_{x \to -3} f(x)$ does not exist since the left and right limits are not equal. (The left limit is -2.)

(iv) $\lim_{x \to 4} f(x) = 2$

(v) $\lim_{x \to 0} f(x) = \infty$ (vi) $\lim_{x \to 2^-} f(x) = -\infty$

(b) The equations of the vertical asymptotes are $x = 0$ and $x = 2$.

(c) f is discontinuous at $x = -3, 0, 2,$ and 4. The discontinuities are jump, infinite, infinite, and removable, respectively.

25. Since the cosine function is continuous on $(-\infty, \infty)$, $\lim_{x \to 0} \cos(x^3 + 3x) = \cos(0^3 + 3 \cdot 0) = \cos 0 = 1$.

27. $\lim_{x \to -3} \dfrac{x^2 - 9}{x^2 + 2x - 3} = \lim_{x \to -3} \dfrac{(x+3)(x-3)}{(x+3)(x-1)} = \lim_{x \to -3} \dfrac{x-3}{x-1} = \dfrac{-3-3}{-3-1} = \dfrac{-6}{-4} = \dfrac{3}{2}$

29. $\lim_{h \to 0} \dfrac{(h-1)^3 + 1}{h} = \lim_{h \to 0} \dfrac{(h^3 - 3h^2 + 3h - 1) + 1}{h} = \lim_{h \to 0} \dfrac{h^3 - 3h^2 + 3h}{h} = \lim_{h \to 0} (h^2 - 3h + 3) = 3$

Another solution: Factor the numerator as a sum of two cubes and then simplify.

$\lim_{h \to 0} \dfrac{(h-1)^3 + 1}{h} = \lim_{h \to 0} \dfrac{(h-1)^3 + 1^3}{h} = \lim_{h \to 0} \dfrac{[(h-1)+1]\left[(h-1)^2 - 1(h-1) + 1^2\right]}{h}$
$= \lim_{h \to 0} \left[(h-1)^2 - h + 2\right] = 1 - 0 + 2 = 3$

31. $\lim_{r \to 9} \dfrac{\sqrt{r}}{(r-9)^4} = \infty$ since $(r-9)^4 \to 0^+$ as $r \to 9$ and $\dfrac{\sqrt{r}}{(r-9)^4} > 0$ for $r \neq 9$.

33. $\lim_{r \to -1} \dfrac{r^2 - 3r - 4}{4r^2 + r - 3} = \lim_{r \to -1} \dfrac{(r-4)(r+1)}{(4r-3)(r+1)} = \lim_{r \to -1} \dfrac{r-4}{4r-3} = \dfrac{-1-4}{4(-1)-3} = \dfrac{-5}{-7} = \dfrac{5}{7}$

35. $\lim_{s \to 16} \dfrac{4 - \sqrt{s}}{s - 16} = \lim_{s \to 16} \dfrac{4 - \sqrt{s}}{(\sqrt{s}+4)(\sqrt{s}-4)} = \lim_{s \to 16} \dfrac{-1}{\sqrt{s}+4} = \dfrac{-1}{\sqrt{16}+4} = -\dfrac{1}{8}$

37. $\lim_{x \to 0} \dfrac{1 - \sqrt{1-x^2}}{x} \cdot \dfrac{1 + \sqrt{1-x^2}}{1 + \sqrt{1-x^2}} = \lim_{x \to 0} \dfrac{1 - (1-x^2)}{x\left(1 + \sqrt{1-x^2}\right)} = \lim_{x \to 0} \dfrac{x^2}{x\left(1 + \sqrt{1-x^2}\right)} = \lim_{x \to 0} \dfrac{x}{1 + \sqrt{1-x^2}} = 0$

39. Since $2x - 1 \le f(x) \le x^2$ for $0 < x < 3$ and $\lim_{x \to 1}(2x - 1) = 1 = \lim_{x \to 1} x^2$, we have $\lim_{x \to 1} f(x) = 1$ by the Squeeze Theorem.

41. Given $\varepsilon > 0$, we need $\delta > 0$ such that if $0 < |x - 2| < \delta$, then $|(14 - 5x) - 4| < \varepsilon$. But $|(14 - 5x) - 4| < \varepsilon \Leftrightarrow$
$|-5x + 10| < \varepsilon \Leftrightarrow |-5||x - 2| < \varepsilon \Leftrightarrow |x - 2| < \varepsilon/5$. So if we choose $\delta = \varepsilon/5$, then $0 < |x - 2| < \delta \Rightarrow$
$|(14 - 5x) - 4| < \varepsilon$. Thus, $\lim_{x \to 2}(14 - 5x) = 4$ by the definition of a limit.

43. Given $\varepsilon > 0$, we need $\delta > 0$ so that if $0 < |x - 2| < \delta$, then $\left|x^2 - 3x - (-2)\right| < \varepsilon$. First, note that if $|x - 2| < 1$, then
$-1 < x - 2 < 1$, so $0 < x - 1 < 2 \Rightarrow |x - 1| < 2$. Now let $\delta = \min\{\varepsilon/2, 1\}$. Then $0 < |x - 2| < \delta \Rightarrow$
$\left|x^2 - 3x - (-2)\right| = |(x - 2)(x - 1)| = |x - 2|\,|x - 1| < (\varepsilon/2)(2) = \varepsilon$.

Thus, $\lim_{x \to 2}(x^2 - 3x) = -2$ by the definition of a limit.

45. (a) $f(x) = \sqrt{-x}$ if $x < 0$, $f(x) = 3 - x$ if $0 \leq x < 3$, $f(x) = (x-3)^2$ if $x > 3$.

(i) $\lim\limits_{x \to 0^+} f(x) = \lim\limits_{x \to 0^+} (3 - x) = 3$

(ii) $\lim\limits_{x \to 0^-} f(x) = \lim\limits_{x \to 0^-} \sqrt{-x} = 0$

(iii) Because of (i) and (ii), $\lim\limits_{x \to 0} f(x)$ does not exist.

(iv) $\lim\limits_{x \to 3^-} f(x) = \lim\limits_{x \to 3^-} (3 - x) = 0$

(v) $\lim\limits_{x \to 3^+} f(x) = \lim\limits_{x \to 3^+} (x - 3)^2 = 0$

(vi) Because of (iv) and (v), $\lim\limits_{x \to 3} f(x) = 0$.

(b) f is discontinuous at 0 since $\lim\limits_{x \to 0} f(x)$ does not exist.

f is discontinuous at 3 since $f(3)$ does not exist.

(c)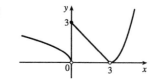

47. x^3 is continuous on $\mathbb{R}$ since it is a polynomial and $\cos x$ is also continuous on $\mathbb{R}$, so the product $x^3 \cos x$ is continuous on $\mathbb{R}$. The root function $\sqrt[4]{x}$ is continuous on its domain, $[0, \infty)$, and so the sum, $h(x) = \sqrt[4]{x} + x^3 \cos x$, is continuous on its domain, $[0, \infty)$.

49. $f(x) = x^5 - x^3 + 3x - 5$ is continuous on the interval $[1, 2]$, $f(1) = -2$, and $f(2) = 25$. Since $-2 < 0 < 25$, there is a number c in $(1, 2)$ such that $f(c) = 0$ by the Intermediate Value Theorem. Thus, there is a solution of the equation $x^5 - x^3 + 3x - 5 = 0$ in the interval $(1, 2)$.

51. $|f(x)| \leq g(x) \iff -g(x) \leq f(x) \leq g(x)$ and $\lim\limits_{x \to a} g(x) = 0 = \lim\limits_{x \to a} -g(x)$.

Thus, by the Squeeze Theorem, $\lim\limits_{x \to a} f(x) = 0$.

PRINCIPLES OF PROBLEM SOLVING

1. $\left|4x - |x+1|\right| = 3 \Rightarrow 4x - |x+1| = -3$ (Equation 1) or $4x - |x+1| = 3$ (Equation 2).

   If $x + 1 < 0$, or $x < -1$, then $|x+1| = -(x+1) = -x-1$. If $x + 1 \geq 0$, or $x \geq -1$, then $|x+1| = x + 1$.

   We thus consider two cases, $x < -1$ (Case 1) and $x \geq -1$ (Case 2), for each of Equations 1 and 2.

   Equation 1, Case 1: $4x - |x+1| = -3 \Rightarrow 4x - (-x-1) = -3 \Rightarrow 5x + 1 = -3 \Rightarrow$
   $5x = -4 \Rightarrow x = -\frac{4}{5}$ which is invalid since $x < -1$.

   Equation 1, Case 2: $4x - |x+1| = -3 \Rightarrow 4x - (x-1) = -3 \Rightarrow 3x - 1 = -3 \Rightarrow$
   $3x = -2 \Rightarrow x = -\frac{2}{3}$, which is valid since $x \geq -1$.

   Equation 2, Case 1: $4x - |x+1| = 3 \Rightarrow 4x - (-x-1) = 3 \Rightarrow 5x + 1 = 3 \Rightarrow$
   $5x = 2 \Rightarrow x = \frac{2}{5}$, which is invalid since $x < -1$.

   Equation 2, Case 2: $4x - |x+1| = 3 \Rightarrow 4x - (x+1) = 3 \Rightarrow 3x - 1 = 3 \Rightarrow$
   $3x = 4 \Rightarrow x = \frac{4}{3}$, which is valid since $x \geq -1$.

   Thus, the solution set is $\left\{-\frac{2}{3}, \frac{4}{3}\right\}$.

3. $f(x) = \left|x^2 - 4|x| + 3\right|$. If $x \geq 0$, then $f(x) = \left|x^2 - 4x + 3\right| = |(x-1)(x-3)|$.

   Case (i): If $0 < x \leq 1$, then $f(x) = x^2 - 4x + 3$.

   Case (ii): If $1 < x \leq 3$, then $f(x) = -(x^2 - 4x + 3) = -x^2 + 4x - 3$.

   Case (iii): If $x > 3$, then $f(x) = x^2 - 4x + 3$.

   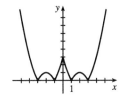

   This enables us to sketch the graph for $x \geq 0$. Then we use the fact that f is an even function to reflect this part of the graph about the y-axis to obtain the entire graph. Or, we could consider also the cases $x < -3$, $-3 \leq x < -1$, and $-1 \leq x < 0$.

5. Remember that $|a| = a$ if $a \geq 0$ and that $|a| = -a$ if $a < 0$. Thus,

   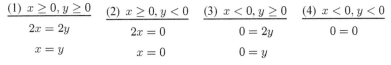

   We will consider the equation $x + |x| = y + |y|$ in four cases.

   (1) $x \geq 0, y \geq 0$ (2) $x \geq 0, y < 0$ (3) $x < 0, y \geq 0$ (4) $x < 0, y < 0$
        $2x = 2y$ $2x = 0$ $0 = 2y$ $0 = 0$
        $x = y$ $x = 0$ $0 = y$

   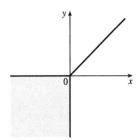

   Case 1 gives us the line $y = x$ with nonnegative x and y.

   Case 2 gives us the portion of the y-axis with y negative.

   Case 3 gives us the portion of the x-axis with x negative.

   Case 4 gives us the entire third quadrant.

7. (a) To sketch the graph of $f(x) = \max\{x, 1/x\}$, we first graph $g(x) = x$ and $h(x) = 1/x$ on the same coordinate axes. Then create the graph of f by plotting the largest y-value of g and h for every value of x.

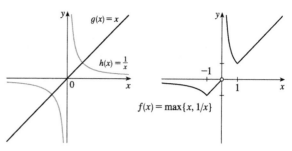

(b)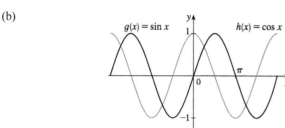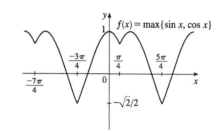

(c)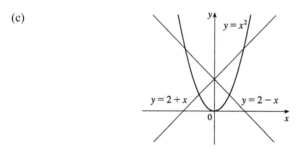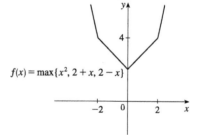

On the TI-84 Plus, max is found under LIST, then under MATH. To graph $f(x) = \max\{x^2, 2+x, 2-x\}$, use $Y = \max(x^2, \max(2+x, 2-x))$.

9. By rearranging terms, we write the given expression as

$$\left(\sin\frac{\pi}{100} + \sin\frac{199\pi}{100}\right) + \left(\sin\frac{2\pi}{100} + \sin\frac{198\pi}{100}\right) + \cdots + \left(\sin\frac{99\pi}{100} + \sin\frac{101\pi}{100}\right) + \sin\frac{100\pi}{100} + \sin\frac{200\pi}{100}$$

Each grouped sum is of the form $\sin x + \sin y$ with $x + y = 2\pi$ so that $\dfrac{x+y}{2} = \dfrac{2\pi}{2} = \pi$. We now derive a useful identity from the product-to-sum identity $\sin x \cos y = \frac{1}{2}[\sin(x+y) + \sin(x-y)]$. If in this identity we replace x with $\dfrac{x+y}{2}$ and y with $\dfrac{x-y}{2}$, we have

$$\sin\left(\frac{x+y}{2}\right)\cos\left(\frac{x-y}{2}\right) = \frac{1}{2}\left[\sin\left(\frac{x+y}{2} + \frac{x-y}{2}\right) + \sin\left(\frac{x+y}{2} - \frac{x-y}{2}\right)\right] = \frac{1}{2}(\sin x + \sin y)$$

Multiplication of the left and right members of this equality by 2 gives the sum-to-product identity

$\sin x + \sin y = 2\sin\left(\dfrac{x+y}{2}\right)\cos\left(\dfrac{x+y}{2}\right)$. Using this sum-to-product identity, we have each grouped sum equal to 0, since

$\sin\left(\dfrac{x+y}{2}\right) = \sin \pi = 0$ is always a factor of the right side. Since $\sin \dfrac{100\pi}{100} = \sin \pi = 0$ and $\sin \dfrac{200\pi}{100} = \sin 2\pi = 0$, the sum of the given expression is 0.

Another approach: Since the sine function is odd, $\sin(-x) = -\sin x$. Because the period of the sine function is 2π, we have $\sin(-x + 2\pi) = -\sin x$. Multiplying each side by -1 and rearranging, we have $\sin x = -\sin(2\pi - x)$. This means that

$\sin \dfrac{\pi}{100} = -\sin\left(2\pi - \dfrac{\pi}{100}\right) = -\sin \dfrac{199\pi}{100}$, $\sin \dfrac{2\pi}{100} = -\sin\left(2\pi - \dfrac{2\pi}{100}\right) = -\sin \dfrac{198\pi}{100}$, and so on, until we have

$\sin \dfrac{99\pi}{100} = -\sin\left(2\pi - \dfrac{99\pi}{100}\right) = -\sin \dfrac{101\pi}{100}$. As before we rearrange terms to write the given expression as

$\left(\sin \dfrac{\pi}{100} + \sin \dfrac{199\pi}{100}\right) + \left(\sin \dfrac{2\pi}{100} + \sin \dfrac{198\pi}{100}\right) + \cdots + \left(\sin \dfrac{99\pi}{100} + \sin \dfrac{101\pi}{100}\right) + \sin \dfrac{100\pi}{100} + \sin \dfrac{200\pi}{100}$

Each sum in parentheses is 0 since the two terms are opposites, and the last two terms again reduce to $\sin \pi$ and $\sin 2\pi$, respectively, each also 0. Thus, the value of the original expression is 0.

11. Let d be the distance traveled on each half of the trip. Let t_1 and t_2 be the times taken for the first and second halves of the trip. For the first half of the trip we have $t_1 = d/30$ and for the second half we have $t_2 = d/60$. Thus, the average speed for the entire trip is $\dfrac{\text{total distance}}{\text{total time}} = \dfrac{2d}{t_1 + t_2} = \dfrac{2d}{\dfrac{d}{30} + \dfrac{d}{60}} \cdot \dfrac{60}{60} = \dfrac{120d}{2d + d} = \dfrac{120d}{3d} = 40$. The average speed for the entire trip is 40 mi/h.

13. Let S_n be the statement that $7^n - 1$ is divisible by 6.

- S_1 is true because $7^1 - 1 = 6$ is divisible by 6.
- Assume S_k is true, that is, $7^k - 1$ is divisible by 6. In other words, $7^k - 1 = 6m$ for some positive integer m. Then $7^{k+1} - 1 = 7^k \cdot 7 - 1 = (6m+1) \cdot 7 - 1 = 42m + 6 = 6(7m+1)$, which is divisible by 6, so S_{k+1} is true.
- Therefore, by mathematical induction, $7^n - 1$ is divisible by 6 for every positive integer n.

15. $f_0(x) = x^2$ and $f_{n+1}(x) = f_0(f_n(x))$ for $n = 0, 1, 2, \ldots$.

$f_1(x) = f_0(f_0(x)) = f_0(x^2) = (x^2)^2 = x^4$, $f_2(x) = f_0(f_1(x)) = f_0(x^4) = (x^4)^2 = x^8$,

$f_3(x) = f_0(f_2(x)) = f_0(x^8) = (x^8)^2 = x^{16}, \ldots$. Thus, a general formula is $f_n(x) = x^{2^{n+1}}$.

17. Let $t = \sqrt[6]{x}$, so $x = t^6$. Then $t \to 1$ as $x \to 1$, so

$\lim\limits_{x \to 1} \dfrac{\sqrt[3]{x} - 1}{\sqrt{x} - 1} = \lim\limits_{t \to 1} \dfrac{t^2 - 1}{t^3 - 1} = \lim\limits_{t \to 1} \dfrac{(t-1)(t+1)}{(t-1)(t^2 + t + 1)} = \lim\limits_{t \to 1} \dfrac{t+1}{t^2 + t + 1} = \dfrac{1+1}{1^2 + 1 + 1} = \dfrac{2}{3}$.

Another method: Multiply both the numerator and the denominator by $(\sqrt{x} + 1)\left(\sqrt[3]{x^2} + \sqrt[3]{x} + 1\right)$.

19. For $-\dfrac{1}{2} < x < \dfrac{1}{2}$, we have $2x - 1 < 0$ and $2x + 1 > 0$, so $|2x - 1| = -(2x - 1)$ and $|2x + 1| = 2x + 1$.

Therefore, $\lim\limits_{x \to 0} \dfrac{|2x-1| - |2x+1|}{x} = \lim\limits_{x \to 0} \dfrac{-(2x-1) - (2x+1)}{x} = \lim\limits_{x \to 0} \dfrac{-4x}{x} = \lim\limits_{x \to 0} (-4) = -4$.

21. (a) For $0 < x < 1$, $[\![x]\!] = 0$, so $\dfrac{[\![x]\!]}{x} = 0$, and $\lim\limits_{x \to 0^+} \dfrac{[\![x]\!]}{x} = 0$. For $-1 < x < 0$, $[\![x]\!] = -1$, so $\dfrac{[\![x]\!]}{x} = \dfrac{-1}{x}$, and

$$\lim_{x \to 0^-} \dfrac{[\![x]\!]}{x} = \lim_{x \to 0^-} \left(\dfrac{-1}{x}\right) = \infty.$$ Since the one-sided limits are not equal, $\lim\limits_{x \to 0} \dfrac{[\![x]\!]}{x}$ does not exist.

(b) For $x > 0$, $1/x - 1 \leq [\![1/x]\!] \leq 1/x \;\Rightarrow\; x(1/x - 1) \leq x[\![1/x]\!] \leq x(1/x) \;\Rightarrow\; 1 - x \leq x[\![1/x]\!] \leq 1$.

As $x \to 0^+$, $1 - x \to 1$, so by the Squeeze Theorem, $\lim\limits_{x \to 0^+} x[\![1/x]\!] = 1$.

For $x < 0$, $1/x - 1 \leq [\![1/x]\!] \leq 1/x \;\Rightarrow\; x(1/x - 1) \geq x[\![1/x]\!] \geq x(1/x) \;\Rightarrow\; 1 - x \geq x[\![1/x]\!] \geq 1$.

As $x \to 0^-$, $1 - x \to 1$, so by the Squeeze Theorem, $\lim\limits_{x \to 0^-} x[\![1/x]\!] = 1$.

Since the one-sided limits are equal, $\lim\limits_{x \to 0} x[\![1/x]\!] = 1$.

23. f is continuous on $(-\infty, a)$ and (a, ∞). To make f continuous on $\mathbb{R}$, we must have continuity at a. Thus,

$$\lim_{x \to a^+} f(x) = \lim_{x \to a^-} f(x) \;\Rightarrow\; \lim_{x \to a^+} x^2 = \lim_{x \to a^-} (x + 1) \;\Rightarrow\; a^2 = a + 1 \;\Rightarrow\; a^2 - a - 1 = 0 \;\Rightarrow$$

[by the quadratic formula] $a = (1 \pm \sqrt{5})/2 \approx 1.618$ or -0.618.

25. $\begin{cases} \lim\limits_{x \to a}[f(x) + g(x)] = 2 \\ \lim\limits_{x \to a}[f(x) - g(x)] = 1 \end{cases} \Rightarrow \begin{cases} \lim\limits_{x \to a} f(x) + \lim\limits_{x \to a} g(x) = 2 \quad (1) \\ \lim\limits_{x \to a} f(x) - \lim\limits_{x \to a} g(x) = 1 \quad (2) \end{cases}$

Adding equations (1) and (2) gives us $2 \lim\limits_{x \to a} f(x) = 3 \;\Rightarrow\; \lim\limits_{x \to a} f(x) = \tfrac{3}{2}$. From equation (1), $\lim\limits_{x \to a} g(x) = \tfrac{1}{2}$. Thus,

$\lim\limits_{x \to a}[f(x)\, g(x)] = \lim\limits_{x \to a} f(x) \cdot \lim\limits_{x \to a} g(x) = \tfrac{3}{2} \cdot \tfrac{1}{2} = \tfrac{3}{4}$.

27. (a) Consider $G(x) = T(x + 180°) - T(x)$. Fix any number a. If $G(a) = 0$, we are done: Temperature at a = Temperature at $a + 180°$. If $G(a) > 0$, then $G(a + 180°) = T(a + 360°) - T(a + 180°) = T(a) - T(a + 180°) = -G(a) < 0$. Also, G is continuous since temperature varies continuously. So, by the Intermediate Value Theorem, G has a zero on the interval $[a, a + 180°]$. If $G(a) < 0$, then a similar argument applies.

(b) Yes. The same argument applies.

(c) The same argument applies for quantities that vary continuously, such as barometric pressure. But one could argue that altitude above sea level is sometimes discontinuous, so the result might not always hold for that quantity.

2 ☐ DERIVATIVES

2.1 Derivatives and Rates of Change

1. (a) This is just the slope of the line through two points: $m_{PQ} = \dfrac{\Delta y}{\Delta x} = \dfrac{f(x) - f(3)}{x - 3}$.

   (b) This is the limit of the slope of the secant line PQ as Q approaches P: $m = \lim\limits_{x \to 3} \dfrac{f(x) - f(3)}{x - 3}$.

3. (a) (i) Using Definition 1 with $f(x) = x^2 + 3x$ and $P(-1, -2)$, the slope of the tangent line is

   $$m = \lim_{x \to a} \frac{f(x) - f(a)}{x - a} = \lim_{x \to -1} \frac{(x^2 + 3x) - (-2)}{x - (-1)} = \lim_{x \to -1} \frac{x^2 + 3x + 2}{x + 1} = \lim_{x \to -1} \frac{(x + 2)(x + 1)}{x + 1}$$

   $$= \lim_{x \to -1} (x + 2) = -1 + 2 = 1$$

   (ii) Using Equation 2 with $f(x) = x^2 + 3x$ and $P(-1, -2)$, the slope of the tangent line is

   $$m = \lim_{h \to 0} \frac{f(a + h) - f(a)}{h} = \lim_{h \to 0} \frac{f(-1 + h) - f(-1)}{h} = \lim_{h \to 0} \frac{[(-1 + h)^2 + 3(-1 + h)] - (-2)}{h}$$

   $$= \lim_{h \to 0} \frac{1 - 2h + h^2 - 3 + 3h + 2}{h} = \lim_{h \to 0} \frac{h^2 + h}{h} = \lim_{h \to 0} \frac{h(h + 1)}{h} = \lim_{h \to 0} (h + 1) = 1$$

   (b) An equation of the tangent line is $y - f(a) = f'(a)(x - a) \Rightarrow y - f(-1) = f'(-1)(x - (-1)) \Rightarrow$
   $y - (-2) = 1(x + 1) \Rightarrow y + 2 = x + 1$, or $y = x - 1$.

   (c) 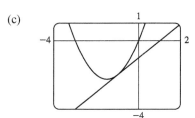 The graph of $y = x - 1$ is tangent to the graph of $y = x^2 + 3x$ at the point $(-1, -2)$. Now zoom in toward the point $(-1, -2)$ until the parabola and the tangent line are indistinguishable.

5. Using (1) with $f(x) = 2x^2 - 5x + 1$ and $P(3, 4)$ [we could also use Equation (2)], the slope of the tangent line is

   $$m = \lim_{x \to a} \frac{f(x) - f(a)}{x - a} = \lim_{x \to 3} \frac{(2x^2 - 5x + 1) - 4}{x - 3} = \lim_{x \to 3} \frac{2x^2 - 5x - 3}{x - 3} = \lim_{x \to 3} \frac{(2x + 1)(x - 3)}{x - 3}$$

   $$= \lim_{x \to 3} (2x + 1) = 2(3) + 1 = 7$$

   Tangent line: $y - 4 = 7(x - 3) \Leftrightarrow y - 4 = 7x - 21 \Leftrightarrow y = 7x - 17$

7. Using (1) with $f(x) = \dfrac{x + 2}{x - 3}$ and $P(2, -4)$, the slope of the tangent line is

   $$m = \lim_{x \to a} \frac{f(x) - f(a)}{x - a} = \lim_{x \to 2} \frac{\dfrac{x + 2}{x - 3} - (-4)}{x - 2} = \lim_{x \to 2} \frac{\dfrac{x + 2 + 4(x - 3)}{x - 3}}{x - 2} = \lim_{x \to 2} \frac{5x - 10}{(x - 2)(x - 3)}$$

   $$= \lim_{x \to 2} \frac{5(x - 2)}{(x - 2)(x - 3)} = \lim_{x \to 2} \frac{5}{x - 3} = \frac{5}{2 - 3} = -5$$

   Tangent line: $y - (-4) = -5(x - 2) \Leftrightarrow y + 4 = -5x + 10 \Leftrightarrow y = -5x + 6$

9. (a) Using (2) with $y = f(x) = 3 + 4x^2 - 2x^3$, the slope of the tangent line is

$$m = \lim_{h \to 0} \frac{f(a+h) - f(a)}{h} = \lim_{h \to 0} \frac{3 + 4(a+h)^2 - 2(a+h)^3 - (3 + 4a^2 - 2a^3)}{h}$$

$$= \lim_{h \to 0} \frac{3 + 4(a^2 + 2ah + h^2) - 2(a^3 + 3a^2h + 3ah^2 + h^3) - 3 - 4a^2 + 2a^3}{h}$$

$$= \lim_{h \to 0} \frac{3 + 4a^2 + 8ah + 4h^2 - 2a^3 - 6a^2h - 6ah^2 - 2h^3 - 3 - 4a^2 + 2a^3}{h}$$

$$= \lim_{h \to 0} \frac{8ah + 4h^2 - 6a^2h - 6ah^2 - 2h^3}{h} = \lim_{h \to 0} \frac{h(8a + 4h - 6a^2 - 6ah - 2h^2)}{h}$$

$$= \lim_{h \to 0} (8a + 4h - 6a^2 - 6ah - 2h^2) = 8a - 6a^2$$

(b) At $(1, 5)$: $m = 8(1) - 6(1)^2 = 2$, so an equation of the tangent line is $y - 5 = 2(x - 1) \Leftrightarrow y = 2x + 3$.

At $(2, 3)$: $m = 8(2) - 6(2)^2 = -8$, so an equation of the tangent line is $y - 3 = -8(x - 2) \Leftrightarrow y = -8x + 19$.

(c)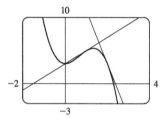

11. (a) We have $d(t) = 16t^2$. The diver will hit the water when $d(t) = 100 \Leftrightarrow 16t^2 = 100 \Leftrightarrow t^2 = \frac{25}{4} \Leftrightarrow t = \frac{5}{2}$ $(t > 0)$. The diver will hit the water after 2.5 seconds.

(b) By Definition 3, the instantaneous velocity of an object with position function $d(t)$ at time $t = 2.5$ is

$$v(2.5) = \lim_{h \to 0} \frac{d(2.5 + h) - d(2.5)}{h} = \lim_{h \to 0} \frac{16(2.5 + h)^2 - 100}{h} = \lim_{h \to 0} \frac{100 + 80h + 16h^2 - 100}{h}$$

$$= \lim_{h \to 0} \frac{80h + 16h^2}{h} = \lim_{h \to 0} \frac{h(80 + 16h)}{h} = \lim_{h \to 0} (80 + 16h) = 80$$

The diver will hit the water with a velocity of 80 ft/s.

13. $v(a) = \lim_{h \to 0} \frac{s(a+h) - s(a)}{h} = \lim_{h \to 0} \frac{\frac{1}{(a+h)^2} - \frac{1}{a^2}}{h} = \lim_{h \to 0} \frac{\frac{a^2 - (a+h)^2}{a^2(a+h)^2}}{h}$

$$= \lim_{h \to 0} \frac{a^2 - (a^2 + 2ah + h^2)}{ha^2(a+h)^2} = \lim_{h \to 0} \frac{-(2ah + h^2)}{ha^2(a+h)^2}$$

$$= \lim_{h \to 0} \frac{-h(2a + h)}{ha^2(a+h)^2} = \lim_{h \to 0} \frac{-(2a + h)}{a^2(a+h)^2} = \frac{-2a}{a^2 \cdot a^2} = \frac{-2}{a^3} \text{ m/s}$$

So $v(1) = \frac{-2}{1^3} = -2$ m/s, $v(2) = \frac{-2}{2^3} = -\frac{1}{4}$ m/s, and $v(3) = \frac{-2}{3^3} = -\frac{2}{27}$ m/s.

15. (a) The particle is moving to the right when s is increasing; that is, on the intervals $(0, 1)$ and $(4, 6)$. The particle is moving to the left when s is decreasing; that is, on the interval $(2, 3)$. The particle is standing still when s is constant; that is, on the intervals $(1, 2)$ and $(3, 4)$.

(b) The velocity of the particle is equal to the slope of the tangent line of the graph. Note that there is no slope at the corner points on the graph. On the interval $(0, 1)$, the slope is $\dfrac{3-0}{1-0} = 3$. On the interval $(2, 3)$, the slope is $\dfrac{1-3}{3-2} = -2$. On the interval $(4, 6)$, the slope is $\dfrac{3-1}{6-4} = 1$.

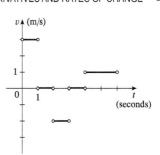

17. $g'(0)$ is the only negative value. The slope at $x = 4$ is smaller than the slope at $x = 2$ and both are smaller than the slope at $x = -2$. Thus, $g'(0) < 0 < g'(4) < g'(2) < g'(-2)$.

19. Using Definition 4 with $f(x) = \sqrt{4x+1}$ and $a = 6$,

$$f'(6) = \lim_{h \to 0} \frac{f(6+h) - f(6)}{h} = \lim_{h \to 0} \frac{\sqrt{4(6+h)+1} - 5}{h} = \lim_{h \to 0} \frac{\sqrt{25+4h} - 5}{h}$$

$$= \lim_{h \to 0} \frac{(\sqrt{25+4h} - 5)(\sqrt{25+4h} + 5)}{h(\sqrt{25+4h} + 5)} = \lim_{h \to 0} \frac{(25+4h) - 25}{h(\sqrt{25+4h} + 5)} = \lim_{h \to 0} \frac{4h}{h(\sqrt{25+4h} + 5)}$$

$$= \lim_{h \to 0} \frac{4}{\sqrt{25+4h} + 5} = \frac{4}{5+5} = \frac{2}{5}$$

21. Using Equation 5 with $f(x) = \dfrac{x^2}{x+6}$ and $a = 3$,

$$f'(3) = \lim_{x \to 3} \frac{f(x) - f(3)}{x - 3} = \lim_{x \to 3} \frac{\dfrac{x^2}{x+6} - 1}{x - 3} = \lim_{x \to 3} \frac{\dfrac{x^2 - (x+6)}{x+6}}{x - 3} = \lim_{x \to 3} \frac{x^2 - x - 6}{(x+6)(x-3)}$$

$$= \lim_{x \to 3} \frac{(x+2)(x-3)}{(x+6)(x-3)} = \lim_{x \to 3} \frac{x+2}{x+6} = \frac{3+2}{3+6} = \frac{5}{9}$$

23. Using Definition 4 with $f(x) = 2x^2 - 5x + 3$,

$$f'(a) = \lim_{h \to 0} \frac{f(a+h) - f(a)}{h} = \lim_{h \to 0} \frac{[2(a+h)^2 - 5(a+h) + 3] - (2a^2 - 5a + 3)}{h}$$

$$= \lim_{h \to 0} \frac{2a^2 + 4ah + 2h^2 - 5a - 5h + 3 - 2a^2 + 5a - 3}{h} = \lim_{h \to 0} \frac{4ah + 2h^2 - 5h}{h}$$

$$= \lim_{h \to 0} \frac{h(4a + 2h - 5)}{h} = \lim_{h \to 0} (4a + 2h - 5) = 4a - 5$$

25. Using Definition 4 with $f(t) = \dfrac{1}{t^2+1}$.

$$f'(a) = \lim_{h \to 0} \frac{f(a+h) - f(a)}{h} = \lim_{h \to 0} \frac{\dfrac{1}{(a+h)^2 + 1} - \dfrac{1}{a^2 + 1}}{h} = \lim_{h \to 0} \frac{\dfrac{(a^2+1) - [(a+h)^2 + 1]}{[(a+h)^2 + 1](a^2 + 1)}}{h}$$

$$= \lim_{h \to 0} \frac{(a^2 + 1) - (a^2 + 2ah + h^2 + 1)}{h[(a+h)^2 + 1](a^2 + 1)} = \lim_{h \to 0} \frac{-(2ah + h^2)}{h[(a+h)^2 + 1](a^2 + 1)} = \lim_{h \to 0} \frac{-h(2a + h)}{h[(a+h)^2 + 1](a^2 + 1)}$$

$$= \lim_{h \to 0} \frac{-(2a + h)}{[(a+h)^2 + 1](a^2 + 1)} = \frac{-2a}{(a^2 + 1)(a^2 + 1)} = -\frac{2a}{(a^2 + 1)^2}$$

27. Since $B(6) = 0$, the point $(6, 0)$ is on the graph of B. Since $B'(6) = -\frac{1}{2}$, the slope of the tangent line at $x = 6$ is $-\frac{1}{2}$. Using the point-slope form of a line gives us $y - 0 = -\frac{1}{2}(x - 6)$, or $y = -\frac{1}{2}x + 3$.

29. Using Definition 4 with $f(x) = 3x^2 - x^3$ and $a = 1$,

$$f'(1) = \lim_{h \to 0} \frac{f(1+h) - f(1)}{h} = \lim_{h \to 0} \frac{[3(1+h)^2 - (1+h)^3] - 2}{h}$$

$$= \lim_{h \to 0} \frac{(3 + 6h + 3h^2) - (1 + 3h + 3h^2 + h^3) - 2}{h} = \lim_{h \to 0} \frac{3h - h^3}{h} = \lim_{h \to 0} \frac{h(3 - h^2)}{h}$$

$$= \lim_{h \to 0} (3 - h^2) = 3 - 0 = 3$$

Tangent line: $y - 2 = 3(x - 1) \Leftrightarrow y - 2 = 3x - 3 \Leftrightarrow y = 3x - 1$

31. (a) Using Definition 4 with $F(x) = 5x/(1 + x^2)$ and the point $(2, 2)$, we have   (b)

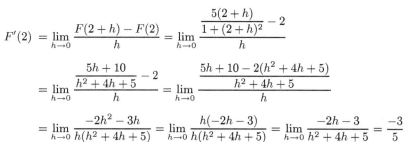

So an equation of the tangent line at $(2, 2)$ is $y - 2 = -\frac{3}{5}(x - 2)$ or $y = -\frac{3}{5}x + \frac{16}{5}$.

33. For the tangent line $y = 4x - 5$: when $x = 2$, $y = 4(2) - 5 = 3$ and its slope is 4 (the coefficient of x). At the point of tangency, these values are shared with the curve $y = f(x)$; that is, $f(2) = 3$ and $f'(2) = 4$.

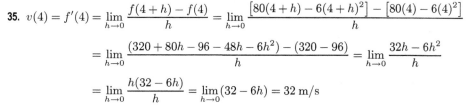

The speed when $t = 4$ is $|32| = 32$ m/s.

37. The sketch shows the graph for a room temperature of $72\,°F$ and a refrigerator temperature of $38\,°F$. The initial rate of change is greater in magnitude than the rate of change after an hour.

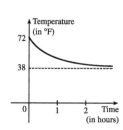

39. We begin by drawing a curve through the origin with a slope of 3 to satisfy $f(0) = 0$ and $f'(0) = 3$. Since $f'(1) = 0$, we will round off our figure so that there is a horizontal tangent directly over $x = 1$. Last, we make sure that the curve has a slope of -1 as we pass over $x = 2$. Two of the many possibilities are shown.

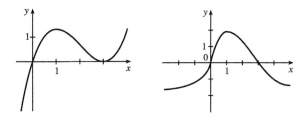

41. We begin by drawing a curve through $(0, 1)$ with a slope of 1 to satisfy $g(0) = 1$ and $g'(0) = 1$. We round off our figure at $x = -2$ to satisfy $g'(-2) = 0$. As $x \to -5^+$, $y \to \infty$, so we draw a vertical asymptote at $x = -5$. As $x \to 5^-$, $y \to 3$, so we draw a dot at $(5, 3)$ [the dot could be open or closed].

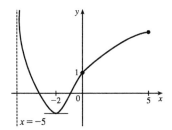

43. By Definition 4, $\lim\limits_{h \to 0} \dfrac{\sqrt{9+h} - 3}{h} = f'(9)$, where $f(x) = \sqrt{x}$ and $a = 9$.

45. By Equation 5, $\lim\limits_{x \to 2} \dfrac{x^6 - 64}{x - 2} = f'(2)$, where $f(x) = x^6$ and $a = 2$.

47. By Definition 4, $\lim\limits_{h \to 0} \dfrac{\tan\left(\frac{\pi}{4} + h\right) - 1}{h} = f'\left(\dfrac{\pi}{4}\right)$, where $f(x) = \tan x$ and $a = \dfrac{\pi}{4}$.

49. (a) (i) $\dfrac{\Delta C}{\Delta x} = \dfrac{C(105) - C(100)}{105 - 100} = \dfrac{6601.25 - 6500}{5} = \$20.25/\text{unit}$.

(ii) $\dfrac{\Delta C}{\Delta x} = \dfrac{C(101) - C(100)}{101 - 100} = \dfrac{6520.05 - 6500}{1} = \$20.05/\text{unit}$.

(b) $\dfrac{C(100+h) - C(100)}{h} = \dfrac{\left[5000 + 10(100+h) + 0.05(100+h)^2\right] - 6500}{h} = \dfrac{20h + 0.05h^2}{h}$
$= 20 + 0.05h$, $h \neq 0$

So the instantaneous rate of change is $\lim\limits_{h \to 0} \dfrac{C(100+h) - C(100)}{h} = \lim\limits_{h \to 0}(20 + 0.05h) = \$20/\text{unit}$.

51. (a) $f'(x)$ is the rate of change of the production cost with respect to the number of ounces of gold produced. Its units are dollars per ounce.

(b) After 800 ounces of gold have been produced, the rate at which the production cost is increasing is $17/ounce. So the cost of producing the 800th (or 801st) ounce is about $17.

(c) In the short term, the values of $f'(x)$ will decrease because more efficient use is made of start-up costs as x increases. But eventually $f'(x)$ might increase due to large-scale operations.

53. (a) $S'(T)$ is the rate at which the oxygen solubility changes with respect to the water temperature. Its units are $(\text{mg/L})/°\text{C}$.

(b) For $T = 16°\text{C}$, it appears that the tangent line to the curve goes through the points $(0, 14)$ and $(32, 6)$. So
$S'(16) \approx \dfrac{6 - 14}{32 - 0} = -\dfrac{8}{32} = -0.25\ (\text{mg/L})/°\text{C}$. This means that as the temperature increases past $16°\text{C}$, the oxygen solubility is decreasing at a rate of $0.25\ (\text{mg/L})/°\text{C}$.

55. (a) (i) $[1.0, 2.0]$: $\dfrac{C(2) - C(1)}{2 - 1} = \dfrac{0.018 - 0.033}{1} = -0.015\ \dfrac{\text{g/dL}}{\text{h}}$

(ii) $[1.5, 2.0]$: $\dfrac{C(2) - C(1.5)}{2 - 1.5} = \dfrac{0.018 - 0.024}{0.5} = \dfrac{-0.006}{0.5} = -0.012\ \dfrac{\text{g/dL}}{\text{h}}$

(iii) $[2.0, 2.5]$: $\dfrac{C(2.5) - C(2)}{2.5 - 2} = \dfrac{0.012 - 0.018}{0.5} = \dfrac{-0.006}{0.5} = -0.012\ \dfrac{\text{g/dL}}{\text{h}}$

(iv) $[2.0, 3.0]$: $\dfrac{C(3) - C(2)}{3 - 2} = \dfrac{0.007 - 0.018}{1} = -0.011\ \dfrac{\text{g/dL}}{\text{h}}$

(b) We estimate the instantaneous rate of change at $t = 2$ by averaging the average rates of change for $[1.5, 2.0]$ and $[2.0, 2.5]$:
$\dfrac{-0.012 + (-0.012)}{2} = -0.012\ \dfrac{\text{g/dL}}{\text{h}}$. After two hours, the BAC is decreasing at a rate of $0.012\ \dfrac{\text{g/dL}}{\text{h}}$.

57. Since $f(x) = x\sin(1/x)$ when $x \neq 0$ and $f(0) = 0$, we have
$f'(0) = \lim\limits_{h \to 0} \dfrac{f(0 + h) - f(0)}{h} = \lim\limits_{h \to 0} \dfrac{h\sin(1/h) - 0}{h} = \lim\limits_{h \to 0} \sin(1/h)$. This limit does not exist since $\sin(1/h)$ takes the values -1 and 1 on any interval containing 0. (Compare with Example 1.5.5.)

59. (a) The slope at the origin appears to be 1.

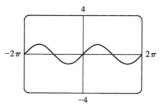

(b) The slope at the origin still appears to be 1.

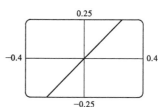

(c) Yes, the slope at the origin now appears to be 0.

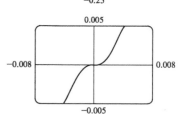

2.2 The Derivative as a Function

1. We estimate the slopes of tangent lines on the graph of f to determine the derivative approximations that follow. Your answers may vary depending on your estimates.

   (a) $f'(0) = \frac{1}{2}$

   (b) $f'(1) \approx 0$

   (c) $f'(2) \approx -1$

   (d) $f'(3) \approx -\frac{3}{2}$

   (e) $f'(4) \approx -1$

   (f) $f'(5) \approx 0$

   (g) $f'(6) \approx 1$

   (h) $f'(7) \approx 1$

3. (a)$' =$ II, since from left to right, the slopes of the tangents to graph (a) start out negative, become 0, then positive, then 0, then negative again. The actual function values in graph II follow the same pattern.

   (b)$' =$ IV, since from left to right, the slopes of the tangents to graph (b) start out at a fixed positive quantity, then suddenly become negative, then positive again. The discontinuities in graph IV indicate sudden changes in the slopes of the tangents.

   (c)$' =$ I, since the slopes of the tangents to graph (c) are negative for $x < 0$ and positive for $x > 0$, as are the function values of graph I.

   (d)$' =$ III, since from left to right, the slopes of the tangents to graph (d) are positive, then 0, then negative, then 0, then positive, then 0, then negative again, and the function values in graph III follow the same pattern.

Hints for Exercises 4–11: First plot x-intercepts on the graph of f' for any horizontal tangents on the graph of f. Look for any corners on the graph of f—there will be a discontinuity on the graph of f'. On any interval where f has a tangent with positive (or negative) slope, the graph of f' will be positive (or negative). If the graph of the function is linear, the graph of f' will be a horizontal line.

5.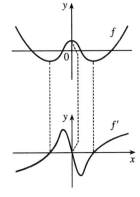

7.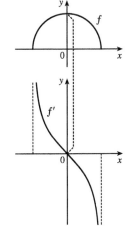

9.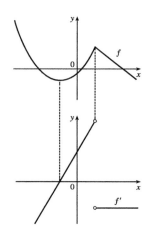

11.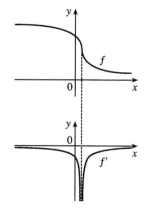

13. (a) $C'(t)$ is the instantaneous rate of change of percentage of full capacity with respect to elapsed time in hours.

(b) The graph of $C'(t)$ tells us that the rate of change of percentage of full capacity is decreasing and approaching 0.

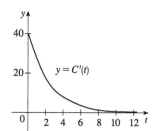

15. It appears that there are horizontal tangents on the graph of f for $t = 2$ and for $t \approx 7.5$. Thus, there are zeros for those values of t on the graph of f'. The derivative is negative for values of t between 0 and 2 and for values of t between approximately 7.5 and 12. The value of $f'(t)$ appears to be largest at $t \approx 5.25$.

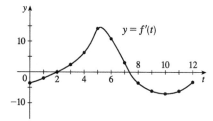

17. (a) By zooming in, we estimate that $f'(0) = 0$, $f'\left(\frac{1}{2}\right) = 1$, $f'(1) = 2$, and $f'(2) = 4$.

(b) By symmetry, $f'(-x) = -f'(x)$. So $f'\left(-\frac{1}{2}\right) = -1$, $f'(-1) = -2$, and $f'(-2) = -4$.

(c) It appears that $f'(x)$ is twice the value of x, so we guess that $f'(x) = 2x$.

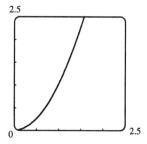

(d) $f'(x) = \lim\limits_{h \to 0} \dfrac{f(x+h) - f(x)}{h} = \lim\limits_{h \to 0} \dfrac{(x+h)^2 - x^2}{h}$

$= \lim\limits_{h \to 0} \dfrac{(x^2 + 2hx + h^2) - x^2}{h} = \lim\limits_{h \to 0} \dfrac{2hx + h^2}{h} = \lim\limits_{h \to 0} \dfrac{h(2x + h)}{h} = \lim\limits_{h \to 0} (2x + h) = 2x$

19. $f'(x) = \lim\limits_{h \to 0} \dfrac{f(x+h) - f(x)}{h} = \lim\limits_{h \to 0} \dfrac{[3(x+h) - 8] - (3x - 8)}{h} = \lim\limits_{h \to 0} \dfrac{3x + 3h - 8 - 3x + 8}{h}$

$= \lim\limits_{h \to 0} \dfrac{3h}{h} = \lim\limits_{h \to 0} 3 = 3$

Domain of f = domain of $f' = \mathbb{R}$.

21. $f'(t) = \lim\limits_{h \to 0} \dfrac{f(t+h) - f(t)}{h} = \lim\limits_{h \to 0} \dfrac{\left[2.5(t+h)^2 + 6(t+h)\right] - \left(2.5t^2 + 6t\right)}{h}$

$= \lim\limits_{h \to 0} \dfrac{2.5(t^2 + 2th + h^2) + 6t + 6h - 2.5t^2 - 6t}{h} = \lim\limits_{h \to 0} \dfrac{2.5t^2 + 5th + 2.5h^2 + 6h - 2.5t^2}{h}$

$= \lim\limits_{h \to 0} \dfrac{5th + 2.5h^2 + 6h}{h} = \lim\limits_{h \to 0} \dfrac{h(5t + 2.5h + 6)}{h} = \lim\limits_{h \to 0} (5t + 2.5h + 6)$

$= 5t + 6$

Domain of f = domain of f' = $\mathbb{R}$.

23. $A'(p) = \lim\limits_{h \to 0} \dfrac{A(p+h) - A(p)}{h} = \lim\limits_{h \to 0} \dfrac{\left[4(p+h)^3 + 3(p+h)\right] - (4p^3 + 3p)}{h}$

$= \lim\limits_{h \to 0} \dfrac{4p^3 + 12p^2h + 12ph^2 + 4h^3 + 3p + 3h - 4p^3 - 3p}{h} = \lim\limits_{h \to 0} \dfrac{12p^2h + 12ph^2 + 4h^3 + 3h}{h}$

$= \lim\limits_{h \to 0} \dfrac{h(12p^2 + 12ph + 4h^2 + 3)}{h} = \lim\limits_{h \to 0} (12p^2 + 12ph + 4h^2 + 3) = 12p^2 + 3$

Domain of A = Domain of A' = $\mathbb{R}$.

25. $f'(x) = \lim\limits_{h \to 0} \dfrac{f(x+h) - f(x)}{h} = \lim\limits_{h \to 0} \dfrac{\dfrac{1}{(x+h)^2 - 4} - \dfrac{1}{x^2 - 4}}{h} = \lim\limits_{h \to 0} \dfrac{\dfrac{(x^2 - 4) - \left[(x+h)^2 - 4\right]}{\left[(x+h)^2 - 4\right](x^2 - 4)}}{h}$

$= \lim\limits_{h \to 0} \dfrac{(x^2 - 4) - (x^2 + 2xh + h^2 - 4)}{h\left[(x+h)^2 - 4\right](x^2 - 4)} = \lim\limits_{h \to 0} \dfrac{x^2 - 4 - x^2 - 2xh - h^2 + 4}{h[(x+h)^2 - 4](x^2 - 4)} = \lim\limits_{h \to 0} \dfrac{-2xh - h^2}{h[(x+h)^2 - 4](x^2 - 4)}$

$= \lim\limits_{h \to 0} \dfrac{h(-2x - h)}{h[(x+h)^2 - 4](x^2 - 4)} = \lim\limits_{h \to 0} \dfrac{-2x - h}{[(x+h)^2 - 4](x^2 - 4)} = \dfrac{-2x}{(x^2 - 4)(x^2 - 4)} = -\dfrac{2x}{(x^2 - 4)^2}$

Domain of f = Domain of f' = $(-\infty, -2) \cup (-2, 2) \cup (2, \infty)$.

27. $g'(u) = \lim\limits_{h \to 0} \dfrac{g(u+h) - g(u)}{h} = \lim\limits_{h \to 0} \dfrac{\dfrac{(u+h)+1}{4(u+h)-1} - \dfrac{u+1}{4u-1}}{h} = \lim\limits_{h \to 0} \dfrac{\dfrac{[(u+h)+1](4u-1) - (u+1)[4(u+h)-1]}{[4(u+h)-1](4u-1)}}{h}$

$= \lim\limits_{h \to 0} \dfrac{\dfrac{(u+h+1)(4u-1) - (u+1)(4u+4h-1)}{[4(u+h)-1](4u-1)}}{h}$

$= \lim\limits_{h \to 0} \dfrac{4u^2 + 4uh + 4u - u - h - 1 - 4u^2 - 4uh + u - 4u - 4h + 1}{h[4(u+h) - 1](4u-1)}$

$= \lim\limits_{h \to 0} \dfrac{-5h}{h[4(u+h)-1](4u-1)} = \lim\limits_{h \to 0} \dfrac{-5}{[4(u+h)-1](4u-1)} = \dfrac{-5}{(4u-1)(4u-1)} = -\dfrac{5}{(4u-1)^2}$

Domain of g = Domain of g' = $\left(-\infty, \tfrac{1}{4}\right) \cup \left(\tfrac{1}{4}, \infty\right)$.

29. $f'(x) = \lim\limits_{h \to 0} \dfrac{f(x+h) - f(x)}{h} = \lim\limits_{h \to 0} \dfrac{\dfrac{1}{\sqrt{1+(x+h)}} - \dfrac{1}{\sqrt{1+x}}}{h}$

$= \lim\limits_{h \to 0} \dfrac{\dfrac{1}{\sqrt{1+(x+h)}} - \dfrac{1}{\sqrt{1+x}}}{h} \cdot \dfrac{\sqrt{1+(x+h)}\,\sqrt{1+x}}{\sqrt{1+(x+h)}\,\sqrt{1+x}}$

$= \lim\limits_{h \to 0} \dfrac{\sqrt{1+x} - \sqrt{1+(x+h)}}{h\sqrt{1+(x+h)}\,\sqrt{1+x}} \cdot \dfrac{\sqrt{1+x} + \sqrt{1+(x+h)}}{\sqrt{1+x} + \sqrt{1+(x+h)}}$

$= \lim\limits_{h \to 0} \dfrac{(1+x) - [1+(x+h)]}{h\sqrt{1+(x+h)}\,\sqrt{1+x}\left(\sqrt{1+x} + \sqrt{1+(x+h)}\right)}$

$= \lim\limits_{h \to 0} \dfrac{-h}{h\sqrt{1+x+h}\,\sqrt{1+x}\left(\sqrt{1+x} + \sqrt{1+x+h}\right)} = \lim\limits_{h \to 0} \dfrac{-1}{\sqrt{1+x+h}\,\sqrt{1+x}\left(\sqrt{1+x} + \sqrt{1+x+h}\right)}$

$= \dfrac{-1}{\sqrt{1+x}\,\sqrt{1+x}\left(\sqrt{1+x} + \sqrt{1+x}\right)} = \dfrac{-1}{(1+x)(2\sqrt{1+x})} = -\dfrac{1}{2(1+x)^{3/2}}$

Domain of f = Domain of f' = $(-1, \infty)$.

31. (a)

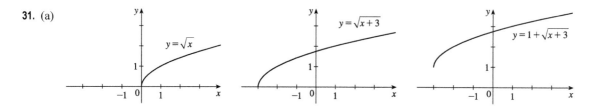

(b) Note that the third graph in part (a) generally has small positive values for its slope, f'; but as $x \to -3^+$, $f' \to \infty$.

See the graph in part (d).

(c) $f'(x) = \lim\limits_{h \to 0} \dfrac{f(x+h) - f(x)}{h} = \lim\limits_{h \to 0} \dfrac{1 + \sqrt{(x+h)+3} - (1 + \sqrt{x+3})}{h}$

$= \lim\limits_{h \to 0} \dfrac{\sqrt{(x+h)+3} - \sqrt{x+3}}{h} = \lim\limits_{h \to 0} \dfrac{\sqrt{(x+h)+3} - \sqrt{x+3}}{h} \left[\dfrac{\sqrt{(x+h)+3} + \sqrt{x+3}}{\sqrt{(x+h)+3} + \sqrt{x+3}} \right]$

$= \lim\limits_{h \to 0} \dfrac{[(x+h)+3] - (x+3)}{h\sqrt{(x+h)+3} + \sqrt{x+3}} = \lim\limits_{h \to 0} \dfrac{x+h+3-x-3}{h\left(\sqrt{(x+h)+3} + \sqrt{x+3}\right)}$

$= \lim\limits_{h \to 0} \dfrac{h}{h\left(\sqrt{(x+h)+3} + \sqrt{x+3}\right)} = \lim\limits_{h \to 0} \dfrac{1}{\sqrt{(x+h)+3} + \sqrt{x+3}} = \dfrac{1}{2\sqrt{x+3}}$

Domain of $f = [-3, \infty)$, Domain of $f' = (-3, \infty)$.

(d)

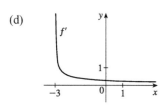

33. (a) $f'(x) = \lim\limits_{h \to 0} \dfrac{f(x+h) - f(x)}{h} = \lim\limits_{h \to 0} \dfrac{[(x+h)^4 + 2(x+h)] - (x^4 + 2x)}{h}$

$= \lim\limits_{h \to 0} \dfrac{x^4 + 4x^3h + 6x^2h^2 + 4xh^3 + h^4 + 2x + 2h - x^4 - 2x}{h}$

$= \lim\limits_{h \to 0} \dfrac{4x^3h + 6x^2h^2 + 4xh^3 + h^4 + 2h}{h} = \lim\limits_{h \to 0} \dfrac{h(4x^3 + 6x^2h + 4xh^2 + h^3 + 2)}{h}$

$= \lim\limits_{h \to 0} (4x^3 + 6x^2h + 4xh^2 + h^3 + 2) = 4x^3 + 2$

(b) Notice that $f'(x) = 0$ when f has a horizontal tangent, $f'(x)$ is positive when the tangents have positive slope, and $f'(x)$ is negative when the tangents have negative slope.

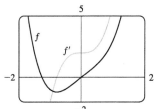

35. As in Exercise 34, we use one-sided difference quotients for the first and last values, and average two difference quotients for all other values.

t	14	21	28	35	42	49
$H(t)$	41	54	64	72	78	83
$H'(t)$	$\frac{13}{7}$	$\frac{23}{14}$	$\frac{18}{14}$	$\frac{14}{14}$	$\frac{11}{14}$	$\frac{5}{7}$

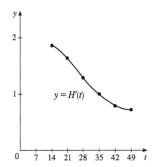

37. (a) dP/dt is the rate at which the percentage of the city's electrical power produced by solar panels changes with respect to time t, measured in percentage points per year.

(b) 2 years after January 1, 2020 (January 1, 2022), the percentage of electrical power produced by solar panels was increasing at a rate of 3.5 percentage points per year.

39. f is not differentiable at $x = -4$, because the graph has a corner there, and at $x = 0$, because there is a discontinuity there.

41. f is not differentiable at $x = 1$, because f is not defined there, and at $x = 5$, because the graph has a vertical tangent there.

43. As we zoom in toward $(-1, 0)$, the curve appears more and more like a straight line, so $f(x) = x + \sqrt{|x|}$ is differentiable at $x = -1$. But no matter how much we zoom in toward the origin, the curve doesn't straighten out—we can't eliminate the sharp point (a cusp). So f is not differentiable at $x = 0$.

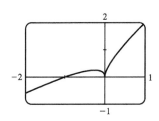

45. Call the curve with the positive y-intercept g and the other curve h. Notice that g has a maximum (horizontal tangent) at $x = 0$, but $h \neq 0$, so h cannot be the derivative of g. Also notice that where g is positive, h is increasing. Thus, $h = f$ and $g = f'$. Now $f'(-1)$ is negative since f' is below the x-axis there and $f''(1)$ is positive since f is concave upward at $x = 1$. Therefore, $f''(1)$ is greater than $f'(-1)$.

47. $a = f$, $b = f'$, $c = f''$. We can see this because where a has a horizontal tangent, $b = 0$, and where b has a horizontal tangent, $c = 0$. We can immediately see that c can be neither f nor f', since at the points where c has a horizontal tangent, neither a nor b is equal to 0.

49. We can immediately see that a is the graph of the acceleration function, since at the points where a has a horizontal tangent, neither c nor b is equal to 0. Next, we note that $a = 0$ at the point where b has a horizontal tangent, so b must be the graph of the velocity function, and hence, $b' = a$. We conclude that c is the graph of the position function.

51. $f'(x) = \lim\limits_{h \to 0} \dfrac{f(x+h) - f(x)}{h} = \lim\limits_{h \to 0} \dfrac{[3(x+h)^2 + 2(x+h) + 1] - (3x^2 + 2x + 1)}{h}$

$= \lim\limits_{h \to 0} \dfrac{(3x^2 + 6xh + 3h^2 + 2x + 2h + 1) - (3x^2 + 2x + 1)}{h} = \lim\limits_{h \to 0} \dfrac{6xh + 3h^2 + 2h}{h}$

$= \lim\limits_{h \to 0} \dfrac{h(6x + 3h + 2)}{h} = \lim\limits_{h \to 0} (6x + 3h + 2) = 6x + 2$

$f''(x) = \lim\limits_{h \to 0} \dfrac{f'(x+h) - f'(x)}{h} = \lim\limits_{h \to 0} \dfrac{[6(x+h) + 2] - (6x + 2)}{h} = \lim\limits_{h \to 0} \dfrac{(6x + 6h + 2) - (6x + 2)}{h}$

$= \lim\limits_{h \to 0} \dfrac{6h}{h} = \lim\limits_{h \to 0} 6 = 6$

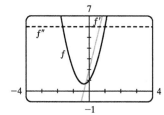

We see from the graph that our answers are reasonable because the graph of f' is that of a linear function and the graph of f'' is that of a constant function.

53. $f'(x) = \lim\limits_{h \to 0} \dfrac{f(x+h) - f(x)}{h} = \lim\limits_{h \to 0} \dfrac{[2(x+h)^2 - (x+h)^3] - (2x^2 - x^3)}{h}$

$= \lim\limits_{h \to 0} \dfrac{h(4x + 2h - 3x^2 - 3xh - h^2)}{h} = \lim\limits_{h \to 0} (4x + 2h - 3x^2 - 3xh - h^2) = 4x - 3x^2$

$f''(x) = \lim\limits_{h \to 0} \dfrac{f'(x+h) - f'(x)}{h} = \lim\limits_{h \to 0} \dfrac{[4(x+h) - 3(x+h)^2] - (4x - 3x^2)}{h} = \lim\limits_{h \to 0} \dfrac{h(4 - 6x - 3h)}{h}$

$= \lim\limits_{h \to 0} (4 - 6x - 3h) = 4 - 6x$

$f'''(x) = \lim\limits_{h \to 0} \dfrac{f''(x+h) - f''(x)}{h} = \lim\limits_{h \to 0} \dfrac{[4 - 6(x+h)] - (4 - 6x)}{h} = \lim\limits_{h \to 0} \dfrac{-6h}{h} = \lim\limits_{h \to 0} (-6) = -6$

$f^{(4)}(x) = \lim\limits_{h \to 0} \dfrac{f'''(x+h) - f'''(x)}{h} = \lim\limits_{h \to 0} \dfrac{-6 - (-6)}{h} = \lim\limits_{h \to 0} \dfrac{0}{h} = \lim\limits_{h \to 0} (0) = 0$

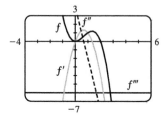

The graphs are consistent with the geometric interpretations of the derivatives because f' has zeros where f has a local minimum and a local maximum, f'' has a zero where f' has a local maximum, and f''' is a constant function equal to the slope of f''.

55. (a) Note that we have factored $x - a$ as the difference of two cubes in the third step.

$$f'(a) = \lim_{x \to a} \frac{f(x) - f(a)}{x - a} = \lim_{x \to a} \frac{x^{1/3} - a^{1/3}}{x - a} = \lim_{x \to a} \frac{x^{1/3} - a^{1/3}}{(x^{1/3} - a^{1/3})(x^{2/3} + x^{1/3}a^{1/3} + a^{2/3})}$$

$$= \lim_{x \to a} \frac{1}{x^{2/3} + x^{1/3}a^{1/3} + a^{2/3}} = \frac{1}{3a^{2/3}} \text{ or } \tfrac{1}{3}a^{-2/3}$$

(b) $f'(0) = \lim_{h \to 0} \frac{f(0 + h) - f(0)}{h} = \lim_{h \to 0} \frac{\sqrt[3]{h} - 0}{h} = \lim_{h \to 0} \frac{1}{h^{2/3}}$. This function increases without bound, so the limit does not exist, and therefore $f'(0)$ does not exist.

(c) $\lim_{x \to 0} |f'(x)| = \lim_{x \to 0} \frac{1}{3x^{2/3}} = \infty$ and f is continuous at $x = 0$ (root function), so f has a vertical tangent at $x = 0$.

57. $f(x) = |x - 6| = \begin{cases} x - 6 & \text{if } x - 6 \geq 6 \\ -(x - 6) & \text{if } x - 6 < 0 \end{cases} = \begin{cases} x - 6 & \text{if } x \geq 6 \\ 6 - x & \text{if } x < 6 \end{cases}$

So the right-hand limit is 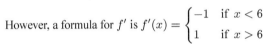 $= \lim_{x \to 6^+} \frac{x - 6}{x - 6} = \lim_{x \to 6^+} 1 = 1$, and the left-hand limit

is $\lim_{x \to 6^-} \frac{f(x) - f(6)}{x - 6} = \lim_{x \to 6^-} \frac{|x - 6| - 0}{x - 6} = \lim_{x \to 6^-} \frac{6 - x}{x - 6} = \lim_{x \to 6^-} (-1) = -1$. Since these limits are not equal,

$f'(6) = \lim_{x \to 6} \frac{f(x) - f(6)}{x - 6}$ does not exist and f is not differentiable at 6.

However, a formula for f' is $f'(x) = \begin{cases} -1 & \text{if } x < 6 \\ 1 & \text{if } x > 6 \end{cases}$

Another way of writing the formula is $f'(x) = \dfrac{x - 6}{|x - 6|}$.

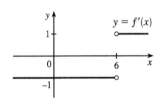

59. (a) $f(x) = x|x| = \begin{cases} x^2 & \text{if } x \geq 0 \\ -x^2 & \text{if } x < 0 \end{cases}$

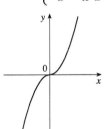

(b) Since $f(x) = x^2$ for $x \geq 0$, we have $f'(x) = 2x$ for $x > 0$. [See Exercise 17(d).] Similarly, since $f(x) = -x^2$ for $x < 0$, we have $f'(x) = -2x$ for $x < 0$. At $x = 0$, we have

$$f'(0) = \lim_{x \to 0} \frac{f(x) - f(0)}{x - 0} = \lim_{x \to 0} \frac{x|x|}{x} = \lim_{x \to 0} |x| = 0.$$

So f is differentiable at 0. Thus, f is differentiable for all x.

(c) From part (b), we have $f'(x) = \begin{cases} 2x & \text{if } x \geq 0 \\ -2x & \text{if } x < 0 \end{cases} = 2|x|$.

61. (a) If f is even, then

$$f'(-x) = \lim_{h \to 0} \frac{f(-x + h) - f(-x)}{h} = \lim_{h \to 0} \frac{f[-(x - h)] - f(-x)}{h}$$

$$= \lim_{h \to 0} \frac{f(x - h) - f(x)}{h} = -\lim_{h \to 0} \frac{f(x - h) - f(x)}{-h} \qquad [\text{let } \Delta x = -h]$$

$$= -\lim_{\Delta x \to 0} \frac{f(x + \Delta x) - f(x)}{\Delta x} = -f'(x)$$

Therefore, f' is odd.

(b) If f is odd, then

$$f'(-x) = \lim_{h \to 0} \frac{f(-x+h) - f(-x)}{h} = \lim_{h \to 0} \frac{f[-(x-h)] - f(-x)}{h}$$

$$= \lim_{h \to 0} \frac{-f(x-h) + f(x)}{h} = \lim_{h \to 0} \frac{f(x-h) - f(x)}{-h} \quad [\text{let } \Delta x = -h]$$

$$= \lim_{\Delta x \to 0} \frac{f(x + \Delta x) - f(x)}{\Delta x} = f'(x)$$

Therefore, f' is even.

63. (a) $f(x) = \begin{cases} 0 & \text{if } x \le 0 \\ 5 - x & \text{if } 0 < x < 4 \\ \dfrac{1}{5 - x} & \text{if } x \ge 4 \end{cases}$

Note that as $h \to 0^-$, $4 + h < 4$, so $f(4+h) = 5 - (4+h)$. As $h \to 0^+$, $4 + h > 4$, so $f(4+h) = \dfrac{1}{5-(4+h)}$.

$$f'_-(4) = \lim_{h \to 0^-} \frac{f(4+h) - f(4)}{h} = \lim_{h \to 0^-} \frac{[5-(4+h)] - (5-4)}{h} = \lim_{h \to 0^-} \frac{(5-4-h) - 1}{h}$$

$$= \lim_{h \to 0^-} \frac{-h}{h} = \lim_{h \to 0^-} (-1) = -1$$

$$f'_+(4) = \lim_{h \to 0^+} \frac{f(4+h) - f(4)}{h} = \lim_{h \to 0^+} \frac{\dfrac{1}{5-(4+h)} - \dfrac{1}{5-4}}{h} = \lim_{h \to 0^+} \frac{\dfrac{1}{1-h} - 1}{h}$$

$$= \lim_{h \to 0^+} \frac{\dfrac{1}{1-h} - \dfrac{1-h}{1-h}}{h} = \lim_{h \to 0^+} \frac{1 - (1-h)}{h(1-h)} = \lim_{h \to 0^+} \frac{h}{h(1-h)} = \lim_{h \to 0^+} \frac{1}{1-h} = 1$$

(b)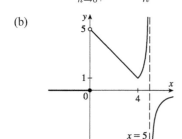

(c) f is discontinuous at $x = 0$ (jump discontinuity) and at $x = 5$ (infinite discontinuity).

(d) f is not differentiable at $x = 0$ [discontinuous, from part (c)], $x = 4$ [one-sided derivatives are not equal, from part (a)], and at $x = 5$ [discontinuous, from part (c)].

65. These graphs are idealizations conveying the spirit of the problem. In reality, changes in speed are not instantaneous, so the graph in (a) would not have corners and the graph in (b) would be continuous.

(a)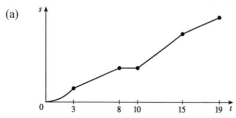

(b)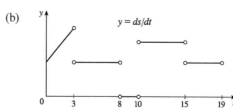

2.3 Differentiation Formulas

1. $g(x) = 4x + 7 \Rightarrow g'(x) = 4(1) + 0 = 4$

3. $f(x) = x^{75} - x + 3 \Rightarrow f'(x) = 75x^{75-1} - 1(1) + 0 = 75x^{74} - 1$

5. $W(v) = 1.8v^{-3} \Rightarrow W'(v) = 1.8(-3v^{-3-1}) = 1.8(-3v^{-4}) = -5.4v^{-4}$

7. $f(x) = x^{3/2} + x^{-3} \Rightarrow f'(x) = \frac{3}{2}x^{1/2} + (-3x^{-4}) = \frac{3}{2}x^{1/2} - 3x^{-4}$

9. $s(t) = \frac{1}{t} + \frac{1}{t^2} = t^{-1} + t^{-2} \Rightarrow s'(t) = -t^{-2} + (-2t^{-3}) = -t^{-2} - 2t^{-3} = -\frac{1}{t^2} - \frac{2}{t^3}$

11. $y = 2x + \sqrt{x} = 2x + x^{1/2} \Rightarrow y' = 2(1) + \frac{1}{2}x^{-1/2} = 2 + \frac{1}{2}x^{-1/2}$ or $2 + \frac{1}{2\sqrt{x}}$

13. $g(x) = \frac{1}{\sqrt{x}} + \sqrt[4]{x} = x^{-1/2} + x^{1/4} \Rightarrow g'(x) = -\frac{1}{2}x^{-3/2} + \frac{1}{4}x^{-3/4}$ or $-\frac{1}{2x\sqrt{x}} + \frac{1}{4\sqrt[4]{x^3}}$

15. $f(x) = x^3(x+3) = x^4 + 3x^3 \Rightarrow f'(x) = 4x^3 + 3(3x^2) = 4x^3 + 9x^2$

17. $f(x) = \frac{3x^2 + x^3}{x} = \frac{3x^2}{x} + \frac{x^3}{x} = 3x + x^2 \Rightarrow f'(x) = 3(1) + 2x = 3 + 2x$

19. $G(q) = (1 + q^{-1})^2 = 1 + 2q^{-1} + q^{-2} \Rightarrow G'(q) = 0 + 2(-1q^{-2}) + (-2q^{-3}) = -2q^{-2} - 2q^{-3}$

21. $G(r) = \frac{3r^{3/2} + r^{5/2}}{r} = \frac{3r^{3/2}}{r} + \frac{r^{5/2}}{r} = 3r^{3/2 - 2/2} + r^{5/2 - 2/2} = 3r^{1/2} + r^{3/2} \Rightarrow$
$G'(r) = 3\left(\frac{1}{2}r^{-1/2}\right) + \frac{3}{2}r^{1/2} = \frac{3}{2}r^{-1/2} + \frac{3}{2}r^{1/2}$ or $\frac{3}{2\sqrt{r}} + \frac{3}{2}\sqrt{r}$

23. $P(w) = \frac{2w^2 - w + 4}{\sqrt{w}} = \frac{2w^2}{\sqrt{w}} - \frac{w}{\sqrt{w}} + \frac{4}{\sqrt{w}} = 2w^{2-1/2} - w^{1-1/2} + 4w^{-1/2} = 2w^{3/2} - w^{1/2} + 4w^{-1/2} \Rightarrow$
$P'(w) = 2\left(\frac{3}{2}w^{1/2}\right) - \frac{1}{2}w^{-1/2} + 4\left(-\frac{1}{2}w^{-3/2}\right) = 3w^{1/2} - \frac{1}{2}w^{-1/2} - 2w^{-3/2}$ or $3\sqrt{w} - \frac{1}{2\sqrt{w}} - \frac{2}{w\sqrt{w}}$

25. $y = tx^2 + t^3 x$.

    To find dy/dx, we treat t as a constant and x as a variable to get $dy/dx = t(2x) + t^3(1) = 2tx + t^3$.

    To find dy/dt, we treat x as a constant and t as a variable to get $dy/dt = (1)x^2 + (3t^2)x = x^2 + 3t^2 x$.

27. Product Rule: $f(x) = (1 + 2x^2)(x - x^2) \Rightarrow$
    $$f'(x) = (1 + 2x^2)(1 - 2x) + (x - x^2)(4x) = 1 - 2x + 2x^2 - 4x^3 + 4x^2 - 4x^3 = 1 - 2x + 6x^2 - 8x^3.$$
    Multiplying first: $f(x) = (1 + 2x^2)(x - x^2) = x - x^2 + 2x^3 - 2x^4 \Rightarrow f'(x) = 1 - 2x + 6x^2 - 8x^3$ (equivalent).

29. By the Product Rule, $f(x) = (3x^2 - 5x)x^2 \Rightarrow$
    $f'(x) = (3x^2 - 5x)(x^2)' + x^2(3x^2 - 5x)' = (3x^2 - 5x)(2x) + x^2(6x - 5)$
    $= 6x^3 - 10x^2 + 6x^3 - 5x^2 = 12x^3 - 15x^2$

31. By the Product Rule, $y = (4x^2 + 3)(2x + 5)$ $\Rightarrow$

$y' = (4x^2 + 3)(2x + 5)' + (2x + 5)(4x^2 + 3)' = (4x^2 + 3)(2) + (2x + 5)(8x)$

$= 8x^2 + 6 + 16x^2 + 40x = 24x^2 + 40x + 6$

33. By the Quotient Rule, $y = \dfrac{5x}{1+x}$ $\Rightarrow$

$y' = \dfrac{(1+x)(5x)' - 5x(1+x)'}{(1+x)^2} = \dfrac{(1+x)5 - 5x(1)}{(1+x)^2} = \dfrac{5 + 5x - 5x}{(1+x)^2} = \dfrac{5}{(1+x)^2}.$

35. $g(t) = \dfrac{3-2t}{5t+1}$ $\overset{QR}{\Rightarrow}$ $g'(t) = \dfrac{(5t+1)(-2) - (3-2t)(5)}{(5t+1)^2} = \dfrac{-10t - 2 - 15 + 10t}{(5t+1)^2} = -\dfrac{17}{(5t+1)^2}$

37. $f(t) = \dfrac{5t}{t^3-t-1}$ $\overset{QR}{\Rightarrow}$ $f'(t) = \dfrac{(t^3-t-1)(5) - (5t)(3t^2-1)}{(t^3-t-1)^2} = \dfrac{5t^3 - 5t - 5 - 15t^3 + 5t}{(t^3-t-1)^2} = -\dfrac{10t^3 + 5}{(t^3-t-1)^2}$

39. $y = \dfrac{s - \sqrt{s}}{s^2} = \dfrac{s}{s^2} - \dfrac{\sqrt{s}}{s^2} = s^{-1} - s^{-3/2}$ $\Rightarrow$ $y' = -s^{-2} + \tfrac{3}{2}s^{-5/2} = \dfrac{-1}{s^2} + \dfrac{3}{2s^{5/2}} = \dfrac{3 - 2\sqrt{s}}{2s^{5/2}}$

41. $F(x) = \dfrac{2x^5 + x^4 - 6x}{x^3} = 2x^2 + x - 6x^{-2}$ $\Rightarrow$ $F'(x) = 4x + 1 + 12x^{-3} = 4x + 1 + \dfrac{12}{x^3}$ or $\dfrac{4x^4 + x^3 + 12}{x^3}$

43. $H(u) = (u - \sqrt{u})(u + \sqrt{u})$ $\overset{PR}{\Rightarrow}$

$H'(u) = \left(u - \sqrt{u}\right)\left(1 + \dfrac{1}{2\sqrt{u}}\right) + \left(u + \sqrt{u}\right)\left(1 - \dfrac{1}{2\sqrt{u}}\right) = u + \tfrac{1}{2}\sqrt{u} - \sqrt{u} - \tfrac{1}{2} + u - \tfrac{1}{2}\sqrt{u} + \sqrt{u} - \tfrac{1}{2} = 2u - 1.$

Alternate solution: An easier method is to simplify first and then differentiate as follows:

$H(u) = (u - \sqrt{u})(u + \sqrt{u}) = u^2 - (\sqrt{u})^2 = u^2 - u$ $\Rightarrow$ $H'(u) = 2u - 1$

45. $J(u) = \left(\dfrac{1}{u} + \dfrac{1}{u^2}\right)\left(u + \dfrac{1}{u}\right) = (u^{-1} + u^{-2})(u + u^{-1})$ $\overset{PR}{\Rightarrow}$

$J'(u) = (u^{-1} + u^{-2})(u + u^{-1})' + (u + u^{-1})(u^{-1} + u^{-2})' = (u^{-1} + u^{-2})(1 - u^{-2}) + (u + u^{-1})(-u^{-2} - 2u^{-3})$

$= u^{-1} - u^{-3} + u^{-2} - u^{-4} - u^{-1} - 2u^{-2} - u^{-3} - 2u^{-4} = -u^{-2} - 2u^{-3} - 3u^{-4} = -\left(\dfrac{1}{u^2} + \dfrac{2}{u^3} + \dfrac{3}{u^4}\right)$

47. $f(t) = \dfrac{\sqrt[3]{t}}{t-3}$ $\overset{QR}{\Rightarrow}$

$f'(t) = \dfrac{(t-3)\left(\tfrac{1}{3}t^{-2/3}\right) - t^{1/3}(1)}{(t-3)^2} = \dfrac{\tfrac{1}{3}t^{1/3} - t^{-2/3} - t^{1/3}}{(t-3)^2} = \dfrac{-\tfrac{2}{3}t^{1/3} - t^{-2/3}}{(t-3)^2}$

$= \dfrac{-\dfrac{2t}{3t^{2/3}} - \dfrac{3}{3t^{2/3}}}{(t-3)^2} = \dfrac{-2t - 3}{3t^{2/3}(t-3)^2}$

49. $G(y) = \dfrac{B}{Ay^3 + B}$ $\overset{QR}{\Rightarrow}$ $G'(y) = \dfrac{(Ay^3 + B)(0) - B(3Ay^2)}{(Ay^3 + B)^2} = -\dfrac{3ABy^2}{(Ay^3 + B)^2}$

51. $f(x) = \dfrac{x}{x + \dfrac{c}{x}}$ $\overset{QR}{\Rightarrow}$ $f'(x) = \dfrac{(x + c/x)(1) - x(1 - c/x^2)}{\left(x + \dfrac{c}{x}\right)^2} = \dfrac{x + c/x - x + c/x}{\left(\dfrac{x^2 + c}{x}\right)^2} = \dfrac{2c/x}{\dfrac{(x^2 + c)^2}{x^2}} \cdot \dfrac{x^2}{x^2} = \dfrac{2cx}{(x^2 + c)^2}$

53. $P(x) = a_n x^n + a_{n-1} x^{n-1} + \cdots + a_2 x^2 + a_1 x + a_0 \Rightarrow P'(x) = n a_n x^{n-1} + (n-1) a_{n-1} x^{n-2} + \cdots + 2 a_2 x + a_1$

55. $f(x) = 3x^{15} - 5x^3 + 3 \Rightarrow f'(x) = 45x^{14} - 15x^2$.

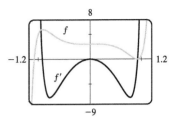

Notice that $f'(x) = 0$ when f has a horizontal tangent, f' is positive when f is increasing, and f' is negative when f is decreasing.

57. (a)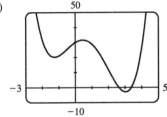

(b) From the graph in part (a), it appears that f' is zero at $x_1 \approx -1.25$, $x_2 \approx 0.5$, and $x_3 \approx 3$. The slopes are negative (so f' is negative) on $(-\infty, x_1)$ and (x_2, x_3). The slopes are positive (so f' is positive) on (x_1, x_2) and (x_3, ∞).

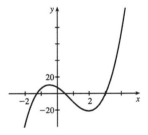

(c) $f(x) = x^4 - 3x^3 - 6x^2 + 7x + 30 \Rightarrow$

$f'(x) = 4x^3 - 9x^2 - 12x + 7$

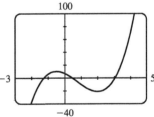

59. $y = \dfrac{2x}{x+1} \Rightarrow y' = \dfrac{(x+1)(2) - (2x)(1)}{(x+1)^2} = \dfrac{2}{(x+1)^2}$.

At $(1, 1)$, $y' = \dfrac{1}{2}$, and an equation of the tangent line is $y - 1 = \dfrac{1}{2}(x - 1)$, or $y = \dfrac{1}{2}x + \dfrac{1}{2}$.

61. $y = x + \sqrt{x} \Rightarrow y' = 1 + \dfrac{1}{2}x^{-1/2} = 1 + 1/(2\sqrt{x})$. At $(1, 2)$, $y' = \dfrac{3}{2}$, and an equation of the tangent line is

$y - 2 = \dfrac{3}{2}(x - 1)$, or $y = \dfrac{3}{2}x + \dfrac{1}{2}$. The slope of the normal line is $-\dfrac{2}{3}$, so an equation of the normal line is

$y - 2 = -\dfrac{2}{3}(x - 1)$, or $y = -\dfrac{2}{3}x + \dfrac{8}{3}$.

63. $y = \dfrac{3x}{1 + 5x^2} \Rightarrow y' = \dfrac{(1 + 5x^2)(3) - 3x(10x)}{(1 + 5x^2)^2} = \dfrac{3 + 15x^2 - 30x^2}{(1 + 5x^2)^2} = \dfrac{3 - 15x^2}{(1 + 5x^2)^2}$

At $\left(1, \dfrac{1}{2}\right)$, $y' = \dfrac{3 - 15(1^2)}{(1 + 5 \cdot 1^2)^2} = \dfrac{-12}{6^2} = -\dfrac{1}{3}$, and an equation of the tangent line is $y - \dfrac{1}{2} = -\dfrac{1}{3}(x - 1)$, or $y = -\dfrac{1}{3}x + \dfrac{5}{6}$.

The slope of the normal line is 3, so an equation of the normal line is $y - \dfrac{1}{2} = 3(x - 1)$, or $y = 3x - \dfrac{5}{2}$.

72 ◻ **CHAPTER 2** DERIVATIVES

65. (a) $y = f(x) = \dfrac{1}{1+x^2}$ $\Rightarrow$

$f'(x) = \dfrac{(1+x^2)(0) - 1(2x)}{(1+x^2)^2} = \dfrac{-2x}{(1+x^2)^2}$. So the slope of the

tangent line at the point $\left(-1, \frac{1}{2}\right)$ is $f'(-1) = \dfrac{2}{2^2} = \frac{1}{2}$ and its

equation is $y - \frac{1}{2} = \frac{1}{2}(x+1)$ or $y = \frac{1}{2}x + 1$.

(b)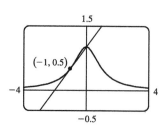

67. $f(x) = 0.001x^5 - 0.02x^3$ $\Rightarrow$ $f'(x) = 0.005x^4 - 0.06x^2$ $\Rightarrow$ $f''(x) = 0.02x^3 - 0.12x$

69. $f(x) = \dfrac{x^2}{1+2x}$ $\Rightarrow$ $f'(x) = \dfrac{(1+2x)(2x) - x^2(2)}{(1+2x)^2} = \dfrac{2x + 4x^2 - 2x^2}{(1+2x)^2} = \dfrac{2x^2 + 2x}{(1+2x)^2}$ $\Rightarrow$

$f''(x) = \dfrac{(1+2x)^2(4x+2) - (2x^2+2x)(1+4x+4x^2)'}{[(1+2x)^2]^2} = \dfrac{2(1+2x)^2(2x+1) - 2x(x+1)(4+8x)}{(1+2x)^4}$

$= \dfrac{2(1+2x)[(1+2x)^2 - 4x(x+1)]}{(1+2x)^4} = \dfrac{2(1+4x+4x^2 - 4x^2 - 4x)}{(1+2x)^3} = \dfrac{2}{(1+2x)^3}$

71. $f(x) = 2x - 5x^{3/4}$ $\Rightarrow$ $f'(x) = 2 - \frac{15}{4}x^{-1/4}$ $\Rightarrow$ $f''(x) = \frac{15}{16}x^{-5/4}$

Note that f' is negative when f is decreasing and positive when f is

increasing. f'' is always positive since f' is always increasing.

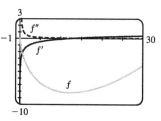

73. (a) $s = t^3 - 3t$ $\Rightarrow$ $v(t) = s'(t) = 3t^2 - 3$ $\Rightarrow$ $a(t) = v'(t) = 6t$

(b) $a(2) = 6(2) = 12 \text{ m/s}^2$

(c) $v(t) = 3t^2 - 3 = 0$ when $t^2 = 1$, that is, $t = 1$ $[t \geq 0]$ and $a(1) = 6 \text{ m/s}^2$.

75. $L = 0.0155A^3 - 0.372A^2 + 3.95A + 1.21$ $\Rightarrow$ $\dfrac{dL}{dA} = 0.0465A^2 - 0.744A + 3.95$, so

$\left.\dfrac{dL}{dA}\right|_{A=12} = 0.0465(12)^2 - 0.744(12) + 3.95 = 1.718$. The derivative is the instantaneous rate of change of the length (in

inches) of an Alaskan rockfish with respect to its age when its age is 12 years. Its units are inches/year.

77. (a) $P = \dfrac{k}{V}$ and $P = 50$ when $V = 0.106$, so $k = PV = 50(0.106) = 5.3$. Thus, $P = \dfrac{5.3}{V}$ and $V = \dfrac{5.3}{P}$.

(b) $V = 5.3P^{-1}$ $\Rightarrow$ $\dfrac{dV}{dP} = 5.3(-1P^{-2}) = -\dfrac{5.3}{P^2}$. When $P = 50$, $\dfrac{dV}{dP} = -\dfrac{5.3}{50^2} = -0.00212$. The derivative is the

instantaneous rate of change of the volume with respect to the pressure at 25°C. Its units are m³/kPa.

79. We are given that $f(5) = 1$, $f'(5) = 6$, $g(5) = -3$, and $g'(5) = 2$.

(a) $(fg)'(5) = f(5)g'(5) + g(5)f'(5) = (1)(2) + (-3)(6) = 2 - 18 = -16$

(b) $\left(\dfrac{f}{g}\right)'(5) = \dfrac{g(5)f'(5) - f(5)g'(5)}{[g(5)]^2} = \dfrac{(-3)(6) - (1)(2)}{(-3)^2} = -\dfrac{20}{9}$

(c) $\left(\dfrac{g}{f}\right)'(5) = \dfrac{f(5)g'(5) - g(5)f'(5)}{[f(5)]^2} = \dfrac{(1)(2) - (-3)(6)}{(1)^2} = 20$

81. $f(x) = \sqrt{x}\, g(x) \;\Rightarrow\; f'(x) = \sqrt{x}\, g'(x) + g(x) \cdot \dfrac{1}{2}x^{-1/2}$, so $f'(4) = \sqrt{4}\, g'(4) + g(4) \cdot \dfrac{1}{2\sqrt{4}} = 2\cdot 7 + 8 \cdot \dfrac{1}{4} = 16$.

83. (a) From the graphs of f and g, we obtain the following values: $f(1) = 2$ since the point $(1, 2)$ is on the graph of f; $g(1) = 3$ since the point $(1, 3)$ is on the graph of g; $f'(1) = \tfrac{1}{3}$ since the slope of the line segment between $(1, 2)$ and $(4, 3)$ is $\dfrac{3-2}{4-1} = \dfrac{1}{3}$; $g'(1) = 1$ since the slope of the line segment between $(-2, 0)$ and $(2, 4)$ is $\dfrac{4-0}{2-(-2)} = \dfrac{4}{4} = 1$.

Now $u(x) = f(x)\,g(x)$, so $u'(1) = f(1)\,g'(1) + g(1)\,f'(1) = 2\cdot 1 + 3\cdot \tfrac{1}{3} = 3$.

(b) From the graphs of f and g, we obtain the following values: $f(4) = 3$ since the point $(4, 3)$ is on the graph of f; $g(4) = 2$ since the point $(4, 2)$ is on the graph of g; $f'(4) = f'(1) = \tfrac{1}{3}$ from the part (a); $g'(4) = 1$ since the slope of the line segment between $(3, 1)$ and $(5, 3)$ is $\dfrac{3-1}{5-3} = \dfrac{2}{2} = 1$.

$v(x) = \dfrac{f(x)}{g(x)}$, so $v'(4) = \dfrac{g(4)f'(4) - f(4)g'(4)}{[g(4)]^2} = \dfrac{2\cdot\tfrac{1}{3} - 3\cdot 1}{2^2} = \dfrac{-\tfrac{7}{3}}{4} = -\dfrac{7}{12}$

85. (a) $y = xg(x) \;\Rightarrow\; y' = xg'(x) + g(x)\cdot 1 = xg'(x) + g(x)$

(b) $y = \dfrac{x}{g(x)} \;\Rightarrow\; y' = \dfrac{g(x)\cdot 1 - xg'(x)}{[g(x)]^2} = \dfrac{g(x) - xg'(x)}{[g(x)]^2}$

(c) $y = \dfrac{g(x)}{x} \;\Rightarrow\; y' = \dfrac{xg'(x) - g(x)\cdot 1}{(x)^2} = \dfrac{xg'(x) - g(x)}{x^2}$

87. $y = x^3 + 3x^2 - 9x + 10 \;\Rightarrow\; y' = 3x^2 + 3(2x) - 9(1) + 0 = 3x^2 + 6x - 9$. Horizontal tangents occur where $y' = 0$. Thus, $3x^2 + 6x - 9 = 0 \;\Rightarrow\; 3(x^2 + 2x - 3) = 0 \;\Rightarrow\; 3(x+3)(x-1) = 0 \;\Rightarrow\; x = -3$ or $x = 1$. The corresponding points are $(-3, 37)$ and $(1, 5)$.

89. $y = 6x^3 + 5x - 3 \;\Rightarrow\; m = y' = 18x^2 + 5$, but $x^2 \geq 0$ for all x, so $m \geq 5$ for all x.

91. The slope of the line $3x - y = 15$ (or $y = 3x - 15$) is 3, so the slope of both tangent lines to the curve is 3. $y = x^3 - 3x^2 + 3x - 3 \;\Rightarrow\; y' = 3x^2 - 6x + 3 = 3(x^2 - 2x + 1) = 3(x-1)^2$. Thus, $3(x-1)^2 = 3 \;\Rightarrow\; (x-1)^2 = 1 \;\Rightarrow\; x - 1 = \pm 1 \;\Rightarrow\; x = 0$ or 2, which are the x-coordinates at which the tangent lines have slope 3. The points on the curve are $(0, -3)$ and $(2, -1)$, so the tangent line equations are $y - (-3) = 3(x - 0)$ or $y = 3x - 3$ and $y - (-1) = 3(x - 2)$ or $y = 3x - 7$.

93. The slope of $y = \sqrt{x}$ is given by $y = \frac{1}{2}x^{-1/2} = \dfrac{1}{2\sqrt{x}}$. The slope of $2x + y = 1$ (or $y = -2x + 1$) is -2, so the desired normal line must have slope -2, and hence, the tangent line to the curve must have slope $\frac{1}{2}$. This occurs if $\dfrac{1}{2\sqrt{x}} = \dfrac{1}{2}$ $\Rightarrow$ $\sqrt{x} = 1$ $\Rightarrow$ $x = 1$. When $x = 1$, $y = \sqrt{1} = 1$, and an equation of the normal line is $y - 1 = -2(x - 1)$ or $y = -2x + 3$.

95. Let (a, a^2) be a point on the parabola at which the tangent line passes through the point $(0, -4)$. The tangent line has slope $2a$ and equation

$y - (-4) = 2a(x - 0)$ $\Leftrightarrow$ $y = 2ax - 4$. Since (a, a^2) also lies on the line, $a^2 = 2a(a) - 4$, or $a^2 = 4$. So $a = \pm 2$ and the points are $(2, 4)$ and $(-2, 4)$.

97. $y = f(x) = ax^2$ $\Rightarrow$ $f'(x) = 2ax$. So the slope of the tangent to the parabola at $x = 2$ is $m = 2a(2) = 4a$. The slope of the given line, $2x + y = b$ $\Leftrightarrow$ $y = -2x + b$, is seen to be -2, so we must have $4a = -2$ $\Leftrightarrow$ $a = -\frac{1}{2}$. So when $x = 2$, the point in question has y-coordinate $-\frac{1}{2} \cdot 2^2 = -2$. Now we simply require that the given line, whose equation is $2x + y = b$, pass through the point $(2, -2)$: $2(2) + (-2) = b$ $\Leftrightarrow$ $b = 2$. So we must have $a = -\frac{1}{2}$ and $b = 2$.

99. Let $P(x) = ax^2 + bx + c$. Then $P'(x) = 2ax + b$ and $P''(x) = 2a$. $P''(2) = 2$ $\Rightarrow$ $2a = 2$ $\Rightarrow$ $a = 1$.
$P'(2) = 3$ $\Rightarrow$ $2(1)(2) + b = 3$ $\Rightarrow$ $4 + b = 3$ $\Rightarrow$ $b = -1$.
$P(2) = 5$ $\Rightarrow$ $1(2)^2 + (-1)(2) + c = 5$ $\Rightarrow$ $2 + c = 5$ $\Rightarrow$ $c = 3$. So $P(x) = x^2 - x + 3$.

101. $y = f(x) = ax^3 + bx^2 + cx + d$ $\Rightarrow$ $f'(x) = 3ax^2 + 2bx + c$. The point $(-2, 6)$ is on f, so $f(-2) = 6$ $\Rightarrow$ $-8a + 4b - 2c + d = 6$ **(1)**. The point $(2, 0)$ is on f, so $f(2) = 0$ $\Rightarrow$ $8a + 4b + 2c + d = 0$ **(2)**. Since there are horizontal tangents at $(-2, 6)$ and $(2, 0)$, $f'(\pm 2) = 0$. $f'(-2) = 0$ $\Rightarrow$ $12a - 4b + c = 0$ **(3)** and $f'(2) = 0$ $\Rightarrow$ $12a + 4b + c = 0$ **(4)**. Subtracting equation **(3)** from **(4)** gives $8b = 0$ $\Rightarrow$ $b = 0$. Adding **(1)** and **(2)** gives $8b + 2d = 6$, so $d = 3$ since $b = 0$. From **(3)** we have $c = -12a$, so **(2)** becomes $8a + 4(0) + 2(-12a) + 3 = 0$ $\Rightarrow$ $3 = 16a$ $\Rightarrow$ $a = \frac{3}{16}$. Now $c = -12a = -12\left(\frac{3}{16}\right) = -\frac{9}{4}$ and the desired cubic function is $y = \frac{3}{16}x^3 - \frac{9}{4}x + 3$.

103. If $P(t)$ denotes the population at time t and $A(t)$ denotes the average annual income, then $T(t) = P(t)\, A(t)$ is the total personal income. The rate at which $T(t)$ is rising is given by $T'(t) = P(t)\, A'(t) + A(t)\, P'(t)$ $\Rightarrow$

$T'(2015) = P(2015)\, A'(2015) + A(2015)\, P'(2015) = (107{,}350)\,(\$2250/\text{year}) + (\$60{,}220)\,(1960/\text{year})$
$= \$241{,}537{,}500/\text{year} + \$118{,}031{,}200/\text{year} = \$359{,}568{,}700/\text{year}$

So the total personal income in Boulder was rising by about $360 million per year in 2015.

The term $P(t)\, A'(t) \approx \$242$ million represents the portion of the rate of change of total income due to the existing population's increasing income. The term $A(t)\, P'(t) \approx \$118$ million represents the portion of the rate of change of total income due to increasing population.

105. $v = \dfrac{0.14[\text{S}]}{0.015 + [\text{S}]} \Rightarrow \dfrac{dv}{d[\text{S}]} = \dfrac{(0.015 + [\text{S}])(0.14) - (0.14[\text{S}])(1)}{(0.015 + [\text{S}])^2} = \dfrac{0.0021}{(0.015 + [\text{S}])^2}$.

$dv/d[\text{S}]$ represents the rate of change of the rate of an enzymatic reaction with respect to the concentration of a substrate S.

107. (a) $(fgh)' = [(fg)h]' = (fg)'h + (fg)h' = (f'g + fg')h + (fg)h' = f'gh + fg'h + fgh'$

(b) Putting $f = g = h$ in part (a), we have $\dfrac{d}{dx}[f(x)]^3 = (fff)' = f'ff + ff'f + fff' = 3fff' = 3[f(x)]^2 f'(x)$.

(c) $y = (x^4 + 3x^3 + 17x + 82)^3 \Rightarrow y' = 3(x^4 + 3x^3 + 17x + 82)^2(4x^3 + 9x^2 + 17)$

109. $F = f/g \Rightarrow f = Fg \Rightarrow f' = F'g + Fg' \Rightarrow F' = \dfrac{f' - Fg'}{g} = \dfrac{f' - (f/g)g'}{g} = \dfrac{f'g - fg'}{g^2}$

111. $f(x) = \begin{cases} x^2 + 1 & \text{if } x < 1 \\ x + 1 & \text{if } x \geq 1 \end{cases}$

Calculate the left- and right-hand derivatives as defined in Exercise 2.2.62:

$f'_-(1) = \lim\limits_{h \to 0^-} \dfrac{f(1+h) - f(1)}{h} = \lim\limits_{h \to 0^-} \dfrac{[(1+h)^2 + 1] - (1+1)}{h} = \lim\limits_{h \to 0^-} \dfrac{h^2 + 2h}{h} = \lim\limits_{h \to 0^-} (h + 2) = 2$ and

$f'_+(1) = \lim\limits_{h \to 0^+} \dfrac{f(1+h) - f(1)}{h} = \lim\limits_{h \to 0^+} \dfrac{[(1+h) + 1] - (1+1)}{h} = \lim\limits_{h \to 0^+} \dfrac{h}{h} = \lim\limits_{h \to 0^+} 1 = 1$.

Since the left and right limits are different,

$\lim\limits_{h \to 0} \dfrac{f(1+h) - f(1)}{h}$ does not exist, that is, $f'(1)$

does not exist. Therefore, f is not differentiable at 1.

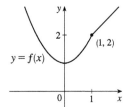
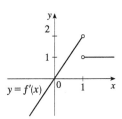

113. (a) Note that $x^2 - 9 < 0$ for $x^2 < 9 \Leftrightarrow |x| < 3 \Leftrightarrow -3 < x < 3$. So

$f(x) = \begin{cases} x^2 - 9 & \text{if } x \leq -3 \\ -x^2 + 9 & \text{if } -3 < x < 3 \\ x^2 - 9 & \text{if } x \geq 3 \end{cases} \Rightarrow f'(x) = \begin{cases} 2x & \text{if } x < -3 \\ -2x & \text{if } -3 < x < 3 \\ 2x & \text{if } x > 3 \end{cases} = \begin{cases} 2x & \text{if } |x| > 3 \\ -2x & \text{if } |x| < 3 \end{cases}$

To show that $f'(3)$ does not exist we investigate $\lim\limits_{h \to 0} \dfrac{f(3+h) - f(3)}{h}$ by computing the left- and right-hand derivatives

defined in Exercise 2.2.62.

$f'_-(3) = \lim\limits_{h \to 0^-} \dfrac{f(3+h) - f(3)}{h} = \lim\limits_{h \to 0^-} \dfrac{[-(3+h)^2 + 9] - 0}{h} = \lim\limits_{h \to 0^-} (-6 - h) = -6$ and

$f'_+(3) = \lim\limits_{h \to 0^+} \dfrac{f(3+h) - f(3)}{h} = \lim\limits_{h \to 0^+} \dfrac{[(3+h)^2 - 9] - 0}{h} = \lim\limits_{h \to 0^+} \dfrac{6h + h^2}{h} = \lim\limits_{h \to 0^+} (6 + h) = 6$.

Since the left and right limits are different,

$\lim\limits_{h \to 0} \dfrac{f(3+h) - f(3)}{h}$ does not exist, that is, $f'(3)$

does not exist. Similarly, $f'(-3)$ does not exist.

Therefore, f is not differentiable at 3 or at -3.

(b)

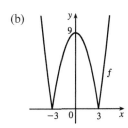
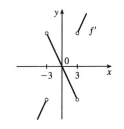

115. f is clearly differentiable for $x < 2$ and for $x > 2$. For $x < 2$, $f'(x) = 2x$, so $f'_-(2) = 4$. For $x > 2$, $f'(x) = m$, so $f'_+(2) = m$. For f to be differentiable at $x = 2$, we need $4 = f'_-(2) = f'_+(2) = m$. So $f(x) = 4x + b$. We must also have continuity at $x = 2$, so $4 = f(2) = \lim\limits_{x \to 2^+} f(x) = \lim\limits_{x \to 2^+} (4x + b) = 8 + b$. Hence, $b = -4$.

117. $y = x^2 \Rightarrow y' = 2x$, so the slope of a tangent line at the point (a, a^2) is $y' = 2a$ and the slope of a normal line is $-1/(2a)$, for $a \neq 0$. The slope of the normal line through the points (a, a^2) and $(0, c)$ is $\dfrac{a^2 - c}{a - 0}$, so $\dfrac{a^2 - c}{a} = -\dfrac{1}{2a} \Rightarrow$
$a^2 - c = -\tfrac{1}{2} \Rightarrow a^2 = c - \tfrac{1}{2}$. The last equation has two solutions if $c > \tfrac{1}{2}$, one solution if $c = \tfrac{1}{2}$, and no solution if $c < \tfrac{1}{2}$. Since the y-axis is normal to $y = x^2$ regardless of the value of c (this is the case for $a = 0$), we have *three* normal lines if $c > \tfrac{1}{2}$ and *one* normal line if $c \leq \tfrac{1}{2}$.

119. *Solution 1:* Let $f(x) = x^{1000}$. Then, by the definition of a derivative, $f'(1) = \lim\limits_{x \to 1} \dfrac{f(x) - f(1)}{x - 1} = \lim\limits_{x \to 1} \dfrac{x^{1000} - 1}{x - 1}$.
But this is just the limit we want to find, and we know (from the Power Rule) that $f'(x) = 1000x^{999}$, so
$f'(1) = 1000(1)^{999} = 1000$. So $\lim\limits_{x \to 1} \dfrac{x^{1000} - 1}{x - 1} = 1000$.

Solution 2: Note that $(x^{1000} - 1) = (x - 1)(x^{999} + x^{998} + x^{997} + \cdots + x^2 + x + 1)$. So
$$\lim_{x \to 1} \frac{x^{1000} - 1}{x - 1} = \lim_{x \to 1} \frac{(x - 1)(x^{999} + x^{998} + x^{997} + \cdots + x^2 + x + 1)}{x - 1} = \lim_{x \to 1} (x^{999} + x^{998} + x^{997} + \cdots + x^2 + x + 1)$$
$$= \underbrace{1 + 1 + 1 + \cdots + 1 + 1 + 1}_{1000 \text{ ones}} = 1000, \text{ as above.}$$

2.4 Derivatives of Trigonometric Functions

1. $f(x) = 3 \sin x - 2 \cos x \Rightarrow f'(x) = 3(\cos x) - 2(-\sin x) = 3 \cos x + 2 \sin x$

3. $y = x^2 + \cot x \Rightarrow y' = 2x + (-\csc^2 x) = 2x - \csc^2 x$

5. $h(\theta) = \theta^2 \sin \theta \overset{\text{PR}}{\Rightarrow} h'(\theta) = \theta^2(\cos \theta) + (\sin \theta)(2\theta) = \theta(\theta \cos \theta + 2 \sin \theta)$

7. $y = \sec \theta \tan \theta \Rightarrow y' = \sec \theta (\sec^2 \theta) + \tan \theta (\sec \theta \tan \theta) = \sec \theta (\sec^2 \theta + \tan^2 \theta)$. Using the identity $1 + \tan^2 \theta = \sec^2 \theta$, we can write alternative forms of the answer as $\sec \theta (1 + 2 \tan^2 \theta)$ or $\sec \theta (2 \sec^2 \theta - 1)$.

9. $f(\theta) = (\theta - \cos \theta) \sin \theta \overset{\text{PR}}{\Rightarrow} f'(\theta) = (\theta - \cos \theta)(\cos \theta) + (\sin \theta)(1 + \sin \theta) = \theta \cos \theta - \cos^2 \theta + \sin \theta + \sin^2 \theta$

11. $H(t) = \cos^2 t = \cos t \cdot \cos t \overset{\text{PR}}{\Rightarrow} H'(t) = \cos t (-\sin t) + \cos t (-\sin t) = -2 \sin t \cos t$. Using the identity $\sin 2t = 2 \sin t \cos t$, we can write an alternative form of the answer as $-\sin 2t$.

13. $f(\theta) = \dfrac{\sin\theta}{1+\cos\theta} \Rightarrow$

$f'(\theta) = \dfrac{(1+\cos\theta)\cos\theta - (\sin\theta)(-\sin\theta)}{(1+\cos\theta)^2} = \dfrac{\cos\theta + \cos^2\theta + \sin^2\theta}{(1+\cos\theta)^2} = \dfrac{\cos\theta + 1}{(1+\cos\theta)^2} = \dfrac{1}{1+\cos\theta}$

15. $y = \dfrac{x}{2-\tan x} \Rightarrow y' = \dfrac{(2-\tan x)(1) - x(-\sec^2 x)}{(2-\tan x)^2} = \dfrac{2 - \tan x + x\sec^2 x}{(2-\tan x)^2}$

17. $f(w) = \dfrac{1+\sec w}{1-\sec w} \quad\overset{QR}{\Rightarrow}$

$f'(w) = \dfrac{(1-\sec w)(\sec w \tan w) - (1+\sec w)(-\sec w \tan w)}{(1-\sec w)^2}$

$= \dfrac{\sec w \tan w - \sec^2 w \tan w + \sec w \tan w + \sec^2 w \tan w}{(1-\sec w)^2} = \dfrac{2\sec w \tan w}{(1-\sec w)^2}$

19. $y = \dfrac{t\sin t}{1+t} \Rightarrow$

$y' = \dfrac{(1+t)(t\cos t + \sin t) - t\sin t(1)}{(1+t)^2} = \dfrac{t\cos t + \sin t + t^2\cos t + t\sin t - t\sin t}{(1+t)^2} = \dfrac{(t^2+t)\cos t + \sin t}{(1+t)^2}$

21. Using Exercise 2.3.107(a), $f(\theta) = \theta \cos\theta \sin\theta \Rightarrow$

$f'(\theta) = 1\cos\theta \sin\theta + \theta(-\sin\theta)\sin\theta + \theta\cos\theta(\cos\theta) = \cos\theta \sin\theta - \theta\sin^2\theta + \theta\cos^2\theta$

$= \sin\theta \cos\theta + \theta(\cos^2\theta - \sin^2\theta) = \tfrac{1}{2}\sin 2\theta + \theta\cos 2\theta$ [using double-angle formulas]

23. $\dfrac{d}{dx}(\csc x) = \dfrac{d}{dx}\left(\dfrac{1}{\sin x}\right) = \dfrac{(\sin x)(0) - 1(\cos x)}{\sin^2 x} = \dfrac{-\cos x}{\sin^2 x} = -\dfrac{1}{\sin x} \cdot \dfrac{\cos x}{\sin x} = -\csc x \cot x$

25. $\dfrac{d}{dx}(\cot x) = \dfrac{d}{dx}\left(\dfrac{\cos x}{\sin x}\right) = \dfrac{(\sin x)(-\sin x) - (\cos x)(\cos x)}{\sin^2 x} = -\dfrac{\sin^2 x + \cos^2 x}{\sin^2 x} = -\dfrac{1}{\sin^2 x} = -\csc^2 x$

27. $y = \sin x + \cos x \Rightarrow y' = \cos x - \sin x$, so $y'(0) = \cos 0 - \sin 0 = 1 - 0 = 1$. An equation of the tangent line to the curve $y = \sin x + \cos x$ at the point $(0,1)$ is $y - 1 = 1(x-0)$ or $y = x + 1$.

29. $y = x + \tan x \Rightarrow y' = 1 + \sec^2 x$, so $y'(\pi) = 1 + (-1)^2 = 2$. An equation of the tangent line to the curve $y = x + \tan x$ at the point (π, π) is $y - \pi = 2(x - \pi)$ or $y = 2x - \pi$.

31. (a) $y = 2x\sin x \Rightarrow y' = 2(x\cos x + \sin x \cdot 1)$. At $\left(\tfrac{\pi}{2}, \pi\right)$,

$y' = 2\left(\tfrac{\pi}{2}\cos\tfrac{\pi}{2} + \sin\tfrac{\pi}{2}\right) = 2(0+1) = 2$, and an equation of the tangent line is $y - \pi = 2\left(x - \tfrac{\pi}{2}\right)$, or $y = 2x$.

(b)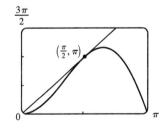

33. (a) $f(x) = \sec x - x \Rightarrow f'(x) = \sec x \tan x - 1$

(b)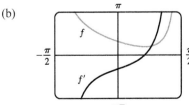

Note that $f' = 0$ where f has a minimum. Also note that f' is negative when f is decreasing and f' is positive when f is increasing.

35. $g(\theta) = \dfrac{\sin\theta}{\theta} \overset{QR}{\Rightarrow} g'(\theta) = \dfrac{\theta(\cos\theta) - (\sin\theta)(1)}{\theta^2} = \dfrac{\theta\cos\theta - \sin\theta}{\theta^2}$

Using the Quotient Rule and $g'(\theta) = \dfrac{\theta\cos\theta - \sin\theta}{\theta^2}$, we get

$g''(\theta) = \dfrac{\theta^2\{[\theta(-\sin\theta) + (\cos\theta)(1)] - \cos\theta\} - (\theta\cos\theta - \sin\theta)(2\theta)}{(\theta^2)^2}$

$= \dfrac{-\theta^3\sin\theta + \theta^2\cos\theta - \theta^2\cos\theta - 2\theta^2\cos\theta + 2\theta\sin\theta}{\theta^4} = \dfrac{\theta(-\theta^2\sin\theta - 2\theta\cos\theta + 2\sin\theta)}{\theta \cdot \theta^3}$

$= \dfrac{-\theta^2\sin\theta - 2\theta\cos\theta + 2\sin\theta}{\theta^3}$

37. (a) $f(x) = \dfrac{\tan x - 1}{\sec x} \Rightarrow$

$f'(x) = \dfrac{\sec x(\sec^2 x) - (\tan x - 1)(\sec x \tan x)}{(\sec x)^2} = \dfrac{\sec x(\sec^2 x - \tan^2 x + \tan x)}{\sec^2 x} = \dfrac{1 + \tan x}{\sec x}$

(b) $f(x) = \dfrac{\tan x - 1}{\sec x} = \dfrac{\dfrac{\sin x}{\cos x} - 1}{\dfrac{1}{\cos x}} = \dfrac{\dfrac{\sin x - \cos x}{\cos x}}{\dfrac{1}{\cos x}} = \sin x - \cos x \Rightarrow f'(x) = \cos x - (-\sin x) = \cos x + \sin x$

(c) From part (a), $f'(x) = \dfrac{1 + \tan x}{\sec x} = \dfrac{1}{\sec x} + \dfrac{\tan x}{\sec x} = \cos x + \sin x$, which is the expression for $f'(x)$ in part (b).

39. $f(x) = x + 2\sin x$ has a horizontal tangent when $f'(x) = 0 \Leftrightarrow 1 + 2\cos x = 0 \Leftrightarrow \cos x = -\tfrac{1}{2} \Leftrightarrow$
$x = \tfrac{2\pi}{3} + 2\pi n$ or $\tfrac{4\pi}{3} + 2\pi n$, where n is an integer. Note that $\tfrac{4\pi}{3}$ and $\tfrac{2\pi}{3}$ are $\pm\tfrac{\pi}{3}$ units from π. This allows us to write the solutions in the more compact equivalent form $(2n + 1)\pi \pm \tfrac{\pi}{3}$, n an integer.

41. (a) $x(t) = 8\sin t \Rightarrow v(t) = x'(t) = 8\cos t \Rightarrow a(t) = x''(t) = -8\sin t$

(b) The mass at time $t = \tfrac{2\pi}{3}$ has position $x\left(\tfrac{2\pi}{3}\right) = 8\sin\tfrac{2\pi}{3} = 8\left(\tfrac{\sqrt{3}}{2}\right) = 4\sqrt{3}$, velocity $v\left(\tfrac{2\pi}{3}\right) = 8\cos\tfrac{2\pi}{3} = 8\left(-\tfrac{1}{2}\right) = -4$,
and acceleration $a\left(\tfrac{2\pi}{3}\right) = -8\sin\tfrac{2\pi}{3} = -8\left(\tfrac{\sqrt{3}}{2}\right) = -4\sqrt{3}$. Since $v\left(\tfrac{2\pi}{3}\right) < 0$, the particle is moving to the left.

43. 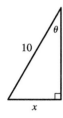 From the diagram we can see that $\sin\theta = x/10 \Leftrightarrow x = 10\sin\theta$. We want to find the rate of change of x with respect to θ, that is, $dx/d\theta$. Taking the derivative of $x = 10\sin\theta$, we get $dx/d\theta = 10(\cos\theta)$. So when $\theta = \tfrac{\pi}{3}$, $\dfrac{dx}{d\theta} = 10\cos\tfrac{\pi}{3} = 10\left(\tfrac{1}{2}\right) = 5$ ft/rad.

45. $\lim\limits_{x\to 0}\dfrac{\sin 5x}{3x} = \lim\limits_{x\to 0}\dfrac{5}{3}\left(\dfrac{\sin 5x}{5x}\right) = \dfrac{5}{3}\lim\limits_{x\to 0}\dfrac{\sin 5x}{5x} = \dfrac{5}{3}\lim\limits_{\theta\to 0}\dfrac{\sin\theta}{\theta}\begin{bmatrix}\text{where }\theta = 5x,\\ \text{using Equation 5}\end{bmatrix} = \dfrac{5}{3}\cdot 1 = \dfrac{5}{3}$

47. $\lim\limits_{t\to 0}\dfrac{\sin 3t}{\sin t} = \lim\limits_{t\to 0}\dfrac{\sin 3t}{3t}\cdot\dfrac{t}{\sin t}\cdot 3 = \lim\limits_{t\to 0}\dfrac{\sin 3t}{3t}\cdot\lim\limits_{t\to 0}\dfrac{t}{\sin t}\cdot\lim\limits_{t\to 0}3 = \lim\limits_{\theta\to 0}\dfrac{\sin\theta}{\theta}\cdot\lim\limits_{t\to 0}\dfrac{1}{\dfrac{\sin t}{t}}\cdot\lim\limits_{t\to 0}3 \quad [\theta = 3t]$

$= 1\cdot 1\cdot 3 = 3$

49. $\lim\limits_{x \to 0} \dfrac{\sin x - \sin x \cos x}{x^2} = \lim\limits_{x \to 0} \dfrac{\sin x\,(1 - \cos x)}{x^2} = \lim\limits_{x \to 0} \dfrac{\sin x}{x} \cdot \lim\limits_{x \to 0} \dfrac{1 - \cos x}{x}$

$= 1 \cdot 0$ [by Equations 5 and 6] $= 0$

51. $\lim\limits_{x \to 0} \dfrac{\tan 2x}{x} = \lim\limits_{x \to 0} \dfrac{\frac{\sin 2x}{\cos 2x}}{x} = \lim\limits_{x \to 0} \dfrac{\sin 2x}{x \cos 2x} = \lim\limits_{x \to 0} \dfrac{\sin 2x}{2x} \cdot \dfrac{2}{\cos 2x}$

$= \lim\limits_{x \to 0} \dfrac{\sin 2x}{2x} \cdot \lim\limits_{x \to 0} \dfrac{2}{\cos 2x} = \lim\limits_{x \to 0} \dfrac{\sin \theta}{\theta} \cdot \lim\limits_{x \to 0} \dfrac{2}{\cos 2x}$ $[\theta = 2x]$

$= 1 \cdot \dfrac{2}{1} = 2$

53. $\lim\limits_{x \to 0} \dfrac{\sin 3x}{5x^3 - 4x} = \lim\limits_{x \to 0} \left(\dfrac{\sin 3x}{3x} \cdot \dfrac{3}{5x^2 - 4} \right) = \lim\limits_{x \to 0} \dfrac{\sin 3x}{3x} \cdot \lim\limits_{x \to 0} \dfrac{3}{5x^2 - 4} = 1 \cdot \left(\dfrac{3}{-4} \right) = -\dfrac{3}{4}$

55. Divide numerator and denominator by θ. [$\sin \theta$ also works.]

$\lim\limits_{\theta \to 0} \dfrac{\sin \theta}{\theta + \tan \theta} = \lim\limits_{\theta \to 0} \dfrac{\frac{\sin \theta}{\theta}}{1 + \frac{\sin \theta}{\theta} \cdot \frac{1}{\cos \theta}} = \dfrac{\lim\limits_{\theta \to 0} \frac{\sin \theta}{\theta}}{1 + \lim\limits_{\theta \to 0} \frac{\sin \theta}{\theta} \lim\limits_{\theta \to 0} \frac{1}{\cos \theta}} = \dfrac{1}{1 + 1 \cdot 1} = \dfrac{1}{2}$

57. $\lim\limits_{\theta \to 0} \dfrac{\cos \theta - 1}{2\theta^2} = \lim\limits_{\theta \to 0} \dfrac{\cos \theta - 1}{2\theta^2} \cdot \dfrac{\cos \theta + 1}{\cos \theta + 1} = \lim\limits_{\theta \to 0} \dfrac{\cos^2 \theta - 1}{2\theta^2 (\cos \theta + 1)} = \lim\limits_{\theta \to 0} \dfrac{-\sin^2 \theta}{2\theta^2 (\cos \theta + 1)}$

$= -\dfrac{1}{2} \lim\limits_{\theta \to 0} \dfrac{\sin \theta}{\theta} \cdot \dfrac{\sin \theta}{\theta} \cdot \dfrac{1}{\cos \theta + 1} = -\dfrac{1}{2} \lim\limits_{\theta \to 0} \dfrac{\sin \theta}{\theta} \cdot \lim\limits_{\theta \to 0} \dfrac{\sin \theta}{\theta} \cdot \lim\limits_{\theta \to 0} \dfrac{1}{\cos \theta + 1}$

$= -\dfrac{1}{2} \cdot 1 \cdot 1 \cdot \dfrac{1}{1+1} = -\dfrac{1}{4}$

59. $\lim\limits_{x \to \pi/4} \dfrac{1 - \tan x}{\sin x - \cos x} = \lim\limits_{x \to \pi/4} \dfrac{\left(1 - \frac{\sin x}{\cos x}\right) \cdot \cos x}{(\sin x - \cos x) \cdot \cos x} = \lim\limits_{x \to \pi/4} \dfrac{\cos x - \sin x}{(\sin x - \cos x) \cos x} = \lim\limits_{x \to \pi/4} \dfrac{-1}{\cos x} = \dfrac{-1}{1/\sqrt{2}} = -\sqrt{2}$

61. $\dfrac{d}{dx}(\sin x) = \cos x \;\Rightarrow\; \dfrac{d^2}{dx^2}(\sin x) = -\sin x \;\Rightarrow\; \dfrac{d^3}{dx^3}(\sin x) = -\cos x \;\Rightarrow\; \dfrac{d^4}{dx^4}(\sin x) = \sin x.$

The derivatives of $\sin x$ occur in a cycle of four. Since $99 = 4(24) + 3$, we have $\dfrac{d^{99}}{dx^{99}}(\sin x) = \dfrac{d^3}{dx^3}(\sin x) = -\cos x$.

63. $y = A \sin x + B \cos x \;\Rightarrow\; y' = A \cos x - B \sin x \;\Rightarrow\; y'' = -A \sin x - B \cos x$. Substituting these expressions for y, y', and y'' into the given differential equation $y'' + y' - 2y = \sin x$ gives us

$(-A \sin x - B \cos x) + (A \cos x - B \sin x) - 2(A \sin x + B \cos x) = \sin x \;\Leftrightarrow\;$

$-3A \sin x - B \sin x + A \cos x - 3B \cos x = \sin x \;\Leftrightarrow\; (-3A - B) \sin x + (A - 3B) \cos x = 1 \sin x$, so we must have $-3A - B = 1$ and $A - 3B = 0$ (since 0 is the coefficient of $\cos x$ on the right side). Solving for A and B, we add the first equation to three times the second to get $B = -\dfrac{1}{10}$ and $A = -\dfrac{3}{10}$.

65. (a) $\dfrac{d}{dx}\tan x = \dfrac{d}{dx}\dfrac{\sin x}{\cos x} \Rightarrow \sec^2 x = \dfrac{\cos x \cos x - \sin x(-\sin x)}{\cos^2 x} = \dfrac{\cos^2 x + \sin^2 x}{\cos^2 x}$. So $\sec^2 x = \dfrac{1}{\cos^2 x}$.

(b) $\dfrac{d}{dx}\sec x = \dfrac{d}{dx}\dfrac{1}{\cos x} \Rightarrow \sec x \tan x = \dfrac{(\cos x)(0) - 1(-\sin x)}{\cos^2 x}$. So $\sec x \tan x = \dfrac{\sin x}{\cos^2 x}$.

(c) $\dfrac{d}{dx}(\sin x + \cos x) = \dfrac{d}{dx}\dfrac{1 + \cot x}{\csc x} \Rightarrow$

$\cos x - \sin x = \dfrac{\csc x(-\csc^2 x) - (1 + \cot x)(-\csc x \cot x)}{\csc^2 x} = \dfrac{\csc x[-\csc^2 x + (1 + \cot x)\cot x]}{\csc^2 x}$

$= \dfrac{-\csc^2 x + \cot^2 x + \cot x}{\csc x} = \dfrac{-1 + \cot x}{\csc x}$

So $\cos x - \sin x = \dfrac{\cot x - 1}{\csc x}$.

67. By the definition of radian measure, $s = r\theta$, where r is the radius of the circle. By drawing the bisector of the angle θ, we can

see that $\sin\dfrac{\theta}{2} = \dfrac{d/2}{r} \Rightarrow d = 2r\sin\dfrac{\theta}{2}$. So $\lim\limits_{\theta \to 0^+}\dfrac{s}{d} = \lim\limits_{\theta \to 0^+}\dfrac{r\theta}{2r\sin(\theta/2)} = \lim\limits_{\theta \to 0^+}\dfrac{2\cdot(\theta/2)}{2\sin(\theta/2)} = \lim\limits_{\theta \to 0}\dfrac{\theta/2}{\sin(\theta/2)} = 1$.

[This is just the reciprocal of the limit $\lim\limits_{x\to 0}\dfrac{\sin x}{x} = 1$ combined with the fact that as $\theta \to 0$, $\dfrac{\theta}{2} \to 0$ also.]

2.5 The Chain Rule

1. Let $u = g(x) = 5 - x^4$ and $y = f(u) = u^3$. Then $\dfrac{dy}{dx} = \dfrac{dy}{du}\dfrac{du}{dx} = (3u^2)(-4x^3) = 3(5 - x^4)^2(-4x^3) = -12x^3(5 - x^4)^2$.

3. Let $u = g(x) = \cos x$ and $y = f(u) = \sin u$. Then

$\dfrac{dy}{dx} = \dfrac{dy}{du}\dfrac{du}{dx} = (\cos u)(-\sin x) = (\cos(\cos x))(-\sin x) = -\sin x\, \cos(\cos x)$.

5. Let $u = g(x) = \sin x$ and $y = f(u) = \sqrt{u}$. Then $\dfrac{dy}{dx} = \dfrac{dy}{du}\dfrac{du}{dx} = \tfrac{1}{2}u^{-1/2}\cos x = \dfrac{\cos x}{2\sqrt{u}} = \dfrac{\cos x}{2\sqrt{\sin x}}$.

7. $f(x) = (2x^3 - 5x^2 + 4)^5 \;\overset{CR}{\Rightarrow}$

$f'(x) = 5(2x^3 - 5x^2 + 4)^4 \cdot \dfrac{d}{dx}(2x^3 - 5x^2 + 4) = 5(2x^3 - 5x^2 + 4)^4(6x^2 - 10x)$

$= 5(2x^3 - 5x^2 + 4)^4 \cdot 2x(3x - 5) = 10x(2x^3 - 5x^2 + 4)^4(3x - 5)$

9. $f(x) = \sqrt{5x + 1} = (5x + 1)^{1/2} \;\overset{CR}{\Rightarrow}\; f'(x) = \tfrac{1}{2}(5x + 1)^{-1/2} \cdot \dfrac{d}{dx}(5x + 1) = \tfrac{1}{2}(5x + 1)^{-1/2}(5) = \dfrac{5}{2\sqrt{5x + 1}}$

11. $g(t) = \dfrac{1}{(2t + 1)^2} = (2t + 1)^{-2} \;\overset{CR}{\Rightarrow}\; g'(t) = -2(2t + 1)^{-3}\cdot\dfrac{d}{dt}(2t + 1) = -2(2t + 1)^{-3}(2) = -\dfrac{4}{(2t + 1)^3}$

13. $A(t) = \dfrac{1}{(\cos t + \tan t)^2} = (\cos t + \tan t)^{-2} \;\overset{CR}{\Rightarrow}\; A'(t) = -2(\cos t + \tan t)^{-3}(-\sin t + \sec^2 t) = \dfrac{2(\sin t - \sec^2 t)}{(\cos t + \tan t)^3}$

15. $f(\theta) = \cos(\theta^2) \;\overset{CR}{\Rightarrow}\; f'(\theta) = -\sin(\theta^2)\cdot\dfrac{d}{d\theta}(\theta^2) = -\sin(\theta^2)\cdot(2\theta) = -2\theta\sin(\theta^2)$

17. $h(v) = v\sqrt[3]{1+v^2} = v(1+v^2)^{1/3}$ $\overset{PR}{\Rightarrow}$

$h'(v) = v \cdot \frac{1}{3}(1+v^2)^{-2/3}(2v) + (1+v^2)^{1/3} \cdot 1 = \frac{1}{3}(1+v^2)^{-2/3}[2v^2 + 3(1+v^2)] = \dfrac{5v^2+3}{3(\sqrt[3]{1+v^2})^2}$

19. $F(x) = (4x+5)^3(x^2-2x+5)^4$ $\Rightarrow$

$F'(x) = (4x+5)^3 \cdot 4(x^2-2x+5)^3(2x-2) + (x^2-2x+5)^4 \cdot 3(4x+5)^2 \cdot 4$

$= 4(4x+5)^2(x^2-2x+5)^3\left[(4x+5)(2x-2) + (x^2-2x+5) \cdot 3\right]$

$= 4(4x+5)^2(x^2-2x+5)^3(8x^2+2x-10+3x^2-6x+15)$

$= 4(4x+5)^2(x^2-2x+5)^3(11x^2-4x+5)$

21. $h(t) = (t+1)^{2/3}(2t^2-1)^3$ $\Rightarrow$

$h'(t) = (t+1)^{2/3} \cdot 3(2t^2-1)^2 \cdot 4t + (2t^2-1)^3 \cdot \frac{2}{3}(t+1)^{-1/3} = \frac{2}{3}(t+1)^{-1/3}(2t^2-1)^2[18t(t+1) + (2t^2-1)]$

$= \frac{2}{3}(t+1)^{-1/3}(2t^2-1)^2(20t^2+18t-1)$

23. $y = \sqrt{\dfrac{x}{x+1}} = \left(\dfrac{x}{x+1}\right)^{1/2}$ $\Rightarrow$

$y' = \dfrac{1}{2}\left(\dfrac{x}{x+1}\right)^{-1/2} \dfrac{d}{dx}\left(\dfrac{x}{x+1}\right) = \dfrac{1}{2}\dfrac{x^{-1/2}}{(x+1)^{-1/2}}\dfrac{(x+1)(1)-x(1)}{(x+1)^2}$

$= \dfrac{1}{2}\dfrac{(x+1)^{1/2}}{x^{1/2}}\dfrac{1}{(x+1)^2} = \dfrac{1}{2\sqrt{x}(x+1)^{3/2}}$

25. $g(u) = \left(\dfrac{u^3-1}{u^3+1}\right)^8$ $\Rightarrow$

$g'(u) = 8\left(\dfrac{u^3-1}{u^3+1}\right)^7 \dfrac{d}{du}\dfrac{u^3-1}{u^3+1} = 8\dfrac{(u^3-1)^7}{(u^3+1)^7}\dfrac{(u^3+1)(3u^2)-(u^3-1)(3u^2)}{(u^3+1)^2}$

$= 8\dfrac{(u^3-1)^7}{(u^3+1)^7}\dfrac{3u^2[(u^3+1)-(u^3-1)]}{(u^3+1)^2} = 8\dfrac{(u^3-1)^7}{(u^3+1)^7}\dfrac{3u^2(2)}{(u^3+1)^2} = \dfrac{48u^2(u^3-1)^7}{(u^3+1)^9}$

27. $H(r) = \dfrac{(r^2-1)^3}{(2r+1)^5}$ $\Rightarrow$

$H'(r) = \dfrac{(2r+1)^5 \cdot 3(r^2-1)^2(2r) - (r^2-1)^3 \cdot 5(2r+1)^4(2)}{[(2r+1)^5]^2} = \dfrac{2(2r+1)^4(r^2-1)^2[3r(2r+1)-5(r^2-1)]}{(2r+1)^{10}}$

$= \dfrac{2(r^2-1)^2(6r^2+3r-5r^2+5)}{(2r+1)^6} = \dfrac{2(r^2-1)^2(r^2+3r+5)}{(2r+1)^6}$

29. $y = \cos(\sec 4x)$ $\Rightarrow$

$y' = -\sin(\sec 4x)\dfrac{d}{dx}\sec 4x = -\sin(\sec 4x) \cdot \sec 4x \tan 4x \cdot 4 = -4\sin(\sec 4x)\sec 4x \tan 4x$

31. $y = \dfrac{\cos x}{\sqrt{1+\sin x}} = (\cos x)(1+\sin x)^{-1/2} \Rightarrow$

$y' = (\cos x) \cdot \left(-\tfrac{1}{2}\right)(1+\sin x)^{-3/2} \cos x + (1+\sin x)^{-1/2}(-\sin x)$

$ = -\tfrac{1}{2}(1+\sin x)^{-3/2}[\cos^2 x + 2(1+\sin x)\sin x] = -\tfrac{1}{2}(1+\sin x)^{-3/2}(\cos^2 x + 2\sin x + 2\sin^2 x)$

$ = -\tfrac{1}{2}(1+\sin x)^{-3/2}(1+2\sin x + \sin^2 x) = -\tfrac{1}{2}(1+\sin x)^{-3/2}(1+\sin x)^2$

$ = -\tfrac{1}{2}(1+\sin x)^{1/2}$ or $-\tfrac{1}{2}\sqrt{1+\sin x}$

33. $y = \left(\dfrac{1-\cos 2x}{1+\cos 2x}\right)^4 \Rightarrow$

$y' = 4\left(\dfrac{1-\cos 2x}{1+\cos 2x}\right)^3 \cdot \dfrac{(1+\cos 2x)(2\sin 2x) + (1-\cos 2x)(-2\sin 2x)}{(1+\cos 2x)^2}$

$ = 4\left(\dfrac{1-\cos 2x}{1+\cos 2x}\right)^3 \cdot \dfrac{2\sin 2x\,(1+\cos 2x + 1 - \cos 2x)}{(1+\cos 2x)^2} = \dfrac{4(1-\cos 2x)^3}{(1+\cos 2x)^3} \dfrac{2\sin 2x\,(2)}{(1+\cos 2x)^2} = \dfrac{16\sin 2x\,(1-\cos 2x)^3}{(1+\cos 2x)^5}$

35. $f(x) = \sin x \cos(1-x^2) \Rightarrow$

$f'(x) = \sin x \left[-\sin(1-x^2)(-2x)\right] + \cos(1-x^2) \cdot \cos x = 2x \sin x \sin(1-x^2) + \cos x \cos(1-x^2)$

37. $F(t) = \tan\sqrt{1+t^2} \Rightarrow F'(t) = \sec^2\sqrt{1+t^2} \cdot \dfrac{1}{2\sqrt{1+t^2}} \cdot 2t = \dfrac{t\sec^2\sqrt{1+t^2}}{\sqrt{1+t^2}}$

39. $y = \sin^2(x^2+1) \Rightarrow y' = 2\sin(x^2+1) \cdot \cos(x^2+1) \cdot 2x = 4x\sin(x^2+1)\cos(x^2+1)$

41. $y = \cos^4(\sin^3 x) = [\cos(\sin^3 x)]^4 \Rightarrow$

$y' = 4[\cos(\sin^3 x)]^3(-\sin(\sin^3 x))\,3\sin^2 x \cos x = -12\cos^3(\sin^3 x)\sin(\sin^3 x)\sin^2 x \cos x$

43. $f(t) = \tan(\sec(\cos t)) \Rightarrow$

$f'(t) = \sec^2(\sec(\cos t))\,\dfrac{d}{dt}\sec(\cos t) = \sec^2(\sec(\cos t))[\sec(\cos t)\tan(\cos t)]\,\dfrac{d}{dt}\cos t$

$ = -\sec^2(\sec(\cos t))\sec(\cos t)\tan(\cos t)\sin t$

45. $g(x) = (2r \sin rx + n)^p \Rightarrow g'(x) = p(2r\sin rx + n)^{p-1}(2r\cos rx \cdot r) = p(2r\sin rx + n)^{p-1}(2r^2 \cos rx)$

47. $y = \cos\sqrt{\sin(\tan \pi x)} = \cos(\sin(\tan \pi x))^{1/2} \Rightarrow$

$y' = -\sin(\sin(\tan \pi x))^{1/2} \cdot \dfrac{d}{dx}(\sin(\tan \pi x))^{1/2} = -\sin(\sin(\tan \pi x))^{1/2} \cdot \tfrac{1}{2}(\sin(\tan \pi x))^{-1/2} \cdot \dfrac{d}{dx}(\sin(\tan \pi x))$

$ = \dfrac{-\sin\sqrt{\sin(\tan \pi x)}}{2\sqrt{\sin(\tan \pi x)}} \cdot \cos(\tan \pi x) \cdot \dfrac{d}{dx}\tan \pi x = \dfrac{-\sin\sqrt{\sin(\tan \pi x)}}{2\sqrt{\sin(\tan \pi x)}} \cdot \cos(\tan \pi x) \cdot \sec^2(\pi x) \cdot \pi$

$ = \dfrac{-\pi \cos(\tan \pi x)\sec^2(\pi x)\sin\sqrt{\sin(\tan \pi x)}}{2\sqrt{\sin(\tan \pi x)}}$

49. $y = \cos(\sin 3\theta) \Rightarrow y' = -\sin(\sin 3\theta) \cdot (\cos 3\theta) \cdot 3 = -3\cos 3\theta \sin(\sin 3\theta) \Rightarrow$

$y'' = -3\,[(\cos 3\theta)\cos(\sin 3\theta)(\cos 3\theta) \cdot 3 + \sin(\sin 3\theta)(-\sin 3\theta) \cdot 3] = -9\cos^2(3\theta)\cos(\sin 3\theta) + 9(\sin 3\theta)\sin(\sin 3\theta)$

51. $y = \sqrt{\cos x}$ $\Rightarrow$ $y' = \dfrac{1}{2\sqrt{\cos x}}(-\sin x) = -\dfrac{\sin x}{2\sqrt{\cos x}}$. With $y' = \dfrac{-\sin x}{2\sqrt{\cos x}}$, we get

$$y'' = \dfrac{2\sqrt{\cos x} \cdot (-\cos x) - (-\sin x)\left(2 \cdot \dfrac{1}{2\sqrt{\cos x}}(-\sin x)\right)}{(2\sqrt{\cos x})^2} = \dfrac{-2\cos x\sqrt{\cos x} - \dfrac{\sin^2 x}{\sqrt{\cos x}}}{4\cos x} \cdot \dfrac{\sqrt{\cos x}}{\sqrt{\cos x}}$$

$$= \dfrac{-2\cos x \cdot \cos x - \sin^2 x}{4\cos x\sqrt{\cos x}} = -\dfrac{2\cos^2 x + \sin^2 x}{4(\cos x)^{3/2}}$$

Using the identity $\sin^2 x + \cos^2 x = 1$, the answer may be written as $-\dfrac{1 + \cos^2 x}{4(\cos x)^{3/2}}$.

53. $y = (3x - 1)^{-6}$ $\Rightarrow$ $y' = -6(3x-1)^{-7} \cdot 3 = -18(3x-1)^{-7}$. At $(0,1)$, $y' = -18(-1)^{-7} = -18(-1) = 18$, and an equation of the tangent line is $y - 1 = 18(x - 0)$, or $y = 18x + 1$.

55. $y = \sin(\sin x)$ $\Rightarrow$ $y' = \cos(\sin x) \cdot \cos x$. At $(\pi, 0)$, $y' = \cos(\sin \pi) \cdot \cos \pi = \cos(0) \cdot (-1) = 1(-1) = -1$, and an equation of the tangent line is $y - 0 = -1(x - \pi)$, or $y = -x + \pi$.

57. (a) $y = f(x) = \tan\left(\tfrac{\pi}{4}x^2\right)$ $\Rightarrow$ $f'(x) = \sec^2\left(\tfrac{\pi}{4}x^2\right)\left(2 \cdot \tfrac{\pi}{4}x\right)$.

The slope of the tangent at $(1, 1)$ is thus

$f'(1) = \sec^2 \tfrac{\pi}{4} \left(\tfrac{\pi}{2}\right) = 2 \cdot \tfrac{\pi}{2} = \pi$, and its equation

is $y - 1 = \pi(x - 1)$ or $y = \pi x - \pi + 1$.

(b)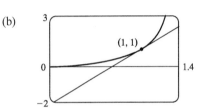

59. (a) $f(x) = x\sqrt{2 - x^2} = x(2 - x^2)^{1/2}$ $\Rightarrow$

$$f'(x) = x \cdot \tfrac{1}{2}(2 - x^2)^{-1/2}(-2x) + (2 - x^2)^{1/2} \cdot 1 = (2 - x^2)^{-1/2}\left[-x^2 + (2 - x^2)\right] = \dfrac{2 - 2x^2}{\sqrt{2 - x^2}}$$

(b)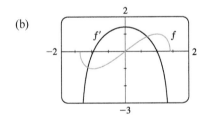

$f' = 0$ when f has a horizontal tangent line, f' is negative when f is decreasing, and f' is positive when f is increasing.

61. For the tangent line to be horizontal, $f'(x) = 0$. $f(x) = 2\sin x + \sin^2 x$ $\Rightarrow$ $f'(x) = 2\cos x + 2\sin x \cos x = 0$ $\Leftrightarrow$ $2\cos x(1 + \sin x) = 0$ $\Leftrightarrow$ $\cos x = 0$ or $\sin x = -1$, so $x = \tfrac{\pi}{2} + 2n\pi$ or $\tfrac{3\pi}{2} + 2n\pi$, where n is any integer. Now $f\left(\tfrac{\pi}{2}\right) = 3$ and $f\left(\tfrac{3\pi}{2}\right) = -1$, so the points on the curve with a horizontal tangent are $\left(\tfrac{\pi}{2} + 2n\pi, 3\right)$ and $\left(\tfrac{3\pi}{2} + 2n\pi, -1\right)$, where n is any integer.

63. $F(x) = f(g(x))$ $\Rightarrow$ $F'(x) = f'(g(x)) \cdot g'(x)$, so $F'(5) = f'(g(5)) \cdot g'(5) = f'(-2) \cdot 6 = 4 \cdot 6 = 24$.

65. (a) $h(x) = f(g(x))$ $\Rightarrow$ $h'(x) = f'(g(x)) \cdot g'(x)$, so $h'(1) = f'(g(1)) \cdot g'(1) = f'(2) \cdot 6 = 5 \cdot 6 = 30$.

(b) $H(x) = g(f(x))$ $\Rightarrow$ $H'(x) = g'(f(x)) \cdot f'(x)$, so $H'(1) = g'(f(1)) \cdot f'(1) = g'(3) \cdot 4 = 9 \cdot 4 = 36$.

67. (a) From the graphs of f and g, we obtain the following values: $g(1) = 4$ since the point $(1, 4)$ is on the graph of g; $f'(4) = -\frac{1}{4}$ since the slope of the line segment between $(2, 4)$ and $(6, 3)$ is $\frac{3-4}{6-2} = -\frac{1}{4}$; and $g'(1) = -1$ since the slope of the line segment between $(0, 5)$ and $(3, 2)$ is $\frac{2-5}{3-0} = -1$. Now $u(x) = f(g(x))$, so

$u'(1) = f'(g(1))\, g'(1) = f'(4)\, g'(1) = -\frac{1}{4}(-1) = \frac{1}{4}$.

(b) From the graphs of f and g, we obtain the following values: $f(1) = 2$ since the point $(1, 2)$ is on the graph of f; $g'(2) = g'(1) = -1$ [see part (a)]; and $f'(1) = 2$ since the slope of the line segment between $(0, 0)$ and $(2, 4)$ is $\frac{4-0}{2-0} = 2$. Now $v(x) = g(f(x))$, so $v'(1) = g'(f(1))\, f'(1) = g'(2)\, f'(1) = -1(2) = -2$.

(c) From part (a), we have $g(1) = 4$ and $g'(1) = -1$. From the graph of g we obtain $g'(4) = \frac{1}{2}$ since the slope of the line segment between $(3, 2)$ and $(7, 4)$ is $\frac{4-2}{7-3} = \frac{1}{2}$. Now $w(x) = g(g(x))$, so

$w'(1) = g'(g(1))\, g'(1) = g'(4)\, g'(1) = \frac{1}{2}(-1) = -\frac{1}{2}$.

69. The point $(3, 2)$ is on the graph of f, so $f(3) = 2$. The tangent line at $(3, 2)$ has slope $\frac{\Delta y}{\Delta x} = \frac{-4}{6} = -\frac{2}{3}$.

$g(x) = \sqrt{f(x)} \;\Rightarrow\; g'(x) = \frac{1}{2}[f(x)]^{-1/2} \cdot f'(x) \;\Rightarrow\;$

$g'(3) = \frac{1}{2}[f(3)]^{-1/2} \cdot f'(3) = \frac{1}{2}(2)^{-1/2}(-\frac{2}{3}) = -\frac{1}{3\sqrt{2}}$ or $-\frac{1}{6}\sqrt{2}$.

71. $r(x) = f(g(h(x))) \;\Rightarrow\; r'(x) = f'(g(h(x))) \cdot g'(h(x)) \cdot h'(x)$, so

$r'(1) = f'(g(h(1))) \cdot g'(h(1)) \cdot h'(1) = f'(g(2)) \cdot g'(2) \cdot 4 = f'(3) \cdot 5 \cdot 4 = 6 \cdot 5 \cdot 4 = 120$

73. $F(x) = f(3f(4f(x))) \;\Rightarrow\;$

$F'(x) = f'(3f(4f(x))) \cdot \frac{d}{dx}(3f(4f(x))) = f'(3f(4f(x))) \cdot 3f'(4f(x)) \cdot \frac{d}{dx}(4f(x))$

$= f'(3f(4f(x))) \cdot 3f'(4f(x)) \cdot 4f'(x)$, so

$F'(0) = f'(3f(4f(0))) \cdot 3f'(4f(0)) \cdot 4f'(0) = f'(3f(4 \cdot 0)) \cdot 3f'(4 \cdot 0) \cdot 4 \cdot 2 = f'(3 \cdot 0) \cdot 3 \cdot 2 \cdot 4 \cdot 2 = 2 \cdot 3 \cdot 2 \cdot 4 \cdot 2 = 96$.

75. Let $f(x) = \cos x$. Then $Df(2x) = 2f'(2x)$, $D^2 f(2x) = 2^2 f''(2x)$, $D^3 f(2x) = 2^3 f'''(2x)$, ..., $D^{(n)} f(2x) = 2^n f^{(n)}(2x)$. Since the derivatives of $\cos x$ occur in a cycle of four, and since $103 = 4(25) + 3$, we have $f^{(103)}(x) = f^{(3)}(x) = \sin x$ and $D^{103} \cos 2x = 2^{103} f^{(103)}(2x) = 2^{103} \sin 2x$.

77. $s(t) = 10 + \frac{1}{4}\sin(10\pi t) \;\Rightarrow\;$ the velocity after t seconds is $v(t) = s'(t) = \frac{1}{4}\cos(10\pi t)(10\pi) = \frac{5\pi}{2}\cos(10\pi t)$ cm/s.

79. (a) $B(t) = 4.0 + 0.35 \sin\left(\frac{2\pi t}{5.4}\right) \;\Rightarrow\; \frac{dB}{dt} = \left(0.35 \cos \frac{2\pi t}{5.4}\right)\left(\frac{2\pi}{5.4}\right) = \frac{0.7\pi}{5.4} \cos \frac{2\pi t}{5.4} = \frac{7\pi}{54} \cos \frac{2\pi t}{5.4}$

(b) At $t = 1$, $\frac{dB}{dt} = \frac{7\pi}{54} \cos \frac{2\pi}{5.4} \approx 0.16$.

81. By the Chain Rule, $a(t) = \dfrac{dv}{dt} = \dfrac{dv}{ds}\dfrac{ds}{dt} = \dfrac{dv}{ds}v(t) = v(t)\dfrac{dv}{ds}$. The derivative dv/dt is the rate of change of the velocity with respect to time (in other words, the acceleration) whereas the derivative dv/ds is the rate of change of the velocity with respect to the displacement.

83. (a) If f is even, then $f(x) = f(-x)$. Using the Chain Rule to differentiate this equation, we get

$$f'(x) = f'(-x)\dfrac{d}{dx}(-x) = -f'(-x). \text{ Thus, } f'(-x) = -f'(x), \text{ so } f' \text{ is odd.}$$

(b) If f is odd, then $f(x) = -f(-x)$. Differentiating this equation, we get $f'(x) = -f'(-x)(-1) = f'(-x)$, so f' is even.

85. Since $\theta° = \left(\dfrac{\pi}{180}\right)\theta$ rad, we have $\dfrac{d}{d\theta}(\sin\theta°) = \dfrac{d}{d\theta}\left(\sin\dfrac{\pi}{180}\theta\right) = \dfrac{\pi}{180}\cos\dfrac{\pi}{180}\theta = \dfrac{\pi}{180}\cos\theta°$.

87. Let $j(x) = g(h(x))$ so that $F(x) = f(g(h(x))) = f(j(x))$. By the Chain Rule, we have $j'(x) = g'(h(x)) \cdot h'(x)$ and, by the Chain Rule and substitution, we have $F'(x) = f'(j(x)) \cdot j'(x) = f'(g(h(x))) \cdot g'(h(x)) \cdot h'(x)$.

2.6 Implicit Differentiation

1. (a) $\dfrac{d}{dx}(5x^2 - y^3) = \dfrac{d}{dx}(7) \;\Rightarrow\; 10x - 3y^2 y' = 0 \;\Rightarrow\; 3y^2 y' = 10x \;\Rightarrow\; y' = \dfrac{10x}{3y^2}$

(b) $5x^2 - y^3 = 7 \;\Rightarrow\; y^3 = 5x^2 - 7 \;\Rightarrow\; y = \sqrt[3]{5x^2 - 7}$, so $y' = \dfrac{1}{3}(5x^2 - 7)^{-2/3}(10x) = \dfrac{10x}{3(5x^2 - 7)^{2/3}}$

(c) From part (a), $y' = \dfrac{10x}{3y^2} = \dfrac{10x}{3(y^3)^{2/3}} = \dfrac{10x}{3(5x^2 - 7)^{2/3}}$, which agrees with part (b).

3. (a) $\dfrac{d}{dx}(\sqrt{x} + \sqrt{y}) = \dfrac{d}{dx}(1) \;\Rightarrow\; \dfrac{1}{2}x^{-1/2} + \dfrac{1}{2}y^{-1/2}y' = 0 \;\Rightarrow\; \dfrac{1}{2\sqrt{y}}y' = -\dfrac{1}{2\sqrt{x}} \;\Rightarrow\; y' = -\dfrac{\sqrt{y}}{\sqrt{x}}$

(b) $\sqrt{x} + \sqrt{y} = 1 \;\Rightarrow\; \sqrt{y} = 1 - \sqrt{x} \;\Rightarrow\; y = (1 - \sqrt{x})^2 \;\Rightarrow\; y = 1 - 2\sqrt{x} + x$, so

$y' = -2 \cdot \dfrac{1}{2}x^{-1/2} + 1 = 1 - \dfrac{1}{\sqrt{x}}$.

(c) From part (a), $y' = -\dfrac{\sqrt{y}}{\sqrt{x}} = -\dfrac{1 - \sqrt{x}}{\sqrt{x}}$ [from part (b)] $= -\dfrac{1}{\sqrt{x}} + 1$, which agrees with part (b).

5. $\dfrac{d}{dx}(x^2 - 4xy + y^2) = \dfrac{d}{dx}(4) \;\Rightarrow\; 2x - 4[xy' + y(1)] + 2y\,y' = 0 \;\Rightarrow\; 2y\,y' - 4xy' = 4y - 2x \;\Rightarrow\;$

$y'(y - 2x) = 2y - x \;\Rightarrow\; y' = \dfrac{2y - x}{y - 2x}$

7. $\dfrac{d}{dx}(x^4 + x^2 y^2 + y^3) = \dfrac{d}{dx}(5) \;\Rightarrow\; 4x^3 + x^2 \cdot 2y\,y' + y^2 \cdot 2x + 3y^2 y' = 0 \;\Rightarrow\; 2x^2 y\,y' + 3y^2 y' = -4x^3 - 2xy^2 \;\Rightarrow\;$

$(2x^2 y + 3y^2)y' = -4x^3 - 2xy^2 \;\Rightarrow\; y' = \dfrac{-4x^3 - 2xy^2}{2x^2 y + 3y^2} = -\dfrac{2x(2x^2 + y^2)}{y(2x^2 + 3y)}$

9. $\dfrac{d}{dx}\left(\dfrac{x^2}{x+y}\right) = \dfrac{d}{dx}(y^2+1) \;\Rightarrow\; \dfrac{(x+y)(2x) - x^2(1+y')}{(x+y)^2} = 2yy' \;\Rightarrow$

$2x^2 + 2xy - x^2 - x^2y' = 2y(x+y)^2 y' \;\Rightarrow\; x^2 + 2xy = 2y(x+y)^2 y' + x^2 y' \;\Rightarrow$

$x(x+2y) = [2y(x^2 + 2xy + y^2) + x^2] y' \;\Rightarrow\; y' = \dfrac{x(x+2y)}{2x^2 y + 4xy^2 + 2y^3 + x^2}$

Or: Start by clearing fractions and then differentiate implicitly.

11. $\dfrac{d}{dx}(\sin x + \cos y) = \dfrac{d}{dx}(2x - 3y) \;\Rightarrow\; \cos x - \sin y \cdot y' = 2 - 3y' \;\Rightarrow\; 3y' - \sin y \cdot y' = 2 - \cos x \;\Rightarrow$

$y'(3 - \sin y) = 2 - \cos x \;\Rightarrow\; y' = \dfrac{2 - \cos x}{3 - \sin y}$

13. $\dfrac{d}{dx}\sin(x+y) = \dfrac{d}{dx}(\cos x + \cos y) \;\Rightarrow\; \cos(x+y)\cdot(1+y') = -\sin x - \sin y \cdot y' \;\Rightarrow$

$\cos(x+y) + y'\cos(x+y) = -\sin x - \sin y \cdot y' \;\Rightarrow\; y'\cos(x+y) + \sin y \cdot y' = -\sin x - \cos(x+y) \;\Rightarrow$

$y'[\cos(x+y) + \sin y] = -[\sin x + \cos(x+y)] \;\Rightarrow\; y' = -\dfrac{\cos(x+y) + \sin x}{\cos(x+y) + \sin y}$

15. $\dfrac{d}{dx}\tan(x/y) = \dfrac{d}{dx}(x+y) \;\Rightarrow\; \sec^2(x/y) \cdot \dfrac{y \cdot 1 - x \cdot y'}{y^2} = 1 + y' \;\Rightarrow$

$y\sec^2(x/y) - x\sec^2(x/y)\cdot y' = y^2 + y^2 y' \;\Rightarrow\; y\sec^2(x/y) - y^2 = y^2 y' + x\sec^2(x/y) \;\Rightarrow$

$y\sec^2(x/y) - y^2 = [y^2 + x\sec^2(x/y)]\cdot y' \;\Rightarrow\; y' = \dfrac{y\sec^2(x/y) - y^2}{y^2 + x\sec^2(x/y)}$

17. $\dfrac{d}{dx}\sqrt{x+y} = \dfrac{d}{dx}(x^4 + y^4) \;\Rightarrow\; \tfrac{1}{2}(x+y)^{-1/2}(1+y') = 4x^3 + 4y^3 y' \;\Rightarrow$

$\dfrac{1}{2\sqrt{x+y}} + \dfrac{1}{2\sqrt{x+y}}y' = 4x^3 + 4y^3 y' \;\Rightarrow\; \dfrac{1}{2\sqrt{x+y}} - 4x^3 = 4y^3 y' - \dfrac{1}{2\sqrt{x+y}}y' \;\Rightarrow$

$\dfrac{1 - 8x^3\sqrt{x+y}}{2\sqrt{x+y}} = \dfrac{8y^3\sqrt{x+y} - 1}{2\sqrt{x+y}}y' \;\Rightarrow\; y' = \dfrac{1 - 8x^3\sqrt{x+y}}{8y^3\sqrt{x+y} - 1}$

19. $\dfrac{d}{dx}\sqrt{xy} = \dfrac{d}{dx}(1 + x^2 y) \;\Rightarrow\; \tfrac{1}{2}(xy)^{-1/2}(xy' + y \cdot 1) = 0 + x^2 y' + y \cdot 2x \;\Rightarrow$

$\dfrac{x}{2\sqrt{xy}}y' + \dfrac{y}{2\sqrt{xy}} = x^2 y' + 2xy \;\Rightarrow\; y'\left(\dfrac{x}{2\sqrt{xy}} - x^2\right) = 2xy - \dfrac{y}{2\sqrt{xy}} \;\Rightarrow$

$y'\left(\dfrac{x - 2x^2\sqrt{xy}}{2\sqrt{xy}}\right) = \dfrac{4xy\sqrt{xy} - y}{2\sqrt{xy}} \;\Rightarrow\; y' = \dfrac{4xy\sqrt{xy} - y}{x - 2x^2\sqrt{xy}}$

21. $\dfrac{d}{dx}\{f(x) + x^2[f(x)]^3\} = \dfrac{d}{dx}(10) \;\Rightarrow\; f'(x) + x^2 \cdot 3[f(x)]^2 \cdot f'(x) + [f(x)]^3 \cdot 2x = 0.$ If $x = 1$, we have

$f'(1) + 1^2 \cdot 3[f(1)]^2 \cdot f'(1) + [f(1)]^3 \cdot 2(1) = 0 \;\Rightarrow\; f'(1) + 1 \cdot 3 \cdot 2^2 \cdot f'(1) + 2^3 \cdot 2 = 0 \;\Rightarrow$

$f'(1) + 12 f'(1) = -16 \;\Rightarrow\; 13 f'(1) = -16 \;\Rightarrow\; f'(1) = -\tfrac{16}{13}.$

23. $\dfrac{d}{dy}(x^4y^2 - x^3y + 2xy^3) = \dfrac{d}{dy}(0) \Rightarrow x^4 \cdot 2y + y^2 \cdot 4x^3\, x' - (x^3 \cdot 1 + y \cdot 3x^2\, x') + 2(x \cdot 3y^2 + y^3 \cdot x') = 0 \Rightarrow$

$4x^3y^2\, x' - 3x^2y\, x' + 2y^3\, x' = -2x^4y + x^3 - 6xy^2 \Rightarrow (4x^3y^2 - 3x^2y + 2y^3)\, x' = -2x^4y + x^3 - 6xy^2 \Rightarrow$

$x' = \dfrac{dx}{dy} = \dfrac{-2x^4y + x^3 - 6xy^2}{4x^3y^2 - 3x^2y + 2y^3}$

25. $y \sin 2x = x \cos 2y \Rightarrow y \cdot \cos 2x \cdot 2 + \sin 2x \cdot y' = x(-\sin 2y \cdot 2y') + \cos(2y) \cdot 1 \Rightarrow$

$\sin 2x \cdot y' + 2x \sin 2y \cdot y' = -2y \cos 2x + \cos 2y \Rightarrow y'(\sin 2x + 2x \sin 2y) = -2y \cos 2x + \cos 2y \Rightarrow$

$y' = \dfrac{-2y \cos 2x + \cos 2y}{\sin 2x + 2x \sin 2y}$. When $x = \dfrac{\pi}{2}$ and $y = \dfrac{\pi}{4}$, we have $y' = \dfrac{(-\pi/2)(-1) + 0}{0 + \pi \cdot 1} = \dfrac{\pi/2}{\pi} = \dfrac{1}{2}$, so an equation of the

tangent line is $y - \dfrac{\pi}{4} = \dfrac{1}{2}(x - \dfrac{\pi}{2})$, or $y = \dfrac{1}{2}x$.

27. $x^{2/3} + y^{2/3} = 4 \Rightarrow \dfrac{2}{3}x^{-1/3} + \dfrac{2}{3}y^{-1/3}y' = 0 \Rightarrow \dfrac{1}{\sqrt[3]{x}} + \dfrac{y'}{\sqrt[3]{y}} = 0 \Rightarrow y' = -\dfrac{\sqrt[3]{y}}{\sqrt[3]{x}}$.

When $x = -3\sqrt{3}$ and $y = 1$, we have $y' = -\dfrac{1}{(-3\sqrt{3})^{1/3}} = -\dfrac{1}{(-3^{3/2})^{1/3}} = \dfrac{1}{3^{1/2}} = \dfrac{1}{\sqrt{3}}$, so an equation of the tangent

line is $y - 1 = \dfrac{1}{\sqrt{3}}\left(x + 3\sqrt{3}\right)$ or $y = \dfrac{1}{\sqrt{3}}x + 4$.

29. $x^2 - xy - y^2 = 1 \Rightarrow 2x - (xy' + y \cdot 1) - 2y\, y' = 0 \Rightarrow 2x - xy' - y - 2y\, y' = 0 \Rightarrow 2x - y = xy' + 2y\, y' \Rightarrow$

$2x - y = (x + 2y)\, y' \Rightarrow y' = \dfrac{2x - y}{x + 2y}$.

When $x = 2$ and $y = 1$, we have $y' = \dfrac{4 - 1}{2 + 2} = \dfrac{3}{4}$, so an equation of the tangent line is $y - 1 = \dfrac{3}{4}(x - 2)$, or $y = \dfrac{3}{4}x - \dfrac{1}{2}$.

31. $x^2 + y^2 = (2x^2 + 2y^2 - x)^2 \Rightarrow 2x + 2y\, y' = 2(2x^2 + 2y^2 - x)(4x + 4y\, y' - 1)$.

When $x = 0$ and $y = \dfrac{1}{2}$, we have $0 + y' = 2(\dfrac{1}{2})(2y' - 1) \Rightarrow y' = 2y' - 1 \Rightarrow y' = 1$, so an equation of the tangent

line is $y - \dfrac{1}{2} = 1(x - 0)$ or $y = x + \dfrac{1}{2}$.

33. $2(x^2 + y^2)^2 = 25(x^2 - y^2) \Rightarrow 4(x^2 + y^2)(2x + 2y\, y') = 25(2x - 2y\, y') \Rightarrow$

$4(x + y\, y')(x^2 + y^2) = 25(x - y\, y') \Rightarrow 4y\, y'(x^2 + y^2) + 25y\, y' = 25x - 4x(x^2 + y^2) \Rightarrow$

$y' = \dfrac{25x - 4x(x^2 + y^2)}{25y + 4y(x^2 + y^2)}$.

When $x = 3$ and $y = 1$, we have $y' = \dfrac{75 - 120}{25 + 40} = -\dfrac{45}{65} = -\dfrac{9}{13}$, so an equation of the tangent line

is $y - 1 = -\dfrac{9}{13}(x - 3)$ or $y = -\dfrac{9}{13}x + \dfrac{40}{13}$.

35. (a) $y^2 = 5x^4 - x^2 \Rightarrow 2y\, y' = 5(4x^3) - 2x \Rightarrow y' = \dfrac{10x^3 - x}{y}$. (b)

So at the point $(1, 2)$ we have $y' = \dfrac{10(1)^3 - 1}{2} = \dfrac{9}{2}$, and an equation

of the tangent line is $y - 2 = \dfrac{9}{2}(x - 1)$ or $y = \dfrac{9}{2}x - \dfrac{5}{2}$.

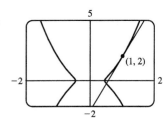

37. $x^2 + 4y^2 = 4 \Rightarrow 2x + 8y\, y' = 0 \Rightarrow y' = -x/(4y) \Rightarrow$

$$y'' = -\frac{1}{4}\frac{y \cdot 1 - x \cdot y'}{y^2} = -\frac{1}{4}\frac{y - x[-x/(4y)]}{y^2} = -\frac{1}{4}\frac{4y^2 + x^2}{4y^3} = -\frac{1}{4}\frac{4}{4y^3} \qquad \left[\begin{array}{l}\text{since } x \text{ and } y \text{ must satisfy the} \\ \text{original equation } x^2 + 4y^2 = 4\end{array}\right]$$

Thus, $y'' = -\dfrac{1}{4y^3}$.

39. $\sin y + \cos x = 1 \Rightarrow \cos y \cdot y' - \sin x = 0 \Rightarrow y' = \dfrac{\sin x}{\cos y} \Rightarrow$

$$y'' = \frac{\cos y \cos x - \sin x (-\sin y)\, y'}{(\cos y)^2} = \frac{\cos y \cos x + \sin x \, \sin y (\sin x / \cos y)}{\cos^2 y}$$

$$= \frac{\cos^2 y \cos x + \sin^2 x \sin y}{\cos^2 y \cos y} = \frac{\cos^2 y \cos x + \sin^2 x \sin y}{\cos^3 y}$$

41. If $x = 0$ in $xy + y^3 = 1$, then we get $y^3 = 1 \Rightarrow y = 1$, so the point where $x = 0$ is $(0, 1)$. Differentiating implicitly with respect to x gives us $xy' + y \cdot 1 + 3y^2 y' = 0$. Substituting 0 for x and 1 for y gives us $1 + 3y' = 0 \Rightarrow y' = -\frac{1}{3}$.
Differentiating $xy' + y + 3y^2 y' = 0$ implicitly with respect to x gives us $xy'' + y' + y' + 3(y^2 y'' + y' \cdot 2y\, y') = 0$. Now substitute 0 for x, 1 for y, and $-\frac{1}{3}$ for y'. $0 - \frac{1}{3} - \frac{1}{3} + 3\left[y'' + \left(-\frac{1}{3}\right) \cdot 2\left(-\frac{1}{3}\right)\right] = 0 \Rightarrow 3\left(y'' + \frac{2}{9}\right) = \frac{2}{3} \Rightarrow$
$y'' + \frac{2}{9} = \frac{2}{9} \Rightarrow y'' = 0$.

43. (a) There are eight points with horizontal tangents: four at $x \approx 1.57735$ and four at $x \approx 0.42265$.

(b) $y' = \dfrac{3x^2 - 6x + 2}{2(2y^3 - 3y^2 - y + 1)} \Rightarrow y' = -1$ at $(0, 1)$ and $y' = \frac{1}{3}$ at $(0, 2)$.

Equations of the tangent lines are $y = -x + 1$ and $y = \frac{1}{3}x + 2$.

(c) $y' = 0 \Rightarrow 3x^2 - 6x + 2 = 0 \Rightarrow x = 1 \pm \frac{1}{3}\sqrt{3}$

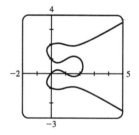

(d) By multiplying the right side of the equation by $x - 3$, we obtain the first graph. By modifying the equation in other ways, we can generate the other graphs.

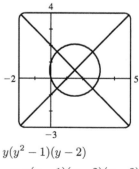

$y(y^2 - 1)(y - 2)$
$= x(x - 1)(x - 2)(x - 3)$

[continued]

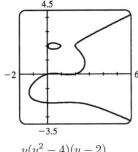
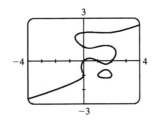
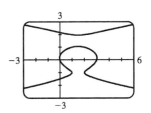

$y(y^2 - 4)(y - 2)$
$= x(x - 1)(x - 2)$

$y(y + 1)(y^2 - 1)(y - 2)$
$= x(x - 1)(x - 2)$

$(y + 1)(y^2 - 1)(y - 2)$
$= (x - 1)(x - 2)$

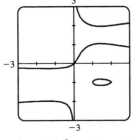
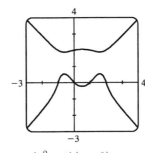
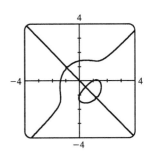

$x(y + 1)(y^2 - 1)(y - 2)$
$= y(x - 1)(x - 2)$

$y(y^2 + 1)(y - 2)$
$= x(x^2 - 1)(x - 2)$

$y(y + 1)(y^2 - 2)$
$= x(x - 1)(x^2 - 2)$

45. From Exercise 33, a tangent to the lemniscate will be horizontal if $y' = 0 \;\Rightarrow\; 25x - 4x(x^2 + y^2) = 0 \;\Rightarrow\;$
$x[25 - 4(x^2 + y^2)] = 0 \;\Rightarrow\; x^2 + y^2 = \frac{25}{4}$ **(1)**. (Note that when x is 0, y is also 0, and there is no horizontal tangent at the origin.) Substituting $\frac{25}{4}$ for $x^2 + y^2$ in the equation of the lemniscate, $2(x^2 + y^2)^2 = 25(x^2 - y^2)$, we get $x^2 - y^2 = \frac{25}{8}$ **(2)**. Solving **(1)** and **(2)**, we have $x^2 = \frac{75}{16}$ and $y^2 = \frac{25}{16}$, so the four points are $\left(\pm\frac{5\sqrt{3}}{4}, \pm\frac{5}{4}\right)$.

47. $\dfrac{x^2}{a^2} - \dfrac{y^2}{b^2} = 1 \;\Rightarrow\; \dfrac{2x}{a^2} - \dfrac{2yy'}{b^2} = 0 \;\Rightarrow\; y' = \dfrac{b^2 x}{a^2 y} \;\Rightarrow\;$ an equation of the tangent line at (x_0, y_0) is

$y - y_0 = \dfrac{b^2 x_0}{a^2 y_0}(x - x_0)$. Multiplying both sides by $\dfrac{y_0}{b^2}$ gives $\dfrac{y_0 y}{b^2} - \dfrac{y_0^2}{b^2} = \dfrac{x_0 x}{a^2} - \dfrac{x_0^2}{a^2}$. Since (x_0, y_0) lies on the hyperbola,

we have $\dfrac{x_0 x}{a^2} - \dfrac{y_0 y}{b^2} = \dfrac{x_0^2}{a^2} - \dfrac{y_0^2}{b^2} = 1$.

49. If the circle has radius r, its equation is $x^2 + y^2 = r^2 \;\Rightarrow\; 2x + 2yy' = 0 \;\Rightarrow\; y' = -\dfrac{x}{y}$, so the slope of the tangent line

at $P(x_0, y_0)$ is $-\dfrac{x_0}{y_0}$. The negative reciprocal of that slope is $\dfrac{-1}{-x_0/y_0} = \dfrac{y_0}{x_0}$, which is the slope of OP, so the tangent line at

P is perpendicular to the radius OP.

51. $x^2 + y^2 = r^2$ is a circle with center O and $ax + by = 0$ is a line through O [assume a and b are not both zero]. $x^2 + y^2 = r^2 \Rightarrow 2x + 2yy' = 0 \Rightarrow y' = -x/y$, so the slope of the tangent line at $P_0(x_0, y_0)$ is $-x_0/y_0$. The slope of the line OP_0 is y_0/x_0, which is the negative reciprocal of $-x_0/y_0$. Hence, the curves are orthogonal, and the families of curves are orthogonal trajectories of each other.

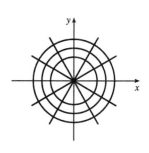

53. $y = cx^2 \Rightarrow y' = 2cx$ and $x^2 + 2y^2 = k$ [assume $k > 0$] $\Rightarrow 2x + 4yy' = 0 \Rightarrow 2yy' = -x \Rightarrow y' = -\dfrac{x}{2(y)} = -\dfrac{x}{2(cx^2)} = -\dfrac{1}{2cx}$, so the curves are orthogonal if $c \neq 0$. If $c = 0$, then the horizontal line $y = cx^2 = 0$ intersects $x^2 + 2y^2 = k$ orthogonally at $\left(\pm\sqrt{k}, 0\right)$, since the ellipse $x^2 + 2y^2 = k$ has vertical tangents at those two points.

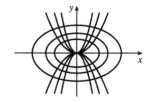

55. Since $A^2 < a^2$, we are assured that there are four points of intersection.

(1) $\dfrac{x^2}{a^2} + \dfrac{y^2}{b^2} = 1 \Rightarrow \dfrac{2x}{a^2} + \dfrac{2yy'}{b^2} = 0 \Rightarrow \dfrac{yy'}{b^2} = -\dfrac{x}{a^2} \Rightarrow$

$y' = m_1 = -\dfrac{xb^2}{ya^2}.$

(2) $\dfrac{x^2}{A^2} - \dfrac{y^2}{B^2} = 1 \Rightarrow \dfrac{2x}{A^2} - \dfrac{2yy'}{B^2} = 0 \Rightarrow \dfrac{yy'}{B^2} = \dfrac{x}{A^2} \Rightarrow$

$y' = m_2 = \dfrac{xB^2}{yA^2}.$

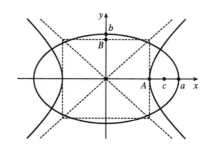

Now $m_1 m_2 = -\dfrac{xb^2}{ya^2} \cdot \dfrac{xB^2}{yA^2} = -\dfrac{b^2 B^2}{a^2 A^2} \cdot \dfrac{x^2}{y^2}$ (3). Subtracting equations, (1) − (2), gives us $\dfrac{x^2}{a^2} + \dfrac{y^2}{b^2} - \dfrac{x^2}{A^2} + \dfrac{y^2}{B^2} = 0 \Rightarrow$

$\dfrac{y^2}{b^2} + \dfrac{y^2}{B^2} = \dfrac{x^2}{A^2} - \dfrac{x^2}{a^2} \Rightarrow \dfrac{y^2 B^2 + y^2 b^2}{b^2 B^2} = \dfrac{x^2 a^2 - x^2 A^2}{A^2 a^2} \Rightarrow \dfrac{y^2(b^2 + B^2)}{b^2 B^2} = \dfrac{x^2(a^2 - A^2)}{a^2 A^2}$ (4). Since

$a^2 - b^2 = A^2 + B^2$, we have $a^2 - A^2 = b^2 + B^2$. Thus, equation (4) becomes $\dfrac{y^2}{b^2 B^2} = \dfrac{x^2}{A^2 a^2} \Rightarrow \dfrac{x^2}{y^2} = \dfrac{A^2 a^2}{b^2 B^2}$, and

substituting for $\dfrac{x^2}{y^2}$ in equation (3) gives us $m_1 m_2 = -\dfrac{b^2 B^2}{a^2 A^2} \cdot \dfrac{a^2 A^2}{b^2 B^2} = -1$. Hence, the ellipse and hyperbola are orthogonal trajectories.

57. (a) $\left(P + \dfrac{n^2 a}{V^2}\right)(V - nb) = nRT \Rightarrow PV - Pnb + \dfrac{n^2 a}{V} - \dfrac{n^3 ab}{V^2} = nRT \Rightarrow$

$\dfrac{d}{dP}(PV - Pnb + n^2 aV^{-1} - n^3 abV^{-2}) = \dfrac{d}{dP}(nRT) \Rightarrow$

$PV' + V \cdot 1 - nb - n^2 aV^{-2} \cdot V' + 2n^3 abV^{-3} \cdot V' = 0 \Rightarrow V'(P - n^2 aV^{-2} + 2n^3 abV^{-3}) = nb - V \Rightarrow$

$V' = \dfrac{nb - V}{P - n^2 aV^{-2} + 2n^3 abV^{-3}}$ or $\dfrac{dV}{dP} = \dfrac{V^3(nb - V)}{PV^3 - n^2 aV + 2n^3 ab}$

(b) Using the last expression for dV/dP from part (a), we get

$$\frac{dV}{dP} = \frac{(10\text{ L})^3[(1\text{ mole})(0.04267\text{ L/mole}) - 10\text{ L}]}{\left[\begin{array}{l}(2.5\text{ atm})(10\text{ L})^3 - (1\text{ mole})^2(3.592\text{ L}^2\text{-atm/mole}^2)(10\text{ L}) \\ \qquad\qquad + 2(1\text{ mole})^3(3.592\text{ L}^2\text{-atm/mole}^2)(0.04267\text{ L/mole})\end{array}\right]}$$

$$= \frac{-9957.33\text{ L}^4}{2464.386541\text{ L}^3\text{-atm}} \approx -4.04\text{ L/atm}.$$

59. To find the points at which the ellipse $x^2 - xy + y^2 = 3$ crosses the x-axis, let $y = 0$ and solve for x.

$y = 0 \;\Rightarrow\; x^2 - x(0) + 0^2 = 3 \;\Leftrightarrow\; x = \pm\sqrt{3}$. So the graph of the ellipse crosses the x-axis at the points $(\pm\sqrt{3}, 0)$.

Using implicit differentiation to find y', we get $2x - xy' - y + 2yy' = 0 \;\Rightarrow\; y'(2y - x) = y - 2x \;\Leftrightarrow\; y' = \dfrac{y - 2x}{2y - x}$.

So y' at $(\sqrt{3}, 0)$ is $\dfrac{0 - 2\sqrt{3}}{2(0) - \sqrt{3}} = 2$ and y' at $(-\sqrt{3}, 0)$ is $\dfrac{0 + 2\sqrt{3}}{2(0) + \sqrt{3}} = 2$. Thus, the tangent lines at these points are parallel.

61. $x^2 y^2 + xy = 2 \;\Rightarrow\; x^2 \cdot 2yy' + y^2 \cdot 2x + x \cdot y' + y \cdot 1 = 0 \;\Leftrightarrow\; y'(2x^2 y + x) = -2xy^2 - y \;\Leftrightarrow$

$y' = -\dfrac{2xy^2 + y}{2x^2 y + x}$. So $-\dfrac{2xy^2 + y}{2x^2 y + x} = -1 \;\Leftrightarrow\; 2xy^2 + y = 2x^2 y + x \;\Leftrightarrow\; y(2xy + 1) = x(2xy + 1) \;\Leftrightarrow$

$y(2xy+1) - x(2xy+1) = 0 \;\Leftrightarrow\; (2xy+1)(y-x) = 0 \;\Leftrightarrow\; xy = -\tfrac{1}{2}$ or $y = x$. But $xy = -\tfrac{1}{2} \;\Rightarrow$

$x^2 y^2 + xy = \tfrac{1}{4} - \tfrac{1}{2} \neq 2$, so we must have $x = y$. Then $x^2 y^2 + xy = 2 \;\Rightarrow\; x^4 + x^2 = 2 \;\Leftrightarrow\; x^4 + x^2 - 2 = 0 \;\Leftrightarrow$

$(x^2 + 2)(x^2 - 1) = 0$. So $x^2 = -2$, which is impossible, or $x^2 = 1 \;\Leftrightarrow\; x = \pm 1$. Since $x = y$, the points on the curve where the tangent line has a slope of -1 are $(-1, -1)$ and $(1, 1)$.

63. For $\dfrac{x}{y} = y^2 + 1$, $y \neq 0$, we have $\dfrac{d}{dx}\left(\dfrac{x}{y}\right) = \dfrac{d}{dx}(y^2 + 1) \;\Rightarrow\; \dfrac{y \cdot 1 - x \cdot y'}{y^2} = 2yy' \;\Rightarrow\; y - xy' = 2y^3 y' \;\Rightarrow$

$2y^3 y' + xy' = y \;\Rightarrow\; y'(2y^3 + x) = y \;\Rightarrow\; y' = \dfrac{y}{2y^3 + x}$.

For $x = y^3 + y$, $y \neq 0$, we have $\dfrac{d}{dx}(x) = \dfrac{d}{dx}(y^3 + y) \;\Rightarrow\; 1 = 3y^2 y' + y' \;\Rightarrow\; 1 = y'(3y^2 + 1) \;\Rightarrow$

$y' = \dfrac{1}{3y^2 + 1}$.

From part (a), $y' = \dfrac{y}{2y^3 + x}$. Since $y \neq 0$, we substitute $y^3 + y$ for x to get

$\dfrac{y}{2y^3 + x} = \dfrac{y}{2y^3 + (y^3 + y)} = \dfrac{y}{3y^3 + y} = \dfrac{y}{y(3y^2 + 1)} = \dfrac{1}{3y^2 + 1}$, which agrees with part (b).

65. $x^2 + 4y^2 = 5 \;\Rightarrow\; 2x + 4(2yy') = 0 \;\Rightarrow\; y' = -\dfrac{x}{4y}$. Now let h be the height of the lamp, and let (a, b) be the point of tangency of the line passing through the points $(3, h)$ and $(-5, 0)$. This line has slope $(h - 0)/[3 - (-5)] = \tfrac{1}{8}h$. But the slope of the tangent line through the point (a, b) can be expressed as $y' = -\dfrac{a}{4b}$, or as $\dfrac{b - 0}{a - (-5)} = \dfrac{b}{a + 5}$ [since the line passes through $(-5, 0)$ and (a, b)], so $-\dfrac{a}{4b} = \dfrac{b}{a+5}$ $\Leftrightarrow$ $4b^2 = -a^2 - 5a$ $\Leftrightarrow$ $a^2 + 4b^2 = -5a$. But $a^2 + 4b^2 = 5$

[since (a, b) is on the ellipse], so $5 = -5a$ $\Leftrightarrow$ $a = -1$. Then $4b^2 = -a^2 - 5a = -1 - 5(-1) = 4$ $\Rightarrow$ $b = 1$, since the

point is on the top half of the ellipse. So $\dfrac{h}{8} = \dfrac{b}{a+5} = \dfrac{1}{-1+5} = \dfrac{1}{4}$ $\Rightarrow$ $h = 2$. So the lamp is located 2 units above the

x-axis.

2.7 Rates of Change in the Natural and Social Sciences

1. (a) $s = f(t) = t^3 - 9t^2 + 24t$ (in feet) $\Rightarrow$ $v(t) = f'(t) = 3t^2 - 18t + 24$ (in ft/s)

(b) $v(1) = 3(1)^2 - 18(1) + 24 = 9$ ft/s

(c) The particle is at rest when $v(t) = 0$: $3t^2 - 18t + 24 = 0$ $\Leftrightarrow$ $3(t-2)(t-4) = 0$ $\Leftrightarrow$ $t = 2$ s or $t = 4$ s.

(d) The particle is moving in the positive direction when $v(t) > 0$. $3(t-2)(t-4) > 0$ $\Rightarrow$ $0 \le t < 2$ or $t > 4$.

(e) Because the particle changes direction when $t = 2$ and

$t = 4$, we need to calculate the distances traveled in the

intervals $[0, 2]$, $[2, 4]$, and $[4, 6]$ separately.

(f)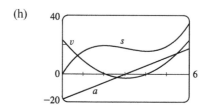

$|f(2) - f(0)| = |20 - 0| = 20$

$|f(4) - f(2)| = |16 - 20| = 4$

$|f(6) - f(4)| = |36 - 16| = 20$

The total distance is $20 + 4 + 20 = 44$ ft.

(g) $v(t) = 3t^2 - 18t + 24$ $\Rightarrow$

$a(t) = v'(t) = 6t - 18$ (in (ft/s)/s or ft/s^2).

$a(1) = 6(1) - 18 = -12$ ft/s^2

(h) [graph showing v, s, a with vertical range -20 to 40 and horizontal range 0 to 6]

(i) The particle is speeding up when v and a have the same sign. This occurs when $2 < t < 3$ (v and a are both negative) and

when $t > 4$ (v and a are both positive). It is slowing down when v and a have opposite signs; that is, when $0 \le t < 2$ and

when $3 < t < 4$.

3. (a) $s = f(t) = \sin(\pi t/2)$ (in feet) $\Rightarrow$ $v(t) = f'(t) = \cos(\pi t/2) \cdot (\pi/2) = \dfrac{\pi}{2} \cos(\pi t/2)$ (in ft/s)

(b) $v(1) = \dfrac{\pi}{2} \cos \dfrac{\pi}{2} = \dfrac{\pi}{2}(0) = 0$ ft/s

(c) The particle is at rest when $v(t) = 0$. $\dfrac{\pi}{2} \cos \dfrac{\pi}{2} t = 0$ $\Leftrightarrow$ $\cos \dfrac{\pi}{2} t = 0$ $\Leftrightarrow$ $\dfrac{\pi}{2} t = \dfrac{\pi}{2} + n\pi$ $\Leftrightarrow$ $t = 1 + 2n$, where n

is a nonnegative integer since $t \ge 0$.

(d) The particle is moving in the positive direction when $v(t) > 0$. From part (c), we see that v changes sign at every positive odd integer. v is positive when $0 < t < 1$, $3 < t < 5$, $7 < t < 9$, and so on.

(e) v changes sign at $t = 1, 3$, and 5 in the interval $[0, 6]$. The total distance traveled during the first 6 seconds is
$$|f(1) - f(0)| + |f(3) - f(1)| + |f(5) - f(3)| + |f(6) - f(5)| = |1 - 0| + |-1 - 1| + |1 - (-1)| + |0 - 1|$$
$$= 1 + 2 + 2 + 1 = 6 \text{ ft}$$

(f)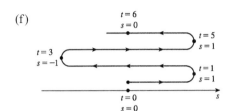

(g) $v(t) = \frac{\pi}{2} \cos(\pi t/2) \Rightarrow$
$$a(t) = v'(t) = \frac{\pi}{2}[-\sin(\pi t/2) \cdot (\pi/2)]$$
$$= (-\pi^2/4)\sin(\pi t/2) \text{ ft/s}^2$$
$$a(1) = (-\pi^2/4)\sin(\pi/2) = -\pi^2/4 \text{ ft/s}^2$$

(h)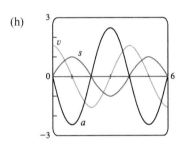

(i) The particle is speeding up when v and a have the same sign. From the figure in part (h), we see that v and a are both positive when $3 < t < 4$ and both negative when $1 < t < 2$ and $5 < t < 6$. Thus, the particle is speeding up when $1 < t < 2$, $3 < t < 4$, and $5 < t < 6$. The particle is slowing down when v and a have opposite signs; that is, when $0 < t < 1$, $2 < t < 3$, and $4 < t < 5$.

5. (a) From the figure, the velocity v is positive on the interval $(0, 2)$ and negative on the interval $(2, 3)$. The acceleration a is positive (negative) when the slope of the tangent line is positive (negative), so the acceleration is positive on the interval $(0, 1)$, and negative on the interval $(1, 3)$. The particle is speeding up when v and a have the same sign, that is, on the interval $(0, 1)$ when $v > 0$ and $a > 0$, and on the interval $(2, 3)$ when $v < 0$ and $a < 0$. The particle is slowing down when v and a have opposite signs, that is, on the interval $(1, 2)$ when $v > 0$ and $a < 0$.

(b) $v > 0$ on $(0, 3)$ and $v < 0$ on $(3, 4)$. $a > 0$ on $(1, 2)$ and $a < 0$ on $(0, 1)$ and $(2, 4)$. The particle is speeding up on $(1, 2)$ $[v > 0, a > 0]$ and on $(3, 4)$ $[v < 0, a < 0]$. The particle is slowing down on $(0, 1)$ and $(2, 3)$ $[v > 0, a < 0]$.

7. The particle is traveling forward when its velocity is positive. From the graph, this occurs when $0 < t < 5$. The particle is traveling backward when its velocity is negative. From the graph, this occurs when $7 < t < 8$. When $5 < t < 7$, its velocity is zero and the particle is not moving.

9. (a) $h(t) = 2 + 24.5t - 4.9t^2 \Rightarrow v(t) = h'(t) = 24.5 - 9.8t$. The velocity after 2 s is $v(2) = 24.5 - 9.8(2) = 4.9$ m/s and after 4 s is $v(4) = 24.5 - 9.8(4) = -14.7$ m/s.

(b) The projectile reaches its maximum height when the velocity is zero. $v(t) = 0 \Leftrightarrow 24.5 - 9.8t = 0 \Leftrightarrow$
$$t = \frac{24.5}{9.8} = 2.5 \text{ s}.$$

(c) The maximum height occurs when $t = 2.5$. $h(2.5) = 2 + 24.5(2.5) - 4.9(2.5)^2 = 32.625$ m [or $32\frac{5}{8}$ m].

(d) The projectile hits the ground when $h = 0$ $\Leftrightarrow$ $2 + 24.5t - 4.9t^2 = 0$ $\Leftrightarrow$

$$t = \frac{-24.5 \pm \sqrt{24.5^2 - 4(-4.9)(2)}}{2(-4.9)} \Rightarrow t = t_f \approx 5.08 \text{ s [since } t \geq 0\text{]}.$$

(e) The projectile hits the ground when $t = t_f$. Its velocity is $v(t_f) = 24.5 - 9.8t_f \approx -25.3$ m/s [downward].

11. (a) $h(t) = 15t - 1.86t^2$ $\Rightarrow$ $v(t) = h'(t) = 15 - 3.72t$. The velocity after 2 s is $v(2) = 15 - 3.72(2) = 7.56$ m/s.

(b) $25 = h$ $\Leftrightarrow$ $1.86t^2 - 15t + 25 = 0$ $\Leftrightarrow$ $t = \dfrac{15 \pm \sqrt{15^2 - 4(1.86)(25)}}{2(1.86)}$ $\Leftrightarrow$ $t = t_1 \approx 2.35$ or $t = t_2 \approx 5.71$.

The velocities are $v(t_1) = 15 - 3.72t_1 \approx 6.24$ m/s [upward] and $v(t_2) = 15 - 3.72t_2 \approx -6.24$ m/s [downward].

13. (a) $A(x) = x^2$ $\Rightarrow$ $A'(x) = 2x$. $A'(15) = 30$ mm^2/mm is the rate at which the area is increasing with respect to the side length as x reaches 15 mm.

(b) The perimeter is $P(x) = 4x$, so $A'(x) = 2x = \frac{1}{2}(4x) = \frac{1}{2}P(x)$. The figure suggests that if Δx is small, then the change in the area of the square is approximately half of its perimeter (2 of the 4 sides) times Δx. From the figure, $\Delta A = 2x(\Delta x) + (\Delta x)^2$. If Δx is small, then $\Delta A \approx 2x(\Delta x)$ and so $\Delta A/\Delta x \approx 2x$.

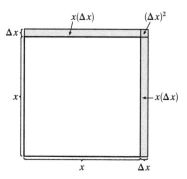

15. (a) Using $A(r) = \pi r^2$, we find that the average rate of change is:

(i) $\dfrac{A(3) - A(2)}{3 - 2} = \dfrac{9\pi - 4\pi}{1} = 5\pi$

(ii) $\dfrac{A(2.5) - A(2)}{2.5 - 2} = \dfrac{6.25\pi - 4\pi}{0.5} = 4.5\pi$

(iii) $\dfrac{A(2.1) - A(2)}{2.1 - 2} = \dfrac{4.41\pi - 4\pi}{0.1} = 4.1\pi$

(b) $A(r) = \pi r^2$ $\Rightarrow$ $A'(r) = 2\pi r$, so $A'(2) = 4\pi$.

(c) The circumference is $C(r) = 2\pi r = A'(r)$. The figure suggests that if Δr is small, then the change in the area of the circle (a ring around the outside) is approximately equal to its circumference times Δr. Straightening out this ring gives us a shape that is approximately rectangular with length $2\pi r$ and width Δr, so $\Delta A \approx 2\pi r(\Delta r)$. Algebraically, $\Delta A = A(r + \Delta r) - A(r) = \pi(r + \Delta r)^2 - \pi r^2 = 2\pi r(\Delta r) + \pi(\Delta r)^2$. So we see that if Δr is small, then $\Delta A \approx 2\pi r(\Delta r)$ and therefore, $\Delta A/\Delta r \approx 2\pi r$.

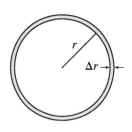

17. $S(r) = 4\pi r^2$ $\Rightarrow$ $S'(r) = 8\pi r$ $\Rightarrow$

(a) $S'(1) = 8\pi$ ft^2/ft

(b) $S'(2) = 16\pi$ ft^2/ft

(c) $S'(3) = 24\pi$ ft^2/ft

As the radius increases, the surface area grows at an increasing rate. In fact, the rate of change is linear with respect to the radius.

19. The mass is $f(x) = 3x^2$, so the linear density at x is $\rho(x) = f'(x) = 6x$.

(a) $\rho(1) = 6$ kg/m (b) $\rho(2) = 12$ kg/m (c) $\rho(3) = 18$ kg/m

Since ρ is an increasing function, the density will be the highest at the right end of the rod and lowest at the left end.

21. The quantity of charge is $Q(t) = t^3 - 2t^2 + 6t + 2$, so the current is $Q'(t) = 3t^2 - 4t + 6$.

(a) $Q'(0.5) = 3(0.5)^2 - 4(0.5) + 6 = 4.75$ A (b) $Q'(1) = 3(1)^2 - 4(1) + 6 = 5$ A

The current is lowest when Q' has a minimum. $Q''(t) = 6t - 4 < 0$ when $t < \frac{2}{3}$. So the current decreases when $t < \frac{2}{3}$ and increases when $t > \frac{2}{3}$. Thus, the current is lowest at $t = \frac{2}{3}$ s.

23. With $m = m_0\left(1 - \dfrac{v^2}{c^2}\right)^{-1/2}$,

$$F = \frac{d}{dt}(mv) = m\frac{d}{dt}(v) + v\frac{d}{dt}(m) = m_0\left(1 - \frac{v^2}{c^2}\right)^{-1/2} \cdot a + v \cdot m_0\left[-\frac{1}{2}\left(1 - \frac{v^2}{c^2}\right)^{-3/2}\right]\left(-\frac{2v}{c^2}\right)\frac{d}{dt}(v)$$

$$= m_0\left(1 - \frac{v^2}{c^2}\right)^{-3/2} \cdot a\left[\left(1 - \frac{v^2}{c^2}\right) + \frac{v^2}{c^2}\right] = \frac{m_0 a}{(1 - v^2/c^2)^{3/2}}$$

Note that we factored out $(1 - v^2/c^2)^{-3/2}$ since $-3/2$ was the lesser exponent. Also note that $\dfrac{d}{dt}(v) = a$.

25. (a) To find the rate of change of volume with respect to pressure, we first solve for V in terms of P.

$$PV = C \;\Rightarrow\; V = \frac{C}{P} \;\Rightarrow\; \frac{dV}{dP} = -\frac{C}{P^2}.$$

(b) From the formula for dV/dP in part (a), we see that as P increases, the absolute value of dV/dP decreases.

Thus, the volume is decreasing more rapidly at the beginning of the 10 minutes.

(c) $\beta = -\dfrac{1}{V}\dfrac{dV}{dP} = -\dfrac{1}{V}\left(-\dfrac{C}{P^2}\right)$ [from part (a)] $= \dfrac{C}{(PV)P} = \dfrac{C}{CP} = \dfrac{1}{P}$

27. (a) **1920:** $m_1 = \dfrac{1860 - 1750}{1920 - 1910} = \dfrac{110}{10} = 11$, $m_2 = \dfrac{2070 - 1860}{1930 - 1920} = \dfrac{210}{10} = 21$,

$(m_1 + m_2)/2 = (11 + 21)/2 = 16$ million/year

1980: $m_1 = \dfrac{4450 - 3710}{1980 - 1970} = \dfrac{740}{10} = 74$, $m_2 = \dfrac{5280 - 4450}{1990 - 1980} = \dfrac{830}{10} = 83$,

$(m_1 + m_2)/2 = (74 + 83)/2 = 78.5$ million/year

(b) $P(t) = at^3 + bt^2 + ct + d$ (in millions of people), where $a \approx -0.000\,284\,900\,3$, $b \approx 0.522\,433\,122\,43$, $c \approx -6.395\,641\,396$, and $d \approx 1720.586\,081$.

(c) $P(t) = at^3 + bt^2 + ct + d \;\Rightarrow\; P'(t) = 3at^2 + 2bt + c$ (in millions of people per year)

(d) 1920 corresponds to $t = 20$ and $P'(20) \approx 14.16$ million/year. 1980 corresponds to $t = 80$ and $P'(80) \approx 71.72$ million/year. These estimates are smaller than the estimates in part (a).

(e) $P'(85) \approx 76.24$ million/year and $f'(85) \approx 64.61$ million/year. The first estimate is probably more accurate.

29. (a) Using $v = \dfrac{P}{4\eta l}(R^2 - r^2)$ with $R = 0.01$, $l = 3$, $P = 3000$, and $\eta = 0.027$, we have v as a function of r:

$$v(r) = \dfrac{3000}{4(0.027)3}(0.01^2 - r^2). \quad v(0) = 0.\overline{925} \text{ cm/s}, \ v(0.005) = 0.69\overline{4} \text{ cm/s}, \ v(0.01) = 0.$$

(b) $v(r) = \dfrac{P}{4\eta l}(R^2 - r^2) \ \Rightarrow \ v'(r) = \dfrac{P}{4\eta l}(-2r) = -\dfrac{Pr}{2\eta l}$. When $l = 3$, $P = 3000$, and $\eta = 0.027$, we have

$$v'(r) = -\dfrac{3000r}{2(0.027)3}. \quad v'(0) = 0, \ v'(0.005) = -92.\overline{592} \text{ (cm/s)/cm, and } v'(0.01) = -185.\overline{185} \text{ (cm/s)/cm}.$$

(c) The velocity is greatest where $r = 0$ (at the center) and the velocity is changing most where $r = R = 0.01$ cm (at the edge).

31. (a) $C(x) = 2000 + 3x + 0.01x^2 + 0.0002x^3 \ \Rightarrow \ C'(x) = 0 + 3(1) + 0.01(2x) + 0.0002(3x^2) = 3 + 0.02x + 0.0006x^2$

(b) $C'(100) = 3 + 0.02(100) + 0.0006(100)^2 = 3 + 2 + 6 = \$11$/pair. $C'(100)$ is the rate at which the cost is increasing as the 100th pair of jeans is produced. It predicts the (approximate) cost of the 101st pair.

(c) The cost of manufacturing the 101st pair of jeans is
$$C(101) - C(100) = 2611.0702 - 2600 = 11.0702 \approx \$11.07.$$ This is close to the marginal cost from part (b).

33. (a) $A(x) = \dfrac{p(x)}{x} \ \Rightarrow \ A'(x) = \dfrac{xp'(x) - p(x) \cdot 1}{x^2} = \dfrac{xp'(x) - p(x)}{x^2}$.

$A'(x) > 0 \ \Rightarrow \ A(x)$ is increasing; that is, the average productivity increases as the size of the workforce increases.

(b) $p'(x)$ is greater than the average productivity $\ \Rightarrow \ p'(x) > A(x) \ \Rightarrow \ p'(x) > \dfrac{p(x)}{x} \ \Rightarrow \ xp'(x) > p(x) \ \Rightarrow $

$xp'(x) - p(x) > 0 \ \Rightarrow \ \dfrac{xp'(x) - p(x)}{x^2} > 0 \ \Rightarrow \ A'(x) > 0.$

35. $PV = nRT \ \Rightarrow \ T = \dfrac{PV}{nR} = \dfrac{PV}{(10)(0.0821)} = \dfrac{1}{0.821}(PV)$. Using the Product Rule, we have

$\dfrac{dT}{dt} = \dfrac{1}{0.821}[P(t)V'(t) + V(t)P'(t)] = \dfrac{1}{0.821}[(8)(-0.15) + (10)(0.10)] \approx -0.2436$ K/min.

37. (a) If the populations are stable, then the growth rates are neither positive nor negative; that is, $\dfrac{dC}{dt} = 0$ and $\dfrac{dW}{dt} = 0$.

(b) "The caribou go extinct" means that the population is zero, or mathematically, $C = 0$.

(c) We have the equations $\dfrac{dC}{dt} = aC - bCW$ and $\dfrac{dW}{dt} = -cW + dCW$. Let $dC/dt = dW/dt = 0$, $a = 0.05$, $b = 0.001$, $c = 0.05$, and $d = 0.0001$ to obtain $0.05C - 0.001CW = 0$ **(1)** and $-0.05W + 0.0001CW = 0$ **(2)**. Adding 10 times **(2)** to **(1)** eliminates the CW-terms and gives us $0.05C - 0.5W = 0 \ \Rightarrow \ C = 10W$. Substituting $C = 10W$ into **(1)** results in $0.05(10W) - 0.001(10W)W = 0 \ \Leftrightarrow \ 0.5W - 0.01W^2 = 0 \ \Leftrightarrow \ 50W - W^2 = 0 \ \Leftrightarrow \ W(50 - W) = 0 \ \Leftrightarrow \ W = 0$ or 50. Since $C = 10W$, $C = 0$ or 500. Thus, the population pairs (C, W) that lead to stable populations are $(0, 0)$ and $(500, 50)$. So it is possible for the two species to live in harmony.

2.8 Related Rates

1. (a) $V = x^3 \;\Rightarrow\; \dfrac{dV}{dt} = \dfrac{dV}{dx}\dfrac{dx}{dt} = 3x^2\dfrac{dx}{dt}$

(b) With $\dfrac{dx}{dt} = 4$ cm/s and $x = 15$ cm, we have $\dfrac{dV}{dt} = 3(15)^2 \cdot 4 = 2700$ cm^3/s.

3. Let s denote the side of a square. The square's area A is given by $A = s^2$. Differentiating with respect to t gives us $\dfrac{dA}{dt} = 2s\dfrac{ds}{dt}$. When $A = 16$, $s = 4$. Substituting 4 for s and 6 for $\dfrac{ds}{dt}$ gives us $\dfrac{dA}{dt} = 2(4)(6) = 48$ cm^2/s.

5. $S = 4\pi r^2 \;\Rightarrow\; \dfrac{dS}{dt} = 4\pi \cdot 2r \dfrac{dr}{dt} \;\Rightarrow\; \dfrac{dS}{dt} = 4\pi \cdot 2 \cdot 8 \cdot 2 = 128\pi$ cm^2/min.

7. $V = \pi r^2 h = \pi(5)^2 h = 25\pi h \;\Rightarrow\; \dfrac{dV}{dt} = 25\pi \dfrac{dh}{dt} \;\Rightarrow\; 3 = 25\pi \dfrac{dh}{dt} \;\Rightarrow\; \dfrac{dh}{dt} = \dfrac{3}{25\pi}$ m/min.

9. (a) $\dfrac{d}{dt}(4x^2 + 9y^2) = \dfrac{d}{dt}(25) \;\Rightarrow\; 8x\dfrac{dx}{dt} + 18y\dfrac{dy}{dt} = 0 \;\Rightarrow\; 4x\dfrac{dx}{dt} + 9y\dfrac{dy}{dt} = 0 \;\Rightarrow$

$4(2)\dfrac{dx}{dt} + 9(1)\cdot\tfrac{1}{3} = 0 \;\Rightarrow\; 8\dfrac{dx}{dt} + 3 = 0 \;\Rightarrow\; \dfrac{dx}{dt} = -\dfrac{3}{8}$

(b) $4x\dfrac{dx}{dt} + 9y\dfrac{dy}{dt} = 0 \;\Rightarrow\; 4(-2)(3) + 9(1)\cdot\dfrac{dy}{dt} = 0 \;\Rightarrow\; -24 + 9\dfrac{dy}{dt} = 0 \;\Rightarrow\; \dfrac{dy}{dt} = \dfrac{24}{9} = \dfrac{8}{3}$

11. $w = w_0\left(\dfrac{3960}{3960+h}\right)^2 = w_0 \cdot 3960^2 (3960+h)^{-2} \;\Rightarrow\; \dfrac{dw}{dt} = w_0 \cdot 3960^2(-2)(3960+h)^{-3}\cdot\dfrac{dh}{dt}$. Then $w_0 = 130$ lb,

$h = 40$ mi, and $dh/dt = 12$ mi/s $\;\Rightarrow\; \dfrac{dw}{dt} = 130\cdot 3960^2(-2)(3960+40)^{-3}(12) = -0.764478 \approx -0.7645$ lb/s.

13. (a) Given: a plane flying horizontally at an altitude of 1 mi and a speed of 500 mi/h passes directly over a radar station. If we let t be time (in hours) and x be the horizontal distance traveled by the plane (in mi), then we are given that $dx/dt = 500$ mi/h.

(b) Unknown: the rate at which the distance from the plane to the station is increasing when it is 2 mi from the station. If we let y be the distance from the plane to the station, then we want to find dy/dt when $y = 2$ mi.

(c)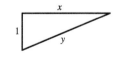

(d) By the Pythagorean Theorem, $y^2 = x^2 + 1 \;\Rightarrow\; 2y\,(dy/dt) = 2x\,(dx/dt)$.

(e) $\dfrac{dy}{dt} = \dfrac{x}{y}\dfrac{dx}{dt} = \dfrac{x}{y}(500)$. Since $y^2 = x^2 + 1$, when $y = 2$, $x = \sqrt{3}$, so $\dfrac{dy}{dt} = \dfrac{\sqrt{3}}{2}(500) = 250\sqrt{3} \approx 433$ mi/h.

15. (a) Given: a man 6 ft tall walks away from a street light mounted on a 15-ft-tall pole at a rate of 5 ft/s. If we let t be time (in s) and x be the distance from the pole to the man (in ft), then we are given that $dx/dt = 5$ ft/s.

(b) Unknown: the rate at which the tip of his shadow is moving when he is 40 ft from the pole. If we let y be the distance from the man to the tip of his shadow (in ft), then we want to find $\dfrac{d}{dt}(x+y)$ when $x=40$ ft.

(c)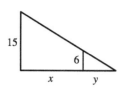

(d) By similar triangles, $\dfrac{15}{6} = \dfrac{x+y}{y}$ $\Rightarrow$ $15y = 6x + 6y$ $\Rightarrow$ $9y = 6x$ $\Rightarrow$ $y = \tfrac{2}{3}x$.

(e) The tip of the shadow moves at a rate of $\dfrac{d}{dt}(x+y) = \dfrac{d}{dt}\left(x + \tfrac{2}{3}x\right) = \dfrac{5}{3}\dfrac{dx}{dt} = \tfrac{5}{3}(5) = \tfrac{25}{3}$ ft/s.

17.
We are given that $\dfrac{dx}{dt} = 60$ mi/h and $\dfrac{dy}{dt} = 25$ mi/h. $z^2 = x^2 + y^2$ $\Rightarrow$
$2z\dfrac{dz}{dt} = 2x\dfrac{dx}{dt} + 2y\dfrac{dy}{dt}$ $\Rightarrow$ $z\dfrac{dz}{dt} = x\dfrac{dx}{dt} + y\dfrac{dy}{dt}$ $\Rightarrow$ $\dfrac{dz}{dt} = \dfrac{1}{z}\left(x\dfrac{dx}{dt} + y\dfrac{dy}{dt}\right)$.
After 2 hours, $x = 2(60) = 120$ and $y = 2(25) = 50$ $\Rightarrow$ $z = \sqrt{120^2 + 50^2} = 130$,
so $\dfrac{dz}{dt} = \dfrac{1}{z}\left(x\dfrac{dx}{dt} + y\dfrac{dy}{dt}\right) = \dfrac{120(60) + 50(25)}{130} = 65$ mi/h.

19.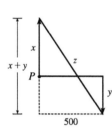
We are given that $\dfrac{dx}{dt} = 4$ ft/s and $\dfrac{dy}{dt} = 5$ ft/s. $z^2 = (x+y)^2 + 500^2$ $\Rightarrow$
$2z\dfrac{dz}{dt} = 2(x+y)\left(\dfrac{dx}{dt} + \dfrac{dy}{dt}\right)$. 15 minutes after the woman starts, we have
$x = (4\text{ ft/s})(20\text{ min})(60\text{ s/min}) = 4800$ ft and $y = 5 \cdot 15 \cdot 60 = 4500$ $\Rightarrow$
$z = \sqrt{(4800 + 4500)^2 + 500^2} = \sqrt{86{,}740{,}000}$, so
$\dfrac{dz}{dt} = \dfrac{x+y}{z}\left(\dfrac{dx}{dt} + \dfrac{dy}{dt}\right) = \dfrac{4800 + 4500}{\sqrt{86{,}740{,}000}}(4+5) = \dfrac{837}{\sqrt{8674}} \approx 8.99$ ft/s.

21. $A = \tfrac{1}{2}bh$, where b is the base and h is the altitude. We are given that $\dfrac{dh}{dt} = 1$ cm/min and $\dfrac{dA}{dt} = 2$ cm^2/min. Using the Product Rule, we have $\dfrac{dA}{dt} = \dfrac{1}{2}\left(b\dfrac{dh}{dt} + h\dfrac{db}{dt}\right)$. When $h = 10$ and $A = 100$, we have $100 = \tfrac{1}{2}b(10)$ $\Rightarrow$ $\tfrac{1}{2}b = 10$ $\Rightarrow$ $b = 20$, so $2 = \dfrac{1}{2}\left(20 \cdot 1 + 10\dfrac{db}{dt}\right)$ $\Rightarrow$ $4 = 20 + 10\dfrac{db}{dt}$ $\Rightarrow$ $\dfrac{db}{dt} = \dfrac{4 - 20}{10} = -1.6$ cm/min.

23. Let x be the distance (in meters) the first dropped stone has traveled, and let y be the distance (in meters) the stone dropped one second later has traveled. Let t be the time (in seconds) since the woman drops the second stone. Using $d = 4.9t^2$, we have $x = 4.9(t+1)^2$ and $y = 4.9t^2$. Let z be the distance between the stones. Then $z = x - y$ and we have
$\dfrac{dz}{dt} = \dfrac{dx}{dt} - \dfrac{dy}{dt}$ $\Rightarrow$ $\dfrac{dz}{dt} = 9.8(t+1) - 9.8t = 9.8$ m/s.

25. If $C =$ the rate at which water is pumped in, then $\dfrac{dV}{dt} = C - 10{,}000$, where
$V = \tfrac{1}{3}\pi r^2 h$ is the volume at time t. By similar triangles, $\dfrac{r}{2} = \dfrac{h}{6}$ $\Rightarrow$ $r = \tfrac{1}{3}h$ $\Rightarrow$
$V = \tfrac{1}{3}\pi\left(\tfrac{1}{3}h\right)^2 h = \dfrac{\pi}{27}h^3$ $\Rightarrow$ $\dfrac{dV}{dt} = \dfrac{\pi}{9}h^2\dfrac{dh}{dt}$. When $h = 200$ cm,

$\dfrac{dh}{dt} = 20$ cm/min, so $C - 10{,}000 = \dfrac{\pi}{9}(200)^2(20)$ $\Rightarrow$ $C = 10{,}000 + \dfrac{800{,}000}{9}\pi \approx 289{,}253$ cm^3/min.

27. 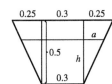 The figure is labeled in meters. The area A of a trapezoid is $\frac{1}{2}(\text{base}_1 + \text{base}_2)(\text{height})$, and the volume V of the 10-meter-long trough is $10A$.

Thus, the volume of the trapezoid with height h is $V = (10)\frac{1}{2}[0.3 + (0.3 + 2a)]h$.

By similar triangles, $\dfrac{a}{h} = \dfrac{0.25}{0.5} = \dfrac{1}{2}$, so $2a = h \;\Rightarrow\; V = 5(0.6 + h)h = 3h + 5h^2$.

Now $\dfrac{dV}{dt} = \dfrac{dV}{dh}\dfrac{dh}{dt} \;\Rightarrow\; 0.2 = (3 + 10h)\dfrac{dh}{dt} \;\Rightarrow\; \dfrac{dh}{dt} = \dfrac{0.2}{3 + 10h}$. When $h = 0.3$,

$\dfrac{dh}{dt} = \dfrac{0.2}{3 + 10(0.3)} = \dfrac{0.2}{6}$ m/min $= \dfrac{1}{30}$ m/min or $\dfrac{10}{3}$ cm/min.

29. We are given that $\dfrac{dV}{dt} = 30$ ft^3/min. $V = \dfrac{1}{3}\pi r^2 h = \dfrac{1}{3}\pi\left(\dfrac{h}{2}\right)^2 h = \dfrac{\pi h^3}{12} \;\Rightarrow\;$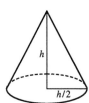

$\dfrac{dV}{dt} = \dfrac{dV}{dh}\dfrac{dh}{dt} \;\Rightarrow\; 30 = \dfrac{\pi h^2}{4}\dfrac{dh}{dt} \;\Rightarrow\; \dfrac{dh}{dt} = \dfrac{120}{\pi h^2}$.

When $h = 10$ ft, $\dfrac{dh}{dt} = \dfrac{120}{10^2\pi} = \dfrac{6}{5\pi} \approx 0.38$ ft/min.

31. The area A of an equilateral triangle with side s is given by $A = \frac{1}{4}\sqrt{3}\,s^2$.

$\dfrac{dA}{dt} = \frac{1}{4}\sqrt{3}\cdot 2s\,\dfrac{ds}{dt} = \frac{1}{4}\sqrt{3}\cdot 2(30)(10) = 150\sqrt{3}$ cm^2/min.

33. 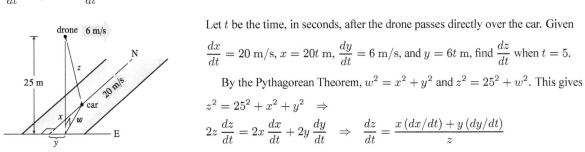 Let t be the time, in seconds, after the drone passes directly over the car. Given $\dfrac{dx}{dt} = 20$ m/s, $x = 20t$ m, $\dfrac{dy}{dt} = 6$ m/s, and $y = 6t$ m, find $\dfrac{dz}{dt}$ when $t = 5$.

By the Pythagorean Theorem, $w^2 = x^2 + y^2$ and $z^2 = 25^2 + w^2$. This gives

$z^2 = 25^2 + x^2 + y^2 \;\Rightarrow\;$

$2z\dfrac{dz}{dt} = 2x\dfrac{dx}{dt} + 2y\dfrac{dy}{dt} \;\Rightarrow\; \dfrac{dz}{dt} = \dfrac{x\,(dx/dt) + y\,(dy/dt)}{z}$

When $t = 5$, $x = 20(5) = 100$ and $y = 6(5) = 30$, so $z^2 = 25^2 + 100^2 + 30^2 \;\Rightarrow\; z = \sqrt{11{,}525}$ m.

$\dfrac{dz}{dt} = \dfrac{100(20) + 30(6)}{\sqrt{11{,}525}} \approx 20.3$ m/s.

35. $\cos\theta = \dfrac{x}{10} \;\Rightarrow\; -\sin\theta\,\dfrac{d\theta}{dt} = \dfrac{1}{10}\dfrac{dx}{dt}$. From Example 2,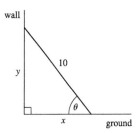

$\dfrac{dx}{dt} = 4$ and when $x = 6$, $y = 8$, so $\sin\theta = \dfrac{8}{10}$.

Thus, $-\dfrac{8}{10}\dfrac{d\theta}{dt} = \dfrac{1}{10}(4) \;\Rightarrow\; \dfrac{d\theta}{dt} = -\dfrac{1}{2}$ rad/s.

37. Differentiating both sides of $PV = C$ with respect to t and using the Product Rule gives us $P\dfrac{dV}{dt} + V\dfrac{dP}{dt} = 0 \;\Rightarrow\;$

$\dfrac{dV}{dt} = -\dfrac{V}{P}\dfrac{dP}{dt}$. When $V = 600$, $P = 150$ and $\dfrac{dP}{dt} = 20$, so we have $\dfrac{dV}{dt} = -\dfrac{600}{150}(20) = -80$. Thus, the volume is decreasing at a rate of 80 cm^3/min.

39. With $R_1 = 80$ and $R_2 = 100$, $\dfrac{1}{R} = \dfrac{1}{R_1} + \dfrac{1}{R_2} = \dfrac{1}{80} + \dfrac{1}{100} = \dfrac{180}{8000} = \dfrac{9}{400}$, so $R = \dfrac{400}{9}$. Differentiating $\dfrac{1}{R} = \dfrac{1}{R_1} + \dfrac{1}{R_2}$

with respect to t, we have $-\dfrac{1}{R^2}\dfrac{dR}{dt} = -\dfrac{1}{R_1^2}\dfrac{dR_1}{dt} - \dfrac{1}{R_2^2}\dfrac{dR_2}{dt} \;\Rightarrow\; \dfrac{dR}{dt} = R^2\left(\dfrac{1}{R_1^2}\dfrac{dR_1}{dt} + \dfrac{1}{R_2^2}\dfrac{dR_2}{dt}\right)$. When $R_1 = 80$ and

$R_2 = 100$, $\dfrac{dR}{dt} = \dfrac{400^2}{9^2}\left[\dfrac{1}{80^2}(0.3) + \dfrac{1}{100^2}(0.2)\right] = \dfrac{107}{810} \approx 0.132\ \Omega/\text{s}$.

41.

We are given that $\dfrac{dx}{dt} = 40$ mi/h and $\dfrac{dy}{dt} = 60$ mi/h. By the Law of Cosines,

$z^2 = x^2 + y^2 - 2xy\cos 60° = x^2 + y^2 - xy \;\Rightarrow\; 2z\dfrac{dz}{dt} = 2x\dfrac{dx}{dt} + 2y\dfrac{dy}{dt} - x\dfrac{dy}{dt} - y\dfrac{dx}{dt}$.

At $t = \tfrac{1}{2}$ h, we have $x = 40\left(\tfrac{1}{2}\right) = 20$ and $y = 60\left(\tfrac{1}{2}\right) = 30 \;\Rightarrow\;$

$z^2 = 20^2 + 30^2 - 20(30) = 700 \;\Rightarrow\; z = \sqrt{700}$ and

$\dfrac{dz}{dt} = \dfrac{2(20)(40) + 2(30)(60) - 20(60) - 30(40)}{2\sqrt{700}} = \dfrac{2800}{2\sqrt{700}} \approx 52.9$ mi/h.

43. We are given $d\theta/dt = 2°/\text{min} = \dfrac{\pi}{90}$ rad/min. By the Law of Cosines,

$x^2 = 12^2 + 15^2 - 2(12)(15)\cos\theta = 369 - 360\cos\theta \;\Rightarrow\;$

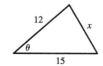

$2x\dfrac{dx}{dt} = 360\sin\theta\dfrac{d\theta}{dt} \;\Rightarrow\; \dfrac{dx}{dt} = \dfrac{180\sin\theta}{x}\dfrac{d\theta}{dt}$. When $\theta = 60°$,

$x = \sqrt{369 - 360\cos 60°} = \sqrt{189} = 3\sqrt{21}$, so $\dfrac{dx}{dt} = \dfrac{180\sin 60°}{3\sqrt{21}}\dfrac{\pi}{90} = \dfrac{\pi\sqrt{3}}{3\sqrt{21}} = \dfrac{\sqrt{7}\pi}{21} \approx 0.396$ m/min.

45. (a) By the Pythagorean Theorem, $4000^2 + y^2 = \ell^2$. Differentiating with respect to t,

we obtain $2y\dfrac{dy}{dt} = 2\ell\dfrac{d\ell}{dt}$. We know that $\dfrac{dy}{dt} = 600$ ft/s, so when $y = 3000$ ft,

$\ell = \sqrt{4000^2 + 3000^2} = \sqrt{25{,}000{,}000} = 5000$ ft

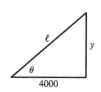

and $\dfrac{d\ell}{dt} = \dfrac{y}{\ell}\dfrac{dy}{dt} = \dfrac{3000}{5000}(600) = \dfrac{1800}{5} = 360$ ft/s.

(b) Here $\tan\theta = \dfrac{y}{4000} \;\Rightarrow\; \dfrac{d}{dt}(\tan\theta) = \dfrac{d}{dt}\left(\dfrac{y}{4000}\right) \;\Rightarrow\; \sec^2\theta\dfrac{d\theta}{dt} = \dfrac{1}{4000}\dfrac{dy}{dt} \;\Rightarrow\; \dfrac{d\theta}{dt} = \dfrac{\cos^2\theta}{4000}\dfrac{dy}{dt}$. When

$y = 3000$ ft, $\dfrac{dy}{dt} = 600$ ft/s, $\ell = 5000$ and $\cos\theta = \dfrac{4000}{\ell} = \dfrac{4000}{5000} = \dfrac{4}{5}$, so $\dfrac{d\theta}{dt} = \dfrac{(4/5)^2}{4000}(600) = 0.096$ rad/s.

47. $\cot\theta = \dfrac{x}{5} \;\Rightarrow\; -\csc^2\theta\dfrac{d\theta}{dt} = \dfrac{1}{5}\dfrac{dx}{dt} \;\Rightarrow\; -\left(\csc\dfrac{\pi}{3}\right)^2\left(-\dfrac{\pi}{6}\right) = \dfrac{1}{5}\dfrac{dx}{dt} \;\Rightarrow\;$

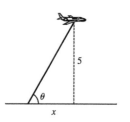

$\dfrac{dx}{dt} = \dfrac{5\pi}{6}\left(\dfrac{2}{\sqrt{3}}\right)^2 = \dfrac{10}{9}\pi$ km/min $[\approx 130$ mi/h$]$

49. We are given that $\dfrac{dx}{dt} = 300$ km/h. By the Law of Cosines,

$y^2 = x^2 + 1^2 - 2(1)(x)\cos 120° = x^2 + 1 - 2x\left(-\tfrac{1}{2}\right) = x^2 + x + 1$, so

$2y\dfrac{dy}{dt} = 2x\dfrac{dx}{dt} + \dfrac{dx}{dt} \;\Rightarrow\; \dfrac{dy}{dt} = \dfrac{2x+1}{2y}\dfrac{dx}{dt}$. After 1 minute, $x = \tfrac{300}{60} = 5$ km $\Rightarrow$

$y = \sqrt{5^2 + 5 + 1} = \sqrt{31}$ km $\;\Rightarrow\; \dfrac{dy}{dt} = \dfrac{2(5)+1}{2\sqrt{31}}(300) = \dfrac{1650}{\sqrt{31}} \approx 296$ km/h.

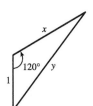

51. Let the distance between the runner and the friend be ℓ. Then by the Law of Cosines,

$\ell^2 = 200^2 + 100^2 - 2 \cdot 200 \cdot 100 \cdot \cos\theta = 50{,}000 - 40{,}000\cos\theta$ $(\star)$. Differentiating

implicitly with respect to t, we obtain $2\ell\dfrac{d\ell}{dt} = -40{,}000(-\sin\theta)\dfrac{d\theta}{dt}$. Now if D is the

distance run when the angle is θ radians, then by the formula for the length of an arc

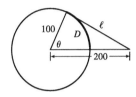

on a circle, $s = r\theta$, we have $D = 100\theta$, so $\theta = \dfrac{1}{100}D \;\Rightarrow\; \dfrac{d\theta}{dt} = \dfrac{1}{100}\dfrac{dD}{dt} = \dfrac{7}{100}$. To substitute into the expression for

$\dfrac{d\ell}{dt}$, we must know $\sin\theta$ at the time when $\ell = 200$, which we find from $(\star)$: $200^2 = 50{,}000 - 40{,}000\cos\theta \;\Leftrightarrow\;$

$\cos\theta = \tfrac{1}{4} \;\Rightarrow\; \sin\theta = \sqrt{1 - \left(\tfrac{1}{4}\right)^2} = \tfrac{\sqrt{15}}{4}$. Substituting, we get $2(200)\dfrac{d\ell}{dt} = 40{,}000\dfrac{\sqrt{15}}{4}\left(\tfrac{7}{100}\right) \;\Rightarrow\;$

$d\ell/dt = \tfrac{7\sqrt{15}}{4} \approx 6.78$ m/s. Whether the distance between them is increasing or decreasing depends on the direction in which

the runner is running.

53. The volume of the snowball is given by $V = \tfrac{4}{3}\pi r^3$, so $\dfrac{dV}{dt} = \tfrac{4}{3}\pi \cdot 3r^2\dfrac{dr}{dt} = 4\pi r^2\dfrac{dr}{dt}$. Since the volume is proportional to

the surface area S, with $S = 4\pi r^2$, we also have $\dfrac{dV}{dt} = k \cdot 4\pi r^2$ for some constant k. Equating the two expressions for $\dfrac{dV}{dt}$

gives $4\pi r^2\dfrac{dr}{dt} = k \cdot 4\pi r^2 \;\Rightarrow\; \dfrac{dr}{dt} = k$, that is, dr/dt is constant.

2.9 Linear Approximations and Differentials

1. $f(x) = x^3 - x^2 + 3 \;\Rightarrow\; f'(x) = 3x^2 - 2x$, so $f(-2) = -9$ and $f'(-2) = 16$. Thus,
$L(x) = f(-2) + f'(-2)(x - (-2)) = -9 + 16(x+2) = 16x + 23$.

3. $f(x) = \sqrt[3]{x} \;\Rightarrow\; f'(x) = \dfrac{1}{3\sqrt[3]{x^2}}$, so $f(8) = 2$ and $f'(8) = \dfrac{1}{12}$. Thus,
$L(x) = f(8) + f'(8)(x - 8) = 2 + \tfrac{1}{12}(x - 8) = \tfrac{1}{12}x + \tfrac{4}{3}$.

5. $f(x) = \sqrt{1 - x} \;\Rightarrow\; f'(x) = \dfrac{-1}{2\sqrt{1-x}}$, so $f(0) = 1$ and $f'(0) = -\tfrac{1}{2}$.

Therefore,

$\sqrt{1-x} = f(x) \approx f(0) + f'(0)(x - 0) = 1 + \left(-\tfrac{1}{2}\right)(x - 0) = 1 - \tfrac{1}{2}x$.

So $\sqrt{0.9} = \sqrt{1 - 0.1} \approx 1 - \tfrac{1}{2}(0.1) = 0.95$

and $\sqrt{0.99} = \sqrt{1 - 0.01} \approx 1 - \tfrac{1}{2}(0.01) = 0.995$.

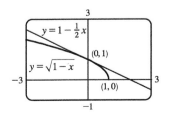

7. $f(x) = \sqrt[4]{1+2x}\ \Rightarrow\ f'(x) = \frac{1}{4}(1+2x)^{-3/4}(2) = \frac{1}{2}(1+2x)^{-3/4}$, so

$f(0) = 1$ and $f'(0) = \frac{1}{2}$. Thus, $f(x) \approx f(0) + f'(0)(x-0) = 1 + \frac{1}{2}x$.

We need $\sqrt[4]{1+2x} - 0.1 < 1 + \frac{1}{2}x < \sqrt[4]{1+2x} + 0.1$, which is true when

$-0.368 < x < 0.677$. Note that to ensure the accuracy, we have rounded the

smaller value up and the larger value down.

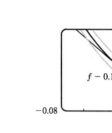

9. $f(x) = \dfrac{1}{(1+2x)^4} = (1+2x)^{-4}\ \Rightarrow$

$f'(x) = -4(1+2x)^{-5}(2) = \dfrac{-8}{(1+2x)^5}$, so $f(0) = 1$ and $f'(0) = -8$.

Thus, $f(x) \approx f(0) + f'(0)(x-0) = 1 + (-8)(x-0) = 1 - 8x$.

We need $\dfrac{1}{(1+2x)^4} - 0.1 < 1 - 8x < \dfrac{1}{(1+2x)^4} + 0.1$, which is true

when $-0.045 < x < 0.055$.

11. The differential dy is defined in terms of dx by the equation $dy = f'(x)\,dx$. For $y = f(x) = (x^2-3)^{-2}$,

$f'(x) = -2(x^2-3)^{-3}(2x) = -\dfrac{4x}{(x^2-3)^3}$, so $dy = -\dfrac{4x}{(x^2-3)^3}\,dx$.

13. For $y = f(u) = \dfrac{1+2u}{1+3u}$, $f'(u) = \dfrac{(1+3u)(2) - (1+2u)(3)}{(1+3u)^2} = \dfrac{-1}{(1+3u)^2}$, so $dy = \dfrac{-1}{(1+3u)^2}\,du$.

15. For $y = f(x) = \dfrac{1}{x^2-3x} = (x^2-3x)^{-1}$, $f'(x) = -(x^2-3x)^{-2}\cdot(2x-3) = -\dfrac{2x-3}{(x^2-3x)^2}$, so $dy = -\dfrac{2x-3}{(x^2-3x)^2}\,dx$.

17. For $y = f(t) = \sqrt{t-\cos t}$, $f'(t) = \frac{1}{2}(t-\cos t)^{-1/2}(1+\sin t) = \dfrac{1+\sin t}{2\sqrt{t-\cos t}}$, so $dy = \dfrac{1+\sin t}{2\sqrt{t-\cos t}}\,dt$.

19. (a) $y = \tan x\ \Rightarrow\ dy = \sec^2 x\,dx$

(b) When $x = \pi/4$ and $dx = -0.1$, $dy = [\sec(\pi/4)]^2(-0.1) = \left(\sqrt{2}\right)^2(-0.1) = -0.2$.

21. (a) $y = \sqrt{3+x^2}\ \Rightarrow\ dy = \dfrac{1}{2}(3+x^2)^{-1/2}(2x)\,dx = \dfrac{x}{\sqrt{3+x^2}}\,dx$

(b) $x = 1$ and $dx = -0.1\ \Rightarrow\ dy = \dfrac{1}{\sqrt{3+1^2}}(-0.1) = \dfrac{1}{2}(-0.1) = -0.05$.

23. $y = f(x) = x^2 - 4x$, $x = 3$, $\Delta x = 0.5\ \Rightarrow$

$\Delta y = f(3.5) - f(3) = -1.75 - (-3) = 1.25$

$dy = f'(x)\,dx = (2x-4)\,dx = (6-4)(0.5) = 1$

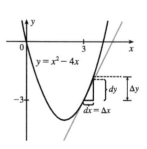

25. $y = f(x) = \sqrt{x-2}$, $x = 3$, $\Delta x = 0.8$ $\Rightarrow$

$\Delta y = f(3.8) - f(3) = \sqrt{1.8} - 1 \approx 0.34$

$dy = f'(x)\,dx = \dfrac{1}{2\sqrt{x-2}}\,dx = \dfrac{1}{2(1)}(0.8) = 0.4$

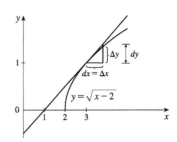

27. $y = f(x) = x^4 - x + 1$. If x changes from 1 to 1.05, $dx = \Delta x = 1.05 - 1 = 0.05$ $\Rightarrow$

$\Delta y = f(1.05) - f(1) = 1.16550625 - 1 \approx 0.1655$ and $dy = (4x^3 - 1)\,dx = (4 \cdot 1^3 - 1)(0.05) = 3(0.05) = 0.15$.

If x changes from 1 to 1.01, $dx = \Delta x = 1.01 - 1 = 0.01$ $\Rightarrow$ $\Delta y = f(1.01) - f(1) = 1.03060401 - 1 \approx 0.0306$

and $dy = (4x^3 - 1)\,dx = 3(0.01) = 0.03$.

With $\Delta x = 0.05$, $|dy - \Delta y| \approx |0.15 - 0.1655| = 0.0155$. With $\Delta x = 0.01$, $|dy - \Delta y| \approx |0.03 - 0.0306| = 0.0006$.

Since $|dy - \Delta y|$ is smaller for $\Delta x = 0.01$ than for $\Delta x = 0.05$, yes, the approximation $\Delta y \approx dy$ becomes better as Δx gets smaller.

29. $y = f(x) = \sqrt{5-x}$. If x changes from 1 to 1.05, $dx = \Delta x = 1.05 - 1 = 0.05$ $\Rightarrow$

$\Delta y = f(1.05) - f(1) = \sqrt{3.95} - 2 \approx -0.012539$ and $dy = -\dfrac{1}{2\sqrt{5-x}}\,dx = -\dfrac{1}{2\sqrt{5-1}}\,dx = -\dfrac{1}{4}(0.05) = -0.0125$.

If x changes from 1 to 1.01, $dx = \Delta x = 1.01 - 1 = 0.01$ $\Rightarrow$ $\Delta y = f(1.01) - f(1) = \sqrt{3.99} - 2 \approx -0.002502$

and $dy = -\dfrac{1}{4}(0.01) = -0.0025$.

With $\Delta x = 0.05$, $|dy - \Delta y| \approx |-0.0125 - (-0.012539)| = 0.000039$. With $\Delta x = 0.01$,

$|dy - \Delta y| \approx |-0.0025 - (-0.002502)| = 0.000002$. Since $|dy - \Delta y|$ is smaller for $\Delta x = 0.01$ than for $\Delta x = 0.05$,

yes, the approximation $\Delta y \approx dy$ becomes better as Δx gets smaller.

31. To estimate $(1.999)^4$, we'll find the linearization of $f(x) = x^4$ at $a = 2$. Since $f'(x) = 4x^3$, $f(2) = 16$, and

$f'(2) = 32$, we have $L(x) = 16 + 32(x-2)$. Thus, $x^4 \approx 16 + 32(x-2)$ when x is near 2, so

$(1.999)^4 \approx 16 + 32(1.999 - 2) = 16 - 0.032 = 15.968$.

33. $y = f(x) = \sqrt[3]{x}$ $\Rightarrow$ $dy = \tfrac{1}{3}x^{-2/3}\,dx$. When $x = 1000$ and $dx = 1$, $dy = \tfrac{1}{3}(1000)^{-2/3}(1) = \tfrac{1}{300}$, so

$\sqrt[3]{1001} = f(1001) \approx f(1000) + dy = 10 + \tfrac{1}{300} = 10.00\overline{3} \approx 10.003$.

35. $y = f(x) = \tan x$ $\Rightarrow$ $dy = \sec^2 x\,dx$. When $x = 0°$ [i.e., 0 radians] and $dx = 2°$ [i.e., $\tfrac{\pi}{90}$ radians],

$dy = (\sec^2 0)\left(\tfrac{\pi}{90}\right) = 1^2\left(\tfrac{\pi}{90}\right) = \tfrac{\pi}{90}$, so $\tan 2° = f(2°) \approx f(0°) + dy = 0 + \tfrac{\pi}{90} = \tfrac{\pi}{90} \approx 0.0349$.

37. $y = f(x) = \sec x$ $\Rightarrow$ $f'(x) = \sec x \tan x$, so $f(0) = 1$ and $f'(0) = 1 \cdot 0 = 0$. The linear approximation of f at 0 is

$f(0) + f'(0)(x - 0) = 1 + 0(x) = 1$. Since 0.08 is close to 0, approximating $\sec 0.08$ with 1 is reasonable.

39. (a) If x is the edge length, then $V = x^3 \;\Rightarrow\; dV = 3x^2\,dx$. When $x = 30$ and $dx = 0.1$, $dV = 3(30)^2(0.1) = 270$, so the maximum possible error in computing the volume of the cube is about 270 cm³. The relative error is calculated by dividing the change in V, ΔV, by V. We approximate ΔV with dV.

$$\text{Relative error} = \frac{\Delta V}{V} \approx \frac{dV}{V} = \frac{3x^2\,dx}{x^3} = 3\frac{dx}{x} = 3\left(\frac{0.1}{30}\right) = 0.01.$$

Percentage error = relative error × 100% = 0.01 × 100% = 1%.

(b) $S = 6x^2 \;\Rightarrow\; dS = 12x\,dx$. When $x = 30$ and $dx = 0.1$, $dS = 12(30)(0.1) = 36$, so the maximum possible error in computing the surface area of the cube is about 36 cm².

$$\text{Relative error} = \frac{\Delta S}{S} \approx \frac{dS}{S} = \frac{12x\,dx}{6x^2} = 2\frac{dx}{x} = 2\left(\frac{0.1}{30}\right) = 0.00\overline{6}.$$

Percentage error = relative error × 100% = $0.00\overline{6}$ × 100% = $0.\overline{6}$%.

41. (a) For a sphere of radius r, the circumference is $C = 2\pi r$ and the surface area is $S = 4\pi r^2$, so

$$r = \frac{C}{2\pi} \;\Rightarrow\; S = 4\pi\left(\frac{C}{2\pi}\right)^2 = \frac{C^2}{\pi} \;\Rightarrow\; dS = \frac{2}{\pi}C\,dC.$$ When $C = 84$ and $dC = 0.5$, $dS = \frac{2}{\pi}(84)(0.5) = \frac{84}{\pi}$,

so the maximum error is about $\dfrac{84}{\pi} \approx 27$ cm². Relative error $\approx \dfrac{dS}{S} = \dfrac{84/\pi}{84^2/\pi} = \dfrac{1}{84} \approx 0.012 = 1.2\%$

(b) $V = \dfrac{4}{3}\pi r^3 = \dfrac{4}{3}\pi\left(\dfrac{C}{2\pi}\right)^3 = \dfrac{C^3}{6\pi^2} \;\Rightarrow\; dV = \dfrac{1}{2\pi^2}C^2\,dC$. When $C = 84$ and $dC = 0.5$,

$dV = \dfrac{1}{2\pi^2}(84)^2(0.5) = \dfrac{1764}{\pi^2}$, so the maximum error is about $\dfrac{1764}{\pi^2} \approx 179$ cm³.

The relative error is approximately $\dfrac{dV}{V} = \dfrac{1764/\pi^2}{(84)^3/(6\pi^2)} = \dfrac{1}{56} \approx 0.018 = 1.8\%$.

43. (a) $V = \pi r^2 h \;\Rightarrow\; \Delta V \approx dV = 2\pi r h\,dr = 2\pi r h\,\Delta r$

(b) The error is

$$\Delta V - dV = [\pi(r+\Delta r)^2 h - \pi r^2 h] - 2\pi r h\,\Delta r = \pi r^2 h + 2\pi r h\,\Delta r + \pi(\Delta r)^2 h - \pi r^2 h - 2\pi r h\,\Delta r = \pi(\Delta r)^2 h.$$

45. $V = RI \;\Rightarrow\; I = \dfrac{V}{R} \;\Rightarrow\; dI = -\dfrac{V}{R^2}\,dR$. The relative error in calculating I is $\dfrac{\Delta I}{I} \approx \dfrac{dI}{I} = \dfrac{-(V/R^2)\,dR}{V/R} = -\dfrac{dR}{R}$.

Hence, the relative error in calculating I is approximately the same (in magnitude) as the relative error in R.

47. (a) $dc = \dfrac{dc}{dx}\,dx = 0\,dx = 0$ \qquad (b) $d(cu) = \dfrac{d}{dx}(cu)\,dx = c\dfrac{du}{dx}\,dx = c\,du$

(c) $d(u+v) = \dfrac{d}{dx}(u+v)\,dx = \left(\dfrac{du}{dx} + \dfrac{dv}{dx}\right)dx = \dfrac{du}{dx}\,dx + \dfrac{dv}{dx}\,dx = du + dv$

(d) $d(uv) = \dfrac{d}{dx}(uv)\,dx = \left(u\dfrac{dv}{dx} + v\dfrac{du}{dx}\right)dx = u\dfrac{dv}{dx}\,dx + v\dfrac{du}{dx}\,dx = u\,dv + v\,du$

(e) $d\left(\dfrac{u}{v}\right) = \dfrac{d}{dx}\left(\dfrac{u}{v}\right)dx = \dfrac{v\dfrac{du}{dx} - u\dfrac{dv}{dx}}{v^2}\,dx = \dfrac{v\dfrac{du}{dx}\,dx - u\dfrac{dv}{dx}\,dx}{v^2} = \dfrac{v\,du - u\,dv}{v^2}$

(f) $d(x^n) = \dfrac{d}{dx}(x^n)\,dx = nx^{n-1}\,dx$

49. (a) The graph shows that $f'(1) = 2$, so $L(x) = f(1) + f'(1)(x-1) = 5 + 2(x-1) = 2x + 3$.

$f(0.9) \approx L(0.9) = 4.8$ and $f(1.1) \approx L(1.1) = 5.2$.

(b) From the graph, we see that $f'(x)$ is positive and decreasing. This means that the slopes of the tangent lines are positive, but the tangents are becoming less steep. So the tangent lines lie *above* the curve. Thus, the estimates in part (a) are too large.

2 Review

TRUE-FALSE QUIZ

1. False. See the note after Theorem 2.2.4.

3. False. See the warning before the Product Rule.

5. True. $\dfrac{d}{dx}\sqrt{f(x)} = \dfrac{d}{dx}[f(x)]^{1/2} = \dfrac{1}{2}[f(x)]^{-1/2}f'(x) = \dfrac{f'(x)}{2\sqrt{f(x)}}$

7. False. $f(x) = |x^2 + x| = x^2 + x$ for $x \geq 0$ or $x \leq -1$ and $|x^2 + x| = -(x^2 + x)$ for $-1 < x < 0$.
So $f'(x) = 2x + 1$ for $x > 0$ or $x < -1$ and $f'(x) = -(2x+1)$ for $-1 < x < 0$. But $|2x+1| = 2x+1$ for $x \geq -\tfrac{1}{2}$ and $|2x+1| = -2x-1$ for $x < -\tfrac{1}{2}$.

9. True. $g(x) = x^5 \;\Rightarrow\; g'(x) = 5x^4 \;\Rightarrow\; g'(2) = 5(2)^4 = 80$, and by the definition of the derivative,
$\displaystyle\lim_{x \to 2}\dfrac{g(x) - g(2)}{x - 2} = g'(2) = 5(2)^4 = 80$.

11. False. A tangent line to the parabola $y = x^2$ has slope $dy/dx = 2x$, so at $(-2, 4)$ the slope of the tangent is $2(-2) = -4$ and an equation of the tangent line is $y - 4 = -4(x+2)$. [The given equation, $y - 4 = 2x(x+2)$, is not even linear!]

13. True. If $p(x) = a_n x^n + a_{n-1}x^{n-1} + \cdots + a_1 x + a_0$, then $p'(x) = na_n x^{n-1} + (n-1)a_{n-1}x^{n-2} + \cdots + a_1$, which is a polynomial.

15. False. For example, let $f(x) = x$ and $a = 0$. Then f is differentiable at a, but $|f| = |x|$ is not.

EXERCISES

1. (a) $s = s(t) = 1 + 2t + t^2/4$. The average velocity over the time interval $[1, 1+h]$ is

$$v_{\text{avg}} = \frac{s(1+h) - s(1)}{(1+h) - 1} = \frac{1 + 2(1+h) + (1+h)^2/4 - 13/4}{h} = \frac{10h + h^2}{4h} = \frac{10 + h}{4}$$

So for the following intervals the average velocities are:

(i) $[1, 3]$: $h = 2$, $v_{\text{avg}} = (10 + 2)/4 = 3$ m/s (ii) $[1, 2]$: $h = 1$, $v_{\text{avg}} = (10 + 1)/4 = 2.75$ m/s

(iii) $[1, 1.5]$: $h = 0.5$, $v_{\text{avg}} = (10 + 0.5)/4 = 2.625$ m/s (iv) $[1, 1.1]$: $h = 0.1$, $v_{\text{avg}} = (10 + 0.1)/4 = 2.525$ m/s

(b) When $t = 1$, the instantaneous velocity is $\lim\limits_{h \to 0} \dfrac{s(1+h) - s(1)}{h} = \lim\limits_{h \to 0} \dfrac{10 + h}{4} = \dfrac{10}{4} = 2.5$ m/s.

3.

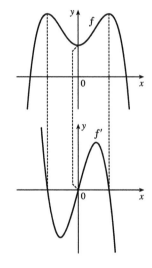

5. The graph of a has tangent lines with positive slope for $x < 0$ and negative slope for $x > 0$, and the values of c fit this pattern, so c must be the graph of the derivative of the function for a. The graph of c has horizontal tangent lines to the left and right of the x-axis and b has zeros at these points. Hence, b is the graph of the derivative of the function for c. Therefore, a is the graph of f, c is the graph of f', and b is the graph of f''.

7. (a) $f'(r)$ is the rate at which the total cost changes with respect to the interest rate. Its units are dollars/(percent per year).

(b) The total cost of paying off the loan is increasing by $1200/(percent per year) as the interest rate reaches 10%. So if the interest rate goes up from 10% to 11%, the cost goes up approximately $1200.

(c) As r increases, C increases. So $f'(r)$ will always be positive.

9. (a) $P'(t)$ is the rate at which the percentage of Americans under the age of 18 is changing with respect to time. Its units are percent per year (%/yr).

(b) To find $P'(t)$, we use $\lim\limits_{h \to 0} \dfrac{P(t+h) - P(t)}{h} \approx \dfrac{P(t+h) - P(t)}{h}$ for small values of h.

For 1950: $P'(1950) \approx \dfrac{P(1960) - P(1950)}{1960 - 1950} = \dfrac{35.7 - 31.1}{10} = 0.46$

For 1960: We estimate $P'(1960)$ by using $h = -10$ and $h = 10$, and then average the two results to obtain a final estimate.

$h = -10 \;\Rightarrow\; P'(1960) \approx \dfrac{P(1950) - P(1960)}{1950 - 1960} = \dfrac{31.1 - 35.7}{-10} = 0.46$

$h = 10 \;\Rightarrow\; P'(1960) \approx \dfrac{P(1970) - P(1960)}{1970 - 1960} = \dfrac{34.0 - 35.7}{10} = -0.17$

So we estimate that $P'(1960) \approx \tfrac{1}{2}[0.46 + (-0.17)] = 0.145$.

t	1950	1960	1970	1980	1990	2000	2010
$P'(t)$	0.460	0.145	-0.385	-0.415	-0.115	-0.085	-0.170

(c)

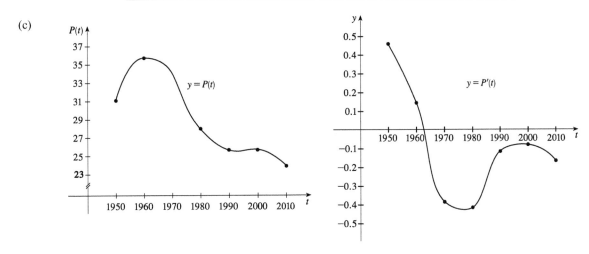

(d) We could get more accurate values for $P'(t)$ by obtaining data for the mid-decade years 1955, 1965, 1975, 1985, 1995, and 2005.

11. $f(x) = x^3 + 5x + 4 \;\Rightarrow\;$

$f'(x) = \lim\limits_{h \to 0} \dfrac{f(x+h) - f(x)}{h} = \lim\limits_{h \to 0} \dfrac{(x+h)^3 + 5(x+h) + 4 - (x^3 + 5x + 4)}{h}$

$= \lim\limits_{h \to 0} \dfrac{3x^2 h + 3xh^2 + h^3 + 5h}{h} = \lim\limits_{h \to 0} (3x^2 + 3xh + h^2 + 5) = 3x^2 + 5$

13. $y = (x^2 + x^3)^4 \;\Rightarrow\; y' = 4(x^2 + x^3)^3(2x + 3x^2) = 4(x^2)^3(1+x)^3 x(2+3x) = 4x^7(x+1)^3(3x+2)$

15. $y = \dfrac{x^2 - x + 2}{\sqrt{x}} = x^{3/2} - x^{1/2} + 2x^{-1/2} \;\Rightarrow\; y' = \dfrac{3}{2}x^{1/2} - \dfrac{1}{2}x^{-1/2} - x^{-3/2} = \dfrac{3}{2}\sqrt{x} - \dfrac{1}{2\sqrt{x}} - \dfrac{1}{\sqrt{x^3}}$

17. $y = x^2 \sin \pi x \;\Rightarrow\; y' = x^2(\cos \pi x)\pi + (\sin \pi x)(2x) = x(\pi x \cos \pi x + 2 \sin \pi x)$

19. $y = \dfrac{t^4 - 1}{t^4 + 1} \Rightarrow y' = \dfrac{(t^4 + 1)4t^3 - (t^4 - 1)4t^3}{(t^4 + 1)^2} = \dfrac{4t^3[(t^4 + 1) - (t^4 - 1)]}{(t^4 + 1)^2} = \dfrac{8t^3}{(t^4 + 1)^2}$

21. $y = \tan\sqrt{1 - x} \Rightarrow y' = \left(\sec^2\sqrt{1 - x}\right)\left(\dfrac{1}{2\sqrt{1 - x}}\right)(-1) = -\dfrac{\sec^2\sqrt{1 - x}}{2\sqrt{1 - x}}$

23. $\dfrac{d}{dx}(xy^4 + x^2 y) = \dfrac{d}{dx}(x + 3y) \Rightarrow x \cdot 4y^3 y' + y^4 \cdot 1 + x^2 \cdot y' + y \cdot 2x = 1 + 3y' \Rightarrow$

$y'(4xy^3 + x^2 - 3) = 1 - y^4 - 2xy \Rightarrow y' = \dfrac{1 - y^4 - 2xy}{4xy^3 + x^2 - 3}$

25. $y = \dfrac{\sec 2\theta}{1 + \tan 2\theta} \Rightarrow$

$y' = \dfrac{(1 + \tan 2\theta)(\sec 2\theta \tan 2\theta \cdot 2) - (\sec 2\theta)(\sec^2 2\theta \cdot 2)}{(1 + \tan 2\theta)^2} = \dfrac{2\sec 2\theta\,[(1 + \tan 2\theta)\tan 2\theta - \sec^2 2\theta]}{(1 + \tan 2\theta)^2}$

$= \dfrac{2\sec 2\theta\,(\tan 2\theta + \tan^2 2\theta - \sec^2 2\theta)}{(1 + \tan 2\theta)^2} = \dfrac{2\sec 2\theta\,(\tan 2\theta - 1)}{(1 + \tan 2\theta)^2} \quad [1 + \tan^2 x = \sec^2 x]$

27. $y = (1 - x^{-1})^{-1} \Rightarrow$

$y' = -1(1 - x^{-1})^{-2}[-(-1x^{-2})] = -(1 - 1/x)^{-2} x^{-2} = -((x - 1)/x)^{-2} x^{-2} = -(x - 1)^{-2}$

29. $\sin(xy) = x^2 - y \Rightarrow \cos(xy)(xy' + y \cdot 1) = 2x - y' \Rightarrow x\cos(xy)y' + y' = 2x - y\cos(xy) \Rightarrow$

$y'[x\cos(xy) + 1] = 2x - y\cos(xy) \Rightarrow y' = \dfrac{2x - y\cos(xy)}{x\cos(xy) + 1}$

31. $y = \cot(3x^2 + 5) \Rightarrow y' = -\csc^2(3x^2 + 5)(6x) = -6x\csc^2(3x^2 + 5)$

33. $y = \sqrt{x}\cos\sqrt{x} \Rightarrow$

$y' = \sqrt{x}\left(\cos\sqrt{x}\right)' + \cos\sqrt{x}\left(\sqrt{x}\right)' = \sqrt{x}\left[-\sin\sqrt{x}\left(\tfrac{1}{2}x^{-1/2}\right)\right] + \cos\sqrt{x}\left(\tfrac{1}{2}x^{-1/2}\right)$

$= \tfrac{1}{2}x^{-1/2}\left(-\sqrt{x}\sin\sqrt{x} + \cos\sqrt{x}\right) = \dfrac{\cos\sqrt{x} - \sqrt{x}\sin\sqrt{x}}{2\sqrt{x}}$

35. $y = \tan^2(\sin\theta) = [\tan(\sin\theta)]^2 \Rightarrow y' = 2[\tan(\sin\theta)] \cdot \sec^2(\sin\theta) \cdot \cos\theta$

37. $y = (x\tan x)^{1/5} \Rightarrow y' = \tfrac{1}{5}(x\tan x)^{-4/5}(\tan x + x\sec^2 x)$

39. $y = \sin\left(\tan\sqrt{1 + x^3}\right) \Rightarrow y' = \cos\left(\tan\sqrt{1 + x^3}\right)\left(\sec^2\sqrt{1 + x^3}\right)\left[3x^2/(2\sqrt{1 + x^3})\right]$

41. $f(t) = \sqrt{4t + 1} \Rightarrow f'(t) = \tfrac{1}{2}(4t + 1)^{-1/2} \cdot 4 = 2(4t + 1)^{-1/2} \Rightarrow$

$f''(t) = 2(-\tfrac{1}{2})(4t + 1)^{-3/2} \cdot 4 = -4/(4t + 1)^{3/2}$, so $f''(2) = -4/9^{3/2} = -\tfrac{4}{27}$.

43. $x^6 + y^6 = 1 \;\Rightarrow\; 6x^5 + 6y^5 y' = 0 \;\Rightarrow\; y' = -x^5/y^5 \;\Rightarrow\;$

$$y'' = -\frac{y^5(5x^4) - x^5(5y^4 y')}{(y^5)^2} = -\frac{5x^4 y^4 \left[y - x(-x^5/y^5)\right]}{y^{10}} = -\frac{5x^4 \left[(y^6 + x^6)/y^5\right]}{y^6} = -\frac{5x^4}{y^{11}}$$

45. $\displaystyle\lim_{x \to 0} \frac{\sec x}{1 - \sin x} = \frac{\sec 0}{1 - \sin 0} = \frac{1}{1 - 0} = 1$

47. $y = 4\sin^2 x \;\Rightarrow\; y' = 4 \cdot 2\sin x \cos x$. At $\left(\frac{\pi}{6}, 1\right)$, $y' = 8 \cdot \frac{1}{2} \cdot \frac{\sqrt{3}}{2} = 2\sqrt{3}$, so an equation of the tangent line is $y - 1 = 2\sqrt{3}\left(x - \frac{\pi}{6}\right)$, or $y = 2\sqrt{3}\,x + 1 - \pi\sqrt{3}/3$.

49. $y = \sqrt{1 + 4\sin x} \;\Rightarrow\; y' = \tfrac{1}{2}(1 + 4\sin x)^{-1/2} \cdot 4\cos x = \dfrac{2\cos x}{\sqrt{1 + 4\sin x}}$.

At $(0, 1)$, $y' = \dfrac{2}{\sqrt{1}} = 2$, so an equation of the tangent line is $y - 1 = 2(x - 0)$, or $y = 2x + 1$.

The slope of the normal line is $-\tfrac{1}{2}$, so an equation of the normal line is $y - 1 = -\tfrac{1}{2}(x - 0)$, or $y = -\tfrac{1}{2}x + 1$.

51. (a) $f(x) = x\sqrt{5 - x} \;\Rightarrow\;$

$$f'(x) = x\left[\tfrac{1}{2}(5-x)^{-1/2}(-1)\right] + \sqrt{5-x} = \frac{-x}{2\sqrt{5-x}} + \sqrt{5-x} \cdot \frac{2\sqrt{5-x}}{2\sqrt{5-x}} = \frac{-x}{2\sqrt{5-x}} + \frac{2(5-x)}{2\sqrt{5-x}}$$

$$= \frac{-x + 10 - 2x}{2\sqrt{5-x}} = \frac{10 - 3x}{2\sqrt{5-x}}$$

(b) At $(1, 2)$: $f'(1) = \tfrac{7}{4}$.

So an equation of the tangent line is $y - 2 = \tfrac{7}{4}(x - 1)$ or $y = \tfrac{7}{4}x + \tfrac{1}{4}$.

At $(4, 4)$: $f'(4) = -\tfrac{2}{2} = -1$.

So an equation of the tangent line is $y - 4 = -1(x - 4)$ or $y = -x + 8$.

(c)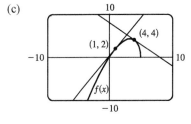

(d)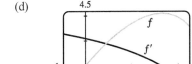

The graphs look reasonable, since f' is positive where f has tangents with positive slope, and f' is negative where f has tangents with negative slope.

53. $y = \sin x + \cos x \;\Rightarrow\; y' = \cos x - \sin x = 0 \;\Leftrightarrow\; \cos x = \sin x$ and $0 \le x \le 2\pi \;\Leftrightarrow\; x = \tfrac{\pi}{4}$ or $\tfrac{5\pi}{4}$, so the points are $\left(\tfrac{\pi}{4}, \sqrt{2}\right)$ and $\left(\tfrac{5\pi}{4}, -\sqrt{2}\right)$.

55. $y = f(x) = ax^2 + bx + c \;\Rightarrow\; f'(x) = 2ax + b$. We know that $f'(-1) = 6$ and $f'(5) = -2$, so $-2a + b = 6$ and $10a + b = -2$. Subtracting the first equation from the second gives $12a = -8 \;\Rightarrow\; a = -\tfrac{2}{3}$. Substituting $-\tfrac{2}{3}$ for a in the first equation gives $b = \tfrac{14}{3}$. Now $f(1) = 4 \;\Rightarrow\; 4 = a + b + c$, so $c = 4 + \tfrac{2}{3} - \tfrac{14}{3} = 0$ and hence, $f(x) = -\tfrac{2}{3}x^2 + \tfrac{14}{3}x$.

57. $f(x) = (x-a)(x-b)(x-c) \Rightarrow f'(x) = (x-b)(x-c) + (x-a)(x-c) + (x-a)(x-b)$.

So $\dfrac{f'(x)}{f(x)} = \dfrac{(x-b)(x-c) + (x-a)(x-c) + (x-a)(x-b)}{(x-a)(x-b)(x-c)} = \dfrac{1}{x-a} + \dfrac{1}{x-b} + \dfrac{1}{x-c}$.

Or: $f(x) = (x-a)(x-b)(x-c) \Rightarrow \ln|f(x)| = \ln|x-a| + \ln|x-b| + \ln|x-c| \Rightarrow$

$\dfrac{f'(x)}{f(x)} = \dfrac{1}{x-a} + \dfrac{1}{x-b} + \dfrac{1}{x-c}$

59. (a) $S(x) = f(x) + g(x) \Rightarrow S'(x) = f'(x) + g'(x) \Rightarrow S'(1) = f'(1) + g'(1) = 3 + 1 = 4$

(b) $P(x) = f(x)\,g(x) \Rightarrow P'(x) = f(x)\,g'(x) + g(x)\,f'(x) \Rightarrow$

$P'(2) = f(2)\,g'(2) + g(2)f'(2) = 1(4) + 1(2) = 4 + 2 = 6$

(c) $Q(x) = \dfrac{f(x)}{g(x)} \Rightarrow Q'(x) = \dfrac{g(x)\,f'(x) - f(x)\,g'(x)}{[g(x)]^2} \Rightarrow$

$Q'(1) = \dfrac{g(1)\,f'(1) - f(1)\,g'(1)}{[g(1)]^2} = \dfrac{3(3) - 2(1)}{3^2} = \dfrac{9-2}{9} = \dfrac{7}{9}$

(d) $C(x) = f(g(x)) \Rightarrow C'(x) = f'(g(x))\,g'(x) \Rightarrow C'(2) = f'(g(2))\,g'(2) = f'(1) \cdot 4 = 3 \cdot 4 = 12$

61. $f(x) = x^2 g(x) \Rightarrow f'(x) = x^2 g'(x) + g(x)(2x)$ or $x[xg'(x) + 2g(x)]$

63. $f(x) = [g(x)]^2 \Rightarrow f'(x) = 2[g(x)] \cdot g'(x) = 2g(x)\,g'(x)$

65. $f(x) = g(g(x)) \Rightarrow f'(x) = g'(g(x))\,g'(x)$

67. $f(x) = g(\sin x) \Rightarrow f'(x) = g'(\sin x) \cdot \cos x$

69. $h(x) = \dfrac{f(x)\,g(x)}{f(x) + g(x)} \Rightarrow$

$h'(x) = \dfrac{[f(x) + g(x)]\,[f(x)\,g'(x) + g(x)\,f'(x)] - f(x)\,g(x)\,[f'(x) + g'(x)]}{[f(x) + g(x)]^2}$

$= \dfrac{[f(x)]^2 g'(x) + f(x)\,g(x)\,f'(x) + f(x)\,g(x)\,g'(x) + [g(x)]^2 f'(x) - f(x)\,g(x)\,f'(x) - f(x)\,g(x)\,g'(x)}{[f(x) + g(x)]^2}$

$= \dfrac{f'(x)\,[g(x)]^2 + g'(x)\,[f(x)]^2}{[f(x) + g(x)]^2}$

71. (a)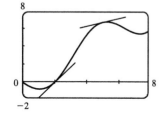

(b) The average rate of change is larger on $[2, 3]$.

(c) The instantaneous rate of change (the slope of the tangent) is larger at $x = 2$.

(d) $f(x) = x - 2\sin x \Rightarrow f'(x) = 1 - 2\cos x$,

so $f'(2) = 1 - 2\cos 2 \approx 1.8323$ and $f'(5) = 1 - 2\cos 5 \approx 0.4327$.

So $f'(2) > f'(5)$, as predicted in part (c).

73. (a) $y = t^3 - 12t + 3 \Rightarrow v(t) = y' = 3t^2 - 12 \Rightarrow a(t) = v'(t) = 6t$

(b) $v(t) = 3(t^2 - 4) > 0$ when $t > 2$, so it moves upward when $t > 2$ and downward when $0 \leq t < 2$.

(c) Distance upward $= y(3) - y(2) = -6 - (-13) = 7$,

Distance downward $= y(0) - y(2) = 3 - (-13) = 16$. Total distance $= 7 + 16 = 23$.

(d)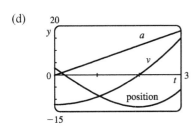

(e) The particle is speeding up when v and a have the same sign, that is, when $t > 2$. The particle is slowing down when v and a have opposite signs; that is, when $0 < t < 2$.

75. The linear density ρ is the rate of change of mass m with respect to length x.

$m = x\left(1 + \sqrt{x}\right) = x + x^{3/2} \Rightarrow \rho = dm/dx = 1 + \tfrac{3}{2}\sqrt{x}$, so the linear density when $x = 4$ is $1 + \tfrac{3}{2}\sqrt{4} = 4$ kg/m.

77. If $x =$ edge length, then $V = x^3 \Rightarrow dV/dt = 3x^2\, dx/dt = 10 \Rightarrow dx/dt = 10/(3x^2)$ and $S = 6x^2 \Rightarrow$
$dS/dt = (12x)\, dx/dt = 12x[10/(3x^2)] = 40/x$. When $x = 30$, $dS/dt = \tfrac{40}{30} = \tfrac{4}{3}$ cm^2/min.

79. Given $dh/dt = 5$ and $dx/dt = 15$, find dz/dt. $z^2 = x^2 + h^2 \Rightarrow$

$2z\dfrac{dz}{dt} = 2x\dfrac{dx}{dt} + 2h\dfrac{dh}{dt} \Rightarrow \dfrac{dz}{dt} = \dfrac{1}{z}(15x + 5h)$. When $t = 3$,

$h = 45 + 3(5) = 60$ and $x = 15(3) = 45 \Rightarrow z = \sqrt{45^2 + 60^2} = 75$,

so $\dfrac{dz}{dt} = \dfrac{1}{75}[15(45) + 5(60)] = 13$ ft/s.

81. We are given $d\theta/dt = -0.25$ rad/h. $\tan\theta = 400/x \Rightarrow$

$x = 400\cot\theta \Rightarrow \dfrac{dx}{dt} = -400\csc^2\theta\, \dfrac{d\theta}{dt}$. When $\theta = \tfrac{\pi}{6}$,

$\dfrac{dx}{dt} = -400(2)^2(-0.25) = 400$ ft/h.

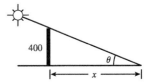

83. (a) $f(x) = \sqrt[3]{1+3x} = (1+3x)^{1/3} \Rightarrow f'(x) = (1+3x)^{-2/3}$, so the linearization of f at $a = 0$ is

$L(x) = f(0) + f'(0)(x-0) = 1^{1/3} + 1^{-2/3}x = 1 + x$. Thus, $\sqrt[3]{1+3x} \approx 1 + x \Rightarrow$

$\sqrt[3]{1.03} = \sqrt[3]{1+3(0.01)} \approx 1 + (0.01) = 1.01$.

(b) The linear approximation is $\sqrt[3]{1+3x} \approx 1 + x$, so for the required accuracy
we want $\sqrt[3]{1+3x} - 0.1 < 1 + x < \sqrt[3]{1+3x} + 0.1$. From the graph,
it appears that this is true when $-0.235 < x < 0.401$.

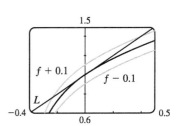

85. $A = x^2 + \frac{1}{2}\pi\left(\frac{1}{2}x\right)^2 = \left(1 + \frac{\pi}{8}\right)x^2 \Rightarrow dA = \left(2 + \frac{\pi}{4}\right)x\,dx$. When $x = 60$ and $dx = 0.1$, $dA = \left(2 + \frac{\pi}{4}\right)60(0.1) = 12 + \frac{3\pi}{2}$, so the maximum error is approximately $12 + \frac{3\pi}{2} \approx 16.7$ cm^2.

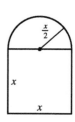

87. $\lim\limits_{h \to 0} \dfrac{\sqrt[4]{16 + h} - 2}{h} = \left[\dfrac{d}{dx}\sqrt[4]{x}\right]_{x=16} = \dfrac{1}{4}x^{-3/4}\Big|_{x=16} = \dfrac{1}{4\left(\sqrt[4]{16}\right)^3} = \dfrac{1}{32}$

89. $\lim\limits_{x \to 0} \dfrac{\sqrt{1 + \tan x} - \sqrt{1 + \sin x}}{x^3} = \lim\limits_{x \to 0} \dfrac{\left(\sqrt{1 + \tan x} - \sqrt{1 + \sin x}\right)\left(\sqrt{1 + \tan x} + \sqrt{1 + \sin x}\right)}{x^3\left(\sqrt{1 + \tan x} + \sqrt{1 + \sin x}\right)}$

$= \lim\limits_{x \to 0} \dfrac{(1 + \tan x) - (1 + \sin x)}{x^3\left(\sqrt{1 + \tan x} + \sqrt{1 + \sin x}\right)} = \lim\limits_{x \to 0} \dfrac{\sin x\,(1/\cos x - 1)}{x^3\left(\sqrt{1 + \tan x} + \sqrt{1 + \sin x}\right)} \cdot \dfrac{\cos x}{\cos x}$

$= \lim\limits_{x \to 0} \dfrac{\sin x\,(1 - \cos x)}{x^3\left(\sqrt{1 + \tan x} + \sqrt{1 + \sin x}\right)\cos x} \cdot \dfrac{1 + \cos x}{1 + \cos x}$

$= \lim\limits_{x \to 0} \dfrac{\sin x \cdot \sin^2 x}{x^3\left(\sqrt{1 + \tan x} + \sqrt{1 + \sin x}\right)\cos x\,(1 + \cos x)}$

$= \left(\lim\limits_{x \to 0}\dfrac{\sin x}{x}\right)^3 \lim\limits_{x \to 0} \dfrac{1}{\left(\sqrt{1 + \tan x} + \sqrt{1 + \sin x}\right)\cos x\,(1 + \cos x)}$

$= 1^3 \cdot \dfrac{1}{\left(\sqrt{1} + \sqrt{1}\right)\cdot 1 \cdot (1 + 1)} = \dfrac{1}{4}$

91. $\dfrac{d}{dx}[f(2x)] = x^2 \Rightarrow f'(2x)\cdot 2 = x^2 \Rightarrow f'(2x) = \frac{1}{2}x^2$. Let $t = 2x$. Then $f'(t) = \frac{1}{2}\left(\frac{1}{2}t\right)^2 = \frac{1}{8}t^2$, so $f'(x) = \frac{1}{8}x^2$.

☐ PROBLEMS PLUS

1. Let a be the x-coordinate of Q. Since the derivative of $y = 1 - x^2$ is $y' = -2x$, the slope at Q is $-2a$. But since the triangle is equilateral, $\overline{AO}/\overline{OC} = \sqrt{3}/1$, so the slope at Q is $-\sqrt{3}$. Therefore, we must have that $-2a = -\sqrt{3} \Rightarrow a = \frac{\sqrt{3}}{2}$.

   Thus, the point Q has coordinates $\left(\frac{\sqrt{3}}{2}, 1 - \left(\frac{\sqrt{3}}{2}\right)^2\right) = \left(\frac{\sqrt{3}}{2}, \frac{1}{4}\right)$ and by symmetry, P has coordinates $\left(-\frac{\sqrt{3}}{2}, \frac{1}{4}\right)$.

3.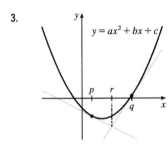

   We must show that r (in the figure) is halfway between p and q, that is, $r = (p+q)/2$. For the parabola $y = ax^2 + bx + c$, the slope of the tangent line is given by $y' = 2ax + b$. An equation of the tangent line at $x = p$ is

   $y - (ap^2 + bp + c) = (2ap + b)(x - p)$. Solving for y gives us

   $$y = (2ap + b)x - 2ap^2 - bp + (ap^2 + bp + c)$$

   or $\quad y = (2ap + b)x + c - ap^2 \quad$ **(1)**

   Similarly, an equation of the tangent line at $x = q$ is

   $$y = (2aq + b)x + c - aq^2 \quad \textbf{(2)}$$

   We can eliminate y and solve for x by subtracting equation **(1)** from equation **(2)**.

   $$[(2aq + b) - (2ap + b)]x - aq^2 + ap^2 = 0$$
   $$(2aq - 2ap)x = aq^2 - ap^2$$
   $$2a(q - p)x = a(q^2 - p^2)$$
   $$x = \frac{a(q+p)(q-p)}{2a(q-p)} = \frac{p+q}{2}$$

   Thus, the x-coordinate of the point of intersection of the two tangent lines, namely r, is $(p+q)/2$.

5. Using $f'(a) = \lim\limits_{x \to a} \dfrac{f(x) - f(a)}{x - a}$, we recognize the given expression, $f(x) = \lim\limits_{t \to x} \dfrac{\sec t - \sec x}{t - x}$, as $g'(x)$
   with $g(x) = \sec x$. Now $f'(\frac{\pi}{4}) = g''(\frac{\pi}{4})$, so we will find $g''(x)$. $g'(x) = \sec x \tan x \Rightarrow$
   $g''(x) = \sec x \sec^2 x + \tan x \sec x \tan x = \sec x (\sec^2 x + \tan^2 x)$, so $g''(\frac{\pi}{4}) = \sqrt{2}(\sqrt{2}^2 + 1^2) = \sqrt{2}(2 + 1) = 3\sqrt{2}$.

7. We use mathematical induction. Let S_n be the statement that $\dfrac{d^n}{dx^n}(\sin^4 x + \cos^4 x) = 4^{n-1}\cos(4x + n\pi/2)$.

   S_1 is true because

   $$\frac{d}{dx}(\sin^4 x + \cos^4 x) = 4\sin^3 x \cos x - 4\cos^3 x \sin x = 4\sin x \cos x \left(\sin^2 x - \cos^2 x\right)$$
   $$= -4\sin x \cos x \cos 2x = -2\sin 2x \cos 2 = -\sin 4x = \sin(-4x)$$
   $$= \cos\left(\tfrac{\pi}{2} - (-4x)\right) = \cos\left(\tfrac{\pi}{2} + 4x\right) = 4^{n-1}\cos\left(4x + n\tfrac{\pi}{2}\right) \text{ when } n = 1$$

   [continued]

Now assume S_k is true, that is, $\dfrac{d^k}{dx^k}(\sin^4 x + \cos^4 x) = 4^{k-1}\cos(4x + k\tfrac{\pi}{2})$. Then

$$\dfrac{d^{k+1}}{dx^{k+1}}(\sin^4 x + \cos^4 x) = \dfrac{d}{dx}\left[\dfrac{d^k}{dx^k}(\sin^4 x + \cos^4 x)\right] = \dfrac{d}{dx}\left[4^{k-1}\cos(4x + k\tfrac{\pi}{2})\right]$$

$$= -4^{k-1}\sin(4x + k\tfrac{\pi}{2}) \cdot \dfrac{d}{dx}(4x + k\tfrac{\pi}{2}) = -4^k \sin(4x + k\tfrac{\pi}{2})$$

$$= 4^k \sin(-4x - k\tfrac{\pi}{2}) = 4^k \cos(\tfrac{\pi}{2} - (-4x - k\tfrac{\pi}{2})) = 4^k \cos(4x + (k+1)\tfrac{\pi}{2})$$

which shows that S_{k+1} is true.

Therefore, $\dfrac{d^n}{dx^n}(\sin^4 x + \cos^4 x) = 4^{n-1}\cos(4x + n\tfrac{\pi}{2})$ for every positive integer n, by mathematical induction.

Another proof: First write

$$\sin^4 x + \cos^4 x = (\sin^2 x + \cos^2 x)^2 - 2\sin^2 x \cos^2 x = 1 - \tfrac{1}{2}\sin^2 2x = 1 - \tfrac{1}{4}(1 - \cos 4x) = \tfrac{3}{4} + \tfrac{1}{4}\cos 4x$$

Then we have $\dfrac{d^n}{dx^n}(\sin^4 x + \cos^4 x) = \dfrac{d^n}{dx^n}\left(\dfrac{3}{4} + \dfrac{1}{4}\cos 4x\right) = \dfrac{1}{4}\cdot 4^n \cos\left(4x + n\dfrac{\pi}{2}\right) = 4^{n-1}\cos\left(4x + n\dfrac{\pi}{2}\right)$.

9. We must find a value x_0 such that the normal lines to the parabola $y = x^2$ at $x = \pm x_0$ intersect at a point one unit from the points $(\pm x_0, x_0^2)$. The normals to $y = x^2$ at $x = \pm x_0$ have slopes $-\dfrac{1}{\pm 2x_0}$ and pass through $(\pm x_0, x_0^2)$ respectively, so the normals have the equations $y - x_0^2 = -\dfrac{1}{2x_0}(x - x_0)$ and $y - x_0^2 = \dfrac{1}{2x_0}(x + x_0)$. The common y-intercept is $x_0^2 + \dfrac{1}{2}$.

We want to find the value of x_0 for which the distance from $(0, x_0^2 + \tfrac{1}{2})$ to (x_0, x_0^2) equals 1. The square of the distance is $(x_0 - 0)^2 + [x_0^2 - (x_0^2 + \tfrac{1}{2})]^2 = x_0^2 + \tfrac{1}{4} = 1 \Leftrightarrow x_0 = \pm\tfrac{\sqrt{3}}{2}$. For these values of x_0, the y-intercept is $x_0^2 + \tfrac{1}{2} = \tfrac{5}{4}$, so the center of the circle is at $(0, \tfrac{5}{4})$.

Another solution: Let the center of the circle be $(0, a)$. Then the equation of the circle is $x^2 + (y - a)^2 = 1$.

Solving with the equation of the parabola, $y = x^2$, we get $x^2 + (x^2 - a)^2 = 1 \Leftrightarrow x^2 + x^4 - 2ax^2 + a^2 = 1 \Leftrightarrow$

$x^4 + (1 - 2a)x^2 + a^2 - 1 = 0$. The parabola and the circle will be tangent to each other when this quadratic equation in x^2 has equal roots; that is, when the discriminant is 0. Thus, $(1 - 2a)^2 - 4(a^2 - 1) = 0 \Leftrightarrow$

$1 - 4a + 4a^2 - 4a^2 + 4 = 0 \Leftrightarrow 4a = 5$, so $a = \tfrac{5}{4}$. The center of the circle is $(0, \tfrac{5}{4})$.

11. See the figure. Clearly, the line $y = 2$ is tangent to both circles at the point $(0, 2)$. We'll look for a tangent line L through the points (a, b) and (c, d), and if such a line exists, then its reflection through the y-axis is another such line. The slope of L is the same at (a, b) and (c, d). Find those slopes: $x^2 + y^2 = 4 \Rightarrow$

$2x + 2y\,y' = 0 \Rightarrow y' = -\dfrac{x}{y} \left[= -\dfrac{a}{b}\right]$ and $x^2 + (y - 3)^2 = 1 \Rightarrow$

$2x + 2(y - 3)y' = 0 \Rightarrow y' = -\dfrac{x}{y - 3} \left[= -\dfrac{c}{d - 3}\right]$.

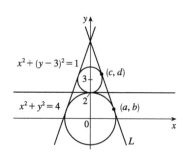

[continued]

Now an equation for L can be written using either point-slope pair, so we get $y - b = -\frac{a}{b}(x - a)$ $\left[\text{or } y = -\frac{a}{b}x + \frac{a^2}{b} + b\right]$

and $y - d = -\frac{c}{d-3}(x - c)$ $\left[\text{or } y = -\frac{c}{d-3}x + \frac{c^2}{d-3} + d\right]$. The slopes are equal, so $-\frac{a}{b} = -\frac{c}{d-3}$ $\Leftrightarrow$

$d - 3 = \frac{bc}{a}$. Since (c, d) is a solution of $x^2 + (y - 3)^2 = 1$, we have $c^2 + (d - 3)^2 = 1$, so $c^2 + \left(\frac{bc}{a}\right)^2 = 1$ $\Rightarrow$

$a^2c^2 + b^2c^2 = a^2$ $\Rightarrow$ $c^2(a^2 + b^2) = a^2$ $\Rightarrow$ $4c^2 = a^2$ [since (a, b) is a solution of $x^2 + y^2 = 4$] $\Rightarrow$ $a = 2c$.

Now $d - 3 = \frac{bc}{a}$ $\Rightarrow$ $d = 3 + \frac{bc}{2c}$, so $d = 3 + \frac{b}{2}$. The y-intercepts are equal, so $\frac{a^2}{b} + b = \frac{c^2}{d-3} + d$ $\Leftrightarrow$

$\frac{a^2}{b} + b = \frac{(a/2)^2}{b/2} + \left(3 + \frac{b}{2}\right)$ $\Leftrightarrow$ $\left[\frac{a^2}{b} + b = \frac{a^2}{2b} + 3 + \frac{b}{2}\right](2b)$ $\Leftrightarrow$ $2a^2 + 2b^2 = a^2 + 6b + b^2$ $\Leftrightarrow$

$a^2 + b^2 = 6b$ $\Leftrightarrow$ $4 = 6b$ $\Leftrightarrow$ $b = \frac{2}{3}$. It follows that $d = 3 + \frac{b}{2} = \frac{10}{3}$, $a^2 = 4 - b^2 = 4 - \frac{4}{9} = \frac{32}{9}$ $\Rightarrow$ $a = \frac{4}{3}\sqrt{2}$,

and $c^2 = 1 - (d - 3)^2 = 1 - \left(\frac{1}{3}\right)^2 = \frac{8}{9}$ $\Rightarrow$ $c = \frac{2}{3}\sqrt{2}$. Thus, L has equation $y - \frac{2}{3} = -\frac{(4/3)\sqrt{2}}{2/3}\left(x - \frac{4}{3}\sqrt{2}\right)$ $\Leftrightarrow$

$y - \frac{2}{3} = -2\sqrt{2}\left(x - \frac{4}{3}\sqrt{2}\right)$ $\Leftrightarrow$ $y = -2\sqrt{2}\,x + 6$. Its reflection has equation $y = 2\sqrt{2}\,x + 6$.

In summary, there are three lines tangent to both circles: $y = 2$ touches at $(0, 2)$, L touches at $\left(\frac{4}{3}\sqrt{2}, \frac{2}{3}\right)$ and $\left(\frac{2}{3}\sqrt{2}, \frac{10}{3}\right)$, and its reflection through the y-axis touches at $\left(-\frac{4}{3}\sqrt{2}, \frac{2}{3}\right)$ and $\left(-\frac{2}{3}\sqrt{2}, \frac{10}{3}\right)$.

13. We can assume without loss of generality that $\theta = 0$ at time $t = 0$, so that $\theta = 12\pi t$ rad. [The angular velocity of the wheel is 360 rpm $= 360 \cdot (2\pi \text{ rad})/(60 \text{ s}) = 12\pi$ rad/s.] Then the position of A as a function of time is

$A = (40\cos\theta, 40\sin\theta) = (40\cos 12\pi t, 40\sin 12\pi t)$, so $\sin\alpha = \frac{y}{1.2 \text{ m}} = \frac{40\sin\theta}{120} = \frac{\sin\theta}{3} = \frac{1}{3}\sin 12\pi t$.

(a) Differentiating the expression for $\sin\alpha$, we get $\cos\alpha \cdot \frac{d\alpha}{dt} = \frac{1}{3} \cdot 12\pi \cdot \cos 12\pi t = 4\pi\cos\theta$. When $\theta = \frac{\pi}{3}$, we have

$\sin\alpha = \frac{1}{3}\sin\theta = \frac{\sqrt{3}}{6}$, so $\cos\alpha = \sqrt{1 - \left(\frac{\sqrt{3}}{6}\right)^2} = \sqrt{\frac{11}{12}}$ and $\frac{d\alpha}{dt} = \frac{4\pi\cos\frac{\pi}{3}}{\cos\alpha} = \frac{2\pi}{\sqrt{11/12}} = \frac{4\pi\sqrt{3}}{\sqrt{11}} \approx 6.56$ rad/s.

(b) By the Law of Cosines, $|AP|^2 = |OA|^2 + |OP|^2 - 2|OA||OP|\cos\theta$ $\Rightarrow$

$120^2 = 40^2 + |OP|^2 - 2 \cdot 40|OP|\cos\theta$ $\Rightarrow$ $|OP|^2 - (80\cos\theta)|OP| - 12{,}800 = 0$ $\Rightarrow$

$|OP| = \frac{1}{2}\left(80\cos\theta \pm \sqrt{6400\cos^2\theta + 51{,}200}\,\right) = 40\cos\theta \pm 40\sqrt{\cos^2\theta + 8} = 40\left(\cos\theta + \sqrt{8 + \cos^2\theta}\,\right)$ cm

[since $|OP| > 0$]. As a check, note that $|OP| = 160$ cm when $\theta = 0$ and $|OP| = 80\sqrt{2}$ cm when $\theta = \frac{\pi}{2}$.

(c) By part (b), the x-coordinate of P is given by $x = 40\left(\cos\theta + \sqrt{8 + \cos^2\theta}\,\right)$, so

$\frac{dx}{dt} = \frac{dx}{d\theta}\frac{d\theta}{dt} = 40\left(-\sin\theta - \frac{2\cos\theta\sin\theta}{2\sqrt{8 + \cos^2\theta}}\right) \cdot 12\pi = -480\pi\sin\theta\left(1 + \frac{\cos\theta}{\sqrt{8 + \cos^2\theta}}\right)$ cm/s.

In particular, $dx/dt = 0$ cm/s when $\theta = 0$ and $dx/dt = -480\pi$ cm/s when $\theta = \frac{\pi}{2}$.

15. It seems from the figure that as P approaches the point $(0, 2)$ from the right, $x_T \to \infty$ and $y_T \to 2^+$. As P approaches the point $(3, 0)$ from the left, it appears that $x_T \to 3^+$ and $y_T \to \infty$. So we guess that $x_T \in (3, \infty)$ and $y_T \in (2, \infty)$. It is more difficult to estimate the range of values for x_N and y_N. We might perhaps guess that $x_N \in (0, 3)$, and $y_N \in (-\infty, 0)$ or $(-2, 0)$.

In order to actually solve the problem, we implicitly differentiate the equation of the ellipse to find the equation of the tangent line: $\dfrac{x^2}{9} + \dfrac{y^2}{4} = 1 \;\Rightarrow\; \dfrac{2x}{9} + \dfrac{2y}{4}y' = 0$, so $y' = -\dfrac{4}{9}\dfrac{x}{y}$. So at the point (x_0, y_0) on the ellipse, an equation of the tangent line is $y - y_0 = -\dfrac{4}{9}\dfrac{x_0}{y_0}(x - x_0)$ or $4x_0 x + 9y_0 y = 4x_0^2 + 9y_0^2$. This can be written as $\dfrac{x_0 x}{9} + \dfrac{y_0 y}{4} = \dfrac{x_0^2}{9} + \dfrac{y_0^2}{4} = 1$, because (x_0, y_0) lies on the ellipse. So an equation of the tangent line is $\dfrac{x_0 x}{9} + \dfrac{y_0 y}{4} = 1$.

Therefore, the x-intercept x_T for the tangent line is given by $\dfrac{x_0 x_T}{9} = 1 \;\Leftrightarrow\; x_T = \dfrac{9}{x_0}$, and the y-intercept y_T is given by $\dfrac{y_0 y_T}{4} = 1 \;\Leftrightarrow\; y_T = \dfrac{4}{y_0}$.

So as x_0 takes on all values in $(0, 3)$, x_T takes on all values in $(3, \infty)$, and as y_0 takes on all values in $(0, 2)$, y_T takes on all values in $(2, \infty)$. At the point (x_0, y_0) on the ellipse, the slope of the normal line is $-\dfrac{1}{y'(x_0, y_0)} = \dfrac{9}{4}\dfrac{y_0}{x_0}$, and its equation is $y - y_0 = \dfrac{9}{4}\dfrac{y_0}{x_0}(x - x_0)$. So the x-intercept x_N for the normal line is given by $0 - y_0 = \dfrac{9}{4}\dfrac{y_0}{x_0}(x_N - x_0) \;\Rightarrow\;$
$x_N = -\dfrac{4x_0}{9} + x_0 = \dfrac{5x_0}{9}$, and the y-intercept y_N is given by $y_N - y_0 = \dfrac{9}{4}\dfrac{y_0}{x_0}(0 - x_0) \;\Rightarrow\; y_N = -\dfrac{9y_0}{4} + y_0 = -\dfrac{5y_0}{4}$.

So as x_0 takes on all values in $(0, 3)$, x_N takes on all values in $\left(0, \tfrac{5}{3}\right)$, and as y_0 takes on all values in $(0, 2)$, y_N takes on all values in $\left(-\tfrac{5}{2}, 0\right)$.

17. (a) If the two lines L_1 and L_2 have slopes m_1 and m_2 and angles of inclination ϕ_1 and ϕ_2, then $m_1 = \tan \phi_1$ and $m_2 = \tan \phi_2$. The triangle in the figure shows that $\phi_1 + \alpha + (180° - \phi_2) = 180°$ and so $\alpha = \phi_2 - \phi_1$. Therefore, using the identity for $\tan(x - y)$, we have
$$\tan \alpha = \tan(\phi_2 - \phi_1) = \dfrac{\tan \phi_2 - \tan \phi_1}{1 + \tan \phi_2 \tan \phi_1} \text{ and so } \tan \alpha = \dfrac{m_2 - m_1}{1 + m_1 m_2}.$$

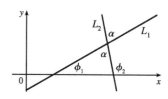

(b) (i) The parabolas intersect when $x^2 = (x - 2)^2 \;\Rightarrow\; x = 1$. If $y = x^2$, then $y' = 2x$, so the slope of the tangent to $y = x^2$ at $(1, 1)$ is $m_1 = 2(1) = 2$. If $y = (x - 2)^2$, then $y' = 2(x - 2)$, so the slope of the tangent to $y = (x - 2)^2$ at $(1, 1)$ is $m_2 = 2(1 - 2) = -2$. Therefore, $\tan \alpha = \dfrac{m_2 - m_1}{1 + m_1 m_2} = \dfrac{-2 - 2}{1 + 2(-2)} = \dfrac{4}{3}$ and so $\alpha = \tan^{-1}\left(\tfrac{4}{3}\right) \approx 53°$ [or $127°$].

(ii) $x^2 - y^2 = 3$ and $x^2 - 4x + y^2 + 3 = 0$ intersect when $x^2 - 4x + (x^2 - 3) + 3 = 0 \;\Leftrightarrow\; 2x(x - 2) = 0 \;\Rightarrow\; x = 0$ or 2, but 0 is extraneous. If $x = 2$, then $y = \pm 1$. If $x^2 - y^2 = 3$ then $2x - 2yy' = 0 \;\Rightarrow\; y' = x/y$ and

$x^2 - 4x + y^2 + 3 = 0 \;\Rightarrow\; 2x - 4 + 2yy' = 0 \;\Rightarrow\; y' = \dfrac{2-x}{y}$. At $(2, 1)$ the slopes are $m_1 = 2$ and $m_2 = 0$, so $\tan \alpha = \dfrac{0-2}{1+2\cdot 0} = -2 \;\Rightarrow\; \alpha \approx 117°$. At $(2, -1)$ the slopes are $m_1 = -2$ and $m_2 = 0$, so $\tan \alpha = \dfrac{0-(-2)}{1+(-2)(0)} = 2 \;\Rightarrow\; \alpha \approx 63°$ [or $117°$].

19. Since $\angle ROQ = \angle OQP = \theta$, the triangle QOR is isosceles, so
$|QR| = |RO| = x$. By the Law of Cosines, $x^2 = x^2 + r^2 - 2rx\cos\theta$. Hence,
$2rx\cos\theta = r^2$, so $x = \dfrac{r^2}{2r\cos\theta} = \dfrac{r}{2\cos\theta}$. Note that as $y \to 0^+, \theta \to 0^+$ (since
$\sin\theta = y/r$), and hence $x \to \dfrac{r}{2\cos 0} = \dfrac{r}{2}$. Thus, as P is taken closer and closer
to the x-axis, the point R approaches the midpoint of the radius AO.

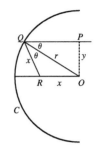

21. $\displaystyle\lim_{x\to 0} \dfrac{\sin(a+2x) - 2\sin(a+x) + \sin a}{x^2}$

$= \displaystyle\lim_{x\to 0} \dfrac{\sin a \cos 2x + \cos a \sin 2x - 2\sin a \cos x - 2\cos a \sin x + \sin a}{x^2}$

$= \displaystyle\lim_{x\to 0} \dfrac{\sin a\,(\cos 2x - 2\cos x + 1) + \cos a\,(\sin 2x - 2\sin x)}{x^2}$

$= \displaystyle\lim_{x\to 0} \dfrac{\sin a\,(2\cos^2 x - 1 - 2\cos x + 1) + \cos a\,(2\sin x \cos x - 2\sin x)}{x^2}$

$= \displaystyle\lim_{x\to 0} \dfrac{\sin a\,(2\cos x)(\cos x - 1) + \cos a\,(2\sin x)(\cos x - 1)}{x^2}$

$= \displaystyle\lim_{x\to 0} \dfrac{2(\cos x - 1)[\sin a \cos x + \cos a \sin x](\cos x + 1)}{x^2(\cos x + 1)}$

$= \displaystyle\lim_{x\to 0} \dfrac{-2\sin^2 x\,[\sin(a+x)]}{x^2(\cos x + 1)} = -2\lim_{x\to 0}\left(\dfrac{\sin x}{x}\right)^2 \cdot \dfrac{\sin(a+x)}{\cos x + 1} = -2(1)^2 \dfrac{\sin(a+0)}{\cos 0 + 1} = -\sin a$

23. $y = x^4 - 2x^2 - x \;\Rightarrow\; y' = 4x^3 - 4x - 1$. The equation of the tangent line at $x = a$ is
$y - (a^4 - 2a^2 - a) = (4a^3 - 4a - 1)(x - a)$ or $y = (4a^3 - 4a - 1)x + (-3a^4 + 2a^2)$ and similarly for $x = b$. So if at $x = a$ and $x = b$ we have the same tangent line, then $4a^3 - 4a - 1 = 4b^3 - 4b - 1$ and $-3a^4 + 2a^2 = -3b^4 + 2b^2$. The first equation gives $a^3 - b^3 = a - b \;\Rightarrow\; (a-b)(a^2 + ab + b^2) = (a - b)$. Assuming $a \ne b$, we have $1 = a^2 + ab + b^2$. The second equation gives $3(a^4 - b^4) = 2(a^2 - b^2) \;\Rightarrow\; 3(a^2 - b^2)(a^2 + b^2) = 2(a^2 - b^2)$ which is true if $a = -b$. Substituting into $1 = a^2 + ab + b^2$ gives $1 = a^2 - a^2 + a^2 \;\Rightarrow\; a = \pm 1$ so that $a = 1$ and $b = -1$ or vice versa. Thus, the points $(1, -2)$ and $(-1, 0)$ have a common tangent line.

As long as there are only two such points, we are done. So we show that these are in fact the only two such points. Suppose that $a^2 - b^2 \ne 0$. Then $3(a^2 - b^2)(a^2 + b^2) = 2(a^2 - b^2)$ gives $3(a^2 + b^2) = 2$ or $a^2 + b^2 = \tfrac{2}{3}$. Thus, $ab = (a^2 + ab + b^2) - (a^2 + b^2) = 1 - \dfrac{2}{3} = \dfrac{1}{3}$, so $b = \dfrac{1}{3a}$. Hence, $a^2 + \dfrac{1}{9a^2} = \dfrac{2}{3}$, so $9a^4 + 1 = 6a^2 \;\Rightarrow\;$

$0 = 9a^4 - 6a^2 + 1 = (3a^2 - 1)^2$. So $3a^2 - 1 = 0 \Rightarrow a^2 = \frac{1}{3} \Rightarrow b^2 = \frac{1}{9a^2} = \frac{1}{3} = a^2$, contradicting our assumption that $a^2 \neq b^2$.

25. Because of the periodic nature of the lattice points, it suffices to consider the points in the 5×2 grid shown. We can see that the minimum value of r occurs when there is a line with slope $\frac{2}{5}$ which touches the circle centered at $(3, 1)$ and the circles centered at $(0, 0)$ and $(5, 2)$.

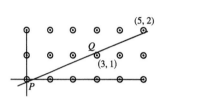
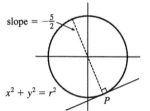

To find P, the point at which the line is tangent to the circle at $(0, 0)$, we simultaneously solve $x^2 + y^2 = r^2$ and $y = -\frac{5}{2}x \Rightarrow x^2 + \frac{25}{4}x^2 = r^2 \Rightarrow x^2 = \frac{4}{29}r^2 \Rightarrow x = \frac{2}{\sqrt{29}}r, y = -\frac{5}{\sqrt{29}}r$. To find Q, we either use symmetry or solve $(x - 3)^2 + (y - 1)^2 = r^2$ and $y - 1 = -\frac{5}{2}(x - 3)$. As above, we get $x = 3 - \frac{2}{\sqrt{29}}r, y = 1 + \frac{5}{\sqrt{29}}r$. Now the slope of the line PQ is $\frac{2}{5}$, so $m_{PQ} = \dfrac{1 + \frac{5}{\sqrt{29}}r - \left(-\frac{5}{\sqrt{29}}r\right)}{3 - \frac{2}{\sqrt{29}}r - \frac{2}{\sqrt{29}}r} = \dfrac{1 + \frac{10}{\sqrt{29}}r}{3 - \frac{4}{\sqrt{29}}r} = \dfrac{\sqrt{29} + 10r}{3\sqrt{29} - 4r} = \dfrac{2}{5} \Rightarrow$

$5\sqrt{29} + 50r = 6\sqrt{29} - 8r \Leftrightarrow 58r = \sqrt{29} \Leftrightarrow r = \frac{\sqrt{29}}{58}$. So the minimum value of r for which any line with slope $\frac{2}{5}$ intersects circles with radius r centered at the lattice points on the plane is $r = \frac{\sqrt{29}}{58} \approx 0.093$.

27. By similar triangles, $\dfrac{r}{5} = \dfrac{h}{16} \Rightarrow r = \dfrac{5h}{16}$. The volume of the cone is

$V = \frac{1}{3}\pi r^2 h = \frac{1}{3}\pi \left(\dfrac{5h}{16}\right)^2 h = \dfrac{25\pi}{768}h^3$, so $\dfrac{dV}{dt} = \dfrac{25\pi}{256}h^2 \dfrac{dh}{dt}$. Now the rate of

change of the volume is also equal to the difference of what is being added

(2 cm^3/min) and what is oozing out ($k\pi rl$, where πrl is the area of the cone and k

is a proportionality constant). Thus, $\dfrac{dV}{dt} = 2 - k\pi rl$.

Equating the two expressions for $\dfrac{dV}{dt}$ and substituting $h = 10, \dfrac{dh}{dt} = -0.3, r = \dfrac{5(10)}{16} = \dfrac{25}{8}$, and $\dfrac{l}{\sqrt{281}} = \dfrac{10}{16} \Leftrightarrow$

$l = \dfrac{5}{8}\sqrt{281}$, we get $\dfrac{25\pi}{256}(10)^2(-0.3) = 2 - k\pi \cdot \dfrac{25}{8} \cdot \dfrac{5}{8}\sqrt{281} \Leftrightarrow \dfrac{125k\pi\sqrt{281}}{64} = 2 + \dfrac{750\pi}{256}$. Solving for k gives us

$k = \dfrac{256 + 375\pi}{250\pi\sqrt{281}}$. To maintain a certain height, the rate of oozing, $k\pi rl$, must equal the rate of the liquid being poured in;

that is, $\dfrac{dV}{dt} = 0$. Thus, the rate at which we should pour the liquid into the container is

$$k\pi rl = \dfrac{256 + 375\pi}{250\pi\sqrt{281}} \cdot \pi \cdot \dfrac{25}{8} \cdot \dfrac{5\sqrt{281}}{8} = \dfrac{256 + 375\pi}{128} \approx 11.204 \text{ cm}^3/\text{min}$$

3 ☐ APPLICATIONS OF DIFFERENTIATION

3.1 Maximum and Minimum Values

1. A function f has an **absolute minimum** at $x = c$ if $f(c)$ is the smallest function value on the entire domain of f, whereas f has a **local minimum** at c if $f(c)$ is the smallest function value when x is near c.

3. Absolute maximum at s, absolute minimum at r, local maximum at c, local minima at b and r, neither a maximum nor a minimum at a and d.

5. Absolute maximum value is $f(4) = 5$; there is no absolute minimum value; local maximum values are $f(4) = 5$ and $f(6) = 4$; local minimum values are $f(2) = 2$ and $f(1) = f(5) = 3$.

7. Absolute maximum at 5, absolute minimum at 2, local maximum at 3, local minima at 2 and 4

9. Absolute minimum at 3, absolute maximum at 4, local maximum at 2

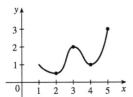
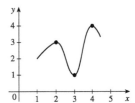

11. (a) 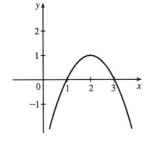 (b) 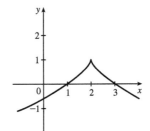 (c)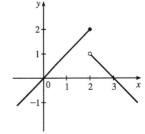

13. (a) *Note:* By the Extreme Value Theorem, f must *not* be continuous; because if it were, it would attain an absolute minimum.

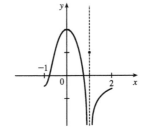

(b)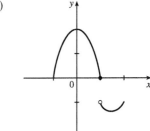

15. $f(x) = 3 - 2x$, $x \geq -1$. Absolute maximum $f(-1) = 5$; no local maximum. No absolute or local minimum.

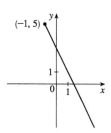

17. $f(x) = 1/x$, $x \geq 1$. Absolute maximum $f(1) = 1$; no local maximum. No absolute or local minimum.

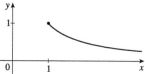

19. $f(x) = \sin x$, $0 \leq x < \pi/2$. No absolute or local maximum. Absolute minimum $f(0) = 0$; no local minimum.

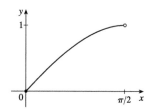

21. $f(x) = \sin x$, $-\pi/2 \leq x \leq \pi/2$. Absolute maximum $f\left(\frac{\pi}{2}\right) = 1$; no local maximum. Absolute minimum $f\left(-\frac{\pi}{2}\right) = -1$; no local minimum.

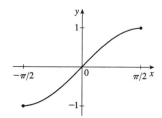

23. $f(x) = 1 + (x+1)^2$, $-2 \leq x < 5$. No absolute or local maximum. Absolute and local minimum $f(-1) = 1$.

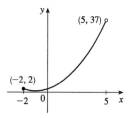

25. $f(x) = 1 - \sqrt{x}$. Absolute maximum $f(0) = 1$; no local maximum. No absolute or local minimum.

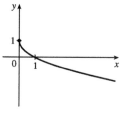

27. $f(x) = \begin{cases} x^2 & \text{if } -1 \leq x \leq 0 \\ 2 - 3x & \text{if } 0 < x \leq 1 \end{cases}$

No absolute or local maximum.
Absolute minimum $f(1) = -1$.
Local minimum $f(0) = 0$.

29. $f(x) = 3x^2 + x - 2 \;\Rightarrow\; f'(x) = 6x + 1$. $f'(x) = 0 \;\Rightarrow\; x = -\frac{1}{6}$. This is the only critical number.

31. $f(x) = 3x^4 + 8x^3 - 48x^2 \;\Rightarrow\; f'(x) = 12x^3 + 24x^2 - 96x = 12x(x^2 + 2x - 8) = 12x(x+4)(x-2)$.
$f'(x) = 0 \;\Rightarrow\; x = -4, 0, 2$. These are the only critical numbers.

33. $g(t) = t^5 + 5t^3 + 50t \;\Rightarrow\; g'(t) = 5t^4 + 15t^2 + 50 = 5(t^4 + 3t^2 + 10)$. Using the quadratic formula to solve for t^2,

$g'(t) = 0 \;\Leftrightarrow\; t^2 = \dfrac{-3 \pm \sqrt{3^2 - 4(1)(10)}}{2(1)} = \dfrac{-3 \pm \sqrt{-31}}{2}$. Since the discriminant, -31, is negative, there are no real solutions, and hence, there are no critical numbers.

35. $g(y) = \dfrac{y-1}{y^2 - y + 1} \;\Rightarrow\;$

$g'(y) = \dfrac{(y^2 - y + 1)(1) - (y-1)(2y-1)}{(y^2 - y + 1)^2} = \dfrac{y^2 - y + 1 - (2y^2 - 3y + 1)}{(y^2 - y + 1)^2} = \dfrac{-y^2 + 2y}{(y^2 - y + 1)^2} = \dfrac{y(2-y)}{(y^2 - y + 1)^2}$.

$g'(y) = 0 \;\Rightarrow\; y = 0, 2$. The expression $y^2 - y + 1$ is never equal to 0, so $g'(y)$ exists for all real numbers. The critical numbers are 0 and 2.

37. $p(x) = \dfrac{x^2 + 2}{2x - 1} \;\Rightarrow\;$

$p'(x) = \dfrac{(2x-1)(2x) - (x^2+2)(2)}{(2x-1)^2} = \dfrac{4x^2 - 2x - 2x^2 - 4}{(2x-1)^2} = \dfrac{2x^2 - 2x - 4}{(2x-1)^2} = \dfrac{2(x^2 - x - 2)}{(2x-1)^2} = \dfrac{2(x-2)(x+1)}{(2x-1)^2}$.

$p'(x) = 0 \;\Rightarrow\; x = -1$ or 2. $p'(x)$ does not exist at $x = \tfrac{1}{2}$, but $\tfrac{1}{2}$ is not in the domain of p, so the critical numbers are -1 and 2.

39. $h(t) = t^{3/4} - 2t^{1/4} \;\Rightarrow\; h'(t) = \tfrac{3}{4} t^{-1/4} - \tfrac{2}{4} t^{-3/4} = \tfrac{1}{4} t^{-3/4}(3t^{1/2} - 2) = \dfrac{3\sqrt{t} - 2}{4\sqrt[4]{t^3}}$.

$h'(t) = 0 \;\Rightarrow\; 3\sqrt{t} = 2 \;\Rightarrow\; \sqrt{t} = \tfrac{2}{3} \;\Rightarrow\; t = \tfrac{4}{9}$. $h'(t)$ does not exist at $t = 0$, so the critical numbers are 0 and $\tfrac{4}{9}$.

41. $F(x) = x^{4/5}(x-4)^2 \;\Rightarrow\;$

$F'(x) = x^{4/5} \cdot 2(x-4) + (x-4)^2 \cdot \tfrac{4}{5} x^{-1/5} = \tfrac{1}{5} x^{-1/5}(x-4)[5 \cdot x \cdot 2 + (x-4) \cdot 4]$

$= \dfrac{(x-4)(14x - 16)}{5x^{1/5}} = \dfrac{2(x-4)(7x-8)}{5x^{1/5}}$

$F'(x) = 0 \;\Rightarrow\; x = 4, \tfrac{8}{7}$. $F'(0)$ does not exist. Thus, the three critical numbers are 0, $\tfrac{8}{7}$, and 4.

43. $f(x) = x^{1/3}(4-x)^{2/3} \;\Rightarrow\;$

$f'(x) = x^{1/3} \cdot \tfrac{2}{3} (4-x)^{-1/3} \cdot (-1) + (4-x)^{2/3} \cdot \tfrac{1}{3} x^{-2/3} = \tfrac{1}{3} x^{-2/3}(4-x)^{-1/3}[-2x + (4-x)] = \dfrac{4 - 3x}{3x^{2/3}(4-x)^{1/3}}$.

$f'(x) = 0 \;\Rightarrow\; 4 - 3x = 0 \;\Rightarrow\; x = \tfrac{4}{3}$. $f'(0)$ and $f'(4)$ are undefined. Thus, the three critical numbers are 0, $\tfrac{4}{3}$, and 4.

45. $f(\theta) = 2\cos\theta + \sin^2\theta \;\Rightarrow\; f'(\theta) = -2\sin\theta + 2\sin\theta\cos\theta$. $f'(\theta) = 0 \;\Rightarrow\; 2\sin\theta(\cos\theta - 1) = 0 \;\Rightarrow\; \sin\theta = 0$ or $\cos\theta = 1 \;\Rightarrow\; \theta = n\pi$ [n an integer] or $\theta = 2n\pi$. The solutions $\theta = n\pi$ include the solutions $\theta = 2n\pi$, so the critical numbers are $\theta = n\pi$.

47. A graph of $f'(x) = 1 + \dfrac{210\sin x}{x^2 - 6x + 10}$ is shown. There are 10 zeros between -25 and 25 (one is approximately -0.05). f' exists everywhere, so f has 10 critical numbers.

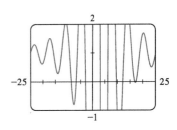

49. $f(x) = 12 + 4x - x^2$, $[0, 5]$. $f'(x) = 4 - 2x = 0 \Leftrightarrow x = 2$. $f(0) = 12$, $f(2) = 16$, and $f(5) = 7$.
So $f(2) = 16$ is the absolute maximum value and $f(5) = 7$ is the absolute minimum value.

51. $f(x) = 2x^3 - 3x^2 - 12x + 1$, $[-2, 3]$. $f'(x) = 6x^2 - 6x - 12 = 6(x^2 - x - 2) = 6(x - 2)(x + 1) = 0 \Leftrightarrow$
$x = 2, -1$. $f(-2) = -3$, $f(-1) = 8$, $f(2) = -19$, and $f(3) = -8$. So $f(-1) = 8$ is the absolute maximum value and
$f(2) = -19$ is the absolute minimum value.

53. $f(x) = 3x^4 - 4x^3 - 12x^2 + 1$, $[-2, 3]$. $f'(x) = 12x^3 - 12x^2 - 24x = 12x(x^2 - x - 2) = 12x(x + 1)(x - 2) = 0 \Leftrightarrow$
$x = -1, 0, 2$. $f(-2) = 33$, $f(-1) = -4$, $f(0) = 1$, $f(2) = -31$, and $f(3) = 28$. So $f(-2) = 33$ is the absolute maximum
value and $f(2) = -31$ is the absolute minimum value.

55. $f(x) = x + \dfrac{1}{x}$, $[0.2, 4]$. $f'(x) = 1 - \dfrac{1}{x^2} = \dfrac{x^2 - 1}{x^2} = \dfrac{(x+1)(x-1)}{x^2} = 0 \Leftrightarrow x = \pm 1$, but $x = -1$ is not in the given
interval, $[0.2, 4]$. $f'(x)$ does not exist when $x = 0$, but 0 is not in the given interval, so 1 is the only critical nuumber.
$f(0.2) = 5.2$, $f(1) = 2$, and $f(4) = 4.25$. So $f(0.2) = 5.2$ is the absolute maximum value and $f(1) = 2$ is the absolute
minimum value.

57. $f(t) = t - \sqrt[3]{t}$, $[-1, 4]$. $f'(t) = 1 - \tfrac{1}{3}t^{-2/3} = 1 - \dfrac{1}{3t^{2/3}}$. $f'(t) = 0 \Leftrightarrow 1 = \dfrac{1}{3t^{2/3}} \Leftrightarrow t^{2/3} = \dfrac{1}{3} \Leftrightarrow$
$t = \pm\left(\dfrac{1}{3}\right)^{3/2} = \pm\sqrt{\dfrac{1}{27}} = \pm\dfrac{1}{3\sqrt{3}} = \pm\dfrac{\sqrt{3}}{9}$. $f'(t)$ does not exist when $t = 0$. $f(-1) = 0$, $f(0) = 0$,
$f\left(\dfrac{-1}{3\sqrt{3}}\right) = \dfrac{-1}{3\sqrt{3}} - \dfrac{-1}{\sqrt{3}} = \dfrac{-1+3}{3\sqrt{3}} = \dfrac{2\sqrt{3}}{9} \approx 0.3849$, $f\left(\dfrac{1}{3\sqrt{3}}\right) = \dfrac{1}{3\sqrt{3}} - \dfrac{1}{\sqrt{3}} = -\dfrac{2\sqrt{3}}{9}$, and
$f(4) = 4 - \sqrt[3]{4} \approx 2.413$. So $f(4) = 4 - \sqrt[3]{4}$ is the absolute maximum value and $f\left(\dfrac{\sqrt{3}}{9}\right) = -\dfrac{2\sqrt{3}}{9}$ is the absolute
minimum value.

59. $f(t) = 2\cos t + \sin 2t$, $[0, \pi/2]$.
$f'(t) = -2\sin t + \cos 2t \cdot 2 = -2\sin t + 2(1 - 2\sin^2 t) = -2(2\sin^2 t + \sin t - 1) = -2(2\sin t - 1)(\sin t + 1)$.
$f'(t) = 0 \Rightarrow \sin t = \tfrac{1}{2}$ or $\sin t = -1 \Rightarrow t = \tfrac{\pi}{6}$. $f(0) = 2$, $f(\tfrac{\pi}{6}) = \sqrt{3} + \tfrac{1}{2}\sqrt{3} = \tfrac{3}{2}\sqrt{3} \approx 2.60$, and $f(\tfrac{\pi}{2}) = 0$.
So $f(\tfrac{\pi}{6}) = \tfrac{3}{2}\sqrt{3}$ is the absolute maximum value and $f(\tfrac{\pi}{2}) = 0$ is the absolute minimum value.

61. $f(x) = x^a(1-x)^b$, $0 \le x \le 1$, $a > 0$, $b > 0$.
$f'(x) = x^a \cdot b(1-x)^{b-1}(-1) + (1-x)^b \cdot ax^{a-1} = x^{a-1}(1-x)^{b-1}[x \cdot b(-1) + (1-x) \cdot a]$
$= x^{a-1}(1-x)^{b-1}(a - ax - bx)$

At the endpoints, we have $f(0) = f(1) = 0$ [the minimum value of f]. In the interval $(0, 1)$, $f'(x) = 0 \Leftrightarrow x = \dfrac{a}{a+b}$.

$f\left(\dfrac{a}{a+b}\right) = \left(\dfrac{a}{a+b}\right)^a\left(1 - \dfrac{a}{a+b}\right)^b = \dfrac{a^a}{(a+b)^a}\left(\dfrac{a+b-a}{a+b}\right)^b = \dfrac{a^a}{(a+b)^a} \cdot \dfrac{b^b}{(a+b)^b} = \dfrac{a^a b^b}{(a+b)^{a+b}}$.

So $f\left(\dfrac{a}{a+b}\right) = \dfrac{a^a b^b}{(a+b)^{a+b}}$ is the absolute maximum value.

63. (a)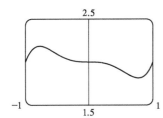

From the graph, it appears that the absolute maximum value is about $f(-0.77) = 2.19$, and the absolute minimum value is about $f(0.77) = 1.81$.

(b) $f(x) = x^5 - x^3 + 2 \Rightarrow f'(x) = 5x^4 - 3x^2 = x^2(5x^2 - 3)$. So $f'(x) = 0 \Rightarrow x = 0, \pm\sqrt{\frac{3}{5}}$.

$$f\left(-\sqrt{\tfrac{3}{5}}\right) = \left(-\sqrt{\tfrac{3}{5}}\right)^5 - \left(-\sqrt{\tfrac{3}{5}}\right)^3 + 2 = -\left(\tfrac{3}{5}\right)^2\sqrt{\tfrac{3}{5}} + \tfrac{3}{5}\sqrt{\tfrac{3}{5}} + 2$$
$$= \left(\tfrac{3}{5} - \tfrac{9}{25}\right)\sqrt{\tfrac{3}{5}} + 2 = \tfrac{6}{25}\sqrt{\tfrac{3}{5}} + 2 \text{ (maximum)}$$

and similarly, $f\left(\sqrt{\tfrac{3}{5}}\right) = -\tfrac{6}{25}\sqrt{\tfrac{3}{5}} + 2$ (minimum).

65. (a)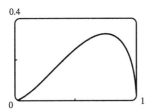

From the graph, it appears that the absolute maximum value is about $f(0.75) = 0.32$, and the absolute minimum value is $f(0) = f(1) = 0$; that is, at both endpoints.

(b) $f(x) = x\sqrt{x - x^2} \Rightarrow f'(x) = x \cdot \dfrac{1 - 2x}{2\sqrt{x - x^2}} + \sqrt{x - x^2} = \dfrac{(x - 2x^2) + (2x - 2x^2)}{2\sqrt{x - x^2}} = \dfrac{3x - 4x^2}{2\sqrt{x - x^2}}$.

So $f'(x) = 0 \Rightarrow 3x - 4x^2 = 0 \Rightarrow x(3 - 4x) = 0 \Rightarrow x = 0$ or $\tfrac{3}{4}$.

$f(0) = f(1) = 0$ (minimum), and $f\left(\tfrac{3}{4}\right) = \tfrac{3}{4}\sqrt{\tfrac{3}{4} - \left(\tfrac{3}{4}\right)^2} = \tfrac{3}{4}\sqrt{\tfrac{3}{16}} = \tfrac{3\sqrt{3}}{16}$ (maximum).

67. The density is defined as $\rho = \dfrac{\text{mass}}{\text{volume}} = \dfrac{1000}{V(T)}$ (in g/cm^3). But a critical point of ρ will also be a critical point of V

$\left[\text{since } \dfrac{d\rho}{dT} = -1000 V^{-2} \dfrac{dV}{dT} \text{ and } V \text{ is never 0}\right]$, and V is easier to differentiate than ρ.

$V(T) = 999.87 - 0.06426T + 0.0085043T^2 - 0.0000679T^3 \Rightarrow V'(T) = -0.06426 + 0.0170086T - 0.0002037T^2$.

Setting this equal to 0 and using the quadratic formula to find T, we get

$T = \dfrac{-0.0170086 \pm \sqrt{0.0170086^2 - 4 \cdot 0.0002037 \cdot 0.06426}}{2(-0.0002037)} \approx 3.9665°\text{C}$ or $79.5318°\text{C}$. Since we are only interested

in the region $0°\text{C} \leq T \leq 30°\text{C}$, we check the density ρ at the endpoints and at $3.9665°\text{C}$: $\rho(0) \approx \dfrac{1000}{999.87} \approx 1.00013$;

$\rho(30) \approx \dfrac{1000}{1003.7628} \approx 0.99625$; $\rho(3.9665) \approx \dfrac{1000}{999.7447} \approx 1.000255$. So water has its maximum density at about $3.9665°\text{C}$.

69. $L(t) = 0.01441t^3 - 0.4177t^2 + 2.703t + 1060.1 \Rightarrow L'(t) = 0.04323t^2 - 0.8354t + 2.703$. Use the quadratic formula

to solve $L'(t) = 0$. $t = \dfrac{0.8354 \pm \sqrt{(0.8354)^2 - 4(0.04323)(2.703)}}{2(0.04323)} \approx 4.1$ or 15.2. For $0 \leq t \leq 12$, we have

71. (a) $v(r) = k(r_0 - r)r^2 = kr_0r^2 - kr^3 \;\Rightarrow\; v'(r) = 2kr_0r - 3kr^2$. $v'(r) = 0 \;\Rightarrow\; kr(2r_0 - 3r) = 0 \;\Rightarrow\;$ $r = 0$ or $\frac{2}{3}r_0$ (but 0 is not in the interval). Evaluating v at $\frac{1}{2}r_0$, $\frac{2}{3}r_0$, and r_0, we get $v(\frac{1}{2}r_0) = \frac{1}{8}kr_0^3$, $v(\frac{2}{3}r_0) = \frac{4}{27}kr_0^3$, and $v(r_0) = 0$. Since $\frac{4}{27} > \frac{1}{8}$, v attains its maximum value at $r = \frac{2}{3}r_0$. This supports the statement in the text.

(b) From part (a), the maximum value of v is $\frac{4}{27}kr_0^3$.

(c)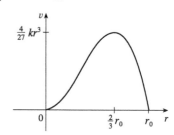

73. (a) Suppose that f has a local minimum value at c, so $f(x) \geq f(c)$ for all x near c. Then $g(x) = -f(x) \leq -f(c) = g(c)$ for all x near c, so $g(x)$ has a local maximum value at c.

(b) If f has a local minimum at c, then $g(x) = -f(x)$ has a local maximum at c, so $g'(c) = 0$ by the case of Fermat's Theorem proved in the text. Thus, $f'(c) = -g'(c) = 0$.

3.2 The Mean Value Theorem

1. **(1)** f is continuous on the closed interval $[0, 8]$.

   (2) f is differentiable on the open interval $(0, 8)$.

   (3) $f(0) = 3$ and $f(8) = 3$

   Thus, f satisfies the hypotheses of Rolle's Theorem. The numbers $c = 1$ and $c = 5$ satisfy the conclusion of Rolle's Theorem since $f'(1) = f'(5) = 0$.

3. (a) **(1)** g is continuous on the closed interval $[0, 8]$.

   (2) g is differentiable on the open interval $(0, 8)$.

(b) $g'(c) = \dfrac{g(8) - g(0)}{8 - 0} = \dfrac{4 - 1}{8} = \dfrac{3}{8}$.

It appears that $g'(c) = \frac{3}{8}$ when $c \approx 2.2$ and 6.4.

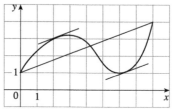

(c) $g'(c) = \dfrac{g(6) - g(2)}{6 - 2} = \dfrac{1 - 3}{4} = -\dfrac{1}{2}$.

It appears that $g'(c) = -\frac{1}{2}$ when $c \approx 3.7$ and 5.5.

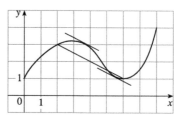

5. **(1)** f is continuous on the closed interval $[0, 5]$.

(2) f is not differentiable on the open interval $(0, 5)$ since f is not differentiable at 3.

Thus, f does not satisfy the hypotheses of the Mean Value Theorem on the interval $[0, 5]$.

7. **(1)** f is continuous on the closed interval $[0, 5]$.

(2) f is differentiable on the open interval $(0, 5)$.

Thus, f satisfies the hypotheses of the Mean Value Theorem on the interval $[0, 5]$.

The line passing through $(0, f(0))$ and $(5, f(5))$ has slope

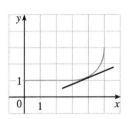

$\dfrac{f(5) - f(0)}{5 - 0} = \dfrac{3 - 1}{5} = \dfrac{2}{5}$. It appears that $f'(c) = \dfrac{2}{5}$ when $c \approx 3.8$.

9. $f(x) = 2x^2 - 4x + 5$, $[-1, 3]$. f is a polynomial, so it's continuous and differentiable on $\mathbb{R}$, and hence, continuous on $[-1, 3]$ and differentiable on $(-1, 3)$. Since $f(-1) = 11$ and $f(3) = 11$, f satisfies all the hypotheses of Rolle's Theorem. $f'(c) = 4c - 4$ and $f'(c) = 0 \iff 4c - 4 = 0 \iff c = 1$. $c = 1$ is in the interval $(-1, 3)$, so 1 satisfies the conclusion of Rolle's Theorem.

11. $f(x) = \sin(x/2)$, $[\pi/2, 3\pi/2]$. f, being the composite of the sine function and the polynomial $x/2$, is continuous and differentiable on $\mathbb{R}$, so it is continuous on $[\pi/2, 3\pi/2]$ and differentiable on $(\pi/2, 3\pi/2)$. Also, $f\left(\frac{\pi}{2}\right) = \frac{1}{2}\sqrt{2} = f\left(\frac{3\pi}{2}\right)$.

$f'(c) = 0 \iff \frac{1}{2}\cos(c/2) = 0 \iff \cos(c/2) = 0 \iff c/2 = \frac{\pi}{2} + n\pi$ [n an integer] $\iff c = \pi + 2n\pi$.

Only $c = \pi$ [when $n = 0$] is in $(\pi/2, 3\pi/2)$, so π satisfies the conclusion of Rolle's Theorem.

13. $f(x) = 1 - x^{2/3}$. $f(-1) = 1 - (-1)^{2/3} = 1 - 1 = 0 = f(1)$. $f'(x) = -\frac{2}{3}x^{-1/3}$, so $f'(c) = 0$ has no solution. This does not contradict Rolle's Theorem, since $f'(0)$ does not exist, and so f is not differentiable on $(-1, 1)$.

15. $f(x) = 2x^2 - 3x + 1$, $[0, 2]$. f is continuous on $[0, 2]$ and differentiable on $(0, 2)$ since polynomials are continuous and differentiable on $\mathbb{R}$. $f'(c) = \dfrac{f(b) - f(a)}{b - a} \iff 4c - 3 = \dfrac{f(2) - f(0)}{2 - 0} = \dfrac{3 - 1}{2} = 1 \iff 4c = 4 \iff c = 1$, which is in $(0, 2)$.

17. $f(x) = \sqrt[3]{x}$, $[0, 1]$. f is continuous on $\mathbb{R}$ and differentiable on $(-\infty, 0) \cup (0, \infty)$, so f is continuous on $[0, 1]$ and differentiable on $(0, 1)$. $f'(c) = \dfrac{f(b) - f(a)}{b - a} \iff \dfrac{1}{3c^{2/3}} = \dfrac{f(1) - f(0)}{1 - 0} \iff \dfrac{1}{3c^{2/3}} = \dfrac{1 - 0}{1} \iff$

$3c^{2/3} = 1 \iff c^{2/3} = \frac{1}{3} \iff c^2 = \left(\frac{1}{3}\right)^3 = \frac{1}{27} \iff c = \pm\sqrt{\frac{1}{27}} = \pm\frac{\sqrt{3}}{9}$, but only $\frac{\sqrt{3}}{9}$ is in $(0, 1)$.

19. $f(x) = \sqrt{x}$, $[0, 4]$. $f'(c) = \dfrac{f(4) - f(0)}{4 - 0} \iff \dfrac{1}{2\sqrt{c}} = \dfrac{2 - 0}{4} \iff$

$\dfrac{1}{2\sqrt{c}} = \dfrac{1}{2} \iff \sqrt{c} = 1 \iff c = 1$. The secant line and the tangent line are parallel.

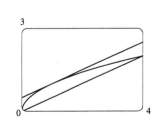

21. $f(x) = (x-3)^{-2}$ $\Rightarrow$ $f'(x) = -2(x-3)^{-3}$. $f(4) - f(1) = f'(c)(4-1)$ $\Rightarrow$ $\dfrac{1}{1^2} - \dfrac{1}{(-2)^2} = \dfrac{-2}{(c-3)^3} \cdot 3$ $\Rightarrow$ $\dfrac{3}{4} = \dfrac{-6}{(c-3)^3}$ $\Rightarrow$ $(c-3)^3 = -8$ $\Rightarrow$ $c - 3 = -2$ $\Rightarrow$ $c = 1$, which is not in the *open* interval $(1, 4)$. This does not contradict the Mean Value Theorem since f is not continuous at $x = 3$, which is in the interval $[1, 4]$.

23. Let $f(x) = 2x + \cos x$. Then $f(-\pi) = -2\pi - 1 < 0$ and $f(0) = 1 > 0$. Since f is the sum of the polynomial $2x$ and the trignometric function $\cos x$, f is continuous and differentiable for all x. By the Intermediate Value Theorem, there is a number c in $(-\pi, 0)$ such that $f(c) = 0$. Thus, the given equation has at least one real solution. If the equation has distinct real solutions a and b with $a < b$, then $f(a) = f(b) = 0$. Since f is continuous on $[a, b]$ and differentiable on (a, b), Rolle's Theorem implies that there is a number r in (a, b) such that $f'(r) = 0$. But $f'(r) = 2 - \sin r > 0$ since $\sin r \leq 1$. This contradiction shows that the given equation can't have two distinct real solutions, so it has exactly one solution.

25. Let $f(x) = x^3 - 15x + c$ for x in $[-2, 2]$. If f has two real solutions a and b in $[-2, 2]$, with $a < b$, then $f(a) = f(b) = 0$. Since the polynomial f is continuous on $[a, b]$ and differentiable on (a, b), Rolle's Theorem implies that there is a number r in (a, b) such that $f'(r) = 0$. Now $f'(r) = 3r^2 - 15$. Since r is in (a, b), which is contained in $[-2, 2]$, we have $|r| < 2$, so $r^2 < 4$. It follows that $3r^2 - 15 < 3 \cdot 4 - 15 = -3 < 0$. This contradicts $f'(r) = 0$, so the given equation, $x^3 - 15x + c = 0$, can't have two real solutions in $[-2, 2]$. Hence, it has at most one real solution in $[-2, 2]$.

27. (a) Suppose that a cubic polynomial $P(x)$ has zeros $a_1 < a_2 < a_3 < a_4$, so $P(a_1) = P(a_2) = P(a_3) = P(a_4)$. By Rolle's Theorem there are numbers c_1, c_2, c_3 with $a_1 < c_1 < a_2$, $a_2 < c_2 < a_3$, and $a_3 < c_3 < a_4$ and $P'(c_1) = P'(c_2) = P'(c_3) = 0$. Thus, the second-degree polynomial $P'(x)$ has three distinct real zeros, which is impossible. This contradiction tells us that a polynomial of degree 3 has at most three real zeros.

(b) We prove by induction that a polynomial of degree n has at most n real zeros. This is certainly true for $n = 1$. Suppose that the result is true for all polynomials of degree n and let $P(x)$ be a polynomial of degree $n + 1$. Suppose that $P(x)$ has more than $n + 1$ real zeros, say $a_1 < a_2 < a_3 < \cdots < a_{n+1} < a_{n+2}$. Then $P(a_1) = P(a_2) = \cdots = P(a_{n+2}) = 0$. By Rolle's Theorem there are real numbers $c_1, \ldots, c_{n+1}$ with $a_1 < c_1 < a_2, \ldots, a_{n+1} < c_{n+1} < a_{n+2}$ and $P'(c_1) = \cdots = P'(c_{n+1}) = 0$. Thus, the nth-degree polynomial $P'(x)$ has at least $n + 1$ zeros. This contradiction shows that $P(x)$ has at most $n + 1$ real zeros and hence, a polynomial of degree n has at most n real zeros.

29. By the Mean Value Theorem, $f(4) - f(1) = f'(c)(4-1)$ for some $c \in (1, 4)$. (f is differentiable for all x, so, in particular, f is differentiable on $(1, 4)$ and continuous on $[1, 4]$. Thus, the hypotheses of the Mean Value Theorem are satisfied.) For every $c \in (1, 4)$, we have $f'(c) \geq 2$. Putting $f'(c) \geq 2$ into the above equation and substituting $f(1) = 10$, we get $f(4) = f(1) + f'(c)(4 - 1) = 10 + 3f'(c) \geq 10 + 3 \cdot 2 = 16$. So the smallest possible value of $f(4)$ is 16.

31. Suppose that such a function f exists. By the Mean Value Theorem, there is a number c such that $0 < c < 2$ with $f'(c) = \dfrac{f(2) - f(0)}{2 - 0} = \dfrac{4 - (-1)}{2 - 0} = \dfrac{5}{2}$. This result, $f'(c) = \dfrac{5}{2}$, is impossible since $f'(x) \leq 2$ for all x, so no such function f exists.

33. Consider the function $f(x) = \sin x$, which is continuous and differentiable on $\mathbb{R}$. Let a be a number such that $0 < a < 2\pi$. Then f is continuous on $[0, a]$ and differentiable on $(0, a)$. By the Mean Value Theorem, there is a number c in $(0, a)$ such that $f(a) - f(0) = f'(c)(a - 0)$; that is, $\sin a - 0 = (\cos c)(a)$. Now $\cos c < 1$ for $0 < c < 2\pi$, so $\sin a < 1 \cdot a = a$. We took a to be an arbitrary number in $(0, 2\pi)$, so $\sin x < x$ for all x satisfying $0 < x < 2\pi$.

35. Let $f(x) = \sin x$ on $[a, b]$. Then f is continuous on $[a, b]$ and differentiable on (a, b). By the Mean Value Theorem, there is a number $c \in (a, b)$ with $f(b) - f(a) = f'(c)(b - a)$ or, equivalently, $\sin b - \sin a = (\cos c)(b - a)$. Taking absolute values, $|\sin b - \sin a| \le |\cos c| \, |b - a|$ or, equivalently, $|\sin a - \sin b| \le 1 \, |b - a|$. If $b < a$, then $|\sin a - \sin b| \le |a - b|$. If $a = b$, both sides of the inequality are 0, which proves the given inequality for all a and b.

37. For $x > 0$, $f(x) = g(x)$, so $f'(x) = g'(x)$. For $x < 0$, $f'(x) = (1/x)' = -1/x^2$ and $g'(x) = (1 + 1/x)' = -1/x^2$, so again $f'(x) = g'(x)$. However, the domain of $g(x)$ is not an interval [it is $(-\infty, 0) \cup (0, \infty)$], so we cannot conclude that $f - g$ is constant (in fact, it is not).

39. Let $g(t)$ and $h(t)$ be the position functions of the two runners and let $f(t) = g(t) - h(t)$. By hypothesis, where b is the finishing time, $f(0) = g(0) - h(0) = 0$ and $f(b) = g(b) - h(b) = 0$. Then by the Mean Value Theorem, there is a time c, with $0 < c < b$, such that $f'(c) = \dfrac{f(b) - f(0)}{b - 0} = \dfrac{0 - 0}{b} = 0$. Since $f'(c) = g'(c) - h'(c) = 0$, we have $g'(c) = h'(c)$ [the velocities are equal]. So at time c, both runners have the same speed.

3.3 What Derivatives Tell Us about the Shape of a Graph

1. (a) f is increasing on $(1, 3)$ and $(4, 6)$.   (b) f is decreasing on $(0, 1)$ and $(3, 4)$.

(c) f is concave upward on $(0, 2)$.   (d) f is concave downward on $(2, 4)$ and $(4, 6)$.

(e) The point of inflection is $(2, 3)$.

3. (a) Use the Increasing/Decreasing (I/D) Test.   (b) Use the Concavity Test.

(c) At any value of x where the concavity changes, we have an inflection point at $(x, f(x))$.

5. (a) Since $f'(x) > 0$ on $(0, 1)$ and $(3, 5)$, f is increasing on these intervals. Since $f'(x) < 0$ on $(1, 3)$ and $(5, 6)$, f is decreasing on these intervals.

(b) Since $f'(x) = 0$ at $x = 1$ and $x = 5$, and f' changes from positive to negative at both values, f changes from increasing to decreasing and has local maxima at $x = 1$ and $x = 5$. Since $f'(x) = 0$ at $x = 3$, and f' changes from negative to positive there, f changes from decreasing to increasing and has a local minimum at $x = 3$.

7. (a) There is an IP at $x = 3$ because the graph of f changes from CD to CU there. There is an IP at $x = 5$ because the graph of f changes from CU to CD there.

(b) There is an IP at $x = 2$ and at $x = 6$ because $f'(x)$ has a maximum value there, and so $f''(x)$ changes from positive to negative there. There is an IP at $x = 4$ because $f'(x)$ has a minimum value there and so $f''(x)$ changes from negative to positive there.

(c) There is an inflection point at $x = 1$ because $f''(x)$ changes from negative to positive there, and so the graph of f changes from concave downward to concave upward. There is an inflection point at $x = 7$ because $f''(x)$ changes from positive to negative there, and so the graph of f changes from concave upward to concave downward.

9. $f(x) = 2x^3 - 15x^2 + 24x - 5 \;\Rightarrow\; f'(x) = 6x^2 - 30x + 24 = 6(x^2 - 5x + 4) = 6(x-1)(x-4)$.

Interval	$x-1$	$x-4$	$f'(x)$	f
$x < 1$	$-$	$-$	$+$	increasing on $(-\infty, 1)$
$1 < x < 4$	$+$	$-$	$-$	decreasing on $(1, 4)$
$x > 4$	$+$	$+$	$+$	increasing on $(4, \infty)$

f changes from increasing to decreasing at $x = 1$ and from decreasing to increasing at $x = 4$. Thus, $f(1) = 6$ is a local maximum value and $f(4) = -21$ is a local minimum value.

11. $f(x) = 6x^4 - 16x^3 + 1 \;\Rightarrow\; f'(x) = 24x^3 - 48x^2 = 24x^2(x-2)$.

Interval	x^2	$x-2$	$f'(x)$	f
$x < 0$	$+$	$-$	$-$	decreasing on $(-\infty, 0)$
$0 < x < 2$	$+$	$-$	$-$	decreasing on $(0, 2)$
$x > 2$	$+$	$+$	$+$	increasing on $(2, \infty)$

Note that f is differentiable and $f'(x) < 0$ on the interval $(-\infty, 2)$ except for the single number $x = 0$. By applying the result of Exercise 3.3.79, we can say that f is decreasing on the entire interval $(-\infty, 2)$. f changes from decreasing to increasing at $x = 2$. Thus, $f(2) = -31$ is a local minimum value.

13. $f(x) = \dfrac{x^2 - 24}{x - 5} \;\Rightarrow\;$

$f'(x) = \dfrac{(x-5)(2x) - (x^2-24)(1)}{(x-5)^2} = \dfrac{2x^2 - 10x - x^2 + 24}{(x-5)^2} = \dfrac{x^2 - 10x + 24}{(x-5)^2} = \dfrac{(x-4)(x-6)}{(x-5)^2}$.

Interval	$x-4$	$x-6$	$(x-5)^2$	$f'(x)$	f
$x < 4$	$-$	$-$	$+$	$+$	increasing on $(-\infty, 4)$
$4 < x < 5$	$+$	$-$	$+$	$-$	decreasing on $(4, 5)$
$5 < x < 6$	$+$	$-$	$+$	$-$	decreasing on $(5, 6)$
$x > 6$	$+$	$+$	$+$	$+$	increasing on $(6, \infty)$

$x = 5$ is not in the domain of f. f changes from increasing to decreasing at $x = 4$ and from decreasing to increasing at $x = 6$. Thus, $f(4) = 8$ is a local maximum value and $f(6) = 12$ is a local minimum value.

15. $f(x) = x^3 - 3x^2 - 9x + 4 \;\Rightarrow\; f'(x) = 3x^2 - 6x - 9 \;\Rightarrow\; f''(x) = 6x - 6 = 6(x-1)$. $f''(x) > 0 \;\Leftrightarrow\; x > 1$ and $f''(x) < 0 \;\Leftrightarrow\; x < 1$. Thus, f is concave upward on $(1, \infty)$ and concave downward on $(-\infty, 1)$. There is an inflection point at $(1, f(1)) = (1, -7)$.

17. $f(x) = \sin^2 x - \cos 2x$, $0 \le x \le \pi$. $f'(x) = 2\sin x \cos x + 2\sin 2x = \sin 2x + 2\sin 2x = 3\sin 2x$ and $f''(x) = 6\cos 2x$. $f''(x) > 0 \Leftrightarrow \cos 2x > 0 \Leftrightarrow 0 < x < \frac{\pi}{4}$ and $\frac{3\pi}{4} < x < \pi$ and $f''(x) < 0 \Leftrightarrow \cos 2x < 0 \Leftrightarrow \frac{\pi}{4} < x < \frac{3\pi}{4}$. Thus, f is concave upward on $\left(0, \frac{\pi}{4}\right)$ and $\left(\frac{3\pi}{4}, \pi\right)$ and concave downward on $\left(\frac{\pi}{4}, \frac{3\pi}{4}\right)$. There are inflection points at $\left(\frac{\pi}{4}, \frac{1}{2}\right)$ and $\left(\frac{3\pi}{4}, \frac{1}{2}\right)$.

19. (a) $f(x) = x^4 - 2x^2 + 3 \Rightarrow f'(x) = 4x^3 - 4x = 4x(x^2 - 1) = 4x(x+1)(x-1)$.

Interval	$x+1$	$4x$	$x-1$	$f'(x)$	f
$x < -1$	$-$	$-$	$-$	$-$	decreasing on $(-\infty, -1)$
$-1 < x < 0$	$+$	$-$	$-$	$+$	increasing on $(-1, 0)$
$0 < x < 1$	$+$	$+$	$-$	$-$	decreasing on $(0, 1)$
$x > 1$	$+$	$+$	$+$	$+$	increasing on $(1, \infty)$

(b) f changes from increasing to decreasing at $x = 0$ and from decreasing to increasing at $x = -1$ and $x = 1$. Thus, $f(0) = 3$ is a local maximum value and $f(\pm 1) = 2$ are local minimum values.

(c) $f''(x) = 12x^2 - 4 = 12\left(x^2 - \frac{1}{3}\right) = 12\left(x + 1/\sqrt{3}\right)\left(x - 1/\sqrt{3}\right)$. $f''(x) > 0 \Leftrightarrow x < -1/\sqrt{3}$ or $x > 1/\sqrt{3}$ and $f''(x) < 0 \Leftrightarrow -1/\sqrt{3} < x < 1/\sqrt{3}$. Thus, f is concave upward on $\left(-\infty, -\sqrt{3}/3\right)$ and $\left(\sqrt{3}/3, \infty\right)$ and concave downward on $\left(-\sqrt{3}/3, \sqrt{3}/3\right)$. There are inflection points at $\left(\pm\sqrt{3}/3, \frac{22}{9}\right)$.

21. (a) $f(x) = \sin x + \cos x$, $0 \le x \le 2\pi$. $f'(x) = \cos x - \sin x = 0 \Rightarrow \cos x = \sin x \Rightarrow 1 = \frac{\sin x}{\cos x} \Rightarrow \tan x = 1 \Rightarrow x = \frac{\pi}{4}$ or $\frac{5\pi}{4}$. Thus, $f'(x) > 0 \Leftrightarrow \cos x - \sin x > 0 \Leftrightarrow \cos x > \sin x \Leftrightarrow 0 < x < \frac{\pi}{4}$ or $\frac{5\pi}{4} < x < 2\pi$ and $f'(x) < 0 \Leftrightarrow \cos x < \sin x \Leftrightarrow \frac{\pi}{4} < x < \frac{5\pi}{4}$. So f is increasing on $\left(0, \frac{\pi}{4}\right)$ and $\left(\frac{5\pi}{4}, 2\pi\right)$ and f is decreasing on $\left(\frac{\pi}{4}, \frac{5\pi}{4}\right)$.

(b) f changes from increasing to decreasing at $x = \frac{\pi}{4}$ and from decreasing to increasing at $x = \frac{5\pi}{4}$. Thus, $f\left(\frac{\pi}{4}\right) = \sqrt{2}$ is a local maximum value and $f\left(\frac{5\pi}{4}\right) = -\sqrt{2}$ is a local minimum value.

(c) $f''(x) = -\sin x - \cos x = 0 \Rightarrow -\sin x = \cos x \Rightarrow \tan x = -1 \Rightarrow x = \frac{3\pi}{4}$ or $\frac{7\pi}{4}$. Divide the interval $(0, 2\pi)$ into subintervals with these numbers as endpoints and complete a second derivative chart.

Interval	$f''(x) = -\sin x - \cos x$	Concavity
$\left(0, \frac{3\pi}{4}\right)$	$f''\left(\frac{\pi}{2}\right) = -1 < 0$	downward
$\left(\frac{3\pi}{4}, \frac{7\pi}{4}\right)$	$f''(\pi) = 1 > 0$	upward
$\left(\frac{7\pi}{4}, 2\pi\right)$	$f''\left(\frac{11\pi}{6}\right) = \frac{1}{2} - \frac{1}{2}\sqrt{3} < 0$	downward

There are inflection points at $\left(\frac{3\pi}{4}, 0\right)$ and $\left(\frac{7\pi}{4}, 0\right)$.

23. $f(x) = 1 + 3x^2 - 2x^3 \Rightarrow f'(x) = 6x - 6x^2 = 6x(1-x)$.

First Derivative Test: $f'(x) > 0 \Rightarrow 0 < x < 1$ and $f'(x) < 0 \Rightarrow x < 0$ or $x > 1$. Since f' changes from negative to positive at $x = 0$, $f(0) = 1$ is a local minimum value; and since f' changes from positive to negative at $x = 1$, $f(1) = 2$ is a local maximum value.

Second Derivative Test: $f''(x) = 6 - 12x$. $f'(x) = 0 \Leftrightarrow x = 0, 1$. $f''(0) = 6 > 0 \Rightarrow f(0) = 1$ is a local minimum value. $f''(1) = -6 < 0 \Rightarrow f(1) = 2$ is a local maximum value.

Preference: For this function, the two tests are equally easy.

25. $f'(x) = (x-4)^2(x+3)^7(x-5)^8$. The factors $(x-4)^2$ and $(x-5)^8$ are nonnegative. Hence, the sign of f' is determined by the sign of $(x+3)^7$, which is positive for $x > -3$. Thus, f increases on the intervals $(-3, 4)$, $(4, 5)$, and $(5, \infty)$. Note that f is differentiable and $f'(x) > 0$ on the interval $(-3, \infty)$ except for the numbers $x = 4$ and $x = 5$. By applying the result of Exercise 3.3.79, we can say that f is increasing on the entire interval $(-3, \infty)$.

27. (a) By the Second Derivative Test, if $f'(2) = 0$ and $f''(2) = -5 < 0$, f has a local maximum at $x = 2$.

(b) If $f'(6) = 0$, we know that f has a horizontal tangent at $x = 6$. Knowing that $f''(6) = 0$ does not provide any additional information since the Second Derivative Test fails. For example, the first and second derivatives of $y = (x-6)^4$, $y = -(x-6)^4$, and $y = (x-6)^3$ all equal zero for $x = 6$, but the first has a local minimum at $x = 6$, the second has a local maximum at $x = 6$, and the third has an inflection point at $x = 6$.

29. (a) $f'(x) > 0$ and $f''(x) < 0$ for all x

The function must be always increasing (since the first derivative is always positive) and concave downward (since the second derivative is always negative).

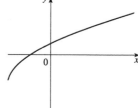

(b) $f'(x) < 0$ and $f''(x) > 0$ for all x

The function must be always decreasing (since the first derivative is always negative) and concave upward (since the second derivative is always positive).

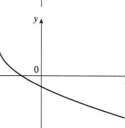

31. $f'(0) = f'(2) = f'(4) = 0 \Rightarrow$ horizontal tangents at $x = 0, 2, 4$.

$f'(x) > 0$ if $x < 0$ or $2 < x < 4 \Rightarrow f$ is increasing on $(-\infty, 0)$ and $(2, 4)$.

$f'(x) < 0$ if $0 < x < 2$ or $x > 4 \Rightarrow f$ is decreasing on $(0, 2)$ and $(4, \infty)$.

$f''(x) > 0$ if $1 < x < 3 \Rightarrow f$ is concave upward on $(1, 3)$.

$f''(x) < 0$ if $x < 1$ or $x > 3 \Rightarrow f$ is concave downward on $(-\infty, 1)$ and $(3, \infty)$. There are inflection points when $x = 1$ and 3.

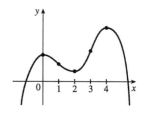

33. $f'(5) = 0 \Rightarrow$ horizontal tangent at $x = 5$.

$f'(x) < 0$ when $x < 5 \Rightarrow f$ is decreasing on $(-\infty, 5)$.

$f'(x) > 0$ when $x > 5 \Rightarrow f$ is increasing on $(5, \infty)$.

$f''(2) = 0$, $f''(8) = 0$, $f''(x) < 0$ when $x < 2$ or $x > 8$,

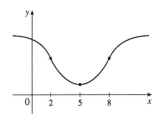

$f''(x) > 0$ for $2 < x < 8 \Rightarrow f$ is concave upward on $(2, 8)$ and concave downward on $(-\infty, 2)$ and $(8, \infty)$. There are inflection points at $x = 2$ and $x = 8$.

35. $f(0) = f'(0) = 0 \Rightarrow$ the graph of f passes through the origin and has a horizontal tangent there. $f'(2) = f'(4) = f'(6) = 0 \Rightarrow$ horizontal tangents at $x = 2, 4, 6$. $f'(x) > 0$ if $0 < x < 2$ or $4 < x < 6 \Rightarrow f$ increasing on $(0, 2)$ and $(4, 6)$. $f'(x) < 0$ if $2 < x < 4$ or $x > 6 \Rightarrow f$ decreasing on $(2, 4)$ and $(6, \infty)$. $f''(x) > 0$ if $0 < x < 1$ or $3 < x < 5 \Rightarrow f$ is CU on $(0, 1)$ and $(3, 5)$.

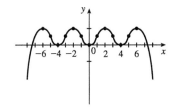

$f''(x) < 0$ if $1 < x < 3$ or $x > 5 \Rightarrow f$ is CD on $(1, 3)$ and $(5, \infty)$. $f(-x) = f(x) \Rightarrow f$ is even and the graph is symmetric about the y-axis.

37. (a) f is increasing where f' is positive, that is, on $(0, 2)$, $(4, 6)$, and $(8, \infty)$; and decreasing where f' is negative, that is, on $(2, 4)$ and $(6, 8)$.

(b) f has local maxima where f' changes from positive to negative, at $x = 2$ and at $x = 6$, and local minima where f' changes from negative to positive, at $x = 4$ and at $x = 8$.

(c) f is concave upward (CU) where f' is increasing, that is, on $(3, 6)$ and $(6, \infty)$, and concave downward (CD) where f' is decreasing, that is, on $(0, 3)$.

(d) There is a point of inflection where f changes from being CD to being CU, that is, at $x = 3$.

(e)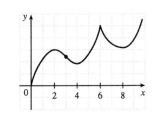

39. (a) $f(x) = x^3 - 3x^2 + 4 \Rightarrow f'(x) = 3x^2 - 6x = 3x(x - 2)$.

Interval	$3x$	$x - 2$	$f'(x)$	f
$x < 0$	−	−	+	increasing on $(-\infty, 0)$
$0 < x < 2$	+	−	−	decreasing on $(0, 2)$
$x > 2$	+	+	+	increasing on $(2, \infty)$

(b) f changes from increasing to decreasing at $x = 0$ and from decreasing to increasing at $x = 2$. Thus, $f(0) = 4$ is a local maximum value and $f(2) = 0$ is a local minimum value.

(c) $f''(x) = 6x - 6 = 6(x-1)$. $f''(x) = 0 \Leftrightarrow x = 1$. $f''(x) > 0$ on $(1, \infty)$ and $f''(x) < 0$ on $(-\infty, 1)$. So f is concave upward on $(1, \infty)$ and f is concave downward on $(-\infty, 1)$. There is an inflection point at $(1, 2)$.

(d)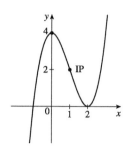

41. (a) $f(x) = \frac{1}{2}x^4 - 4x^2 + 3 \Rightarrow f'(x) = 2x^3 - 8x = 2x(x^2 - 4) = 2x(x+2)(x-2)$. $f'(x) > 0 \Leftrightarrow -2 < x < 0$ or $x > 2$, and $f'(x) < 0 \Leftrightarrow x < -2$ or $0 < x < 2$. So f is increasing on $(-2, 0)$ and $(2, \infty)$ and f is decreasing on $(-\infty, -2)$ and $(0, 2)$.

(b) f changes from increasing to decreasing at $x = 0$, so $f(0) = 3$ is a local maximum value.

f changes from decreasing to increasing at $x = \pm 2$, so $f(\pm 2) = -5$ is a local minimum value.

(c) $f''(x) = 6x^2 - 8 = 6\left(x^2 - \frac{4}{3}\right) = 6\left(x + \frac{2}{\sqrt{3}}\right)\left(x - \frac{2}{\sqrt{3}}\right)$.

$f''(x) = 0 \Leftrightarrow x = \pm\frac{2}{\sqrt{3}}$. $f''(x) > 0$ on $\left(-\infty, -\frac{2}{\sqrt{3}}\right)$ and $\left(\frac{2}{\sqrt{3}}, \infty\right)$ and $f''(x) < 0$ on $\left(-\frac{2}{\sqrt{3}}, \frac{2}{\sqrt{3}}\right)$. So f is CU on $\left(-\infty, -\frac{2}{\sqrt{3}}\right)$ and $\left(\frac{2}{\sqrt{3}}, \infty\right)$, and f is CD on $\left(-\frac{2}{\sqrt{3}}, \frac{2}{\sqrt{3}}\right)$. There are inflection points at $\left(\pm\frac{2}{\sqrt{3}}, -\frac{13}{9}\right)$.

(d)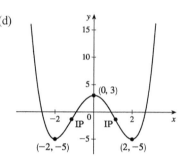

43. (a) $g(t) = 3t^4 - 8t^3 + 12 \Rightarrow g'(t) = 12t^3 - 24t^2 = 12t^2(t - 2)$.

Interval	$12t^2$	$t - 2$	$g'(t)$	g
$t < 0$	$+$	$-$	$-$	decreasing on $(-\infty, 0)$
$0 < t < 2$	$+$	$-$	$-$	decreasing on $(0, 2)$
$t > 2$	$+$	$+$	$+$	increasing on $(2, \infty)$

So f is increasing on $(2, \infty)$, and f is decreasing on $(-\infty, 0)$ and $(0, 2)$. By Exercise 3.3.79, we can say that f increasing on $(2, \infty)$ and decreasing on $(-\infty, 2)$.

(b) g changes from decreasing to increasing at $x = 2$. Thus, $g(2) = -4$ is a local minimum value.

(c) $g''(t) = 36t^2 - 48t = 12t(3t - 4)$. $g''(t) = 0 \Leftrightarrow t = 0$ or $t = \frac{4}{3}$.

(d)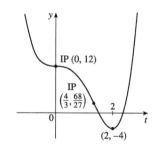

Interval	$12t$	$3t - 4$	$g''(t)$	g
$t < 0$	$-$	$-$	$+$	concave up on $(-\infty, 0)$
$0 < t < \frac{4}{3}$	$+$	$-$	$-$	concave down on $\left(0, \frac{4}{3}\right)$
$t > \frac{4}{3}$	$+$	$+$	$+$	concave up on $\left(\frac{4}{3}, \infty\right)$

There are inflection points at $(0, 12)$ and $\left(\frac{4}{3}, \frac{68}{27}\right)$.

45. (a) $f(z) = z^7 - 112z^2 \Rightarrow f'(z) = 7z^6 - 224z = 7z(z^5 - 32)$. $f'(z) = 0 \Rightarrow z = 0, 2$.

Interval	$7z$	$z^5 - 32$	$f'(z)$	f
$z < 0$	−	−	+	increasing on $(-\infty, 0)$
$0 < z < 2$	+	−	−	decreasing on $(0, 2)$
$z > 2$	+	+	+	increasing on $(2, \infty)$

(b) f changes from increasing to decreasing at $x = 0$ and from decreasing to increasing at $x = 2$. Thus, $f(0) = 0$ is a local maximum value and $f(2) = -320$ is a local minimum value.

(c) $f''(z) = 42z^5 - 224 = 14(3z^5 - 16)$. $f''(z) = 0 \Leftrightarrow 3z^5 = 16 \Leftrightarrow z^5 = \frac{16}{3} \Leftrightarrow z = \sqrt[5]{\frac{16}{3}}$ [call this value a]. $f''(z) > 0 \Leftrightarrow z > a$ and $f''(z) < 0 \Leftrightarrow z < a$. So, f is concave up on (a, ∞) and concave down on $(-\infty, a)$. There is an inflection point at $(a, f(a)) = \left(\sqrt[5]{\frac{16}{3}}, -\frac{320}{3}\sqrt[5]{\frac{256}{9}}\right) \approx (1.398, -208.4)$.

(d)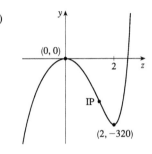

47. (a) $F(x) = x\sqrt{6 - x} \Rightarrow$

$F'(x) = x \cdot \frac{1}{2}(6-x)^{-1/2}(-1) + (6-x)^{1/2}(1) = \frac{1}{2}(6-x)^{-1/2}[-x + 2(6-x)] = \frac{-3x + 12}{2\sqrt{6-x}}$.

$F'(x) > 0 \Leftrightarrow -3x + 12 > 0 \Leftrightarrow x < 4$ and $F'(x) < 0 \Leftrightarrow 4 < x < 6$. So F is increasing on $(-\infty, 4)$ and F is decreasing on $(4, 6)$.

(b) F changes from increasing to decreasing at $x = 4$, so $F(4) = 4\sqrt{2}$ is a local maximum value. There is no local minimum value.

(c) $F'(x) = -\frac{3}{2}(x - 4)(6 - x)^{-1/2} \Rightarrow$

$F''(x) = -\frac{3}{2}\left[(x-4)\left(-\frac{1}{2}(6-x)^{-3/2}(-1)\right) + (6-x)^{-1/2}(1)\right]$

$= -\frac{3}{2} \cdot \frac{1}{2}(6-x)^{-3/2}[(x-4) + 2(6-x)] = \frac{3(x-8)}{4(6-x)^{3/2}}$

$F''(x) < 0$ on $(-\infty, 6)$, so F is CD on $(-\infty, 6)$. There is no inflection point.

(d)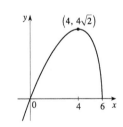

49. (a) $C(x) = x^{1/3}(x + 4) = x^{4/3} + 4x^{1/3} \Rightarrow C'(x) = \frac{4}{3}x^{1/3} + \frac{4}{3}x^{-2/3} = \frac{4}{3}x^{-2/3}(x + 1) = \frac{4(x+1)}{3\sqrt[3]{x^2}}$. $C'(x) > 0$ if $-1 < x < 0$ or $x > 0$ and $C'(x) < 0$ for $x < -1$, so C is increasing on $(-1, \infty)$ and C is decreasing on $(-\infty, -1)$.

(b) $C(-1) = -3$ is a local minimum value.

(c) $C''(x) = \frac{4}{9}x^{-2/3} - \frac{8}{9}x^{-5/3} = \frac{4}{9}x^{-5/3}(x - 2) = \frac{4(x-2)}{9\sqrt[3]{x^5}}$.

$C''(x) < 0$ for $0 < x < 2$ and $C''(x) > 0$ for $x < 0$ and $x > 2$, so C is concave downward on $(0, 2)$ and concave upward on $(-\infty, 0)$ and $(2, \infty)$. There are inflection points at $(0, 0)$ and $\left(2, 6\sqrt[3]{2}\right) \approx (2, 7.56)$.

(d)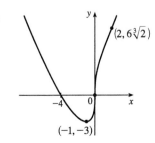

51. (a) $f(\theta) = 2\cos\theta + \cos^2\theta$, $0 \leq \theta \leq 2\pi$ $\Rightarrow$ $f'(\theta) = -2\sin\theta + 2\cos\theta(-\sin\theta) = -2\sin\theta(1+\cos\theta)$.
$f'(\theta) = 0 \Leftrightarrow \theta = 0, \pi,$ and 2π. $f'(\theta) > 0 \Leftrightarrow \pi < \theta < 2\pi$ and $f'(\theta) < 0 \Leftrightarrow 0 < \theta < \pi$. So f is increasing on $(\pi, 2\pi)$ and f is decreasing on $(0, \pi)$.

(b) $f(\pi) = -1$ is a local minimum value.

(c) $f'(\theta) = -2\sin\theta(1+\cos\theta)$ $\Rightarrow$

$$f''(\theta) = -2\sin\theta(-\sin\theta) + (1+\cos\theta)(-2\cos\theta) = 2\sin^2\theta - 2\cos\theta - 2\cos^2\theta$$
$$= 2(1-\cos^2\theta) - 2\cos\theta - 2\cos^2\theta = -4\cos^2\theta - 2\cos\theta + 2$$
$$= -2(2\cos^2\theta + \cos\theta - 1) = -2(2\cos\theta - 1)(\cos\theta + 1)$$

Since $-2(\cos\theta + 1) < 0$ [for $\theta \neq \pi$], $f''(\theta) > 0 \Rightarrow 2\cos\theta - 1 < 0 \Rightarrow \cos\theta < \frac{1}{2} \Rightarrow \frac{\pi}{3} < \theta < \frac{5\pi}{3}$ and $f''(\theta) < 0 \Rightarrow \cos\theta > \frac{1}{2} \Rightarrow 0 < \theta < \frac{\pi}{3}$ or $\frac{5\pi}{3} < \theta < 2\pi$. So f is CU on $\left(\frac{\pi}{3}, \frac{5\pi}{3}\right)$ and f is CD on $\left(0, \frac{\pi}{3}\right)$ and $\left(\frac{5\pi}{3}, 2\pi\right)$. There are points of inflection at $\left(\frac{\pi}{3}, f\left(\frac{\pi}{3}\right)\right) = \left(\frac{\pi}{3}, \frac{5}{4}\right)$ and $\left(\frac{5\pi}{3}, f\left(\frac{5\pi}{3}\right)\right) = \left(\frac{5\pi}{3}, \frac{5}{4}\right)$.

(d)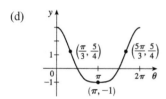

53. $f(x) = x^4 - cx$, $c > 0$ $\Rightarrow$ $f'(x) = 4x^3 - c$ $\Rightarrow$ $f'(x) = 0 \Leftrightarrow x = \sqrt[3]{c/4}$. $f'(x) > 0 \Leftrightarrow x > \sqrt[3]{c/4}$ and $f'(x) < 0 \Leftrightarrow x < \sqrt[3]{c/4}$. Thus, f is increasing on $\left(\sqrt[3]{c/4}, \infty\right)$ and decreasing on $\left(-\infty, \sqrt[3]{c/4}\right)$. f changes from decreasing to increasing at $x = \sqrt[3]{c/4}$. Thus, $f\left(\sqrt[3]{c/4}\right)$ is a local minimum value. $f''(x) = 12x^2$ is positive except at $x = 0$, so f is concave up on $(-\infty, 0)$ and $(0, \infty)$. There are no inflection points.

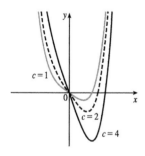

The members of this family have one local minimum point with increasing x-coordinate and decreasing y-coordinate as c increases. The graphs are concave up on $(-\infty, 0)$ and $(0, \infty)$. Since the graphs are continuous at $x = 0$, and the graphs lie above their tangents, we can say that the graphs are concave up on $(-\infty, \infty)$. There is no inflection point.

55. (a)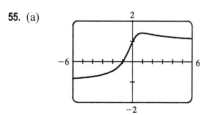

From the graph, we get an estimate of $f(1) \approx 1.41$ as a local maximum value, and no local minimum value.

$$f(x) = \frac{x+1}{\sqrt{x^2+1}} \Rightarrow f'(x) = \frac{1-x}{(x^2+1)^{3/2}}.$$

$f'(x) = 0 \Leftrightarrow x = 1$. $f(1) = \frac{2}{\sqrt{2}} = \sqrt{2}$ is the exact value.

(b) From the graph in part (a), f increases most rapidly somewhere between $x = -\frac{1}{2}$ and $x = -\frac{1}{4}$. To find the exact value, we need to find the maximum value of f', which we can do by finding the critical numbers of f'.

$f''(x) = \dfrac{2x^2 - 3x - 1}{(x^2+1)^{5/2}} = 0 \Leftrightarrow x = \dfrac{3 \pm \sqrt{17}}{4}$. By the First Derivative Test applied to f', $x = \dfrac{3+\sqrt{17}}{4}$ corresponds to the *minimum* value of f' and the maximum value of f' occurs at $x = \dfrac{3-\sqrt{17}}{4} \approx -0.28$.

57. $f(x) = \sin 2x + \sin 4x \Rightarrow f'(x) = 2\cos 2x + 4\cos 4x \Rightarrow f''(x) = -4\sin 2x - 16\sin 4x$

(a) From the graph of f, it seems that f is CD on $(0, 0.8)$, CU on $(0.8, 1.6)$, CD on $(1.6, 2.3)$, and CU on $(2.3, \pi)$. The inflection points appear to be at $(0.8, 0.7)$, $(1.6, 0)$, and $(2.3, -0.7)$.

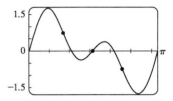

(b) From the graph of f'' (and zooming in near the zeros), it seems that f is CD on $(0, 0.85)$, CU on $(0.85, 1.57)$, CD on $(1.57, 2.29)$, and CU on $(2.29, \pi)$. Refined estimates of the inflection points are $(0.85, 0.74)$, $(1.57, 0)$, and $(2.29, -0.74)$.

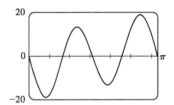

59. $f(x) = \dfrac{x^4 + x^3 + 1}{\sqrt{x^2 + x + 1}}$. In Maple, we define f and then use the command `plot(diff(diff(f,x),x),x=-2..2);`. In Mathematica, we define f and then use `Plot[Dt[Dt[f,x],x],{x,-2,2}]`. We see that $f'' > 0$ for $x < -0.6$ and $x > 0.0$ [≈ 0.03] and $f'' < 0$ for $-0.6 < x < 0.0$. So f is CU on $(-\infty, -0.6)$ and $(0.0, \infty)$ and CD on $(-0.6, 0.0)$.

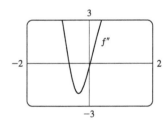

61. (a) The rate of increase of the population is initially very small, then gets larger until it reaches a maximum at about $t = 8$ hours, and decreases toward 0 as the population begins to level off.

(b) The rate of increase has its maximum value at $t = 8$ hours.

(c) The population function is concave upward on $(0, 8)$ and concave downward on $(8, 18)$.

(d) At $t = 8$, the population is about 350, so the inflection point is about $(8, 350)$.

63. If $D(t)$ is the size of the national deficit as a function of time t, then at the time of the speech $D'(t) > 0$ (since the deficit is increasing), and $D''(t) < 0$ (since the rate of increase of the deficit is decreasing).

65. Most students learn more in the third hour of studying than in the eighth hour, so $K(3) - K(2)$ is larger than $K(8) - K(7)$. In other words, as you begin studying for a test, the rate of knowledge gain is large and then starts to taper off, so $K'(t)$ decreases and the graph of K is concave downward.

67. $f(x) = ax^3 + bx^2 + cx + d \Rightarrow f'(x) = 3ax^2 + 2bx + c$.

We are given that $f(1) = 0$ and $f(-2) = 3$, so $f(1) = a + b + c + d = 0$ and

$f(-2) = -8a + 4b - 2c + d = 3$. Also $f'(1) = 3a + 2b + c = 0$ and

$f'(-2) = 12a - 4b + c = 0$ by Fermat's Theorem. Solving these four equations, we get

$a = \frac{2}{9}, b = \frac{1}{3}, c = -\frac{4}{3}, d = \frac{7}{9}$, so the function is $f(x) = \frac{1}{9}(2x^3 + 3x^2 - 12x + 7)$.

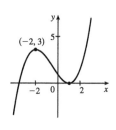

69. $y = x \sin x \Rightarrow y' = x \cos x + \sin x \Rightarrow y'' = -x \sin x + 2\cos x$. $y'' = 0 \Rightarrow 2\cos x = x \sin x$ [which is y] $\Rightarrow$

$(2\cos x)^2 = (x \sin x)^2 \Rightarrow 4\cos^2 x = x^2 \sin^2 x \Rightarrow 4\cos^2 x = x^2(1 - \cos^2 x) \Rightarrow 4\cos^2 x + x^2 \cos^2 x = x^2 \Rightarrow$

$\cos^2 x (4 + x^2) = x^2 \Rightarrow 4\cos^2 x (x^2 + 4) = 4x^2 \Rightarrow y^2(x^2 + 4) = 4x^2$ since $y = 2\cos x$ when $y'' = 0$.

71. (a) Since f and g are positive, increasing, and CU on I with f'' and g'' never equal to 0, we have $f > 0$, $f' \geq 0$, $f'' > 0$,

$g > 0$, $g' \geq 0$, $g'' > 0$ on I. Then $(fg)' = f'g + fg' \Rightarrow (fg)'' = f''g + 2f'g' + fg'' \geq f''g + fg'' > 0$ on $I \Rightarrow$

fg is CU on I.

(b) In part (a), if f and g are both decreasing instead of increasing, then $f' \leq 0$ and $g' \leq 0$ on I, so we still have $2f'g' \geq 0$

on I. Thus, $(fg)'' = f''g + 2f'g' + fg'' \geq f''g + fg'' > 0$ on $I \Rightarrow fg$ is CU on I as in part (a).

(c) Suppose f is increasing and g is decreasing [with f and g positive and CU]. Then $f' \geq 0$ and $g' \leq 0$ on I, so $2f'g' \leq 0$

on I and the argument in parts (a) and (b) fails.

Example 1. $I = (0, \infty)$, $f(x) = x^3$, $g(x) = 1/x$. Then $(fg)(x) = x^2$, so $(fg)'(x) = 2x$ and

$(fg)''(x) = 2 > 0$ on I. Thus, fg is CU on I.

Example 2. $I = (0, \infty)$, $f(x) = 4x\sqrt{x}$, $g(x) = 1/x$. Then $(fg)(x) = 4\sqrt{x}$, so $(fg)'(x) = 2/\sqrt{x}$ and

$(fg)''(x) = -1/\sqrt{x^3} < 0$ on I. Thus, fg is CD on I.

Example 3. $I = (0, \infty)$, $f(x) = x^2$, $g(x) = 1/x$. Thus, $(fg)(x) = x$, so fg is linear on I.

73. Let the cubic function be $f(x) = ax^3 + bx^2 + cx + d \Rightarrow f'(x) = 3ax^2 + 2bx + c \Rightarrow f''(x) = 6ax + 2b$.

So f is CU when $6ax + 2b > 0 \Leftrightarrow x > -b/(3a)$, CD when $x < -b/(3a)$, and so the only point of inflection occurs

when $x = -b/(3a)$. If the graph has three x-intercepts x_1, x_2 and x_3, then the expression for $f(x)$ must factor as

$f(x) = a(x - x_1)(x - x_2)(x - x_3)$. Multiplying these factors together gives us

$$f(x) = a[x^3 - (x_1 + x_2 + x_3)x^2 + (x_1x_2 + x_1x_3 + x_2x_3)x - x_1x_2x_3]$$

Equating the coefficients of the x^2-terms for the two forms of f gives us $b = -a(x_1 + x_2 + x_3)$. Hence, the x-coordinate of

the point of inflection is $-\dfrac{b}{3a} = -\dfrac{-a(x_1 + x_2 + x_3)}{3a} = \dfrac{x_1 + x_2 + x_3}{3}$.

75. By hypothesis $g = f'$ is differentiable on an open interval containing c. Since $(c, f(c))$ is a point of inflection, the concavity

changes at $x = c$, so $f''(x)$ changes signs at $x = c$. Hence, by the First Derivative Test, f' has a local extremum at $x = c$.

Thus, by Fermat's Theorem $f''(c) = 0$.

77. Using the fact that $|x| = \sqrt{x^2}$, we have that $g(x) = x\,|x| = x\sqrt{x^2}\;\Rightarrow\;g'(x) = \sqrt{x^2} + \sqrt{x^2} = 2\sqrt{x^2} = 2\,|x|\;\Rightarrow$
$g''(x) = 2x(x^2)^{-1/2} = \dfrac{2x}{|x|} < 0$ for $x < 0$ and $g''(x) > 0$ for $x > 0$, so $(0,0)$ is an inflection point. But $g''(0)$ does not exist.

79. Suppose that f is differentiable on an interval I and $f'(x) > 0$ for all x in I except $x = c$. To show that f is increasing on I, let x_1, x_2 be two numbers in I with $x_1 < x_2$.

    Case 1 $x_1 < x_2 < c$. Let J be the interval $\{x \in I \mid x < c\}$. By applying the Increasing/Decreasing Test to f on J, we see that f is increasing on J, so $f(x_1) < f(x_2)$.

    Case 2 $c < x_1 < x_2$. Apply the Increasing/Decreasing Test to f on $K = \{x \in I \mid x > c\}$.

    Case 3 $x_1 < x_2 = c$. Apply the proof of the Increasing/Decreasing Test, using the Mean Value Theorem (MVT) on the interval $[x_1, x_2]$ and noting that the MVT does not require f to be differentiable at the endpoints of $[x_1, x_2]$.

    Case 4 $c = x_1 < x_2$. Same proof as in Case 3.

    Case 5 $x_1 < c < x_2$. By Cases 3 and 4, f is increasing on $[x_1, c]$ and on $[c, x_2]$, so $f(x_1) < f(c) < f(x_2)$.

In all cases, we have shown that $f(x_1) < f(x_2)$. Since x_1, x_2 were any numbers in I with $x_1 < x_2$, we have shown that f is increasing on I.

81. (a) $f(x) = x^4 \sin\dfrac{1}{x}\;\Rightarrow\;f'(x) = x^4 \cos\dfrac{1}{x}\left(-\dfrac{1}{x^2}\right) + \sin\dfrac{1}{x}(4x^3) = 4x^3 \sin\dfrac{1}{x} - x^2 \cos\dfrac{1}{x}$.

$g(x) = x^4\left(2 + \sin\dfrac{1}{x}\right) = 2x^4 + f(x)\;\Rightarrow\;g'(x) = 8x^3 + f'(x)$.

$h(x) = x^4\left(-2 + \sin\dfrac{1}{x}\right) = -2x^4 + f(x)\;\Rightarrow\;h'(x) = -8x^3 + f'(x)$.

It is given that $f(0) = 0$, so $f'(0) = \lim\limits_{x\to 0}\dfrac{f(x) - f(0)}{x - 0} = \lim\limits_{x\to 0}\dfrac{x^4 \sin\frac{1}{x} - 0}{x} = \lim\limits_{x\to 0} x^3 \sin\dfrac{1}{x}$. Since

$-|x^3| \le x^3 \sin\dfrac{1}{x} \le |x^3|$ and $\lim\limits_{x\to 0}|x^3| = 0$, we see that $f'(0) = 0$ by the Squeeze Theorem. Also,

$g'(0) = 8(0)^3 + f'(0) = 0$ and $h'(0) = -8(0)^3 + f'(0) = 0$, so 0 is a critical number of f, g, and h.

For $x_{2n} = \dfrac{1}{2n\pi}$ [n a nonzero integer], $\sin\dfrac{1}{x_{2n}} = \sin 2n\pi = 0$ and $\cos\dfrac{1}{x_{2n}} = \cos 2n\pi = 1$, so $f'(x_{2n}) = -x_{2n}^2 < 0$.

For $x_{2n+1} = \dfrac{1}{(2n+1)\pi}$, $\sin\dfrac{1}{x_{2n+1}} = \sin(2n+1)\pi = 0$ and $\cos\dfrac{1}{x_{2n+1}} = \cos(2n+1)\pi = -1$, so

$f'(x_{2n+1}) = x_{2n+1}^2 > 0$. Thus, f' changes sign infinitely often on both sides of 0.

Next, $g'(x_{2n}) = 8x_{2n}^3 + f'(x_{2n}) = 8x_{2n}^3 - x_{2n}^2 = x_{2n}^2(8x_{2n} - 1) < 0$ for $x_{2n} < \frac{1}{8}$, but

$g'(x_{2n+1}) = 8x_{2n+1}^3 + x_{2n+1}^2 = x_{2n+1}^2(8x_{2n+1} + 1) > 0$ for $x_{2n+1} > -\frac{1}{8}$, so g' changes sign infinitely often on both sides of 0.

[continued]

Last, $h'(x_{2n}) = -8x_{2n}^3 + f'(x_{2n}) = -8x_{2n}^3 - x_{2n}^2 = -x_{2n}^2(8x_{2n} + 1) < 0$ for $x_{2n} > -\frac{1}{8}$ and $h'(x_{2n+1}) = -8x_{2n+1}^3 + x_{2n+1}^2 = x_{2n+1}^2(-8x_{2n+1} + 1) > 0$ for $x_{2n+1} < \frac{1}{8}$, so h' changes sign infinitely often on both sides of 0.

(b) $f(0) = 0$ and since $\sin\frac{1}{x}$ and hence $x^4 \sin\frac{1}{x}$ is both positive and negative inifinitely often on both sides of 0, and arbitrarily close to 0, f has neither a local maximum nor a local minimum at 0.

Since $2 + \sin\frac{1}{x} \geq 1$, $g(x) = x^4\left(2 + \sin\frac{1}{x}\right) > 0$ for $x \neq 0$, so $g(0) = 0$ is a local minimum.

Since $-2 + \sin\frac{1}{x} \leq -1$, $h(x) = x^4\left(-2 + \sin\frac{1}{x}\right) < 0$ for $x \neq 0$, so $h(0) = 0$ is a local maximum.

3.4 Limits at Infinity; Horizontal Asymptotes

1. (a) As x becomes large, the values of $f(x)$ approach 5.

(b) As x becomes large negative, the values of $f(x)$ approach 3.

3. (a) $\lim\limits_{x \to \infty} f(x) = -2$ (b) $\lim\limits_{x \to -\infty} f(x) = 2$ (c) $\lim\limits_{x \to 1} f(x) = \infty$

(d) $\lim\limits_{x \to 3} f(x) = -\infty$ (e) Vertical: $x = 1$, $x = 3$; horizontal: $y = -2$, $y = 2$

5. If $f(x) = x^2/2^x$, then a calculator gives $f(0) = 0$, $f(1) = 0.5$, $f(2) = 1$, $f(3) = 1.125$, $f(4) = 1$, $f(5) = 0.78125$, $f(6) = 0.5625$, $f(7) = 0.3828125$, $f(8) = 0.25$, $f(9) = 0.158203125$, $f(10) = 0.09765625$, $f(20) \approx 0.00038147$, $f(50) \approx 2.2204 \times 10^{-12}$, $f(100) \approx 7.8886 \times 10^{-27}$. It appears that $\lim\limits_{x \to \infty}\left(x^2/2^x\right) = 0$.

7. $\lim\limits_{x \to \infty} \dfrac{2x^2 - 7}{5x^2 + x - 3} = \lim\limits_{x \to \infty} \dfrac{(2x^2 - 7)/x^2}{(5x^2 + x - 3)/x^2}$ [Divide both the numerator and denominator by x^2 (the highest power of x that appears in the denominator)]

$= \dfrac{\lim\limits_{x \to \infty}(2 - 7/x^2)}{\lim\limits_{x \to \infty}(5 + 1/x - 3/x^2)}$ [Limit Law 5]

$= \dfrac{\lim\limits_{x \to \infty} 2 - \lim\limits_{x \to \infty}(7/x^2)}{\lim\limits_{x \to \infty} 5 + \lim\limits_{x \to \infty}(1/x) - \lim\limits_{x \to \infty}(3/x^2)}$ [Limit Laws 1 and 2]

$= \dfrac{2 - 7\lim\limits_{x \to \infty}(1/x^2)}{5 + \lim\limits_{x \to \infty}(1/x) - 3\lim\limits_{x \to \infty}(1/x^2)}$ [Limit Laws 8 and 3]

$= \dfrac{2 - 7(0)}{5 + 0 + 3(0)}$ [Theorem 4]

$= \dfrac{2}{5}$

9. $\lim\limits_{x \to \infty} \dfrac{4x+3}{5x-1} = \lim\limits_{x \to \infty} \dfrac{(4x+3)/x}{(5x-1)/x} = \lim\limits_{x \to \infty} \dfrac{4+3/x}{5-1/x} = \dfrac{\lim\limits_{x\to\infty} 4 + 3\lim\limits_{x\to\infty}(1/x)}{\lim\limits_{x\to\infty} 5 - \lim\limits_{x\to\infty}(1/x)} = \dfrac{4+3(0)}{5-0} = \dfrac{4}{5}$

11. $\lim\limits_{t \to -\infty} \dfrac{3t^2+t}{t^3-4t+1} = \lim\limits_{t \to -\infty} \dfrac{(3t^2+t)/t^3}{(t^3-4t+1)/t^3} = \lim\limits_{t \to -\infty} \dfrac{3/t + 1/t^2}{1 - 4/t^2 + 1/t^3}$

$= \dfrac{3\lim\limits_{t\to-\infty}(1/t) + \lim\limits_{t\to-\infty}(1/t^2)}{\lim\limits_{t\to-\infty} 1 - 4\lim\limits_{t\to-\infty}(1/t^2) + \lim\limits_{t\to-\infty}(1/t^3)} = \dfrac{3(0)+0}{1-4(0)+0} = 0$

13. $\lim\limits_{r \to \infty} \dfrac{r-r^3}{2-r^2+3r^3} = \lim\limits_{r \to \infty} \dfrac{(r-r^3)/r^3}{(2-r^2+3r^3)/r^3} = \lim\limits_{r \to \infty} \dfrac{1/r^2 - 1}{2/r^3 - 1/r + 3} = \dfrac{0-1}{0-0+3} = -\dfrac{1}{3}$

15. $\lim\limits_{x \to \infty} \dfrac{4-\sqrt{x}}{2+\sqrt{x}} = \lim\limits_{x \to \infty} \dfrac{(4-\sqrt{x})/\sqrt{x}}{(2+\sqrt{x})/\sqrt{x}} = \lim\limits_{x \to \infty} \dfrac{4/\sqrt{x}-1}{2/\sqrt{x}+1} = \dfrac{0-1}{0+1} = -1$

17. $\lim\limits_{x \to \infty} \dfrac{\sqrt{x+3x^2}}{4x-1} = \lim\limits_{x \to \infty} \dfrac{\sqrt{x+3x^2}/x}{(4x-1)/x} = \dfrac{\lim\limits_{x\to\infty}\sqrt{(x+3x^2)/x^2}}{\lim\limits_{x\to\infty}(4-1/x)}$ [since $x = \sqrt{x^2}$ for $x > 0$]

$= \dfrac{\lim\limits_{x\to\infty}\sqrt{1/x+3}}{\lim\limits_{x\to\infty} 4 - \lim\limits_{x\to\infty}(1/x)} = \dfrac{\sqrt{\lim\limits_{x\to\infty}(1/x)+\lim\limits_{x\to\infty} 3}}{4-0} = \dfrac{\sqrt{0+3}}{4} = \dfrac{\sqrt{3}}{4}$

19. $\lim\limits_{x \to \infty} \dfrac{\sqrt{1+4x^6}}{2-x^3} = \lim\limits_{x \to \infty} \dfrac{\sqrt{1+4x^6}/x^3}{(2-x^3)/x^3} = \dfrac{\lim\limits_{x\to\infty}\sqrt{(1+4x^6)/x^6}}{\lim\limits_{x\to\infty}(2/x^3-1)}$ $\left[\text{since } x^3 = \sqrt{x^6} \text{ for } x > 0\right]$

$= \dfrac{\lim\limits_{x\to\infty}\sqrt{1/x^6+4}}{\lim\limits_{x\to\infty}(2/x^3) - \lim\limits_{x\to\infty} 1} = \dfrac{\sqrt{\lim\limits_{x\to\infty}(1/x^6)+\lim\limits_{x\to\infty} 4}}{0-1}$

$= \dfrac{\sqrt{0+4}}{-1} = \dfrac{2}{-1} = -2$

21. $\lim\limits_{x \to -\infty} \dfrac{2x^5-x}{x^4+3} = \lim\limits_{x \to -\infty} \dfrac{(2x^5-x)/x^4}{(x^4+3)/x^4} = \lim\limits_{x \to -\infty} \dfrac{2x-1/x^3}{1+3/x^4}$

$= -\infty$ since $2x - 1/x^3 \to -\infty$ and $1+3/x^4 \to 1$ as $x \to -\infty$

23. $\lim\limits_{x \to \infty} \cos x$ does not exist because as x increases $\cos x$ does not approach any one value, but oscillates between 1 and -1.

25. $\lim\limits_{t \to \infty}\left(\sqrt{25t^2+2}-5t\right) = \lim\limits_{t \to \infty}\left(\sqrt{25t^2+2}-5t\right)\left(\dfrac{\sqrt{25t^2+2}+5t}{\sqrt{25t^2+2}+5t}\right) = \lim\limits_{t \to \infty} \dfrac{(25t^2+2)-(5t)^2}{\sqrt{25t^2+2}+5t}$

$= \lim\limits_{t \to \infty} \dfrac{2}{\sqrt{25t^2+2}+5t} = \lim\limits_{t \to \infty} \dfrac{2/t}{(\sqrt{25t^2+2}+5t)/t}$

$= \lim\limits_{t \to \infty} \dfrac{2/t}{\sqrt{25+2/t^2}+5}$ [since $t = \sqrt{t^2}$ for $t > 0$]

$= \dfrac{0}{\sqrt{25+0}+5} = 0$

27. $\lim\limits_{x\to\infty} \left(\sqrt{x^2+ax} - \sqrt{x^2+bx}\right) = \lim\limits_{x\to\infty} \dfrac{\left(\sqrt{x^2+ax} - \sqrt{x^2+bx}\right)\left(\sqrt{x^2+ax} + \sqrt{x^2+bx}\right)}{\sqrt{x^2+ax} + \sqrt{x^2+bx}}$

$= \lim\limits_{x\to\infty} \dfrac{(x^2+ax) - (x^2+bx)}{\sqrt{x^2+ax} + \sqrt{x^2+bx}} = \lim\limits_{x\to\infty} \dfrac{[(a-b)x]/x}{\left(\sqrt{x^2+ax} + \sqrt{x^2+bx}\right)/\sqrt{x^2}}$

$= \lim\limits_{x\to\infty} \dfrac{a-b}{\sqrt{1+a/x} + \sqrt{1+b/x}} = \dfrac{a-b}{\sqrt{1+0} + \sqrt{1+0}} = \dfrac{a-b}{2}$

29. $\lim\limits_{x\to-\infty} (x^2 + 2x^7) = \lim\limits_{x\to-\infty} x^7\left(\dfrac{1}{x^5} + 2\right)$ [factor out the largest power of x] $= -\infty$ because $x^7 \to -\infty$ and $1/x^5 + 2 \to 2$ as $x \to -\infty$.

Or: $\lim\limits_{x\to-\infty} (x^2 + 2x^7) = \lim\limits_{x\to-\infty} x^2(1 + 2x^5) = -\infty$.

31. If $t = \dfrac{1}{x}$, then $\lim\limits_{x\to\infty} x \sin\dfrac{1}{x} = \lim\limits_{t\to 0^+} \dfrac{1}{t}\sin t = \lim\limits_{t\to 0^+} \dfrac{\sin t}{t} = 1$.

33. (a)

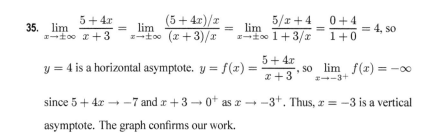

From the graph of $f(x) = \sqrt{x^2 + x + 1} + x$, we estimate the value of $\lim\limits_{x\to-\infty} f(x)$ to be -0.5.

(b)

x	$f(x)$
$-10{,}000$	$-0.499\,962\,5$
$-100{,}000$	$-0.499\,996\,2$
$-1{,}000{,}000$	$-0.499\,999\,6$

From the table, we estimate the limit to be -0.5.

(c) $\lim\limits_{x\to-\infty}\left(\sqrt{x^2+x+1}+x\right) = \lim\limits_{x\to-\infty}\left(\sqrt{x^2+x+1}+x\right)\left[\dfrac{\sqrt{x^2+x+1}-x}{\sqrt{x^2+x+1}-x}\right] = \lim\limits_{x\to-\infty}\dfrac{(x^2+x+1)-x^2}{\sqrt{x^2+x+1}-x}$

$= \lim\limits_{x\to-\infty}\dfrac{(x+1)(1/x)}{(\sqrt{x^2+x+1}-x)(1/x)} = \lim\limits_{x\to-\infty}\dfrac{1+(1/x)}{-\sqrt{1+(1/x)+(1/x^2)}-1}$

$= \dfrac{1+0}{-\sqrt{1+0+0}-1} = -\dfrac{1}{2}$

Note that for $x < 0$, we have $\sqrt{x^2} = |x| = -x$, so when we divide the radical by x, with $x < 0$, we get

$\dfrac{1}{x}\sqrt{x^2+x+1} = -\dfrac{1}{\sqrt{x^2}}\sqrt{x^2+x+1} = -\sqrt{1+(1/x)+(1/x^2)}$.

35. $\lim\limits_{x\to\pm\infty}\dfrac{5+4x}{x+3} = \lim\limits_{x\to\pm\infty}\dfrac{(5+4x)/x}{(x+3)/x} = \lim\limits_{x\to\pm\infty}\dfrac{5/x+4}{1+3/x} = \dfrac{0+4}{1+0} = 4$, so

$y = 4$ is a horizontal asymptote. $y = f(x) = \dfrac{5+4x}{x+3}$, so $\lim\limits_{x\to-3^+} f(x) = -\infty$

since $5+4x \to -7$ and $x+3 \to 0^+$ as $x \to -3^+$. Thus, $x = -3$ is a vertical asymptote. The graph confirms our work.

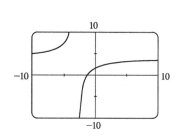

37. $\lim\limits_{x\to\pm\infty} \dfrac{2x^2+x-1}{x^2+x-2} = \lim\limits_{x\to\pm\infty} \dfrac{\dfrac{2x^2+x-1}{x^2}}{\dfrac{x^2+x-2}{x^2}} = \lim\limits_{x\to\pm\infty} \dfrac{2+\dfrac{1}{x}-\dfrac{1}{x^2}}{1+\dfrac{1}{x}-\dfrac{2}{x^2}} = \dfrac{\lim\limits_{x\to\pm\infty}\left(2+\dfrac{1}{x}-\dfrac{1}{x^2}\right)}{\lim\limits_{x\to\pm\infty}\left(1+\dfrac{1}{x}-\dfrac{2}{x^2}\right)}$

$= \dfrac{\lim\limits_{x\to\pm\infty} 2 + \lim\limits_{x\to\pm\infty}\dfrac{1}{x} - \lim\limits_{x\to\pm\infty}\dfrac{1}{x^2}}{\lim\limits_{x\to\pm\infty} 1 + \lim\limits_{x\to\pm\infty}\dfrac{1}{x} - 2\lim\limits_{x\to\pm\infty}\dfrac{1}{x^2}} = \dfrac{2+0-0}{1+0-2(0)} = 2$, so $y=2$ is a horizontal asymptote.

$y = f(x) = \dfrac{2x^2+x-1}{x^2+x-2} = \dfrac{(2x-1)(x+1)}{(x+2)(x-1)}$, so $\lim\limits_{x\to -2^-} f(x) = \infty$,

$\lim\limits_{x\to -2^+} f(x) = -\infty$, $\lim\limits_{x\to 1^-} f(x) = -\infty$, and $\lim\limits_{x\to 1^+} f(x) = \infty$. Thus, $x=-2$

and $x=1$ are vertical asymptotes. The graph confirms our work.

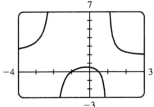

39. $y = f(x) = \dfrac{x^3-x}{x^2-6x+5} = \dfrac{x(x^2-1)}{(x-1)(x-5)} = \dfrac{x(x+1)(x-1)}{(x-1)(x-5)} = \dfrac{x(x+1)}{x-5} = g(x)$ for $x \ne 1$.

The graph of g is the same as the graph of f with the exception of a hole in the

graph of f at $x=1$. By long division, $g(x) = \dfrac{x^2+x}{x-5} = x+6 + \dfrac{30}{x-5}$.

As $x \to \pm\infty$, $g(x) \to \pm\infty$, so there is no horizontal asymptote. The denominator

of g is zero when $x=5$. $\lim\limits_{x\to 5^-} g(x) = -\infty$ and $\lim\limits_{x\to 5^+} g(x) = \infty$, so $x=5$ is a

vertical asymptote. The graph confirms our work.

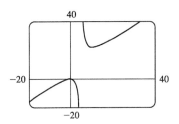

41. From the graph, it appears $y=1$ is a horizontal asymptote.

$\lim\limits_{x\to\pm\infty} \dfrac{3x^3+500x^2}{x^3+500x^2+100x+2000} = \lim\limits_{x\to\pm\infty} \dfrac{\dfrac{3x^3+500x^2}{x^3}}{\dfrac{x^3+500x^2+100x+2000}{x^3}}$

$= \lim\limits_{x\to\pm\infty} \dfrac{3+(500/x)}{1+(500/x)+(100/x^2)+(2000/x^3)}$

$= \dfrac{3+0}{1+0+0+0} = 3$, so $y=3$ is a horizontal asymptote.

The discrepancy can be explained by the choice of the viewing window. Try

$[-100{,}000, 100{,}000]$ by $[-1,4]$ to get a graph that lends credibility to our

calculation that $y=3$ is a horizontal asymptote.

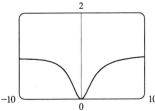

43. Divide the numerator and the denominator by the highest power of x in $Q(x)$.

(a) If $\deg P < \deg Q$, then the numerator $\to 0$ but the denominator doesn't. So $\lim\limits_{x\to\infty} [P(x)/Q(x)] = 0$.

(b) If $\deg P > \deg Q$, then the numerator $\to \pm\infty$ but the denominator doesn't, so $\lim\limits_{x\to\infty} [P(x)/Q(x)] = \pm\infty$

(depending on the ratio of the leading coefficients of P and Q).

45. Let's look for a rational function.

(1) $\lim\limits_{x \to \pm\infty} f(x) = 0 \;\Rightarrow\;$ degree of numerator $<$ degree of denominator

(2) $\lim\limits_{x \to 0} f(x) = -\infty \;\Rightarrow\;$ there is a factor of x^2 in the denominator (not just x, since that would produce a sign change at $x = 0$), and the function is negative near $x = 0$.

(3) $\lim\limits_{x \to 3^-} f(x) = \infty$ and $\lim\limits_{x \to 3^+} f(x) = -\infty \;\Rightarrow\;$ vertical asymptote at $x = 3$; there is a factor of $(x - 3)$ in the denominator.

(4) $f(2) = 0 \;\Rightarrow\;$ 2 is an x-intercept; there is at least one factor of $(x - 2)$ in the numerator.

Combining all of this information and putting in a negative sign to give us the desired left- and right-hand limits gives us

$f(x) = \dfrac{2 - x}{x^2(x - 3)}$ as one possibility.

47. (a) We must first find the function f. Since f has a vertical asymptote $x = 4$ and x-intercept $x = 1$, $x - 4$ is a factor of the denominator and $x - 1$ is a factor of the numerator. There is a removable discontinuity at $x = -1$, so $x - (-1) = x + 1$ is a factor of both the numerator and denominator. Thus, f now looks like this: $f(x) = \dfrac{a(x - 1)(x + 1)}{(x - 4)(x + 1)}$, where a is still to be determined. Then $\lim\limits_{x \to -1} f(x) = \lim\limits_{x \to -1} \dfrac{a(x - 1)(x + 1)}{(x - 4)(x + 1)} = \lim\limits_{x \to -1} \dfrac{a(x - 1)}{x - 4} = \dfrac{a(-1 - 1)}{(-1 - 4)} = \dfrac{2}{5}a$, so $\dfrac{2}{5}a = 2$, and $a = 5$. Thus $f(x) = \dfrac{5(x - 1)(x + 1)}{(x - 4)(x + 1)}$ is a ratio of quadratic functions satisfying all the given conditions and $f(0) = \dfrac{5(-1)(1)}{(-4)(1)} = \dfrac{5}{4}$.

(b) $\lim\limits_{x \to \infty} f(x) = 5 \lim\limits_{x \to \infty} \dfrac{x^2 - 1}{x^2 - 3x - 4} = 5 \lim\limits_{x \to \infty} \dfrac{(x^2/x^2) - (1/x^2)}{(x^2/x^2) - (3x/x^2) - (4/x^2)} = 5 \dfrac{1 - 0}{1 - 0 - 0} = 5(1) = 5$

49. $y = \dfrac{1 - x}{1 + x}$ has domain $(-\infty, -1) \cup (-1, \infty)$.

$\lim\limits_{x \to \pm\infty} \dfrac{1 - x}{1 + x} = \lim\limits_{x \to \pm\infty} \dfrac{1/x - 1}{1/x + 1} = \dfrac{0 - 1}{0 + 1} = -1$, so $y = -1$ is a HA.

The line $x = -1$ is a VA.

$y' = \dfrac{(1 + x)(-1) - (1 - x)(1)}{(1 + x)^2} = \dfrac{-2}{(1 + x)^2} < 0$ for $x \neq 1$. Thus,

$(-\infty, -1)$ and $(-1, \infty)$ are intervals of decrease.

$y'' = -2 \cdot \dfrac{-2(1 + x)}{[(1 + x)^2]^2} = \dfrac{4}{(1 + x)^3} < 0$ for $x < -1$ and $y'' > 0$ for $x > -1$, so the curve is CD on $(-\infty, -1)$ and CU on

$(-1, \infty)$. Since $x = -1$ is not in the domain, there is no IP.

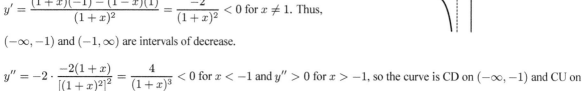

51. $\lim\limits_{x\to\pm\infty} \dfrac{x}{x^2+1} = \lim\limits_{x\to\pm\infty} \dfrac{1/x}{1+1/x^2} = \dfrac{0}{1+0} = 0$, so $y = 0$ is a

horizontal asymptote.

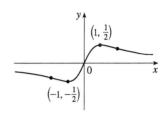

$y' = \dfrac{x^2+1 - x(2x)}{(x^2+1)^2} = \dfrac{1-x^2}{(x^2+1)^2} = 0$ when $x = \pm 1$ and $y' > 0 \Leftrightarrow$

$x^2 < 1 \Leftrightarrow -1 < x < 1$, so y is increasing on $(-1, 1)$ and decreasing

on $(-\infty, -1)$ and $(1, \infty)$.

$y'' = \dfrac{(1+x^2)^2(-2x) - (1-x^2)2(x^2+1)2x}{(1+x^2)^4} = \dfrac{2x(x^2-3)}{(1+x^2)^3} > 0 \Leftrightarrow x > \sqrt{3}$ or $-\sqrt{3} < x < 0$, so y is CU on

$(\sqrt{3}, \infty)$ and $(-\sqrt{3}, 0)$ and CD on $(-\infty, -\sqrt{3})$ and $(0, \sqrt{3})$.

53. $y = f(x) = x^4 - x^6 = x^4(1-x^2) = x^4(1+x)(1-x)$. The y-intercept is

$f(0) = 0$. The x-intercepts are $0, -1$, and 1 [found by solving $f(x) = 0$ for x].

Since $x^4 > 0$ for $x \neq 0$, f doesn't change sign at $x = 0$. The function does change

sign at $x = -1$ and $x = 1$. As $x \to \pm\infty$, $f(x) = x^4(1-x^2)$ approaches $-\infty$

because $x^4 \to \infty$ and $(1 - x^2) \to -\infty$.

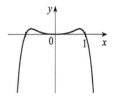

55. $y = f(x) = (3-x)(1+x)^2(1-x)^4$. The y-intercept is $f(0) = 3(1)^2(1)^4 = 3$.

The x-intercepts are $3, -1$, and 1. There is a sign change at 3, but not at -1 and 1.

When x is large positive, $3 - x$ is negative and the other factors are positive, so

$\lim\limits_{x\to\infty} f(x) = -\infty$. When x is large negative, $3 - x$ is positive, so

$\lim\limits_{x\to-\infty} f(x) = \infty$.

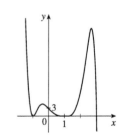

57. $f(2) = 4$, $f(-2) = -4$, $\lim\limits_{x\to-\infty} f(x) = 0$, $\lim\limits_{x\to\infty} f(x) = 2$

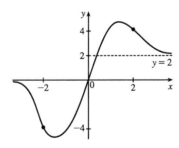

59. First we plot the points which are known to be on the graph: $(2, -1)$ and

$(0, 0)$. We can also draw a short line segment of slope 0 at $x = 2$, since

we are given that $f'(2) = 0$. Now we know that $f'(x) < 0$ (that is, the

function is decreasing) on $(0, 2)$, and that $f''(x) < 0$ on $(0, 1)$ and

$f''(x) > 0$ on $(1, 2)$. So we must join the points $(0, 0)$ and $(2, -1)$ in

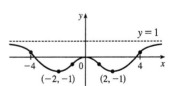

such a way that the curve is concave down on $(0, 1)$ and concave up on $(1, 2)$. The curve must be concave up and increasing on $(2, 4)$ and concave down and increasing toward $y = 1$ on $(4, \infty)$. Now we just need to reflect the curve in the y-axis, since we are given that f is an even function [the condition that $f(-x) = f(x)$ for all x].

61. We are given that $f(1) = f'(1) = 0$. So we can draw a short horizontal line at the point $(1, 0)$ to represent this situation. We are given that $x = 0$ and $x = 2$ are vertical asymptotes, with $\lim_{x \to 0} f(x) = -\infty$, $\lim_{x \to 2^+} f(x) = \infty$ and $\lim_{x \to 2^-} f(x) = -\infty$, so we can draw the parts of the curve which approach these asymptotes. On the interval $(-\infty, 0)$, the graph is concave down, and $f(x) \to \infty$ as $x \to -\infty$. Between the asymptotes the graph is concave down. On the interval $(2, \infty)$ the graph is concave up, and $f(x) \to 0$ as $x \to \infty$, so $y = 0$ is a horizontal asymptote. The diagram shows one possible function satisfying all of the given conditions.

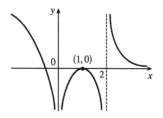

63. (a) Since $-1 \leq \sin x \leq 1$ for all x, $-\dfrac{1}{x} \leq \dfrac{\sin x}{x} \leq \dfrac{1}{x}$ for $x > 0$. As $x \to \infty$, $-1/x \to 0$ and $1/x \to 0$, so by the Squeeze Theorem, $(\sin x)/x \to 0$. Thus, $\lim_{x \to \infty} \dfrac{\sin x}{x} = 0$.

(b) From part (a), the horizontal asymptote is $y = 0$. The function $y = (\sin x)/x$ crosses the horizontal asymptote whenever $\sin x = 0$; that is, at $x = \pi n$ for every integer n. Thus, the graph crosses the asymptote *an infinite number of times*.

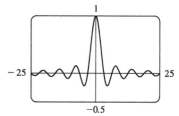

65. $\lim_{x \to \infty} \dfrac{4x - 1}{x} = \lim_{x \to \infty} \left(4 - \dfrac{1}{x}\right) = 4$ and $\lim_{x \to \infty} \dfrac{4x^2 + 3x}{x^2} = \lim_{x \to \infty} \left(4 + \dfrac{3}{x}\right) = 4$. Therefore, by the Squeeze Theorem, $\lim_{x \to \infty} f(x) = 4$.

67. Let $g(x) = \dfrac{3x^2 + 1}{2x^2 + x + 1}$ and $f(x) = |g(x) - 1.5|$. Note that $\lim_{x \to \infty} g(x) = \tfrac{3}{2}$ and $\lim_{x \to \infty} f(x) = 0$. We are interested in finding the x-value at which $f(x) < 0.05$. From the graph, we find that $x \approx 14.804$, so we choose $N = 15$ (or any larger number).

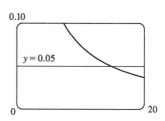

69. We want a value of N such that $x < N \Rightarrow \left| \dfrac{1 - 3x}{\sqrt{x^2 + 1}} - 3 \right| < \varepsilon$, or equivalently, $3 - \varepsilon < \dfrac{1 - 3x}{\sqrt{x^2 + 1}} < 3 + \varepsilon$. When $\varepsilon = 0.1$, we graph $y = f(x) = \dfrac{1 - 3x}{\sqrt{x^2 + 1}}$, $y = 3.1$, and $y = 2.9$. From the graph, we find that $f(x) = 3.1$ at about $x = -8.092$, so we choose $N=-9$ (or any lesser number). Similarly for $\varepsilon=0.05$, we find that $f(x)=3.05$ at about $x=-18.338$, so we choose $N=-19$ (or any lesser number).

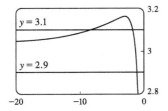
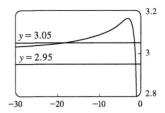

71. (a) $1/x^2 < 0.0001 \Leftrightarrow x^2 > 1/0.0001 = 10\,000 \Leftrightarrow x > 100 \quad (x>0)$

(b) If $\varepsilon > 0$ is given, then $1/x^2 < \varepsilon \Leftrightarrow x^2 > 1/\varepsilon \Leftrightarrow x > 1/\sqrt{\varepsilon}$. Let $N = 1/\sqrt{\varepsilon}$.

Then $x > N \Rightarrow x > \dfrac{1}{\sqrt{\varepsilon}} \Rightarrow \left|\dfrac{1}{x^2} - 0\right| = \dfrac{1}{x^2} < \varepsilon$, so $\lim\limits_{x\to\infty} \dfrac{1}{x^2} = 0$.

73. For $x < 0$, $|1/x - 0| = -1/x$. If $\varepsilon > 0$ is given, then $-1/x < \varepsilon \Leftrightarrow x < -1/\varepsilon$.

Take $N = -1/\varepsilon$. Then $x < N \Rightarrow x < -1/\varepsilon \Rightarrow |(1/x) - 0| = -1/x < \varepsilon$, so $\lim\limits_{x\to -\infty} (1/x) = 0$.

75. (a) Suppose that $\lim\limits_{x\to\infty} f(x) = L$. Then for every $\varepsilon > 0$ there is a corresponding positive number N such that $|f(x) - L| < \varepsilon$

whenever $x > N$. If $t = 1/x$, then $x > N \Leftrightarrow 0 < 1/x < 1/N \Leftrightarrow 0 < t < 1/N$. Thus, for every

$\varepsilon > 0$ there is a corresponding $\delta > 0$ (namely $1/N$) such that $|f(1/t) - L| < \varepsilon$ whenever $0 < t < \delta$. This proves that

$\lim\limits_{t\to 0^+} f(1/t) = L = \lim\limits_{x\to\infty} f(x)$.

Now suppose that $\lim\limits_{x\to -\infty} f(x) = L$. Then for every $\varepsilon > 0$ there is a corresponding negative number N such that

$|f(x) - L| < \varepsilon$ whenever $x < N$. If $t = 1/x$, then $x < N \Leftrightarrow 1/N < 1/x < 0 \Leftrightarrow 1/N < t < 0$. Thus, for every

$\varepsilon > 0$ there is a corresponding $\delta > 0$ (namely $-1/N$) such that $|f(1/t) - L| < \varepsilon$ whenever $-\delta < t < 0$. This proves that

$\lim\limits_{t\to 0^-} f(1/t) = L = \lim\limits_{x\to -\infty} f(x)$.

(b) $\lim\limits_{x\to 0^+} x \sin\dfrac{1}{x} = \lim\limits_{t\to 0^+} t \sin\dfrac{1}{t}$ [let $x = t$]

$\qquad\qquad\qquad = \lim\limits_{y\to\infty} \dfrac{1}{y} \sin y$ [part (a) with $y = 1/t$]

$\qquad\qquad\qquad = \lim\limits_{x\to\infty} \dfrac{\sin x}{x}$ [let $y = x$]

$\qquad\qquad\qquad = 0$ [by Exercise 63]

3.5 Summary of Curve Sketching

1. $y = f(x) = x^3 + 3x^2 = x^2(x+3)$ **A.** f is a polynomial, so $D = \mathbb{R}$.
B. y-intercept $= f(0) = 0$, x-intercepts are 0 and -3 **C.** No symmetry
D. No asymptote   **E.** $f'(x) = 3x^2 + 6x = 3x(x+2) > 0 \Leftrightarrow x < -2$ or $x > 0$, so f is increasing on $(-\infty, -2)$ and $(0, \infty)$, and decreasing on $(-2, 0)$.
F. Local maximum value $f(-2) = 4$, local minimum value $f(0) = 0$
G. $f''(x) = 6x + 6 = 6(x+1) > 0 \Leftrightarrow x > -1$, so f is CU on $(-1, \infty)$ and CD on $(-\infty, -1)$. IP at $(-1, 2)$

H.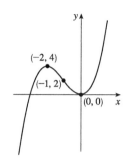

3. $y = f(x) = x^4 - 4x = x(x^3 - 4)$ **A.** $D = \mathbb{R}$ **B.** x-intercepts are 0 and $\sqrt[3]{4}$, y-intercept $= f(0) = 0$ **C.** No symmetry   **D.** No asymptote
E. $f'(x) = 4x^3 - 4 = 4(x^3 - 1) = 4(x-1)(x^2+x+1) > 0 \Leftrightarrow x > 1$, so f is increasing on $(1, \infty)$ and decreasing on $(-\infty, 1)$.   **F.** Local minimum value $f(1) = -3$, no local maximum   **G.** $f''(x) = 12x^2 > 0$ for all x, so f is CU on $(-\infty, \infty)$. No IP

H.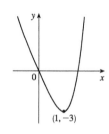

5. $y = f(x) = x(x-4)^3$ **A.** $D = \mathbb{R}$ **B.** x-intercepts are 0 and 4, y-intercept $f(0) = 0$ **C.** No symmetry
D. No asymptote
E. $f'(x) = x \cdot 3(x-4)^2 + (x-4)^3 \cdot 1 = (x-4)^2[3x + (x-4)]$
$= (x-4)^2(4x-4) = 4(x-1)(x-4)^2 > 0 \Leftrightarrow$
$x > 1$, so f is increasing on $(1, \infty)$ and decreasing on $(-\infty, 1)$.
F. Local minimum value $f(1) = -27$, no local maximum value
G. $f''(x) = 4[(x-1) \cdot 2(x-4) + (x-4)^2 \cdot 1] = 4(x-4)[2(x-1) + (x-4)]$
$= 4(x-4)(3x-6) = 12(x-4)(x-2) < 0 \Leftrightarrow$
$2 < x < 4$, so f is CD on $(2, 4)$ and CU on $(-\infty, 2)$ and $(4, \infty)$. IPs at $(2, -16)$ and $(4, 0)$

H.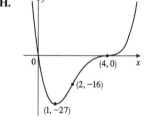

7. $y = f(x) = \frac{1}{5}x^5 - \frac{8}{3}x^3 + 16x = x(\frac{1}{5}x^4 - \frac{8}{3}x^2 + 16)$ **A.** $D = \mathbb{R}$ **B.** x-intercept 0, y-intercept $= f(0) = 0$
C. $f(-x) = -f(x)$, so f is odd; the curve is symmetric about the origin.   **D.** No asymptote
E. $f'(x) = x^4 - 8x^2 + 16 = (x^2 - 4)^2 = (x+2)^2(x-2)^2 > 0$ for all x except ± 2, so f is increasing on $\mathbb{R}$.   **F.** No extreme values
G. $f''(x) = 4x^3 - 16x = 4x(x^2 - 4) = 4x(x+2)(x-2) > 0 \Leftrightarrow$
$-2 < x < 0$ or $x > 2$, so f is CU on $(-2, 0)$ and $(2, \infty)$, and f is CD on $(-\infty, -2)$ and $(0, 2)$. IP at $\left(-2, -\frac{256}{15}\right)$, $(0, 0)$, and $\left(2, \frac{256}{15}\right)$

H.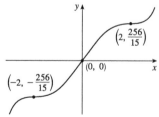

9. $y = f(x) = \dfrac{2x+3}{x+2}$ **A.** $D = \{x \mid x \neq -2\} = (-\infty, -2) \cup (-2, \infty)$ **B.** x-intercept $= -\dfrac{3}{2}$, y-intercept $= f(0) = \dfrac{3}{2}$

C. No symmetry **D.** $\lim\limits_{x \to \pm\infty} \dfrac{2x+3}{x+2} = 2$, so $y = 2$ is a HA. $\lim\limits_{x \to -2^-} \dfrac{2x+3}{x+2} = \infty$, $\lim\limits_{x \to -2^+} \dfrac{2x+3}{x+2} = -\infty$, so $x = -2$

is a VA. **E.** $f'(x) = \dfrac{(x+2)\cdot 2 - (2x+3)\cdot 1}{(x+2)^2} = \dfrac{1}{(x+2)^2} > 0$ for **H.**

$x \neq -2$, so f is increasing on $(-\infty, -2)$ and $(-2, \infty)$. **F.** No extreme values

G. $f''(x) = \dfrac{-2}{(x+2)^3} > 0 \Leftrightarrow x < -2$, so f is CU on $(-\infty, -2)$ and CD

on $(-2, \infty)$. No IP

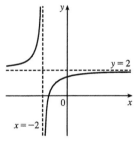

11. $y = f(x) = \dfrac{x - x^2}{2 - 3x + x^2} = \dfrac{x(1-x)}{(1-x)(2-x)} = \dfrac{x}{2-x}$ for $x \neq 1$. There is a hole in the graph at $(1, 1)$.

A. $D = \{x \mid x \neq 1, 2\} = (-\infty, 1) \cup (1, 2) \cup (2, \infty)$ **B.** x-intercept $= 0$, y-intercept $= f(0) = 0$ **C.** No symmetry

D. $\lim\limits_{x \to \pm\infty} \dfrac{x}{2-x} = -1$, so $y = -1$ is a HA. $\lim\limits_{x \to 2^-} \dfrac{x}{2-x} = \infty$, $\lim\limits_{x \to 2^+} \dfrac{x}{2-x} = -\infty$, so $x = 2$ is a VA.

E. $f'(x) = \dfrac{(2-x)(1) - x(-1)}{(2-x)^2} = \dfrac{2}{(2-x)^2} > 0$ $[x \neq 1, 2]$, so f is **H.**

increasing on $(-\infty, 1)$, $(1, 2)$, and $(2, \infty)$. **F.** No extreme values

G. $f'(x) = 2(2-x)^{-2} \Rightarrow$

$f''(x) = -4(2-x)^{-3}(-1) = \dfrac{4}{(2-x)^3} > 0 \Leftrightarrow x < 2$, so f is CU on

$(-\infty, 1)$ and $(1, 2)$, and f is CD on $(2, \infty)$. No IP

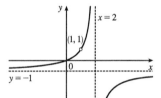

13. $y = f(x) = \dfrac{x}{x^2 - 4} = \dfrac{x}{(x+2)(x-2)}$ **A.** $D = (-\infty, -2) \cup (-2, 2) \cup (2, \infty)$ **B.** x-intercept $= 0$,

y-intercept $= f(0) = 0$ **C.** $f(-x) = -f(x)$, so f is odd; the graph is symmetric about the origin.

D. $\lim\limits_{x \to 2^+} \dfrac{x}{x^2 - 4} = \infty$, $\lim\limits_{x \to 2^-} f(x) = -\infty$, $\lim\limits_{x \to -2^+} f(x) = \infty$, $\lim\limits_{x \to -2^-} f(x) = -\infty$, so $x = \pm 2$ are VAs.

$\lim\limits_{x \to \pm\infty} \dfrac{x}{x^2 - 4} = 0$, so $y = 0$ is a HA. **E.** $f'(x) = \dfrac{(x^2-4)(1) - x(2x)}{(x^2-4)^2} = -\dfrac{x^2+4}{(x^2-4)^2} < 0$ for all x in D, so f is

decreasing on $(-\infty, -2)$, $(-2, 2)$, and $(2, \infty)$. **F.** No extreme values

G. $f''(x) = -\dfrac{(x^2-4)^2(2x) - (x^2+4)2(x^2-4)(2x)}{[(x^2-4)^2]^2}$ **H.**

$= -\dfrac{2x(x^2-4)[(x^2-4) - 2(x^2+4)]}{(x^2-4)^4}$

$= -\dfrac{2x(-x^2-12)}{(x^2-4)^3} = \dfrac{2x(x^2+12)}{(x+2)^3(x-2)^3}$.

$f''(x) < 0$ if $x < -2$ or $0 < x < 2$, so f is CD on $(-\infty, -2)$ and $(0, 2)$, and CU

on $(-2, 0)$ and $(2, \infty)$. IP at $(0, 0)$

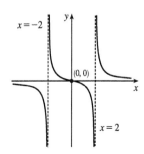

15. $y = f(x) = \dfrac{x^2}{x^2+3} = \dfrac{(x^2+3)-3}{x^2+3} = 1 - \dfrac{3}{x^2+3}$ **A.** $D = \mathbb{R}$ **B.** y-intercept: $f(0) = 0$;

x-intercept: $f(x) = 0 \Leftrightarrow x = 0$ **C.** $f(-x) = f(x)$, so f is even; the graph is symmetric about the y-axis.

D. $\lim\limits_{x \to \pm\infty} \dfrac{x^2}{x^2+3} = 1$, so $y = 1$ is a HA. No VA. **E.** Using the Reciprocal Rule, $f'(x) = -3 \cdot \dfrac{-2x}{(x^2+3)^2} = \dfrac{6x}{(x^2+3)^2}$.

$f'(x) > 0 \Leftrightarrow x > 0$ and $f'(x) < 0 \Leftrightarrow x < 0$, so f is decreasing on $(-\infty, 0)$ and increasing on $(0, \infty)$.

F. Local minimum value $f(0) = 0$, no local maximum.

G. $f''(x) = \dfrac{(x^2+3)^2 \cdot 6 - 6x \cdot 2(x^2+3) \cdot 2x}{[(x^2+3)^2]^2}$

$= \dfrac{6(x^2+3)[(x^2+3) - 4x^2]}{(x^2+3)^4} = \dfrac{6(3-3x^2)}{(x^2+3)^3} = \dfrac{-18(x+1)(x-1)}{(x^2+3)^3}$

$f''(x)$ is negative on $(-\infty, -1)$ and $(1, \infty)$ and positive on $(-1, 1)$,

so f is CD on $(-\infty, -1)$ and $(1, \infty)$ and CU on $(-1, 1)$. IP at $\left(\pm 1, \tfrac{1}{4}\right)$

H.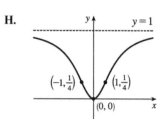

17. $y = f(x) = \dfrac{x-1}{x^2}$ **A.** $D = \{x \mid x \ne 0\} = (-\infty, 0) \cup (0, \infty)$ **B.** No y-intercept; x-intercept: $f(x) = 0 \Leftrightarrow x = 1$

C. No symmetry **D.** $\lim\limits_{x \to \pm\infty} \dfrac{x-1}{x^2} = 0$, so $y = 0$ is a HA. $\lim\limits_{x \to 0} \dfrac{x-1}{x^2} = -\infty$, so $x = 0$ is a VA.

E. $f'(x) = \dfrac{x^2 \cdot 1 - (x-1) \cdot 2x}{(x^2)^2} = \dfrac{-x^2 + 2x}{x^4} = \dfrac{-(x-2)}{x^3}$, so $f'(x) > 0 \Leftrightarrow 0 < x < 2$ and $f'(x) < 0 \Leftrightarrow$

$x < 0$ or $x > 2$. Thus, f is increasing on $(0, 2)$ and decreasing on $(-\infty, 0)$

and $(2, \infty)$. **F.** No local minimum, local maximum value $f(2) = \tfrac{1}{4}$.

G. $f''(x) = \dfrac{x^3 \cdot (-1) - [-(x-2)] \cdot 3x^2}{(x^3)^2} = \dfrac{2x^3 - 6x^2}{x^6} = \dfrac{2(x-3)}{x^4}$.

$f''(x)$ is negative on $(-\infty, 0)$ and $(0, 3)$ and positive on $(3, \infty)$, so f is CD

on $(-\infty, 0)$ and $(0, 3)$ and CU on $(3, \infty)$. IP at $\left(3, \tfrac{2}{9}\right)$

H.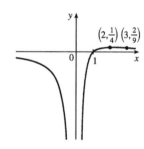

19. $y = f(x) = \dfrac{x^3}{x^3+1} = \dfrac{x^3}{(x+1)(x^2-x+1)}$ **A.** $D = (-\infty, -1) \cup (-1, \infty)$ **B.** y-intercept: $f(0) = 0$; x-intercept:

$f(x) = 0 \Leftrightarrow x = 0$ **C.** No symmetry **D.** $\lim\limits_{x \to \pm\infty} \dfrac{x^3}{x^3+1} = \dfrac{1}{1+1/x^3} = 1$, so $y = 1$ is a HA. $\lim\limits_{x \to -1^-} f(x) = \infty$ and

$\lim\limits_{x \to -1^+} f(x) = -\infty$, so $x = -1$ is a VA. **E.** $f'(x) = \dfrac{(x^3+1)(3x^2) - x^3(3x^2)}{(x^3+1)^2} = \dfrac{3x^2}{(x^3+1)^2}$. $f'(x) > 0$ for $x \ne -1$

(not in the domain) and $x \ne 0$ ($f' = 0$), so f is increasing on $(-\infty, -1)$, $(-1, 0)$, and $(0, \infty)$, and furthermore, by

Exercise 3.3.79, f is increasing on $(-\infty, -1)$, and $(-1, \infty)$. **F.** No extreme values

[continued]

G. $f''(x) = \dfrac{(x^3+1)^2(6x) - 3x^2[2(x^3+1)(3x^2)]}{[(x^3+1)^2]^2}$

$= \dfrac{(x^3+1)(6x)[(x^3+1) - 3x^3]}{(x^3+1)^4} = \dfrac{6x(1-2x^3)}{(x^3+1)^3}$

$f''(x) > 0 \Leftrightarrow x < -1$ or $0 < x < \sqrt[3]{\tfrac{1}{2}}$ [≈ 0.79], so f is CU on $(-\infty, -1)$ and $\left(0, \sqrt[3]{\tfrac{1}{2}}\right)$ and CD on $(-1, 0)$ and $\left(\sqrt[3]{\tfrac{1}{2}}, \infty\right)$. There are IPs at $(0, 0)$ and $\left(\sqrt[3]{\tfrac{1}{2}}, \tfrac{1}{3}\right)$.

H.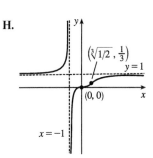

21. $y = f(x) = (x-3)\sqrt{x} = x^{3/2} - 3x^{1/2}$ **A.** $D = [0, \infty)$ **B.** x-intercepts: $0, 3$; y-intercept $= f(0) = 0$ **C.** No symmetry **D.** No asymptote **E.** $f'(x) = \tfrac{3}{2}x^{1/2} - \tfrac{3}{2}x^{-1/2} = \tfrac{3}{2}x^{-1/2}(x-1) = \dfrac{3(x-1)}{2\sqrt{x}} > 0 \Leftrightarrow x > 1$,

so f is increasing on $(1, \infty)$ and decreasing on $(0, 1)$.

F. Local minimum value $f(1) = -2$, no local maximum value

G. $f''(x) = \tfrac{3}{4}x^{-1/2} + \tfrac{3}{4}x^{-3/2} = \tfrac{3}{4}x^{-3/2}(x+1) = \dfrac{3(x+1)}{4x^{3/2}} > 0$ for $x > 0$,

so f is CU on $(0, \infty)$. No IP

H.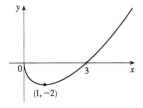

23. $y = f(x) = \sqrt{x^2 + x - 2} = \sqrt{(x+2)(x-1)}$ **A.** $D = \{x \mid (x+2)(x-1) \geq 0\} = (-\infty, -2] \cup [1, \infty)$

B. y-intercept: none; x-intercepts: -2 and 1 **C.** No symmetry **D.** No asymptote

E. $f'(x) = \tfrac{1}{2}(x^2 + x - 2)^{-1/2}(2x+1) = \dfrac{2x+1}{2\sqrt{x^2+x-2}}$, $f'(x) = 0$ if $x = -\tfrac{1}{2}$, but $-\tfrac{1}{2}$ is not in the domain.

$f'(x) > 0 \Rightarrow x > -\tfrac{1}{2}$ and $f'(x) < 0 \Rightarrow x < -\tfrac{1}{2}$, so (considering the domain) f is increasing on $(1, \infty)$ and

f is decreasing on $(-\infty, -2)$. **F.** No extreme values

G. $f''(x) = \dfrac{2(x^2+x-2)^{1/2}(2) - (2x+1) \cdot 2 \cdot \tfrac{1}{2}(x^2+x-2)^{-1/2}(2x+1)}{\left(2\sqrt{x^2+x-2}\right)^2}$

$= \dfrac{(x^2+x-2)^{-1/2}\left[4(x^2+x-2) - (4x^2+4x+1)\right]}{4(x^2+x-2)}$

$= \dfrac{-9}{4(x^2+x-2)^{3/2}} < 0$

so f is CD on $(-\infty, -2)$ and $(1, \infty)$. No IP

H.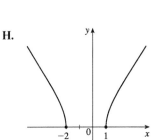

25. $y = f(x) = x/\sqrt{x^2+1}$ **A.** $D = \mathbb{R}$ **B.** y-intercept: $f(0) = 0$; x-intercepts: $f(x) = 0 \Rightarrow x = 0$

C. $f(-x) = -f(x)$, so f is odd; the graph is symmetric about the origin.

D. $\lim_{x \to \infty} f(x) = \lim_{x \to \infty} \dfrac{x}{\sqrt{x^2+1}} = \lim_{x \to \infty} \dfrac{x/x}{\sqrt{x^2+1}/x} = \lim_{x \to \infty} \dfrac{x/x}{\sqrt{x^2+1}/\sqrt{x^2}} = \lim_{x \to \infty} \dfrac{1}{\sqrt{1+1/x^2}} = \dfrac{1}{\sqrt{1+0}} = 1$

[continued]

and

$$\lim_{x\to-\infty} f(x) = \lim_{x\to-\infty} \frac{x}{\sqrt{x^2+1}} = \lim_{x\to-\infty} \frac{x/x}{\sqrt{x^2+1}/x} = \lim_{x\to-\infty} \frac{x/x}{\sqrt{x^2+1}/\left(-\sqrt{x^2}\right)} = \lim_{x\to-\infty} \frac{1}{-\sqrt{1+1/x^2}}$$

$$= \frac{1}{-\sqrt{1+0}} = -1 \text{ so } y = \pm 1 \text{ are HA. No VA}$$

E. $f'(x) = \dfrac{\sqrt{x^2+1} - x \cdot \dfrac{2x}{2\sqrt{x^2+1}}}{[(x^2+1)^{1/2}]^2} = \dfrac{x^2+1-x^2}{(x^2+1)^{3/2}} = \dfrac{1}{(x^2+1)^{3/2}} > 0$ for all x, so f is increasing on $\mathbb{R}$.

F. No extreme values G. $f''(x) = -\frac{3}{2}(x^2+1)^{-5/2} \cdot 2x = \dfrac{-3x}{(x^2+1)^{5/2}}$, so   H.

$f''(x) > 0$ for $x < 0$ and $f''(x) < 0$ for $x > 0$. Thus, f is CU on $(-\infty, 0)$ and CD on $(0, \infty)$. IP at $(0, 0)$

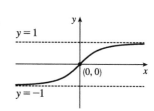

27. $y = f(x) = \sqrt{1-x^2}/x$ A. $D = \{x \mid |x| \le 1, x \ne 0\} = [-1, 0) \cup (0, 1]$ B. x-intercepts ± 1, no y-intercept

C. $f(-x) = -f(x)$, so the curve is symmetric about $(0, 0)$. D. $\lim_{x\to 0^+} \dfrac{\sqrt{1-x^2}}{x} = \infty$, $\lim_{x\to 0^-} \dfrac{\sqrt{1-x^2}}{x} = -\infty$,

so $x = 0$ is a VA. E. $f'(x) = \dfrac{\left(-x^2/\sqrt{1-x^2}\right) - \sqrt{1-x^2}}{x^2} = -\dfrac{1}{x^2\sqrt{1-x^2}} < 0$, so f is decreasing

on $(-1, 0)$ and $(0, 1)$. F. No extreme values   H.

G. $f''(x) = \dfrac{2-3x^2}{x^3(1-x^2)^{3/2}} > 0 \Leftrightarrow -1 < x < -\sqrt{\frac{2}{3}}$ or $0 < x < \sqrt{\frac{2}{3}}$, so

f is CU on $\left(-1, -\sqrt{\frac{2}{3}}\right)$ and $\left(0, \sqrt{\frac{2}{3}}\right)$ and CD on $\left(-\sqrt{\frac{2}{3}}, 0\right)$ and $\left(\sqrt{\frac{2}{3}}, 1\right)$.

IP at $\left(\pm\sqrt{\frac{2}{3}}, \pm\frac{1}{\sqrt{2}}\right)$

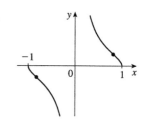

29. $y = f(x) = x - 3x^{1/3}$ A. $D = \mathbb{R}$ B. y-intercept: $f(0) = 0$; x-intercepts: $f(x) = 0 \Rightarrow x = 3x^{1/3} \Rightarrow$

$x^3 = 27x \Rightarrow x^3 - 27x = 0 \Rightarrow x(x^2 - 27) = 0 \Rightarrow x = 0, \pm 3\sqrt{3}$ C. $f(-x) = -f(x)$, so f is odd;

the graph is symmetric about the origin. D. No asymptote E. $f'(x) = 1 - x^{-2/3} = 1 - \dfrac{1}{x^{2/3}} = \dfrac{x^{2/3} - 1}{x^{2/3}}$.

$f'(x) > 0$ when $|x| > 1$ and $f'(x) < 0$ when $0 < |x| < 1$, so f is increasing on $(-\infty, -1)$ and $(1, \infty)$, and

decreasing on $(-1, 0)$ and $(0, 1)$ [hence decreasing on $(-1, 1)$ since f is   H.

continuous on $(-1, 1)$]. F. Local maximum value $f(-1) = 2$, local minimum

value $f(1) = -2$ G. $f''(x) = \frac{2}{3}x^{-5/3} < 0$ when $x < 0$ and $f''(x) > 0$

when $x > 0$, so f is CD on $(-\infty, 0)$ and CU on $(0, \infty)$. IP at $(0, 0)$

31. $y = f(x) = \sqrt[3]{x^2 - 1}$ **A.** $D = \mathbb{R}$ **B.** y-intercept: $f(0) = -1$; x-intercepts: $f(x) = 0 \Leftrightarrow x^2 - 1 = 0 \Leftrightarrow x = \pm 1$ **C.** $f(-x) = f(x)$, so the curve is symmetric about the y-axis. **D.** No asymptote

E. $f'(x) = \frac{1}{3}(x^2 - 1)^{-2/3}(2x) = \dfrac{2x}{3\sqrt[3]{(x^2-1)^2}}$. $f'(x) > 0 \Leftrightarrow x > 0$ and $f'(x) < 0 \Leftrightarrow x < 0$, so f is increasing on $(0, \infty)$ and decreasing on $(-\infty, 0)$. **F.** Local minimum value $f(0) = -1$

G. $f''(x) = \dfrac{2}{3} \cdot \dfrac{(x^2-1)^{2/3}(1) - x \cdot \frac{2}{3}(x^2-1)^{-1/3}(2x)}{[(x^2-1)^{2/3}]^2}$

$= \dfrac{2}{9} \cdot \dfrac{(x^2-1)^{-1/3}[3(x^2-1) - 4x^2]}{(x^2-1)^{4/3}} = -\dfrac{2(x^2+3)}{9(x^2-1)^{5/3}}$

H.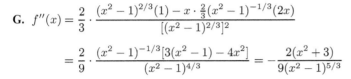

$f''(x) > 0 \Leftrightarrow -1 < x < 1$ and $f''(x) < 0 \Leftrightarrow x < -1$ or $x > 1$, so f is CU on $(-1, 1)$ and f is CD on $(-\infty, -1)$ and $(1, \infty)$. IP at $(\pm 1, 0)$

33. $y = f(x) = \sin^3 x$ **A.** $D = \mathbb{R}$ **B.** x-intercepts: $f(x) = 0 \Leftrightarrow x = n\pi$, n an integer; y-intercept $= f(0) = 0$

C. $f(-x) = -f(x)$, so f is odd and the curve is symmetric about the origin. Also, $f(x + 2\pi) = f(x)$, so f is periodic with period 2π, and we determine **E–G** for $0 \le x \le \pi$. Since f is odd, we can reflect the graph of f on $[0, \pi]$ about the origin to obtain the graph of f on $[-\pi, \pi]$, and then since f has period 2π, we can extend the graph of f for all real numbers.

D. No asymptote **E.** $f'(x) = 3\sin^2 x \cos x > 0 \Leftrightarrow \cos x > 0$ and $\sin x \ne 0 \Leftrightarrow 0 < x < \frac{\pi}{2}$, so f is increasing on $(0, \frac{\pi}{2})$ and f is decreasing on $(\frac{\pi}{2}, \pi)$. **F.** Local maximum value $f(\frac{\pi}{2}) = 1$ $\left[\text{local minimum value } f(-\frac{\pi}{2}) = -1\right]$

G. $f''(x) = 3\sin^2 x (-\sin x) + 3\cos x (2\sin x \cos x) = 3\sin x (2\cos^2 x - \sin^2 x)$
$= 3\sin x [2(1 - \sin^2 x) - \sin^2 x] = 3\sin x (2 - 3\sin^2 x) > 0 \Leftrightarrow$

$\sin x > 0$ and $\sin^2 x < \frac{2}{3} \Leftrightarrow 0 < x < \pi$ and $0 < \sin x < \sqrt{\frac{2}{3}} \Leftrightarrow 0 < x < \sin^{-1}\sqrt{\frac{2}{3}}$ $\left[\text{let } \alpha = \sin^{-1}\sqrt{\frac{2}{3}}\right]$ or $\pi - \alpha < x < \pi$, so f is CU on $(0, \alpha)$ and $(\pi - \alpha, \pi)$, and f is CD on $(\alpha, \pi - \alpha)$. There are inflection points at $x = 0, \pi, \alpha$, and $x = \pi - \alpha$.

H.

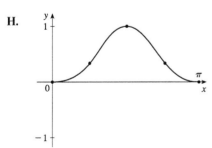 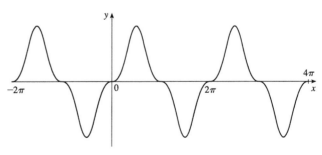

35. $y = f(x) = x \tan x$, $-\frac{\pi}{2} < x < \frac{\pi}{2}$ **A.** $D = \left(-\frac{\pi}{2}, \frac{\pi}{2}\right)$ **B.** Intercepts are 0 **C.** $f(-x) = f(x)$, so the curve is symmetric about the y-axis. **D.** $\lim\limits_{x \to (\pi/2)^-} x \tan x = \infty$ and $\lim\limits_{x \to -(\pi/2)^+} x \tan x = \infty$, so $x = \frac{\pi}{2}$ and $x = -\frac{\pi}{2}$ are VA.

[continued]

E. $f'(x) = \tan x + x\sec^2 x > 0 \Leftrightarrow 0 < x < \frac{\pi}{2}$, so f increases on $\left(0, \frac{\pi}{2}\right)$ and decreases on $\left(-\frac{\pi}{2}, 0\right)$. **F.** Absolute and local minimum value $f(0) = 0$.

G. $y'' = 2\sec^2 x + 2x\tan x \sec^2 x > 0$ for $-\frac{\pi}{2} < x < \frac{\pi}{2}$, so f is CU on $\left(-\frac{\pi}{2}, \frac{\pi}{2}\right)$. No IP

H.

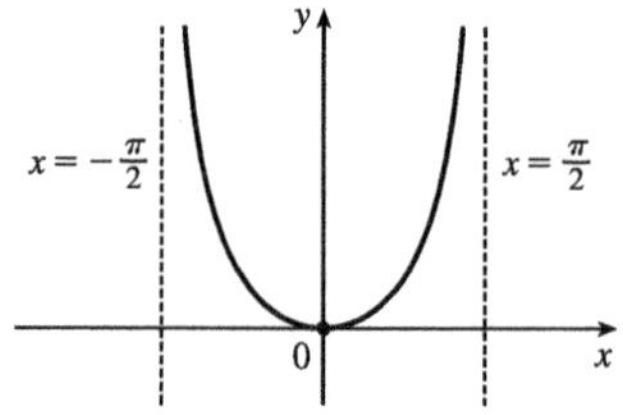

37. $y = f(x) = \sin x + \sqrt{3}\cos x$, $-2\pi \leq x \leq 2\pi$ **A.** $D = [-2\pi, 2\pi]$ **B.** y-intercept: $f(0) = \sqrt{3}$; x-intercepts: $f(x) = 0 \Leftrightarrow \sin x = -\sqrt{3}\cos x \Leftrightarrow \tan x = -\sqrt{3} \Leftrightarrow x = -\frac{4\pi}{3}, -\frac{\pi}{3}, \frac{2\pi}{3}$, or $\frac{5\pi}{3}$ **C.** f is periodic with period 2π. **D.** No asymptote **E.** $f'(x) = \cos x - \sqrt{3}\sin x$. $f'(x) = 0 \Leftrightarrow \cos x = \sqrt{3}\sin x \Leftrightarrow \tan x = \dfrac{1}{\sqrt{3}} \Leftrightarrow$ $x = -\frac{11\pi}{6}, -\frac{5\pi}{6}, \frac{\pi}{6}$, or $\frac{7\pi}{6}$. $f'(x) < 0 \Leftrightarrow -\frac{11\pi}{6} < x < -\frac{5\pi}{6}$ or $\frac{\pi}{6} < x < \frac{7\pi}{6}$, so f is decreasing on $\left(-\frac{11\pi}{6}, -\frac{5\pi}{6}\right)$ and $\left(\frac{\pi}{6}, \frac{7\pi}{6}\right)$, and f is increasing on $\left(-2\pi, -\frac{11\pi}{6}\right)$, $\left(-\frac{5\pi}{6}, \frac{\pi}{6}\right)$, and $\left(\frac{7\pi}{6}, 2\pi\right)$. **F.** Local maximum value $f\left(-\frac{11\pi}{6}\right) = f\left(\frac{\pi}{6}\right) = \frac{1}{2} + \sqrt{3}\left(\frac{1}{2}\sqrt{3}\right) = 2$, local minimum value $f\left(-\frac{5\pi}{6}\right) = f\left(\frac{7\pi}{6}\right) = -\frac{1}{2} + \sqrt{3}\left(-\frac{1}{2}\sqrt{3}\right) = -2$

G. $f''(x) = -\sin x - \sqrt{3}\cos x$. $f''(x) = 0 \Leftrightarrow \sin x = -\sqrt{3}\cos x \Leftrightarrow$ $\tan x = -\dfrac{1}{\sqrt{3}} \Leftrightarrow x = -\frac{4\pi}{3}, -\frac{\pi}{3}, \frac{2\pi}{3}$, or $\frac{5\pi}{3}$. $f''(x) > 0 \Leftrightarrow$ $-\frac{4\pi}{3} < x < -\frac{\pi}{3}$ or $\frac{2\pi}{3} < x < \frac{5\pi}{3}$, so f is CU on $\left(-\frac{4\pi}{3}, -\frac{\pi}{3}\right)$ and $\left(\frac{2\pi}{3}, \frac{5\pi}{3}\right)$, and f is CD on $\left(-2\pi, -\frac{4\pi}{3}\right)$, $\left(-\frac{\pi}{3}, \frac{2\pi}{3}\right)$, and $\left(\frac{5\pi}{3}, 2\pi\right)$. There are IPs at $\left(-\frac{4\pi}{3}, 0\right)$, $\left(-\frac{\pi}{3}, 0\right)$, $\left(\frac{2\pi}{3}, 0\right)$, and $\left(\frac{5\pi}{3}, 0\right)$.

H.

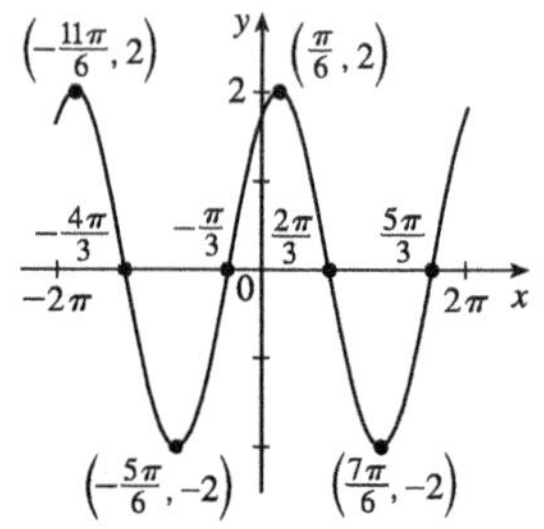

39. $y = f(x) = \dfrac{\sin x}{1 + \cos x}$ $\left[\overset{\text{when } \cos x \neq 1}{=} \dfrac{\sin x}{1 + \cos x} \cdot \dfrac{1 - \cos x}{1 - \cos x} = \dfrac{\sin x\,(1 - \cos x)}{\sin^2 x} = \dfrac{1 - \cos x}{\sin x} = \csc x - \cot x\right]$

A. The domain of f is the set of all real numbers except odd integer multiples of π; that is, all reals except $(2n+1)\pi$, where n is an integer. **B.** y-intercept: $f(0) = 0$; x-intercepts: $x = 2n\pi$, n an integer. **C.** $f(-x) = -f(x)$, so f is an odd function; the graph is symmetric about the origin and has period 2π. **D.** When n is an odd integer, $\lim\limits_{x \to (n\pi)^-} f(x) = \infty$ and $\lim\limits_{x \to (n\pi)^+} f(x) = -\infty$, so $x = n\pi$ is a VA for each odd integer n (or, equivalently, $x = (2n+1)\pi$ for every integer n).

No HA. **E.** $f'(x) = \dfrac{(1 + \cos x) \cdot \cos x - \sin x(-\sin x)}{(1 + \cos x)^2} = \dfrac{1 + \cos x}{(1 + \cos x)^2} = \dfrac{1}{1 + \cos x}$. $f'(x) > 0$ for all x except odd multiples of π, so f is increasing on $((2k-1)\pi, (2k+1)\pi)$ for each integer k. **F.** No extreme values

[continued]

G. $f''(x) = \dfrac{\sin x}{(1+\cos x)^2} > 0 \Rightarrow \sin x > 0 \Rightarrow$

$x \in (2k\pi, (2k+1)\pi)$ and $f''(x) < 0$ on $((2k-1)\pi, 2k\pi)$ for each integer k. f is CU on $(2k\pi, (2k+1)\pi)$ and CD on $((2k-1)\pi, 2k\pi)$ for each integer k. f has IPs at $(2k\pi, 0)$ for each integer k.

H.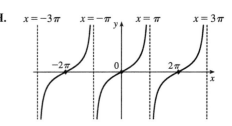

41. $g(x) = \sqrt{f(x)}$

(a) The domain of g consists of all x such that $f(x) \geq 0$, so g has domain $(-\infty, 7]$. $g'(x) = \dfrac{1}{2\sqrt{f(x)}} \cdot f'(x) = \dfrac{f'(x)}{2\sqrt{f(x)}}$.

Since $f'(3)$ does not exist, $g'(3)$ does not exist. (Note that $f(7) = 0$, but 7 is an endpoint of the domain of g.) The domain of g' is $(-\infty, 3) \cup (3, 7)$.

(b) $g'(x) = 0 \Rightarrow f'(x) = 0$ on the domain of $g \Rightarrow x = 5$ [there is a horizontal tangent line there]. From part (a), $g'(3)$ does not exist. So the critical numbers of g are 3 and 5.

(c) From part (a), $g'(x) = \dfrac{f'(x)}{2\sqrt{f(x)}}$. $g'(6) = \dfrac{f'(6)}{2\sqrt{f(6)}} \approx \dfrac{-2}{2\sqrt{3}} = -\dfrac{1}{\sqrt{3}} \approx -0.58$.

(d) $\lim\limits_{x \to -\infty} g(x) = \lim\limits_{x \to -\infty} \sqrt{f(x)} = \sqrt{2}$, so $y = \sqrt{2}$ is a horizontal asymptote. No VA

43. $g(x) = |f(x)|$

(a) Since the absolute-value function is defined for all reals, the domain of g equals the domain of f, $(-\infty, \infty)$. The domain of g' equals the domain of f' except for any values of x such that both $f(x) = 0$ and $f'(x) \neq 0$. $f'(3)$ does not exist, $f(7) = 0$, and $f'(7) \neq 0$. Thus, the domain of g' is $(-\infty, 3) \cup (3, 7) \cup (7, \infty)$.

(b) $g'(x) = 0 \Rightarrow f'(x) = 0 \Rightarrow x = 5$ or $x = 9$ [there are horizontal tangent lines there]. From part (a), g' does not exist at $x = 3$ and $x = 7$. So the critical numbers of g are 3, 5, 7, and 9.

(c) Since f is positive near $x = 6$, $g(x) = |f(x)| = f(x)$ near 6, so $g'(6) = f'(6) \approx -2$.

(d) $\lim\limits_{x \to -\infty} g(x) = \lim\limits_{x \to -\infty} |f(x)| = |2| = 2$ and $\lim\limits_{x \to \infty} g(x) = \lim\limits_{x \to \infty} |f(x)| = |-1| = 1$, so $y = 2$ and $y = 1$ are horizontal asymptotes. No VA

45. $m = f(v) = \dfrac{m_0}{\sqrt{1 - v^2/c^2}}$. The m-intercept is $f(0) = m_0$. There are no v-intercepts. $\lim\limits_{v \to c^-} f(v) = \infty$, so $v = c$ is a VA.

$f'(v) = -\tfrac{1}{2} m_0 (1 - v^2/c^2)^{-3/2} (-2v/c^2) = \dfrac{m_0 v}{c^2(1 - v^2/c^2)^{3/2}} = \dfrac{m_0 v}{\dfrac{c^2(c^2 - v^2)^{3/2}}{c^3}} = \dfrac{m_0 c v}{(c^2 - v^2)^{3/2}} > 0$, so f is

increasing on $(0, c)$. There are no local extreme values.

$f''(v) = \dfrac{(c^2 - v^2)^{3/2}(m_0 c) - m_0 c v \cdot \tfrac{3}{2}(c^2 - v^2)^{1/2}(-2v)}{[(c^2 - v^2)^{3/2}]^2}$

$= \dfrac{m_0 c (c^2 - v^2)^{1/2}[(c^2 - v^2) + 3v^2]}{(c^2 - v^2)^3} = \dfrac{m_0 c (c^2 + 2v^2)}{(c^2 - v^2)^{5/2}} > 0$,

so f is CU on $(0, c)$. There are no inflection points.

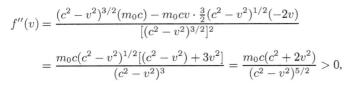

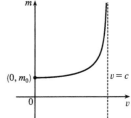

47. $y = -\dfrac{W}{24EI}x^4 + \dfrac{WL}{12EI}x^3 - \dfrac{WL^2}{24EI}x^2 = -\dfrac{W}{24EI}x^2(x^2 - 2Lx + L^2)$

$= \dfrac{-W}{24EI}x^2(x-L)^2 = cx^2(x-L)^2$

where $c = -\dfrac{W}{24EI}$ is a negative constant and $0 \le x \le L$. We sketch

$f(x) = cx^2(x-L)^2$ for $c = -1$. $f(0) = f(L) = 0$.

$f'(x) = cx^2[2(x-L)] + (x-L)^2(2cx) = 2cx(x-L)[x+(x-L)] = 2cx(x-L)(2x-L)$. So for $0 < x < L$,

$f'(x) > 0 \Leftrightarrow x(x-L)(2x-L) < 0$ [since $c < 0$] $\Leftrightarrow L/2 < x < L$ and $f'(x) < 0 \Leftrightarrow 0 < x < L/2$.

Thus, f is increasing on $(L/2, L)$ and decreasing on $(0, L/2)$, and there is a local and absolute

minimum at the point $(L/2, f(L/2)) = (L/2, cL^4/16)$. $f'(x) = 2c[x(x-L)(2x-L)] \Rightarrow$

$f''(x) = 2c[1(x-L)(2x-L) + x(1)(2x-L) + x(x-L)(2)] = 2c(6x^2 - 6Lx + L^2) = 0 \Leftrightarrow$

$x = \dfrac{6L \pm \sqrt{12L^2}}{12} = \tfrac{1}{2}L \pm \tfrac{\sqrt{3}}{6}L$, and these are the x-coordinates of the two inflection points.

49. $y = \dfrac{x^2+1}{x+1}$. Long division gives us:

$$\begin{array}{r} x - 1 \\ x+1 \overline{\smash{\big)}\, x^2 \phantom{{}+x} + 1} \\ \underline{x^2 + x \phantom{{}+1}} \\ -x + 1 \\ \underline{-x - 1} \\ 2 \end{array}$$

Thus, $y = f(x) = \dfrac{x^2+1}{x+1} = x - 1 + \dfrac{2}{x+1}$ and $f(x) - (x-1) = \dfrac{2}{x+1} = \dfrac{\frac{2}{x}}{1+\frac{1}{x}}$ [for $x \ne 0$] $\to 0$ as $x \to \pm\infty$.

So the line $y = x - 1$ is a slant asymptote (SA).

51. $y = \dfrac{2x^3 - 5x^2 + 3x}{x^2 - x - 2}$. Long division gives us:

$$\begin{array}{r} 2x - 3 \\ x^2 - x - 2 \overline{\smash{\big)}\, 2x^3 - 5x^2 + 3x} \\ \underline{2x^3 - 2x^2 - 4x} \\ -3x^2 + 7x \\ \underline{-3x^2 + 3x + 6} \\ 4x - 6 \end{array}$$

Thus, $y = f(x) = \dfrac{2x^3 - 5x^2 + 3x}{x^2 - x - 2} = 2x - 3 + \dfrac{4x-6}{x^2-x-2}$ and $f(x) - (2x-3) = \dfrac{4x-6}{x^2-x-2} = \dfrac{\frac{4}{x} - \frac{6}{x^2}}{1 - \frac{1}{x} - \frac{1}{x^2}}$

[for $x \ne 0$] $\to \tfrac{0}{1} = 0$ as $x \to \pm\infty$. So the line $y = 2x - 3$ is a slant asymptote (SA).

53. $y = f(x) = \dfrac{x^2}{x-1} = x+1 + \dfrac{1}{x-1}$ **A.** $D = (-\infty, 1) \cup (1, \infty)$ **B.** x-intercept: $f(x) = 0 \Leftrightarrow x = 0$;

y-intercept: $f(0) = 0$ **C.** No symmetry **D.** $\lim\limits_{x \to 1^-} f(x) = -\infty$ and $\lim\limits_{x \to 1^+} f(x) = \infty$, so $x = 1$ is a VA.

$\lim\limits_{x \to \pm\infty} [f(x) - (x+1)] = \lim\limits_{x \to \pm\infty} \dfrac{1}{x-1} = 0$, so the line $y = x+1$ is a SA.

E. $f'(x) = 1 - \dfrac{1}{(x-1)^2} = \dfrac{(x-1)^2 - 1}{(x-1)^2} = \dfrac{x^2 - 2x}{(x-1)^2} = \dfrac{x(x-2)}{(x-1)^2} > 0$ for **H.**

$x < 0$ or $x > 2$, so f is increasing on $(-\infty, 0)$ and $(2, \infty)$, and f is decreasing

on $(0, 1)$ and $(1, 2)$. **F.** Local maximum value $f(0) = 0$, local minimum value

$f(2) = 4$ **G.** $f''(x) = \dfrac{2}{(x-1)^3} > 0$ for $x > 1$, so f is CU on $(1, \infty)$ and f

is CD on $(-\infty, 1)$. No IP

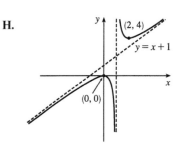

55. $y = f(x) = \dfrac{x^3 + 4}{x^2} = x + \dfrac{4}{x^2}$ **A.** $D = (-\infty, 0) \cup (0, \infty)$ **B.** x-intercept: $f(x) = 0 \Leftrightarrow x = -\sqrt[3]{4}$; no y-intercept

C. No symmetry **D.** $\lim\limits_{x \to 0} f(x) = \infty$, so $x = 0$ is a VA. $\lim\limits_{x \to \pm\infty} [f(x) - x] = \lim\limits_{x \to \pm\infty} \dfrac{4}{x^2} = 0$, so $y = x$ is a SA.

E. $f'(x) = 1 - \dfrac{8}{x^3} = \dfrac{x^3 - 8}{x^3} > 0$ for $x < 0$ or $x > 2$, so f is increasing on **H.**

$(-\infty, 0)$ and $(2, \infty)$, and f is decreasing on $(0, 2)$. **F.** Local minimum value

$f(2) = 3$, no local maximum value **G.** $f''(x) = \dfrac{24}{x^4} > 0$ for $x \ne 0$, so f is CU

on $(-\infty, 0)$ and $(0, \infty)$. No IP

57. $y = f(x) = \dfrac{2x^3 + x^2 + 1}{x^2 + 1} = 2x + 1 + \dfrac{-2x}{x^2 + 1}$ **A.** $D = \mathbb{R}$ **B.** y-intercept: $f(0) = 1$; x-intercept: $f(x) = 0 \Rightarrow$

$0 = 2x^3 + x^2 + 1 = (x+1)(2x^2 - x + 1) \Rightarrow x = -1$ **C.** No symmetry **D.** No VA

$\lim\limits_{x \to \pm\infty} [f(x) - (2x+1)] = \lim\limits_{x \to \pm\infty} \dfrac{-2x}{x^2 + 1} = \lim\limits_{x \to \pm\infty} \dfrac{-2/x}{1 + 1/x^2} = 0$, so the line $y = 2x + 1$ is a slant asymptote.

E. $f'(x) = 2 + \dfrac{(x^2+1)(-2) - (-2x)(2x)}{(x^2+1)^2} = \dfrac{2(x^4 + 2x^2 + 1) - 2x^2 - 2 + 4x^2}{(x^2+1)^2} = \dfrac{2x^4 + 6x^2}{(x^2+1)^2} = \dfrac{2x^2(x^2+3)}{(x^2+1)^2}$

so $f'(x) > 0$ if $x \ne 0$. Thus, f is increasing on $(-\infty, 0)$ and $(0, \infty)$. Since f is continuous at 0, f is increasing on $\mathbb{R}$.

F. No extreme values

G. $f''(x) = \dfrac{(x^2+1)^2 \cdot (8x^3 + 12x) - (2x^4 + 6x^2) \cdot 2(x^2+1)(2x)}{[(x^2+1)^2]^2}$

$= \dfrac{4x(x^2+1)[(x^2+1)(2x^2+3) - 2x^4 - 6x^2]}{(x^2+1)^4} = \dfrac{4x(-x^2+3)}{(x^2+1)^3}$

[continued]

so $f''(x) > 0$ for $x < -\sqrt{3}$ and $0 < x < \sqrt{3}$, and $f''(x) < 0$ for $-\sqrt{3} < x < 0$ and $x > \sqrt{3}$. f is CU on $(-\infty, -\sqrt{3})$ and $(0, \sqrt{3})$, and CD on $(-\sqrt{3}, 0)$ and $(\sqrt{3}, \infty)$. There are three IPs: $(0, 1)$, $(-\sqrt{3}, -\tfrac{3}{2}\sqrt{3} + 1) \approx (-1.73, -1.60)$, and $(\sqrt{3}, \tfrac{3}{2}\sqrt{3} + 1) \approx (1.73, 3.60)$.

H.
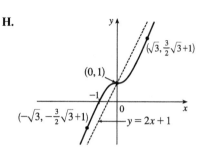

59. $y = f(x) = \sqrt{4x^2 + 9} \;\Rightarrow\; f'(x) = \dfrac{4x}{\sqrt{4x^2 + 9}} \;\Rightarrow\;$

$f''(x) = \dfrac{\sqrt{4x^2+9}\cdot 4 - 4x\cdot 4x/\sqrt{4x^2+9}}{4x^2+9} = \dfrac{4(4x^2+9) - 16x^2}{(4x^2+9)^{3/2}} = \dfrac{36}{(4x^2+9)^{3/2}}$. f is defined on $(-\infty, \infty)$.

$f(-x) = f(x)$, so f is even, which means its graph is symmetric about the y-axis. The y-intercept is $f(0) = 3$. There are no x-intercepts since $f(x) > 0$ for all x.

$\lim\limits_{x\to\infty}\left(\sqrt{4x^2+9} - 2x\right) = \lim\limits_{x\to\infty}\dfrac{\left(\sqrt{4x^2+9} - 2x\right)\left(\sqrt{4x^2+9} + 2x\right)}{\sqrt{4x^2+9} + 2x}$

$= \lim\limits_{x\to\infty}\dfrac{(4x^2+9) - 4x^2}{\sqrt{4x^2+9} + 2x} = \lim\limits_{x\to\infty}\dfrac{9}{\sqrt{4x^2+9} + 2x} = 0$

and, similarly, $\lim\limits_{x\to-\infty}\left(\sqrt{4x^2+9} + 2x\right) = \lim\limits_{x\to-\infty}\dfrac{9}{\sqrt{4x^2+9} - 2x} = 0$,

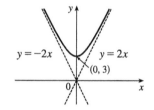

so $y = \pm 2x$ are slant asymptotes. f is decreasing on $(-\infty, 0)$ and increasing on $(0, \infty)$ with local minimum $f(0) = 3$. $f''(x) > 0$ for all x, so f is CU on $\mathbb{R}$.

61. $\dfrac{x^2}{a^2} - \dfrac{y^2}{b^2} = 1 \;\Rightarrow\; y = \pm\dfrac{b}{a}\sqrt{x^2 - a^2}$. Now

$\lim\limits_{x\to\infty}\left[\dfrac{b}{a}\sqrt{x^2 - a^2} - \dfrac{b}{a}x\right] = \dfrac{b}{a}\cdot\lim\limits_{x\to\infty}\left(\sqrt{x^2 - a^2} - x\right)\dfrac{\sqrt{x^2 - a^2} + x}{\sqrt{x^2 - a^2} + x} = \dfrac{b}{a}\cdot\lim\limits_{x\to\infty}\dfrac{-a^2}{\sqrt{x^2 - a^2} + x} = 0$,

which shows that $y = \dfrac{b}{a}x$ is a slant asymptote. Similarly,

$\lim\limits_{x\to\infty}\left[-\dfrac{b}{a}\sqrt{x^2 - a^2} - \left(-\dfrac{b}{a}x\right)\right] = -\dfrac{b}{a}\cdot\lim\limits_{x\to\infty}\dfrac{-a^2}{\sqrt{x^2 - a^2} + x} = 0$, so $y = -\dfrac{b}{a}x$ is a slant asymptote.

63. $\lim\limits_{x\to\pm\infty}\left[f(x) - x^3\right] = \lim\limits_{x\to\pm\infty}\dfrac{x^4 + 1}{x} - \dfrac{x^4}{x} = \lim\limits_{x\to\pm\infty}\dfrac{1}{x} = 0$, so the graph of f is asymptotic to that of $y = x^3$.

A. $D = \{x \mid x \neq 0\}$ **B.** No intercept **C.** f is symmetric about the origin. **D.** $\lim\limits_{x\to 0^-}\left(x^3 + \dfrac{1}{x}\right) = -\infty$ and

$\lim\limits_{x\to 0^+}\left(x^3 + \dfrac{1}{x}\right) = \infty$, so $x = 0$ is a vertical asymptote, and as shown above, the graph of f is asymptotic to that of $y = x^3$.

E. $f'(x) = 3x^2 - 1/x^2 > 0 \;\Leftrightarrow\; x^4 > \tfrac{1}{3} \;\Leftrightarrow\; |x| > \dfrac{1}{\sqrt[4]{3}}$, so f is increasing on $\left(-\infty, -\dfrac{1}{\sqrt[4]{3}}\right)$ and $\left(\dfrac{1}{\sqrt[4]{3}}, \infty\right)$ and decreasing on $\left(-\dfrac{1}{\sqrt[4]{3}}, 0\right)$ and $\left(0, \dfrac{1}{\sqrt[4]{3}}\right)$. **F.** Local maximum value

$f\left(-\dfrac{1}{\sqrt[4]{3}}\right) = -4 \cdot 3^{-5/4}$, local minimum value $f\left(\dfrac{1}{\sqrt[4]{3}}\right) = 4 \cdot 3^{-5/4}$

G. $f''(x) = 6x + 2/x^3 > 0 \Leftrightarrow x > 0$, so f is CU on $(0, \infty)$ and CD on $(-\infty, 0)$. No IP

H.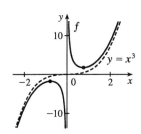

3.6 Graphing with Calculus and Technology

1. $f(x) = x^5 - 5x^4 - x^3 + 28x^2 - 2x \;\Rightarrow\; f'(x) = 5x^4 - 20x^3 - 3x^2 + 56x - 2 \;\Rightarrow\; f''(x) = 20x^3 - 60x^2 - 6x + 56$.
$f(x) = 0 \;\Leftrightarrow\; x = 0$ or $x \approx -2.09, 0.07$; $f'(x) = 0 \;\Leftrightarrow\; x \approx -1.50, 0.04, 2.62, 2.84$; $f''(x) = 0 \;\Leftrightarrow\; x \approx -0.89, 1.15, 2.74$.

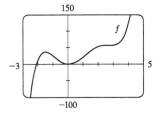 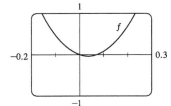 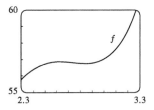

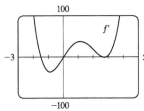 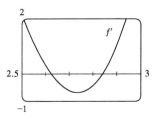

From the graphs of f', we estimate that $f' < 0$ and that f is decreasing on $(-1.50, 0.04)$ and $(2.62, 2.84)$, and that $f' > 0$ and f is increasing on $(-\infty, -1.50)$, $(0.04, 2.62)$, and $(2.84, \infty)$ with local minimum values $f(0.04) \approx -0.04$ and $f(2.84) \approx 56.73$ and local maximum values $f(-1.50) \approx 36.47$ and $f(2.62) \approx 56.83$.

From the graph of f'', we estimate that $f'' > 0$ and that f is CU on $(-0.89, 1.15)$ and $(2.74, \infty)$, and that $f'' < 0$ and f is CD on $(-\infty, -0.89)$ and $(1.15, 2.74)$. There are inflection points at about $(-0.89, 20.90)$, $(1.15, 26.57)$, and $(2.74, 56.78)$.

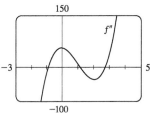

3. $f(x) = x^6 - 5x^5 + 25x^3 - 6x^2 - 48x \;\Rightarrow\;$
$f'(x) = 6x^5 - 25x^4 + 75x^2 - 12x - 48 \;\Rightarrow\;$
$f''(x) = 30x^4 - 100x^3 + 150x - 12$. $f(x) = 0 \;\Leftrightarrow\; x = 0$ or $x \approx 3.20$;
$f'(x) = 0 \;\Leftrightarrow\; x \approx -1.31, -0.84, 1.06, 2.50, 2.75$; $f''(x) = 0 \;\Leftrightarrow\;$
$x \approx -1.10, 0.08, 1.72, 2.64$.

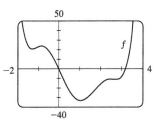

[continued]

From the graph of f', we estimate that f is decreasing on $(-\infty, -1.31)$, increasing on $(-1.31, -0.84)$, decreasing on $(-0.84, 1.06)$, increasing on $(1.06, 2.50)$, decreasing on $(2.50, 2.75)$, and increasing on $(2.75, \infty)$. f has local minimum values $f(-1.31) \approx 20.72$, $f(1.06) \approx -33.12$, and $f(2.75) \approx -11.33$. f has local maximum values $f(-0.84) \approx 23.71$ and $f(2.50) \approx -11.02$.

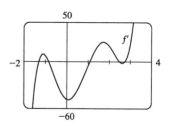

From the graph of f'', we estimate that f is CU on $(-\infty, -1.10)$, CD on $(-1.10, 0.08)$, CU on $(0.08, 1.72)$, CD on $(1.72, 2.64)$, and CU on $(2.64, \infty)$. There are inflection points at about $(-1.10, 22.09)$, $(0.08, -3.88)$, $(1.72, -22.53)$, and $(2.64, -11.18)$.

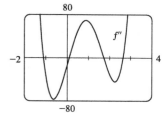

5. $f(x) = \dfrac{x}{x^3 + x^2 + 1} \;\Rightarrow\; f'(x) = -\dfrac{2x^3 + x^2 - 1}{(x^3 + x^2 + 1)^2} \;\Rightarrow\; f''(x) = \dfrac{2x(3x^4 + 3x^3 + x^2 - 6x - 3)}{(x^3 + x^2 + 1)^3}$

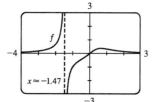 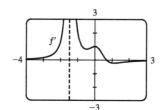 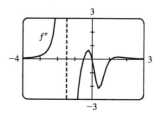

From the graph of f, we see that there is a VA at $x \approx -1.47$. From the graph of f', we estimate that f is increasing on $(-\infty, -1.47)$, increasing on $(-1.47, 0.66)$, and decreasing on $(0.66, \infty)$, with local maximum value $f(0.66) \approx 0.38$.

From the graph of f'', we estimate that f is CU on $(-\infty, -1.47)$, CD on $(-1.47, -0.49)$, CU on $(-0.49, 0)$, CD on $(0, 1.10)$, and CU on $(1.10, \infty)$. There is an inflection point at $(0, 0)$ and at about $(-0.49, -0.44)$ and $(1.10, 0.31)$.

7. $f(x) = 6 \sin x + \cot x$, $-\pi \le x \le \pi \;\Rightarrow\; f'(x) = 6\cos x - \csc^2 x \;\Rightarrow\; f''(x) = -6\sin x + 2\csc^2 x \cot x$

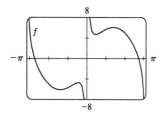 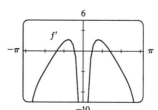 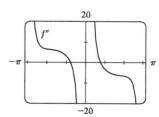

From the graph of f, we see that there are VAs at $x = 0$ and $x = \pm\pi$. f is an odd function, so its graph is symmetric about the origin. From the graph of f', we estimate that f is decreasing on $(-\pi, -1.40)$, increasing on $(-1.40, -0.44)$, decreasing on $(-0.44, 0)$, decreasing on $(0, 0.44)$, increasing on $(0.44, 1.40)$, and decreasing on $(1.40, \pi)$, with local minimum values $f(-1.40) \approx -6.09$ and $f(0.44) \approx 4.68$, and local maximum values $f(-0.44) \approx -4.68$ and $f(1.40) \approx 6.09$.

From the graph of f'', we estimate that f is CU on $(-\pi, -0.77)$, CD on $(-0.77, 0)$, CU on $(0, 0.77)$, and CD on $(0.77, \pi)$. There are IPs at about $(-0.77, -5.22)$ and $(0.77, 5.22)$.

9. $f(x) = 1 + \dfrac{1}{x} + \dfrac{8}{x^2} + \dfrac{1}{x^3}$ $\Rightarrow$ $f'(x) = -\dfrac{1}{x^2} - \dfrac{16}{x^3} - \dfrac{3}{x^4} = -\dfrac{1}{x^4}(x^2 + 16x + 3)$ $\Rightarrow$

$f''(x) = \dfrac{2}{x^3} + \dfrac{48}{x^4} + \dfrac{12}{x^5} = \dfrac{2}{x^5}(x^2 + 24x + 6)$.

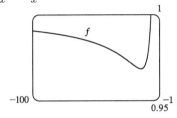
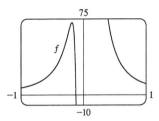

From the graphs, it appears that f increases on $(-15.8, -0.2)$ and decreases on $(-\infty, -15.8)$, $(-0.2, 0)$, and $(0, \infty)$; that f has a local minimum value of $f(-15.8) \approx 0.97$ and a local maximum value of $f(-0.2) \approx 72$; that f is CD on $(-\infty, -24)$ and $(-0.25, 0)$ and is CU on $(-24, -0.25)$ and $(0, \infty)$; and that f has IPs at $(-24, 0.97)$ and $(-0.25, 60)$.

To find the exact values, note that $f' = 0$ $\Rightarrow$ $x = \dfrac{-16 \pm \sqrt{256 - 12}}{2} = -8 \pm \sqrt{61}$ [≈ -0.19 and -15.81].

f' is positive (f is increasing) on $(-8 - \sqrt{61}, -8 + \sqrt{61})$ and f' is negative (f is decreasing) on $(-\infty, -8 - \sqrt{61})$,

$(-8 + \sqrt{61}, 0)$, and $(0, \infty)$. $f'' = 0$ $\Rightarrow$ $x = \dfrac{-24 \pm \sqrt{576 - 24}}{2} = -12 \pm \sqrt{138}$ [≈ -0.25 and -23.75]. f'' is

positive (f is CU) on $(-12 - \sqrt{138}, -12 + \sqrt{138})$ and $(0, \infty)$ and f'' is negative (f is CD) on $(-\infty, -12 - \sqrt{138})$ and $(-12 + \sqrt{138}, 0)$.

11.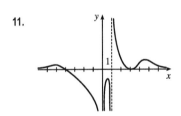

$f(x) = \dfrac{(x+4)(x-3)^2}{x^4(x-1)}$ has VA at $x = 0$ and at $x = 1$ since $\lim_{x \to 0} f(x) = -\infty$,

$\lim_{x \to 1^-} f(x) = -\infty$ and $\lim_{x \to 1^+} f(x) = \infty$.

$f(x) = \dfrac{\dfrac{x+4}{x} \cdot \dfrac{(x-3)^2}{x^2}}{\dfrac{x^4}{x^3} \cdot (x-1)}$ $\begin{bmatrix} \text{dividing numerator} \\ \text{and denominator by } x^3 \end{bmatrix} = \dfrac{(1 + 4/x)(1 - 3/x)^2}{x(x-1)} \to 0$

as $x \to \pm\infty$, so f is asymptotic to the x-axis.

Since f is undefined at $x = 0$, it has no y-intercept. $f(x) = 0$ $\Rightarrow$ $(x+4)(x-3)^2 = 0$ $\Rightarrow$ $x = -4$ or $x = 3$, so f has x-intercepts -4 and 3. Note, however, that the graph of f is only tangent to the x-axis and does not cross it at $x = 3$, since f is positive as $x \to 3^-$ and as $x \to 3^+$.

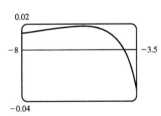
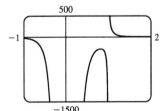
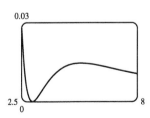

From these graphs, it appears that f has three maximum values and one minimum value. The maximum values are approximately $f(-5.6) = 0.0182$, $f(0.82) = -281.5$ and $f(5.2) = 0.0145$ and we know (since the graph is tangent to the x-axis at $x = 3$) that the minimum value is $f(3) = 0$.

13. $f(x) = \dfrac{x^2(x+1)^3}{(x-2)^2(x-4)^4} \quad \Rightarrow \quad f'(x) = -\dfrac{x(x+1)^2(x^3 + 18x^2 - 44x - 16)}{(x-2)^3(x-4)^5}$ [from CAS].

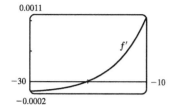 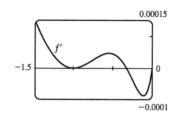 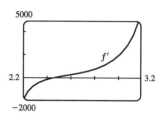

From the graphs of f', it seems that the critical points which indicate extrema occur at $x \approx -20, -0.3$, and 2.5, as estimated in Example 3. (There is another critical point at $x = -1$, but the sign of f' does not change there.) We differentiate again,

obtaining $f''(x) = 2\dfrac{(x+1)(x^6 + 36x^5 + 6x^4 - 628x^3 + 684x^2 + 672x + 64)}{(x-2)^4(x-4)^6}$.

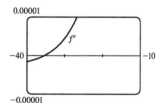 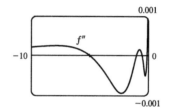 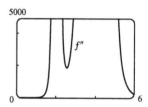

From the graphs of f'', it appears that f is CU on $(-35.3, -5.0)$, $(-1, -0.5)$, $(-0.1, 2)$, $(2, 4)$, and $(4, \infty)$ and CD on $(-\infty, -35.3)$, $(-5.0, -1)$ and $(-0.5, -0.1)$. We check back on the graphs of f to find the y-coordinates of the inflection points, and find that these points are approximately $(-35.3, -0.015)$, $(-5.0, -0.005)$, $(-1, 0)$, $(-0.5, 0.00001)$, and $(-0.1, 0.0000066)$.

15. $f(x) = \dfrac{x^3 + 5x^2 + 1}{x^4 + x^3 - x^2 + 2}$. From a CAS, $f'(x) = \dfrac{-x(x^5 + 10x^4 + 6x^3 + 4x^2 - 3x - 22)}{(x^4 + x^3 - x^2 + 2)^2}$ and

$f''(x) = \dfrac{2(x^9 + 15x^8 + 18x^7 + 21x^6 - 9x^5 - 135x^4 - 76x^3 + 21x^2 + 6x + 22)}{(x^4 + x^3 - x^2 + 2)^3}$

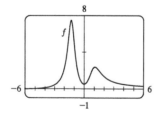 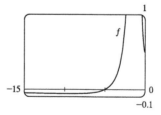

The first graph of f shows that $y = 0$ is a HA. As $x \to \infty$, $f(x) \to 0$ through positive values. As $x \to -\infty$, it is not clear if $f(x) \to 0$ through positive or negative values. The second graph of f shows that f has an x-intercept near -5, and will have a

local minimum and inflection point to the left of -5.

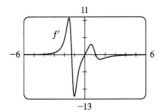 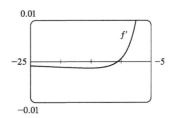

From the two graphs of f', we see that f' has four zeros. We conclude that f is decreasing on $(-\infty, -9.41)$, increasing on $(-9.41, -1.29)$, decreasing on $(-1.29, 0)$, increasing on $(0, 1.05)$, and decreasing on $(1.05, \infty)$. We have local minimum values $f(-9.41) \approx -0.056$ and $f(0) = 0.5$, and local maximum values $f(-1.29) \approx 7.49$ and $f(1.05) \approx 2.35$.

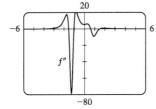 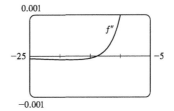

From the two graphs of f'', we see that f'' has five zeros. We conclude that f is CD on $(-\infty, -13.81)$, CU on $(-13.81, -1.55)$, CD on $(-1.55, -1.03)$, CU on $(-1.03, 0.60)$, CD on $(0.60, 1.48)$, and CU on $(1.48, \infty)$. There are five inflection points: $(-13.81, -0.05)$, $(-1.55, 5.64)$, $(-1.03, 5.39)$, $(0.60, 1.52)$, and $(1.48, 1.93)$.

17. $y = f(x) = \sqrt{x + 5\sin x}$, $x \le 20$.

From a CAS, $y' = \dfrac{5\cos x + 1}{2\sqrt{x + 5\sin x}}$ and $y'' = -\dfrac{10\cos x + 25\sin^2 x + 10x\sin x + 26}{4(x + 5\sin x)^{3/2}}$.

We'll start with a graph of $g(x) = x + 5\sin x$. Note that $f(x) = \sqrt{g(x)}$ is only defined if $g(x) \ge 0$. $g(x) = 0 \Leftrightarrow x = 0$ or $x \approx -4.91, -4.10, 4.10,$ and 4.91. Thus, the domain of f is $[-4.91, -4.10] \cup [0, 4.10] \cup [4.91, 20]$.

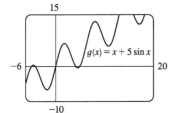 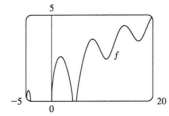 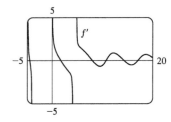

From the expression for y', we see that $y' = 0 \Leftrightarrow 5\cos x + 1 = 0 \Rightarrow x_1 = \cos^{-1}(-\tfrac{1}{5}) \approx 1.77$ and $x_2 = 2\pi - x_1 \approx -4.51$ (not in the domain of f). The leftmost zero of f' is $x_1 - 2\pi \approx -4.51$. Moving to the right, the zeros of f' are $x_1, x_1 + 2\pi, x_2 + 2\pi, x_1 + 4\pi,$ and $x_2 + 4\pi$. Thus, f is increasing on $(-4.91, -4.51)$, decreasing on $(-4.51, -4.10)$, increasing on $(0, 1.77)$, decreasing on $(1.77, 4.10)$, increasing on $(4.91, 8.06)$, decreasing on $(8.06, 10.79)$, increasing on $(10.79, 14.34)$, decreasing on $(14.34, 17.08)$, and increasing on $(17.08, 20)$. The local maximum values are

$f(-4.51) \approx 0.62$, $f(1.77) \approx 2.58$, $f(8.06) \approx 3.60$, and $f(14.34) \approx 4.39$. The local minimum values are $f(10.79) \approx 2.43$ and $f(17.08) \approx 3.49$.

f is CD on $(-4.91, -4.10)$, $(0, 4.10)$, $(4.91, 9.60)$, CU on $(9.60, 12.25)$, CD on $(12.25, 15.81)$, CU on $(15.81, 18.65)$, and CD on $(18.65, 20)$. There are inflection points at $(9.60, 2.95)$, $(12.25, 3.27)$, $(15.81, 3.91)$, and $(18.65, 4.20)$.

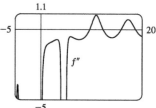

19.

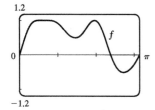 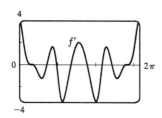 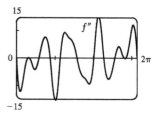

From the graph of $f(x) = \sin(x + \sin 3x)$ in the viewing rectangle $[0, \pi]$ by $[-1.2, 1.2]$, it looks like f has two maxima and two minima. If we calculate and graph $f'(x) = [\cos(x + \sin 3x)](1 + 3\cos 3x)$ on $[0, 2\pi]$, we see that the graph of f' appears to be almost tangent to the x-axis at about $x = 0.7$. The graph of

$$f'' = -[\sin(x + \sin 3x)](1 + 3\cos 3x)^2 + \cos(x + \sin 3x)(-9\sin 3x)$$

is even more interesting near this x-value: it seems to just touch the x-axis.

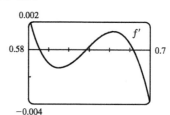

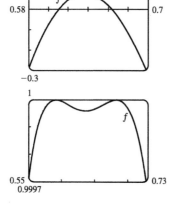 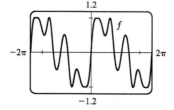

If we zoom in on this place on the graph of f'', we see that f'' actually does cross the axis twice near $x = 0.65$, indicating a change in concavity for a very short interval. If we look at the graph of f' on the same interval, we see that it changes sign three times near $x = 0.65$, indicating that what we had thought was a broad extremum at about $x = 0.7$ actually consists of three extrema (two maxima and a minimum). These maximum values are roughly $f(0.59) = 1$ and $f(0.68) = 1$, and the minimum value is roughly $f(0.64) = 0.99996$. There are also a maximum value of about $f(1.96) = 1$ and minimum values of about $f(1.46) = 0.49$ and $f(2.73) = -0.51$. The points of inflection on $(0, \pi)$ are about $(0.61, 0.99998)$, $(0.66, 0.99998)$, $(1.17, 0.72)$, $(1.75, 0.77)$, and $(2.28, 0.34)$. On $(\pi, 2\pi)$, they are about $(4.01, -0.34)$, $(4.54, -0.77)$, $(5.11, -0.72)$, $(5.62, -0.99998)$, and $(5.67, -0.99998)$. There are also IP at $(0, 0)$ and $(\pi, 0)$. Note that the function is odd and periodic with period 2π, and it is also rotationally symmetric about all points of the form $((2n + 1)\pi, 0)$, n an integer.

21. $f(x) = x^2 + 6x + c/x$ $\Rightarrow$ $f'(x) = 2x + 6 - c/x^2$ $\Rightarrow$ $f''(x) = 2 + 2c/x^3$

$c = 0$: The graph is the parabola $y = x^2 + 6x$, which has x-intercepts -6 and 0, vertex $(-3, -9)$, and opens upward.

$c \neq 0$: The parabola $y = x^2 + 6x$ is an asymptote that the graph of f approaches as $x \to \pm\infty$. The y-axis is a vertical asymptote.

$c < 0$: The x-intercepts are found by solving $f(x) = 0$ $\Leftrightarrow$ $x^3 + 6x^2 + c = g(x) = 0$. Now $g'(x) = 0$ $\Leftrightarrow$ $x = -4$ or 0, and g (not f) has a local maximum at $x = -4$. $g(-4) = 32 + c$, so if $c < -32$, the maximum is negative and there are no negative x-intercepts; if $c = -32$, the maximum is 0 and there is one negative x-intercept; if $-32 < c < 0$, the maximum is positive and there are two negative x-intercepts. In all cases, there is one positive x-intercept.

As $c \to 0^-$, the local minimum point moves down and right, approaching $(-3, -9)$. [Note that since $f'(x) = \dfrac{2x^3 + 6x^2 - c}{x^2}$, Descartes' Rule of Signs implies that f' has no positive solutions and one negative solution when $c < 0$. $f''(x) = \dfrac{2(x^3 + c)}{x^3} > 0$ at that negative solution, so that critical point yields a local minimum value. This tells us that there are no local maximums when $c < 0$.] $f'(x) > 0$ for $x > 0$, so f is increasing on $(0, \infty)$. From $f''(x) = \dfrac{2(x^3 + c)}{x^3}$, we see that f has an inflection point at $\left(\sqrt[3]{-c}, 6\sqrt[3]{-c}\right)$. This inflection point moves down and left, approaching the origin as $c \to 0^-$.

f is CU on $(-\infty, 0)$, CD on $\left(0, \sqrt[3]{-c}\right)$, and CU on $\left(\sqrt[3]{-c}, \infty\right)$.

$c > 0$: The inflection point $\left(\sqrt[3]{-c}, 6\sqrt[3]{-c}\right)$ is now in the third quadrant and moves up and right, approaching the origin as $c \to 0^+$. f is CU on $\left(-\infty, \sqrt[3]{-c}\right)$, CD on $\left(\sqrt[3]{-c}, 0\right)$, and CU on $(0, \infty)$. f has a local minimum point in the first quadrant. It moves down and left, approaching the origin as $c \to 0^+$. $f'(x) = 0$ $\Leftrightarrow$ $2x^3 + 6x^2 - c = h(x) = 0$. Now $h'(x) = 0$ $\Leftrightarrow$ $x = -2$ or 0, and h (not f) has a local maximum at $x = -2$. $h(-2) = 8 - c$, so $c = 8$ makes $h(x) = 0$, and hence, $f'(x) = 0$. When $c > 8$, $f'(x) < 0$ and f is decreasing on $(-\infty, 0)$. For $0 < c < 8$, there is a local minimum that moves toward $(-3, -9)$ and a local maximum that moves toward the origin as c decreases.

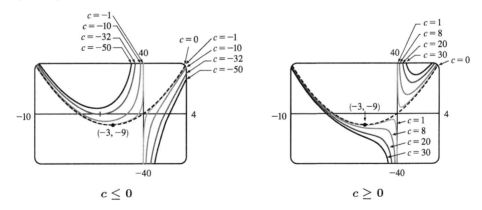

23. Note that $c = 0$ is a transitional value at which the graph consists of the x-axis. Also, we can see that if we substitute $-c$ for c, the function $f(x) = \dfrac{cx}{1 + c^2 x^2}$ will be reflected in the x-axis, so we investigate only positive values of c (except $c = -1$, as

a demonstration of this reflective property). Also, f is an odd function. $\lim_{x \to \pm\infty} f(x) = 0$, so $y = 0$ is a horizontal asymptote

for all c. We calculate $f'(x) = \dfrac{(1+c^2x^2)c - cx(2c^2x)}{(1+c^2x^2)^2} = -\dfrac{c(c^2x^2-1)}{(1+c^2x^2)^2}$. $f'(x) = 0 \Leftrightarrow c^2x^2 - 1 = 0 \Leftrightarrow$

$x = \pm 1/c$. So there is an absolute maximum value of $f(1/c) = \frac{1}{2}$ and an absolute minimum value of $f(-1/c) = -\frac{1}{2}$.

These extrema have the same value regardless of c, but the maximum points move closer to the y-axis as c increases.

$f''(x) = \dfrac{(-2c^3x)(1+c^2x^2)^2 - (-c^3x^2+c)[2(1+c^2x^2)(2c^2x)]}{(1+c^2x^2)^4}$

$= \dfrac{(-2c^3x)(1+c^2x^2) + (c^3x^2-c)(4c^2x)}{(1+c^2x^2)^3} = \dfrac{2c^3x(c^2x^2-3)}{(1+c^2x^2)^3}$

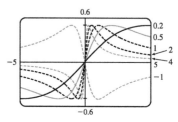

$f''(x) = 0 \Leftrightarrow x = 0$ or $\pm\sqrt{3}/c$, so there are inflection points at $(0, 0)$ and

at $(\pm\sqrt{3}/c, \pm\sqrt{3}/4)$. Again, the y-coordinate of the inflection points does not depend on c, but as c increases, both inflection points approach the y-axis.

25. $f(x) = cx + \sin x \Rightarrow f'(x) = c + \cos x \Rightarrow f''(x) = -\sin x$

$f(-x) = -f(x)$, so f is an odd function and its graph is symmetric with respect to the origin.

$f(x) = 0 \Leftrightarrow \sin x = -cx$, so 0 is always an x-intercept.

$f'(x) = 0 \Leftrightarrow \cos x = -c$, so there is no critical number when $|c| > 1$. If $|c| \le 1$, then there are infinitely many critical numbers. If x_1 is the unique solution of $\cos x = -c$ in the interval $[0, \pi]$, then the critical numbers are $2n\pi \pm x_1$, where n ranges over the integers. (Special cases: When $c = -1$, $x_1 = 0$; when $c = 0$, $x = \frac{\pi}{2}$; and when $c = 1$, $x_1 = \pi$.)

$f''(x) < 0 \Leftrightarrow \sin x > 0$, so f is CD on intervals of the form $(2n\pi, (2n+1)\pi)$. f is CU on intervals of the form $((2n-1)\pi, 2n\pi)$. The inflection points of f are the points $(n\pi, n\pi c)$, where n is an integer.

If $c \ge 1$, then $f'(x) \ge 0$ for all x, so f is increasing and has no extremum. If $c \le -1$, then $f'(x) \le 0$ for all x, so f is decreasing and has no extremum. If $|c| < 1$, then $f'(x) > 0 \Leftrightarrow \cos x > -c \Leftrightarrow x$ is in an interval of the form $(2n\pi - x_1, 2n\pi + x_1)$ for some integer n. These are the intervals on which f is increasing. Similarly, we find that f is decreasing on the intervals of the form $(2n\pi + x_1, 2(n+1)\pi - x_1)$. Thus, f has local maxima at the points $2n\pi + x_1$, where f has the values $c(2n\pi + x_1) + \sin x_1 = c(2n\pi + x_1) + \sqrt{1-c^2}$, and f has local minima at the points $2n\pi - x_1$, where we have $f(2n\pi - x_1) = c(2n\pi - x_1) - \sin x_1 = c(2n\pi - x_1) - \sqrt{1-c^2}$.

The transitional values of c are -1 and 1. The inflection points move vertically, but not horizontally, when c changes.

When $|c| \ge 1$, there is no extremum. For $|c| < 1$, the maxima are spaced 2π apart horizontally, as are the minima. The horizontal spacing between maxima and adjacent minima is regular (and equals π) when $c = 0$, but the horizontal space between a local maximum and the nearest local minimum shrinks as $|c|$ approaches 1.

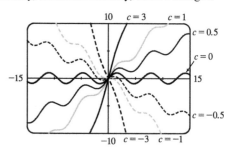

27. For $c = 0$, there is no inflection point; the curve is CU everywhere. If c increases, the curve simply becomes steeper, and there are still no inflection points. If c starts at 0 and decreases, a slight upward bulge appears near $x = 0$, so that there are two inflection points for any $c < 0$. This can be seen algebraically by calculating the second derivative:

$f(x) = x^4 + cx^2 + x \Rightarrow f'(x) = 4x^3 + 2cx + 1 \Rightarrow f''(x) = 12x^2 + 2c$. Thus, $f''(x) > 0$ when $c > 0$. For $c < 0$, there are inflection points when $x = \pm\sqrt{-\frac{1}{6}c}$. For $c = 0$, the graph has one critical number, at the absolute minimum somewhere around $x = -0.6$. As c increases, the number of critical points does not change. If c instead decreases from 0, we see that the graph eventually sprouts another local minimum, to the right of the origin, somewhere between $x = 1$ and $x = 2$. Consequently, there is also a maximum near $x = 0$.

After a bit of experimentation, we find that at $c = -1.5$, there appear to be two critical numbers: the absolute minimum at about $x = -1$, and a horizontal tangent with no extremum at about $x = 0.5$. For any c smaller than this there will be 3 critical points, as shown in the graphs with $c = -3$ and with $c = -5$.

To prove this algebraically, we calculate $f'(x) = 4x^3 + 2cx + 1$. Now if we substitute our value of $c = -1.5$, the formula for $f'(x)$ becomes $4x^3 - 3x + 1 = (x+1)(2x-1)^2$. This has a double solution at $x = \frac{1}{2}$, indicating that the function has two critical points: $x = -1$ and $x = \frac{1}{2}$, just as we had guessed from the graph.

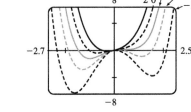

3.7 Optimization Problems

1. (a)

First Number	Second Number	Product
1	22	22
2	21	42
3	20	60
4	19	76
5	18	90
6	17	102
7	16	112
8	15	120
9	14	126
10	13	130
11	12	132

We needn't consider pairs where the first number is larger than the second, since we can just interchange the numbers in such cases. The answer appears to be 11 and 12, but we have considered only integers in the table.

(b) Call the two numbers x and y. Then $x + y = 23$, so $y = 23 - x$. Call the product P. Then

$P = xy = x(23 - x) = 23x - x^2$, so we wish to maximize the function $P(x) = 23x - x^2$. Since $P'(x) = 23 - 2x$, we see that $P'(x) = 0 \Leftrightarrow x = \frac{23}{2} = 11.5$. Thus, the maximum value of P is $P(11.5) = (11.5)^2 = 132.25$ and it occurs when $x = y = 11.5$.

Or: Note that $P''(x) = -2 < 0$ for all x, so P is everywhere concave downward and the local maximum at $x = 11.5$ must be an absolute maximum.

3. The two numbers are x and $\dfrac{100}{x}$, where $x > 0$. Minimize $f(x) = x + \dfrac{100}{x}$. $f'(x) = 1 - \dfrac{100}{x^2} = \dfrac{x^2 - 100}{x^2}$. The critical number is $x = 10$. Since $f'(x) < 0$ for $0 < x < 10$ and $f'(x) > 0$ for $x > 10$, there is an absolute minimum at $x = 10$. The numbers are 10 and 10.

5. Let the vertical distance be given by $v(x) = (x+2) - x^2$, $-1 \leq x \leq 2$.
$v'(x) = 1 - 2x = 0 \Leftrightarrow x = \tfrac{1}{2}$. $v(-1) = 0$, $v(\tfrac{1}{2}) = \tfrac{9}{4}$, and $v(2) = 0$, so there is an absolute maximum at $x = \tfrac{1}{2}$. The maximum distance is
$v(\tfrac{1}{2}) = \tfrac{1}{2} + 2 - \tfrac{1}{4} = \tfrac{9}{4}$.

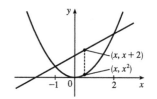

7. If the rectangle has dimensions x and y, then its perimeter is $2x + 2y = 100$ m, so $y = 50 - x$. Thus, the area is
$A = xy = x(50 - x)$. We wish to maximize the function $A(x) = x(50 - x) = 50x - x^2$, where $0 < x < 50$. Since
$A'(x) = 50 - 2x = -2(x - 25)$, $A'(x) > 0$ for $0 < x < 25$ and $A'(x) < 0$ for $25 < x < 50$. Thus, A has an absolute
maximum at $x = 25$, and $A(25) = 25^2 = 625$ m². The dimensions of the rectangle that maximize its area are $x = y = 25$ m.
(The rectangle is a square.)

9. We need to maximize Y for $N \geq 0$. $Y(N) = \dfrac{kN}{1+N^2} \Rightarrow$

$Y'(N) = \dfrac{(1+N^2)k - kN(2N)}{(1+N^2)^2} = \dfrac{k(1-N^2)}{(1+N^2)^2} = \dfrac{k(1+N)(1-N)}{(1+N^2)^2}$. $Y'(N) > 0$ for $0 < N < 1$ and $Y'(N) < 0$

for $N > 1$. Thus, Y has an absolute maximum of $Y(1) = \tfrac{1}{2}k$ at $N = 1$.

11. (a)

 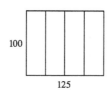 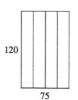

The areas of the three figures are 12,500, 12,500, and 9000 ft². There appears to be a maximum area of at least 12,500 ft².

(b) Let x denote the length of each of two sides and three dividers.

Let y denote the length of the other two sides.

(c) Area $A = $ length $\times$ width $= y \cdot x$

(d) Length of fencing $= 750 \Rightarrow 5x + 2y = 750$

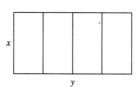

(e) $5x + 2y = 750 \Rightarrow y = 375 - \tfrac{5}{2}x \Rightarrow A(x) = (375 - \tfrac{5}{2}x)x = 375x - \tfrac{5}{2}x^2$

(f) $A'(x) = 375 - 5x = 0 \Rightarrow x = 75$. Since $A''(x) = -5 < 0$ there is an absolute maximum when $x = 75$. Then
$y = \tfrac{375}{2} = 187.5$. The largest area is $75\left(\tfrac{375}{2}\right) = 14{,}062.5$ ft². These values of x and y are between the values in the first
and second figures in part (a). Our original estimate was low.

13. 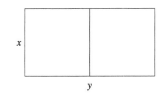 $xy = 1.5 \times 10^6$, so $y = 1.5 \times 10^6/x$. Minimize the amount of fencing, which is
$3x + 2y = 3x + 2(1.5 \times 10^6/x) = 3x + 3 \times 10^6/x = F(x)$.
$F'(x) = 3 - 3 \times 10^6/x^2 = 3(x^2 - 10^6)/x^2$. The critical number is $x = 10^3$ and
$F'(x) < 0$ for $0 < x < 10^3$ and $F'(x) > 0$ if $x > 10^3$, so the absolute minimum
occurs when $x = 10^3$ and $y = 1.5 \times 10^3$.

The field should be 1000 feet by 1500 feet with the middle fence parallel to the short side of the field.

15. 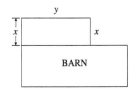 See the figure. The fencing cost $20 per linear foot to install and the cost of the fencing on the west side will be split with the neighbor, so the farmer's cost C will be $C = \frac{1}{2}(20x) + 20y + 20x = 20y + 30x$. The area A will be maximized when $C = 5000$, so $5000 = 20y + 30x \Leftrightarrow 20y = 5000 - 30x \Leftrightarrow$
$y = 250 - \frac{3}{2}x$. Now $A = xy = x\left(250 - \frac{3}{2}x\right) = 250x - \frac{3}{2}x^2 \Rightarrow A' = 250 - 3x$. $A' = 0 \Leftrightarrow x = \frac{250}{3}$ and since $A'' = -3 < 0$, we have a maximum for A when $x = \frac{250}{3}$ ft and $y = 250 - \frac{3}{2}\left(\frac{250}{3}\right) = 125$ ft. [The maximum area is $125\left(\frac{250}{3}\right) = 10{,}416.\overline{6}$ ft^2.]

17. (a) Let the rectangle have sides x and y and area A, so $A = xy$ or $y = A/x$. The problem is to minimize the perimeter $= 2x + 2y = 2x + 2A/x = P(x)$. Now $P'(x) = 2 - 2A/x^2 = 2(x^2 - A)/x^2$. So the critical number is $x = \sqrt{A}$. Since $P'(x) < 0$ for $0 < x < \sqrt{A}$ and $P'(x) > 0$ for $x > \sqrt{A}$, there is an absolute minimum at $x = \sqrt{A}$. The sides of the rectangle are $\sqrt{A}$ and $A/\sqrt{A} = \sqrt{A}$, so the rectangle is a square.

(b) Let p be the perimeter and x and y the lengths of the sides, so $p = 2x + 2y \Rightarrow 2y = p - 2x \Rightarrow y = \frac{1}{2}p - x$.
The area is $A(x) = x\left(\frac{1}{2}p - x\right) = \frac{1}{2}px - x^2$. Now $A'(x) = 0 \Rightarrow \frac{1}{2}p - 2x = 0 \Rightarrow 2x = \frac{1}{2}p \Rightarrow x = \frac{1}{4}p$. Since $A''(x) = -2 < 0$, there is an absolute maximum for A when $x = \frac{1}{4}p$ by the Second Derivative Test. The sides of the rectangle are $\frac{1}{4}p$ and $\frac{1}{2}p - \frac{1}{4}p = \frac{1}{4}p$, so the rectangle is a square.

19. Let b be the length of the base of the box and h the height. The surface area is $1200 = b^2 + 4hb \Rightarrow h = (1200 - b^2)/(4b)$.
The volume is $V = b^2 h = b^2(1200 - b^2)/4b = 300b - b^3/4 \Rightarrow V'(b) = 300 - \frac{3}{4}b^2$.
$V'(b) = 0 \Rightarrow 300 = \frac{3}{4}b^2 \Rightarrow b^2 = 400 \Rightarrow b = \sqrt{400} = 20$. Since $V'(b) > 0$ for $0 < b < 20$ and $V'(b) < 0$ for $b > 20$, there is an absolute maximum when $b = 20$ by the First Derivative Test for Absolute Extreme Values. If $b = 20$, then $h = (1200 - 20^2)/(4 \cdot 20) = 10$, so the largest possible volume is $b^2 h = (20)^2(10) = 4000$ cm^3.

21. 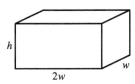 $V = lwh \Rightarrow 10 = (2w)(w)h = 2w^2 h$, so $h = 5/w^2$.
The cost is $10(2w^2) + 6[2(2wh) + 2(hw)] = 20w^2 + 36wh$, so
$C(w) = 20w^2 + 36w(5/w^2) = 20w^2 + 180/w$.

$C'(w) = 40w - 180/w^2 = (40w^3 - 180)/w^2 = 40\left(w^3 - \frac{9}{2}\right)/w^2 \Rightarrow w = \sqrt[3]{\frac{9}{2}}$ is the critical number. There is an absolute minimum for C when $w = \sqrt[3]{\frac{9}{2}}$ since $C'(w) < 0$ for $0 < w < \sqrt[3]{\frac{9}{2}}$ and $C'(w) > 0$ for $w > \sqrt[3]{\frac{9}{2}}$. The minimum cost is $C\left(\sqrt[3]{\frac{9}{2}}\right) = 20\left(\sqrt[3]{\frac{9}{2}}\right)^2 + \dfrac{180}{\sqrt[3]{9/2}} \approx \$163.54$.

23. Let $x > 0$ be the length of the package and $y > 0$ be the length of the sides of the square base. We have $x + 4y = 108 \Rightarrow x = 108 - 4y$. The volume is $V = xy^2 = (108 - 4y)y^2 = 108y^2 - 4y^3$ [for $0 < y < 27$].

$V'(y) = 216y - 12y^2 = 12y(18 - y) = 0 \Rightarrow y = 18$ [since $y \neq 0$]. Since $V'(y) > 0$ for $0 < y < 18$ and $V'(y) < 0$ for $y > 18$, there is an absolute maximum when $y = 18$ by the First Derivative Test for Absolute Extreme Values. If $y = 18$, then $x = 108 - 4(18) = 36$, so the dimensions that give the greatest volume are 18 in $\times$ 18 in $\times$ 36 in, giving a greatest possible volume of 11,664 in^3.

25. The distance d from the origin $(0, 0)$ to a point $(x, 2x + 3)$ on the line is given by $d = \sqrt{(x - 0)^2 + (2x + 3 - 0)^2}$ and the square of the distance is $S = d^2 = x^2 + (2x + 3)^2$. $S' = 2x + 2(2x + 3)2 = 10x + 12$ and $S' = 0 \Leftrightarrow x = -\frac{6}{5}$. Now $S'' = 10 > 0$, so we know that S has a minimum at $x = -\frac{6}{5}$. Thus, the y-value is $2\left(-\frac{6}{5}\right) + 3 = \frac{3}{5}$ and the point is $\left(-\frac{6}{5}, \frac{3}{5}\right)$.

27.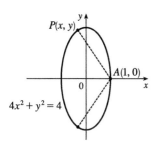

From the figure, we see that there are two points that are farthest away from $A(1, 0)$. The distance d from A to an arbitrary point $P(x, y)$ on the ellipse is $d = \sqrt{(x - 1)^2 + (y - 0)^2}$ and the square of the distance is $S = d^2 = x^2 - 2x + 1 + y^2 = x^2 - 2x + 1 + (4 - 4x^2) = -3x^2 - 2x + 5$. $S' = -6x - 2$ and $S' = 0 \Rightarrow x = -\frac{1}{3}$. Now $S'' = -6 < 0$, so we know that S has a maximum at $x = -\frac{1}{3}$. Since $-1 \le x \le 1$, $S(-1) = 4$,

$S\left(-\frac{1}{3}\right) = \frac{16}{3}$, and $S(1) = 0$, we see that the maximum distance is $\sqrt{\frac{16}{3}}$. The corresponding y-values are

$y = \pm\sqrt{4 - 4\left(-\frac{1}{3}\right)^2} = \pm\sqrt{\frac{32}{9}} = \pm\frac{4}{3}\sqrt{2} \approx \pm 1.89$. The points are $\left(-\frac{1}{3}, \pm\frac{4}{3}\sqrt{2}\right)$.

29.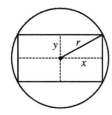

The area of the rectangle is $(2x)(2y) = 4xy$. Also $r^2 = x^2 + y^2$ so $y = \sqrt{r^2 - x^2}$, so the area is $A(x) = 4x\sqrt{r^2 - x^2}$. Now

$A'(x) = 4\left(\sqrt{r^2 - x^2} - \dfrac{x^2}{\sqrt{r^2 - x^2}}\right) = 4\dfrac{r^2 - 2x^2}{\sqrt{r^2 - x^2}}$. The critical number is

$x = \frac{1}{\sqrt{2}}r$. Clearly this gives a maximum.

$y = \sqrt{r^2 - \left(\frac{1}{\sqrt{2}}r\right)^2} = \sqrt{\frac{1}{2}r^2} = \frac{1}{\sqrt{2}}r = x$, which tells us that the rectangle is a square. The dimensions are $2x = \sqrt{2}\,r$ and $2y = \sqrt{2}\,r$.

31. 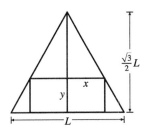 The height h of the equilateral triangle with sides of length L is $\frac{\sqrt{3}}{2}L$, since $h^2 + (L/2)^2 = L^2 \Rightarrow h^2 = L^2 - \frac{1}{4}L^2 = \frac{3}{4}L^2 \Rightarrow$

$h = \frac{\sqrt{3}}{2}L$. Using similar triangles, $\dfrac{\frac{\sqrt{3}}{2}L - y}{x} = \dfrac{\frac{\sqrt{3}}{2}L}{L/2} = \sqrt{3} \Rightarrow$

$\sqrt{3}\,x = \frac{\sqrt{3}}{2}L - y \Rightarrow y = \frac{\sqrt{3}}{2}L - \sqrt{3}\,x \Rightarrow y = \frac{\sqrt{3}}{2}(L - 2x)$.

The area of the inscribed rectangle is $A(x) = (2x)y = \sqrt{3}\,x(L - 2x) = \sqrt{3}\,Lx - 2\sqrt{3}\,x^2$, where $0 \le x \le L/2$. Now $0 = A'(x) = \sqrt{3}\,L - 4\sqrt{3}\,x \Rightarrow x = \sqrt{3}\,L/(4\sqrt{3}) = L/4$. Since $A(0) = A(L/2) = 0$, the maximum occurs when $x = L/4$, and $y = \frac{\sqrt{3}}{2}L - \frac{\sqrt{3}}{4}L = \frac{\sqrt{3}}{4}L$, so the dimensions are $L/2$ and $\frac{\sqrt{3}}{4}L$.

33. 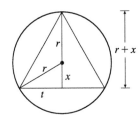 The area of the triangle is
$A(x) = \frac{1}{2}(2t)(r + x) = t(r + x) = \sqrt{r^2 - x^2}\,(r + x)$. Then

$0 = A'(x) = r\dfrac{-2x}{2\sqrt{r^2 - x^2}} + \sqrt{r^2 - x^2} + x\dfrac{-2x}{2\sqrt{r^2 - x^2}}$

$= -\dfrac{x^2 + rx}{\sqrt{r^2 - x^2}} + \sqrt{r^2 - x^2} \Rightarrow$

$\dfrac{x^2 + rx}{\sqrt{r^2 - x^2}} = \sqrt{r^2 - x^2} \Rightarrow x^2 + rx = r^2 - x^2 \Rightarrow 0 = 2x^2 + rx - r^2 = (2x - r)(x + r) \Rightarrow$

$x = \frac{1}{2}r$ or $x = -r$. Now $A(r) = 0 = A(-r) \Rightarrow$ the maximum occurs where $x = \frac{1}{2}r$, so the triangle has height $r + \frac{1}{2}r = \frac{3}{2}r$ and base $2\sqrt{r^2 - \left(\frac{1}{2}r\right)^2} = 2\sqrt{\frac{3}{4}r^2} = \sqrt{3}\,r$.

35. The area of the triangle is $A = \frac{1}{2}a(2a)\sin\theta$ for $0 < \theta < \pi$. $A'(\theta) = a^2\cos\theta = 0 \Rightarrow \cos\theta = 0 \Rightarrow \theta = \frac{\pi}{2}$. Since $A'(\theta) > 0$ for $0 < \theta < \frac{\pi}{2}$ and $A'(\theta) < 0$ for $\frac{\pi}{2} < \theta < \pi$, there is an absolute maximum when $\theta = \frac{\pi}{2}$ by the First Derivative Test for Absolute Extreme Values. (The maximum area of $\frac{1}{2}a(2a)\sin\frac{\pi}{2} = a^2$ results from the triangle being a right triangle.)

37. 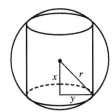 The cylinder has volume $V = \pi y^2(2x)$. Also $x^2 + y^2 = r^2 \Rightarrow y^2 = r^2 - x^2$, so $V(x) = \pi(r^2 - x^2)(2x) = 2\pi(r^2x - x^3)$, where $0 \le x \le r$.

$V'(x) = 2\pi(r^2 - 3x^2) = 0 \Rightarrow x = r/\sqrt{3}$. Now $V(0) = V(r) = 0$, so there is a maximum when $x = r/\sqrt{3}$ and $V\left(r/\sqrt{3}\right) = \pi(r^2 - r^2/3)(2r/\sqrt{3}) = 4\pi r^3/(3\sqrt{3})$.

39. 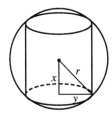 The cylinder has surface area

$2(\text{area of the base}) + (\text{lateral surface area}) = 2\pi(\text{radius})^2 + 2\pi(\text{radius})(\text{height})$
$= 2\pi y^2 + 2\pi y(2x)$

Now $x^2 + y^2 = r^2 \Rightarrow y^2 = r^2 - x^2 \Rightarrow y = \sqrt{r^2 - x^2}$, so the surface area is

$S(x) = 2\pi(r^2 - x^2) + 4\pi x\sqrt{r^2 - x^2}$, $0 \le x \le r$
$= 2\pi r^2 - 2\pi x^2 + 4\pi\left(x\sqrt{r^2 - x^2}\right)$

[continued]

Thus,
$$S'(x) = 0 - 4\pi x + 4\pi\left[x \cdot \tfrac{1}{2}(r^2 - x^2)^{-1/2}(-2x) + (r^2 - x^2)^{1/2} \cdot 1\right]$$
$$= 4\pi\left[-x - \frac{x^2}{\sqrt{r^2 - x^2}} + \sqrt{r^2 - x^2}\right] = 4\pi \cdot \frac{-x\sqrt{r^2 - x^2} - x^2 + r^2 - x^2}{\sqrt{r^2 - x^2}}$$

$S'(x) = 0 \Rightarrow x\sqrt{r^2 - x^2} = r^2 - 2x^2$ $(\star)$ $\Rightarrow \left(x\sqrt{r^2 - x^2}\right)^2 = (r^2 - 2x^2)^2 \Rightarrow$
$x^2(r^2 - x^2) = r^4 - 4r^2x^2 + 4x^4 \Rightarrow r^2x^2 - x^4 = r^4 - 4r^2x^2 + 4x^4 \Rightarrow 5x^4 - 5r^2x^2 + r^4 = 0$.

This is a quadratic equation in x^2. By the quadratic formula, $x^2 = \frac{5 \pm \sqrt{5}}{10}r^2$, but we reject the solution with the $+$ sign since it doesn't satisfy $(\star)$. [The right side is negative and the left side is positive.] So $x = \sqrt{\frac{5 - \sqrt{5}}{10}}\, r$. Since $S(0) = S(r) = 0$, the maximum surface area occurs at the critical number and $x^2 = \frac{5 - \sqrt{5}}{10}r^2 \Rightarrow y^2 = r^2 - \frac{5 - \sqrt{5}}{10}r^2 = \frac{5 + \sqrt{5}}{10}r^2 \Rightarrow$ the surface area is

$$2\pi\left(\tfrac{5 + \sqrt{5}}{10}\right)r^2 + 4\pi\sqrt{\tfrac{5 - \sqrt{5}}{10}}\sqrt{\tfrac{5 + \sqrt{5}}{10}}r^2 = \pi r^2\left[2 \cdot \tfrac{5 + \sqrt{5}}{10} + 4\tfrac{\sqrt{(5 - \sqrt{5})(5 + \sqrt{5})}}{10}\right] = \pi r^2\left[\tfrac{5 + \sqrt{5}}{5} + \tfrac{2\sqrt{20}}{5}\right]$$
$$= \pi r^2\left[\tfrac{5 + \sqrt{5} + 2 \cdot 2\sqrt{5}}{5}\right] = \pi r^2\left[\tfrac{5 + 5\sqrt{5}}{5}\right] = \pi r^2(1 + \sqrt{5}).$$

41.

$xy = 384 \Rightarrow y = 384/x$. Total area is
$A(x) = (8 + x)(12 + 384/x) = 12(40 + x + 256/x)$, so
$A'(x) = 12(1 - 256/x^2) = 0 \Rightarrow x = 16$. There is an absolute minimum when $x = 16$ since $A'(x) < 0$ for $0 < x < 16$ and $A'(x) > 0$ for $x > 16$.
When $x = 16$, $y = 384/16 = 24$, so the dimensions are 24 cm and 36 cm.

43. Let x be the length of the wire used for the square. The total area is
$$A(x) = \left(\tfrac{x}{4}\right)^2 + \tfrac{1}{2}\left(\tfrac{10 - x}{3}\right)\tfrac{\sqrt{3}}{2}\left(\tfrac{10 - x}{3}\right)$$
$$= \tfrac{1}{16}x^2 + \tfrac{\sqrt{3}}{36}(10 - x)^2, \; 0 \le x \le 10$$

$A'(x) = \tfrac{1}{8}x - \tfrac{\sqrt{3}}{18}(10 - x) = 0 \; \Leftrightarrow \; \tfrac{9}{72}x + \tfrac{4\sqrt{3}}{72}x - \tfrac{40\sqrt{3}}{72} = 0 \; \Leftrightarrow \; x = \tfrac{40\sqrt{3}}{9 + 4\sqrt{3}}$.

Now $A(0) = \left(\tfrac{\sqrt{3}}{36}\right)100 \approx 4.81$, $A(10) = \tfrac{100}{16} = 6.25$ and $A\left(\tfrac{40\sqrt{3}}{9 + 4\sqrt{3}}\right) \approx 2.72$, so

(a) The maximum area occurs when $x = 10$ m, and all the wire is used for the square.

(b) The minimum area occurs when $x = \tfrac{40\sqrt{3}}{9 + 4\sqrt{3}} \approx 4.35$ m.

45. From the figure, the perimeter of the slice is $2r + r\theta = 32$, so $\theta = \dfrac{32 - 2r}{r}$. The area A of the slice is $A = \tfrac{1}{2}r^2\theta = \tfrac{1}{2}r^2\left(\dfrac{32 - 2r}{r}\right) = r(16 - r) = 16r - r^2$ for $0 \le r \le 16$. $A'(r) = 16 - 2r$, so $A' = 0$ when $r = 8$. Since $A(0) = 0$, $A(16) = 0$, and $A(8) = 64$ in.^{2}, the largest piece comes from a pizza with radius 8 in. and diameter 16 in. Note that $\theta = 2$ radians $\approx 114.6°$, which is about 32% of the whole pizza.

47.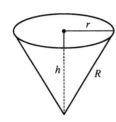

$h^2 + r^2 = R^2 \Rightarrow V = \frac{\pi}{3}r^2h = \frac{\pi}{3}(R^2 - h^2)h = \frac{\pi}{3}(R^2h - h^3)$.

$V'(h) = \frac{\pi}{3}(R^2 - 3h^2) = 0$ when $h = \frac{1}{\sqrt{3}}R$. This gives an absolute maximum, since

$V'(h) > 0$ for $0 < h < \frac{1}{\sqrt{3}}R$ and $V'(h) < 0$ for $h > \frac{1}{\sqrt{3}}R$. The maximum volume is

$V\left(\frac{1}{\sqrt{3}}R\right) = \frac{\pi}{3}\left(\frac{1}{\sqrt{3}}R^3 - \frac{1}{3\sqrt{3}}R^3\right) = \frac{2}{9\sqrt{3}}\pi R^3$.

49.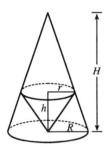

By similar triangles, $\dfrac{H}{R} = \dfrac{H-h}{r}$ (1). The volume of the inner cone is $V = \frac{1}{3}\pi r^2 h$,

so we'll solve (1) for h. $\dfrac{Hr}{R} = H - h \Rightarrow$

$h = H - \dfrac{Hr}{R} = \dfrac{HR - Hr}{R} = \dfrac{H}{R}(R - r)$ (2).

Thus, $V(r) = \dfrac{\pi}{3}r^2 \cdot \dfrac{H}{R}(R - r) = \dfrac{\pi H}{3R}(Rr^2 - r^3) \Rightarrow$

$V'(r) = \dfrac{\pi H}{3R}(2Rr - 3r^2) = \dfrac{\pi H}{3R}r(2R - 3r)$.

$V'(r) = 0 \Rightarrow r = 0$ or $2R = 3r \Rightarrow r = \frac{2}{3}R$ and from (2), $h = \dfrac{H}{R}\left(R - \frac{2}{3}R\right) = \dfrac{H}{R}\left(\frac{1}{3}R\right) = \frac{1}{3}H$.

$V'(r)$ changes from positive to negative at $r = \frac{2}{3}R$, so the inner cone has a maximum volume of

$V = \frac{1}{3}\pi r^2 h = \frac{1}{3}\pi\left(\frac{2}{3}R\right)^2\left(\frac{1}{3}H\right) = \frac{4}{27} \cdot \frac{1}{3}\pi R^2 H$, which is approximately 15% of the volume of the larger cone.

51. $P(R) = \dfrac{E^2 R}{(R+r)^2} \Rightarrow$

$P'(R) = \dfrac{(R+r)^2 \cdot E^2 - E^2 R \cdot 2(R+r)}{[(R+r)^2]^2} = \dfrac{(R^2 + 2Rr + r^2)E^2 - 2E^2R^2 - 2E^2Rr}{(R+r)^4}$

$= \dfrac{E^2 r^2 - E^2 R^2}{(R+r)^4} = \dfrac{E^2(r^2 - R^2)}{(R+r)^4} = \dfrac{E^2(r+R)(r-R)}{(R+r)^4} = \dfrac{E^2(r-R)}{(R+r)^3}$

$P'(R) = 0 \Rightarrow R = r \Rightarrow P(r) = \dfrac{E^2 r}{(r+r)^2} = \dfrac{E^2 r}{4r^2} = \dfrac{E^2}{4r}$.

The expression for $P'(R)$ shows that $P'(R) > 0$ for $R < r$ and $P'(R) < 0$ for $R > r$. Thus, the maximum value of the power is $E^2/(4r)$, and this occurs when $R = r$.

53. $S = 6sh - \frac{3}{2}s^2 \cot\theta + \left(\frac{3}{2}\sqrt{3}\,s^2\right)\csc\theta$

(a) $\dfrac{dS}{d\theta} = \frac{3}{2}s^2 \csc^2\theta - \left(\frac{3}{2}\sqrt{3}\,s^2\right)\csc\theta\cot\theta$ or $\frac{3}{2}s^2\csc\theta\left(\csc\theta - \sqrt{3}\cot\theta\right)$.

(b) $\dfrac{dS}{d\theta} = 0$ when $\csc\theta - \sqrt{3}\cot\theta = 0 \Rightarrow \dfrac{1}{\sin\theta} - \sqrt{3}\,\dfrac{\cos\theta}{\sin\theta} = 0 \Rightarrow \cos\theta = \frac{1}{\sqrt{3}}$. The First Derivative Test shows that the minimum surface area occurs when $\theta = \cos^{-1}\left(\frac{1}{\sqrt{3}}\right) \approx 55°$.

(c)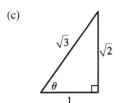

If $\cos\theta = \frac{1}{\sqrt{3}}$, then $\cot\theta = \frac{1}{\sqrt{2}}$ and $\csc\theta = \frac{\sqrt{3}}{\sqrt{2}}$, so the surface area is

$S = 6sh - \frac{3}{2}s^2\,\frac{1}{\sqrt{2}} + 3s^2\,\frac{\sqrt{3}}{2}\,\frac{\sqrt{3}}{\sqrt{2}} = 6sh - \dfrac{3}{2\sqrt{2}}s^2 + \dfrac{9}{2\sqrt{2}}s^2$

$= 6sh + \dfrac{6}{2\sqrt{2}}s^2 = 6s\left(h + \dfrac{1}{2\sqrt{2}}s\right)$

55. Here $T(x) = \dfrac{\sqrt{x^2+25}}{6} + \dfrac{5-x}{8}$, $0 \le x \le 5 \Rightarrow T'(x) = \dfrac{x}{6\sqrt{x^2+25}} - \dfrac{1}{8} = 0 \Leftrightarrow 8x = 6\sqrt{x^2+25} \Leftrightarrow$ $16x^2 = 9(x^2+25) \Leftrightarrow x = \dfrac{15}{\sqrt{7}}$. But $\dfrac{15}{\sqrt{7}} > 5$, so T has no critical number. Since $T(0) \approx 1.46$ and $T(5) \approx 1.18$, she should row directly to B.

57. There are $(6-x)$ km over land and $\sqrt{x^2+4}$ km under the river. We need to minimize the cost C (measured in $100,000) of the pipeline.

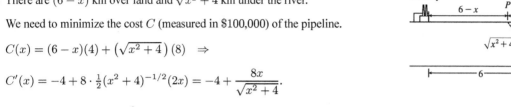

$C(x) = (6-x)(4) + \left(\sqrt{x^2+4}\right)(8) \Rightarrow$

$C'(x) = -4 + 8 \cdot \tfrac{1}{2}(x^2+4)^{-1/2}(2x) = -4 + \dfrac{8x}{\sqrt{x^2+4}}.$

$C'(x) = 0 \Rightarrow 4 = \dfrac{8x}{\sqrt{x^2+4}} \Rightarrow \sqrt{x^2+4} = 2x \Rightarrow x^2+4 = 4x^2 \Rightarrow 4 = 3x^2 \Rightarrow x^2 = \tfrac{4}{3} \Rightarrow$

$x = 2/\sqrt{3}$ $[0 \le x \le 6]$. Compare the costs for $x = 0, 2/\sqrt{3}$, and 6. $C(0) = 24 + 16 = 40$,

$C(2/\sqrt{3}) = 24 - 8/\sqrt{3} + 32/\sqrt{3} = 24 + 24/\sqrt{3} \approx 37.9$, and $C(6) = 0 + 8\sqrt{40} \approx 50.6$. So the minimum cost is about $3.79 million when P is $6 - 2/\sqrt{3} \approx 4.85$ km east of the refinery.

59.

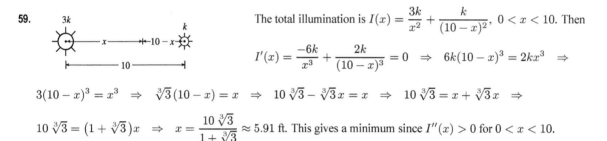

The total illumination is $I(x) = \dfrac{3k}{x^2} + \dfrac{k}{(10-x)^2}$, $0 < x < 10$. Then

$I'(x) = \dfrac{-6k}{x^3} + \dfrac{2k}{(10-x)^3} = 0 \Rightarrow 6k(10-x)^3 = 2kx^3 \Rightarrow$

$3(10-x)^3 = x^3 \Rightarrow \sqrt[3]{3}(10-x) = x \Rightarrow 10\sqrt[3]{3} - \sqrt[3]{3}\,x = x \Rightarrow 10\sqrt[3]{3} = x + \sqrt[3]{3}\,x \Rightarrow$

$10\sqrt[3]{3} = (1+\sqrt[3]{3})x \Rightarrow x = \dfrac{10\sqrt[3]{3}}{1+\sqrt[3]{3}} \approx 5.91$ ft. This gives a minimum since $I''(x) > 0$ for $0 < x < 10$.

61. Every line segment in the first quadrant passing through (a,b) with endpoints on the x- and y-axes satisfies an equation of the form $y - b = m(x-a)$, where $m < 0$. By setting $x = 0$ and then $y = 0$, we find its endpoints, $A(0, b-am)$ and $B\left(a - \dfrac{b}{m}, 0\right)$. The distance d from A to B is given by $d = \sqrt{\left[\left(a - \dfrac{b}{m}\right) - 0\right]^2 + [0 - (b-am)]^2}$.

It follows that the square of the length of the line segment, as a function of m, is given by

$S(m) = \left(a - \dfrac{b}{m}\right)^2 + (am-b)^2 = a^2 - \dfrac{2ab}{m} + \dfrac{b^2}{m^2} + a^2m^2 - 2abm + b^2.$ Thus,

$S'(m) = \dfrac{2ab}{m^2} - \dfrac{2b^2}{m^3} + 2a^2m - 2ab = \dfrac{2}{m^3}(abm - b^2 + a^2m^4 - abm^3)$

$= \dfrac{2}{m^3}[b(am-b) + am^3(am-b)] = \dfrac{2}{m^3}(am-b)(b+am^3)$

Thus, $S'(m) = 0 \Leftrightarrow m = b/a$ or $m = -\sqrt[3]{\dfrac{b}{a}}$. Since $b/a > 0$ and $m < 0$, m must equal $-\sqrt[3]{\dfrac{b}{a}}$. Since $\dfrac{2}{m^3} < 0$, we see that $S'(m) < 0$ for $m < -\sqrt[3]{\dfrac{b}{a}}$ and $S'(m) > 0$ for $m > -\sqrt[3]{\dfrac{b}{a}}$. Thus, S has its absolute minimum value when $m = -\sqrt[3]{\dfrac{b}{a}}$.

[continued]

That value is

$$S\left(-\sqrt[3]{\frac{b}{a}}\right) = \left(a + b\sqrt[3]{\frac{a}{b}}\right)^2 + \left(-a\sqrt[3]{\frac{b}{a}} - b\right)^2 = \left(a + \sqrt[3]{ab^2}\right)^2 + \left(\sqrt[3]{a^2b} + b\right)^2$$

$$= a^2 + 2a^{4/3}b^{2/3} + a^{2/3}b^{4/3} + a^{4/3}b^{2/3} + 2a^{2/3}b^{4/3} + b^2 = a^2 + 3a^{4/3}b^{2/3} + 3a^{2/3}b^{4/3} + b^2$$

The last expression is of the form $x^3 + 3x^2y + 3xy^2 + y^3$ $[= (x+y)^3]$ with $x = a^{2/3}$ and $y = b^{2/3}$,
so we can write it as $(a^{2/3} + b^{2/3})^3$ and the shortest such line segment has length $\sqrt{S} = (a^{2/3} + b^{2/3})^{3/2}$.

63. $y = \dfrac{3}{x}$ $\Rightarrow$ $y' = -\dfrac{3}{x^2}$, so an equation of the tangent line at the point $\left(a, \dfrac{3}{a}\right)$ is

$y - \dfrac{3}{a} = -\dfrac{3}{a^2}(x - a)$, or $y = -\dfrac{3}{a^2}x + \dfrac{6}{a}$. The y-intercept $[x = 0]$ is $6/a$. The

x-intercept $[y = 0]$ is $2a$. The distance d of the line segment that has endpoints at the

intercepts is $d = \sqrt{(2a - 0)^2 + (0 - 6/a)^2}$. Let $S = d^2$, so $S = 4a^2 + \dfrac{36}{a^2}$ $\Rightarrow$

$S' = 8a - \dfrac{72}{a^3}$. $S' = 0$ $\Leftrightarrow$ $\dfrac{72}{a^3} = 8a$ $\Leftrightarrow$ $a^4 = 9$ $\Leftrightarrow$ $a^2 = 3$ $\Rightarrow$ $a = \sqrt{3}$.

$S'' = 8 + \dfrac{216}{a^4} > 0$, so there is an absolute minimum at $a = \sqrt{3}$. Thus, $S = 4(3) + \dfrac{36}{3} = 12 + 12 = 24$ and
hence, $d = \sqrt{24} = 2\sqrt{6}$.

65. (a) If $c(x) = \dfrac{C(x)}{x}$, then, by the Quotient Rule, we have $c'(x) = \dfrac{xC'(x) - C(x)}{x^2}$. Now $c'(x) = 0$ when

$xC'(x) - C(x) = 0$ and this gives $C'(x) = \dfrac{C(x)}{x} = c(x)$. Therefore, the marginal cost equals the average cost.

(b) (i) $C(x) = 16{,}000 + 200x + 4x^{3/2}$, $C(1000) = 16{,}000 + 200{,}000 + 40{,}000\sqrt{10} \approx 216{,}000 + 126{,}491$, so

$C(1000) \approx \$342{,}491$. $c(x) = C(x)/x = \dfrac{16{,}000}{x} + 200 + 4x^{1/2}$, $c(1000) \approx \$342.49/\text{unit}$. $C'(x) = 200 + 6x^{1/2}$,

$C'(1000) = 200 + 60\sqrt{10} \approx \$389.74/\text{unit}$.

(ii) We must have $C'(x) = c(x)$ $\Leftrightarrow$ $200 + 6x^{1/2} = \dfrac{16{,}000}{x} + 200 + 4x^{1/2}$ $\Leftrightarrow$ $2x^{3/2} = 16{,}000$ $\Leftrightarrow$

$x = (8{,}000)^{2/3} = 400$ units. To check that this is a minimum, we calculate

$c'(x) = \dfrac{-16{,}000}{x^2} + \dfrac{2}{\sqrt{x}} = \dfrac{2}{x^2}(x^{3/2} - 8000)$. This is negative for $x < (8000)^{2/3} = 400$, zero at $x = 400$,

and positive for $x > 400$, so c is decreasing on $(0, 400)$ and increasing on $(400, \infty)$. Thus, c has an absolute minimum
at $x = 400$. [Note: $c''(x)$ is *not* positive for all $x > 0$.]

(iii) The minimum average cost is $c(400) = 40 + 200 + 80 = \$320/\text{unit}$.

67. (a) We are given that the demand function p is linear and $p(27{,}000) = 10$, $p(33{,}000) = 8$, so the slope is

$\dfrac{10 - 8}{27{,}000 - 33{,}000} = -\dfrac{1}{3000}$ and an equation of the line is $y - 10 = \left(-\dfrac{1}{3000}\right)(x - 27{,}000)$ $\Rightarrow$

$y = p(x) = -\dfrac{1}{3000}x + 19 = 19 - (x/3000)$.

(b) The revenue is $R(x) = xp(x) = 19x - (x^2/3000)$ $\Rightarrow$ $R'(x) = 19 - (x/1500) = 0$ when $x = 28{,}500$. Since $R''(x) = -1/1500 < 0$, the maximum revenue occurs when $x = 28{,}500$ $\Rightarrow$ the price is $p(28{,}500) = \$9.50$.

69. (a) As in Example 6, we see that the demand function p is linear. We are given that $p(1200) = 350$ and deduce that $p(1280) = 340$, since a \$10 reduction in price increases sales by 80 per week. The slope for p is $\dfrac{340 - 350}{1280 - 1200} = -\dfrac{1}{8}$, so an equation is $p - 350 = -\tfrac{1}{8}(x - 1200)$ or $p(x) = -\tfrac{1}{8}x + 500$, where $x \geq 1200$.

(b) $R(x) = x\, p(x) = -\tfrac{1}{8}x^2 + 500x$. $R'(x) = -\tfrac{1}{4}x + 500 = 0$ when $x = 4(500) = 2000$. $p(2000) = 250$, so the price should be set at \$250 to maximize revenue.

(c) $C(x) = 35{,}000 + 120x$ $\Rightarrow$ $P(x) = R(x) - C(x) = -\tfrac{1}{8}x^2 + 500x - 35{,}000 - 120x = -\tfrac{1}{8}x^2 + 380x - 35{,}000$. $P'(x) = -\tfrac{1}{4}x + 380 = 0$ when $x = 4(380) = 1520$. $p(1520) = 310$, so the price should be set at \$310 to maximize profit.

71. 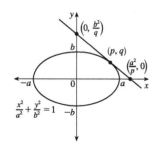 Here $s^2 = h^2 + b^2/4$, so $h^2 = s^2 - b^2/4$. The area is $A = \tfrac{1}{2}b\sqrt{s^2 - b^2/4}$.

Let the perimeter be p, so $2s + b = p$ or $s = (p-b)/2$ $\Rightarrow$
$A(b) = \tfrac{1}{2}b\sqrt{(p-b)^2/4 - b^2/4} = b\sqrt{p^2 - 2pb}/4$. Now
$$A'(b) = \frac{\sqrt{p^2 - 2pb}}{4} - \frac{bp/4}{\sqrt{p^2 - 2pb}} = \frac{-3pb + p^2}{4\sqrt{p^2 - 2pb}}.$$

Therefore, $A'(b) = 0$ $\Rightarrow$ $-3pb + p^2 = 0$ $\Rightarrow$ $b = p/3$. Since $A'(b) > 0$ for $b < p/3$ and $A'(b) < 0$ for $b > p/3$, there is an absolute maximum when $b = p/3$. But then $2s + p/3 = p$, so $s = p/3$ $\Rightarrow$ $s = b$ $\Rightarrow$ the triangle is equilateral.

73. (a) Using implicit differentiation, $\dfrac{x^2}{a^2} + \dfrac{y^2}{b^2} = 1$ $\Rightarrow$ $\dfrac{2x}{a^2} + \dfrac{2y\, y'}{b^2} = 0$ $\Rightarrow$
$\dfrac{2y\, y'}{b^2} = -\dfrac{2x}{a^2}$ $\Rightarrow$ $y' = -\dfrac{b^2 x}{a^2 y}$. At (p, q), $y' = -\dfrac{b^2 p}{a^2 q}$, and an equation of the
tangent line is $y - q = -\dfrac{b^2 p}{a^2 q}(x - p)$ $\Leftrightarrow$ $y = -\dfrac{b^2 p}{a^2 q}x + \dfrac{b^2 p^2}{a^2 q} + q$ $\Leftrightarrow$
$y = -\dfrac{b^2 p}{a^2 q}x + \dfrac{b^2 p^2 + a^2 q^2}{a^2 q}$. The last term is the y-intercept, but not the term we

want, namely b^2/q. Since (p, q) is on the ellipse, we know $\dfrac{p^2}{a^2} + \dfrac{q^2}{b^2} = 1$. To use that relationship we must divide $b^2 p^2$ in

the y-intercept by $a^2 b^2$, so divide all terms by $a^2 b^2$. $\dfrac{(b^2 p^2 + a^2 q^2)/a^2 b^2}{(a^2 q)/a^2 b^2} = \dfrac{p^2/a^2 + q^2/b^2}{q/b^2} = \dfrac{1}{q/b^2} = \dfrac{b^2}{q}$. So the

tangent line has equation $y = -\dfrac{b^2 p}{a^2 q}x + \dfrac{b^2}{q}$. Let $y = 0$ and solve for x to find that x-intercept: $\dfrac{b^2 p}{a^2 q}x = \dfrac{b^2}{q}$ $\Leftrightarrow$
$x = \dfrac{b^2 a^2 q}{q b^2 p} = \dfrac{a^2}{p}$.

(b) The portion of the tangent line cut off by the coordinate axes is the distance between the intercepts, $(a^2/p, 0)$ and $(0, b^2/q)$: $\sqrt{\left(\dfrac{a^2}{p}\right)^2 + \left(-\dfrac{b^2}{q}\right)^2} = \sqrt{\dfrac{a^4}{p^2} + \dfrac{b^4}{q^2}}$. To eliminate p or q, we turn to the relationship $\dfrac{p^2}{a^2} + \dfrac{q^2}{b^2} = 1 \Leftrightarrow$

$\dfrac{q^2}{b^2} = 1 - \dfrac{p^2}{a^2} \Leftrightarrow q^2 = b^2 - \dfrac{b^2 p^2}{a^2} \Leftrightarrow q^2 = \dfrac{b^2(a^2 - p^2)}{a^2}$. Now substitute for q^2 and use the square S of the

distance. $S(p) = \dfrac{a^4}{p^2} + \dfrac{b^4 a^2}{b^2(a^2 - p^2)} = \dfrac{a^4}{p^2} + \dfrac{a^2 b^2}{a^2 - p^2}$ for $0 < p < a$. Note that as $p \to 0$ or $p \to a$, $S(p) \to \infty$,

so the minimum value of S must occur at a critical number. Now $S'(p) = -\dfrac{2a^4}{p^3} + \dfrac{2a^2 b^2 p}{(a^2 - p^2)^2}$ and $S'(p) = 0 \Leftrightarrow$

$\dfrac{2a^4}{p^3} = \dfrac{2a^2 b^2 p}{(a^2 - p^2)^2} \Leftrightarrow a^2(a^2 - p^2)^2 = b^2 p^4 \Rightarrow a(a^2 - p^2) = bp^2 \Leftrightarrow a^3 = (a + b)p^2 \Leftrightarrow p^2 = \dfrac{a^3}{a + b}$.

Substitute for p^2 in $S(p)$:

$$\dfrac{a^4}{\dfrac{a^3}{a+b}} + \dfrac{a^2 b^2}{a^2 - \dfrac{a^3}{a+b}} = \dfrac{a^4(a+b)}{a^3} + \dfrac{a^2 b^2(a+b)}{a^2(a+b) - a^3} = \dfrac{a(a+b)}{1} + \dfrac{a^2 b^2(a+b)}{a^2 b}$$

$$= a(a+b) + b(a+b) = (a+b)(a+b) = (a+b)^2$$

Taking the square root gives us the desired minimum length of $a + b$.

(c) The triangle formed by the tangent line and the coordinate axes has area $A = \dfrac{1}{2}\left(\dfrac{a^2}{p}\right)\left(\dfrac{b^2}{q}\right)$. As in part (b), we'll use the

square of the area and substitute for q^2. $S = \dfrac{a^4 b^4}{4p^2 q^2} = \dfrac{a^4 b^4 a^2}{4p^2 b^2(a^2 - p^2)} = \dfrac{a^6 b^2}{4p^2(a^2 - p^2)}$. Minimizing S (and hence A)

is equivalent to maximizing $p^2(a^2 - p^2)$. Let $f(p) = p^2(a^2 - p^2) = a^2 p^2 - p^4$ for $0 < p < a$. As in part (b), the

minimum value of S must occur at a critical number. Now $f'(p) = 2a^2 p - 4p^3 = 2p(a^2 - 2p^2)$. $f'(p) = 0 \Rightarrow$

$p^2 = a^2/2 \Rightarrow p = a/\sqrt{2}$ $[p > 0]$. Substitute for p^2 in $S(p)$: $\dfrac{a^6 b^2}{4\left(\dfrac{a^2}{2}\right)\left(a^2 - \dfrac{a^2}{2}\right)} = \dfrac{a^6 b^2}{a^4} = a^2 b^2 = (ab)^2$.

Taking the square root gives us the desired minimum area of ab.

75. Note that $|AD| = |AP| + |PD| \Rightarrow 5 = x + |PD| \Rightarrow |PD| = 5 - x$.

Using the Pythagorean Theorem for $\triangle PDB$ and $\triangle PDC$ gives us

$L(x) = |AP| + |BP| + |CP| = x + \sqrt{(5-x)^2 + 2^2} + \sqrt{(5-x)^2 + 3^2}$

$= x + \sqrt{x^2 - 10x + 29} + \sqrt{x^2 - 10x + 34} \Rightarrow$

$L'(x) = 1 + \dfrac{x - 5}{\sqrt{x^2 - 10x + 29}} + \dfrac{x - 5}{\sqrt{x^2 - 10x + 34}}$. From the graphs of L

and L', it seems that the minimum value of L is about $L(3.59) = 9.35$ m.

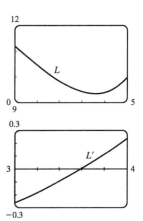

77.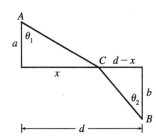

The total time is

$$T(x) = \text{(time from } A \text{ to } C) + \text{(time from } C \text{ to } B)$$
$$= \frac{\sqrt{a^2 + x^2}}{v_1} + \frac{\sqrt{b^2 + (d-x)^2}}{v_2}, \quad 0 < x < d$$

$$T'(x) = \frac{x}{v_1\sqrt{a^2 + x^2}} - \frac{d-x}{v_2\sqrt{b^2 + (d-x)^2}} = \frac{\sin\theta_1}{v_1} - \frac{\sin\theta_2}{v_2}$$

The minimum occurs when $T'(x) = 0 \Rightarrow \dfrac{\sin\theta_1}{v_1} = \dfrac{\sin\theta_2}{v_2}$, or,

equivalently, $\dfrac{\sin\theta_1}{\sin\theta_2} = \dfrac{v_1}{v_2}$. [Note: $T''(x) > 0$]

79.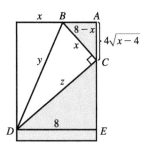

$y^2 = x^2 + z^2$, but triangles CDE and BCA are similar, so

$z/8 = x/(4\sqrt{x-4}) \Rightarrow z = 2x/\sqrt{x-4}$. Thus, we minimize

$f(x) = y^2 = x^2 + 4x^2/(x-4) = x^3/(x-4), \ 4 < x \leq 8$.

$f'(x) = \dfrac{(x-4)(3x^2) - x^3}{(x-4)^2} = \dfrac{x^2[3(x-4) - x]}{(x-4)^2} = \dfrac{2x^2(x-6)}{(x-4)^2} = 0$

when $x = 6$. $f'(x) < 0$ when $x < 6$, $f'(x) > 0$ when $x > 6$, so the minimum

occurs when $x = 6$ in.

81.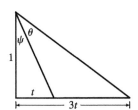

It suffices to maximize $\tan\theta$. Now

$\dfrac{3t}{1} = \tan(\psi + \theta) = \dfrac{\tan\psi + \tan\theta}{1 - \tan\psi\tan\theta} = \dfrac{t + \tan\theta}{1 - t\tan\theta}$. So

$3t(1 - t\tan\theta) = t + \tan\theta \Rightarrow 2t = (1 + 3t^2)\tan\theta \Rightarrow \tan\theta = \dfrac{2t}{1 + 3t^2}$.

Let $f(t) = \tan\theta = \dfrac{2t}{1 + 3t^2} \Rightarrow f'(t) = \dfrac{2(1 + 3t^2) - 2t(6t)}{(1 + 3t^2)^2} = \dfrac{2(1 - 3t^2)}{(1 + 3t^2)^2} = 0 \Leftrightarrow 1 - 3t^2 = 0 \Leftrightarrow$

$t = \dfrac{1}{\sqrt{3}}$ since $t \geq 0$. Now $f'(t) > 0$ for $0 \leq t < \dfrac{1}{\sqrt{3}}$ and $f'(t) < 0$ for $t > \dfrac{1}{\sqrt{3}}$, so f has an absolute maximum when $t = \dfrac{1}{\sqrt{3}}$

and $\tan\theta = \dfrac{2(1/\sqrt{3})}{1 + 3(1/\sqrt{3})^2} = \dfrac{1}{\sqrt{3}} \Rightarrow \theta = \dfrac{\pi}{6}$. Substituting for t and θ in $3t = \tan(\psi + \theta)$ gives us

$\sqrt{3} = \tan(\psi + \dfrac{\pi}{6}) \Rightarrow \psi = \dfrac{\pi}{6}$.

83.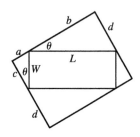

In the small triangle with sides a and c and hypotenuse W, $\sin\theta = \dfrac{a}{W}$ and

$\cos\theta = \dfrac{c}{W}$. In the triangle with sides b and d and hypotenuse L, $\sin\theta = \dfrac{d}{L}$ and

$\cos\theta = \dfrac{b}{L}$. Thus, $a = W\sin\theta$, $c = W\cos\theta$, $d = L\sin\theta$, and $b = L\cos\theta$, so the

area of the circumscribed rectangle is

$$A(\theta) = (a+b)(c+d) = (W\sin\theta + L\cos\theta)(W\cos\theta + L\sin\theta)$$
$$= W^2 \sin\theta \cos\theta + WL\sin^2\theta + LW\cos^2\theta + L^2 \sin\theta \cos\theta$$
$$= LW \sin^2\theta + LW \cos^2\theta + (L^2 + W^2) \sin\theta \cos\theta$$
$$= LW(\sin^2\theta + \cos^2\theta) + (L^2 + W^2) \cdot \tfrac{1}{2} \cdot 2\sin\theta \cos\theta = LW + \tfrac{1}{2}(L^2 + W^2)\sin 2\theta, \ 0 \le \theta \le \tfrac{\pi}{2}$$

This expression shows, without calculus, that the maximum value of $A(\theta)$ occurs when $\sin 2\theta = 1 \ \Leftrightarrow \ 2\theta = \tfrac{\pi}{2} \ \Rightarrow \ \theta = \tfrac{\pi}{4}$. So the maximum area is $A\left(\tfrac{\pi}{4}\right) = LW + \tfrac{1}{2}(L^2 + W^2) = \tfrac{1}{2}(L^2 + 2LW + W^2) = \tfrac{1}{2}(L+W)^2$.

85. (a)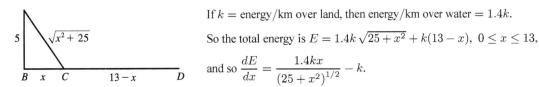

If k = energy/km over land, then energy/km over water = $1.4k$.
So the total energy is $E = 1.4k\sqrt{25 + x^2} + k(13 - x), \ 0 \le x \le 13$,
and so $\dfrac{dE}{dx} = \dfrac{1.4kx}{(25 + x^2)^{1/2}} - k$.

Set $\dfrac{dE}{dx} = 0$: $1.4kx = k(25 + x^2)^{1/2} \ \Rightarrow \ 1.96x^2 = x^2 + 25 \ \Rightarrow \ 0.96x^2 = 25 \ \Rightarrow \ x = \dfrac{5}{\sqrt{0.96}} \approx 5.1$.

Testing against the value of E at the endpoints: $E(0) = 1.4k(5) + 13k = 20k$, $E(5.1) \approx 17.9k$, $E(13) \approx 19.5k$.
Thus, to minimize energy, the bird should fly to a point about 5.1 km from B.

(b) If W/L is large, the bird would fly to a point C that is closer to B than to D to minimize the energy used flying over water.
If W/L is small, the bird would fly to a point C that is closer to D than to B to minimize the distance of the flight.

$E = W\sqrt{25 + x^2} + L(13 - x) \ \Rightarrow \ \dfrac{dE}{dx} = \dfrac{Wx}{\sqrt{25 + x^2}} - L = 0$ when $\dfrac{W}{L} = \dfrac{\sqrt{25 + x^2}}{x}$. By the same sort of

argument as in part (a), this ratio will give the minimal expenditure of energy if the bird heads for the point x km from B.

(c) For flight direct to D, $x = 13$, so from part (b), $W/L = \dfrac{\sqrt{25 + 13^2}}{13} \approx 1.07$. There is no value of W/L for which the bird should fly directly to B. But note that $\lim\limits_{x \to 0^+}(W/L) = \infty$, so if the point at which E is a minimum is close to B, then W/L is large.

(d) Assuming that the birds instinctively choose the path that minimizes the energy expenditure, we can use the equation for $dE/dx = 0$ from part (a) with $1.4k = c$, $x = 4$, and $k = 1$: $c(4) = 1 \cdot (25 + 4^2)^{1/2} \ \Rightarrow \ c = \sqrt{41}/4 \approx 1.6$.

3.8 Newton's Method

1. (a)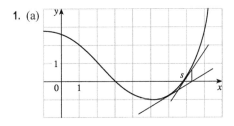

The tangent line at $x_1 = 6$ intersects the x-axis at $x \approx 7.3$, so $x_2 = 7.3$.
The tangent line at $x = 7.3$ intersects the x-axis at $x \approx 6.8$, so $x_3 \approx 6.8$.

(b) $x_1 = 8$ would be a better first approximation because the tangent line at $x = 8$ intersects the x-axis closer to s than does the first approximation $x_1 = 6$.

3. Since the tangent line $y = 9 - 2x$ is tangent to the curve $y = f(x)$ at the point $(2, 5)$, we have $x_1 = 2$, $f(x_1) = 5$, and $f'(x_1) = -2$ [the slope of the tangent line]. Thus, by Equation 2,

$$x_2 = x_1 - \frac{f(x_1)}{f'(x_1)} = 2 - \frac{5}{-2} = \frac{9}{2}$$

Note that geometrically $\frac{9}{2}$ represents the x-intercept of the tangent line $y = 9 - 2x$.

5. The initial approximations $x_1 = a, b$, and c will work, resulting in a second approximation closer to the origin, and lead to the solution of the equation $f(x) = 0$, namely, $x = 0$. The initial approximation $x_1 = d$ will not work because it will result in successive approximations farther and farther from the origin.

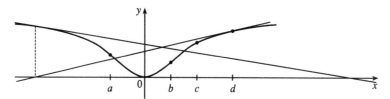

7. $f(x) = \dfrac{2}{x} - x^2 + 1 \;\Rightarrow\; f'(x) = -\dfrac{2}{x^2} - 2x$, so $x_{n+1} = x_n - \dfrac{2/x_n - x_n^2 + 1}{-2/x_n^2 - 2x_n}$. Now $x_1 = 2 \;\Rightarrow$

$x_2 = 2 - \dfrac{1 - 4 + 1}{-1/2 - 4} = 2 - \dfrac{-2}{-9/2} = \dfrac{14}{9} \;\Rightarrow\; x_3 = \dfrac{14}{9} - \dfrac{2/(14/9) - (14/9)^2 + 1}{-2(14/9)^2 - 2(14/9)} \approx 1.5215$.

9. $f(x) = x^3 + x + 3 \;\Rightarrow\; f'(x) = 3x^2 + 1$, so $x_{n+1} = x_n - \dfrac{x_n^3 + x_n + 3}{3x_n^2 + 1}$.

Now $x_1 = -1 \;\Rightarrow$

$x_2 = -1 - \dfrac{(-1)^3 + (-1) + 3}{3(-1)^2 + 1} = -1 - \dfrac{-1 - 1 + 3}{3 + 1} = -1 - \dfrac{1}{4} = -1.25$.

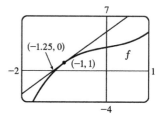

Newton's method follows the tangent line at $(-1, 1)$ down to its intersection with the x-axis at $(-1.25, 0)$, giving the second approximation $x_2 = -1.25$.

11. To approximate $x = \sqrt[4]{75}$ (so that $x^4 = 75$), we can take $f(x) = x^4 - 75$. So $f'(x) = 4x^3$, and thus,

$x_{n+1} = x_n - \dfrac{x_n^4 - 75}{4x_n^3}$. Since $\sqrt[4]{81} = 3$ and 81 is reasonably close to 75, we'll use $x_1 = 3$. We need to find approximations

until they agree to eight decimal places. $x_1 = 3 \;\Rightarrow\; x_2 = 2.9\overline{4}$, $x_3 \approx 2.94283228$, $x_4 \approx 2.94283096 \approx x_5$. So $\sqrt[4]{75} \approx 2.94283096$, to eight decimal places.

To use Newton's method on a calculator, assign f to Y_1 and f' to Y_2. Then store x_1 in X and enter $X - Y_1/Y_2 \to X$ to get x_2 and further approximations (repeatedly press ENTER).

13. (a) Let $f(x) = 3x^4 - 8x^3 + 2$. The polynomial f is continuous on $[2, 3]$, $f(2) = -14 < 0$, and $f(3) = 29 > 0$, so by the Intermediate Value Theorem, there is a number c in $(2, 3)$ such that $f(c) = 0$. In other words, the equation $3x^4 - 8x^3 + 2 = 0$ has a solution in $[2, 3]$.

(b) $f'(x) = 12x^3 - 24x^2 \Rightarrow x_{n+1} = x_n - \dfrac{3x_n^4 - 8x_n^3 + 2}{12x_n^3 - 24x_n^2}$. Taking $x_1 = 2.5$, we get $x_2 = 2.655$, $x_3 \approx 2.630725$, $x_4 \approx 2.630021$, $x_5 \approx 2.630020 \approx x_6$. To six decimal places, the solution is 2.630020. Note that taking $x_1 = 2$ is not allowed since $f'(2) = 0$.

15.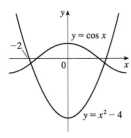

From the graph, we see that the negative solution of $\cos x = x^2 - 4$ is near $x = -2$.
Solving $\cos x = x^2 - 4$ is the same as solving $f(x) = \cos x - x^2 + 4 = 0$.
$f'(x) = -\sin x - 2x$, so $x_{n+1} = x_n - \dfrac{\cos x_n - x_n^2 + 4}{-\sin x_n - 2x_n}$. $x_1 = -2 \Rightarrow$
$x_2 \approx -1.915233$, $x_3 \approx -1.914021 \approx x_4$. Thus, the negative solution is -1.914021, to six decimal places.

17.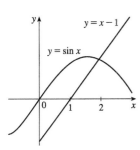

From the graph, we see that there appears to be a point of intersection near $x = 2$.
Solving $\sin x = x - 1$ is the same as solving $f(x) = \sin x - x + 1 = 0$.
$f'(x) = \cos x - 1$, so $x_{n+1} = x_n - \dfrac{\sin x_n - x_n + 1}{\cos x_n - 1}$. $x_1 = 2 \Rightarrow$
$x_2 \approx 1.935951$, $x_3 \approx 1.934564$, $x_4 \approx 1.934563 \approx x_5$. Thus, the solution is 1.934563, to six decimal places.

19.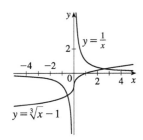

From the graph, we see that there appear to be points of intersection near
$x = -0.5$ and $x = 2.5$. Solving $\dfrac{1}{x} = \sqrt[3]{x} - 1$ is the same as solving
$f(x) = \dfrac{1}{x} - \sqrt[3]{x} + 1 = 0$. $f'(x) = -\dfrac{1}{x^2} - \dfrac{1}{3x^{2/3}}$, so
$x_{n+1} = x_n - \dfrac{1/x_n - \sqrt[3]{x_n} + 1}{-1/x_n^2 - 1/(3x_n^{2/3})}$.

$x_1 = -0.5$	$x_1 = 2.5$
$x_2 \approx -0.545549$	$x_2 \approx 2.625502$
$x_3 \approx -0.549672$	$x_3 \approx 2.629654$
$x_4 \approx -0.549700 \approx x_5$	$x_4 \approx 2.629658 \approx x_5$

To six decimal places, the roots of the equation are -0.549700 and 2.629658.

21.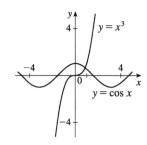
From the graph, we see that there appears to be a point of intersection near $x = 1$. Solving $x^3 = \cos x$ is the same as solving $f(x) = x^3 - \cos x = 0$.
$f'(x) = 3x^2 + \sin x$, so $x_{n+1} = x_n - \dfrac{x_n^3 - \cos x_n}{3x_n^2 + \sin x_n}$.

$x_1 = 1 \Rightarrow x_2 \approx 0.880333$, $x_3 \approx 0.865684$, $x_4 = 0.865474 \approx x_5$.

To six decimal places, the only solution is 0.865474.

23.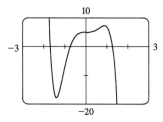
$f(x) = -2x^7 - 5x^4 + 9x^3 + 5 \Rightarrow f'(x) = -14x^6 - 20x^3 + 27x^2 \Rightarrow$
$x_{n+1} = x_n - \dfrac{-2x_n^7 - 5x_n^4 + 9x_n^3 + 5}{-14x_n^6 - 20x_n^3 + 27x_n^2}$.

From the graph of f, there appear to be solutions near -1.7, -0.7, and 1.3.

$x_1 = -1.7$	$x_1 = -0.7$	$x_1 = 1.3$
$x_2 = -1.693255$	$x_2 \approx -0.74756345$	$x_2 = 1.268776$
$x_3 \approx -1.69312035$	$x_3 \approx -0.74467752$	$x_3 \approx 1.26589387$
$x_4 \approx -1.69312029 \approx x_5$	$x_4 \approx -0.74466668 \approx x_5$	$x_4 \approx 1.26587094 \approx x_5$

To eight decimal places, the solutions of the equation are -1.69312029, -0.74466668, and 1.26587094.

25.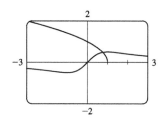
Solving $\dfrac{x}{x^2+1} = \sqrt{1-x}$ is the same as solving

$f(x) = \dfrac{x}{x^2+1} - \sqrt{1-x} = 0$. $f'(x) = \dfrac{1-x^2}{(x^2+1)^2} + \dfrac{1}{2\sqrt{1-x}} \Rightarrow$

$x_{n+1} = x_n - \dfrac{\dfrac{x_n}{x_n^2+1} - \sqrt{1-x_n}}{\dfrac{1-x_n^2}{(x_n^2+1)^2} + \dfrac{1}{2\sqrt{1-x_n}}}$.

From the graph, we see that the curves intersect at about 0.8. $x_1 = 0.8 \Rightarrow x_2 \approx 0.76757581$, $x_3 \approx 0.76682610$, $x_4 \approx 0.76682579 \approx x_5$. To eight decimal places, the solution of the equation is 0.76682579.

27. (a) $f(x) = x^2 - a \Rightarrow f'(x) = 2x$, so Newton's method gives

$x_{n+1} = x_n - \dfrac{x_n^2 - a}{2x_n} = x_n - \dfrac{1}{2}x_n + \dfrac{a}{2x_n} = \dfrac{1}{2}x_n + \dfrac{a}{2x_n} = \dfrac{1}{2}\left(x_n + \dfrac{a}{x_n}\right)$.

(b) Using (a) with $a = 1000$ and $x_1 = \sqrt{900} = 30$, we get $x_2 \approx 31.666667$, $x_3 \approx 31.622807$, and $x_4 \approx 31.622777 \approx x_5$. So $\sqrt{1000} \approx 31.622777$.

29. $f(x) = x^3 - 3x + 6 \Rightarrow f'(x) = 3x^2 - 3$. If $x_1 = 1$, then $f'(x_1) = 0$ and the tangent line used for approximating x_2 is horizontal. Attempting to find x_2 results in trying to divide by zero.

31. For $f(x) = x^{1/3}$, $f'(x) = \frac{1}{3}x^{-2/3}$ and

$$x_{n+1} = x_n - \frac{f(x_n)}{f'(x_n)} = x_n - \frac{x_n^{1/3}}{\frac{1}{3}x_n^{-2/3}} = x_n - 3x_n = -2x_n.$$

Therefore, each successive approximation becomes twice as large as the previous one in absolute value, so the sequence of approximations fails to converge to the solution, which is 0. In the figure, we have $x_1 = 0.5$, $x_2 = -2(0.5) = -1$, and $x_3 = -2(-1) = 2$.

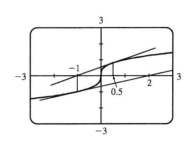

33. (a) $f(x) = x^6 - x^4 + 3x^3 - 2x \Rightarrow f'(x) = 6x^5 - 4x^3 + 9x^2 - 2 \Rightarrow$
$f''(x) = 30x^4 - 12x^2 + 18x$. To find the critical numbers of f, we'll find the zeros of f'. From the graph of f', it appears there are zeros at approximately $x = -1.3, -0.4,$ and 0.5. Try $x_1 = -1.3 \Rightarrow$

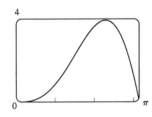

$$x_2 = x_1 - \frac{f'(x_1)}{f''(x_1)} \approx -1.293344 \Rightarrow x_3 \approx -1.293227 \approx x_4.$$

Now try $x_1 = -0.4 \Rightarrow x_2 \approx -0.443755 \Rightarrow x_3 \approx -0.441735 \Rightarrow x_4 \approx -0.441731 \approx x_5$. Finally try $x_1 = 0.5 \Rightarrow x_2 \approx 0.507937 \Rightarrow x_3 \approx 0.507854 \approx x_4$. Therefore, $x = -1.293227, -0.441731,$ and 0.507854 are all the critical numbers correct to six decimal places.

(b) There are two critical numbers where f' changes from negative to positive, so f changes from decreasing to increasing. $f(-1.293227) \approx -2.0212$ and $f(0.507854) \approx -0.6721$, so -2.0212 is the absolute minimum value of f correct to four decimal places.

35.

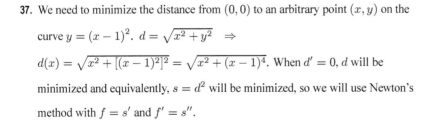

From the graph of $y = x^2 \sin x$, we see that $x = 1.5$ is a reasonable guess for the x-coordinate of the inflection point. Using Newton's method with $g(x) = y''$ and $g'(x) = y'''$, we get $x_1 = 1.5 \Rightarrow x_2 \approx 1.520092, x_3 \approx 1.519855 \approx x_4$. The inflection point is about $(1.519855, 2.306964)$.

37. We need to minimize the distance from $(0,0)$ to an arbitrary point (x, y) on the curve $y = (x-1)^2$. $d = \sqrt{x^2 + y^2} \Rightarrow$

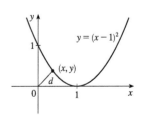

$d(x) = \sqrt{x^2 + [(x-1)^2]^2} = \sqrt{x^2 + (x-1)^4}$. When $d' = 0$, d will be minimized and equivalently, $s = d^2$ will be minimized, so we will use Newton's method with $f = s'$ and $f' = s''$.

[continued]

$f(x) = 2x + 4(x-1)^3 \Rightarrow f'(x) = 2 + 12(x-1)^2$, so $x_{n+1} = x_n - \dfrac{2x_n + 4(x_n-1)^3}{2 + 12(x_n-1)^2}$. Try $x_1 = 0.5 \Rightarrow$

$x_2 = 0.4$, $x_3 \approx 0.410127$, $x_4 \approx 0.410245 \approx x_5$. Now $d(0.410245) \approx 0.537841$ is the minimum distance and the point on the parabola is $(0.410245, 0.347810)$, correct to six decimal places.

39. In this case, $A = 18{,}000$, $R = 375$, and $n = 5(12) = 60$. So the formula $A = \dfrac{R}{i}[1 - (1+i)^{-n}]$ becomes

$18{,}000 = \dfrac{375}{x}[1 - (1+x)^{-60}] \Leftrightarrow 48x = 1 - (1+x)^{-60}$ [multiply each term by $(1+x)^{60}$] $\Leftrightarrow$

$48x(1+x)^{60} - (1+x)^{60} + 1 = 0$. Let the LHS be called $f(x)$, so that

$$f'(x) = 48x(60)(1+x)^{59} + 48(1+x)^{60} - 60(1+x)^{59}$$
$$= 12(1+x)^{59}[4x(60) + 4(1+x) - 5] = 12(1+x)^{59}(244x - 1)$$

$x_{n+1} = x_n - \dfrac{48x_n(1+x_n)^{60} - (1+x_n)^{60} + 1}{12(1+x_n)^{59}(244x_n - 1)}$. An interest rate of 1% per month seems like a reasonable estimate for

$x = i$. So let $x_1 = 1\% = 0.01$, and we get $x_2 \approx 0.0082202$, $x_3 \approx 0.0076802$, $x_4 \approx 0.0076291$, $x_5 \approx 0.0076286 \approx x_6$. Thus, the dealer is charging a monthly interest rate of 0.76286% (or 9.55% per year, compounded monthly).

3.9 Antiderivatives

1. (a) $f(x) = 6 \Rightarrow F(x) = 6x$ is an antiderivative.

   (b) $g(t) = 3t^2 \Rightarrow G(t) = 3\dfrac{t^{2+1}}{2+1} = t^3$ is an antiderivative.

3. (a) $h(q) = \cos q \Rightarrow H(q) = \sin q$ is an antiderivative.

   (b) $f(x) = \sec x \tan x \Rightarrow F(x) = \sec x$ is an antiderivative.

5. $f(x) = 4x + 7 = 4x^1 + 7 \Rightarrow F(x) = 4\dfrac{x^{1+1}}{1+1} + 7x + C = 2x^2 + 7x + C$

Check: $F'(x) = 2(2x) + 7 + 0 = 4x + 7 = f(x)$

7. $f(x) = 2x^3 - \tfrac{2}{3}x^2 + 5x \Rightarrow F(x) = 2\dfrac{x^{3+1}}{3+1} - \dfrac{2}{3}\dfrac{x^{2+1}}{2+1} + 5\dfrac{x^{1+1}}{1+1} = \tfrac{1}{2}x^4 - \tfrac{2}{9}x^3 + \tfrac{5}{2}x^2 + C$

Check: $F'(x) = \tfrac{1}{2}(4x^3) - \tfrac{2}{9}(3x^2) + \tfrac{5}{2}(2x) + 0 = 2x^3 - \tfrac{2}{3}x^2 + 5x = f(x)$

9. $f(x) = x(12x + 8) = 12x^2 + 8x \Rightarrow F(x) = 12\dfrac{x^3}{3} + 8\dfrac{x^2}{2} + C = 4x^3 + 4x^2 + C$

11. $g(x) = 4x^{-2/3} - 2x^{5/3} \Rightarrow G(x) = 4(3x^{1/3}) - 2\left(\tfrac{3}{8}x^{8/3}\right) + C = 12x^{1/3} - \tfrac{3}{4}x^{8/3} + C$

13. $f(x) = 3\sqrt{x} - 2\sqrt[3]{x} = 3x^{1/2} - 2x^{1/3} \Rightarrow F(x) = 3\left(\tfrac{2}{3}x^{3/2}\right) - 2\left(\tfrac{3}{4}x^{4/3}\right) + C = 2x^{3/2} - \tfrac{3}{2}x^{4/3} + C$

15. $f(t) = \dfrac{2t - 4 + 3\sqrt{t}}{\sqrt{t}} = 2t^{1/2} - 4t^{-1/2} + 3 \Rightarrow F(t) = 2\dfrac{t^{3/2}}{3/2} - 4\dfrac{t^{1/2}}{1/2} + 3t + C = \tfrac{4}{3}t^{3/2} - 8\sqrt{t} + 3t + C$

17. $f(x) = \dfrac{10}{x^9} = 10x^{-9}$ has domain $(-\infty, 0) \cup (0, \infty)$, so $F(x) = \begin{cases} \dfrac{10x^{-8}}{-8} + C_1 = -\dfrac{5}{4x^8} + C_1 & \text{if } x < 0 \\ -\dfrac{5}{4x^8} + C_2 & \text{if } x > 0 \end{cases}$

See Example 1(c) for a similar problem.

19. $f(\theta) = 2\sin\theta - 3\sec\theta\tan\theta \;\Rightarrow\; F(\theta) = -2\cos\theta - 3\sec\theta + C_n$ on the interval $\left(n\pi - \tfrac{\pi}{2}, n\pi + \tfrac{\pi}{2}\right)$, n an integer.

21. $h(\theta) = 2\sin\theta - \sec^2\theta \;\Rightarrow\; H(\theta) = -2\cos\theta - \tan\theta + C_n$ on the interval $\left(n\pi - \tfrac{\pi}{2}, n\pi + \tfrac{\pi}{2}\right)$, n an integer.

23. $g(v) = \sqrt[3]{v^2} - 2\sec^2 v \;\Rightarrow\; G(v) = \tfrac{3}{5}v^{5/3} - 2\tan v + C_n$ on $\left(n\pi - \tfrac{\pi}{2}, n\pi + \tfrac{\pi}{2}\right)$, n an integer.

25. $f(x) = 5x^4 - 2x^5 \;\Rightarrow\; F(x) = 5 \cdot \dfrac{x^5}{5} - 2 \cdot \dfrac{x^6}{6} + C = x^5 - \tfrac{1}{3}x^6 + C$.

$F(0) = 4 \;\Rightarrow\; 0^5 - \tfrac{1}{3} \cdot 0^6 + C = 4 \;\Rightarrow\; C = 4$, so $F(x) = x^5 - \tfrac{1}{3}x^6 + 4$.

The graph confirms our answer since $f(x) = 0$ when F has a local maximum, f is positive when F is increasing, and f is negative when F is decreasing.

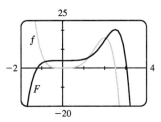

27. $f''(x) = 24x \;\Rightarrow\; f'(x) = 24\left(\dfrac{x^2}{2}\right) + C = 12x^2 + C \;\Rightarrow\; f(x) = 12\left(\dfrac{x^3}{3}\right) + Cx + D = 4x^3 + Cx + D$

29. $f''(x) = 4x^3 + 24x - 1 \;\Rightarrow\; f'(x) = 4\left(\dfrac{x^4}{4}\right) + 24\left(\dfrac{x^2}{2}\right) - x + C = x^4 + 12x^2 - x + C \;\Rightarrow\;$

$f(x) = \dfrac{x^5}{5} + 12\left(\dfrac{x^3}{3}\right) - \dfrac{x^2}{2} + Cx + D = \tfrac{1}{5}x^5 + 4x^3 - \tfrac{1}{2}x^2 + Cx + D$

31. $f''(x) = 4 - \sqrt[3]{x} \;\Rightarrow\; f'(x) = 4x - \tfrac{3}{4}x^{4/3} + C \;\Rightarrow\; f(x) = 4 \cdot \tfrac{1}{2}x^2 - \tfrac{3}{4} \cdot \tfrac{3}{7}x^{7/3} + Cx + D = 2x^2 - \tfrac{9}{28}x^{7/3} + Cx + D$

33. $f'''(t) = 12 + \sin t \;\Rightarrow\; f''(t) = 12t - \cos t + C_1 \;\Rightarrow\; f'(t) = 6t^2 - \sin t + C_1 t + D \;\Rightarrow\;$
$f(t) = 2t^3 + \cos t + Ct^2 + Dt + E$, where $C = \tfrac{1}{2}C_1$.

35. $f'(x) = 5x^4 - 3x^2 + 4 \;\Rightarrow\; f(x) = x^5 - x^3 + 4x + C.$ $f(-1) = -1 + 1 - 4 + C$ and $f(-1) = 2 \;\Rightarrow\;$
$-4 + C = 2 \;\Rightarrow\; C = 6$, so $f(x) = x^5 - x^3 + 4x + 6$.

37. $f'(x) = 5x^{2/3} \;\Rightarrow\; f(x) = 5\left(\tfrac{3}{5}x^{5/3}\right) + C = 3x^{5/3} + C$.

$f(8) = 3 \cdot 32 + C$ and $f(8) = 21 \;\Rightarrow\; 96 + C = 21 \;\Rightarrow\; C = -75$, so $f(x) = 3x^{5/3} - 75$.

39. $f'(t) = \sec t(\sec t + \tan t) = \sec^2 t + \sec t \tan t$, $-\tfrac{\pi}{2} < t < \tfrac{\pi}{2} \;\Rightarrow\; f(t) = \tan t + \sec t + C$. $f\left(\tfrac{\pi}{4}\right) = 1 + \sqrt{2} + C$
and $f\left(\tfrac{\pi}{4}\right) = -1 \;\Rightarrow\; 1 + \sqrt{2} + C = -1 \;\Rightarrow\; C = -2 - \sqrt{2}$, so $f(t) = \tan t + \sec t - 2 - \sqrt{2}$.

Note: The fact that f is defined and continuous on $\left(-\tfrac{\pi}{2}, \tfrac{\pi}{2}\right)$ means that we have only one constant of integration.

41. $f''(x) = -2 + 12x - 12x^2 \;\Rightarrow\; f'(x) = -2x + 6x^2 - 4x^3 + C.$ $f'(0) = C$ and $f'(0) = 12 \;\Rightarrow\; C = 12$, so
$f'(x) = -2x + 6x^2 - 4x^3 + 12$ and hence, $f(x) = -x^2 + 2x^3 - x^4 + 12x + D$. $f(0) = D$ and $f(0) = 4 \;\Rightarrow\; D = 4$,
so $f(x) = -x^2 + 2x^3 - x^4 + 12x + 4$.

43. $f''(\theta) = \sin\theta + \cos\theta \;\Rightarrow\; f'(\theta) = -\cos\theta + \sin\theta + C.\; f'(0) = -1 + C$ and $f'(0) = 4 \;\Rightarrow\; C = 5$, so $f'(\theta) = -\cos\theta + \sin\theta + 5$ and hence, $f(\theta) = -\sin\theta - \cos\theta + 5\theta + D.\; f(0) = -1 + D$ and $f(0) = 3 \;\Rightarrow\; D = 4$, so $f(\theta) = -\sin\theta - \cos\theta + 5\theta + 4$.

45. $f''(x) = 4 + 6x + 24x^2 \;\Rightarrow\; f'(x) = 4x + 3x^2 + 8x^3 + C \;\Rightarrow\; f(x) = 2x^2 + x^3 + 2x^4 + Cx + D.\; f(0) = D$ and $f(0) = 3 \;\Rightarrow\; D = 3$, so $f(x) = 2x^2 + x^3 + 2x^4 + Cx + 3.\; f(1) = 8 + C$ and $f(1) = 10 \;\Rightarrow\; C = 2$, so $f(x) = 2x^2 + x^3 + 2x^4 + 2x + 3$.

47. $f''(t) = \sqrt[3]{t} - \cos t = t^{1/3} - \cos t \;\Rightarrow\; f'(t) = \tfrac{3}{4}t^{4/3} - \sin t + C \;\Rightarrow\; f(t) = \tfrac{9}{28}t^{7/3} + \cos t + Ct + D.$
$f(0) = 0 + 1 + 0 + D$ and $f(0) = 2 \;\Rightarrow\; D = 1$, so $f(t) = \tfrac{9}{28}t^{7/3} + \cos t + Ct + 1.\; f(1) = \tfrac{9}{28} + \cos 1 + C + 1$ and $f(1) = 2 \;\Rightarrow\; C = 2 - \tfrac{9}{28} - \cos 1 - 1 = \tfrac{19}{28} - \cos 1$, so $f(t) = \tfrac{9}{28}t^{7/3} + \cos t + \left(\tfrac{19}{28} - \cos 1\right)t + 1.$

49. "The slope of its tangent line at $(x, f(x))$ is $3 - 4x$" means that $f'(x) = 3 - 4x$, so $f(x) = 3x - 2x^2 + C$.

"The graph of f passes through the point $(2, 5)$" means that $f(2) = 5$, but $f(2) = 3(2) - 2(2)^2 + C$, so $5 = 6 - 8 + C \;\Rightarrow\; C = 7.$ Thus, $f(x) = 3x - 2x^2 + 7$ and $f(1) = 3 - 2 + 7 = 8$.

51. b is the antiderivative of f. For small x, f is negative, so the graph of its antiderivative must be decreasing. But both a and c are increasing for small x, so only b can be f's antiderivative. Also, f is positive where b is increasing, which supports our conclusion.

53.

The graph of F must start at $(0, 1)$. Where the given graph, $y = f(x)$, has a local minimum or maximum, the graph of F will have an inflection point.

Where f is negative (positive), F is decreasing (increasing).

Where f changes from negative to positive, F will have a minimum.

Where f changes from positive to negative, F will have a maximum.

Where f is decreasing (increasing), F is concave downward (upward).

55.

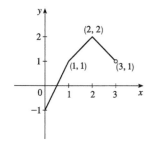

$f'(x) = \begin{cases} 2 & \text{if } 0 \le x < 1 \\ 1 & \text{if } 1 < x < 2 \\ -1 & \text{if } 2 < x < 3 \end{cases} \;\Rightarrow\; f(x) = \begin{cases} 2x + C & \text{if } 0 \le x < 1 \\ x + D & \text{if } 1 < x < 2 \\ -x + E & \text{if } 2 < x < 3 \end{cases}$

$f(0) = -1 \;\Rightarrow\; 2(0) + C = -1 \;\Rightarrow\; C = -1.$ Starting at the point $(0, -1)$ and moving to the right on a line with slope 2 gets us to the point $(1, 1)$.

The slope for $1 < x < 2$ is 1, so we get to the point $(2, 2)$. Here we have used the fact that f is continuous. We can include the point $x = 1$ on either the first or the second part of f. The line connecting $(1, 1)$ to $(2, 2)$ is $y = x$, so $D = 0$. The slope for

$2 < x < 3$ is -1, so we get to $(3, 1)$. $f(2) = 2 \Rightarrow -2 + E = 2 \Rightarrow E = 4$. Thus,

$$f(x) = \begin{cases} 2x - 1 & \text{if } 0 \leq x \leq 1 \\ x & \text{if } 1 < x < 2 \\ -x + 4 & \text{if } 2 \leq x < 3 \end{cases}$$

Note that $f'(x)$ does not exist at $x = 1, 2,$ or 3.

57. $f(x) = \dfrac{\sin x}{1 + x^2}$, $-2\pi \leq x \leq 2\pi$

Note that the graph of f is one of an odd function, so the graph of F will be one of an even function.

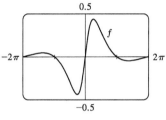

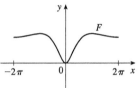

59. $v(t) = s'(t) = 2\cos t + 4\sin t \Rightarrow s(t) = 2\sin t - 4\cos t + C$. $s(0) = -4 + C$ and $s(0) = 3 \Rightarrow C = 7$, so $s(t) = 2\sin t - 4\cos t + 7$.

61. $a(t) = v'(t) = 2t + 1 \Rightarrow v(t) = t^2 + t + C$. $v(0) = C$ and $v(0) = -2 \Rightarrow C = -2$, so $v(t) = t^2 + t - 2$ and $s(t) = \frac{1}{3}t^3 + \frac{1}{2}t^2 - 2t + D$. $s(0) = D$ and $s(0) = 3 \Rightarrow D = 3$, so $s(t) = \frac{1}{3}t^3 + \frac{1}{2}t^2 - 2t + 3$.

63. $a(t) = v'(t) = \sin t - \cos t \Rightarrow v(t) = s'(t) = -\cos t - \sin t + C \Rightarrow s(t) = -\sin t + \cos t + Ct + D$.
$s(0) = 1 + D = 0$ and $s(\pi) = -1 + C\pi + D = 6 \Rightarrow D = -1$ and $C = \dfrac{8}{\pi}$. Thus, $s(t) = -\sin t + \cos t + \dfrac{8}{\pi}t - 1$.

65. (a) We first observe that since the stone is dropped 450 m above the ground, $v(0) = 0$ and $s(0) = 450$.
$v'(t) = a(t) = -9.8 \Rightarrow v(t) = -9.8t + C$. Now $v(0) = 0 \Rightarrow C = 0$, so $v(t) = -9.8t \Rightarrow$
$s(t) = -4.9t^2 + D$. Last, $s(0) = 450 \Rightarrow D = 450 \Rightarrow s(t) = 450 - 4.9t^2$.

(b) The stone reaches the ground when $s(t) = 0$. $450 - 4.9t^2 = 0 \Rightarrow t^2 = 450/4.9 \Rightarrow t_1 = \sqrt{450/4.9} \approx 9.58$ s.

(c) The velocity with which the stone strikes the ground is $v(t_1) = -9.8\sqrt{450/4.9} \approx -93.9$ m/s.

(d) This is just reworking parts (a) and (b) with $v(0) = -5$. Using $v(t) = -9.8t + C$, $v(0) = -5 \Rightarrow 0 + C = -5 \Rightarrow$
$v(t) = -9.8t - 5$. So $s(t) = -4.9t^2 - 5t + D$ and $s(0) = 450 \Rightarrow D = 450 \Rightarrow s(t) = -4.9t^2 - 5t + 450$.
Solving $s(t) = 0$ by using the quadratic formula gives us $t = \left(5 \pm \sqrt{8845}\right)/(-9.8) \Rightarrow t_1 \approx 9.09$ s.

67. By Exercise 66 with $a = -9.8$, $s(t) = -4.9t^2 + v_0 t + s_0$ and $v(t) = s'(t) = -9.8t + v_0$. So
$[v(t)]^2 = (-9.8t + v_0)^2 = (9.8)^2 t^2 - 19.6 v_0 t + v_0^2 = v_0^2 + 96.04 t^2 - 19.6 v_0 t = v_0^2 - 19.6\left(-4.9t^2 + v_0 t\right)$.
But $-4.9t^2 + v_0 t$ is just $s(t)$ without the s_0 term; that is, $s(t) - s_0$. Thus, $[v(t)]^2 = v_0^2 - 19.6\left[s(t) - s_0\right]$.

69. Using Exercise 66 with $a = -32$, $v_0 = 0$, and $s_0 = h$ (the height of the cliff), we know that the height at time t is

$s(t) = -16t^2 + h$. $v(t) = s'(t) = -32t$ and $v(t) = -120 \Rightarrow -32t = -120 \Rightarrow t = 3.75$, so

$0 = s(3.75) = -16(3.75)^2 + h \Rightarrow h = 16(3.75)^2 = 225$ ft.

71. Marginal cost $= 1.92 - 0.002x = C'(x) \Rightarrow C(x) = 1.92x - 0.001x^2 + K$. But $C(1) = 1.92 - 0.001 + K = 562 \Rightarrow$
$K = 560.081$. Therefore, $C(x) = 1.92x - 0.001x^2 + 560.081 \Rightarrow C(100) = 742.081$, so the cost of producing 100 items is \$742.08.

73. Taking the upward direction to be positive we have that for $0 \le t \le 10$ (using the subscript 1 to refer to $0 \le t \le 10$),

$a_1(t) = -(9 - 0.9t) = v_1'(t) \Rightarrow v_1(t) = -9t + 0.45t^2 + v_0$, but $v_1(0) = v_0 = -10 \Rightarrow$

$v_1(t) = -9t + 0.45t^2 - 10 = s_1'(t) \Rightarrow s_1(t) = -\frac{9}{2}t^2 + 0.15t^3 - 10t + s_0$. But $s_1(0) = 500 = s_0 \Rightarrow$

$s_1(t) = -\frac{9}{2}t^2 + 0.15t^3 - 10t + 500$. $s_1(10) = -450 + 150 - 100 + 500 = 100$, so it takes

more than 10 seconds for the raindrop to fall. Now for $t > 10$, $a(t) = 0 = v'(t) \Rightarrow$

$v(t) = \text{constant} = v_1(10) = -9(10) + 0.45(10)^2 - 10 = -55 \Rightarrow v(t) = -55$.

At 55 m/s, it will take $100/55 \approx 1.8$ s to fall the last 100 m. Hence, the total time is $10 + \frac{100}{55} = \frac{130}{11} \approx 11.8$ s.

75. $a(t) = k$, the initial velocity is 30 mi/h $= 30 \cdot \frac{5280}{3600} = 44$ ft/s, and the final velocity (after 5 seconds) is

50 mi/h $= 50 \cdot \frac{5280}{3600} = \frac{220}{3}$ ft/s. So $v(t) = kt + C$ and $v(0) = 44 \Rightarrow C = 44$. Thus, $v(t) = kt + 44 \Rightarrow$

$v(5) = 5k + 44$. But $v(5) = \frac{220}{3}$, so $5k + 44 = \frac{220}{3} \Rightarrow 5k = \frac{88}{3} \Rightarrow k = \frac{88}{15} \approx 5.87$ ft/s^2.

77. Let the acceleration be $a(t) = k$ km/h^2. We have $v(0) = 100$ km/h and we can take the initial position $s(0)$ to be 0.

We want the time t_f for which $v(t) = 0$ to satisfy $s(t) < 0.08$ km. In general, $v'(t) = a(t) = k$, so $v(t) = kt + C$,

where $C = v(0) = 100$. Now $s'(t) = v(t) = kt + 100$, so $s(t) = \frac{1}{2}kt^2 + 100t + D$, where $D = s(0) = 0$.

Thus, $s(t) = \frac{1}{2}kt^2 + 100t$. Since $v(t_f) = 0$, we have $kt_f + 100 = 0$ or $t_f = -100/k$, so

$s(t_f) = \frac{1}{2}k\left(-\frac{100}{k}\right)^2 + 100\left(-\frac{100}{k}\right) = 10{,}000\left(\frac{1}{2k} - \frac{1}{k}\right) = -\frac{5{,}000}{k}$. The condition $s(t_f)$ must satisfy is

$-\frac{5{,}000}{k} < 0.08 \Rightarrow -\frac{5{,}000}{0.08} > k$ [k is negative] $\Rightarrow k < -62{,}500$ km/h^2, or equivalently,

$k < -\frac{3125}{648} \approx -4.82$ m/s^2.

79. (a) First note that 180 mi/h $= 180 \times \frac{5280}{3600}$ ft/s $= 264$ ft/s. Then $a(t) = 2.4$ ft/s$^2 \Rightarrow v(t) = 2.4t + C$, but

$v(0) = 0 \Rightarrow C = 0$. Now $2.4t = 264$ when $t = \frac{264}{2.4} = 110$ s, so it takes 110 s to reach 264 ft/s. Therefore, taking

$s(0) = 0$, we have $s(t) = 1.2t^2$, $0 \le t \le 110$. So $s(110) = 14{,}520$ ft. 20 minutes $= 20(60) = 1200$ s, so for

$110 \le t \le 1310$ we have $v(t) = 264$ ft/s $\Rightarrow s(1310) = 264(1200) + 14{,}520 = 331{,}320$ ft $= 62.75$ mi.

(b) As in part (a), the train accelerates for 110 s and travels 14,520 ft while doing so. Similarly, it decelerates for 110 s and travels 14,520 ft at the end of its trip. During the remaining $1200 - 2(110) = 980$ s it travels at 264 ft/s, so the distance traveled is $264 \cdot 980 = 258{,}720$ ft. Thus, the total distance is $14{,}520 + 258{,}720 + 14{,}520 = 287{,}760$ ft $= 54.5$ mi.

(c) 60 mi $= 60(5280) = 316{,}800$ ft. Subtract $2(14{,}520)$ to take care of the speeding up and slowing down, and we have 287,760 ft at 264 ft/s for a trip of $287{,}760/264 = 1090$ s at 180 mi/h. The total time is
$1090 + 2(110) = 1310$ s $= 21$ min 50 s $= 21.8\overline{3}$ min.

(d) $37.5(60) = 2250$ s. Then $2250 - 2(110) = 2030$ s at maximum speed. $2030(264) + 2(14{,}520) = 564{,}960$ total feet or $564{,}960/5280 = 107$ mi.

3 Review

TRUE-FALSE QUIZ

1. False. For example, take $f(x) = x^3$, then $f'(x) = 3x^2$ and $f'(0) = 0$, but $f(0) = 0$ is not a maximum or minimum; $(0, 0)$ is an inflection point.

3. False. For example, $f(x) = x$ is continuous on $(0, 1)$ but attains neither a maximum nor a minimum value on $(0, 1)$. Don't confuse this with f being continuous on the *closed* interval $[a, b]$, which would make the statement true.

5. True. This is an example of part (b) of the Increasing/Decreasing Test.

7. False. $f'(x) = g'(x) \;\Rightarrow\; f(x) = g(x) + C$. For example, if $f(x) = x + 2$ and $g(x) = x + 1$, then $f'(x) = g'(x) = 1$, but $f(x) \neq g(x)$.

9. True. The graph of one such function is sketched.

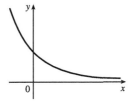

11. True. Let $x_1 < x_2$ where $x_1, x_2 \in I$. Then $f(x_1) < f(x_2)$ and $g(x_1) < g(x_2)$ [since f and g are increasing on I], so $(f + g)(x_1) = f(x_1) + g(x_1) < f(x_2) + g(x_2) = (f + g)(x_2)$.

13. False. Take $f(x) = x$ and $g(x) = x - 1$. Then both f and g are increasing on $(0, 1)$. But $f(x)\,g(x) = x(x - 1)$ is not increasing on $(0, 1)$.

15. True. Let $x_1, x_2 \in I$ and $x_1 < x_2$. Then $f(x_1) < f(x_2)$ [f is increasing] $\;\Rightarrow\; \dfrac{1}{f(x_1)} > \dfrac{1}{f(x_2)}$ [f is positive] $\;\Rightarrow\;$
$g(x_1) > g(x_2) \;\Rightarrow\; g(x) = 1/f(x)$ is decreasing on I.

17. True. If f is periodic, then there is a number p such that $f(x + p) = f(p)$ for all x. Differentiating gives
$f'(x) = f'(x + p) \cdot (x + p)' = f'(x + p) \cdot 1 = f'(x + p)$, so f' is periodic.

19. True. By the Mean Value Theorem, there exists a number c in $(0, 1)$ such that $f(1) - f(0) = f'(c)(1 - 0) = f'(c)$. Since $f'(c)$ is nonzero, $f(1) - f(0) \neq 0$, so $f(1) \neq f(0)$.

EXERCISES

1. $f(x) = x^3 - 9x^2 + 24x - 2$, $[0, 5]$. $f'(x) = 3x^2 - 18x + 24 = 3(x^2 - 6x + 8) = 3(x - 2)(x - 4)$. $f'(x) = 0 \Leftrightarrow x = 2$ or $x = 4$. $f'(x) > 0$ for $0 < x < 2$, $f'(x) < 0$ for $2 < x < 4$, and $f'(x) > 0$ for $4 < x < 5$, so $f(2) = 18$ is a local maximum value and $f(4) = 14$ is a local minimum value. Checking the endpoints, we find $f(0) = -2$ and $f(5) = 18$. Thus, $f(0) = -2$ is the absolute minimum value and $f(2) = f(5) = 18$ is the absolute maximum value.

3. $f(x) = \dfrac{3x - 4}{x^2 + 1}$, $[-2, 2]$. $f'(x) = \dfrac{(x^2 + 1)(3) - (3x - 4)(2x)}{(x^2 + 1)^2} = \dfrac{-(3x^2 - 8x - 3)}{(x^2 + 1)^2} = \dfrac{-(3x + 1)(x - 3)}{(x^2 + 1)^2}$.

$f'(x) = 0 \Rightarrow x = -\frac{1}{3}$ or $x = 3$, but 3 is not in the interval. $f'(x) > 0$ for $-\frac{1}{3} < x < 2$ and $f'(x) < 0$ for $-2 < x < -\frac{1}{3}$, so $f(-\frac{1}{3}) = \dfrac{-5}{10/9} = -\frac{9}{2}$ is a local minimum value. Checking the endpoints, we find $f(-2) = -2$ and $f(2) = \frac{2}{5}$. Thus, $f(-\frac{1}{3}) = -\frac{9}{2}$ is the absolute minimum value and $f(2) = \frac{2}{5}$ is the absolute maximum value.

5. $f(x) = x + 2\cos x$, $[-\pi, \pi]$. $f'(x) = 1 - 2\sin x$. $f'(x) = 0 \Rightarrow \sin x = \frac{1}{2} \Rightarrow x = \frac{\pi}{6}, \frac{5\pi}{6}$. $f'(x) > 0$ for $\left(-\pi, \frac{\pi}{6}\right)$ and $\left(\frac{5\pi}{6}, \pi\right)$, and $f'(x) < 0$ for $\left(\frac{\pi}{6}, \frac{5\pi}{6}\right)$, so $f\left(\frac{\pi}{6}\right) = \frac{\pi}{6} + \sqrt{3} \approx 2.26$ is a local maximum value and $f\left(\frac{5\pi}{6}\right) = \frac{5\pi}{6} - \sqrt{3} \approx 0.89$ is a local minimum value. Checking the endpoints, we find $f(-\pi) = -\pi - 2 \approx -5.14$ and $f(\pi) = \pi - 2 \approx 1.14$. Thus, $f(-\pi) = -\pi - 2$ is the absolute minimum value and $f\left(\frac{\pi}{6}\right) = \frac{\pi}{6} + \sqrt{3}$ is the absolute maximum value.

7. $\displaystyle\lim_{x \to \infty} \dfrac{3x^4 + x - 5}{6x^4 - 2x^2 + 1} = \lim_{x \to \infty} \dfrac{3 + \dfrac{1}{x^3} - \dfrac{5}{x^4}}{6 - \dfrac{2}{x^2} + \dfrac{1}{x^4}} = \dfrac{3 + 0 + 0}{6 - 0 + 0} = \dfrac{1}{2}$

9. $\displaystyle\lim_{x \to -\infty} \dfrac{\sqrt{4x^2 + 1}}{3x - 1} = \lim_{x \to -\infty} \dfrac{\sqrt{4x^2 + 1}/\sqrt{x^2}}{(3x - 1)/\sqrt{x^2}} = \lim_{x \to -\infty} \dfrac{\sqrt{4 + 1/x^2}}{-3 + 1/x}$ [since $-x = |x| = \sqrt{x^2}$ for $x < 0$]

$= \dfrac{2}{-3 + 0} = -\dfrac{2}{3}$

11. $\displaystyle\lim_{x \to \infty} \left(\sqrt{4x^2 + 3x} - 2x\right) = \lim_{x \to \infty} \dfrac{\sqrt{4x^2 + 3x} - 2x}{1} \cdot \dfrac{\sqrt{4x^2 + 3x} + 2x}{\sqrt{4x^2 + 3x} + 2x} = \lim_{x \to \infty} \dfrac{(4x^2 + 3x) - 4x^2}{\sqrt{4x^2 + 3x} + 2x}$

$= \displaystyle\lim_{x \to \infty} \dfrac{3x}{\sqrt{4x^2 + 3x} + 2x} = \lim_{x \to \infty} \dfrac{3x/\sqrt{x^2}}{\left(\sqrt{4x^2 + 3x} + 2x\right)/\sqrt{x^2}}$

$= \displaystyle\lim_{x \to \infty} \dfrac{3}{\sqrt{4 + 3/x} + 2}$ [since $x = |x| = \sqrt{x^2}$ for $x > 0$]

$= \dfrac{3}{2 + 2} = \dfrac{3}{4}$

13. $f(0) = 0$, $f'(-2) = f'(1) = f'(9) = 0$, $\lim_{x \to \infty} f(x) = 0$, $\lim_{x \to 6} f(x) = -\infty$,

$f'(x) < 0$ on $(-\infty, -2)$, $(1, 6)$, and $(9, \infty)$, $f'(x) > 0$ on $(-2, 1)$ and $(6, 9)$,

$f''(x) > 0$ on $(-\infty, 0)$ and $(12, \infty)$, $f''(x) < 0$ on $(0, 6)$ and $(6, 12)$

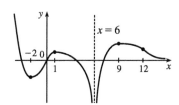

15. f is odd, $f'(x) < 0$ for $0 < x < 2$, $f'(x) > 0$ for $x > 2$,

$f''(x) > 0$ for $0 < x < 3$, $f''(x) < 0$ for $x > 3$,

$\lim_{x \to \infty} f(x) = -2$

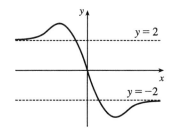

17. $y = f(x) = 2 - 2x - x^3$ **A.** $D = \mathbb{R}$ **B.** y-intercept: $f(0) = 2$.
The x-intercept (approximately 0.770917) can be found using Newton's
Method. **C.** No symmetry **D.** No asymptote
E. $f'(x) = -2 - 3x^2 = -(3x^2 + 2) < 0$, so f is decreasing on $\mathbb{R}$.
F. No extreme values **G.** $f''(x) = -6x < 0$ on $(0, \infty)$ and $f''(x) > 0$ on
$(-\infty, 0)$, so f is CD on $(0, \infty)$ and CU on $(-\infty, 0)$. There is an IP at $(0, 2)$.

H.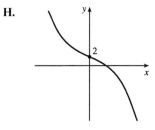

19. $y = f(x) = 3x^4 - 4x^3 + 2$ **A.** $D = \mathbb{R}$ **B.** y-intercept: $f(0) = 2$; no x-intercept **C.** No symmetry **D.** No asymptote
E. $f'(x) = 12x^3 - 12x^2 = 12x^2(x - 1)$. $f'(x) > 0$ for $x > 1$, so f is
increasing on $(1, \infty)$ and decreasing on $(-\infty, 1)$. **F.** $f'(x)$ does not change
sign at $x = 0$, so there is no local extremum there. $f(1) = 1$ is a local minimum
value. **G.** $f''(x) = 36x^2 - 24x = 12x(3x - 2)$. $f''(x) < 0$ for $0 < x < \frac{2}{3}$,
so f is CD on $\left(0, \frac{2}{3}\right)$ and f is CU on $(-\infty, 0)$ and $\left(\frac{2}{3}, \infty\right)$. There are inflection
points at $(0, 2)$ and $\left(\frac{2}{3}, \frac{38}{27}\right)$.

H.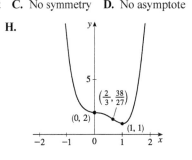

21. $y = f(x) = \dfrac{1}{x(x-3)^2}$ **A.** $D = \{x \mid x \neq 0, 3\} = (-\infty, 0) \cup (0, 3) \cup (3, \infty)$ **B.** No intercepts. **C.** No symmetry.

D. $\lim_{x \to \pm\infty} \dfrac{1}{x(x-3)^2} = 0$, so $y = 0$ is a HA. $\lim_{x \to 0^+} \dfrac{1}{x(x-3)^2} = \infty$, $\lim_{x \to 0^-} \dfrac{1}{x(x-3)^2} = -\infty$, $\lim_{x \to 3} \dfrac{1}{x(x-3)^2} = \infty$,

so $x = 0$ and $x = 3$ are VA. **E.** $f'(x) = -\dfrac{(x-3)^2 + 2x(x-3)}{x^2(x-3)^4} = \dfrac{3(1-x)}{x^2(x-3)^3}$ $\Rightarrow$ $f'(x) > 0$ $\Leftrightarrow$ $1 < x < 3$,

so f is increasing on $(1, 3)$ and decreasing on $(-\infty, 0)$, $(0, 1)$, and $(3, \infty)$.

F. Local minimum value $f(1) = \frac{1}{4}$ **G.** $f''(x) = \dfrac{6(2x^2 - 4x + 3)}{x^3(x-3)^4}$.

Note that $2x^2 - 4x + 3 > 0$ for all x since it has negative discriminant.
So $f''(x) > 0$ $\Leftrightarrow$ $x > 0$ $\Rightarrow$ f is CU on $(0, 3)$ and $(3, \infty)$ and
CD on $(-\infty, 0)$. No IP

H.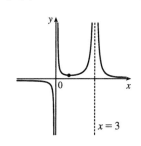

23. $y = f(x) = \dfrac{(x-1)^3}{x^2} = \dfrac{x^3 - 3x^2 + 3x - 1}{x^2} = x - 3 + \dfrac{3x-1}{x^2}$ **A.** $D = \{x \mid x \neq 0\} = (-\infty, 0) \cup (0, \infty)$

B. y-intercept: none; x-intercept: $f(x) = 0 \Leftrightarrow x = 1$ **C.** No symmetry **D.** $\lim\limits_{x \to 0^-} \dfrac{(x-1)^3}{x^2} = -\infty$ and

$\lim\limits_{x \to 0^+} f(x) = -\infty$, so $x = 0$ is a VA. $f(x) - (x-3) = \dfrac{3x-1}{x^2} \to 0$ as $x \to \pm\infty$, so $y = x - 3$ is a SA.

E. $f'(x) = \dfrac{x^2 \cdot 3(x-1)^2 - (x-1)^3(2x)}{(x^2)^2} = \dfrac{x(x-1)^2[3x - 2(x-1)]}{x^4} = \dfrac{(x-1)^2(x+2)}{x^3}$. $f'(x) < 0$ for $-2 < x < 0$,

so f is increasing on $(-\infty, -2)$, decreasing on $(-2, 0)$, and increasing on $(0, \infty)$. **H.**

F. Local maximum value $f(-2) = -\tfrac{27}{4}$ **G.** $f(x) = x - 3 + \dfrac{3}{x} - \dfrac{1}{x^2} \Rightarrow$

$f'(x) = 1 - \dfrac{3}{x^2} + \dfrac{2}{x^3} \Rightarrow f''(x) = \dfrac{6}{x^3} - \dfrac{6}{x^4} = \dfrac{6x - 6}{x^4} = \dfrac{6(x-1)}{x^4}$.

$f''(x) > 0$ for $x > 1$, so f is CD on $(-\infty, 0)$ and $(0, 1)$, and f is CU on $(1, \infty)$.

There is an inflection point at $(1, 0)$.

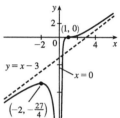

25. $y = f(x) = x\sqrt{2+x}$ **A.** $D = [-2, \infty)$ **B.** y-intercept: $f(0) = 0$; x-intercepts: -2 and 0 **C.** No symmetry

D. No asymptote **E.** $f'(x) = \dfrac{x}{2\sqrt{2+x}} + \sqrt{2+x} = \dfrac{1}{2\sqrt{2+x}}[x + 2(2+x)] = \dfrac{3x+4}{2\sqrt{2+x}} = 0$ when $x = -\tfrac{4}{3}$, so f is

decreasing on $\left(-2, -\tfrac{4}{3}\right)$ and increasing on $\left(-\tfrac{4}{3}, \infty\right)$. **F.** Local minimum value $f\left(-\tfrac{4}{3}\right) = -\tfrac{4}{3}\sqrt{\tfrac{2}{3}} = -\tfrac{4\sqrt{6}}{9} \approx -1.09$,

no local maximum **H.**

G. $f''(x) = \dfrac{2\sqrt{2+x} \cdot 3 - (3x+4)\dfrac{1}{\sqrt{2+x}}}{4(2+x)} = \dfrac{6(2+x) - (3x+4)}{4(2+x)^{3/2}}$

$= \dfrac{3x+8}{4(2+x)^{3/2}}$

$f''(x) > 0$ for $x > -2$, so f is CU on $(-2, \infty)$. No IP

27. $y = f(x) = \sin^2 x - 2\cos x$ **A.** $D = \mathbb{R}$ **B.** y-intercept: $f(0) = -2$ **C.** $f(-x) = f(x)$, so f is symmetric with respect

to the y-axis. f has period 2π. **D.** No asymptote **E.** $y' = 2\sin x \cos x + 2\sin x = 2\sin x(\cos x + 1)$. $y' = 0 \Leftrightarrow$

$\sin x = 0$ or $\cos x = -1 \Leftrightarrow x = n\pi$ or $x = (2n+1)\pi$. $y' > 0$ when $\sin x > 0$, since $\cos x + 1 \geq 0$ for all x.

Therefore, $y' > 0$ [and so f is increasing] on $(2n\pi, (2n+1)\pi)$; $y' < 0$ [and so f is decreasing] on $((2n-1)\pi, 2n\pi)$.

F. Local maximum values are $f((2n+1)\pi) = 2$; local minimum values are $f(2n\pi) = -2$.

G. $y' = \sin 2x + 2\sin x \Rightarrow y'' = 2\cos 2x + 2\cos x = 2(2\cos^2 x - 1) + 2\cos x = 4\cos^2 x + 2\cos x - 2$

$\qquad = 2(2\cos^2 x + \cos x - 1) = 2(2\cos x - 1)(\cos x + 1)$

$y'' = 0 \Leftrightarrow \cos x = \tfrac{1}{2}$ or $-1 \Leftrightarrow x = 2n\pi \pm \tfrac{\pi}{3}$ or $x = (2n+1)\pi$. **H.**

$y'' > 0$ [and so f is CU] on $\left(2n\pi - \tfrac{\pi}{3}, 2n\pi + \tfrac{\pi}{3}\right)$; $y'' \leq 0$ [and so f is CD]

on $\left(2n\pi + \tfrac{\pi}{3}, 2n\pi + \tfrac{5\pi}{3}\right)$. There are inflection points at $\left(2n\pi \pm \tfrac{\pi}{3}, -\tfrac{1}{4}\right)$.

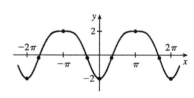

29. $f(x) = \dfrac{x^2-1}{x^3} \Rightarrow f'(x) = \dfrac{x^3(2x) - (x^2-1)3x^2}{x^6} = \dfrac{3-x^2}{x^4} \Rightarrow$

$f''(x) = \dfrac{x^4(-2x) - (3-x^2)4x^3}{x^8} = \dfrac{2x^2 - 12}{x^5}$

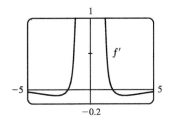

Estimates: From the graphs of f' and f'', it appears that f is increasing on $(-1.73, 0)$ and $(0, 1.73)$ and decreasing on $(-\infty, -1.73)$ and $(1.73, \infty)$; f has a local maximum of about $f(1.73) = 0.38$ and a local minimum of about $f(-1.7) = -0.38$; f is CU on $(-2.45, 0)$ and $(2.45, \infty)$, and CD on $(-\infty, -2.45)$ and $(0, 2.45)$; and f has inflection points at about $(-2.45, -0.34)$ and $(2.45, 0.34)$.

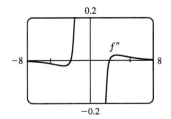

Exact: Now $f'(x) = \dfrac{3-x^2}{x^4}$ is positive for $0 < x^2 < 3$, that is, f is increasing on $(-\sqrt{3}, 0)$ and $(0, \sqrt{3})$; and $f'(x)$ is negative (and so f is decreasing) on $(-\infty, -\sqrt{3})$ and $(\sqrt{3}, \infty)$. $f'(x) = 0$ when $x = \pm\sqrt{3}$.

f' goes from positive to negative at $x = \sqrt{3}$, so f has a local maximum of

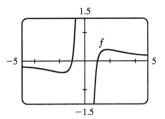

$f(\sqrt{3}) = \dfrac{(\sqrt{3})^2 - 1}{(\sqrt{3})^3} = \dfrac{2\sqrt{3}}{9}$; and since f is odd, we know that maxima on the interval $(0, \infty)$ correspond to minima on $(-\infty, 0)$, so f has a local minimum of $f(-\sqrt{3}) = -\dfrac{2\sqrt{3}}{9}$. Also, $f''(x) = \dfrac{2x^2 - 12}{x^5}$ is positive (so f is CU) on $(-\sqrt{6}, 0)$ and $(\sqrt{6}, \infty)$, and negative (so f is CD) on $(-\infty, -\sqrt{6})$ and $(0, \sqrt{6})$. There are IP at $\left(\sqrt{6}, \dfrac{5\sqrt{6}}{36}\right)$ and $\left(-\sqrt{6}, -\dfrac{5\sqrt{6}}{36}\right)$.

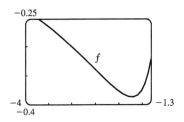

31. $f(x) = 3x^6 - 5x^5 + x^4 - 5x^3 - 2x^2 + 2 \Rightarrow f'(x) = 18x^5 - 25x^4 + 4x^3 - 15x^2 - 4x \Rightarrow$

$f''(x) = 90x^4 - 100x^3 + 12x^2 - 30x - 4$

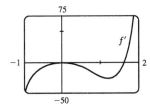 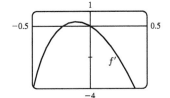 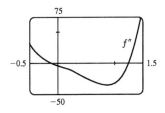

From the graphs of f' and f'', it appears that f is increasing on $(-0.23, 0)$ and $(1.62, \infty)$ and decreasing on $(-\infty, -0.23)$ and $(0, 1.62)$; f has a local maximum of $f(0) = 2$ and local minima of about $f(-0.23) = 1.96$ and $f(1.62) = -19.2$;

f is CU on $(-\infty, -0.12)$ and $(1.24, \infty)$ and CD on $(-0.12, 1.24)$; and f has inflection points at about $(-0.12, 1.98)$ and $(1.24, -12.1)$.

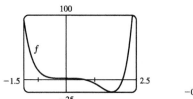
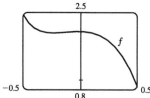

33. Let $f(x) = 3x + 2\cos x + 5$. Then $f(0) = 7 > 0$ and $f(-\pi) = -3\pi - 2 + 5 = -3\pi + 3 = -3(\pi - 1) < 0$, and since f is continuous on $\mathbb{R}$ (hence on $[-\pi, 0]$), the Intermediate Value Theorem assures us that there is at least one solution of f in $[-\pi, 0]$. Now $f'(x) = 3 - 2\sin x > 0$ implies that f is increasing on $\mathbb{R}$, so there is exactly one solution of f, and hence, exactly one real solution of the equation $3x + 2\cos x + 5 = 0$.

35. Since f is continuous on $[32, 33]$ and differentiable on $(32, 33)$, then by the Mean Value Theorem there exists a number c in $(32, 33)$ such that $f'(c) = \frac{1}{5}c^{-4/5} = \frac{\sqrt[5]{33} - \sqrt[5]{32}}{33 - 32} = \sqrt[5]{33} - 2$, but $\frac{1}{5}c^{-4/5} > 0 \ \Rightarrow\ \sqrt[5]{33} - 2 > 0 \ \Rightarrow\ \sqrt[5]{33} > 2$. Also f' is decreasing, so that $f'(c) < f'(32) = \frac{1}{5}(32)^{-4/5} = 0.0125 \ \Rightarrow\ 0.0125 > f'(c) = \sqrt[5]{33} - 2 \ \Rightarrow\ \sqrt[5]{33} < 2.0125$. Therefore, $2 < \sqrt[5]{33} < 2.0125$.

37. (a) $g(x) = f(x^2) \ \Rightarrow\ g'(x) = 2xf'(x^2)$ by the Chain Rule. Since $f'(x) > 0$ for all $x \neq 0$, we must have $f'(x^2) > 0$ for $x \neq 0$, so $g'(x) = 0 \ \Leftrightarrow\ x = 0$. Now $g'(x)$ changes sign (from negative to positive) at $x = 0$, since one of its factors, $f'(x^2)$, is positive for all x, and its other factor, $2x$, changes from negative to positive at this point, so by the First Derivative Test, f has a local and absolute minimum at $x = 0$.

(b) $g'(x) = 2xf'(x^2) \ \Rightarrow\ g''(x) = 2[xf''(x^2)(2x) + f'(x^2)] = 4x^2 f''(x^2) + 2f'(x^2)$ by the Product Rule and the Chain Rule. But $x^2 > 0$ for all $x \neq 0$, $f''(x^2) > 0$ [since f is CU for $x > 0$], and $f'(x^2) > 0$ for all $x \neq 0$, so since all of its factors are positive, $g''(x) > 0$ for $x \neq 0$. Whether $g''(0)$ is positive or 0 doesn't matter [since the sign of g'' does not change there]; g is concave upward on $\mathbb{R}$.

39. If $B = 0$, the line is vertical and the distance from $x = -\frac{C}{A}$ to (x_1, y_1) is $\left|x_1 + \frac{C}{A}\right| = \frac{|Ax_1 + By_1 + C|}{\sqrt{A^2 + B^2}}$, so assume $B \neq 0$. The square of the distance from (x_1, y_1) to the line is $f(x) = (x - x_1)^2 + (y - y_1)^2$ where $Ax + By + C = 0$, so we minimize $f(x) = (x - x_1)^2 + \left(-\frac{A}{B}x - \frac{C}{B} - y_1\right)^2 \ \Rightarrow\ f'(x) = 2(x - x_1) + 2\left(-\frac{A}{B}x - \frac{C}{B} - y_1\right)\left(-\frac{A}{B}\right)$.

$f'(x) = 0 \ \Rightarrow\ x = \frac{B^2 x_1 - ABy_1 - AC}{A^2 + B^2}$ and this gives a minimum since $f''(x) = 2\left(1 + \frac{A^2}{B^2}\right) > 0$. Substituting this value of x into $f(x)$ and simplifying gives $f(x) = \frac{(Ax_1 + By_1 + C)^2}{A^2 + B^2}$, so the minimum distance is

$\sqrt{f(x)} = \frac{|Ax_1 + By_1 + C|}{\sqrt{A^2 + B^2}}$.

41. 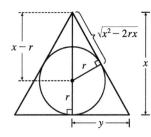 By similar triangles, $\dfrac{y}{x} = \dfrac{r}{\sqrt{x^2 - 2rx}}$, so the area of the triangle is

$$A(x) = \tfrac{1}{2}(2y)x = xy = \dfrac{rx^2}{\sqrt{x^2 - 2rx}} \quad \Rightarrow$$

$$A'(x) = \dfrac{2rx\sqrt{x^2 - 2rx} - rx^2(x - r)/\sqrt{x^2 - 2rx}}{x^2 - 2rx} = \dfrac{rx^2(x - 3r)}{(x^2 - 2rx)^{3/2}} = 0$$

when $x = 3r$.

$A'(x) < 0$ when $2r < x < 3r$, $A'(x) > 0$ when $x > 3r$. So $x = 3r$ gives a minimum and $A(3r) = \dfrac{r(9r^2)}{\sqrt{3}\,r} = 3\sqrt{3}\,r^2$.

43. 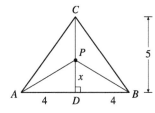 We minimize $L(x) = |PA| + |PB| + |PC| = 2\sqrt{x^2 + 16} + (5 - x)$,

$0 \le x \le 5$. $L'(x) = 2x/\sqrt{x^2 + 16} - 1 = 0 \Leftrightarrow 2x = \sqrt{x^2 + 16} \Leftrightarrow$

$4x^2 = x^2 + 16 \Leftrightarrow x = \tfrac{4}{\sqrt{3}}$. $L(0) = 13$, $L\left(\tfrac{4}{\sqrt{3}}\right) \approx 11.9$, $L(5) \approx 12.8$, so the

minimum occurs when $x = \tfrac{4}{\sqrt{3}} \approx 2.3$.

45. $v = K\sqrt{\dfrac{L}{C} + \dfrac{C}{L}} \Rightarrow \dfrac{dv}{dL} = \dfrac{K}{2\sqrt{(L/C) + (C/L)}}\left(\dfrac{1}{C} - \dfrac{C}{L^2}\right) = 0 \Leftrightarrow \dfrac{1}{C} = \dfrac{C}{L^2} \Leftrightarrow L^2 = C^2 \Leftrightarrow L = C$.

This gives the minimum velocity since $v' < 0$ for $0 < L < C$ and $v' > 0$ for $L > C$.

47. Let x denote the number of $1 decreases in ticket price. Then the ticket price is $12 − $1($x$), and the average attendance is $11{,}000 + 1000(x)$. Now the revenue per game is

$$R(x) = (\text{price per person}) \times (\text{number of people per game})$$

$$= (12 - x)(11{,}000 + 1000x) = -1000x^2 + 1000x + 132{,}000$$

for $0 \le x \le 4$ [since the seating capacity is 15,000] $\Rightarrow R'(x) = -2000x + 1000 = 0 \Leftrightarrow x = 0.5$. This is a maximum since $R''(x) = -2000 < 0$ for all x. Now we must check the value of $R(x) = (12 - x)(11{,}000 + 1000x)$ at $x = 0.5$ and at the endpoints of the domain to see which value of x gives the maximum value of R.

$R(0) = (12)(11{,}000) = 132{,}000$, $R(0.5) = (11.5)(11{,}500) = 132{,}250$, and $R(4) = (8)(15{,}000) = 120{,}000$. Thus, the maximum revenue of $132,250 per game occurs when the average attendance is 11,500 and the ticket price is $11.50.

49. $f(x) = x^5 - x^4 + 3x^2 - 3x - 2 \Rightarrow f'(x) = 5x^4 - 4x^3 + 6x - 3$, so $x_{n+1} = x_n - \dfrac{x_n^5 - x_n^4 + 3x_n^2 - 3x_n - 2}{5x_n^4 - 4x_n^3 + 6x_n - 3}$.

Now $x_1 = 1 \Rightarrow x_2 = 1.5 \Rightarrow x_3 \approx 1.343860 \Rightarrow x_4 \approx 1.300320 \Rightarrow x_5 \approx 1.297396 \Rightarrow$

$x_6 \approx 1.297383 \approx x_7$, so the solution in $[1, 2]$ is 1.297383, to six decimal places.

51. $f(t) = \cos t + t - t^2 \Rightarrow f'(t) = -\sin t + 1 - 2t$. $f'(t)$ exists for all t, so to find the maximum of f, we can examine the zeros of f'. From the graph of f', we see that a good choice for t_1 is $t_1 = 0.3$. Use $g(t) = -\sin t + 1 - 2t$ and $g'(t) = -\cos t - 2$ to obtain $t_2 \approx 0.33535293$, $t_3 \approx 0.33541803 \approx t_4$. Since $f''(t) = -\cos t - 2 < 0$ for all t, $f(0.33541803) \approx 1.16718557$ is the absolute maximum.

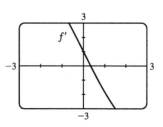

53. $f(x) = 4\sqrt{x} - 6x^2 + 3 = 4x^{1/2} - 6x^2 + 3 \Rightarrow F(x) = 4\left(\tfrac{2}{3}x^{3/2}\right) - 6\left(\tfrac{1}{3}x^3\right) + 3x + C = \tfrac{8}{3}x^{3/2} - 2x^3 + 3x + C$

55. $h(t) = t^{-3} + 5\sin t \Rightarrow H(t) = \begin{cases} -\tfrac{1}{2}t^{-2} - 5\cos t + C_1 & \text{if } t < 0 \\ -\tfrac{1}{2}t^{-2} - 5\cos t + C_2 & \text{if } t > 0 \end{cases}$.

See Example 3.9.1(c) for a similar problem.

57. $f'(t) = 2t - 3\sin t \Rightarrow f(t) = t^2 + 3\cos t + C$.
$f(0) = 3 + C$ and $f(0) = 5 \Rightarrow C = 2$, so $f(t) = t^2 + 3\cos t + 2$.

59. $f''(x) = 1 - 6x + 48x^2 \Rightarrow f'(x) = x - 3x^2 + 16x^3 + C$. $f'(0) = C$ and $f'(0) = 2 \Rightarrow C = 2$, so $f'(x) = x - 3x^2 + 16x^3 + 2$ and hence, $f(x) = \tfrac{1}{2}x^2 - x^3 + 4x^4 + 2x + D$.
$f(0) = D$ and $f(0) = 1 \Rightarrow D = 1$, so $f(x) = \tfrac{1}{2}x^2 - x^3 + 4x^4 + 2x + 1$.

61. $v(t) = s'(t) = 2t - \sin t \Rightarrow s(t) = t^2 + \cos t + C$.
$s(0) = 0 + 1 + C = C + 1$ and $s(0) = 3 \Rightarrow C + 1 = 3 \Rightarrow C = 2$, so $s(t) = t^2 + \cos t + 2$.

63. $f(x) = x^2 \sin(x^2)$, $0 \le x \le \pi$

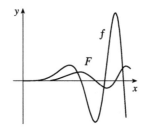

65. Choosing the positive direction to be upward, we have $a(t) = -9.8 \Rightarrow v(t) = -9.8t + v_0$, but $v(0) = 0 = v_0 \Rightarrow v(t) = -9.8t = s'(t) \Rightarrow s(t) = -4.9t^2 + s_0$, but $s(0) = s_0 = 500 \Rightarrow s(t) = -4.9t^2 + 500$. When $s = 0$, $-4.9t^2 + 500 = 0 \Rightarrow t_1 = \sqrt{\tfrac{500}{4.9}} \approx 10.1 \Rightarrow v(t_1) = -9.8\sqrt{\tfrac{500}{4.9}} \approx -98.995$ m/s. Since the canister has been designed to withstand an impact velocity of 100 m/s, the canister will *not burst*.

67. (a)

The cross-sectional area of the rectangular beam is
$A = 2x \cdot 2y = 4xy = 4x\sqrt{100 - x^2}$, $0 \le x \le 10$, so
$\dfrac{dA}{dx} = 4x\left(\tfrac{1}{2}\right)(100 - x^2)^{-1/2}(-2x) + (100 - x^2)^{1/2} \cdot 4$
$= \dfrac{-4x^2}{(100 - x^2)^{1/2}} + 4(100 - x^2)^{1/2} = \dfrac{4[-x^2 + (100 - x^2)]}{(100 - x^2)^{1/2}}$.

$\dfrac{dA}{dx} = 0$ when $-x^2 + (100 - x^2) = 0 \Rightarrow x^2 = 50 \Rightarrow x = \sqrt{50} \approx 7.07 \Rightarrow y = \sqrt{100 - (\sqrt{50})^2} = \sqrt{50}$.

Since $A(0) = A(10) = 0$, the rectangle of maximum area is a square.

(b) 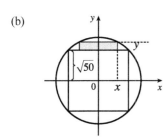 The cross-sectional area of each rectangular plank (shaded in the figure) is

$A = 2x(y - \sqrt{50}) = 2x[\sqrt{100 - x^2} - \sqrt{50}]$, $0 \le x \le \sqrt{50}$, so

$\dfrac{dA}{dx} = 2(\sqrt{100 - x^2} - \sqrt{50}) + 2x(\tfrac{1}{2})(100 - x^2)^{-1/2}(-2x)$

$= 2(100 - x^2)^{1/2} - 2\sqrt{50} - \dfrac{2x^2}{(100 - x^2)^{1/2}}$

Set $\dfrac{dA}{dx} = 0$: $(100 - x^2) - \sqrt{50}(100 - x^2)^{1/2} - x^2 = 0 \Rightarrow 100 - 2x^2 = \sqrt{50}(100 - x^2)^{1/2} \Rightarrow$

$10{,}000 - 400x^2 + 4x^4 = 50(100 - x^2) \Rightarrow 4x^4 - 350x^2 + 5000 = 0 \Rightarrow 2x^4 - 175x^2 + 2500 = 0 \Rightarrow$

$x^2 = \dfrac{175 \pm \sqrt{10{,}625}}{4} \approx 69.52$ or $17.98 \Rightarrow x \approx 8.34$ or 4.24. But $8.34 > \sqrt{50}$, so $x_1 \approx 4.24 \Rightarrow$

$y - \sqrt{50} = \sqrt{100 - x_1^2} - \sqrt{50} \approx 1.99$. Each plank should have dimensions about $8\tfrac{1}{2}$ inches by 2 inches.

(c) From the figure in part (a), the width is $2x$ and the depth is $2y$, so the strength is

$S = k(2x)(2y)^2 = 8kxy^2 = 8kx(100 - x^2) = 800kx - 8kx^3$, $0 \le x \le 10$. $dS/dx = 800k - 24kx^2 = 0$ when

$24kx^2 = 800k \Rightarrow x^2 = \tfrac{100}{3} \Rightarrow x = \tfrac{10}{\sqrt{3}} \Rightarrow y = \sqrt{\tfrac{200}{3}} = \tfrac{10\sqrt{2}}{\sqrt{3}} = \sqrt{2}\,x$. Since $S(0) = S(10) = 0$, the

maximum strength occurs when $x = \tfrac{10}{\sqrt{3}}$. The dimensions should be $\tfrac{20}{\sqrt{3}} \approx 11.55$ inches by $\tfrac{20\sqrt{2}}{\sqrt{3}} \approx 16.33$ inches.

69. Let c be the diameter of the semicircle. Then, from the given figure and the Law of Cosines, $c^2 = a^2 + a^2 - 2a \cdot a \cdot \cos\theta$.

The radius of the semicircle is $\tfrac{1}{2}c$, or $\tfrac{1}{2}\sqrt{2a^2 - 2a^2\cos\theta}$. The area of the figure is given by

$$A(\theta) = \text{area of triangle} + \text{area of circle}$$

$$= \tfrac{1}{2}a \cdot a \cdot \sin\theta + \tfrac{1}{2}\pi\left(\tfrac{1}{2}\sqrt{2a^2 - 2a^2\cos\theta}\right)^2 = \tfrac{1}{2}a^2\sin\theta + \tfrac{1}{8}\pi(2a^2 - 2a^2\cos\theta)$$

$A'(\theta) = \tfrac{1}{2}a^2\cos\theta + \tfrac{1}{4}\pi a^2 \sin\theta = 0 \Rightarrow \tfrac{1}{4}\pi a^2 \sin\theta = -\tfrac{1}{2}a^2\cos\theta \Rightarrow \tan\theta = -\dfrac{2}{\pi} \Rightarrow$

$\theta = \tan^{-1}\left(-\dfrac{2}{\pi}\right) + n\pi$ (n an integer). We let $n = 1$ so that $0 \le \theta \le \pi$, giving

$\theta = \tan^{-1}\left(-\dfrac{2}{\pi}\right) + \pi = \tan^{-1}\left(-\dfrac{2}{\pi}\right) + 180° \approx 147.5°$.

$A''(\theta) = -\tfrac{1}{2}a^2\sin\theta + \tfrac{1}{4}\pi a^2 \cos\theta < 0$ [θ. is a second-quadrant angle, so $\sin\theta > 0$ and $\cos\theta < 0$], so this value of θ gives a maximum.

71. (a) $I = \dfrac{k\cos\theta}{d^2} = \dfrac{k(h/d)}{d^2} = k\dfrac{h}{d^3} = k\dfrac{h}{\left(\sqrt{40^2+h^2}\right)^3} = k\dfrac{h}{(1600+h^2)^{3/2}} \Rightarrow$

$$\dfrac{dI}{dh} = k\dfrac{(1600+h^2)^{3/2} - h\frac{3}{2}(1600+h^2)^{1/2}\cdot 2h}{[(1600+h^2)^{3/2}]^2} = \dfrac{k(1600+h^2)^{1/2}(1600+h^2-3h^2)}{(1600+h^2)^3}$$

$$= \dfrac{k(1600-2h^2)}{(1600+h^2)^{5/2}} \quad \text{[k is the constant of proportionality]}$$

Set $dI/dh = 0$: $1600 - 2h^2 = 0 \Rightarrow h^2 = 800 \Rightarrow h = \sqrt{800} = 20\sqrt{2}$. By the First Derivative Test, I has a local maximum at $h = 20\sqrt{2} \approx 28$ ft.

(b)

$\dfrac{dx}{dt} = 4$ ft/s

$I = \dfrac{k\cos\theta}{d^2} = \dfrac{k[(h-4)/d]}{d^2} = \dfrac{k(h-4)}{d^3}$

$= \dfrac{k(h-4)}{[(h-4)^2+x^2]^{3/2}} = k(h-4)[(h-4)^2+x^2]^{-3/2}$

$\dfrac{dI}{dt} = \dfrac{dI}{dx}\cdot\dfrac{dx}{dt} = k(h-4)\left(-\tfrac{3}{2}\right)[(h-4)^2+x^2]^{-5/2}\cdot 2x\cdot\dfrac{dx}{dt}$

$= k(h-4)(-3x)[(h-4)^2+x^2]^{-5/2}\cdot 4 = \dfrac{-12xk(h-4)}{[(h-4)^2+x^2]^{5/2}}$

$\left.\dfrac{dI}{dt}\right|_{x=40} = -\dfrac{480k(h-4)}{[(h-4)^2+1600]^{5/2}}$

PROBLEMS PLUS

1. Let $f(x) = \sin x - \cos x$ on $[0, 2\pi]$ since f has period 2π. $f'(x) = \cos x + \sin x = 0 \Leftrightarrow \cos x = -\sin x \Leftrightarrow \tan x = -1 \Leftrightarrow x = \frac{3\pi}{4}$ or $\frac{7\pi}{4}$. Evaluating f at its critical numbers and endpoints, we get $f(0) = -1$, $f\left(\frac{3\pi}{4}\right) = \sqrt{2}$, $f\left(\frac{7\pi}{4}\right) = -\sqrt{2}$, and $f(2\pi) = -1$. So f has absolute maximum value $\sqrt{2}$ and absolute minimum value $-\sqrt{2}$. Thus, $-\sqrt{2} \leq \sin x - \cos x \leq \sqrt{2} \Rightarrow |\sin x - \cos x| \leq \sqrt{2}$.

3. $y = \dfrac{\sin x}{x} \Rightarrow y' = \dfrac{x \cos x - \sin x}{x^2} \Rightarrow y'' = \dfrac{-x^2 \sin x - 2x \cos x + 2 \sin x}{x^3}$. If (x, y) is an inflection point, then $y'' = 0 \Rightarrow (2 - x^2) \sin x = 2x \cos x \Rightarrow (2 - x^2)^2 \sin^2 x = 4x^2 \cos^2 x \Rightarrow$
$(2 - x^2)^2 \sin^2 x = 4x^2 (1 - \sin^2 x) \Rightarrow (4 - 4x^2 + x^4) \sin^2 x = 4x^2 - 4x^2 \sin^2 x \Rightarrow$
$(4 + x^4) \sin^2 x = 4x^2 \Rightarrow (x^4 + 4) \dfrac{\sin^2 x}{x^2} = 4 \Rightarrow y^2 (x^4 + 4) = 4$ since $y = \dfrac{\sin x}{x}$.

5. Differentiating $x^2 + xy + y^2 = 12$ implicitly with respect to x gives $2x + y + x \dfrac{dy}{dx} + 2y \dfrac{dy}{dx} = 0$, so $\dfrac{dy}{dx} = -\dfrac{2x + y}{x + 2y}$.

At a highest or lowest point, $\dfrac{dy}{dx} = 0 \Leftrightarrow y = -2x$. Substituting $-2x$ for y in the original equation gives
$x^2 + x(-2x) + (-2x)^2 = 12$, so $3x^2 = 12$ and $x = \pm 2$. If $x = 2$, then $y = -2x = -4$, and if $x = -2$ then $y = 4$. Thus, the highest and lowest points are $(-2, 4)$ and $(2, -4)$.

7. Since
$$\dfrac{f(x + n) - f(x)}{n} = f'(x) \quad (1)$$

holds for all real numbers x and all positive integers n, we have
$$\dfrac{f(x + n) - f(x)}{n} = \dfrac{f(x + 2n) - f(x)}{2n}$$

for every real number x. It follows that
$$f(x + 2n) - 2f(x + n) = -f(x) \quad (2)$$

Now, again from (1), we can write
$$n f'(x) = f(x + n) - f(x)$$

The right-hand side of this equation is differentiable by hypothesis, so the left-hand side is also differentiable. Differentiating and then using (1) again—twice this time—we get
$$n f''(x) = f'(x + n) - f'(x)$$
$$= \dfrac{f(x + 2n) - f(x + n)}{n} - \dfrac{f(x + n) - f(x)}{n} = \dfrac{f(x + 2n) - 2f(x + n) + f(x)}{n}$$

[continued]

Rearranging this last equation and simplifying using **(2)**, we get

$$n^2 f''(x) = f(x+2n) - 2f(x+n) + f(x) = -f(x) + f(x) = 0$$

Thus, $f''(x) = 0$ for all x, so f is a linear function.

9. (a) $y = x^2 \;\Rightarrow\; y' = 2x$, so the slope of the tangent line at $P(a, a^2)$ is $2a$ and the slope of the normal line is $-\dfrac{1}{2a}$ for $a \neq 0$. An equation of the normal line is $y - a^2 = -\dfrac{1}{2a}(x-a)$. Substitute x^2 for y to find the x-coordinates of the two points of intersection of the parabola and the normal line. $x^2 - a^2 = -\dfrac{x}{2a} + \dfrac{1}{2} \;\Leftrightarrow\; x^2 + \left(\dfrac{1}{2a}\right)x - \dfrac{1}{2} - a^2 = 0$. We know that a is a root of this quadratic equation, so $x - a$ is a factor, and we have $(x-a)\left(x + \dfrac{1}{2a} + a\right) = 0$, and hence, $x = -a - \dfrac{1}{2a}$ is the x-coordinate of the point Q. We want to minimize the y-coordinate of Q, which is

$\left(-a - \dfrac{1}{2a}\right)^2 = a^2 + 1 + \dfrac{1}{4a^2} = y(a)$. Now $y'(a) = 2a - \dfrac{1}{2a^3} = \dfrac{4a^4 - 1}{2a^3} = \dfrac{(2a^2+1)(2a^2-1)}{2a^3} = 0 \;\Rightarrow\;$
$a = \dfrac{1}{\sqrt{2}}$ for $a > 0$. Since $y''(a) = 2 + \dfrac{3}{2a^4} > 0$, we see that $a = \dfrac{1}{\sqrt{2}}$ gives us the minimum value of the y-coordinate of Q.

(b) The square S of the distance from $P(a, a^2)$ to $Q\left(-a - \dfrac{1}{2a}, \left(-a - \dfrac{1}{2a}\right)^2\right)$ is given by

$$S = \left(-a - \dfrac{1}{2a} - a\right)^2 + \left[\left(-a - \dfrac{1}{2a}\right)^2 - a^2\right]^2 = \left(-2a - \dfrac{1}{2a}\right)^2 + \left[\left(a^2 + 1 + \dfrac{1}{4a^2}\right) - a^2\right]^2$$

$$= \left(4a^2 + 2 + \dfrac{1}{4a^2}\right) + \left(1 + \dfrac{1}{4a^2}\right)^2 = \left(4a^2 + 2 + \dfrac{1}{4a^2}\right) + 1 + \dfrac{2}{4a^2} + \dfrac{1}{16a^4}$$

$$= 4a^2 + 3 + \dfrac{3}{4a^2} + \dfrac{1}{16a^4}$$

$S' = 8a - \dfrac{6}{4a^3} - \dfrac{4}{16a^5} = 8a - \dfrac{3}{2a^3} - \dfrac{1}{4a^5} = \dfrac{32a^6 - 6a^2 - 1}{4a^5} = \dfrac{(2a^2-1)(4a^2+1)^2}{4a^5}$. The only real positive zero of the equation $S' = 0$ is $a = \dfrac{1}{\sqrt{2}}$. Since $S'' = 8 + \dfrac{9}{2a^4} + \dfrac{5}{4a^6} > 0$, $a = \dfrac{1}{\sqrt{2}}$ corresponds to the shortest possible length of the line segment PQ.

11. $A = (x_1, x_1^2)$ and $B = (x_2, x_2^2)$, where x_1 and x_2 are the solutions of the quadratic equation $x^2 = mx + b$. Let $P = (x, x^2)$ and set $A_1 = (x_1, 0)$, $B_1 = (x_2, 0)$, and $P_1 = (x, 0)$. Let $f(x)$ denote the area of triangle PAB. Then $f(x)$ can be expressed in terms of the areas of three trapezoids as follows:

$$f(x) = \text{area}\,(A_1ABB_1) - \text{area}\,(A_1APP_1) - \text{area}\,(B_1BPP_1)$$
$$= \tfrac{1}{2}(x_1^2 + x_2^2)(x_2 - x_1) - \tfrac{1}{2}(x_1^2 + x^2)(x - x_1) - \tfrac{1}{2}(x^2 + x_2^2)(x_2 - x)$$

[continued]

After expanding and canceling terms, we get

$f(x) = \frac{1}{2}(x_2x_1^2 - x_1x_2^2 - xx_1^2 + x_1x^2 - x_2x^2 + xx_2^2) = \frac{1}{2}[x_1^2(x_2 - x) + x_2^2(x - x_1) + x^2(x_1 - x_2)]$

$f'(x) = \frac{1}{2}[-x_1^2 + x_2^2 + 2x(x_1 - x_2)]. \quad f''(x) = \frac{1}{2}[2(x_1 - x_2)] = x_1 - x_2 < 0$ since $x_2 > x_1$.

$f'(x) = 0 \;\Rightarrow\; 2x(x_1 - x_2) = x_1^2 - x_2^2 \;\Rightarrow\; x_P = \frac{1}{2}(x_1 + x_2)$.

$f(x_P) = \frac{1}{2}\big(x_1^2\big[\frac{1}{2}(x_2 - x_1)\big] + x_2^2\big[\frac{1}{2}(x_2 - x_1)\big] + \frac{1}{4}(x_1 + x_2)^2(x_1 - x_2)\big)$

$= \frac{1}{2}\big[\frac{1}{2}(x_2 - x_1)(x_1^2 + x_2^2) - \frac{1}{4}(x_2 - x_1)(x_1 + x_2)^2\big] = \frac{1}{8}(x_2 - x_1)\big[2(x_1^2 + x_2^2) - (x_1^2 + 2x_1x_2 + x_2^2)\big]$

$= \frac{1}{8}(x_2 - x_1)(x_1^2 - 2x_1x_2 + x_2^2) = \frac{1}{8}(x_2 - x_1)(x_1 - x_2)^2 = \frac{1}{8}(x_2 - x_1)(x_2 - x_1)^2 = \frac{1}{8}(x_2 - x_1)^3$

To put this in terms of m and b, we solve the system $y = x_1^2$ and $y = mx_1 + b$, giving us $x_1^2 - mx_1 - b = 0 \;\Rightarrow\;$
$x_1 = \frac{1}{2}(m - \sqrt{m^2 + 4b})$. Similarly, $x_2 = \frac{1}{2}(m + \sqrt{m^2 + 4b})$. The area is then $\frac{1}{8}(x_2 - x_1)^3 = \frac{1}{8}(\sqrt{m^2 + 4b})^3$,
and is attained at the point $P(x_P, x_P^2) = P(\frac{1}{2}m, \frac{1}{4}m^2)$.

Note: Another way to get an expression for $f(x)$ is to use the formula for an area of a triangle in terms of the coordinates of the vertices: $f(x) = \frac{1}{2}\big[(x_2x_1^2 - x_1x_2^2) + (x_1x^2 - xx_1^2) + (xx_2^2 - x_2x^2)\big]$.

13. $f(x) = (a^2 + a - 6)\cos 2x + (a - 2)x + \cos 1 \;\Rightarrow\; f'(x) = -(a^2 + a - 6)\sin 2x \,(2) + (a - 2)$. The derivative exists for all x, so the only possible critical points will occur where $f'(x) = 0 \;\Leftrightarrow\; 2(a - 2)(a + 3)\sin 2x = a - 2 \;\Leftrightarrow\;$
either $a = 2$ or $2(a + 3)\sin 2x = 1$, with the latter implying that $\sin 2x = \dfrac{1}{2(a + 3)}$. Since the range of $\sin 2x$ is $[-1, 1]$,
this equation has no solution whenever either $\dfrac{1}{2(a + 3)} < -1$ or $\dfrac{1}{2(a + 3)} > 1$. Solving these inequalities, we get
$-\frac{7}{2} < a < -\frac{5}{2}$.

15. (a) Let $y = |AD|$, $x = |AB|$, and $1/x = |AC|$, so that $|AB| \cdot |AC| = 1$. We compute the area $\mathcal{A}$ of $\triangle ABC$ in two ways.

First, $\mathcal{A} = \frac{1}{2}|AB||AC|\sin\frac{2\pi}{3} = \frac{1}{2} \cdot 1 \cdot \frac{\sqrt{3}}{2} = \frac{\sqrt{3}}{4}$.

Second,

$\mathcal{A} = (\text{area of } \triangle ABD) + (\text{area of } \triangle ACD) = \frac{1}{2}|AB||AD|\sin\frac{\pi}{3} + \frac{1}{2}|AD||AC|\sin\frac{\pi}{3}$

$= \frac{1}{2}xy\frac{\sqrt{3}}{2} + \frac{1}{2}y(1/x)\frac{\sqrt{3}}{2} = \frac{\sqrt{3}}{4}y(x + 1/x)$

Equating the two expressions for the area, we get $\frac{\sqrt{3}}{4}y\left(x + \dfrac{1}{x}\right) = \frac{\sqrt{3}}{4} \;\Leftrightarrow\; y = \dfrac{1}{x + 1/x} = \dfrac{x}{x^2 + 1}$, $x > 0$.

Another method: Use the Law of Sines on the triangles ABD and ABC. In $\triangle ABD$, we have
$\angle A + \angle B + \angle D = 180° \;\Leftrightarrow\; 60° + \alpha + \angle D = 180° \;\Leftrightarrow\; \angle D = 120° - \alpha$. Thus,

$\dfrac{x}{y} = \dfrac{\sin(120° - \alpha)}{\sin\alpha} = \dfrac{\sin 120° \cos\alpha - \cos 120° \sin\alpha}{\sin\alpha} = \dfrac{\frac{\sqrt{3}}{2}\cos\alpha + \frac{1}{2}\sin\alpha}{\sin\alpha} \;\Rightarrow\; \dfrac{x}{y} = \dfrac{\sqrt{3}}{2}\cot\alpha + \frac{1}{2}$, and by a similar argument with $\triangle ABC$, $\frac{\sqrt{3}}{2}\cot\alpha = x^2 + \frac{1}{2}$. Eliminating $\cot\alpha$ gives $\frac{x}{y} = (x^2 + \frac{1}{2}) + \frac{1}{2}$ $\Rightarrow$

$$y = \frac{x}{x^2 + 1}, \quad x > 0.$$

(b) We differentiate our expression for y with respect to x to find the maximum:

$$\frac{dy}{dx} = \frac{(x^2 + 1) - x(2x)}{(x^2 + 1)^2} = \frac{1 - x^2}{(x^2 + 1)^2} = 0 \text{ when } x = 1.$$ This indicates a maximum by the First Derivative Test, since

$y'(x) > 0$ for $0 < x < 1$ and $y'(x) < 0$ for $x > 1$, so the maximum value of y is $y(1) = \frac{1}{2}$.

17. (a) $A = \frac{1}{2}bh$ with $\sin\theta = h/c$, so $A = \frac{1}{2}bc\sin\theta$. But A is a constant,

so differentiating this equation with respect to t, we get

$$\frac{dA}{dt} = 0 = \frac{1}{2}\left[bc\cos\theta\,\frac{d\theta}{dt} + b\frac{dc}{dt}\sin\theta + \frac{db}{dt}c\sin\theta\right] \Rightarrow$$

$$bc\cos\theta\,\frac{d\theta}{dt} = -\sin\theta\left[b\frac{dc}{dt} + c\frac{db}{dt}\right] \Rightarrow \frac{d\theta}{dt} = -\tan\theta\left[\frac{1}{c}\frac{dc}{dt} + \frac{1}{b}\frac{db}{dt}\right].$$

(b) We use the Law of Cosines to get the length of side a in terms of those of b and c, and then we differentiate implicitly with respect to t: $a^2 = b^2 + c^2 - 2bc\cos\theta$ $\Rightarrow$

$$2a\frac{da}{dt} = 2b\frac{db}{dt} + 2c\frac{dc}{dt} - 2\left[bc(-\sin\theta)\frac{d\theta}{dt} + b\frac{dc}{dt}\cos\theta + \frac{db}{dt}c\cos\theta\right] \Rightarrow$$

$$\frac{da}{dt} = \frac{1}{a}\left(b\frac{db}{dt} + c\frac{dc}{dt} + bc\sin\theta\,\frac{d\theta}{dt} - b\frac{dc}{dt}\cos\theta - c\frac{db}{dt}\cos\theta\right).$$ Now we substitute our value of a from the Law of

Cosines and the value of $d\theta/dt$ from part (a), and simplify (primes signify differentiation by t):

$$\frac{da}{dt} = \frac{bb' + cc' + bc\sin\theta\left[-\tan\theta(c'/c + b'/b)\right] - (bc' + cb')(\cos\theta)}{\sqrt{b^2 + c^2 - 2bc\cos\theta}}$$

$$= \frac{bb' + cc' - [\sin^2\theta\,(bc' + cb') + \cos^2\theta\,(bc' + cb')]/\cos\theta}{\sqrt{b^2 + c^2 - 2bc\cos\theta}} = \frac{bb' + cc' - (bc' + cb')\sec\theta}{\sqrt{b^2 + c^2 - 2bc\cos\theta}}$$

or $\dfrac{da}{dt} = \dfrac{b\dfrac{db}{dt} + c\dfrac{dc}{dt} - \left(b\dfrac{dc}{dt} + c\dfrac{db}{dt}\right)\sec\theta}{\sqrt{b^2 + c^2 - 2bc\cos\theta}}$.

19. (a) Distance = rate × time, so time = distance/rate. $T_1 = \dfrac{D}{c_1}$, $T_2 = \dfrac{2|PR|}{c_1} + \dfrac{|RS|}{c_2} = \dfrac{2h\sec\theta}{c_1} + \dfrac{D - 2h\tan\theta}{c_2}$,

$$T_3 = \frac{2\sqrt{h^2 + D^2/4}}{c_1} = \frac{\sqrt{4h^2 + D^2}}{c_1}.$$

(b) $\dfrac{dT_2}{d\theta} = \dfrac{2h}{c_1}\cdot\sec\theta\tan\theta - \dfrac{2h}{c_2}\sec^2\theta = 0$ when $2h\sec\theta\left(\dfrac{1}{c_1}\tan\theta - \dfrac{1}{c_2}\sec\theta\right) = 0$ $\Rightarrow$

$\dfrac{1}{c_1}\dfrac{\sin\theta}{\cos\theta} - \dfrac{1}{c_2}\dfrac{1}{\cos\theta} = 0$ $\Rightarrow$ $\dfrac{\sin\theta}{c_1\cos\theta} = \dfrac{1}{c_2\cos\theta}$ $\Rightarrow$ $\sin\theta = \dfrac{c_1}{c_2}$. The First Derivative Test shows that this gives

a minimum.

(c) Using part (a) with $D = 1$ and $T_1 = 0.26$, we have $T_1 = \dfrac{D}{c_1}$ $\Rightarrow$ $c_1 = \dfrac{1}{0.26} \approx 3.85$ km/s. $T_3 = \dfrac{\sqrt{4h^2 + D^2}}{c_1}$ $\Rightarrow$

$4h^2 + D^2 = T_3^2 c_1^2$ $\Rightarrow$ $h = \tfrac{1}{2}\sqrt{T_3^2 c_1^2 - D^2} = \tfrac{1}{2}\sqrt{(0.34)^2(1/0.26)^2 - 1^2} \approx 0.42$ km. To find c_2, we use $\sin\theta = \dfrac{c_1}{c_2}$

from part (b) and $T_2 = \dfrac{2h\sec\theta}{c_1} + \dfrac{D - 2h\tan\theta}{c_2}$ from part (a). From the figure,

$\sin\theta = \dfrac{c_1}{c_2}$ $\Rightarrow$ $\sec\theta = \dfrac{c_2}{\sqrt{c_2^2 - c_1^2}}$ and $\tan\theta = \dfrac{c_1}{\sqrt{c_2^2 - c_1^2}}$, so

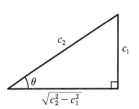

$T_2 = \dfrac{2hc_2}{c_1\sqrt{c_2^2 - c_1^2}} + \dfrac{D\sqrt{c_2^2 - c_1^2} - 2hc_1}{c_2\sqrt{c_2^2 - c_1^2}}$. Using the values for T_2 [given as 0.32],

h, c_1, and D, we can graph $Y_1 = T_2$ and $Y_2 = \dfrac{2hc_2}{c_1\sqrt{c_2^2 - c_1^2}} + \dfrac{D\sqrt{c_2^2 - c_1^2} - 2hc_1}{c_2\sqrt{c_2^2 - c_1^2}}$ and find their intersection

points. Doing so gives us $c_2 \approx 4.10$ and 7.66, but if $c_2 = 4.10$, then $\theta = \sin^{-1}(c_1/c_2) \approx 69.6°$, which implies that point S is to the left of point R in the diagram. So $c_2 = 7.66$ km/s.

21.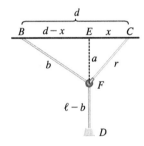

Let $a = |EF|$ and $b = |BF|$ as shown in the figure.
Since $\ell = |BF| + |FD|$, $|FD| = \ell - b$. Now

$$|ED| = |EF| + |FD| = a + \ell - b$$
$$\sqrt{r^2 - x^2} + \ell - \sqrt{(d-x)^2 + a^2}$$
$$= \sqrt{r^2 - x^2} + \ell - \sqrt{(d-x)^2 + \left(\sqrt{r^2 - x^2}\right)^2}$$
$$= \sqrt{r^2 - x^2} + \ell - \sqrt{d^2 - 2dx + x^2 + r^2 - x^2}$$

Let $f(x) = \sqrt{r^2 - x^2} + \ell - \sqrt{d^2 + r^2 - 2dx}$.

$f'(x) = \tfrac{1}{2}(r^2 - x^2)^{-1/2}(-2x) - \tfrac{1}{2}(d^2 + r^2 - 2dx)^{-1/2}(-2d) = \dfrac{-x}{\sqrt{r^2 - x^2}} + \dfrac{d}{\sqrt{d^2 + r^2 - 2dx}}$.

$f'(x) = 0$ $\Rightarrow$ $\dfrac{x}{\sqrt{r^2 - x^2}} = \dfrac{d}{\sqrt{d^2 + r^2 - 2dx}}$ $\Rightarrow$ $\dfrac{x^2}{r^2 - x^2} = \dfrac{d^2}{d^2 + r^2 - 2dx}$ $\Rightarrow$

$d^2 x^2 + r^2 x^2 - 2dx^3 = d^2 r^2 - d^2 x^2$ $\Rightarrow$ $0 = 2dx^3 - 2d^2 x^2 - r^2 x^2 + d^2 r^2$ $\Rightarrow$

$0 = 2dx^2(x - d) - r^2(x^2 - d^2)$ $\Rightarrow$ $0 = 2dx^2(x - d) - r^2(x + d)(x - d)$ $\Rightarrow$ $0 = (x - d)\left[2dx^2 - r^2(x + d)\right]$

But $d > r > x$, so $x \neq d$. Thus, we solve $2dx^2 - r^2 x - dr^2 = 0$ for x:

$x = \dfrac{-(-r^2) \pm \sqrt{(-r^2)^2 - 4(2d)(-dr^2)}}{2(2d)} = \dfrac{r^2 \pm \sqrt{r^4 + 8d^2 r^2}}{4d}$. Because $\sqrt{r^4 + 8d^2 r^2} > r^2$, the "negative" can be

discarded. Thus, $x = \dfrac{r^2 + \sqrt{r^2}\sqrt{r^2 + 8d^2}}{4d} = \dfrac{r^2 + r\sqrt{r^2 + 8d^2}}{4d}$ $[r > 0]$ $= \dfrac{r}{4d}\left(r + \sqrt{r^2 + 8d^2}\right)$. The maximum

value of $|ED|$ occurs at this value of x.

23. $V = \frac{4}{3}\pi r^3$ $\Rightarrow$ $\frac{dV}{dt} = 4\pi r^2 \frac{dr}{dt}$. But $\frac{dV}{dt}$ is proportional to the surface area, so $\frac{dV}{dt} = k \cdot 4\pi r^2$ for some constant k.

Therefore, $4\pi r^2 \frac{dr}{dt} = k \cdot 4\pi r^2$ $\Leftrightarrow$ $\frac{dr}{dt} = k =$ constant. An antiderivative of k with respect to t is kt, so $r = kt + C$.

When $t = 0$, the radius r must equal the original radius r_0, so $C = r_0$, and $r = kt + r_0$. To find k we use the fact that when $t = 3$, $r = 3k + r_0$ and $V = \frac{1}{2}V_0$ $\Rightarrow$ $\frac{4}{3}\pi(3k + r_0)^3 = \frac{1}{2} \cdot \frac{4}{3}\pi r_0^3$ $\Rightarrow$ $(3k + r_0)^3 = \frac{1}{2}r_0^3$ $\Rightarrow$

$3k + r_0 = \frac{1}{\sqrt[3]{2}}r_0$ $\Rightarrow$ $k = \frac{1}{3}r_0\left(\frac{1}{\sqrt[3]{2}} - 1\right)$. Since $r = kt + r_0$, $r = \frac{1}{3}r_0\left(\frac{1}{\sqrt[3]{2}} - 1\right)t + r_0$. When the snowball

has melted completely we have $r = 0$ $\Rightarrow$ $\frac{1}{3}r_0\left(\frac{1}{\sqrt[3]{2}} - 1\right)t + r_0 = 0$ which gives $t = \frac{3\sqrt[3]{2}}{\sqrt[3]{2} - 1}$. Hence, it takes

$\frac{3\sqrt[3]{2}}{\sqrt[3]{2} - 1} - 3 = \frac{3}{\sqrt[3]{2} - 1} \approx 11$ h 33 min longer.

4 ☐ INTEGRALS

4.1 The Area and Distance Problems

1. (a) Since f is *decreasing*, we can obtain a *lower* estimate by using *right* endpoints. We are instructed to use five rectangles, so $n = 5$.

$$R_5 = \sum_{i=1}^{5} f(x_i)\,\Delta x \quad \left[\Delta x = \frac{b-a}{n} = \frac{10-0}{5} = 2\right]$$

$$= f(x_1) \cdot 2 + f(x_2) \cdot 2 + f(x_3) \cdot 2 + f(x_4) \cdot 2 + f(x_5) \cdot 2$$

$$= 2[f(2) + f(4) + f(6) + f(8) + f(10)]$$

$$\approx 2(3.2 + 1.8 + 0.8 + 0.2 + 0)$$

$$= 2(6) = 12$$

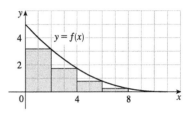

Since f is *decreasing*, we can obtain an *upper* estimate by using *left* endpoints.

$$L_5 = \sum_{i=1}^{5} f(x_{i-1})\,\Delta x$$

$$= f(x_0) \cdot 2 + f(x_1) \cdot 2 + f(x_2) \cdot 2 + f(x_3) \cdot 2 + f(x_4) \cdot 2$$

$$= 2[f(0) + f(2) + f(4) + f(6) + f(8)]$$

$$\approx 2(5 + 3.2 + 1.8 + 0.8 + 0.2)$$

$$= 2(11) = 22$$

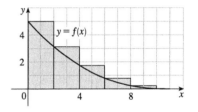

(b) $R_{10} = \sum_{i=1}^{10} f(x_i)\,\Delta x \quad \left[\Delta x = \frac{10-0}{10} = 1\right]$

$$= 1[f(x_1) + f(x_2) + \cdots + f(x_{10})]$$

$$= f(1) + f(2) + \cdots + f(10)$$

$$\approx 4 + 3.2 + 2.5 + 1.8 + 1.3 + 0.8 + 0.5 + 0.2 + 0.1 + 0$$

$$= 14.4$$

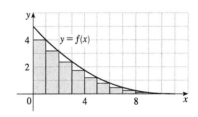

$L_{10} = \sum_{i=1}^{10} f(x_{i-1})\,\Delta x$

$$= f(0) + f(1) + \cdots + f(9)$$

$$= R_{10} + 1 \cdot f(0) - 1 \cdot f(10) \quad \left[\begin{array}{l}\text{add leftmost upper rectangle,} \\ \text{subtract rightmost lower rectangle}\end{array}\right]$$

$$= 14.4 + 5 - 0$$

$$= 19.4$$

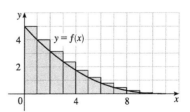

3. (a) $R_4 = \sum_{i=1}^{4} f(x_i)\,\Delta x \quad \left[\Delta x = \dfrac{2-1}{4} = \dfrac{1}{4}\right] = \left[\sum_{i=1}^{4} f(x_i)\right]\Delta x$

$= [f(x_1) + f(x_2) + f(x_3) + f(x_4)]\,\Delta x$

$= \left[\dfrac{1}{5/4} + \dfrac{1}{6/4} + \dfrac{1}{7/4} + \dfrac{1}{8/4}\right]\dfrac{1}{4} = \left[\dfrac{4}{5} + \dfrac{2}{3} + \dfrac{4}{7} + \dfrac{1}{2}\right]\dfrac{1}{4} \approx 0.6345$

Since f is *decreasing* on $[1,2]$, an *underestimate* is obtained by using the *right* endpoint approximation, R_4.

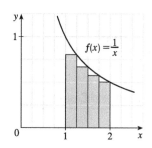

(b) $L_4 = \sum_{i=1}^{4} f(x_{i-1})\,\Delta x = \left[\sum_{i=1}^{4} f(x_{i-1})\right]\Delta x$

$= [f(x_0) + f(x_1) + f(x_2) + f(x_3)]\,\Delta x$

$= \left[\dfrac{1}{1} + \dfrac{1}{5/4} + \dfrac{1}{6/4} + \dfrac{1}{7/4}\right]\dfrac{1}{4} = \left[1 + \dfrac{4}{5} + \dfrac{2}{3} + \dfrac{4}{7}\right]\dfrac{1}{4} \approx 0.7595$

L_4 is an overestimate. Alternatively, we could just add the area of the leftmost upper rectangle and subtract the area of the rightmost lower rectangle; that is, $L_4 = R_4 + f(1)\cdot\dfrac{1}{4} - f(2)\cdot\dfrac{1}{4}$.

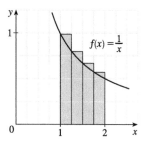

5. (a) $f(x) = 1 + x^2$ and $\Delta x = \dfrac{2-(-1)}{3} = 1 \;\Rightarrow$

$R_3 = 1\cdot f(0) + 1\cdot f(1) + 1\cdot f(2) = 1\cdot 1 + 1\cdot 2 + 1\cdot 5 = 8.$

$\Delta x = \dfrac{2-(-1)}{6} = 0.5 \;\Rightarrow$

$R_6 = 0.5[f(-0.5) + f(0) + f(0.5) + f(1) + f(1.5) + f(2)]$

$= 0.5(1.25 + 1 + 1.25 + 2 + 3.25 + 5)$

$= 0.5(13.75) = 6.875$

(b) $L_3 = 1\cdot f(-1) + 1\cdot f(0) + 1\cdot f(1) = 1\cdot 2 + 1\cdot 1 + 1\cdot 2 = 5$

$L_6 = 0.5[f(-1) + f(-0.5) + f(0) + f(0.5) + f(1) + f(1.5)]$

$= 0.5(2 + 1.25 + 1 + 1.25 + 2 + 3.25)$

$= 0.5(10.75) = 5.375$

(c) $M_3 = 1\cdot f(-0.5) + 1\cdot f(0.5) + 1\cdot f(1.5)$

$= 1\cdot 1.25 + 1\cdot 1.25 + 1\cdot 3.25 = 5.75$

$M_6 = 0.5[f(-0.75) + f(-0.25) + f(0.25)$
$\qquad + f(0.75) + f(1.25) + f(1.75)]$

$= 0.5(1.5625 + 1.0625 + 1.0625 + 1.5625 + 2.5625 + 4.0625)$

$= 0.5(11.875) = 5.9375$

(d) M_6 appears to be the best estimate.

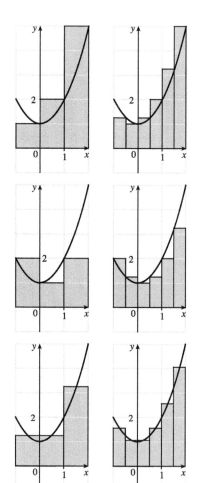

7. $f(x) = 6 - x^2$, $-2 \leq x \leq 2$, $\Delta x = \dfrac{2-(-2)}{n} = \dfrac{4}{n}$.

$n = 2$:

The maximum values of f on both subintervals occur at $x = 0$, so

upper sum $= f(0) \cdot 2 + f(0) \cdot 2 = 6 \cdot 2 + 6 \cdot 2 = 24$.

The minimum values of f on the subintervals occur at $x = -2$ and $x = 2$, so

lower sum $= f(-2) \cdot 2 + f(2) \cdot 2 = 2 \cdot 2 + 2 \cdot 2 = 8$.

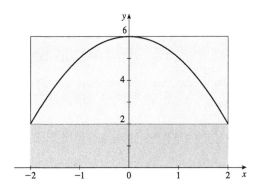

$n = 4$:

upper sum $= [f(-1) + f(0) + f(0) + f(1)](1)$
$= [5 + 6 + 6 + 5](1)$
$= 22$

lower sum $= [f(-2) + f(-1) + f(1) + f(2)](1)$
$= [2 + 5 + 5 + 2](1)$
$= 14$

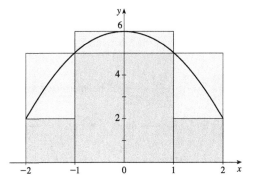

$n = 8$:

upper sum $= [f(-1.5) + f(-1) + f(-0.5) + f(0)$
$\quad + f(0) + f(0.5) + f(1) + f(1.5)](0.5)$
$= 20.5$

lower sum $= [f(-2) + f(-1.5) + f(-1) + f(-0.5)$
$\quad + f(0.5) + f(1) + f(1.5) + f(2)](0.5)$
$= 16.5$

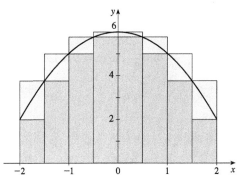

9. Since v is an increasing function, L_6 will give us a lower estimate and R_6 will give us an upper estimate.

$L_6 = (0 \text{ ft/s})(0.5 \text{ s}) + (6.2)(0.5) + (10.8)(0.5) + (14.9)(0.5) + (18.1)(0.5) + (19.4)(0.5) = 0.5(69.4) = 34.7$ ft

$R_6 = 0.5(6.2 + 10.8 + 14.9 + 18.1 + 19.4 + 20.2) = 0.5(89.6) = 44.8$ ft

11. Lower estimate for oil leakage: $R_5 = (7.6 + 6.8 + 6.2 + 5.7 + 5.3)(2) = (31.6)(2) = 63.2$ L.

Upper estimate for oil leakage: $L_5 = (8.7 + 7.6 + 6.8 + 6.2 + 5.7)(2) = (35)(2) = 70$ L.

13. For a decreasing function, using left endpoints gives us an overestimate and using right endpoints results in an underestimate. We will use M_6 to get an estimate. $\Delta t = 1$, so

$M_6 = 1[v(0.5) + v(1.5) + v(2.5) + v(3.5) + v(4.5) + v(5.5)] \approx 55 + 40 + 28 + 18 + 10 + 4 = 155$ ft

For a very rough check on the above calculation, we can draw a line from $(0, 70)$ to $(6, 0)$ and calculate the area of the triangle: $\frac{1}{2}(70)(6) = 210$. This is clearly an overestimate, so our midpoint estimate of 155 is reasonable.

15. $f(t) = -t(t-21)(t+1)$ and $\Delta t = \frac{12-0}{6} = 2$

$$M_6 = 2 \cdot f(1) + 2 \cdot f(3) + 2 \cdot f(5) + 2 \cdot f(7) + 2 \cdot f(9) + 2 \cdot f(11)$$
$$= 2 \cdot 40 + 2 \cdot 216 + 2 \cdot 480 + 2 \cdot 784 + 2 \cdot 1080 + 2 \cdot 1320$$
$$= 7840 \text{ (infected cells/mL)} \cdot \text{days}$$

Thus, the total amount of infection needed to develop symptoms of measles is about 7840 infected cells per mL of blood plasma.

17. $f(x) = 2 + \sin^2 x$, $0 \le x \le \pi$. $\Delta x = (\pi - 0)/n = \pi/n$ and $x_i = 0 + i\Delta x = \pi i/n$.

$$A = \lim_{n \to \infty} R_n = \lim_{n \to \infty} \sum_{i=1}^n f(x_i)\Delta x = \lim_{n \to \infty} \sum_{i=1}^n \left[2 + \sin^2(\pi i/n)\right] \cdot \frac{\pi}{n}.$$

19. $f(x) = x\sqrt{x^3 + 8}$, $1 \le x \le 5$. $\Delta x = (5-1)/n = 4/n$ and $x_i = 1 + i\Delta x = 1 + 4i/n$.

$$A = \lim_{n \to \infty} R_n = \lim_{n \to \infty} \sum_{i=1}^n f(x_i)\Delta x = \lim_{n \to \infty} \sum_{i=1}^n \left[(1 + 4i/n)\sqrt{(1 + 4i/n)^3 + 8}\right] \cdot \frac{4}{n}.$$

21. $\lim_{n \to \infty} \sum_{i=1}^n \frac{2}{n} \frac{1}{1 + (2i/n)}$ can be interpreted as the area of the region lying under the graph of $y = \frac{1}{1+x}$ on the interval $[0, 2]$,

since for $y = \frac{1}{1+x}$ on $[0, 2]$ with $\Delta x = \frac{2-0}{n} = \frac{2}{n}$, $x_i = 0 + i\Delta x = \frac{2i}{n}$, and $x_i^* = x_i$, the expression for area is

$A = \lim_{n \to \infty} \sum_{i=1}^n f(x_i^*)\Delta x = \lim_{n \to \infty} \sum_{i=1}^n \left(\frac{1}{1+(2i/n)}\right) \cdot \frac{2}{n}$. Note that this answer is not unique. For example, we could also use

$y = \frac{1}{x}$ on $[1, 3]$.

23. $\lim_{n \to \infty} \sum_{i=1}^n \frac{\pi}{4n} \tan \frac{i\pi}{4n}$ can be interpreted as the area of the region lying under the graph of $y = \tan x$ on the interval $\left[0, \frac{\pi}{4}\right]$,

since for $y = \tan x$ on $\left[0, \frac{\pi}{4}\right]$ with $\Delta x = \frac{\pi/4 - 0}{n} = \frac{\pi}{4n}$, $x_i = 0 + i\Delta x = \frac{i\pi}{4n}$, and $x_i^* = x_i$, the expression for the area is

$A = \lim_{n \to \infty} \sum_{i=1}^n f(x_i^*)\Delta x = \lim_{n \to \infty} \sum_{i=1}^n \tan\left(\frac{i\pi}{4n}\right)\frac{\pi}{4n}$. Note that this answer is not unique. For example, we could also use

$y = \tan(x - \pi)$ on $\left[\pi, \frac{5\pi}{4}\right]$.

25. (a) Since f is an increasing function, L_n is an underestimate of A [lower sum] and R_n is an overestimate of A [upper sum].

Thus, A, L_n, and R_n are related by the inequality $L_n < A < R_n$.

(b) $R_n = f(x_1)\Delta x + f(x_2)\Delta x + \cdots + f(x_n)\Delta x$

$L_n = f(x_0)\Delta x + f(x_1)\Delta x + \cdots + f(x_{n-1})\Delta x$

$R_n - L_n = f(x_n)\Delta x - f(x_0)\Delta x$

$\quad = \Delta x[f(x_n) - f(x_0)]$

$\quad = \frac{b-a}{n}[f(b) - f(a)]$

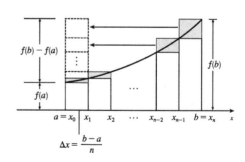

[continued]

In the diagram, $R_n - L_n$ is the sum of the areas of the shaded rectangles. By sliding the shaded rectangles to the left so that they stack on top of the leftmost shaded rectangle, we form a rectangle of height $f(b) - f(a)$ and width $\dfrac{b-a}{n}$.

(c) $A > L_n$, so $R_n - A < R_n - L_n$; that is, $R_n - A < \dfrac{b-a}{n}[f(b) - f(a)]$.

27. Here is one possible algorithm (ordered sequence of operations) for calculating the sums:

    1 Let SUM = 0, X_MIN = 0, X_MAX = 1, N = 10 (depending on which sum we are calculating),

    DELTA_X = (X_MAX - X_MIN)/N, and RIGHT_ENDPOINT = X_MIN + DELTA_X.

    2 Repeat steps 2a, 2b in sequence until RIGHT_ENDPOINT > X_MAX.

    2a Add (RIGHT_ENDPOINT)^4 to SUM.

    2b Add DELTA_X to RIGHT_ENDPOINT.

    At the end of this procedure, (DELTA_X)·(SUM) is equal to the answer we are looking for. We find that

    $R_{10} = \dfrac{1}{10} \sum_{i=1}^{10} \left(\dfrac{i}{10}\right)^4 \approx 0.2533$, $R_{30} = \dfrac{1}{30} \sum_{i=1}^{30} \left(\dfrac{i}{30}\right)^4 \approx 0.2170$, $R_{50} = \dfrac{1}{50} \sum_{i=1}^{50} \left(\dfrac{i}{50}\right)^4 \approx 0.2101$, and

    $R_{100} = \dfrac{1}{100} \sum_{i=1}^{100} \left(\dfrac{i}{100}\right)^4 \approx 0.2050$. It appears that the exact area is 0.2. The following display shows the program SUMRIGHT and its output from a TI-83/4 Plus calculator. To generalize the program, we have input (rather than assign) values for Xmin, Xmax, and N. Also, the function, x^4, is assigned to Y_1, enabling us to evaluate any right sum merely by changing Y_1 and running the program.

29. In Maple, we have to perform a number of steps before getting a numerical answer. After loading the student package [command: `with(student);`] we use the command `left_sum:=leftsum(1/(x^2+1),x=0..1,10 [or 30, or 50]);` which gives us the expression in summation notation. To get a numerical approximation to the sum, we use `evalf(left_sum);`. Mathematica does not have a special command for these sums, so we must type them in manually. For example, the first left sum is given by `(1/10)*Sum[1/(((i-1)/10)^2+1)],{i,1,10}]`, and we use the N command on the resulting output to get a numerical approximation.

[continued]

(a) With $f(x) = \dfrac{1}{x^2+1}$, $0 \le x \le 1$, the left sums are of the form $L_n = \dfrac{1}{n}\sum_{i=1}^{n}\dfrac{1}{\left(\frac{i-1}{n}\right)^2+1}$. Specifically, $L_{10} \approx 0.8100$, $L_{30} \approx 0.7937$, and $L_{50} \approx 0.7904$. The right sums are of the form $R_n = \dfrac{1}{n}\sum_{i=1}^{n}\dfrac{1}{\left(\frac{i}{n}\right)^2+1}$. Specifically, $R_{10} \approx 0.7600$, $R_{30} \approx 0.7770$, and $R_{50} \approx 0.7804$.

(b) In Maple, we use the leftbox (with the same arguments as left_sum) and rightbox commands to generate the graphs.

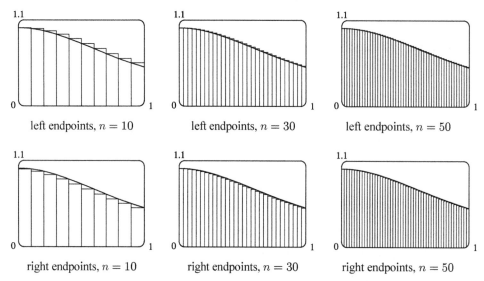

left endpoints, $n = 10$ left endpoints, $n = 30$ left endpoints, $n = 50$

right endpoints, $n = 10$ right endpoints, $n = 30$ right endpoints, $n = 50$

(c) We know that since $y = 1/(x^2+1)$ is a decreasing function on $(0,1)$, all of the left sums are larger than the actual area, and all of the right sums are smaller than the actual area. Since the left sum with $n = 50$ is about $0.7904 < 0.791$ and the right sum with $n = 50$ is about $0.7804 > 0.780$, we conclude that $0.780 < R_{50} <$ exact area $< L_{50} < 0.791$, so the exact area is between 0.780 and 0.791.

31. (a) $y = f(x) = x^5$. $\Delta x = \dfrac{2-0}{n} = \dfrac{2}{n}$ and $x_i = 0 + i\,\Delta x = \dfrac{2i}{n}$.

$$A = \lim_{n\to\infty} R_n = \lim_{n\to\infty}\sum_{i=1}^{n} f(x_i)\,\Delta x = \lim_{n\to\infty}\sum_{i=1}^{n}\left(\dfrac{2i}{n}\right)^5 \cdot \dfrac{2}{n} = \lim_{n\to\infty}\sum_{i=1}^{n}\dfrac{32i^5}{n^5}\cdot\dfrac{2}{n} = \lim_{n\to\infty}\dfrac{64}{n^6}\sum_{i=1}^{n} i^5.$$

(b) $\sum_{i=1}^{n} i^5 \stackrel{\text{CAS}}{=} \dfrac{n^2(n+1)^2(2n^2+2n-1)}{12}$

(c) $\lim_{n\to\infty}\dfrac{64}{n^6}\cdot\dfrac{n^2(n+1)^2(2n^2+2n-1)}{12} = \dfrac{64}{12}\lim_{n\to\infty}\dfrac{(n^2+2n+1)(2n^2+2n-1)}{n^2\cdot n^2}$

$$= \dfrac{16}{3}\lim_{n\to\infty}\left(1+\dfrac{2}{n}+\dfrac{1}{n^2}\right)\left(2+\dfrac{2}{n}-\dfrac{1}{n^2}\right) = \dfrac{16}{3}\cdot 1 \cdot 2 = \dfrac{32}{3}$$

33. $y = f(x) = \cos x.$ $\Delta x = \dfrac{b-0}{n} = \dfrac{b}{n}$ and $x_i = 0 + i\,\Delta x = \dfrac{bi}{n}.$

$$A = \lim_{n\to\infty} R_n = \lim_{n\to\infty} \sum_{i=1}^{n} f(x_i)\,\Delta x = \lim_{n\to\infty} \sum_{i=1}^{n} \cos\left(\dfrac{bi}{n}\right)\cdot \dfrac{b}{n}$$

$$\stackrel{\text{CAS}}{=} \lim_{n\to\infty}\left[\dfrac{b\sin\left(b\left(\dfrac{1}{2n}+1\right)\right)}{2n\sin\left(\dfrac{b}{2n}\right)} - \dfrac{b}{2n}\right] \stackrel{\text{CAS}}{=} \sin b$$

If $b = \dfrac{\pi}{2}$, then $A = \sin \dfrac{\pi}{2} = 1.$

4.2 The Definite Integral

1. $f(x) = x - 1,\ -6 \le x \le 4.\ \Delta x = \dfrac{b-a}{n} = \dfrac{4-(-6)}{5} = 2.$

Since we are using right endpoints, $x_i^* = x_i.$

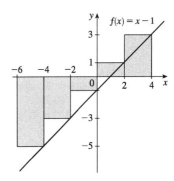

$R_5 = \sum_{i=1}^{5} f(x_i)\,\Delta x$

$ = (\Delta x)[f(x_1) + f(x_2) + f(x_3) + f(x_4) + f(x_5) + f(x_6)]$

$ = 2[f(-4) + f(-2) + f(0) + f(2) + f(4)]$

$ = 2[-5 + (-3) + (-1) + 1 + 3]$

$ = 2(-5) = -10$

The Riemann sum represents the sum of the areas of the two rectangles above the x-axis minus the sum of the areas of the three rectangles below the x-axis; that is, the *net area* of the rectangles with respect to the x-axis.

3. $f(x) = x^2 - 4,\ 0 \le x \le 3.\ \Delta x = \dfrac{b-a}{n} = \dfrac{3-0}{6} = \dfrac{1}{2}.$

Since we are using midpoints, $x_i^* = \overline{x}_i = \tfrac{1}{2}(x_{i-1} + x_i).$

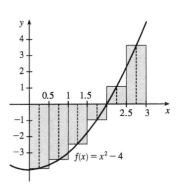

$M_6 = \sum_{i=1}^{6} f(\overline{x}_i)\,\Delta x$

$ = (\Delta x)[f(\overline{x}_1) + f(\overline{x}_2) + f(\overline{x}_3) + f(\overline{x}_4) + f(\overline{x}_5) + f(\overline{x}_6)]$

$ = \tfrac{1}{2}\left[f\!\left(\tfrac{1}{4}\right) + f\!\left(\tfrac{3}{4}\right) + f\!\left(\tfrac{5}{4}\right) + f\!\left(\tfrac{7}{4}\right) + f\!\left(\tfrac{9}{4}\right) + f\!\left(\tfrac{11}{4}\right)\right]$

$ = \tfrac{1}{2}\left(-\tfrac{63}{16} - \tfrac{55}{16} - \tfrac{39}{16} - \tfrac{15}{16} + \tfrac{17}{16} + \tfrac{57}{16}\right) = \tfrac{1}{2}\left(-\tfrac{98}{16}\right) = -\tfrac{49}{16}$

The Riemann sum represents the sum of the areas of the two rectangles above the x-axis minus the sum of the areas of the four rectangles below the x-axis; that is, the *net area* of the rectangles with respect to the x-axis.

5. (a) $\int_0^{10} f(x)\,dx \approx R_5 = [f(2) + f(4) + f(6) + f(8) + f(10)]\,\Delta x$

$\phantom{\text{(a) }\int_0^{10} f(x)\,dx} = [-1 + 1 + 3 + (-1) + 0](2) = 2(2) = 4$

(b) $\int_0^{10} f(x)\,dx \approx L_5 = [f(0) + f(2) + f(4) + f(6) + f(8)]\,\Delta x$

$\phantom{\text{(b) }\int_0^{10} f(x)\,dx} = [-1 + (-1) + 1 + 3 + (-1)](2) = 1(2) = 2$

(c) $\int_0^{10} f(x)\,dx \approx M_5 = [f(1) + f(3) + f(5) + f(7) + f(9)]\,\Delta x$

$\qquad\qquad\qquad\qquad = [2 + 0 + 2 + 1 + (-2)](2) = 3(2) = 6$

7. Since f is increasing, $L_5 \leq \int_{10}^{30} f(x)\,dx \leq R_5$.

Lower estimate $= L_5 = \displaystyle\sum_{i=1}^{5} f(x_{i-1})\Delta x = 4[f(10) + f(14) + f(18) + f(22) + f(26)]$

$\qquad\qquad\qquad\qquad = 4[-12 + (-6) + (-2) + 1 + 3] = 4(-16) = -64$

Upper estimate $= R_5 = \displaystyle\sum_{i=1}^{5} f(x)\Delta x = 4[f(14) + f(18) + f(22) + f(26) + f(30)]$

$\qquad\qquad\qquad\qquad = 4[-6 + (-2) + 1 + 3 + 8] = 4(4) = 16$

9. $\Delta x = (8-0)/4 = 2$, so the endpoints are 0, 2, 4, 6, and 8, and the midpoints are 1, 3, 5, and 7. The Midpoint Rule gives $\displaystyle\int_0^8 x^2\,dx \approx \sum_{i=1}^{4} f(\overline{x}_i)\,\Delta x = 2(1^2 + 3^2 + 5^2 + 7^2) = 2(84) = 168$.

11. $\Delta x = (8-0)/4 = 2$, so the endpoints are 0, 2, 4, 6, and 8, and the midpoints are 1, 3, 5, and 7. The Midpoint Rule gives $\displaystyle\int_0^8 \sin\sqrt{x}\,dx \approx \sum_{i=1}^{4} f(\overline{x}_i)\,\Delta x = 2\left(\sin\sqrt{1} + \sin\sqrt{3} + \sin\sqrt{5} + \sin\sqrt{7}\right) \approx 2(3.0910) = 6.1820$.

13. $\Delta x = (3-1)/5 = \frac{2}{5}$, so the endpoints are 1, $\frac{7}{5}$, $\frac{9}{5}$, $\frac{11}{5}$, $\frac{13}{5}$, and 3, and the midpoints are $\frac{6}{5}$, $\frac{8}{5}$, 2, $\frac{12}{5}$, and $\frac{14}{5}$. The Midpoint Rule gives

$$\int_1^3 \frac{x}{x^2+8}\,dx \approx \sum_{i=1}^{5} f(\overline{x}_i)\,\Delta x = \left(\frac{6/5}{(6/5)^2+8} + \frac{8/5}{(8/5)^2+8} + \frac{2}{2^2+8} + \frac{12/5}{(12/5)^2+8} + \frac{14/5}{(14/5)^2+8}\right)\left(\frac{2}{5}\right)$$

$\qquad\qquad\qquad \approx 0.3186$

15. Using Mathematica and the Riemann Sum notebook from MathWorld, we obtain the following for $f(x) = x/(x^2 + 8)$:

$M_5 \approx 0.318595$ $\qquad\qquad\qquad\qquad$ $M_{10} \approx 0.318144$ $\qquad\qquad\qquad\qquad$ $M_{20} \approx 0.318032$

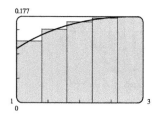 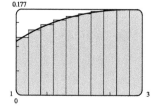 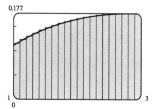

17. We'll create the table of values to approximate $\int_0^\pi \sin x\,dx$ by using the program in the solution to Exercise 4.1.27 with $Y_1 = \sin x$, Xmin $= 0$, Xmax $= \pi$, and $n = 5, 10, 50$, and 100.

The values of R_n appear to be approaching 2.

n	R_n
5	1.933766
10	1.983524
50	1.999342
100	1.999836

19. On $[0, \pi]$, $\lim\limits_{n \to \infty} \sum\limits_{i=1}^{n} \dfrac{\sin x_i}{1 + x_i} \Delta x = \int_0^{\pi} \dfrac{\sin x}{1 + x} \, dx$.

21. On $[2, 7]$, $\lim\limits_{n \to \infty} \sum\limits_{i=1}^{n} [5(x_i^*)^3 - 4x_i^*] \Delta x = \int_2^7 (5x^3 - 4x) \, dx$.

23. For $\int_0^4 (x - x^2) \, dx$, $\Delta x = \dfrac{4 - 0}{n} = \dfrac{4}{n}$, and $x_i = 0 + i\Delta x = \dfrac{4i}{n}$. Then

$$\int_0^4 (x - x^2) \, dx = \lim_{n \to \infty} \sum_{i=1}^{n} f\left(\dfrac{4i}{n}\right) \dfrac{4}{n} = \lim_{n \to \infty} \sum_{i=1}^{n} \left[\left(\dfrac{4i}{n}\right) - \left(\dfrac{4i}{n}\right)^2\right] \dfrac{4}{n} = \lim_{n \to \infty} \dfrac{4}{n} \sum_{i=1}^{n} \left[\dfrac{4i}{n} - \dfrac{16i^2}{n^2}\right] = \lim_{n \to \infty} R_n.$$

$$\lim_{n \to \infty} \dfrac{4}{n} \sum_{i=1}^{n} \left[\dfrac{4i}{n} - \dfrac{16i^2}{n^2}\right] = \lim_{n \to \infty} \dfrac{4}{n} \left[\dfrac{4}{n} \sum_{i=1}^{n} i - \dfrac{16}{n^2} \sum_{i=1}^{n} i^2\right] = \lim_{n \to \infty} \left[\dfrac{16}{n^2} \dfrac{n(n+1)}{2} - \dfrac{64}{n^3} \dfrac{n(n+1)(2n+1)}{6}\right]$$

$$= \lim_{n \to \infty} \left[\dfrac{8}{n}(n+1) - \dfrac{32}{3n^2}(n+1)(2n+1)\right]$$

$$= \lim_{n \to \infty} \left[8\left(1 + \dfrac{1}{n}\right) - \dfrac{32}{3}\left(1 + \dfrac{1}{n}\right)\left(2 + \dfrac{1}{n}\right)\right] = 8(1) - \dfrac{32}{3}(1)(2) = -\dfrac{40}{3}$$

25. $f(x) = \sqrt{4 + x^2}$, $a = 1$, $b = 3$, and $\Delta x = \dfrac{3 - 1}{n} = \dfrac{2}{n}$. Using Theorem 4, we get $x_i^* = x_i = 1 + i\Delta x = 1 + \dfrac{2i}{n}$, so

$$\int_1^3 \sqrt{4 + x^2} \, dx = \lim_{n \to \infty} R_n = \lim_{n \to \infty} \sum_{i=1}^{n} \sqrt{4 + \left(1 + \dfrac{2i}{n}\right)^2} \cdot \dfrac{2}{n}.$$

27. Note that $\Delta x = \dfrac{2 - 0}{n} = \dfrac{2}{n}$ and $x_i = 0 + i\Delta x = \dfrac{2i}{n}$.

$$\int_0^2 3x \, dx = \lim_{n \to \infty} \sum_{i=1}^{n} f(x_i) \Delta x = \lim_{n \to \infty} \sum_{i=1}^{n} f\left(\dfrac{2i}{n}\right) \dfrac{2}{n} = \lim_{n \to \infty} \dfrac{2}{n} \sum_{i=1}^{n} 3\left(\dfrac{2i}{n}\right) = \lim_{n \to \infty} \dfrac{2}{n} \sum_{i=1}^{n} \dfrac{6i}{n}$$

$$= \lim_{n \to \infty} \dfrac{2}{n} \left(\dfrac{6}{n}\right) \sum_{i=1}^{n} i = \lim_{n \to \infty} \dfrac{2}{n} \left(\dfrac{6}{n}\right) \left[\dfrac{n(n+1)}{2}\right] = \lim_{n \to \infty} 6\left(\dfrac{n+1}{n}\right)$$

$$= \lim_{n \to \infty} 6\left(1 + \dfrac{1}{n}\right) = 6(1) = 6$$

29. Note that $\Delta x = \dfrac{3 - 0}{n} = \dfrac{3}{n}$ and $x_i = 0 + i\Delta x = \dfrac{3i}{n}$.

$$\int_0^3 (5x + 2) \, dx = \lim_{n \to \infty} \sum_{i=1}^{n} f(x_i) \Delta x = \lim_{n \to \infty} \sum_{i=1}^{n} \left[5\left(\dfrac{3i}{n}\right) + 2\right] \dfrac{3}{n} = \lim_{n \to \infty} \dfrac{3}{n} \sum_{i=1}^{n} \left(\dfrac{15i}{n} + 2\right)$$

$$= \lim_{n \to \infty} \dfrac{3}{n} \left(\dfrac{15}{n} \sum_{i=1}^{n} i + \sum_{i=1}^{n} 2\right) = \lim_{n \to \infty} \left[\dfrac{45}{n^2} \dfrac{n(n+1)}{2} + \dfrac{3}{n} \cdot n(2)\right]$$

$$= \lim_{n \to \infty} \left[\dfrac{45}{2n}(n+1) + 3(2)\right] = \lim_{n \to \infty} \left[\dfrac{45}{2}\left(1 + \dfrac{1}{n}\right) + 6\right] = \dfrac{45}{2}(1) + 6 = \dfrac{57}{2}$$

31. Note that $\Delta x = \dfrac{5-1}{n} = \dfrac{4}{n}$ and $x_i = 1 + i\,\Delta x = 1 + \dfrac{4i}{n}$.

$$\int_1^5 (3x^2 + 7x)\,dx = \lim_{n\to\infty} \sum_{i=1}^n f(x_i)\,\Delta x = \lim_{n\to\infty} \sum_{i=1}^n f\!\left(1 + \dfrac{4i}{n}\right)\dfrac{4}{n}$$

$$= \lim_{n\to\infty} \dfrac{4}{n} \sum_{i=1}^n \left[3\!\left(1+\dfrac{4i}{n}\right)^2 + 7\!\left(1+\dfrac{4i}{n}\right)\right]$$

$$= \lim_{n\to\infty} \dfrac{4}{n} \sum_{i=1}^n \left[3\!\left(1+\dfrac{8i}{n}+\dfrac{16i^2}{n^2}\right) + 7\!\left(1+\dfrac{4i}{n}\right)\right]$$

$$= \lim_{n\to\infty} \dfrac{4}{n} \sum_{i=1}^n \left[3 + \dfrac{24i}{n} + \dfrac{48i^2}{n^2} + 7 + \dfrac{28i}{n}\right] = \lim_{n\to\infty} \dfrac{4}{n} \sum_{i=1}^n \left[10 + \dfrac{52i}{n} + \dfrac{48i^2}{n^2}\right]$$

$$= \lim_{n\to\infty} \dfrac{4}{n} \left[\sum_{i=1}^n 10 + \dfrac{52}{n}\sum_{i=1}^n i + \dfrac{48}{n^2}\sum_{i=1}^n i^2\right]$$

$$= \lim_{n\to\infty} \left[\dfrac{4}{n}\cdot n(10) + \dfrac{208}{n^2}\dfrac{n(n+1)}{2} + \dfrac{192}{n^3}\dfrac{n(n+1)(2n+1)}{6}\right]$$

$$= \lim_{n\to\infty} \left[4(10) + \dfrac{104}{n}(n+1) + \dfrac{32}{n^2}(n+1)(2n+1)\right]$$

$$= \lim_{n\to\infty} \left[40 + 104\!\left(1+\dfrac{1}{n}\right) + 32\!\left(1+\dfrac{1}{n}\right)\!\left(2+\dfrac{1}{n}\right)\right]$$

$$= 40 + 104(1) + 32(1)(2) = 208$$

33. Note that $\Delta x = \dfrac{1-0}{n} = \dfrac{1}{n}$ and $x_i = 0 + i\,\Delta x = \dfrac{i}{n}$.

$$\int_0^1 (x^3 - 3x^2)\,dx = \lim_{n\to\infty}\sum_{i=1}^n f(x_i)\,\Delta x = \lim_{n\to\infty}\sum_{i=1}^n f\!\left(\dfrac{i}{n}\right)\Delta x = \lim_{n\to\infty}\sum_{i=1}^n \left[\left(\dfrac{i}{n}\right)^3 - 3\!\left(\dfrac{i}{n}\right)^2\right]\dfrac{1}{n}$$

$$= \lim_{n\to\infty} \dfrac{1}{n}\sum_{i=1}^n \left[\dfrac{i^3}{n^3} - \dfrac{3i^2}{n^2}\right] = \lim_{n\to\infty} \dfrac{1}{n}\left[\dfrac{1}{n^3}\sum_{i=1}^n i^3 - \dfrac{3}{n^2}\sum_{i=1}^n i^2\right]$$

$$= \lim_{n\to\infty} \left\{\dfrac{1}{n^4}\!\left[\dfrac{n(n+1)}{2}\right]^2 - \dfrac{3}{n^3}\dfrac{n(n+1)(2n+1)}{6}\right\} = \lim_{n\to\infty}\left[\dfrac{1}{4}\dfrac{n+1}{n}\dfrac{n+1}{n} - \dfrac{1}{2}\dfrac{n+1}{n}\dfrac{2n+1}{n}\right]$$

$$= \lim_{n\to\infty} \left[\dfrac{1}{4}\!\left(1+\dfrac{1}{n}\right)\!\left(1+\dfrac{1}{n}\right) - \dfrac{1}{2}\!\left(1+\dfrac{1}{n}\right)\!\left(2+\dfrac{1}{n}\right)\right] = \dfrac{1}{4}(1)(1) - \dfrac{1}{2}(1)(2) = -\dfrac{3}{4}$$

35. (a) Think of $\int_0^2 f(x)\,dx$ as the area of a trapezoid with bases 1 and 3 and height 2. The area of a trapezoid is $A = \tfrac{1}{2}(b+B)h$, so $\int_0^2 f(x)\,dx = \tfrac{1}{2}(1+3)2 = 4$.

(b) $\int_0^5 f(x)\,dx = \int_0^2 f(x)\,dx + \int_2^3 f(x)\,dx + \int_3^5 f(x)\,dx$
    trapezoid   rectangle   triangle
$= \tfrac{1}{2}(1+3)2 + \quad 3\cdot 1 \quad + \quad \tfrac{1}{2}\cdot 2\cdot 3 \;=\; 4+3+3 = 10$

(c) $\int_5^7 f(x)\,dx$ is the negative of the area of the triangle with base 2 and height 3. $\int_5^7 f(x)\,dx = -\tfrac{1}{2}\cdot 2 \cdot 3 = -3$.

(d) $\int_3^7 f(x)\,dx = \int_3^5 f(x)\,dx + \int_5^7 f(x)\,dx$. $\int_3^5 f(x)\,dx$ is the area of the triangle with base 2 and height 3.

$\int_3^5 f(x)\,dx = \frac{1}{2} \cdot 2 \cdot 3 = 3$. From part (c), $\int_5^7 f(x)\,dx = -3$. Thus, $\int_3^7 f(x)\,dx = 3 + (-3) = 0$.

Or: Since $\int_3^5 f(x)\,dx$ is the same figure as in part (c), but with opposite sign, it has value 3. Thus,

$\int_3^7 f(x)\,dx = 3 + (-3) = 0$.

(e) $\int_3^7 |f(x)|\,dx = \int_3^5 |f(x)|\,dx + \int_5^7 |f(x)|\,dx = \int_3^5 f(x)\,dx + \int_5^7 [-f(x)]\,dx$. From part (d), $\int_3^5 f(x)\,dx = 3$.

From part (c), $\int_5^7 f(x)\,dx = -3$, so $\int_5^7 [-f(x)]\,dx = -(-3) = 3$. Thus, $\int_3^7 |f(x)|\,dx = 3 + 3 = 6$.

(f) $\int_2^0 f(x)\,dx = -\int_0^2 f(x)\,dx$. From part (a), $\int_0^2 f(x)\,dx = 4$, so $\int_2^0 f(x)\,dx = -4$.

37. (a) Note that $\Delta x = \dfrac{3-0}{n} = \dfrac{3}{n}$ and $x_i = 0 + i\,\Delta x = \dfrac{3i}{n}$.

$$\int_0^3 4x\,dx = \lim_{n\to\infty} \sum_{i=1}^n f(x_i)\,\Delta x = \lim_{n\to\infty} \sum_{i=1}^n f\left(\frac{3i}{n}\right)\frac{3}{n} = \lim_{n\to\infty} \frac{3}{n}\sum_{i=1}^n 4\left(\frac{3i}{n}\right)$$

$$= \lim_{n\to\infty} \frac{3}{n}\sum_{i=1}^n \frac{12i}{n} = \lim_{n\to\infty} \frac{3}{n}\left(\frac{12}{n}\right)\sum_{i=1}^n i = \lim_{n\to\infty} \frac{36}{n^2}\left[\frac{n(n+1)}{2}\right]$$

$$= \lim_{n\to\infty} 18\left(\frac{n+1}{n}\right) = \lim_{n\to\infty} 18\left(1 + \frac{1}{n}\right) = 18(1) = 18$$

(b) $\int_0^3 4x\,dx$ can be interpreted as the area of the shaded triangle; that is, $\frac{1}{2}(3)(12) = 18$.

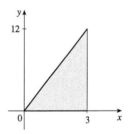

39. (a) $\Delta x = (8-0)/8 = 1$ and $x_i^* = x_i = 0 + 1i = i$.

$\int_0^8 (3-2x)\,dx$

$\approx \sum_{i=1}^8 f(x_i^*)\,\Delta x$

$= 1\{[3 - 2(1)] + [3 - 2(2)] + \cdots + [3 - 2(8)]\}$

$= 1[1 + (-1) + (-3) + (-5) + (-7) + (-9) + (-11) + (-13)]$

$= -48$

(b)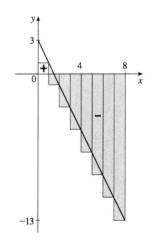

(c) Note that $\Delta x = \dfrac{8-0}{n} = \dfrac{8}{n}$ and $x_i = 0 + i\,\Delta x = \dfrac{8i}{n}$.

$$\int_0^8 (3-2x)\,dx = \lim_{n\to\infty}\sum_{i=1}^n f(x_i)\,\Delta x = \lim_{n\to\infty}\sum_{i=1}^n\left[3-2\left(\dfrac{8i}{n}\right)\right]\left(\dfrac{8}{n}\right) = \lim_{n\to\infty}\dfrac{8}{n}\sum_{i=1}^n\left[3-\dfrac{16i}{n}\right]$$

$$= \lim_{n\to\infty}\left[\dfrac{8}{n}\sum_{i=1}^n 3 - \dfrac{128}{n^2}\sum_{i=1}^n i\right] = \lim_{n\to\infty}\left[\dfrac{8}{n}\cdot n(3) - \dfrac{128}{n^2}\dfrac{n(n+1)}{2}\right]$$

$$= \lim_{n\to\infty}\left[8(3) - \dfrac{64}{n}(n+1)\right] = \lim_{n\to\infty}\left[24 - 64\left(1+\dfrac{1}{n}\right)\right]$$

$$= 24 - 64(1) = -40$$

(d) $\int_0^8 (3-2x)\,dx = A_1 - A_2$, where A_1 is the area marked $+$ and A_2 is the area marked $-$.

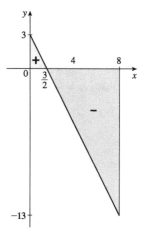

41. $\int_{-2}^{5}(10-5x)\,dx$ can be interpreted as the difference of the areas of the two shaded triangles; that is, $\tfrac{1}{2}(4)(20) - \tfrac{1}{2}(3)(15) = 40 - \tfrac{45}{2} = \tfrac{35}{2}$.

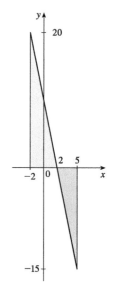

43. $\int_{-4}^{3}\left|\tfrac{1}{2}x\right|\,dx$ can be interpreted as the sum of the areas of the two shaded triangles; that is, $\tfrac{1}{2}(4)(2) + \tfrac{1}{2}(3)\left(\tfrac{3}{2}\right) = 4 + \tfrac{9}{4} = \tfrac{25}{4}$.

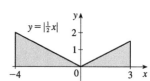

45. $\int_{-3}^{0} \left(1 + \sqrt{9-x^2}\right) dx$ can be interpreted as the area under the graph of

$f(x) = 1 + \sqrt{9-x^2}$ between $x = -3$ and $x = 0$. This is equal to one-quarter

the area of the circle with radius 3, plus the area of the rectangle, so

$\int_{-3}^{0} \left(1 + \sqrt{9-x^2}\right) dx = \frac{1}{4}\pi \cdot 3^2 + 1 \cdot 3 = 3 + \frac{9}{4}\pi$.

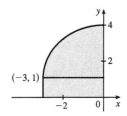

47. $\displaystyle\int_a^b x \, dx = \lim_{n \to \infty} \frac{b-a}{n} \sum_{i=1}^{n} \left[a + \frac{b-a}{n}i\right] = \lim_{n \to \infty} \left[\frac{a(b-a)}{n} \sum_{i=1}^{n} 1 + \frac{(b-a)^2}{n^2} \sum_{i=1}^{n} i\right]$

$\displaystyle = \lim_{n \to \infty} \left[\frac{a(b-a)}{n} n + \frac{(b-a)^2}{n^2} \cdot \frac{n(n+1)}{2}\right] = a(b-a) + \lim_{n \to \infty} \frac{(b-a)^2}{2} \left(1 + \frac{1}{n}\right)$

$\displaystyle = a(b-a) + \frac{1}{2}(b-a)^2 = (b-a)\left(a + \frac{1}{2}b - \frac{1}{2}a\right) = (b-a)\frac{1}{2}(b+a) = \frac{1}{2}(b^2 - a^2)$

49. $\Delta x = (\pi - 0)/n = \pi/n$ and $x_i^* = x_i = \pi i/n$.

$\displaystyle\int_0^{\pi} \sin 5x \, dx = \lim_{n \to \infty} \sum_{i=1}^{n} (\sin 5x_i)\left(\frac{\pi}{n}\right) = \lim_{n \to \infty} \sum_{i=1}^{n} \sin \frac{5\pi i}{n} \cdot \frac{\pi}{n} \stackrel{CAS}{=} \pi \lim_{n \to \infty} \frac{1}{n} \cot\left(\frac{5\pi}{2n}\right) \stackrel{CAS}{=} \pi\left(\frac{2}{5\pi}\right) = \frac{2}{5}$

51. $\int_1^1 \sqrt{1 + x^4} \, dx = 0$ since the limits of integration are equal.

53. $\int_0^1 (5 - 6x^2) \, dx = \int_0^1 5 \, dx - 6\int_0^1 x^2 \, dx = 5(1 - 0) - 6\left(\frac{1}{3}\right) = 5 - 2 = 3$

55. $\int_1^4 (2x^2 - 3x + 1) \, dx = 2\int_1^4 x^2 \, dx - 3\int_1^4 x \, dx + \int_1^4 1 \, dx$

$\displaystyle = 2 \cdot \frac{1}{3}(4^3 - 1^3) - 3 \cdot \frac{1}{2}(4^2 - 1^2) + 1(4 - 1) = \frac{45}{2} = 22.5$

57. $\int_{-2}^{2} f(x) \, dx + \int_{2}^{5} f(x) \, dx - \int_{-2}^{-1} f(x) \, dx = \int_{-2}^{5} f(x) \, dx + \int_{-1}^{-2} f(x) \, dx$ [by Property 5 and reversing limits]

$\displaystyle \hspace{3.5cm} = \int_{-1}^{5} f(x) \, dx$ \hspace{1cm} [Property 5]

59. $\int_0^9 [2f(x) + 3g(x)] \, dx = 2\int_0^9 f(x) \, dx + 3\int_0^9 g(x) \, dx = 2(37) + 3(16) = 122$

61. $\int_0^3 f(x) \, dx$ is clearly less than -1 and has the smallest value. The slope of the tangent line of f at $x = 1$, $f'(1)$, has a value

between -1 and 0, so it has the next smallest value. The largest value is $\int_3^8 f(x) \, dx$, followed by $\int_4^8 f(x) \, dx$, which has a

value about 1 unit less than $\int_3^8 f(x) \, dx$. Still positive, but with a smaller value than $\int_4^8 f(x) \, dx$, is $\int_0^8 f(x) \, dx$. Ordering these

quantities from smallest to largest gives us

$\int_0^3 f(x) \, dx < f'(1) < \int_0^8 f(x) \, dx < \int_4^8 f(x) \, dx < \int_3^8 f(x) \, dx$ or $B < E < A < D < C$

63. $I = \int_{-4}^{2}[f(x) + 2x + 5]\,dx = \int_{-4}^{2} f(x)\,dx + 2\int_{-4}^{2} x\,dx + \int_{-4}^{2} 5\,dx = I_1 + 2I_2 + I_3$

$I_1 = -3$ [area below x-axis] $\quad +3 - 3 = -3$

$I_2 = -\frac{1}{2}(4)(4)$ [area of triangle, see figure] $\quad +\frac{1}{2}(2)(2)$

$= -8 + 2 = -6$

$I_3 = 5[2 - (-4)] = 5(6) = 30$

Thus, $I = -3 + 2(-6) + 30 = 15$.

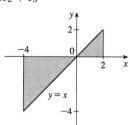

65. $x^2 - 4x + 4 = (x-2)^2 \geq 0$ on $[0, 4]$, so $\int_0^4 (x^2 - 4x + 4)\,dx \geq 0$ [Property 6].

67. If $-1 \leq x \leq 1$, then $0 \leq x^2 \leq 1$ and $1 \leq 1 + x^2 \leq 2$, so $1 \leq \sqrt{1+x^2} \leq \sqrt{2}$ and
$1[1-(-1)] \leq \int_{-1}^{1} \sqrt{1+x^2}\,dx \leq \sqrt{2}\,[1-(-1)]$ [Property 8]; that is, $2 \leq \int_{-1}^{1} \sqrt{1+x^2}\,dx \leq 2\sqrt{2}$.

69. If $0 \leq x \leq 1$, then $0 \leq x^3 \leq 1$, so $0(1-0) \leq \int_0^1 x^3\,dx \leq 1(1-0)$ [Property 8]; that is, $0 \leq \int_0^1 x^3\,dx \leq 1$.

71. If $\frac{\pi}{4} \leq x \leq \frac{\pi}{3}$, then $1 \leq \tan x \leq \sqrt{3}$, so $1\left(\frac{\pi}{3} - \frac{\pi}{4}\right) \leq \int_{\pi/4}^{\pi/3} \tan x\,dx \leq \sqrt{3}\left(\frac{\pi}{3} - \frac{\pi}{4}\right)$ or $\frac{\pi}{12} \leq \int_{\pi/4}^{\pi/3} \tan x\,dx \leq \frac{\pi}{12}\sqrt{3}$.

73. For $-1 \leq x \leq 1$, $0 \leq x^4 \leq 1$ and $1 \leq \sqrt{1+x^4} \leq \sqrt{2}$, so $1[1-(-1)] \leq \int_{-1}^{1} \sqrt{1+x^4}\,dx \leq \sqrt{2}\,[1-(-1)]$
or $2 \leq \int_{-1}^{1} \sqrt{1+x^4}\,dx \leq 2\sqrt{2}$.

75. $\sqrt{x^4+1} \geq \sqrt{x^4} = x^2$, so $\int_1^3 \sqrt{x^4+1}\,dx \geq \int_1^3 x^2\,dx = \frac{1}{3}(3^3 - 1^3) = \frac{26}{3}$.

77. $1/x < \sqrt{x} < x$ for $1 < x \leq 2$ and $\sqrt{x}$ is an increasing function, so $\sqrt{1/x} < \sqrt{\sqrt{x}} < \sqrt{x}$, and hence
$\int_1^2 \sqrt{1/x}\,dx < \int_1^2 \sqrt{\sqrt{x}}\,dx < \int_1^2 \sqrt{x}\,dx$. Thus, $\int_1^2 \sqrt{x}\,dx$ has the largest value.

79. Using right endpoints as in the proof of Property 2, we calculate

$\int_a^b cf(x)\,dx = \lim\limits_{n \to \infty} \sum\limits_{i=1}^{n} cf(x_i)\,\Delta x = \lim\limits_{n \to \infty} c \sum\limits_{i=1}^{n} f(x_i)\,\Delta x = c \lim\limits_{n \to \infty} \sum\limits_{i=1}^{n} f(x_i)\,\Delta x = c \int_a^b f(x)\,dx$.

81. Suppose that f is integrable on $[0, 1]$, that is, $\lim\limits_{n \to \infty} \sum\limits_{i=1}^{n} f(x_i^*)\,\Delta x$ exists for any choice of x_i^* in $[x_{i-1}, x_i]$. Let n denote a positive integer and divide the interval $[0, 1]$ into n equal subintervals $\left[0, \frac{1}{n}\right], \left[\frac{1}{n}, \frac{2}{n}\right], \ldots, \left[\frac{n-1}{n}, 1\right]$. If we choose x_i^* to be a rational number in the ith subinterval, then we obtain the Riemann sum $\sum\limits_{i=1}^{n} f(x_i^*) \cdot \frac{1}{n} = 0$, so

$\lim\limits_{n \to \infty} \sum\limits_{i=1}^{n} f(x_i^*) \cdot \frac{1}{n} = \lim\limits_{n \to \infty} 0 = 0$. Now suppose we choose x_i^* to be an irrational number. Then we get

$\sum\limits_{i=1}^{n} f(x_i^*) \cdot \frac{1}{n} = \sum\limits_{i=1}^{n} 1 \cdot \frac{1}{n} = n \cdot \frac{1}{n} = 1$ for each n, so $\lim\limits_{n \to \infty} \sum\limits_{i=1}^{n} f(x_i^*) \cdot \frac{1}{n} = \lim\limits_{n \to \infty} 1 = 1$. Since the value of

$\lim\limits_{n \to \infty} \sum\limits_{i=1}^{n} f(x_i^*)\,\Delta x$ depends on the choice of the sample points x_i^*, the limit does not exist, and f is not integrable on $[0, 1]$.

83. $\lim\limits_{n\to\infty} \sum\limits_{i=1}^{n} \dfrac{i^4}{n^5} = \lim\limits_{n\to\infty} \sum\limits_{i=1}^{n} \dfrac{i^4}{n^4} \cdot \dfrac{1}{n} = \lim\limits_{n\to\infty} \sum\limits_{i=1}^{n} \left(\dfrac{i}{n}\right)^4 \dfrac{1}{n}$. At this point, we need to recognize the limit as being of the form

$\lim\limits_{n\to\infty} \sum\limits_{i=1}^{n} f(x_i)\,\Delta x$, where $\Delta x = (1-0)/n = 1/n$, $x_i = 0 + i\,\Delta x = i/n$, and $f(x) = x^4$. Thus, the definite integral

is $\int_0^1 x^4\,dx$.

85. Choose $x_i = 1 + \dfrac{i}{n}$ and $x_i^* = \sqrt{x_{i-1}x_i} = \sqrt{\left(1 + \dfrac{i-1}{n}\right)\left(1 + \dfrac{i}{n}\right)}$. Then

$$\int_1^2 x^{-2}\,dx = \lim_{n\to\infty} \dfrac{1}{n} \sum_{i=1}^{n} \dfrac{1}{\left(1 + \tfrac{i-1}{n}\right)\left(1 + \tfrac{i}{n}\right)} = \lim_{n\to\infty} n \sum_{i=1}^{n} \dfrac{1}{(n+i-1)(n+i)}$$

$$= \lim_{n\to\infty} n \sum_{i=1}^{n} \left(\dfrac{1}{n+i-1} - \dfrac{1}{n+i}\right) \quad \text{[by the hint]} \quad = \lim_{n\to\infty} n\left(\sum_{i=0}^{n-1} \dfrac{1}{n+i} - \sum_{i=1}^{n} \dfrac{1}{n+i}\right)$$

$$= \lim_{n\to\infty} n\left(\left[\dfrac{1}{n} + \dfrac{1}{n+1} + \cdots + \dfrac{1}{2n-1}\right] - \left[\dfrac{1}{n+1} + \cdots + \dfrac{1}{2n-1} + \dfrac{1}{2n}\right]\right)$$

$$= \lim_{n\to\infty} n\left(\dfrac{1}{n} - \dfrac{1}{2n}\right) = \lim_{n\to\infty} \left(1 - \tfrac{1}{2}\right) = \tfrac{1}{2}$$

4.3 The Fundamental Theorem of Calculus

1. One process undoes what the other one does. The precise version of this statement is given by the Fundamental Theorem of Calculus. See the statement of this theorem and the paragraph that follows it.

3. (a) $g(x) = \int_0^x f(t)\,dt$.

$g(0) = \int_0^0 f(t)\,dt = 0$

$g(1) = \int_0^1 f(t)\,dt = 1 \cdot 2 = 2$ [rectangle],

$g(2) = \int_0^2 f(t)\,dt = \int_0^1 f(t)\,dt + \int_1^2 f(t)\,dt = g(1) + \int_1^2 f(t)\,dt$

$\quad = 2 + 1\cdot 2 + \tfrac{1}{2}\cdot 1\cdot 2 = 5$ [rectangle plus triangle],

$g(3) = \int_0^3 f(t)\,dt = g(2) + \int_2^3 f(t)\,dt = 5 + \tfrac{1}{2}\cdot 1\cdot 4 = 7$,

$g(6) = g(3) + \int_3^6 f(t)\,dt$ [the integral is negative since f lies under the t-axis]

$\quad = 7 + \left[-\left(\tfrac{1}{2}\cdot 2\cdot 2 + 1\cdot 2\right)\right] = 7 - 4 = 3$

(b) g is increasing on $(0, 3)$ because as x increases from 0 to 3, we keep adding more area.

(c) g has a maximum value when we start subtracting area; that is, at $x = 3$.

(d)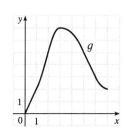

5. (a) $g(x) = \int_0^x f(t)\,dt = 3x$ [area of a rectangle with base x and height 3]

(b) $g(x) = 3x \;\Rightarrow\; g'(x) = 3 = f(x)$, so g is an antiderivative of f. Since $f(t) = 3$, 3 is the integrand in part (a), and 3 is the integrand evaluated at upper limit of integration x (since 3 is constant), verifying FTC1.

7.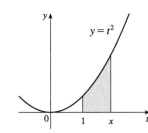

(a) By FTC1 with $f(t) = t^2$ and $a = 1$, $g(x) = \int_1^x t^2 \, dt \Rightarrow g'(x) = f(x) = x^2$.

(b) Using FTC2, $g(x) = \int_1^x t^2 \, dt = \left[\frac{1}{3}t^3\right]_1^x = \frac{1}{3}x^3 - \frac{1}{3} \Rightarrow g'(x) = x^2$.

9. $f(t) = \sqrt{t + t^3}$ and $g(x) = \int_0^x \sqrt{t + t^3} \, dt$, so by FTC1, $g'(x) = f(x) = \sqrt{x + x^3}$.

11. $f(t) = \sin(1 + t^3)$ and $g(w) = \int_0^w \sin(1 + t^3) \, dt$, so by FTC1, $g'(w) = f(w) = \sin(1 + w^3)$.

13. $F(x) = \int_x^0 \sqrt{1 + \sec t} \, dt = -\int_0^x \sqrt{1 + \sec t} \, dt \overset{\text{FTC1}}{\Rightarrow} F'(x) = -\dfrac{d}{dx}\int_0^x \sqrt{1 + \sec t} \, dt = -\sqrt{1 + \sec x}$

15. Let $u = \dfrac{1}{x}$. Then $\dfrac{du}{dx} = -\dfrac{1}{x^2}$. Also, $\dfrac{dh}{dx} = \dfrac{dh}{du}\dfrac{du}{dx}$, so

$h'(x) = \dfrac{d}{dx}\int_2^{1/x} \sin^4 t \, dt = \dfrac{d}{du}\int_2^u \sin^4 t \, dt \cdot \dfrac{du}{dx} = \sin^4 u \, \dfrac{du}{dx} = \dfrac{-\sin^4(1/x)}{x^2}$.

17. Let $u = 3x + 2$. Then $\dfrac{du}{dx} = 3$. Also, $\dfrac{dy}{dx} = \dfrac{dy}{du}\dfrac{du}{dx}$, so by FTC1,

$y' = \dfrac{d}{dx}\int_1^{3x+2} \dfrac{t}{1 + t^3} \, dt = \dfrac{d}{du}\int_1^u \dfrac{t}{1 + t^3} \, dt \cdot \dfrac{du}{dx} = \dfrac{u}{1 + u^3}\dfrac{du}{dx} = \dfrac{3x + 2}{1 + (3x + 2)^3} \cdot 3 = \dfrac{3(3x + 2)}{1 + (3x + 2)^3}$

19. Let $u = \sqrt{x}$. Then $\dfrac{du}{dx} = \dfrac{1}{2\sqrt{x}}$. Also, $\dfrac{dy}{dx} = \dfrac{dy}{du}\dfrac{du}{dx}$, so by FTC1,

$y' = \dfrac{d}{dx}\int_{\sqrt{x}}^{\pi/4} \theta \tan \theta \, d\theta = -\dfrac{d}{du}\int_{\pi/4}^{\sqrt{x}} \theta \tan \theta \, d\theta \cdot \dfrac{du}{dx} = -u \tan u \, \dfrac{du}{dx} = -\sqrt{x}\tan\sqrt{x} \cdot \dfrac{1}{2\sqrt{x}} = -\dfrac{1}{2}\tan\sqrt{x}$

21. $\int_{-1}^{2} x^3 \, dx = \left[\frac{1}{4}x^4\right]_{-1}^{2} = 4 - \frac{1}{4} = \frac{15}{4} = 3.75$

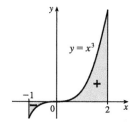

23. $\int_{\pi/2}^{2\pi}(2 \sin x) \, dx = \left[-2\cos x\right]_{\pi/2}^{2\pi}$

$= (-2\cos 2\pi) - \left(-2\cos \frac{\pi}{2}\right)$

$= -2(1) + 2(0) = -2$

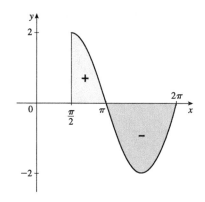

25. $\int_1^3 (x^2 + 2x - 4)\, dx = \left[\tfrac{1}{3}x^3 + x^2 - 4x\right]_1^3 = (9 + 9 - 12) - \left(\tfrac{1}{3} + 1 - 4\right) = 6 + \tfrac{8}{3} = \tfrac{26}{3}$

27. $\int_0^2 \left(\tfrac{4}{5}t^3 - \tfrac{3}{4}t^2 + \tfrac{2}{5}t\right) dt = \left[\tfrac{1}{5}t^4 - \tfrac{1}{4}t^3 + \tfrac{1}{5}t^2\right]_0^2 = \left(\tfrac{16}{5} - 2 + \tfrac{4}{5}\right) - 0 = 2$

29. $\int_1^9 \sqrt{x}\, dx = \int_1^9 x^{1/2}\, dx = \left[\dfrac{x^{3/2}}{3/2}\right]_1^9 = \tfrac{2}{3}\left[x^{3/2}\right]_1^9 = \tfrac{2}{3}(9^{3/2} - 1^{3/2}) = \tfrac{2}{3}(27 - 1) = \tfrac{52}{3}$

31. $\int_0^4 (t^2 + t^{3/2})\, dt = \left[\tfrac{1}{3}t^3 + \tfrac{2}{5}t^{5/2}\right]_0^4 = \left(\tfrac{64}{3} + \tfrac{64}{5}\right) - 0 = \tfrac{512}{15}$

33. $\int_{\pi/2}^0 \cos\theta\, d\theta = \left[\sin\theta\right]_{\pi/2}^0 = \sin 0 - \sin\tfrac{\pi}{2} = 0 - 1 = -1$

35. $\int_0^1 (u + 2)(u - 3)\, du = \int_0^1 (u^2 - u - 6)\, du = \left[\tfrac{1}{3}u^3 - \tfrac{1}{2}u^2 - 6u\right]_0^1 = \left(\tfrac{1}{3} - \tfrac{1}{2} - 6\right) - 0 = -\tfrac{37}{6}$

37. $\displaystyle\int_1^4 \dfrac{2 + x^2}{\sqrt{x}}\, dx = \int_1^4 \left(\dfrac{2}{\sqrt{x}} + \dfrac{x^2}{\sqrt{x}}\right) dx = \int_1^4 (2x^{-1/2} + x^{3/2})\, dx$

$\quad = \left[4x^{1/2} + \tfrac{2}{5}x^{5/2}\right]_1^4 = \left[4(2) + \tfrac{2}{5}(32)\right] - \left(4 + \tfrac{2}{5}\right) = 8 + \tfrac{64}{5} - 4 - \tfrac{2}{5} = \tfrac{82}{5}$

39. $\int_1^2 \dfrac{s^4 + 1}{s^2}\, ds = \int_1^2 (s^2 + s^{-2})\, ds = \left[\tfrac{1}{3}s^3 - \tfrac{1}{s}\right]_1^2 = \left(\tfrac{8}{3} - \tfrac{1}{2}\right) - \left(\tfrac{1}{3} - 1\right) = \tfrac{7}{3} + \tfrac{1}{2} = \tfrac{17}{6}$

41. $\int_0^{\pi/3} \sec\theta\tan\theta\, d\theta = \left[\sec\theta\right]_0^{\pi/3} = \sec\tfrac{\pi}{3} - \sec 0 = 2 - 1 = 1$

43. $\int_0^1 (1 + r)^3\, dr = \int_0^1 (1 + 3r + 3r^2 + r^3)\, dr = \left[r + \tfrac{3}{2}r^2 + r^3 + \tfrac{1}{4}r^4\right]_0^1 = \left(1 + \tfrac{3}{2} + 1 + \tfrac{1}{4}\right) - 0 = \tfrac{15}{4}$

45. If $f(x) = \begin{cases} \sin x & \text{if } 0 \le x < \pi/2 \\ \cos x & \text{if } \pi/2 \le x \le \pi \end{cases}$ then

$\int_0^\pi f(x)\, dx = \int_0^{\pi/2} \sin x\, dx + \int_{\pi/2}^\pi \cos x\, dx = \left[-\cos x\right]_0^{\pi/2} + \left[\sin x\right]_{\pi/2}^\pi = -\cos\tfrac{\pi}{2} + \cos 0 + \sin\pi - \sin\tfrac{\pi}{2}$

$= -0 + 1 + 0 - 1 = 0$

Note that f is integrable by Theorem 3 in Section 4.2.

47. Area $= \int_0^4 \sqrt{x}\, dx = \int_0^4 x^{1/2}\, dx = \left[\tfrac{2}{3}x^{3/2}\right]_0^4 = \tfrac{2}{3}(8) - 0 = \tfrac{16}{3}$

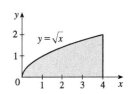

49. Area $= \int_{-2}^2 (4 - x^2)\, dx = \left[4x - \tfrac{1}{3}x^3\right]_{-2}^2 = \left(8 - \tfrac{8}{3}\right) - \left(-8 + \tfrac{8}{3}\right) = \tfrac{32}{3}$

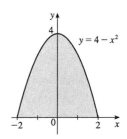

51. From the graph, it appears that the area is about 60. The actual area is

$\int_0^{27} x^{1/3}\,dx = \left[\tfrac{3}{4}x^{4/3}\right]_0^{27} = \tfrac{3}{4}\cdot 81 - 0 = \tfrac{243}{4} = 60.75$. This is $\tfrac{3}{4}$ of the

area of the viewing rectangle.

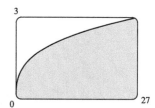

53. It appears that the area under the graph is about $\tfrac{2}{3}$ of the area of the viewing

rectangle, or about $\tfrac{2}{3}\pi \approx 2.1$. The actual area is

$\int_0^{\pi}\sin x\,dx = [-\cos x]_0^{\pi} = (-\cos\pi) - (-\cos 0) = -(-1) + 1 = 2$.

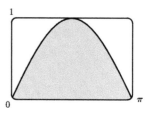

55. $f(x) = x^{-4}$ is not continuous on the interval $[-2, 1]$, so FTC2 cannot be applied. In fact, f has an infinite discontinuity at $x = 0$, so $\int_{-2}^{1} x^{-4}\,dx$ does not exist.

57. $f(\theta) = \sec\theta\tan\theta$ is not continuous on the interval $[\pi/3, \pi]$, so FTC2 cannot be applied. In fact, f has an infinite discontinuity at $x = \pi/2$, so $\int_{\pi/3}^{\pi}\sec\theta\tan\theta\,d\theta$ does not exist.

59. $g(x) = \int_{2x}^{3x}\dfrac{u^2-1}{u^2+1}\,du = \int_{2x}^{0}\dfrac{u^2-1}{u^2+1}\,du + \int_{0}^{3x}\dfrac{u^2-1}{u^2+1}\,du = -\int_{0}^{2x}\dfrac{u^2-1}{u^2+1}\,du + \int_{0}^{3x}\dfrac{u^2-1}{u^2+1}\,du \Rightarrow$

$g'(x) = -\dfrac{(2x)^2-1}{(2x)^2+1}\cdot\dfrac{d}{dx}(2x) + \dfrac{(3x)^2-1}{(3x)^2+1}\cdot\dfrac{d}{dx}(3x) = -2\cdot\dfrac{4x^2-1}{4x^2+1} + 3\cdot\dfrac{9x^2-1}{9x^2+1}$

61. $h(x) = \int_{\sqrt{x}}^{x^3}\cos(t^2)\,dt = \int_{\sqrt{x}}^{0}\cos(t^2)\,dt + \int_{0}^{x^3}\cos(t^2)\,dt = -\int_{0}^{\sqrt{x}}\cos(t^2)\,dt + \int_{0}^{x^3}\cos(t^2)\,dt \Rightarrow$

$h'(x) = -\cos\left((\sqrt{x})^2\right)\cdot\dfrac{d}{dx}(\sqrt{x}) + [\cos(x^3)^2]\cdot\dfrac{d}{dx}(x^3) = -\dfrac{1}{2\sqrt{x}}\cos x + 3x^2\cos(x^6)$

63. $F(x) = \int_{\pi}^{x}\dfrac{\cos t}{t}\,dt \Rightarrow F'(x) = \dfrac{\cos x}{x}$, so the slope at $x = \pi$ is $\dfrac{\cos\pi}{\pi} = -\dfrac{1}{\pi}$. The y-coordinate of the point on F at

$x = \pi$ is $F(\pi) = \int_{\pi}^{\pi}\dfrac{\cos t}{t}\,dt = 0$ since the limits are equal. An equation of the tangent line is $y - 0 = -\dfrac{1}{\pi}(x - \pi)$,

or $y = -\dfrac{1}{\pi}x + 1$.

65. $y = \int_{0}^{x}\dfrac{t^2}{t^2+t+2}\,dt \Rightarrow y' = \dfrac{x^2}{x^2+x+2} \Rightarrow$

$y'' = \dfrac{(x^2+x+2)(2x) - x^2(2x+1)}{(x^2+x+2)^2} = \dfrac{2x^3+2x^2+4x-2x^3-x^2}{(x^2+x+2)^2} = \dfrac{x^2+4x}{(x^2+x+2)^2} = \dfrac{x(x+4)}{(x^2+x+2)^2}$.

The curve y is concave downward when $y'' < 0$; that is, on the interval $(-4, 0)$.

67. By FTC2, $\int_{1}^{4} f'(x)\,dx = f(4) - f(1)$, so $17 = f(4) - 12 \Rightarrow f(4) = 17 + 12 = 29$.

69. (a) The Fresnel function $S(x) = \int_0^x \sin\left(\frac{\pi}{2}t^2\right) dt$ has local maximum values where $0 = S'(x) = \sin\left(\frac{\pi}{2}t^2\right)$ and S' changes from positive to negative. For $x > 0$, this happens when $\frac{\pi}{2}x^2 = (2n-1)\pi$ [odd multiples of π] $\Leftrightarrow$ $x^2 = 2(2n-1)$ $\Leftrightarrow$ $x = \sqrt{4n-2}$, n any positive integer. For $x < 0$, S' changes from positive to negative where $\frac{\pi}{2}x^2 = 2n\pi$ [even multiples of π] $\Leftrightarrow$ $x^2 = 4n$ $\Leftrightarrow$ $x = -2\sqrt{n}$. S' does not change sign at $x = 0$.

(b) S is concave upward on those intervals where $S''(x) > 0$. Differentiating our expression for $S'(x)$, we get $S''(x) = \cos\left(\frac{\pi}{2}x^2\right)\left(2\frac{\pi}{2}x\right) = \pi x \cos\left(\frac{\pi}{2}x^2\right)$. For $x > 0$, $S''(x) > 0$ where $\cos\left(\frac{\pi}{2}x^2\right) > 0$ $\Leftrightarrow$ $0 < \frac{\pi}{2}x^2 < \frac{\pi}{2}$ or $\left(2n - \frac{1}{2}\right)\pi < \frac{\pi}{2}x^2 < \left(2n + \frac{1}{2}\right)\pi$, n any integer $\Leftrightarrow$ $0 < x < 1$ or $\sqrt{4n-1} < x < \sqrt{4n+1}$, n any positive integer. For $x < 0$, $S''(x) > 0$ where $\cos\left(\frac{\pi}{2}x^2\right) < 0$ $\Leftrightarrow$ $\left(2n - \frac{3}{2}\right)\pi < \frac{\pi}{2}x^2 < \left(2n - \frac{1}{2}\right)\pi$, n any integer $\Leftrightarrow$ $4n - 3 < x^2 < 4n - 1$ $\Leftrightarrow$ $\sqrt{4n-3} < |x| < \sqrt{4n-1}$ $\Rightarrow$ $\sqrt{4n-3} < -x < \sqrt{4n-1}$ $\Rightarrow$ $-\sqrt{4n-3} > x > -\sqrt{4n-1}$, so the intervals of upward concavity for $x < 0$ are $\left(-\sqrt{4n-1}, -\sqrt{4n-3}\right)$, n any positive integer. To summarize: S is concave upward on the intervals $(0, 1)$, $(-\sqrt{3}, -1)$, $(\sqrt{3}, \sqrt{5})$, $(-\sqrt{7}, -\sqrt{5})$, $(\sqrt{7}, 3), \ldots$.

(c) In Maple, we use `plot({int(sin(Pi*t^2/2),t=0..x),0.2},x=0..2);`. Note that Maple recognizes the Fresnel function, calling it `FresnelS(x)`. In Mathematica, we use `Plot[{Integrate[Sin[Pi*t^2/2],{t,0,x}],0.2},{x,0,2}]`. In Derive, we load the utility file FRESNEL and plot FRESNEL_SIN(x). From the graphs, we see that $\int_0^x \sin\left(\frac{\pi}{2}t^2\right) dt = 0.2$ at $x \approx 0.74$.

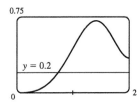 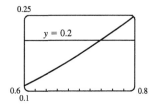

71. (a) By FTC1, $g'(x) = f(x)$. So $g'(x) = f(x) = 0$ at $x = 1, 3, 5, 7,$ and 9. g has local maxima at $x = 1$ and 5 (since $f = g'$ changes from positive to negative there) and local minima at $x = 3$ and 7. There is no local maximum or minimum at $x = 9$, since f is not defined for $x > 9$.

(b) We can see from the graph that $\left|\int_0^1 f\, dt\right| < \left|\int_1^3 f\, dt\right| < \left|\int_3^5 f\, dt\right| < \left|\int_5^7 f\, dt\right| < \left|\int_7^9 f\, dt\right|$. So $g(1) = \left|\int_0^1 f\, dt\right|$, $g(5) = \int_0^5 f\, dt = g(1) - \left|\int_1^3 f\, dt\right| + \left|\int_3^5 f\, dt\right|$, and $g(9) = \int_0^9 f\, dt = g(5) - \left|\int_5^7 f\, dt\right| + \left|\int_7^9 f\, dt\right|$. Thus, $g(1) < g(5) < g(9)$, and so the absolute maximum of $g(x)$ occurs at $x = 9$.

(c) g is concave downward on those intervals where $g'' < 0$. But $g'(x) = f(x)$, so $g''(x) = f'(x)$, which is negative on (approximately) $\left(\frac{1}{2}, 2\right)$, $(4, 6)$ and $(8, 9)$. So g is concave downward on these intervals.

(d)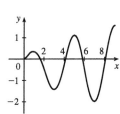

73. $\lim\limits_{n\to\infty} \sum\limits_{i=1}^{n} \left(\dfrac{i^4}{n^5} + \dfrac{i}{n^2}\right) = \lim\limits_{n\to\infty} \sum\limits_{i=1}^{n} \left(\dfrac{i^4}{n^4} + \dfrac{i}{n}\right)\dfrac{1}{n} = \lim\limits_{n\to\infty} \dfrac{1-0}{n} \sum\limits_{i=1}^{n}\left[\left(\dfrac{i}{n}\right)^4 + \dfrac{i}{n}\right] = \int_0^1 (x^4 + x)\,dx$

$= \left[\tfrac{1}{5}x^5 + \tfrac{1}{2}x^2\right]_0^1 = \left(\tfrac{1}{5} + \tfrac{1}{2}\right) - 0 = \tfrac{7}{10}$

75. Suppose $h < 0$. Since f is continuous on $[x+h, x]$, the Extreme Value Theorem says that there are numbers u and v in $[x+h, x]$ such that $f(u) = m$ and $f(v) = M$, where m and M are the absolute minimum and maximum values of f on $[x+h, x]$. By Integral Property 4.2.8, $m(-h) \le \int_{x+h}^{x} f(t)\,dt \le M(-h)$; that is, $f(u)(-h) \le -\int_{x}^{x+h} f(t)\,dt \le f(v)(-h)$.

Since $-h > 0$, we can divide this inequality by $-h$: $f(u) \le \dfrac{1}{h}\int_{x}^{x+h} f(t)\,dt \le f(v)$. By Equation 2,

$\dfrac{g(x+h) - g(x)}{h} = \dfrac{1}{h}\int_{x}^{x+h} f(t)\,dt$ for $h \ne 0$, and hence $f(u) \le \dfrac{g(x+h) - g(x)}{h} \le f(v)$, which is Equation 3 in the case where $h < 0$.

77. (a) Let $f(x) = \sqrt{x}$ $\Rightarrow$ $f'(x) = 1/(2\sqrt{x}) > 0$ for $x > 0$ $\Rightarrow$ f is increasing on $(0, \infty)$. If $x \ge 0$, then $x^3 \ge 0$, so $1 + x^3 \ge 1$ and since f is increasing, this means that $f(1+x^3) \ge f(1)$ $\Rightarrow$ $\sqrt{1+x^3} \ge 1$ for $x \ge 0$. Next let $g(t) = t^2 - t$ $\Rightarrow$ $g'(t) = 2t - 1$ $\Rightarrow$ $g'(t) > 0$ when $t \ge 1$. Thus, g is increasing on $(1, \infty)$. And since $g(1) = 0$, $g(t) \ge 0$ when $t \ge 1$. Now let $t = \sqrt{1+x^3}$, where $x \ge 0$. $\sqrt{1+x^3} \ge 1$ (from above) $\Rightarrow$ $t \ge 1$ $\Rightarrow$ $g(t) \ge 0$ $\Rightarrow$ $(1+x^3) - \sqrt{1+x^3} \ge 0$ for $x \ge 0$. Therefore, $1 \le \sqrt{1+x^3} \le 1 + x^3$ for $x \ge 0$.

(b) From part (a) and Integral Property 4.2.7: $\int_0^1 1\,dx \le \int_0^1 \sqrt{1+x^3}\,dx \le \int_0^1 (1+x^3)\,dx$ $\Leftrightarrow$

$[x]_0^1 \le \int_0^1 \sqrt{1+x^3}\,dx \le \left[x + \tfrac{1}{4}x^4\right]_0^1$ $\Leftrightarrow$ $1 \le \int_0^1 \sqrt{1+x^3}\,dx \le 1 + \tfrac{1}{4} = 1.25$.

79. $0 < \dfrac{x^2}{x^4 + x^2 + 1} < \dfrac{x^2}{x^4} = \dfrac{1}{x^2}$ on $[5, 10]$, so

$0 \le \int_5^{10} \dfrac{x^2}{x^4+x^2+1}\,dx < \int_5^{10} \dfrac{1}{x^2}\,dx = \left[-\dfrac{1}{x}\right]_5^{10} = -\dfrac{1}{10} - \left(-\dfrac{1}{5}\right) = \dfrac{1}{10} = 0.1$.

81. Using FTC1, we differentiate both sides of $6 + \int_a^x \dfrac{f(t)}{t^2}\,dt = 2\sqrt{x}$ to get $\dfrac{f(x)}{x^2} = 2\dfrac{1}{2\sqrt{x}}$ $\Rightarrow$ $f(x) = x^{3/2}$.

To find a, we substitute $x = a$ in the original equation to obtain $6 + \int_a^a \dfrac{f(t)}{t^2}\,dt = 2\sqrt{a}$ $\Rightarrow$ $6 + 0 = 2\sqrt{a}$ $\Rightarrow$ $3 = \sqrt{a}$ $\Rightarrow$ $a = 9$.

83. (a) Let $F(t) = \int_0^t f(s)\,ds$. Then, by FTC1, $F'(t) = f(t) =$ rate of depreciation, so $F(t)$ represents the loss in value over the interval $[0, t]$.

(b) $C(t) = \dfrac{1}{t}\left[A + \int_0^t f(s)\,ds\right] = \dfrac{A + F(t)}{t}$ represents the average expenditure per unit of t during the interval $[0, t]$, assuming that there has been only one overhaul during that time period. The company wants to minimize average expenditure.

(c) $C(t) = \dfrac{1}{t}\left[A + \int_0^t f(s)\,ds\right]$. Using FTC1, we have $C'(t) = -\dfrac{1}{t^2}\left[A + \int_0^t f(s)\,ds\right] + \dfrac{1}{t}f(t)$.

$C'(t) = 0 \;\Rightarrow\; tf(t) = A + \int_0^t f(s)\,ds \;\Rightarrow\; f(t) = \dfrac{1}{t}\left[A + \int_0^t f(s)\,ds\right] = C(t)$.

85. $\displaystyle\int_1^9 \dfrac{1}{2x}\,dx = \dfrac{1}{2}\int_1^9 \dfrac{1}{x}\,dx = \dfrac{1}{2}\Big[\ln|x|\Big]_1^9 = \dfrac{1}{2}(\ln 9 - \ln 1) = \dfrac{1}{2}\ln 9 - 0 = \ln 9^{1/2} = \ln 3$

87. $\displaystyle\int_{1/2}^{1/\sqrt{2}} \dfrac{4}{\sqrt{1-x^2}}\,dx = \Big[4\arcsin x\Big]_{1/2}^{1/\sqrt{2}} = 4\left(\dfrac{\pi}{4} - \dfrac{\pi}{6}\right) = 4\left(\dfrac{\pi}{12}\right) = \dfrac{\pi}{3}$

89. $\displaystyle\int_{-1}^{1} e^{u+1}\,du = \Big[e^{u+1}\Big]_{-1}^{1} = e^2 - e^0 = e^2 - 1$ [or start with $e^{u+1} = e^u e^1$]

4.4 Indefinite Integrals and the Net Change Theorem

1. $\dfrac{d}{dx}\left(\dfrac{1}{2}x + \dfrac{1}{4}\sin 2x + C\right) = \dfrac{1}{2} + \dfrac{1}{4}\cos 2x \cdot 2 + 0 = \dfrac{1}{2} + \dfrac{1}{2}\cos 2x$

$\qquad = \dfrac{1}{2} + \dfrac{1}{2}(2\cos^2 x - 1) = \dfrac{1}{2} + \cos^2 x - \dfrac{1}{2} = \cos^2 x$

3. $\dfrac{d}{dx}\left[-\dfrac{\sqrt{1+x^2}}{x} + C\right] = \dfrac{d}{dx}\left[-\dfrac{(1+x^2)^{1/2}}{x} + C\right] = -\dfrac{x \cdot \frac{1}{2}(1+x^2)^{-1/2}(2x) - (1+x^2)^{1/2} \cdot 1}{(x)^2} + 0$

$\qquad = -\dfrac{(1+x^2)^{-1/2}\left[x^2 - (1+x^2)\right]}{x^2} = -\dfrac{-1}{(1+x^2)^{1/2}x^2} = \dfrac{1}{x^2\sqrt{1+x^2}}$

5. $\displaystyle\int (3x^2 + 4x + 1)\,dx = 3 \cdot \dfrac{1}{3}x^3 + 4 \cdot \dfrac{1}{2}x^2 + x + C = x^3 + 2x^2 + x + C$

7. $\displaystyle\int (x + \cos x)\,dx = \dfrac{1}{2}x^2 + \sin x + C$

9. $\displaystyle\int (x^{1.3} + 7x^{2.5})\,dx = \dfrac{1}{2.3}x^{2.3} + \dfrac{7}{3.5}x^{3.5} + C = \dfrac{1}{2.3}x^{2.3} + 2x^{3.5} + C$

11. $\displaystyle\int \left(5 + \dfrac{2}{3}x^2 + \dfrac{3}{4}x^3\right)dx = 5x + \dfrac{2}{3} \cdot \dfrac{1}{3}x^3 + \dfrac{3}{4} \cdot \dfrac{1}{4}x^4 + C = 5x + \dfrac{2}{9}x^3 + \dfrac{3}{16}x^4 + C$

13. $\displaystyle\int (u+4)(2u+1)\,du = \int (2u^2 + 9u + 4)\,du = 2\dfrac{u^3}{3} + 9\dfrac{u^2}{2} + 4u + C = \dfrac{2}{3}u^3 + \dfrac{9}{2}u^2 + 4u + C$

15. $\displaystyle\int \dfrac{1+\sqrt{x}+x}{\sqrt{x}}\,dx = \int \left(\dfrac{1}{\sqrt{x}} + 1 + \sqrt{x}\right)dx = \int (x^{-1/2} + 1 + x^{1/2})\,dx$

$\qquad = 2x^{1/2} + x + \dfrac{2}{3}x^{3/2} + C = 2\sqrt{x} + x + \dfrac{2}{3}x^{3/2} + C$

17. $\displaystyle\int (2 + \tan^2\theta)\,d\theta = \int [2 + (\sec^2\theta - 1)]\,d\theta = \int (1 + \sec^2\theta)\,d\theta = \theta + \tan\theta + C$

19. $\displaystyle\int 3\csc^2 t\,dt = -3\cot t + C$

21. $\int \left(\cos x + \frac{1}{2}x\right) dx = \sin x + \frac{1}{4}x^2 + C$. The members of the family in the figure correspond to $C = -5, 0, 5$, and 10.

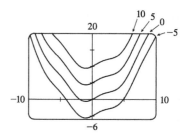

23. $\int_{-2}^{3}(x^2 - 3)\,dx = \left[\frac{1}{3}x^3 - 3x\right]_{-2}^{3} = (9 - 9) - \left(-\frac{8}{3} + 6\right) = \frac{8}{3} - \frac{18}{3} = -\frac{10}{3}$

25. $\int_{1}^{4}(8t^3 - 6t^{-2})\,dt = \left[2t^4 + \frac{6}{t}\right]_{1}^{4} = \left(2 \cdot 4^4 + \frac{6}{4}\right) - \left(2 \cdot 1^4 + \frac{6}{1}\right) = \left(512 + \frac{3}{2}\right) - (2 + 6) = \frac{1011}{2} = 505.5$

27. $\int_{0}^{2}(2x - 3)(4x^2 + 1)\,dx = \int_{0}^{2}(8x^3 - 12x^2 + 2x - 3)\,dx = \left[2x^4 - 4x^3 + x^2 - 3x\right]_{0}^{2} = (32 - 32 + 4 - 6) - 0 = -2$

29. $\int_{0}^{\pi}(4\sin\theta - 3\cos\theta)\,d\theta = \left[-4\cos\theta - 3\sin\theta\right]_{0}^{\pi} = (4 - 0) - (-4 - 0) = 8$

31. $\int_{1}^{4}\left(\frac{4 + 6u}{\sqrt{u}}\right) du = \int_{1}^{4}\left(\frac{4}{\sqrt{u}} + \frac{6u}{\sqrt{u}}\right) du = \int_{1}^{4}(4u^{-1/2} + 6u^{1/2})\,du = \left[8u^{1/2} + 4u^{3/2}\right]_{1}^{4} = (16 + 32) - (8 + 4) = 36$

33. $\int_{\pi/6}^{\pi/3}(4\sec^2 y)\,dy = \left[4\tan y\right]_{\pi/6}^{\pi/3} = 4\tan\frac{\pi}{3} - 4\tan\frac{\pi}{6} = 4 \cdot \sqrt{3} - 4 \cdot \frac{1}{\sqrt{3}} = 4\sqrt{3} - \frac{4\sqrt{3}}{3} = \frac{8\sqrt{3}}{3}$ or $\frac{8}{\sqrt{3}}$

35. $\int_{0}^{1} x\left(\sqrt[3]{x} + \sqrt[4]{x}\right) dx = \int_{0}^{1}(x^{4/3} + x^{5/4})\,dx = \left[\frac{3}{7}x^{7/3} + \frac{4}{9}x^{9/4}\right]_{0}^{1} = \left(\frac{3}{7} + \frac{4}{9}\right) - 0 = \frac{55}{63}$

37. $\int_{1}^{4}\sqrt{5/x}\,dx = \sqrt{5}\int_{1}^{4} x^{-1/2}\,dx = \sqrt{5}\left[2\sqrt{x}\right]_{1}^{4} = \sqrt{5}(2 \cdot 2 - 2 \cdot 1) = 2\sqrt{5}$

39. $\int_{0}^{\pi/4} \frac{1 + \cos^2\theta}{\cos^2\theta}\,d\theta = \int_{0}^{\pi/4}\left(\frac{1}{\cos^2\theta} + \frac{\cos^2\theta}{\cos^2\theta}\right) d\theta = \int_{0}^{\pi/4}(\sec^2\theta + 1)\,d\theta$
$= \left[\tan\theta + \theta\right]_{0}^{\pi/4} = \left(\tan\frac{\pi}{4} + \frac{\pi}{4}\right) - (0 + 0) = 1 + \frac{\pi}{4}$

41. $\int_{0}^{64} \sqrt{u}\left(u - \sqrt[3]{u}\right) du = \int_{0}^{64}(u^{3/2} - u^{5/6})\,du = \left[\frac{2}{5}u^{5/2} - \frac{6}{11}u^{11/6}\right]_{0}^{64} = \left(\frac{65{,}536}{5} - \frac{12{,}288}{11}\right) - 0 = \frac{659{,}456}{55}$

43. $|x - 3| = \begin{cases} x - 3 & \text{if } x - 3 \geq 0 \\ -(x - 3) & \text{if } x - 3 < 0 \end{cases} = \begin{cases} x - 3 & \text{if } x \geq 3 \\ 3 - x & \text{if } x < 3 \end{cases}$

Thus, $\int_{2}^{5}|x - 3|\,dx = \int_{2}^{3}(3 - x)\,dx + \int_{3}^{5}(x - 3)\,dx = \left[3x - \frac{1}{2}x^2\right]_{2}^{3} + \left[\frac{1}{2}x^2 - 3x\right]_{3}^{5}$
$= \left(9 - \frac{9}{2}\right) - (6 - 2) + \left(\frac{25}{2} - 15\right) - \left(\frac{9}{2} - 9\right) = \frac{5}{2}$

45. $\int_{-1}^{2}(x - 2|x|)\,dx = \int_{-1}^{0}[x - 2(-x)]\,dx + \int_{0}^{2}[x - 2(x)]\,dx = \int_{-1}^{0} 3x\,dx + \int_{0}^{2}(-x)\,dx = 3\left[\frac{1}{2}x^2\right]_{-1}^{0} - \left[\frac{1}{2}x^2\right]_{0}^{2}$
$= 3\left(0 - \frac{1}{2}\right) - (2 - 0) = -\frac{7}{2} = -3.5$

47. The graph shows that $y = 1 - 2x - 5x^4$ has x-intercepts at $x = a \approx -0.86$ and at $x = b \approx 0.42$. So the area of the region that lies under the curve and above the x-axis is
$\int_a^b (1 - 2x - 5x^4)\, dx = \left[x - x^2 - x^5 \right]_a^b$
$= (b - b^2 - b^5) - (a - a^2 - a^5) \approx 1.36$

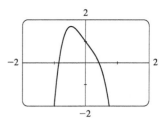

49. $A = \int_0^2 (2y - y^2)\, dy = \left[y^2 - \tfrac{1}{3}y^3 \right]_0^2 = \left(4 - \tfrac{8}{3} \right) - 0 = \tfrac{4}{3}$

51. If $w'(t)$ is the rate of change of weight in pounds per year, then $w(t)$ represents the weight in pounds of the child at age t. We know from the Net Change Theorem that $\int_5^{10} w'(t)\, dt = w(10) - w(5)$, so the integral represents the increase in the child's weight (in pounds) between the ages of 5 and 10.

53. Since $r(t)$ is the rate at which oil leaks, we can write $r(t) = -V'(t)$, where $V(t)$ is the volume of oil at time t. [Note that the minus sign is needed because V is decreasing, so $V'(t)$ is negative, but $r(t)$ is positive.] Thus, by the Net Change Theorem,
$\int_0^{120} r(t)\, dt = -\int_0^{120} V'(t)\, dt = -[V(120) - V(0)] = V(0) - V(120)$, which is the number of gallons of oil that leaked from the tank in the first two hours (120 minutes).

55. By the Net Change Theorem, $\int_{1000}^{5000} R'(x)\, dx = R(5000) - R(1000)$, so it represents the increase in revenue when production is increased from 1000 units to 5000 units.

57. The function h gives the rate of change in the total number of heartbeats H, so $h(t) = H'(t)$. By the Net Change Theorem, $\int_0^{30} h(t)\, dt = H(30) - H(0) = H(30) - 0 = H(30)$ represents the total number of heartbeats during the first 30 minutes of an exercise session.

59. In general, the unit of measurement for $\int_a^b f(x)\, dx$ is the product of the unit for $f(x)$ and the unit for x. Since $f(x)$ is measured in newtons and x is measured in meters, the units for $\int_0^{100} f(x)\, dx$ are newton-meters (or joules). (A newton-meter is abbreviated N·m.)

61. (a) Displacement $= \int_0^3 (3t - 5)\, dt = \left[\tfrac{3}{2}t^2 - 5t \right]_0^3 = \tfrac{27}{2} - 15 = -\tfrac{3}{2}$ m

   (b) Distance traveled $= \int_0^3 |3t - 5|\, dt = \int_0^{5/3} (5 - 3t)\, dt + \int_{5/3}^3 (3t - 5)\, dt$
   $= \left[5t - \tfrac{3}{2}t^2 \right]_0^{5/3} + \left[\tfrac{3}{2}t^2 - 5t \right]_{5/3}^3 = \tfrac{25}{3} - \tfrac{3}{2} \cdot \tfrac{25}{9} + \tfrac{27}{2} - 15 - \left(\tfrac{3}{2} \cdot \tfrac{25}{9} - \tfrac{25}{3} \right) = \tfrac{41}{6}$ m

63. (a) $v'(t) = a(t) = t + 4 \;\Rightarrow\; v(t) = \tfrac{1}{2}t^2 + 4t + C \;\Rightarrow\; v(0) = C = 5 \;\Rightarrow\; v(t) = \tfrac{1}{2}t^2 + 4t + 5$ m/s

   (b) Distance traveled $= \int_0^{10} |v(t)|\, dt = \int_0^{10} \left| \tfrac{1}{2}t^2 + 4t + 5 \right| dt = \int_0^{10} \left(\tfrac{1}{2}t^2 + 4t + 5 \right) dt = \left[\tfrac{1}{6}t^3 + 2t^2 + 5t \right]_0^{10}$
   $= \tfrac{500}{3} + 200 + 50 = 416\tfrac{2}{3}$ m

65. Since $m'(x) = \rho(x)$, $m = \int_0^4 \rho(x)\, dx = \int_0^4 \left(9 + 2\sqrt{x} \right) dx = \left[9x + \tfrac{4}{3}x^{3/2} \right]_0^4 = 36 + \tfrac{32}{3} - 0 = \tfrac{140}{3} = 46\tfrac{2}{3}$ kg.

67. Let s be the position of the car. We know from Equation 2 that $s(100) - s(0) = \int_0^{100} v(t)\,dt$. We use the Midpoint Rule for $0 \le t \le 100$ with $n = 5$. Note that the length of each of the five time intervals is 20 seconds $= \frac{20}{3600}$ hour $= \frac{1}{180}$ hour. So the distance traveled is

$\int_0^{100} v(t)\,dt \approx \frac{1}{180}[v(10) + v(30) + v(50) + v(70) + v(90)] = \frac{1}{180}(38 + 58 + 51 + 53 + 47) = \frac{247}{180} \approx 1.37$ miles.

69. Use the midpoint of each of four 2-day intervals. Let $t = 0$ correspond to July 18 and note that the inflow rate, $r(t)$, is in ft^3/s.

Amount of water $= \int_0^8 r(t)\,dt \approx [r(1) + r(3) + r(5) + r(7)]\dfrac{8-0}{4} \approx [6401 + 4249 + 3821 + 2628](2) = 34{,}198$.

Now multiply by the number of seconds in a day, $24 \cdot 60^2$, to get 2,954,707,200 ft^3.

71. To use the Midpoint Rule, we'll use the midpoint of each of three 2-second intervals.

$v(6) - v(0) = \int_0^6 a(t)\,dt \approx [a(1) + a(3) + a(5)]\dfrac{6-0}{3} \approx (0.6 + 10 + 9.3)(2) = 39.8$ ft/s

73. Power is the rate of change of energy with respect to time; that is, $P(t) = E'(t)$. By the Net Change Theorem and the Midpoint Rule, the electric energy used on that day is

$$E(24) - E(0) = \int_0^{24} P(t)\,dt \approx \frac{24-0}{12}[P(1) + P(3) + P(5) + \cdots + P(21) + P(23)]$$

$$\approx 2(11 + 10.5 + 11 + 14 + 15.1 + 15.5 + 15.1 + 15 + 15 + 16.1 + 15 + 13)$$

$$= 2(166.3) = 332.6 \text{ gigawatt-hours}$$

75. $\int (\sin x + \sinh x)\,dx = -\cos x + \cosh x + C$

77. $\int \left(x^2 + 1 + \dfrac{1}{x^2+1}\right) dx = \dfrac{x^3}{3} + x + \tan^{-1} x + C$

79. $\displaystyle\int_0^{1/\sqrt{3}} \dfrac{t^2-1}{t^4-1}\,dt = \int_0^{1/\sqrt{3}} \dfrac{t^2-1}{(t^2+1)(t^2-1)}\,dt = \int_0^{1/\sqrt{3}} \dfrac{1}{t^2+1}\,dt = \big[\arctan t\big]_0^{1/\sqrt{3}} = \arctan\left(1/\sqrt{3}\right) - \arctan 0$

$= \dfrac{\pi}{6} - 0 = \dfrac{\pi}{6}$

4.5 The Substitution Rule

1. Let $u = 2x$. Then $du = 2\,dx$ and $dx = \frac{1}{2}\,du$, so $\int \cos 2x\,dx = \int \cos u\,\left(\frac{1}{2}\,du\right) = \frac{1}{2}\sin u + C = \frac{1}{2}\sin 2x + C$.

3. Let $u = x^3 + 1$. Then $du = 3x^2\,dx$ and $x^2\,dx = \frac{1}{3}\,du$, so

$\int x^2\sqrt{x^3+1}\,dx = \int \sqrt{u}\,\left(\tfrac{1}{3}\,du\right) = \dfrac{1}{3}\dfrac{u^{3/2}}{3/2} + C = \dfrac{1}{3}\cdot\dfrac{2}{3}u^{3/2} + C = \tfrac{2}{9}(x^3+1)^{3/2} + C.$

5. Let $u = x^4 - 5$. Then $du = 4x^3\,dx$ and $x^3\,dx = \frac{1}{4}\,du$, so

$\int \dfrac{x^3}{(x^4-5)^2}\,dx = \int \dfrac{1}{u^2}\left(\tfrac{1}{4}\,du\right) = \dfrac{1}{4}\int u^{-2}\,du = \dfrac{1}{4}\dfrac{u^{-1}}{-1} + C = -\dfrac{1}{4u} + C = -\dfrac{1}{4(x^4-5)} + C.$

7. Let $u = \sqrt{t}$. Then $du = \dfrac{1}{2\sqrt{t}} dt$ and $\dfrac{1}{\sqrt{t}} dt = 2\, du$, so $\displaystyle\int \dfrac{\cos\sqrt{t}}{\sqrt{t}} dt = \int \cos u\, (2\, du) = 2\sin u + C = 2\sin\sqrt{t} + C$.

9. Let $u = 1 - x^2$. Then $du = -2x\, dx$ and $x\, dx = -\tfrac{1}{2} du$, so

$\int x\sqrt{1-x^2}\, dx = \int \sqrt{u}\, (-\tfrac{1}{2} du) = -\tfrac{1}{2} \cdot \tfrac{2}{3} u^{3/2} + C = -\tfrac{1}{3}(1-x^2)^{3/2} + C$.

11. Let $u = x^3$. Then $du = 3x^2\, dx$ and $x^2\, dx = \tfrac{1}{3} du$, so $\int x^2 \cos(x^3)\, dx = \int \cos u\, (\tfrac{1}{3} du) = \tfrac{1}{3} \sin u + C = \tfrac{1}{3}\sin(x^3) + C$.

13. Let $u = \dfrac{\pi}{3} t$. Then $du = \dfrac{\pi}{3} dt$ and $dt = \dfrac{3}{\pi} du$, so

$\displaystyle\int \sin\left(\dfrac{\pi t}{3}\right) dt = \int \sin u \left(\dfrac{3}{\pi} du\right) = \dfrac{3}{\pi} \cdot (-\cos u) + C = -\dfrac{3}{\pi} \cos\left(\dfrac{\pi}{3} t\right) + C$.

15. Let $u = 3t$. Then $du = 3\, dt$ and $dt = \tfrac{1}{3} du$, so $\int \sec 3t \tan 3t\, dt = \int \sec u \tan u\, (\tfrac{1}{3} du) = \tfrac{1}{3} \sec u + C = \tfrac{1}{3} \sec 3t + C$.

17. Let $u = 1 + 5t$. Then $du = 5\, dt$ and $dt = \tfrac{1}{5} du$, so

$\int \cos(1+5t)\, dt = \int \cos u\, (\tfrac{1}{5} du) = \tfrac{1}{5} \sin u + C = \tfrac{1}{5} \sin(1+5t) + C$.

19. Let $u = \cos\theta$. Then $du = -\sin\theta\, d\theta$ and $\sin\theta\, d\theta = -du$, so

$\int \cos^3\theta \sin\theta\, d\theta = \int u^3(-du) = -\tfrac{1}{4} u^4 + C = -\tfrac{1}{4} \cos^4\theta + C$.

21. Let $u = x^2 + \dfrac{2}{x}$. Then $du = \left(2x - \dfrac{2}{x^2}\right) dx = 2\left(x - \dfrac{1}{x^2}\right) dx$ and $\left(x - \dfrac{1}{x^2}\right) dx = \dfrac{1}{2} du$, so

$\displaystyle\int \left(x - \dfrac{1}{x^2}\right)\left(x^2 + \dfrac{2}{x}\right)^5 dx = \int u^5 \left(\dfrac{1}{2} du\right) = \dfrac{1}{2} \cdot \dfrac{1}{6} u^6 + C = \dfrac{1}{12}\left(x^2 + \dfrac{2}{x}\right)^6 + C$.

23. Let $u = 3ax + bx^3$. Then $du = (3a + 3bx^2)\, dx = 3(a + bx^2)\, dx$, so

$\displaystyle\int \dfrac{a+bx^2}{\sqrt{3ax+bx^3}}\, dx = \int \dfrac{\tfrac{1}{3} du}{u^{1/2}} = \dfrac{1}{3} \int u^{-1/2}\, du = \dfrac{1}{3} \cdot 2u^{1/2} + C = \tfrac{2}{3}\sqrt{3ax+bx^3} + C$.

25. Let $u = 1 + z^3$. Then $du = 3z^2\, dz$ and $z^2\, dz = \tfrac{1}{3} du$, so

$\displaystyle\int \dfrac{z^2}{\sqrt[3]{1+z^3}}\, dz = \int u^{-1/3}(\tfrac{1}{3} du) = \tfrac{1}{3} \cdot \tfrac{3}{2} u^{2/3} + C = \tfrac{1}{2}(1+z^3)^{2/3} + C$.

27. Let $u = \cot x$. Then $du = -\csc^2 x\, dx$ and $\csc^2 x\, dx = -du$, so

$\displaystyle\int \sqrt{\cot x}\, \csc^2 x\, dx = \int \sqrt{u}\, (-du) = -\dfrac{u^{3/2}}{3/2} + C = -\tfrac{2}{3}(\cot x)^{3/2} + C$.

29. Let $u = \sec x$. Then $du = \sec x \tan x\, dx$, so

$\int \sec^3 x \tan x\, dx = \int \sec^2 x\, (\sec x \tan x)\, dx = \int u^2\, du = \tfrac{1}{3} u^3 + C = \tfrac{1}{3} \sec^3 x + C$.

31. Let $u = 2x + 5$. Then $du = 2\, dx$ and $x = \tfrac{1}{2}(u - 5)$, so

$\int x(2x+5)^8\, dx = \int \tfrac{1}{2}(u-5) u^8 (\tfrac{1}{2} du) = \tfrac{1}{4} \int (u^9 - 5u^8)\, du$

$= \tfrac{1}{4}\left(\tfrac{1}{10} u^{10} - \tfrac{5}{9} u^9\right) + C = \tfrac{1}{40}(2x+5)^{10} - \tfrac{5}{36}(2x+5)^9 + C$

33. $f(x) = x(x^2 - 1)^3$. $u = x^2 - 1 \Rightarrow du = 2x\,dx$, so

$$\int x(x^2 - 1)^3\,dx = \int u^3\left(\tfrac{1}{2}\,du\right) = \tfrac{1}{8}u^4 + C = \tfrac{1}{8}(x^2 - 1)^4 + C$$

Where f is positive (negative), F is increasing (decreasing). Where f changes from negative to positive (positive to negative), F has a local minimum (maximum).

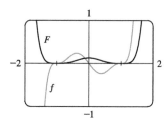

35. $f(x) = \sin^3 x \cos x$. $u = \sin x \Rightarrow du = \cos x\,dx$, so

$$\int \sin^3 x \cos x\,dx = \int u^3\,du = \tfrac{1}{4}u^4 + C = \tfrac{1}{4}\sin^4 x + C$$

Note that at $x = \tfrac{\pi}{2}$, f changes from positive to negative and F has a local maximum. Also, both f and F are periodic with period π, so at $x = 0$ and at $x = \pi$, f changes from negative to positive and F has local minima.

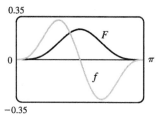

37. Let $u = \tfrac{\pi}{2}t$, so $du = \tfrac{\pi}{2}\,dt$. When $t = 0$, $u = 0$; when $t = 1$, $u = \tfrac{\pi}{2}$. Thus,

$$\int_0^1 \cos(\pi t/2)\,dt = \int_0^{\pi/2} \cos u \left(\tfrac{2}{\pi}\,du\right) = \tfrac{2}{\pi}[\sin u]_0^{\pi/2} = \tfrac{2}{\pi}\left(\sin \tfrac{\pi}{2} - \sin 0\right) = \tfrac{2}{\pi}(1 - 0) = \tfrac{2}{\pi}$$

39. Let $u = 1 + 7x$, so $du = 7\,dx$. When $x = 0$, $u = 1$; when $x = 1$, $u = 8$. Thus,

$$\int_0^1 \sqrt[3]{1 + 7x}\,dx = \int_1^8 u^{1/3}\left(\tfrac{1}{7}\,du\right) = \tfrac{1}{7}\left[\tfrac{3}{4}u^{4/3}\right]_1^8 = \tfrac{3}{28}(8^{4/3} - 1^{4/3}) = \tfrac{3}{28}(16 - 1) = \tfrac{45}{28}$$

41. Let $u = \cos t$, so $du = -\sin t\,dt$. When $t = 0$, $u = 1$; when $t = \tfrac{\pi}{6}$, $u = \sqrt{3}/2$. Thus,

$$\int_0^{\pi/6} \frac{\sin t}{\cos^2 t}\,dt = \int_1^{\sqrt{3}/2} \frac{1}{u^2}(-du) = \left[\frac{1}{u}\right]_1^{\sqrt{3}/2} = \frac{2}{\sqrt{3}} - 1.$$

43. $\int_{-\pi/4}^{\pi/4}(x^3 + x^4 \tan x)\,dx = 0$ by Theorem 6(b), since $f(x) = x^3 + x^4 \tan x$ is an odd function.

45. Let $u = 1 + 2x$, so $du = 2\,dx$. When $x = 0$, $u = 1$; when $x = 13$, $u = 27$. Thus,

$$\int_0^{13} \frac{dx}{\sqrt[3]{(1 + 2x)^2}} = \int_1^{27} u^{-2/3}\left(\tfrac{1}{2}\,du\right) = \left[\tfrac{1}{2} \cdot 3u^{1/3}\right]_1^{27} = \tfrac{3}{2}(3 - 1) = 3.$$

47. Let $u = x^2 + a^2$, so $du = 2x\,dx$ and $x\,dx = \tfrac{1}{2}\,du$. When $x = 0$, $u = a^2$; when $x = a$, $u = 2a^2$. Thus,

$$\int_0^a x\sqrt{x^2 + a^2}\,dx = \int_{a^2}^{2a^2} u^{1/2}\left(\tfrac{1}{2}\,du\right) = \tfrac{1}{2}\left[\tfrac{2}{3}u^{3/2}\right]_{a^2}^{2a^2} = \left[\tfrac{1}{3}u^{3/2}\right]_{a^2}^{2a^2} = \tfrac{1}{3}\left[(2a^2)^{3/2} - (a^2)^{3/2}\right] = \tfrac{1}{3}\left(2\sqrt{2} - 1\right)a^3$$

49. Let $u = x - 1$, so $u + 1 = x$ and $du = dx$. When $x = 1$, $u = 0$; when $x = 2$, $u = 1$. Thus,

$$\int_1^2 x\sqrt{x - 1}\,dx = \int_0^1 (u + 1)\sqrt{u}\,du = \int_0^1 (u^{3/2} + u^{1/2})\,du = \left[\tfrac{2}{5}u^{5/2} + \tfrac{2}{3}u^{3/2}\right]_0^1 = \tfrac{2}{5} + \tfrac{2}{3} = \tfrac{16}{15}.$$

51. Let $u = x^{-2}$, so $du = -2x^{-3}\,dx$. When $x = \tfrac{1}{2}$, $u = 4$; when $x = 1$, $u = 1$. Thus,

$$\int_{1/2}^1 \frac{\cos(x^{-2})}{x^3}\,dx = \int_4^1 \cos u\left(\frac{du}{-2}\right) = \frac{1}{2}\int_1^4 \cos u\,du = \frac{1}{2}[\sin u]_1^4 = \frac{1}{2}(\sin 4 - \sin 1).$$

53. From the graph, it appears that the area under the curve is about $1 + \left(\text{a little more than } \tfrac{1}{2} \cdot 1 \cdot 0.7\right)$, or about 1.4. The exact area is given by $A = \int_0^1 \sqrt{2x+1}\, dx$. Let $u = 2x + 1$, so $du = 2\, dx$. The limits change to $2 \cdot 0 + 1 = 1$ and $2 \cdot 1 + 1 = 3$, and

$$A = \int_1^3 \sqrt{u}\left(\tfrac{1}{2}\, du\right) = \tfrac{1}{2}\left[\tfrac{2}{3}u^{3/2}\right]_1^3 = \tfrac{1}{3}(3\sqrt{3} - 1) = \sqrt{3} - \tfrac{1}{3} \approx 1.399.$$

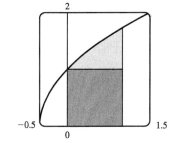

55. First write the integral as a sum of two integrals:

$I = \int_{-2}^{2}(x+3)\sqrt{4-x^2}\, dx = I_1 + I_2 = \int_{-2}^{2} x\sqrt{4-x^2}\, dx + \int_{-2}^{2} 3\sqrt{4-x^2}\, dx$. $I_1 = 0$ by Theorem 6(b), since $f(x) = x\sqrt{4-x^2}$ is an odd function and we are integrating from $x = -2$ to $x = 2$. We interpret I_2 as three times the area of a semicircle with radius 2, so $I = 0 + 3 \cdot \tfrac{1}{2}(\pi \cdot 2^2) = 6\pi$.

57. The volume of inhaled air in the lungs at time t is

$$V(t) = \int_0^t f(u)\, du = \int_0^t \tfrac{1}{2}\sin\!\left(\tfrac{2\pi}{5}u\right) du = \int_0^{2\pi t/5} \tfrac{1}{2}\sin v\left(\tfrac{5}{2\pi}\, dv\right) \quad [\text{substitute } v = \tfrac{2\pi}{5}u,\ dv = \tfrac{2\pi}{5}\, du]$$

$$= \tfrac{5}{4\pi}\bigl[-\cos v\bigr]_0^{2\pi t/5} = \tfrac{5}{4\pi}\left[-\cos\!\left(\tfrac{2\pi}{5}t\right) + 1\right] = \tfrac{5}{4\pi}\left[1 - \cos\!\left(\tfrac{2\pi}{5}t\right)\right] \text{ liters}$$

59. Let $u = 2x$. Then $du = 2\, dx$, so $\int_0^2 f(2x)\, dx = \int_0^4 f(u)\left(\tfrac{1}{2}\, du\right) = \tfrac{1}{2}\int_0^4 f(u)\, du = \tfrac{1}{2}(10) = 5$.

61. Let $u = -x$. Then $du = -dx$, so

$$\int_a^b f(-x)\, dx = \int_{-a}^{-b} f(u)(-du) = \int_{-b}^{-a} f(u)\, du = \int_{-b}^{-a} f(x)\, dx$$

From the diagram, we see that the equality follows from the fact that we are reflecting the graph of f, and the limits of integration, about the y-axis.

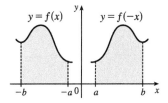

63. Let $u = 1 - x$. Then $x = 1 - u$ and $dx = -du$, so

$$\int_0^1 x^a(1-x)^b\, dx = \int_1^0 (1-u)^a u^b(-du) = \int_0^1 u^b(1-u)^a\, du = \int_0^1 x^b(1-x)^a\, dx.$$

65. $\int_0^{\pi/2} f(\cos x)\, dx = \int_0^{\pi/2} f\!\left[\sin\!\left(\tfrac{\pi}{2} - x\right)\right] dx \quad [u = \tfrac{\pi}{2} - x,\ du = -dx]$

$= \int_{\pi/2}^0 f(\sin u)(-du) = \int_0^{\pi/2} f(\sin u)\, du = \int_0^{\pi/2} f(\sin x)\, dx$

Continuity of f is needed in order to apply the substitution rule for definite integrals.

67. Let $u = 4x + 7$. Then $du = 4\, dx$ and $dx = \tfrac{1}{4}\, du$, so $\int \dfrac{dx}{4x+7} = \int \dfrac{1}{u}\left(\tfrac{1}{4}\, du\right) = \tfrac{1}{4}\ln|u| + C = \tfrac{1}{4}\ln|4x+7| + C$.

69. Let $u = \ln x$. Then $du = \dfrac{dx}{x}$, so $\int \dfrac{(\ln x)^2}{x}\, dx = \int u^2\, du = \tfrac{1}{3}u^3 + C = \tfrac{1}{3}(\ln x)^3 + C$.

71. Let $u = 2 + 3e^r$. Then $du = 3e^r\, dr$ and $e^r\, dr = \tfrac{1}{3}\, du$, so

$\int e^r(2 + 3e^r)^{3/2}\, dr = \int u^{3/2}\left(\tfrac{1}{3}\, du\right) = \tfrac{1}{3} \cdot \tfrac{2}{5}u^{5/2} + C = \tfrac{2}{15}(2 + 3e^r)^{5/2} + C.$

73. Let $u = \arctan x$. Then $du = \dfrac{1}{x^2+1}\,dx$, so $\displaystyle\int \dfrac{(\arctan x)^2}{x^2+1}\,dx = \int u^2\,du = \tfrac{1}{3}u^3 + C = \tfrac{1}{3}(\arctan x)^3 + C$.

75. Let $u = 1 + x^2$. Then $du = 2x\,dx$, so

$$\int \dfrac{1+x}{1+x^2}\,dx = \int \dfrac{1}{1+x^2}\,dx + \int \dfrac{x}{1+x^2}\,dx = \tan^{-1}x + \int \dfrac{\tfrac{1}{2}du}{u} = \tan^{-1}x + \tfrac{1}{2}\ln|u| + C$$

$$= \tan^{-1}x + \tfrac{1}{2}\ln\left|1+x^2\right| + C = \tan^{-1}x + \tfrac{1}{2}\ln(1+x^2) + C \quad [\text{since } 1+x^2 > 0].$$

77. $\displaystyle\int \dfrac{\sin 2x}{1+\cos^2 x}\,dx = 2\int \dfrac{\sin x \cos x}{1+\cos^2 x}\,dx = 2I$. Let $u = \cos x$. Then $du = -\sin x\,dx$, so

$$2I = -2\int \dfrac{u\,du}{1+u^2} = -2 \cdot \tfrac{1}{2}\ln(1+u^2) + C = -\ln(1+u^2) + C = -\ln(1+\cos^2 x) + C.$$

Or: Let $u = 1 + \cos^2 x$.

79. $\displaystyle\int \cot x\,dx = \int \dfrac{\cos x}{\sin x}\,dx$. Let $u = \sin x$. Then $du = \cos x\,dx$, so $\displaystyle\int \cot x\,dx = \int \dfrac{1}{u}\,du = \ln|u| + C = \ln|\sin x| + C$.

81. Let $u = \ln x$, so $du = \dfrac{dx}{x}$. When $x = e$, $u = 1$; when $x = e^4$; $u = 4$. Thus,

$$\int_e^{e^4} \dfrac{dx}{x\sqrt{\ln x}} = \int_1^4 u^{-1/2}\,du = 2\left[u^{1/2}\right]_1^4 = 2(2-1) = 2.$$

83. Let $u = e^z + z$, so $du = (e^z + 1)\,dz$. When $z = 0$, $u = 1$; when $z = 1$, $u = e + 1$. Thus,

$$\int_0^1 \dfrac{e^z+1}{e^z+z}\,dz = \int_1^{e+1} \dfrac{1}{u}\,du = \Big[\ln|u|\Big]_1^{e+1} = \ln|e+1| - \ln|1| = \ln(e+1).$$

85. $\dfrac{x\sin x}{1+\cos^2 x} = x \cdot \dfrac{\sin x}{2 - \sin^2 x} = x\,f(\sin x)$, where $f(t) = \dfrac{t}{2-t^2}$. By Exercise 64,

$$\int_0^\pi \dfrac{x\sin x}{1+\cos^2 x}\,dx = \int_0^\pi x\,f(\sin x)\,dx = \dfrac{\pi}{2}\int_0^\pi f(\sin x)\,dx = \dfrac{\pi}{2}\int_0^\pi \dfrac{\sin x}{1+\cos^2 x}\,dx$$

Let $u = \cos x$. Then $du = -\sin x\,dx$. When $x = \pi$, $u = -1$ and when $x = 0$, $u = 1$. So

$$\dfrac{\pi}{2}\int_0^\pi \dfrac{\sin x}{1+\cos^2 x}\,dx = -\dfrac{\pi}{2}\int_1^{-1} \dfrac{du}{1+u^2} = \dfrac{\pi}{2}\int_{-1}^1 \dfrac{du}{1+u^2} = \dfrac{\pi}{2}\Big[\tan^{-1}u\Big]_{-1}^1$$

$$= \dfrac{\pi}{2}[\tan^{-1}1 - \tan^{-1}(-1)] = \dfrac{\pi}{2}\left[\dfrac{\pi}{4} - \left(-\dfrac{\pi}{4}\right)\right] = \dfrac{\pi^2}{4}$$

4 Review

TRUE-FALSE QUIZ

1. True by Property 2 of the Integral in Section 4.2.

3. True by Property 3 of the Integral in Section 4.2.

5. False. For example, let $f(x) = x^2$. Then $\int_0^1 \sqrt{x^2}\,dx = \int_0^1 x\,dx = \tfrac{1}{2}$, but $\sqrt{\int_0^1 x^2\,dx} = \sqrt{\tfrac{1}{3}} = \tfrac{1}{\sqrt{3}}$.

7. True by the Net Change Theorem.

9. False. $\int_a^b f'(x)\,[f(x)]^4\,dx$ is a definite integral and, thus, is a number; $\frac{1}{5}[f(x)]^5 + C$ is a family of functions. The statement would be true without the limits of integration.

11. True by Comparison Property 7 of the Integral in Section 4.2.

13. False. For example, the function $y = |x|$ is continuous on $\mathbb{R}$, but has no derivative at $x = 0$.

15. True. By Property 5 in Section 4.2, $\int_\pi^{3\pi} \frac{\sin x}{x}\,dx = \int_\pi^{2\pi} \frac{\sin x}{x}\,dx + \int_{2\pi}^{3\pi} \frac{\sin x}{x}\,dx \Rightarrow$

$\int_\pi^{2\pi} \frac{\sin x}{x}\,dx = \int_\pi^{3\pi} \frac{\sin x}{x}\,dx - \int_{2\pi}^{3\pi} \frac{\sin x}{x}\,dx \Rightarrow \int_\pi^{2\pi} \frac{\sin x}{x}\,dx = \int_\pi^{3\pi} \frac{\sin x}{x}\,dx + \int_{3\pi}^{2\pi} \frac{\sin x}{x}\,dx$

[by reversing limits].

17. False. $\int_a^b f(x)\,dx$ is a constant, so $\frac{d}{dx}\left(\int_a^b f(x)\,dx\right) = 0$, not $f(x)$ [unless $f(x) = 0$]. Compare the given statement carefully with FTC1, in which the upper limit in the integral is x.

19. False. The function $f(x) = 1/x^4$ is not bounded on the interval $[-2, 1]$. It has an infinite discontinuity at $x = 0$, so it is not integrable on the interval. (If the integral were to exist, a positive value would be expected, by Comparison Property 4.2.6 of Integrals.)

EXERCISES

1. (a)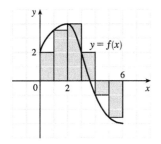

$L_6 = \sum_{i=1}^{6} f(x_{i-1})\,\Delta x \quad [\Delta x = \frac{6-0}{6} = 1]$

$= f(x_0) \cdot 1 + f(x_1) \cdot 1 + f(x_2) \cdot 1 + f(x_3) \cdot 1 + f(x_4) \cdot 1 + f(x_5) \cdot 1$

$\approx 2 + 3.5 + 4 + 2 + (-1) + (-2.5) = 8$

The Riemann sum represents the sum of the areas of the four rectangles above the x-axis minus the sum of the areas of the two rectangles below the x-axis.

(b)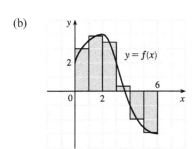

$M_6 = \sum_{i=1}^{6} f(\bar{x}_i)\,\Delta x \quad [\Delta x = \frac{6-0}{6} = 1]$

$= f(\bar{x}_1) \cdot 1 + f(\bar{x}_2) \cdot 1 + f(\bar{x}_3) \cdot 1 + f(\bar{x}_4) \cdot 1 + f(\bar{x}_5) \cdot 1 + f(\bar{x}_6) \cdot 1$

$= f(0.5) + f(1.5) + f(2.5) + f(3.5) + f(4.5) + f(5.5)$

$\approx 3 + 3.9 + 3.4 + 0.3 + (-2) + (-2.9) = 5.7$

The Riemann sum represents the sum of the areas of the four rectangles above the x-axis minus the sum of the areas of the two rectangles below the x-axis.

3. $\int_0^1 \left(x + \sqrt{1-x^2}\right) dx = \int_0^1 x\, dx + \int_0^1 \sqrt{1-x^2}\, dx = I_1 + I_2$.

I_1 can be interpreted as the area of the triangle shown in the figure and I_2 can be interpreted as the area of the quarter-circle.

Area $= \frac{1}{2}(1)(1) + \frac{1}{4}(\pi)(1)^2 = \frac{1}{2} + \frac{\pi}{4}$.

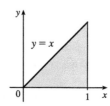
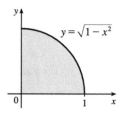

5. $\int_0^6 f(x)\, dx = \int_0^4 f(x)\, dx + \int_4^6 f(x)\, dx \;\Rightarrow\; 10 = 7 + \int_4^6 f(x)\, dx \;\Rightarrow\; \int_4^6 f(x)\, dx = 10 - 7 = 3$

7. First note that either a or b must be the graph of $\int_0^x f(t)\, dt$, since $\int_0^0 f(t)\, dt = 0$, and $c(0) \neq 0$. Now notice that $b > 0$ when c is increasing, and that $c > 0$ when a is increasing. It follows that c is the graph of $f(x)$, b is the graph of $f'(x)$, and a is the graph of $\int_0^x f(t)\, dt$.

9. $g(4) = \int_0^4 f(t)\, dt = \int_0^1 f(t)\, dt + \int_1^2 f(t)\, dt + \int_2^3 f(t)\, dt + \int_3^4 f(t)\, dt$

$= -\frac{1}{2} \cdot 1 \cdot 2 \begin{bmatrix} \text{area of triangle,} \\ \text{below } t\text{-axis} \end{bmatrix} + \frac{1}{2} \cdot 1 \cdot 2 + 1 \cdot 2 + \frac{1}{2} \cdot 1 \cdot 2 = 3$

By FTC1, $g'(x) = f(x)$, so $g'(4) = f(4) = 0$.

11. $\int_{-1}^0 (x^2 + 5x)\, dx = \left[\frac{1}{3}x^3 + \frac{5}{2}x^2\right]_{-1}^0 = 0 - \left(-\frac{1}{3} + \frac{5}{2}\right) = -\frac{13}{6}$

13. $\int_0^1 (1 - x^9)\, dx = \left[x - \frac{1}{10}x^{10}\right]_0^1 = \left(1 - \frac{1}{10}\right) - 0 = \frac{9}{10}$

15. $\int_1^9 \frac{\sqrt{u} - 2u^2}{u}\, du = \int_1^9 (u^{-1/2} - 2u)\, du = \left[2u^{1/2} - u^2\right]_1^9 = (6 - 81) - (2 - 1) = -76$

17. Let $u = y^2 + 1$, so $du = 2y\, dy$ and $y\, dy = \frac{1}{2}\, du$. When $y = 0$, $u = 1$; when $y = 1$, $u = 2$. Thus,

$\int_0^1 y(y^2 + 1)^5\, dy = \int_1^2 u^5 \left(\frac{1}{2}\, du\right) = \frac{1}{2}\left[\frac{1}{6}u^6\right]_1^2 = \frac{1}{12}(64 - 1) = \frac{63}{12} = \frac{21}{4}$.

19. $\int_1^5 \frac{dt}{(t-4)^2}$ does not exist because the function $f(t) = \frac{1}{(t-4)^2}$ has an infinite discontinuity at $t = 4$; that is, f is discontinuous on the interval $[1, 5]$.

21. Let $u = v^3$, so $du = 3v^2\, dv$. When $v = 0$, $u = 0$; when $v = 1$, $u = 1$. Thus,

$\int_0^1 v^2 \cos(v^3)\, dv = \int_0^1 \cos u \left(\frac{1}{3}\, du\right) = \frac{1}{3}[\sin u]_0^1 = \frac{1}{3}(\sin 1 - 0) = \frac{1}{3} \sin 1$.

23. $\int_{-\pi/4}^{\pi/4} \frac{t^4 \tan t}{2 + \cos t}\, dt = 0$ by Theorem 4.5.6, since $f(t) = \frac{t^4 \tan t}{2 + \cos t}$ is an odd function.

25. Let $u = \sin \pi t$. Then $du = \pi \cos \pi t\, dt$, so $\int \sin \pi t \cos \pi t\, dt = \int u\left(\frac{1}{\pi}\, du\right) = \frac{1}{\pi} \cdot \frac{1}{2}u^2 + C = \frac{1}{2\pi}(\sin \pi t)^2 + C$.

27. Let $u = 2\theta$. Then $du = 2\, d\theta$, so

$\int_0^{\pi/8} \sec 2\theta \tan 2\theta\, d\theta = \int_0^{\pi/4} \sec u \tan u \left(\frac{1}{2}\, du\right) = \frac{1}{2}[\sec u]_0^{\pi/4} = \frac{1}{2}\left(\sec \frac{\pi}{4} - \sec 0\right) = \frac{1}{2}(\sqrt{2} - 1) = \frac{1}{2}\sqrt{2} - \frac{1}{2}$.

29. Let $u = 1 - x$. Then $x = 1 - u$, $du = -dx$, and $dx = -du$, so

$$\int x(1-x)^{2/3}\, dx = \int (1-u) \cdot u^{2/3}(-du) = \int (u^{5/3} - u^{2/3})\, du = \tfrac{3}{8}u^{8/3} - \tfrac{3}{5}u^{5/3} + C$$
$$= \tfrac{3}{8}(1-x)^{8/3} - \tfrac{3}{5}(1-x)^{5/3} + C$$

31. Since $x^2 - 4 < 0$ for $0 \le x < 2$ and $x^2 - 4 > 0$ for $2 < x \le 3$, we have $|x^2 - 4| = -(x^2 - 4) = 4 - x^2$ for $0 \le x < 2$ and $|x^2 - 4| = x^2 - 4$ for $2 < x \le 3$. Thus,

$$\int_0^3 |x^2 - 4|\, dx = \int_0^2 (4 - x^2)\, dx + \int_2^3 (x^2 - 4)\, dx = \left[4x - \frac{x^3}{3}\right]_0^2 + \left[\frac{x^3}{3} - 4x\right]_2^3$$
$$= (8 - \tfrac{8}{3}) - 0 + (9 - 12) - (\tfrac{8}{3} - 8) = \tfrac{16}{3} - 3 + \tfrac{16}{3} = \tfrac{32}{3} - \tfrac{9}{3} = \tfrac{23}{3}$$

33. Let $u = 1 + \sin x$. Then $du = \cos x\, dx$, so

$$\int \frac{\cos x\, dx}{\sqrt{1 + \sin x}} = \int u^{-1/2}\, du = 2u^{1/2} + C = 2\sqrt{1 + \sin x} + C.$$

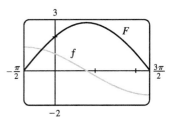

35. From the graph, it appears that the area under the curve $y = x\sqrt{x}$ between $x = 0$ and $x = 4$ is somewhat less than half the area of an 8×4 rectangle, so perhaps about 13 or 14. To find the exact value, we evaluate

$$\int_0^4 x\sqrt{x}\, dx = \int_0^4 x^{3/2}\, dx = \left[\tfrac{2}{5}x^{5/2}\right]_0^4 = \tfrac{2}{5}(4)^{5/2} = \tfrac{64}{5} = 12.8.$$

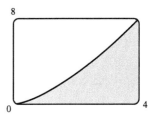

37. Area $= \int_0^4 (x^2 + 5)\, dx = \left[\tfrac{1}{3}x^3 + 5x\right]_0^4 = \left(\tfrac{64}{3} + 20\right) - 0 = \tfrac{124}{3}$

39. (a) $\int_1^5 f(x)\, dx = \int_1^3 f(x)\, dx + \int_3^4 f(x)\, dx + \int_4^5 f(x)\, dx = 3 - 2 + 1 = 2$

(b) $\int_1^5 |f(x)|\, dx = \int_1^3 f(x)\, dx + \int_3^4 [-f(x)]\, dx + \int_4^5 f(x)\, dx = 3 + 2 + 1 = 6$

41. $F(x) = \int_0^x \frac{t^2}{1 + t^3}\, dt \;\Rightarrow\; F'(x) = \frac{d}{dx}\int_0^x \frac{t^2}{1 + t^3}\, dt = \frac{x^2}{1 + x^3}$

43. Let $u = x^4$. Then $\dfrac{du}{dx} = 4x^3$. Also, $\dfrac{dg}{dx} = \dfrac{dg}{du}\dfrac{du}{dx}$, so

$$g'(x) = \frac{d}{dx}\int_0^{x^4} \cos(t^2)\, dt = \frac{d}{du}\int_0^u \cos(t^2)\, dt \cdot \frac{du}{dx} = \cos(u^2)\frac{du}{dx} = 4x^3 \cos(x^8).$$

45. $y = \displaystyle\int_{\sqrt{x}}^x \frac{\cos \theta}{\theta}\, d\theta = \int_1^x \frac{\cos \theta}{\theta}\, d\theta + \int_{\sqrt{x}}^1 \frac{\cos \theta}{\theta}\, d\theta = \int_1^x \frac{\cos \theta}{\theta}\, d\theta - \int_1^{\sqrt{x}} \frac{\cos \theta}{\theta}\, d\theta \;\Rightarrow\;$

$y' = \dfrac{\cos x}{x} - \dfrac{\cos \sqrt{x}}{\sqrt{x}}\dfrac{1}{2\sqrt{x}} = \dfrac{2\cos x - \cos \sqrt{x}}{2x}$

47. If $1 \le x \le 3$, then $\sqrt{1^2 + 3} \le \sqrt{x^2 + 3} \le \sqrt{3^2 + 3} \;\Rightarrow\; 2 \le \sqrt{x^2 + 3} \le 2\sqrt{3}$, so $2(3 - 1) \le \int_1^3 \sqrt{x^2 + 3}\, dx \le 2\sqrt{3}(3 - 1)$; that is, $4 \le \int_1^3 \sqrt{x^2 + 3}\, dx \le 4\sqrt{3}$.

49. $0 \le x \le 1 \;\Rightarrow\; 0 \le \cos x \le 1 \;\Rightarrow\; x^2 \cos x \le x^2 \;\Rightarrow\; \int_0^1 x^2 \cos x\, dx \le \int_0^1 x^2\, dx = \frac{1}{3}\left[x^3\right]_0^1 = \frac{1}{3}$ [Property 4.2.7].

51. $\Delta x = (3-0)/6 = \frac{1}{2}$, so the endpoints are $0, \frac{1}{2}, 1, \frac{3}{2}, 2, \frac{5}{2}$, and 3, and the midpoints are $\frac{1}{4}, \frac{3}{4}, \frac{5}{4}, \frac{7}{4}, \frac{9}{4}$, and $\frac{11}{4}$.

The Midpoint Rule gives

$$\int_0^3 \sin(x^3)\, dx \approx \sum_{i=1}^{6} f(\overline{x}_i)\,\Delta x = \frac{1}{2}\left[\sin\left(\tfrac{1}{4}\right)^3 + \sin\left(\tfrac{3}{4}\right)^3 + \sin\left(\tfrac{5}{4}\right)^3 + \sin\left(\tfrac{7}{4}\right)^3 + \sin\left(\tfrac{9}{4}\right)^3 + \sin\left(\tfrac{11}{4}\right)^3\right] \approx 0.2810.$$

53. Note that $r(t) = b'(t)$, where $b(t) =$ the number of barrels of oil consumed up to time t. So, by the Net Change Theorem, $\int_{15}^{20} r(t)\, dt = b(20) - b(15)$ represents the number of barrels of oil consumed from Jan. 1, 2015, through Jan. 1, 2020.

55. We use the Midpoint Rule with $n = 6$ and $\Delta t = \frac{24-0}{6} = 4$. The increase in the bee population was

$$\int_0^{24} r(t)\, dt \approx M_6 = 4[r(2) + r(6) + r(10) + r(14) + r(18) + r(22)]$$
$$\approx 4[50 + 1000 + 7000 + 8550 + 1350 + 150] = 4(18{,}100) = 72{,}400$$

57. Let $u = 2\sin\theta$. Then $du = 2\cos\theta\, d\theta$ and when $\theta = 0$, $u = 0$; when $\theta = \frac{\pi}{2}$, $u = 2$. Thus,

$$\int_0^{\pi/2} f(2\sin\theta)\cos\theta\, d\theta = \int_0^2 f(u)\left(\tfrac{1}{2}\,du\right) = \tfrac{1}{2}\int_0^2 f(u)\, du = \tfrac{1}{2}\int_0^2 f(x)\, dx = \tfrac{1}{2}(6) = 3.$$

59. $\displaystyle\int_0^x f(t)\, dt = x\sin x + \int_0^x \frac{f(t)}{1+t^2}\, dt \;\Rightarrow\; f(x) = x\cos x + \sin x + \frac{f(x)}{1+x^2}$ [by differentiation] $\Rightarrow$

$f(x)\left(1 - \dfrac{1}{1+x^2}\right) = x\cos x + \sin x \;\Rightarrow\; f(x)\left(\dfrac{x^2}{1+x^2}\right) = x\cos x + \sin x \;\Rightarrow\; f(x) = \dfrac{1+x^2}{x^2}(x\cos x + \sin x)$

61. Let $u = f(x)$ and $du = f'(x)\, dx$. So $2\int_a^b f(x)f'(x)\, dx = 2\int_{f(a)}^{f(b)} u\, du = \left[u^2\right]_{f(a)}^{f(b)} = [f(b)]^2 - [f(a)]^2$.

63. Let $u = 1 - x$. Then $du = -dx$, so $\int_0^1 f(1-x)\, dx = \int_1^0 f(u)(-du) = \int_0^1 f(u)\, du = \int_0^1 f(x)\, dx$.

☐ PROBLEMS PLUS

1. Differentiating both sides of the equation $x \sin \pi x = \int_0^{x^2} f(t)\, dt$ (using FTC1 and the Chain Rule for the right side) gives $\sin \pi x + \pi x \cos \pi x = 2x f(x^2)$. Letting $x = 2$ so that $f(x^2) = f(4)$, we obtain $\sin 2\pi + 2\pi \cos 2\pi = 4f(4)$, so $f(4) = \frac{1}{4}(0 + 2\pi \cdot 1) = \frac{\pi}{2}$.

3. Differentiating the given equation, $\int_0^x f(t)\, dt = [f(x)]^2$, using FTC1 gives $f(x) = 2f(x)f'(x) \Rightarrow$ $f(x)[2f'(x) - 1] = 0$, so $f(x) = 0$ or $f'(x) = \frac{1}{2}$. Since $f(x)$ is never 0, we must have $f'(x) = \frac{1}{2}$ and $f'(x) = \frac{1}{2} \Rightarrow$ $f(x) = \frac{1}{2}x + C$. To find C, we substitute into the given equation to get $\int_0^x \left(\frac{1}{2}t + C\right) dt = \left(\frac{1}{2}x + C\right)^2 \Leftrightarrow$ $\frac{1}{4}x^2 + Cx = \frac{1}{4}x^2 + Cx + C^2$. It follows that $C^2 = 0$, so $C = 0$, and $f(x) = \frac{1}{2}x$.

5. $f(x) = \int_0^{g(x)} \dfrac{1}{\sqrt{1+t^3}}\, dt$, where $g(x) = \int_0^{\cos x} [1 + \sin(t^2)]\, dt$. Using FTC1 and the Chain Rule (twice) we have

$f'(x) = \dfrac{1}{\sqrt{1 + [g(x)]^3}}\, g'(x) = \dfrac{1}{\sqrt{1 + [g(x)]^3}} [1 + \sin(\cos^2 x)](-\sin x)$. Now $g\left(\frac{\pi}{2}\right) = \int_0^0 [1 + \sin(t^2)]\, dt = 0$, so

$f'\left(\frac{\pi}{2}\right) = \dfrac{1}{\sqrt{1 + 0}} (1 + \sin 0)(-1) = 1 \cdot 1 \cdot (-1) = -1$.

7. $f(x) = 2 + x - x^2 = (-x + 2)(x + 1) = 0 \Leftrightarrow x = 2$ or $x = -1$. $f(x) \geq 0$ for $x \in [-1, 2]$ and $f(x) < 0$ everywhere else. The integral $\int_a^b (2 + x - x^2)\, dx$ has a maximum on the interval where the integrand is positive, which is $[-1, 2]$. So $a = -1, b = 2$. (Any larger interval gives a smaller integral since $f(x) < 0$ outside $[-1, 2]$. Any smaller interval also gives a smaller integral since $f(x) \geq 0$ in $[-1, 2]$.)

9. (a) We can split the integral $\int_0^n [\![x]\!]\, dx$ into the sum $\sum\limits_{i=1}^{n} \left[\int_{i-1}^{i} [\![x]\!]\, dx\right]$. But on each of the intervals $[i-1, i)$ of integration, $[\![x]\!]$ is a constant function, namely $i - 1$. So the ith integral in the sum is equal to $(i-1)[i - (i-1)] = (i-1)$. So the original integral is equal to $\sum\limits_{i=1}^{n}(i-1) = \sum\limits_{i=1}^{n-1} i = \dfrac{(n-1)n}{2}$.

(b) We can write $\int_a^b [\![x]\!]\, dx = \int_0^b [\![x]\!]\, dx - \int_0^a [\![x]\!]\, dx$.

Now $\int_0^b [\![x]\!]\, dx = \int_0^{[\![b]\!]} [\![x]\!]\, dx + \int_{[\![b]\!]}^b [\![x]\!]\, dx$. The first of these integrals is equal to $\frac{1}{2}([\![b]\!] - 1)[\![b]\!]$, by part (a), and since $[\![x]\!] = [\![b]\!]$ on $[[\![b]\!], b]$, the second integral is just $[\![b]\!](b - [\![b]\!])$. So

$\int_0^b [\![x]\!]\, dx = \frac{1}{2}([\![b]\!] - 1)[\![b]\!] + [\![b]\!](b - [\![b]\!]) = \frac{1}{2}[\![b]\!](2b - [\![b]\!] - 1)$ and similarly $\int_0^a [\![x]\!]\, dx = \frac{1}{2}[\![a]\!](2a - [\![a]\!] - 1)$.

Therefore, $\int_a^b [\![x]\!]\, dx = \frac{1}{2}[\![b]\!](2b - [\![b]\!] - 1) - \frac{1}{2}[\![a]\!](2a - [\![a]\!] - 1)$.

11. Let $Q(x) = \int_0^x P(t)\,dt = \left[at + \dfrac{b}{2}t^2 + \dfrac{c}{3}t^3 + \dfrac{d}{4}t^4\right]_0^x = ax + \dfrac{b}{2}x^2 + \dfrac{c}{3}x^3 + \dfrac{d}{4}x^4$. Then $Q(0) = 0$, and $Q(1) = 0$ by the given condition, $a + \dfrac{b}{2} + \dfrac{c}{3} + \dfrac{d}{4} = 0$. Also, $Q'(x) = P(x) = a + bx + cx^2 + dx^3$ by FTC1. By Rolle's Theorem, applied to Q on $[0, 1]$, there is a number r in $(0, 1)$ such that $Q'(r) = 0$, that is, such that $P(r) = 0$. Thus, the equation $P(x) = 0$ has a solution between 0 and 1.

More generally, if $P(x) = a_0 + a_1 x + a_2 x^2 + \cdots + a_n x^n$ and if $a_0 + \dfrac{a_1}{2} + \dfrac{a_2}{3} + \cdots + \dfrac{a_n}{n+1} = 0$, then the equation $P(x) = 0$ has a solution between 0 and 1. The proof is the same as before:

Let $Q(x) = \int_0^x P(t)\,dt = a_0 x + \dfrac{a_1}{2}x^2 + \dfrac{a_2}{3}x^3 + \cdots + \dfrac{a_n}{n+1}x^n$. Then $Q(0) = Q(1) = 0$ and $Q'(x) = P(x)$. By Rolle's Theorem applied to Q on $[0, 1]$, there is a number r in $(0, 1)$ such that $Q'(r) = 0$, that is, such that $P(r) = 0$.

13. Note that $\dfrac{d}{dx}\left(\int_0^x \left[\int_0^u f(t)\,dt\right] du\right) = \int_0^x f(t)\,dt$ by FTC1, while

$$\dfrac{d}{dx}\left[\int_0^x f(u)(x-u)\,du\right] = \dfrac{d}{dx}\left[x\int_0^x f(u)\,du\right] - \dfrac{d}{dx}\left[\int_0^x f(u)u\,du\right]$$
$$= \int_0^x f(u)\,du + xf(x) - f(x)x = \int_0^x f(u)\,du$$

Hence, $\int_0^x f(u)(x-u)\,du = \int_0^x \left[\int_0^u f(t)\,dt\right] du + C$. Setting $x = 0$ gives $C = 0$.

15.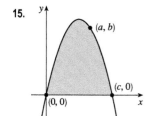

Let c be the nonzero x-intercept so that the parabola has equation $f(x) = kx(x - c)$, or $y = kx^2 - ckx$, where $k < 0$. The area A under the parabola is
$$A = \int_0^c kx(x-c)\,dx = k\int_0^c (x^2 - cx)\,dx = k\left[\tfrac{1}{3}x^3 - \tfrac{1}{2}cx^2\right]_0^c$$
$$= k\left(\tfrac{1}{3}c^3 - \tfrac{1}{2}c^3\right) = -\tfrac{1}{6}kc^3$$

The point (a, b) is on the parabola, so $f(a) = b \Rightarrow b = ka(a - c) \Rightarrow$

$k = \dfrac{b}{a(a-c)}$. Substituting for k in A gives $A(c) = -\dfrac{b}{6a}\cdot\dfrac{c^3}{a-c} \Rightarrow$

$$A'(c) = -\dfrac{b}{6a}\cdot\dfrac{(a-c)(3c^2) - c^3(-1)}{(a-c)^2} = -\dfrac{b}{6a}\cdot\dfrac{c^2[3(a-c)+c]}{(a-c)^2} = -\dfrac{bc^2(3a-2c)}{6a(a-c)^2}$$

Now $A' = 0 \Rightarrow c = \tfrac{3}{2}a$. Since $A'(c) < 0$ for $a < c < \tfrac{3}{2}a$ and $A'(c) > 0$ for $c > \tfrac{3}{2}a$, so A has an absolute minimum when $c = \tfrac{3}{2}a$. Substituting for c in k gives us $k = \dfrac{b}{a(a - \tfrac{3}{2}a)} = -\dfrac{2b}{a^2}$, so $f(x) = -\dfrac{2b}{a^2}x(x - \tfrac{3}{2}a)$, or

$f(x) = -\dfrac{2b}{a^2}x^2 + \dfrac{3b}{a}x$. Note that the vertex of the parabola is $\left(\tfrac{3}{4}a, \tfrac{9}{8}b\right)$ and the minimal area under the parabola is $A\left(\tfrac{3}{2}a\right) = \tfrac{9}{8}ab$.

17. $\displaystyle\lim_{n\to\infty}\left(\dfrac{1}{\sqrt{n}\sqrt{n+1}}+\dfrac{1}{\sqrt{n}\sqrt{n+2}}+\cdots+\dfrac{1}{\sqrt{n}\sqrt{n+n}}\right)$

$\displaystyle =\lim_{n\to\infty}\dfrac{1}{n}\left(\sqrt{\dfrac{n}{n+1}}+\sqrt{\dfrac{n}{n+2}}+\cdots+\sqrt{\dfrac{n}{n+n}}\right)$

$\displaystyle =\lim_{n\to\infty}\dfrac{1}{n}\left(\dfrac{1}{\sqrt{1+1/n}}+\dfrac{1}{\sqrt{1+2/n}}+\cdots+\dfrac{1}{\sqrt{1+1}}\right)$

$\displaystyle =\lim_{n\to\infty}\dfrac{1}{n}\sum_{i=1}^{n}f\left(\dfrac{i}{n}\right)\quad\left[\text{where }f(x)=\dfrac{1}{\sqrt{1+x}}\right]$

$\displaystyle =\int_{0}^{1}\dfrac{1}{\sqrt{1+x}}\,dx=\left[2\sqrt{1+x}\,\right]_{0}^{1}=2\left(\sqrt{2}-1\right)$

5 APPLICATIONS OF INTEGRATION

5.1 Areas Between Curves

1. (a) $A = \int_{x=0}^{x=2} (y_T - y_B)\, dx = \int_0^2 [(3x - x^2) - x]\, dx = \int_0^2 (2x - x^2)\, dx$

   (b) $\int_0^2 (2x - x^2)\, dx = \left[x^2 - \frac{1}{3}x^3\right]_0^2 = \left(4 - \frac{8}{3}\right) - 0 = \frac{4}{3}$

3. (a) $A = \int_{y=0}^{y=1} (x_R - x_L)\, dy = \int_0^1 \left[\sqrt{y} - (y^2 - 1)\right] dy = \int_0^1 \left(\sqrt{y} - y^2 + 1\right) dy$

   (b) $\int_0^1 (y^{1/2} - y^2 + 1)\, dy = \left[\frac{2}{3}y^{3/2} - \frac{1}{3}y^3 + y\right]_0^1 = \left(\frac{2}{3} - \frac{1}{3} + 1\right) - 0 = \frac{4}{3}$

5. $A = \int_{-2}^{0} [(x^3 - 3x) - x]\, dx + \int_0^2 [x - (x^3 - 3x)]\, dx \stackrel{\text{symmetry}}{=} 2\int_0^2 [x - (x^3 - 3x)]\, dx$

   $= 2\int_0^2 (4x - x^3)\, dx = 2\left[2x^2 - \frac{1}{4}x^4\right]_0^2 = 2[(8 - 4) - 0] = 8$

7. The curves intersect when $\frac{1}{x} = \frac{1}{x^2}$ $\Rightarrow$ $x^2 = x$ $\Leftrightarrow$

   $x^2 - x = 0$ $\Leftrightarrow$ $x(x-1) = 0$ $\Leftrightarrow$ $x = 0$ or $x = 1$.

   $A = \int_1^2 \left(\frac{1}{x} - \frac{1}{x^2}\right) dx$

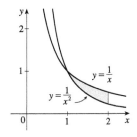

9. The curves intersect when $2 - x = 2x - x^2$ $\Leftrightarrow$

   $x^2 - 3x + 2 = 0$ $\Leftrightarrow$ $(x-2)(x-1) = 0$ $\Leftrightarrow$

   $x = 1$ or $x = 2$.

   $A = \int_1^2 [(2x - x^2) - (2 - x)]\, dx$ or $\int_1^2 (-x^2 + 3x - 2)\, dx$

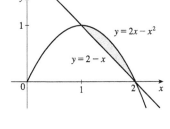

11. $A = \int_0^1 [(x^2 + 2) - (-x - 1)]\, dx$

    $= \int_0^1 (x^2 + x + 3)\, dx$

    $= \left[\frac{1}{3}x^3 + \frac{1}{2}x^2 + 3x\right]_0^1 = \left(\frac{1}{3} + \frac{1}{2} + 3\right) - 0$

    $= \frac{23}{6}$

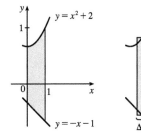

13. The curves intersect when $(x-2)^2 = x \Leftrightarrow x^2 - 4x + 4 = x \Leftrightarrow x^2 - 5x + 4 = 0 \Leftrightarrow$
$(x-1)(x-4) = 0 \Leftrightarrow x = 1$ or 4.

$$A = \int_1^4 [x - (x-2)^2]\,dx = \int_1^4 (-x^2 + 5x - 4)\,dx$$
$$= \left[-\tfrac{1}{3}x^3 + \tfrac{5}{2}x^2 - 4x\right]_1^4$$
$$= \left(-\tfrac{64}{3} + 40 - 16\right) - \left(-\tfrac{1}{3} + \tfrac{5}{2} - 4\right)$$
$$= \tfrac{9}{2}$$

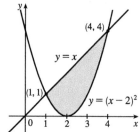 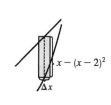

15. First find the points of intersection: $\sqrt{x+3} = \dfrac{x+3}{2} \Rightarrow (\sqrt{x+3})^2 = \left(\dfrac{x+3}{2}\right)^2 \Rightarrow x+3 = \tfrac{1}{4}(x+3)^2 \Rightarrow$
$4(x+3) - (x+3)^2 = 0 \Rightarrow (x+3)[4 - (x+3)] = 0 \Rightarrow (x+3)(1-x) = 0 \Rightarrow x = -3$ or 1. So

$$A = \int_{-3}^1 \left(\sqrt{x+3} - \dfrac{x+3}{2}\right)dx$$
$$= \left[\tfrac{2}{3}(x+3)^{3/2} - \dfrac{(x+3)^2}{4}\right]_{-3}^1$$
$$= \left(\tfrac{16}{3} - 4\right) - (0 - 0) = \tfrac{4}{3}$$

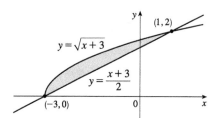 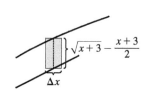

17. The curves intersect when $1 - y^2 = y^2 - 1 \Leftrightarrow 2 = 2y^2 \Leftrightarrow y^2 = 1 \Leftrightarrow y = \pm 1$.

$$A = \int_{-1}^1 [(1-y^2) - (y^2 - 1)]\,dy$$
$$= \int_{-1}^1 2(1-y^2)\,dy = 2 \cdot 2 \int_0^1 (1-y^2)\,dy$$
$$= 4\left[y - \tfrac{1}{3}y^3\right]_0^1 = 4\left(1 - \tfrac{1}{3}\right) = \tfrac{8}{3}$$

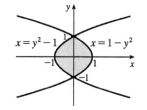 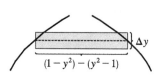

19. $12 - x^2 = x^2 - 6 \Leftrightarrow 2x^2 = 18 \Leftrightarrow$
$x^2 = 9 \Leftrightarrow x = \pm 3$, so

$$A = \int_{-3}^3 [(12 - x^2) - (x^2 - 6)]\,dx$$
$$= 2\int_0^3 (18 - 2x^2)\,dx \quad \text{[by symmetry]}$$
$$= 2\left[18x - \tfrac{2}{3}x^3\right]_0^3 = 2[(54 - 18) - 0]$$
$$= 2(36) = 72$$

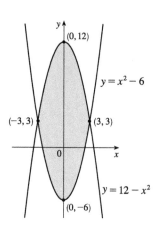

21. $2y^2 = 4 + y^2 \Leftrightarrow y^2 = 4 \Leftrightarrow y = \pm 2$, so

$$A = \int_{-2}^{2} [(4+y^2) - 2y^2]\, dy$$
$$= 2\int_{0}^{2} (4-y^2)\, dy \quad \text{[by symmetry]}$$
$$= 2\left[4y - \tfrac{1}{3}y^3\right]_0^2 = 2\left(8 - \tfrac{8}{3}\right) = \tfrac{32}{3}$$

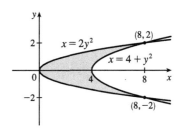

23. $\sqrt[3]{2x} = \tfrac{1}{2}x \Leftrightarrow 2x = \left(\tfrac{1}{2}x\right)^3 = \tfrac{1}{8}x^3$
$\Leftrightarrow 16x = x^3 \Leftrightarrow x^3 - 16x = 0$
$\Leftrightarrow x(x^2 - 16) = 0 \Leftrightarrow x = -4, 0, \text{ and } 4$

By symmetry,

$A = 2\int_0^4 \left(\sqrt[3]{2x} - \tfrac{1}{2}x\right) dx = 2\left[\tfrac{3}{8}(2x)^{4/3} - \tfrac{1}{4}x^2\right]_0^4$
$= 2[(6 - 4) - 0] = 4$

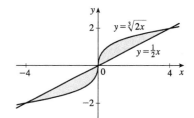

25. $\sqrt{x} = \tfrac{1}{3}x \Rightarrow x = \tfrac{1}{9}x^2 \Leftrightarrow \tfrac{1}{9}x^2 - x = 0 \Leftrightarrow$
$\tfrac{1}{9}x(x-9) = 0 \Leftrightarrow x = 0 \text{ or } x = 9$

$$A = \int_0^9 \left(\sqrt{x} - \tfrac{1}{3}x\right) dx + \int_9^{16}\left(\tfrac{1}{3}x - \sqrt{x}\right) dx$$
$$= \left[\tfrac{2}{3}x^{3/2} - \tfrac{1}{6}x^2\right]_0^9 + \left[\tfrac{1}{6}x^2 - \tfrac{2}{3}x^{3/2}\right]_9^{16}$$
$$= \left[\left(18 - \tfrac{27}{2}\right) - 0\right] + \left[\left(\tfrac{128}{3} - \tfrac{128}{3}\right) - \left(\tfrac{27}{2} - 18\right)\right] = 9$$

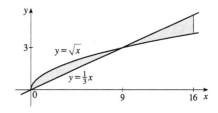

27. $\cos x = \sin 2x = 2\sin x \cos x \Leftrightarrow 2\sin x \cos x - \cos x = 0 \Leftrightarrow \cos x (2\sin x - 1) = 0 \Leftrightarrow$
$\cos x = 0 \text{ or } \sin x = \tfrac{1}{2} \Leftrightarrow x = \tfrac{\pi}{2} \text{ or } x = \tfrac{\pi}{6} \text{ on } \left[0, \tfrac{\pi}{2}\right]$

$A = \int_0^{\pi/6}(\cos x - \sin 2x)\, dx + \int_{\pi/6}^{\pi/2}(\sin 2x - \cos x)\, dx$

$= \left[\sin x + \tfrac{1}{2}\cos 2x\right]_0^{\pi/6} + \left[-\tfrac{1}{2}\cos 2x - \sin x\right]_{\pi/6}^{\pi/2}$

$= \left[\left(\sin \tfrac{\pi}{6} + \tfrac{1}{2}\cos\tfrac{\pi}{3}\right) - \left(\sin 0 + \tfrac{1}{2}\cos 0\right)\right]$
$\quad + \left[\left(-\tfrac{1}{2}\cos \pi - \sin\tfrac{\pi}{2}\right) - \left(-\tfrac{1}{2}\cos\tfrac{\pi}{3} - \sin\tfrac{\pi}{6}\right)\right]$

$= \left[\left(\tfrac{1}{2} + \tfrac{1}{4}\right) - \left(0 + \tfrac{1}{2}\right)\right] + \left[\left(\tfrac{1}{2} - 1\right) - \left(-\tfrac{1}{4} - \tfrac{1}{2}\right)\right] = \tfrac{1}{2}$

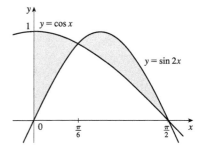

29. The curves intersect when $8\cos x = \sec^2 x \;\Rightarrow\; 8\cos^3 x = 1 \;\Rightarrow\; \cos^3 x = \frac{1}{8} \;\Rightarrow\; \cos x = \frac{1}{2} \;\Rightarrow\;$ $x = \frac{\pi}{3}$ for $0 < x < \frac{\pi}{2}$. By symmetry,

$$A = 2\int_0^{\pi/3} (8\cos x - \sec^2 x)\, dx$$

$$= 2\left[8\sin x - \tan x\right]_0^{\pi/3}$$

$$= 2\left(8 \cdot \tfrac{\sqrt{3}}{2} - \sqrt{3}\right) = 2(3\sqrt{3})$$

$$= 6\sqrt{3}$$

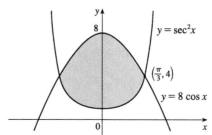

31. By inspection, we see that the curves intersect at $x = \pm 1$ and that the area of the region enclosed by the curves is twice the area enclosed in the first quadrant.

$$A = 2\int_0^1 [(2-x) - x^4]\, dx = 2\left[2x - \tfrac{1}{2}x^2 - \tfrac{1}{5}x^5\right]_0^1$$

$$= 2\left[\left(2 - \tfrac{1}{2} - \tfrac{1}{5}\right) - 0\right] = 2\left(\tfrac{13}{10}\right) = \tfrac{13}{5}$$

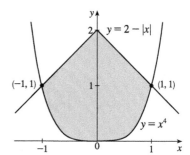

33. By inspection, we see that the curves intersect at $x = -1, 0,$ and 1. By symmetry,

$$A = 2\int_0^1 \left[\sin\!\left(\tfrac{\pi x}{2}\right) - x^3\right] dx = 2\left[-\tfrac{2}{\pi}\cos\!\left(\tfrac{\pi x}{2}\right) - \tfrac{1}{4}x^4\right]_0^1$$

$$= 2\left[\left(-\tfrac{2}{\pi}\cos\tfrac{\pi}{2} - \tfrac{1}{4}\right) - \left(-\tfrac{2}{\pi}\cos 0 - 0\right)\right]$$

$$= 2\left[\left(0 - \tfrac{1}{4}\right) - \left(-\tfrac{2}{\pi} - 0\right)\right] = 2\left(\tfrac{2}{\pi} - \tfrac{1}{4}\right) = \tfrac{4}{\pi} - \tfrac{1}{2}$$

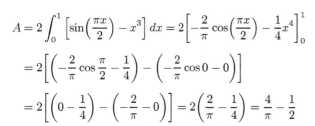

35. (a) Total area $= 12 + 27 = 39$.

(b) $f(x) \le g(x)$ for $0 \le x \le 2$ and $f(x) \ge g(x)$ for $2 \le x \le 5$, so

$$\int_0^5 [f(x) - g(x)]\, dx = \int_0^2 [f(x) - g(x)]\, dx + \int_2^5 [f(x) - g(x)]\, dx$$

$$= -\int_0^2 [g(x) - f(x)]\, dx + \int_2^5 [f(x) - g(x)]\, dx$$

$$= -(12) + 27 = 15$$

37. $\cos^2 x \sin x = \sin x \;\Leftrightarrow\; \cos^2 x \sin x - \sin x = 0 \;\Leftrightarrow\; \sin x(\cos^2 x - 1) = 0 \;\Leftrightarrow\; \sin x(-\sin^2 x) = 0 \;\Leftrightarrow\;$ $\sin x = 0 \;\Leftrightarrow\; x = 0$ or π.

$$A = \int_0^\pi (\sin x - \cos^2 x \sin x)\, dx$$

$$= \left[-\cos x + \tfrac{1}{3}\cos^3 x\right]_0^\pi$$

$$= \left(1 - \tfrac{1}{3}\right) - \left(-1 + \tfrac{1}{3}\right) = \tfrac{4}{3}$$

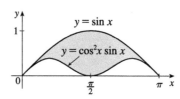

39. An equation of the line through $(0, 0)$ and $(3, 1)$ is $y = \frac{1}{3}x$; through $(0, 0)$ and $(1, 2)$ is $y = 2x$; through $(3, 1)$ and $(1, 2)$ is $y = -\frac{1}{2}x + \frac{5}{2}$.

$$A = \int_0^1 \left(2x - \tfrac{1}{3}x\right) dx + \int_1^3 \left[\left(-\tfrac{1}{2}x + \tfrac{5}{2}\right) - \tfrac{1}{3}x\right] dx$$

$$= \int_0^1 \tfrac{5}{3}x \, dx + \int_1^3 \left(-\tfrac{5}{6}x + \tfrac{5}{2}\right) dx = \left[\tfrac{5}{6}x^2\right]_0^1 + \left[-\tfrac{5}{12}x^2 + \tfrac{5}{2}x\right]_1^3$$

$$= \tfrac{5}{6} + \left(-\tfrac{15}{4} + \tfrac{15}{2}\right) - \left(-\tfrac{5}{12} + \tfrac{5}{2}\right) = \tfrac{5}{2}$$

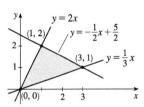

41. The curves intersect when $\sin x = \cos 2x$ (on $[0, \pi/2]$) $\Leftrightarrow$ $\sin x = 1 - 2\sin^2 x$ $\Leftrightarrow$ $2\sin^2 x + \sin x - 1 = 0$ $\Leftrightarrow$ $(2\sin x - 1)(\sin x + 1) = 0$ $\Rightarrow$ $\sin x = \tfrac{1}{2}$ $\Rightarrow$ $x = \tfrac{\pi}{6}$.

$$A = \int_0^{\pi/2} |\sin x - \cos 2x| \, dx$$

$$= \int_0^{\pi/6} (\cos 2x - \sin x) \, dx + \int_{\pi/6}^{\pi/2} (\sin x - \cos 2x) \, dx$$

$$= \left[\tfrac{1}{2}\sin 2x + \cos x\right]_0^{\pi/6} + \left[-\cos x - \tfrac{1}{2}\sin 2x\right]_{\pi/6}^{\pi/2}$$

$$= \left(\tfrac{1}{4}\sqrt{3} + \tfrac{1}{2}\sqrt{3}\right) - (0 + 1) + (0 - 0) - \left(-\tfrac{1}{2}\sqrt{3} - \tfrac{1}{4}\sqrt{3}\right)$$

$$= \tfrac{3}{2}\sqrt{3} - 1$$

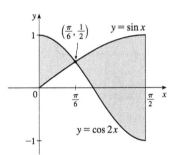

43.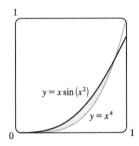

From the graph, we see that the curves intersect at $x = 0$ and $x = a \approx 0.896$, with $x \sin(x^2) > x^4$ on $(0, a)$. So the area A of the region bounded by the curves is

$$A = \int_0^a \left[x \sin(x^2) - x^4\right] dx = \left[-\tfrac{1}{2}\cos(x^2) - \tfrac{1}{5}x^5\right]_0^a$$

$$= -\tfrac{1}{2}\cos(a^2) - \tfrac{1}{5}a^5 + \tfrac{1}{2} \approx 0.037$$

45.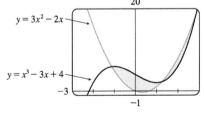

From the graph, we see that the curves intersect at

$x = a \approx -1.11$, $x = b \approx 1.25$, and $x = c \approx 2.86$, with

$x^3 - 3x + 4 > 3x^2 - 2x$ on (a, b) and $3x^2 - 2x > x^3 - 3x + 4$

on (b, c). So the area of the region bounded by the curves is

$$A = \int_a^b \left[(x^3 - 3x + 4) - (3x^2 - 2x)\right] dx + \int_b^c \left[(3x^2 - 2x) - (x^3 - 3x + 4)\right] dx$$

$$= \int_a^b (x^3 - 3x^2 - x + 4) \, dx + \int_b^c (-x^3 + 3x^2 + x - 4) \, dx$$

$$= \left[\tfrac{1}{4}x^4 - x^3 - \tfrac{1}{2}x^2 + 4x\right]_a^b + \left[-\tfrac{1}{4}x^4 + x^3 + \tfrac{1}{2}x^2 - 4x\right]_b^c \approx 8.38$$

47.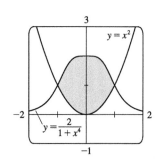

Graph $Y_1=2/(1+x^{\wedge}4)$ and $Y_2=x^{\wedge}2$. We see that $Y_1 > Y_2$ on $(-1, 1)$, so the area is given by $\int_{-1}^{1}\left(\dfrac{2}{1+x^4}-x^2\right)dx$. Evaluate the integral with a command such as $\text{fnInt}(Y_1-Y_2, x, -1, 1)$ to get 2.80123 to five decimal places.

Another method: Graph $f(x) = Y_1=2/(1+x^{\wedge}4)-x^{\wedge}2$ and from the graph evaluate $\int f(x)\,dx$ from -1 to 1.

49.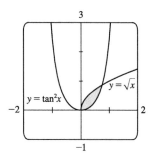

The curves intersect at $x = 0$ and $x = a \approx 0.749363$.
$$A = \int_0^a \left(\sqrt{x} - \tan^2 x\right)dx \approx 0.25142$$

51. As the figure illustrates, the curves $y = x$ and $y = x^5 - 6x^3 + 4x$ enclose a four-part region symmetric about the origin (since $x^5 - 6x^3 + 4x$ and x are odd functions of x). The curves intersect at values of x where $x^5 - 6x^3 + 4x = x$; that is, where $x(x^4 - 6x^2 + 3) = 0$. That happens at $x = 0$ and where

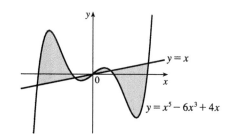

$x^2 = \dfrac{6 \pm \sqrt{36-12}}{2} = 3 \pm \sqrt{6}$; that is, at $x = -\sqrt{3+\sqrt{6}}, -\sqrt{3-\sqrt{6}}, 0, \sqrt{3-\sqrt{6}},$ and $\sqrt{3+\sqrt{6}}$. The exact area is

$2\displaystyle\int_0^{\sqrt{3+\sqrt{6}}} \left|(x^5 - 6x^3 + 4x) - x\right| dx = 2\int_0^{\sqrt{3+\sqrt{6}}} \left|x^5 - 6x^3 + 3x\right| dx$

$\qquad = 2\displaystyle\int_0^{\sqrt{3-\sqrt{6}}} (x^5 - 6x^3 + 3x)\,dx + 2\int_{\sqrt{3-\sqrt{6}}}^{\sqrt{3+\sqrt{6}}} (-x^5 + 6x^3 - 3x)\,dx$

$\qquad \overset{\text{CAS}}{=} 12\sqrt{6} - 9$

53. 1 second = $\dfrac{1}{3600}$ hour, so 10 s = $\dfrac{1}{360}$ h. With the given data, we can take $n = 5$ to use the Midpoint Rule. $\Delta t = \dfrac{1/360 - 0}{5} = \dfrac{1}{1800}$, so

distance $_{\text{Kelly}}$ − distance $_{\text{Chris}} = \int_0^{1/360} v_K\,dt - \int_0^{1/360} v_C\,dt = \int_0^{1/360}(v_K - v_C)\,dt$

$\qquad \approx M_5 = \dfrac{1}{1800}\left[(v_K - v_C)(1) + (v_K - v_C)(3) + (v_K - v_C)(5)\right.$
$\qquad\qquad\qquad\qquad \left. + (v_K - v_C)(7) + (v_K - v_C)(9)\right]$

$\qquad = \dfrac{1}{1800}[(22 - 20) + (52 - 46) + (71 - 62) + (86 - 75) + (98 - 86)]$

$\qquad = \dfrac{1}{1800}(2 + 6 + 9 + 11 + 12) = \dfrac{1}{1800}(40) = \dfrac{1}{45}$ mile, or $117\dfrac{1}{3}$ feet

55. Let $h(x)$ denote the height of the wing at x cm from the left end.

$$A \approx M_5 = \frac{200 - 0}{5} [h(20) + h(60) + h(100) + h(140) + h(180)]$$
$$= 40(20.3 + 29.0 + 27.3 + 20.5 + 8.7) = 40(105.8) = 4232 \text{ cm}^2$$

57. (a) From Example 8(a), the infectiousness concentration is 1210 cells/mL. $g(t) = 1210 \Leftrightarrow 0.9f(t) = 1210 \Leftrightarrow$ $0.9(-t)(t-21)(t+1) = 1210$. Using a calculator to solve the last equation for $t > 0$ gives us two solutions with the lesser being $t = t_3 \approx 11.26$ days, or the 12th day.

(b) From Example 8(b), the slope of the line through P_1 and P_2 is -23. From part (a), $P_3 = (t_3, 1210)$. An equation of the line through P_3 that is parallel to $\overline{P_1 P_2}$ is $N - 1210 = -23(t - t_3)$, or $N = -23t + 23t_3 + 1210$. Using a calculator, we find that this line intersects g at $t = t_4 \approx 17.18$, or the 18th day. So in the patient with some immunity, the infection lasts about 2 days less than in the patient without immunity.

(c) The level of infectiousness for this patient is the area between the graph of g and the line in part (b). This area is

$$\int_{t_3}^{t_4} \left[g(t) - (-23t + 23t_3 + 1210) \right] dt \approx \int_{11.26}^{17.18} (-0.9t^3 + 18t^2 + 41.9t - 1468.94) \, dt$$
$$= \left[-0.225t^4 + 6t^3 + 20.95t^2 - 1468.94t \right]_{11.26}^{17.18} \approx 706 \text{ (cells/mL)} \cdot \text{days}$$

59. We know that the area under curve A between $t = 0$ and $t = x$ is $\int_0^x v_A(t) \, dt = s_A(x)$, where $v_A(t)$ is the velocity of car A and s_A is its displacement. Similarly, the area under curve B between $t = 0$ and $t = x$ is $\int_0^x v_B(t) \, dt = s_B(x)$.

(a) After one minute, the area under curve A is greater than the area under curve B. So car A is ahead after one minute.

(b) The area of the shaded region has numerical value $s_A(1) - s_B(1)$, which is the distance by which A is ahead of B after 1 minute.

(c) After two minutes, car B is traveling faster than car A and has gained some ground, but the area under curve A from $t = 0$ to $t = 2$ is still greater than the corresponding area for curve B, so car A is still ahead.

(d) From the graph, it appears that the area between curves A and B for $0 \le t \le 1$ (when car A is going faster), which corresponds to the distance by which car A is ahead, seems to be about 3 squares. Therefore, the cars will be side by side at the time x where the area between the curves for $1 \le t \le x$ (when car B is going faster) is the same as the area for $0 \le t \le 1$. From the graph, it appears that this time is $x \approx 2.2$. So the cars are side by side when $t \approx 2.2$ minutes.

61. 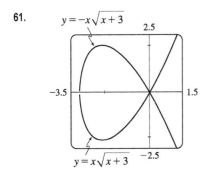 To graph this function, we must first express it as a combination of explicit functions of y; namely, $y = \pm x \sqrt{x + 3}$. We can see from the graph that the loop extends from $x = -3$ to $x = 0$, and that by symmetry, the area we seek is just twice the area under the top half of the curve on this interval, the equation of the top half being $y = -x \sqrt{x + 3}$. So the area is $A = 2 \int_{-3}^{0} \left(-x \sqrt{x + 3} \right) dx$. We substitute $u = x + 3$, so $du = dx$ and the limits change to 0 and 3, and we get

$$A = -2 \int_0^3 [(u - 3)\sqrt{u}] \, du = -2 \int_0^3 (u^{3/2} - 3u^{1/2}) \, du$$
$$= -2 \left[\tfrac{2}{5} u^{5/2} - 2u^{3/2} \right]_0^3 = -2 \left[\tfrac{2}{5} (3^2 \sqrt{3}) - 2(3\sqrt{3}) \right] = \tfrac{24}{5} \sqrt{3}$$

63.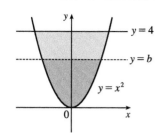

By the symmetry of the problem, we consider only the first quadrant, where $y = x^2 \Rightarrow x = \sqrt{y}$. We are looking for a number b such that

$$\int_0^b \sqrt{y}\,dy = \int_b^4 \sqrt{y}\,dy \Rightarrow \tfrac{2}{3}\left[y^{3/2}\right]_0^b = \tfrac{2}{3}\left[y^{3/2}\right]_b^4 \Rightarrow$$

$b^{3/2} = 4^{3/2} - b^{3/2} \Rightarrow 2b^{3/2} = 8 \Rightarrow b^{3/2} = 4 \Rightarrow b = 4^{2/3} \approx 2.52$.

65. We first assume that $c > 0$, since c can be replaced by $-c$ in both equations without changing the graphs, and if $c = 0$ the curves do not enclose a region. We see from the graph that the enclosed area A lies between $x = -c$ and $x = c$, and by symmetry, it is equal to four times the area in the first quadrant. The enclosed area is

$A = 4\int_0^c (c^2 - x^2)\,dx = 4\left[c^2 x - \tfrac{1}{3}x^3\right]_0^c = 4\left(c^3 - \tfrac{1}{3}c^3\right) = 4\left(\tfrac{2}{3}c^3\right) = \tfrac{8}{3}c^3$

So $A = 576 \Leftrightarrow \tfrac{8}{3}c^3 = 576 \Leftrightarrow c^3 = 216 \Leftrightarrow c = \sqrt[3]{216} = 6$.

Note that $c = -6$ is another solution, since the graphs are the same.

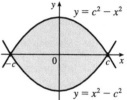

67. Let a and b be the x-coordinates of the points where the line intersects the curve. From the figure, $R_1 = R_2 \Rightarrow$

$\int_0^a \left[c - (8x - 27x^3)\right]dx = \int_a^b \left[(8x - 27x^3) - c\right]dx$

$\left[cx - 4x^2 + \tfrac{27}{4}x^4\right]_0^a = \left[4x^2 - \tfrac{27}{4}x^4 - cx\right]_a^b$

$ac - 4a^2 + \tfrac{27}{4}a^4 = \left(4b^2 - \tfrac{27}{4}b^4 - bc\right) - \left(4a^2 - \tfrac{27}{4}a^4 - ac\right)$

$0 = 4b^2 - \tfrac{27}{4}b^4 - bc = 4b^2 - \tfrac{27}{4}b^4 - b(8b - 27b^3)$

$= 4b^2 - \tfrac{27}{4}b^4 - 8b^2 + 27b^4 = \tfrac{81}{4}b^4 - 4b^2$

$= b^2\left(\tfrac{81}{4}b^2 - 4\right)$

So for $b > 0$, $b^2 = \tfrac{16}{81} \Rightarrow b = \tfrac{4}{9}$. Thus, $c = 8b - 27b^3 = 8\left(\tfrac{4}{9}\right) - 27\left(\tfrac{64}{729}\right) = \tfrac{32}{9} - \tfrac{64}{27} = \tfrac{32}{27}$.

69. The curves intersect when $\tan x = 2 \sin x$ (on $[-\pi/3, \pi/3]$) $\Leftrightarrow \sin x = 2 \sin x \cos x \Leftrightarrow$

$2 \sin x \cos x - \sin x = 0 \Leftrightarrow \sin x (2 \cos x - 1) = 0 \Leftrightarrow \sin x = 0$ or $\cos x = \tfrac{1}{2} \Leftrightarrow x = 0$ or $x = \pm\tfrac{\pi}{3}$.

$A = 2\int_0^{\pi/3} (2\sin x - \tan x)\,dx$ [by symmetry]

$= 2\left[-2\cos x - \ln|\sec x|\right]_0^{\pi/3}$

$= 2\left[(-1 - \ln 2) - (-2 - 0)\right]$

$= 2(1 - \ln 2) = 2 - 2\ln 2$

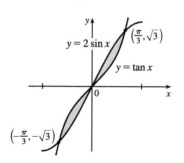

71. $1/x = x \Leftrightarrow 1 = x^2 \Leftrightarrow x = \pm 1$ and $1/x = \tfrac{1}{4}x \Leftrightarrow$
$4 = x^2 \Leftrightarrow x = \pm 2$, so for $x > 0$,

$$A = \int_0^1 \left(x - \tfrac{1}{4}x\right) dx + \int_1^2 \left(\tfrac{1}{x} - \tfrac{1}{4}x\right) dx$$

$$= \int_0^1 \left(\tfrac{3}{4}x\right) dx + \int_1^2 \left(\tfrac{1}{x} - \tfrac{1}{4}x\right) dx$$

$$= \left[\tfrac{3}{8}x^2\right]_0^1 + \left[\ln|x| - \tfrac{1}{8}x^2\right]_1^2$$

$$= \tfrac{3}{8} + \left(\ln 2 - \tfrac{1}{2}\right) - \left(0 - \tfrac{1}{8}\right) = \ln 2$$

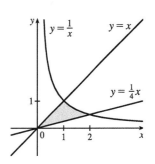

5.2 Volumes

1. (a)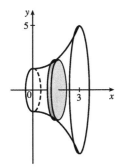

(b) A cross-section is a disk with radius $x^2 + 5$, so its area is
$$A(x) = \pi(x^2 + 5)^2 = \pi(x^4 + 10x^2 + 25).$$
$$V = \int_0^3 A(x)\,dx = \int_0^3 \pi(x^4 + 10x^2 + 25)\,dx$$

(c) $\displaystyle\int_0^3 \pi(x^4 + 10x^2 + 25)\,dx = \pi\left[\tfrac{1}{5}x^5 + \tfrac{10}{3}x^3 + 25x\right]_0^3 = \pi\left(\tfrac{243}{5} + 90 + 75\right) = \tfrac{1068}{5}\pi$

3. (a)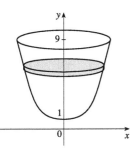

(b) $y = x^3 + 1 \;\Rightarrow\; y - 1 = x^3 \;\Rightarrow\; x = \sqrt[3]{y-1}$. Therefore, a cross-section is a disk with radius $\sqrt[3]{y-1}$, so its area is
$$A(y) = \pi\left(\sqrt[3]{y-1}\right)^2 = \pi(y-1)^{2/3}.$$
$$V = \int_1^9 A(y)\,dy = \int_1^9 \pi(y-1)^{2/3}\,dy$$

(c) $\displaystyle\int_1^9 \pi(y-1)^{2/3}\,dy = \pi\left[\tfrac{3}{5}(y-1)^{5/3}\right]_1^9 = \tfrac{3}{5}\pi(32 - 0) = \tfrac{96}{5}\pi$

5. $\displaystyle V = \int_1^3 \pi\left(1 - \tfrac{1}{x}\right)^2 dx$

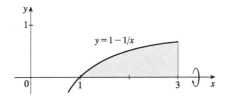

248 ☐ **CHAPTER 5** APPLICATIONS OF INTEGRATION

7. $8y = x^2 \Rightarrow x = \sqrt{8y}$ for $x \geq 0$; $y = \sqrt{x} \Rightarrow x = y^2$ for $y \geq 0$.

$y^2 = \sqrt{8y} \Rightarrow y^4 = 8y \Leftrightarrow y^4 - 8y = 0 \Leftrightarrow y(y^3 - 8) = 0 \Leftrightarrow$
$y = 0$ or $y = 2$.

$V = \int_0^2 \pi\left[\left(\sqrt{8y}\right)^2 - (y^2)^2\right] dy$

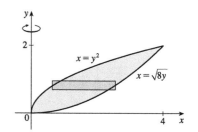

9. $V = \pi \int_0^\pi \left\{[\sin x - (-2)]^2 - [0 - (-2)]^2\right\} dx$

$= \pi \int_0^\pi \left[(\sin x + 2)^2 - 2^2\right] dx$

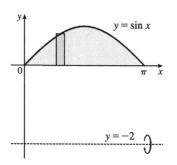

11. A cross-section is a disk with radius $x + 1$, so its area is

$A(x) = \pi(x+1)^2 = \pi(x^2 + 2x + 1)$.

$V = \int_0^2 A(x)\,dx = \int_0^2 \pi(x^2 + 2x + 1)\,dx$

$= \pi\left[\tfrac{1}{3}x^3 + x^2 + x\right]_0^2$

$= \pi\left(\tfrac{8}{3} + 4 + 2\right) = \tfrac{26\pi}{3}$

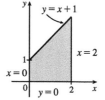
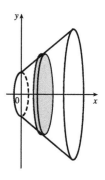

13. A cross-section is a disk with radius $\sqrt{x-1}$, so its area is $A(x) = \pi\left(\sqrt{x-1}\right)^2 = \pi(x-1)$.

$V = \int_1^5 A(x)\,dx = \int_1^5 \pi(x-1)\,dx = \pi\left[\tfrac{1}{2}x^2 - x\right]_1^5 = \pi\left[\left(\tfrac{25}{2} - 5\right) - \left(\tfrac{1}{2} - 1\right)\right] = 8\pi$

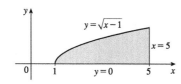
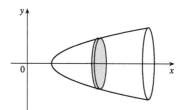

15. A cross-section is a disk with radius $2\sqrt{y}$, so its area is $A(y) = \pi\left(2\sqrt{y}\right)^2$.

$V = \int_0^9 A(y)\,dy = \int_0^9 \pi\left(2\sqrt{y}\right)^2 dy = 4\pi \int_0^9 y\,dy$

$= 4\pi\left[\tfrac{1}{2}y^2\right]_0^9 = 2\pi(81) = 162\pi$

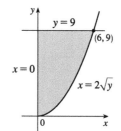
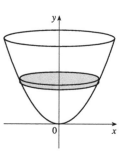

17. A cross-section is a washer with inner radius $\frac{1}{2}y$ and

outer radius $\sqrt{y}$, so its area is

$A(y) = \pi\left[\left(\sqrt{y}\right)^2 - \left(\frac{1}{2}y\right)^2\right] = \pi\left(y - \frac{1}{4}y^2\right)$.

$V = \int_0^4 \pi\left(y - \frac{1}{4}y^2\right)dy = \pi\left[\frac{1}{2}y^2 - \frac{1}{12}y^3\right]_0^4$

$= \pi\left[\left(8 - \frac{16}{3}\right) - 0\right] = \frac{8}{3}\pi$

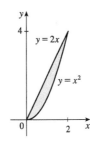

19. A cross-section is a washer with inner radius x^3 and outer radius $\sqrt{x}$, so its area is

$A(x) = \pi\left[\left(\sqrt{x}\right)^2 - (x^3)^2\right] = \pi(x - x^6)$.

$V = \int_0^1 \pi(x - x^6)\,dx = \pi\left[\frac{1}{2}x^2 - \frac{1}{7}x^7\right]_0^1$

$= \pi\left[\left(\frac{1}{2} - \frac{1}{7}\right) - 0\right] = \frac{5}{14}\pi$

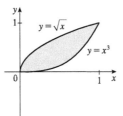
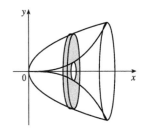

21. A cross-section is a washer with inner radius $1 - \sqrt{x}$ and outer radius $1 - x^2$, so its area is

$A(x) = \pi\left[(1-x^2)^2 - (1-\sqrt{x})^2\right]$

$= \pi\left[(1 - 2x^2 + x^4) - (1 - 2\sqrt{x} + x)\right]$

$= \pi\left(x^4 - 2x^2 + 2\sqrt{x} - x\right)$.

$V = \int_0^1 A(x)\,dx = \int_0^1 \pi(x^4 - 2x^2 + 2x^{1/2} - x)\,dx$

$= \pi\left[\frac{1}{5}x^5 - \frac{2}{3}x^3 + \frac{4}{3}x^{3/2} - \frac{1}{2}x^2\right]_0^1$

$= \pi\left(\frac{1}{5} - \frac{2}{3} + \frac{4}{3} - \frac{1}{2}\right) = \frac{11}{30}\pi$

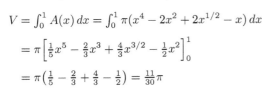

23. A cross-section is a washer with inner radius $(1 + \sec x) - 1 = \sec x$ and outer radius $3 - 1 = 2$, so its area is

$A(x) = \pi\left[2^2 - (\sec x)^2\right] = \pi(4 - \sec^2 x)$.

$V = \int_{-\pi/3}^{\pi/3} A(x)\,dx = \int_{-\pi/3}^{\pi/3} \pi(4 - \sec^2 x)\,dx$

$= 2\pi\int_0^{\pi/3}(4 - \sec^2 x)\,dx$ [by symmetry]

$= 2\pi\left[4x - \tan x\right]_0^{\pi/3} = 2\pi\left[\left(\frac{4\pi}{3} - \sqrt{3}\right) - 0\right]$

$= 2\pi\left(\frac{4\pi}{3} - \sqrt{3}\right)$

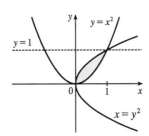
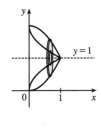

25. A cross-section is a washer with inner radius $2 - 1$ and outer radius $2 - \sqrt[3]{y}$, so its area is

$A(y) = \pi\left[(2 - \sqrt[3]{y})^2 - (2-1)^2\right] = \pi\left[4 - 4\sqrt[3]{y} + \sqrt[3]{y^2} - 1\right]$.

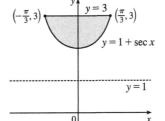
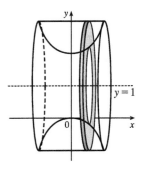

[continued]

$$V = \int_0^1 A(y)\,dy = \int_0^1 \pi(3 - 4y^{1/3} + y^{2/3})\,dy = \pi\left[3y - 3y^{4/3} + \tfrac{3}{5}y^{5/3}\right]_0^1 = \pi\left(3 - 3 + \tfrac{3}{5}\right) = \tfrac{3}{5}\pi.$$

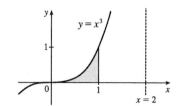
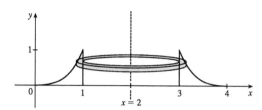

27. From the symmetry of the curves, we see they intersect at $x = \tfrac{1}{2}$ and so $y^2 = \tfrac{1}{2}$ $\Leftrightarrow$ $y = \pm\sqrt{\tfrac{1}{2}}$. A cross-section is a washer with inner radius $3 - (1 - y^2)$ and outer radius $3 - y^2$, so its area is

$$A(y) = \pi\left[(3 - y^2)^2 - (2 + y^2)^2\right]$$
$$= \pi\left[(9 - 6y^2 + y^4) - (4 + 4y^2 + y^4)\right]$$
$$= \pi(5 - 10y^2).$$

$$V = \int_{-\sqrt{1/2}}^{\sqrt{1/2}} A(y)\,dy$$

$$= 2\int_0^{\sqrt{1/2}} 5\pi(1 - 2y^2)\,dy \quad \text{[by symmetry]}$$

$$= 10\pi\left[y - \tfrac{2}{3}y^3\right]_0^{\sqrt{2}/2} = 10\pi\left(\tfrac{\sqrt{2}}{2} - \tfrac{\sqrt{2}}{6}\right)$$

$$= 10\pi\left(\tfrac{\sqrt{2}}{3}\right) = \tfrac{10}{3}\sqrt{2}\,\pi$$

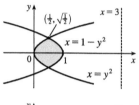
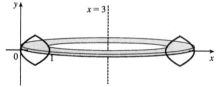

29. $\mathcal{R}_1$ about OA (the line $y = 0$):

$$V = \int_0^1 A(x)\,dx = \int_0^1 \pi(x)^2\,dx = \pi\left[\tfrac{1}{3}x^3\right]_0^1 = \tfrac{1}{3}\pi$$

31. $\mathcal{R}_1$ about AB (the line $x = 1$):

$$V = \int_0^1 A(y)\,dy = \int_0^1 \pi(1 - y)^2\,dy = \pi\int_0^1 (1 - 2y + y^2)\,dy = \pi\left[y - y^2 + \tfrac{1}{3}y^3\right]_0^1 = \tfrac{1}{3}\pi$$

33. $\mathcal{R}_2$ about OA (the line $y = 0$):

$$V = \int_0^1 A(x)\,dx = \int_0^1 \pi\left[1^2 - (\sqrt[4]{x})^2\right]dx = \pi\int_0^1 (1 - x^{1/2})\,dx = \pi\left[x - \tfrac{2}{3}x^{3/2}\right]_0^1 = \pi\left(1 - \tfrac{2}{3}\right) = \tfrac{1}{3}\pi$$

35. $\mathcal{R}_2$ about AB (the line $x = 1$):

$$V = \int_0^1 A(y)\,dy = \int_0^1 \pi[1^2 - (1 - y^4)^2]\,dy = \pi\int_0^1 [1 - (1 - 2y^4 + y^8)]\,dy$$

$$= \pi\int_0^1 (2y^4 - y^8)\,dy = \pi\left[\tfrac{2}{5}y^5 - \tfrac{1}{9}y^9\right]_0^1 = \pi\left(\tfrac{2}{5} - \tfrac{1}{9}\right) = \tfrac{13}{45}\pi$$

37. $\mathcal{R}_3$ about OA (the line $y = 0$):

$$V = \int_0^1 A(x)\,dx = \int_0^1 \pi\left[(\sqrt[4]{x})^2 - x^2\right]dx = \pi\int_0^1 (x^{1/2} - x^2)\,dx = \pi\left[\tfrac{2}{3}x^{3/2} - \tfrac{1}{3}x^3\right]_0^1 = \pi\left(\tfrac{2}{3} - \tfrac{1}{3}\right) = \tfrac{1}{3}\pi$$

[continued]

Note: Let $\mathcal{R} = \mathcal{R}_1 \cup \mathcal{R}_2 \cup \mathcal{R}_3$. If we rotate $\mathcal{R}$ about any of the segments OA, OC, AB, or BC, we obtain a right circular cylinder of height 1 and radius 1. Its volume is $\pi r^2 h = \pi(1)^2 \cdot 1 = \pi$. As a check for Exercises 29, 33, and 37, we can add the answers, and that sum must equal π. Thus, $\frac{1}{3}\pi + \frac{1}{3}\pi + \frac{1}{3}\pi = \pi$.

39. $\mathcal{R}_3$ about AB (the line $x = 1$):

$$V = \int_0^1 A(y)\, dy = \int_0^1 \pi[(1-y^4)^2 - (1-y)^2]\, dy = \pi \int_0^1 [(1 - 2y^4 + y^8) - (1 - 2y + y^2)]\, dy$$

$$= \pi \int_0^1 (y^8 - 2y^4 - y^2 + 2y)\, dy = \pi \left[\tfrac{1}{9}y^9 - \tfrac{2}{5}y^5 - \tfrac{1}{3}y^3 + y^2\right]_0^1 = \pi\left(\tfrac{1}{9} - \tfrac{2}{5} - \tfrac{1}{3} + 1\right) = \tfrac{17}{45}\pi$$

Note: See the note in the solution to Exercise 37. For Exercises 31, 35, and 39, we have $\frac{1}{3}\pi + \frac{13}{45}\pi + \frac{17}{45}\pi = \pi$.

41. (a) About the x-axis:

$$V = \int_0^{\pi/4} \pi(\tan x)^2\, dx = \pi \int_0^{\pi/4} \tan^2 x\, dx$$

$$\approx 0.67419$$

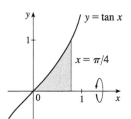

(b) About $y = -1$:

$$V = \int_0^{\pi/4} \pi\left\{[\tan x - (-1)]^2 - [0 - (-1)]^2\right\}\, dx$$

$$= \pi \int_0^{\pi/4} [(\tan x + 1)^2 - 1^2]\, dx$$

$$= \pi \int_0^{\pi/4} (\tan^2 x + 2\tan x)\, dx$$

$$\approx 2.85178$$

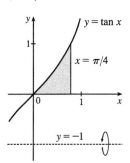

43. (a) About $y = 2$:

$$x^2 + 4y^2 = 4 \quad\Rightarrow\quad 4y^2 = 4 - x^2 \quad\Rightarrow\quad y^2 = 1 - x^2/4 \quad\Rightarrow$$

$$y = \pm\sqrt{1 - x^2/4}$$

$$V = \int_{-2}^{2} \pi\left\{\left[2 - \left(-\sqrt{1 - x^2/4}\right)\right]^2 - \left(2 - \sqrt{1 - x^2/4}\right)^2\right\}\, dx$$

$$= 2\pi \int_0^2 8\sqrt{1 - x^2/4}\, dx \approx 78.95684$$

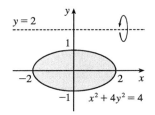

(b) About $x = 2$:

$$x^2 + 4y^2 = 4 \quad\Rightarrow\quad x^2 = 4 - 4y^2 \quad\Rightarrow\quad x = \pm\sqrt{4 - 4y^2}$$

$$V = \int_{-1}^{1} \pi\left\{\left[2 - \left(-\sqrt{4 - 4y^2}\right)\right]^2 - \left(2 - \sqrt{4 - 4y^2}\right)^2\right\}\, dy$$

$$= 2\pi \int_0^1 8\sqrt{4 - 4y^2}\, dy \approx 78.95684$$

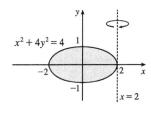

[Notice that this is the same approximation as in part (a). This can be explained by Pappus's Theorem in Section 8.3.]

45. $y = 1 + x^4$ and $y = \sqrt{3 - x^3}$ intersect at $x = -1$ and $x = a \approx 0.857$.

$$V = \pi \int_{-1}^{a} \left[\left(\sqrt{3 - x^3}\right)^2 - \left(1 + x^4\right)^2 \right] dx$$

$$\approx 9.756$$

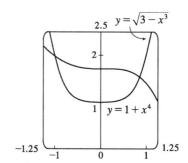

47. $V = \pi \int_{0}^{\pi} \left\{ \left[\sin^2 x - (-1)\right]^2 - \left[0 - (-1)\right]^2 \right\} dx$

$\stackrel{\text{CAS}}{=} \frac{11}{8}\pi^2$

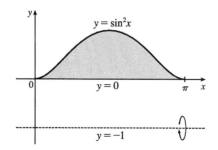

49. $\pi \int_{0}^{\pi/2} \sin^2 x \, dx = \pi \int_{0}^{\pi/2} (\sin x)^2 \, dx$ describes the volume of the solid obtained by rotating the region
$\mathcal{R} = \{(x, y) \mid 0 \leq x \leq \pi/2,\, 0 \leq y \leq \sin x\}$ of the xy-plane about the x-axis.

51. $\pi \int_{0}^{1} (x^4 - x^6) \, dx = \pi \int_{0}^{1} [(x^2)^2 - (x^3)^2] \, dx$ describes the volume of the solid obtained by rotating the region
$\mathcal{R} = \{(x, y) \mid 0 \leq x \leq 1,\, x^3 \leq y \leq x^2\}$ of the xy-plane about the x-axis.

53. $\pi \int_{0}^{4} y \, dy = \pi \int_{0}^{4} \left(\sqrt{y}\right)^2 dy$ describes the volume of the solid obtained by rotating the region
$\mathcal{R} = \left\{ (x, y) \mid 0 \leq y \leq 4,\, 0 \leq x \leq \sqrt{y} \right\}$ of the xy-plane about the y-axis.

55. There are 10 subintervals over the 15-cm length, so we'll use $n = 10/2 = 5$ for the Midpoint Rule.

$V = \int_{0}^{15} A(x) \, dx \approx M_5 = \frac{15 - 0}{5}[A(1.5) + A(4.5) + A(7.5) + A(10.5) + A(13.5)]$

$ = 3(18 + 79 + 106 + 128 + 39) = 3 \cdot 370 = 1110 \text{ cm}^3$

57. (a) $V = \int_{2}^{10} \pi [f(x)]^2 \, dx \approx \pi \frac{10 - 2}{4} \left\{ [f(3)]^2 + [f(5)]^2 + [f(7)]^2 + [f(9)]^2 \right\}$

$ \approx 2\pi \left[(1.5)^2 + (2.2)^2 + (3.8)^2 + (3.1)^2 \right] \approx 196 \text{ units}^3$

(b) $V = \int_{0}^{4} \pi \left[(\text{outer radius})^2 - (\text{inner radius})^2 \right] dy$

$ \approx \pi \frac{4 - 0}{4} \left\{ [(9.9)^2 - (2.2)^2] + [(9.7)^2 - (3.0)^2] + [(9.3)^2 - (5.6)^2] + [(8.7)^2 - (6.5)^2] \right\}$

$ \approx 838 \text{ units}^3$

59. We'll form a right circular cone with height h and base radius r by revolving the line $y = \frac{r}{h}x$ about the x-axis.

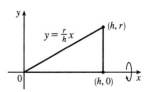

$$V = \pi \int_0^h \left(\frac{r}{h}x\right)^2 dx = \pi \int_0^h \frac{r^2}{h^2}x^2 dx = \pi \frac{r^2}{h^2}\left[\frac{1}{3}x^3\right]_0^h$$

$$= \pi \frac{r^2}{h^2}\left(\frac{1}{3}h^3\right) = \frac{1}{3}\pi r^2 h$$

Another solution: Revolve $x = -\frac{r}{h}y + r$ about the y-axis.

$$V = \pi \int_0^h \left(-\frac{r}{h}y + r\right)^2 dy \overset{*}{=} \pi \int_0^h \left[\frac{r^2}{h^2}y^2 - \frac{2r^2}{h}y + r^2\right] dy$$

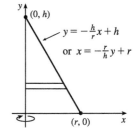

$$= \pi \left[\frac{r^2}{3h^2}y^3 - \frac{r^2}{h}y^2 + r^2 y\right]_0^h = \pi\left(\frac{1}{3}r^2 h - r^2 h + r^2 h\right) = \frac{1}{3}\pi r^2 h$$

* Or use substitution with $u = r - \frac{r}{h}y$ and $du = -\frac{r}{h}dy$ to get

$$\pi \int_r^0 u^2 \left(-\frac{h}{r}du\right) = -\pi \frac{h}{r}\left[\frac{1}{3}u^3\right]_r^0 = -\pi\frac{h}{r}\left(-\frac{1}{3}r^3\right) = \frac{1}{3}\pi r^2 h.$$

61. $x^2 + y^2 = r^2 \Leftrightarrow x^2 = r^2 - y^2$

$$V = \pi \int_{r-h}^r (r^2 - y^2) dy = \pi \left[r^2 y - \frac{y^3}{3}\right]_{r-h}^r = \pi\left\{\left[r^3 - \frac{r^3}{3}\right] - \left[r^2(r-h) - \frac{(r-h)^3}{3}\right]\right\}$$

$$= \pi\left\{\tfrac{2}{3}r^3 - \tfrac{1}{3}(r-h)\left[3r^2 - (r-h)^2\right]\right\}$$

$$= \tfrac{1}{3}\pi\left\{2r^3 - (r-h)\left[3r^2 - (r^2 - 2rh + h^2)\right]\right\}$$

$$= \tfrac{1}{3}\pi\left\{2r^3 - (r-h)\left[2r^2 + 2rh - h^2\right]\right\}$$

$$= \tfrac{1}{3}\pi\left(2r^3 - 2r^3 - 2r^2 h + rh^2 + 2r^2 h + 2rh^2 - h^3\right)$$

$$= \tfrac{1}{3}\pi(3rh^2 - h^3) = \tfrac{1}{3}\pi h^2(3r - h), \text{ or, equivalently, } \pi h^2\left(r - \frac{h}{3}\right)$$

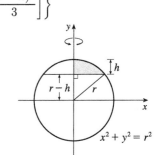

63. For a cross-section at height y, we see from similar triangles that $\dfrac{\alpha/2}{b/2} = \dfrac{h-y}{h}$, so $\alpha = b\left(1 - \dfrac{y}{h}\right)$.

Similarly, for cross-sections having $2b$ as their base and β replacing α, $\beta = 2b\left(1 - \dfrac{y}{h}\right)$. So

$$V = \int_0^h A(y)\, dy = \int_0^h \left[b\left(1 - \frac{y}{h}\right)\right]\left[2b\left(1 - \frac{y}{h}\right)\right] dy$$

$$= \int_0^h 2b^2\left(1 - \frac{y}{h}\right)^2 dy = 2b^2 \int_0^h \left(1 - \frac{2y}{h} + \frac{y^2}{h^2}\right) dy$$

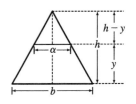

$$= 2b^2 \left[y - \frac{y^2}{h} + \frac{y^3}{3h^2}\right]_0^h = 2b^2\left[h - h + \tfrac{1}{3}h\right]$$

$$= \tfrac{2}{3}b^2 h \quad [= \tfrac{1}{3}Bh \text{ where } B \text{ is the area of the base, as with any pyramid.}]$$

65. A cross-section at height z is a triangle similar to the base, so we'll multiply the legs of the base triangle, 3 and 4, by a proportionality factor of $(5-z)/5$. Thus, the triangle at height z has area

$$A(z) = \frac{1}{2} \cdot 3\left(\frac{5-z}{5}\right) \cdot 4\left(\frac{5-z}{5}\right) = 6\left(1-\frac{z}{5}\right)^2, \text{ so}$$

$$V = \int_0^5 A(z)\,dz = 6\int_0^5 \left(1-\frac{z}{5}\right)^2 dz = 6\int_1^0 u^2(-5\,du) \quad \begin{bmatrix} u = 1 - z/5, \\ du = -\frac{1}{5}\,dz \end{bmatrix}$$

$$= -30\left[\tfrac{1}{3}u^3\right]_1^0 = -30(-\tfrac{1}{3}) = 10 \text{ cm}^3$$

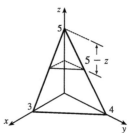

67. If l is a leg of the isosceles right triangle and $2y$ is the hypotenuse,

then $l^2 + l^2 = (2y)^2 \;\Rightarrow\; 2l^2 = 4y^2 \;\Rightarrow\; l^2 = 2y^2$.

$V = \int_{-2}^{2} A(x)\,dx = 2\int_0^2 A(x)\,dx = 2\int_0^2 \tfrac{1}{2}(l)(l)\,dx = 2\int_0^2 y^2\,dx$

$= 2\int_0^2 \tfrac{1}{4}(36 - 9x^2)\,dx = \tfrac{9}{2}\int_0^2 (4 - x^2)\,dx$

$= \tfrac{9}{2}\left[4x - \tfrac{1}{3}x^3\right]_0^2 = \tfrac{9}{2}\left(8 - \tfrac{8}{3}\right) = 24$

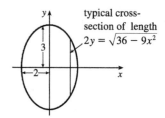

typical cross-section of length $2y = \sqrt{36 - 9x^2}$

69. The cross-section of the base corresponding to the coordinate x has length $y = 1 - x$. The corresponding square with side s has area

$A(x) = s^2 = (1-x)^2 = 1 - 2x + x^2$. Therefore,

$V = \int_0^1 A(x)\,dx = \int_0^1 (1 - 2x + x^2)\,dx$

$= \left[x - x^2 + \tfrac{1}{3}x^3\right]_0^1 = (1 - 1 + \tfrac{1}{3}) - 0 = \tfrac{1}{3}$

Or: $\int_0^1 (1-x)^2\,dx = \int_1^0 u^2(-du) \quad [u = 1 - x] = \left[\tfrac{1}{3}u^3\right]_0^1 = \tfrac{1}{3}$

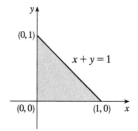

71. The cross-section of the base b corresponding to the coordinate x has length $1 - x^2$. The height h also has length $1 - x^2$, so the corresponding isosceles triangle has area $A(x) = \tfrac{1}{2}bh = \tfrac{1}{2}(1-x^2)^2$. Therefore,

$V = \int_{-1}^{1} A(x)\,dx = \int_{-1}^{1} \tfrac{1}{2}(1-x^2)^2\,dx$

$= 2 \cdot \tfrac{1}{2} \int_0^1 (1 - 2x^2 + x^4)\,dx$ [by symmetry]

$= \left[x - \tfrac{2}{3}x^3 + \tfrac{1}{5}x^5\right]_0^1 = (1 - \tfrac{2}{3} + \tfrac{1}{5}) - 0 = \tfrac{8}{15}$

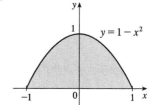

73. The cross-section of S at coordinate x, $-1 \leq x \leq 1$, is a circle centered at the point $(x, \tfrac{1}{2}(1-x^2))$ with radius $\tfrac{1}{2}(1-x^2)$.

The area of the cross-section is

$$A(x) = \pi\left[\tfrac{1}{2}(1-x^2)\right]^2 = \tfrac{\pi}{4}(1 - 2x^2 + x^4)$$

The volume of S is

$$V = \int_{-1}^{1} A(x)\,dx = 2\int_0^1 \tfrac{\pi}{4}(1 - 2x^2 + x^4)\,dx = \tfrac{\pi}{2}\left[x - \tfrac{2}{3}x^3 + \tfrac{1}{5}x^5\right]_0^1 = \tfrac{\pi}{2}\left(1 - \tfrac{2}{3} + \tfrac{1}{5}\right) = \tfrac{\pi}{2}\left(\tfrac{8}{15}\right) = \tfrac{4\pi}{15}$$

75. (a) The torus is obtained by rotating the circle $(x - R)^2 + y^2 = r^2$ about the y-axis. Solving for x, we see that the right half of the circle is given by $x = R + \sqrt{r^2 - y^2} = f(y)$ and the left half by $x = R - \sqrt{r^2 - y^2} = g(y)$. So

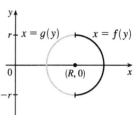

$$V = \pi \int_{-r}^{r} \left\{ [f(y)]^2 - [g(y)]^2 \right\} dy$$

$$= 2\pi \int_0^r \left[\left(R^2 + 2R\sqrt{r^2 - y^2} + r^2 - y^2 \right) - \left(R^2 - 2R\sqrt{r^2 - y^2} + r^2 - y^2 \right) \right] dy$$

$$= 2\pi \int_0^r 4R\sqrt{r^2 - y^2}\, dy = 8\pi R \int_0^r \sqrt{r^2 - y^2}\, dy$$

(b) Observe that the integral represents a quarter of the area of a circle with radius r, so

$$8\pi R \int_0^r \sqrt{r^2 - y^2}\, dy = 8\pi R \cdot \tfrac{1}{4}\pi r^2 = 2\pi^2 r^2 R.$$

77. The cross-sections perpendicular to the y-axis in Figure 17 are rectangles. The rectangle corresponding to the coordinate y has a base of length $2\sqrt{16 - y^2}$ in the xy-plane and a height of $\tfrac{1}{\sqrt{3}} y$, since $\angle BAC = 30°$ and $|BC| = \tfrac{1}{\sqrt{3}} |AB|$. Thus, $A(y) = \tfrac{2}{\sqrt{3}} y \sqrt{16 - y^2}$ and

$$V = \int_0^4 A(y)\, dy = \tfrac{2}{\sqrt{3}} \int_0^4 \sqrt{16 - y^2}\, y\, dy = \tfrac{2}{\sqrt{3}} \int_{16}^0 u^{1/2} \left(-\tfrac{1}{2}\, du \right) \quad \text{[Put } u = 16 - y^2 \text{, so } du = -2y\, dy\text{]}$$

$$= \tfrac{1}{\sqrt{3}} \int_0^{16} u^{1/2}\, du = \tfrac{1}{\sqrt{3}} \tfrac{2}{3} \left[u^{3/2} \right]_0^{16} = \tfrac{2}{3\sqrt{3}} (64) = \tfrac{128}{3\sqrt{3}}$$

79.

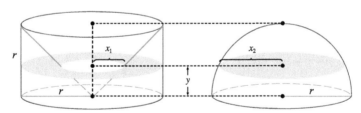

By similar triangles, the radius x_1 of the cross-section at height y of the cone removed from the cylinder satisfies $\dfrac{x_1}{y} = \dfrac{r}{r} \Rightarrow x_1 = y$. Thus, the area of the annular cross-section at height y remaining once the cone is removed from the cylinder is $\pi(r^2 - y^2)$.

The radius x_2 of the cross-section at height y of the hemisphere satisfies $x_2^2 + y^2 = r^2 \Rightarrow x_2 = \sqrt{r^2 - y^2}$.

The area of the circular cross-section at height y is then $\pi \left(\sqrt{r^2 - y^2} \right)^2 = \pi(r^2 - y^2)$.

Each cross-section at height y of the cylinder with cone removed has area equal to that of the corresponding cross-section at height y of the hemisphere. By Cavalieri's Principle, the volumes of the solids are then equal.

81. The volume is obtained by rotating the area common to two circles of radius r, as shown. The volume of the right half is

$$V_{\text{right}} = \pi \int_0^{r/2} y^2\, dx = \pi \int_0^{r/2} \left[r^2 - \left(\tfrac{1}{2}r + x\right)^2\right] dx$$

$$= \pi\left[r^2 x - \tfrac{1}{3}\left(\tfrac{1}{2}r + x\right)^3\right]_0^{r/2} = \pi\left[\left(\tfrac{1}{2}r^3 - \tfrac{1}{3}r^3\right) - \left(0 - \tfrac{1}{24}r^3\right)\right] = \tfrac{5}{24}\pi r^3$$

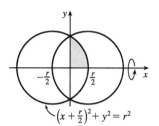

So by symmetry, the total volume is twice this, or $\tfrac{5}{12}\pi r^3$.

Another solution: We observe that the volume is the twice the volume of a cap of a sphere, so we can use the formula from Exercise 61 with $h = \tfrac{1}{2}r$: $V = 2 \cdot \tfrac{1}{3}\pi h^2 (3r - h) = \tfrac{2}{3}\pi\left(\tfrac{1}{2}r\right)^2\left(3r - \tfrac{1}{2}r\right) = \tfrac{5}{12}\pi r^3$.

83. Take the x-axis to be the axis of the cylindrical hole of radius r. A quarter of the cross-section through y, perpendicular to the y-axis, is the rectangle shown. Using the Pythagorean Theorem twice, we see that the dimensions of this rectangle are
$x = \sqrt{R^2 - y^2}$ and $z = \sqrt{r^2 - y^2}$, so

$\tfrac{1}{4}A(y) = xz = \sqrt{r^2 - y^2}\sqrt{R^2 - y^2}$, and

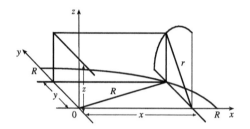

$$V = \int_{-r}^{r} A(y)\, dy = \int_{-r}^{r} 4\sqrt{r^2 - y^2}\sqrt{R^2 - y^2}\, dy = 8\int_0^r \sqrt{r^2 - y^2}\sqrt{R^2 - y^2}\, dy$$

85. (a) The radius of the barrel is the same at each end by symmetry, since the function $y = R - cx^2$ is even. Since the barrel is obtained by rotating the graph of the function y about the x-axis, this radius is equal to the value of y at $x = \tfrac{1}{2}h$, which is $R - c\left(\tfrac{1}{2}h\right)^2 = R - d = r$.

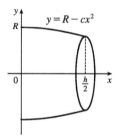

(b) The barrel is symmetric about the y-axis, so its volume is twice the volume of that part of the barrel for $x > 0$. Also, the barrel is a volume of rotation, so

$$V = 2\int_0^{h/2} \pi y^2\, dx = 2\pi \int_0^{h/2} (R - cx^2)^2\, dx = 2\pi\left[R^2 x - \tfrac{2}{3} Rcx^3 + \tfrac{1}{5} c^2 x^5\right]_0^{h/2}$$

$$= 2\pi\left(\tfrac{1}{2}R^2 h - \tfrac{1}{12} Rch^3 + \tfrac{1}{160} c^2 h^5\right)$$

Trying to make this look more like the expression we want, we rewrite it as $V = \tfrac{1}{3}\pi h\left[2R^2 + \left(R^2 - \tfrac{1}{2} Rch^2 + \tfrac{3}{80} c^2 h^4\right)\right]$.

But $R^2 - \tfrac{1}{2}Rch^2 + \tfrac{3}{80}c^2 h^4 = \left(R - \tfrac{1}{4}ch^2\right)^2 - \tfrac{1}{40}c^2 h^4 = (R - d)^2 - \tfrac{2}{5}\left(\tfrac{1}{4}ch^2\right)^2 = r^2 - \tfrac{2}{5}d^2$.

Substituting this back into V, we see that $V = \tfrac{1}{3}\pi h\left(2R^2 + r^2 - \tfrac{2}{5}d^2\right)$, as required.

87. (a) $y = x^3 \Rightarrow x = \sqrt[3]{y}$. $V_1 = \int_1^8 \pi \left(\sqrt[3]{y}\right)^2 dy = \int_1^8 \pi y^{2/3} dy = \frac{3}{5}\pi \left[y^{5/3}\right]_1^8 = \frac{3}{5}\pi(32 - 1) = \frac{93}{5}\pi$

(b) If each y is replaced with cy, then $y = 1$ will be mapped to $y = 1c = c$, and $y = 8$ will be mapped to $y = 8c$, each as shown. If each x is replaced with cx, then $y = (cx)^3/c^2 = c^3 x^3/c^2 = cx^3$, so y has again been mapped to cy. A dilation with scaling factor c therefore transforms the region R_1 into R_2.

(c) $y = x^3/c^2 \Rightarrow c^2 y = x^3 \Rightarrow x = \sqrt[3]{c^2 y}$. Then

$V_2 = \int_c^{8c} \pi \left(\sqrt[3]{c^2 y}\right)^2 dy = \int_c^{8c} \pi \cdot c^{4/3} y^{2/3} dy = \frac{3}{5}\pi \cdot c^{4/3} \left[y^{5/3}\right]_c^{8c} = \frac{3}{5}\pi \cdot c^{4/3}(32 c^{5/3} - c^{5/3})$

$= \frac{3}{5}\pi \cdot c^{4/3} \cdot 31 c^{5/3} = \frac{3}{5}\pi \cdot 31 c^3 = \frac{93}{5}\pi c^3 = c^3 V_1$

(d) $V_2 = 5$ L $\Rightarrow \frac{93}{5}\pi c^3 = 5000$ cm^3 $\Rightarrow c^3 = \frac{25{,}000}{93\pi} \Rightarrow c = \sqrt[3]{\frac{25{,}000}{93\pi}} \approx 4.41$

5.3 Volumes by Cylindrical Shells

1.

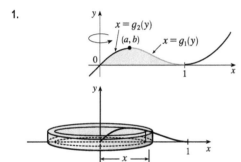

If we were to use the "washer" method, we would first have to locate the local maximum point (a, b) of $y = x(x - 1)^2$ using the methods of Chapter 3. Then we would have to solve the equation $y = x(x - 1)^2$ for x in terms of y to obtain the functions $x = g_1(y)$ and $x = g_2(y)$ shown in the first figure. This step would be difficult because it involves the cubic formula. Finally we would find the volume using $V = \pi \int_0^b \{[g_1(y)]^2 - [g_2(y)]^2\} dy$.

Using shells, we find that a typical approximating shell has radius x, so its circumference is $2\pi x$. Its height is y, that is, $x(x - 1)^2$. So the total volume is

$$V = \int_0^1 2\pi x \left[x(x-1)^2\right] dx = 2\pi \int_0^1 (x^4 - 2x^3 + x^2) dx = 2\pi \left[\frac{x^5}{5} - 2\frac{x^4}{4} + \frac{x^3}{3}\right]_0^1 = \frac{\pi}{15}$$

3. (a)

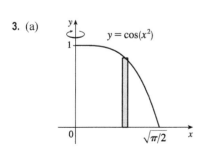
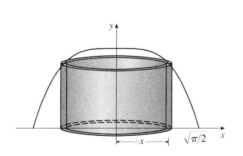

$$V = \int_0^{\sqrt{\pi/2}} 2\pi x \cos(x^2) \, dx$$

(b) $V = \int_0^{\sqrt{\pi/2}} 2\pi x \cos(x^2) \, dx = 2\pi \int_0^{\sqrt{\pi/2}} x \cos(x^2) \, dx = 2\pi \cdot \frac{1}{2}\left[\sin(x^2)\right]_0^{\sqrt{\pi/2}} = \pi\left[\sin\left(\frac{\pi}{2}\right) - \sin 0\right]$

$= \pi(1 - 0) = \pi$

5. $V = \int_0^2 2\pi x \sqrt[4]{x}\, dx$

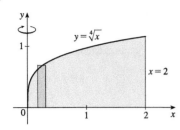

7. $y = \sqrt{x+4}\ \Rightarrow\ y^2 = x+4\ \Leftrightarrow\ x = y^2 - 4$

$V = \int_0^2 2\pi(3-y)[0 - (y^2 - 4)]\, dy$

$ = \int_0^2 2\pi(3-y)(4 - y^2)\, dy$

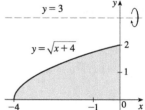

9. The shell has radius x, circumference $2\pi x$, and height $\sqrt{x}$, so

$V = \int_0^4 2\pi x \sqrt{x}\, dx = \int_0^4 2\pi x^{3/2}\, dx = 2\pi\left[\tfrac{2}{5}x^{5/2}\right]_0^4 = 2\pi \cdot \tfrac{2}{5}(32 - 0) = \tfrac{128}{5}\pi.$

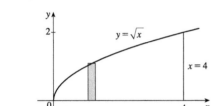
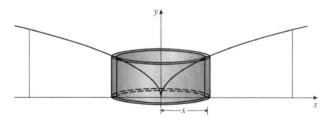

11. The shell has radius x, circumference $2\pi x$, and height $1/x$, so

$V = \int_1^4 2\pi x\left(\dfrac{1}{x}\right) dx = \int_1^4 2\pi\, dx = 2\pi\Big[x\Big]_1^4 = 2\pi(4 - 1) = 6\pi.$

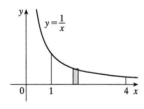

13. The shell has radius x, circumference $2\pi x$, and height $\sqrt{5 + x^2}$, so

$V = \displaystyle\int_0^2 2\pi x\sqrt{5 + x^2}\, dx\quad [u = 5 + x^2,\ du = 2x\, dx]$

$ = \displaystyle\int_5^9 2\pi \cdot \tfrac{1}{2} u^{1/2}\, du = \pi \cdot \tfrac{2}{3}\Big[u^{3/2}\Big]_5^9$

$ = \tfrac{2}{3}\pi(27 - 5^{3/2})$

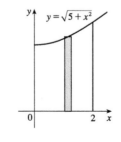
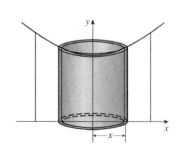

15. $xy = 1 \Rightarrow x = \frac{1}{y}$. The shell has radius y, circumference $2\pi y$, and height $1/y$, so

$$V = \int_1^3 2\pi y\left(\frac{1}{y}\right) dy$$

$$= 2\pi \int_1^3 dy = 2\pi \left[y\right]_1^3$$

$$= 2\pi(3-1) = 4\pi$$

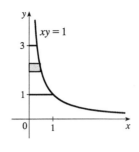
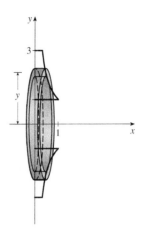

17. $y = x^{3/2} \Rightarrow x = y^{2/3}$. The shell has radius y, circumference $2\pi y$, and height $y^{2/3}$, so

$$V = \int_0^8 2\pi y(y^{2/3})\, dy = 2\pi \int_0^8 y^{5/3}\, dy$$

$$= 2\pi \left[\tfrac{3}{8} y^{8/3}\right]_0^8$$

$$= 2\pi \cdot \tfrac{3}{8} \cdot 256 = 192\pi$$

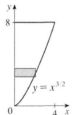
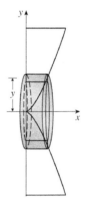

19. The shell has radius y, circumference $2\pi y$, and height

$2 - \left[1 + (y-2)^2\right] = 1 - (y-2)^2 = 1 - \left(y^2 - 4y + 4\right)$
$\qquad\qquad = -y^2 + 4y - 3$, so

$V = \int_1^3 2\pi y(-y^2 + 4y - 3)\, dy$

$\quad = 2\pi \int_1^3 (-y^3 + 4y^2 - 3y)\, dy$

$\quad = 2\pi \left[-\tfrac{1}{4}y^4 + \tfrac{4}{3}y^3 - \tfrac{3}{2}y^2\right]_1^3$

$\quad = 2\pi \left[\left(-\tfrac{81}{4} + 36 - \tfrac{27}{2}\right) - \left(-\tfrac{1}{4} + \tfrac{4}{3} - \tfrac{3}{2}\right)\right]$

$\quad = 2\pi \left(\tfrac{8}{3}\right) = \tfrac{16}{3}\pi$

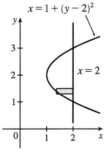
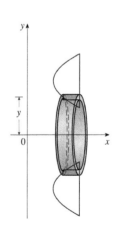

21. $x^2 = 8\sqrt{x} \Rightarrow x^4 = 64x \Rightarrow x^4 - 64x = 0 \Rightarrow$
$x(x^3 - 64) = 0 \Rightarrow x = 0 \text{ or } x = 4$

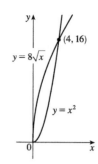

(a) By cylindrical shells:

$V = \int_0^4 2\pi x \left(8\sqrt{x} - x^2\right) dx$

$= \int_0^4 2\pi (8x^{3/2} - x^3) \, dx$

$= 2\pi \left[\frac{16}{5} x^{5/2} - \frac{1}{4} x^4 \right]_0^4$

$= 2\pi \left[\left(\frac{512}{5} - 64 \right) - 0 \right] = \frac{384}{5} \pi$

(b) By washers:

$V = \int_0^{16} \pi \left[\left(\sqrt{y} \right)^2 - \left(\frac{1}{64} y^2 \right)^2 \right] dy$

$= \int_0^{16} \pi \left(y - \frac{1}{4096} y^4 \right) dy = \pi \left[\frac{1}{2} y^2 - \frac{1}{20{,}480} y^5 \right]_0^{16}$

$= \pi \left(128 - \frac{256}{5} \right) = \frac{384}{5} \pi$

23. (a) The shell has radius $x - (-2) = x + 2$,

circumference $2\pi(x + 2)$, and height $4x - x^2$.

(b) $V = \int_0^4 2\pi(x+2)(4x - x^2) \, dx$

(c) $V = \int_0^4 2\pi(x+2)(4x - x^2) \, dx$

$= \int_0^4 2\pi(2x^2 - x^3 + 8x) \, dx = 2\pi \left[\frac{2}{3} x^3 - \frac{1}{4} x^4 + 4x^2 \right]_0^4$

$= 2\pi \left[\left(\frac{128}{3} - 64 + 64 \right) - 0 \right] = \frac{256}{3} \pi$

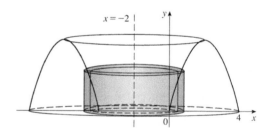

25. The shell has radius $3 - x$, circumference

$2\pi(3 - x)$, and height $8 - x^3$.

$V = \int_0^2 2\pi(3 - x)(8 - x^3) \, dx$

$= 2\pi \int_0^2 (x^4 - 3x^3 - 8x + 24) \, dx$

$= 2\pi \left[\frac{1}{5} x^5 - \frac{3}{4} x^4 - 4x^2 + 24x \right]_0^2$

$= 2\pi \left(\frac{32}{5} - 12 - 16 + 48 \right) = 2\pi \left(\frac{132}{5} \right) = \frac{264\pi}{5}$

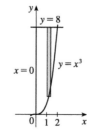 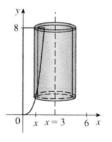

27. The shell has radius $x - 1$, circumference $2\pi(x - 1)$, and height $(4x - x^2) - 3 = -x^2 + 4x - 3$.

$V = \int_1^3 2\pi(x - 1)(-x^2 + 4x - 3)\,dx$

$= 2\pi \int_1^3 (-x^3 + 5x^2 - 7x + 3)\,dx$

$= 2\pi \left[-\frac{1}{4}x^4 + \frac{5}{3}x^3 - \frac{7}{2}x^2 + 3x \right]_1^3$

$= 2\pi \left[\left(-\frac{81}{4} + 45 - \frac{63}{2} + 9 \right) - \left(-\frac{1}{4} + \frac{5}{3} - \frac{7}{2} + 3 \right) \right]$

$= 2\pi \left(\frac{4}{3} \right) = \frac{8}{3}\pi$

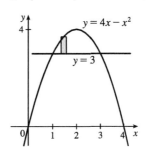
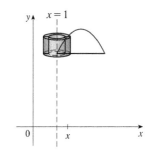

29. The shell has radius $2 - y$, circumference $2\pi(2 - y)$, and height $2 - 2y^2$.

$V = \int_0^1 2\pi(2 - y)(2 - 2y^2)\,dy$

$= 4\pi \int_0^1 (2 - y)(1 - y^2)\,dy$

$= 4\pi \int_0^1 (y^3 - 2y^2 - y + 2)\,dy$

$= 4\pi \left[\frac{1}{4}y^4 - \frac{2}{3}y^3 - \frac{1}{2}y^2 + 2y \right]_0^1$

$= 4\pi \left(\frac{1}{4} - \frac{2}{3} - \frac{1}{2} + 2 \right)$

$= 4\pi \left(\frac{13}{12} \right) = \frac{13\pi}{3}$

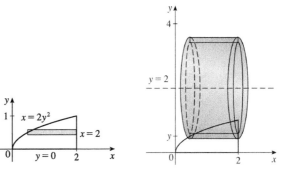

31. (a) $V = \int_{2\pi}^{3\pi} 2\pi x \sin x\,dx$

(b) $V \approx 98.69604$

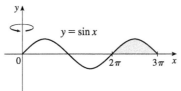

33. (a) $V = 2\pi \int_{-\pi/2}^{\pi/2} (\pi - x)[\cos^4 x - (-\cos^4 x)]\,dx$

$= 4\pi \int_{-\pi/2}^{\pi/2} (\pi - x) \cos^4 x\,dx$

[or $8\pi^2 \int_0^{\pi/2} \cos^4 x\,dx$ using Theorem 4.5.6]

(b) $V \approx 46.50942$

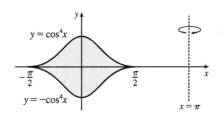

35. (a) $V = \int_0^\pi 2\pi(4 - y)\sqrt{\sin y}\,dy$ (b) $V \approx 36.57476$

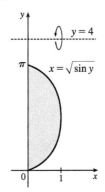

37. $V = \int_0^1 2\pi x\sqrt{1+x^3}\,dx$. Let $f(x) = x\sqrt{1+x^3}$.

Then the Midpoint Rule with $n = 5$ gives

$\int_0^1 f(x)\,dx \approx \frac{1-0}{5}[f(0.1) + f(0.3) + f(0.5) + f(0.7) + f(0.9)]$

$\approx 0.2(2.9290)$

Multiplying by 2π gives $V \approx 3.68$.

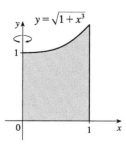

39. $\int_0^3 2\pi x^5\,dx = 2\pi \int_0^3 x(x^4)\,dx$. By the method of cylindrical shells, this integral represents the volume of the solid obtained by rotating the region $0 \le y \le x^4$, $0 \le x \le 3$ about the y-axis ($x = 0$).

41. $2\pi \int_1^4 \frac{y+2}{y^2}\,dy = 2\pi \int_1^4 (y+2)\left(\frac{1}{y^2}\right)dy$. By the method of cylindrical shells, this integral represents the volume of the solid obtained by rotating the region $0 \le x \le 1/y^2$, $1 \le y \le 4$ about the line $y = -2$.

43. From the graph, the curves intersect at $x = 0$ and $x = a \approx 2.175$, with

$\frac{x}{x^2+1} > x^2 - 2x$ on the interval $(0, a)$. So the volume of the solid

obtained by rotating the region about the y-axis is

$V = 2\pi \int_0^a x\left[\frac{x}{x^2+1} - (x^2 - 2x)\right]dx \approx 14.450$

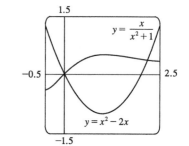

45. $V = 2\pi \int_0^{\pi/2} \left[\left(\frac{\pi}{2} - x\right)(\sin^2 x - \sin^4 x)\right] dx$

$\stackrel{\text{CAS}}{=} \frac{1}{32}\pi^3$

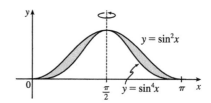

47. (a) Use shells. Each shell has radius x, circumference $2\pi x$, and height $\left[\left(1 - \sqrt[3]{x}\right) - (x-1)\right] = 2 - \sqrt[3]{x} - x$.

$V = \int_0^1 2\pi x\left(2 - \sqrt[3]{x} - x\right) dx$

(b) $V = \int_0^1 2\pi x\left(2 - \sqrt[3]{x} - x\right) dx = \int_0^1 2\pi(2x - x^{4/3} - x^2)\,dx$

$= 2\pi\left[x^2 - \frac{3}{7}x^{7/3} - \frac{1}{3}x^3\right]_0^1 = 2\pi\left[\left(1 - \frac{3}{7} - \frac{1}{3}\right) - 0\right] = 2\pi\left(\frac{5}{21}\right) = \frac{10}{21}\pi$

49. (a) Use disks. $V = \int_0^\pi \pi\left(\sqrt{\sin x}\right)^2 dx$

(b) $V = \int_0^\pi \pi\left(\sqrt{\sin x}\right)^2 dx = \int_0^\pi \pi \sin x\,dx = \pi\Big[-\cos x\Big]_0^\pi = \pi[-\cos\pi - (-\cos 0)] = \pi(1+1) = 2\pi$

51. (a) Use shells. Each shell has radius $x - (-2) = x + 2$, circumference $2\pi(x+2)$, and height $x^2 - x^3$.

$V = \int_0^{1/2} 2\pi(x+2)(x^2 - x^3)\,dx$

(b) $V = \int_0^{1/2} 2\pi(x+2)(x^2 - x^3)\,dx = \int_0^{1/2} 2\pi(-x^3 - x^4 + 2x^2)\,dx = 2\pi\left[-\frac{1}{4}x^4 - \frac{1}{5}x^5 + \frac{2}{3}x^3\right]_0^{1/2}$

$= 2\pi\left[\left(-\frac{1}{64} - \frac{1}{160} + \frac{1}{12}\right) - 0\right] = \frac{59}{480}\pi$

53. Use shells:

$V = \int_2^4 2\pi x(-x^2 + 6x - 8)\,dx = 2\pi \int_2^4 (-x^3 + 6x^2 - 8x)\,dx$

$= 2\pi\left[-\frac{1}{4}x^4 + 2x^3 - 4x^2\right]_2^4$

$= 2\pi[(-64 + 128 - 64) - (-4 + 16 - 16)]$

$= 2\pi(4) = 8\pi$

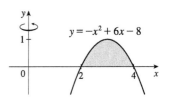

55. Use washers: $y^2 - x^2 = 1 \Rightarrow y = \pm\sqrt{x^2 \pm 1}$

$V = \int_{-\sqrt{3}}^{\sqrt{3}} \pi\left[(2-0)^2 - \left(\sqrt{x^2+1} - 0\right)^2\right] dx$

$= 2\pi \int_0^{\sqrt{3}} [4 - (x^2 + 1)]\,dx \quad$ [by symmetry]

$= 2\pi \int_0^{\sqrt{3}} (3 - x^2)\,dx = 2\pi\left[3x - \frac{1}{3}x^3\right]_0^{\sqrt{3}}$

$= 2\pi(3\sqrt{3} - \sqrt{3}) = 4\sqrt{3}\,\pi$

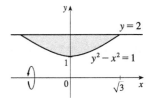

57. Use disks: $x^2 + (y-1)^2 = 1 \Leftrightarrow x = \pm\sqrt{1 - (y-1)^2}$

$V = \pi \int_0^2 \left[\sqrt{1 - (y-1)^2}\right]^2 dy = \pi \int_0^2 (2y - y^2)\,dy$

$= \pi\left[y^2 - \frac{1}{3}y^3\right]_0^2 = \pi\left(4 - \frac{8}{3}\right) = \frac{4}{3}\pi$

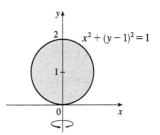

59. $y + 1 = (y-1)^2 \Leftrightarrow y + 1 = y^2 - 2y + 1 \Leftrightarrow 0 = y^2 - 3y \Leftrightarrow$

$0 = y(y-3) \Leftrightarrow y = 0$ or 3.

Use washers:

$V = \pi \int_0^3 \left\{[(y+1) - (-1)]^2 - [(y-1)^2 - (-1)]^2\right\} dy$

$= \pi \int_0^3 [(y+2)^2 - (y^2 - 2y + 2)^2]\,dy$

$= \pi \int_0^3 [(y^2 + 4y + 4) - (y^4 - 4y^3 + 8y^2 - 8y + 4)]\,dy = \pi \int_0^3 (-y^4 + 4y^3 - 7y^2 + 12y)\,dy$

$= \pi\left[-\frac{1}{5}y^5 + y^4 - \frac{7}{3}y^3 + 6y^2\right]_0^3 = \pi\left(-\frac{243}{5} + 81 - 63 + 54\right) = \frac{117}{5}\pi$

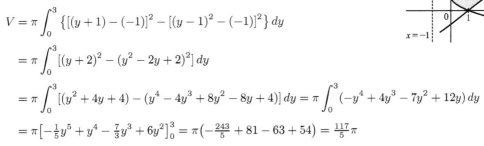

61. Use shells:

$V = 2\int_0^r 2\pi x\sqrt{r^2 - x^2}\,dx = -2\pi \int_0^r (r^2 - x^2)^{1/2}(-2x)\,dx$

$= \left[-2\pi \cdot \frac{2}{3}(r^2 - x^2)^{3/2}\right]_0^r = -\frac{4}{3}\pi(0 - r^3) = \frac{4}{3}\pi r^3$

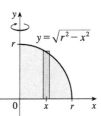

63. $V = 2\pi \int_0^r x\left(-\frac{h}{r}x + h\right)dx = 2\pi h \int_0^r \left(-\frac{x^2}{r} + x\right)dx$

$= 2\pi h \left[-\frac{x^3}{3r} + \frac{x^2}{2}\right]_0^r = 2\pi h \frac{r^2}{6} = \frac{\pi r^2 h}{3}$

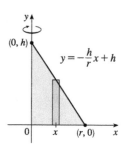

5.4 Work

1. The force exerted by the weight lifter is $F = mg = (200 \text{ kg})(9.8 \text{ m/s}^2) = 1960$ N. The work done by the weight lifter in lifting the weight from 1.5 m to 2.0 m above the ground is then

$$W = Fd = (1960 \text{ N})(2.0 \text{ m} - 1.5 \text{ m}) = (1960 \text{ N})(0.5 \text{ m}) = 980 \text{ N-m} = 980 \text{ J}$$

3. $W = \int_a^b f(x)\,dx = \int_1^{10} 5x^{-2}\,dx = 5\left[-x^{-1}\right]_1^{10} = 5\left(-\frac{1}{10} + 1\right) = 4.5$ ft-lb

5. The force function is given by $F(x)$ (in newtons) and the work (in joules) is the area under the curve, given by

$\int_0^8 F(x)\,dx = \int_0^4 F(x)\,dx + \int_4^8 F(x)\,dx = \frac{1}{2}(4)(30) + (4)(30) = 180$ J.

7. According to Hooke's Law, the force required to maintain a spring stretched x units beyond its natural length (or compressed x units less than its natural length) is proportional to x, that is, $f(x) = kx$. Here, the amount stretched is 4 in. $= \frac{1}{3}$ ft and the force is 10 lb. Thus, $10 = k\left(\frac{1}{3}\right)$ $\Rightarrow$ $k = 30$ lb/ft, and $f(x) = 30x$. The work done in stretching the spring from its natural length to 6 in. $= \frac{1}{2}$ ft beyond its natural length is $W = \int_0^{1/2} 30x\,dx = \left[15x^2\right]_0^{1/2} = \frac{15}{4}$ ft-lb.

9. (a) If $\int_0^{0.12} kx\,dx = 2$ J, then $2 = \left[\frac{1}{2}kx^2\right]_0^{0.12} = \frac{1}{2}k(0.0144) = 0.0072k$ and $k = \frac{2}{0.0072} = \frac{2500}{9} \approx 277.78$ N/m.

Thus, the work needed to stretch the spring from 35 cm to 40 cm is

$\int_{0.05}^{0.10} \frac{2500}{9}x\,dx = \left[\frac{1250}{9}x^2\right]_{1/20}^{1/10} = \frac{1250}{9}\left(\frac{1}{100} - \frac{1}{400}\right) = \frac{25}{24} \approx 1.04$ J.

(b) $f(x) = kx$, so $30 = \frac{2500}{9}x$ and $x = \frac{270}{2500}$ m $= 10.8$ cm

11. The distance from 20 cm to 30 cm is 0.1 m, so with $f(x) = kx$, we get $W_1 = \int_0^{0.1} kx\,dx = k\left[\frac{1}{2}x^2\right]_0^{0.1} = \frac{1}{200}k$.

Now $W_2 = \int_{0.1}^{0.2} kx\,dx = k\left[\frac{1}{2}x^2\right]_{0.1}^{0.2} = k\left(\frac{4}{200} - \frac{1}{200}\right) = \frac{3}{200}k$. Thus, $W_2 = 3W_1$.

In Exercises 13 – 22, n is the number of subintervals of length Δx, and x_i^* is a sample point in the ith subinterval $[x_{i-1}, x_i]$.

13. (a) The portion of the rope from x ft to $(x + \Delta x)$ ft below the top of the building weighs $\frac{1}{2}\Delta x$ lb and must be lifted x_i^* ft, so its contribution to the total work is $\frac{1}{2}x_i^* \Delta x$ ft-lb. The total work is

$$W = \lim_{n\to\infty} \sum_{i=1}^n \frac{1}{2}x_i^* \Delta x = \int_0^{50} \frac{1}{2}x\,dx = \left[\frac{1}{4}x^2\right]_0^{50} = \frac{2500}{4} = 625 \text{ ft-lb}$$

Notice that the exact height of the building does not matter (as long as it is more than 50 ft).

(b) When half the rope is pulled to the top of the building, the work to lift the top half of the rope is

$W_1 = \int_0^{25} \frac{1}{2}x\, dx = \left[\frac{1}{4}x^2\right]_0^{25} = \frac{625}{4}$ ft-lb. The bottom half of the rope is lifted 25 ft and the work needed to accomplish

that is $W_2 = \int_{25}^{50} \frac{1}{2} \cdot 25\, dx = \frac{25}{2}\left[x\right]_{25}^{50} = \frac{625}{2}$ ft-lb. The total work done in pulling half the rope to the top of the building

is $W = W_1 + W_2 = \frac{625}{2} + \frac{625}{4} = \frac{3}{4} \cdot 625 = \frac{1875}{4}$ ft-lb.

15. The work needed to lift the cable is $\lim_{n \to \infty} \sum_{i=1}^{n} 2x_i^* \, \Delta x = \int_0^{500} 2x\, dx = \left[x^2\right]_0^{500} = 250{,}000$ ft-lb. The work needed to lift

the coal is 800 lb $\cdot$ 500 ft $= 400{,}000$ ft-lb. Thus, the total work required is $250{,}000 + 400{,}000 = 650{,}000$ ft-lb.

17. The chain's weight density is $\dfrac{25 \text{ lb}}{10 \text{ ft}} = 2.5$ lb/ft. The part of the chain x ft below the ceiling (for $5 \leq x \leq 10$) has to be lifted

$2(x - 5)$ ft, so the work needed to lift the ith subinterval of the chain is $2(x_i^* - 5)(2.5\, \Delta x)$. The total work needed is

$$W = \lim_{n \to \infty} \sum_{i=1}^{n} 2(x_i^* - 5)(2.5)\, \Delta x = \int_5^{10}[2(x-5)(2.5)]\, dx = 5\int_5^{10}(x - 5)\, dx$$

$$= 5\left[\tfrac{1}{2}x^2 - 5x\right]_5^{10} = 5\left[(50 - 50) - \left(\tfrac{25}{2} - 25\right)\right] = 5\left(\tfrac{25}{2}\right) = 62.5 \text{ ft-lb}$$

19. At a height of x meters ($0 \leq x \leq 12$), the mass of the rope is $(0.8 \text{ kg/m})(12 - x \text{ m}) = (9.6 - 0.8x)$ kg and the mass of the

water is $\left(\tfrac{36}{12} \text{ kg/m}\right)(12 - x \text{ m}) = (36 - 3x)$ kg. The mass of the bucket is 10 kg, so the total mass is

$(9.6 - 0.8x) + (36 - 3x) + 10 = (55.6 - 3.8x)$ kg, and hence, the total force is $9.8(55.6 - 3.8x)$ N. The work needed to lift

the bucket Δx m through the ith subinterval of $[0, 12]$ is $9.8(55.6 - 3.8x_i^*)\Delta x$, so the total work is

$$W = \lim_{n \to \infty} \sum_{i=1}^{n} 9.8(55.6 - 3.8x_i^*)\, \Delta x = \int_0^{12}(9.8)(55.6 - 3.8x)\, dx = 9.8\left[55.6x - 1.9x^2\right]_0^{12} = 9.8(393.6) \approx 3857 \text{ J}$$

21. A "slice" of water Δx m thick and lying at a depth of x_i^* m (where $0 \leq x_i^* \leq \tfrac{1}{2}$) has volume $(2 \times 1 \times \Delta x)$ m^3, a mass of

$2000\, \Delta x$ kg, weighs about $(9.8)(2000\, \Delta x) = 19{,}600\, \Delta x$ N, and thus requires about $19{,}600 x_i^* \, \Delta x$ J of work for its removal.

So $W = \lim_{n \to \infty} \sum_{i=1}^{n} 19{,}600 x_i^*\, \Delta x = \int_0^{1/2} 19{,}600x\, dx = \left[9800 x^2\right]_0^{1/2} = 2450$ J.

23. A rectangular "slice" of water Δx m thick and lying x m above the bottom has width x m and volume $8x\, \Delta x$ m^3. It weighs

about $(9.8 \times 1000)(8x\, \Delta x)$ N, and must be lifted $(5 - x)$ m by the pump, so the work needed is about

$(9.8 \times 10^3)(5 - x)(8x\, \Delta x)$ J. The total work required is

$$W \approx \int_0^3 (9.8 \times 10^3)(5 - x)8x\, dx = (9.8 \times 10^3)\int_0^3 (40x - 8x^2)\, dx = (9.8 \times 10^3)\left[20x^2 - \tfrac{8}{3}x^3\right]_0^3$$

$$= (9.8 \times 10^3)(180 - 72) = (9.8 \times 10^3)(108) = 1058.4 \times 10^3 \approx 1.06 \times 10^6 \text{ J}$$

25. Let x measure depth (in feet) below the spout at the top of the tank. A horizontal disk-shaped "slice" of water Δx ft thick and lying at coordinate x has radius $\frac{3}{8}(16-x)$ ft $(\star)$ and volume $\pi r^2 \Delta x = \pi \cdot \frac{9}{64}(16-x)^2 \Delta x$ ft^3. It weighs about $(62.5)\frac{9\pi}{64}(16-x)^2 \Delta x$ lb and must be lifted x ft by the pump, so the work needed to pump it out is about $(62.5)x\frac{9\pi}{64}(16-x)^2 \Delta x$ ft-lb. The total work required is

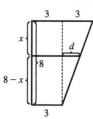

$W \approx \int_0^8 (62.5)x \frac{9\pi}{64}(16-x)^2\,dx = (62.5)\frac{9\pi}{64}\int_0^8 x(256 - 32x + x^2)\,dx$

$= (62.5)\frac{9\pi}{64}\int_0^8 (256x - 32x^2 + x^3)\,dx = (62.5)\frac{9\pi}{64}\left[128x^2 - \frac{32}{3}x^3 + \frac{1}{4}x^4\right]_0^8$

$= (62.5)\frac{9\pi}{64}\left(\frac{11{,}264}{3}\right) = 33{,}000\pi \approx 1.04 \times 10^5$ ft-lb

$(\star)$ From similar triangles, $\dfrac{d}{8-x} = \dfrac{3}{8}$.

So $r = 3 + d = 3 + \frac{3}{8}(8-x)$

$= \dfrac{3(8)}{8} + \dfrac{3}{8}(8-x)$

$= \frac{3}{8}(16-x)$

27. If only 4.7×10^5 J of work is done, then only the water above a certain level (call it h) will be pumped out. So we use the same formula as in Exercise 23, except that the work is fixed, and we are trying to find the lower limit of integration:

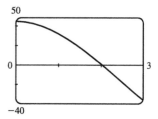

$4.7 \times 10^5 \approx \int_h^3 (9.8 \times 10^3)(5-x)8x\,dx = (9.8 \times 10^3)\left[20x^2 - \frac{8}{3}x^3\right]_h^3 \Leftrightarrow$

$\frac{4.7}{9.8} \times 10^2 \approx 48 = \left(20 \cdot 3^2 - \frac{8}{3} \cdot 3^3\right) - \left(20h^2 - \frac{8}{3}h^3\right) \Leftrightarrow$

$2h^3 - 15h^2 + 45 = 0$. To find the solution of this equation, we plot $2h^3 - 15h^2 + 45$ between $h = 0$ and $h = 3$. We see that the equation is satisfied for $h \approx 2.0$. So the depth of water remaining in the tank is about 2.0 m.

29. $V = \pi r^2 x$, so V is a function of x and P can also be regarded as a function of x. If $V_1 = \pi r^2 x_1$ and $V_2 = \pi r^2 x_2$, then

$W = \int_{x_1}^{x_2} F(x)\,dx = \int_{x_1}^{x_2} \pi r^2 P(V(x))\,dx = \int_{x_1}^{x_2} P(V(x))\,dV(x)$ [Let $V(x) = \pi r^2 x$, so $dV(x) = \pi r^2\,dx$.]

$= \int_{V_1}^{V_2} P(V)\,dV$ by the Substitution Rule.

31. $W = \int_{x_1}^{x_2} f(x)\,dx = \int_{t_1}^{t_2} f(s(t))v(t)\,dt \quad \begin{bmatrix} x = s(t), \\ dx = v(t)\,dt \end{bmatrix}$

$= \int_{t_1}^{t_2} m\,a(t)\,v(t)\,dt = \int_{v_1}^{v_2} m\,u\,du \quad \begin{bmatrix} u = v(t), \\ du = a(t)\,dt \end{bmatrix}$

$= \left[\frac{1}{2}mu^2\right]_{v_1}^{v_2} = \frac{1}{2}mv_2^2 - \frac{1}{2}mv_1^2$

33. The work required to move the 800-kg roller coaster car is

$W = \int_0^{60} (5.7x^2 + 1.5x)\,dx = \left[1.9x^3 + 0.75x^2\right]_0^{60} = 410{,}400 + 2700 = 413{,}100$ J.

Using Exercise 31 with $v_1 = 0$, we get $W = \frac{1}{2}mv_2^2 \Rightarrow v_2 = \sqrt{\dfrac{2W}{m}} = \sqrt{\dfrac{2(413{,}100)}{800}} \approx 32.14$ m/s.

35. (a) $W = \int_a^b F(r)\,dr = \int_a^b G\dfrac{m_1 m_2}{r^2}\,dr = Gm_1 m_2 \left[\dfrac{-1}{r}\right]_a^b = Gm_1 m_2\left(\dfrac{1}{a} - \dfrac{1}{b}\right)$

(b) By part (a), $W = GMm\left(\dfrac{1}{R} - \dfrac{1}{R + 1{,}000{,}000}\right)$ where M = mass of the earth in kg, R = radius of the earth in m,

and m = mass of satellite in kg. (Note that 1000 km = 1,000,000 m.) Thus,

$$W = (6.67 \times 10^{-11})(5.98 \times 10^{24})(1000) \times \left(\dfrac{1}{6.37 \times 10^6} - \dfrac{1}{7.37 \times 10^6}\right) \approx 8.50 \times 10^9 \text{ J}$$

5.5 Average Value of a Function

1. $f_{\text{avg}} = \dfrac{1}{b-a}\int_a^b f(x)\,dx = \dfrac{1}{2-(-1)}\int_{-1}^{2}(3x^2 + 8x)\,dx = \tfrac{1}{3}[x^3 + 4x^2]_{-1}^{2} = \tfrac{1}{3}[(8 + 16) - (-1 + 4)] = 7$

3. $g_{\text{avg}} = \dfrac{1}{b-a}\int_a^b g(x)\,dx = \dfrac{1}{\pi/2 - (-\pi/2)}\int_{-\pi/2}^{\pi/2} 3\cos x\,dx = \dfrac{3 \cdot 2}{\pi}\int_0^{\pi/2}\cos x\,dx$ [by Theorem 4.5.6]

$= \dfrac{6}{\pi}\bigl[\sin x\bigr]_0^{\pi/2} = \dfrac{6}{\pi}(1 - 0) = \dfrac{6}{\pi}$

5. $f_{\text{avg}} = \dfrac{1}{b-a}\int_a^b f(t)\,dt = \dfrac{1}{2-0}\int_0^2 t^2(1 + t^3)^4\,dt = \tfrac{1}{2}\int_1^9 u^4\left(\tfrac{1}{3}\,du\right)$ $[u = 1 + t^3,\ du = 3t^2\,dx]$

$= \tfrac{1}{6}\bigl[\tfrac{1}{5}u^5\bigr]_1^9 = \tfrac{1}{30}(9^5 - 1) = \dfrac{29{,}524}{15} = 1968.2\overline{6}$

7. $h_{\text{avg}} = \dfrac{1}{b-a}\int_a^b h(x)\,dx = \dfrac{1}{\pi - 0}\int_0^{\pi}\cos^4 x\,\sin x\,dx = \tfrac{1}{\pi}\int_1^{-1} u^4(-du)$ $[u = \cos x,\ du = -\sin x\,dx]$

$= \tfrac{1}{\pi}\int_{-1}^{1} u^4\,du = \tfrac{1}{\pi}\cdot 2\int_0^1 u^4\,du$ [by Theorem 4.5.6] $= \tfrac{2}{\pi}\bigl[\tfrac{1}{5}u^5\bigr]_0^1 = \dfrac{2}{5\pi}$

9. (a) $f_{\text{avg}} = \dfrac{1}{3-1}\int_1^3 \dfrac{1}{t^2}\,dt = \tfrac{1}{2}\left[-\dfrac{1}{t}\right]_1^3$

$= \tfrac{1}{2}\left[-\tfrac{1}{3} - (-1)\right] = \tfrac{1}{3}$

(b) $f(c) = f_{\text{avg}} \Leftrightarrow \dfrac{1}{c^2} = \dfrac{1}{3} \Leftrightarrow c^2 = 3 \Rightarrow c = \sqrt{3}$

since $-\sqrt{3}$ is not in $[1, 3]$.

(c)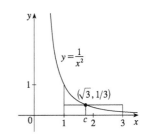

11. (a) $f_{\text{avg}} = \dfrac{1}{\pi - 0}\int_0^{\pi}(2\sin x - \sin 2x)\,dx$

$= \tfrac{1}{\pi}\bigl[-2\cos x + \tfrac{1}{2}\cos 2x\bigr]_0^{\pi}$

$= \tfrac{1}{\pi}\bigl[(2 + \tfrac{1}{2}) - (-2 + \tfrac{1}{2})\bigr] = \dfrac{4}{\pi}$

(c) $f(c) = f_{\text{avg}} \Leftrightarrow 2\sin c - \sin 2c = \dfrac{4}{\pi} \Leftrightarrow$

$c = c_1 \approx 1.238$ or $c = c_2 \approx 2.808$

(b)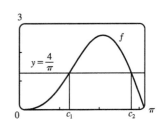

13. f is continuous on $[1, 3]$, so by the Mean Value Theorem for Integrals there exists a number c in $[1, 3]$ such that

$\int_1^3 f(x)\,dx = f(c)(3 - 1) \Rightarrow 8 = 2f(c)$; that is, there is a number c such that $f(c) = \tfrac{8}{2} = 4$.

15. Use geometric interpretations to find the values of the integrals.

$\int_0^8 f(x)\,dx = \int_0^1 f(x)\,dx + \int_1^2 f(x)\,dx + \int_2^3 f(x)\,dx + \int_3^4 f(x)\,dx + \int_4^6 f(x)\,dx + \int_6^7 f(x)\,dx + \int_7^8 f(x)\,dx$

$= -\tfrac{1}{2} + \tfrac{1}{2} + \tfrac{1}{2} + 1 + 4 + \tfrac{3}{2} + 2 = 9$

Thus, the average value of f on $[0, 8] = f_{\text{avg}} = \tfrac{1}{8-0}\int_0^8 f(x)\,dx = \tfrac{1}{8}(9) = \tfrac{9}{8}$.

17. Let $t = 0$ and $t = 12$ correspond to 9 AM and 9 PM, respectively.

$T_{\text{avg}} = \tfrac{1}{12-0}\int_0^{12}\left[50 + 14\sin\tfrac{1}{12}\pi t\right]dt = \tfrac{1}{12}\left[50t - 14\cdot\tfrac{12}{\pi}\cos\tfrac{1}{12}\pi t\right]_0^{12}$

$= \tfrac{1}{12}\left[50\cdot 12 + 14\cdot\tfrac{12}{\pi} + 14\cdot\tfrac{12}{\pi}\right] = 50 + \tfrac{28}{\pi} \approx 59°\text{F}$

19. $\rho_{\text{avg}} = \dfrac{1}{8}\displaystyle\int_0^8 \dfrac{12}{\sqrt{x+1}}\,dx = \dfrac{3}{2}\int_0^8 (x+1)^{-1/2}\,dx = \left[3\sqrt{x+1}\right]_0^8 = 9 - 3 = 6\text{ kg/m}$

21. $V_{\text{avg}} = \tfrac{1}{5}\int_0^5 V(t)\,dt = \tfrac{1}{5}\int_0^5 \tfrac{5}{4\pi}\left[1 - \cos\left(\tfrac{2}{5}\pi t\right)\right]dt = \tfrac{1}{4\pi}\int_0^5 \left[1 - \cos\left(\tfrac{2}{5}\pi t\right)\right]dt$

$= \tfrac{1}{4\pi}\left[t - \tfrac{5}{2\pi}\sin\left(\tfrac{2}{5}\pi t\right)\right]_0^5 = \tfrac{1}{4\pi}[(5-0) - 0] = \tfrac{5}{4\pi} \approx 0.4\text{ L}$

23. $f_{\text{avg}} = \dfrac{1}{b-a}\displaystyle\int_a^b f(x)\,dx$

$> \dfrac{1}{b-a}$ (area of trapezoid $ABDF$)

$= \dfrac{1}{b-a}$ (area of rectangle $ACEF$)

$= \dfrac{1}{b-a}\left[f\left(\tfrac{a+b}{2}\right)\cdot(b-a)\right]$

$= f\left(\tfrac{a+b}{2}\right)$

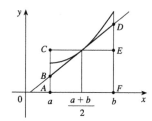

25. Since f is continuous on $[a, a+h]$, the Fundamental Theorem of Calculus can be applied.

$\displaystyle\lim_{h\to 0^+} f_{\text{avg}}[a, a+h] = \lim_{h\to 0^+} \dfrac{1}{(a+h)-a}\int_a^{a+h} f(t)\,dt \stackrel{\text{FTC2}}{=} \lim_{h\to 0^+} \dfrac{F(a+h)-F(a)}{h}$, where $F(x) = \int_c^x f(t)\,dt$. But we

recognize this limit as being $F'(a)$ by the definition of a derivative. Therefore, $\displaystyle\lim_{h\to 0^+} f_{\text{avg}}[a, a+h] = F'(a) = f(a)$

by FTC1.

5 Review

TRUE-FALSE QUIZ

1. False. For example, let $f(x) = x$, $g(x) = 2x$, $a = 1$, and $b = 2$. The area between the curves for $a \le x \le b$ is $A = \int_a^b [g(x) - f(x)]\,dx$. The given integral represents area when $f(x) \ge g(x)$ for $a \le x \le b$.

3. False. Cross-sections perpendicular to the x-axis are washers, and we find cross-sectional area by subtracting the area of the inner circle from the area of the outer circle. Thus, $A(x) = \pi\left(\sqrt{x}\right)^2 - \pi(x)^2 = \pi\left[\left(\sqrt{x}\right)^2 - x^2\right]$, and the volume of the resulting solid is $V = \int_0^1 A(x)\,dx = \int_0^1 \pi\left[\left(\sqrt{x}\right)^2 - x^2\right]dx$.

5. True. See "Volumes of Solids of Revolution" in Section 5.2.

CHAPTER 5 REVIEW ☐ 269

7. False.   Cross-sections perpendicular to the y-axis are washers.

9. True.   A cross-section of S perpendicular to the x-axis is a square with side length $f(x)$, so each cross-section has area
$A(x) = [f(x)]^2$ and volume $V = \int_a^b A(x)\, dx = \int_a^b [f(x)]^2\, dx$.

11. True.   By definition of the average value of f on the interval $[a, b]$, the average value of f on $[2, 5]$ is
$\frac{1}{5-2} \int_2^5 f(x)\, dx = \frac{1}{3}(12) = 4$.

EXERCISES

1. The curves intersect when $x^2 = 8x - x^2 \;\Leftrightarrow\; 2x^2 - 8x = 0 \;\Leftrightarrow\;$
$2x(x-4) = 0 \;\Leftrightarrow\; x = 0$ or 4.

$A = \int_0^4 \left[(8x - x^2) - x^2\right] dx = \int_0^4 (8x - 2x^2)\, dx$

$= \left[4x^2 - \tfrac{2}{3}x^3\right]_0^4 = \left[\left(64 - \tfrac{128}{3}\right) - 0\right] = \tfrac{64}{3}$

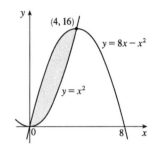

3. If $x \geq 0$, then $|x| = x$, and the graphs intersect when $x = 1 - 2x^2 \;\Leftrightarrow\; 2x^2 + x - 1 = 0 \;\Leftrightarrow\; (2x-1)(x+1) = 0 \;\Leftrightarrow\;$
$x = \tfrac{1}{2}$ or -1, but $-1 < 0$. By symmetry, we can double the area from $x = 0$ to $x = \tfrac{1}{2}$.

$A = 2 \int_0^{1/2} \left[(1 - 2x^2) - x\right] dx = 2 \int_0^{1/2} (-2x^2 - x + 1)\, dx$

$= 2\left[-\tfrac{2}{3}x^3 - \tfrac{1}{2}x^2 + x\right]_0^{1/2} = 2\left[\left(-\tfrac{1}{12} - \tfrac{1}{8} + \tfrac{1}{2}\right) - 0\right]$

$= 2\left(\tfrac{7}{24}\right) = \tfrac{7}{12}$

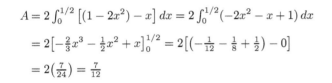

5. $A = \int_0^2 \left[\sin\left(\tfrac{\pi x}{2}\right) - (x^2 - 2x)\right] dx$

$= \left[-\tfrac{2}{\pi}\cos\left(\tfrac{\pi x}{2}\right) - \tfrac{1}{3}x^3 + x^2\right]_0^2$

$= \left(\tfrac{2}{\pi} - \tfrac{8}{3} + 4\right) - \left(-\tfrac{2}{\pi} - 0 + 0\right) = \tfrac{4}{3} + \tfrac{4}{\pi}$

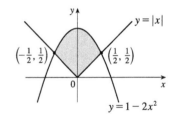

7. Using washers with inner radius x^2 and outer radius $2x$, we have

$V = \pi \int_0^2 \left[(2x)^2 - (x^2)^2\right] dx = \pi \int_0^2 (4x^2 - x^4)\, dx$

$= \pi \left[\tfrac{4}{3}x^3 - \tfrac{1}{5}x^5\right]_0^2 = \pi \left(\tfrac{32}{3} - \tfrac{32}{5}\right)$

$= 32\pi \cdot \tfrac{2}{15} = \tfrac{64}{15}\pi$

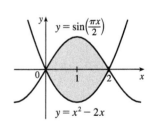

© 2021 Cengage Learning. All Rights Reserved. May not be scanned, copied, or duplicated, or posted to a publicly accessible website, in whole or in part.

9. $V = \pi \int_{-3}^{3} \left\{ [(9-y^2)-(-1)]^2 - [0-(-1)]^2 \right\} dy$

$= 2\pi \int_0^3 \left[(10-y^2)^2 - 1 \right] dy = 2\pi \int_0^3 (100 - 20y^2 + y^4 - 1) \, dy$

$= 2\pi \int_0^3 (99 - 20y^2 + y^4) \, dy = 2\pi \left[99y - \frac{20}{3}y^3 + \frac{1}{5}y^5 \right]_0^3$

$= 2\pi \left(297 - 180 + \frac{243}{5} \right) = \frac{1656}{5}\pi$

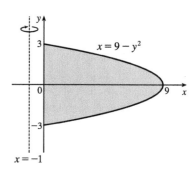

11. The graph of $x^2 - y^2 = a^2$ is a hyperbola with right and left branches.

Solving for y gives us $y^2 = x^2 - a^2 \Rightarrow y = \pm\sqrt{x^2 - a^2}$.

We'll use shells and the height of each shell is

$\sqrt{x^2 - a^2} - (-\sqrt{x^2 - a^2}) = 2\sqrt{x^2 - a^2}$.

The volume is $V = \int_a^{a+h} 2\pi x \cdot 2\sqrt{x^2 - a^2} \, dx$. To evaluate, let $u = x^2 - a^2$,

so $du = 2x\,dx$ and $x\,dx = \frac{1}{2}du$. When $x = a$, $u = 0$, and when $x = a+h$,

$u = (a+h)^2 - a^2 = a^2 + 2ah + h^2 - a^2 = 2ah + h^2$.

Thus, $V = 4\pi \int_0^{2ah+h^2} \sqrt{u} \left(\frac{1}{2} du \right) = 2\pi \left[\frac{2}{3} u^{3/2} \right]_0^{2ah+h^2} = \frac{4}{3}\pi (2ah + h^2)^{3/2}$.

13. A shell has radius $\frac{\pi}{2} - x$, circumference $2\pi\left(\frac{\pi}{2} - x\right)$, and height $\cos^2 x - \frac{1}{4}$.

$y = \cos^2 x$ intersects $y = \frac{1}{4}$ when $\cos^2 x = \frac{1}{4} \Leftrightarrow$

$\cos x = \pm\frac{1}{2} \quad [|x| \leq \pi/2] \Leftrightarrow x = \pm\frac{\pi}{3}$.

$V = \int_{-\pi/3}^{\pi/3} 2\pi \left(\frac{\pi}{2} - x \right) \left(\cos^2 x - \frac{1}{4} \right) dx$

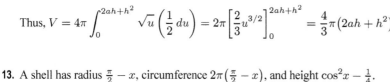

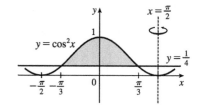

15. $3x^2 = x^3 \Leftrightarrow x^3 - 3x^2 = 0 \Leftrightarrow$

$x^2(x-3) = 0 \Leftrightarrow x = 0 \text{ or } x = 3$

$y = 3x^2 \Rightarrow x = \sqrt{\frac{y}{3}} \quad (x > 0)$

$y = x^3 \Rightarrow x = \sqrt[3]{y}$

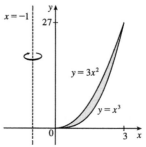

(a) With x as the variable of integration, we use the method of cylindrical shells.

$V = \int_0^3 2\pi(x+1)(3x^2 - x^3) \, dx = \int_0^3 2\pi(2x^3 - x^4 + 3x^2) \, dx = 2\pi \left[\frac{1}{2}x^4 - \frac{1}{5}x^5 + x^3 \right]_0^3$

$= 2\pi \left[\left(\frac{81}{2} - \frac{243}{5} + 27 \right) - 0 \right] = 2\pi \cdot \frac{189}{10} = \frac{189}{5}\pi$

(b) With y as the variable of integration, we use washers with inner radius $\sqrt{\dfrac{y}{3}}+1$ and outer radius $\sqrt[3]{y}+1$.

The area of a cross-section is

$$\pi\left(\sqrt[3]{y}+1\right)^2 - \pi\left(\sqrt{\dfrac{y}{3}}+1\right)^2 = \pi\left[\left(y^{2/3}+2y^{1/3}+1\right) - \left(\dfrac{y}{3}+\dfrac{2}{\sqrt{3}}y^{1/2}+1\right)\right]$$

$$= \pi\left(y^{2/3}+2y^{1/3}-\dfrac{y}{3}-\dfrac{2}{\sqrt{3}}y^{1/2}\right)$$

$$V = \int_0^{27} \pi\left(y^{2/3}+2y^{1/3}-\dfrac{y}{3}-\dfrac{2}{\sqrt{3}}y^{1/2}\right)dy = \pi\left[\dfrac{3}{5}y^{5/3}+\dfrac{3}{2}y^{4/3}-\dfrac{1}{6}y^2-\dfrac{4}{3\sqrt{3}}y^{3/2}\right]_0^{27}$$

$$= \pi\left[\left(\dfrac{729}{5}+\dfrac{243}{2}-\dfrac{243}{2}-108\right)-0\right] = \dfrac{189}{5}\pi$$

17. (a) A cross-section is a washer with inner radius x^2 and outer radius x.

$$V = \int_0^1 \pi\left[(x)^2-(x^2)^2\right]dx = \int_0^1 \pi(x^2-x^4)\,dx = \pi\left[\dfrac{1}{3}x^3-\dfrac{1}{5}x^5\right]_0^1 = \pi\left[\dfrac{1}{3}-\dfrac{1}{5}\right] = \dfrac{2}{15}\pi$$

(b) A cross-section is a washer with inner radius y and outer radius $\sqrt{y}$.

$$V = \int_0^1 \pi\left[\left(\sqrt{y}\right)^2-y^2\right]dy = \int_0^1 \pi(y-y^2)\,dy = \pi\left[\dfrac{1}{2}y^2-\dfrac{1}{3}y^3\right]_0^1 = \pi\left[\dfrac{1}{2}-\dfrac{1}{3}\right] = \dfrac{\pi}{6}$$

(c) A cross-section is a washer with inner radius $2-x$ and outer radius $2-x^2$.

$$V = \int_0^1 \pi\left[(2-x^2)^2-(2-x)^2\right]dx = \int_0^1 \pi(x^4-5x^2+4x)\,dx = \pi\left[\dfrac{1}{5}x^5-\dfrac{5}{3}x^3+2x^2\right]_0^1 = \pi\left[\dfrac{1}{5}-\dfrac{5}{3}+2\right] = \dfrac{8}{15}\pi$$

19. (a) Using the Midpoint Rule on $[0,1]$ with $f(x)=\tan(x^2)$ and $n=4$, we estimate

$$A = \int_0^1 \tan(x^2)\,dx \approx \dfrac{1}{4}\left[\tan\left(\left(\dfrac{1}{8}\right)^2\right)+\tan\left(\left(\dfrac{3}{8}\right)^2\right)+\tan\left(\left(\dfrac{5}{8}\right)^2\right)+\tan\left(\left(\dfrac{7}{8}\right)^2\right)\right] \approx \dfrac{1}{4}(1.53) \approx 0.38$$

(b) Using the Midpoint Rule on $[0,1]$ with $f(x)=\pi\tan^2(x^2)$ (for disks) and $n=4$, we estimate

$$V = \int_0^1 f(x)\,dx \approx \dfrac{1}{4}\pi\left[\tan^2\left(\left(\dfrac{1}{8}\right)^2\right)+\tan^2\left(\left(\dfrac{3}{8}\right)^2\right)+\tan^2\left(\left(\dfrac{5}{8}\right)^2\right)+\tan^2\left(\left(\dfrac{7}{8}\right)^2\right)\right] \approx \dfrac{\pi}{4}(1.114) \approx 0.87$$

21. $\int_0^{\pi/2} 2\pi x \cos x\,dx = \int_0^{\pi/2}(2\pi x)\cos x\,dx$

The solid is obtained by rotating the region $\mathcal{R} = \{(x,y) \mid 0 \le x \le \dfrac{\pi}{2}, 0 \le y \le \cos x\}$ about the y-axis.

23. $\int_0^{\pi} \pi(2-\sin x)^2\,dx$

The solid is obtained by rotating the region $\mathcal{R} = \{(x,y) \mid 0 \le x \le \pi, 0 \le y \le 2-\sin x\}$ about the x-axis.

25. Take the base to be the disk $x^2+y^2 \le 9$. Then $V = \int_{-3}^{3} A(x)\,dx$, where $A(x_0)$ is the area of the isosceles right triangle whose hypotenuse lies along the line $x=x_0$ in the xy-plane. The length of the hypotenuse is $2\sqrt{9-x^2}$ and the length of each leg is $\sqrt{2}\sqrt{9-x^2}$. $A(x) = \dfrac{1}{2}\left(\sqrt{2}\sqrt{9-x^2}\right)^2 = 9-x^2$, so

$$V = 2\int_0^3 A(x)\,dx = 2\int_0^3 (9-x^2)\,dx = 2\left[9x-\dfrac{1}{3}x^3\right]_0^3 = 2(27-9) = 36$$

27. Equilateral triangles with sides measuring $\frac{1}{4}x$ meters have height $\frac{1}{4}x \sin 60° = \frac{\sqrt{3}}{8}x$. Therefore,

$A(x) = \frac{1}{2} \cdot \frac{1}{4}x \cdot \frac{\sqrt{3}}{8}x = \frac{\sqrt{3}}{64}x^2$. $V = \int_0^{20} A(x)\,dx = \frac{\sqrt{3}}{64} \int_0^{20} x^2\,dx = \frac{\sqrt{3}}{64}\left[\frac{1}{3}x^3\right]_0^{20} = \frac{8000\sqrt{3}}{64 \cdot 3} = \frac{125\sqrt{3}}{3}$ m^3.

29. $f(x) = kx \Rightarrow 30\text{ N} = k(15-12)$ cm $\Rightarrow k = 10$ N/cm $= 1000$ N/m. 20 cm -12 cm $= 0.08$ m $\Rightarrow$
$W = \int_0^{0.08} kx\,dx = 1000 \int_0^{0.08} x\,dx = 500\left[x^2\right]_0^{0.08} = 500(0.08)^2 = 3.2$ N·m $= 3.2$ J.

31. (a) The parabola has equation $y = ax^2$ with vertex at the origin and passing through

$(4, 4)$. $4 = a \cdot 4^2 \Rightarrow a = \frac{1}{4} \Rightarrow y = \frac{1}{4}x^2 \Rightarrow x^2 = 4y \Rightarrow$

$x = 2\sqrt{y}$. Each circular disk has radius $2\sqrt{y}$ and is moved $4 - y$ ft.

$W = \int_0^4 \pi \left(2\sqrt{y}\right)^2 62.5(4-y)\,dy = 250\pi \int_0^4 y(4-y)\,dy$

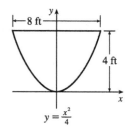

$= 250\pi \left[2y^2 - \frac{1}{3}y^3\right]_0^4 = 250\pi \left(32 - \frac{64}{3}\right) = \frac{8000\pi}{3} \approx 8378$ ft-lb

(b) In part (a) we knew the final water level (0) but not the amount of work done. Here we use the same equation, except with the work fixed, and the lower limit of integration (that is, the final water level — call it h) unknown: $W = 4000 \Leftrightarrow$

$250\pi \left[2y^2 - \frac{1}{3}y^3\right]_h^4 = 4000 \Leftrightarrow \frac{16}{\pi} = \left[\left(32 - \frac{64}{3}\right) - \left(2h^2 - \frac{1}{3}h^3\right)\right] \Leftrightarrow$

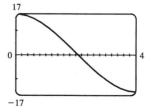

$h^3 - 6h^2 + 32 - \frac{48}{\pi} = 0$. We graph the function $f(h) = h^3 - 6h^2 + 32 - \frac{48}{\pi}$
on the interval $[0, 4]$ to see where it is 0. From the graph, $f(h) = 0$ for $h \approx 2.1$.
So the depth of water remaining is about 2.1 ft.

33. $f_{\text{avg}} = \frac{1}{b-a} \int_a^b f(t)\,dt = \frac{1}{\pi/4 - 0} \int_0^{\pi/4} \sec^2 t\,dt = \frac{4}{\pi}\left[\tan t\right]_0^{\pi/4} = \frac{4}{\pi}(1-0) = \frac{4}{\pi}$

35. (a) The regions $\mathcal{R}_1$ and $\mathcal{R}_2$ are shown in the figure.

The area of $\mathcal{R}_1$ is $A_1 = \int_0^b x^2\,dx = \left[\frac{1}{3}x^3\right]_0^b = \frac{1}{3}b^3$, and the area of $\mathcal{R}_2$ is

$A_2 = \int_0^{b^2} \sqrt{y}\,dy = \left[\frac{2}{3}y^{3/2}\right]_0^{b^2} = \frac{2}{3}b^3$. So there is no solution to $A_1 = A_2$
for $b \neq 0$.

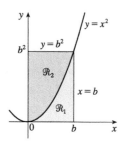

(b) Using disks, we calculate the volume of rotation of $\mathcal{R}_1$ about the x-axis to be $V_{1,x} = \pi \int_0^b (x^2)^2\,dx = \frac{1}{5}\pi b^5$.

Using cylindrical shells, we calculate the volume of rotation of $\mathcal{R}_1$ about the y-axis to be

$V_{1,y} = 2\pi \int_0^b x(x^2)\,dx = 2\pi \left[\frac{1}{4}x^4\right]_0^b = \frac{1}{2}\pi b^4$. So $V_{1,x} = V_{1,y} \Leftrightarrow \frac{1}{5}\pi b^5 = \frac{1}{2}\pi b^4 \Leftrightarrow 2b = 5 \Leftrightarrow b = \frac{5}{2}$.

So the volumes of rotation about the x- and y-axes are the same for $b = \frac{5}{2}$.

(c) We use cylindrical shells to calculate the volume of rotation of $\mathcal{R}_2$ about the x-axis:

$\mathcal{R}_{2,x} = 2\pi \int_0^{b^2} y\left(\sqrt{y}\right) dy = 2\pi \left[\frac{2}{5} y^{5/2}\right]_0^{b^2} = \frac{4}{5}\pi b^5$. We already know the volume of rotation of $\mathcal{R}_1$ about the x-axis from part (b), and $\mathcal{R}_{1,x} = \mathcal{R}_{2,x} \Leftrightarrow \frac{1}{5}\pi b^5 = \frac{4}{5}\pi b^5$, which has no solution for $b \neq 0$.

(d) We use disks to calculate the volume of rotation of $\mathcal{R}_2$ about the y-axis: $\mathcal{R}_{2,y} = \pi \int_0^{b^2} \left(\sqrt{y}\right)^2 dy = \pi \left[\frac{1}{2} y^2\right]_0^{b^2} = \frac{1}{2}\pi b^4$.

We know the volume of rotation of $\mathcal{R}_1$ about the y-axis from part (b), and $\mathcal{R}_{1,y} = \mathcal{R}_{2,y} \Leftrightarrow \frac{1}{2}\pi b^4 = \frac{1}{2}\pi b^4$. But this equation is true for all b, so the volumes of rotation about the y-axis are equal for all values of b.

PROBLEMS PLUS

1. The volume generated from $x = 0$ to $x = b$ is $\int_0^b \pi[f(x)]^2\,dx$. Hence, we are given that $b^2 = \int_0^b \pi[f(x)]^2\,dx$ for all $b > 0$. Differentiating both sides of this equation with respect to b using the Fundamental Theorem of Calculus gives $2b = \pi[f(b)]^2 \Rightarrow f(b) = \sqrt{2b/\pi}$, since f is positive. Therefore, $f(x) = \sqrt{2x/\pi}$.

3. We must find expressions for the areas A and B, and then set them equal and see what this says about the curve C. If $P = (a, 2a^2)$, then area A is just $\int_0^a (2x^2 - x^2)\,dx = \int_0^a x^2\,dx = \tfrac{1}{3}a^3$. To find area B, we use y as the variable of integration. So we find the equation of the middle curve as a function of y: $y = 2x^2 \Leftrightarrow x = \sqrt{y/2}$, since we are concerned with the first quadrant only. We can express area B as

$$\int_0^{2a^2} \left[\sqrt{y/2} - C(y)\right] dy = \left[\tfrac{4}{3}(y/2)^{3/2}\right]_0^{2a^2} - \int_0^{2a^2} C(y)\,dy = \tfrac{4}{3}a^3 - \int_0^{2a^2} C(y)\,dy$$

where $C(y)$ is the function with graph C. Setting $A = B$, we get $\tfrac{1}{3}a^3 = \tfrac{4}{3}a^3 - \int_0^{2a^2} C(y)\,dy \Leftrightarrow \int_0^{2a^2} C(y)\,dy = a^3$. Now we differentiate this equation with respect to a using the Chain Rule and the Fundamental Theorem:

$C(2a^2)(4a) = 3a^2 \Rightarrow C(y) = \tfrac{3}{4}\sqrt{y/2}$, where $y = 2a^2$. Now we can solve for y: $x = \tfrac{3}{4}\sqrt{y/2} \Rightarrow x^2 = \tfrac{9}{16}(y/2) \Rightarrow y = \tfrac{32}{9}x^2$.

5. We are given that the rate of change of the volume of water is $\dfrac{dV}{dt} = -kA(x)$, where k is some positive constant and $A(x)$ is the area of the surface when the water has depth x. Now we are concerned with the rate of change of the depth of the water with respect to time, that is, $\dfrac{dx}{dt}$. But by the Chain Rule, $\dfrac{dV}{dt} = \dfrac{dV}{dx}\dfrac{dx}{dt}$, so the first equation can be written

$\dfrac{dV}{dx}\dfrac{dx}{dt} = -kA(x)$ $(\star)$. Also, we know that the total volume of water up to a depth x is $V(x) = \int_0^x A(s)\,ds$, where $A(s)$ is the area of a cross-section of the water at a depth s. Differentiating this equation with respect to x, we get $dV/dx = A(x)$. Substituting this into equation $\star$, we get $A(x)(dx/dt) = -kA(x) \Rightarrow dx/dt = -k$, a constant.

7. A typical sphere of radius r is shown in the figure. We wish to maximize the shaded volume V, which can be thought of as the volume of a hemisphere of radius r minus the volume of the spherical cap with height $h = 1 - \sqrt{1-r^2}$ and radius 1.

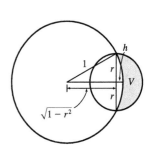

$V = \tfrac{1}{2} \cdot \tfrac{4}{3}\pi r^3 - \tfrac{1}{3}\pi\left(1 - \sqrt{1-r^2}\right)^2 \left[3(1) - \left(1 - \sqrt{1-r^2}\right)\right]$ [by Exercise 5.2.61]

$= \tfrac{1}{3}\pi\left[2r^3 - \left(2 - 2\sqrt{1-r^2} - r^2\right)\left(2 + \sqrt{1-r^2}\right)\right]$

$= \tfrac{1}{3}\pi\left[2r^3 - 2 + (r^2 + 2)\sqrt{1-r^2}\right]$

[continued]

$$V' = \tfrac{1}{3}\pi\left[6r^2 + \frac{(r^2+2)(-r)}{\sqrt{1-r^2}} + \sqrt{1-r^2}(2r)\right] = \tfrac{1}{3}\pi\left[\frac{6r^2\sqrt{1-r^2} - r(r^2+2) + 2r(1-r^2)}{\sqrt{1-r^2}}\right]$$

$$= \tfrac{1}{3}\pi\left(\frac{6r^2\sqrt{1-r^2} - 3r^3}{\sqrt{1-r^2}}\right) = \frac{\pi r^2(2\sqrt{1-r^2} - r)}{\sqrt{1-r^2}}$$

$V'(r) = 0 \Leftrightarrow 2\sqrt{1-r^2} = r \Leftrightarrow 4 - 4r^2 = r^2 \Leftrightarrow r^2 = \tfrac{4}{5} \Leftrightarrow r = \tfrac{2}{\sqrt{5}} \approx 0.89$.

Since $V'(r) > 0$ for $0 < r < \tfrac{2}{\sqrt{5}}$ and $V'(r) < 0$ for $\tfrac{2}{\sqrt{5}} < r < 1$, we know that V attains a maximum at $r = \tfrac{2}{\sqrt{5}}$.

9. (a) Stacking disks along the y-axis gives us $V = \int_0^h \pi[f(y)]^2\,dy$.

(b) Using the Chain Rule, $\dfrac{dV}{dt} = \dfrac{dV}{dh} \cdot \dfrac{dh}{dt} = \pi[f(h)]^2\dfrac{dh}{dt}$.

(c) $kA\sqrt{h} = \pi[f(h)]^2\dfrac{dh}{dt}$. Set $\dfrac{dh}{dt} = C$: $\pi[f(h)]^2 C = kA\sqrt{h} \Rightarrow [f(h)]^2 = \dfrac{kA}{\pi C}\sqrt{h} \Rightarrow f(h) = \sqrt{\dfrac{kA}{\pi C}}h^{1/4}$; that

is, $f(y) = \sqrt{\dfrac{kA}{\pi C}}y^{1/4}$. The advantage of having $\dfrac{dh}{dt} = C$ is that the markings on the container are equally spaced.

11. The cubic polynomial passes through the origin, so let its equation be
$y = px^3 + qx^2 + rx$. The curves intersect when $px^3 + qx^2 + rx = x^2 \Leftrightarrow$
$px^3 + (q-1)x^2 + rx = 0$. Call the left side $f(x)$. Since $f(a) = f(b) = 0$,
another form of f is

$$f(x) = px(x-a)(x-b) = px[x^2 - (a+b)x + ab]$$
$$= p[x^3 - (a+b)x^2 + abx]$$

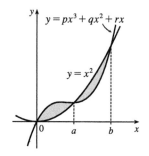

Since the two areas are equal, we must have $\int_0^a f(x)\,dx = -\int_a^b f(x)\,dx \Rightarrow$
$[F(x)]_0^a = [F(x)]_b^a \Rightarrow F(a) - F(0) = F(a) - F(b) \Rightarrow F(0) = F(b)$, where F is an antiderivative of f.
Now $F(x) = \int f(x)\,dx = \int p[x^3 - (a+b)x^2 + abx]\,dx = p\left[\tfrac{1}{4}x^4 - \tfrac{1}{3}(a+b)x^3 + \tfrac{1}{2}abx^2\right] + C$, so
$F(0) = F(b) \Rightarrow C = p\left[\tfrac{1}{4}b^4 - \tfrac{1}{3}(a+b)b^3 + \tfrac{1}{2}ab^3\right] + C \Rightarrow 0 = p\left[\tfrac{1}{4}b^4 - \tfrac{1}{3}(a+b)b^3 + \tfrac{1}{2}ab^3\right] \Rightarrow$
$0 = 3b - 4(a+b) + 6a$ [multiply by $12/(pb^3)$, $b \neq 0$] $\Rightarrow 0 = 3b - 4a - 4b + 6a \Rightarrow b = 2a$.
Hence, b is twice the value of a.

13. We assume that P lies in the region of positive x. Since $y = x^3$ is an odd
function, this assumption will not affect the result of the calculation. Let
$P = (a, a^3)$. The slope of the tangent to the curve $y = x^3$ at P is $3a^2$,
and so the equation of the tangent is $y - a^3 = 3a^2(x - a) \Leftrightarrow$
$y = 3a^2 x - 2a^3$.

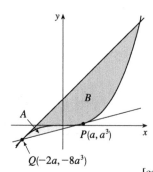

[continued]

We solve this simultaneously with $y = x^3$ to find the other point of intersection: $x^3 = 3a^2x - 2a^3$ $\Leftrightarrow$ $(x-a)^2(x+2a) = 0$. So $Q = (-2a, -8a^3)$ is the other point of intersection. The equation of the tangent at Q is $y - (-8a^3) = 12a^2[x - (-2a)]$ $\Leftrightarrow$ $y = 12a^2x + 16a^3$. By symmetry, this tangent will intersect the curve again at $x = -2(-2a) = 4a$. The curve lies above the first tangent, and below the second, so we are looking for a relationship between $A = \int_{-2a}^{a} \left[x^3 - (3a^2x - 2a^3)\right] dx$ and $B = \int_{-2a}^{4a} \left[(12a^2x + 16a^3) - x^3\right] dx$. We calculate $A = \left[\frac{1}{4}x^4 - \frac{3}{2}a^2x^2 + 2a^3x\right]_{-2a}^{a} = \frac{3}{4}a^4 - (-6a^4) = \frac{27}{4}a^4$, and $B = \left[6a^2x^2 + 16a^3x - \frac{1}{4}x^4\right]_{-2a}^{4a} = 96a^4 - (-12a^4) = 108a^4$. We see that $B = 16A = 2^4 A$. This is because our calculation of area B was essentially the same as that of area A, with a replaced by $-2a$, so if we replace a with $-2a$ in our expression for A, we get $\frac{27}{4}(-2a)^4 = 108a^4 = B$.

6 □ INVERSE FUNCTIONS:
Exponential, Logarithmic, and Inverse Trigonometric Functions

6.1 Inverse Functions and Their Derivatives

1. (a) See Definition 1.

   (b) It must pass the Horizontal Line Test.

3. f is not one-to-one because $2 \neq 6$, but $f(2) = 2.0 = f(6)$.

5. We could draw a horizontal line that intersects the graph in more than one point. Thus, by the Horizontal Line Test, the function is not one-to-one.

7. No horizontal line intersects the graph more than once. Thus, by the Horizontal Line Test, the function is one-to-one.

9. The graph of $f(x) = 2x - 3$ is a line with slope 2. It passes the Horizontal Line Test, so f is one-to-one.

   Algebraic solution: If $x_1 \neq x_2$, then $2x_1 \neq 2x_2 \Rightarrow 2x_1 - 3 \neq 2x_2 - 3 \Rightarrow f(x_1) \neq f(x_2)$, so f is one-to-one.

11. No horizontal line intersects the graph of $r(t) = t^3 + 4$ more than once. Thus, by the Horizontal Line Test, the function is one-to-one.

    Algebraic solution: If $t_1 \neq t_2$, then $t_1^3 \neq t_2^3 \Rightarrow t_1^3 + 4 \neq t_2^3 + 4 \Rightarrow r(t_1) \neq r(t_2)$, so r is one-to-one.

13. $g(x) = 1 - \sin x$. $g(0) = 1$ and $g(\pi) = 1$, so g is not one-to-one.

15. A football will attain every height h up to its maximum height twice: once on the way up, and again on the way down. Thus, even if t_1 does not equal t_2, $f(t_1)$ may equal $f(t_2)$, so f is not 1-1.

17. (a) Since f is 1-1, $f(6) = 17 \Leftrightarrow f^{-1}(17) = 6$.

    (b) Since f is 1-1, $f^{-1}(3) = 2 \Leftrightarrow f(2) = 3$.

19. $h(x) = x + \sqrt{x} \Rightarrow h'(x) = 1 + 1/(2\sqrt{x}) > 0$ on $(0, \infty)$. So h is increasing and hence, 1-1. By inspection, $h(4) = 4 + \sqrt{4} = 6$, so $h^{-1}(6) = 4$.

21. We solve $C = \frac{5}{9}(F - 32)$ for F: $\frac{9}{5}C = F - 32 \Rightarrow F = \frac{9}{5}C + 32$. This gives us a formula for the inverse function, that is, the Fahrenheit temperature F as a function of the Celsius temperature C. $F \geq -459.67 \Rightarrow \frac{9}{5}C + 32 \geq -459.67 \Rightarrow \frac{9}{5}C \geq -491.67 \Rightarrow C \geq -273.15$, the domain of the inverse function.

23. $y = f(x) = 5 - 4x \Rightarrow 4x = 5 - y \Rightarrow x = \frac{1}{4}(5 - y)$. Interchange x and y: $y = \frac{1}{4}(5 - x)$. So $f^{-1}(x) = \frac{1}{4}(5 - x) = \frac{5}{4} - \frac{1}{4}x$.

25. First note that $f(x) = 1 - x^2$, $x \geq 0$, is one-to-one. We first write $y = 1 - x^2$, $x \geq 0$, and solve for x:
    $x^2 = 1 - y \Rightarrow x = \sqrt{1-y}$ (since $x \geq 0$). Interchanging x and y gives $y = \sqrt{1-x}$, so the inverse function is $f^{-1}(x) = \sqrt{1-x}$.

27. First write $y = g(x) = 2 + \sqrt{x+1}$ and note that $y \geq 2$. Solve for x: $y - 2 = \sqrt{x+1} \Rightarrow (y-2)^2 = x+1 \Rightarrow x = (y-2)^2 - 1 \, (y \geq 2)$. Interchanging x and y gives $y = (x-2)^2 - 1$, so $g^{-1}(x) = (x-2)^2 - 1$ with domain $x \geq 2$.

29. We solve $y = \left(2 + \sqrt[3]{x}\right)^5$ for x: $\sqrt[5]{y} = 2 + \sqrt[3]{x} \Rightarrow \sqrt[3]{x} = \sqrt[5]{y} - 2 \Rightarrow x = \left(\sqrt[5]{y} - 2\right)^3$. Interchanging x and y gives the inverse function $y = \left(\sqrt[5]{x} - 2\right)^3$.

31. $y = f(x) = \sqrt{4x+3} \, (y \geq 0) \Rightarrow y^2 = 4x+3 \Rightarrow x = \dfrac{y^2 - 3}{4}$.

Interchange x and y: $y = \dfrac{x^2 - 3}{4}$. So $f^{-1}(x) = \dfrac{x^2 - 3}{4} \, (x \geq 0)$. From the graph, we see that f and f^{-1} are reflections about the line $y = x$.

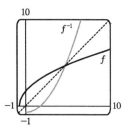

33. Reflect the graph of f about the line $y = x$. The points $(-1, -2)$, $(1, -1)$, $(2, 2)$, and $(3, 3)$ on f are reflected to $(-2, -1)$, $(-1, 1)$, $(2, 2)$, and $(3, 3)$ on f^{-1}.

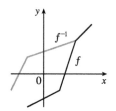

35. (a) $y = f(x) = \sqrt{1 - x^2} \, (0 \leq x \leq 1$ and note that $y \geq 0) \Rightarrow y^2 = 1 - x^2 \Rightarrow x^2 = 1 - y^2 \Rightarrow x = \sqrt{1 - y^2}$. So $f^{-1}(x) = \sqrt{1 - x^2}$, $0 \leq x \leq 1$. We see that f^{-1} and f are the same function.

(b) The graph of f is the portion of the circle $x^2 + y^2 = 1$ with $0 \leq x \leq 1$ and $0 \leq y \leq 1$ (quarter-circle in the first quadrant). The graph of f is symmetric with respect to the line $y = x$, so its reflection about $y = x$ is itself, that is, $f^{-1} = f$.

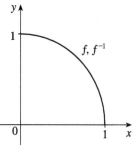

37. (a) $x_1 \neq x_2 \Rightarrow x_1^3 \neq x_2^3 \Rightarrow f(x_1) \neq f(x_2)$, so f is one-to-one.

(b) $f'(x) = 3x^2$ and $f(2) = 8 \Rightarrow f^{-1}(8) = 2$, so $(f^{-1})'(8) = 1/f'(f^{-1}(8)) = 1/f'(2) = \frac{1}{12}$.

(c) $y = x^3 \Rightarrow x = y^{1/3}$. Interchanging x and y gives $y = x^{1/3}$, so $f^{-1}(x) = x^{1/3}$. Domain$(f^{-1}) = $ range$(f) = \mathbb{R}$. Range$(f^{-1}) = $ domain$(f) = \mathbb{R}$.

(e)

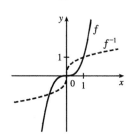

(d) $f^{-1}(x) = x^{1/3} \Rightarrow (f^{-1})'(x) = \frac{1}{3}x^{-2/3} \Rightarrow (f^{-1})'(8) = \frac{1}{3}\left(\frac{1}{4}\right) = \frac{1}{12}$ as in part (b).

39. (a) Since $x \geq 0$, $x_1 \neq x_2$ $\Rightarrow$ $x_1^2 \neq x_2^2$ $\Rightarrow$ $9 - x_1^2 \neq 9 - x_2^2$ $\Rightarrow$ $f(x_1) \neq f(x_2)$, so f is 1-1.

(b) $f'(x) = -2x$ and $f(1) = 8$ $\Rightarrow$ $f^{-1}(8) = 1$, so $(f^{-1})'(8) = \dfrac{1}{f'(f^{-1}(8))} = \dfrac{1}{f'(1)} = \dfrac{1}{-2} = -\dfrac{1}{2}$.

(c) $y = 9 - x^2$ $\Rightarrow$ $x^2 = 9 - y$ $\Rightarrow$ $x = \sqrt{9-y}$.
Interchange x and y: $y = \sqrt{9-x}$, so $f^{-1}(x) = \sqrt{9-x}$.
Domain(f^{-1}) = range $(f) = [0, 9]$.
Range(f^{-1}) = domain $(f) = [0, 3]$.

(e)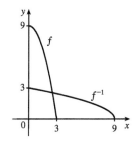

(d) $(f^{-1})'(x) = -1/(2\sqrt{9-x})$ $\Rightarrow$ $(f^{-1})'(8) = -\frac{1}{2}$ as in part (b).

41. $f(0) = 5$ $\Rightarrow$ $f^{-1}(5) = 0$, and $f(x) = 3x^3 + 4x^2 + 6x + 5$ $\Rightarrow$ $f'(x) = 9x^2 + 8x + 6$ and $f'(0) = 6$. Thus,
$(f^{-1})'(5) = \dfrac{1}{f'(f^{-1}(5))} = \dfrac{1}{f'(0)} = \dfrac{1}{6}$.

43. $f(0) = 3$ $\Rightarrow$ $f^{-1}(3) = 0$, and $f(x) = 3 + x^2 + \tan(\pi x/2)$ $\Rightarrow$ $f'(x) = 2x + \frac{\pi}{2}\sec^2(\pi x/2)$ and
$f'(0) = \frac{\pi}{2} \cdot 1 = \frac{\pi}{2}$. Thus, $(f^{-1})'(3) = 1/f'(f^{-1}(3)) = 1/f'(0) = 2/\pi$.

45. $f(4) = 5$ $\Rightarrow$ $f^{-1}(5) = 4$. Thus, $(f^{-1})'(5) = \dfrac{1}{f'(f^{-1}(5))} = \dfrac{1}{f'(4)} = \dfrac{1}{2/3} = \dfrac{3}{2}$.

47. $f(x) = \int_3^x \sqrt{1+t^3}\,dt$ $\Rightarrow$ $f'(x) = \sqrt{1+x^3} > 0$, so f is an increasing function and it has an inverse. Since
$f(3) = \int_3^3 \sqrt{1+t^3}\,dt = 0$, $f^{-1}(0) = 3$. Thus, $(f^{-1})'(0) = \dfrac{1}{f'(f^{-1}(0))} = \dfrac{1}{f'(3)} = \dfrac{1}{\sqrt{1+3^3}} = \dfrac{1}{\sqrt{28}}$.

49. We see that the graph of $y = f(x) = \sqrt{x^3 + x^2 + x + 1}$ is increasing, so f is 1-1.
(We could verify this by showing that $f'(x) > 0$.) Enter $x = \sqrt{y^3 + y^2 + y + 1}$
and use your CAS to solve the equation for y. You will likely get two (irrelevant)
solutions involving imaginary expressions, as well as one which can be simplified
to
$y = f^{-1}(x) = -\dfrac{\sqrt[3]{4}}{6}\left(\sqrt[3]{D - 27x^2 + 20} - \sqrt[3]{D + 27x^2 - 20} + \sqrt[3]{2}\right)$
where $D = 3\sqrt{3}\sqrt{27x^4 - 40x^2 + 16}$ or, equivalently, $\dfrac{1}{6}\dfrac{M^{2/3} - 8 - 2M^{1/3}}{2M^{1/3}}$,
where $M = 108x^2 + 12\sqrt{48 - 120x^2 + 81x^4} - 80$.

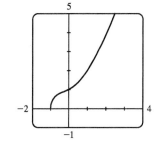

51. (a) If the point (x, y) is on the graph of $y = f(x)$, then the point $(x - c, y)$ is that point shifted c units to the left. Since f is 1-1, the point (y, x) is on the graph of $y = f^{-1}(x)$ and the point corresponding to $(x - c, y)$ on the graph of f is $(y, x - c)$ on the graph of f^{-1}. Thus, the curve's reflection is shifted *down* the same number of units as the curve itself is shifted to the left. So an expression for the inverse function is $g^{-1}(x) = f^{-1}(x) - c$.

(b) If we compress (or stretch) a curve horizontally, the curve's reflection in the line $y = x$ is compressed (or stretched) *vertically* by the same factor. Using this geometric principle, we see that the inverse of $h(x) = f(cx)$ can be expressed as $h^{-1}(x) = (1/c) f^{-1}(x)$.

6.2 Exponential Functions and Their Derivatives

1. (a) $f(x) = b^x$, $b > 0$ (b) $\mathbb{R}$ (c) $(0, \infty)$ (d) See Figures 4(c), 4(b), and 4(a), respectively.

3. All of these graphs approach 0 as $x \to -\infty$, all of them pass through the point $(0, 1)$, and all of them are increasing and approach ∞ as $x \to \infty$. The larger the base, the faster the function increases for $x > 0$, and the faster it approaches 0 as $x \to -\infty$.

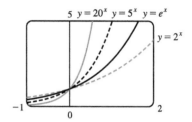

5. The functions with base greater than 1 (3^x and 10^x) are increasing, while those with base less than 1 $\left[\left(\frac{1}{3}\right)^x \text{ and } \left(\frac{1}{10}\right)^x\right]$ are decreasing. The graph of $\left(\frac{1}{3}\right)^x$ is the reflection of that of 3^x about the y-axis, and the graph of $\left(\frac{1}{10}\right)^x$ is the reflection of that of 10^x about the y-axis. The graph of 10^x increases more quickly than that of 3^x for $x > 0$, and approaches 0 faster as $x \to -\infty$.

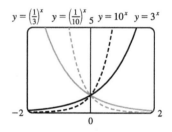

7. We start with the graph of $y = 3^x$ (Figure 12) and shift 1 unit upward to get the graph of $g(x) = 3^x + 1$.

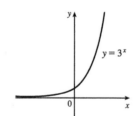
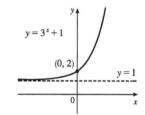

9. We start with the graph of $y = e^x$ (Figure 14) and reflect about the y-axis to get the graph of $y = e^{-x}$. Then we reflect the graph about the x-axis to get the graph of $y = -e^{-x}$.

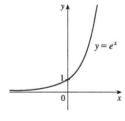
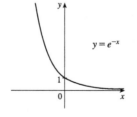
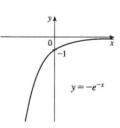

11. We start with the graph of $y = e^x$ (Figure 14) and reflect about the y-axis to get the graph of $y = e^{-x}$. Then we compress the graph vertically by a factor of 2 to obtain the graph of $y = \frac{1}{2}e^{-x}$ and then reflect about the x-axis to get the graph of $y = -\frac{1}{2}e^{-x}$. Finally, we shift the graph one unit upward to get the graph of $y = 1 - \frac{1}{2}e^{-x}$.

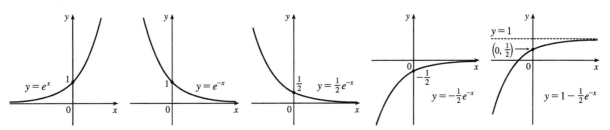

13. (a) To find the equation of the graph that results from shifting the graph of $y = e^x$ two units downward, we subtract 2 from the original function to get $y = e^x - 2$.

(b) To find the equation of the graph that results from shifting the graph of $y = e^x$ two units to the right, we replace x with $x - 2$ in the original function to get $y = e^{x-2}$.

(c) To find the equation of the graph that results from reflecting the graph of $y = e^x$ about the x-axis, we multiply the original function by -1 to get $y = -e^x$.

(d) To find the equation of the graph that results from reflecting the graph of $y = e^x$ about the y-axis, we replace x with $-x$ in the original function to get $y = e^{-x}$.

(e) To find the equation of the graph that results from reflecting the graph of $y = e^x$ about the x-axis and then about the y-axis, we first multiply the original function by -1 (to get $y = -e^x$) and then replace x with $-x$ in this equation to get $y = -e^{-x}$.

15. (a) The denominator is zero when $1 - e^{1-x^2} = 0$ $\Leftrightarrow$ $e^{1-x^2} = 1$ $\Leftrightarrow$ $1 - x^2 = 0$ $\Leftrightarrow$ $x = \pm 1$. Thus, the function $f(x) = \dfrac{1 - e^{x^2}}{1 - e^{1-x^2}}$ has domain $\{x \mid x \neq \pm 1\} = (-\infty, -1) \cup (-1, 1) \cup (1, \infty)$.

(b) The denominator is never equal to zero, so the function $f(x) = \dfrac{1+x}{e^{\cos x}}$ has domain $\mathbb{R}$, or $(-\infty, \infty)$.

17. Use $y = Cb^x$ with the points $(1, 6)$ and $(3, 24)$. $6 = Cb^1$ $[C = \frac{6}{b}]$ and $24 = Cb^3$ $\Rightarrow$ $24 = \left(\dfrac{6}{b}\right)b^3$ $\Rightarrow$ $4 = b^2$ $\Rightarrow$ $b = 2$ [since $b > 0$] and $C = \frac{6}{2} = 3$. The function is $f(x) = 3 \cdot 2^x$.

19. 2 ft = 24 in, $f(24) = 24^2$ in = 576 in = 48 ft. $g(24) = 2^{24}$ in = $2^{24}/(12 \cdot 5280)$ mi ≈ 265 mi

21. The graph of g finally surpasses that of f at $x \approx 35.8$.

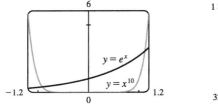 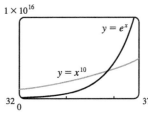

284 ☐ **CHAPTER 6** INVERSE FUNCTIONS

23. $\lim\limits_{x \to \infty} (1.001)^x = \infty$ by (3), since $1.001 > 1$.

25. Divide numerator and denominator by e^{3x}: $\lim\limits_{x \to \infty} \dfrac{e^{3x} - e^{-3x}}{e^{3x} + e^{-3x}} = \lim\limits_{x \to \infty} \dfrac{1 - e^{-6x}}{1 + e^{-6x}} = \dfrac{1 - 0}{1 + 0} = 1$

27. Let $t = 3/(2-x)$. As $x \to 2^+$, $t \to -\infty$. So $\lim\limits_{x \to 2^+} e^{3/(2-x)} = \lim\limits_{t \to -\infty} e^t = 0$ by (11).

29. Since $-1 \le \cos x \le 1$ and $e^{-2x} > 0$, we have $-e^{-2x} \le e^{-2x} \cos x \le e^{-2x}$. We know that $\lim\limits_{x \to \infty} (-e^{-2x}) = 0$ and $\lim\limits_{x \to \infty} (e^{-2x}) = 0$, so by the Squeeze Theorem, $\lim\limits_{x \to \infty} (e^{-2x} \cos x) = 0$.

31. $f(t) = -2e^t \;\Rightarrow\; f'(t) = -2(e^t) = -2e^t$

33. $f(x) = (3x^2 - 5x)e^x \;\overset{\text{PR}}{\Rightarrow}$

$f'(x) = (3x^2 - 5x)(e^x)' + e^x(3x^2 - 5x)' = (3x^2 - 5x)e^x + e^x(6x - 5)$
$= e^x[(3x^2 - 5x) + (6x - 5)] = e^x(3x^2 + x - 5)$

35. By (9), $y = e^{ax^3} \;\Rightarrow\; y' = e^{ax^3} \dfrac{d}{dx}(ax^3) = 3ax^2 e^{ax^3}$.

37. $y = e^{\tan\theta} \;\Rightarrow\; y' = e^{\tan\theta} \dfrac{d}{d\theta}(\tan\theta) = (\sec^2\theta)e^{\tan\theta}$

39. $f(x) = \dfrac{x^2 e^x}{x^2 + e^x} \;\overset{\text{QR}}{\Rightarrow}$

$f'(x) = \dfrac{(x^2 + e^x)[x^2 e^x + e^x(2x)] - x^2 e^x(2x + e^x)}{(x^2 + e^x)^2} = \dfrac{x^4 e^x + 2x^3 e^x + x^2 e^{2x} + 2xe^{2x} - 2x^3 e^x - x^2 e^{2x}}{(x^2 + e^x)^2}$

$= \dfrac{x^4 e^x + 2xe^{2x}}{(x^2 + e^x)^2} = \dfrac{xe^x(x^3 + 2e^x)}{(x^2 + e^x)^2}$

41. Using the Product Rule and the Chain Rule, $y = x^2 e^{-3x} \;\Rightarrow\;$
$y' = x^2 e^{-3x}(-3) + e^{-3x}(2x) = e^{-3x}(-3x^2 + 2x) = xe^{-3x}(2 - 3x)$.

43. $f(t) = e^{at} \sin bt \;\Rightarrow\; f'(t) = e^{at}(\cos bt) \cdot b + (\sin bt)e^{at} \cdot a = e^{at}(b \cos bt + a \sin bt)$

45. By (9), $F(t) = e^{t \sin 2t} \;\Rightarrow\;$
$F'(t) = e^{t \sin 2t}(t \sin 2t)' = e^{t \sin 2t}(t \cdot 2\cos 2t + \sin 2t \cdot 1) = e^{t \sin 2t}(2t \cos 2t + \sin 2t)$

47. $g(u) = e^{\sqrt{\sec u^2}} \;\Rightarrow\;$

$g'(u) = e^{\sqrt{\sec u^2}} \dfrac{d}{du}\sqrt{\sec u^2} = e^{\sqrt{\sec u^2}} \dfrac{1}{2}(\sec u^2)^{-1/2}\dfrac{d}{du}\sec u^2$

$= e^{\sqrt{\sec u^2}} \dfrac{1}{2\sqrt{\sec u^2}} \cdot \sec u^2 \tan u^2 \cdot 2u = u\sqrt{\sec u^2}\, \tan u^2\, e^{\sqrt{\sec u^2}}$

© 2021 Cengage Learning. All Rights Reserved. May not be scanned, copied, or duplicated, or posted to a publicly accessible website, in whole or in part.

49. $g(x) = \sin\left(\dfrac{e^x}{1+e^x}\right) \Rightarrow$

$g'(x) = \cos\left(\dfrac{e^x}{1+e^x}\right) \cdot \dfrac{(1+e^x)e^x - e^x(e^x)}{(1+e^x)^2} = \cos\left(\dfrac{e^x}{1+e^x}\right) \cdot \dfrac{e^x(1+e^x - e^x)}{(1+e^x)^2} = \dfrac{e^x}{(1+e^x)^2} \cos\left(\dfrac{e^x}{1+e^x}\right)$

51. $y = e^x \cos x + \sin x \Rightarrow y' = e^x(-\sin x) + (\cos x)(e^x) + \cos x = e^x(\cos x - \sin x) + \cos x$, so

$y'(0) = e^0(\cos 0 - \sin 0) + \cos 0 = 1(1-0) + 1 = 2$. An equation of the tangent line to the curve $y = e^x \cos x + \sin x$ at

the point $(0, 1)$ is $y - 1 = 2(x - 0)$ or $y = 2x + 1$.

53. $\dfrac{d}{dx}(e^{x/y}) = \dfrac{d}{dx}(x - y) \Rightarrow e^{x/y} \cdot \dfrac{d}{dx}\left(\dfrac{x}{y}\right) = 1 - y' \Rightarrow e^{x/y} \cdot \dfrac{y \cdot 1 - x \cdot y'}{y^2} = 1 - y' \Rightarrow$

$e^{x/y} \cdot \dfrac{1}{y} - \dfrac{xe^{x/y}}{y^2} \cdot y' = 1 - y' \Rightarrow y' - \dfrac{xe^{x/y}}{y^2} \cdot y' = 1 - \dfrac{e^{x/y}}{y} \Rightarrow y'\left(1 - \dfrac{xe^{x/y}}{y^2}\right) = \dfrac{y - e^{x/y}}{y} \Rightarrow$

$y' = \dfrac{\dfrac{y - e^{x/y}}{y}}{\dfrac{y^2 - xe^{x/y}}{y^2}} = \dfrac{y(y - e^{x/y})}{y^2 - xe^{x/y}}$

55. $y = e^x + e^{-x/2} \Rightarrow y' = e^x - \tfrac{1}{2}e^{-x/2} \Rightarrow y'' = e^x + \tfrac{1}{4}e^{-x/2}$, so

$2y'' - y' - y = 2\left(e^x + \tfrac{1}{4}e^{-x/2}\right) - \left(e^x - \tfrac{1}{2}e^{-x/2}\right) - \left(e^x + e^{-x/2}\right) = 0$.

57. $y = e^{rx} \Rightarrow y' = re^{rx} \Rightarrow y'' = r^2 e^{rx}$, so if $y = e^{rx}$ satisfies the differential equation $y'' + 6y' + 8y = 0$,

then $r^2 e^{rx} + 6re^{rx} + 8e^{rx} = 0$; that is, $e^{rx}(r^2 + 6r + 8) = 0$. Since $e^{rx} > 0$ for all x, we must have $r^2 + 6r + 8 = 0$,

or $(r+2)(r+4) = 0$, so $r = -2$ or -4.

59. $f(x) = e^{2x} \Rightarrow f'(x) = 2e^{2x} \Rightarrow f''(x) = 2 \cdot 2e^{2x} = 2^2 e^{2x} \Rightarrow$

$f'''(x) = 2^2 \cdot 2e^{2x} = 2^3 e^{2x} \Rightarrow \cdots \Rightarrow f^{(n)}(x) = 2^n e^{2x}$

61. (a) $f(x) = e^x + x$ is continuous on $\mathbb{R}$ and $f(-1) = e^{-1} - 1 < 0 < 1 = f(0)$, so by the Intermediate Value Theorem,

$e^x + x = 0$ has a solution in $(-1, 0)$.

(b) $f(x) = e^x + x \Rightarrow f'(x) = e^x + 1$, so $x_{n+1} = x_n - \dfrac{e^{x_n} + x_n}{e^{x_n} + 1}$. Using $x_1 = -0.5$, we get $x_2 \approx -0.566311$,

$x_3 \approx -0.567143 \approx x_4$, so the solution is -0.567143 to six decimal places.

63. Half of 76.0 RNA copies per mL, corresponding to $t = 1$, is 38.0 RNA copies per mL. Using the graph of V in Figure 11, we

estimate that it takes about 3.5 additional days for the patient's viral load to decrease to 38 RNA copies per mL.

65. (a) $\lim\limits_{t \to \infty} p(t) = \lim\limits_{t \to \infty} \dfrac{1}{1 + ae^{-kt}} = \dfrac{1}{1 + a \cdot 0} = 1$, since $k > 0 \Rightarrow -kt \to -\infty \Rightarrow e^{-kt} \to 0$. As time increases, the

proportion of the population that has heard the rumor approaches 1; that is, everyone in the population has heard the rumor.

(b) $p(t) = (1 + ae^{-kt})^{-1} \Rightarrow \dfrac{dp}{dt} = -(1 + ae^{-kt})^{-2}(-kae^{-kt}) = \dfrac{kae^{-kt}}{(1 + ae^{-kt})^2}$

(c)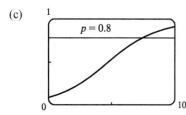

From the graph of $p(t) = (1 + 10e^{-0.5t})^{-1}$, it seems that $p(t) = 0.8$ (indicating that 80% of the population has heard the rumor) when $t \approx 7.4$ hours.

67. $f(x) = \dfrac{e^x}{1 + x^2}$, $[0, 3]$. $f'(x) = \dfrac{(1 + x^2)e^x - e^x(2x)}{(1 + x^2)^2} = \dfrac{e^x(x^2 - 2x + 1)}{(1 + x^2)^2} = \dfrac{e^x(x - 1)^2}{(1 + x^2)^2}$. $f'(x) = 0 \Rightarrow$

$(x - 1)^2 = 0 \Leftrightarrow x = 1$. $f'(x)$ exists for all real numbers since $1 + x^2$ is never equal to 0. $f(0) = 1$,

$f(1) = e/2 \approx 1.359$, and $f(3) = e^3/10 \approx 2.009$. So $f(3) = e^3/10$ is the absolute maximum value and $f(0) = 1$ is the absolute minimum value.

69. $f(x) = x - e^x \Rightarrow f'(x) = 1 - e^x = 0 \Leftrightarrow e^x = 1 \Leftrightarrow x = 0$. Now $f'(x) > 0$ for all $x < 0$ and $f'(x) < 0$ for all $x > 0$, so the absolute maximum value is $f(0) = 0 - 1 = -1$.

71. (a) $f(x) = xe^{2x} \Rightarrow f'(x) = x(2e^{2x}) + e^{2x}(1) = e^{2x}(2x + 1)$. Thus, $f'(x) > 0$ if $x > -\tfrac{1}{2}$ and $f'(x) < 0$ if $x < -\tfrac{1}{2}$.

So f is increasing on $\left(-\tfrac{1}{2}, \infty\right)$ and f is decreasing on $\left(-\infty, -\tfrac{1}{2}\right)$.

(b) $f''(x) = e^{2x}(2) + (2x + 1) \cdot 2e^{2x} = 2e^{2x}[1 + (2x + 1)] = 2e^{2x}(2x + 2) = 4e^{2x}(x + 1)$. $f''(x) > 0 \Leftrightarrow x > -1$ and $f''(x) < 0 \Leftrightarrow x < -1$. Thus, f is concave upward on $(-1, \infty)$ and f is concave downward on $(-\infty, -1)$.

(c) There is an inflection point at $\left(-1, -e^{-2}\right)$, or $\left(-1, -1/e^2\right)$.

73. $y = f(x) = e^{-1/(x+1)}$ **A.** $D = \{x \mid x \neq -1\} = (-\infty, -1) \cup (-1, \infty)$ **B.** No x-intercept; y-intercept $= f(0) = e^{-1}$

C. No symmetry **D.** $\lim\limits_{x \to \pm\infty} e^{-1/(x+1)} = 1$ since $-1/(x+1) \to 0$, so $y = 1$ is a HA. $\lim\limits_{x \to -1^+} e^{-1/(x+1)} = 0$ since

$-1/(x+1) \to -\infty$, $\lim\limits_{x \to -1^-} e^{-1/(x+1)} = \infty$ since $-1/(x+1) \to \infty$, so $x = -1$ is a VA.

E. $f'(x) = e^{-1/(x+1)}/(x + 1)^2 \Rightarrow f'(x) > 0$ for all x except 1, so

f is increasing on $(-\infty, -1)$ and $(-1, \infty)$. **F.** No extreme values **H.**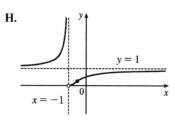

G. $f''(x) = \dfrac{e^{-1/(x+1)}}{(x + 1)^4} + \dfrac{e^{-1/(x+1)}(-2)}{(x + 1)^3} = -\dfrac{e^{-1/(x+1)}(2x + 1)}{(x + 1)^4} \Rightarrow$

$f''(x) > 0 \Leftrightarrow 2x + 1 < 0 \Leftrightarrow x < -\tfrac{1}{2}$, so f is CU on $(-\infty, -1)$

and $\left(-1, -\tfrac{1}{2}\right)$, and CD on $\left(-\tfrac{1}{2}, \infty\right)$. f has an IP at $\left(-\tfrac{1}{2}, e^{-2}\right)$.

75. $y = 1/(1 + e^{-x})$ **A.** $D = \mathbb{R}$ **B.** No x-intercept; y-intercept $= f(0) = \tfrac{1}{2}$. **C.** No symmetry

D. $\lim\limits_{x \to \infty} 1/(1 + e^{-x}) = \tfrac{1}{1+0} = 1$ and $\lim\limits_{x \to -\infty} 1/(1 + e^{-x}) = 0$ since $\lim\limits_{x \to -\infty} e^{-x} = \infty$, so f has horizontal asymptotes

$y = 0$ and $y = 1$.   **E.** $f'(x) = -(1+e^{-x})^{-2}(-e^{-x}) = e^{-x}/(1+e^{-x})^2$. This is positive for all x, so f is increasing on $\mathbb{R}$.

F. No extreme values   **G.** $f''(x) = \dfrac{(1+e^{-x})^2(-e^{-x}) - e^{-x}(2)(1+e^{-x})(-e^{-x})}{(1+e^{-x})^4} = \dfrac{e^{-x}(e^{-x}-1)}{(1+e^{-x})^3}$

The second factor in the numerator is negative for $x > 0$ and positive for $x < 0$,   **H.**
and the other factors are always positive, so f is CU on $(-\infty, 0)$ and CD
on $(0, \infty)$. IP at $\left(0, \tfrac{1}{2}\right)$

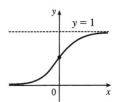

77. $f(x) = e^{x^3-x} \to 0$ as $x \to -\infty$, and
$f(x) \to \infty$ as $x \to \infty$. From the graph,
it appears that f has a local minimum of
about $f(0.58) = 0.68$, and a local
maximum of about $f(-0.58) = 1.47$.
To find the exact values, we calculate

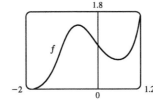
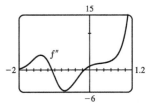

$f'(x) = (3x^2 - 1)e^{x^3-x}$, which is 0 when $3x^2 - 1 = 0 \Leftrightarrow x = \pm\tfrac{1}{\sqrt{3}}$. The negative solution corresponds to the local

maximum $f\left(-\tfrac{1}{\sqrt{3}}\right) = e^{(-1/\sqrt{3})^3 - (-1/\sqrt{3})} = e^{2\sqrt{3}/9}$, and the positive solution corresponds to the local minimum

$f\left(\tfrac{1}{\sqrt{3}}\right) = e^{(1/\sqrt{3})^3 - (1/\sqrt{3})} = e^{-2\sqrt{3}/9}$. To estimate the inflection points, we calculate and graph

$f''(x) = \dfrac{d}{dx}\left[(3x^2-1)e^{x^3-x}\right] = (3x^2-1)e^{x^3-x}(3x^2-1) + e^{x^3-x}(6x) = e^{x^3-x}(9x^4 - 6x^2 + 6x + 1)$.

From the graph, it appears that $f''(x)$ changes sign (and thus f has inflection points) at $x \approx -0.15$ and $x \approx -1.09$. From the
graph of f, we see that these x-values correspond to inflection points at about $(-0.15, 1.15)$ and $(-1.09, 0.82)$.

79. Let $a = 0.135$ and $b = -2.802$. Then $C(t) = ate^{bt} \Rightarrow C'(t) = a(t \cdot e^{bt} \cdot b + e^{bt} \cdot 1) = ae^{bt}(bt + 1)$. $C'(t) = 0 \Leftrightarrow$

$bt + 1 = 0 \Leftrightarrow t = -\dfrac{1}{b} \approx 0.36$ h. $C(0) = 0, C(-1/b) = -\dfrac{a}{b}e^{-1} = -\dfrac{a}{be} \approx 0.0177$, and $C(3) = 3ae^{3b} \approx 0.00009$.

The maximum average BAC during the first three hours is about 0.0177 g/dL and it occurs at approximately 0.36 h
(21.4 min).

81. $\displaystyle\int_0^1 (x^e + e^x)\,dx = \left[\dfrac{x^{e+1}}{e+1} + e^x\right]_0^1 = \left(\dfrac{1}{e+1} + e\right) - (0+1) = \dfrac{1}{e+1} + e - 1$

83. $\displaystyle\int_0^2 \dfrac{dx}{e^{\pi x}} = \int_0^2 e^{-\pi x}\,dx = \left[-\dfrac{1}{\pi}e^{-\pi x}\right]_0^2 = -\dfrac{1}{\pi}e^{-2\pi} + \dfrac{1}{\pi}e^0 = \dfrac{1}{\pi}(1 - e^{-2\pi})$

85. Let $u = 1 + e^x$. Then $du = e^x\,dx$, so $\int e^x\sqrt{1+e^x}\,dx = \int \sqrt{u}\,du = \tfrac{2}{3}u^{3/2} + C = \tfrac{2}{3}(1+e^x)^{3/2} + C$.

87. $\int (e^x + e^{-x})^2\,dx = \int (e^{2x} + 2 + e^{-2x})\,dx = \tfrac{1}{2}e^{2x} + 2x - \tfrac{1}{2}e^{-2x} + C$

89. Let $x = 1 - e^u$. Then $dx = -e^u\,du$ and $e^u\,du = -dx$, so

$\displaystyle\int \dfrac{e^u}{(1-e^u)^2}\,du = \int \dfrac{1}{x^2}(-dx) = -\int x^{-2}\,dx = -(-x^{-1}) + C = \dfrac{1}{x} + C = \dfrac{1}{1-e^u} + C$.

91. Let $u = 1/x$, so $du = -1/x^2\,dx$. When $x = 1$, $u = 1$; when $x = 2$, $u = \frac{1}{2}$. Thus,
$$\int_1^2 \frac{e^{1/x}}{x^2}\,dx = \int_1^{1/2} e^u\,(-du) = -\left[e^u\right]_1^{1/2} = -(e^{1/2} - e) = e - \sqrt{e}.$$

93. $f_{\text{avg}} = \frac{1}{2-0}\int_0^2 2xe^{-x^2}\,dx$
$= \frac{1}{2}\left[-e^{-x^2}\right]_0^2 = \frac{1}{2}(-e^{-4} + 1)$

95. Area $= \int_0^1 \left(e^{3x} - e^x\right) dx = \left[\frac{1}{3}e^{3x} - e^x\right]_0^1 = \left(\frac{1}{3}e^3 - e\right) - \left(\frac{1}{3} - 1\right) = \frac{1}{3}e^3 - e + \frac{2}{3} \approx 4.644$

97. $V = \int_0^1 \pi(e^x)^2\,dx = \pi\int_0^1 e^{2x}\,dx = \frac{1}{2}\pi\left[e^{2x}\right]_0^1 = \frac{\pi}{2}\left(e^2 - 1\right)$

99. First Figure Let $u = \sqrt{x}$, so $x = u^2$ and $dx = 2u\,du$. When $x = 0$, $u = 0$; when $x = 1$, $u = 1$. Thus,
$A_1 = \int_0^1 e^{\sqrt{x}}\,dx = \int_0^1 e^u\,(2u\,du) = 2\int_0^1 ue^u\,du.$

Second Figure $A_2 = \int_0^1 2xe^x\,dx = 2\int_0^1 ue^u\,du.$

Third Figure Let $u = \sin x$, so $du = \cos x\,dx$. When $x = 0$, $u = 0$; when $x = \frac{\pi}{2}$, $u = 1$. Thus,
$A_3 = \int_0^{\pi/2} e^{\sin x} \sin 2x\,dx = \int_0^{\pi/2} e^{\sin x}(2\sin x\,\cos x)\,dx = \int_0^1 e^u\,(2u\,du) = 2\int_0^1 ue^u\,du.$

Since $A_1 = A_2 = A_3$, all three areas are equal.

101. The rate is measured in liters per minute. Integrating from $t = 0$ minutes to $t = 60$ minutes will give us the total amount of oil that leaks out (in liters) during the first hour.
$\int_0^{60} r(t)\,dt = \int_0^{60} 100e^{-0.01t}\,dt \quad [u = -0.01t, du = -0.01dt]$
$= 100\int_0^{-0.6} e^u\,(-100\,du) = -10{,}000\left[e^u\right]_0^{-0.6} = -10{,}000(e^{-0.6} - 1) \approx 4511.9 \approx 4512\,\text{L}$

103. $\int_0^{30} u(t)\,dt = \int_0^{30} \frac{r}{V}C_0 e^{-rt/V}\,dt = C_0\int_1^{e^{-30r/V}}(-dx) \quad \left[\begin{array}{l}x = e^{-rt/V},\\ dx = -\frac{r}{V}e^{-rt/V}\,dt\end{array}\right]$
$= C_0\left[-x\right]_1^{e^{-30r/V}} = C_0(-e^{-30r/V} + 1)$

The integral $\int_0^{30} u(t)\,dt$ represents the total amount of urea removed from the blood in the first 30 minutes of dialysis.

105. We use Theorem 6.1.7. Note that $f(0) = 3 + 0 + e^0 = 4$, so $f^{-1}(4) = 0$. Also $f'(x) = 1 + e^x$. Therefore,
$\left(f^{-1}\right)'(4) = \frac{1}{f'(f^{-1}(4))} = \frac{1}{f'(0)} = \frac{1}{1 + e^0} = \frac{1}{2}.$

107.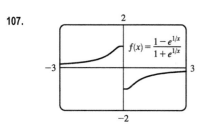

From the graph, it appears that f is an odd function (f is undefined for $x = 0$).
To prove this, we must show that $f(-x) = -f(x)$.

$f(-x) = \dfrac{1 - e^{1/(-x)}}{1 + e^{1/(-x)}} = \dfrac{1 - e^{(-1/x)}}{1 + e^{(-1/x)}} = \dfrac{1 - \dfrac{1}{e^{1/x}}}{1 + \dfrac{1}{e^{1/x}}} \cdot \dfrac{e^{1/x}}{e^{1/x}} = \dfrac{e^{1/x} - 1}{e^{1/x} + 1}$

$= -\dfrac{1 - e^{1/x}}{1 + e^{1/x}} = -f(x)$

so f is an odd function.

6.3 Logarithmic Functions

1. (a) It is defined as the inverse of the exponential function with base b, that is, $\log_b x = y \Leftrightarrow b^y = x$.

(b) $(0, \infty)$ (c) $\mathbb{R}$ (d) See Figure 1.

3. (a) $\log_3 81 = \log_3 3^4 = 4$ (b) $\log_3\left(\frac{1}{81}\right) = \log_3 3^{-4} = -4$ (c) $\log_9 3 = \log_9 9^{1/2} = \frac{1}{2}$

5. (a) $\log_2 30 - \log_2 15 = \log_2\left(\frac{30}{15}\right) = \log_2 2 = 1$

(b) $\log_3 10 - \log_3 5 - \log_3 18 = \log_3\left(\frac{10}{5}\right) - \log_3 18 = \log_3 2 - \log_3 18 = \log_3\left(\frac{2}{18}\right) = \log_3\left(\frac{1}{9}\right)$
$= \log_3 3^{-2} = -2$

(c) $2\log_5 100 - 4\log_5 50 = \log_5 100^2 - \log_5 50^4 = \log_5\left(\frac{100^2}{50^4}\right) = \log_5\left(\frac{10^4}{5^4 \cdot 10^4}\right) = \log_5 5^{-4} = -4$

7. (a) $\log_{10}\left(x^2 y^3 z\right) = \log_{10} x^2 + \log_{10} y^3 + \log_{10} z$ [Law 1]
$\qquad\qquad\qquad = 2\log_{10} x + 3\log_{10} y + \log_{10} z$ [Law 3]

(b) $\ln\left(\dfrac{x^4}{\sqrt{x^2-4}}\right) = \ln x^4 - \ln(x^2-4)^{1/2}$ [Law 2]
$\qquad\qquad\qquad = 4\ln x - \frac{1}{2}\ln[(x+2)(x-2)]$ [Law 3]
$\qquad\qquad\qquad = 4\ln x - \frac{1}{2}[\ln(x+2) + \ln(x-2)]$ [Law 1]
$\qquad\qquad\qquad = 4\ln x - \frac{1}{2}\ln(x+2) - \frac{1}{2}\ln(x-2)$

9. (a) $\log_{10} 20 - \frac{1}{3}\log_{10} 1000 = \log_{10} 20 - \log_{10} 1000^{1/3} = \log_{10} 20 - \log_{10} \sqrt[3]{1000}$
$= \log_{10} 20 - \log_{10} 10 = \log_{10}\left(\frac{20}{10}\right) = \log_{10} 2$

(b) $\ln a - 2\ln b + 3\ln c = \ln a - \ln b^2 + \ln c^3 = \ln \dfrac{a}{b^2} + \ln c^3 = \ln \dfrac{ac^3}{b^2}$

11. (a) $3\ln(x-2) - \ln(x^2-5x+6) + 2\ln(x-3) = \ln(x-2)^3 - \ln[(x-2)(x-3)] + \ln(x-3)^2$
$= \ln\left[\dfrac{(x-2)^3(x-3)^2}{(x-2)(x-3)}\right] = \ln[(x-2)^2(x-3)]$

(b) $c\log_a x - d\log_a y + \log_a z = \log_a x^c - \log_a y^d + \log_a z = \log_a\left(\dfrac{x^c z}{y^d}\right)$

13. (a) $\log_3 12 = \dfrac{\ln 12}{\ln 3} \approx 2.261860$

(b) $\log_{12} 6 = \dfrac{\ln 6}{\ln 12} \approx 0.721057$

15. To graph these functions, we use $\log_{1.5} x = \dfrac{\ln x}{\ln 1.5}$ and $\log_{50} x = \dfrac{\ln x}{\ln 50}$.
These graphs all approach $-\infty$ as $x \to 0^+$, and they all pass through the point $(1, 0)$. Also, they are all increasing, and all approach ∞ as $x \to \infty$. The functions with larger bases increase extremely slowly, and the ones with smaller bases do so somewhat more quickly. The functions with large bases approach the y-axis more closely as $x \to 0^+$.

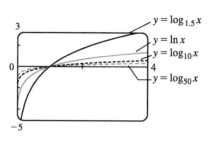

17. 3 ft = 36 in, so we need x such that $\log_2 x = 36 \;\Leftrightarrow\; x = 2^{36} = 68{,}719{,}476{,}736$. In miles, this is
$68{,}719{,}476{,}736 \text{ in} \cdot \dfrac{1 \text{ ft}}{12 \text{ in}} \cdot \dfrac{1 \text{ mi}}{5280 \text{ ft}} \approx 1{,}084{,}587.7 \text{ mi}$.

19. (a) Shift the graph of $y = \log_{10} x$ five units to the left to obtain the graph of $y = \log_{10}(x+5)$. Note the vertical asymptote of $x = -5$.

(b) Reflect the graph of $y = \ln x$ about the x-axis to obtain the graph of $y = -\ln x$.

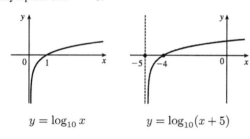
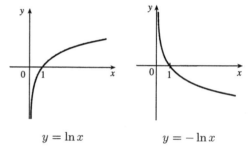

21. (a) The domain of $f(x) = \ln x + 2$ is $x > 0$ and the range is $\mathbb{R}$.

(b) $y = 0 \;\Rightarrow\; \ln x + 2 = 0 \;\Rightarrow\; \ln x = -2 \;\Rightarrow\; x = e^{-2}$

(c) We shift the graph of $y = \ln x$ two units upward.

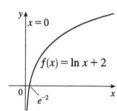

23. (a) $\ln(4x+2) = 3 \;\Rightarrow\; e^{\ln(4x+2)} = e^3 \;\Rightarrow\; 4x + 2 = e^3 \;\Rightarrow\; 4x = e^3 - 2 \;\Rightarrow\; x = \tfrac{1}{4}(e^3 - 2) \approx 4.521$

(b) $e^{2x-3} = 12 \;\Rightarrow\; \ln e^{2x-3} = \ln 12 \;\Rightarrow\; 2x - 3 = \ln 12 \;\Rightarrow\; 2x = 3 + \ln 12 \;\Rightarrow\; x = \tfrac{1}{2}(3 + \ln 12) \approx 2.742$

25. (a) $\ln x + \ln(x-1) = 0 \;\Rightarrow\; \ln[x(x-1)] = 0 \;\Rightarrow\; e^{\ln[x^2 - x]} = e^0 \;\Rightarrow\; x^2 - x = 1 \;\Rightarrow\; x^2 - x - 1 = 0$. The quadratic formula gives $x = \dfrac{1 \pm \sqrt{(-1)^2 - 4(1)(-1)}}{2(1)} = \dfrac{1 \pm \sqrt{5}}{2}$, but we note that $\ln \dfrac{1 - \sqrt{5}}{2}$ is undefined because $\dfrac{1 - \sqrt{5}}{2} < 0$. Thus, $x = \dfrac{1 + \sqrt{5}}{2} \approx 1.618$.

(b) $5^{1-2x} = 9 \;\Rightarrow\; \ln 5^{1-2x} = \ln 9 \;\Rightarrow\; (1 - 2x)\ln 5 = \ln 9 \;\Rightarrow\; 1 - 2x = \dfrac{\ln 9}{\ln 5} \;\Rightarrow\; x = \dfrac{1}{2} - \dfrac{\ln 9}{2 \ln 5} \approx -0.183$

27. (a) $e^{2x} - 3e^x + 2 = 0 \;\Leftrightarrow\; (e^x - 1)(e^x - 2) = 0 \;\Leftrightarrow\; e^x = 1$ or $e^x = 2 \;\Leftrightarrow\; x = \ln 1$ or $x = \ln 2$, so $x = 0$ or $\ln 2$.

(b) $e^{e^x} = 10 \;\Leftrightarrow\; \ln\left(e^{e^x}\right) = \ln 10 \;\Leftrightarrow\; e^x \ln e = e^x = \ln 10 \;\Leftrightarrow\; \ln e^x = \ln(\ln 10) \;\Leftrightarrow\; x = \ln \ln 10$

29. (a) $\ln(1+x^3) - 4 = 0 \;\Leftrightarrow\; \ln(1+x^3) = 4 \;\Leftrightarrow\; 1+x^3 = e^4 \;\Leftrightarrow\; x^3 = e^4 - 1 \;\Leftrightarrow\; x = \sqrt[3]{e^4 - 1} \approx 3.7704$.

(b) $2e^{1/x} = 42 \;\Leftrightarrow\; e^{1/x} = 21 \;\Leftrightarrow\; \dfrac{1}{x} = \ln 21 \;\Leftrightarrow\; x = \dfrac{1}{\ln 21} \approx 0.3285$.

31. (a) $\ln x < 0 \;\Rightarrow\; x < e^0 \;\Rightarrow\; x < 1$. Since the domain of $f(x) = \ln x$ is $x > 0$, the solution of the original inequality is $0 < x < 1$.

(b) $e^x > 5 \;\Rightarrow\; \ln e^x > \ln 5 \;\Rightarrow\; x > \ln 5$

33. If I is the intensity of the 1989 San Francisco earthquake, then $\log_{10}(I/S) = 7.1 \;\Rightarrow\;$

$\log_{10}(16I/S) = \log_{10} 16 + \log_{10}(I/S) = \log_{10} 16 + 7.1 \approx 8.3$.

35. (a) $n = f(t) = 100 \cdot 2^{t/3} \;\Rightarrow\; \dfrac{n}{100} = 2^{t/3} \;\Rightarrow\; \log_2\left(\dfrac{n}{100}\right) = \dfrac{t}{3} \;\Rightarrow\; t = 3\log_2\left(\dfrac{n}{100}\right)$. Using the Change of Base Formula, we can write this as $t = f^{-1}(n) = 3 \cdot \dfrac{\ln(n/100)}{\ln 2}$. This function tells us how long it will take to obtain n bacteria (given the number n).

(b) $n = 50{,}000 \;\Rightarrow\; t = f^{-1}(50{,}000) = 3 \cdot \dfrac{\ln\left(\frac{50{,}000}{100}\right)}{\ln 2} = 3\left(\dfrac{\ln 500}{\ln 2}\right) \approx 26.9$ hours

37. $\lim\limits_{x \to 1^+} \ln(\sqrt{x} - 1) = -\infty$ since $\sqrt{x} - 1 \to 0^+$ as $x \to 1^+$.

39. $\lim\limits_{x \to 0} \ln(\cos x) = \ln 1 = 0$. [$\ln(\cos x)$ is continuous at $x = 0$ since it is the composite of two continuous functions.]

41. $\lim\limits_{x \to \infty} [\ln(1+x^2) - \ln(1+x)] = \lim\limits_{x \to \infty} \ln \dfrac{1+x^2}{1+x} = \ln\left(\lim\limits_{x \to \infty} \dfrac{1+x^2}{1+x}\right) = \ln\left(\lim\limits_{x \to \infty} \dfrac{\frac{1}{x} + x}{\frac{1}{x} + 1}\right) = \infty$, since the limit in parentheses is ∞.

43. $f(x) = \ln(4 - x^2)$.

$D_f = \{x \mid 4 - x^2 > 0\} = \{x \mid x^2 < 4\} = \{x \mid |x| < 2\} = (-2, 2)$

45. (a) For $f(x) = \sqrt{3 - e^{2x}}$, we must have $3 - e^{2x} \geq 0 \;\Rightarrow\; e^{2x} \leq 3 \;\Rightarrow\; 2x \leq \ln 3 \;\Rightarrow\; x \leq \tfrac{1}{2} \ln 3$.

Thus, the domain of f is $(-\infty, \tfrac{1}{2} \ln 3]$.

(b) $y = f(x) = \sqrt{3 - e^{2x}}$ [note that $y \geq 0$] $\;\Rightarrow\; y^2 = 3 - e^{2x} \;\Rightarrow\; e^{2x} = 3 - y^2 \;\Rightarrow\; 2x = \ln(3 - y^2) \;\Rightarrow\;$

$x = \tfrac{1}{2} \ln(3 - y^2)$. Interchange x and y: $y = \tfrac{1}{2} \ln(3 - x^2)$. So $f^{-1}(x) = \tfrac{1}{2} \ln(3 - x^2)$. For the domain of f^{-1},

we must have $3 - x^2 > 0 \;\Rightarrow\; x^2 < 3 \;\Rightarrow\; |x| < \sqrt{3} \;\Rightarrow\; -\sqrt{3} < x < \sqrt{3} \;\Rightarrow\; 0 \leq x < \sqrt{3}$ since $x \geq 0$. Note that the domain of f^{-1}, $[0, \sqrt{3})$, equals the range of f.

47. (a) We must have $e^x - 3 > 0 \;\Leftrightarrow\; e^x > 3 \;\Leftrightarrow\; x > \ln 3$. Thus, the domain of $f(x) = \ln(e^x - 3)$ is $(\ln 3, \infty)$.

(b) $y = \ln(e^x - 3) \;\Rightarrow\; e^y = e^x - 3 \;\Rightarrow\; e^x = e^y + 3 \;\Rightarrow\; x = \ln(e^y + 3)$, so $f^{-1}(x) = \ln(e^x + 3)$.

Now $e^x + 3 > 0 \;\Rightarrow\; e^x > -3$, which is true for any real x, so the domain of f^{-1} is $\mathbb{R}$.

49. We solve $y = 3\ln(x-2)$ for x: $y/3 = \ln(x-2) \Rightarrow e^{y/3} = x - 2 \Rightarrow x = 2 + e^{y/3}$. Interchanging x and y gives the inverse function $y = 2 + e^{x/3}$.

51. We solve $y = e^{1-x}$ for x: $\ln y = \ln e^{1-x} \Rightarrow \ln y = 1 - x \Rightarrow x = 1 - \ln y$. Interchanging x and y gives the inverse function $y = 1 - \ln x$.

53. $y = f(x) = 3^{2x-4} \Rightarrow \log_3 y = 2x - 4 \Rightarrow 2x = \log_3 y + 4 \Rightarrow x = \frac{1}{2}\log_3 y + 2$. Interchange x and y: $y = \frac{1}{2}\log_3 x + 2$. So $f^{-1}(x) = \frac{1}{2}\log_3 x + 2$.

55. $f(x) = e^{3x} - e^x \Rightarrow f'(x) = 3e^{3x} - e^x$. Thus, $f'(x) > 0 \Leftrightarrow 3e^{3x} > e^x \Leftrightarrow \dfrac{3e^{3x}}{e^x} > \dfrac{e^x}{e^x} \Leftrightarrow 3e^{2x} > 1 \Leftrightarrow e^{2x} > \frac{1}{3} \Leftrightarrow 2x > \ln\left(\frac{1}{3}\right) = -\ln 3 \Leftrightarrow x > -\frac{1}{2}\ln 3$, so f is increasing on $\left(-\frac{1}{2}\ln 3, \infty\right)$.

57. (a) We have to show that $-f(x) = f(-x)$.

$$-f(x) = -\ln\left(x + \sqrt{x^2+1}\right) = \ln\left(\left(x + \sqrt{x^2+1}\right)^{-1}\right) = \ln\dfrac{1}{x + \sqrt{x^2+1}}$$

$$= \ln\left(\dfrac{1}{x+\sqrt{x^2+1}} \cdot \dfrac{x - \sqrt{x^2+1}}{x - \sqrt{x^2+1}}\right) = \ln\dfrac{x - \sqrt{x^2+1}}{x^2 - x^2 - 1} = \ln\left(\sqrt{x^2+1} - x\right) = f(-x)$$

Thus, f is an odd function.

(b) Let $y = \ln\left(x + \sqrt{x^2+1}\right)$. Then $e^y = x + \sqrt{x^2+1} \Leftrightarrow (e^y - x)^2 = x^2 + 1 \Leftrightarrow e^{2y} - 2xe^y + x^2 = x^2 + 1 \Leftrightarrow 2xe^y = e^{2y} - 1 \Leftrightarrow x = \dfrac{e^{2y}-1}{2e^y} = \frac{1}{2}(e^y - e^{-y})$. Thus, the inverse function is $f^{-1}(x) = \frac{1}{2}(e^x - e^{-x})$.

59. $x^{1/\ln x} = 2 \Rightarrow \ln(x^{1/\ln x}) = \ln(2) \Rightarrow \dfrac{1}{\ln x} \cdot \ln x = \ln 2 \Rightarrow 1 = \ln 2$, a contradiction, so the given equation has no solution. The function $f(x) = x^{1/\ln x} = (e^{\ln x})^{1/\ln x} = e^1 = e$ for all $x > 0$, so the function $f(x) = x^{1/\ln x}$ is the constant function $f(x) = e$.

61. (a) Let $\varepsilon > 0$ be given. We need N such that $|b^x - 0| < \varepsilon$ when $x < N$. But $b^x < \varepsilon \Leftrightarrow x < \log_b \varepsilon$. Let $N = \log_b \varepsilon$. Then $x < N \Rightarrow x < \log_b \varepsilon \Rightarrow |b^x - 0| = b^x < \varepsilon$, so $\lim\limits_{x \to -\infty} b^x = 0$.

(b) Let $M > 0$ be given. We need N such that $b^x > M$ when $x > N$. But $b^x > M \Leftrightarrow x > \log_b M$. Let $N = \log_b M$. Then $x > N \Rightarrow x > \log_b M \Rightarrow b^x > M$, so $\lim\limits_{x \to \infty} b^x = \infty$.

63. $\ln(x^2 - 2x - 2) \leq 0 \Rightarrow 0 < x^2 - 2x - 2 \leq 1$. Now $x^2 - 2x - 2 \leq 1$ gives $x^2 - 2x - 3 \leq 0$ and hence $(x-3)(x+1) \leq 0$. So $-1 \leq x \leq 3$. Now $0 < x^2 - 2x - 2 \Rightarrow x < 1 - \sqrt{3}$ or $x > 1 + \sqrt{3}$. Therefore, $\ln(x^2 - 2x - 2) \leq 0 \Leftrightarrow -1 \leq x < 1 - \sqrt{3}$ or $1 + \sqrt{3} < x \leq 3$.

6.4 Derivatives of Logarithmic Functions

1. The differentiation formula for logarithmic functions, $\frac{d}{dx}(\log_b x) = \frac{1}{x \ln b}$, is simplest when $b = e$ because $\ln e = 1$.

3. $f(x) = \ln(x^2 + 3x + 5)$ $\Rightarrow$ $f'(x) = \frac{1}{x^2 + 3x + 5} \cdot \frac{d}{dx}(x^2 + 3x + 5) = \frac{1}{x^2 + 3x + 5} \cdot (2x + 3) = \frac{2x + 3}{x^2 + 3x + 5}$

5. $f(x) = \sin(\ln x)$ $\Rightarrow$ $f'(x) = \cos(\ln x) \cdot \frac{d}{dx} \ln x = \cos(\ln x) \cdot \frac{1}{x} = \frac{\cos(\ln x)}{x}$

7. $f(x) = \ln \frac{1}{x}$ $\Rightarrow$ $f'(x) = \frac{1}{1/x} \frac{d}{dx}\left(\frac{1}{x}\right) = x\left(-\frac{1}{x^2}\right) = -\frac{1}{x}$.

Another solution: $f(x) = \ln \frac{1}{x} = \ln 1 - \ln x = -\ln x$ $\Rightarrow$ $f'(x) = -\frac{1}{x}$.

9. $g(x) = \ln(xe^{-2x}) = \ln x + \ln e^{-2x} = \ln x - 2x$ $\Rightarrow$ $g'(x) = \frac{1}{x} - 2$

11. $F(t) = (\ln t)^2 \sin t$ $\Rightarrow$ $F'(t) = (\ln t)^2 \cos t + \sin t \cdot 2 \ln t \cdot \frac{1}{t} = \ln t \left(\ln t \cos t + \frac{2 \sin t}{t}\right)$

13. $y = \log_8(x^2 + 3x)$ $\Rightarrow$ $y' = \frac{1}{(x^2 + 3x) \ln 8} \cdot \frac{d}{dx}(x^2 + 3x) = \frac{1}{(x^2 + 3x) \ln 8} \cdot (2x + 3) = \frac{2x + 3}{(x^2 + 3x) \ln 8}$

15. $f(u) = \frac{\ln u}{1 + \ln(2u)}$ $\Rightarrow$

$f'(u) = \frac{[1 + \ln(2u)] \cdot \frac{1}{u} - \ln u \cdot \frac{1}{2u} \cdot 2}{[1 + \ln(2u)]^2} = \frac{\frac{1}{u}[1 + \ln(2u) - \ln u]}{[1 + \ln(2u)]^2} = \frac{1 + (\ln 2 + \ln u) - \ln u}{u[1 + \ln(2u)]^2} = \frac{1 + \ln 2}{u[1 + \ln(2u)]^2}$

17. $f(x) = x^5 + 5^x$ $\Rightarrow$ $f'(x) = 5x^4 + 5^x \ln 5$

19. $T(z) = 2^z \log_2 z$ $\Rightarrow$ $T'(z) = 2^z \frac{1}{z \ln 2} + \log_2 z \cdot 2^z \ln 2 = 2^z \left(\frac{1}{z \ln 2} + \log_2 z \, (\ln 2)\right)$.

Note that $\log_2 z \, (\ln 2) = \frac{\ln z}{\ln 2}(\ln 2) = \ln z$ by the change of base formula. Thus, $T'(z) = 2^z \left(\frac{1}{z \ln 2} + \ln z\right)$.

21. $g(t) = \ln \frac{t(t^2 + 1)^4}{\sqrt[3]{2t - 1}} = \ln t + \ln(t^2 + 1)^4 - \ln \sqrt[3]{2t - 1} = \ln t + 4\ln(t^2 + 1) - \frac{1}{3}\ln(2t - 1)$ $\Rightarrow$

$g'(t) = \frac{1}{t} + 4 \cdot \frac{1}{t^2 + 1} \cdot 2t - \frac{1}{3} \cdot \frac{1}{2t - 1} \cdot 2 = \frac{1}{t} + \frac{8t}{t^2 + 1} - \frac{2}{3(2t - 1)}$

23. $y = \ln|3 - 2x^5|$ $\Rightarrow$ $y' = \frac{1}{3 - 2x^5} \cdot (-10x^4) = \frac{-10x^4}{3 - 2x^5}$

25. $y = \tan[\ln(ax + b)]$ $\Rightarrow$ $y' = \sec^2[\ln(ax + b)] \cdot \frac{1}{ax + b} \cdot a = \sec^2[\ln(ax + b)]\frac{a}{ax + b}$

27. Using Formula 7 and the Chain Rule, $G(x) = 4^{C/x}$ $\Rightarrow$

$G'(x) = 4^{C/x} (\ln 4) \frac{d}{dx} \frac{C}{x} \quad \left[\frac{C}{x} = Cx^{-1}\right] = 4^{C/x} (\ln 4)\left(-Cx^{-2}\right) = -C(\ln 4)\frac{4^{C/x}}{x^2}$.

29. $\dfrac{d}{dx}\ln\left(x+\sqrt{x^2+1}\right) = \dfrac{1}{x+\sqrt{x^2+1}}\cdot\dfrac{d}{dx}\left(x+\sqrt{x^2+1}\right) = \dfrac{1}{x+\sqrt{x^2+1}}\cdot\left(1+\dfrac{2x}{2\sqrt{x^2+1}}\right)$

$= \dfrac{1}{x+\sqrt{x^2+1}}\cdot\left(\dfrac{\sqrt{x^2+1}}{\sqrt{x^2+1}}+\dfrac{x}{\sqrt{x^2+1}}\right) = \dfrac{1}{x+\sqrt{x^2+1}}\cdot\left(\dfrac{x+\sqrt{x^2+1}}{\sqrt{x^2+1}}\right) = \dfrac{1}{\sqrt{x^2+1}}$

31. $y = \sqrt{x}\ln x \;\Rightarrow\; y' = \sqrt{x}\cdot\dfrac{1}{x}+(\ln x)\dfrac{1}{2\sqrt{x}} = \dfrac{2+\ln x}{2\sqrt{x}} \;\Rightarrow\;$

$y'' = \dfrac{2\sqrt{x}\,(1/x)-(2+\ln x)(1/\sqrt{x})}{(2\sqrt{x})^2} = \dfrac{2/\sqrt{x}-(2+\ln x)(1/\sqrt{x})}{4x} = \dfrac{2-(2+\ln x)}{\sqrt{x}(4x)} = -\dfrac{\ln x}{4x\sqrt{x}}$

33. $y = \ln|\sec x| \;\Rightarrow\; y' = \dfrac{1}{\sec x}\dfrac{d}{dx}\sec x = \dfrac{1}{\sec x}\sec x\tan x = \tan x \;\Rightarrow\; y'' = \sec^2 x$

35. $f(x) = \dfrac{x}{1-\ln(x-1)} \;\Rightarrow\;$

$f'(x) = \dfrac{[1-\ln(x-1)]\cdot 1 - x\cdot\dfrac{-1}{x-1}}{[1-\ln(x-1)]^2} = \dfrac{\dfrac{(x-1)[1-\ln(x-1)]+x}{x-1}}{[1-\ln(x-1)]^2} = \dfrac{x-1-(x-1)\ln(x-1)+x}{(x-1)[1-\ln(x-1)]^2}$

$= \dfrac{2x-1-(x-1)\ln(x-1)}{(x-1)[1-\ln(x-1)]^2}$

$\text{Dom}(f) = \{x \mid x-1 > 0 \text{ and } 1-\ln(x-1)\neq 0\} = \{x \mid x > 1 \text{ and } \ln(x-1)\neq 1\}$

$= \{x \mid x > 1 \text{ and } x-1\neq e^1\} = \{x \mid x > 1 \text{ and } x\neq 1+e\} = (1, 1+e)\cup(1+e,\infty)$

37. $f(x) = \ln(x^2-2x) \;\Rightarrow\; f'(x) = \dfrac{1}{x^2-2x}(2x-2) = \dfrac{2(x-1)}{x(x-2)}$.

$\text{Dom}(f) = \{x \mid x(x-2) > 0\} = (-\infty, 0)\cup(2,\infty)$.

39. $f(x) = \ln(x+\ln x) \;\Rightarrow\; f'(x) = \dfrac{1}{x+\ln x}\dfrac{d}{dx}(x+\ln x) = \dfrac{1}{x+\ln x}\left(1+\dfrac{1}{x}\right)$.

Substitute 1 for x to get $f'(1) = \dfrac{1}{1+\ln 1}\left(1+\dfrac{1}{1}\right) = \dfrac{1}{1+0}(1+1) = 1\cdot 2 = 2$.

41. $y = \ln(x^2-3x+1) \;\Rightarrow\; y' = \dfrac{1}{x^2-3x+1}\cdot(2x-3) \;\Rightarrow\; y'(3) = \tfrac{1}{1}\cdot 3 = 3$, so an equation of a tangent line at $(3,0)$ is $y-0 = 3(x-3)$, or $y = 3x-9$.

43. $f(x) = \sin x + \ln x \;\Rightarrow\; f'(x) = \cos x + 1/x$.

This is reasonable, because the graph shows that f increases when f' is positive, and $f'(x) = 0$ when f has a horizontal tangent.

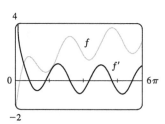

45. $f(x) = cx + \ln(\cos x) \;\Rightarrow\; f'(x) = c + \dfrac{1}{\cos x}\cdot(-\sin x) = c - \tan x$.

$f'(\tfrac{\pi}{4}) = 6 \;\Rightarrow\; c - \tan\tfrac{\pi}{4} = 6 \;\Rightarrow\; c - 1 = 6 \;\Rightarrow\; c = 7$.

47. $y = (x^2+2)^2(x^4+4)^4 \;\Rightarrow\; \ln y = \ln[(x^2+2)^2(x^4+4)^4] \;\Rightarrow\; \ln y = 2\ln(x^2+2) + 4\ln(x^4+4) \;\Rightarrow\;$

$\dfrac{1}{y}y' = 2 \cdot \dfrac{1}{x^2+2} \cdot 2x + 4 \cdot \dfrac{1}{x^4+4} \cdot 4x^3 \;\Rightarrow\; y' = y\left(\dfrac{4x}{x^2+2} + \dfrac{16x^3}{x^4+4}\right) \;\Rightarrow\;$

$y' = (x^2+2)^2(x^4+4)^4\left(\dfrac{4x}{x^2+2} + \dfrac{16x^3}{x^4+4}\right)$

49. $y = \sqrt{\dfrac{x-1}{x^4+1}} \;\Rightarrow\; \ln y = \ln\left(\dfrac{x-1}{x^4+1}\right)^{1/2} \;\Rightarrow\; \ln y = \dfrac{1}{2}\ln(x-1) - \dfrac{1}{2}\ln(x^4+1) \;\Rightarrow\;$

$\dfrac{1}{y}y' = \dfrac{1}{2}\dfrac{1}{x-1} - \dfrac{1}{2}\dfrac{1}{x^4+1}\cdot 4x^3 \;\Rightarrow\; y' = y\left(\dfrac{1}{2(x-1)} - \dfrac{2x^3}{x^4+1}\right) \;\Rightarrow\; y' = \sqrt{\dfrac{x-1}{x^4+1}}\left(\dfrac{1}{2x-2} - \dfrac{2x^3}{x^4+1}\right)$

51. $y = x^x \;\Rightarrow\; \ln y = \ln x^x \;\Rightarrow\; \ln y = x\ln x \;\Rightarrow\; y'/y = x(1/x) + (\ln x)\cdot 1 \;\Rightarrow\; y' = y(1 + \ln x) \;\Rightarrow\;$

$y' = x^x(1 + \ln x)$

53. $y = x^{\sin x} \;\Rightarrow\; \ln y = \ln x^{\sin x} \;\Rightarrow\; \ln y = \sin x \ln x \;\Rightarrow\; \dfrac{y'}{y} = (\sin x)\cdot\dfrac{1}{x} + (\ln x)(\cos x) \;\Rightarrow\;$

$y' = y\left(\dfrac{\sin x}{x} + \ln x \cos x\right) \;\Rightarrow\; y' = x^{\sin x}\left(\dfrac{\sin x}{x} + \ln x \cos x\right)$

55. $y = (\cos x)^x \;\Rightarrow\; \ln y = \ln(\cos x)^x \;\Rightarrow\; \ln y = x\ln\cos x \;\Rightarrow\; \dfrac{1}{y}y' = x\cdot\dfrac{1}{\cos x}\cdot(-\sin x) + \ln\cos x\cdot 1 \;\Rightarrow\;$

$y' = y\left(\ln\cos x - \dfrac{x\sin x}{\cos x}\right) \;\Rightarrow\; y' = (\cos x)^x(\ln\cos x - x\tan x)$

57. $y = x^{\ln x} \;\Rightarrow\; \ln y = \ln x \ln x = (\ln x)^2 \;\Rightarrow\; \dfrac{y'}{y} = 2\ln x\left(\dfrac{1}{x}\right) \;\Rightarrow\; y' = x^{\ln x}\left(\dfrac{2\ln x}{x}\right)$

59. $y = \ln(x^2 + y^2) \;\Rightarrow\; y' = \dfrac{1}{x^2+y^2}\dfrac{d}{dx}(x^2+y^2) \;\Rightarrow\; y' = \dfrac{2x + 2yy'}{x^2+y^2} \;\Rightarrow\; x^2y' + y^2y' = 2x + 2yy' \;\Rightarrow\;$

$x^2y' + y^2y' - 2yy' = 2x \;\Rightarrow\; (x^2+y^2-2y)y' = 2x \;\Rightarrow\; y' = \dfrac{2x}{x^2+y^2-2y}$

61. $f(x) = \ln(x-1) \;\Rightarrow\; f'(x) = \dfrac{1}{(x-1)} = (x-1)^{-1} \;\Rightarrow\; f''(x) = -(x-1)^{-2} \;\Rightarrow\; f'''(x) = 2(x-1)^{-3} \;\Rightarrow\;$

$f^{(4)}(x) = -2\cdot 3(x-1)^{-4} \;\Rightarrow\; \cdots \;\Rightarrow\; f^{(n)}(x) = (-1)^{n-1}\cdot 2\cdot 3\cdot 4 \cdots (n-1)(x-1)^{-n} = (-1)^{n-1}\dfrac{(n-1)!}{(x-1)^n}$

63. $f(x) = \dfrac{\ln x}{\sqrt{x}} \;\Rightarrow\; f'(x) = \dfrac{\sqrt{x}(1/x) - (\ln x)[1/(2\sqrt{x})]}{x} = \dfrac{2 - \ln x}{2x^{3/2}} \;\Rightarrow\;$

$f''(x) = \dfrac{2x^{3/2}(-1/x) - (2 - \ln x)(3x^{1/2})}{4x^3} = \dfrac{3\ln x - 8}{4x^{5/2}} > 0 \;\Leftrightarrow\; \ln x > \tfrac{8}{3} \;\Leftrightarrow\; x > e^{8/3}$, so f is CU on $(e^{8/3}, \infty)$

and CD on $(0, e^{8/3})$. The inflection point is $\left(e^{8/3}, \tfrac{8}{3}e^{-4/3}\right)$.

65. $y = f(x) = \ln(\sin x)$

A. $D = \{x \text{ in } \mathbb{R} \mid \sin x > 0\} = \bigcup_{n=-\infty}^{\infty} (2n\pi, (2n+1)\pi) = \cdots \cup (-4\pi, -3\pi) \cup (-2\pi, -\pi) \cup (0, \pi) \cup (2\pi, 3\pi) \cup \cdots$

B. No y-intercept; x-intercepts: $f(x) = 0 \Leftrightarrow \ln(\sin x) = 0 \Leftrightarrow \sin x = e^0 = 1 \Leftrightarrow x = 2n\pi + \frac{\pi}{2}$ for each integer n. **C.** f is periodic with period 2π. **D.** $\lim\limits_{x \to (2n\pi)^+} f(x) = -\infty$ and $\lim\limits_{x \to [(2n+1)\pi]^-} f(x) = -\infty$, so the lines $x = n\pi$ are VAs for all integers n. **E.** $f'(x) = \dfrac{\cos x}{\sin x} = \cot x$, so $f'(x) > 0$ when $2n\pi < x < 2n\pi + \frac{\pi}{2}$ for each integer n, and $f'(x) < 0$ when $2n\pi + \frac{\pi}{2} < x < (2n+1)\pi$. Thus, f is increasing on $(2n\pi, 2n\pi + \frac{\pi}{2})$ and decreasing on $(2n\pi + \frac{\pi}{2}, (2n+1)\pi)$ for each integer n.

F. Local maximum values $f(2n\pi + \frac{\pi}{2}) = 0$, no local minimum.

G. $f''(x) = -\csc^2 x < 0$, so f is CD on $(2n\pi, (2n+1)\pi)$ for each integer n. No IP

H.

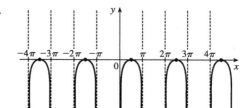

67. $y = f(x) = \ln(1 + x^2)$ **A.** $D = \mathbb{R}$ **B.** Both intercepts are 0. **C.** $f(-x) = f(x)$, so the curve is symmetric about the y-axis. **D.** $\lim\limits_{x \to \pm\infty} \ln(1 + x^2) = \infty$, no asymptotes. **E.** $f'(x) = \dfrac{2x}{1 + x^2} > 0 \Leftrightarrow x > 0$, so f is increasing on $(0, \infty)$ and decreasing on $(-\infty, 0)$.

F. $f(0) = 0$ is a local and absolute minimum.

G. $f''(x) = \dfrac{2(1 + x^2) - 2x(2x)}{(1 + x^2)^2} = \dfrac{2(1 - x^2)}{(1 + x^2)^2} > 0 \Leftrightarrow$
$|x| < 1$, so f is CU on $(-1, 1)$, CD on $(-\infty, -1)$ and $(1, \infty)$.
IP $(1, \ln 2)$ and $(-1, \ln 2)$.

H.

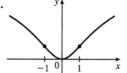

69. We use the CAS to calculate $f'(x) = \dfrac{2 + \sin x + x \cos x}{2x + x \sin x}$ and

$f''(x) = \dfrac{2x^2 \sin x + 4\sin x - \cos^2 x + x^2 + 5}{x^2(\cos^2 x - 4\sin x - 5)}$. From the graphs, it

seems that $f' > 0$ (and so f is increasing) on approximately the intervals $(0, 2.7)$, $(4.5, 8.2)$ and $(10.9, 14.3)$. It seems that f'' changes sign (indicating inflection points) at $x \approx 3.8, 5.7, 10.0$ and 12.0.

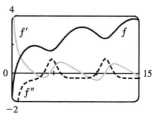

Looking back at the graph of $f(x) = \ln(2x + x\sin x)$, this implies that the inflection points have approximate coordinates $(3.8, 1.7)$, $(5.7, 2.1)$, $(10.0, 2.7)$, and $(12.0, 2.9)$.

71. 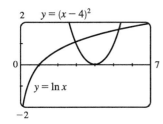 From the graph, it appears that the curves $y = (x-4)^2$ and $y = \ln x$ intersect just to the left of $x = 3$ and to the right of $x = 5$, at about $x = 5.3$. Let $f(x) = \ln x - (x-4)^2$. Then $f'(x) = 1/x - 2(x-4)$, so Newton's method says that $x_{n+1} = x_n - f(x_n)/f'(x_n) = x_n - \dfrac{\ln x_n - (x_n - 4)^2}{1/x_n - 2(x_n - 4)}$. Taking $x_0 = 3$, we get $x_1 \approx 2.957738$, $x_2 \approx 2.958516 \approx x_3$, so the first solution is 2.958516, to six decimal places. Taking $x_0 = 5$, we get $x_1 \approx 5.290755$, $x_2 \approx 5.290718 \approx x_3$, so the second (and final) solution is 5.290718, to six decimal places.

73. (a) Let $f(x) = \ln x \Rightarrow f'(x) = 1/x \Rightarrow f''(x) = -1/x^2$. The linear approximation to $\ln x$ near 1 is
$$\ln x \approx f(1) + f'(1)(x-1) = \ln 1 + \tfrac{1}{1}(x-1) = x - 1.$$

(b) 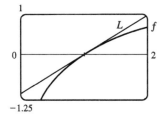   (c)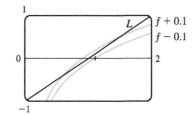

From the graph, it appears that the linear approximation is accurate to within 0.1 for x between about 0.62 and 1.51.

75. $\displaystyle\int_2^4 \dfrac{3}{x}\,dx = 3\int_2^4 \dfrac{1}{x}\,dx = 3\Big[\ln|x|\Big]_2^4 = 3(\ln 4 - \ln 2) = 3\ln\dfrac{4}{2} = 3\ln 2$

77. $\displaystyle\int_1^2 \dfrac{dt}{8-3t} = \left[-\dfrac{1}{3}\ln|8-3t|\right]_1^2 = -\dfrac{1}{3}\ln 2 - \left(-\dfrac{1}{3}\ln 5\right) = \dfrac{1}{3}(\ln 5 - \ln 2) = \dfrac{1}{3}\ln\dfrac{5}{2}$

Or: Let $u = 8 - 3t$. Then $du = -3\,dt$, so
$$\int_1^2 \dfrac{dt}{8-3t} = \int_5^2 \dfrac{-\tfrac{1}{3}\,du}{u} = \left[-\dfrac{1}{3}\ln|u|\right]_5^2 = -\dfrac{1}{3}\ln 2 - \left(-\dfrac{1}{3}\ln 5\right) = \dfrac{1}{3}(\ln 5 - \ln 2) = \dfrac{1}{3}\ln\dfrac{5}{2}.$$

79. $\displaystyle\int_1^3 \left(\dfrac{3x^2 + 4x + 1}{x}\right) dx = \int_1^3 \left(3x + 4 + \dfrac{1}{x}\right) dx = \left[\dfrac{3}{2}x^2 + 4x + \ln|x|\right]_1^3 = \left(\dfrac{27}{2} + 12 + \ln 3\right) - \left(\dfrac{3}{2} + 4 + \ln 1\right)$
$= 20 + \ln 3$

81. Let $u = \ln x$. Then $du = \dfrac{dx}{x} \Rightarrow \displaystyle\int \dfrac{(\ln x)^2}{x}\,dx = \int u^2\,du = \dfrac{1}{3}u^3 + C = \dfrac{1}{3}(\ln x)^3 + C.$

83. $\displaystyle\int \dfrac{\sin 2x}{1 + \cos^2 x}\,dx = 2\int \dfrac{\sin x \cos x}{1 + \cos^2 x}\,dx = 2I$. Let $u = \cos x$. Then $du = -\sin x\,dx$, so
$$2I = -2\int \dfrac{u\,du}{1+u^2} = -2 \cdot \tfrac{1}{2}\ln(1+u^2) + C = -\ln(1+u^2) + C = -\ln(1+\cos^2 x) + C.$$

Or: Let $u = 1 + \cos^2 x$.

85. $\int_0^4 2^s\, ds = \left[\dfrac{1}{\ln 2} 2^s\right]_0^4 = \dfrac{16}{\ln 2} - \dfrac{1}{\ln 2} = \dfrac{15}{\ln 2}$

87. (a) $\dfrac{d}{dx}(\ln|\sin x| + C) = \dfrac{1}{\sin x} \cos x = \cot x$

(b) Let $u = \sin x$. Then $du = \cos x\, dx$, so $\displaystyle\int \cot x\, dx = \int \dfrac{\cos x}{\sin x}\, dx = \int \dfrac{du}{u} = \ln|u| + C = \ln|\sin x| + C$.

89. The cross-sectional area is $\pi\left(1/\sqrt{x+1}\right)^2 = \pi/(x+1)$. Therefore, the volume is

$$\int_0^1 \dfrac{\pi}{x+1}\, dx = \pi[\ln(x+1)]_0^1 = \pi(\ln 2 - \ln 1) = \pi \ln 2.$$

91. $W = \int_{V_1}^{V_2} P\, dV = \int_{600}^{1000} \dfrac{C}{V}\, dV = C\int_{600}^{1000} \dfrac{1}{V}\, dV = C\bigl[\ln|V|\bigr]_{600}^{1000} = C(\ln 1000 - \ln 600) = C\ln \dfrac{1000}{600} = C\ln \dfrac{5}{3}$.

Initially, $PV = C$, where $P = 150$ kPa and $V = 600$ cm^3, so $C = (150)(600) = 90{,}000$ kPa $\cdot$ cm^3. Thus,

$$W = 90{,}000 \ln \tfrac{5}{3}\ \text{kPa}\cdot\text{cm}^3 = 90{,}000\ln\tfrac{5}{3}\left(1000\,\dfrac{\text{N}}{\text{m}^2}\right)\left(\dfrac{1}{100}\,\text{m}\right)^3$$

$$= 90\ln\tfrac{5}{3}\ \text{N}\cdot\text{m} \approx 45.974\ \text{J}$$

93. $f(x) = 2x + \ln x \;\Rightarrow\; f'(x) = 2 + 1/x$. If $g = f^{-1}$, then $f(1) = 2 \;\Rightarrow\; g(2) = 1$, so $g'(2) = 1/f'(g(2)) = 1/f'(1) = \tfrac{1}{3}$.

95. The curve and the line will determine a region when they intersect at two or more points. So we solve the equation $x/(x^2+1) = mx \;\Rightarrow\; x = 0$ or

$mx^2 + m - 1 = 0 \;\Rightarrow\; x = 0$ or $x = \dfrac{\pm\sqrt{-4(m)(m-1)}}{2m} = \pm\sqrt{\dfrac{1}{m} - 1}$.

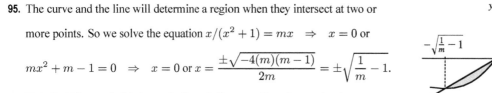

Note that if $m = 1$, this has only the solution $x = 0$, and no region is determined. But if $1/m - 1 > 0 \;\Leftrightarrow\; 1/m > 1 \;\Leftrightarrow\; 0 < m < 1$, then there are two solutions. [Another way of seeing this is to observe that the slope of the tangent to $y = x/(x^2+1)$ at the origin is $y' = 1$ and therefore we must have $0 < m < 1$.] Note that we cannot just integrate between the positive and negative roots, since the curve and the line cross at the origin. Since mx and $x/(x^2+1)$ are both odd functions, the total area is twice the area between the curves on the interval $\left[0, \sqrt{1/m - 1}\right]$. So the total area enclosed is

$$2\int_0^{\sqrt{1/m-1}}\left[\dfrac{x}{x^2+1} - mx\right]dx = 2\bigl[\tfrac{1}{2}\ln(x^2+1) - \tfrac{1}{2}mx^2\bigr]_0^{\sqrt{1/m-1}}$$

$$= \left[\ln\left(\dfrac{1}{m} - 1 + 1\right) - m\left(\dfrac{1}{m} - 1\right)\right] - (\ln 1 - 0)$$

$$= \ln\left(\dfrac{1}{m}\right) + m - 1 = m - \ln m - 1$$

97. If $f(x) = \ln(1+x)$, then $f'(x) = \dfrac{1}{1+x}$, so $f'(0) = 1$.

Thus, $\displaystyle\lim_{x \to 0} \dfrac{\ln(1+x)}{x} = \lim_{x \to 0} \dfrac{f(x)}{x} = \lim_{x \to 0} \dfrac{f(x) - f(0)}{x - 0} = f'(0) = 1$.

99. $y = b^x \;\Rightarrow\; y' = b^x \ln b$, so the slope of the tangent line to the curve $y = b^x$ at the point (a, b^a) is $b^a \ln b$. An equation of this tangent line is then $y - b^a = b^a \ln b \, (x - a)$. If c is the x-intercept of this tangent line, then $0 - b^a = b^a \ln b \, (c - a) \;\Rightarrow\;$ $-1 = \ln b \, (c - a) \;\Rightarrow\; \dfrac{-1}{\ln b} = c - a \;\Rightarrow\; |c - a| = \left|\dfrac{-1}{\ln b}\right| = \dfrac{1}{|\ln b|}$. The distance between $(a, 0)$ and $(c, 0)$ is $|c - a|$, and this distance is the constant $\dfrac{1}{|\ln b|}$ for any a. [*Note*: The absolute value is needed for the case $0 < b < 1$ because $\ln b$ is negative there. If $b > 1$, we can write $a - c = 1/(\ln b)$ as the constant distance between $(a, 0)$ and $(c, 0)$.]

6.2* The Natural Logarithmic Function

1. (a) $\ln \sqrt{ab} = \ln(ab)^{1/2} = \tfrac{1}{2}\ln(ab) = \tfrac{1}{2}(\ln a + \ln b) = \tfrac{1}{2}\ln a + \tfrac{1}{2}\ln b$ [assuming that the variables are positive]

(b) $\ln\left(\dfrac{x^4}{\sqrt{x^2 - 4}}\right) = \ln x^4 - \ln(x^2 - 4)^{1/2}$ [Law 2]

$= 4\ln x - \tfrac{1}{2}\ln[(x+2)(x-2)]$ [Law 3]

$= 4\ln x - \tfrac{1}{2}[\ln(x+2) + \ln(x-2)]$ [Law 1]

$= 4\ln x - \tfrac{1}{2}\ln(x+2) - \tfrac{1}{2}\ln(x-2)$

3. (a) $\ln a - 2\ln b + 3\ln c = \ln a - \ln b^2 + \ln c^3 = \ln\dfrac{a}{b^2} + \ln c^3 = \ln\dfrac{ac^3}{b^2}$

(b) $\ln 4 + \ln a - \tfrac{1}{3}\ln(a+1) = \ln(4 \cdot a) - \ln(a+1)^{1/3} = \ln(4a) - \ln\sqrt[3]{a+1} = \ln\dfrac{4a}{\sqrt[3]{a+1}}$

5. (a) $\ln 3 + \tfrac{1}{3}\ln 8 = \ln 3 + \ln 8^{1/3} = \ln 3 + \ln 2 = \ln(3 \cdot 2) = \ln 6$

(b) $\tfrac{1}{3}\ln(x+2)^3 + \tfrac{1}{2}\left[\ln x - \ln(x^2 + 3x + 2)^2\right] = \ln[(x+2)^3]^{1/3} + \tfrac{1}{2}\ln\dfrac{x}{(x^2 + 3x + 2)^2}$ [by Laws 3, 2]

$= \ln(x+2) + \ln\dfrac{\sqrt{x}}{x^2 + 3x + 2}$ [by Law 3]

$= \ln\dfrac{(x+2)\sqrt{x}}{(x+1)(x+2)}$ [by Law 1]

$= \ln\dfrac{\sqrt{x}}{x+1}$

Note that since $\ln x$ is defined for $x > 0$, we have $x + 1$, $x + 2$, and $x^2 + 3x + 2$ all positive, and hence their logarithms are defined.

7. Reflect the graph of $y = \ln x$ about the x-axis to obtain the graph of $y = -\ln x$.

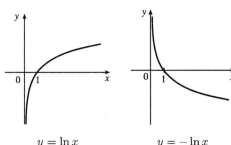

9.

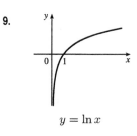
$y = \ln x$

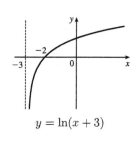
$y = \ln(x+3)$

11. $\lim\limits_{x \to 0} \ln(\cos x) = \ln 1 = 0$. [$\ln(\cos x)$ is continuous at $x = 0$ since it is the composite of two continuous functions.]

13. $\lim\limits_{x \to \infty} [\ln(1+x^2) - \ln(1+x)] = \lim\limits_{x \to \infty} \ln \dfrac{1+x^2}{1+x} = \ln \left(\lim\limits_{x \to \infty} \dfrac{1+x^2}{1+x} \right) = \ln \left(\lim\limits_{x \to \infty} \dfrac{\frac{1}{x}+x}{\frac{1}{x}+1} \right) = \infty$, since the limit in parentheses is ∞.

15. $f(x) = x^3 \ln x \;\Rightarrow\; f'(x) = x^3 \cdot \dfrac{1}{x} + (\ln x) \cdot 3x^2 = x^2 + 3x^2 \ln x = x^2(1 + 3\ln x)$

17. $f(x) = \ln(x^2 + 3x + 5) \;\Rightarrow\; f'(x) = \dfrac{1}{x^2+3x+5} \cdot \dfrac{d}{dx}(x^2+3x+5) = \dfrac{1}{x^2+3x+5} \cdot (2x+3) = \dfrac{2x+3}{x^2+3x+5}$

19. $f(x) = \sin(\ln x) \;\Rightarrow\; f'(x) = \cos(\ln x) \cdot \dfrac{d}{dx} \ln x = \cos(\ln x) \cdot \dfrac{1}{x} = \dfrac{\cos(\ln x)}{x}$

21. $f(x) = \ln \dfrac{1}{x} \;\Rightarrow\; f'(x) = \dfrac{1}{1/x} \dfrac{d}{dx} \left(\dfrac{1}{x} \right) = x \left(-\dfrac{1}{x^2} \right) = -\dfrac{1}{x}$.

Another solution: $f(x) = \ln \dfrac{1}{x} = \ln 1 - \ln x = -\ln x \;\Rightarrow\; f'(x) = -\dfrac{1}{x}$.

23. $f(x) = \sin x \ln(5x) \;\Rightarrow\; f'(x) = \sin x \cdot \dfrac{1}{5x} \cdot \dfrac{d}{dx}(5x) + \ln(5x) \cdot \cos x = \dfrac{\sin x \cdot 5}{5x} + \cos x \ln(5x) = \dfrac{\sin x}{x} + \cos x \ln(5x)$

25. $F(t) = (\ln t)^2 \sin t \;\Rightarrow\; F'(t) = (\ln t)^2 \cos t + \sin t \cdot 2 \ln t \cdot \dfrac{1}{t} = \ln t \left(\ln t \cos t + \dfrac{2 \sin t}{t} \right)$

27. $y = (\ln \tan x)^2 \;\Rightarrow\; y' = 2(\ln \tan x) \cdot \dfrac{1}{\tan x} \cdot \sec^2 x = \dfrac{2(\ln \tan x) \sec^2 x}{\tan x}$ or $y' = \dfrac{2 \ln \tan x}{\sin x \cos x}$

29. $f(u) = \dfrac{\ln u}{1 + \ln(2u)} \;\Rightarrow\;$

$f'(u) = \dfrac{[1 + \ln(2u)] \cdot \frac{1}{u} - \ln u \cdot \frac{1}{2u} \cdot 2}{[1 + \ln(2u)]^2} = \dfrac{\frac{1}{u}[1 + \ln(2u) - \ln u]}{[1 + \ln(2u)]^2} = \dfrac{1 + (\ln 2 + \ln u) - \ln u}{u[1 + \ln(2u)]^2} = \dfrac{1 + \ln 2}{u[1 + \ln(2u)]^2}$

31. $g(t) = \ln \dfrac{t(t^2+1)^4}{\sqrt[3]{2t-1}} = \ln t + \ln(t^2+1)^4 - \ln \sqrt[3]{2t-1} = \ln t + 4\ln(t^2+1) - \tfrac{1}{3} \ln(2t-1) \;\Rightarrow\;$

$g'(t) = \dfrac{1}{t} + 4 \cdot \dfrac{1}{t^2+1} \cdot 2t - \dfrac{1}{3} \cdot \dfrac{1}{2t-1} \cdot 2 = \dfrac{1}{t} + \dfrac{8t}{t^2+1} - \dfrac{2}{3(2t-1)}$

33. $y = \ln|3 - 2x^5| \;\Rightarrow\; y' = \dfrac{1}{3-2x^5} \cdot (-10x^4) = \dfrac{-10x^4}{3-2x^5}$

35. $\dfrac{d}{dx} \ln\left(x + \sqrt{x^2+1}\right) = \dfrac{1}{x+\sqrt{x^2+1}} \cdot \dfrac{d}{dx}\left(x+\sqrt{x^2+1}\right) = \dfrac{1}{x+\sqrt{x^2+1}} \cdot \left(1 + \dfrac{2x}{2\sqrt{x^2+1}}\right)$

$= \dfrac{1}{x+\sqrt{x^2+1}} \cdot \left(\dfrac{\sqrt{x^2+1}}{\sqrt{x^2+1}} + \dfrac{x}{\sqrt{x^2+1}}\right) = \dfrac{1}{x+\sqrt{x^2+1}} \cdot \left(\dfrac{x+\sqrt{x^2+1}}{\sqrt{x^2+1}}\right) = \dfrac{1}{\sqrt{x^2+1}}$

37. $y = \sqrt{x}\,\ln x \;\Rightarrow\; y' = \sqrt{x} \cdot \dfrac{1}{x} + (\ln x)\dfrac{1}{2\sqrt{x}} = \dfrac{2 + \ln x}{2\sqrt{x}} \;\Rightarrow\;$

$y'' = \dfrac{2\sqrt{x}\,(1/x) - (2+\ln x)(1/\sqrt{x})}{(2\sqrt{x})^2} = \dfrac{2/\sqrt{x} - (2+\ln x)(1/\sqrt{x})}{4x} = \dfrac{2-(2+\ln x)}{\sqrt{x}(4x)} = -\dfrac{\ln x}{4x\sqrt{x}}$

39. $y = \ln|\sec x| \;\Rightarrow\; y' = \dfrac{1}{\sec x}\dfrac{d}{dx}\sec x = \dfrac{1}{\sec x}\sec x\,\tan x = \tan x \;\Rightarrow\; y'' = \sec^2 x$

41. $f(x) = \dfrac{x}{1 - \ln(x-1)} \;\Rightarrow\;$

$f'(x) = \dfrac{[1-\ln(x-1)]\cdot 1 - x \cdot \dfrac{-1}{x-1}}{[1-\ln(x-1)]^2} = \dfrac{\dfrac{(x-1)[1-\ln(x-1)] + x}{x-1}}{[1-\ln(x-1)]^2} = \dfrac{x - 1 - (x-1)\ln(x-1) + x}{(x-1)[1-\ln(x-1)]^2}$

$= \dfrac{2x - 1 - (x-1)\ln(x-1)}{(x-1)[1-\ln(x-1)]^2}$

$\text{Dom}(f) = \{x \mid x - 1 > 0 \text{ and } 1 - \ln(x-1) \neq 0\} = \{x \mid x > 1 \text{ and } \ln(x-1) \neq 1\}$
$= \{x \mid x > 1 \text{ and } x - 1 \neq e^1\} = \{x \mid x > 1 \text{ and } x \neq 1 + e\} = (1, 1+e) \cup (1+e, \infty)$

43. $f(x) = \ln(x^2 - 2x) \;\Rightarrow\; f'(x) = \dfrac{1}{x^2-2x}(2x-2) = \dfrac{2(x-1)}{x(x-2)}.$

$\text{Dom}(f) = \{x \mid x(x-2) > 0\} = (-\infty, 0) \cup (2, \infty).$

45. $f(x) = \ln(x + \ln x) \;\Rightarrow\; f'(x) = \dfrac{1}{x + \ln x}\dfrac{d}{dx}(x+\ln x) = \dfrac{1}{x+\ln x}\left(1 + \dfrac{1}{x}\right).$

Substitute 1 for x to get $f'(1) = \dfrac{1}{1 + \ln 1}\left(1 + \dfrac{1}{1}\right) = \dfrac{1}{1+0}(1+1) = 1 \cdot 2 = 2.$

47. $f(x) = \sin x + \ln x \;\Rightarrow\; f'(x) = \cos x + 1/x.$

This is reasonable, because the graph shows that f increases when f' is positive, and $f'(x) = 0$ when f has a horizontal tangent.

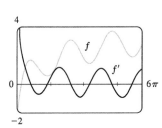

49. $y = \sin(2\ln x) \;\Rightarrow\; y' = \cos(2\ln x) \cdot \dfrac{2}{x}.$ At $(1, 0)$, $y' = \cos 0 \cdot \dfrac{2}{1} = 2$, so an equation of the tangent line is

$y - 0 = 2 \cdot (x-1)$, or $y = 2x - 2$.

51. $y = \ln(x^2 + y^2)$ $\Rightarrow$ $y' = \dfrac{1}{x^2 + y^2} \dfrac{d}{dx}(x^2 + y^2)$ $\Rightarrow$ $y' = \dfrac{2x + 2yy'}{x^2 + y^2}$ $\Rightarrow$ $x^2 y' + y^2 y' = 2x + 2yy'$ $\Rightarrow$

$x^2 y' + y^2 y' - 2yy' = 2x$ $\Rightarrow$ $(x^2 + y^2 - 2y)y' = 2x$ $\Rightarrow$ $y' = \dfrac{2x}{x^2 + y^2 - 2y}$

53. $f(x) = \ln(x - 1)$ $\Rightarrow$ $f'(x) = \dfrac{1}{(x-1)} = (x-1)^{-1}$ $\Rightarrow$ $f''(x) = -(x-1)^{-2}$ $\Rightarrow$ $f'''(x) = 2(x-1)^{-3}$ $\Rightarrow$

$f^{(4)}(x) = -2 \cdot 3(x-1)^{-4}$ $\Rightarrow$ $\cdots$ $\Rightarrow$ $f^{(n)}(x) = (-1)^{n-1} \cdot 2 \cdot 3 \cdot 4 \cdot \cdots \cdot (n-1)(x-1)^{-n} = (-1)^{n-1} \dfrac{(n-1)!}{(x-1)^n}$

55. $y = f(x) = \ln(\sin x)$

A. $D = \{x \text{ in } \mathbb{R} \mid \sin x > 0\} = \bigcup\limits_{n=-\infty}^{\infty} (2n\pi, (2n+1)\pi) = \cdots \cup (-4\pi, -3\pi) \cup (-2\pi, -\pi) \cup (0, \pi) \cup (2\pi, 3\pi) \cup \cdots$

B. No y-intercept; x-intercepts: $f(x) = 0$ $\Leftrightarrow$ $\ln(\sin x) = 0$ $\Leftrightarrow$ $\sin x = e^0 = 1$ $\Leftrightarrow$ $x = 2n\pi + \frac{\pi}{2}$ for each integer n. **C.** f is periodic with period 2π. **D.** $\lim\limits_{x \to (2n\pi)^+} f(x) = -\infty$ and $\lim\limits_{x \to [(2n+1)\pi]^-} f(x) = -\infty$, so the lines $x = n\pi$ are VAs for all integers n. **E.** $f'(x) = \dfrac{\cos x}{\sin x} = \cot x$, so $f'(x) > 0$ when $2n\pi < x < 2n\pi + \frac{\pi}{2}$ for each integer n, and $f'(x) < 0$ when $2n\pi + \frac{\pi}{2} < x < (2n+1)\pi$. Thus, f is increasing on $(2n\pi, 2n\pi + \frac{\pi}{2})$ and decreasing on $(2n\pi + \frac{\pi}{2}, (2n+1)\pi)$ for each integer n. **H.**

F. Local maximum values $f(2n\pi + \frac{\pi}{2}) = 0$, no local minimum.

G. $f''(x) = -\csc^2 x < 0$, so f is CD on $(2n\pi, (2n+1)\pi)$ for each integer n. No IP

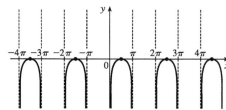

57. $y = f(x) = \ln(1 + x^2)$ **A.** $D = \mathbb{R}$ **B.** Both intercepts are 0. **C.** $f(-x) = f(x)$, so the curve is symmetric about the y-axis. **D.** $\lim\limits_{x \to \pm\infty} \ln(1 + x^2) = \infty$, no asymptotes. **E.** $f'(x) = \dfrac{2x}{1 + x^2} > 0$ $\Leftrightarrow$

$x > 0$, so f is increasing on $(0, \infty)$ and decreasing on $(-\infty, 0)$. **H.**

F. $f(0) = 0$ is a local and absolute minimum.

G. $f''(x) = \dfrac{2(1 + x^2) - 2x(2x)}{(1 + x^2)^2} = \dfrac{2(1 - x^2)}{(1 + x^2)^2} > 0$ $\Leftrightarrow$

$|x| < 1$, so f is CU on $(-1, 1)$, CD on $(-\infty, -1)$ and $(1, \infty)$.
IP $(1, \ln 2)$ and $(-1, \ln 2)$.

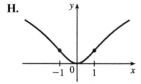

59. We use the CAS to calculate $f'(x) = \dfrac{2 + \sin x + x \cos x}{2x + x \sin x}$ and

$f''(x) = \dfrac{2x^2 \sin x + 4\sin x - \cos^2 x + x^2 + 5}{x^2(\cos^2 x - 4\sin x - 5)}$. From the graphs, it

seems that $f' > 0$ (and so f is increasing) on approximately the intervals

$(0, 2.7)$, $(4.5, 8.2)$ and $(10.9, 14.3)$. It seems that f'' changes sign
(indicating inflection points) at $x \approx 3.8, 5.7, 10.0$ and 12.0.

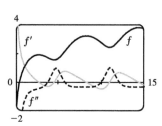

[continued]

Looking back at the graph of $f(x) = \ln(2x + x\sin x)$, this implies that the inflection points have approximate coordinates $(3.8, 1.7)$, $(5.7, 2.1)$, $(10.0, 2.7)$, and $(12.0, 2.9)$.

61.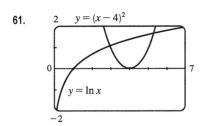

From the graph, it appears that the curves $y = (x-4)^2$ and $y = \ln x$ intersect just to the left of $x = 3$ and to the right of $x = 5$, at about $x = 5.3$. Let $f(x) = \ln x - (x-4)^2$. Then $f'(x) = 1/x - 2(x-4)$, so Newton's method says that $x_{n+1} = x_n - f(x_n)/f'(x_n) = x_n - \dfrac{\ln x_n - (x_n - 4)^2}{1/x_n - 2(x_n - 4)}$. Taking $x_0 = 3$, we get $x_1 \approx 2.957738$, $x_2 \approx 2.958516 \approx x_3$, so the first solution is 2.958516, to six decimal places. Taking $x_0 = 5$, we get $x_1 \approx 5.290755$, $x_2 \approx 5.290718 \approx x_3$, so the second (and final) solution is 5.290718, to six decimal places.

63. $y = (x^2 + 2)^2(x^4 + 4)^4 \;\Rightarrow\; \ln y = \ln[(x^2+2)^2(x^4+4)^4] \;\Rightarrow\; \ln y = 2\ln(x^2+2) + 4\ln(x^4+4) \;\Rightarrow\;$

$\dfrac{1}{y}y' = 2 \cdot \dfrac{1}{x^2+2} \cdot 2x + 4 \cdot \dfrac{1}{x^4+4} \cdot 4x^3 \;\Rightarrow\; y' = y\left(\dfrac{4x}{x^2+2} + \dfrac{16x^3}{x^4+4}\right) \;\Rightarrow\;$

$y' = (x^2+2)^2(x^4+4)^4\left(\dfrac{4x}{x^2+2} + \dfrac{16x^3}{x^4+4}\right)$

65. $y = \sqrt{\dfrac{x-1}{x^4+1}} \;\Rightarrow\; \ln y = \ln\left(\dfrac{x-1}{x^4+1}\right)^{1/2} \;\Rightarrow\; \ln y = \dfrac{1}{2}\ln(x-1) - \dfrac{1}{2}\ln(x^4+1) \;\Rightarrow\;$

$\dfrac{1}{y}y' = \dfrac{1}{2}\dfrac{1}{x-1} - \dfrac{1}{2}\dfrac{1}{x^4+1} \cdot 4x^3 \;\Rightarrow\; y' = y\left(\dfrac{1}{2(x-1)} - \dfrac{2x^3}{x^4+1}\right) \;\Rightarrow\; y' = \sqrt{\dfrac{x-1}{x^4+1}}\left(\dfrac{1}{2x-2} - \dfrac{2x^3}{x^4+1}\right)$

67. $\displaystyle\int_2^4 \dfrac{3}{x}\,dx = 3\int_2^4 \dfrac{1}{x}\,dx = 3\Big[\ln|x|\Big]_2^4 = 3(\ln 4 - \ln 2) = 3\ln\dfrac{4}{2} = 3\ln 2$

69. $\displaystyle\int_1^2 \dfrac{dt}{8-3t} = \left[-\dfrac{1}{3}\ln|8-3t|\right]_1^2 = -\dfrac{1}{3}\ln 2 - \left(-\dfrac{1}{3}\ln 5\right) = \dfrac{1}{3}(\ln 5 - \ln 2) = \dfrac{1}{3}\ln\dfrac{5}{2}$

Or: Let $u = 8 - 3t$. Then $du = -3\,dt$, so

$\displaystyle\int_1^2 \dfrac{dt}{8-3t} = \int_5^2 \dfrac{-\frac{1}{3}\,du}{u} = \left[-\dfrac{1}{3}\ln|u|\right]_5^2 = -\dfrac{1}{3}\ln 2 - \left(-\dfrac{1}{3}\ln 5\right) = \dfrac{1}{3}(\ln 5 - \ln 2) = \dfrac{1}{3}\ln\dfrac{5}{2}.$

71. $\displaystyle\int_1^3 \left(\dfrac{3x^2 + 4x + 1}{x}\right)dx = \int_1^3\left(3x + 4 + \dfrac{1}{x}\right)dx = \left[\dfrac{3}{2}x^2 + 4x + \ln|x|\right]_1^3 = \left(\dfrac{27}{2} + 12 + \ln 3\right) - \left(\dfrac{3}{2} + 4 + \ln 1\right)$

$= 20 + \ln 3$

73. Let $u = \ln x$. Then $du = \dfrac{dx}{x} \;\Rightarrow\; \displaystyle\int \dfrac{(\ln x)^2}{x}\,dx = \int u^2\,du = \dfrac{1}{3}u^3 + C = \dfrac{1}{3}(\ln x)^3 + C.$

75. $\displaystyle\int \dfrac{\sin 2x}{1+\cos^2 x}\,dx = 2\int \dfrac{\sin x\cos x}{1+\cos^2 x}\,dx = 2I.$ Let $u = \cos x$. Then $du = -\sin x\,dx$, so

$2I = -2\displaystyle\int \dfrac{u\,du}{1+u^2} = -2 \cdot \dfrac{1}{2}\ln(1+u^2) + C = -\ln(1+u^2) + C = -\ln(1+\cos^2 x) + C.$

Or: Let $u = 1 + \cos^2 x$.

77. (a) $\dfrac{d}{dx}(\ln|\sin x| + C) = \dfrac{1}{\sin x}\cos x = \cot x$

(b) Let $u = \sin x$. Then $du = \cos x\, dx$, so $\displaystyle\int \cot x\, dx = \int \dfrac{\cos x}{\sin x}\, dx = \int \dfrac{du}{u} = \ln|u| + C = \ln|\sin x| + C.$

79. The cross-sectional area is $\pi\left(1/\sqrt{x+1}\right)^2 = \pi/(x+1)$. Therefore, the volume is

$$\int_0^1 \dfrac{\pi}{x+1}\, dx = \pi[\ln(x+1)]_0^1 = \pi(\ln 2 - \ln 1) = \pi \ln 2.$$

81. $W = \displaystyle\int_{V_1}^{V_2} P\, dV = \int_{600}^{1000} \dfrac{C}{V}\, dV = C\int_{600}^{1000} \dfrac{1}{V}\, dV = C\Big[\ln|V|\Big]_{600}^{1000} = C(\ln 1000 - \ln 600) = C\ln \dfrac{1000}{600} = C\ln \dfrac{5}{3}.$

Initially, $PV = C$, where $P = 150$ kPa and $V = 600$ cm^3, so $C = (150)(600) = 90{,}000$ kPa·cm^3. Thus,

$$W = 90{,}000\ln \tfrac{5}{3}\ \text{kPa}\cdot\text{cm}^3 = 90{,}000\ln \tfrac{5}{3}\left(1000\ \dfrac{\text{N}}{\text{m}^2}\right)\left(\dfrac{1}{100}\ \text{m}\right)^3$$

$$= 90\ln \tfrac{5}{3}\ \text{N·m} \approx 45.974\ \text{J}$$

83. $f(x) = 2x + \ln x\ \Rightarrow\ f'(x) = 2 + 1/x$. If $g = f^{-1}$, then $f(1) = 2\ \Rightarrow\ g(2) = 1$, so $g'(2) = 1/f'(g(2)) = 1/f'(1) = \tfrac{1}{3}.$

85. (a)

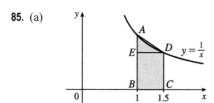

We interpret $\ln 1.5$ as the area under the curve $y = 1/x$ from $x = 1$ to $x = 1.5$. The area of the rectangle $BCDE$ is $\tfrac{1}{2}\cdot\tfrac{2}{3} = \tfrac{1}{3}$. The area of the trapezoid $ABCD$ is $\tfrac{1}{2}\cdot\tfrac{1}{2}\left(1+\tfrac{2}{3}\right) = \tfrac{5}{12}$. Thus, by comparing areas, we observe that $\tfrac{1}{3} < \ln 1.5 < \tfrac{5}{12}.$

(b) With $f(t) = 1/t$, $n = 10$, and $\Delta t = 0.05$, we have

$$\ln 1.5 = \int_1^{1.5}(1/t)\, dt \approx (0.05)[f(1.025) + f(1.075) + \cdots + f(1.475)]$$
$$= (0.05)\left[\tfrac{1}{1.025} + \tfrac{1}{1.075} + \cdots + \tfrac{1}{1.475}\right] \approx 0.4054$$

87.

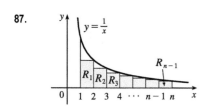

The area of R_i is $\dfrac{1}{i+1}$ and so $\dfrac{1}{2} + \dfrac{1}{3} + \cdots + \dfrac{1}{n} < \displaystyle\int_1^n \dfrac{1}{t}\, dt = \ln n.$

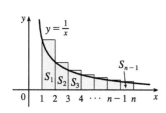

The area of S_i is $\dfrac{1}{i}$ and so $1 + \dfrac{1}{2} + \cdots + \dfrac{1}{n-1} > \displaystyle\int_1^n \dfrac{1}{t}\, dt = \ln n.$

89. The curve and the line will determine a region when they intersect at two or more points. So we solve the equation $x/(x^2+1) = mx \Rightarrow x = 0$ or

$mx^2 + m - 1 = 0 \Rightarrow x = 0$ or $x = \dfrac{\pm\sqrt{-4(m)(m-1)}}{2m} = \pm\sqrt{\dfrac{1}{m} - 1}$.

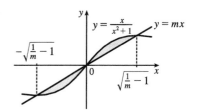

Note that if $m = 1$, this has only the solution $x = 0$, and no region is determined. But if $1/m - 1 > 0 \Leftrightarrow 1/m > 1 \Leftrightarrow 0 < m < 1$, then there are two solutions. [Another way of seeing this is to observe that the slope of the tangent to $y = x/(x^2+1)$ at the origin is $y' = 1$ and therefore we must have $0 < m < 1$.] Note that we cannot just integrate between the positive and negative roots, since the curve and the line cross at the origin. Since mx and $x/(x^2+1)$ are both odd functions, the total area is twice the area between the curves on the interval $\left[0, \sqrt{1/m - 1}\right]$. So the total area enclosed is

$$2\int_0^{\sqrt{1/m-1}} \left[\dfrac{x}{x^2+1} - mx\right] dx = 2\left[\tfrac{1}{2}\ln(x^2+1) - \tfrac{1}{2}mx^2\right]_0^{\sqrt{1/m-1}}$$

$$= \left[\ln\left(\dfrac{1}{m} - 1 + 1\right) - m\left(\dfrac{1}{m} - 1\right)\right] - (\ln 1 - 0)$$

$$= \ln\left(\dfrac{1}{m}\right) + m - 1 = m - \ln m - 1$$

91. If $f(x) = \ln(1+x)$, then $f'(x) = \dfrac{1}{1+x}$, so $f'(0) = 1$.

Thus, $\lim\limits_{x \to 0} \dfrac{\ln(1+x)}{x} = \lim\limits_{x \to 0} \dfrac{f(x)}{x} = \lim\limits_{x \to 0} \dfrac{f(x) - f(0)}{x - 0} = f'(0) = 1$.

6.3* The Natural Exponential Function

1. The function value at $x = 0$ is 1 and the slope at $x = 0$ is 1; that is, if $f(x) = e^x$, then $f'(0) = 1$.

3. (a) $\ln \dfrac{1}{e^2} = \ln e^{-2} = -2$ (b) $\ln \sqrt{e} = \ln e^{1/2} = \tfrac{1}{2}$ (c) $\ln e^{\sin x} = \sin x$

5. (a) $\ln(4x+2) = 3 \Rightarrow e^{\ln(4x+2)} = e^3 \Rightarrow 4x + 2 = e^3 \Rightarrow 4x = e^3 - 2 \Rightarrow x = \tfrac{1}{4}(e^3 - 2) \approx 4.521$

(b) $e^{2x-3} = 12 \Rightarrow \ln e^{2x-3} = \ln 12 \Rightarrow 2x - 3 = \ln 12 \Rightarrow 2x = 3 + \ln 12 \Rightarrow x = \tfrac{1}{2}(3 + \ln 12) \approx 2.742$

7. (a) $\ln x + \ln(x-1) = 0 \Rightarrow \ln[x(x-1)] = 0 \Rightarrow e^{\ln[x^2-x]} = e^0 \Rightarrow x^2 - x = 1 \Rightarrow x^2 - x - 1 = 0$. The quadratic formula gives $x = \dfrac{1 \pm \sqrt{(-1)^2 - 4(1)(-1)}}{2(1)} = \dfrac{1 \pm \sqrt{5}}{2}$, but we note that $\ln \dfrac{1 - \sqrt{5}}{2}$ is undefined because $\dfrac{1 - \sqrt{5}}{2} < 0$. Thus, $x = \dfrac{1 + \sqrt{5}}{2} \approx 1.618$.

(b) $e - e^{-2x} = 1 \Rightarrow e - 1 = e^{-2x} \Rightarrow \ln(e-1) = \ln e^{-2x} \Rightarrow \ln(e-1) = -2x \Rightarrow$
$x = -\frac{1}{2}\ln(e-1) \approx -0.271$

9. (a) $e^{2x} - 3e^x + 2 = 0 \Leftrightarrow (e^x - 1)(e^x - 2) = 0 \Leftrightarrow e^x = 1$ or $e^x = 2 \Leftrightarrow x = \ln 1$ or $x = \ln 2$, so $x = 0$ or $\ln 2$.

(b) $e^{e^x} = 10 \Leftrightarrow \ln\left(e^{e^x}\right) = \ln 10 \Leftrightarrow e^x \ln e = e^x = \ln 10 \Leftrightarrow \ln e^x = \ln(\ln 10) \Leftrightarrow x = \ln \ln 10$

11. (a) $\ln(1 + x^3) - 4 = 0 \Leftrightarrow \ln(1 + x^3) = 4 \Leftrightarrow 1 + x^3 = e^4 \Leftrightarrow x^3 = e^4 - 1 \Leftrightarrow x = \sqrt[3]{e^4 - 1} \approx 3.7704$.

(b) $2e^{1/x} = 42 \Leftrightarrow e^{1/x} = 21 \Leftrightarrow \frac{1}{x} = \ln 21 \Leftrightarrow x = \frac{1}{\ln 21} \approx 0.3285$.

13. (a) $\ln x < 0 \Rightarrow x < e^0 \Rightarrow x < 1$. Since the domain of $f(x) = \ln x$ is $x > 0$, the solution of the original inequality is $0 < x < 1$.

(b) $e^x > 5 \Rightarrow \ln e^x > \ln 5 \Rightarrow x > \ln 5$

15. We start with the graph of $y = e^x$ (Figure 2) and reflect about the y-axis to get the graph of $y = e^{-x}$. Then we reflect the graph about the x-axis to get the graph of $y = -e^{-x}$.

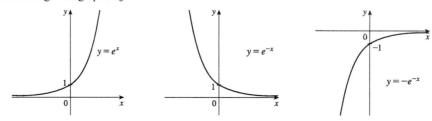

17. We start with the graph of $y = e^x$ (Figure 2) and reflect about the y-axis to get the graph of $y = e^{-x}$. Then we compress the graph vertically by a factor of 2 to obtain the graph of $y = \frac{1}{2}e^{-x}$ and then reflect about the x-axis to get the graph of $y = -\frac{1}{2}e^{-x}$. Finally, we shift the graph one unit upward to get the graph of $y = 1 - \frac{1}{2}e^{-x}$.

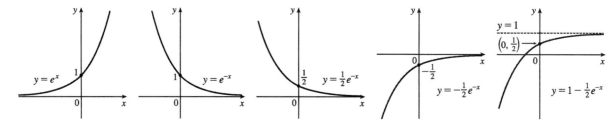

19. (a) For $f(x) = \sqrt{3 - e^{2x}}$, we must have $3 - e^{2x} \geq 0 \Rightarrow e^{2x} \leq 3 \Rightarrow 2x \leq \ln 3 \Rightarrow x \leq \frac{1}{2}\ln 3$.
Thus, the domain of f is $(-\infty, \frac{1}{2}\ln 3]$.

(b) $y = f(x) = \sqrt{3 - e^{2x}}$ [note that $y \geq 0$] $\Rightarrow y^2 = 3 - e^{2x} \Rightarrow e^{2x} = 3 - y^2 \Rightarrow 2x = \ln(3 - y^2) \Rightarrow$
$x = \frac{1}{2}\ln(3 - y^2)$. Interchange x and y: $y = \frac{1}{2}\ln(3 - x^2)$. So $f^{-1}(x) = \frac{1}{2}\ln(3 - x^2)$. For the domain of f^{-1},
we must have $3 - x^2 > 0 \Rightarrow x^2 < 3 \Rightarrow |x| < \sqrt{3} \Rightarrow -\sqrt{3} < x < \sqrt{3} \Rightarrow 0 \leq x < \sqrt{3}$ since $x \geq 0$. Note
that the domain of f^{-1}, $[0, \sqrt{3})$, equals the range of f.

21. We solve $y = 3\ln(x-2)$ for x: $y/3 = \ln(x-2) \;\Rightarrow\; e^{y/3} = x - 2 \;\Rightarrow\; x = 2 + e^{y/3}$. Interchanging x and y gives the inverse function $y = 2 + e^{x/3}$.

23. We solve $y = e^{1-x}$ for x: $\ln y = \ln e^{1-x} \;\Rightarrow\; \ln y = 1 - x \;\Rightarrow\; x = 1 - \ln y$. Interchanging x and y gives the inverse function $y = 1 - \ln x$.

25. Divide numerator and denominator by e^{3x}: $\displaystyle\lim_{x \to \infty} \frac{e^{3x} - e^{-3x}}{e^{3x} + e^{-3x}} = \lim_{x \to \infty} \frac{1 - e^{-6x}}{1 + e^{-6x}} = \frac{1 - 0}{1 + 0} = 1$

27. Let $t = 3/(2-x)$. As $x \to 2^+$, $t \to -\infty$. So $\displaystyle\lim_{x \to 2^+} e^{3/(2-x)} = \lim_{t \to -\infty} e^t = 0$ by (6).

29. Since $-1 \leq \cos x \leq 1$ and $e^{-2x} > 0$, we have $-e^{-2x} \leq e^{-2x} \cos x \leq e^{-2x}$. We know that $\displaystyle\lim_{x \to \infty}(-e^{-2x}) = 0$ and $\displaystyle\lim_{x \to \infty}(e^{-2x}) = 0$, so by the Squeeze Theorem, $\displaystyle\lim_{x \to \infty}(e^{-2x} \cos x) = 0$.

31. $f(t) = -2e^t \;\Rightarrow\; f'(t) = -2(e^t) = -2e^t$

33. $f(x) = (3x^2 - 5x)e^x \;\overset{PR}{\Rightarrow}$
$f'(x) = (3x^2 - 5x)(e^x)' + e^x(3x^2 - 5x)' = (3x^2 - 5x)e^x + e^x(6x - 5)$
$= e^x[(3x^2 - 5x) + (6x - 5)] = e^x(3x^2 + x - 5)$

35. By (9), $y = e^{ax^3} \;\Rightarrow\; y' = e^{ax^3} \dfrac{d}{dx}(ax^3) = 3ax^2 e^{ax^3}$.

37. $y = e^{\tan\theta} \;\Rightarrow\; y' = e^{\tan\theta} \dfrac{d}{d\theta}(\tan\theta) = (\sec^2\theta)e^{\tan\theta}$

39. $f(x) = \dfrac{x^2 e^x}{x^2 + e^x} \;\overset{QR}{\Rightarrow}$
$f'(x) = \dfrac{(x^2 + e^x)[x^2 e^x + e^x(2x)] - x^2 e^x(2x + e^x)}{(x^2 + e^x)^2} = \dfrac{x^4 e^x + 2x^3 e^x + x^2 e^{2x} + 2xe^{2x} - 2x^3 e^x - x^2 e^{2x}}{(x^2 + e^x)^2}$
$= \dfrac{x^4 e^x + 2xe^{2x}}{(x^2 + e^x)^2} = \dfrac{xe^x(x^3 + 2e^x)}{(x^2 + e^x)^2}$

41. Using the Product Rule and the Chain Rule, $y = x^2 e^{-3x} \;\Rightarrow$
$y' = x^2 e^{-3x}(-3) + e^{-3x}(2x) = e^{-3x}(-3x^2 + 2x) = xe^{-3x}(2 - 3x)$.

43. $f(t) = e^{at}\sin bt \;\Rightarrow\; f'(t) = e^{at}(\cos bt)\cdot b + (\sin bt)e^{at}\cdot a = e^{at}(b\cos bt + a\sin bt)$

45. By (9), $F(t) = e^{t\sin 2t} \;\Rightarrow$
$F'(t) = e^{t\sin 2t}(t\sin 2t)' = e^{t\sin 2t}(t\cdot 2\cos 2t + \sin 2t \cdot 1) = e^{t\sin 2t}(2t\cos 2t + \sin 2t)$

47. $g(u) = e^{\sqrt{\sec u^2}} \;\Rightarrow$
$g'(u) = e^{\sqrt{\sec u^2}}\dfrac{d}{du}\sqrt{\sec u^2} = e^{\sqrt{\sec u^2}}\dfrac{1}{2}(\sec u^2)^{-1/2}\dfrac{d}{du}\sec u^2$
$= e^{\sqrt{\sec u^2}}\dfrac{1}{2\sqrt{\sec u^2}}\cdot \sec u^2 \tan u^2 \cdot 2u = u\sqrt{\sec u^2}\tan u^2\, e^{\sqrt{\sec u^2}}$

49. $g(x) = \sin\left(\dfrac{e^x}{1+e^x}\right) \Rightarrow$

$g'(x) = \cos\left(\dfrac{e^x}{1+e^x}\right) \cdot \dfrac{(1+e^x)e^x - e^x(e^x)}{(1+e^x)^2} = \cos\left(\dfrac{e^x}{1+e^x}\right) \cdot \dfrac{e^x(1+e^x - e^x)}{(1+e^x)^2} = \dfrac{e^x}{(1+e^x)^2} \cos\left(\dfrac{e^x}{1+e^x}\right)$

51. $y = e^x \cos x + \sin x \;\Rightarrow\; y' = e^x(-\sin x) + (\cos x)(e^x) + \cos x = e^x(\cos x - \sin x) + \cos x$, so

$y'(0) = e^0(\cos 0 - \sin 0) + \cos 0 = 1(1 - 0) + 1 = 2$. An equation of the tangent line to the curve $y = e^x \cos x + \sin x$ at the point $(0, 1)$ is $y - 1 = 2(x - 0)$ or $y = 2x + 1$.

53. $\dfrac{d}{dx}(e^{x/y}) = \dfrac{d}{dx}(x - y) \;\Rightarrow\; e^{x/y} \cdot \dfrac{d}{dx}\left(\dfrac{x}{y}\right) = 1 - y' \;\Rightarrow\; e^{x/y} \cdot \dfrac{y \cdot 1 - x \cdot y'}{y^2} = 1 - y' \;\Rightarrow$

$e^{x/y} \cdot \dfrac{1}{y} - \dfrac{xe^{x/y}}{y^2} \cdot y' = 1 - y' \;\Rightarrow\; y' - \dfrac{xe^{x/y}}{y^2} \cdot y' = 1 - \dfrac{e^{x/y}}{y} \;\Rightarrow\; y'\left(1 - \dfrac{xe^{x/y}}{y^2}\right) = \dfrac{y - e^{x/y}}{y} \;\Rightarrow$

$y' = \dfrac{\dfrac{y - e^{x/y}}{y}}{\dfrac{y^2 - xe^{x/y}}{y^2}} = \dfrac{y(y - e^{x/y})}{y^2 - xe^{x/y}}$

55. $y = e^x + e^{-x/2} \;\Rightarrow\; y' = e^x - \tfrac{1}{2}e^{-x/2} \;\Rightarrow\; y'' = e^x + \tfrac{1}{4}e^{-x/2}$, so

$2y'' - y' - y = 2\left(e^x + \tfrac{1}{4}e^{-x/2}\right) - \left(e^x - \tfrac{1}{2}e^{-x/2}\right) - \left(e^x + e^{-x/2}\right) = 0$.

57. $y = e^{rx} \;\Rightarrow\; y' = re^{rx} \;\Rightarrow\; y'' = r^2 e^{rx}$, so if $y = e^{rx}$ satisfies the differential equation $y'' + 6y' + 8y = 0$,

then $r^2 e^{rx} + 6re^{rx} + 8e^{rx} = 0$; that is, $e^{rx}(r^2 + 6r + 8) = 0$. Since $e^{rx} > 0$ for all x, we must have $r^2 + 6r + 8 = 0$,

or $(r + 2)(r + 4) = 0$, so $r = -2$ or -4.

59. $f(x) = e^{2x} \;\Rightarrow\; f'(x) = 2e^{2x} \;\Rightarrow\; f''(x) = 2 \cdot 2e^{2x} = 2^2 e^{2x} \;\Rightarrow$

$f'''(x) = 2^2 \cdot 2e^{2x} = 2^3 e^{2x} \;\Rightarrow\; \cdots \;\Rightarrow\; f^{(n)}(x) = 2^n e^{2x}$

61. (a) $f(x) = e^x + x$ is continuous on $\mathbb{R}$ and $f(-1) = e^{-1} - 1 < 0 < 1 = f(0)$, so by the Intermediate Value Theorem, $e^x + x = 0$ has a solution in $(-1, 0)$.

(b) $f(x) = e^x + x \;\Rightarrow\; f'(x) = e^x + 1$, so $x_{n+1} = x_n - \dfrac{e^{x_n} + x_n}{e^{x_n} + 1}$. Using $x_1 = -0.5$, we get $x_2 \approx -0.566311$,

$x_3 \approx -0.567143 \approx x_4$, so the solution is -0.567143 to six decimal places.

63. (a) $\lim\limits_{t \to \infty} p(t) = \lim\limits_{t \to \infty} \dfrac{1}{1 + ae^{-kt}} = \dfrac{1}{1 + a \cdot 0} = 1$, since $k > 0 \;\Rightarrow\; -kt \to -\infty \;\Rightarrow\; e^{-kt} \to 0$. As time increases, the proportion of the population that has heard the rumor approaches 1; that is, everyone in the population has heard the rumor.

(b) $p(t) = (1 + ae^{-kt})^{-1} \;\Rightarrow\; \dfrac{dp}{dt} = -(1 + ae^{-kt})^{-2}(-kae^{-kt}) = \dfrac{kae^{-kt}}{(1 + ae^{-kt})^2}$

(c)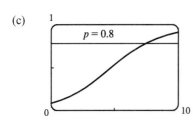

From the graph of $p(t) = (1 + 10e^{-0.5t})^{-1}$, it seems that $p(t) = 0.8$ (indicating that 80% of the population has heard the rumor) when $t \approx 7.4$ hours.

65. $f(x) = \dfrac{e^x}{1+x^2}$, $[0, 3]$. $f'(x) = \dfrac{(1+x^2)e^x - e^x(2x)}{(1+x^2)^2} = \dfrac{e^x(x^2 - 2x + 1)}{(1+x^2)^2} = \dfrac{e^x(x-1)^2}{(1+x^2)^2}$. $f'(x) = 0 \Rightarrow$ $(x-1)^2 = 0 \Leftrightarrow x = 1$. $f'(x)$ exists for all real numbers since $1 + x^2$ is never equal to 0. $f(0) = 1$, $f(1) = e/2 \approx 1.359$, and $f(3) = e^3/10 \approx 2.009$. So $f(3) = e^3/10$ is the absolute maximum value and $f(0) = 1$ is the absolute minimum value.

67. $f(x) = x - e^x \Rightarrow f'(x) = 1 - e^x = 0 \Leftrightarrow e^x = 1 \Leftrightarrow x = 0$. Now $f'(x) > 0$ for all $x < 0$ and $f'(x) < 0$ for all $x > 0$, so the absolute maximum value is $f(0) = 0 - 1 = -1$.

69. (a) $f(x) = xe^{2x} \Rightarrow f'(x) = x(2e^{2x}) + e^{2x}(1) = e^{2x}(2x + 1)$. Thus, $f'(x) > 0$ if $x > -\frac{1}{2}$ and $f'(x) < 0$ if $x < -\frac{1}{2}$. So f is increasing on $\left(-\frac{1}{2}, \infty\right)$ and f is decreasing on $\left(-\infty, -\frac{1}{2}\right)$.

(b) $f''(x) = e^{2x}(2) + (2x+1) \cdot 2e^{2x} = 2e^{2x}[1 + (2x+1)] = 2e^{2x}(2x+2) = 4e^{2x}(x+1)$. $f''(x) > 0 \Leftrightarrow x > -1$ and $f''(x) < 0 \Leftrightarrow x < -1$. Thus, f is concave upward on $(-1, \infty)$ and f is concave downward on $(-\infty, -1)$.

(c) There is an inflection point at $\left(-1, -e^{-2}\right)$, or $\left(-1, -1/e^2\right)$.

71. $y = f(x) = e^{-1/(x+1)}$ **A.** $D = \{x \mid x \neq -1\} = (-\infty, -1) \cup (-1, \infty)$ **B.** No x-intercept; y-intercept $= f(0) = e^{-1}$
C. No symmetry **D.** $\lim\limits_{x \to \pm\infty} e^{-1/(x+1)} = 1$ since $-1/(x+1) \to 0$, so $y = 1$ is a HA. $\lim\limits_{x \to -1^+} e^{-1/(x+1)} = 0$ since $-1/(x+1) \to -\infty$, $\lim\limits_{x \to -1^-} e^{-1/(x+1)} = \infty$ since $-1/(x+1) \to \infty$, so $x = -1$ is a VA.
E. $f'(x) = e^{-1/(x+1)}/(x+1)^2 \Rightarrow f'(x) > 0$ for all x except 1, so f is increasing on $(-\infty, -1)$ and $(-1, \infty)$. **F.** No extreme values **H.**
G. $f''(x) = \dfrac{e^{-1/(x+1)}}{(x+1)^4} + \dfrac{e^{-1/(x+1)}(-2)}{(x+1)^3} = -\dfrac{e^{-1/(x+1)}(2x+1)}{(x+1)^4} \Rightarrow$
$f''(x) > 0 \Leftrightarrow 2x + 1 < 0 \Leftrightarrow x < -\frac{1}{2}$, so f is CU on $(-\infty, -1)$ and $\left(-1, -\frac{1}{2}\right)$, and CD on $\left(-\frac{1}{2}, \infty\right)$. f has an IP at $\left(-\frac{1}{2}, e^{-2}\right)$.

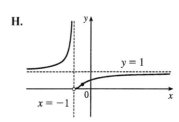

73. $y = 1/(1 + e^{-x})$ **A.** $D = \mathbb{R}$ **B.** No x-intercept; y-intercept $= f(0) = \frac{1}{2}$. **C.** No symmetry
D. $\lim\limits_{x \to \infty} 1/(1+e^{-x}) = \frac{1}{1+0} = 1$ and $\lim\limits_{x \to -\infty} 1/(1+e^{-x}) = 0$ since $\lim\limits_{x \to -\infty} e^{-x} = \infty$, so f has horizontal asymptotes $y = 0$ and $y = 1$. **E.** $f'(x) = -(1+e^{-x})^{-2}(-e^{-x}) = e^{-x}/(1+e^{-x})^2$. This is positive for all x, so f is increasing on $\mathbb{R}$.

[continued]

F. No extreme values   **G.** $f''(x) = \dfrac{(1+e^{-x})^2(-e^{-x}) - e^{-x}(2)(1+e^{-x})(-e^{-x})}{(1+e^{-x})^4} = \dfrac{e^{-x}(e^{-x}-1)}{(1+e^{-x})^3}$

The second factor in the numerator is negative for $x > 0$ and positive for $x < 0$,   **H.**

and the other factors are always positive, so f is CU on $(-\infty, 0)$ and CD

on $(0, \infty)$. IP at $\left(0, \tfrac{1}{2}\right)$

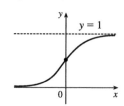

75. $f(x) = e^{x^3 - x} \to 0$ as $x \to -\infty$, and

$f(x) \to \infty$ as $x \to \infty$. From the graph,

it appears that f has a local minimum of

about $f(0.58) = 0.68$, and a local

maximum of about $f(-0.58) = 1.47$.

To find the exact values, we calculate

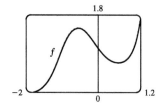 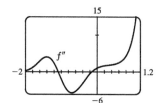

$f'(x) = (3x^2 - 1)e^{x^3 - x}$, which is 0 when $3x^2 - 1 = 0 \iff x = \pm\tfrac{1}{\sqrt{3}}$. The negative solution corresponds to the local

maximum $f\left(-\tfrac{1}{\sqrt{3}}\right) = e^{(-1/\sqrt{3})^3 - (-1/\sqrt{3})} = e^{2\sqrt{3}/9}$, and the positive solution corresponds to the local minimum

$f\left(\tfrac{1}{\sqrt{3}}\right) = e^{(1/\sqrt{3})^3 - (1/\sqrt{3})} = e^{-2\sqrt{3}/9}$. To estimate the inflection points, we calculate and graph

$f''(x) = \dfrac{d}{dx}\left[(3x^2 - 1)e^{x^3 - x}\right] = (3x^2 - 1)e^{x^3 - x}(3x^2 - 1) + e^{x^3 - x}(6x) = e^{x^3 - x}(9x^4 - 6x^2 + 6x + 1)$.

From the graph, it appears that $f''(x)$ changes sign (and thus f has inflection points) at $x \approx -0.15$ and $x \approx -1.09$. From the

graph of f, we see that these x-values correspond to inflection points at about $(-0.15, 1.15)$ and $(-1.09, 0.82)$.

77. Let $a = 0.135$ and $b = -2.802$. Then $C(t) = ate^{bt} \Rightarrow C'(t) = a(t \cdot e^{bt} \cdot b + e^{bt} \cdot 1) = ae^{bt}(bt + 1)$. $C'(t) = 0 \iff$

$bt + 1 = 0 \iff t = -\dfrac{1}{b} \approx 0.36$ h. $C(0) = 0$, $C(-1/b) = -\dfrac{a}{b}e^{-1} = -\dfrac{a}{be} \approx 0.0177$, and $C(3) = 3ae^{3b} \approx 0.00009$.

The maximum average BAC during the first three hours is about 0.0177 g/dL and it occurs at approximately 0.36 h

(21.4 min).

79. $\displaystyle\int_0^1 (x^e + e^x)\, dx = \left[\dfrac{x^{e+1}}{e+1} + e^x\right]_0^1 = \left(\dfrac{1}{e+1} + e\right) - (0 + 1) = \dfrac{1}{e+1} + e - 1$

81. $\displaystyle\int_0^2 \dfrac{dx}{e^{\pi x}} = \int_0^2 e^{-\pi x}\, dx = \left[-\dfrac{1}{\pi}e^{-\pi x}\right]_0^2 = -\dfrac{1}{\pi}e^{-2\pi} + \dfrac{1}{\pi}e^0 = \dfrac{1}{\pi}(1 - e^{-2\pi})$

83. Let $u = 1 + e^x$. Then $du = e^x\, dx$, so $\int e^x\sqrt{1 + e^x}\, dx = \int \sqrt{u}\, du = \tfrac{2}{3}u^{3/2} + C = \tfrac{2}{3}(1 + e^x)^{3/2} + C$.

85. $\int (e^x + e^{-x})^2\, dx = \int (e^{2x} + 2 + e^{-2x})\, dx = \tfrac{1}{2}e^{2x} + 2x - \tfrac{1}{2}e^{-2x} + C$

87. Let $x = 1 - e^u$. Then $dx = -e^u\, du$ and $e^u\, du = -dx$, so

$\displaystyle\int \dfrac{e^u}{(1 - e^u)^2}\, du = \int \dfrac{1}{x^2}(-dx) = -\int x^{-2}\, dx = -(-x^{-1}) + C = \dfrac{1}{x} + C = \dfrac{1}{1 - e^u} + C$.

89. Let $u = 1/x$, so $du = -1/x^2\, dx$. When $x = 1$, $u = 1$; when $x = 2$, $u = \frac{1}{2}$. Thus,

$$\int_1^2 \frac{e^{1/x}}{x^2}\, dx = \int_1^{1/2} e^u\, (-du) = -\bigl[e^u\bigr]_1^{1/2} = -(e^{1/2} - e) = e - \sqrt{e}.$$

91. $f_{\text{avg}} = \frac{1}{2-0} \int_0^2 2xe^{-x^2}\, dx$

$= \frac{1}{2}\Bigl[-e^{-x^2}\Bigr]_0^2 = \frac{1}{2}(-e^{-4} + 1)$

93. Area $= \int_0^1 (e^{3x} - e^x)\, dx = \bigl[\tfrac{1}{3}e^{3x} - e^x\bigr]_0^1 = \bigl(\tfrac{1}{3}e^3 - e\bigr) - \bigl(\tfrac{1}{3} - 1\bigr) = \tfrac{1}{3}e^3 - e + \tfrac{2}{3} \approx 4.644$

95. $V = \int_0^1 \pi(e^x)^2\, dx = \pi \int_0^1 e^{2x}\, dx = \tfrac{1}{2}\pi\bigl[e^{2x}\bigr]_0^1 = \tfrac{\pi}{2}\bigl(e^2 - 1\bigr)$

97. First Figure Let $u = \sqrt{x}$, so $x = u^2$ and $dx = 2u\, du$. When $x = 0$, $u = 0$; when $x = 1$, $u = 1$. Thus,

$A_1 = \int_0^1 e^{\sqrt{x}}\, dx = \int_0^1 e^u (2u\, du) = 2\int_0^1 ue^u\, du.$

Second Figure $A_2 = \int_0^1 2xe^x\, dx = 2\int_0^1 ue^u\, du.$

Third Figure Let $u = \sin x$, so $du = \cos x\, dx$. When $x = 0$, $u = 0$; when $x = \tfrac{\pi}{2}$, $u = 1$. Thus,

$A_3 = \int_0^{\pi/2} e^{\sin x} \sin 2x\, dx = \int_0^{\pi/2} e^{\sin x}(2\sin x \cos x)\, dx = \int_0^1 e^u(2u\, du) = 2\int_0^1 ue^u\, du.$

Since $A_1 = A_2 = A_3$, all three areas are equal.

99. The rate is measured in liters per minute. Integrating from $t = 0$ minutes to $t = 60$ minutes will give us the total amount of oil that leaks out (in liters) during the first hour.

$\int_0^{60} r(t)\, dt = \int_0^{60} 100 e^{-0.01t}\, dt \qquad [u = -0.01t,\, du = -0.01\, dt]$

$= 100 \int_0^{-0.6} e^u (-100\, du) = -10{,}000\bigl[e^u\bigr]_0^{-0.6} = -10{,}000(e^{-0.6} - 1) \approx 4511.9 \approx 4512\, \text{L}$

101. $\displaystyle\int_0^{30} u(t)\, dt = \int_0^{30} \frac{r}{V} C_0 e^{-rt/V}\, dt = C_0 \int_1^{e^{-30r/V}} (-dx) \qquad \begin{bmatrix} x = e^{-rt/V}, \\ dx = -\frac{r}{V}e^{-rt/V}\, dt \end{bmatrix}$

$= C_0 \bigl[-x\bigr]_1^{e^{-30r/V}} = C_0(-e^{-30r/V} + 1)$

The integral $\int_0^{30} u(t)\, dt$ represents the total amount of urea removed from the blood in the first 30 minutes of dialysis.

103.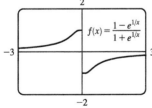

From the graph, it appears that f is an odd function (f is undefined for $x = 0$). To prove this, we must show that $f(-x) = -f(x)$.

$f(-x) = \dfrac{1 - e^{1/(-x)}}{1 + e^{1/(-x)}} = \dfrac{1 - e^{(-1/x)}}{1 + e^{(-1/x)}} = \dfrac{1 - \dfrac{1}{e^{1/x}}}{1 + \dfrac{1}{e^{1/x}}} \cdot \dfrac{e^{1/x}}{e^{1/x}} = \dfrac{e^{1/x} - 1}{e^{1/x} + 1}$

$= -\dfrac{1 - e^{1/x}}{1 + e^{1/x}} = -f(x)$

so f is an odd function.

105. Using the second law of logarithms and Equation 5, we have $\ln(e^x/e^y) = \ln e^x - \ln e^y = x - y = \ln(e^{x-y})$. Since $\ln$ is a one-to-one function, it follows that $e^x/e^y = e^{x-y}$.

6.4* General Logarithmic and Exponential Functions

1. (a) $b^x = e^{x \ln b}$ (b) The domain of $f(x) = b^x$ is $\mathbb{R}$.

    (c) The range of $f(x) = b^x$ $[b \neq 1]$ is $(0, \infty)$.

    (d) (i) See Figure 1.   (ii) See Figure 3.   (iii) See Figure 2.

3. Since $b^x = e^{x \ln b}$, $4^{-\pi} = e^{-\pi \ln 4}$.

5. Since $b^x = e^{x \ln b}$, $10^{x^2} = e^{x^2 \ln 10}$.

7. (a) $\log_3 81 = \log_3 3^4 = 4$ (b) $\log_3 \left(\frac{1}{81}\right) = \log_3 3^{-4} = -4$ (c) $\log_9 3 = \log_9 9^{1/2} = \frac{1}{2}$

9. (a) $\log_3 10 - \log_3 5 - \log_3 18 = \log_3 \left(\frac{10}{5}\right) - \log_3 18 = \log_3 2 - \log_3 18 = \log_3 \left(\frac{2}{18}\right) = \log_3 \left(\frac{1}{9}\right)$
$= \log_3 3^{-2} = -2$

    (b) $2 \log_5 100 - 4 \log_5 50 = \log_5 100^2 - \log_5 50^4 = \log_5 \left(\frac{100^2}{50^4}\right) = \log_5 \left(\frac{10^4}{5^4 \cdot 10^4}\right) = \log_5 5^{-4} = -4$

11. All of these graphs approach 0 as $x \to -\infty$, all of them pass through the point $(0, 1)$, and all of them are increasing and approach ∞ as $x \to \infty$. The larger the base, the faster the function increases for $x > 0$, and the faster it approaches 0 as $x \to -\infty$.

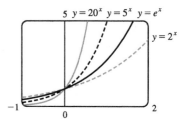

13. (a) $\log_5 10 = \dfrac{\ln 10}{\ln 5} \approx 1.430677$ (b) $\log_3 12 = \dfrac{\ln 12}{\ln 3} \approx 2.261860$ (c) $\log_{12} 6 = \dfrac{\ln 6}{\ln 12} \approx 0.721057$

15. To graph these functions, we use $\log_{1.5} x = \dfrac{\ln x}{\ln 1.5}$ and $\log_{50} x = \dfrac{\ln x}{\ln 50}$.
These graphs all approach $-\infty$ as $x \to 0^+$, and they all pass through the point $(1, 0)$. Also, they are all increasing, and all approach ∞ as $x \to \infty$. The functions with larger bases increase extremely slowly, and the ones with smaller bases do so somewhat more quickly. The functions with large bases approach the y-axis more closely as $x \to 0^+$.

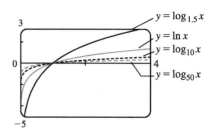

17. Use $y = Cb^x$ with the points $(1, 6)$ and $(3, 24)$. $6 = Cb^1$ $[C = \frac{6}{b}]$ and $24 = Cb^3$ $\Rightarrow$ $24 = \left(\frac{6}{b}\right)b^3$ $\Rightarrow$
$4 = b^2$ $\Rightarrow$ $b = 2$ [since $b > 0$] and $C = \frac{6}{2} = 3$. The function is $f(x) = 3 \cdot 2^x$.

19. (a) 2 ft = 24 in, $f(24) = 24^2$ in = 576 in = 48 ft. $g(24) = 2^{24}$ in = $2^{24}/(12 \cdot 5280)$ mi ≈ 265 mi

(b) 3 ft = 36 in, so we need x such that $\log_2 x = 36 \Leftrightarrow x = 2^{36} = 68{,}719{,}476{,}736$. In miles, this is

$$68{,}719{,}476{,}736 \text{ in} \cdot \frac{1 \text{ ft}}{12 \text{ in}} \cdot \frac{1 \text{ mi}}{5280 \text{ ft}} \approx 1{,}084{,}587.7 \text{ mi}.$$

21. $\lim_{x \to \infty} (1.001)^x = \infty$ by Figure 1, since $1.001 > 1$.

23. $\lim_{t \to \infty} 2^{-t^2} = \lim_{u \to -\infty} 2^u$ [where $u = -t^2$] $= 0$

25. $f(x) = x^5 + 5^x \Rightarrow f'(x) = 5x^4 + 5^x \ln 5$

27. Using Formula 4 and the Chain Rule, $G(x) = 4^{C/x} \Rightarrow$

$$G'(x) = 4^{C/x} (\ln 4) \frac{d}{dx} \frac{C}{x} \quad \left[\frac{C}{x} = Cx^{-1}\right] = 4^{C/x} (\ln 4)(-Cx^{-2}) = -C (\ln 4) \frac{4^{C/x}}{x^2}.$$

29. $L(v) = \tan(4^{v^2}) \Rightarrow L'(v) = \sec^2\left(4^{v^2}\right) \frac{d}{dv}\left(4^{v^2}\right) = \sec^2\left(4^{v^2}\right) \cdot 4^{v^2} \ln 4 \frac{d}{dv}(v^2) = 2v \ln 4 \sec^2\left(4^{v^2}\right) \cdot 4^{v^2}$

31. $y = \log_8(x^2 + 3x) \Rightarrow y' = \frac{1}{(x^2 + 3x) \ln 8} \cdot \frac{d}{dx}(x^2 + 3x) = \frac{1}{(x^2 + 3x) \ln 8} \cdot (2x + 3) = \frac{2x + 3}{(x^2 + 3x) \ln 8}$

33. $y = x \log_4 \sin x \Rightarrow y' = x \cdot \frac{1}{\sin x \ln 4} \cdot \cos x + \log_4 \sin x \cdot 1 = \frac{x \cot x}{\ln 4} + \log_4 \sin x$

35. $y = x^x \Rightarrow \ln y = \ln x^x \Rightarrow \ln y = x \ln x \Rightarrow y'/y = x(1/x) + (\ln x) \cdot 1 \Rightarrow y' = y(1 + \ln x) \Rightarrow$
$y' = x^x(1 + \ln x)$

37. $y = x^{\sin x} \Rightarrow \ln y = \ln x^{\sin x} \Rightarrow \ln y = \sin x \ln x \Rightarrow \frac{y'}{y} = (\sin x) \cdot \frac{1}{x} + (\ln x)(\cos x) \Rightarrow$
$y' = y\left(\frac{\sin x}{x} + \ln x \cos x\right) \Rightarrow y' = x^{\sin x}\left(\frac{\sin x}{x} + \ln x \cos x\right)$

39. $y = (\cos x)^x \Rightarrow \ln y = \ln(\cos x)^x \Rightarrow \ln y = x \ln \cos x \Rightarrow \frac{1}{y} y' = x \cdot \frac{1}{\cos x} \cdot (-\sin x) + \ln \cos x \cdot 1 \Rightarrow$
$y' = y\left(\ln \cos x - \frac{x \sin x}{\cos x}\right) \Rightarrow y' = (\cos x)^x (\ln \cos x - x \tan x)$

41. $y = x^{\ln x} \Rightarrow \ln y = \ln x \ln x = (\ln x)^2 \Rightarrow \frac{y'}{y} = 2 \ln x \left(\frac{1}{x}\right) \Rightarrow y' = x^{\ln x}\left(\frac{2 \ln x}{x}\right)$

43. $y = 10^x \Rightarrow y' = 10^x \ln 10$, so at $(1, 10)$, the slope of the tangent line is $10^1 \ln 10 = 10 \ln 10$, and its equation is
$y - 10 = 10 \ln 10 (x - 1)$, or $y = (10 \ln 10)x + 10(1 - \ln 10)$.

45. $\int_0^4 2^s \, ds = \left[\frac{1}{\ln 2} 2^s\right]_0^4 = \frac{16}{\ln 2} - \frac{1}{\ln 2} = \frac{15}{\ln 2}$

47. $\int \dfrac{\log_{10} x}{x}\, dx = \int \dfrac{(\ln x)/(\ln 10)}{x}\, dx = \dfrac{1}{\ln 10} \int \dfrac{\ln x}{x}\, dx$. Now put $u = \ln x$, so $du = \dfrac{1}{x}\, dx$, and the expression becomes

$\dfrac{1}{\ln 10} \int u\, du = \dfrac{1}{\ln 10}\left(\tfrac{1}{2} u^2 + C_1\right) = \dfrac{1}{2 \ln 10}(\ln x)^2 + C$.

Or: The substitution $u = \log_{10} x$ gives $du = \dfrac{dx}{x \ln 10}$ and we get $\int \dfrac{\log_{10} x}{x}\, dx = \tfrac{1}{2} \ln 10 (\log_{10} x)^2 + C$.

49. Let $u = \sin \theta$. Then $du = \cos \theta\, d\theta$ and $\int 3^{\sin \theta} \cos \theta\, d\theta = \int 3^u\, du = \dfrac{3^u}{\ln 3} + C = \dfrac{1}{\ln 3} 3^{\sin \theta} + C$.

51. $A = \displaystyle\int_{-1}^{0} (2^x - 5^x)\, dx + \int_{0}^{1} (5^x - 2^x)\, dx = \left[\dfrac{2^x}{\ln 2} - \dfrac{5^x}{\ln 5}\right]_{-1}^{0} + \left[\dfrac{5^x}{\ln 5} - \dfrac{2^x}{\ln 2}\right]_{0}^{1}$

$= \left(\dfrac{1}{\ln 2} - \dfrac{1}{\ln 5}\right) - \left(\dfrac{1/2}{\ln 2} - \dfrac{1/5}{\ln 5}\right) + \left(\dfrac{5}{\ln 5} - \dfrac{2}{\ln 2}\right) - \left(\dfrac{1}{\ln 5} - \dfrac{1}{\ln 2}\right)$

$= \dfrac{16}{5 \ln 5} - \dfrac{1}{2 \ln 2}$

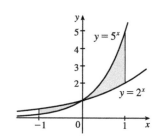

53. We see that the graphs of $y = 2^x$ and $y = 1 + 3^{-x}$ intersect at $x \approx 0.6$. We let $f(x) = 2^x - 1 - 3^{-x}$ and calculate $f'(x) = 2^x \ln 2 + 3^{-x} \ln 3$, and using the formula $x_{n+1} = x_n - f(x_n)/f'(x_n)$ (Newton's method), we get $x_1 = 0.6$, $x_2 \approx x_3 \approx 0.600967$. So, correct to six decimal places, the solution occurs at $x = 0.600967$.

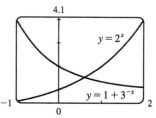

55. $y = g(x) = \log_4(x^3 + 2) \;\Rightarrow\; 4^y = x^3 + 2 \;\Rightarrow\; x^3 = 4^y - 2 \;\Rightarrow\; x = \sqrt[3]{4^y - 2}$. Interchange x and y: $y = \sqrt[3]{4^x - 2}$. So $g^{-1}(x) = \sqrt[3]{4^x - 2}$.

57. If I is the intensity of the 1989 San Francisco earthquake, then $\log_{10}(I/S) = 7.1 \;\Rightarrow\;$
$\log_{10}(16I/S) = \log_{10} 16 + \log_{10}(I/S) = \log_{10} 16 + 7.1 \approx 8.3$.

59. We find I with the loudness formula from Exercise 58, substituting $I_0 = 10^{-12}$ and $L = 50$:

$50 = 10 \log_{10} \dfrac{I}{10^{-12}} \;\Leftrightarrow\; 5 = \log_{10} \dfrac{I}{10^{-12}} \;\Leftrightarrow\; 10^5 = \dfrac{I}{10^{-12}} \;\Leftrightarrow\; I = 10^{-7}$ watt/m^2. Now we differentiate L with respect to I: $L = 10 \log_{10} \dfrac{I}{I_0} \;\Rightarrow\; \dfrac{dL}{dI} = 10 \dfrac{1}{(I/I_0) \ln 10}\left(\dfrac{1}{I_0}\right) = \dfrac{10}{\ln 10}\left(\dfrac{1}{I}\right)$. Substituting $I = 10^{-7}$, we get

$L'(50) = \dfrac{10}{\ln 10}\left(\dfrac{1}{10^{-7}}\right) = \dfrac{10^8}{\ln 10} \approx 4.34 \times 10^7 \dfrac{\text{dB}}{\text{watt/m}^2}$.

61. (a) $I = \log_2\left(\dfrac{2D}{W}\right) \;\Rightarrow\; \dfrac{dI}{dD}$ [W constant] $= \dfrac{1}{\left(\dfrac{2D}{W}\right) \ln 2} \cdot \dfrac{2}{W} = \dfrac{1}{D \ln 2}$

As D increases, the rate of change of difficulty decreases.

(b) $I = \log_2\left(\dfrac{2D}{W}\right) \Rightarrow \dfrac{dI}{dW}$ [D constant] $= \dfrac{1}{\left(\dfrac{2D}{W}\right)\ln 2} \cdot (-2DW^{-2}) = \dfrac{W}{2D\ln 2} \cdot \dfrac{-2D}{W^2} = -\dfrac{1}{W\ln 2}$

The negative sign indicates that difficulty decreases with increasing width. While the magnitude of the rate of change decreases with increasing width $\left(\text{that is, } \left|-\dfrac{1}{W\ln 2}\right| = \dfrac{1}{W\ln 2} \text{ decreases as } W \text{ increases}\right)$, the rate of change itself increases (gets closer to zero from the negative side) with increasing values of W.

(c) The answers to (a) and (b) agree with intuition. For fixed width, the difficulty of acquiring a target increases, but less and less so, as the distance to the target increases. Similarly, for a fixed distance to a target, the difficulty of acquiring the target decreases, but less and less so, as the width of the target increases.

63. Half of 76.0 RNA copies per mL, corresponding to $t = 1$, is 38.0 RNA copies per mL. Using the graph of V in Figure 9, we estimate that it takes about 3.5 additional days for the patient's viral load to decrease to 38 RNA copies per mL.

65. (a) $P = ab^t$ with $a = 4.502714 \times 10^{-20}$ and $b = 1.029953851$, where P is measured in thousands of people. The fit appears to be very good.

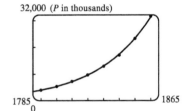

(b) **For 1800:** $m_1 = \dfrac{5308 - 3929}{1800 - 1790} = 137.9$, $m_2 = \dfrac{7240 - 5308}{1810 - 1800} = 193.2$.

So $P'(1800) \approx (m_1 + m_2)/2 = 165.55$ thousand people/year.

For 1850: $m_1 = \dfrac{23{,}192 - 17{,}063}{1850 - 1840} = 612.9$, $m_2 = \dfrac{31{,}443 - 23{,}192}{1860 - 1850} = 825.1$.

So $P'(1850) \approx (m_1 + m_2)/2 = 719$ thousand people/year.

(c) Using $P'(t) = ab^t \ln b$ (from Formula 4) with the values of a and b from part (a), we get $P'(1800) \approx 156.85$ thousand people/year and $P'(1850) \approx 686.07$. These estimates are somewhat less than the ones in part (b).

(d) $P(1870) \approx 41{,}946.56$. The difference of 3.4 million people is most likely due to the Civil War (1861–1865).

67. $y = b^x \Rightarrow y' = b^x \ln b$, so the slope of the tangent line to the curve $y = b^x$ at the point (a, b^a) is $b^a \ln b$. An equation of this tangent line is then $y - b^a = b^a \ln b\,(x - a)$. If c is the x-intercept of this tangent line, then $0 - b^a = b^a \ln b\,(c - a) \Rightarrow$ $-1 = \ln b\,(c - a) \Rightarrow \dfrac{-1}{\ln b} = c - a \Rightarrow |c - a| = \left|\dfrac{-1}{\ln b}\right| = \dfrac{1}{|\ln b|}$. The distance between $(a, 0)$ and $(c, 0)$ is $|c - a|$, and this distance is the constant $\dfrac{1}{|\ln b|}$ for any a. [*Note:* The absolute value is needed for the case $0 < b < 1$ because $\ln b$ is negative there. If $b > 1$, we can write $a - c = 1/(\ln b)$ as the constant distance between $(a, 0)$ and $(c, 0)$.]

69. Using Definition 1 and the second law of exponents for e^x, we have $b^{x-y} = e^{(x-y)\ln b} = e^{x\ln b - y\ln b} = \dfrac{e^{x\ln b}}{e^{y\ln b}} = \dfrac{b^x}{b^y}$.

71. Let $\log_b x = r$ and $\log_b y = s$. Then $b^r = x$ and $b^s = y$.

(a) $xy = b^r b^s = b^{r+s}$ $\Rightarrow$ $\log_b(xy) = r + s = \log_b x + \log_b y$

(b) $\dfrac{x}{y} = \dfrac{b^r}{b^s} = b^{r-s}$ $\Rightarrow$ $\log_b \dfrac{x}{y} = r - s = \log_b x - \log_b y$

(c) $x^y = (b^r)^y = b^{ry}$ $\Rightarrow$ $\log_b(x^y) = ry = y \log_b x$

6.5 Exponential Growth and Decay

1. The relative growth rate is $\dfrac{1}{P}\dfrac{dP}{dt} = 0.4159$, so $\dfrac{dP}{dt} = 0.4159P$ and by Theorem 2,

$P(t) = P(0)e^{0.4159t} = 3.8e^{0.4159t}$ million cells. Thus, $P(2) = 3.8e^{0.4159(2)} \approx 8.7$ million cells.

3. (a) By Theorem 2, $P(t) = P(0)e^{kt} = 50e^{kt}$. Now $P(1.5) = 50e^{k(1.5)} = 975$ $\Rightarrow$ $e^{1.5k} = \dfrac{975}{50}$ $\Rightarrow$

$1.5k = \ln 19.5$ $\Rightarrow$ $k = \dfrac{1}{1.5}\ln 19.5 \approx 1.9803$. So $P(t) \approx 50e^{1.9803t}$ cells.

(b) Using 1.9803 for k, we get $P(3) = 50e^{1.9803(3)} = 19{,}013.85 \approx 19{,}014$ cells.

(c) $\dfrac{dP}{dt} = kP$ $\Rightarrow$ $P'(3) = k \cdot P(3) = 1.9803 \cdot 19{,}014$ [from parts (a) and (b)] $= 37{,}653.4 \approx 37{,}653$ cells/h

(d) $P(t) = 50e^{1.9803t} = 250{,}000$ $\Rightarrow$ $e^{1.9803t} = \dfrac{250{,}000}{50}$ $\Rightarrow$ $e^{1.9803t} = 5000$ $\Rightarrow$ $1.9803t = \ln 5000$ $\Rightarrow$

$t = \dfrac{\ln 5000}{1.9803} \approx 4.30$ h

5. (a) Let the population (in millions) in the year t be $P(t)$. Since the initial time is the year 1750, we substitute $t - 1750$ for t in Theorem 2, so the exponential model gives $P(t) = P(1750)e^{k(t-1750)}$. Then $P(1800) = 980 = 790e^{k(1800-1750)}$ $\Rightarrow$

$\dfrac{980}{790} = e^{k(50)}$ $\Rightarrow$ $\ln \dfrac{980}{790} = 50k$ $\Rightarrow$ $k = \dfrac{1}{50}\ln \dfrac{980}{790} \approx 0.0043104$. So with this model, we have

$P(1900) = 790e^{k(1900-1750)} \approx 1508$ million, and $P(1950) = 790e^{k(1950-1750)} \approx 1871$ million. Both of these estimates are much too low.

(b) In this case, the exponential model gives $P(t) = P(1850)e^{k(t-1850)}$ $\Rightarrow$ $P(1900) = 1650 = 1260e^{k(1900-1850)}$ $\Rightarrow$

$\ln \dfrac{1650}{1260} = k(50)$ $\Rightarrow$ $k = \dfrac{1}{50}\ln \dfrac{1650}{1260} \approx 0.005393$. So with this model, we estimate

$P(1950) = 1260e^{k(1950-1850)} \approx 2161$ million. This is still too low, but closer than the estimate of $P(1950)$ in part (a).

(c) The exponential model gives $P(t) = P(1900)e^{k(t-1900)}$ $\Rightarrow$ $P(1950) = 2560 = 1650e^{k(1950-1900)}$ $\Rightarrow$

$\ln \dfrac{2560}{1650} = k(50)$ $\Rightarrow$ $k = \dfrac{1}{50}\ln \dfrac{2560}{1650} \approx 0.008785$. With this model, we estimate

$P(2000) = 1650e^{k(2000-1900)} \approx 3972$ million. This is much too low. The discrepancy is explained by the fact that the world birth rate (average yearly number of births per person) is about the same as always, whereas the mortality rate (especially the infant mortality rate) is much lower, owing mostly to advances in medical science and to the wars in the first part of the twentieth century. The exponential model assumes, among other things, that the birth and mortality rates will remain constant.

7. (a) If $y = [\text{N}_2\text{O}_5]$ then by Theorem 2, $\dfrac{dy}{dt} = -0.0005y \;\Rightarrow\; y(t) = y(0)e^{-0.0005t} = Ce^{-0.0005t}$.

(b) $y(t) = Ce^{-0.0005t} = 0.9C \;\Rightarrow\; e^{-0.0005t} = 0.9 \;\Rightarrow\; -0.0005t = \ln 0.9 \;\Rightarrow\; t = -2000 \ln 0.9 \approx 211$ s

9. (a) If $y(t)$ is the mass (in mg) remaining after t years, then $y(t) = y(0)e^{kt} = 100e^{kt}$.

$y(30) = 100e^{30k} = \tfrac{1}{2}(100) \;\Rightarrow\; e^{30k} = \tfrac{1}{2} \;\Rightarrow\; k = -(\ln 2)/30 \;\Rightarrow\; y(t) = 100e^{-(\ln 2)t/30} = 100 \cdot 2^{-t/30}$

(b) $y(100) = 100 \cdot 2^{-100/30} \approx 9.92$ mg

(c) $100e^{-(\ln 2)t/30} = 1 \;\Rightarrow\; -(\ln 2)t/30 = \ln \tfrac{1}{100} \;\Rightarrow\; t = -30 \, \dfrac{\ln 0.01}{\ln 2} \approx 199.3$ years

11. Let $y(t)$ be the level of radioactivity. Thus, $y(t) = y(0)e^{-kt}$ and k is determined by using the half-life:

$y(5730) = \tfrac{1}{2} y(0) \;\Rightarrow\; y(0)e^{-k(5730)} = \tfrac{1}{2} y(0) \;\Rightarrow\; e^{-5730k} = \tfrac{1}{2} \;\Rightarrow\; -5730k = \ln \tfrac{1}{2} \;\Rightarrow\; k = -\dfrac{\ln \tfrac{1}{2}}{5730} = \dfrac{\ln 2}{5730}$.

If 74% of the ^{14}C remains, then we know that $y(t) = 0.74y(0) \;\Rightarrow\; 0.74 = e^{-t(\ln 2)/5730} \;\Rightarrow\; \ln 0.74 = -\dfrac{t \ln 2}{5730} \;\Rightarrow\;$

$t = -\dfrac{5730(\ln 0.74)}{\ln 2} \approx 2489 \approx 2500$ years.

13. Let t measure time since a dinosaur died in millions of years, and let $y(t)$ be the amount of ^{40}K in the dinosaur's bones at time t. Then $y(t) = y(0)e^{-kt}$ and k is determined by the half-life: $y(1250) = \tfrac{1}{2} y(0) \;\Rightarrow\; y(0)e^{-k(1250)} = \tfrac{1}{2} y(0) \;\Rightarrow\;$

$e^{-1250k} = \tfrac{1}{2} \;\Rightarrow\; -1250k = \ln \tfrac{1}{2} \;\Rightarrow\; k = -\dfrac{\ln \tfrac{1}{2}}{1250} = \dfrac{\ln 2}{1250}$. To determine if a dinosaur dating of 68 million years is possible, we find that $y(68) = y(0)e^{-k(68)} \approx 0.963y(0)$, indicating that about 96% of the ^{40}K is remaining, which is clearly detectable. To determine the maximum age of a fossil by using ^{40}K, we solve $y(t) = 0.1\%y(0)$ for t.

$y(0)e^{-kt} = 0.001y(0) \;\Leftrightarrow\; e^{-kt} = 0.001 \;\Leftrightarrow\; -kt = \ln 0.001 \;\Leftrightarrow\; t = \dfrac{\ln 0.001}{-(\ln 2)/1250} \approx 12{,}457$ million, or 12.457 billion years.

15. (a) Using Newton's Law of Cooling, $\dfrac{dT}{dt} = k(T - T_s)$, we have $\dfrac{dT}{dt} = k(T - 75)$. Now let $y = T - 75$, so

$y(0) = T(0) - 75 = 185 - 75 = 110$, so y is a solution of the initial-value problem $dy/dt = ky$ with $y(0) = 110$ and by Theorem 2 we have $y(t) = y(0)e^{kt} = 110e^{kt}$.

$y(30) = 110e^{30k} = 150 - 75 \;\Rightarrow\; e^{30k} = \tfrac{75}{110} = \tfrac{15}{22} \;\Rightarrow\; k = \tfrac{1}{30} \ln \tfrac{15}{22}$, so $y(t) = 110e^{\tfrac{1}{30} t \ln\left(\tfrac{15}{22}\right)}$ and

$y(45) = 110e^{\tfrac{45}{30} \ln\left(\tfrac{15}{22}\right)} \approx 62°$F. Thus, $T(45) \approx 62 + 75 = 137°$F.

(b) $T(t) = 100 \implies y(t) = 25$. $y(t) = 110e^{\frac{1}{30}t\ln\left(\frac{15}{22}\right)} = 25 \implies e^{\frac{1}{30}t\ln\left(\frac{15}{22}\right)} = \frac{25}{110} \implies \frac{1}{30}t\ln\frac{15}{22} = \ln\frac{25}{110} \implies$

$t = \dfrac{30\ln\frac{25}{110}}{\ln\frac{15}{22}} \approx 116$ min.

17. $\dfrac{dT}{dt} = k(T - 20)$. Letting $y = T - 20$, we get $\dfrac{dy}{dt} = ky$, so $y(t) = y(0)e^{kt}$. $y(0) = T(0) - 20 = 5 - 20 = -15$, so

$y(25) = y(0)e^{25k} = -15e^{25k}$, and $y(25) = T(25) - 20 = 10 - 20 = -10$, so $-15e^{25k} = -10 \implies e^{25k} = \frac{2}{3}$. Thus,

$25k = \ln\left(\frac{2}{3}\right)$ and $k = \frac{1}{25}\ln\left(\frac{2}{3}\right)$, so $y(t) = y(0)e^{kt} = -15e^{(1/25)\ln(2/3)t}$. More simply, $e^{25k} = \frac{2}{3} \implies e^k = \left(\frac{2}{3}\right)^{1/25} \implies$

$e^{kt} = \left(\frac{2}{3}\right)^{t/25} \implies y(t) = -15 \cdot \left(\frac{2}{3}\right)^{t/25}$.

(a) $T(50) = 20 + y(50) = 20 - 15 \cdot \left(\frac{2}{3}\right)^{50/25} = 20 - 15 \cdot \left(\frac{2}{3}\right)^2 = 20 - \frac{20}{3} = 13.\overline{3}\,°C$

(b) $15 = T(t) = 20 + y(t) = 20 - 15 \cdot \left(\frac{2}{3}\right)^{t/25} \implies 15 \cdot \left(\frac{2}{3}\right)^{t/25} = 5 \implies \left(\frac{2}{3}\right)^{t/25} = \frac{1}{3} \implies$

$(t/25)\ln\left(\frac{2}{3}\right) = \ln\left(\frac{1}{3}\right) \implies t = 25\ln\left(\frac{1}{3}\right)/\ln\left(\frac{2}{3}\right) \approx 67.74$ min.

19. (a) Let $P(h)$ be the pressure at altitude h. Then $dP/dh = kP \implies P(h) = P(0)e^{kh} = 101.3e^{kh}$.

$P(1000) = 101.3e^{1000k} = 87.14 \implies 1000k = \ln\left(\frac{87.14}{101.3}\right) \implies k = \frac{1}{1000}\ln\left(\frac{87.14}{101.3}\right) \implies$

$P(h) = 101.3\, e^{\frac{1}{1000}h\ln\left(\frac{87.14}{101.3}\right)}$, so $P(3000) = 101.3e^{3\ln\left(\frac{87.14}{101.3}\right)} \approx 64.5$ kPa.

(b) $P(6187) = 101.3\, e^{\frac{6187}{1000}\ln\left(\frac{87.14}{101.3}\right)} \approx 39.9$ kPa

21. (a) Using $A = A_0\left(1 + \dfrac{r}{n}\right)^{nt}$ with $A_0 = 4000$, $r = 0.0175$, and $t = 5$, we have:

(i) Annually: $n = 1$ $A = 4000\left(1 + \dfrac{0.0175}{1}\right)^{1\cdot 5} = \$4362.47$

(ii) Semiannually: $n = 2$ $A = 4000\left(1 + \dfrac{0.0175}{2}\right)^{2\cdot 5} = \$4364.11$

(iii) Monthly: $n = 12$ $A = 4000\left(1 + \dfrac{0.0175}{12}\right)^{12\cdot 5} = \$4365.49$

(iv) Weekly: $n = 52$ $A = 4000\left(1 + \dfrac{0.0175}{52}\right)^{52\cdot 5} = \$4365.70$

(v) Daily: $n = 365$ $A = 4000\left(1 + \dfrac{0.0175}{365}\right)^{365\cdot 5} = \$4365.76$

(vi) Continuously: $A = 4000e^{(0.0175)5} = \$4365.77$

(b) $dA/dt = 0.0175A$ and $A(0) = 4000$.

6.6 Inverse Trigonometric Functions

1. (a) $\sin^{-1}(0.5) = \frac{\pi}{6}$ because $\sin\frac{\pi}{6} = 0.5$ and $\frac{\pi}{6}$ is in the interval $\left[-\frac{\pi}{2}, \frac{\pi}{2}\right]$ (the range of $\sin^{-1}$).

(b) $\cos^{-1}(-1) = \pi$ because $\cos\pi = -1$ and π is in the interval $[0, \pi]$ (the range of $\cos^{-1}$).

3. (a) $\csc^{-1}\sqrt{2} = \frac{\pi}{4}$ because $\csc\frac{\pi}{4} = \sqrt{2}$ and $\frac{\pi}{4}$ is in $\left(0, \frac{\pi}{2}\right] \cup \left(\pi, \frac{3\pi}{2}\right]$ (the range of $\csc^{-1}$).

(b) $\cos^{-1}(\sqrt{3}/2) = \frac{\pi}{6}$ because $\cos\frac{\pi}{6} = \sqrt{3}/2$ and $\frac{\pi}{6}$ is in $[0, \pi]$.

5. (a) In general, $\tan(\arctan x) = x$ for any real number x. Thus, $\tan(\arctan 10) = 10$.

(b) $\arcsin(\sin(5\pi/4)) = \arcsin(-1/\sqrt{2}) = -\frac{\pi}{4}$ because $\sin\left(-\frac{\pi}{4}\right) = -1/\sqrt{2}$ and $-\frac{\pi}{4}$ is in $\left[-\frac{\pi}{2}, \frac{\pi}{2}\right]$.

7. Let $\theta = \sin^{-1}\left(\frac{2}{3}\right)$ [see the figure].

Then $\tan\left(\sin^{-1}\left(\frac{2}{3}\right)\right) = \tan\theta = \dfrac{2}{\sqrt{5}}$.

9. Let $\theta = \sin^{-1}\left(\frac{5}{13}\right)$ [see the figure].

$\cos\left(2\sin^{-1}\left(\frac{5}{13}\right)\right) = \cos 2\theta = \cos^2\theta - \sin^2\theta$

$= \left(\frac{12}{13}\right)^2 - \left(\frac{5}{13}\right)^2 = \frac{144}{169} - \frac{25}{169} = \frac{119}{169}$

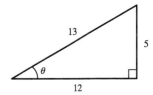

11. Let $y = \sin^{-1}x$. Then $-\frac{\pi}{2} \le y \le \frac{\pi}{2}$ $\Rightarrow$ $\cos y \ge 0$, so $\cos(\sin^{-1}x) = \cos y = \sqrt{1 - \sin^2 y} = \sqrt{1 - x^2}$.

13. Let $y = \tan^{-1}x$. Then $\tan y = x$, so from the triangle (which illustrates the case $y > 0$), we see that

$\sin(\tan^{-1}x) = \sin y = \dfrac{x}{\sqrt{1 + x^2}}$.

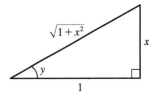

15.

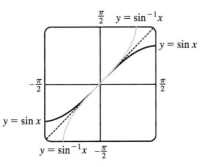

The graph of $\sin^{-1}x$ is the reflection of the graph of $\sin x$ about the line $y = x$.

17. Let $y = \cos^{-1}x$. Then $\cos y = x$ and $0 \le y \le \pi$ $\Rightarrow$ $-\sin y \dfrac{dy}{dx} = 1$ $\Rightarrow$

$\dfrac{dy}{dx} = -\dfrac{1}{\sin y} = -\dfrac{1}{\sqrt{1 - \cos^2 y}} = -\dfrac{1}{\sqrt{1 - x^2}}$. [Note that $\sin y \ge 0$ for $0 \le y \le \pi$.]

320 ☐ **CHAPTER 6** INVERSE FUNCTIONS

19. Let $y = \cot^{-1} x$. Then $\cot y = x \;\Rightarrow\; -\csc^2 y \dfrac{dy}{dx} = 1 \;\Rightarrow\; \dfrac{dy}{dx} = -\dfrac{1}{\csc^2 y} = -\dfrac{1}{1+\cot^2 y} = -\dfrac{1}{1+x^2}$.

21. Let $y = \csc^{-1} x$. Then $\csc y = x \;\Rightarrow\; -\csc y \cot y \dfrac{dy}{dx} = 1 \;\Rightarrow\;$

$\dfrac{dy}{dx} = -\dfrac{1}{\csc y \cot y} = -\dfrac{1}{\csc y \sqrt{\csc^2 y - 1}} = -\dfrac{1}{x\sqrt{x^2-1}}$. Note that $\cot y \geq 0$ on the domain of $\csc^{-1} x$.

23. $f(x) = \sin^{-1}(5x) \Rightarrow f'(x) = \dfrac{1}{\sqrt{1-(5x)^2}} \cdot \dfrac{d}{dx}(5x) = \dfrac{5}{\sqrt{1-25x^2}}$

25. $y = (\tan^{-1} x)^2 \;\Rightarrow\; y' = 2(\tan^{-1} x)^1 \cdot \dfrac{d}{dx}(\tan^{-1} x) = 2\tan^{-1} x \cdot \dfrac{1}{1+x^2} = \dfrac{2\tan^{-1} x}{1+x^2}$

27. $y = \tan^{-1}\sqrt{x-1} \;\Rightarrow\;$

$y' = \dfrac{1}{1+\left(\sqrt{x-1}\right)^2} \cdot \dfrac{d}{dt}\left(\sqrt{x-1}\right) = \dfrac{1}{1+(x-1)} \cdot \dfrac{1}{2\sqrt{x-1}} = \dfrac{1}{x} \cdot \dfrac{1}{2\sqrt{x-1}} = \dfrac{1}{2x\sqrt{x-1}}$

29. $y = \arctan(\cos\theta) \;\Rightarrow\; y' = \dfrac{1}{1+(\cos\theta)^2}(-\sin\theta) = -\dfrac{\sin\theta}{1+\cos^2\theta}$

31. $f(z) = e^{\arcsin(z^2)} \;\Rightarrow\; f'(z) = e^{\arcsin(z^2)} \cdot \dfrac{d}{dz}[\arcsin(z^2)] = e^{\arcsin(z^2)} \cdot \dfrac{1}{\sqrt{1-(z^2)^2}} \cdot 2z = \dfrac{2z e^{\arcsin(z^2)}}{\sqrt{1-z^4}}$

33. $h(t) = \cot^{-1}(t) + \cot^{-1}(1/t) \;\Rightarrow\;$

$h'(t) = -\dfrac{1}{1+t^2} - \dfrac{1}{1+(1/t)^2} \cdot \dfrac{d}{dt}\dfrac{1}{t} = -\dfrac{1}{1+t^2} - \dfrac{t^2}{t^2+1} \cdot \left(-\dfrac{1}{t^2}\right) = -\dfrac{1}{1+t^2} + \dfrac{1}{t^2+1} = 0$.

Note that this makes sense because $h(t) = \dfrac{\pi}{2}$ for $t > 0$ and $h(t) = \dfrac{3\pi}{2}$ for $t < 0$.

35. $y = x\sin^{-1} x + \sqrt{1-x^2} \;\Rightarrow\;$

$y' = x \cdot \dfrac{1}{\sqrt{1-x^2}} + (\sin^{-1} x)(1) + \dfrac{1}{2}(1-x^2)^{-1/2}(-2x) = \dfrac{x}{\sqrt{1-x^2}} + \sin^{-1} x - \dfrac{x}{\sqrt{1-x^2}} = \sin^{-1} x$

37. $y = \tan^{-1}\left(\dfrac{x}{a}\right) + \ln\sqrt{\dfrac{x-a}{x+a}} = \tan^{-1}\left(\dfrac{x}{a}\right) + \dfrac{1}{2}\ln\left(\dfrac{x-a}{x+a}\right) \;\Rightarrow\;$

$y' = \dfrac{1}{1+\left(\dfrac{x}{a}\right)^2} \cdot \dfrac{1}{a} + \dfrac{1}{2} \cdot \dfrac{1}{\dfrac{x-a}{x+a}} \cdot \dfrac{(x+a)\cdot 1 - (x-a)\cdot 1}{(x+a)^2} = \dfrac{1}{a+\dfrac{x^2}{a}} + \dfrac{1}{2} \cdot \dfrac{x+a}{x-a} \cdot \dfrac{2a}{(x+a)^2}$

$= \dfrac{1}{a+\dfrac{x^2}{a}} \cdot \dfrac{a}{a} + \dfrac{a}{(x-a)(x+a)} = \dfrac{a}{x^2+a^2} + \dfrac{a}{x^2-a^2}$

39. $g(x) = \cos^{-1}(3-2x)$ $\Rightarrow$ $g'(x) = -\dfrac{1}{\sqrt{1-(3-2x)^2}}(-2) = \dfrac{2}{\sqrt{1-(3-2x)^2}}$.

Domain$(g) = \{x \mid -1 \le 3 - 2x \le 1\} = \{x \mid -4 \le -2x \le -2\} = \{x \mid 2 \ge x \ge 1\} = [1, 2]$.

Domain$(g') = \{x \mid 1 - (3-2x)^2 > 0\} = \{x \mid (3-2x)^2 < 1\} = \{x \mid |3-2x| < 1\}$
$= \{x \mid -1 < 3 - 2x < 1\} = \{x \mid -4 < -2x < -2\} = \{x \mid 2 > x > 1\} = (1, 2)$

41. $g(x) = x\sin^{-1}\left(\dfrac{x}{4}\right) + \sqrt{16 - x^2}$ $\Rightarrow$ $g'(x) = \sin^{-1}\left(\dfrac{x}{4}\right) + \dfrac{x}{4\sqrt{1-(x/4)^2}} - \dfrac{x}{\sqrt{16-x^2}} = \sin^{-1}\left(\dfrac{x}{4}\right)$ $\Rightarrow$

$g'(2) = \sin^{-1}\left(\tfrac{1}{2}\right) = \tfrac{\pi}{6}$

43. $f(x) = \sqrt{1-x^2}\arcsin x$ $\Rightarrow$ $f'(x) = \sqrt{1-x^2}\cdot \dfrac{1}{\sqrt{1-x^2}} + \arcsin x \cdot \dfrac{1}{2}(1-x^2)^{-1/2}(-2x) = 1 - \dfrac{x\arcsin x}{\sqrt{1-x^2}}$

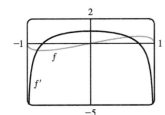

Note that $f' = 0$ where the graph of f has a horizontal tangent. Also note that f' is negative when f is decreasing and f' is positive when f is increasing.

45. $\lim\limits_{x\to -1^+} \sin^{-1} x = \sin^{-1}(-1) = -\tfrac{\pi}{2}$

47. Let $t = e^x$. As $x \to \infty$, $t \to \infty$. $\lim\limits_{x\to\infty} \arctan(e^x) = \lim\limits_{t\to\infty} \arctan t = \tfrac{\pi}{2}$ by (8).

49.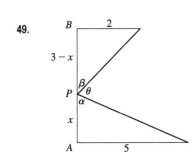

From the figure, $\tan\alpha = \dfrac{5}{x}$ and $\tan\beta = \dfrac{2}{3-x}$. Since

$\alpha + \beta + \theta = 180° = \pi$, $\theta = \pi - \tan^{-1}\left(\dfrac{5}{x}\right) - \tan^{-1}\left(\dfrac{2}{3-x}\right)$ $\Rightarrow$

$\dfrac{d\theta}{dx} = -\dfrac{1}{1+\left(\dfrac{5}{x}\right)^2}\left(-\dfrac{5}{x^2}\right) - \dfrac{1}{1+\left(\dfrac{2}{3-x}\right)^2}\left[\dfrac{2}{(3-x)^2}\right]$

$= \dfrac{x^2}{x^2+25}\cdot\dfrac{5}{x^2} - \dfrac{(3-x)^2}{(3-x)^2+4}\cdot\dfrac{2}{(3-x)^2}$.

Now $\dfrac{d\theta}{dx} = 0$ $\Rightarrow$ $\dfrac{5}{x^2+25} = \dfrac{2}{x^2-6x+13}$ $\Rightarrow$ $2x^2 + 50 = 5x^2 - 30x + 65$ $\Rightarrow$

$3x^2 - 30x + 15 = 0$ $\Rightarrow$ $x^2 - 10x + 5 = 0$ $\Rightarrow$ $x = 5 \pm 2\sqrt{5}$. We reject the solution with the $+$ sign, since it is larger than 3. $d\theta/dx > 0$ for $x < 5 - 2\sqrt{5}$ and $d\theta/dx < 0$ for $x > 5 - 2\sqrt{5}$, so θ is maximized when $|AP| = x = 5 - 2\sqrt{5} \approx 0.53$.

51. $\dfrac{dx}{dt} = 2$ ft/s, $\sin\theta = \dfrac{x}{10}$ $\Rightarrow$ $\theta = \sin^{-1}\left(\dfrac{x}{10}\right)$, $\dfrac{d\theta}{dx} = \dfrac{1/10}{\sqrt{1-(x/10)^2}}$,

$\dfrac{d\theta}{dt} = \dfrac{d\theta}{dx}\dfrac{dx}{dt} = \dfrac{1/10}{\sqrt{1-(x/10)^2}}(2)$ rad/s, $\left.\dfrac{d\theta}{dt}\right]_{x=6} = \dfrac{2/10}{\sqrt{1-(6/10)^2}}$ rad/s $= \tfrac{1}{4}$ rad/s

53. $y = f(x) = \sin^{-1}(x/(x+1))$ **A.** $D = \{x \mid -1 \le x/(x+1) \le 1\}$. For $x > -1$ we have $-x - 1 \le x \le x + 1$ $\Leftrightarrow$ $2x \ge -1$ $\Leftrightarrow$ $x \ge -\frac{1}{2}$, so $D = [-\frac{1}{2}, \infty)$.   **B.** Intercepts are 0   **C.** No symmetry

D. $\lim\limits_{x \to \infty} \sin^{-1}\left(\dfrac{x}{x+1}\right) = \lim\limits_{x \to \infty} \sin^{-1}\left(\dfrac{1}{1+1/x}\right) = \sin^{-1} 1 = \frac{\pi}{2}$, so $y = \frac{\pi}{2}$ is a HA.

E. $f'(x) = \dfrac{1}{\sqrt{1 - [x/(x+1)]^2}} \cdot \dfrac{(x+1) - x}{(x+1)^2} = \dfrac{1}{(x+1)\sqrt{2x+1}} > 0$,   **H.**

so f is increasing on $(-\frac{1}{2}, \infty)$.   **F.** No local maximum or minimum,

$f(-\frac{1}{2}) = \sin^{-1}(-1) = -\frac{\pi}{2}$ is an absolute minimum

G. $f''(x) = -\dfrac{\sqrt{2x+1} + (x+1)/\sqrt{2x+1}}{(x+1)^2(2x+1)}$

$= -\dfrac{3x+2}{(x+1)^2(2x+1)^{3/2}} < 0$ on D, so f is CD on $(-\frac{1}{2}, \infty)$.

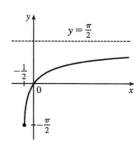

55. $y = f(x) = x - \tan^{-1} x$ **A.** $D = \mathbb{R}$ **B.** Intercepts are 0   **C.** $f(-x) = -f(x)$, so the curve is symmetric about the origin.   **D.** $\lim\limits_{x \to \infty} (x - \tan^{-1} x) = \infty$ and $\lim\limits_{x \to -\infty} (x - \tan^{-1} x) = -\infty$, no HA.

But $f(x) - (x - \frac{\pi}{2}) = -\tan^{-1} x + \frac{\pi}{2} \to 0$ as $x \to \infty$, and

$f(x) - (x + \frac{\pi}{2}) = -\tan^{-1} x - \frac{\pi}{2} \to 0$ as $x \to -\infty$, so $y = x \pm \frac{\pi}{2}$ are

slant asymptotes.   **E.** $f'(x) = 1 - \dfrac{1}{x^2+1} = \dfrac{x^2}{x^2+1} > 0$, so f is

increasing on $\mathbb{R}$.   **F.** No extrema

H.

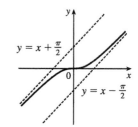

G. $f''(x) = \dfrac{(1+x^2)(2x) - x^2(2x)}{(1+x^2)^2} = \dfrac{2x}{(1+x^2)^2} > 0 \Leftrightarrow x > 0$, so

f is CU on $(0, \infty)$, CD on $(-\infty, 0)$. IP at $(0, 0)$.

57. $f(x) = \arctan(\cos(3 \arcsin x))$. We use a CAS to compute f' and f'', and to graph f, f', and f'':

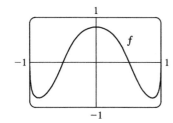 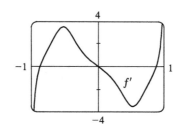 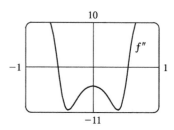

From the graph of f', it appears that the only maximum occurs at $x = 0$ and there are minima at $x = \pm 0.87$. From the graph of f'', it appears that there are inflection points at $x = \pm 0.52$.

59. $f(x) = \dfrac{2x^2 + 5}{x^2 + 1} = \dfrac{2(x^2+1) + 3}{x^2+1} = 2 + \dfrac{3}{x^2+1}$ $\Rightarrow$ $F(x) = 2x + 3\tan^{-1} x + C$

61. $\displaystyle\int_{1/\sqrt{3}}^{\sqrt{3}} \dfrac{8}{1+x^2}\, dx = \Big[8\arctan x\Big]_{1/\sqrt{3}}^{\sqrt{3}} = 8\left(\dfrac{\pi}{3} - \dfrac{\pi}{6}\right) = 8\left(\dfrac{\pi}{6}\right) = \dfrac{4\pi}{3}$

63. Let $u = \sin^{-1}x$, so $du = \dfrac{dx}{\sqrt{1-x^2}}$. When $x=0$, $u=0$; when $x=\tfrac{1}{2}$, $u=\tfrac{\pi}{6}$. Thus,

$$\int_0^{1/2} \dfrac{\sin^{-1}x}{\sqrt{1-x^2}}\, dx = \int_0^{\pi/6} u\, du = \left[\dfrac{u^2}{2}\right]_0^{\pi/6} = \dfrac{\pi^2}{72}.$$

65. Let $u = 1 + x^2$. Then $du = 2x\, dx$, so

$$\int \dfrac{1+x}{1+x^2}\, dx = \int \dfrac{1}{1+x^2}\, dx + \int \dfrac{x}{1+x^2}\, dx = \tan^{-1}x + \int \dfrac{\tfrac{1}{2}\, du}{u} = \tan^{-1}x + \tfrac{1}{2}\ln|u| + C$$

$$= \tan^{-1}x + \tfrac{1}{2}\ln\left|1+x^2\right| + C = \tan^{-1}x + \tfrac{1}{2}\ln(1+x^2) + C \quad [\text{since } 1+x^2 > 0].$$

67. Let $u = \arctan x$. Then $du = \dfrac{1}{x^2+1}\, dx$, so $\displaystyle\int \dfrac{(\arctan x)^2}{x^2+1}\, dx = \int u^2\, du = \tfrac{1}{3}u^3 + C = \tfrac{1}{3}(\arctan x)^3 + C$.

69. Let $u = \arcsin x$. Then $du = \dfrac{1}{\sqrt{1-x^2}}\, dx$, so $\displaystyle\int \dfrac{e^{\arcsin x}}{\sqrt{1-x^2}}\, dx = \int e^u\, du = e^u + C = e^{\arcsin x} + C$.

71. Let $u = t^3$. Then $du = 3t^2\, dt$ and $\displaystyle\int \dfrac{t^2}{\sqrt{1-t^6}}\, dt = \int \dfrac{\tfrac{1}{3}\, du}{\sqrt{1-u^2}} = \tfrac{1}{3}\sin^{-1}u + C = \tfrac{1}{3}\sin^{-1}(t^3) + C$.

73. Let $u = \sqrt{x}$. Then $du = \dfrac{dx}{2\sqrt{x}}$ and $\displaystyle\int \dfrac{dx}{\sqrt{x}\,(1+x)} = \int \dfrac{2\, du}{1+u^2} = 2\tan^{-1}u + C = 2\tan^{-1}\sqrt{x} + C$.

75. Let $u = x/a$. Then $du = dx/a$, so

$$\int \dfrac{1}{\sqrt{a^2-x^2}}\, dx = \int \dfrac{1}{a\sqrt{1-(x/a)^2}}\, dx = \int \dfrac{du}{\sqrt{1-u^2}} = \sin^{-1}u + C = \sin^{-1}\left(\dfrac{x}{a}\right) + C.$$

77.

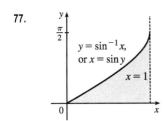

The integral represents the area below the curve $y = \sin^{-1}x$ on the interval $x \in [0,1]$. The bounding curves are $y = \sin^{-1}x \iff x = \sin y$, $y = 0$ and $x = 1$. We see that y ranges between $\sin^{-1}0 = 0$ and $\sin^{-1}1 = \tfrac{\pi}{2}$. So we have to integrate the function $x = 1 - \sin y$ between $y = 0$ and $y = \tfrac{\pi}{2}$:

$$\int_0^1 \sin^{-1}x\, dx = \int_0^{\pi/2} (1 - \sin y)\, dy = \left(\tfrac{\pi}{2} + \cos\tfrac{\pi}{2}\right) - (0 + \cos 0) = \tfrac{\pi}{2} - 1.$$

79. (a) $\arctan\tfrac{1}{2} + \arctan\tfrac{1}{3} = \arctan\left(\dfrac{\tfrac{1}{2}+\tfrac{1}{3}}{1-\tfrac{1}{2}\cdot\tfrac{1}{3}}\right) = \arctan 1 = \dfrac{\pi}{4}$

(b) $2\arctan\tfrac{1}{3} + \arctan\tfrac{1}{7} = \left(\arctan\tfrac{1}{3} + \arctan\tfrac{1}{3}\right) + \arctan\tfrac{1}{7} = \arctan\left(\dfrac{\tfrac{1}{3}+\tfrac{1}{3}}{1-\tfrac{1}{3}\cdot\tfrac{1}{3}}\right) + \arctan\tfrac{1}{7}$

$= \arctan\tfrac{3}{4} + \arctan\tfrac{1}{7} = \arctan\left(\dfrac{\tfrac{3}{4}+\tfrac{1}{7}}{1-\tfrac{3}{4}\cdot\tfrac{1}{7}}\right) = \arctan 1 = \dfrac{\pi}{4}$

81. Let $f(x) = 2\sin^{-1}x - \cos^{-1}(1-2x^2)$. Then $f'(x) = \dfrac{2}{\sqrt{1-x^2}} - \dfrac{4x}{\sqrt{1-(1-2x^2)^2}} = \dfrac{2}{\sqrt{1-x^2}} - \dfrac{4x}{2x\sqrt{1-x^2}} = 0$

[since $x \geq 0$]. Thus $f'(x) = 0$ for all $x \in [0,1)$. Thus $f(x) = C$. To find C let $x = 0$. Thus $2\sin^{-1}(0) - \cos^{-1}(1) = 0 = C$. Therefore we see that $f(x) = 2\sin^{-1}x - \cos^{-1}(1-2x^2) = 0 \Rightarrow 2\sin^{-1}x = \cos^{-1}(1-2x^2)$.

83. $y = \sec^{-1} x \Rightarrow \sec y = x \Rightarrow \sec y \tan y \dfrac{dy}{dx} = 1 \Rightarrow \dfrac{dy}{dx} = \dfrac{1}{\sec y \tan y}$. Now $\tan^2 y = \sec^2 y - 1 = x^2 - 1$, so $\tan y = \pm\sqrt{x^2 - 1}$. For $y \in [0, \frac{\pi}{2})$, $x \geq 1$, so $\sec y = x = |x|$ and $\tan y \geq 0 \Rightarrow \dfrac{dy}{dx} = \dfrac{1}{x\sqrt{x^2 - 1}} = \dfrac{1}{|x|\sqrt{x^2 - 1}}$.

For $y \in (\frac{\pi}{2}, \pi]$, $x \leq -1$, so $|x| = -x$ and $\tan y = -\sqrt{x^2 - 1} \Rightarrow$

$$\dfrac{dy}{dx} = \dfrac{1}{\sec y \tan y} = \dfrac{1}{x(-\sqrt{x^2 - 1})} = \dfrac{1}{(-x)\sqrt{x^2 - 1}} = \dfrac{1}{|x|\sqrt{x^2 - 1}}$$

6.7 Hyperbolic Functions

1. (a) $\sinh 0 = \frac{1}{2}(e^0 - e^{-0}) = 0$ (b) $\cosh 0 = \frac{1}{2}(e^0 + e^{-0}) = \frac{1}{2}(1 + 1) = 1$

3. (a) $\cosh(\ln 5) = \frac{1}{2}(e^{\ln 5} + e^{-\ln 5}) = \frac{1}{2}(5 + (e^{\ln 5})^{-1}) = \frac{1}{2}(5 + 5^{-1}) = \frac{1}{2}(5 + \frac{1}{5}) = \frac{13}{5}$

(b) $\cosh 5 = \frac{1}{2}(e^5 + e^{-5}) \approx 74.20995$

5. (a) $\operatorname{sech} 0 = \dfrac{1}{\cosh 0} = \dfrac{1}{1} = 1$ (b) $\cosh^{-1} 1 = 0$ because $\cosh 0 = 1$.

7. $8\sinh x + 5\cosh x = 8\left(\dfrac{e^x - e^{-x}}{2}\right) + 5\left(\dfrac{e^x + e^{-x}}{2}\right) = \dfrac{8}{2}e^x - \dfrac{8}{2}e^{-x} + \dfrac{5}{2}e^x + \dfrac{5}{2}e^{-x} = \dfrac{13}{2}e^x - \dfrac{3}{2}e^{-x}$

9. $\sinh(\ln x) = \dfrac{1}{2}(e^{\ln x} - e^{-\ln x}) = \dfrac{1}{2}\left(x - e^{\ln x^{-1}}\right) = \dfrac{1}{2}(x - x^{-1}) = \dfrac{1}{2}\left(x - \dfrac{1}{x}\right) = \dfrac{1}{2}\left(\dfrac{x^2 - 1}{x}\right) = \dfrac{x^2 - 1}{2x}$

11. $\sinh(-x) = \frac{1}{2}[e^{-x} - e^{-(-x)}] = \frac{1}{2}(e^{-x} - e^x) = -\frac{1}{2}(e^x - e^{-x}) = -\sinh x$

13. $\cosh x + \sinh x = \frac{1}{2}(e^x + e^{-x}) + \frac{1}{2}(e^x - e^{-x}) = \frac{1}{2}(2e^x) = e^x$

15. $\sinh x \cosh y + \cosh x \sinh y = \left[\frac{1}{2}(e^x - e^{-x})\right]\left[\frac{1}{2}(e^y + e^{-y})\right] + \left[\frac{1}{2}(e^x + e^{-x})\right]\left[\frac{1}{2}(e^y - e^{-y})\right]$

$= \frac{1}{4}[(e^{x+y} + e^{x-y} - e^{-x+y} - e^{-x-y}) + (e^{x+y} - e^{x-y} + e^{-x+y} - e^{-x-y})]$

$= \frac{1}{4}(2e^{x+y} - 2e^{-x-y}) = \frac{1}{2}[e^{x+y} - e^{-(x+y)}] = \sinh(x+y)$

17. Divide both sides of the identity $\cosh^2 x - \sinh^2 x = 1$ by $\sinh^2 x$:

$\dfrac{\cosh^2 x}{\sinh^2 x} - \dfrac{\sinh^2 x}{\sinh^2 x} = \dfrac{1}{\sinh^2 x} \Leftrightarrow \coth^2 x - 1 = \operatorname{csch}^2 x$.

19. Putting $y = x$ in the result from Exercise 15, we have

$\sinh 2x = \sinh(x + x) = \sinh x \cosh x + \cosh x \sinh x = 2\sinh x \cosh x$.

21. $\tanh(\ln x) = \dfrac{\sinh(\ln x)}{\cosh(\ln x)} = \dfrac{(e^{\ln x} - e^{-\ln x})/2}{(e^{\ln x} + e^{-\ln x})/2} = \dfrac{x - (e^{\ln x})^{-1}}{x + (e^{\ln x})^{-1}} = \dfrac{x - x^{-1}}{x + x^{-1}} = \dfrac{x - 1/x}{x + 1/x} = \dfrac{(x^2 - 1)/x}{(x^2 + 1)/x} = \dfrac{x^2 - 1}{x^2 + 1}$

23. By Exercise 13, $(\cosh x + \sinh x)^n = (e^x)^n = e^{nx} = \cosh nx + \sinh nx$.

25. $\operatorname{sech} x = \dfrac{1}{\cosh x} \;\Rightarrow\; \operatorname{sech} x = \dfrac{1}{5/3} = \dfrac{3}{5}$.

$\cosh^2 x - \sinh^2 x = 1 \;\Rightarrow\; \sinh^2 x = \cosh^2 x - 1 = \left(\tfrac{5}{3}\right)^2 - 1 = \tfrac{16}{9} \;\Rightarrow\; \sinh x = \tfrac{4}{3}$ [because $x > 0$].

$\operatorname{csch} x = \dfrac{1}{\sinh x} \;\Rightarrow\; \operatorname{csch} x = \dfrac{1}{4/3} = \dfrac{3}{4}$.

$\tanh x = \dfrac{\sinh x}{\cosh x} \;\Rightarrow\; \tanh x = \dfrac{4/3}{5/3} = \dfrac{4}{5}$.

$\coth x = \dfrac{1}{\tanh x} \;\Rightarrow\; \coth x = \dfrac{1}{4/5} = \dfrac{5}{4}$.

27. (a) $\displaystyle\lim_{x\to\infty} \tanh x = \lim_{x\to\infty} \dfrac{e^x - e^{-x}}{e^x + e^{-x}} \cdot \dfrac{e^{-x}}{e^{-x}} = \lim_{x\to\infty} \dfrac{1 - e^{-2x}}{1 + e^{-2x}} = \dfrac{1-0}{1+0} = 1$

(b) $\displaystyle\lim_{x\to-\infty} \tanh x = \lim_{x\to-\infty} \dfrac{e^x - e^{-x}}{e^x + e^{-x}} \cdot \dfrac{e^x}{e^x} = \lim_{x\to-\infty} \dfrac{e^{2x} - 1}{e^{2x} + 1} = \dfrac{0-1}{0+1} = -1$

(c) $\displaystyle\lim_{x\to\infty} \sinh x = \lim_{x\to\infty} \dfrac{e^x - e^{-x}}{2} = \infty$

(d) $\displaystyle\lim_{x\to-\infty} \sinh x = \lim_{x\to-\infty} \dfrac{e^x - e^{-x}}{2} = -\infty$

(e) $\displaystyle\lim_{x\to\infty} \operatorname{sech} x = \lim_{x\to\infty} \dfrac{2}{e^x + e^{-x}} = 0$

(f) $\displaystyle\lim_{x\to\infty} \coth x = \lim_{x\to\infty} \dfrac{e^x + e^{-x}}{e^x - e^{-x}} \cdot \dfrac{e^{-x}}{e^{-x}} = \lim_{x\to\infty} \dfrac{1 + e^{-2x}}{1 - e^{-2x}} = \dfrac{1+0}{1-0} = 1$ [Or: Use part (a).]

(g) $\displaystyle\lim_{x\to 0^+} \coth x = \lim_{x\to 0^+} \dfrac{\cosh x}{\sinh x} = \infty$, since $\sinh x \to 0$ through positive values and $\cosh x \to 1$.

(h) $\displaystyle\lim_{x\to 0^-} \coth x = \lim_{x\to 0^-} \dfrac{\cosh x}{\sinh x} = -\infty$, since $\sinh x \to 0$ through negative values and $\cosh x \to 1$.

(i) $\displaystyle\lim_{x\to-\infty} \operatorname{csch} x = \lim_{x\to-\infty} \dfrac{2}{e^x - e^{-x}} = 0$

(j) $\displaystyle\lim_{x\to\infty} \dfrac{\sinh x}{e^x} = \lim_{x\to\infty} \dfrac{e^x - e^{-x}}{2e^x} = \lim_{x\to\infty} \dfrac{1 - e^{-2x}}{2} = \dfrac{1-0}{2} = \dfrac{1}{2}$

29. Let $y = \sinh^{-1} x$. Then $\sinh y = x$ and, by Example 1(a), $\cosh^2 y - \sinh^2 y = 1 \;\Rightarrow\;$ [with $\cosh y > 0$] $\cosh y = \sqrt{1 + \sinh^2 y} = \sqrt{1 + x^2}$. So by Exercise 13, $e^y = \sinh y + \cosh y = x + \sqrt{1 + x^2} \;\Rightarrow\; y = \ln(x + \sqrt{1 + x^2})$.

31. (a) Let $y = \tanh^{-1} x$. Then $x = \tanh y = \dfrac{\sinh y}{\cosh y} = \dfrac{(e^y - e^{-y})/2}{(e^y + e^{-y})/2} \cdot \dfrac{e^y}{e^y} = \dfrac{e^{2y} - 1}{e^{2y} + 1} \;\Rightarrow\; xe^{2y} + x = e^{2y} - 1 \;\Rightarrow\;$

$1 + x = e^{2y} - xe^{2y} \;\Rightarrow\; 1 + x = e^{2y}(1 - x) \;\Rightarrow\; e^{2y} = \dfrac{1+x}{1-x} \;\Rightarrow\; 2y = \ln\left(\dfrac{1+x}{1-x}\right) \;\Rightarrow\;$

$y = \dfrac{1}{2}\ln\left(\dfrac{1+x}{1-x}\right)$.

(b) Let $y = \tanh^{-1} x$. Then $x = \tanh y$, so from Exercise 22 we have

$$e^{2y} = \frac{1 + \tanh y}{1 - \tanh y} = \frac{1 + x}{1 - x} \quad \Rightarrow \quad 2y = \ln\left(\frac{1+x}{1-x}\right) \quad \Rightarrow \quad y = \frac{1}{2}\ln\left(\frac{1+x}{1-x}\right).$$

33. (a) Let $y = \cosh^{-1} x$. Then $\cosh y = x$ and $y \geq 0 \quad \Rightarrow \quad \sinh y \dfrac{dy}{dx} = 1 \quad \Rightarrow$

$$\frac{dy}{dx} = \frac{1}{\sinh y} = \frac{1}{\sqrt{\cosh^2 y - 1}} = \frac{1}{\sqrt{x^2 - 1}} \quad \text{[since } \sinh y \geq 0 \text{ for } y \geq 0\text{]}. \quad \textit{Or:} \text{ Use Formula 4.}$$

(b) Let $y = \tanh^{-1} x$. Then $\tanh y = x \quad \Rightarrow \quad \operatorname{sech}^2 y \dfrac{dy}{dx} = 1 \quad \Rightarrow \quad \dfrac{dy}{dx} = \dfrac{1}{\operatorname{sech}^2 y} = \dfrac{1}{1 - \tanh^2 y} = \dfrac{1}{1 - x^2}$.

Or: Use Formula 5.

(c) Let $y = \coth^{-1} x$. Then $\coth y = x \quad \Rightarrow \quad -\operatorname{csch}^2 y \dfrac{dy}{dx} = 1 \quad \Rightarrow \quad \dfrac{dy}{dx} = -\dfrac{1}{\operatorname{csch}^2 y} = \dfrac{1}{1 - \coth^2 y} = \dfrac{1}{1 - x^2}$

by Exercise 17.

35. $f(x) = \cosh 3x \quad \Rightarrow \quad f'(x) = \sinh(3x) \cdot \dfrac{d}{dx}(3x) = \sinh(3x) \cdot 3 = 3\sinh 3x$

37. $h(x) = \sinh(x^2) \quad \Rightarrow \quad h'(x) = \cosh(x^2) \dfrac{d}{dx}(x^2) = 2x \cosh(x^2)$

39. $G(t) = \sinh(\ln t) \quad \Rightarrow \quad G'(t) = \cosh(\ln t) \dfrac{d}{dt} \ln t = \dfrac{1}{2}\left(e^{\ln t} + e^{-\ln t}\right)\left(\dfrac{1}{t}\right) = \dfrac{1}{2t}\left(t + \dfrac{1}{t}\right) = \dfrac{1}{2t}\left(\dfrac{t^2+1}{t}\right) = \dfrac{t^2+1}{2t^2}$

Or: $G(t) = \sinh(\ln t) = \dfrac{1}{2}(e^{\ln t} - e^{-\ln t}) = \dfrac{1}{2}\left(t - \dfrac{1}{t}\right) \quad \Rightarrow \quad G'(t) = \dfrac{1}{2}\left(1 + \dfrac{1}{t^2}\right) = \dfrac{t^2+1}{2t^2}$

41. $f(x) = \tanh \sqrt{x} \quad \Rightarrow \quad f'(x) = \operatorname{sech}^2 \sqrt{x} \dfrac{d}{dx}\sqrt{x} = \operatorname{sech}^2 \sqrt{x} \left(\dfrac{1}{2\sqrt{x}}\right) = \dfrac{\operatorname{sech}^2 \sqrt{x}}{2\sqrt{x}}$

43. $y = \operatorname{sech} x \tanh x \quad \overset{\text{PR}}{\Rightarrow} \quad y' = \operatorname{sech} x \cdot \operatorname{sech}^2 x + \tanh x \cdot (-\operatorname{sech} x \tanh x) = \operatorname{sech}^3 x - \operatorname{sech} x \tanh^2 x$

45. $g(t) = t \coth\sqrt{t^2+1} \quad \overset{\text{PR}}{\Rightarrow}$

$g'(t) = t\left[-\operatorname{csch}^2 \sqrt{t^2+1}\left(\tfrac{1}{2}(t^2+1)^{-1/2} \cdot 2t\right)\right] + \left(\coth\sqrt{t^2+1}\right)(1) = \coth\sqrt{t^2+1} - \dfrac{t^2}{\sqrt{t^2+1}}\operatorname{csch}^2\sqrt{t^2+1}$

47. $f(x) = \sinh^{-1}(-2x) \quad \Rightarrow \quad f'(x) = \dfrac{1}{\sqrt{1+(-2x)^2}} \cdot \dfrac{d}{dx}(-2x) = -\dfrac{2}{\sqrt{1+4x^2}}$

49. $y = \cosh^{-1}(\sec\theta) \quad \Rightarrow$

$y' = \dfrac{1}{\sqrt{\sec^2\theta - 1}} \cdot \dfrac{d}{d\theta}(\sec\theta) = \dfrac{1}{\sqrt{\tan^2\theta}} \cdot \sec\theta \tan\theta = \dfrac{1}{\tan\theta} \cdot \sec\theta \tan\theta \quad \text{[since } 0 \leq \theta < \pi/2\text{]} = \sec\theta$

51. $G(u) = \cosh^{-1}\sqrt{1+u^2} \quad \Rightarrow$

$G'(u) = \dfrac{1}{\sqrt{\left(\sqrt{1+u^2}\right)^2 - 1}} \cdot \dfrac{d}{du}\left(\sqrt{1+u^2}\right) = \dfrac{1}{\sqrt{(1+u^2)-1}} \cdot \dfrac{2u}{2\sqrt{1+u^2}} = \dfrac{u}{\sqrt{u^2}\cdot\sqrt{1+u^2}}$

$= \dfrac{u}{u\sqrt{1+u^2}} \quad \text{[since } u > 0\text{]} = \dfrac{1}{\sqrt{1+u^2}}$

53. $y = x\sinh^{-1}(x/3) - \sqrt{9+x^2} \;\Rightarrow$

$y' = \sinh^{-1}\left(\dfrac{x}{3}\right) + x \cdot \dfrac{1/3}{\sqrt{1+(x/3)^2}} - \dfrac{2x}{2\sqrt{9+x^2}} = \sinh^{-1}\left(\dfrac{x}{3}\right) + \dfrac{x}{\sqrt{9+x^2}} - \dfrac{x}{\sqrt{9+x^2}} = \sinh^{-1}\left(\dfrac{x}{3}\right)$

55. $\dfrac{d}{dx}\arctan(\tanh x) = \dfrac{1}{1+(\tanh x)^2}\dfrac{d}{dx}(\tanh x) = \dfrac{\text{sech}^2 x}{1+\tanh^2 x} = \dfrac{1/\cosh^2 x}{1+(\sinh^2 x)/\cosh^2 x}$

$= \dfrac{1}{\cosh^2 x + \sinh^2 x} = \dfrac{1}{\cosh 2x}$ [by Exercise 20] $= \text{sech}\, 2x$

57. As the depth d of the water gets large, the fraction $\dfrac{2\pi d}{L}$ gets large, and from Figure 5 or Exercise 27(a), $\tanh\left(\dfrac{2\pi d}{L}\right)$

approaches 1. Thus, $v = \sqrt{\dfrac{gL}{2\pi}\tanh\left(\dfrac{2\pi d}{L}\right)} \approx \sqrt{\dfrac{gL}{2\pi}(1)} = \sqrt{\dfrac{gL}{2\pi}}$.

59. (a) $y = 20\cosh(x/20) - 15 \;\Rightarrow\; y' = 20\sinh(x/20) \cdot \tfrac{1}{20} = \sinh(x/20)$. Since the right pole is positioned at $x = 7$,

we have $y'(7) = \sinh\tfrac{7}{20} \approx 0.3572$.

(b) If α is the angle between the tangent line and the x-axis, then $\tan\alpha =$ slope of the line $= \sinh\tfrac{7}{20}$, so

$\alpha = \tan^{-1}\left(\sinh\tfrac{7}{20}\right) \approx 0.343$ rad $\approx 19.66°$. Thus, the angle between the line and the pole is $\theta = 90° - \alpha \approx 70.34°$.

61. (a) From Exercise 60, the shape of the cable is given by $y = f(x) = \dfrac{T}{\rho g}\cosh\left(\dfrac{\rho g x}{T}\right)$. The shape is symmetric about the

y-axis, so the lowest point is $(0, f(0)) = \left(0, \dfrac{T}{\rho g}\right)$ and the poles are at $x = \pm 100$. We want to find T when the lowest

point is 60 m, so $\dfrac{T}{\rho g} = 60 \;\Rightarrow\; T = 60\rho g = (60\text{ m})(2\text{ kg/m})(9.8\text{ m/s}^2) = 1176\,\dfrac{\text{kg-m}}{\text{s}^2}$, or 1176 N (newtons).

The height of each pole is $f(100) = \dfrac{T}{\rho g}\cosh\left(\dfrac{\rho g \cdot 100}{T}\right) = 60\cosh\left(\dfrac{100}{60}\right) \approx 164.50$ m.

(b) If the tension is doubled from T to $2T$, then the low point is doubled since $\dfrac{T}{\rho g} = 60 \;\Rightarrow\; \dfrac{2T}{\rho g} = 120$. The height of the

poles is now $f(100) = \dfrac{2T}{\rho g}\cosh\left(\dfrac{\rho g \cdot 100}{2T}\right) = 120\cosh\left(\dfrac{100}{120}\right) \approx 164.13$ m, just a slight decrease.

63. (a) $y = A\sinh mx + B\cosh mx \;\Rightarrow\; y' = mA\cosh mx + mB\sinh mx \;\Rightarrow$

$y'' = m^2 A\sinh mx + m^2 B\cosh mx = m^2(A\sinh mx + B\cosh mx) = m^2 y$

(b) From part (a), a solution of $y'' = 9y$ is $y(x) = A\sinh 3x + B\cosh 3x$. Now $-4 = y(0) = A\sinh 0 + B\cosh 0 = B$, so

$B = -4$. Also, $y'(x) = 3A\cosh 3x - 12\sinh 3x$, so $6 = y'(0) = 3A \;\Rightarrow\; A = 2$. Thus, $y = 2\sinh 3x - 4\cosh 3x$.

65. The tangent to $y = \cosh x$ has slope 1 when $y' = \sinh x = 1 \;\Rightarrow\; x = \sinh^{-1} 1 = \ln(1+\sqrt{2})$, by Equation 3.

Since $\sinh x = 1$ and $y = \cosh x = \sqrt{1+\sinh^2 x}$, we have $\cosh x = \sqrt{2}$. The point is $\left(\ln(1+\sqrt{2}),\, \sqrt{2}\right)$.

67. Let $u = \cosh x$. Then $du = \sinh x\, dx$, so $\int \sinh x\, \cosh^2 x\, dx = \int u^2\, du = \tfrac{1}{3}u^3 + C = \tfrac{1}{3}\cosh^3 x + C$.

69. Let $u = \sqrt{x}$. Then $du = \dfrac{dx}{2\sqrt{x}}$ and $\displaystyle\int \dfrac{\sinh\sqrt{x}}{\sqrt{x}}\,dx = \int \sinh u \cdot 2\,du = 2\cosh u + C = 2\cosh\sqrt{x} + C$.

71. $\displaystyle\int \dfrac{\cosh x}{\cosh^2 x - 1}\,dx = \int \dfrac{\cosh x}{\sinh^2 x}\,dx = \int \dfrac{\cosh x}{\sinh x} \cdot \dfrac{1}{\sinh x}\,dx = \int \coth x\,\operatorname{csch} x\,dx = -\operatorname{csch} x + C$

73. Let $t = 3u$. Then $dt = 3\,du$ and

$$\int_4^6 \dfrac{1}{\sqrt{t^2-9}}\,dt = \int_{4/3}^2 \dfrac{1}{\sqrt{9u^2-9}}\,3\,du = \int_{4/3}^2 \dfrac{du}{\sqrt{u^2-1}} = \left[\cosh^{-1} u\right]_{4/3}^2 = \cosh^{-1} 2 - \cosh^{-1}\left(\tfrac{4}{3}\right) \text{ or}$$

$$= \left[\cosh^{-1} u\right]_{4/3}^2 = \left[\ln\left(u + \sqrt{u^2-1}\right)\right]_{4/3}^2 = \ln(2+\sqrt{3}) - \ln\left(\dfrac{4+\sqrt{7}}{3}\right) = \ln\left(\dfrac{6+3\sqrt{3}}{4+\sqrt{7}}\right)$$

75. Let $u = e^x$. Then $du = e^x\,dx$ and $\displaystyle\int \dfrac{e^x}{1-e^{2x}}\,dx = \int \dfrac{du}{1-u^2} = \tanh^{-1} u + C = \tanh^{-1} e^x + C$

$\left[\text{or } \tfrac{1}{2}\ln\left(\dfrac{1+e^x}{1-e^x}\right) + C\right]$.

77. (a) From the graphs, we estimate that the two curves $y = \cosh 2x$ and $y = 1 + \sinh x$ intersect at $x = 0$ and at $x = a \approx 0.481$.

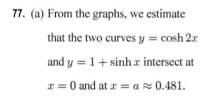
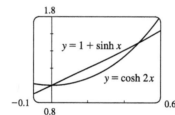
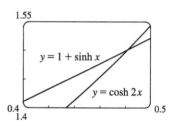

(b) We have found the two solutions of the equation $\cosh 2x = 1 + \sinh x$ to be $x = 0$ and $x = a \approx 0.481$. Note from the first graph that $1 + \sinh x > \cosh 2x$ on the interval $(0, a)$, so the area between the two curves is

$$A = \int_0^a (1 + \sinh x - \cosh 2x)\,dx = \left[x + \cosh x - \tfrac{1}{2}\sinh 2x\right]_0^a$$
$$= \left[a + \cosh a - \tfrac{1}{2}\sinh 2a\right] - \left[0 + \cosh 0 - \tfrac{1}{2}\sinh 0\right] \approx 0.0402$$

79. If $ae^x + be^{-x} = \alpha\cosh(x + \beta)$ [or $\alpha\sinh(x+\beta)$], then

$ae^x + be^{-x} = \tfrac{\alpha}{2}\left(e^{x+\beta} \pm e^{-x-\beta}\right) = \tfrac{\alpha}{2}\left(e^x e^\beta \pm e^{-x}e^{-\beta}\right) = \left(\tfrac{\alpha}{2}e^\beta\right)e^x \pm \left(\tfrac{\alpha}{2}e^{-\beta}\right)e^{-x}$. Comparing coefficients of e^x and e^{-x}, we have $a = \tfrac{\alpha}{2}e^\beta$ **(1)** and $b = \pm\tfrac{\alpha}{2}e^{-\beta}$ **(2)**. We need to find α and β. Dividing equation **(1)** by equation **(2)** gives us $\tfrac{a}{b} = \pm e^{2\beta}$ $\Rightarrow$ $(\star)$ $2\beta = \ln\left(\pm\tfrac{a}{b}\right)$ $\Rightarrow$ $\beta = \tfrac{1}{2}\ln\left(\pm\tfrac{a}{b}\right)$. Solving equations **(1)** and **(2)** for e^β gives us

$e^\beta = \dfrac{2a}{\alpha}$ and $e^\beta = \pm\dfrac{\alpha}{2b}$, so $\dfrac{2a}{\alpha} = \pm\dfrac{\alpha}{2b}$ $\Rightarrow$ $\alpha^2 = \pm 4ab$ $\Rightarrow$ $\alpha = 2\sqrt{\pm ab}$.

$(\star)$ If $\tfrac{a}{b} > 0$, we use the $+$ sign and obtain a cosh function, whereas if $\tfrac{a}{b} < 0$, we use the $-$ sign and obtain a sinh function.

In summary, if a and b have the same sign, we have $ae^x + be^{-x} = 2\sqrt{ab}\cosh\left(x + \tfrac{1}{2}\ln\tfrac{a}{b}\right)$, whereas if a and b have the opposite sign, then $ae^x + be^{-x} = 2\sqrt{-ab}\sinh\left(x + \tfrac{1}{2}\ln\left(-\tfrac{a}{b}\right)\right)$.

6.8 Indeterminate Forms and l'Hospital's Rule

Note: The use of l'Hospital's Rule is indicated by an H above the equal sign: $\stackrel{H}{=}$

1. (a) $\lim\limits_{x \to a} \dfrac{f(x)}{g(x)}$ is an indeterminate form of type $\dfrac{0}{0}$.

(b) $\lim\limits_{x \to a} \dfrac{f(x)}{p(x)} = 0$ because the numerator approaches 0 while the denominator becomes large.

(c) $\lim\limits_{x \to a} \dfrac{h(x)}{p(x)} = 0$ because the numerator approaches a finite number while the denominator becomes large.

(d) If $\lim\limits_{x \to a} p(x) = \infty$ and $f(x) \to 0$ through positive values, then $\lim\limits_{x \to a} \dfrac{p(x)}{f(x)} = \infty$. [For example, take $a = 0$, $p(x) = 1/x^2$, and $f(x) = x^2$.] If $f(x) \to 0$ through negative values, then $\lim\limits_{x \to a} \dfrac{p(x)}{f(x)} = -\infty$. [For example, take $a = 0$, $p(x) = 1/x^2$, and $f(x) = -x^2$.] If $f(x) \to 0$ through both positive and negative values, then the limit might not exist. [For example, take $a = 0$, $p(x) = 1/x^2$, and $f(x) = x$.]

(e) $\lim\limits_{x \to a} \dfrac{p(x)}{q(x)}$ is an indeterminate form of type $\dfrac{\infty}{\infty}$.

3. (a) When x is near a, $f(x)$ is near 0 and $p(x)$ is large, so $f(x) - p(x)$ is large negative. Thus, $\lim\limits_{x \to a} [f(x) - p(x)] = -\infty$.

(b) $\lim\limits_{x \to a} [p(x) - q(x)]$ is an indeterminate form of type $\infty - \infty$.

(c) When x is near a, $p(x)$ and $q(x)$ are both large, so $p(x) + q(x)$ is large. Thus, $\lim\limits_{x \to a} [p(x) + q(x)] = \infty$.

5. From the graphs of f and g, we see that $\lim\limits_{x \to 2} f(x) = 0$ and $\lim\limits_{x \to 2} g(x) = 0$, so l'Hospital's Rule applies.

$$\lim_{x \to 2} \frac{f(x)}{g(x)} \stackrel{H}{=} \lim_{x \to 2} \frac{f'(x)}{g'(x)} = \frac{\lim\limits_{x \to 2} f'(x)}{\lim\limits_{x \to 2} g'(x)} = \frac{f'(2)}{g'(2)} = \frac{1.8}{\frac{4}{5}} = \frac{9}{4}$$

7. f and $g = e^x - 1$ are differentiable and $g' = e^x \neq 0$ on an open interval that contains 0. $\lim\limits_{x \to 0} f(x) = 0$ and $\lim\limits_{x \to 0} g(x) = 0$, so we have the indeterminate form $\frac{0}{0}$ and can apply l'Hospital's Rule.

$$\lim_{x \to 0} \frac{f(x)}{e^x - 1} \stackrel{H}{=} \lim_{x \to 0} \frac{f'(x)}{e^x} = \frac{1}{1} = 1$$

Note that $\lim\limits_{x \to 0} f'(x) = 1$ since the graph of f has the same slope as the line $y = x$ at $x = 0$.

9. This limit has the form $\frac{0}{0}$. $\lim\limits_{x \to 4} \dfrac{x^2 - 2x - 8}{x - 4} = \lim\limits_{x \to 4} \dfrac{(x-4)(x+2)}{x-4} = \lim\limits_{x \to 4} (x+2) = 4 + 2 = 6$

Note: Alternatively, we could apply l'Hospital's Rule.

11. This limit has the form $\frac{0}{0}$. $\lim\limits_{x \to 1} \dfrac{x^7 - 1}{x^3 - 1} \stackrel{H}{=} \lim\limits_{x \to 1} \dfrac{7x^6}{3x^2} = \dfrac{7}{3}$

Note: Alternatively, we could factor and simplify.

13. This limit has the form $\frac{0}{0}$. $\lim\limits_{x \to \pi/4} \dfrac{\sin x - \cos x}{\tan x - 1} \overset{\text{H}}{=} \lim\limits_{x \to \pi/4} \dfrac{\cos x + \sin x}{\sec^2 x} = \lim\limits_{x \to 4} \dfrac{\frac{\sqrt{2}}{2} + \frac{\sqrt{2}}{2}}{(\sqrt{2})^2} = \dfrac{\sqrt{2}}{2}$

15. This limit has the form $\frac{0}{0}$. $\lim\limits_{t \to 0} \dfrac{e^{2t} - 1}{\sin t} \overset{\text{H}}{=} \lim\limits_{t \to 0} \dfrac{2e^{2t}}{\cos t} = \dfrac{2(1)}{1} = 2$

17. This limit has the form $\frac{0}{0}$. $\lim\limits_{x \to 1} \dfrac{\sin(x-1)}{x^3 + x - 2} \overset{\text{H}}{=} \lim\limits_{x \to 1} \dfrac{\cos(x-1)}{3x^2 + 1} = \dfrac{\cos 0}{3(1)^2 + 1} = \dfrac{1}{4}$

19. This limit has the form $\frac{\infty}{\infty}$. $\lim\limits_{x \to \infty} \dfrac{\sqrt{x}}{1 + e^x} \overset{\text{H}}{=} \lim\limits_{x \to \infty} \dfrac{\frac{1}{2\sqrt{x}}}{e^x} = \lim\limits_{x \to \infty} \dfrac{1}{e^x \cdot 2\sqrt{x}} = 0$

21. $\lim\limits_{x \to 0^+} [(\ln x)/x] = -\infty$ since $\ln x \to -\infty$ as $x \to 0^+$ and dividing by small values of x just increases the magnitude of the quotient $(\ln x)/x$. L'Hospital's Rule does not apply.

23. This limit has the form $\frac{0}{0}$. $\lim\limits_{x \to 3} \dfrac{\ln(x/3)}{3 - x} \overset{\text{H}}{=} \lim\limits_{x \to 3} \dfrac{1/(x/3) \cdot (1/3)}{-1} = \lim\limits_{x \to 3} \left(-\dfrac{1}{x}\right) = -\dfrac{1}{3}$

25. This limit has the form $\frac{0}{0}$.

$\lim\limits_{x \to 0} \dfrac{\sqrt{1 + 2x} - \sqrt{1 - 4x}}{x} \overset{\text{H}}{=} \lim\limits_{x \to 0} \dfrac{\frac{1}{2}(1 + 2x)^{-1/2} \cdot 2 - \frac{1}{2}(1 - 4x)^{-1/2}(-4)}{1}$

$= \lim\limits_{x \to 0} \left(\dfrac{1}{\sqrt{1 + 2x}} + \dfrac{2}{\sqrt{1 - 4x}}\right) = \dfrac{1}{\sqrt{1}} + \dfrac{2}{\sqrt{1}} = 3$

27. This limit has the form $\frac{0}{0}$. $\lim\limits_{x \to 0} \dfrac{e^x + e^{-x} - 2}{e^x - x - 1} \overset{\text{H}}{=} \lim\limits_{x \to 0} \dfrac{e^x - e^{-x}}{e^x - 1} \overset{\text{H}}{=} \lim\limits_{x \to 0} \dfrac{e^x + e^{-x}}{e^x} = \dfrac{1 + 1}{1} = 2$

29. This limit has the form $\frac{0}{0}$. $\lim\limits_{x \to 0} \dfrac{\tanh x}{\tan x} \overset{\text{H}}{=} \lim\limits_{x \to 0} \dfrac{\text{sech}^2 x}{\sec^2 x} = \dfrac{\text{sech}^2 0}{\sec^2 0} = \dfrac{1}{1} = 1$

31. This limit has the form $\frac{0}{0}$. $\lim\limits_{x \to 0} \dfrac{\sin^{-1} x}{x} \overset{\text{H}}{=} \lim\limits_{x \to 0} \dfrac{1/\sqrt{1 - x^2}}{1} = \lim\limits_{x \to 0} \dfrac{1}{\sqrt{1 - x^2}} = \dfrac{1}{1} = 1$

33. This limit has the form $\frac{0}{0}$. $\lim\limits_{x \to 0} \dfrac{x 3^x}{3^x - 1} \overset{\text{H}}{=} \lim\limits_{x \to 0} \dfrac{x 3^x \ln 3 + 3^x}{3^x \ln 3} = \lim\limits_{x \to 0} \dfrac{3^x (x \ln 3 + 1)}{3^x \ln 3} = \lim\limits_{x \to 0} \dfrac{x \ln 3 + 1}{\ln 3} = \dfrac{1}{\ln 3}$

35. This limit can be evaluated by substituting 0 for x. $\lim\limits_{x \to 0} \dfrac{\ln(1 + x)}{\cos x + e^x - 1} = \dfrac{\ln 1}{1 + 1 - 1} = \dfrac{0}{1} = 0$

37. This limit has the form $\frac{0}{\infty}$, so l'Hospital's Rule doesn't apply. As $x \to 0^+$, $\arctan 2x \to 0$ and $\ln x \to -\infty$,

so $\lim\limits_{x \to 0^+} \dfrac{\arctan 2x}{\ln x} = 0$.

39. This limit has the form $\frac{0}{0}$. $\lim\limits_{x \to 1} \dfrac{x^a - 1}{x^b - 1}$ [for $b \neq 0$] $\overset{\text{H}}{=} \lim\limits_{x \to 1} \dfrac{ax^{a-1}}{bx^{b-1}} = \dfrac{a(1)}{b(1)} = \dfrac{a}{b}$

41. This limit has the form $\frac{0}{0}$.

$$\lim_{x\to 0}\frac{\cos x-1+\frac{1}{2}x^2}{x^4}\stackrel{\text{H}}{=}\lim_{x\to 0}\frac{-\sin x+x}{4x^3}\stackrel{\text{H}}{=}\lim_{x\to 0}\frac{-\cos x+1}{12x^2}\stackrel{\text{H}}{=}\lim_{x\to 0}\frac{\sin x}{24x}\stackrel{\text{H}}{=}\lim_{x\to 0}\frac{\cos x}{24}=\frac{1}{24}$$

43. This limit has the form $\infty\cdot 0$. We'll change it to the form $\frac{0}{0}$.

$$\lim_{x\to\infty}x\sin(\pi/x)=\lim_{x\to\infty}\frac{\sin(\pi/x)}{1/x}\stackrel{\text{H}}{=}\lim_{x\to\infty}\frac{\cos(\pi/x)(-\pi/x^2)}{-1/x^2}=\pi\lim_{x\to\infty}\cos(\pi/x)=\pi(1)=\pi$$

45. This limit has the form $0\cdot\infty$. We'll change it to the form $\frac{0}{0}$. $\lim_{x\to 0}\sin 5x\csc 3x=\lim_{x\to 0}\frac{\sin 5x}{\sin 3x}\stackrel{\text{H}}{=}\lim_{x\to 0}\frac{5\cos 5x}{3\cos 3x}=\frac{5\cdot 1}{3\cdot 1}=\frac{5}{3}$

47. This limit has the form $\infty\cdot 0$. $\lim_{x\to\infty}x^3e^{-x^2}=\lim_{x\to\infty}\frac{x^3}{e^{x^2}}\stackrel{\text{H}}{=}\lim_{x\to\infty}\frac{3x^2}{2xe^{x^2}}=\lim_{x\to\infty}\frac{3x}{2e^{x^2}}\stackrel{\text{H}}{=}\lim_{x\to\infty}\frac{3}{4xe^{x^2}}=0$

49. This limit has the form $0\cdot(-\infty)$.

$$\lim_{x\to 1^+}\ln x\tan(\pi x/2)=\lim_{x\to 1^+}\frac{\ln x}{\cot(\pi x/2)}\stackrel{\text{H}}{=}\lim_{x\to 1^+}\frac{1/x}{(-\pi/2)\csc^2(\pi x/2)}=\frac{1}{(-\pi/2)(1)^2}=-\frac{2}{\pi}$$

51. This limit has the form $\infty-\infty$.

$$\lim_{x\to 1}\left(\frac{x}{x-1}-\frac{1}{\ln x}\right)=\lim_{x\to 1}\frac{x\ln x-(x-1)}{(x-1)\ln x}\stackrel{\text{H}}{=}\lim_{x\to 1}\frac{x(1/x)+\ln x-1}{(x-1)(1/x)+\ln x}=\lim_{x\to 1}\frac{\ln x}{1-(1/x)+\ln x}$$

$$\stackrel{\text{H}}{=}\lim_{x\to 1}\frac{1/x}{1/x^2+1/x}\cdot\frac{x^2}{x^2}=\lim_{x\to 1}\frac{x}{1+x}=\frac{1}{1+1}=\frac{1}{2}$$

53. This limit has the form $\infty-\infty$.

$$\lim_{x\to 0^+}\left(\frac{1}{x}-\frac{1}{e^x-1}\right)=\lim_{x\to 0^+}\frac{e^x-1-x}{x(e^x-1)}\stackrel{\text{H}}{=}\lim_{x\to 0^+}\frac{e^x-1}{xe^x+e^x-1}\stackrel{\text{H}}{=}\lim_{x\to 0^+}\frac{e^x}{xe^x+e^x+e^x}=\frac{1}{0+1+1}=\frac{1}{2}$$

55. This limit has the form $\infty-\infty$.

$$\lim_{x\to 0^+}\left(\frac{1}{x}-\frac{1}{\tan x}\right)=\lim_{x\to 0^+}\frac{\tan x-x}{x\tan x}\stackrel{\text{H}}{=}\lim_{x\to 0^+}\frac{\sec^2 x-1}{x\sec^2 x+\tan x}\stackrel{\text{H}}{=}\lim_{x\to 0^+}\frac{2\sec x\cdot\sec x\tan x}{x\cdot 2\sec x\cdot\sec x\tan x+\sec^2 x+\sec^2 x}$$

$$=\frac{0}{0+1+1}=0$$

57. $y=x^{\sqrt{x}}\;\Rightarrow\;\ln y=\sqrt{x}\ln x$, so

$$\lim_{x\to 0^+}\ln y=\lim_{x\to 0^+}\sqrt{x}\ln x=\lim_{x\to 0^+}\frac{\ln x}{x^{-1/2}}\stackrel{\text{H}}{=}\lim_{x\to 0^+}\frac{1/x}{-\frac{1}{2}x^{-3/2}}=-2\lim_{x\to 0^+}\sqrt{x}=0\;\Rightarrow$$

$$\lim_{x\to 0^+}x^{\sqrt{x}}=\lim_{x\to 0^+}e^{\ln y}=e^0=1.$$

59. $y=(1-2x)^{1/x}\;\Rightarrow\;\ln y=\frac{1}{x}\ln(1-2x)$, so $\lim_{x\to 0}\ln y=\lim_{x\to 0}\frac{\ln(1-2x)}{x}\stackrel{\text{H}}{=}\lim_{x\to 0}\frac{-2/(1-2x)}{1}=-2\;\Rightarrow$

$$\lim_{x\to 0}(1-2x)^{1/x}=\lim_{x\to 0}e^{\ln y}=e^{-2}.$$

61. $y = x^{1/(1-x)} \;\Rightarrow\; \ln y = \dfrac{1}{1-x}\ln x$, so $\displaystyle\lim_{x\to 1^+}\ln y = \lim_{x\to 1^+}\dfrac{1}{1-x}\ln x = \lim_{x\to 1^+}\dfrac{\ln x}{1-x} \stackrel{\mathrm{H}}{=} \lim_{x\to 1^+}\dfrac{1/x}{-1} = -1 \;\Rightarrow\;$
$\displaystyle\lim_{x\to 1^+} x^{1/(1-x)} = \lim_{x\to 1^+} e^{\ln y} = e^{-1} = \dfrac{1}{e}$.

63. $y = x^{1/x} \;\Rightarrow\; \ln y = (1/x)\ln x \;\Rightarrow\; \displaystyle\lim_{x\to\infty}\ln y = \lim_{x\to\infty}\dfrac{\ln x}{x} \stackrel{\mathrm{H}}{=} \lim_{x\to\infty}\dfrac{1/x}{1} = 0 \;\Rightarrow\;$
$\displaystyle\lim_{x\to\infty} x^{1/x} = \lim_{x\to\infty} e^{\ln y} = e^0 = 1$

65. $y = (4x+1)^{\cot x} \;\Rightarrow\; \ln y = \cot x \,\ln(4x+1)$, so $\displaystyle\lim_{x\to 0^+}\ln y = \lim_{x\to 0^+}\dfrac{\ln(4x+1)}{\tan x} \stackrel{\mathrm{H}}{=} \lim_{x\to 0^+}\dfrac{\dfrac{4}{4x+1}}{\sec^2 x} = 4 \;\Rightarrow\;$
$\displaystyle\lim_{x\to 0^+}(4x+1)^{\cot x} = \lim_{x\to 0^+} e^{\ln y} = e^4$.

67. $y = (1+\sin 3x)^{1/x} \;\Rightarrow\; \ln y = \dfrac{1}{x}\ln(1+\sin 3x) \;\Rightarrow\;$

$\displaystyle\lim_{x\to 0^+}\ln y = \lim_{x\to 0^+}\dfrac{\ln(1+\sin 3x)}{x} \stackrel{\mathrm{H}}{=} \lim_{x\to 0^+}\dfrac{[1/(1+\sin 3x)]\cdot 3\cos 3x}{1} = \lim_{x\to 0^+}\dfrac{3\cos 3x}{1+\sin 3x} = \dfrac{3\cdot 1}{1+0} = 3 \;\Rightarrow\;$
$\displaystyle\lim_{x\to 0^+}(1+\sin 3x)^{1/x} = \lim_{x\to 0^+} e^{\ln y} = e^3$

69. The given limit is $\displaystyle\lim_{x\to 0^+}\dfrac{x^x - 1}{\ln x + x - 1}$. Note that $y = x^x \;\Rightarrow\; \ln y = x\ln x$, so

$\displaystyle\lim_{x\to 0^+}\ln y = \lim_{x\to 0^+} x\ln x = \lim_{x\to 0^+}\dfrac{\ln x}{1/x} \stackrel{\mathrm{H}}{=} \lim_{x\to 0^+}\dfrac{\dfrac{1}{x}}{-\dfrac{1}{x^2}} = \lim_{x\to 0^+}(-x) = 0 \;\Rightarrow\; \displaystyle\lim_{x\to 0^+} x^x = \lim_{x\to 0^+} e^{\ln y} = e^0 = 1$.

Therefore, the numerator of the given limit has limit $1 - 1 = 0$ as $x \to 0^+$. The denominator of the given limit $\to -\infty$ as $x \to 0^+$ since $\ln x \to -\infty$ as $x \to 0^+$. Thus, $\displaystyle\lim_{x\to 0^+}\dfrac{x^x - 1}{\ln x + x - 1} = 0$.

71.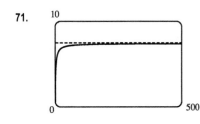

From the graph, if $x = 500$, $y \approx 7.36$. The limit has the form 1^∞.

Now $y = \left(1+\dfrac{2}{x}\right)^x \;\Rightarrow\; \ln y = x\ln\left(1+\dfrac{2}{x}\right) \;\Rightarrow\;$

$\displaystyle\lim_{x\to\infty}\ln y = \lim_{x\to\infty}\dfrac{\ln(1+2/x)}{1/x} \stackrel{\mathrm{H}}{=} \lim_{x\to\infty}\dfrac{\dfrac{1}{1+2/x}\left(-\dfrac{2}{x^2}\right)}{-1/x^2}$

$= 2\displaystyle\lim_{x\to\infty}\dfrac{1}{1+2/x} = 2(1) = 2 \;\Rightarrow\;$

$\displaystyle\lim_{x\to\infty}\left(1+\dfrac{2}{x}\right)^x = \lim_{x\to\infty} e^{\ln y} = e^2 \;[\approx 7.39]$

73.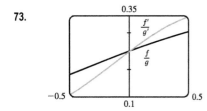

From the graph, it appears that $\displaystyle\lim_{x\to 0}\dfrac{f(x)}{g(x)} = \lim_{x\to 0}\dfrac{f'(x)}{g'(x)} = 0.25$.

We calculate $\displaystyle\lim_{x\to 0}\dfrac{f(x)}{g(x)} = \lim_{x\to 0}\dfrac{e^x - 1}{x^3 + 4x} \stackrel{\mathrm{H}}{=} \lim_{x\to 0}\dfrac{e^x}{3x^2 + 4} = \dfrac{1}{4}$.

75. $\lim\limits_{x\to\infty} \dfrac{e^x}{x^n} \stackrel{H}{=} \lim\limits_{x\to\infty} \dfrac{e^x}{nx^{n-1}} \stackrel{H}{=} \lim\limits_{x\to\infty} \dfrac{e^x}{n(n-1)x^{n-2}} \stackrel{H}{=} \cdots \stackrel{H}{=} \lim\limits_{x\to\infty} \dfrac{e^x}{n!} = \infty$

77. $\lim\limits_{x\to\infty} \dfrac{x}{\sqrt{x^2+1}} \stackrel{H}{=} \lim\limits_{x\to\infty} \dfrac{1}{\frac{1}{2}(x^2+1)^{-1/2}(2x)} = \lim\limits_{x\to\infty} \dfrac{\sqrt{x^2+1}}{x}$. Repeated applications of l'Hospital's Rule result in the

original limit or the limit of the reciprocal of the function. Another method is to try dividing the numerator and denominator

by x: $\lim\limits_{x\to\infty} \dfrac{x}{\sqrt{x^2+1}} = \lim\limits_{x\to\infty} \dfrac{x/x}{\sqrt{x^2/x^2+1/x^2}} = \lim\limits_{x\to\infty} \dfrac{1}{\sqrt{1+1/x^2}} = \dfrac{1}{1} = 1$

79. $y = f(x) = xe^{-x}$ **A.** $D = \mathbb{R}$ **B.** Intercepts are 0   **C.** No symmetry

D. $\lim\limits_{x\to\infty} xe^{-x} = \lim\limits_{x\to\infty} \dfrac{x}{e^x} \stackrel{H}{=} \lim\limits_{x\to\infty} \dfrac{1}{e^x} = 0$, so $y = 0$ is a HA. $\lim\limits_{x\to-\infty} xe^{-x} = -\infty$

E. $f'(x) = e^{-x} - xe^{-x} = e^{-x}(1-x) > 0 \Leftrightarrow x < 1$, so f is increasing

on $(-\infty, 1)$ and decreasing on $(1, \infty)$.   **F.** Absolute and local maximum

value $f(1) = 1/e$.   **G.** $f''(x) = e^{-x}(x-2) > 0 \Leftrightarrow x > 2$, so f is CU

on $(2, \infty)$ and CD on $(-\infty, 2)$.   IP at $\left(2, 2/e^2\right)$

H.
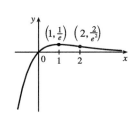

81. $y = f(x) = xe^{-x^2}$ **A.** $D = \mathbb{R}$ **B.** Intercepts are 0   **C.** $f(-x) = -f(x)$, so the curve is symmetric

about the origin.   **D.** $\lim\limits_{x\to\pm\infty} xe^{-x^2} = \lim\limits_{x\to\pm\infty} \dfrac{x}{e^{x^2}} \stackrel{H}{=} \lim\limits_{x\to\pm\infty} \dfrac{1}{2xe^{x^2}} = 0$, so $y = 0$ is a HA.

E. $f'(x) = e^{-x^2} - 2x^2 e^{-x^2} = e^{-x^2}(1-2x^2) > 0 \Leftrightarrow x^2 < \tfrac{1}{2} \Leftrightarrow |x| < \tfrac{1}{\sqrt{2}}$, so f is increasing on $\left(-\tfrac{1}{\sqrt{2}}, \tfrac{1}{\sqrt{2}}\right)$

and decreasing on $\left(-\infty, -\tfrac{1}{\sqrt{2}}\right)$ and $\left(\tfrac{1}{\sqrt{2}}, \infty\right)$.   **F.** Local maximum value $f\left(\tfrac{1}{\sqrt{2}}\right) = 1/\sqrt{2e}$, local minimum

value $f\left(-\tfrac{1}{\sqrt{2}}\right) = -1/\sqrt{2e}$ **G.** $f''(x) = -2xe^{-x^2}(1-2x^2) - 4xe^{-x^2} = 2xe^{-x^2}(2x^2-3) > 0 \Leftrightarrow$

$x > \sqrt{\tfrac{3}{2}}$ or $-\sqrt{\tfrac{3}{2}} < x < 0$, so f is CU on $\left(\sqrt{\tfrac{3}{2}}, \infty\right)$ and

$\left(-\sqrt{\tfrac{3}{2}}, 0\right)$ and CD on $\left(-\infty, -\sqrt{\tfrac{3}{2}}\right)$ and $\left(0, \sqrt{\tfrac{3}{2}}\right)$.

IP are $(0, 0)$ and $\left(\pm\sqrt{\tfrac{3}{2}}, \pm\sqrt{\tfrac{3}{2}}\, e^{-3/2}\right)$.

H.
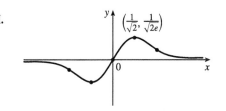

83. $y = f(x) = \dfrac{1}{x} + \ln x$ **A.** $D = (0, \infty)$ [same as $\ln x$]   **B.** No y-intercept; no x-intercept $[1/x > |\ln x|$ on $(0, 1)$, and $1/x$

and $\ln x$ are both positive on $(1, \infty)$]   **C.** No symmetry   **D.** $\lim\limits_{x\to 0^+} f(x) = \infty$, so $x = 0$ is a VA.

E. $f'(x) = -\dfrac{1}{x^2} + \dfrac{1}{x} = \dfrac{x-1}{x^2}$. $f'(x) > 0$ for $x > 1$, so f is increasing on

$(1, \infty)$ and f is decreasing on $(0, 1)$.

F. Local minimum value $f(1) = 1$ **G.** $f''(x) = \dfrac{2}{x^3} - \dfrac{1}{x^2} = \dfrac{2-x}{x^3}$.

$f''(x) > 0$ for $0 < x < 2$, so f is CU on $(0, 2)$, and f is CD on $(2, \infty)$.

IP at $\left(2, \tfrac{1}{2} + \ln 2\right)$

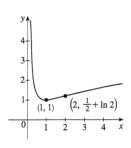

85. (a) $f(x) = x^{-x}$

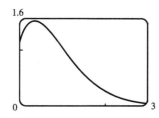

(b) $y = f(x) = x^{-x}$. We note that $\ln f(x) = \ln x^{-x} = -x \ln x = -\dfrac{\ln x}{1/x}$, so

$$\lim_{x \to 0^+} \ln f(x) \stackrel{H}{=} \lim_{x \to 0^+} -\dfrac{1/x}{-x^{-2}} = \lim_{x \to 0^+} x = 0. \text{ Thus } \lim_{x \to 0^+} f(x) = \lim_{x \to 0^+} e^{\ln f(x)} = e^0 = 1.$$

(c) From the graph, it appears that there is a local and absolute maximum of about $f(0.37) \approx 1.44$. To find the exact value, we differentiate: $f(x) = x^{-x} = e^{-x \ln x} \Rightarrow f'(x) = e^{-x \ln x}\left[-x\left(\dfrac{1}{x}\right) + \ln x(-1)\right] = -x^{-x}(1 + \ln x)$. This is 0 only when $1 + \ln x = 0 \Leftrightarrow x = e^{-1}$. Also $f'(x)$ changes from positive to negative at e^{-1}. So the maximum value is $f(1/e) = (1/e)^{-1/e} = e^{1/e}$.

(d)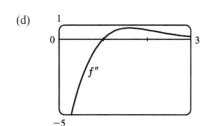

We differentiate again to get

$$f''(x) = -x^{-x}(1/x) + (1 + \ln x)^2(x^{-x}) = x^{-x}[(1 + \ln x)^2 - 1/x]$$

From the graph of $f''(x)$, it seems that $f''(x)$ changes from negative to positive at $x = 1$, so we estimate that f has an IP at $x = 1$.

87. (a) $f(x) = x^{1/x}$

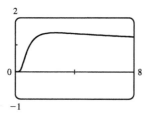

(b) Recall that $a^b = e^{b \ln a}$. $\lim_{x \to 0^+} x^{1/x} = \lim_{x \to 0^+} e^{(1/x) \ln x}$. As $x \to 0^+$, $\dfrac{\ln x}{x} \to -\infty$, so $x^{1/x} = e^{(1/x) \ln x} \to 0$. This indicates that there is a hole at $(0, 0)$. As $x \to \infty$, we have the indeterminate form ∞^0. $\lim_{x \to \infty} x^{1/x} = \lim_{x \to \infty} e^{(1/x) \ln x}$, but $\lim_{x \to \infty} \dfrac{\ln x}{x} \stackrel{H}{=} \lim_{x \to \infty} \dfrac{1/x}{1} = 0$, so $\lim_{x \to \infty} x^{1/x} = e^0 = 1$. This indicates that $y = 1$ is a HA.

(c) Estimated maximum: $(2.72, 1.45)$. No estimated minimum. We use logarithmic differentiation to find any critical numbers. $y = x^{1/x} \Rightarrow \ln y = \dfrac{1}{x} \ln x \Rightarrow \dfrac{y'}{y} = \dfrac{1}{x} \cdot \dfrac{1}{x} + (\ln x)\left(-\dfrac{1}{x^2}\right) \Rightarrow y' = x^{1/x}\left(\dfrac{1 - \ln x}{x^2}\right) = 0 \Rightarrow \ln x = 1 \Rightarrow x = e$. For $0 < x < e$, $y' > 0$ and for $x > e$, $y' < 0$, so $f(e) = e^{1/e}$ is a local maximum value. This point is approximately $(2.7183, 1.4447)$, which agrees with our estimate.

(d)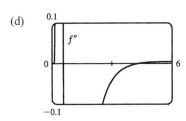

From the graph, we see that $f''(x) = 0$ at $x \approx 0.58$ and $x \approx 4.37$. Since f'' changes sign at these values, they are x-coordinates of inflection points.

89. $f(x) = e^x - cx \;\Rightarrow\; f'(x) = e^x - c = 0 \;\Leftrightarrow\; e^x = c \;\Leftrightarrow\; x = \ln c,\, c > 0.\; f''(x) = e^x > 0$, so f is CU on $(-\infty, \infty)$. $\lim_{x\to\infty}(e^x - cx) = \lim_{x\to\infty}\left[x\left(\dfrac{e^x}{x} - c\right)\right] = L_1$. Now $\lim_{x\to\infty}\dfrac{e^x}{x} \stackrel{H}{=} \lim_{x\to\infty}\dfrac{e^x}{1} = \infty$, so $L_1 = \infty$, regardless of the value of c. For $L = \lim_{x\to-\infty}(e^x - cx)$, $e^x \to 0$, so L is determined by $-cx$. If $c > 0$, $-cx \to \infty$, and $L = \infty$. If $c < 0$, $-cx \to -\infty$, and $L = -\infty$. Thus, f has an absolute minimum for $c > 0$. As c increases, the minimum points $(\ln c,\, c - c\ln c)$ get farther away from the origin.

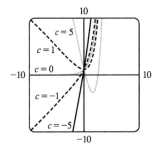

91. First we will find $\lim_{n\to\infty}\left(1 + \dfrac{r}{n}\right)^{nt}$, which is of the form 1^∞. $y = \left(1 + \dfrac{r}{n}\right)^{nt} \;\Rightarrow\; \ln y = nt\ln\left(1 + \dfrac{r}{n}\right)$, so

$$\lim_{n\to\infty}\ln y = \lim_{n\to\infty} nt\ln\left(1 + \dfrac{r}{n}\right) = t\lim_{n\to\infty}\dfrac{\ln(1 + r/n)}{1/n} \stackrel{H}{=} t\lim_{n\to\infty}\dfrac{(-r/n^2)}{(1 + r/n)(-1/n^2)} = t\lim_{n\to\infty}\dfrac{r}{1 + i/n} = tr \;\Rightarrow$$

$\lim_{n\to\infty} y = e^{rt}$. Thus, as $n \to \infty$, $A = A_0\left(1 + \dfrac{r}{n}\right)^{nt} \to A_0 e^{rt}$.

93. (a) $\lim_{t\to\infty} P(t) = \lim_{t\to\infty}\dfrac{M}{1 + Ae^{-kt}} = \dfrac{M}{1 + A\cdot 0} = M$

It is to be expected that a population that is growing will eventually reach the maximum population size that can be supported.

(b) $\lim_{M\to\infty} P(t) = \lim_{M\to\infty}\dfrac{M}{1 + \dfrac{M - P_0}{P_0}e^{-kt}} = \lim_{M\to\infty}\dfrac{M}{1 + \left(\dfrac{M}{P_0} - 1\right)e^{-kt}} \stackrel{H}{=} \lim_{M\to\infty}\dfrac{1}{\dfrac{1}{P_0}e^{-kt}} = P_0 e^{kt}$

$P_0 e^{kt}$ is an exponential function.

95. $\lim_{x\to 0}\dfrac{1}{x^2}\int_0^x \dfrac{2t}{\sqrt{t^3+1}}\,dt = \lim_{x\to 0}\dfrac{\int_0^x \dfrac{2t}{\sqrt{t^3+1}}\,dt}{x^2}\;\left[\text{form }\tfrac{0}{0}\right] \stackrel{H}{=} \lim_{x\to 0}\dfrac{\dfrac{2x}{\sqrt{x^3+1}}}{2x} = \lim_{x\to 0}\dfrac{1}{\sqrt{x^3+1}} = 1$

97. Both numerator and denominator approach 0 as $x \to 0$, so we use l'Hospital's Rule (and FTC1):

$$\lim_{x\to 0}\dfrac{S(x)}{x^3} = \lim_{x\to 0}\dfrac{\int_0^x \sin(\pi t^2/2)\,dt}{x^3} \stackrel{H}{=} \lim_{x\to 0}\dfrac{\sin(\pi x^2/2)}{3x^2} \stackrel{H}{=} \lim_{x\to 0}\dfrac{\pi x\cos(\pi x^2/2)}{6x} = \dfrac{\pi}{6}\cdot\cos 0 = \dfrac{\pi}{6}$$

99. We see that both numerator and denominator approach 0, so we can use l'Hospital's Rule:

$$\lim_{x \to a} \frac{\sqrt{2a^3x - x^4} - a\sqrt[3]{aax}}{a - \sqrt[4]{ax^3}} \stackrel{H}{=} \lim_{x \to a} \frac{\frac{1}{2}(2a^3x - x^4)^{-1/2}(2a^3 - 4x^3) - a\left(\frac{1}{3}\right)(aax)^{-2/3}a^2}{-\frac{1}{4}(ax^3)^{-3/4}(3ax^2)}$$

$$= \frac{\frac{1}{2}(2a^3 a - a^4)^{-1/2}(2a^3 - 4a^3) - \frac{1}{3}a^3(a^2 a)^{-2/3}}{-\frac{1}{4}(aa^3)^{-3/4}(3aa^2)}$$

$$= \frac{(a^4)^{-1/2}(-a^3) - \frac{1}{3}a^3(a^3)^{-2/3}}{-\frac{3}{4}a^3(a^4)^{-3/4}} = \frac{-a - \frac{1}{3}a}{-\frac{3}{4}} = \frac{4}{3}\left(\frac{4}{3}a\right) = \frac{16}{9}a$$

101. The limit, $L = \lim_{x \to \infty}\left[x - x^2 \ln\left(\frac{1+x}{x}\right)\right] = \lim_{x \to \infty}\left[x - x^2 \ln\left(\frac{1}{x} + 1\right)\right]$. Let $t = 1/x$, so as $x \to \infty$, $t \to 0^+$.

$$L = \lim_{t \to 0^+}\left[\frac{1}{t} - \frac{1}{t^2}\ln(t+1)\right] = \lim_{t \to 0^+}\frac{t - \ln(t+1)}{t^2} \stackrel{H}{=} \lim_{t \to 0^+}\frac{1 - \frac{1}{t+1}}{2t} = \lim_{t \to 0^+}\frac{t/(t+1)}{2t} = \lim_{t \to 0^+}\frac{1}{2(t+1)} = \frac{1}{2}$$

Note: Starting the solution by factoring out x or x^2 leads to a more complicated solution.

103. (a) We look for functions f and g whose individual limits are ∞ as $x \to 0$, but whose quotient has a limit of 7 as $x \to 0$.

One such pair of functions is $f(x) = \dfrac{7}{x^2}$ and $g(x) = \dfrac{1}{x^2}$. We have $\lim_{x \to 0} f(x) = \lim_{x \to 0} g(x) = \infty$, and

$$\lim_{x \to 0}\frac{f(x)}{g(x)} = \lim_{x \to 0}\frac{7/x^2}{1/x^2} = \lim_{x \to 0} 7 = 7.$$

(b) We look for functions f and g whose individual limits are ∞ as $x \to 0$, but whose difference has a limit of 7 as $x \to 0$.

One such pair of functions is $f(x) = \dfrac{1}{x^2} + 7$ and $g(x) = \dfrac{1}{x^2}$. We have $\lim_{x \to 0} f(x) = \lim_{x \to 0} g(x) = \infty$, and

$$\lim_{x \to 0}[f(x) - g(x)] = \lim_{x \to 0}\left[\left(\frac{1}{x^2} + 7\right) - \frac{1}{x^2}\right] = \lim_{x \to 0} 7 = 7.$$

105. (a) We show that $\lim_{x \to 0}\dfrac{f(x)}{x^n} = 0$ for every integer $n \geq 0$. Let $y = \dfrac{1}{x^2}$. Then

$$\lim_{x \to 0}\frac{f(x)}{x^{2n}} = \lim_{x \to 0}\frac{e^{-1/x^2}}{(x^2)^n} = \lim_{y \to \infty}\frac{y^n}{e^y} \stackrel{H}{=} \lim_{y \to \infty}\frac{ny^{n-1}}{e^y} \stackrel{H}{=} \cdots \stackrel{H}{=} \lim_{y \to \infty}\frac{n!}{e^y} = 0 \Rightarrow$$

$$\lim_{x \to 0}\frac{f(x)}{x^n} = \lim_{x \to 0} x^n\frac{f(x)}{x^{2n}} = \lim_{x \to 0} x^n \lim_{x \to 0}\frac{f(x)}{x^{2n}} = 0. \text{ Thus, } f'(0) = \lim_{x \to 0}\frac{f(x) - f(0)}{x - 0} = \lim_{x \to 0}\frac{f(x)}{x} = 0.$$

(b) Using the Chain Rule and the Quotient Rule we see that $f^{(n)}(x)$ exists for $x \neq 0$. In fact, we prove by induction that for each $n \geq 0$, there is a polynomial p_n and a non-negative integer k_n with $f^{(n)}(x) = p_n(x)f(x)/x^{k_n}$ for $x \neq 0$. This is true for $n = 0$; suppose it is true for the nth derivative. Then $f'(x) = f(x)(2/x^3)$, so

$$f^{(n+1)}(x) = \left[x^{k_n}[p'_n(x)f(x) + p_n(x)f'(x)] - k_n x^{k_n-1} p_n(x) f(x)\right] x^{-2k_n}$$

$$= \left[x^{k_n} p'_n(x) + p_n(x)(2/x^3) - k_n x^{k_n-1} p_n(x)\right] f(x) x^{-2k_n}$$

$$= \left[x^{k_n+3} p'_n(x) + 2 p_n(x) - k_n x^{k_n+2} p_n(x)\right] f(x) x^{-(2k_n+3)}$$

which has the desired form.

[continued]

Now we show by induction that $f^{(n)}(0) = 0$ for all n. By part (a), $f'(0) = 0$. Suppose that $f^{(n)}(0) = 0$. Then

$$f^{(n+1)}(0) = \lim_{x \to 0} \frac{f^{(n)}(x) - f^{(n)}(0)}{x - 0} = \lim_{x \to 0} \frac{f^{(n)}(x)}{x} = \lim_{x \to 0} \frac{p_n(x) f(x)/x^{k_n}}{x} = \lim_{x \to 0} \frac{p_n(x) f(x)}{x^{k_n+1}}$$

$$= \lim_{x \to 0} p_n(x) \lim_{x \to 0} \frac{f(x)}{x^{k_n+1}} = p_n(0) \cdot 0 = 0 \qquad \text{[by part (a)]}$$

6 Review

TRUE-FALSE QUIZ

1. True. If f is one-to-one, with domain $\mathbb{R}$, then $f^{-1}(f(6)) = 6$ by the first cancellation equation [see (6.1.4)].

3. False. For example, $\cos \frac{\pi}{2} = \cos\left(-\frac{\pi}{2}\right)$, so $\cos x$ is not 1-1.

5. True. The function $y = \ln x$ is increasing on $(0, \infty)$, so if $0 < a < b$, then $\ln a < \ln b$.

7. True. We can divide by e^x since $e^x \neq 0$ for every x.

9. False. Let $x = e$. Then $(\ln x)^6 = (\ln e)^6 = 1^6 = 1$, but $6 \ln x = 6 \ln e = 6 \cdot 1 = 6 \neq 1 = (\ln x)^6$. What *is* true, however, is that $\ln(x^6) = 6 \ln x$ for $x > 0$.

11. False. $\ln 10$ is a constant, so its derivative, $\dfrac{d}{dx}(\ln 10)$, is 0, not $\frac{1}{10}$.

13. False. The "-1" is not an exponent; it is an indication of an inverse function.

15. True. See Figure 2 in Section 6.7.

17. True. $\displaystyle\int_2^{16} \frac{dx}{x} = \Big[\ln |x|\Big]_2^{16} = \ln 16 - \ln 2 = \ln \frac{16}{2} = \ln 8 = \ln 2^3 = 3 \ln 2$

19. False. Let $f(x) = 1 + \dfrac{1}{x}$ and $g(x) = x$. Then $\displaystyle\lim_{x \to \infty} f(x) = 1$ and $\displaystyle\lim_{x \to \infty} g(x) = \infty$, but

$$\lim_{x \to \infty} [f(x)]^{g(x)} = \lim_{x \to \infty} \left(1 + \frac{1}{x}\right)^x = e, \text{ not } 1.$$

EXERCISES

1. No. f is not 1-1 because the graph of f fails the Horizontal Line Test.

3. (a) $f^{-1}(3) = 7$ since $f(7) = 3$. **(b)** $(f^{-1})'(3) = \dfrac{1}{f'(f^{-1}(3))} = \dfrac{1}{f'(7)} = \dfrac{1}{8}$

5.

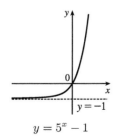

$y = 5^x - 1$

7. Reflect the graph of $y = \ln x$ about the x-axis to obtain the graph of $y = -\ln x$.

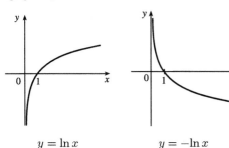

9.

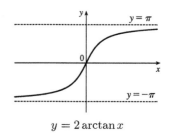

$y = 2\arctan x$

11. (a) $e^{2\ln 5} = e^{\ln 5^2} = 5^2 = 25$

(b) $\log_6 4 + \log_6 54 = \log_6(4 \cdot 54) = \log_6 216 = \log_6 6^3 = 3$

(c) Let $\theta = \arcsin\frac{4}{5}$, so $\sin\theta = \frac{4}{5}$. Draw a right triangle with angle θ as shown in the figure. By the Pythagorean Theorem, the adjacent side has length 3, and $\tan\left(\arcsin\frac{4}{5}\right) = \tan\theta = \frac{\text{opp}}{\text{adj}} = \frac{4}{3}$.

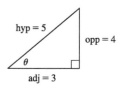

13. $e^{2x} = 3 \;\Rightarrow\; \ln(e^{2x}) = \ln 3 \;\Rightarrow\; 2x = \ln 3 \;\Rightarrow\; x = \frac{1}{2}\ln 3 \approx 0.549$

15. $e^{e^x} = 10 \;\Rightarrow\; \ln\left(e^{e^x}\right) = \ln 10 \;\Rightarrow\; e^x = \ln 10 \;\Rightarrow\; \ln e^x = \ln(\ln 10) \;\Rightarrow\; x = \ln(\ln 10) \approx 0.834$

17. $\tan^{-1}(3x^2) = \frac{\pi}{4} \;\Rightarrow\; \tan(\tan^{-1}(3x^2)) = \tan\frac{\pi}{4} \;\Rightarrow\; 3x^2 = 1 \;\Rightarrow\; x^2 = \frac{1}{3} \;\Rightarrow\; x = \pm\frac{1}{\sqrt{3}} \approx \pm 0.577$

19. $\ln x - 1 = \ln(5+x) - 4 \;\Rightarrow\; \ln x - \ln(5+x) = -4+1 \;\Rightarrow\; \ln\frac{x}{5+x} = -3 \;\Rightarrow\; e^{\ln(x/(5+x))} = e^{-3} \;\Rightarrow\;$

$\frac{x}{5+x} = e^{-3} \;\Rightarrow\; x = 5e^{-3} + xe^{-3} \;\Rightarrow\; x - xe^{-3} = 5e^{-3} \;\Rightarrow\; x(1-e^{-3}) = 5e^{-3} \;\Rightarrow\; x = \frac{5e^{-3}}{1-e^{-3}}$

or, multiplying by $\frac{e^3}{e^3}$, we have $x = \frac{5}{e^3 - 1} \approx 0.262$.

21. $f(t) = t^2 \ln t \;\Rightarrow\; f'(t) = t^2 \cdot \frac{1}{t} + (\ln t)(2t) = t + 2t\ln t$ or $t(1 + 2\ln t)$

23. $h(\theta) = e^{\tan 2\theta} \;\Rightarrow\; h'(\theta) = e^{\tan 2\theta} \cdot \sec^2 2\theta \cdot 2 = 2\sec^2(2\theta)\, e^{\tan 2\theta}$

25. $y = \ln|\sec 5x + \tan 5x| \;\Rightarrow\;$

$y' = \frac{1}{\sec 5x + \tan 5x}(\sec 5x \tan 5x \cdot 5 + \sec^2 5x \cdot 5) = \frac{5\sec 5x\,(\tan 5x + \sec 5x)}{\sec 5x + \tan 5x} = 5\sec 5x$

27. $y = \ln(\sec^2 x) = 2\ln|\sec x| \;\Rightarrow\; y' = (2/\sec x)(\sec x \tan x) = 2\tan x$

29. $y = \frac{e^{1/x}}{x^2} \;\Rightarrow\; y' = \frac{x^2(e^{1/x})' - e^{1/x}(x^2)'}{(x^2)^2} = \frac{x^2(e^{1/x})(-1/x^2) - e^{1/x}(2x)}{x^4} = \frac{-e^{1/x}(1+2x)}{x^4}$

31. $y = 5\arctan\dfrac{1}{x}$ $\Rightarrow$ $y' = 5 \cdot \dfrac{1}{1+\left(\dfrac{1}{x}\right)^2} \cdot \dfrac{d}{dx}\left(\dfrac{1}{x}\right) = \dfrac{5}{1+\dfrac{1}{x^2}}\left(-\dfrac{1}{x^2}\right) = -\dfrac{5}{x^2+1}$

33. $y = 3^{x\ln x}$ $\Rightarrow$ $y' = 3^{x\ln x}(\ln 3)\dfrac{d}{dx}(x\ln x) = 3^{x\ln x}(\ln 3)\left(x\cdot\dfrac{1}{x}+\ln x \cdot 1\right) = 3^{x\ln x}(\ln 3)(1+\ln x)$

35. $y = x\tan^{-1}x - \tfrac{1}{2}\ln(1+x^2)$ $\Rightarrow$

$y' = x\cdot\dfrac{1}{1+x^2}+(\tan^{-1}x)\cdot 1 - \dfrac{1}{2}\left(\dfrac{2x}{1+x^2}\right) = \dfrac{x}{1+x^2}+\tan^{-1}x - \dfrac{x}{1+x^2} = \tan^{-1}x$

37. $y = x\sinh(x^2)$ $\Rightarrow$ $y' = x\cosh(x^2)\cdot 2x + \sinh(x^2)\cdot 1 = 2x^2\cosh(x^2)+\sinh(x^2)$

39. $y = \ln(\arcsin x^2)$ $\Rightarrow$ $y' = \dfrac{1}{\arcsin x^2}\cdot\dfrac{d}{dx}(\arcsin x^2) = \dfrac{1}{\arcsin x^2}\cdot\dfrac{1}{\sqrt{1-(x^2)^2}}\cdot 2x = \dfrac{2x}{(\arcsin x^2)\sqrt{1-x^4}}$

41. $y = \ln\left(\dfrac{1}{x}\right)+\dfrac{1}{\ln x} = \ln x^{-1}+(\ln x)^{-1} = -\ln x + (\ln x)^{-1}$ $\Rightarrow$ $y' = -1\cdot\dfrac{1}{x}+(-1)(\ln x)^{-2}\cdot\dfrac{1}{x} = -\dfrac{1}{x}-\dfrac{1}{x(\ln x)^2}$

43. $y = \ln(\cosh 3x)$ $\Rightarrow$ $y' = (1/\cosh 3x)(\sinh 3x)(3) = 3\tanh 3x$

45. $y = \cosh^{-1}(\sinh x)$ $\Rightarrow$ $y' = (\cosh x)/\sqrt{\sinh^2 x - 1}$

47. $y = \cos\left(e^{\sqrt{\tan 3x}}\right)$ $\Rightarrow$

$y' = -\sin\left(e^{\sqrt{\tan 3x}}\right)\cdot\left(e^{\sqrt{\tan 3x}}\right)' = -\sin\left(e^{\sqrt{\tan 3x}}\right)e^{\sqrt{\tan 3x}}\cdot\tfrac{1}{2}(\tan 3x)^{-1/2}\cdot\sec^2(3x)\cdot 3$

$= \dfrac{-3\sin\left(e^{\sqrt{\tan 3x}}\right)e^{\sqrt{\tan 3x}}\sec^2(3x)}{2\sqrt{\tan 3x}}$

49. $f(x) = e^{g(x)}$ $\Rightarrow$ $f'(x) = e^{g(x)}g'(x)$

51. $f(x) = \ln|g(x)|$ $\Rightarrow$ $f'(x) = \dfrac{1}{g(x)}g'(x) = \dfrac{g'(x)}{g(x)}$

53. $f(x) = 2^x$ $\Rightarrow$ $f'(x) = 2^x\ln 2$ $\Rightarrow$ $f''(x) = 2^x(\ln 2)^2$ $\Rightarrow$ $\cdots$ $\Rightarrow$ $f^{(n)}(x) = 2^x(\ln 2)^n$

55. We first show it is true for $n = 1$: $f'(x) = e^x + xe^x = (x+1)e^x$. We now assume it is true for $n = k$:

$f^{(k)}(x) = (x+k)e^x$. With this assumption, we must show it is true for $n = k+1$:

$f^{(k+1)}(x) = \dfrac{d}{dx}\left[f^{(k)}(x)\right] = \dfrac{d}{dx}[(x+k)e^x] = e^x + (x+k)e^x = [x+(k+1)]e^x$.

Therefore, $f^{(n)}(x) = (x+n)e^x$ by mathematical induction.

57. $y = (2+x)e^{-x}$ $\Rightarrow$ $y' = (2+x)(-e^{-x})+e^{-x}\cdot 1 = e^{-x}[-(2+x)+1] = e^{-x}(-x-1)$. At $(0,2)$, $y' = 1(-1) = -1$, so an equation of the tangent line is $y - 2 = -1(x-0)$, or $y = -x + 2$.

59. $y = [\ln(x+4)]^2 \;\Rightarrow\; y' = 2[\ln(x+4)]^1 \cdot \dfrac{1}{x+4} \cdot 1 = 2\,\dfrac{\ln(x+4)}{x+4}$ and $y' = 0 \;\Leftrightarrow\; \ln(x+4) = 0 \;\Leftrightarrow\;$
$x+4 = e^0 \;\Rightarrow\; x+4 = 1 \;\Leftrightarrow\; x = -3$, so the tangent is horizontal at the point $(-3, 0)$.

61. (a) The line $x - 4y = 1$ has slope $\tfrac{1}{4}$. A tangent to $y = e^x$ has slope $\tfrac{1}{4}$ when $y' = e^x = \tfrac{1}{4} \;\Rightarrow\; x = \ln\tfrac{1}{4} = -\ln 4$.

Since $y = e^x$, the y-coordinate is $\tfrac{1}{4}$ and the point of tangency is $(-\ln 4, \tfrac{1}{4})$. Thus, an equation of the tangent line is $y - \tfrac{1}{4} = \tfrac{1}{4}(x + \ln 4)$ or $y = \tfrac{1}{4}x + \tfrac{1}{4}(\ln 4 + 1)$.

(b) The slope of the tangent at the point (a, e^a) is $\left.\dfrac{d}{dx} e^x\right|_{x=a} = e^a$. Thus, an equation of the tangent line is

$y - e^a = e^a(x - a)$. We substitute $x = 0, y = 0$ into this equation, since we want the line to pass through the origin:

$0 - e^a = e^a(0 - a) \;\Leftrightarrow\; -e^a = e^a(-a) \;\Leftrightarrow\; a = 1$. So an equation of the tangent line at the point $(a, e^a) = (1, e)$ is $y - e = e(x - 1)$ or $y = ex$.

63. $\lim\limits_{x \to \infty} e^{-3x} = 0$ since $-3x \to -\infty$ as $x \to \infty$ and $\lim\limits_{t \to -\infty} e^t = 0$.

65. Let $t = 2/(x-3)$. As $x \to 3^-$, $t \to -\infty$. $\quad \lim\limits_{x \to 3^-} e^{2/(x-3)} = \lim\limits_{t \to -\infty} e^t = 0$

67. As $x \to 0$, $\cosh x \to 1$, so $\lim\limits_{x \to 0} \ln(\cosh x) = 0$.

69. $\lim\limits_{x \to \infty} \dfrac{1 + 2^x}{1 - 2^x} = \lim\limits_{x \to \infty} \dfrac{(1 + 2^x)/2^x}{(1 - 2^x)/2^x} = \lim\limits_{x \to \infty} \dfrac{1/2^x + 1}{1/2^x - 1} = \dfrac{0 + 1}{0 - 1} = -1$

71. This limit has the form $\tfrac{0}{0}$. $\quad \lim\limits_{x \to 0} \dfrac{e^x - 1}{\tan x} \stackrel{\text{H}}{=} \lim\limits_{x \to 0} \dfrac{e^x}{\sec^2 x} = \dfrac{1}{1} = 1$

73. This limit has the form $\tfrac{0}{0}$. $\quad \lim\limits_{x \to 0} \dfrac{e^{2x} - e^{-2x}}{\ln(x+1)} \stackrel{\text{H}}{=} \lim\limits_{x \to 0} \dfrac{2e^{2x} + 2e^{-2x}}{1/(x+1)} = \dfrac{2 + 2}{1} = 4$

75. This limit has the form $\infty \cdot 0$.

$$\lim_{x \to -\infty} (x^2 - x^3)e^{2x} = \lim_{x \to -\infty} \dfrac{x^2 - x^3}{e^{-2x}} \; \left[\tfrac{\infty}{\infty} \text{ form}\right] \stackrel{\text{H}}{=} \lim_{x \to -\infty} \dfrac{2x - 3x^2}{-2e^{-2x}} \; \left[\tfrac{\infty}{\infty} \text{ form}\right]$$

$$\stackrel{\text{H}}{=} \lim_{x \to -\infty} \dfrac{2 - 6x}{4e^{-2x}} \; \left[\tfrac{\infty}{\infty} \text{ form}\right] \stackrel{\text{H}}{=} \lim_{x \to -\infty} \dfrac{-6}{-8e^{-2x}} = 0$$

77. This limit has the form $\infty - \infty$.

$$\lim_{x \to 1^+} \left(\dfrac{x}{x-1} - \dfrac{1}{\ln x} \right) = \lim_{x \to 1^+} \dfrac{x \ln x - x + 1}{(x-1)\ln x} \stackrel{\text{H}}{=} \lim_{x \to 1^+} \dfrac{x \cdot (1/x) + \ln x - 1}{(x-1)\cdot(1/x) + \ln x} = \lim_{x \to 1^+} \dfrac{\ln x}{1 - 1/x + \ln x}$$

$$\stackrel{\text{H}}{=} \lim_{x \to 1^+} \dfrac{1/x}{1/x^2 + 1/x} = \dfrac{1}{1+1} = \dfrac{1}{2}$$

79. $y = f(x) = e^x \sin x$, $-\pi \le x \le \pi$ **A.** $D = [-\pi, \pi]$ **B.** y-intercept: $f(0) = 0$; $f(x) = 0 \;\Leftrightarrow\; \sin x = 0 \;\Rightarrow\;$
$x = -\pi, 0, \pi$. **C.** No symmetry **D.** No asymptote **E.** $f'(x) = e^x \cos x + \sin x \cdot e^x = e^x(\cos x + \sin x)$.

[continued]

$f'(x) = 0 \Leftrightarrow -\cos x = \sin x \Leftrightarrow -1 = \tan x \Rightarrow x = -\frac{\pi}{4}, \frac{3\pi}{4}$. $f'(x) > 0$ for $-\frac{\pi}{4} < x < \frac{3\pi}{4}$ and $f'(x) < 0$ for $-\pi < x < -\frac{\pi}{4}$ and $\frac{3\pi}{4} < x < \pi$, so f is increasing on $\left(-\frac{\pi}{4}, \frac{3\pi}{4}\right)$ and f is decreasing on $\left(-\pi, -\frac{\pi}{4}\right)$ and $\left(\frac{3\pi}{4}, \pi\right)$.

F. Local minimum value $f\left(-\frac{\pi}{4}\right) = (-\sqrt{2}/2)e^{-\pi/4} \approx -0.32$ and local maximum value $f\left(\frac{3\pi}{4}\right) = (\sqrt{2}/2)\,e^{3\pi/4} \approx 7.46$

G. $f''(x) = e^x(-\sin x + \cos x) + (\cos x + \sin x)e^x = e^x(2\cos x) > 0 \Rightarrow -\frac{\pi}{2} < x < \frac{\pi}{2}$ and $f''(x) < 0 \Rightarrow -\pi < x < -\frac{\pi}{2}$ and $\frac{\pi}{2} < x < \pi$, so f is CU on $\left(-\frac{\pi}{2}, \frac{\pi}{2}\right)$, and f is CD on $\left(-\pi, -\frac{\pi}{2}\right)$ and $\left(\frac{\pi}{2}, \pi\right)$. There are inflection points at $\left(-\frac{\pi}{2}, -e^{-\pi/2}\right)$ and $\left(\frac{\pi}{2}, e^{\pi/2}\right)$.

H.

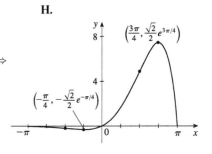

81. $y = f(x) = x \ln x$ **A.** $D = (0, \infty)$ **B.** No y-intercept; x-intercept 1.   **C.** No symmetry   **D.** No asymptote [Note that the graph approaches the point $(0,0)$ as $x \to 0^+$.]

E. $f'(x) = x(1/x) + (\ln x)(1) = 1 + \ln x$, so $f'(x) \to -\infty$ as $x \to 0^+$ and $f'(x) \to \infty$ as $x \to \infty$. $f'(x) = 0 \Leftrightarrow \ln x = -1 \Leftrightarrow x = e^{-1} = 1/e$. $f'(x) > 0$ for $x > 1/e$, so f is decreasing on $(0, 1/e)$ and increasing on $(1/e, \infty)$.   **F.** Local minimum: $f(1/e) = -1/e$. No local maximum.

G. $f''(x) = 1/x$, so $f''(x) > 0$ for $x > 0$. The graph is CU on $(0, \infty)$ and there is no IP.

H.

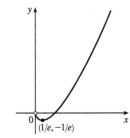

83. $y = f(x) = (x-2)e^{-x}$ **A.** $D = \mathbb{R}$ **B.** y-intercept: $f(0) = -2$; x-intercept: $f(x) = 0 \Leftrightarrow x = 2$

C. No symmetry   **D.** $\lim\limits_{x \to \infty} \dfrac{x-2}{e^x} \stackrel{H}{=} \lim\limits_{x \to \infty} \dfrac{1}{e^x} = 0$, so $y = 0$ is a HA. No VA

E. $f'(x) = (x-2)(-e^{-x}) + e^{-x}(1) = e^{-x}[-(x-2)+1] = (3-x)e^{-x}$. $f'(x) > 0$ for $x < 3$, so f is increasing on $(-\infty, 3)$ and decreasing on $(3, \infty)$.

F. Local maximum value $f(3) = e^{-3}$, no local minimum value

G. $f''(x) = (3-x)(-e^{-x}) + e^{-x}(-1) = e^{-x}[-(3-x)+(-1)]$
$= (x-4)e^{-x} > 0$
for $x > 4$, so f is CU on $(4, \infty)$ and CD on $(-\infty, 4)$. IP at $(4, 2e^{-4})$

H.

85. If $c < 0$, then $\lim\limits_{x \to -\infty} f(x) = \lim\limits_{x \to -\infty} xe^{-cx} = \lim\limits_{x \to -\infty} \dfrac{x}{e^{cx}} \stackrel{H}{=} \lim\limits_{x \to -\infty} \dfrac{1}{ce^{cx}} = 0$, and $\lim\limits_{x \to \infty} f(x) = \infty$.

If $c > 0$, then $\lim\limits_{x \to -\infty} f(x) = -\infty$, and $\lim\limits_{x \to \infty} f(x) \stackrel{H}{=} \lim\limits_{x \to \infty} \dfrac{1}{ce^{cx}} = 0$.

If $c = 0$, then $f(x) = x$, so $\lim\limits_{x \to \pm\infty} f(x) = \pm\infty$, respectively.

So we see that $c = 0$ is a transitional value. We now exclude the case $c = 0$, since we know how the function behaves in that case. To find the maxima and minima of f, we differentiate: $f(x) = xe^{-cx} \Rightarrow$

$f'(x) = x(-ce^{-cx}) + e^{-cx} = (1-cx)e^{-cx}$. This is 0 when $1 - cx = 0 \Leftrightarrow x = 1/c$. If $c < 0$ then this represents a minimum value of $f(1/c) = 1/(ce)$, since $f'(x)$ changes from negative to positive at $x = 1/c$; and if $c > 0$, it represents a maximum value. As $|c|$ increases, the maximum or minimum point gets closer to the origin. To find the inflection points, we differentiate again: $f'(x) = e^{-cx}(1 - cx)$ $\Rightarrow$

$f''(x) = e^{-cx}(-c) + (1 - cx)(-ce^{-cx}) = (cx - 2)ce^{-cx}$. This changes sign when $cx - 2 = 0$ $\Leftrightarrow$ $x = 2/c$. So as $|c|$ increases, the points of inflection get closer to the origin.

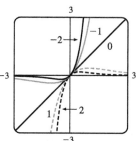

87. $s(t) = Ae^{-ct}\cos(\omega t + \delta)$ $\Rightarrow$

$v(t) = s'(t) = A\{e^{-ct}[-\omega\sin(\omega t + \delta)] + \cos(\omega t + \delta)(-ce^{-ct})\} = -Ae^{-ct}[\omega\sin(\omega t + \delta) + c\cos(\omega t + \delta)]$ $\Rightarrow$

$a(t) = v'(t) = -A\{e^{-ct}[\omega^2\cos(\omega t + \delta) - c\omega\sin(\omega t + \delta)] + [\omega\sin(\omega t + \delta) + c\cos(\omega t + \delta)](-ce^{-ct})\}$

$= -Ae^{-ct}[\omega^2\cos(\omega t + \delta) - c\omega\sin(\omega t + \delta) - c\omega\sin(\omega t + \delta) - c^2\cos(\omega t + \delta)]$

$= -Ae^{-ct}[(\omega^2 - c^2)\cos(\omega t + \delta) - 2c\omega\sin(\omega t + \delta)] = Ae^{-ct}[(c^2 - \omega^2)\cos(\omega t + \delta) + 2c\omega\sin(\omega t + \delta)]$

89. (a) $y(t) = y(0)e^{kt} = 200e^{kt}$ $\Rightarrow$ $y(0.5) = 200e^{0.5k} = 360$ $\Rightarrow$ $e^{0.5k} = 1.8$ $\Rightarrow$ $0.5k = \ln 1.8$ $\Rightarrow$

$k = 2\ln 1.8 = \ln(1.8)^2 = \ln 3.24$ $\Rightarrow$ $y(t) = 200e^{(\ln 3.24)t} = 200(3.24)^t$

(b) $y(4) = 200(3.24)^4 \approx 22{,}040$ cells

(c) $y'(t) = 200(3.24)^t \cdot \ln 3.24$, so $y'(4) = 200(3.24)^4 \cdot \ln 3.24 \approx 25{,}910$ cells per hour

(d) $200(3.24)^t = 10{,}000$ $\Rightarrow$ $(3.24)^t = 50$ $\Rightarrow$ $t\ln 3.24 = \ln 50$ $\Rightarrow$ $t = (\ln 50)/(\ln 3.24) \approx 3.33$ hours

91. Let $P(t) = \dfrac{64}{1 + 31e^{-0.7944t}} = \dfrac{A}{1 + Be^{ct}} = A(1 + Be^{ct})^{-1}$, where $A = 64$, $B = 31$, and $c = -0.7944$.

$P'(t) = -A(1 + Be^{ct})^{-2}(Bce^{ct}) = -ABce^{ct}(1 + Be^{ct})^{-2}$

$P''(t) = -ABce^{ct}[-2(1 + Be^{ct})^{-3}(Bce^{ct})] + (1 + Be^{ct})^{-2}(-ABc^2e^{ct})$

$= -ABc^2e^{ct}(1 + Be^{ct})^{-3}[-2Be^{ct} + (1 + Be^{ct})] = -\dfrac{ABc^2e^{ct}(1 - Be^{ct})}{(1 + Be^{ct})^3}$

The population is increasing most rapidly when its graph changes from CU to CD; that is, when $P''(t) = 0$ in this case.

$P''(t) = 0$ $\Rightarrow$ $Be^{ct} = 1$ $\Rightarrow$ $e^{ct} = \dfrac{1}{B}$ $\Rightarrow$ $ct = \ln\dfrac{1}{B}$ $\Rightarrow$ $t = \dfrac{\ln(1/B)}{c} = \dfrac{\ln(1/31)}{-0.7944} \approx 4.32$ days. Note that

$P\left(\dfrac{1}{c}\ln\dfrac{1}{B}\right) = \dfrac{A}{1 + Be^{c(1/c)\ln(1/B)}} = \dfrac{A}{1 + Be^{\ln(1/B)}} = \dfrac{A}{1 + B(1/B)} = \dfrac{A}{1 + 1} = \dfrac{A}{2}$, one-half the limit of P as $t \to \infty$.

93. Let $u = -2y^2$. Then $du = -4y\,dy$ and $\int_0^1 ye^{-2y^2}\,dy = \int_0^{-2} e^u\left(-\tfrac{1}{4}du\right) = -\tfrac{1}{4}[e^u]_0^{-2} = -\tfrac{1}{4}(e^{-2} - 1) = \tfrac{1}{4}(1 - e^{-2})$.

95. Let $u = e^x$, so $du = e^x\,dx$. When $x = 0, u = 1$; when $x = 1, u = e$. Thus,

$$\int_0^1 \frac{e^x}{1+e^{2x}}\,dx = \int_1^e \frac{1}{1+u^2}\,du = \big[\arctan u\big]_1^e = \arctan e - \arctan 1 = \arctan e - \tfrac{\pi}{4}.$$

97. Let $u = \sqrt{x}$. Then $du = \dfrac{dx}{2\sqrt{x}} \;\Rightarrow\; \displaystyle\int \frac{e^{\sqrt{x}}}{\sqrt{x}}\,dx = 2\int e^u\,du = 2e^u + C = 2e^{\sqrt{x}} + C.$

99. Let $u = x^2 + 2x$. Then $du = (2x+2)\,dx = 2(x+1)\,dx$ and

$$\int \frac{x+1}{x^2+2x}\,dx = \int \frac{\tfrac{1}{2}\,du}{u} = \tfrac{1}{2}\ln|u| + C = \tfrac{1}{2}\ln\left|x^2+2x\right| + C.$$

101. Let $u = \ln(\cos x)$. Then $du = \dfrac{-\sin x}{\cos x}\,dx = -\tan x\,dx \;\Rightarrow\;$

$\int \tan x \ln(\cos x)\,dx = -\int u\,du = -\tfrac{1}{2}u^2 + C = -\tfrac{1}{2}[\ln(\cos x)]^2 + C.$

103. Let $u = \tan\theta$. Then $du = \sec^2\theta\,d\theta$ and $\displaystyle\int 2^{\tan\theta}\sec^2\theta\,d\theta = \int 2^u\,du = \dfrac{2^u}{\ln 2} + C = \dfrac{2^{\tan\theta}}{\ln 2} + C.$

105. $\displaystyle\int_{-2}^{-1}\frac{z^2+1}{z}\,dz = \int_{-2}^{-1}\left(z + \frac{1}{z}\right)dz = \left[\tfrac{1}{2}z^2 + \ln|z|\right]_{-2}^{-1} = \left(\tfrac{1}{2} + \ln|-1|\right) - \left(2 + \ln|-2|\right) = -\tfrac{3}{2} - \ln 2$

107. $\cos x \le 1 \;\Rightarrow\; e^x \cos x \le e^x \;\Rightarrow\; \int_0^1 e^x \cos x\,dx \le \int_0^1 e^x\,dx = e^x\big]_0^1 = e - 1$

109. $f(x) = \displaystyle\int_1^{\sqrt{x}} \frac{e^s}{s}\,ds \;\Rightarrow\; f'(x) = \frac{d}{dx}\int_1^{\sqrt{x}} \frac{e^s}{s}\,ds = \frac{e^{\sqrt{x}}}{\sqrt{x}}\frac{d}{dx}\sqrt{x} = \frac{e^{\sqrt{x}}}{\sqrt{x}}\cdot\frac{1}{2\sqrt{x}} = \frac{e^{\sqrt{x}}}{2x}$

111. (a) $f_{\text{avg}} = \dfrac{1}{b-a}\displaystyle\int_a^b f(x)\,dx = \dfrac{1}{5-1}\int_1^5 \frac{\ln x}{x}\,dx = \dfrac{1}{4}\int_0^{\ln 5} u\,du \quad \begin{bmatrix} u = \ln x, \\ du = (1/x)\,dx \end{bmatrix}$

$= \tfrac{1}{4}\left[\tfrac{1}{2}u^2\right]_0^{\ln 5} = \tfrac{1}{8}(\ln 5)^2$

(b) $f(x) = \dfrac{\ln x}{x}$, $[1,5]$. $f'(x) = \dfrac{x(1/x) - (\ln x)\cdot 1}{x^2} = \dfrac{1 - \ln x}{x^2} = 0 \;\Leftrightarrow\; 1 - \ln x = 0 \;\Leftrightarrow\; \ln x = 1 \;\Leftrightarrow\; x = e.$

$f(1) = 0$, $f(e) = 1/e \approx 0.37$, and $f(5) = \tfrac{1}{5}\ln 5 \approx 0.32$. So $f(e) = 1/e$ is the absolute maximum value and $f(1) = 0$ is the absolute minimum value.

113. $V = \displaystyle\int_0^1 \frac{2\pi x}{1+x^4}\,dx$ by cylindrical shells. Let $u = x^2 \;\Rightarrow\; du = 2x\,dx$. Then

$V = \displaystyle\int_0^1 \frac{\pi}{1+u^2}\,du = \pi\big[\tan^{-1} u\big]_0^1 = \pi(\tan^{-1} 1 - \tan^{-1} 0) = \pi\left(\dfrac{\pi}{4}\right) = \dfrac{\pi^2}{4}.$

115. $f(x) = \ln x + \tan^{-1} x \;\Rightarrow\; f(1) = \ln 1 + \tan^{-1} 1 = \tfrac{\pi}{4} \;\Rightarrow\; g\!\left(\tfrac{\pi}{4}\right) = 1$ [where $g = f^{-1}$].

$f'(x) = \dfrac{1}{x} + \dfrac{1}{1+x^2}$, so $g'\!\left(\dfrac{\pi}{4}\right) = \dfrac{1}{f'(1)} = \dfrac{1}{3/2} = \dfrac{2}{3}.$

117. 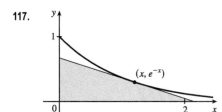 We find the equation of a tangent to the curve $y = e^{-x}$, so that we can find the x- and y-intercepts of this tangent, and then we can find the area of the triangle.

The slope of the tangent at the point (a, e^{-a}) is given by $\dfrac{d}{dx}e^{-x}\bigg]_{x=a} = -e^{-a}$,

and so the equation of the tangent is $y - e^{-a} = -e^{-a}(x - a) \Leftrightarrow$

$y = e^{-a}(a - x + 1)$.

The y-intercept of this line is $y = e^{-a}(a - 0 + 1) = e^{-a}(a + 1)$. To find the x-intercept we set $y = 0 \Rightarrow$

$e^{-a}(a - x + 1) = 0 \Rightarrow x = a + 1$. So the area of the triangle is $A(a) = \frac{1}{2}[e^{-a}(a+1)](a+1) = \frac{1}{2}e^{-a}(a+1)^2$. We

differentiate this with respect to a: $A'(a) = \frac{1}{2}[e^{-a}(2)(a+1) + (a+1)^2 e^{-a}(-1)] = \frac{1}{2}e^{-a}(1 - a^2)$. This is 0

at $a = \pm 1$, and the root $a = 1$ gives a maximum, by the First Derivative Test. So the maximum area of the triangle is

$A(1) = \frac{1}{2}e^{-1}(1+1)^2 = 2e^{-1} = 2/e$.

119. $\lim\limits_{x \to -1} F(x) = \lim\limits_{x \to -1} \dfrac{b^{x+1} - a^{x+1}}{x+1} \stackrel{H}{=} \lim\limits_{x \to -1} \dfrac{b^{x+1} \ln b - a^{x+1} \ln a}{1} = \ln b - \ln a = F(-1)$, so F is continuous at -1.

121. Using FTC1, we differentiate both sides of the given equation, $\int_1^x f(t)\,dt = (x-1)e^{2x} + \int_1^x e^{-t}f(t)\,dt$, and get

$f(x) = (x-1) \cdot e^{2x} \cdot 2 + e^{2x} + e^{-x}f(x) \Rightarrow f(x)(1 - e^{-x}) = 2(x-1)e^{2x} + e^{2x} \Rightarrow$

$f(x) = \dfrac{e^{2x}[1 + 2(x-1)]}{1 - e^{-x}} = \dfrac{e^{2x}(2x - 1)}{1 - e^{-x}}$.

PROBLEMS PLUS

1. Let $y = f(x) = e^{-x^2}$. The area of the rectangle under the curve from $-x$ to x is $A(x) = 2xe^{-x^2}$ where $x \geq 0$. We maximize $A(x)$: $A'(x) = 2e^{-x^2} - 4x^2 e^{-x^2} = 2e^{-x^2}(1 - 2x^2) = 0 \Rightarrow x = \frac{1}{\sqrt{2}}$. This gives a maximum since $A'(x) > 0$ for $0 \leq x < \frac{1}{\sqrt{2}}$ and $A'(x) < 0$ for $x > \frac{1}{\sqrt{2}}$. We next determine the points of inflection of $f(x)$. Notice that $f'(x) = -2xe^{-x^2} = -A(x)$. So $f''(x) = -A'(x)$ and hence, $f''(x) < 0$ for $-\frac{1}{\sqrt{2}} < x < \frac{1}{\sqrt{2}}$ and $f''(x) > 0$ for $x < -\frac{1}{\sqrt{2}}$ and $x > \frac{1}{\sqrt{2}}$. So $f(x)$ changes concavity at $x = \pm\frac{1}{\sqrt{2}}$, and the two vertices of the rectangle of largest area are at the inflection points.

3. $\ln(x^2 - 2x - 2) \leq 0 \Rightarrow x^2 - 2x - 2 \leq e^0 = 1 \Rightarrow x^2 - 2x - 3 \leq 0 \Rightarrow (x-3)(x+1) \leq 0 \Rightarrow x \in [-1, 3]$. Since the argument must be positive, $x^2 - 2x - 2 > 0 \Rightarrow \left[x - (1 - \sqrt{3})\right]\left[x - (1 + \sqrt{3})\right] > 0 \Rightarrow x \in (-\infty, 1 - \sqrt{3}) \cup (1 + \sqrt{3}, \infty)$. The intersection of these intervals is $[-1, 1 - \sqrt{3}) \cup (1 + \sqrt{3}, 3]$.

5. $f(x)$ has the form $e^{g(x)}$, so it will have an absolute maximum (minimum) where g has an absolute maximum (minimum).

$g(x) = 10|x - 2| - x^2 = \begin{cases} 10(x - 2) - x^2 & \text{if } x - 2 > 0 \\ 10[-(x - 2)] - x^2 & \text{if } x - 2 < 0 \end{cases} = \begin{cases} -x^2 + 10x - 20 & \text{if } x > 2 \\ -x^2 - 10x + 20 & \text{if } x < 2 \end{cases} \Rightarrow$

$g'(x) = \begin{cases} -2x + 10 & \text{if } x > 2 \\ -2x - 10 & \text{if } x < 2 \end{cases}$

$g'(x) = 0$ if $x = -5$ or $x = 5$, and $g'(2)$ does not exist, so the critical numbers of g are $-5, 2,$ and 5. Since $g''(x) = -2$ for all $x \neq 2$, g is concave downward on $(-\infty, 2)$ and $(2, \infty)$, and g will attain its absolute maximum at one of the critical numbers. Since $g(-5) = 45$, $g(2) = -4$, and $g(5) = 5$, we see that $f(-5) = e^{45}$ is the absolute maximum value of f. Also, $\lim_{x \to \infty} g(x) = -\infty$, so $\lim_{x \to \infty} f(x) = \lim_{x \to \infty} e^{g(x)} = 0$. But $f(x) > 0$ for all x, so there is no absolute minimum value of f.

7. Consider the statement that $\dfrac{d^n}{dx^n}(e^{ax} \sin bx) = r^n e^{ax} \sin(bx + n\theta)$. For $n = 1$,

$\dfrac{d}{dx}(e^{ax} \sin bx) = ae^{ax} \sin bx + be^{ax} \cos bx$, and

$re^{ax} \sin(bx + \theta) = re^{ax}[\sin bx \cos \theta + \cos bx \sin \theta] = re^{ax}\left(\dfrac{a}{r} \sin bx + \dfrac{b}{r} \cos bx\right) = ae^{ax} \sin bx + be^{ax} \cos bx$

since $\tan \theta = \dfrac{b}{a} \Rightarrow \sin \theta = \dfrac{b}{r}$ and $\cos \theta = \dfrac{a}{r}$. So the statement is true for $n = 1$.

Assume it is true for $n = k$. Then

$\dfrac{d^{k+1}}{dx^{k+1}}(e^{ax} \sin bx) = \dfrac{d}{dx}\left[r^k e^{ax} \sin(bx + k\theta)\right] = r^k a e^{ax} \sin(bx + k\theta) + r^k e^{ax} b \cos(bx + k\theta)$

$= r^k e^{ax}[a \sin(bx + k\theta) + b \cos(bx + k\theta)]$

[continued]

But

$\sin[bx + (k+1)\theta] = \sin[(bx + k\theta) + \theta] = \sin(bx + k\theta)\cos\theta + \sin\theta\cos(bx + k\theta) = \frac{a}{r}\sin(bx + k\theta) + \frac{b}{r}\cos(bx + k\theta).$

Hence, $a\sin(bx + k\theta) + b\cos(bx + k\theta) = r\sin[bx + (k+1)\theta]$. So

$\frac{d^{k+1}}{dx^{k+1}}(e^{ax}\sin bx) = r^k e^{ax}[a\sin(bx + k\theta) + b\cos(bx + k\theta)] = r^k e^{ax}[r\sin(bx + (k+1)\theta)] = r^{k+1}e^{ax}[\sin(bx + (k+1)\theta)].$

Therefore, the statement is true for all n by mathematical induction.

9. We first show that $\frac{x}{1+x^2} < \tan^{-1}x$ for $x > 0$. Let $f(x) = \tan^{-1}x - \frac{x}{1+x^2}$. Then

$f'(x) = \frac{1}{1+x^2} - \frac{1(1+x^2) - x(2x)}{(1+x^2)^2} = \frac{(1+x^2) - (1-x^2)}{(1+x^2)^2} = \frac{2x^2}{(1+x^2)^2} > 0$ for $x > 0$. So $f(x)$ is increasing

on $(0, \infty)$. Hence, $0 < x \;\Rightarrow\; 0 = f(0) < f(x) = \tan^{-1}x - \frac{x}{1+x^2}$. So $\frac{x}{1+x^2} < \tan^{-1}x$ for $0 < x$. We next show that

$\tan^{-1}x < x$ for $x > 0$. Let $h(x) = x - \tan^{-1}x$. Then $h'(x) = 1 - \frac{1}{1+x^2} = \frac{x^2}{1+x^2} > 0$. Hence, $h(x)$ is increasing

on $(0, \infty)$. So for $0 < x$, $0 = h(0) < h(x) = x - \tan^{-1}x$. Hence, $\tan^{-1}x < x$ for $x > 0$, and we conclude that

$\frac{x}{1+x^2} < \tan^{-1}x < x$ for $x > 0$.

11. By the Fundamental Theorem of Calculus, $f(x) = \int_1^x \sqrt{1+t^3}\,dt \;\Rightarrow\; f'(x) = \sqrt{1+x^3} > 0$ for $x > -1$.

So f is increasing on $(-1, \infty)$ and hence is one-to-one. Note that $f(1) = 0$, so $f^{-1}(1) = 0 \;\Rightarrow$

$(f^{-1})'(0) = 1/f'(1) = \frac{1}{\sqrt{2}}$.

13. If $L = \lim\limits_{x\to\infty}\left(\frac{x+a}{x-a}\right)^x$, then L has the indeterminate form 1^∞, so

$\ln L = \lim\limits_{x\to\infty}\ln\left(\frac{x+a}{x-a}\right)^x = \lim\limits_{x\to\infty} x\ln\left(\frac{x+a}{x-a}\right) = \lim\limits_{x\to\infty}\frac{\ln(x+a) - \ln(x-a)}{1/x} \stackrel{H}{=} \lim\limits_{x\to\infty}\frac{\frac{1}{x+a} - \frac{1}{x-a}}{-1/x^2}$

$= \lim\limits_{x\to\infty}\left[\frac{(x-a)-(x+a)}{(x+a)(x-a)}\cdot\frac{-x^2}{1}\right] = \lim\limits_{x\to\infty}\frac{2ax^2}{x^2 - a^2} = \lim\limits_{x\to\infty}\frac{2a}{1 - a^2/x^2} = 2a$

Hence, $\ln L = 2a$, so $L = e^{2a}$. From the original equation, we want $L = e^1 \;\Rightarrow\; 2a = 1 \;\Rightarrow\; a = \frac{1}{2}$.

15. As in Exercise 4.3.70, assume that the integrand is defined at $t = 0$ so that it is continuous there. By l'Hospital's Rule and the Fundamental Theorem, using the notation $\exp(y) = e^y$,

$\lim\limits_{x\to 0}\frac{\int_0^x (1 - \tan 2t)^{1/t}\,dt}{x} \stackrel{H}{=} \lim\limits_{x\to 0}\frac{(1 - \tan 2x)^{1/x}}{1} = \exp\left[\ln\left(\lim\limits_{x\to 0}(1 - \tan 2x)^{1/x}\right)\right] = \exp\left(\lim\limits_{x\to 0}\frac{\ln(1 - \tan 2x)}{x}\right)$

$\stackrel{H}{=} \exp\left(\lim\limits_{x\to 0}\frac{-2\sec^2 2x}{1 - \tan 2x}\right) = \exp\left(\frac{-2\cdot 1^2}{1 - 0}\right) = e^{-2}$

17. Both sides of the inequality are positive, so $\cosh(\sinh x) < \sinh(\cosh x)$

$\Leftrightarrow \quad \cosh^2(\sinh x) < \sinh^2(\cosh x) \quad \Leftrightarrow \quad \sinh^2(\sinh x) + 1 < \sinh^2(\cosh x)$

$\Leftrightarrow \quad 1 < [\sinh(\cosh x) - \sinh(\sinh x)][\sinh(\cosh x) + \sinh(\sinh x)]$

$\Leftrightarrow \quad 1 < \left[\sinh\left(\dfrac{e^x + e^{-x}}{2}\right) - \sinh\left(\dfrac{e^x - e^{-x}}{2}\right)\right]\left[\sinh\left(\dfrac{e^x + e^{-x}}{2}\right) + \sinh\left(\dfrac{e^x - e^{-x}}{2}\right)\right]$

$\Leftrightarrow \quad 1 < [2\cosh(e^x/2)\sinh(e^{-x}/2)][2\sinh(e^x/2)\cosh(e^{-x}/2)]$ [use the addition formulas and cancel]

$\Leftrightarrow \quad 1 < [2\sinh(e^x/2)\cosh(e^x/2)][2\sinh(e^{-x}/2)\cosh(e^{-x}/2)] \quad \Leftrightarrow \quad 1 < \sinh e^x \sinh e^{-x}$,

by the half-angle formula. Now both e^x and e^{-x} are positive, and $\sinh y > y$ for $y > 0$, since $\sinh 0 = 0$ and $(\sinh y - y)' = \cosh y - 1 > 0$ for $x > 0$, so $1 = e^x e^{-x} < \sinh e^x \sinh e^{-x}$. So, following this chain of reasoning backward, we arrive at the desired result.

19. 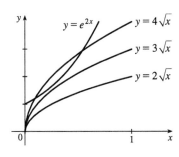 Let $f(x) = e^{2x}$ and $g(x) = k\sqrt{x}$ $[k > 0]$. From the graphs of f and g, we see that f will intersect g exactly once when f and g share a tangent line. Thus, we must have $f = g$ and $f' = g'$ at $x = a$.

$$f(a) = g(a) \quad \Rightarrow \quad e^{2a} = k\sqrt{a} \quad (\star)$$

and $\quad f'(a) = g'(a) \quad \Rightarrow \quad 2e^{2a} = \dfrac{k}{2\sqrt{a}} \quad \Rightarrow \quad e^{2a} = \dfrac{k}{4\sqrt{a}}$.

So we must have $k\sqrt{a} = \dfrac{k}{4\sqrt{a}} \quad \Rightarrow \quad \left(\sqrt{a}\right)^2 = \dfrac{k}{4k} \quad \Rightarrow \quad a = \tfrac{1}{4}$. From $(\star)$, $e^{2(1/4)} = k\sqrt{1/4} \quad \Rightarrow$

$k = 2e^{1/2} = 2\sqrt{e} \approx 3.297$.

21. Suppose that the curve $y = a^x$ intersects the line $y = x$. Then $a^{x_0} = x_0$ for some $x_0 > 0$, and hence $a = x_0^{1/x_0}$. We find the maximum value of $g(x) = x^{1/x}$, $x > 0$, because if a is larger than the maximum value of this function, then the curve $y = a^x$ does not intersect the line $y = x$. $g'(x) = e^{(1/x)\ln x}\left(-\dfrac{1}{x^2}\ln x + \dfrac{1}{x} \cdot \dfrac{1}{x}\right) = x^{1/x}\left(\dfrac{1}{x^2}\right)(1 - \ln x)$. This is 0 only where $x = e$, and for $0 < x < e$, $f'(x) > 0$, while for $x > e$, $f'(x) < 0$, so g has an absolute maximum of $g(e) = e^{1/e}$. So if $y = a^x$ intersects $y = x$, we must have $0 < a \leq e^{1/e}$. Conversely, suppose that $0 < a \leq e^{1/e}$. Then $a^e \leq e$, so the graph of $y = a^x$ lies below or touches the graph of $y = x$ at $x = e$. Also $a^0 = 1 > 0$, so the graph of $y = a^x$ lies above that of $y = x$ at $x = 0$. Therefore, by the Intermediate Value Theorem, the graphs of $y = a^x$ and $y = x$ must intersect somewhere between $x = 0$ and $x = e$.

7 ☐ TECHNIQUES OF INTEGRATION

7.1 Integration by Parts

1. Let $u = x$, $dv = e^{2x}\,dx$ $\Rightarrow$ $du = dx$, $v = \frac{1}{2}e^{2x}$. Then by Equation 2,
$\int xe^{2x}\,dx = \frac{1}{2}xe^{2x} - \int \frac{1}{2}e^{2x}\,dx = \frac{1}{2}xe^{2x} - \frac{1}{4}e^{2x} + C$.

3. Let $u = x$, $dv = \cos 4x\,dx$ $\Rightarrow$ $du = dx$, $v = \frac{1}{4}\sin 4x$. Then by Equation 2,
$\int x\cos 4x\,dx = \frac{1}{4}x\sin 4x - \int \frac{1}{4}\sin 4x\,dx = \frac{1}{4}x\sin 4x + \frac{1}{16}\cos 4x + C$.

Note: A mnemonic device which is helpful for selecting u when using integration by parts is the LIATE principle of precedence for u:

Logarithmic
Inverse trigonometric
Algebraic
Trigonometric
Exponential

If the integrand has several factors, then we try to choose among them a u which appears as high as possible on the list. For example, in $\int xe^{2x}\,dx$ the integrand is xe^{2x}, which is the product of an algebraic function (x) and an exponential function (e^{2x}). Since Algebraic appears before Exponential, we choose $u = x$. Sometimes the integration turns out to be similar regardless of the selection of u and dv, but it is advisable to refer to LIATE when in doubt.

5. Let $u = t$, $dv = e^{2t}\,dt$ $\Rightarrow$ $du = dt$, $v = \frac{1}{2}e^{2t}$. Then by Equation 2,
$\int te^{2t}\,dt = \frac{1}{2}te^{2t} - \int \frac{1}{2}e^{2t}\,dt = \frac{1}{2}te^{2t} - \frac{1}{4}e^{2t} + C$.

7. Let $u = x$, $dv = \sin 10x\,dx$ $\Rightarrow$ $du = dx$, $v = -\frac{1}{10}\cos 10x$. Then by Equation 2,
$$\int x\sin 10x\,dx = -\frac{1}{10}x\cos 10x - \int -\frac{1}{10}\cos 10x\,dx = -\frac{1}{10}x\cos 10x + \frac{1}{10}\int \cos 10x\,dx$$
$$= -\frac{1}{10}x\cos 10x + \frac{1}{100}\sin 10x + C$$

9. Let $u = \ln w$, $dv = w\,dw$ $\Rightarrow$ $du = \frac{1}{w}\,dw$, $v = \frac{1}{2}w^2$. Then by Equation 2,
$$\int w\ln w\,dw = \frac{1}{2}w^2 \ln w - \int \frac{1}{2}w^2 \cdot \frac{1}{w}\,dw = \frac{1}{2}w^2 \ln w - \frac{1}{2}\int w\,dw = \frac{1}{2}w^2 \ln w - \frac{1}{4}w^2 + C.$$

11. First let $u = x^2 + 2x$, $dv = \cos x\,dx$ $\Rightarrow$ $du = (2x+2)\,dx$, $v = \sin x$. Then by Equation 2,
$I = \int (x^2+2x)\cos x\,dx = (x^2+2x)\sin x - \int (2x+2)\sin x\,dx$. Next let $U = 2x+2$, $dV = \sin x\,dx$ $\Rightarrow$ $dU = 2\,dx$, $V = -\cos x$, so $\int (2x+2)\sin x\,dx = -(2x+2)\cos x - \int -2\cos x\,dx = -(2x+2)\cos x + 2\sin x$.
Thus, $I = (x^2+2x)\sin x + (2x+2)\cos x - 2\sin x + C$.

13. Let $u = \cos^{-1}x$, $dv = dx$ $\Rightarrow$ $du = \dfrac{-1}{\sqrt{1-x^2}}\,dx$, $v = x$. Then by Equation 2,
$$\int \cos^{-1}x\,dx = x\cos^{-1}x - \int \frac{-x}{\sqrt{1-x^2}}\,dx = x\cos^{-1}x - \int \frac{1}{\sqrt{t}}\left(\frac{1}{2}\,dt\right) \quad \begin{bmatrix} t = 1 - x^2, \\ dt = -2x\,dx \end{bmatrix}$$
$$= x\cos^{-1}x - \frac{1}{2}\cdot 2t^{1/2} + C = x\cos^{-1}x - \sqrt{1-x^2} + C$$

350 ◻ **CHAPTER 7** TECHNIQUES OF INTEGRATION

15. Let $u = \ln t$, $dv = t^4\, dt$ $\Rightarrow$ $du = \dfrac{1}{t}\, dt$, $v = \dfrac{1}{5}t^5$. Then by Equation 2,

$$\int t^4 \ln t\, dt = \tfrac{1}{5}t^5 \ln t - \int \tfrac{1}{5}t^5 \cdot \tfrac{1}{t}\, dt = \tfrac{1}{5}t^5 \ln t - \int \tfrac{1}{5}t^4\, dt = \tfrac{1}{5}t^5 \ln t - \tfrac{1}{25}t^5 + C.$$

17. Let $u = t$, $dv = \csc^2 t\, dt$ $\Rightarrow$ $du = dt$, $v = -\cot t$. Then by Equation 2,

$$\int t \csc^2 t\, dt = -t \cot t - \int -\cot t\, dt = -t \cot t + \int \frac{\cos t}{\sin t}\, dt = -t \cot t + \int \frac{1}{z}\, dz \quad \begin{bmatrix} z = \sin t, \\ dz = \cos t\, dt \end{bmatrix}$$

$$= -t \cot t + \ln|z| + C = -t \cot t + \ln|\sin t| + C$$

19. First let $u = (\ln x)^2$, $dv = dx$ $\Rightarrow$ $du = 2 \ln x \cdot \tfrac{1}{x}\, dx$, $v = x$. Then by Equation 2,

$I = \int (\ln x)^2\, dx = x(\ln x)^2 - 2 \int x \ln x \cdot \tfrac{1}{x}\, dx = x(\ln x)^2 - 2 \int \ln x\, dx$. Next let $U = \ln x$, $dV = dx$ $\Rightarrow$

$dU = 1/x\, dx$, $V = x$ to get $\int \ln x\, dx = x \ln x - \int x \cdot (1/x)\, dx = x \ln x - \int dx = x \ln x - x + C_1$. Thus,

$I = x(\ln x)^2 - 2(x \ln x - x + C_1) = x(\ln x)^2 - 2x \ln x + 2x + C$, where $C = -2C_1$.

21. First let $u = e^{3x}$, $dv = \cos x\, dx$ $\Rightarrow$ $du = 3e^{3x}\, dx$, $v = \sin x$. Then

$I = \int e^{3x} \cos x\, dx = e^{3x} \sin x - 3 \int e^{3x} \sin x\, dx$. Next, let $U = e^{3x}$, $dV = \sin x\, dx$ $\Rightarrow$ $dU = 3e^{3x}\, dx$, $V = -\cos x$,

so $\int e^{3x} \sin x\, dx = -e^{3x} \cos x + 3 \int e^{3x} \cos x\, dx$. Substituting in the previous formula gives

$I = e^{3x} \sin x - 3(-e^{3x} \cos x + 3I) = e^{3x} \sin x + 3e^{3x} \cos x - 9I$ $\Rightarrow$ $10I = e^{3x} \sin x + 3e^{3x} \cos x + C_1$ $\Rightarrow$

$I = \tfrac{1}{10}e^{3x} \sin x + \tfrac{3}{10}e^{3x} \cos x + C$, where $C = \tfrac{1}{10}C_1$.

23. First let $u = \sin 3\theta$, $dv = e^{2\theta}\, d\theta$ $\Rightarrow$ $du = 3 \cos 3\theta\, d\theta$, $v = \tfrac{1}{2}e^{2\theta}$. Then

$I = \int e^{2\theta} \sin 3\theta\, d\theta = \tfrac{1}{2}e^{2\theta} \sin 3\theta - \tfrac{3}{2} \int e^{2\theta} \cos 3\theta\, d\theta$. Next let $U = \cos 3\theta$, $dV = e^{2\theta}\, d\theta$ $\Rightarrow$ $dU = -3 \sin 3\theta\, d\theta$,

$V = \tfrac{1}{2}e^{2\theta}$ to get $\int e^{2\theta} \cos 3\theta\, d\theta = \tfrac{1}{2}e^{2\theta} \cos 3\theta + \tfrac{3}{2} \int e^{2\theta} \sin 3\theta\, d\theta$. Substituting in the previous formula gives

$I = \tfrac{1}{2}e^{2\theta} \sin 3\theta - \tfrac{3}{4}e^{2\theta} \cos 3\theta - \tfrac{9}{4} \int e^{2\theta} \sin 3\theta\, d\theta = \tfrac{1}{2}e^{2\theta} \sin 3\theta - \tfrac{3}{4}e^{2\theta} \cos 3\theta - \tfrac{9}{4}I$ $\Rightarrow$

$\tfrac{13}{4}I = \tfrac{1}{2}e^{2\theta} \sin 3\theta - \tfrac{3}{4}e^{2\theta} \cos 3\theta + C_1$. Hence, $I = \tfrac{1}{13}e^{2\theta}(2 \sin 3\theta - 3 \cos 3\theta) + C$, where $C = \tfrac{4}{13}C_1$.

25. First let $u = z^3$, $dv = e^z\, dz$ $\Rightarrow$ $du = 3z^2\, dz$, $v = e^z$. Then $I_1 = \int z^3 e^z\, dz = z^3 e^z - 3 \int z^2 e^z\, dz$. Next let $u_1 = z^2$,

$dv_1 = e^z\, dz$ $\Rightarrow$ $du_1 = 2z\, dz$, $v_1 = e^z$. Then $I_2 = z^2 e^z - 2 \int z e^z\, dz$. Finally, let $u_2 = z$, $dv_2 = e^z\, dz$ $\Rightarrow$

$du_2 = dz$, $v_2 = e^z$. Then $\int z e^z\, dz = z e^z - \int e^z\, dz = z e^z - e^z + C_1$. Substituting in the expression for I_2, we get

$I_2 = z^2 e^z - 2(z e^z - e^z + C_1) = z^2 e^z - 2z e^z + 2e^z - 2C_1$. Substituting the last expression for I_2 into I_1 gives

$I_1 = z^3 e^z - 3(z^2 e^z - 2z e^z + 2e^z - 2C_1) = z^3 e^z - 3z^2 e^z + 6z e^z - 6e^z + C$, where $C = 6C_1$.

27. First let $u = 1 + x^2$, $dv = e^{3x}\, dx$ $\Rightarrow$ $du = 2x\, dx$, $v = \tfrac{1}{3}e^{3x}$. Then

$I = \int (1 + x^2) e^{3x}\, dx = \tfrac{1}{3}e^{3x}(1 + x^2) - \tfrac{2}{3} \int x e^{3x}\, dx$. Next, let $U = x$, $dV = e^{3x}\, dx$ $\Rightarrow$ $dU = dx$, $V = \tfrac{1}{3}e^{3x}$, so

$\int xe^{3x}\,dx = \frac{1}{3}xe^{3x} - \frac{1}{3}\int e^{3x}\,dx = \frac{1}{3}xe^{3x} - \frac{1}{9}e^{3x} + C_1$. Substituting in the previous formula gives

$$I = \tfrac{1}{3}e^{3x}(1+x^2) - \tfrac{2}{3}\left(\tfrac{1}{3}xe^{3x} - \tfrac{1}{9}e^{3x} + C_1\right) = \tfrac{1}{3}e^{3x} + \tfrac{1}{3}x^2 e^{3x} - \tfrac{2}{9}xe^{3x} + \tfrac{2}{27}e^{3x} - \tfrac{2}{3}C_1$$
$$= \tfrac{11}{27}e^{3x} - \tfrac{2}{9}xe^{3x} + \tfrac{1}{3}x^2 e^{3x} + C, \text{ where } C = -\tfrac{2}{3}C_1$$

29. Let $u = x$, $dv = 3^x\,dx$ $\Rightarrow$ $du = dx$, $v = \dfrac{1}{\ln 3}3^x$. By (6),

$$\int_0^1 x3^x\,dx = \left[\frac{1}{\ln 3}x3^x\right]_0^1 - \frac{1}{\ln 3}\int_0^1 3^x\,dx = \left(\frac{3}{\ln 3} - 0\right) - \frac{1}{\ln 3}\left[\frac{1}{\ln 3}3^x\right]_0^1 = \frac{3}{\ln 3} - \frac{1}{(\ln 3)^2}(3-1)$$
$$= \frac{3}{\ln 3} - \frac{2}{(\ln 3)^2}$$

31. Let $u = y$, $dv = \sinh y\,dy$ $\Rightarrow$ $du = dy$, $v = \cosh y$. By (6),

$$\int_0^2 y \sinh y\,dy = \left[y \cosh y\right]_0^2 - \int_0^2 \cosh y\,dy = 2\cosh 2 - 0 - \left[\sinh y\right]_0^2 = 2\cosh 2 - \sinh 2.$$

33. Let $u = \ln R$, $dv = \dfrac{1}{R^2}\,dR$ $\Rightarrow$ $du = \dfrac{1}{R}\,dR$, $v = -\dfrac{1}{R}$. By (6),

$$\int_1^5 \frac{\ln R}{R^2}\,dR = \left[-\frac{1}{R}\ln R\right]_1^5 - \int_1^5 -\frac{1}{R^2}\,dR = -\tfrac{1}{5}\ln 5 - 0 - \left[\frac{1}{R}\right]_1^5 = -\tfrac{1}{5}\ln 5 - \left(\tfrac{1}{5} - 1\right) = \tfrac{4}{5} - \tfrac{1}{5}\ln 5.$$

35. $\sin 2x = 2\sin x \cos x$, so $\int_0^\pi x \sin x \cos x\,dx = \tfrac{1}{2}\int_0^\pi x \sin 2x\,dx$. Let $u = x$, $dv = \sin 2x\,dx$ $\Rightarrow$ $du = dx$, $v = -\tfrac{1}{2}\cos 2x$. By (6),

$$\tfrac{1}{2}\int_0^\pi x \sin 2x\,dx = \tfrac{1}{2}\left[-\tfrac{1}{2}x \cos 2x\right]_0^\pi - \tfrac{1}{2}\int_0^\pi -\tfrac{1}{2}\cos 2x\,dx = -\tfrac{1}{4}\pi - 0 + \tfrac{1}{4}\left[\tfrac{1}{2}\sin 2x\right]_0^\pi = -\frac{\pi}{4}$$

37. Let $u = M$, $dv = e^{-M}\,dM$ $\Rightarrow$ $du = dM$, $v = -e^{-M}$. By (6),

$$\int_1^5 \frac{M}{e^M}\,dM = \int_1^5 Me^{-M}\,dM = \left[-Me^{-M}\right]_1^5 - \int_1^5 -e^{-M}\,dM = -5e^{-5} + e^{-1} - \left[e^{-M}\right]_1^5$$
$$= -5e^{-5} + e^{-1} - (e^{-5} - e^{-1}) = 2e^{-1} - 6e^{-5}$$

39. Let $u = \ln(\cos x)$, $dv = \sin x\,dx$ $\Rightarrow$ $du = \dfrac{1}{\cos x}(-\sin x)\,dx$, $v = -\cos x$. By (6),

$$\int_0^{\pi/3} \sin x \ln(\cos x)\,dx = \left[-\cos x \ln(\cos x)\right]_0^{\pi/3} - \int_0^{\pi/3} \sin x\,dx = -\tfrac{1}{2}\ln\tfrac{1}{2} - 0 - \left[-\cos x\right]_0^{\pi/3}$$
$$= -\tfrac{1}{2}\ln\tfrac{1}{2} + \left(\tfrac{1}{2} - 1\right) = \tfrac{1}{2}\ln 2 - \tfrac{1}{2}$$

41. Let $u = \cos x$, $dv = \sinh x\,dx$ $\Rightarrow$ $du = -\sin x\,dx$, $v = \cosh x$. By (6),

$$I = \int_0^\pi \cos x \sinh x\,dx = \left[\cos x \cosh x\right]_0^\pi - \int_0^\pi -\sin x \cosh x\,dx = -\cosh \pi - 1 + \int_0^\pi \sin x \cosh x\,dx.$$

Now let $U = \sin x$, $dV = \cosh x\,dx$ $\Rightarrow$ $dU = \cos x\,dx$, $V = \sinh x$. Then

$$\int_0^\pi \sin x \cosh x\,dx = \left[\sin x \sinh x\right]_0^\pi - \int_0^\pi \cos x \sinh x\,dx = (0 - 0) - \int_0^\pi \cos x \sinh x\,dx = -I.$$

Substituting in the previous formula gives $I = -\cosh \pi - 1 - I$ $\Rightarrow$ $2I = -(\cosh \pi + 1)$ $\Rightarrow$ $I = -\dfrac{\cosh \pi + 1}{2}$.

[We could also write the answer as $I = -\tfrac{1}{4}(2 + e^\pi + e^{-\pi})$.]

43. Let $t = \sqrt{x}$, so that $t^2 = x$ and $2t\, dt = dx$. Thus, $\int e^{\sqrt{x}}\, dx = \int e^t (2t)\, dt$. Now use parts with $u = t$, $dv = e^t\, dt$, $du = dt$, and $v = e^t$ to get $2 \int t e^t\, dt = 2t e^t - 2 \int e^t\, dt = 2t e^t - 2 e^t + C = 2\sqrt{x}\, e^{\sqrt{x}} - 2 e^{\sqrt{x}} + C$.

45. Let $x = \theta^2$, so that $dx = 2\theta\, d\theta$. Thus, $\displaystyle\int_{\sqrt{\pi/2}}^{\sqrt{\pi}} \theta^3 \cos(\theta^2)\, d\theta = \int_{\sqrt{\pi/2}}^{\sqrt{\pi}} \theta^2 \cos(\theta^2) \cdot \tfrac{1}{2}(2\theta\, d\theta) = \tfrac{1}{2}\int_{\pi/2}^{\pi} x \cos x\, dx$. Now use parts with $u = x$, $dv = \cos x\, dx$, $du = dx$, $v = \sin x$ to get

$$\tfrac{1}{2}\int_{\pi/2}^{\pi} x \cos x\, dx = \tfrac{1}{2}\left([x \sin x]_{\pi/2}^{\pi} - \int_{\pi/2}^{\pi} \sin x\, dx \right) = \tfrac{1}{2}[x \sin x + \cos x]_{\pi/2}^{\pi}$$
$$= \tfrac{1}{2}(\pi \sin \pi + \cos \pi) - \tfrac{1}{2}\left(\tfrac{\pi}{2}\sin\tfrac{\pi}{2} + \cos\tfrac{\pi}{2}\right) = \tfrac{1}{2}(\pi \cdot 0 - 1) - \tfrac{1}{2}\left(\tfrac{\pi}{2}\cdot 1 + 0\right) = -\tfrac{1}{2} - \tfrac{\pi}{4}$$

47. Let $y = 1 + x$, so that $dy = dx$. Thus, $\int x \ln(1+x)\, dx = \int (y-1) \ln y\, dy$. Now use parts with $u = \ln y$, $dv = (y-1)\, dy$, $du = \tfrac{1}{y}\, dy$, $v = \tfrac{1}{2}y^2 - y$ to get

$$\int (y-1) \ln y\, dy = \left(\tfrac{1}{2}y^2 - y\right) \ln y - \int \left(\tfrac{1}{2}y - 1\right) dy = \tfrac{1}{2}y(y-2) \ln y - \tfrac{1}{4}y^2 + y + C$$
$$= \tfrac{1}{2}(1+x)(x-1)\ln(1+x) - \tfrac{1}{4}(1+x)^2 + 1 + x + C,$$

which can be written as $\tfrac{1}{2}(x^2 - 1)\ln(1+x) - \tfrac{1}{4}x^2 + \tfrac{1}{2}x + \tfrac{3}{4} + C$.

49. Let $u = x$, $dv = e^{-2x}\, dx$ $\Rightarrow$ $du = dx$, $v = -\tfrac{1}{2}e^{-2x}$. Then
$\int xe^{-2x}\, dx = -\tfrac{1}{2}xe^{-2x} + \int \tfrac{1}{2}e^{-2x}\, dx = -\tfrac{1}{2}xe^{-2x} - \tfrac{1}{4}e^{-2x} + C$. We see from the graph that this is reasonable, since F has a minimum where f changes from negative to positive. Also, F increases where f is positive and F decreases where f is negative.

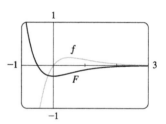

51. Let $u = \tfrac{1}{2}x^2$, $dv = 2x\sqrt{1+x^2}\, dx$ $\Rightarrow$ $du = x\, dx$, $v = \tfrac{2}{3}(1+x^2)^{3/2}$. Then

$$\int x^3\sqrt{1+x^2}\, dx = \tfrac{1}{2}x^2\left[\tfrac{2}{3}(1+x^2)^{3/2}\right] - \tfrac{2}{3}\int x(1+x^2)^{3/2}\, dx$$
$$= \tfrac{1}{3}x^2(1+x^2)^{3/2} - \tfrac{2}{3}\cdot\tfrac{2}{5}\cdot\tfrac{1}{2}(1+x^2)^{5/2} + C$$
$$= \tfrac{1}{3}x^2(1+x^2)^{3/2} - \tfrac{2}{15}(1+x^2)^{5/2} + C$$

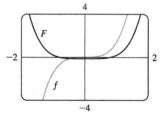

We see from the graph that this is reasonable, since F increases where f is positive and F decreases where f is negative. Note also that f is an odd function and F is an even function.

Another method: Use substitution with $u = 1 + x^2$ to get $\tfrac{1}{5}(1+x^2)^{5/2} - \tfrac{1}{3}(1+x^2)^{3/2} + C$.

53. (a) Take $n = 2$ in Example 6 to get $\displaystyle\int \sin^2 x\, dx = -\tfrac{1}{2}\cos x \sin x + \tfrac{1}{2}\int 1\, dx = \tfrac{x}{2} - \tfrac{\sin 2x}{4} + C$.

(b) $\int \sin^4 x\, dx = -\tfrac{1}{4}\cos x \sin^3 x + \tfrac{3}{4}\int \sin^2 x\, dx = -\tfrac{1}{4}\cos x \sin^3 x + \tfrac{3}{8}x - \tfrac{3}{16}\sin 2x + C$.

55. (a) From Example 6, $\int \sin^n x\, dx = -\dfrac{1}{n}\cos x \sin^{n-1} x + \dfrac{n-1}{n}\int \sin^{n-2} x\, dx$. Using (6),

$$\int_0^{\pi/2} \sin^n x\, dx = \left[-\dfrac{\cos x \sin^{n-1} x}{n}\right]_0^{\pi/2} + \dfrac{n-1}{n}\int_0^{\pi/2} \sin^{n-2} x\, dx$$

$$= (0-0) + \dfrac{n-1}{n}\int_0^{\pi/2} \sin^{n-2} x\, dx = \dfrac{n-1}{n}\int_0^{\pi/2} \sin^{n-2} x\, dx$$

(b) Using $n=3$ in part (a), we have $\int_0^{\pi/2} \sin^3 x\, dx = \tfrac{2}{3}\int_0^{\pi/2} \sin x\, dx = \left[-\tfrac{2}{3}\cos x\right]_0^{\pi/2} = \tfrac{2}{3}$.

Using $n=5$ in part (a), we have $\int_0^{\pi/2} \sin^5 x\, dx = \tfrac{4}{5}\int_0^{\pi/2} \sin^3 x\, dx = \tfrac{4}{5} \cdot \tfrac{2}{3} = \tfrac{8}{15}$.

(c) The formula holds for $n=1$ (that is, $2n+1=3$) by (b). Assume it holds for some $k \geq 1$. Then

$$\int_0^{\pi/2} \sin^{2k+1} x\, dx = \dfrac{2 \cdot 4 \cdot 6 \cdots (2k)}{3 \cdot 5 \cdot 7 \cdots (2k+1)}.$$ By Example 6,

$$\int_0^{\pi/2} \sin^{2k+3} x\, dx = \dfrac{2k+2}{2k+3}\int_0^{\pi/2} \sin^{2k+1} x\, dx = \dfrac{2k+2}{2k+3} \cdot \dfrac{2 \cdot 4 \cdot 6 \cdots (2k)}{3 \cdot 5 \cdot 7 \cdots (2k+1)}$$

$$= \dfrac{2 \cdot 4 \cdot 6 \cdots (2k)[2(k+1)]}{3 \cdot 5 \cdot 7 \cdots (2k+1)[2(k+1)+1]},$$

so the formula holds for $n = k+1$. By induction, the formula holds for all $n \geq 1$.

57. Let $u = (\ln x)^n$, $dv = dx$ $\;\Rightarrow\;$ $du = n(\ln x)^{n-1}(dx/x)$, $v = x$. By Equation 2,

$\int (\ln x)^n\, dx = x(\ln x)^n - \int nx(\ln x)^{n-1}(dx/x) = x(\ln x)^n - n\int (\ln x)^{n-1}\, dx$.

59. $\int \tan^n x\, dx = \int \tan^{n-2} x\, \tan^2 x\, dx = \int \tan^{n-2} x\,(\sec^2 x - 1)\, dx = \int \tan^{n-2} x\, \sec^2 x\, dx - \int \tan^{n-2} x\, dx$

$= I - \int \tan^{n-2} x\, dx$.

Let $u = \tan^{n-2} x$, $dv = \sec^2 x\, dx$ $\;\Rightarrow\;$ $du = (n-2)\tan^{n-3} x\, \sec^2 x\, dx$, $v = \tan x$. Then, by Equation 2,

$$I = \tan^{n-1} x - (n-2)\int \tan^{n-2} x\, \sec^2 x\, dx$$

$$1I = \tan^{n-1} x - (n-2)I$$

$$(n-1)I = \tan^{n-1} x$$

$$I = \dfrac{\tan^{n-1} x}{n-1}$$

Returning to the original integral, $\int \tan^n x\, dx = \dfrac{\tan^{n-1} x}{n-1} - \int \tan^{n-2} x\, dx$.

61. By repeated applications of the reduction formula in Exercise 57,

$$\int (\ln x)^3\, dx = x\,(\ln x)^3 - 3\int (\ln x)^2\, dx = x(\ln x)^3 - 3\bigl[x(\ln x)^2 - 2\int (\ln x)^1\, dx\bigr]$$

$$= x\,(\ln x)^3 - 3x(\ln x)^2 + 6\bigl[x(\ln x)^1 - 1\int (\ln x)^0\, dx\bigr]$$

$$= x\,(\ln x)^3 - 3x(\ln x)^2 + 6x\ln x - 6\int 1\, dx = x\,(\ln x)^3 - 3x(\ln x)^2 + 6x\ln x - 6x + C$$

63. The curves $y = x^2 \ln x$ and $y = 4\ln x$ intersect when $x^2 \ln x = 4\ln x \iff$
$x^2 \ln x - 4\ln x = 0 \iff (x^2 - 4)\ln x = 0 \iff$
$x = 1$ or 2 [since $x > 0$]. For $1 < x < 2$, $4\ln x > x^2 \ln x$. Thus,
area $= \int_1^2 (4\ln x - x^2 \ln x)\,dx = \int_1^2 [(4 - x^2)\ln x]\,dx$. Let $u = \ln x$,
$dv = (4 - x^2)\,dx \;\Rightarrow\; du = \frac{1}{x}\,dx$, $v = 4x - \frac{1}{3}x^3$. Then

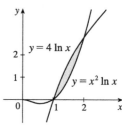

$$\text{area} = \left[(\ln x)\left(4x - \tfrac{1}{3}x^3\right)\right]_1^2 - \int_1^2 \left[\left(4x - \tfrac{1}{3}x^3\right)\tfrac{1}{x}\right]dx = (\ln 2)\left(\tfrac{16}{3}\right) - 0 - \int_1^2 \left(4 - \tfrac{1}{3}x^2\right)dx$$
$$= \tfrac{16}{3}\ln 2 - \left[4x - \tfrac{1}{9}x^3\right]_1^2 = \tfrac{16}{3}\ln 2 - \left(\tfrac{64}{9} - \tfrac{35}{9}\right) = \tfrac{16}{3}\ln 2 - \tfrac{29}{9}$$

65. The curves $y = \arcsin\left(\tfrac{1}{2}x\right)$ and $y = 2 - x^2$ intersect at
$x = a \approx -1.75119$ and $x = b \approx 1.17210$. From the figure, the area
bounded by the curves is given by
$A = \int_a^b [(2 - x^2) - \arcsin\left(\tfrac{1}{2}x\right)]\,dx = \left[2x - \tfrac{1}{3}x^3\right]_a^b - \int_a^b \arcsin\left(\tfrac{1}{2}x\right)dx$.
Let $u = \arcsin\left(\tfrac{1}{2}x\right)$, $dv = dx \;\Rightarrow\; du = \dfrac{1}{\sqrt{1 - \left(\tfrac{1}{2}x\right)^2}} \cdot \tfrac{1}{2}\,dx$, $v = x$.

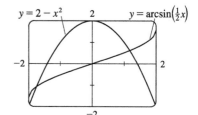

Then
$$A = \left[2x - \tfrac{1}{3}x^3\right]_a^b - \left\{\left[x\arcsin\left(\tfrac{1}{2}x\right)\right]_a^b - \int_a^b \dfrac{x}{2\sqrt{1 - \tfrac{1}{4}x^2}}\,dx\right\}$$
$$= \left[2x - \tfrac{1}{3}x^3 - x\arcsin\left(\tfrac{1}{2}x\right) - 2\sqrt{1 - \tfrac{1}{4}x^2}\right]_a^b \approx 3.99926$$

67. Volume $= \int_0^1 2\pi x \cos(\pi x/2)\,dx$. Let $u = x$, $dv = \cos(\pi x/2)\,dx \;\Rightarrow\; du = dx$, $v = \tfrac{2}{\pi}\sin(\pi x/2)$.
$$V = 2\pi\left[\tfrac{2}{\pi}x\sin\left(\tfrac{\pi x}{2}\right)\right]_0^1 - 2\pi \cdot \tfrac{2}{\pi}\int_0^1 \sin\left(\tfrac{\pi x}{2}\right)dx = 2\pi\left(\tfrac{2}{\pi} - 0\right) - 4\left[-\tfrac{2}{\pi}\cos\left(\tfrac{\pi x}{2}\right)\right]_0^1 = 4 + \tfrac{8}{\pi}(0 - 1) = 4 - \tfrac{8}{\pi}.$$

69. Volume $= \int_{-1}^0 2\pi(1 - x)e^{-x}\,dx$. Let $u = 1 - x$, $dv = e^{-x}\,dx \;\Rightarrow\; du = -dx$, $v = -e^{-x}$.
$$V = 2\pi\left[(1 - x)(-e^{-x})\right]_{-1}^0 - 2\pi\int_{-1}^0 e^{-x}\,dx = 2\pi\left[(x - 1)(e^{-x}) + e^{-x}\right]_{-1}^0 = 2\pi\left[xe^{-x}\right]_{-1}^0 = 2\pi(0 + e) = 2\pi e.$$

71. (a) Use shells about the y-axis:
$$V = \int_1^2 2\pi x \ln x\,dx \qquad \left[\begin{array}{l} u = \ln x,\; dv = x\,dx \\ du = \tfrac{1}{x}\,dx,\; v = \tfrac{1}{2}x^2 \end{array}\right]$$
$$= 2\pi\left\{\left[\tfrac{1}{2}x^2\ln x\right]_1^2 - \int_1^2 \tfrac{1}{2}x\,dx\right\} = 2\pi\left\{(2\ln 2 - 0) - \left[\tfrac{1}{4}x^2\right]_1^2\right\} = 2\pi\left(2\ln 2 - \tfrac{3}{4}\right)$$

(b) Use disks about the x-axis:
$$V = \int_1^2 \pi(\ln x)^2\,dx \qquad \left[\begin{array}{l} u = (\ln x)^2,\; dv = dx \\ du = 2\ln x \cdot \tfrac{1}{x}\,dx,\; v = x \end{array}\right]$$
$$= \pi\left\{\left[x(\ln x)^2\right]_1^2 - \int_1^2 2\ln x\,dx\right\} \qquad \left[\begin{array}{l} u = \ln x,\; dv = dx \\ du = \tfrac{1}{x}\,dx,\; v = x \end{array}\right]$$
$$= \pi\left\{2(\ln 2)^2 - 2\left(\left[x\ln x\right]_1^2 - \int_1^2 dx\right)\right\} = \pi\left\{2(\ln 2)^2 - 4\ln 2 + 2\left[x\right]_1^2\right\}$$
$$= \pi[2(\ln 2)^2 - 4\ln 2 + 2] = 2\pi[(\ln 2)^2 - 2\ln 2 + 1]$$

73. $S(x) = \int_0^x \sin\left(\tfrac{1}{2}\pi t^2\right) dt \;\Rightarrow\; \int S(x)\, dx = \int \left[\int_0^x \sin\left(\tfrac{1}{2}\pi t^2\right) dt\right] dx.$

Let $u = \int_0^x \sin\left(\tfrac{1}{2}\pi t^2\right) dt = S(x)$, $dv = dx \;\Rightarrow\; du = \sin\left(\tfrac{1}{2}\pi x^2\right) dx$, $v = x$. Thus,

$$\int S(x)\, dx = xS(x) - \int x \sin\left(\tfrac{1}{2}\pi x^2\right) dx = xS(x) - \int \sin y \left(\tfrac{1}{\pi}\, dy\right) \quad \begin{bmatrix} u = \tfrac{1}{2}\pi x^2, \\ du = \pi x\, dx \end{bmatrix}$$

$$= xS(x) + \tfrac{1}{\pi}\cos y + C = xS(x) + \tfrac{1}{\pi}\cos\left(\tfrac{1}{2}\pi x^2\right) + C$$

75. Since $v(t) > 0$ for all t, the desired distance is $s(t) = \int_0^t v(w)\, dw = \int_0^t w^2 e^{-w}\, dw$.

First let $u = w^2$, $dv = e^{-w}\, dw \;\Rightarrow\; du = 2w\, dw$, $v = -e^{-w}$. Then $s(t) = \left[-w^2 e^{-w}\right]_0^t + 2\int_0^t we^{-w}\, dw$.

Next let $U = w$, $dV = e^{-w}\, dw \;\Rightarrow\; dU = dw$, $V = -e^{-w}$. Then

$$s(t) = -t^2 e^{-t} + 2\left(\left[-we^{-w}\right]_0^t + \int_0^t e^{-w}\, dw\right) = -t^2 e^{-t} + 2\left(-te^{-t} + 0 + \left[-e^{-w}\right]_0^t\right)$$

$$= -t^2 e^{-t} + 2(-te^{-t} - e^{-t} + 1) = -t^2 e^{-t} - 2te^{-t} - 2e^{-t} + 2 = 2 - e^{-t}(t^2 + 2t + 2) \text{ meters}$$

77. For $I = \int_1^4 x f''(x)\, dx$, let $u = x$, $dv = f''(x)\, dx \;\Rightarrow\; du = dx$, $v = f'(x)$. Then

$$I = \left[xf'(x)\right]_1^4 - \int_1^4 f'(x)\, dx = 4f'(4) - 1 \cdot f'(1) - [f(4) - f(1)] = 4 \cdot 3 - 1 \cdot 5 - (7 - 2) = 12 - 5 - 5 = 2.$$

We used the fact that f'' is continuous to guarantee that I exists.

79. (a) Assuming $f(x)$ and $g(x)$ are differentiable functions, the Quotient Rule for differentiation states

$$\frac{d}{dx}\left[\frac{f(x)}{g(x)}\right] = \frac{g(x)f'(x) - f(x)g'(x)}{[g(x)]^2}.$$ Writing in integral form gives

$$\frac{f(x)}{g(x)} = \int \frac{g(x)f'(x) - f(x)g'(x)}{[g(x)]^2}\, dx = \int \frac{1}{g(x)} f'(x)\, dx - \int \frac{f(x)}{[g(x)]^2} g'(x)\, dx.$$ Now let $u = f(x)$ and $v = g(x)$ so

that $du = f'(x)\, dx$ and $dv = g'(x)\, dx$. Substituting into the above equation gives $\dfrac{u}{v} = \int \dfrac{1}{v}\, du - \int \dfrac{u}{v^2}\, dv \;\Rightarrow\;$

$$\int \frac{u}{v^2}\, dv = -\frac{u}{v} + \int \frac{1}{v}\, du.$$

(b) Let $u = \ln x$, $v = x \;\Rightarrow\; du = \dfrac{1}{x}\, dx$. Then, using the formula from part (a), we get

$$\int \frac{\ln x}{x^2}\, dx = -\frac{\ln x}{x} + \int \frac{1}{x}\left(\frac{1}{x}\, dx\right) = -\frac{\ln x}{x} + \int \frac{1}{x^2}\, dx = -\frac{\ln x}{x} - \frac{1}{x} + C.$$

81. Using the formula for volumes of rotation and the figure, we see that

Volume $= \int_0^d \pi b^2\, dy - \int_0^c \pi a^2\, dy - \int_c^d \pi [g(y)]^2\, dy = \pi b^2 d - \pi a^2 c - \int_c^d \pi [g(y)]^2\, dy$. Let $y = f(x)$,

which gives $dy = f'(x)\, dx$ and $g(y) = x$, so that $V = \pi b^2 d - \pi a^2 c - \pi \int_a^b x^2 f'(x)\, dx$.

Now integrate by parts with $u = x^2$, and $dv = f'(x)\, dx \;\Rightarrow\; du = 2x\, dx$, $v = f(x)$, and

$\int_a^b x^2 f'(x)\, dx = \left[x^2 f(x)\right]_a^b - \int_a^b 2x f(x)\, dx = b^2 f(b) - a^2 f(a) - \int_a^b 2x f(x)\, dx$, but $f(a) = c$ and $f(b) = d \;\Rightarrow\;$

$V = \pi b^2 d - \pi a^2 c - \pi \left[b^2 d - a^2 c - \int_a^b 2x f(x)\, dx\right] = \int_a^b 2\pi x f(x)\, dx.$

7.2 Trigonometric Integrals

The symbols $\stackrel{s}{=}$ and $\stackrel{c}{=}$ indicate the use of the substitutions $\{u = \sin x, du = \cos x \, dx\}$ and $\{u = \cos x, du = -\sin x \, dx\}$, respectively.

1. $\int \sin^3 x \cos^2 x \, dx = \int \sin^2 x \cos^2 x \sin x \, dx = \int (1 - \cos^2 x) \cos^2 x \sin x \, dx \stackrel{c}{=} \int (1 - u^2) u^2 \, (-du)$

   $= \int (u^4 - u^2) \, du = \tfrac{1}{5} u^5 - \tfrac{1}{3} u^3 + C = \tfrac{1}{5} \cos^5 x - \tfrac{1}{3} \cos^3 x + C$

3. $\int_0^{\pi/2} \cos^9 x \sin^5 x \, dx = \int_0^{\pi/2} \cos^9 x \, (\sin^2 x)^2 \sin x \, dx = \int_0^{\pi/2} \cos^9 x \, (1 - \cos^2 x)^2 \sin x \, dx$

   $\stackrel{c}{=} \int_1^0 u^9 (1 - u^2)^2 \, (-du) = \int_0^1 u^9 (1 - 2u^2 + u^4) \, du = \int_0^1 (u^9 - 2u^{11} + u^{13}) \, du$

   $= \left[\tfrac{1}{10} u^{10} - \tfrac{1}{6} u^{12} + \tfrac{1}{14} u^{14} \right]_0^1 = \left(\tfrac{1}{10} - \tfrac{1}{6} + \tfrac{1}{14} \right) - 0 = \tfrac{1}{210}$

5. $\int \sin^5(2t) \cos^2(2t) \, dt = \int \sin^4(2t) \cos^2(2t) \sin(2t) \, dt = \int [1 - \cos^2(2t)]^2 \cos^2(2t) \sin(2t) \, dt$

   $= \int (1 - u^2)^2 u^2 \left(-\tfrac{1}{2} du \right) \qquad [u = \cos(2t), \, du = -2 \sin(2t) \, dt]$

   $= -\tfrac{1}{2} \int (u^4 - 2u^2 + 1) u^2 \, du = -\tfrac{1}{2} \int (u^6 - 2u^4 + u^2) \, du$

   $= -\tfrac{1}{2} \left(\tfrac{1}{7} u^7 - \tfrac{2}{5} u^5 + \tfrac{1}{3} u^3 \right) + C = -\tfrac{1}{14} \cos^7(2t) + \tfrac{1}{5} \cos^5(2t) - \tfrac{1}{6} \cos^3(2t) + C$

7. $\int_0^{\pi/2} \cos^2 \theta \, d\theta = \int_0^{\pi/2} \tfrac{1}{2} (1 + \cos 2\theta) \, d\theta \qquad$ [half-angle identity]

   $= \tfrac{1}{2} \left[\theta + \tfrac{1}{2} \sin 2\theta \right]_0^{\pi/2} = \tfrac{1}{2} \left[\left(\tfrac{\pi}{2} + 0 \right) - (0 + 0) \right] = \tfrac{\pi}{4}$

9. $\int_0^{\pi} \cos^4(2t) \, dt = \int_0^{\pi} [\cos^2(2t)]^2 \, dt = \int_0^{\pi} \left[\tfrac{1}{2} (1 + \cos(2 \cdot 2t)) \right]^2 dt \qquad$ [half-angle identity]

   $= \tfrac{1}{4} \int_0^{\pi} [1 + 2 \cos 4t + \cos^2(4t)] \, dt = \tfrac{1}{4} \int_0^{\pi} \left[1 + 2 \cos 4t + \tfrac{1}{2} (1 + \cos 8t) \right] dt$

   $= \tfrac{1}{4} \int_0^{\pi} \left(\tfrac{3}{2} + 2 \cos 4t + \tfrac{1}{2} \cos 8t \right) dt = \tfrac{1}{4} \left[\tfrac{3}{2} t + \tfrac{1}{2} \sin 4t + \tfrac{1}{16} \sin 8t \right]_0^{\pi} = \tfrac{1}{4} \left[\left(\tfrac{3}{2} \pi + 0 + 0 \right) - 0 \right] = \tfrac{3}{8} \pi$

11. $\int_0^{\pi/2} \sin^2 x \cos^2 x \, dx = \int_0^{\pi/2} \tfrac{1}{4} (4 \sin^2 x \cos^2 x) \, dx = \int_0^{\pi/2} \tfrac{1}{4} (2 \sin x \cos x)^2 \, dx = \tfrac{1}{4} \int_0^{\pi/2} \sin^2 2x \, dx$

    $= \tfrac{1}{4} \int_0^{\pi/2} \tfrac{1}{2} (1 - \cos 4x) \, dx = \tfrac{1}{8} \int_0^{\pi/2} (1 - \cos 4x) \, dx = \tfrac{1}{8} \left[x - \tfrac{1}{4} \sin 4x \right]_0^{\pi/2} = \tfrac{1}{8} \left(\tfrac{\pi}{2} \right) = \tfrac{\pi}{16}$

13. $\int \sqrt{\cos \theta} \, \sin^3 \theta \, d\theta = \int \sqrt{\cos \theta} \, \sin^2 \theta \sin \theta \, d\theta = \int (\cos \theta)^{1/2} (1 - \cos^2 \theta) \sin \theta \, d\theta$

    $\stackrel{c}{=} \int u^{1/2} (1 - u^2) (-du) = \int (u^{5/2} - u^{1/2}) \, du$

    $= \tfrac{2}{7} u^{7/2} - \tfrac{2}{3} u^{3/2} + C = \tfrac{2}{7} (\cos \theta)^{7/2} - \tfrac{2}{3} (\cos \theta)^{3/2} + C$

15. $\int \sin x \sec^5 x \, dx = \int \dfrac{\sin x}{\cos^5 x} \, dx \stackrel{c}{=} \int \dfrac{1}{u^5} (-du) = \dfrac{1}{4u^4} + C = \dfrac{1}{4 \cos^4 x} + C = \tfrac{1}{4} \sec^4 x + C$

17. $\int \cot x \cos^2 x \, dx = \int \dfrac{\cos x}{\sin x} (1 - \sin^2 x) \, dx$

    $\stackrel{s}{=} \int \dfrac{1 - u^2}{u} \, du = \int \left(\dfrac{1}{u} - u \right) du = \ln |u| - \tfrac{1}{2} u^2 + C = \ln |\sin x| - \tfrac{1}{2} \sin^2 x + C$

19. $\int \sin^2 x \sin 2x \, dx = \int \sin^2 x \, (2 \sin x \cos x) \, dx \stackrel{s}{=} \int 2u^3 \, du = \tfrac{1}{2} u^4 + C = \tfrac{1}{2} \sin^4 x + C$

21. $\int \tan x \sec^3 x \, dx = \int \tan x \sec x \sec^2 x \, dx = \int u^2 \, du \qquad [u = \sec x, \, du = \sec x \tan x \, dx]$

    $= \tfrac{1}{3} u^3 + C = \tfrac{1}{3} \sec^3 x + C$

SECTION 7.2 TRIGONOMETRIC INTEGRALS 357

23. $\int \tan^2 x \, dx = \int (\sec^2 x - 1) \, dx = \tan x - x + C$

25. Let $u = \tan x$. Then $du = \sec^2 x \, dx$, so

$$\int \tan^4 x \sec^6 x \, dx = \int \tan^4 x \sec^4 x (\sec^2 x \, dx) = \int \tan^4 x (1 + \tan^2 x)^2 (\sec^2 x \, dx)$$
$$= \int u^4 (1 + u^2)^2 \, du = \int (u^8 + 2u^6 + u^4) \, du$$
$$= \tfrac{1}{9} u^9 + \tfrac{2}{7} u^7 + \tfrac{1}{5} u^5 + C = \tfrac{1}{9} \tan^9 x + \tfrac{2}{7} \tan^7 x + \tfrac{1}{5} \tan^5 x + C$$

27. $\int \tan^3 x \sec x \, dx = \int \tan^2 x \sec x \tan x \, dx = \int (\sec^2 x - 1) \sec x \tan x \, dx$
$$= \int (u^2 - 1) \, du \quad [u = \sec x, du = \sec x \tan x \, dx] \quad = \tfrac{1}{3} u^3 - u + C = \tfrac{1}{3} \sec^3 x - \sec x + C$$

29. $\int \tan^3 x \sec^6 x \, dx = \int \tan^3 x \sec^4 x \sec^2 x \, dx = \int \tan^3 x (1 + \tan^2 x)^2 \sec^2 x \, dx$
$$= \int u^3 (1 + u^2)^2 \, du \quad \begin{bmatrix} u = \tan x, \\ du = \sec^2 x \, dx \end{bmatrix}$$
$$= \int u^3 (u^4 + 2u^2 + 1) \, du = \int (u^7 + 2u^5 + u^3) \, du$$
$$= \tfrac{1}{8} u^8 + \tfrac{1}{3} u^6 + \tfrac{1}{4} u^4 + C = \tfrac{1}{8} \tan^8 x + \tfrac{1}{3} \tan^6 x + \tfrac{1}{4} \tan^4 x + C$$

31. $\int \tan^5 x \, dx = \int (\sec^2 x - 1)^2 \tan x \, dx = \int \sec^4 x \tan x \, dx - 2 \int \sec^2 x \tan x \, dx + \int \tan x \, dx$
$$= \int \sec^3 x \sec x \tan x \, dx - 2 \int \tan x \sec^2 x \, dx + \int \tan x \, dx$$
$$= \tfrac{1}{4} \sec^4 x - \tan^2 x + \ln|\sec x| + C \quad [\text{or } \tfrac{1}{4} \sec^4 x - \sec^2 x + \ln|\sec x| + C]$$

33. $\int \dfrac{1 - \tan^2 x}{\sec^2 x} \, dx = \int (\cos^2 x - \sin^2 x) \, dx = \int \cos 2x \, dx = \tfrac{1}{2} \sin 2x + C$

35. $\int_0^{\pi/4} \dfrac{\sin^3 x}{\cos x} \, dx = \int_0^{\pi/4} \dfrac{\sin^2 x}{\cos x} \sin x \, dx = \int_0^{\pi/4} \dfrac{1 - \cos^2 x}{\cos x} \sin x \, dx \overset{c}{=} \int_1^{1/\sqrt{2}} \dfrac{1 - u^2}{u} (-du)$
$$= \int_{1/\sqrt{2}}^1 \left(\dfrac{1}{u} - u \right) du = \left[\ln|u| - \tfrac{1}{2} u^2 \right]_{1/\sqrt{2}}^1 = \left(\ln 1 - \tfrac{1}{2} \right) - \left(\ln \dfrac{1}{\sqrt{2}} - \tfrac{1}{4} \right) = -\tfrac{1}{4} - \ln \dfrac{\sqrt{2}}{2}$$

37. $\int_{\pi/6}^{\pi/2} \cot^2 x \, dx = \int_{\pi/6}^{\pi/2} (\csc^2 x - 1) \, dx = \left[-\cot x - x \right]_{\pi/6}^{\pi/2} = \left(0 - \tfrac{\pi}{2} \right) - \left(-\sqrt{3} - \tfrac{\pi}{6} \right) = \sqrt{3} - \tfrac{\pi}{3}$

39. $\int_{\pi/4}^{\pi/2} \cot^5 \phi \csc^3 \phi \, d\phi = \int_{\pi/4}^{\pi/2} \cot^4 \phi \csc^2 \phi \csc \phi \cot \phi \, d\phi = \int_{\pi/4}^{\pi/2} (\csc^2 \phi - 1)^2 \csc^2 \phi \csc \phi \cot \phi \, d\phi$
$$= \int_{\sqrt{2}}^1 (u^2 - 1)^2 u^2 (-du) \quad [u = \csc \phi, du = -\csc \phi \cot \phi \, d\phi]$$
$$= \int_1^{\sqrt{2}} (u^6 - 2u^4 + u^2) \, du = \left[\tfrac{1}{7} u^7 - \tfrac{2}{5} u^5 + \tfrac{1}{3} u^3 \right]_1^{\sqrt{2}} = \left(\tfrac{8}{7} \sqrt{2} - \tfrac{8}{5} \sqrt{2} + \tfrac{2}{3} \sqrt{2} \right) - \left(\tfrac{1}{7} - \tfrac{2}{5} + \tfrac{1}{3} \right)$$
$$= \dfrac{120 - 168 + 70}{105} \sqrt{2} - \dfrac{15 - 42 + 35}{105} = \dfrac{22}{105} \sqrt{2} - \dfrac{8}{105}$$

41. $I = \int \csc x \, dx = \int \dfrac{\csc x (\csc x - \cot x)}{\csc x - \cot x} \, dx = \int \dfrac{-\csc x \cot x + \csc^2 x}{\csc x - \cot x} \, dx$. Let $u = \csc x - \cot x \Rightarrow$
$du = (-\csc x \cot x + \csc^2 x) \, dx$. Then $I = \int du/u = \ln|u| = \ln|\csc x - \cot x| + C$.

43. $\int \sin 8x \cos 5x \, dx \overset{2a}{=} \int \tfrac{1}{2} [\sin(8x - 5x) + \sin(8x + 5x)] \, dx = \tfrac{1}{2} \int (\sin 3x + \sin 13x) \, dx$
$$= \tfrac{1}{2} \left(-\tfrac{1}{3} \cos 3x - \tfrac{1}{13} \cos 13x \right) + C = -\tfrac{1}{6} \cos 3x - \tfrac{1}{26} \cos 13x + C$$

45. $\int_0^{\pi/2} \cos 5t \cos 10t \, dt \stackrel{2c}{=} \int_0^{\pi/2} \frac{1}{2}[\cos(5t - 10t) + \cos(5t + 10t)] \, dt$

$= \frac{1}{2} \int_0^{\pi/2} [\cos(-5t) + \cos 15t] \, dt = \frac{1}{2} \int_0^{\pi/2} (\cos 5t + \cos 15t) \, dt$

$= \frac{1}{2} \left[\frac{1}{5} \sin 5t + \frac{1}{15} \sin 15t \right]_0^{\pi/2} = \frac{1}{2} \left(\frac{1}{5} - \frac{1}{15} \right) = \frac{1}{15}$

47. $\int \frac{\sin^2(1/t)}{t^2} \, dt = \int \sin^2 u \, (-du) \quad [u = \frac{1}{t}, \, du = -\frac{1}{t^2} \, dt]$

$= -\int \frac{1}{2}(1 - \cos 2u) \, du = -\frac{1}{2}\left(u - \frac{1}{2} \sin 2u \right) + C = -\frac{1}{2t} + \frac{1}{4} \sin\left(\frac{2}{t}\right) + C$

49. $\int_0^{\pi/6} \sqrt{1 + \cos 2x} \, dx = \int_0^{\pi/6} \sqrt{1 + (2\cos^2 x - 1)} \, dx = \int_0^{\pi/6} \sqrt{2\cos^2 x} \, dx = \sqrt{2} \int_0^{\pi/6} \sqrt{\cos^2 x} \, dx$

$= \sqrt{2} \int_0^{\pi/6} |\cos x| \, dx = \sqrt{2} \int_0^{\pi/6} \cos x \, dx \qquad [\text{since } \cos x > 0 \text{ for } 0 \leq x \leq \pi/6]$

$= \sqrt{2} \left[\sin x\right]_0^{\pi/6} = \sqrt{2}\left(\frac{1}{2} - 0\right) = \frac{1}{2}\sqrt{2}$

51. $\int t \sin^2 t \, dt = \int t \left[\frac{1}{2}(1 - \cos 2t)\right] dt = \frac{1}{2} \int (t - t \cos 2t) \, dt = \frac{1}{2} \int t \, dt - \frac{1}{2} \int t \cos 2t \, dt$

$= \frac{1}{2}\left(\frac{1}{2}t^2\right) - \frac{1}{2}\left(\frac{1}{2}t \sin 2t - \int \frac{1}{2} \sin 2t \, dt\right) \qquad \begin{bmatrix} u = t, & dv = \cos 2t \, dt \\ du = dt, & v = \frac{1}{2} \sin 2t \end{bmatrix}$

$= \frac{1}{4}t^2 - \frac{1}{4}t \sin 2t + \frac{1}{2}\left(-\frac{1}{4}\cos 2t\right) + C = \frac{1}{4}t^2 - \frac{1}{4}t \sin 2t - \frac{1}{8}\cos 2t + C$

53. $\int x \tan^2 x \, dx = \int x(\sec^2 x - 1) \, dx = \int x \sec^2 x \, dx - \int x \, dx$

$= x \tan x - \int \tan x \, dx - \frac{1}{2}x^2 \qquad \begin{bmatrix} u = x, & dv = \sec^2 x \, dx \\ du = dx, & v = \tan x \end{bmatrix}$

$= x \tan x - \ln|\sec x| - \frac{1}{2}x^2 + C$

55. $\int \frac{dx}{\cos x - 1} = \int \frac{1}{\cos x - 1} \cdot \frac{\cos x + 1}{\cos x + 1} \, dx = \int \frac{\cos x + 1}{\cos^2 x - 1} \, dx = \int \frac{\cos x + 1}{-\sin^2 x} \, dx$

$= \int \left(-\cot x \csc x - \csc^2 x\right) dx = \csc x + \cot x + C$

In Exercises 57–60, let $f(x)$ denote the integrand and $F(x)$ its antiderivative (with $C = 0$).

57. Let $u = x^2$, so that $du = 2x \, dx$. Then

$\int x \sin^2(x^2) \, dx = \int \sin^2 u \left(\frac{1}{2} du\right) = \frac{1}{2} \int \frac{1}{2}(1 - \cos 2u) \, du$

$= \frac{1}{4}\left(u - \frac{1}{2}\sin 2u\right) + C = \frac{1}{4}u - \frac{1}{4}\left(\frac{1}{2} \cdot 2 \sin u \cos u\right) + C$

$= \frac{1}{4}x^2 - \frac{1}{4}\sin(x^2)\cos(x^2) + C$

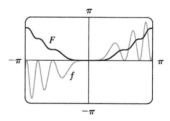

We see from the graph that this is reasonable, since F increases where f is positive and F decreases where f is negative. Note also that f is an odd function and F is an even function.

59. $\int \sin 3x \sin 6x \, dx = \int \frac{1}{2}[\cos(3x - 6x) - \cos(3x + 6x)] \, dx$

$= \frac{1}{2} \int (\cos 3x - \cos 9x) \, dx$

$= \frac{1}{6} \sin 3x - \frac{1}{18} \sin 9x + C$

Notice that $f(x) = 0$ whenever F has a horizontal tangent.

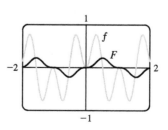

61. Let $u = \tan^7 x$, $dv = \sec x \tan x \, dx$ $\Rightarrow$ $du = 7\tan^6 x \sec^2 x \, dx$, $v = \sec x$. Then

$$\int \tan^8 x \sec x \, dx = \int \tan^7 x \cdot \sec x \tan x \, dx = \tan^7 x \sec x - \int 7 \tan^6 x \sec^2 x \sec x \, dx$$

$$= \tan^7 x \sec x - 7 \int \tan^6 x \, (\tan^2 x + 1) \sec x \, dx$$

$$= \tan^7 x \sec x - 7 \int \tan^8 x \sec x \, dx - 7 \int \tan^6 x \sec x \, dx$$

Thus, $8 \int \tan^8 x \sec x \, dx = \tan^7 x \sec x - 7 \int \tan^6 x \sec x \, dx$ and

$$\int_0^{\pi/4} \tan^8 x \sec x \, dx = \frac{1}{8}\left[\tan^7 x \sec x\right]_0^{\pi/4} - \frac{7}{8}\int_0^{\pi/4} \tan^6 x \sec x \, dx = \frac{\sqrt{2}}{8} - \frac{7}{8}I.$$

63. $f_{\text{avg}} = \frac{1}{2\pi}\int_{-\pi}^{\pi} \sin^2 x \, \cos^3 x \, dx = \frac{1}{2\pi}\int_{-\pi}^{\pi} \sin^2 x \, (1 - \sin^2 x) \, \cos x \, dx$

$= \frac{1}{2\pi}\int_0^0 u^2(1 - u^2) \, du$ [where $u = \sin x$] $= 0$

65. $A = \int_0^\pi (\sin^2 x - \sin^3 x) \, dx = \int_0^\pi \left[\frac{1}{2}(1 - \cos 2x) - \sin x \, (1 - \cos^2 x)\right] dx$

$= \int_0^\pi \left(\frac{1}{2} - \frac{1}{2}\cos 2x\right) dx + \int_1^{-1}(1 - u^2) \, du$ $\begin{bmatrix} u = \cos x, \\ du = -\sin x \, dx \end{bmatrix}$

$= \left[\frac{1}{2}x - \frac{1}{4}\sin 2x\right]_0^\pi + 2\int_0^1 (u^2 - 1) \, du$

$= \left(\frac{1}{2}\pi - 0\right) - (0 - 0) + 2\left[\frac{1}{3}u^3 - u\right]_0^1$

$= \frac{1}{2}\pi + 2\left(\frac{1}{3} - 1\right) = \frac{1}{2}\pi - \frac{4}{3}$

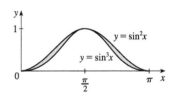

67.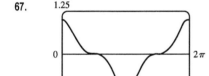

It seems from the graph that $\int_0^{2\pi} \cos^3 x \, dx = 0$, since the area below the x-axis and above the graph looks about equal to the area above the axis and below the graph. By Example 1, the integral is $\left[\sin x - \frac{1}{3}\sin^3 x\right]_0^{2\pi} = 0$. Note that due to symmetry, the integral of any odd power of $\sin x$ or $\cos x$ between limits which differ by $2n\pi$ (n any integer) is 0.

69. Using disks, $V = \int_{\pi/2}^{\pi} \pi \sin^2 x \, dx = \pi \int_{\pi/2}^{\pi} \frac{1}{2}(1 - \cos 2x) \, dx = \pi\left[\frac{1}{2}x - \frac{1}{4}\sin 2x\right]_{\pi/2}^{\pi} = \pi\left(\frac{\pi}{2} - 0 - \frac{\pi}{4} + 0\right) = \frac{\pi^2}{4}$

71. Using washers,

$V = \int_0^{\pi/4} \pi\left[(1 - \sin x)^2 - (1 - \cos x)^2\right] dx$

$= \pi \int_0^{\pi/4} \left[(1 - 2\sin x + \sin^2 x) - (1 - 2\cos x + \cos^2 x)\right] dx$

$= \pi \int_0^{\pi/4} (2\cos x - 2\sin x + \sin^2 x - \cos^2 x) \, dx$

$= \pi \int_0^{\pi/4} (2\cos x - 2\sin x - \cos 2x) \, dx = \pi \left[2\sin x + 2\cos x - \frac{1}{2}\sin 2x\right]_0^{\pi/4}$

$= \pi\left[\left(\sqrt{2} + \sqrt{2} - \frac{1}{2}\right) - (0 + 2 - 0)\right] = \pi\left(2\sqrt{2} - \frac{5}{2}\right)$

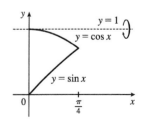

73. $s = f(t) = \int_0^t \sin \omega u \, \cos^2 \omega u \, du$. Let $y = \cos \omega u$ $\Rightarrow$ $dy = -\omega \sin \omega u \, du$. Then

$s = -\frac{1}{\omega}\int_1^{\cos \omega t} y^2 \, dy = -\frac{1}{\omega}\left[\frac{1}{3}y^3\right]_1^{\cos \omega t} = \frac{1}{3\omega}(1 - \cos^3 \omega t)$.

75. Just note that the integrand is odd [$f(-x) = -f(x)$].

Or: If $m \neq n$, calculate

$$\int_{-\pi}^{\pi} \sin mx \cos nx \, dx = \int_{-\pi}^{\pi} \tfrac{1}{2}[\sin(m-n)x + \sin(m+n)x] \, dx = \tfrac{1}{2}\left[-\frac{\cos(m-n)x}{m-n} - \frac{\cos(m+n)x}{m+n}\right]_{-\pi}^{\pi} = 0$$

If $m = n$, then the first term in each set of brackets is zero.

77. $\int_{-\pi}^{\pi} \cos mx \cos nx \, dx = \int_{-\pi}^{\pi} \tfrac{1}{2}[\cos(m-n)x + \cos(m+n)x] \, dx.$

If $m \neq n$, this is equal to $\dfrac{1}{2}\left[\dfrac{\sin(m-n)x}{m-n} + \dfrac{\sin(m+n)x}{m+n}\right]_{-\pi}^{\pi} = 0.$

If $m = n$, we get $\int_{-\pi}^{\pi} \tfrac{1}{2}[1 + \cos(m+n)x] \, dx = \left[\tfrac{1}{2}x\right]_{-\pi}^{\pi} + \left[\dfrac{\sin(m+n)x}{2(m+n)}\right]_{-\pi}^{\pi} = \pi + 0 = \pi.$

7.3 Trigonometric Substitution

1. (a) Use $x = \tan\theta$, where $-\pi/2 < \theta < \pi/2$, since the integrand contains the expression $\sqrt{1+x^2}$.

(b) $x = \tan\theta \;\Rightarrow\; dx = \sec^2\theta \, d\theta$ and $\sqrt{1+x^2} = \sqrt{1+\tan^2\theta} = \sqrt{\sec^2\theta} = |\sec\theta| = \sec\theta.$

Then $\displaystyle\int \frac{x^3}{\sqrt{1+x^2}} \, dx = \int \frac{\tan^3\theta}{\sec\theta} \sec^2\theta \, d\theta = \int \tan^3\theta \sec\theta \, d\theta.$

3. (a) Use $x = \sqrt{2}\sec\theta$, where $0 < \theta < \pi/2$ or $\pi < \theta < 3\pi/2$, since the integrand contains the expression $\sqrt{x^2-2}$.

(b) $x = \sqrt{2}\sec\theta \;\Rightarrow\; dx = \sqrt{2}\sec\theta\tan\theta \, d\theta$ and

$\sqrt{x^2-2} = \sqrt{2\sec^2\theta - 2} = \sqrt{2(\sec^2\theta - 1)} = \sqrt{2\tan^2\theta} = \sqrt{2}\,|\tan\theta| = \sqrt{2}\tan\theta.$

Then $\displaystyle\int \frac{x^2}{\sqrt{x^2-2}} \, dx = \int \frac{2\sec^2\theta}{\sqrt{2}\tan\theta} \sqrt{2}\sec\theta\tan\theta \, d\theta = \int 2\sec^3\theta \, d\theta.$

5. Let $x = \sin\theta$, where $-\pi/2 \leq \theta \leq \pi/2$. Then $dx = \cos\theta \, d\theta$ and

$\sqrt{1-x^2} = \sqrt{1-\sin^2\theta} = \sqrt{\cos^2\theta} = |\cos\theta| = \cos\theta.$ Thus,

$\displaystyle\int \frac{x^3}{\sqrt{1-x^2}} \, dx = \int \frac{\sin^3\theta}{\cos\theta} \cos\theta \, d\theta = \int (1-\cos^2\theta)\sin\theta \, d\theta$

$\overset{c}{=} \displaystyle\int (1-u^2)(-du) = \int (-1+u^2) \, du = -u + \tfrac{1}{3}u^3 + C$

$= -\cos\theta + \tfrac{1}{3}\cos^3\theta + C = -\sqrt{1-x^2} + \tfrac{1}{3}\left(\sqrt{1-x^2}\right)^3 + C$

7. Let $x = \tfrac{5}{2}\sec\theta$, where $0 \leq \theta \leq \tfrac{\pi}{2}$ or $\pi \leq \theta < \tfrac{3\pi}{2}$. Then $dx = \tfrac{5}{2}\sec\theta\tan\theta \, d\theta$

and $\sqrt{4x^2-25} = \sqrt{25\sec^2\theta - 25} = \sqrt{25\tan^2\theta} = 5\,|\tan\theta| = 5\tan\theta$ for

the relevant values of θ, so

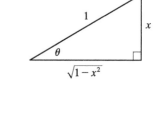

$\displaystyle\int \frac{\sqrt{4x^2-25}}{x} \, dx = \int \frac{5\tan\theta}{\tfrac{5}{2}\sec\theta}\left(\tfrac{5}{2}\sec\theta\tan\theta \, d\theta\right) = 5\int \tan^2\theta \, d\theta$

$= 5(\tan\theta - \theta) + C \quad$ [by Exercise 7.1.59 or integration by parts]

$= 5\left(\dfrac{\sqrt{4x^2-25}}{5} - \sec^{-1}\left(\dfrac{2x}{5}\right)\right) + C = \sqrt{4x^2-25} - 5\sec^{-1}\left(\tfrac{2}{5}x\right) + C$

9. Let $x = 4\tan\theta$, where $-\pi/2 < \theta < \pi/2$. Then $dx = 4\sec^2\theta\,d\theta$ and

$\sqrt{16+x^2} = \sqrt{16+16\tan^2\theta} = \sqrt{16\sec^2\theta} = 4|\sec\theta| = 4\sec\theta$. Thus,

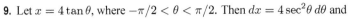

$\int x^3\sqrt{16+x^2}\,dx = \int 64\tan^3\theta\,(4\sec\theta)(4\sec^2\theta\,d\theta) = 1024\int \tan^3\theta\,\sec^3\theta\,d\theta$

$\quad = 1024\int \tan^2\theta\,\sec^2\theta\,\sec\theta\tan\theta\,d\theta = 1024\int(\sec^2\theta - 1)\sec^2\theta\,\sec\theta\tan\theta\,d\theta$

$\quad = 1024\int (u^2 - 1)u^2\,du \quad [u = \sec\theta,\,du = \sec\theta\tan\theta\,d\theta]$

$\quad = 1024\left(\dfrac{1}{5}u^5 - \dfrac{1}{3}u^3\right) + C = \dfrac{1024}{5}\left(\dfrac{\sqrt{16+x^2}}{4}\right)^5 - \dfrac{1024}{3}\left(\dfrac{\sqrt{16+x^2}}{4}\right)^3 + C$

$\quad = (16+x^2)^{3/2}\left(\dfrac{1}{5}(16+x^2) - \dfrac{16}{3}\right) + C = (16+x^2)^{3/2}\left(\dfrac{1}{5}x^2 - \dfrac{32}{15}\right) + C$

11. Let $x = \sec\theta$, where $0 \leq \theta \leq \frac{\pi}{2}$ or $\pi \leq \theta < \frac{3\pi}{2}$. Then $dx = \sec\theta\tan\theta\,d\theta$

and $\sqrt{x^2-1} = \sqrt{\sec^2\theta - 1} = \sqrt{\tan^2\theta} = |\tan\theta| = \tan\theta$ for the relevant

values of θ, so

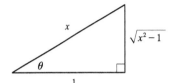

$\displaystyle\int \dfrac{\sqrt{x^2-1}}{x^4}\,dx = \int \dfrac{\tan\theta}{\sec^4\theta}\sec\theta\tan\theta\,d\theta = \int \tan^2\theta\,\cos^3\theta\,d\theta$

$\quad = \displaystyle\int \sin^2\theta\,\cos\theta\,d\theta \stackrel{s}{=} \int u^2\,du = \tfrac{1}{3}u^3 + C = \tfrac{1}{3}\sin^3\theta + C$

$\quad = \dfrac{1}{3}\left(\dfrac{\sqrt{x^2-1}}{x}\right)^3 + C = \dfrac{1}{3}\dfrac{(x^2-1)^{3/2}}{x^3} + C$

13. Let $x = a\tan\theta$, where $a > 0$ and $-\frac{\pi}{2} < \theta < \frac{\pi}{2}$. Then $dx = a\sec^2\theta\,d\theta$, $x = 0 \Rightarrow \theta = 0$, and $x = a \Rightarrow$

$\theta = \frac{\pi}{4}$. Thus,

$\displaystyle\int_0^a \dfrac{dx}{(a^2+x^2)^{3/2}} = \int_0^{\pi/4} \dfrac{a\sec^2\theta\,d\theta}{[a^2(1+\tan^2\theta)]^{3/2}} = \int_0^{\pi/4} \dfrac{a\sec^2\theta\,d\theta}{a^3\sec^3\theta} = \dfrac{1}{a^2}\int_0^{\pi/4} \cos\theta\,d\theta$

$\quad = \dfrac{1}{a^2}\Big[\sin\theta\Big]_0^{\pi/4} = \dfrac{1}{a^2}\left(\dfrac{\sqrt{2}}{2} - 0\right) = \dfrac{1}{\sqrt{2}\,a^2}$.

15. Let $x = \sec\theta$, so $dx = \sec\theta\tan\theta\,d\theta$, $x = 2 \Rightarrow \theta = \frac{\pi}{3}$, and

$x = 3 \Rightarrow \theta = \sec^{-1}3$. Then

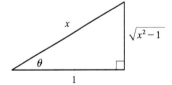

$\displaystyle\int_2^3 \dfrac{dx}{(x^2-1)^{3/2}} = \int_{\pi/3}^{\sec^{-1}3} \dfrac{\sec\theta\tan\theta\,d\theta}{\tan^3\theta} = \int_{\pi/3}^{\sec^{-1}3} \dfrac{\cos\theta}{\sin^2\theta}\,d\theta$

$\quad \stackrel{s}{=} \displaystyle\int_{\sqrt{3}/2}^{\sqrt{8}/3} \dfrac{1}{u^2}\,du = \left[-\dfrac{1}{u}\right]_{\sqrt{3}/2}^{\sqrt{8}/3} = \dfrac{-3}{\sqrt{8}} + \dfrac{2}{\sqrt{3}} = -\dfrac{3}{4}\sqrt{2} + \dfrac{2}{3}\sqrt{3}$

17. $\displaystyle\int_0^{1/2} x\sqrt{1-4x^2}\,dx = \int_1^0 u^{1/2}\left(-\dfrac{1}{8}\,du\right)\quad \begin{bmatrix} u = 1 - 4x^2, \\ du = -8x\,dx \end{bmatrix}$

$\quad = \dfrac{1}{8}\left[\dfrac{2}{3}u^{3/2}\right]_0^1 = \dfrac{1}{12}(1-0) = \dfrac{1}{12}$

19. Let $x = 3\sec\theta$, where $0 \le \theta < \frac{\pi}{2}$ or $\pi \le \theta < \frac{3\pi}{2}$. Then
$dx = 3\sec\theta \tan\theta \, d\theta$ and $\sqrt{x^2 - 9} = 3\tan\theta$, so

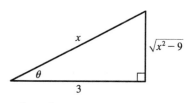

$$\int \frac{\sqrt{x^2-9}}{x^3}\,dx = \int \frac{3\tan\theta}{27\sec^3\theta}\, 3\sec\theta\tan\theta\, d\theta = \frac{1}{3}\int \frac{\tan^2\theta}{\sec^2\theta}\, d\theta$$

$$= \frac{1}{3}\int \sin^2\theta\, d\theta = \frac{1}{3}\int \frac{1}{2}(1-\cos 2\theta)\, d\theta = \frac{1}{6}\theta - \frac{1}{12}\sin 2\theta + C = \frac{1}{6}\theta - \frac{1}{6}\sin\theta\cos\theta + C$$

$$= \frac{1}{6}\sec^{-1}\!\left(\frac{x}{3}\right) - \frac{1}{6}\frac{\sqrt{x^2-9}}{x}\frac{3}{x} + C = \frac{1}{6}\sec^{-1}\!\left(\frac{x}{3}\right) - \frac{\sqrt{x^2-9}}{2x^2} + C$$

21. Let $x = a\sin\theta$, $dx = a\cos\theta\, d\theta$, $x = 0 \;\Rightarrow\; \theta = 0$ and $x = a \;\Rightarrow\; \theta = \frac{\pi}{2}$. Then

$$\int_0^a x^2\sqrt{a^2-x^2}\, dx = \int_0^{\pi/2} a^2\sin^2\theta\,(a\cos\theta)\,a\cos\theta\, d\theta = a^4\int_0^{\pi/2}\sin^2\theta\cos^2\theta\, d\theta$$

$$= a^4\int_0^{\pi/2}\left[\tfrac{1}{2}(2\sin\theta\cos\theta)\right]^2 d\theta = \frac{a^4}{4}\int_0^{\pi/2}\sin^2 2\theta\, d\theta = \frac{a^4}{4}\int_0^{\pi/2}\tfrac{1}{2}(1-\cos 4\theta)\, d\theta$$

$$= \frac{a^4}{8}\left[\theta - \tfrac{1}{4}\sin 4\theta\right]_0^{\pi/2} = \frac{a^4}{8}\left[\left(\tfrac{\pi}{2} - 0\right) - 0\right] = \frac{\pi}{16}a^4$$

23. Let $u = x^2 - 7$, so $du = 2x\, dx$. Then $\int \frac{x}{\sqrt{x^2-7}}\, dx = \frac{1}{2}\int \frac{1}{\sqrt{u}}\, du = \frac{1}{2}\cdot 2\sqrt{u} + C = \sqrt{x^2-7} + C$.

25. Let $x = \tan\theta$, where $-\frac{\pi}{2} < \theta < \frac{\pi}{2}$. Then $dx = \sec^2\theta\, d\theta$
and $\sqrt{1+x^2} = \sec\theta$, so

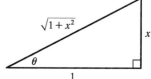

$$\int \frac{\sqrt{1+x^2}}{x}\, dx = \int \frac{\sec\theta}{\tan\theta}\sec^2\theta\, d\theta = \int \frac{\sec\theta}{\tan\theta}(1+\tan^2\theta)\, d\theta$$

$$= \int(\csc\theta + \sec\theta\tan\theta)\, d\theta$$

$$= \ln|\csc\theta - \cot\theta| + \sec\theta + C \qquad \text{[by Exercise 7.2.41]}$$

$$= \ln\left|\frac{\sqrt{1+x^2}}{x} - \frac{1}{x}\right| + \frac{\sqrt{1+x^2}}{1} + C = \ln\left|\frac{\sqrt{1+x^2}-1}{x}\right| + \sqrt{1+x^2} + C$$

27. Let $x = \frac{3}{5}\sin\theta$, so $dx = \frac{3}{5}\cos\theta\, d\theta$, $x = 0 \;\Rightarrow\; \theta = 0$, and $x = 0.6 \;\Rightarrow\; \theta = \frac{\pi}{2}$. Then

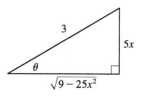

$$\int_0^{0.6} \frac{x^2}{\sqrt{9-25x^2}}\, dx = \int_0^{\pi/2} \frac{\left(\frac{3}{5}\right)^2\sin^2\theta}{3\cos\theta}\left(\tfrac{3}{5}\cos\theta\, d\theta\right) = \frac{9}{125}\int_0^{\pi/2}\sin^2\theta\, d\theta$$

$$= \frac{9}{125}\int_0^{\pi/2}\tfrac{1}{2}(1-\cos 2\theta)\, d\theta = \frac{9}{250}\left[\theta - \tfrac{1}{2}\sin 2\theta\right]_0^{\pi/2}$$

$$= \frac{9}{250}\left[\left(\tfrac{\pi}{2} - 0\right) - 0\right] = \frac{9}{500}\pi$$

29. $\displaystyle\int \frac{dx}{\sqrt{x^2+2x+5}} = \int \frac{dx}{\sqrt{(x+1)^2+4}} = \int \frac{2\sec^2\theta\, d\theta}{\sqrt{4\tan^2\theta+4}} \quad \begin{bmatrix} x+1 = 2\tan\theta,\\ dx = 2\sec^2\theta\, d\theta \end{bmatrix}$

$$= \int \frac{2\sec^2\theta\, d\theta}{2\sec\theta} = \int \sec\theta\, d\theta = \ln|\sec\theta + \tan\theta| + C_1$$

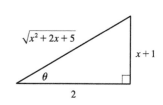

$$= \ln\left|\frac{\sqrt{x^2+2x+5}}{2} + \frac{x+1}{2}\right| + C_1,$$

or $\ln\left|\sqrt{x^2+2x+5} + x+1\right| + C$, where $C = C_1 - \ln 2$.

31. $\int x^2\sqrt{3+2x-x^2}\,dx = \int x^2\sqrt{4-(x^2+2x+1)}\,dx = \int x^2\sqrt{2^2-(x-1)^2}\,dx$

$\quad = \int(1+2\sin\theta)^2\sqrt{4\cos^2\theta}\,2\cos\theta\,d\theta \quad \begin{bmatrix} x-1=2\sin\theta, \\ dx = 2\cos\theta\,d\theta \end{bmatrix}$

$\quad = \int(1+4\sin\theta+4\sin^2\theta)\,4\cos^2\theta\,d\theta$

$\quad = 4\int(\cos^2\theta + 4\sin\theta\cos^2\theta + 4\sin^2\theta\cos^2\theta)\,d\theta$

$\quad = 4\int\tfrac{1}{2}(1+\cos 2\theta)\,d\theta + 4\int 4\sin\theta\cos^2\theta\,d\theta + 4\int(2\sin\theta\cos\theta)^2\,d\theta$

$\quad = 2\int(1+\cos 2\theta)\,d\theta + 16\int\sin\theta\cos^2\theta\,d\theta + 4\int\sin^2 2\theta\,d\theta$

$\quad = 2\left(\theta + \tfrac{1}{2}\sin 2\theta\right) + 16\left(-\tfrac{1}{3}\cos^3\theta\right) + 4\int\tfrac{1}{2}(1-\cos 4\theta)\,d\theta$

$\quad = 2\theta + \sin 2\theta - \tfrac{16}{3}\cos^3\theta + 2\left(\theta - \tfrac{1}{4}\sin 4\theta\right) + C$

$\quad = 4\theta - \tfrac{1}{2}\sin 4\theta + \sin 2\theta - \tfrac{16}{3}\cos^3\theta + C$

$\quad = 4\theta - \tfrac{1}{2}(2\sin 2\theta\cos 2\theta) + \sin 2\theta - \tfrac{16}{3}\cos^3\theta + C$

$\quad = 4\theta + \sin 2\theta(1 - \cos 2\theta) - \tfrac{16}{3}\cos^3\theta + C$

$\quad = 4\theta + (2\sin\theta\cos\theta)(2\sin^2\theta) - \tfrac{16}{3}\cos^3\theta + C$

$\quad = 4\theta + 4\sin^3\theta\cos\theta - \tfrac{16}{3}\cos^3\theta + C$

$\quad = 4\sin^{-1}\!\left(\dfrac{x-1}{2}\right) + 4\left(\dfrac{x-1}{2}\right)^3\dfrac{\sqrt{3+2x-x^2}}{2} - \dfrac{16}{3}\dfrac{(3+2x-x^2)^{3/2}}{2^3} + C$

$\quad = 4\sin^{-1}\!\left(\dfrac{x-1}{2}\right) + \tfrac{1}{4}(x-1)^3\sqrt{3+2x-x^2} - \tfrac{2}{3}(3+2x-x^2)^{3/2} + C$

33. $x^2 + 2x = (x^2+2x+1) - 1 = (x+1)^2 - 1$. Let $x+1 = 1\sec\theta$, so $dx = \sec\theta\tan\theta\,d\theta$ and $\sqrt{x^2+2x} = \tan\theta$. Then

$\int\sqrt{x^2+2x}\,dx = \int\tan\theta(\sec\theta\tan\theta\,d\theta) = \int\tan^2\theta\sec\theta\,d\theta$

$\quad = \int(\sec^2\theta - 1)\sec\theta\,d\theta = \int\sec^3\theta\,d\theta - \int\sec\theta\,d\theta$

$\quad = \tfrac{1}{2}\sec\theta\tan\theta + \tfrac{1}{2}\ln|\sec\theta+\tan\theta| - \ln|\sec\theta+\tan\theta| + C$

$\quad = \tfrac{1}{2}\sec\theta\tan\theta - \tfrac{1}{2}\ln|\sec\theta+\tan\theta| + C = \tfrac{1}{2}(x+1)\sqrt{x^2+2x} - \tfrac{1}{2}\ln\left|x+1+\sqrt{x^2+2x}\right| + C$

35. Let $u = x^2$, $du = 2x\,dx$. Then

$\int x\sqrt{1-x^4}\,dx = \int\sqrt{1-u^2}\left(\tfrac{1}{2}du\right) = \tfrac{1}{2}\int\cos\theta\cdot\cos\theta\,d\theta \quad \begin{bmatrix} \text{where } u = \sin\theta,\ du = \cos\theta\,d\theta, \\ \text{and } \sqrt{1-u^2} = \cos\theta \end{bmatrix}$

$\quad = \tfrac{1}{2}\int\tfrac{1}{2}(1+\cos 2\theta)\,d\theta = \tfrac{1}{4}\theta + \tfrac{1}{8}\sin 2\theta + C = \tfrac{1}{4}\theta + \tfrac{1}{4}\sin\theta\cos\theta + C$

$\quad = \tfrac{1}{4}\sin^{-1}u + \tfrac{1}{4}u\sqrt{1-u^2} + C = \tfrac{1}{4}\sin^{-1}(x^2) + \tfrac{1}{4}x^2\sqrt{1-x^4} + C$

37. (a) Let $x = a\tan\theta$, where $-\frac{\pi}{2} < \theta < \frac{\pi}{2}$. Then $\sqrt{x^2+a^2} = a\sec\theta$ and

$$\int \frac{dx}{\sqrt{x^2+a^2}} = \int \frac{a\sec^2\theta\, d\theta}{a\sec\theta} = \int \sec\theta\, d\theta = \ln|\sec\theta + \tan\theta| + C_1 = \ln\left|\frac{\sqrt{x^2+a^2}}{a} + \frac{x}{a}\right| + C_1$$

$$= \ln\left(x + \sqrt{x^2+a^2}\right) + C \quad \text{where } C = C_1 - \ln|a|$$

(b) Let $x = a\sinh t$, so that $dx = a\cosh t\, dt$ and $\sqrt{x^2+a^2} = a\cosh t$. Then

$$\int \frac{dx}{\sqrt{x^2+a^2}} = \int \frac{a\cosh t\, dt}{a\cosh t} = t + C = \sinh^{-1}\frac{x}{a} + C.$$

39. The average value of $f(x) = \sqrt{x^2-1}/x$ on the interval $[1, 7]$ is

$$\frac{1}{7-1}\int_1^7 \frac{\sqrt{x^2-1}}{x}\, dx = \frac{1}{6}\int_0^\alpha \frac{\tan\theta}{\sec\theta}\cdot\sec\theta\tan\theta\, d\theta \quad \left[\begin{array}{l}\text{where } x = \sec\theta,\, dx = \sec\theta\tan\theta\, d\theta, \\ \sqrt{x^2-1} = \tan\theta,\text{ and } \alpha = \sec^{-1}7\end{array}\right]$$

$$= \tfrac{1}{6}\int_0^\alpha \tan^2\theta\, d\theta = \tfrac{1}{6}\int_0^\alpha (\sec^2\theta - 1)\, d\theta = \tfrac{1}{6}\left[\tan\theta - \theta\right]_0^\alpha$$

$$= \tfrac{1}{6}(\tan\alpha - \alpha) = \tfrac{1}{6}\left(\sqrt{48} - \sec^{-1}7\right)$$

41. Area of $\triangle POQ = \tfrac{1}{2}(r\cos\theta)(r\sin\theta) = \tfrac{1}{2}r^2\sin\theta\cos\theta$. Area of region $PQR = \int_{r\cos\theta}^r \sqrt{r^2-x^2}\, dx$.

Let $x = r\cos u \Rightarrow dx = -r\sin u\, du$ for $0 \le u \le \tfrac{\pi}{2}$. Then we obtain

$$\int \sqrt{r^2-x^2}\, dx = \int r\sin u\,(-r\sin u)\, du = -r^2\int \sin^2 u\, du = -\tfrac{1}{2}r^2(u - \sin u\cos u) + C$$

$$= -\tfrac{1}{2}r^2\cos^{-1}(x/r) + \tfrac{1}{2}x\sqrt{r^2-x^2} + C$$

so

$$\text{area of region } PQR = \tfrac{1}{2}\left[-r^2\cos^{-1}(x/r) + x\sqrt{r^2-x^2}\right]_{r\cos\theta}^r$$

$$= \tfrac{1}{2}\left[0 - (-r^2\theta + r\cos\theta\, r\sin\theta)\right] = \tfrac{1}{2}r^2\theta - \tfrac{1}{2}r^2\sin\theta\cos\theta$$

and thus, (area of sector POR) = (area of $\triangle POQ$) + (area of region PQR) = $\tfrac{1}{2}r^2\theta$.

43. Use disks about the x-axis:

$$V = \int_0^3 \pi\left(\frac{9}{x^2+9}\right)^2 dx = 81\pi\int_0^3 \frac{1}{(x^2+9)^2}\, dx$$

Let $x = 3\tan\theta$, so $dx = 3\sec^2\theta\, d\theta$, $x = 0 \Rightarrow \theta = 0$ and $x = 3 \Rightarrow \theta = \tfrac{\pi}{4}$. Thus,

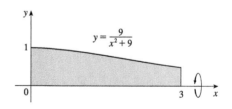

$$V = 81\pi\int_0^{\pi/4} \frac{1}{(9\sec^2\theta)^2}\, 3\sec^2\theta\, d\theta = 3\pi\int_0^{\pi/4} \cos^2\theta\, d\theta = 3\pi\int_0^{\pi/4} \tfrac{1}{2}(1 + \cos 2\theta)\, d\theta$$

$$= \tfrac{3\pi}{2}\left[\theta + \tfrac{1}{2}\sin 2\theta\right]_0^{\pi/4} = \tfrac{3\pi}{2}\left[\left(\tfrac{\pi}{4} + \tfrac{1}{2}\right) - 0\right] = \tfrac{3}{8}\pi^2 + \tfrac{3}{4}\pi$$

45. (a) Let $t = a \sin \theta$, $dt = a \cos \theta \, d\theta$, $t = 0 \;\Rightarrow\; \theta = 0$ and $t = x \;\Rightarrow\;$
$\theta = \sin^{-1}(x/a)$. Then

$$\int_0^x \sqrt{a^2 - t^2}\, dt = \int_0^{\sin^{-1}(x/a)} a \cos\theta \, (a\cos\theta \, d\theta) = a^2 \int_0^{\sin^{-1}(x/a)} \cos^2\theta \, d\theta$$

$$= \frac{a^2}{2} \int_0^{\sin^{-1}(x/a)} (1 + \cos 2\theta)\, d\theta = \frac{a^2}{2}\Big[\theta + \tfrac{1}{2}\sin 2\theta\Big]_0^{\sin^{-1}(x/a)} = \frac{a^2}{2}\Big[\theta + \sin\theta \cos\theta\Big]_0^{\sin^{-1}(x/a)}$$

$$= \frac{a^2}{2}\left[\left(\sin^{-1}\!\left(\frac{x}{a}\right) + \frac{x}{a} \cdot \frac{\sqrt{a^2 - x^2}}{a}\right) - 0\right] = \tfrac{1}{2}a^2 \sin^{-1}(x/a) + \tfrac{1}{2}x\sqrt{a^2 - x^2}$$

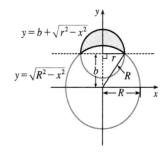

(b) The integral $\int_0^x \sqrt{a^2 - t^2}\, dt$ represents the area under the curve $y = \sqrt{a^2 - t^2}$ between the vertical lines $t = 0$ and $t = x$. The figure shows that this area consists of a triangular region and a sector of the circle $t^2 + y^2 = a^2$. The triangular region has base x and height $\sqrt{a^2 - x^2}$, so its area is $\tfrac{1}{2}x\sqrt{a^2 - x^2}$. The sector has area $\tfrac{1}{2}a^2\theta = \tfrac{1}{2}a^2 \sin^{-1}(x/a)$.

47. We use cylindrical shells and assume that $R > r$. $x^2 = r^2 - (y - R)^2 \;\Rightarrow\; x = \pm\sqrt{r^2 - (y-R)^2}$,
so $g(y) = 2\sqrt{r^2 - (y-R)^2}$ and

$$V = \int_{R-r}^{R+r} 2\pi y \cdot 2\sqrt{r^2 - (y-R)^2}\, dy = \int_{-r}^{r} 4\pi (u + R)\sqrt{r^2 - u^2}\, du \quad \text{[where } u = y - R\text{]}$$

$$= 4\pi \int_{-r}^{r} u\sqrt{r^2 - u^2}\, du + 4\pi R \int_{-r}^{r} \sqrt{r^2 - u^2}\, du \quad \begin{bmatrix}\text{where } u = r\sin\theta,\, du = r\cos\theta \, d\theta \\ \text{in the second integral}\end{bmatrix}$$

$$= 4\pi\left[-\tfrac{1}{3}(r^2 - u^2)^{3/2}\right]_{-r}^{r} + 4\pi R \int_{-\pi/2}^{\pi/2} r^2 \cos^2\theta \, d\theta = -\tfrac{4\pi}{3}(0 - 0) + 4\pi Rr^2 \int_{-\pi/2}^{\pi/2} \cos^2\theta \, d\theta$$

$$= 2\pi Rr^2 \int_{-\pi/2}^{\pi/2} (1 + \cos 2\theta)\, d\theta = 2\pi Rr^2 \Big[\theta + \tfrac{1}{2}\sin 2\theta\Big]_{-\pi/2}^{\pi/2} = 2\pi^2 Rr^2$$

Another method: Use washers instead of shells, so $V = 8\pi R \int_0^r \sqrt{r^2 - y^2}\, dy$ as in Exercise 6.2.75(a), but evaluate the integral using $y = r\sin\theta$.

49. Let the equation of the large circle be $x^2 + y^2 = R^2$. Then the equation of the small circle is $x^2 + (y-b)^2 = r^2$, where $b = \sqrt{R^2 - r^2}$ is the distance between the centers of the circles. The desired area is

$$A = \int_{-r}^{r}\left[\left(b + \sqrt{r^2 - x^2}\right) - \sqrt{R^2 - x^2}\right] dx$$

$$= 2\int_0^r \left(b + \sqrt{r^2 - x^2} - \sqrt{R^2 - x^2}\right) dx$$

$$= 2\int_0^r b\, dx + 2\int_0^r \sqrt{r^2 - x^2}\, dx - 2\int_0^r \sqrt{R^2 - x^2}\, dx$$

The first integral is just $2br = 2r\sqrt{R^2 - r^2}$. The second integral represents the area of a quarter-circle of radius r, so its value is $\tfrac{1}{4}\pi r^2$. To evaluate the other integral, note that

$$\int \sqrt{a^2 - x^2}\, dx = \int a^2 \cos^2\theta \, d\theta \quad [x = a\sin\theta,\, dx = a\cos\theta \, d\theta] = \left(\tfrac{1}{2}a^2\right)\int (1 + \cos 2\theta)\, d\theta$$

$$= \tfrac{1}{2}a^2\left(\theta + \tfrac{1}{2}\sin 2\theta\right) + C = \tfrac{1}{2}a^2(\theta + \sin\theta \cos\theta) + C$$

$$= \frac{a^2}{2}\arcsin\!\left(\frac{x}{a}\right) + \frac{a^2}{2}\left(\frac{x}{a}\right)\frac{\sqrt{a^2 - x^2}}{a} + C = \frac{a^2}{2}\arcsin\!\left(\frac{x}{a}\right) + \frac{x}{2}\sqrt{a^2 - x^2} + C$$

[continued]

Thus, the desired area is

$$A = 2r\sqrt{R^2 - r^2} + 2\left(\tfrac{1}{4}\pi r^2\right) - \left[R^2 \arcsin(x/R) + x\sqrt{R^2 - x^2}\right]_0^r$$
$$= 2r\sqrt{R^2 - r^2} + \tfrac{1}{2}\pi r^2 - \left[R^2 \arcsin(r/R) + r\sqrt{R^2 - r^2}\right] = r\sqrt{R^2 - r^2} + \tfrac{\pi}{2}r^2 - R^2 \arcsin(r/R)$$

7.4 Integration of Rational Functions by Partial Fractions

1. (a) $\dfrac{1}{(x-3)(x+5)} = \dfrac{A}{x-3} + \dfrac{B}{x+5}$

(b) $\dfrac{2x+5}{(x-2)^2(x^2+2)} = \dfrac{A}{x-2} + \dfrac{B}{(x-2)^2} + \dfrac{Cx+D}{x^2+2}$

3. (a) $\dfrac{x^2+4}{x^3-3x^2+2x} = \dfrac{x^2+4}{x(x^2-3x+2)} = \dfrac{x^2+4}{x(x-1)(x-2)} = \dfrac{A}{x} + \dfrac{B}{x-1} + \dfrac{C}{x-2}$

(b) $\dfrac{x^3+x}{x(2x-1)^2(x^2+3)^2} = \dfrac{A}{x} + \dfrac{B}{2x-1} + \dfrac{C}{(2x-1)^2} + \dfrac{Dx+E}{x^2+3} + \dfrac{Fx+G}{(x^2+3)^2}$

5. (a) $\dfrac{x^5+1}{(x^2-x)(x^4+2x^2+1)} = \dfrac{x^5+1}{x(x-1)(x^2+1)^2} = \dfrac{A}{x} + \dfrac{B}{x-1} + \dfrac{Cx+D}{x^2+1} + \dfrac{Ex+F}{(x^2+1)^2}$

(b) $\dfrac{x^2}{x^2+x-6} = 1 + \dfrac{-x+6}{x^2+x-6} = 1 + \dfrac{-x+6}{(x-2)(x+3)} = 1 + \dfrac{A}{x-2} + \dfrac{B}{x+3}$

7. $\dfrac{5}{(x-1)(x+4)} = \dfrac{A}{x-1} + \dfrac{B}{x+4}$. Multiply both sides by $(x-1)(x+4)$ to get $5 = A(x+4) + B(x-1)$ $\Rightarrow$

$5 = (A+B)x + (4A-B)$. The coefficients of x must be equal and the constant terms are also equal, so $A+B = 0$ and $4A - B = 5$. Adding the equations together gives $5A = 5$ $\Leftrightarrow$ $A = 1$, and hence $B = -1$. Thus,

$$\int \dfrac{5}{(x-1)(x+4)}\, dx = \int \left(\dfrac{1}{x-1} - \dfrac{1}{x+4}\right) dx = \ln|x-1| - \ln|x+4| + C.$$

9. $\dfrac{5x+1}{(2x+1)(x-1)} = \dfrac{A}{2x+1} + \dfrac{B}{x-1}$. Multiply both sides by $(2x+1)(x-1)$ to get $5x+1 = A(x-1) + B(2x+1)$ $\Rightarrow$

$5x + 1 = Ax - A + 2Bx + B$ $\Rightarrow$ $5x+1 = (A+2B)x + (-A+B)$.

The coefficients of x must be equal and the constant terms are also equal, so $A + 2B = 5$ and $-A + B = 1$. Adding these equations gives us $3B = 6$ $\Leftrightarrow$ $B = 2$, and hence $A = 1$. Thus,

$$\int \dfrac{5x+1}{(2x+1)(x-1)}\, dx = \int \left(\dfrac{1}{2x+1} + \dfrac{2}{x-1}\right) dx = \tfrac{1}{2}\ln|2x+1| + 2\ln|x-1| + C.$$

Another method: Substituting 1 for x in the equation $5x + 1 = A(x-1) + B(2x+1)$ gives $6 = 3B$ $\Leftrightarrow$ $B = 2$. Substituting $-\tfrac{1}{2}$ for x gives $-\tfrac{3}{2} = -\tfrac{3}{2}A$ $\Leftrightarrow$ $A = 1$.

11. $\dfrac{2}{2x^2+3x+1} = \dfrac{2}{(2x+1)(x+1)} = \dfrac{A}{2x+1} + \dfrac{B}{x+1}$. Multiply both sides by $(2x+1)(x+1)$ to get

$2 = A(x+1) + B(2x+1)$. The coefficients of x must be equal and the constant terms are also equal, so $A + 2B = 0$ and

SECTION 7.4 INTEGRATION OF RATIONAL FUNCTIONS BY PARTIAL FRACTIONS □ 367

$A + B = 2$. Subtracting the second equation from the first gives $B = -2$, and hence, $A = 4$. Thus,

$$\int_0^1 \frac{2}{2x^2 + 3x + 1} \, dx = \int_0^1 \left(\frac{4}{2x+1} - \frac{2}{x+1} \right) dx = \left[\frac{4}{2} \ln|2x+1| - 2\ln|x+1| \right]_0^1 = (2\ln 3 - 2\ln 2) - 0 = 2\ln \frac{3}{2}.$$

Another method: Substituting -1 for x in the equation $2 = A(x+1) + B(2x+1)$ gives $2 = -B \iff B = -2$.
Substituting $-\frac{1}{2}$ for x gives $2 = \frac{1}{2}A \iff A = 4$.

13. $\dfrac{1}{x(x-a)} = \dfrac{A}{x} + \dfrac{B}{x-a}$. Multiply both sides by $x(x-a)$ to get $1 = A(x-a) + Bx \Rightarrow 1 = (A+B)x + (-aA)$.

The coefficients of x must be equal and the constant terms are also equal, so $A + B = 0$ and $-aA = 1$. The second equation gives $A = -1/a$, which after substituting in the first equation gives $B = 1/a$. Thus,

$$\int \frac{1}{x(x-a)} \, dx = \int \left(-\frac{1/a}{x} + \frac{1/a}{x-a} \right) dx = -\frac{1}{a} \ln|x| + \frac{1}{a} \ln|x-a| + C.$$

15. $\dfrac{x^2}{x-1} = \dfrac{(x^2 - 1) + 1}{x - 1} = \dfrac{(x+1)(x-1) + 1}{x-1} = x + 1 + \dfrac{1}{x-1}$. [This result can also be obtained using long division.]

Thus, $\displaystyle\int \frac{x^2}{x-1} \, dx = \int \left(x + 1 + \frac{1}{x-1} \right) dx = \frac{1}{2} x^2 + x + \ln|x-1| + C$.

17. $\dfrac{4y^2 - 7y - 12}{y(y+2)(y-3)} = \dfrac{A}{y} + \dfrac{B}{y+2} + \dfrac{C}{y-3} \Rightarrow 4y^2 - 7y - 12 = A(y+2)(y-3) + By(y-3) + Cy(y+2)$. Setting

$y = 0$ gives $-12 = -6A$, so $A = 2$. Setting $y = -2$ gives $18 = 10B$, so $B = \frac{9}{5}$. Setting $y = 3$ gives $3 = 15C$, so $C = \frac{1}{5}$.

Now

$$\int_1^2 \frac{4y^2 - 7y - 12}{y(y+2)(y-3)} \, dy = \int_1^2 \left(\frac{2}{y} + \frac{9/5}{y+2} + \frac{1/5}{y-3} \right) dy = \left[2\ln|y| + \frac{9}{5} \ln|y+2| + \frac{1}{5} \ln|y-3| \right]_1^2$$

$$= 2\ln 2 + \frac{9}{5} \ln 4 + \frac{1}{5} \ln 1 - 2\ln 1 - \frac{9}{5} \ln 3 - \frac{1}{5} \ln 2$$

$$= 2\ln 2 + \frac{18}{5} \ln 2 - \frac{1}{5} \ln 2 - \frac{9}{5} \ln 3 = \frac{27}{5} \ln 2 - \frac{9}{5} \ln 3 = \frac{9}{5}(3\ln 2 - \ln 3) = \frac{9}{5} \ln \frac{8}{3}$$

19. $\dfrac{x^2 + x + 1}{(x+1)^2(x+2)} = \dfrac{A}{x+1} + \dfrac{B}{(x+1)^2} + \dfrac{C}{x+2}$. Multiplying both sides by $(x+1)^2(x+2)$ gives

$x^2 + x + 1 = A(x+1)(x+2) + B(x+2) + C(x+1)^2$. Substituting -1 for x gives $1 = B$. Substituting -2 for x gives

$3 = C$. Equating coefficients of x^2 gives $1 = A + C = A + 3$, so $A = -2$. Thus,

$$\int_0^1 \frac{x^2 + x + 1}{(x+1)^2(x+2)} \, dx = \int_0^1 \left(\frac{-2}{x+1} + \frac{1}{(x+1)^2} + \frac{3}{x+2} \right) dx = \left[-2\ln|x+1| - \frac{1}{x+1} + 3\ln|x+2| \right]_0^1$$

$$= (-2\ln 2 - \frac{1}{2} + 3\ln 3) - (0 - 1 + 3\ln 2) = \frac{1}{2} - 5\ln 2 + 3\ln 3, \text{ or } \frac{1}{2} + \ln \frac{27}{32}$$

21. $\dfrac{1}{(t^2 - 1)^2} = \dfrac{1}{(t+1)^2(t-1)^2} = \dfrac{A}{t+1} + \dfrac{B}{(t+1)^2} + \dfrac{C}{t-1} + \dfrac{D}{(t-1)^2}$. Multiplying both sides by $(t+1)^2(t-1)^2$ gives

$1 = A(t+1)(t-1)^2 + B(t-1)^2 + C(t-1)(t+1)^2 + D(t+1)^2$. Substituting 1 for t gives $1 = 4D \iff D = \frac{1}{4}$.

Substituting -1 for t gives $1 = 4B \iff B = \frac{1}{4}$. Substituting 0 for t gives $1 = A + B - C + D = A + \frac{1}{4} - C + \frac{1}{4}$, so

$\frac{1}{2} = A - C$. Equating coefficients of t^3 gives $0 = A + C$. Adding the last two equations gives $2A = \frac{1}{2} \Leftrightarrow A = \frac{1}{4}$, and so $C = -\frac{1}{4}$. Thus,

$$\int \frac{dt}{(t^2-1)^2} = \int \left[\frac{1/4}{t+1} + \frac{1/4}{(t+1)^2} - \frac{1/4}{t-1} + \frac{1/4}{(t-1)^2} \right] dt$$

$$= \frac{1}{4} \left[\ln|t+1| - \frac{1}{t+1} - \ln|t-1| - \frac{1}{t-1} \right] + C, \text{ or } \frac{1}{4} \left(\ln\left|\frac{t+1}{t-1}\right| + \frac{2t}{1-t^2} \right) + C$$

23. $\dfrac{10}{(x-1)(x^2+9)} = \dfrac{A}{x-1} + \dfrac{Bx+C}{x^2+9}$. Multiply both sides by $(x-1)(x^2+9)$ to get

$10 = A(x^2+9) + (Bx+C)(x-1)$ ($\star$). Substituting 1 for x gives $10 = 10A \Leftrightarrow A = 1$. Substituting 0 for x gives $10 = 9A - C \Rightarrow C = 9(1) - 10 = -1$. The coefficients of the x^2-terms in ($\star$) must be equal, so $0 = A + B \Rightarrow B = -1$. Thus,

$$\int \frac{10}{(x-1)(x^2+9)} dx = \int \left(\frac{1}{x-1} + \frac{-x-1}{x^2+9} \right) dx = \int \left(\frac{1}{x-1} - \frac{x}{x^2+9} - \frac{1}{x^2+9} \right) dx$$

$$= \ln|x-1| - \tfrac{1}{2}\ln(x^2+9) - \tfrac{1}{3}\tan^{-1}\left(\tfrac{x}{3}\right) + C$$

In the second term we used the substitution $u = x^2 + 9$ and in the last term we used Formula 10.

25. $\dfrac{x^3 - 4x + 1}{x^2 - 3x + 2} = x + 3 + \dfrac{3x-5}{(x-1)(x-2)}$. Write $\dfrac{3x-5}{(x-1)(x-2)} = \dfrac{A}{x-1} + \dfrac{B}{x-2}$. Multiplying both sides by $(x-1)(x-2)$ gives $3x - 5 = A(x-2) + B(x-1)$. Substituting 2 for x gives $1 = B$. Substituting 1 for x gives $-2 = -A \Leftrightarrow A = 2$. Thus,

$$\int_{-1}^{0} \frac{x^3 - 4x + 1}{x^2 - 3x + 2} dx = \int_{-1}^{0} \left(x + 3 + \frac{2}{x-1} + \frac{1}{x-2} \right) dx = \left[\tfrac{1}{2}x^2 + 3x + 2\ln|x-1| + \ln|x-2| \right]_{-1}^{0}$$

$$= (0 + 0 + 0 + \ln 2) - (\tfrac{1}{2} - 3 + 2\ln 2 + \ln 3) = \tfrac{5}{2} - \ln 2 - \ln 3, \text{ or } \tfrac{5}{2} - \ln 6$$

27. $\dfrac{4x}{x^3 + x^2 + x + 1} = \dfrac{4x}{x^2(x+1) + 1(x+1)} = \dfrac{4x}{(x+1)(x^2+1)} = \dfrac{A}{x+1} + \dfrac{Bx+C}{x^2+1}$. Multiply both sides by $(x+1)(x^2+1)$ to get $4x = A(x^2+1) + (Bx+C)(x+1) \Leftrightarrow 4x = Ax^2 + A + Bx^2 + Bx + Cx + C \Leftrightarrow 4x = (A+B)x^2 + (B+C)x + (A+C)$. Comparing coefficients gives us the following system of equations:

$$A + B = 0 \quad \textbf{(1)} \qquad B + C = 4 \quad \textbf{(2)} \qquad A + C = 0 \quad \textbf{(3)}$$

Subtracting equation **(1)** from equation **(2)** gives us $-A + C = 4$, and adding that equation to equation **(3)** gives us $2C = 4 \Leftrightarrow C = 2$, and hence $A = -2$ and $B = 2$. Thus,

$$\int \frac{4x}{x^3+x^2+x+1} dx = \int \left(\frac{-2}{x+1} + \frac{2x+2}{x^2+1} \right) dx = \int \left(\frac{-2}{x+1} + \frac{2x}{x^2+1} + \frac{2}{x^2+1} \right) dx$$

$$= -2\ln|x+1| + \ln(x^2+1) + 2\tan^{-1}x + C$$

29. $\dfrac{x^3 + 4x + 3}{x^4 + 5x^2 + 4} = \dfrac{x^3 + 4x + 3}{(x^2 + 1)(x^2 + 4)} = \dfrac{Ax + B}{x^2 + 1} + \dfrac{Cx + D}{x^2 + 4}$. Multiply both sides by $(x^2 + 1)(x^2 + 4)$

to get $x^3 + 4x + 3 = (Ax + B)(x^2 + 4) + (Cx + D)(x^2 + 1)$ $\Leftrightarrow$

$x^3 + 4x + 3 = Ax^3 + Bx^2 + 4Ax + 4B + Cx^3 + Dx^2 + Cx + D$ $\Leftrightarrow$

$x^3 + 4x + 3 = (A + C)x^3 + (B + D)x^2 + (4A + C)x + (4B + D)$. Comparing coefficients gives us the following system of equations:

$\qquad A + C = 1 \quad \text{(1)} \qquad\qquad B + D = 0 \quad \text{(2)} \qquad\qquad 4A + C = 4 \quad \text{(3)} \qquad\qquad 4B + D = 3 \quad \text{(4)}$

Subtracting equation **(1)** from equation **(3)** gives us $A = 1$ and hence, $C = 0$. Subtracting equation **(2)** from equation **(4)** gives us $B = 1$ and hence, $D = -1$. Thus,

$$\int \dfrac{x^3 + 4x + 3}{x^4 + 5x^2 + 4}\, dx = \int \left(\dfrac{x + 1}{x^2 + 1} + \dfrac{-1}{x^2 + 4} \right) dx = \int \left(\dfrac{x}{x^2 + 1} + \dfrac{1}{x^2 + 1} - \dfrac{1}{x^2 + 4} \right) dx$$

$$= \dfrac{1}{2}\ln(x^2 + 1) + \tan^{-1} x - \dfrac{1}{2}\tan^{-1}\left(\dfrac{x}{2}\right) + C$$

31. $\displaystyle\int \dfrac{x + 4}{x^2 + 2x + 5}\, dx = \int \dfrac{x + 1}{x^2 + 2x + 5}\, dx + \int \dfrac{3}{x^2 + 2x + 5}\, dx = \dfrac{1}{2}\int \dfrac{(2x + 2)\, dx}{x^2 + 2x + 5} + \int \dfrac{3\, dx}{(x + 1)^2 + 4}$

$= \dfrac{1}{2}\ln|x^2 + 2x + 5| + 3\displaystyle\int \dfrac{2\, du}{4(u^2 + 1)} \qquad \left[\begin{array}{l}\text{where } x + 1 = 2u, \\ \text{and } dx = 2\, du\end{array}\right]$

$= \dfrac{1}{2}\ln(x^2 + 2x + 5) + \dfrac{3}{2}\tan^{-1} u + C = \dfrac{1}{2}\ln(x^2 + 2x + 5) + \dfrac{3}{2}\tan^{-1}\left(\dfrac{x + 1}{2}\right) + C$

33. $\dfrac{1}{x^3 - 1} = \dfrac{1}{(x - 1)(x^2 + x + 1)} = \dfrac{A}{x - 1} + \dfrac{Bx + C}{x^2 + x + 1}$ $\Rightarrow$ $1 = A(x^2 + x + 1) + (Bx + C)(x - 1)$.

Take $x = 1$ to get $A = \dfrac{1}{3}$. Equating coefficients of x^2 and then comparing the constant terms, we get $0 = \dfrac{1}{3} + B$, $1 = \dfrac{1}{3} - C$, so $B = -\dfrac{1}{3}$, $C = -\dfrac{2}{3}$ $\Rightarrow$

$$\int \dfrac{1}{x^3 - 1}\, dx = \int \dfrac{\frac{1}{3}}{x - 1}\, dx + \int \dfrac{-\frac{1}{3}x - \frac{2}{3}}{x^2 + x + 1}\, dx = \dfrac{1}{3}\ln|x - 1| - \dfrac{1}{3}\int \dfrac{x + 2}{x^2 + x + 1}\, dx$$

$$= \dfrac{1}{3}\ln|x - 1| - \dfrac{1}{3}\int \dfrac{x + 1/2}{x^2 + x + 1}\, dx - \dfrac{1}{3}\int \dfrac{(3/2)\, dx}{(x + 1/2)^2 + 3/4}$$

$$= \dfrac{1}{3}\ln|x - 1| - \dfrac{1}{6}\ln(x^2 + x + 1) - \dfrac{1}{2}\left(\dfrac{2}{\sqrt{3}}\right)\tan^{-1}\left(\dfrac{x + \frac{1}{2}}{\sqrt{3}/2}\right) + K$$

$$= \dfrac{1}{3}\ln|x - 1| - \dfrac{1}{6}\ln(x^2 + x + 1) - \dfrac{1}{\sqrt{3}}\tan^{-1}\left(\dfrac{1}{\sqrt{3}}(2x + 1)\right) + K$$

35. Let $u = x^4 + 4x^2 + 3$, so that $du = (4x^3 + 8x)\, dx = 4(x^3 + 2x)\, dx$, $x = 0$ $\Rightarrow$ $u = 3$, and $x = 1$ $\Rightarrow$ $u = 8$.

Then $\displaystyle\int_0^1 \dfrac{x^3 + 2x}{x^4 + 4x^2 + 3}\, dx = \int_3^8 \dfrac{1}{u}\left(\dfrac{1}{4}\, du\right) = \dfrac{1}{4}\left[\ln|u|\right]_3^8 = \dfrac{1}{4}(\ln 8 - \ln 3) = \dfrac{1}{4}\ln\dfrac{8}{3}$.

37. $\dfrac{5x^4 + 7x^2 + x + 2}{x(x^2+1)^2} = \dfrac{A}{x} + \dfrac{Bx+C}{x^2+1} + \dfrac{Dx+E}{(x^2+1)^2}$. Multiply by $x(x^2+1)^2$ to get

$5x^4 + 7x^2 + x + 2 = A(x^2+1)^2 + (Bx+C)x(x^2+1) + (Dx+E)x \iff$

$5x^4 + 7x^2 + x + 2 = A(x^4 + 2x^2 + 1) + (Bx^2 + Cx)(x^2+1) + Dx^2 + Ex \iff$

$5x^4 + 7x^2 + x + 2 = Ax^4 + 2Ax^2 + A + Bx^4 + Cx^3 + Bx^2 + Cx + Dx^2 + Ex \iff$

$5x^4 + 7x^2 + x + 2 = (A+B)x^4 + Cx^3 + (2A+B+D)x^2 + (C+E)x + A$. Equating coefficients gives us $C = 0$,
$A = 2$, $A + B = 5 \;\Rightarrow\; B = 3$, $C + E = 1 \;\Rightarrow\; E = 1$, and $2A + B + D = 7 \;\Rightarrow\; D = 0$. Thus,

$\displaystyle\int \dfrac{5x^4 + 7x^2 + x + 2}{x(x^2+1)^2}\,dx = \int\left[\dfrac{2}{x} + \dfrac{3x}{x^2+1} + \dfrac{1}{(x^2+1)^2}\right]dx = I$. Now

$\displaystyle\int \dfrac{dx}{(x^2+1)^2} = \int \dfrac{\sec^2\theta\,d\theta}{(\tan^2\theta+1)^2}\quad \begin{bmatrix} x = \tan\theta, \\ dx = \sec^2\theta\,d\theta\end{bmatrix}$

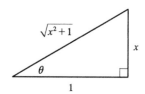

$= \displaystyle\int \dfrac{\sec^2\theta}{\sec^4\theta}\,d\theta = \int \cos^2\theta\,d\theta = \int \tfrac{1}{2}(1 + \cos 2\theta)\,d\theta$

$= \tfrac{1}{2}\theta + \tfrac{1}{4}\sin 2\theta + C = \tfrac{1}{2}\theta + \tfrac{1}{2}\sin\theta\cos\theta + C$

$= \dfrac{1}{2}\tan^{-1}x + \dfrac{1}{2}\dfrac{x}{\sqrt{x^2+1}}\dfrac{1}{\sqrt{x^2+1}} + C$

Therefore, $I = 2\ln|x| + \tfrac{3}{2}\ln(x^2+1) + \tfrac{1}{2}\tan^{-1}x + \dfrac{x}{2(x^2+1)} + C$.

39. $\dfrac{x^2 - 3x + 7}{(x^2 - 4x + 6)^2} = \dfrac{Ax + B}{x^2 - 4x + 6} + \dfrac{Cx + D}{(x^2 - 4x + 6)^2} \;\Rightarrow\; x^2 - 3x + 7 = (Ax+B)(x^2 - 4x + 6) + Cx + D \;\Rightarrow\;$

$x^2 - 3x + 7 = Ax^3 + (-4A+B)x^2 + (6A - 4B + C)x + (6B + D)$. So $A = 0$, $-4A + B = 1 \;\Rightarrow\; B = 1$,
$6A - 4B + C = -3 \;\Rightarrow\; C = 1$, $6B + D = 7 \;\Rightarrow\; D = 1$. Thus,

$I = \displaystyle\int \dfrac{x^2 - 3x + 7}{(x^2 - 4x + 6)^2}\,dx = \int\left(\dfrac{1}{x^2 - 4x + 6} + \dfrac{x + 1}{(x^2 - 4x + 6)^2}\right)dx$

$= \displaystyle\int \dfrac{1}{(x-2)^2 + 2}\,dx + \int \dfrac{x - 2}{(x^2 - 4x + 6)^2}\,dx + \int \dfrac{3}{(x^2 - 4x + 6)^2}\,dx$

$= I_1 + I_2 + I_3$.

$I_1 = \displaystyle\int \dfrac{1}{(x-2)^2 + \left(\sqrt{2}\right)^2}\,dx = \dfrac{1}{\sqrt{2}}\tan^{-1}\!\left(\dfrac{x-2}{\sqrt{2}}\right) + C_1$

$I_2 = \dfrac{1}{2}\displaystyle\int \dfrac{2x - 4}{(x^2 - 4x + 6)^2}\,dx = \dfrac{1}{2}\int \dfrac{1}{u^2}\,du = \dfrac{1}{2}\!\left(-\dfrac{1}{u}\right) + C_2 = -\dfrac{1}{2(x^2 - 4x + 6)} + C_2$

[continued]

SECTION 7.4 INTEGRATION OF RATIONAL FUNCTIONS BY PARTIAL FRACTIONS 371

$$I_3 = 3\int \frac{1}{\left[(x-2)^2 + (\sqrt{2})^2\right]^2}\,dx = 3\int \frac{1}{[2(\tan^2\theta + 1)]^2}\sqrt{2}\sec^2\theta\,d\theta \qquad \begin{bmatrix} x - 2 = \sqrt{2}\tan\theta, \\ dx = \sqrt{2}\sec^2\theta\,d\theta \end{bmatrix}$$

$$= \frac{3\sqrt{2}}{4}\int \frac{\sec^2\theta}{\sec^4\theta}\,d\theta = \frac{3\sqrt{2}}{4}\int \cos^2\theta\,d\theta = \frac{3\sqrt{2}}{4}\int \tfrac{1}{2}(1 + \cos 2\theta)\,d\theta$$

$$= \frac{3\sqrt{2}}{8}\left(\theta + \tfrac{1}{2}\sin 2\theta\right) + C_3 = \frac{3\sqrt{2}}{8}\tan^{-1}\!\left(\frac{x-2}{\sqrt{2}}\right) + \frac{3\sqrt{2}}{8}\left(\tfrac{1}{2}\cdot 2\sin\theta\cos\theta\right) + C_3$$

$$= \frac{3\sqrt{2}}{8}\tan^{-1}\!\left(\frac{x-2}{\sqrt{2}}\right) + \frac{3\sqrt{2}}{8}\cdot\frac{x-2}{\sqrt{x^2-4x+6}}\cdot\frac{\sqrt{2}}{\sqrt{x^2-4x+6}} + C_3$$

$$= \frac{3\sqrt{2}}{8}\tan^{-1}\!\left(\frac{x-2}{\sqrt{2}}\right) + \frac{3(x-2)}{4(x^2-4x+6)} + C_3$$

So $I = I_1 + I_2 + I_3 \qquad [C = C_1 + C_2 + C_3]$

$$= \frac{1}{\sqrt{2}}\tan^{-1}\!\left(\frac{x-2}{\sqrt{2}}\right) + \frac{-1}{2(x^2-4x+6)} + \frac{3\sqrt{2}}{8}\tan^{-1}\!\left(\frac{x-2}{\sqrt{2}}\right) + \frac{3(x-2)}{4(x^2-4x+6)} + C$$

$$= \left(\frac{4\sqrt{2}}{8} + \frac{3\sqrt{2}}{8}\right)\tan^{-1}\!\left(\frac{x-2}{\sqrt{2}}\right) + \frac{3(x-2)-2}{4(x^2-4x+6)} + C = \frac{7\sqrt{2}}{8}\tan^{-1}\!\left(\frac{x-2}{\sqrt{2}}\right) + \frac{3x-8}{4(x^2-4x+6)} + C$$

41. $\displaystyle\int \frac{dx}{x\sqrt{x-1}} = \int \frac{2u}{u(u^2+1)}\,du \qquad \begin{bmatrix} u = \sqrt{x-1},\ x = u^2 + 1 \\ u^2 = x - 1,\ dx = 2u\,du \end{bmatrix}$

$$= 2\int \frac{1}{u^2+1}\,du = 2\tan^{-1}u + C = 2\tan^{-1}\sqrt{x-1} + C$$

43. Let $u = \sqrt{x}$, so $u^2 = x$ and $2u\,du = dx$. Then $\displaystyle\int \frac{dx}{x^2 + x\sqrt{x}} = \int \frac{2u\,du}{u^4 + u^3} = \int \frac{2\,du}{u^3 + u^2} = \int \frac{2\,du}{u^2(u+1)}$.

$$\frac{2}{u^2(u+1)} = \frac{A}{u} + \frac{B}{u^2} + \frac{C}{u+1} \;\Rightarrow\; 2 = Au(u+1) + B(u+1) + Cu^2.\ \text{Setting } u = 0 \text{ gives } B = 2.\ \text{Setting } u = -1$$

gives $C = 2$. Equating coefficients of u^2, we get $0 = A + C$, so $A = -2$. Thus,

$$\int \frac{2\,du}{u^2(u+1)} = \int \left(\frac{-2}{u} + \frac{2}{u^2} + \frac{2}{u+1}\right)du = -2\ln|u| - \frac{2}{u} + 2\ln|u+1| + C = -2\ln\sqrt{x} - \frac{2}{\sqrt{x}} + 2\ln(\sqrt{x}+1) + C.$$

45. Let $u = \sqrt[3]{x^2+1}$. Then $x^2 = u^3 - 1$, $2x\,dx = 3u^2\,du \;\Rightarrow\;$

$$\int \frac{x^3}{\sqrt[3]{x^2+1}}\,dx = \int \frac{(u^3-1)\tfrac{3}{2}u^2\,du}{u} = \frac{3}{2}\int(u^4 - u)\,du$$

$$= \tfrac{3}{10}u^5 - \tfrac{3}{4}u^2 + C = \tfrac{3}{10}(x^2+1)^{5/3} - \tfrac{3}{4}(x^2+1)^{2/3} + C$$

47. If we were to substitute $u = \sqrt{x}$, then the square root would disappear but a cube root would remain. On the other hand, the substitution $u = \sqrt[3]{x}$ would eliminate the cube root but leave a square root. We can eliminate both roots by means of the substitution $u = \sqrt[6]{x}$. (Note that 6 is the least common multiple of 2 and 3.)

[continued]

Let $u = \sqrt[6]{x}$. Then $x = u^6$, so $dx = 6u^5\,du$ and $\sqrt{x} = u^3$, $\sqrt[3]{x} = u^2$. Thus,

$$\int \frac{1}{\sqrt{x} - \sqrt[3]{x}}\,dx = \int \frac{6u^5\,du}{u^3 - u^2} = 6\int \frac{u^5}{u^2(u-1)}\,du = 6\int \frac{u^3}{u-1}\,du$$

$$= 6\int \left(u^2 + u + 1 + \frac{1}{u-1}\right)du \quad \text{[by long division]}$$

$$= 6\left(\tfrac{1}{3}u^3 + \tfrac{1}{2}u^2 + u + \ln|u-1|\right) + C = 2\sqrt{x} + 3\sqrt[3]{x} + 6\sqrt[6]{x} + 6\ln\left|\sqrt[6]{x} - 1\right| + C$$

49. Let $u = \sqrt{x} \Rightarrow x = u^2$, so $dx = 2u\,du$. This substitution gives $I = \int \frac{1}{x - 3\sqrt{x} + 2}\,dx = \int \frac{2u}{u^2 - 3u + 2}\,du$. Now

$$\frac{2u}{u^2 - 3u + 2} = \frac{2u}{(u-2)(u-1)} = \frac{A}{u-2} + \frac{B}{u-1}.$$ Multiply both sides by $(u-2)(u-1)$ to get

$2u = A(u-1) + B(u-2)$. Setting $u = 1$ gives $2 = -B$ or $B = -2$, and setting $u = 2$ gives $A = 4$. Thus,

$$I = \int \left(\frac{4}{u-2} - \frac{2}{u-1}\right)du = 4\ln|u-2| - 2\ln|u-1| + C = 4\ln\left|\sqrt{x} - 2\right| - 2\ln\left|\sqrt{x} - 1\right| + C.$$

51. Let $u = e^x$. Then $x = \ln u$, $dx = \dfrac{du}{u} \Rightarrow$

$$\int \frac{e^{2x}\,dx}{e^{2x} + 3e^x + 2} = \int \frac{u^2\,(du/u)}{u^2 + 3u + 2} = \int \frac{u\,du}{(u+1)(u+2)} = \int \left[\frac{-1}{u+1} + \frac{2}{u+2}\right]du$$

$$= 2\ln|u+2| - \ln|u+1| + C = \ln\frac{(e^x + 2)^2}{e^x + 1} + C$$

53. Let $u = \tan t$, so that $du = \sec^2 t\,dt$. Then $\displaystyle\int \frac{\sec^2 t}{\tan^2 t + 3\tan t + 2}\,dt = \int \frac{1}{u^2 + 3u + 2}\,du = \int \frac{1}{(u+1)(u+2)}\,du.$

Now $\dfrac{1}{(u+1)(u+2)} = \dfrac{A}{u+1} + \dfrac{B}{u+2} \Rightarrow 1 = A(u+2) + B(u+1).$

Setting $u = -2$ gives $1 = -B$, so $B = -1$. Setting $u = -1$ gives $1 = A$.

Thus, $\displaystyle\int \frac{1}{(u+1)(u+2)}\,du = \int \left(\frac{1}{u+1} - \frac{1}{u+2}\right)du = \ln|u+1| - \ln|u+2| + C = \ln|\tan t + 1| - \ln|\tan t + 2| + C.$

55. Let $u = e^x$, so that $du = e^x\,dx$ and $dx = \dfrac{du}{u}$. Then $\displaystyle\int \frac{dx}{1 + e^x} = \int \frac{du}{(1+u)u} \cdot \frac{1}{u(u+1)} = \frac{A}{u} + \frac{B}{u+1} \Rightarrow$

$1 = A(u+1) + Bu$. Setting $u = -1$ gives $B = -1$. Setting $u = 0$ gives $A = 1$. Thus,

$$\int \frac{du}{u(u+1)} = \int \left(\frac{1}{u} - \frac{1}{u+1}\right)du = \ln|u| - \ln|u+1| + C = \ln e^x - \ln(e^x + 1) + C = x - \ln(e^x + 1) + C.$$

57. Let $u = \ln(x^2 - x + 2)$, $dv = dx$. Then $du = \dfrac{2x-1}{x^2-x+2}\,dx$, $v = x$, and (by integration by parts)

$$\int \ln(x^2 - x + 2)\,dx = x\ln(x^2 - x + 2) - \int \frac{2x^2 - x}{x^2 - x + 2}\,dx = x\ln(x^2 - x + 2) - \int \left(2 + \frac{x-4}{x^2-x+2}\right)dx$$

$$= x\ln(x^2-x+2) - 2x - \int \frac{\tfrac{1}{2}(2x-1)}{x^2-x+2}\,dx + \frac{7}{2}\int \frac{dx}{(x-\tfrac{1}{2})^2 + \tfrac{7}{4}}$$

$$= x\ln(x^2-x+2) - 2x - \tfrac{1}{2}\ln(x^2-x+2) + \frac{7}{2}\int \frac{\tfrac{\sqrt{7}}{2}\,du}{\tfrac{7}{4}(u^2+1)} \quad \left[\begin{array}{l}\text{where } x - \tfrac{1}{2} = \tfrac{\sqrt{7}}{2}u,\\ dx = \tfrac{\sqrt{7}}{2}\,du,\\ (x-\tfrac{1}{2})^2 + \tfrac{7}{4} = \tfrac{7}{4}(u^2+1)\end{array}\right]$$

$$= (x - \tfrac{1}{2})\ln(x^2-x+2) - 2x + \sqrt{7}\tan^{-1}u + C$$

$$= (x - \tfrac{1}{2})\ln(x^2-x+2) - 2x + \sqrt{7}\tan^{-1}\frac{2x-1}{\sqrt{7}} + C$$

59.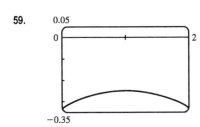

From the graph, we see that the integral will be negative, and we guess that the area is about the same as that of a rectangle with width 2 and height 0.3, so we estimate the integral to be $-(2 \cdot 0.3) = -0.6$. Now

$$\frac{1}{x^2 - 2x - 3} = \frac{1}{(x-3)(x+1)} = \frac{A}{x-3} + \frac{B}{x+1} \iff$$

$1 = (A+B)x + A - 3B$, so $A = -B$ and $A - 3B = 1 \iff A = \tfrac{1}{4}$

and $B = -\tfrac{1}{4}$, so the integral becomes

$$\int_0^2 \frac{dx}{x^2 - 2x - 3} = \frac{1}{4}\int_0^2 \frac{dx}{x-3} - \frac{1}{4}\int_0^2 \frac{dx}{x+1} = \frac{1}{4}\Big[\ln|x-3| - \ln|x+1|\Big]_0^2 = \frac{1}{4}\left[\ln\left|\frac{x-3}{x+1}\right|\right]_0^2$$

$$= \tfrac{1}{4}\left(\ln\tfrac{1}{3} - \ln 3\right) = -\tfrac{1}{2}\ln 3 \approx -0.55$$

61. $\displaystyle\int \frac{dx}{x^2 - 2x} = \int \frac{dx}{(x-1)^2 - 1} = \int \frac{du}{u^2 - 1}$ [put $u = x - 1$]

$= \dfrac{1}{2}\ln\left|\dfrac{u-1}{u+1}\right| + C$ [by Equation 6] $= \dfrac{1}{2}\ln\left|\dfrac{x-2}{x}\right| + C$

63. (a) If $t = \tan\left(\dfrac{x}{2}\right)$, then $\dfrac{x}{2} = \tan^{-1}t$. The figure gives

$\cos\left(\dfrac{x}{2}\right) = \dfrac{1}{\sqrt{1+t^2}}$ and $\sin\left(\dfrac{x}{2}\right) = \dfrac{t}{\sqrt{1+t^2}}$.

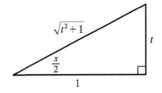

(b) $\cos x = \cos\left(2 \cdot \dfrac{x}{2}\right) = 2\cos^2\left(\dfrac{x}{2}\right) - 1$

$= 2\left(\dfrac{1}{\sqrt{1+t^2}}\right)^2 - 1 = \dfrac{2}{1+t^2} - 1 = \dfrac{1-t^2}{1+t^2}$

$\sin x = \sin\left(2 \cdot \dfrac{x}{2}\right) = 2\sin\left(\dfrac{x}{2}\right)\cos\left(\dfrac{x}{2}\right) = 2\dfrac{t}{\sqrt{t^2+1}} \cdot \dfrac{1}{\sqrt{t^2+1}} = \dfrac{2t}{t^2+1}$

(c) $\dfrac{x}{2} = \arctan t \Rightarrow x = 2\arctan t \Rightarrow dx = \dfrac{2}{1+t^2}\,dt$

65. Let $t = \tan(x/2)$. Then, using the expressions in Exercise 63, we have

$$\int \frac{1}{3\sin x - 4\cos x}\,dx = \int \frac{1}{3\left(\dfrac{2t}{1+t^2}\right) - 4\left(\dfrac{1-t^2}{1+t^2}\right)} \cdot \frac{2\,dt}{1+t^2} = 2\int \frac{dt}{3(2t) - 4(1-t^2)} = \int \frac{dt}{2t^2 + 3t - 2}$$

$$= \int \frac{dt}{(2t-1)(t+2)} = \int \left[\frac{2}{5}\frac{1}{2t-1} - \frac{1}{5}\frac{1}{t+2}\right] dt \quad \text{[using partial fractions]}$$

$$= \tfrac{1}{5}\Big[\ln|2t-1| - \ln|t+2|\Big] + C = \tfrac{1}{5}\ln\left|\dfrac{2t-1}{t+2}\right| + C = \tfrac{1}{5}\ln\left|\dfrac{2\tan(x/2) - 1}{\tan(x/2) + 2}\right| + C$$

67. Let $t = \tan(x/2)$. Then, by Exercise 63,

$$\int_0^{\pi/2} \frac{\sin 2x}{2 + \cos x}\,dx = \int_0^{\pi/2} \frac{2\sin x \cos x}{2 + \cos x}\,dx = \int_0^1 \frac{2 \cdot \dfrac{2t}{1+t^2} \cdot \dfrac{1-t^2}{1+t^2}}{2 + \dfrac{1-t^2}{1+t^2}} \cdot \frac{2}{1+t^2}\,dt = \int_0^1 \frac{\dfrac{8t(1-t^2)}{(1+t^2)^2}}{2(1+t^2) + (1-t^2)}\,dt$$

$$= \int_0^1 8t \cdot \frac{1-t^2}{(t^2+3)(t^2+1)^2}\,dt = I$$

If we now let $u = t^2$, then $\dfrac{1-t^2}{(t^2+3)(t^2+1)^2} = \dfrac{1-u}{(u+3)(u+1)^2} = \dfrac{A}{u+3} + \dfrac{B}{u+1} + \dfrac{C}{(u+1)^2} \Rightarrow$

$1 - u = A(u+1)^2 + B(u+3)(u+1) + C(u+3)$. Set $u = -1$ to get $2 = 2C$, so $C = 1$. Set $u = -3$ to get $4 = 4A$, so $A = 1$. Set $u = 0$ to get $1 = 1 + 3B + 3$, so $B = -1$. So

$$I = \int_0^1 \left[\frac{8t}{t^2+3} - \frac{8t}{t^2+1} + \frac{8t}{(t^2+1)^2}\right] dt = \left[4\ln(t^2+3) - 4\ln(t^2+1) - \frac{4}{t^2+1}\right]_0^1$$

$$= (4\ln 4 - 4\ln 2 - 2) - (4\ln 3 - 0 - 4) = 8\ln 2 - 4\ln 2 - 4\ln 3 + 2 = 4\ln \tfrac{2}{3} + 2$$

69. By long division, $\dfrac{x^2+1}{3x-x^2} = -1 + \dfrac{3x+1}{3x-x^2}$. Now

$\dfrac{3x+1}{3x-x^2} = \dfrac{3x+1}{x(3-x)} = \dfrac{A}{x} + \dfrac{B}{3-x} \Rightarrow 3x+1 = A(3-x) + Bx$. Set $x = 3$ to get $10 = 3B$, so $B = \tfrac{10}{3}$. Set $x = 0$ to get $1 = 3A$, so $A = \tfrac{1}{3}$. Thus, the area is

$$\int_1^2 \frac{x^2+1}{3x-x^2}\,dx = \int_1^2 \left(-1 + \frac{\tfrac{1}{3}}{x} + \frac{\tfrac{10}{3}}{3-x}\right) dx = \left[-x + \tfrac{1}{3}\ln|x| - \tfrac{10}{3}\ln|3-x|\right]_1^2$$

$$= \left(-2 + \tfrac{1}{3}\ln 2 - 0\right) - \left(-1 + 0 - \tfrac{10}{3}\ln 2\right) = -1 + \tfrac{11}{3}\ln 2$$

71. $t = \displaystyle\int \frac{P+S}{P[(r-1)P - S]}\,dP = \int \frac{P+S}{P(0.1P - S)}\,dP$ $[r = 1.1]$. Now $\dfrac{P+S}{P(0.1P - S)} = \dfrac{A}{P} + \dfrac{B}{0.1P - S} \Rightarrow$

$P + S = A(0.1P - S) + BP$. Substituting 0 for P gives $S = -AS \Rightarrow A = -1$. Substituting $10S$ for P gives

$11S = 10BS \Rightarrow B = \tfrac{11}{10}$. Thus, $t = \displaystyle\int \left(\frac{-1}{P} + \frac{11/10}{0.1P - S}\right) dP \Rightarrow t = -\ln P + 11\ln(0.1P - S) + C$.

[continued]

When $t = 0$, $P = 10{,}000$ and $S = 900$, so $0 = -\ln 10{,}000 + 11 \ln(1000 - 900) + C \;\Rightarrow\;$
$C = \ln 10{,}000 - 11 \ln 100 \quad [= \ln 10^{-18} \approx -41.45]$.

Therefore, $t = -\ln P + 11 \ln\left(\tfrac{1}{10}P - 900\right) + \ln 10{,}000 - 11 \ln 100 \;\Rightarrow\; t = \ln\dfrac{10{,}000}{P} + 11 \ln\dfrac{P - 9000}{1000}$.

73. (a) In Maple, we define $f(x)$, and then use `convert(f,parfrac,x);` to obtain

$$f(x) = \frac{24{,}110/4879}{5x + 2} - \frac{668/323}{2x + 1} - \frac{9438/80{,}155}{3x - 7} + \frac{(22{,}098x + 48{,}935)/260{,}015}{x^2 + x + 5}$$

In Mathematica, we use the command `Apart`.

(b) $\displaystyle\int f(x)\,dx = \tfrac{24{,}110}{4879} \cdot \tfrac{1}{5}\ln|5x + 2| - \tfrac{668}{323} \cdot \tfrac{1}{2}\ln|2x + 1| - \tfrac{9438}{80{,}155} \cdot \tfrac{1}{3}\ln|3x - 7|$

$\qquad + \dfrac{1}{260{,}015} \displaystyle\int \dfrac{22{,}098\left(x + \tfrac{1}{2}\right) + 37{,}886}{\left(x + \tfrac{1}{2}\right)^2 + \tfrac{19}{4}}\,dx + C$

$= \tfrac{24{,}110}{4879} \cdot \tfrac{1}{5}\ln|5x + 2| - \tfrac{668}{323} \cdot \tfrac{1}{2}\ln|2x + 1| - \tfrac{9438}{80{,}155} \cdot \tfrac{1}{3}\ln|3x - 7|$

$\qquad + \dfrac{1}{260{,}015}\left[22{,}098 \cdot \tfrac{1}{2}\ln(x^2 + x + 5) + 37{,}886 \cdot \sqrt{\tfrac{4}{19}}\tan^{-1}\left(\tfrac{1}{\sqrt{19/4}}\left(x + \tfrac{1}{2}\right)\right)\right] + C$

$= \tfrac{4822}{4879}\ln|5x + 2| - \tfrac{334}{323}\ln|2x + 1| - \tfrac{3146}{80{,}155}\ln|3x - 7| + \tfrac{11{,}049}{260{,}015}\ln(x^2 + x + 5)$

$\qquad + \dfrac{75{,}772}{260{,}015\sqrt{19}}\tan^{-1}\left[\tfrac{1}{\sqrt{19}}(2x + 1)\right] + C$

Using a CAS, we get

$$\frac{4822 \ln(5x + 2)}{4879} - \frac{334 \ln(2x + 1)}{323} - \frac{3146 \ln(3x - 7)}{80{,}155}$$
$$+ \frac{11{,}049 \ln(x^2 + x + 5)}{260{,}015} + \frac{3988\sqrt{19}}{260{,}015}\tan^{-1}\left[\frac{\sqrt{19}}{19}(2x + 1)\right]$$

The main difference in this answer is that the absolute value signs and the constant of integration have been omitted. Also, the fractions have been reduced and the denominators rationalized.

75. There are only finitely many values of x where $Q(x) = 0$ (assuming that Q is not the zero polynomial). At all other values of x, $F(x)/Q(x) = G(x)/Q(x)$, so $F(x) = G(x)$. In other words, the values of F and G agree at all except perhaps finitely many values of x. By continuity of F and G, the polynomials F and G must agree at those values of x too.

More explicitly: if a is a value of x such that $Q(a) = 0$, then $Q(x) \neq 0$ for all x sufficiently close to a. Thus,

$$F(a) = \lim_{x \to a} F(x) \quad \text{[by continuity of } F\text{]}$$
$$= \lim_{x \to a} G(x) \quad \text{[whenever } Q(x) \neq 0\text{]}$$
$$= G(a) \quad \text{[by continuity of } G\text{]}$$

77. If $a \neq 0$ and n is a positive integer, then $f(x) = \dfrac{1}{x^n(x - a)} = \dfrac{A_1}{x} + \dfrac{A_2}{x^2} + \cdots + \dfrac{A_n}{x^n} + \dfrac{B}{x - a}$. Multiply both sides by $x^n(x - a)$ to get $1 = A_1 x^{n-1}(x - a) + A_2 x^{n-2}(x - a) + \cdots + A_n(x - a) + Bx^n$. Let $x = a$ in the last equation to get $1 = Ba^n \;\Rightarrow\; B = 1/a^n$. So

$$f(x) - \frac{B}{x-a} = \frac{1}{x^n(x-a)} - \frac{1}{a^n(x-a)} = \frac{a^n - x^n}{x^n a^n (x-a)} = -\frac{x^n - a^n}{a^n x^n (x-a)}$$

$$= -\frac{(x-a)(x^{n-1} + x^{n-2}a + x^{n-3}a^2 + \cdots + xa^{n-2} + a^{n-1})}{a^n x^n (x-a)}$$

$$= -\left(\frac{x^{n-1}}{a^n x^n} + \frac{x^{n-2}a}{a^n x^n} + \frac{x^{n-3}a^2}{a^n x^n} + \cdots + \frac{xa^{n-2}}{a^n x^n} + \frac{a^{n-1}}{a^n x^n}\right)$$

$$= -\frac{1}{a^n x} - \frac{1}{a^{n-1}x^2} - \frac{1}{a^{n-2}x^3} - \cdots - \frac{1}{a^2 x^{n-1}} - \frac{1}{ax^n}$$

Thus, $f(x) = \dfrac{1}{x^n(x-a)} = -\dfrac{1}{a^n x} - \dfrac{1}{a^{n-1}x^2} - \cdots - \dfrac{1}{ax^n} + \dfrac{1}{a^n(x-a)}$.

7.5 Strategy for Integration

1. (a) Let $u = 1 + x^2$, so that $du = 2x\,dx \;\Rightarrow\; \frac{1}{2}du = x\,dx$. Then,

$$\int \frac{x}{1+x^2}\,dx = \frac{1}{2}\int \frac{1}{u}\,du = \frac{1}{2}\ln|u| + C = \frac{1}{2}\ln(1+x^2) + C$$

Note the absolute value has been omitted in the last step since $1 + x^2 > 0$ for all $x \in \mathbb{R}$.

(b) $\displaystyle\int \frac{1}{1+x^2}\,dx = \tan^{-1}x + C$

(c) $\displaystyle\int \frac{1}{1-x^2}\,dx = \int \frac{1}{(1+x)(1-x)}\,dx = \frac{1}{2}\int\left(\frac{1}{1+x} + \frac{1}{1-x}\right)dx$ [by partial fractions]

$\qquad = \frac{1}{2}\ln|1+x| - \frac{1}{2}\ln|1-x| + C$

3. (a) Let $u = \ln x$, so that $du = \dfrac{1}{x}\,dx$. Then $\displaystyle\int \frac{\ln x}{x}\,dx = \int u\,du = \frac{1}{2}u^2 + C = \frac{1}{2}(\ln x)^2 + C$.

(b) Use integration by parts with $u = \ln(2x)$, $dv = dx \;\Rightarrow\; du = \dfrac{1}{2x}(2)\,dx = \dfrac{1}{x}\,dx$, $v = x$. Then

$$\int \ln(2x)\,dx = x\ln(2x) - \int x\left(\frac{1}{x}\,dx\right) = x\ln(2x) - \int dx = x\ln(2x) - x + C.$$

(c) Use integration by parts with $u = \ln x$, $dv = x\,dx \;\Rightarrow\; du = \dfrac{1}{x}\,dx$, $v = \dfrac{1}{2}x^2$. Then

$$\int x\ln x\,dx = \frac{1}{2}x^2 \ln x - \int \frac{1}{2}x^2\left(\frac{1}{x}\,dx\right) = \frac{1}{2}x^2 \ln x - \frac{1}{2}\int x\,dx = \frac{1}{2}x^2 \ln x - \frac{1}{4}x^2 + C.$$

5. (a) $\dfrac{1}{x^2 - 4x + 3} = \dfrac{1}{(x-3)(x-1)} = \dfrac{A}{x-3} + \dfrac{B}{x-1}$. Multiply both sides by $(x-3)(x-1)$ to get

$1 = A(x-1) + B(x-3)$. Setting $x = 3$ gives $1 = 2A$, so $A = \frac{1}{2}$. Now setting $x = 1$ gives $1 = -2B$, so $B = -\frac{1}{2}$.

Thus, $\displaystyle\int \frac{1}{x^2 - 4x + 3}\,dx = \frac{1}{2}\int\left(\frac{1}{x-3} - \frac{1}{x-1}\right)dx = \frac{1}{2}\ln|x-3| - \frac{1}{2}\ln|x-1| + C$.

(b) $\dfrac{1}{x^2 - 4x + 4} = \dfrac{1}{(x-2)^2}$. Let $u = x - 2$, so that $du = dx$. Thus,

$$\int \dfrac{1}{x^2 - 4x + 4}\, dx = \int \dfrac{1}{(x-2)^2}\, dx = \int \dfrac{1}{u^2}\, du = \int u^{-2}\, du = -u^{-1} + C = -\dfrac{1}{x-2} + C.$$

(c) $x^2 - 4x + 5$ is an irreducible quadratic, so it cannot be factored. Completing the square gives

$x^2 - 4x + 5 = (x^2 - 4x + 4) - 4 + 5 = (x-2)^2 + 1$. Now, use the substitution $u = x - 2$, so that $du = dx$. Thus,

$$\int \dfrac{1}{x^2 - 4x + 5}\, dx = \int \dfrac{1}{(x-2)^2 + 1}\, dx = \int \dfrac{1}{u^2 + 1}\, du = \tan^{-1} u + C = \tan^{-1}(x-2) + C.$$

7. (a) Let $u = x^3$, so that $du = 3x^2\, dx$ $\Rightarrow$ $\tfrac{1}{3}\, du = x^2\, dx$. Thus, $\displaystyle\int x^2 e^{x^3}\, dx = \dfrac{1}{3}\int e^u\, du = \dfrac{1}{3} e^u + C = \dfrac{1}{3} e^{x^3} + C$.

(b) First, use integration by parts with $u = x^2$, $dv = e^x\, dx$ $\Rightarrow$ $du = 2x\, dx$, $v = e^x$. This gives

$I = \int x^2 e^x\, dx = x^2 e^x - \int 2x e^x\, dx$. Next, use integration by parts for the remaining integral with $U = 2x$,

$dV = e^x\, dx$ $\Rightarrow$ $dU = 2\, dx$, $V = e^x$. Thus, $I = x^2 e^x - \left(2x e^x - \int 2 e^x\, dx\right) = x^2 e^x - 2x e^x + 2 e^x + C$.

(c) Let $y = x^2$, so that $dy = 2x\, dx$. Thus, $\int x^3 e^{x^2}\, dx = \int x^2 e^{x^2} x\, dx = \tfrac{1}{2} \int y e^y\, dy$. Now use integration by parts

with $u = y$, $dv = e^y$ $\Rightarrow$ $du = dy$, $v = e^y$. This gives

$\int x^3 e^{x^2}\, dx = \tfrac{1}{2}\left(y e^y - \int e^y\, dy\right) = \tfrac{1}{2} y e^y - \tfrac{1}{2} e^y + C = \tfrac{1}{2} x^2 e^{x^2} - \tfrac{1}{2} e^{x^2} + C$.

9. Let $u = 1 - \sin x$. Then $du = -\cos x\, dx$ $\Rightarrow$

$$\int \dfrac{\cos x}{1 - \sin x}\, dx = \int \dfrac{1}{u}(-du) = -\ln|u| + C = -\ln|1 - \sin x| + C = -\ln(1 - \sin x) + C.$$

11. Let $u = \ln y$, $dv = \sqrt{y}\, dy$ $\Rightarrow$ $du = \dfrac{1}{y}\, dy$, $v = \dfrac{2}{3} y^{3/2}$. Then

$$\int_1^4 \sqrt{y}\, \ln y\, dy = \left[\dfrac{2}{3} y^{3/2} \ln y\right]_1^4 - \int_1^4 \dfrac{2}{3} y^{1/2}\, dy = \dfrac{2}{3}\cdot 8 \ln 4 - 0 - \left[\dfrac{4}{9} y^{3/2}\right]_1^4$$

$$= \tfrac{16}{3}(2 \ln 2) - \left(\tfrac{4}{9} \cdot 8 - \tfrac{4}{9}\right) = \tfrac{32}{3} \ln 2 - \tfrac{28}{9}$$

13. Let $x = \ln y$, so that $dx = \dfrac{1}{y}\, dy$. Thus, $I = \displaystyle\int \dfrac{\ln(\ln y)}{y}\, dy = \int \ln x\, dx$. Now use integration by parts with

$u = \ln x$, $dv = dx$ $\Rightarrow$ $du = dx/x$, $v = x$. This gives

$$I = x \ln x - \int \dfrac{x}{x}\, dx = x \ln x - \int dx = x \ln x - x + C = \ln y\, [\ln(\ln y)] - \ln y + C.$$

15. Let $u = x^2$, so that $du = 2x\, dx$. Thus,

$$\int \dfrac{x}{x^4 + 9}\, dx = \dfrac{1}{2}\int \dfrac{1}{u^2 + 9}\, du = \dfrac{1}{2}\int \dfrac{(1/3)^2}{(u/3)^2 + 1}\, du = \dfrac{1}{2}\left(\dfrac{1}{3}\tan^{-1}\left(\dfrac{u}{3}\right)\right) + C = \dfrac{1}{6}\tan^{-1}\left(\dfrac{x^2}{3}\right) + C.$$

17. $\dfrac{x+2}{x^2+3x-4} = \dfrac{x+2}{(x+4)(x-1)} = \dfrac{A}{x+4} + \dfrac{B}{x-1}$. Multiply by $(x+4)(x-1)$ to get $x+2 = A(x-1) + B(x+4)$.

Substituting 1 for x gives $3 = 5B \Leftrightarrow B = \frac{3}{5}$. Substituting -4 for x gives $-2 = -5A \Leftrightarrow A = \frac{2}{5}$. Thus,

$$\int_2^4 \dfrac{x+2}{x^2+3x-4}\,dx = \int_2^4 \left(\dfrac{2/5}{x+4} + \dfrac{3/5}{x-1}\right) dx = \left[\dfrac{2}{5}\ln|x+4| + \dfrac{3}{5}\ln|x-1|\right]_2^4$$

$$= \left(\tfrac{2}{5}\ln 8 + \tfrac{3}{5}\ln 3\right) - \left(\tfrac{2}{5}\ln 6 + 0\right) = \tfrac{2}{5}(3\ln 2) + \tfrac{3}{5}\ln 3 - \tfrac{2}{5}(\ln 2 + \ln 3)$$

$$= \tfrac{4}{5}\ln 2 + \tfrac{1}{5}\ln 3, \text{ or } \tfrac{1}{5}\ln 48$$

19. Let $x = \sec\theta$, where $0 \le \theta \le \frac{\pi}{2}$ or $\pi \le \theta < \frac{3\pi}{2}$. Then $dx = \sec\theta\tan\theta\,d\theta$ and

$\sqrt{x^2-1} = \sqrt{\sec^2\theta - 1} = \sqrt{\tan^2\theta} = |\tan\theta| = \tan\theta$ for the relevant values of θ, so

$$\int \dfrac{1}{x^3\sqrt{x^2-1}}\,dx = \int \dfrac{\sec\theta\tan\theta}{\sec^3\theta\tan\theta}\,d\theta = \int \cos^2\theta\,d\theta = \int \tfrac{1}{2}(1+\cos 2\theta)\,d\theta$$

$$= \tfrac{1}{2}\theta + \tfrac{1}{4}\sin 2\theta + C = \tfrac{1}{2}\theta + \tfrac{1}{2}\sin\theta\cos\theta + C$$

$$= \dfrac{1}{2}\sec^{-1}x + \dfrac{1}{2}\dfrac{\sqrt{x^2-1}}{x}\dfrac{1}{x} + C = \dfrac{1}{2}\sec^{-1}x + \dfrac{\sqrt{x^2-1}}{2x^2} + C$$

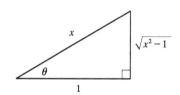

21. $\int \dfrac{\cos^3 x}{\csc x}\,dx = \int \cos^3 x \sin x\,dx \stackrel{c}{=} \int u^3(-du) = -\dfrac{1}{4}u^4 + C = -\dfrac{1}{4}\cos^4 x + C$

23. Let $u = x$, $dv = \sec x \tan x\,dx \Rightarrow du = dx$, $v = \sec x$. Then

$\int x \sec x \tan x\,dx = x\sec x - \int \sec x\,dx = x\sec x - \ln|\sec x + \tan x| + C$.

25. $\int_0^\pi t\cos^2 t\,dt = \int_0^\pi t\left[\tfrac{1}{2}(1+\cos 2t)\right]dt = \tfrac{1}{2}\int_0^\pi t\,dt + \tfrac{1}{2}\int_0^\pi t\cos 2t\,dt$

$$= \tfrac{1}{2}\left[\tfrac{1}{2}t^2\right]_0^\pi + \tfrac{1}{2}\left[\tfrac{1}{2}t\sin 2t\right]_0^\pi - \tfrac{1}{2}\int_0^\pi \tfrac{1}{2}\sin 2t\,dt \qquad \begin{bmatrix} u = t, & dv = \cos 2t\,dt \\ du = dt, & v = \tfrac{1}{2}\sin 2t \end{bmatrix}$$

$$= \tfrac{1}{4}\pi^2 + 0 - \tfrac{1}{4}\left[-\tfrac{1}{2}\cos 2t\right]_0^\pi = \tfrac{1}{4}\pi^2 + \tfrac{1}{8}(1-1) = \tfrac{1}{4}\pi^2$$

27. Let $u = e^x$. Then $\int e^{x+e^x}\,dx = \int e^{e^x}e^x\,dx = \int e^u\,du = e^u + C = e^{e^x} + C$.

29. Let $t = \sqrt{x}$, so that $t^2 = x$ and $2t\,dt = dx$. Then $\int \arctan\sqrt{x}\,dx = \int \arctan t\,(2t\,dt) = I$. Now use parts with

$u = \arctan t$, $dv = 2t\,dt \Rightarrow du = \dfrac{1}{1+t^2}\,dt$, $v = t^2$. Thus,

$$I = t^2 \arctan t - \int \dfrac{t^2}{1+t^2}\,dt = t^2\arctan t - \int\left(1 - \dfrac{1}{1+t^2}\right)dt = t^2\arctan t - t + \arctan t + C$$

$$= x\arctan\sqrt{x} - \sqrt{x} + \arctan\sqrt{x} + C \quad \left[\text{or } (x+1)\arctan\sqrt{x} - \sqrt{x} + C\right]$$

31. Let $u = 1 + \sqrt{x}$. Then $x = (u-1)^2$, $dx = 2(u-1)\,du \Rightarrow$

$$\int_0^1 \left(1 + \sqrt{x}\right)^8 dx = \int_1^2 u^8 \cdot 2(u-1)\,du = 2\int_1^2 (u^9 - u^8)\,du$$

$$= \left[\tfrac{1}{5}u^{10} - 2 \cdot \tfrac{1}{9}u^9\right]_1^2 = \tfrac{1024}{5} - \tfrac{1024}{9} - \tfrac{1}{5} + \tfrac{2}{9} = \tfrac{4097}{45}$$

33. $\displaystyle\int_0^1 \frac{1 + 12t}{1 + 3t}\,dt = \int_0^1 \frac{(12t+4) - 3}{3t+1}\,dt = \int_0^1 \left(4 - \frac{3}{3t+1}\right)dt = \Big[4t - \ln|3t+1|\Big]_0^1$

$= (4 - \ln 4) - (0 - 0) = 4 - \ln 4$

35. Let $u = 1 + e^x$, so that $du = e^x\,dx = (u-1)\,dx$. Then $\displaystyle\int \frac{1}{1+e^x}\,dx = \int \frac{1}{u} \cdot \frac{du}{u-1} = \int \frac{1}{u(u-1)}\,du = I$. Now

$\dfrac{1}{u(u-1)} = \dfrac{A}{u} + \dfrac{B}{u-1} \Rightarrow 1 = A(u-1) + Bu$. Set $u = 1$ to get $1 = B$. Set $u = 0$ to get $1 = -A$, so $A = -1$.

Thus, $I = \displaystyle\int \left(\frac{-1}{u} + \frac{1}{u-1}\right)du = -\ln|u| + \ln|u-1| + C = -\ln(1+e^x) + \ln e^x + C = x - \ln(1+e^x) + C$.

Another method: Multiply numerator and denominator by e^{-x} and let $u = e^{-x} + 1$. This gives the answer in the form $-\ln(e^{-x} + 1) + C$.

37. Use integration by parts with $u = \ln\left(x + \sqrt{x^2 - 1}\right)$, $dv = dx \Rightarrow$

$du = \dfrac{1}{x + \sqrt{x^2-1}}\left(1 + \dfrac{x}{\sqrt{x^2-1}}\right)dx = \dfrac{1}{x + \sqrt{x^2-1}}\left(\dfrac{\sqrt{x^2-1} + x}{\sqrt{x^2-1}}\right)dx = \dfrac{1}{\sqrt{x^2-1}}\,dx$, $v = x$. Then

$\displaystyle\int \ln\left(x + \sqrt{x^2-1}\right)dx = x\ln\left(x + \sqrt{x^2-1}\right) - \int \frac{x}{\sqrt{x^2-1}}\,dx = x\ln\left(x + \sqrt{x^2-1}\right) - \sqrt{x^2-1} + C.$

39. As in Example 5,

$\displaystyle\int \sqrt{\frac{1+x}{1-x}}\,dx = \int \frac{\sqrt{1+x}}{\sqrt{1-x}} \cdot \frac{\sqrt{1+x}}{\sqrt{1+x}}\,dx = \int \frac{1+x}{\sqrt{1-x^2}}\,dx = \int \frac{dx}{\sqrt{1-x^2}} + \int \frac{x\,dx}{\sqrt{1-x^2}} = \sin^{-1}x - \sqrt{1-x^2} + C.$

Another method: Substitute $u = \sqrt{(1+x)/(1-x)}$.

41. $3 - 2x - x^2 = -(x^2 + 2x + 1) + 4 = 4 - (x+1)^2$. Let $x + 1 = 2\sin\theta$,

where $-\tfrac{\pi}{2} \le \theta \le \tfrac{\pi}{2}$. Then $dx = 2\cos\theta\,d\theta$ and

$\displaystyle\int \sqrt{3 - 2x - x^2}\,dx = \int \sqrt{4 - (x+1)^2}\,dx = \int \sqrt{4 - 4\sin^2\theta}\,2\cos\theta\,d\theta$

$= 4\int \cos^2\theta\,d\theta = 2\int(1 + \cos 2\theta)\,d\theta$

$= 2\theta + \sin 2\theta + C = 2\theta + 2\sin\theta\cos\theta + C$

$= 2\sin^{-1}\!\left(\dfrac{x+1}{2}\right) + 2 \cdot \dfrac{x+1}{2} \cdot \dfrac{\sqrt{3 - 2x - x^2}}{2} + C$

$= 2\sin^{-1}\!\left(\dfrac{x+1}{2}\right) + \dfrac{x+1}{2}\sqrt{3 - 2x - x^2} + C$

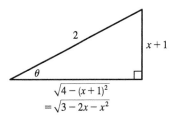

43. The integrand is an odd function, so $\int_{-\pi/2}^{\pi/2} \dfrac{x}{1+\cos^2 x}\, dx = 0$ [by 4.5.6].

45. Let $u = \tan\theta$. Then $du = \sec^2\theta\, d\theta \;\Rightarrow\; \int_0^{\pi/4} \tan^3\theta \sec^2\theta\, d\theta = \int_0^1 u^3\, du = \left[\tfrac{1}{4}u^4\right]_0^1 = \tfrac{1}{4}$.

47. Let $u = \sec\theta$, so that $du = \sec\theta\tan\theta\, d\theta$. Then $\displaystyle\int \dfrac{\sec\theta\tan\theta}{\sec^2\theta - \sec\theta}\, d\theta = \int \dfrac{1}{u^2 - u}\, du = \int \dfrac{1}{u(u-1)}\, du = I$. Now

$$\dfrac{1}{u(u-1)} = \dfrac{A}{u} + \dfrac{B}{u-1} \;\Rightarrow\; 1 = A(u-1) + Bu.\; \text{Set } u=1 \text{ to get } 1=B.\; \text{Set } u=0 \text{ to get } 1=-A,\; \text{so } A=-1.$$

Thus, $I = \displaystyle\int \left(\dfrac{-1}{u} + \dfrac{1}{u-1}\right) du = -\ln|u| + \ln|u-1| + C = \ln|\sec\theta - 1| - \ln|\sec\theta| + C$ [or $\ln|1 - \cos\theta| + C$].

49. Let $u = \theta$, $dv = \tan^2\theta\, d\theta = (\sec^2\theta - 1)\, d\theta \;\Rightarrow\; du = d\theta$ and $v = \tan\theta - \theta$. So

$$\int \theta \tan^2\theta\, d\theta = \theta(\tan\theta - \theta) - \int (\tan\theta - \theta)\, d\theta = \theta\tan\theta - \theta^2 - \ln|\sec\theta| + \tfrac{1}{2}\theta^2 + C$$
$$= \theta\tan\theta - \tfrac{1}{2}\theta^2 - \ln|\sec\theta| + C$$

51. Let $u = \sqrt{x}$, so that $du = \dfrac{1}{2\sqrt{x}}\, dx$. Then

$$\int \dfrac{\sqrt{x}}{1+x^3}\, dx = \int \dfrac{u}{1+u^6}(2u\, du) = 2\int \dfrac{u^2}{1+(u^3)^2}\, du = 2\int \dfrac{1}{1+t^2}\left(\tfrac{1}{3}\, dt\right) \quad \begin{bmatrix} t = u^3 \\ dt = 3u^2\, du \end{bmatrix}$$
$$= \tfrac{2}{3}\tan^{-1} t + C = \tfrac{2}{3}\tan^{-1} u^3 + C = \tfrac{2}{3}\tan^{-1}(x^{3/2}) + C$$

Another method: Let $u = x^{3/2}$ so that $u^2 = x^3$ and $du = \tfrac{3}{2}x^{1/2}\, dx \;\Rightarrow\; \sqrt{x}\, dx = \tfrac{2}{3}\, du$. Then

$$\int \dfrac{\sqrt{x}}{1+x^3}\, dx = \int \dfrac{\tfrac{2}{3}}{1+u^2}\, du = \tfrac{2}{3}\tan^{-1} u + C = \tfrac{2}{3}\tan^{-1}(x^{3/2}) + C.$$

53. Let $u = \sqrt{x}$, so that $x = u^2$ and $dx = 2u\, du$. Thus,

$$\int \dfrac{x}{1+\sqrt{x}}\, dx = \int \dfrac{u^2}{1+u}(2u\, du) = 2\int \dfrac{u^3}{1+u}\, du = 2\int \dfrac{(u^3+1) - 1}{u+1}\, du \quad \text{[or use long division]}$$
$$= 2\int \dfrac{(u+1)(u^2 - u + 1) - 1}{u+1}\, du = 2\int \left(u^2 - u + 1 - \dfrac{1}{u+1}\right) du$$
$$= 2\left(\tfrac{1}{3}u^3 - \tfrac{1}{2}u^2 + u - \ln|u+1|\right) + C = \tfrac{2}{3}u^3 - u^2 + 2u - 2\ln|u+1| + C$$
$$= \tfrac{2}{3}x^{3/2} - x + 2\sqrt{x} - 2\ln\left(\sqrt{x} + 1\right) + C$$

55. Let $u = x - 1$, so that $du = dx$. Then

$$\int x^3(x-1)^{-4}\, dx = \int (u+1)^3 u^{-4}\, du = \int (u^3 + 3u^2 + 3u + 1)u^{-4}\, du = \int (u^{-1} + 3u^{-2} + 3u^{-3} + u^{-4})\, du$$
$$= \ln|u| - 3u^{-1} - \tfrac{3}{2}u^{-2} - \tfrac{1}{3}u^{-3} + C = \ln|x-1| - 3(x-1)^{-1} - \tfrac{3}{2}(x-1)^{-2} - \tfrac{1}{3}(x-1)^{-3} + C$$

57. Let $u = \sqrt{4x+1} \;\Rightarrow\; u^2 = 4x + 1 \;\Rightarrow\; 2u\, du = 4\, dx \;\Rightarrow\; dx = \tfrac{1}{2}u\, du$. So

$$\int \dfrac{1}{x\sqrt{4x+1}}\, dx = \int \dfrac{\tfrac{1}{2}u\, du}{\tfrac{1}{4}(u^2 - 1)u} = 2\int \dfrac{du}{u^2 - 1} = 2\left(\tfrac{1}{2}\right)\ln\left|\dfrac{u-1}{u+1}\right| + C \quad \text{[by Formula 19]}$$
$$= \ln\left|\dfrac{\sqrt{4x+1} - 1}{\sqrt{4x+1} + 1}\right| + C$$

59. Let $2x = \tan\theta \Rightarrow x = \frac{1}{2}\tan\theta$, $dx = \frac{1}{2}\sec^2\theta\,d\theta$, $\sqrt{4x^2+1} = \sec\theta$, so

$$\int \frac{dx}{x\sqrt{4x^2+1}} = \int \frac{\frac{1}{2}\sec^2\theta\,d\theta}{\frac{1}{2}\tan\theta\,\sec\theta} = \int \frac{\sec\theta}{\tan\theta}\,d\theta = \int \csc\theta\,d\theta$$

$$= -\ln|\csc\theta + \cot\theta| + C \quad [\text{or } \ln|\csc\theta - \cot\theta| + C]$$

$$= -\ln\left|\frac{\sqrt{4x^2+1}}{2x} + \frac{1}{2x}\right| + C \quad \left[\text{or } \ln\left|\frac{\sqrt{4x^2+1}}{2x} - \frac{1}{2x}\right| + C\right]$$

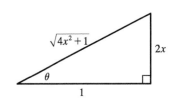

61. $\displaystyle\int x^2 \sinh mx\,dx = \frac{1}{m}x^2 \cosh mx - \frac{2}{m}\int x\cosh mx\,dx \qquad \begin{bmatrix} u = x^2, & dv = \sinh mx\,dx, \\ du = 2x\,dx & v = \frac{1}{m}\cosh mx \end{bmatrix}$

$$= \frac{1}{m}x^2\cosh mx - \frac{2}{m}\left(\frac{1}{m}x\sinh mx - \frac{1}{m}\int \sinh mx\,dx\right) \qquad \begin{bmatrix} U = x, & dV = \cosh mx\,dx, \\ dU = dx & V = \frac{1}{m}\sinh mx \end{bmatrix}$$

$$= \frac{1}{m}x^2\cosh mx - \frac{2}{m^2}x\sinh mx + \frac{2}{m^3}\cosh mx + C$$

63. Let $u = \sqrt{x}$, so that $x = u^2$ and $dx = 2u\,du$. Then $\displaystyle\int \frac{dx}{x + x\sqrt{x}} = \int \frac{2u\,du}{u^2 + u^2 \cdot u} = \int \frac{2}{u(1+u)}\,du = I$.

Now $\dfrac{2}{u(1+u)} = \dfrac{A}{u} + \dfrac{B}{1+u} \Rightarrow 2 = A(1+u) + Bu$. Set $u = -1$ to get $2 = -B$, so $B = -2$. Set $u = 0$ to get $2 = A$.

Thus, $I = \displaystyle\int \left(\frac{2}{u} - \frac{2}{1+u}\right) du = 2\ln|u| - 2\ln|1+u| + C = 2\ln\sqrt{x} - 2\ln\left(1 + \sqrt{x}\right) + C$.

65. Let $u = \sqrt[3]{x+c}$. Then $x = u^3 - c \Rightarrow$

$\int x\sqrt[3]{x+c}\,dx = \int (u^3 - c)u \cdot 3u^2\,du = 3\int (u^6 - cu^3)\,du = \frac{3}{7}u^7 - \frac{3}{4}cu^4 + C = \frac{3}{7}(x+c)^{7/3} - \frac{3}{4}c(x+c)^{4/3} + C$

67. $\dfrac{1}{x^4 - 16} = \dfrac{1}{(x^2-4)(x^2+4)} = \dfrac{1}{(x-2)(x+2)(x^2+4)} = \dfrac{A}{x-2} + \dfrac{B}{x+2} + \dfrac{Cx+D}{x^2+4}$. Multiply by

$(x-2)(x+2)(x^2+4)$ to get $1 = A(x+2)(x^2+4) + B(x-2)(x^2+4) + (Cx+D)(x-2)(x+2)$. Substituting 2 for x

gives $1 = 32A \Leftrightarrow A = \frac{1}{32}$. Substituting -2 for x gives $1 = -32B \Leftrightarrow B = -\frac{1}{32}$. Equating coefficients of x^3 gives

$0 = A + B + C = \frac{1}{32} - \frac{1}{32} + C$, so $C = 0$. Equating constant terms gives $1 = 8A - 8B - 4D = \frac{1}{4} + \frac{1}{4} - 4D$, so

$\frac{1}{2} = -4D \Leftrightarrow D = -\frac{1}{8}$. Thus,

$$\int \frac{dx}{x^4 - 16} = \int \left(\frac{1/32}{x-2} - \frac{1/32}{x+2} - \frac{1/8}{x^2+4}\right) dx = \frac{1}{32}\ln|x-2| - \frac{1}{32}\ln|x+2| - \frac{1}{8} \cdot \frac{1}{2}\tan^{-1}\left(\frac{x}{2}\right) + C$$

$$= \frac{1}{32}\ln\left|\frac{x-2}{x+2}\right| - \frac{1}{16}\tan^{-1}\left(\frac{x}{2}\right) + C$$

69. $\displaystyle\int \frac{d\theta}{1+\cos\theta} = \int\left(\frac{1}{1+\cos\theta}\cdot\frac{1-\cos\theta}{1-\cos\theta}\right)d\theta = \int\frac{1-\cos\theta}{1-\cos^2\theta}\,d\theta = \int\frac{1-\cos\theta}{\sin^2\theta}\,d\theta = \int\left(\frac{1}{\sin^2\theta} - \frac{\cos\theta}{\sin^2\theta}\right)d\theta$

$\qquad = \int(\csc^2\theta - \cot\theta\,\csc\theta)\,d\theta = -\cot\theta + \csc\theta + C$

Another method: Use the substitutions in Exercise 7.4.63.

$\displaystyle\int\frac{d\theta}{1+\cos\theta} = \int\frac{2/(1+t^2)\,dt}{1+(1-t^2)/(1+t^2)} = \int\frac{2\,dt}{(1+t^2)+(1-t^2)} = \int dt = t + C = \tan\!\left(\frac{\theta}{2}\right) + C$

71. Let $y = \sqrt{x}$ so that $dy = \dfrac{1}{2\sqrt{x}}\,dx\;\Rightarrow\;dx = 2\sqrt{x}\,dy = 2y\,dy$. Then

$\displaystyle\int\sqrt{x}\,e^{\sqrt{x}}\,dx = \int ye^y(2y\,dy) = \int 2y^2 e^y\,dy\qquad\begin{bmatrix}u = 2y^2, & dv = e^y\,dy, \\ du = 4y\,dy & v = e^y\end{bmatrix}$

$\qquad = 2y^2 e^y - \int 4y e^y\,dy\qquad\begin{bmatrix}U = 4y, & dV = e^y\,dy, \\ dU = 4\,dy & V = e^y\end{bmatrix}$

$\qquad = 2y^2 e^y - \left(4y e^y - \int 4e^y\,dy\right) = 2y^2 e^y - 4y e^y + 4e^y + C$

$\qquad = 2(y^2 - 2y + 2)e^y + C = 2\!\left(x - 2\sqrt{x} + 2\right)e^{\sqrt{x}} + C$

73. Let $u = \cos^2 x$, so that $du = 2\cos x\,(-\sin x)\,dx$. Then

$\displaystyle\int\frac{\sin 2x}{1+\cos^4 x}\,dx = \int\frac{2\sin x\cos x}{1+(\cos^2 x)^2}\,dx = \int\frac{1}{1+u^2}(-du) = -\tan^{-1}u + C = -\tan^{-1}(\cos^2 x) + C$

75. $\displaystyle\int\frac{1}{\sqrt{x+1}+\sqrt{x}}\,dx = \int\left(\frac{1}{\sqrt{x+1}+\sqrt{x}}\cdot\frac{\sqrt{x+1}-\sqrt{x}}{\sqrt{x+1}-\sqrt{x}}\right)dx = \int\left(\sqrt{x+1}-\sqrt{x}\right)dx$

$\qquad = \tfrac{2}{3}\!\left[(x+1)^{3/2} - x^{3/2}\right] + C$

77. Let $x = \tan\theta$, so that $dx = \sec^2\theta\,d\theta$, $x = \sqrt{3}\;\Rightarrow\;\theta = \tfrac{\pi}{3}$, and $x = 1\;\Rightarrow\;\theta = \tfrac{\pi}{4}$. Then

$\displaystyle\int_1^{\sqrt{3}}\frac{\sqrt{1+x^2}}{x^2}\,dx = \int_{\pi/4}^{\pi/3}\frac{\sec\theta}{\tan^2\theta}\sec^2\theta\,d\theta = \int_{\pi/4}^{\pi/3}\frac{\sec\theta(\tan^2\theta+1)}{\tan^2\theta}\,d\theta = \int_{\pi/4}^{\pi/3}\left(\frac{\sec\theta\,\tan^2\theta}{\tan^2\theta} + \frac{\sec\theta}{\tan^2\theta}\right)d\theta$

$\qquad = \int_{\pi/4}^{\pi/3}(\sec\theta + \csc\theta\,\cot\theta)\,d\theta = \Big[\ln|\sec\theta + \tan\theta| - \csc\theta\Big]_{\pi/4}^{\pi/3}$

$\qquad = \left(\ln|2+\sqrt{3}| - \tfrac{2}{\sqrt{3}}\right) - \left(\ln|\sqrt{2}+1| - \sqrt{2}\right) = \sqrt{2} - \tfrac{2}{\sqrt{3}} + \ln(2+\sqrt{3}) - \ln(1+\sqrt{2})$

79. Let $u = e^x$. Then $x = \ln u$, $dx = du/u\;\Rightarrow$

$\displaystyle\int\frac{e^{2x}}{1+e^x}\,dx = \int\frac{u^2}{1+u}\frac{du}{u} = \int\frac{u}{1+u}\,du = \int\left(1 - \frac{1}{1+u}\right)du = u - \ln|1+u| + C = e^x - \ln(1+e^x) + C.$

81. Let $\theta = \arcsin x$, so that $d\theta = \dfrac{1}{\sqrt{1-x^2}}\,dx$ and $x = \sin\theta$. Then

$\displaystyle\int\frac{x + \arcsin x}{\sqrt{1-x^2}}\,dx = \int(\sin\theta + \theta)\,d\theta = -\cos\theta + \tfrac{1}{2}\theta^2 + C$

$\qquad = -\sqrt{1-x^2} + \tfrac{1}{2}(\arcsin x)^2 + C$

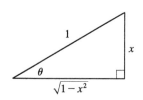

83. $\displaystyle\int \frac{dx}{x\ln x - x} = \int \frac{dx}{x(\ln x - 1)} = \int \frac{du}{u} \quad \begin{bmatrix} u = \ln x - 1, \\ du = (1/x)\, dx \end{bmatrix}$

$\quad = \ln|u| + C = \ln|\ln x - 1| + C$

85. Let $y = \sqrt{1 + e^x}$, so that $y^2 = 1 + e^x$, $2y\, dy = e^x\, dx$, $e^x = y^2 - 1$, and $x = \ln(y^2 - 1)$. Then

$\displaystyle\int \frac{xe^x}{\sqrt{1 + e^x}}\, dx = \int \frac{\ln(y^2 - 1)}{y}(2y\, dy) = 2\int [\ln(y + 1) + \ln(y - 1)]\, dy$

$\quad = 2[(y + 1)\ln(y + 1) - (y + 1) + (y - 1)\ln(y - 1) - (y - 1)] + C \quad$ [by Example 7.1.2]

$\quad = 2[y\ln(y + 1) + \ln(y + 1) - y - 1 + y\ln(y - 1) - \ln(y - 1) - y + 1] + C$

$\quad = 2[y(\ln(y + 1) + \ln(y - 1)) + \ln(y + 1) - \ln(y - 1) - 2y] + C$

$\quad = 2\left[y\ln(y^2 - 1) + \ln\dfrac{y + 1}{y - 1} - 2y\right] + C = 2\left[\sqrt{1 + e^x}\,\ln(e^x) + \ln\dfrac{\sqrt{1 + e^x} + 1}{\sqrt{1 + e^x} - 1} - 2\sqrt{1 + e^x}\right] + C$

$\quad = 2x\sqrt{1 + e^x} + 2\ln\dfrac{\sqrt{1 + e^x} + 1}{\sqrt{1 + e^x} - 1} - 4\sqrt{1 + e^x} + C = 2(x - 2)\sqrt{1 + e^x} + 2\ln\dfrac{\sqrt{1 + e^x} + 1}{\sqrt{1 + e^x} - 1} + C$

87. Let $u = x$, $dv = \sin^2 x \cos x\, dx \ \Rightarrow\ du = dx$, $v = \tfrac{1}{3}\sin^3 x$. Then

$\int x \sin^2 x \cos x\, dx = \tfrac{1}{3}x\sin^3 x - \int \tfrac{1}{3}\sin^3 x\, dx = \tfrac{1}{3}x\sin^3 x - \tfrac{1}{3}\int (1 - \cos^2 x)\sin x\, dx$

$\quad = \dfrac{1}{3}x\sin^3 x + \dfrac{1}{3}\int (1 - y^2)\, dy \quad \begin{bmatrix} u = \cos x, \\ du = -\sin x\, dx \end{bmatrix}$

$\quad = \tfrac{1}{3}x\sin^3 x + \tfrac{1}{3}y - \tfrac{1}{9}y^3 + C = \tfrac{1}{3}x\sin^3 x + \tfrac{1}{3}\cos x - \tfrac{1}{9}\cos^3 x + C$

89. $\displaystyle\int \sqrt{1 - \sin x}\, dx = \int \sqrt{\dfrac{1 - \sin x}{1} \cdot \dfrac{1 + \sin x}{1 + \sin x}}\, dx = \int \sqrt{\dfrac{1 - \sin^2 x}{1 + \sin x}}\, dx$

$\quad = \displaystyle\int \sqrt{\dfrac{\cos^2 x}{1 + \sin x}}\, dx = \int \dfrac{\cos x\, dx}{\sqrt{1 + \sin x}} \quad [\text{assume } \cos x > 0]$

$\quad = \displaystyle\int \dfrac{du}{\sqrt{u}} \quad \begin{bmatrix} u = 1 + \sin x, \\ du = \cos x\, dx \end{bmatrix}$

$\quad = 2\sqrt{u} + C = 2\sqrt{1 + \sin x} + C$

Another method: Let $u = \sin x$ so that $du = \cos x\, dx = \sqrt{1 - \sin^2 x}\, dx = \sqrt{1 - u^2}\, dx$. Then

$\displaystyle\int \sqrt{1 - \sin x}\, dx = \int \sqrt{1 - u}\left(\dfrac{du}{\sqrt{1 - u^2}}\right) = \int \dfrac{1}{\sqrt{1 + u}}\, du = 2\sqrt{1 + u} + C = 2\sqrt{1 + \sin x} + C.$

91. $\displaystyle\int_1^3 \left(\sqrt{\dfrac{9 - x}{x}} - \sqrt{\dfrac{x}{9 - x}}\right) dx = \int_1^3 \left(\dfrac{\sqrt{9 - x}}{\sqrt{x}} - \dfrac{\sqrt{x}}{\sqrt{9 - x}}\right) dx = \int_1^3 \left(\dfrac{9 - x - x}{\sqrt{x}\sqrt{9 - x}}\right) dx = \int_1^3 \left(\dfrac{9 - 2x}{\sqrt{9x - x^2}}\right) dx$

$\quad = \displaystyle\int_8^{18} \left(\dfrac{1}{\sqrt{u}}\right) du \quad \begin{bmatrix} u = 9x - x^2 \\ du = (9 - 2x)\, dx \end{bmatrix} = \int_8^{18} u^{-1/2}\, du = \left[2u^{1/2}\right]_8^{18}$

$\quad = 2\sqrt{18} - 2\sqrt{8} = 6\sqrt{2} - 4\sqrt{2} = 2\sqrt{2}$

93. $\displaystyle\int_0^{\pi/6} \sqrt{1+\sin 2\theta}\, d\theta = \int_0^{\pi/6} \sqrt{(\sin^2\theta + \cos^2\theta) + 2\sin\theta\cos\theta}\, d\theta = \int_0^{\pi/6} \sqrt{(\sin\theta + \cos\theta)^2}\, d\theta$

$\displaystyle = \int_0^{\pi/6} |\sin\theta + \cos\theta|\, d\theta = \int_0^{\pi/6} (\sin\theta + \cos\theta)\, d\theta \quad \begin{bmatrix}\text{since integrand is}\\\text{positive on }[0,\pi/6]\end{bmatrix}$

$\displaystyle = \Big[-\cos\theta + \sin\theta\Big]_0^{\pi/6} = \left(-\frac{\sqrt{3}}{2} + \frac{1}{2}\right) - (-1 + 0) = \frac{3-\sqrt{3}}{2}$

Alternate solution:

$\displaystyle\int_0^{\pi/6} \sqrt{1+\sin 2\theta}\, d\theta = \int_0^{\pi/6} \sqrt{1+\sin 2\theta} \cdot \frac{\sqrt{1-\sin 2\theta}}{\sqrt{1-\sin 2\theta}}\, d\theta = \int_0^{\pi/6} \frac{\sqrt{1-\sin^2 2\theta}}{\sqrt{1-\sin 2\theta}}\, d\theta$

$\displaystyle = \int_0^{\pi/6} \frac{\sqrt{\cos^2 2\theta}}{\sqrt{1-\sin 2\theta}}\, d\theta = \int_0^{\pi/6} \frac{|\cos 2\theta|}{\sqrt{1-\sin 2\theta}}\, d\theta = \int_0^{\pi/6} \frac{\cos 2\theta}{\sqrt{1-\sin 2\theta}}\, d\theta$

$\displaystyle = -\frac{1}{2}\int_1^{1-\sqrt{3}/2} u^{-1/2}\, du \quad [u = 1 - \sin 2\theta,\, du = -2\cos 2\theta\, d\theta]$

$\displaystyle = -\tfrac{1}{2}\Big[2u^{1/2}\Big]_1^{1-\sqrt{3}/2} = 1 - \sqrt{1 - (\sqrt{3}/2)}$

95. The function $y = 2xe^{x^2}$ does have an elementary antiderivative, so we'll use this fact to help evaluate the integral.

$\displaystyle\int (2x^2 + 1)e^{x^2}\, dx = \int 2x^2 e^{x^2}\, dx + \int e^{x^2}\, dx = \int x\left(2xe^{x^2}\right) dx + \int e^{x^2}\, dx$

$\displaystyle = xe^{x^2} - \int e^{x^2}\, dx + \int e^{x^2}\, dx \quad \begin{bmatrix} u = x, & dv = 2xe^{x^2}\, dx,\\ du = dx & v = e^{x^2}\end{bmatrix} = xe^{x^2} + C$

7.6 Integration Using Tables and Technology

Keep in mind that there are several ways to approach many of these exercises, and different methods can lead to different forms of the answer.

1. $\displaystyle\int_0^{\pi/2} \cos 5x \cos 2x\, dx \stackrel{80}{=} \left[\frac{\sin(5-2)x}{2(5-2)} + \frac{\sin(5+2)x}{2(5+2)}\right]_0^{\pi/2} \quad \begin{bmatrix}a=5,\\b=2\end{bmatrix}$

$\displaystyle = \left[\frac{\sin 3x}{6} + \frac{\sin 7x}{14}\right]_0^{\pi/2} = \left(-\frac{1}{6} - \frac{1}{14}\right) - 0 = \frac{-7-3}{42} = -\frac{5}{21}$

3. Let $u = x^2$, so that $du = 2x\, dx$. Thus,

$\displaystyle\int x \arcsin(x^2)\, dx = \tfrac{1}{2}\int \sin^{-1} u\, du \stackrel{87}{=} \tfrac{1}{2}\Big[u\sin^{-1} u + \sqrt{1-u^2}\Big] + C = \tfrac{1}{2}x^2 \sin^{-1}(x^2) + \tfrac{1}{2}\sqrt{1-x^4} + C$.

5. Let $u = y^2$, so that $du = 2y\, dy$. Then,

$\displaystyle\int \frac{y^5}{\sqrt{4+y^4}}\, dy = \int \frac{y^4}{\sqrt{4+y^4}} y\, dy = \frac{1}{2}\int \frac{u^2}{\sqrt{4+u^2}}\, du \stackrel{26}{=} \frac{1}{2}\left[\frac{u}{2}\sqrt{4+u^2} - \frac{4}{2}\ln\left(u + \sqrt{4+u^2}\right)\right] + C$

$\displaystyle = \frac{y^2}{4}\sqrt{4+y^4} - \ln\left(y^2 + \sqrt{4+y^4}\right) + C$

7. $\int_0^{\pi/8} \arctan 2x \, dx = \frac{1}{2} \int_0^{\pi/4} \arctan u \, du \quad [u = 2x, \ du = 2\, dx]$

$\stackrel{89}{=} \frac{1}{2} \left[u \arctan u - \frac{1}{2} \ln(1+u^2) \right]_0^{\pi/4} = \frac{1}{2} \left\{ \left[\frac{\pi}{4} \arctan \frac{\pi}{4} - \frac{1}{2} \ln\left(1 + \frac{\pi^2}{16}\right) \right] - 0 \right\}$

$= \frac{\pi}{8} \arctan \frac{\pi}{4} - \frac{1}{4} \ln\left(1 + \frac{\pi^2}{16}\right)$

9. $\displaystyle\int \frac{\cos x}{\sin^2 x - 9} \, dx = \int \frac{1}{u^2 - 9} \, du \quad \begin{bmatrix} u = \sin x, \\ du = \cos x \, dx \end{bmatrix} \stackrel{20}{=} \frac{1}{2(3)} \ln\left|\frac{u-3}{u+3}\right| + C = \frac{1}{6} \ln\left|\frac{\sin x - 3}{\sin x + 3}\right| + C$

11. $\displaystyle\int \frac{\sqrt{9x^2+4}}{x^2} \, dx = \int \frac{\sqrt{u^2+4}}{u^2/9} \left(\frac{1}{3} du\right) \quad \begin{bmatrix} u = 3x, \\ du = 3\, dx \end{bmatrix}$

$= 3 \int \frac{\sqrt{4+u^2}}{u^2} \, du \stackrel{24}{=} 3 \left[-\frac{\sqrt{4+u^2}}{u} + \ln(u + \sqrt{4+u^2}) \right] + C$

$= -\frac{3\sqrt{4+9x^2}}{3x} + 3 \ln(3x + \sqrt{4+9x^2}) + C = -\frac{\sqrt{9x^2+4}}{x} + 3 \ln(3x + \sqrt{9x^2+4}) + C$

13. $\int_0^\pi \cos^6 \theta \, d\theta \stackrel{74}{=} \left[\frac{1}{6} \cos^5 \theta \sin \theta \right]_0^\pi + \frac{5}{6} \int_0^\pi \cos^4 \theta \, d\theta \stackrel{74}{=} 0 + \frac{5}{6} \left\{ \left[\frac{1}{4} \cos^3 \theta \sin \theta \right]_0^\pi + \frac{3}{4} \int_0^\pi \cos^2 \theta \, d\theta \right\}$

$\stackrel{64}{=} \frac{5}{6} \left\{ 0 + \frac{3}{4} \left[\frac{1}{2}\theta + \frac{1}{4} \sin 2\theta \right]_0^\pi \right\} = \frac{5}{6} \cdot \frac{3}{4} \cdot \frac{\pi}{2} = \frac{5\pi}{16}$

15. $\displaystyle\int \frac{\arctan \sqrt{x}}{\sqrt{x}} \, dx = \int \arctan u \, (2\, du) \quad \begin{bmatrix} u = \sqrt{x}, \\ du = 1/(2\sqrt{x})\, dx \end{bmatrix}$

$\stackrel{89}{=} 2\left[u \arctan u - \frac{1}{2} \ln(1+u^2) \right] + C = 2\sqrt{x} \arctan \sqrt{x} - \ln(1+x) + C$

17. $\displaystyle\int \frac{\coth(1/y)}{y^2} \, dy = \int \coth u \, (-du) \quad \begin{bmatrix} u = 1/y, \\ du = -1/y^2 \, dy \end{bmatrix}$

$\stackrel{106}{=} -\ln|\sinh u| + C = -\ln|\sinh(1/y)| + C$

19. Let $z = 6 + 4y - 4y^2 = 6 - (4y^2 - 4y + 1) + 1 = 7 - (2y-1)^2$, $u = 2y - 1$, and $a = \sqrt{7}$.

Then $z = a^2 - u^2$, $du = 2\, dy$, and

$\int y \sqrt{6 + 4y - 4y^2}\, dy = \int y \sqrt{z}\, dy = \int \frac{1}{2}(u+1) \sqrt{a^2 - u^2}\, \frac{1}{2}\, du = \frac{1}{4} \int u \sqrt{a^2 - u^2}\, du + \frac{1}{4} \int \sqrt{a^2 - u^2}\, du$

$= \frac{1}{4} \int \sqrt{a^2 - u^2}\, du - \frac{1}{8} \int (-2u) \sqrt{a^2 - u^2}\, du$

$\stackrel{30}{=} \frac{u}{8} \sqrt{a^2 - u^2} + \frac{a^2}{8} \sin^{-1}\left(\frac{u}{a}\right) - \frac{1}{8} \int \sqrt{w}\, dw \quad \begin{bmatrix} w = a^2 - u^2, \\ dw = -2u\, du \end{bmatrix}$

$= \frac{2y-1}{8} \sqrt{6 + 4y - 4y^2} + \frac{7}{8} \sin^{-1} \frac{2y-1}{\sqrt{7}} - \frac{1}{8} \cdot \frac{2}{3} w^{3/2} + C$

$= \frac{2y-1}{8} \sqrt{6 + 4y - 4y^2} + \frac{7}{8} \sin^{-1} \frac{2y-1}{\sqrt{7}} - \frac{1}{12}(6 + 4y - 4y^2)^{3/2} + C$

[continued]

This can be rewritten as

$$\sqrt{6+4y-4y^2}\left[\frac{1}{8}(2y-1)-\frac{1}{12}(6+4y-4y^2)\right]+\frac{7}{8}\sin^{-1}\frac{2y-1}{\sqrt{7}}+C$$

$$=\left(\frac{1}{3}y^2-\frac{1}{12}y-\frac{5}{8}\right)\sqrt{6+4y-4y^2}+\frac{7}{8}\sin^{-1}\left(\frac{2y-1}{\sqrt{7}}\right)+C$$

$$=\frac{1}{24}(8y^2-2y-15)\sqrt{6+4y-4y^2}+\frac{7}{8}\sin^{-1}\left(\frac{2y-1}{\sqrt{7}}\right)+C$$

21. Let $u=\sin x$. Then $du=\cos x\,dx$, so

$$\int \sin^2 x\,\cos x\,\ln(\sin x)\,dx=\int u^2\ln u\,du\stackrel{101}{=}\frac{u^{2+1}}{(2+1)^2}[(2+1)\ln u-1]+C=\tfrac{1}{9}u^3(3\ln u-1)+C$$

$$=\tfrac{1}{9}\sin^3 x\,[3\ln(\sin x)-1]+C$$

23. $\displaystyle\int\frac{\sin 2\theta}{\sqrt{\cos^4\theta+4}}\,d\theta=\int\frac{2\sin\theta\cos\theta}{\sqrt{\cos^4\theta+4}}\,d\theta=-\int\frac{1}{\sqrt{u^2+4}}\,du\quad\begin{bmatrix}u=\cos^2\theta\\du=-2\sin\theta\cos\theta\,d\theta\end{bmatrix}$

$$\stackrel{25}{=}\ln(u+\sqrt{u^2+4})+C=-\ln(\cos^2\theta+\sqrt{\cos^4\theta+4})+C$$

25. $\displaystyle\int x^3 e^{2x}\,dx\stackrel{97}{=}\tfrac{1}{2}x^3 e^{2x}-\tfrac{3}{2}\int x^2 e^{2x}\,dx\stackrel{97}{=}\tfrac{1}{2}x^3 e^{2x}-\tfrac{3}{2}\left(\tfrac{1}{2}x^2 e^{2x}-\int xe^{2x}\,dx\right)$

$$\stackrel{96}{=}\tfrac{1}{2}x^3 e^{2x}-\tfrac{3}{2}\left(\tfrac{1}{2}x^2 e^{2x}-\tfrac{1}{4}(2x-1)e^{2x}\right)+C$$

$$=\tfrac{1}{2}x^3 e^{2x}-\tfrac{3}{4}x^2 e^{2x}+\tfrac{3}{4}xe^{2x}-\tfrac{3}{8}e^{2x}+C$$

$$=\tfrac{1}{2}e^{2x}\left(x^3-\tfrac{3}{2}x^2+\tfrac{3}{2}x-\tfrac{3}{4}\right)+C$$

27. $\displaystyle\int\cos^5 y\,dy\stackrel{74}{=}\tfrac{1}{5}\cos^4 y\sin y+\tfrac{4}{5}\int\cos^3 y\,dy\stackrel{68}{=}\tfrac{1}{5}\cos^4 y\sin y+\tfrac{4}{5}\left[\tfrac{1}{3}(2+\cos^2 y)\sin y\right]+C$

$$=\tfrac{1}{5}\cos^4 y\sin y+\tfrac{8}{15}\sin y+\tfrac{4}{15}\cos^2 y\sin y+C=\tfrac{1}{5}\sin y\left(\cos^4 y+\tfrac{4}{3}\cos^2 y+\tfrac{8}{3}\right)+C$$

29. $\displaystyle\int\frac{\cos^{-1}(x^{-2})}{x^3}\,dx=-\frac{1}{2}\int\cos^{-1}u\,du\quad\begin{bmatrix}u=x^{-2},\\du=-2x^{-3}\,dx\end{bmatrix}$

$$\stackrel{88}{=}-\tfrac{1}{2}\left(u\cos^{-1}u-\sqrt{1-u^2}\right)+C=-\tfrac{1}{2}x^{-2}\cos^{-1}(x^{-2})+\tfrac{1}{2}\sqrt{1-x^{-4}}+C$$

31. Let $u=e^x$. Then $x=\ln u$, $dx=du/u$, so

$$\int\sqrt{e^{2x}-1}\,dx=\int\frac{\sqrt{u^2-1}}{u}\,du\stackrel{41}{=}\sqrt{u^2-1}-\cos^{-1}(1/u)+C=\sqrt{e^{2x}-1}-\cos^{-1}(e^{-x})+C.$$

33. $\displaystyle\int\frac{x^4}{\sqrt{x^{10}-2}}\,dx=\int\frac{x^4}{\sqrt{(x^5)^2-2}}\,dx=\frac{1}{5}\int\frac{1}{\sqrt{u^2-2}}\,du\quad\begin{bmatrix}u=x^5,\\du=5x^4\,dx\end{bmatrix}$

$$\stackrel{43}{=}\tfrac{1}{5}\ln\left|u+\sqrt{u^2-2}\right|+C=\tfrac{1}{5}\ln\left|x^5+\sqrt{x^{10}-2}\right|+C$$

35. Use disks about the x-axis:

$$V=\int_0^\pi\pi(\sin^2 x)^2\,dx=\pi\int_0^\pi\sin^4 x\,dx\stackrel{73}{=}\pi\left\{\left[-\tfrac{1}{4}\sin^3 x\cos x\right]_0^\pi+\tfrac{3}{4}\int_0^\pi\sin^2 x\,dx\right\}$$

$$\stackrel{63}{=}\pi\left\{0+\tfrac{3}{4}\left[\tfrac{1}{2}x-\tfrac{1}{4}\sin 2x\right]_0^\pi\right\}=\pi\left[\tfrac{3}{4}\left(\tfrac{1}{2}\pi-0\right)\right]=\tfrac{3}{8}\pi^2$$

37. (a) $\dfrac{d}{du}\left[\dfrac{1}{b^3}\left(a+bu-\dfrac{a^2}{a+bu}-2a\ln|a+bu|\right)+C\right]=\dfrac{1}{b^3}\left[b+\dfrac{ba^2}{(a+bu)^2}-\dfrac{2ab}{(a+bu)}\right]$

$=\dfrac{1}{b^3}\left[\dfrac{b(a+bu)^2+ba^2-(a+bu)2ab}{(a+bu)^2}\right]$

$=\dfrac{1}{b^3}\left[\dfrac{b^3u^2}{(a+bu)^2}\right]=\dfrac{u^2}{(a+bu)^2}$

(b) Let $t=a+bu \;\Rightarrow\; dt=b\,du$. Note that $u=\dfrac{t-a}{b}$ and $du=\dfrac{1}{b}\,dt$.

$\displaystyle\int\dfrac{u^2\,du}{(a+bu)^2}=\dfrac{1}{b^3}\int\dfrac{(t-a)^2}{t^2}\,dt=\dfrac{1}{b^3}\int\dfrac{t^2-2at+a^2}{t^2}\,dt=\dfrac{1}{b^3}\int\left(1-\dfrac{2a}{t}+\dfrac{a^2}{t^2}\right)dt$

$=\dfrac{1}{b^3}\left(t-2a\ln|t|-\dfrac{a^2}{t}\right)+C=\dfrac{1}{b^3}\left(a+bu-\dfrac{a^2}{a+bu}-2a\ln|a+bu|\right)+C$

39. Maple and Mathematica both give $\int\sec^4 x\,dx=\tfrac{2}{3}\tan x+\tfrac{1}{3}\tan x\sec^2 x$. Using Formula 77, we get
$\int\sec^4 x\,dx=\tfrac{1}{3}\tan x\sec^2 x+\tfrac{2}{3}\int\sec^2 x\,dx=\tfrac{1}{3}\tan x\sec^2 x+\tfrac{2}{3}\tan x+C$.

41. Maple gives $\displaystyle\int x^2\sqrt{2^2+x^2}\,dx=\tfrac{1}{4}x(x^2+4)^{3/2}-\tfrac{1}{2}x\sqrt{x^2+4}-2\,\text{arcsinh}\left(\tfrac{1}{2}x\right)$. Applying the command `convert(%,ln);` yields

$\tfrac{1}{4}x(x^2+4)^{3/2}-\tfrac{1}{2}x\sqrt{x^2+4}-2\ln\left(\tfrac{1}{2}x+\tfrac{1}{2}\sqrt{x^2+4}\right)=\tfrac{1}{4}x(x^2+4)^{1/2}\left[(x^2+4)-2\right]-2\ln\left[(x+\sqrt{x^2+4})/2\right]$

$=\tfrac{1}{4}x(x^2+2)\sqrt{x^2+4}-2\ln\left(\sqrt{x^2+4}+x\right)+2\ln 2$

Mathematica gives $\tfrac{1}{4}x(2+x^2)\sqrt{3+x^2}-2\,\text{arcsinh}(x/2)$. Applying the `TrigToExp` and `Simplify` commands gives
$\tfrac{1}{4}\left[x(2+x^2)\sqrt{4+x^2}-8\log\left(\tfrac{1}{2}\left(x+\sqrt{4+x^2}\right)\right)\right]=\tfrac{1}{4}x(x^2+2)\sqrt{x^2+4}-2\ln\left(x+\sqrt{4+x^2}\right)+2\ln 2$, so all are equivalent (without constant).

Now use Formula 22 to get

$\displaystyle\int x^2\sqrt{2^2+x^2}\,dx=\dfrac{x}{8}(2^2+2x^2)\sqrt{2^2+x^2}-\dfrac{2^4}{8}\ln\left(x+\sqrt{2^2+x^2}\right)+C$

$=\dfrac{x}{8}(2)(2+x^2)\sqrt{4+x^2}-2\ln\left(x+\sqrt{4+x^2}\right)+C$

$=\tfrac{1}{4}x(x^2+2)\sqrt{x^2+4}-2\ln\left(\sqrt{x^2+4}+x\right)+C$

43. Maple gives $\displaystyle\int\cos^4 x\,dx=\dfrac{\sin x\cos^3 x}{4}+\dfrac{3\sin x\cos x}{8}+\dfrac{3x}{8}$, whereas Mathematica gives

$\dfrac{3x}{8}+\dfrac{1}{4}\sin(2x)+\dfrac{1}{32}\sin(4x)=\dfrac{3x}{8}+\dfrac{1}{4}(2\sin x\cos x)+\dfrac{1}{32}(2\sin 2x\cos 2x)$

$=\dfrac{3x}{8}+\dfrac{1}{2}\sin x\cos x+\dfrac{1}{16}[2\sin x\cos x(2\cos^2 x-1)]$

$=\dfrac{3x}{8}+\dfrac{1}{2}\sin x\cos x+\dfrac{1}{4}\sin x\cos^3 x-\dfrac{1}{8}\sin x\cos x$,

so both are equivalent.

[continued]

Using tables,

$$\int \cos^4 x\, dx \stackrel{74}{=} \tfrac{1}{4}\cos^3 x \sin x + \tfrac{3}{4}\int \cos^2 x\, dx \stackrel{64}{=} \tfrac{1}{4}\cos^3 x \sin x + \tfrac{3}{4}\left(\tfrac{1}{2}x + \tfrac{1}{4}\sin 2x\right) + C$$

$$= \tfrac{1}{4}\cos^3 x \sin x + \tfrac{3}{8}x + \tfrac{3}{16}(2\sin x \cos x) + C = \tfrac{1}{4}\cos^3 x \sin x + \tfrac{3}{8}x + \tfrac{3}{8}\sin x \cos x + C$$

45. Maple gives $\int \tan^5 x\, dx = \tfrac{1}{4}\tan^4 x - \tfrac{1}{2}\tan^2 x + \tfrac{1}{2}\ln(1 + \tan^2 x)$, and Mathematica gives

$\int \tan^5 x\, dx = \tfrac{1}{4}[-1 - 2\cos(2x)]\sec^4 x - \ln(\cos x)$. These expressions are equivalent, and neither includes absolute value bars or a constant of integration. Note that Mathematica's expression suggests that the integral is undefined where $\cos x < 0$, which is not the case. Using Formula 75, $\int \tan^5 x\, dx = \tfrac{1}{5-1}\tan^{5-1}x - \int \tan^{5-2}x\, dx = \tfrac{1}{4}\tan^4 x - \int \tan^3 x\, dx$. Using Formula 69, $\int \tan^3 x\, dx = \tfrac{1}{2}\tan^2 x + \ln|\cos x| + C$, so $\int \tan^5 x\, dx = \tfrac{1}{4}\tan^4 x - \tfrac{1}{2}\tan^2 x - \ln|\cos x| + C$.

47. (a) $F(x) = \displaystyle\int f(x)\, dx = \int \dfrac{1}{x\sqrt{1-x^2}}\, dx \stackrel{35}{=} -\dfrac{1}{1}\ln\left|\dfrac{1+\sqrt{1-x^2}}{x}\right| + C = -\ln\left|\dfrac{1+\sqrt{1-x^2}}{x}\right| + C$.

f has domain $\{x \mid x \neq 0,\, 1 - x^2 > 0\} = \{x \mid x \neq 0,\, |x| < 1\} = (-1, 0) \cup (0, 1)$. F has the same domain.

(b) Mathematica gives $F(x) = \ln x - \ln\left(1 + \sqrt{1-x^2}\right)$. Maple gives $F(x) = -\operatorname{arctanh}\left(1/\sqrt{1-x^2}\right)$. This function has domain $\{x \mid |x| < 1,\, -1 < 1/\sqrt{1-x^2} < 1\} = \{x \mid |x| < 1,\, 1/\sqrt{1-x^2} < 1\} = \{x \mid |x| < 1,\, \sqrt{1-x^2} > 1\} = \emptyset$, the empty set! If we apply the command convert(%,ln); to Maple's answer, we get

$-\dfrac{1}{2}\ln\left(\dfrac{1}{\sqrt{1-x^2}} + 1\right) + \dfrac{1}{2}\ln\left(1 - \dfrac{1}{\sqrt{1-x^2}}\right)$, which has the same domain, $\emptyset$.

7.7 Approximate Integration

1. (a) $\Delta x = (b-a)/n = (4-0)/2 = 2$

$L_2 = \displaystyle\sum_{i=1}^{2} f(x_{i-1})\,\Delta x = f(x_0)\cdot 2 + f(x_1)\cdot 2 = 2\,[f(0) + f(2)] = 2(0.5 + 2.5) = 6$

$R_2 = \displaystyle\sum_{i=1}^{2} f(x_i)\,\Delta x = f(x_1)\cdot 2 + f(x_2)\cdot 2 = 2\,[f(2) + f(4)] = 2(2.5 + 3.5) = 12$

$M_2 = \displaystyle\sum_{i=1}^{2} f(\overline{x}_i)\,\Delta x = f(\overline{x}_1)\cdot 2 + f(\overline{x}_2)\cdot 2 = 2\,[f(1) + f(3)] \approx 2(1.6 + 3.2) = 9.6$

(b)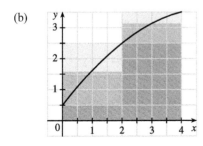

L_2 is an underestimate, since the area under the small rectangles is less than the area under the curve, and R_2 is an overestimate, since the area under the large rectangles is greater than the area under the curve. It appears that M_2 is an overestimate, though it is fairly close to I. See the solution to Exercise 47 for a proof of the fact that if f is concave down on $[a, b]$, then the Midpoint Rule is an overestimate of $\int_a^b f(x)\, dx$.

(c) $T_2 = \left(\frac{1}{2}\Delta x\right)[f(x_0) + 2f(x_1) + f(x_2)] = \frac{2}{2}[f(0) + 2f(2) + f(4)] = 0.5 + 2(2.5) + 3.5 = 9$.

This approximation is an underestimate, since the graph is concave down. Thus, $T_2 = 9 < I$. See the solution to Exercise 47 for a general proof of this conclusion.

(d) For any n, we will have $L_n < T_n < I < M_n < R_n$.

3. $f(x) = \cos(x^2)$, $\Delta x = \frac{1-0}{4} = \frac{1}{4}$

(a) $T_4 = \frac{1}{4 \cdot 2}\left[f(0) + 2f\left(\frac{1}{4}\right) + 2f\left(\frac{2}{4}\right) + 2f\left(\frac{3}{4}\right) + f(1)\right] \approx 0.895759$

(b) $M_4 = \frac{1}{4}\left[f\left(\frac{1}{8}\right) + f\left(\frac{3}{8}\right) + f\left(\frac{5}{8}\right) + f\left(\frac{7}{8}\right)\right] \approx 0.908907$

The graph shows that f is concave down on $[0, 1]$. So T_4 is an underestimate and M_4 is an overestimate. We can conclude that $0.895759 < \int_0^1 \cos(x^2)\,dx < 0.908907$.

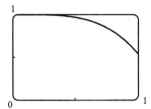

5. (a) $f(x) = x\sin x$, $\Delta x = \dfrac{b-a}{n} = \dfrac{\pi - 0}{6} = \dfrac{\pi}{6}$

$M_6 = \dfrac{\pi}{6}\left[f\left(\dfrac{\pi}{12}\right) + f\left(\dfrac{3\pi}{12}\right) + f\left(\dfrac{5\pi}{12}\right) + f\left(\dfrac{7\pi}{12}\right) + f\left(\dfrac{9\pi}{12}\right) + f\left(\dfrac{11\pi}{12}\right)\right] \approx 3.177769$

(b) $S_6 = \dfrac{\pi}{6 \cdot 3}\left[f(0) + 4f\left(\dfrac{\pi}{6}\right) + 2f\left(\dfrac{2\pi}{6}\right) + 4f\left(\dfrac{3\pi}{6}\right) + 2f\left(\dfrac{4\pi}{6}\right) + 4f\left(\dfrac{5\pi}{6}\right) + f\left(\dfrac{6\pi}{6}\right)\right] \approx 3.142949$

Actual: $I = \int_0^\pi x\sin x\,dx = \left[-x\cos x + \sin x\right]_0^\pi$ [use parts with $u = x$ and $dv = \sin x\,dx$]

$= (-\pi(-1) - 0) - (0 + 0) = \pi \approx 3.141593$

Errors: $E_M = $ actual $- M_6 \approx 3.141593 - 3.177769 \approx -0.036176$

$E_S = $ actual $- S_6 \approx 3.141593 - 3.142949 \approx -0.001356$

7. $f(x) = \sqrt{1+x^3}$, $\Delta x = \dfrac{b-a}{n} = \dfrac{1-0}{4} = \dfrac{1}{4}$

(a) $T_4 = \frac{1}{4 \cdot 2}[f(0) + 2f(0.25) + 2f(0.5) + 2f(0.75) + f(1)] \approx 1.116993$

(b) $M_4 = \frac{1}{4}[f(0.125) + f(0.375) + f(0.625) + f(0.875)] \approx 1.108667$

(c) $S_4 = \frac{1}{4 \cdot 3}[f(0) + 4f(0.25) + 2f(0.5) + 4f(0.75) + f(1)] \approx 1.111363$

9. $f(x) = \sqrt{e^x - 1}$, $\Delta x = \dfrac{b-a}{n} = \dfrac{1-0}{10} = \dfrac{1}{10}$

(a) $T_{10} = \frac{1}{10 \cdot 2}[f(0) + 2f(0.1) + 2f(0.2) + 2f(0.3) + 2f(0.4) + 2f(0.5) + 2f(0.6)$
$\quad + 2f(0.7) + 2f(0.8) + 2f(0.9) + f(1)]$

≈ 0.777722

(b) $M_{10} = \frac{1}{10}[f(0.05) + f(0.15) + f(0.25) + f(0.35) + f(0.45) + f(0.55)$
$\quad + f(0.65) + f(0.75) + f(0.85) + f(0.95)]$

≈ 0.784958

(c) $S_{10} = \frac{1}{10 \cdot 3}[f(0) + 4f(0.1) + 2f(0.2) + 4f(0.3) + 2f(0.4) + 4f(0.5) + 2f(0.6)$
$\qquad + 4f(0.7) + 2f(0.8) + 4f(0.9) + f(1)]$
≈ 0.780895

11. $f(x) = e^{x+\cos x}$, $\Delta x = \frac{2-(-1)}{6} = \frac{1}{2}$

(a) $T_6 = \frac{1}{2}[f(-1.0) + 2f(-0.5) + 2f(0) + 2f(0.5) + 2f(1) + 2f(1.5) + f(2.0)] \approx 10.185560$

(b) $M_6 = \frac{1}{2}[f(-0.75) + f(-0.25) + f(0.25) + f(0.75) + f(1.25) + f(1.75)] \approx 10.208618$

(c) $S_6 = \frac{1}{2 \cdot 3}[f(-1.0) + 4f(-0.5) + 2f(0) + 4f(0.5) + 2f(1.0) + 4f(1.5) + f(2.0)] \approx 10.201790$

13. $f(y) = \sqrt{y} \cos y$, $\Delta y = \frac{4-0}{8} = \frac{1}{2}$

(a) $T_8 = \frac{1}{2 \cdot 2}\left[f(0) + 2f\left(\frac{1}{2}\right) + 2f(1) + 2f\left(\frac{3}{2}\right) + 2f(2) + 2f\left(\frac{5}{2}\right) + 2f(3) + 2f\left(\frac{7}{2}\right) + f(4)\right] \approx -2.364034$

(b) $M_8 = \frac{1}{2}\left[f\left(\frac{1}{4}\right) + f\left(\frac{3}{4}\right) + f\left(\frac{5}{4}\right) + f\left(\frac{7}{4}\right) + f\left(\frac{9}{4}\right) + f\left(\frac{11}{4}\right) + f\left(\frac{13}{4}\right) + f\left(\frac{15}{4}\right)\right] \approx -2.310690$

(c) $S_8 = \frac{1}{2 \cdot 3}\left[f(0) + 4f\left(\frac{1}{2}\right) + 2f(1) + 4f\left(\frac{3}{2}\right) + 2f(2) + 4f\left(\frac{5}{2}\right) + 2f(3) + 4f\left(\frac{7}{2}\right) + f(4)\right] \approx -2.346520$

15. $f(x) = \frac{x^2}{1+x^4}$, $\Delta x = \frac{1-0}{10} = \frac{1}{10}$

(a) $T_{10} = \frac{1}{10 \cdot 2}\{f(0) + 2[f(0.1) + f(0.2) + \cdots + f(0.9)] + f(1)\} \approx 0.243747$

(b) $M_{10} = \frac{1}{10}[f(0.05) + f(0.15) + \cdots + f(0.85) + f(0.95)] \approx 0.243748$

(c) $S_{10} = \frac{1}{10 \cdot 3}[f(0) + 4f(0.1) + 2f(0.2) + 4f(0.3) + 2f(0.4) + 4f(0.5) + 2f(0.6)$
$\qquad + 4f(0.7) + 2f(0.8) + 4f(0.9) + f(1)] \approx 0.243751$

Note: $\int_0^1 f(x)\,dx \approx 0.24374775$. This is a rare case where the Trapezoidal and Midpoint Rules give better approximations than Simpson's Rule.

17. $f(x) = \ln(1+e^x)$, $\Delta x = \frac{4-0}{8} = \frac{1}{2}$

(a) $T_8 = \frac{1}{2 \cdot 2}\{f(0) + 2[f(0.5) + f(1) + \cdots + f(3) + f(3.5)] + f(4)\} \approx 8.814278$

(b) $M_8 = \frac{1}{2}[f(0.25) + f(0.75) + \cdots + f(3.25) + f(3.75)] \approx 8.799212$

(c) $S_8 = \frac{1}{2 \cdot 3}[f(0) + 4f(0.5) + 2f(1) + 4f(1.5) + 2f(2) + 4f(2.5) + 2f(3) + 4f(3.5) + f(4)] \approx 8.804229$

19. $f(x) = \cos(x^2)$, $\Delta x = \frac{1-0}{8} = \frac{1}{8}$

(a) $T_8 = \frac{1}{8 \cdot 2}\{f(0) + 2[f(\frac{1}{8}) + f(\frac{2}{8}) + \cdots + f(\frac{7}{8})] + f(1)\} \approx 0.902333$

$M_8 = \frac{1}{8}[f(\frac{1}{16}) + f(\frac{3}{16}) + f(\frac{5}{16}) + \cdots + f(\frac{15}{16})] = 0.905620$

(b) $f(x) = \cos(x^2)$, $f'(x) = -2x\sin(x^2)$, $f''(x) = -2\sin(x^2) - 4x^2\cos(x^2)$. For $0 \le x \le 1$, sin and cos are positive, so $|f''(x)| = 2\sin(x^2) + 4x^2\cos(x^2) \le 2 \cdot 1 + 4 \cdot 1 \cdot 1 = 6$ since $\sin(x^2) \le 1$ and $\cos(x^2) \le 1$ for all x, and $x^2 \le 1$ for $0 \le x \le 1$. So for $n = 8$, we take $K = 6$, $a = 0$, and $b = 1$ in Theorem 3, to get

$|E_T| \le 6 \cdot 1^3/(12 \cdot 8^2) = \frac{1}{128} = 0.0078125$ and $|E_M| \le \frac{1}{256} = 0.00390625$. [A better estimate is obtained by noting from a graph of f'' that $|f''(x)| \le 4$ for $0 \le x \le 1$.]

(c) Take $K = 6$ [as in part (b)] in Theorem 3. $|E_T| \leq \dfrac{K(b-a)^3}{12n^2} \leq 0.0001 \Leftrightarrow \dfrac{6(1-0)^3}{12n^2} \leq 10^{-4} \Leftrightarrow$

$\dfrac{1}{2n^2} \leq \dfrac{1}{10^4} \Leftrightarrow 2n^2 \geq 10^4 \Leftrightarrow n^2 \geq 5000 \Leftrightarrow n \geq 71$. Take $n = 71$ for T_n. For E_M, again take $K = 6$ in Theorem 3 to get $|E_M| \leq 10^{-4} \Leftrightarrow 4n^2 \geq 10^4 \Leftrightarrow n^2 \geq 2500 \Leftrightarrow n \geq 50$. Take $n = 50$ for M_n.

21. $f(x) = \sin x$, $\Delta x = \dfrac{\pi - 0}{10} = \dfrac{\pi}{10}$

(a) $T_{10} = \dfrac{\pi}{10 \cdot 2}\left[f(0) + 2f\left(\dfrac{\pi}{10}\right) + 2f\left(\dfrac{2\pi}{10}\right) + \cdots + 2f\left(\dfrac{9\pi}{10}\right) + f(\pi)\right] \approx 1.983524$

$M_{10} = \dfrac{\pi}{10}\left[f\left(\dfrac{\pi}{20}\right) + f\left(\dfrac{3\pi}{20}\right) + f\left(\dfrac{5\pi}{20}\right) + \cdots + f\left(\dfrac{19\pi}{20}\right)\right] \approx 2.008248$

$S_{10} = \dfrac{\pi}{10 \cdot 3}\left[f(0) + 4f\left(\dfrac{\pi}{10}\right) + 2f\left(\dfrac{2\pi}{10}\right) + 4f\left(\dfrac{3\pi}{10}\right) + \cdots + 4f\left(\dfrac{9\pi}{10}\right) + f(\pi)\right] \approx 2.000110$

Since $I = \int_0^\pi \sin x \, dx = \left[-\cos x\right]_0^\pi = 1 - (-1) = 2$, $E_T = I - T_{10} \approx 0.016476$, $E_M = I - M_{10} \approx -0.008248$,

and $E_S = I - S_{10} \approx -0.000110$.

(b) $f(x) = \sin x \Rightarrow \left|f^{(n)}(x)\right| \leq 1$, so take $K = 1$ for all error estimates.

$|E_T| \leq \dfrac{K(b-a)^3}{12n^2} = \dfrac{1(\pi-0)^3}{12(10)^2} = \dfrac{\pi^3}{1200} \approx 0.025839$. $|E_M| \leq \dfrac{|E_T|}{2} = \dfrac{\pi^3}{2400} \approx 0.012919$.

$|E_S| \leq \dfrac{K(b-a)^5}{180n^4} = \dfrac{1(\pi-0)^5}{180(10)^4} = \dfrac{\pi^5}{1,800,000} \approx 0.000170$.

The actual error is about 64% of the error estimate in all three cases.

(c) $|E_T| \leq 0.00001 \Leftrightarrow \dfrac{\pi^3}{12n^2} \leq \dfrac{1}{10^5} \Leftrightarrow n^2 \geq \dfrac{10^5\pi^3}{12} \Rightarrow n \geq 508.3$. Take $n = 509$ for T_n.

$|E_M| \leq 0.00001 \Leftrightarrow \dfrac{\pi^3}{24n^2} \leq \dfrac{1}{10^5} \Leftrightarrow n^2 \geq \dfrac{10^5\pi^3}{24} \Rightarrow n \geq 359.4$. Take $n = 360$ for M_n.

$|E_S| \leq 0.00001 \Leftrightarrow \dfrac{\pi^5}{180n^4} \leq \dfrac{1}{10^5} \Leftrightarrow n^4 \geq \dfrac{10^5\pi^5}{180} \Rightarrow n \geq 20.3$.

Take $n = 22$ for S_n (since n must be even).

23. (a) Using a CAS, we differentiate $f(x) = e^{\cos x}$ twice, and find that

$f''(x) = e^{\cos x}(\sin^2 x - \cos x)$. From the graph, we see that the maximum

value of $|f''(x)|$ occurs at the endpoints of the interval $[0, 2\pi]$.

Since $f''(0) = -e$, we can use $K = e$ or $K = 2.8$.

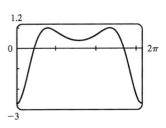

(b) A CAS gives $M_{10} \approx 7.954926518$. (In Maple, use Student[Calculus1][RiemannSum] or

Student[Calculus1][ApproximateInt].)

(c) Using Theorem 3 for the Midpoint Rule, with $K = e$, we get $|E_M| \leq \dfrac{e(2\pi - 0)^3}{24 \cdot 10^2} \approx 0.280945995$.

With $K = 2.8$, we get $|E_M| \leq \dfrac{2.8(2\pi - 0)^3}{24 \cdot 10^2} = 0.289391916$.

(d) A CAS gives $I \approx 7.954926521$.

(e) The actual error is only about 3×10^{-9}, much less than the estimate in part (c).

(f) We use the CAS to differentiate twice more, and then graph
$$f^{(4)}(x) = e^{\cos x}(\sin^4 x - 6\sin^2 x \cos x + 3 - 7\sin^2 x + \cos x).$$
From the graph, we see that the maximum value of $\left|f^{(4)}(x)\right|$ occurs at the endpoints of the interval $[0, 2\pi]$. Since $f^{(4)}(0) = 4e$, we can use $K = 4e$ or $K = 10.9$.

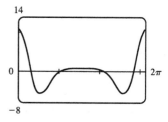

(g) A CAS gives $S_{10} \approx 7.953789422$. (In Maple, use `Student[Calculus1][ApproximateInt]`.)

(h) Using Theorem 4 with $K = 4e$, we get $|E_S| \leq \dfrac{4e(2\pi - 0)^5}{180 \cdot 10^4} \approx 0.059153618$.

With $K = 10.9$, we get $|E_S| \leq \dfrac{10.9(2\pi - 0)^5}{180 \cdot 10^4} \approx 0.059299814$.

(i) The actual error is about $7.954926521 - 7.953789422 \approx 0.00114$. This is quite a bit smaller than the estimate in part (h), though the difference is not nearly as great as it was in the case of the Midpoint Rule.

(j) To ensure that $|E_S| \leq 0.0001$, we use Theorem 4: $|E_S| \leq \dfrac{4e(2\pi)^5}{180 \cdot n^4} \leq 0.0001 \;\Rightarrow\; \dfrac{4e(2\pi)^5}{180 \cdot 0.0001} \leq n^4 \;\Rightarrow\;$
$n^4 \geq 5{,}915{,}362 \;\Leftrightarrow\; n \geq 49.3$. So we must take $n \geq 50$ to ensure that $|I - S_n| \leq 0.0001$.
($K = 10.9$ leads to the same value of n.)

25. $I = \int_0^1 xe^x\,dx = \left[(x-1)e^x\right]_0^1$ [parts or Formula 96] $= 0 - (-1) = 1$, $f(x) = xe^x$, $\Delta x = 1/n$

$n = 5$: $L_5 = \tfrac{1}{5}[f(0) + f(0.2) + f(0.4) + f(0.6) + f(0.8)] \approx 0.742943$

           $R_5 = \tfrac{1}{5}[f(0.2) + f(0.4) + f(0.6) + f(0.8) + f(1)] \approx 1.286599$

           $T_5 = \tfrac{1}{5 \cdot 2}[f(0) + 2f(0.2) + 2f(0.4) + 2f(0.6) + 2f(0.8) + f(1)] \approx 1.014771$

           $M_5 = \tfrac{1}{5}[f(0.1) + f(0.3) + f(0.5) + f(0.7) + f(0.9)] \approx 0.992621$

           $E_L = I - L_5 \approx 1 - 0.742943 = 0.257057$

           $E_R \approx 1 - 1.286599 = -0.286599$

           $E_T \approx 1 - 1.014771 = -0.014771$

           $E_M \approx 1 - 0.992621 = 0.007379$

$n = 10$: $L_{10} = \tfrac{1}{10}[f(0) + f(0.1) + f(0.2) + \cdots + f(0.9)] \approx 0.867782$

            $R_{10} = \tfrac{1}{10}[f(0.1) + f(0.2) + \cdots + f(0.9) + f(1)] \approx 1.139610$

            $T_{10} = \tfrac{1}{10 \cdot 2}\{f(0) + 2[f(0.1) + f(0.2) + \cdots + f(0.9)] + f(1)\} \approx 1.003696$

            $M_{10} = \tfrac{1}{10}[f(0.05) + f(0.15) + \cdots + f(0.85) + f(0.95)] \approx 0.998152$

            $E_L = I - L_{10} \approx 1 - 0.867782 = 0.132218$

            $E_R \approx 1 - 1.139610 = -0.139610$

            $E_T \approx 1 - 1.003696 = -0.003696$

            $E_M \approx 1 - 0.998152 = 0.001848$ [continued]

$n = 20$: $\quad L_{20} = \frac{1}{20}[f(0) + f(0.05) + f(0.10) + \cdots + f(0.95)] \approx 0.932967$

$R_{20} = \frac{1}{20}[f(0.05) + f(0.10) + \cdots + f(0.95) + f(1)] \approx 1.068881$

$T_{20} = \frac{1}{20 \cdot 2}\{f(0) + 2[f(0.05) + f(0.10) + \cdots + f(0.95)] + f(1)\} \approx 1.000924$

$M_{20} = \frac{1}{20}[f(0.025) + f(0.075) + f(0.125) + \cdots + f(0.975)] \approx 0.999538$

$E_L = I - L_{20} \approx 1 - 0.932967 = 0.067033$

$E_R \approx 1 - 1.068881 = -0.068881$

$E_T \approx 1 - 1.000924 = -0.000924$

$E_M \approx 1 - 0.999538 = 0.000462$

n	L_n	R_n	T_n	M_n
5	0.742943	1.286599	1.014771	0.992621
10	0.867782	1.139610	1.003696	0.998152
20	0.932967	1.068881	1.000924	0.999538

n	E_L	E_R	E_T	E_M
5	0.257057	−0.286599	−0.014771	0.007379
10	0.132218	−0.139610	−0.003696	0.001848
20	0.067033	−0.068881	−0.000924	0.000462

Observations:

1. E_L and E_R are always opposite in sign, as are E_T and E_M.
2. As n is doubled, E_L and E_R are decreased by about a factor of 2, and E_T and E_M are decreased by a factor of about 4.
3. The Midpoint approximation is about twice as accurate as the Trapezoidal approximation.
4. All the approximations become more accurate as the value of n increases.
5. The Midpoint and Trapezoidal approximations are much more accurate than the endpoint approximations.

27. $I = \int_0^2 x^4 \, dx = \left[\frac{1}{5}x^5\right]_0^2 = \frac{32}{5} - 0 = 6.4$, $f(x) = x^4$, $\Delta x = \frac{2-0}{n} = \frac{2}{n}$

$n = 6$: $\quad T_6 = \frac{2}{6 \cdot 2}\{f(0) + 2[f(\frac{1}{3}) + f(\frac{2}{3}) + f(\frac{3}{3}) + f(\frac{4}{3}) + f(\frac{5}{3})] + f(2)\} \approx 6.695473$

$M_6 = \frac{2}{6}[f(\frac{1}{6}) + f(\frac{3}{6}) + f(\frac{5}{6}) + f(\frac{7}{6}) + f(\frac{9}{6}) + f(\frac{11}{6})] \approx 6.252572$

$S_6 = \frac{2}{6 \cdot 3}[f(0) + 4f(\frac{1}{3}) + 2f(\frac{2}{3}) + 4f(\frac{3}{3}) + 2f(\frac{4}{3}) + 4f(\frac{5}{3}) + f(2)] \approx 6.403292$

$E_T = I - T_6 \approx 6.4 - 6.695473 = -0.295473$

$E_M \approx 6.4 - 6.252572 = 0.147428$

$E_S \approx 6.4 - 6.403292 = -0.003292$

$n = 12$: $\quad T_{12} = \frac{2}{12 \cdot 2}\{f(0) + 2[f(\frac{1}{6}) + f(\frac{2}{6}) + f(\frac{3}{6}) + \cdots + f(\frac{11}{6})] + f(2)\} \approx 6.474023$

$M_6 = \frac{2}{12}[f(\frac{1}{12}) + f(\frac{3}{12}) + f(\frac{5}{12}) + \cdots + f(\frac{23}{12})] \approx 6.363008$

$S_6 = \frac{2}{12 \cdot 3}[f(0) + 4f(\frac{1}{6}) + 2f(\frac{2}{6}) + 4f(\frac{3}{6}) + 2f(\frac{4}{6}) + \cdots + 4f(\frac{11}{6}) + f(2)] \approx 6.400206$

$E_T = I - T_{12} \approx 6.4 - 6.474023 = -0.074023$

$E_M \approx 6.4 - 6.363008 = 0.036992$

$E_S \approx 6.4 - 6.400206 = -0.000206$

n	T_n	M_n	S_n
6	6.695473	6.252572	6.403292
12	6.474023	6.363008	6.400206

n	E_T	E_M	E_S
6	−0.295473	0.147428	−0.003292
12	−0.074023	0.036992	−0.000206

[continued]

Observations:

1. E_T and E_M are opposite in sign and decrease by a factor of about 4 as n is doubled.
2. The Simpson's approximation is much more accurate than the Midpoint and Trapezoidal approximations, and E_S seems to decrease by a factor of about 16 as n is doubled.

29. (a) $\Delta x = (b-a)/n = (6-0)/6 = 1$

$$T_6 = \tfrac{1}{2}[f(0) + 2f(1) + 2f(2) + 2f(3) + 2f(4) + 2f(5) + f(6)]$$
$$\approx \tfrac{1}{2}[2 + 2(1) + 2(3) + 2(5) + 2(4) + 2(3) + 4] = \tfrac{1}{2}(38) = 19$$

(b) $M_6 = 1[f(0.5) + f(1.5) + f(2.5) + f(3.5) + f(4.5) + f(5.5)] \approx 1.3 + 1.5 + 4.6 + 4.7 + 3.3 + 3.2 = 18.6$

(c) $S_6 = \tfrac{1}{3}[f(0) + 4f(1) + 2f(2) + 4f(3) + 2f(4) + 4f(5) + f(6)]$
$$\approx \tfrac{1}{3}[2 + 4(1) + 2(3) + 4(5) + 2(4) + 4(3) + 4] = \tfrac{1}{3}(56) = 18.\overline{6}$$

31. (a) $\int_1^5 f(x)\, dx \approx M_4 = \tfrac{5-1}{4}[f(1.5) + f(2.5) + f(3.5) + f(4.5)] = 1(2.9 + 3.6 + 4.0 + 3.9) = 14.4$

(b) $-2 \le f''(x) \le 3 \;\Rightarrow\; |f''(x)| \le 3 \;\Rightarrow\; K = 3$, since $|f''(x)| \le K$. The error estimate for the Midpoint Rule is

$$|E_M| \le \frac{K(b-a)^3}{24n^2} = \frac{3(5-1)^3}{24(4)^2} = \frac{1}{2}.$$

33. We use Simpson's Rule with $n = 12$ and $\Delta t = \frac{24-0}{12} = 2$.

$$S_{12} = \tfrac{2}{3}[T(0) + 4T(2) + 2T(4) + 4T(6) + 2T(8) + 4T(10) + 2T(12)$$
$$\qquad + 4T(14) + 2T(16) + 4T(18) + 2T(20) + 4T(22) + T(24)]$$
$$\approx \tfrac{2}{3}[66.6 + 4(65.4) + 2(64.4) + 4(61.7) + 2(67.3) + 4(72.1) + 2(74.9)$$
$$\qquad + 4(77.4) + 2(79.1) + 4(75.4) + 2(75.6) + 4(71.4) + 67.5] = \tfrac{2}{3}(2550.3) = 1700.2$$

Thus, $\int_0^{24} T(t)\, dt \approx S_{12}$ and $T_{\text{avg}} = \frac{1}{24-0}\int_0^{24} T(t)\, dt \approx 70.84°\text{F}$.

35. By the Net Change Theorem, the increase in velocity is equal to $\int_0^6 a(t)\, dt$. We use Simpson's Rule with $n = 6$ and $\Delta t = (6-0)/6 = 1$ to estimate this integral:

$$\int_0^6 a(t)\, dt \approx S_6 = \tfrac{1}{3}[a(0) + 4a(1) + 2a(2) + 4a(3) + 2a(4) + 4a(5) + a(6)]$$
$$\approx \tfrac{1}{3}[0 + 4(0.5) + 2(4.1) + 4(9.8) + 2(12.9) + 4(9.5) + 0] = \tfrac{1}{3}(113.2) = 37.7\overline{3} \text{ ft/s}$$

37. By the Net Change Theorem, the energy used is equal to $\int_0^6 P(t)\, dt$. We use Simpson's Rule with $n = 12$ and $\Delta t = \frac{6-0}{12} = \tfrac{1}{2}$ to estimate this integral:

$$\int_0^6 P(t)\, dt \approx S_{12} = \tfrac{1/2}{3}[P(0) + 4P(0.5) + 2P(1) + 4P(1.5) + 2P(2) + 4P(2.5) + 2P(3)$$
$$\qquad + 4P(3.5) + 2P(4) + 4P(4.5) + 2P(5) + 4P(5.5) + P(6)]$$
$$= \tfrac{1}{6}[1814 + 4(1735) + 2(1686) + 4(1646) + 2(1637) + 4(1609) + 2(1604)$$
$$\qquad + 4(1611) + 2(1621) + 4(1666) + 2(1745) + 4(1886) + 2052]$$
$$= \tfrac{1}{6}(61{,}064) = 10{,}177.\overline{3} \text{ megawatt-hours}$$

39. (a) Let $y = f(x)$ denote the curve. Using disks, $V = \int_2^{10} \pi [f(x)]^2 \, dx = \pi \int_2^{10} g(x) \, dx = \pi I_1$.

Now use Simpson's Rule to approximate I_1:

$$I_1 \approx S_8 = \tfrac{10-2}{3(8)}[g(2) + 4g(3) + 2g(4) + 4g(5) + 2g(6) + 4g(7) + g(8)]$$
$$\approx \tfrac{1}{3}[0^2 + 4(1.5)^2 + 2(1.9)^2 + 4(2.2)^2 + 2(3.0)^2 + 4(3.8)^2 + 2(4.0)^2 + 4(3.1)^2 + 0^2]$$
$$= \tfrac{1}{3}(181.78)$$

Thus, $V \approx \pi \cdot \tfrac{1}{3}(181.78) \approx 190.4$ or 190 cubic units.

(b) Using cylindrical shells, $V = \int_2^{10} 2\pi x f(x) \, dx = 2\pi \int_2^{10} x f(x) \, dx = 2\pi I_1$.

Now use Simpson's Rule to approximate I_1:

$$I_1 \approx S_8 = \tfrac{10-2}{3(8)}[2f(2) + 4 \cdot 3f(3) + 2 \cdot 4f(4) + 4 \cdot 5f(5) + 2 \cdot 6f(6)$$
$$+ 4 \cdot 7f(7) + 2 \cdot 8f(8) + 4 \cdot 9f(9) + 10f(10)]$$
$$\approx \tfrac{1}{3}[2(0) + 12(1.5) + 8(1.9) + 20(2.2) + 12(3.0) + 28(3.8) + 16(4.0) + 36(3.1) + 10(0)]$$
$$= \tfrac{1}{3}(395.2)$$

Thus, $V \approx 2\pi \cdot \tfrac{1}{3}(395.2) \approx 827.7$ or 828 cubic units.

41. The curve is $y = f(x) = 1/(1 + e^{-x})$. Using disks, $V = \int_0^{10} \pi [f(x)]^2 \, dx = \pi \int_0^{10} g(x) \, dx = \pi I_1$. Now use Simpson's Rule to approximate I_1:

$$I_1 \approx S_{10} = \tfrac{10-0}{10 \cdot 3}[g(0) + 4g(1) + 2g(2) + 4g(3) + 2g(4) + 4g(5) + 2g(6) + 4g(7) + 2g(8) + 4g(9) + g(10)]$$
$$\approx 8.80825$$

Thus, $V \approx \pi I_1 \approx 27.7$ or 28 cubic units.

43. $I(\theta) = \dfrac{N^2 \sin^2 k}{k^2}$, where $k = \dfrac{\pi N d \sin \theta}{\lambda}$, $N = 10{,}000$, $d = 10^{-4}$, and $\lambda = 632.8 \times 10^{-9}$. So $I(\theta) = \dfrac{(10^4)^2 \sin^2 k}{k^2}$,

where $k = \dfrac{\pi (10^4)(10^{-4}) \sin \theta}{632.8 \times 10^{-9}}$. Now $n = 10$ and $\Delta \theta = \dfrac{10^{-6} - (-10^{-6})}{10} = 2 \times 10^{-7}$, so

$M_{10} = 2 \times 10^{-7}[I(-0.0000009) + I(-0.0000007) + \cdots + I(0.0000009)] \approx 59.4$.

45. Consider the function f whose graph is shown. The area $\int_0^2 f(x) \, dx$ is close to 2. The Trapezoidal Rule gives

$T_2 = \tfrac{2-0}{2 \cdot 2}[f(0) + 2f(1) + f(2)] = \tfrac{1}{2}[1 + 2 \cdot 1 + 1] = 2$.

The Midpoint Rule gives $M_2 = \tfrac{2-0}{2}[f(0.5) + f(1.5)] = 1[0 + 0] = 0$,

so the Trapezoidal Rule is more accurate.

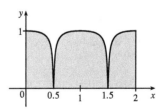

47. Since the Trapezoidal and Midpoint approximations on the interval $[a, b]$ are the sums of the Trapezoidal and Midpoint approximations on the subintervals $[x_{i-1}, x_i]$, $i = 1, 2, \ldots, n$, we can focus our attention on one such interval. The condition $f''(x) < 0$ for $a \leq x \leq b$ means that the graph of f is concave down as in Figure 5. In that figure, T_n is the area of the trapezoid $AQRD$, $\int_a^b f(x)\,dx$ is the area of the region $AQPRD$, and M_n is the area of the trapezoid $ABCD$, so $T_n < \int_a^b f(x)\,dx < M_n$. In general, the condition $f'' < 0$ implies that the graph of f on $[a, b]$ lies above the chord joining the points $(a, f(a))$ and $(b, f(b))$. Thus, $\int_a^b f(x)\,dx > T_n$. Since M_n is the area under a tangent to the graph, and since $f'' < 0$ implies that the tangent lies above the graph, we also have $M_n > \int_a^b f(x)\,dx$. Thus, $T_n < \int_a^b f(x)\,dx < M_n$.

49. $T_n = \frac{1}{2}\Delta x\,[f(x_0) + 2f(x_1) + \cdots + 2f(x_{n-1}) + f(x_n)]$ and

$M_n = \Delta x\,[f(\overline{x}_1) + f(\overline{x}_2) + \cdots + f(\overline{x}_{n-1}) + f(\overline{x}_n)]$, where $\overline{x}_i = \frac{1}{2}(x_{i-1} + x_i)$. Now

$T_{2n} = \frac{1}{2}\left(\frac{1}{2}\Delta x\right)[f(x_0) + 2f(\overline{x}_1) + 2f(x_1) + 2f(\overline{x}_2) + 2f(x_2) + \cdots + 2f(\overline{x}_{n-1}) + 2f(x_{n-1}) + 2f(\overline{x}_n) + f(x_n)]$, so

$\frac{1}{2}(T_n + M_n) = \frac{1}{2}T_n + \frac{1}{2}M_n$

$\qquad = \frac{1}{4}\Delta x[f(x_0) + 2f(x_1) + \cdots + 2f(x_{n-1}) + f(x_n)]$

$\qquad\quad + \frac{1}{4}\Delta x[2f(\overline{x}_1) + 2f(\overline{x}_2) + \cdots + 2f(\overline{x}_{n-1}) + 2f(\overline{x}_n)]$

$\qquad = T_{2n}$

7.8 Improper Integrals

1. (a) Since $y = \dfrac{1}{x-3}$ has an infinite discontinuity at $x = 3$, $\displaystyle\int_1^4 \dfrac{dx}{x-3}$ is a Type 2 improper integral.

(b) Since $\displaystyle\int_3^\infty \dfrac{dx}{x^2 - 4}$ has an infinite interval of integration, it is an improper integral of Type 1.

(c) Since $y = \tan \pi x$ has an infinite discontinuity at $x = \frac{1}{2}$, $\int_0^1 \tan \pi x\,dx$ is a Type 2 improper integral.

(d) Since $\displaystyle\int_{-\infty}^{-1} \dfrac{e^x}{x}\,dx$ has an infinite interval of integration, it is an improper integral of Type 1.

3. The area under the graph of $y = 1/x^3 = x^{-3}$ between $x = 1$ and $x = t$ is

$A(t) = \int_1^t x^{-3}\,dx = \left[-\frac{1}{2}x^{-2}\right]_1^t = -\frac{1}{2}t^{-2} - \left(-\frac{1}{2}\right) = \frac{1}{2} - 1/(2t^2)$. So the area for $1 \le x \le 10$ is

$A(10) = 0.5 - 0.005 = 0.495$, the area for $1 \le x \le 100$ is $A(100) = 0.5 - 0.00005 = 0.49995$, and the area for

$1 \le x \le 1000$ is $A(1000) = 0.5 - 0.0000005 = 0.4999995$. The total area under the curve for $x \ge 1$ is

$\displaystyle\lim_{t\to\infty} A(t) = \lim_{t\to\infty}\left[\frac{1}{2} - 1/(2t^2)\right] = \frac{1}{2}$.

5. $\displaystyle\int_1^\infty 2x^{-3}\,dx = \lim_{t\to\infty}\int_1^t 2x^{-3}\,dx = \lim_{t\to\infty}\left[-x^{-2}\right]_1^t = \lim_{t\to\infty}\left[-t^{-2} + 1\right] = 0 + 1 = 1.$ Convergent

7. $\displaystyle\int_0^\infty e^{-2x}\,dx = \lim_{t\to\infty}\int_0^t e^{-2x}\,dx = \lim_{t\to\infty}\left[-\frac{1}{2}e^{-2x}\right]_0^t = \lim_{t\to\infty}\left[-\frac{1}{2}e^{-2t} + \frac{1}{2}\right] = 0 + \frac{1}{2} = \frac{1}{2}.$ Convergent

9. $\displaystyle\int_{-2}^\infty \dfrac{1}{x+4}\,dx = \lim_{t\to\infty}\int_{-2}^t \dfrac{1}{x+4}\,dx = \lim_{t\to\infty}\Big[\ln|x+4|\Big]_{-2}^t = \lim_{t\to\infty}\Big[\ln|t+4| - \ln 2\Big] = \infty$ since $\displaystyle\lim_{t\to\infty}\ln|x+4| = \infty$.

Divergent

11. $\int_3^\infty \dfrac{1}{(x-2)^{3/2}}\,dx = \lim\limits_{t\to\infty} \int_3^t (x-2)^{-3/2}\,dx = \lim\limits_{t\to\infty} \left[-2(x-2)^{-1/2}\right]_3^t \quad [u = x-2,\, du = dx]$

$= \lim\limits_{t\to\infty} \left(\dfrac{-2}{\sqrt{t-2}} + \dfrac{2}{\sqrt{1}}\right) = 0 + 2 = 2.$ Convergent

13. $\int_{-\infty}^0 \dfrac{x}{(x^2+1)^3}\,dx = \lim\limits_{t\to-\infty} \int_t^0 \dfrac{x}{(x^2+1)^3}\,dx = \lim\limits_{t\to-\infty}\left[-\tfrac{1}{4}(x^2+1)^{-2}\right]_t^0 = \lim\limits_{t\to-\infty}\left[-\dfrac{1}{4(x^2+1)^2}\right]_t^0$

$= \lim\limits_{t\to-\infty}\left[-\dfrac{1}{4} + \dfrac{1}{4(t^2+1)^2}\right] = -\dfrac{1}{4} + 0 = -\dfrac{1}{4}.$ Convergent

15. $\int_1^\infty \dfrac{x^2+x+1}{x^4}\,dx = \lim\limits_{t\to\infty}\int_1^t (x^{-2}+x^{-3}+x^{-4})\,dx$

$= \lim\limits_{t\to\infty}\left[-x^{-1} - \tfrac{1}{2}x^{-2} - \tfrac{1}{3}x^{-3}\right]_1^t = \lim\limits_{t\to\infty}\left[-\dfrac{1}{x} - \dfrac{1}{2x^2} - \dfrac{1}{3x^3}\right]_1^t$

$= \lim\limits_{t\to\infty}\left[\left(-\dfrac{1}{t} - \dfrac{1}{2t^2} - \dfrac{1}{3t^3}\right) - \left(-1 - \dfrac{1}{2} - \dfrac{1}{3}\right)\right] = 0 + \dfrac{11}{6} = \dfrac{11}{6}.$ Convergent

17. $\int_0^\infty \dfrac{e^x}{(1+e^x)^2}\,dx = \lim\limits_{t\to\infty}\int_0^t \dfrac{e^x}{(1+e^x)^2}\,dx = \lim\limits_{t\to\infty}\left[-(1+e^x)^{-1}\right]_0^t = \lim\limits_{t\to\infty}\left[-\dfrac{1}{1+e^x}\right]_0^t$

$= \lim\limits_{t\to\infty}\left[-\dfrac{1}{1+e^t} + \dfrac{1}{2}\right] = 0 + \dfrac{1}{2} = \dfrac{1}{2}.$ Convergent

19. $\int_{-\infty}^\infty xe^{-x^2}\,dx = \int_{-\infty}^0 xe^{-x^2}\,dx + \int_0^\infty xe^{-x^2}\,dx.$

$\int_{-\infty}^0 xe^{-x^2}\,dx = \lim\limits_{t\to-\infty}\left(-\tfrac{1}{2}\right)\left[e^{-x^2}\right]_t^0 = \lim\limits_{t\to-\infty}\left(-\tfrac{1}{2}\right)\left(1 - e^{-t^2}\right) = -\tfrac{1}{2}\cdot(1-0) = -\tfrac{1}{2},$ and

$\int_0^\infty xe^{-x^2}\,dx = \lim\limits_{t\to\infty}\left(-\tfrac{1}{2}\right)\left[e^{-x^2}\right]_0^t = \lim\limits_{t\to\infty}\left(-\tfrac{1}{2}\right)\left(e^{-t^2} - 1\right) = -\tfrac{1}{2}\cdot(0-1) = \tfrac{1}{2}.$

Therefore, $\int_{-\infty}^\infty xe^{-x^2}\,dx = -\tfrac{1}{2} + \tfrac{1}{2} = 0.$ Convergent

21. $I = \int_{-\infty}^\infty \cos 2t\,dt = \int_{-\infty}^0 \cos 2t\,dt + \int_0^\infty \cos 2t\,dt = I_1 + I_2,$ but $I_1 = \lim\limits_{s\to-\infty}\left[\tfrac{1}{2}\sin 2t\right]_s^0 = \lim\limits_{s\to-\infty}\left(-\tfrac{1}{2}\sin 2s\right),$ and this limit does not exist. Since I_1 is divergent, I is divergent, and there is no need to evaluate I_2. Divergent

23. $\int_0^\infty \sin^2\alpha\,d\alpha = \lim\limits_{t\to\infty}\int_0^t \tfrac{1}{2}(1-\cos 2\alpha)\,d\alpha = \lim\limits_{t\to\infty}\left[\tfrac{1}{2}(\alpha - \tfrac{1}{2}\sin 2\alpha)\right]_0^t = \lim\limits_{t\to\infty}\left[\tfrac{1}{2}(t - \tfrac{1}{2}\sin 2t) - 0\right] = \infty.$
Divergent

25. $\int_1^\infty \dfrac{1}{x^2+x}\,dx = \lim\limits_{t\to\infty}\int_1^t \dfrac{1}{x(x+1)}\,dx = \lim\limits_{t\to\infty}\int_1^t \left(\dfrac{1}{x} - \dfrac{1}{x+1}\right)dx$ [partial fractions]

$= \lim\limits_{t\to\infty}\left[\ln|x| - \ln|x+1|\right]_1^t = \lim\limits_{t\to\infty}\left[\ln\left|\dfrac{x}{x+1}\right|\right]_1^t = \lim\limits_{t\to\infty}\left(\ln\dfrac{t}{t+1} - \ln\dfrac{1}{2}\right) = 0 - \ln\dfrac{1}{2} = \ln 2.$

Convergent

27. $\int_{-\infty}^0 ze^{2z}\,dz = \lim\limits_{t\to-\infty}\int_t^0 ze^{2z}\,dz = \lim\limits_{t\to-\infty}\left[\tfrac{1}{2}ze^{2z} - \tfrac{1}{4}e^{2z}\right]_t^0 \quad \begin{bmatrix}\text{integration by parts with}\\ u = z,\, dv = e^{2z}\,dz\end{bmatrix}$

$= \lim\limits_{t\to-\infty}\left[\left(0 - \tfrac{1}{4}\right) - \left(\tfrac{1}{2}te^{2t} - \tfrac{1}{4}e^{2t}\right)\right] = -\tfrac{1}{4} - 0 + 0$ [by l'Hospital's Rule] $= -\tfrac{1}{4}.$ Convergent

29. $\int_1^\infty \dfrac{\ln x}{x}\,dx = \lim\limits_{t\to\infty} \left[\dfrac{(\ln x)^2}{2}\right]_1^t \quad \begin{bmatrix}\text{by substitution with}\\ u=\ln x,\, du = dx/x\end{bmatrix} = \lim\limits_{t\to\infty} \dfrac{(\ln t)^2}{2} = \infty.$ Divergent

31. $\int_{-\infty}^0 \dfrac{z}{z^4+4}\,dz = \lim\limits_{t\to-\infty} \int_t^0 \dfrac{z}{z^4+4}\,dz = \lim\limits_{t\to-\infty} \dfrac{1}{2}\left[\dfrac{1}{2}\tan^{-1}\!\left(\dfrac{z^2}{2}\right)\right]_t^0 \quad \begin{bmatrix}u = z^2,\\ du = 2z\,dz\end{bmatrix}$

$= \lim\limits_{t\to-\infty}\left[0 - \dfrac{1}{4}\tan^{-1}\!\left(\dfrac{t^2}{2}\right)\right] = -\dfrac{1}{4}\!\left(\dfrac{\pi}{2}\right) = -\dfrac{\pi}{8}.$ Convergent

33. $\int_0^\infty e^{-\sqrt{y}}\,dy = \lim\limits_{t\to\infty}\int_0^t e^{-\sqrt{y}}\,dy = \lim\limits_{t\to\infty}\int_0^{\sqrt{t}} e^{-x}(2x\,dx) \quad \begin{bmatrix}x = \sqrt{y},\\ dx = 1/(2\sqrt{y})\,dy\end{bmatrix}$

$= \lim\limits_{t\to\infty}\left\{\left[-2xe^{-x}\right]_0^{\sqrt{t}} + \int_0^{\sqrt{t}} 2e^{-x}\,dx\right\} \quad \begin{bmatrix}u = 2x, & dv = e^{-x}\,dx\\ du = 2\,dx, & v = -e^{-x}\end{bmatrix}$

$= \lim\limits_{t\to\infty}\left(-2\sqrt{t}\,e^{-\sqrt{t}} + \left[-2e^{-x}\right]_0^{\sqrt{t}}\right) = \lim\limits_{t\to\infty}\left(\dfrac{-2\sqrt{t}}{e^{\sqrt{t}}} - \dfrac{2}{e^{\sqrt{t}}} + 2\right) = 0 - 0 + 2 = 2.$

Convergent

Note: $\lim\limits_{t\to\infty}\dfrac{\sqrt{t}}{e^{\sqrt{t}}} \stackrel{H}{=} \lim\limits_{t\to\infty}\dfrac{2\sqrt{t}}{2\sqrt{t}\,e^{\sqrt{t}}} = \lim\limits_{t\to\infty}\dfrac{1}{e^{\sqrt{t}}} = 0$

35. $\int_0^1 \dfrac{1}{x}\,dx = \lim\limits_{t\to 0^+}\int_t^1 \dfrac{1}{x}\,dx = \lim\limits_{t\to 0^+}\Big[\ln|x|\Big]_t^1 = \lim\limits_{t\to 0^+}(-\ln t) = \infty.$ Divergent

37. $\int_{-2}^{14}\dfrac{dx}{\sqrt[4]{x+2}} = \lim\limits_{t\to -2^+}\int_t^{14}(x+2)^{-1/4}\,dx = \lim\limits_{t\to -2^+}\left[\dfrac{4}{3}(x+2)^{3/4}\right]_t^{14} = \dfrac{4}{3}\lim\limits_{t\to -2^+}\left[16^{3/4} - (t+2)^{3/4}\right]$

$= \tfrac{4}{3}(8-0) = \tfrac{32}{3}.$ Convergent

39. $\int_{-2}^3 \dfrac{1}{x^4}\,dx = \int_{-2}^0 \dfrac{dx}{x^4} + \int_0^3 \dfrac{dx}{x^4},$ but $\int_{-2}^0 \dfrac{dx}{x^4} = \lim\limits_{t\to 0^-}\left[-\dfrac{x^{-3}}{3}\right]_{-2}^t = \lim\limits_{t\to 0^-}\left[-\dfrac{1}{3t^3}-\dfrac{1}{24}\right] = \infty.$ Divergent

41. There is an infinite discontinuity at $x = 1$. $\int_0^9 \dfrac{1}{\sqrt[3]{x-1}}\,dx = \int_0^1 (x-1)^{-1/3}\,dx + \int_1^9 (x-1)^{-1/3}\,dx.$

Here $\int_0^1 (x-1)^{-1/3}\,dx = \lim\limits_{t\to 1^-}\int_0^t (x-1)^{-1/3}\,dx = \lim\limits_{t\to 1^-}\left[\tfrac{3}{2}(x-1)^{2/3}\right]_0^t = \lim\limits_{t\to 1^-}\left[\tfrac{3}{2}(t-1)^{2/3} - \tfrac{3}{2}\right] = -\tfrac{3}{2}$

and $\int_1^9 (x-1)^{-1/3}\,dx = \lim\limits_{t\to 1^+}\int_t^9 (x-1)^{-1/3}\,dx = \lim\limits_{t\to 1^+}\left[\tfrac{3}{2}(x-1)^{2/3}\right]_t^9 = \lim\limits_{t\to 1^+}\left[6 - \tfrac{3}{2}(t-1)^{2/3}\right] = 6.$ Thus,

$\int_0^9 \dfrac{1}{\sqrt[3]{x-1}}\,dx = -\dfrac{3}{2} + 6 = \dfrac{9}{2}.$ Convergent

43. $\int_0^{\pi/2}\tan^2\theta\,d\theta = \lim\limits_{t\to (\pi/2)^-}\int_0^t \tan^2\theta\,d\theta = \lim\limits_{t\to (\pi/2)^-}\int_0^t(\sec^2\theta - 1)\,d\theta = \lim\limits_{t\to (\pi/2)^-}\Big[\tan\theta - \theta\Big]_0^t$

$= \lim\limits_{t\to (\pi/2)^-}(\tan t - t) = \infty$ since $\tan t \to \infty$ as $t \to \tfrac{\pi}{2}^-$.   Divergent

45. $\int_0^1 r \ln r \, dr = \lim\limits_{t \to 0^+} \int_t^1 r \ln r \, dr = \lim\limits_{t \to 0^+} \left[\frac{1}{2}r^2 \ln r - \frac{1}{4}r^2\right]_t^1 \quad \begin{bmatrix} u = \ln r, & dv = r \, dr \\ du = (1/r) \, dr, & v = \frac{1}{2}r^2 \end{bmatrix}$

$= \lim\limits_{t \to 0^+} \left[\left(0 - \frac{1}{4}\right) - \left(\frac{1}{2}t^2 \ln t - \frac{1}{4}t^2\right)\right] = -\frac{1}{4} - 0 = -\frac{1}{4}$

since $\lim\limits_{t \to 0^+} t^2 \ln t = \lim\limits_{t \to 0^+} \frac{\ln t}{1/t^2} \overset{\text{H}}{=} \lim\limits_{t \to 0^+} \frac{1/t}{-2/t^3} = \lim\limits_{t \to 0^+} \left(-\frac{1}{2}t^2\right) = 0.$ Convergent

47. $\int_{-1}^0 \frac{e^{1/x}}{x^3} \, dx = \lim\limits_{t \to 0^-} \int_{-1}^t \frac{1}{x} e^{1/x} \cdot \frac{1}{x^2} \, dx = \lim\limits_{t \to 0^-} \int_{-1}^{1/t} u e^u \, (-du) \quad \begin{bmatrix} u = 1/x, \\ du = -dx/x^2 \end{bmatrix}$

$= \lim\limits_{t \to 0^-} \left[(u-1)e^u\right]_{1/t}^{-1} \quad \begin{bmatrix} \text{use parts} \\ \text{or Formula 96} \end{bmatrix} = \lim\limits_{t \to 0^-} \left[-2e^{-1} - \left(\frac{1}{t} - 1\right)e^{1/t}\right]$

$= -\frac{2}{e} - \lim\limits_{s \to -\infty} (s-1)e^s \quad [s = 1/t] \quad = -\frac{2}{e} - \lim\limits_{s \to -\infty} \frac{s-1}{e^{-s}} \overset{\text{H}}{=} -\frac{2}{e} - \lim\limits_{s \to -\infty} \frac{1}{-e^{-s}}$

$= -\frac{2}{e} - 0 = -\frac{2}{e}.$ Convergent

49.

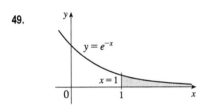

Area $= \int_1^\infty e^{-x} \, dx = \lim\limits_{t \to \infty} \int_1^t e^{-x} \, dx = \lim\limits_{t \to \infty} \left[-e^{-x}\right]_1^t$

$= \lim\limits_{t \to \infty} (-e^{-t} + e^{-1}) = 0 + e^{-1} = 1/e$

51.

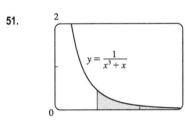

Area $= \int_1^\infty \frac{1}{x^3 + x} \, dx = \lim\limits_{t \to \infty} \int_1^t \frac{1}{x(x^2+1)} \, dx$

$= \lim\limits_{t \to \infty} \int_1^t \left(\frac{1}{x} - \frac{x}{x^2+1}\right) dx$ [partial fractions]

$= \lim\limits_{t \to \infty} \left[\ln|x| - \frac{1}{2}\ln|x^2+1|\right]_1^t = \lim\limits_{t \to \infty} \left[\ln \frac{x}{\sqrt{x^2+1}}\right]_1^t$

$= \lim\limits_{t \to \infty} \left(\ln \frac{t}{\sqrt{t^2+1}} - \ln \frac{1}{\sqrt{2}}\right) = \ln 1 - \ln 2^{-1/2} = \frac{1}{2}\ln 2$

53.

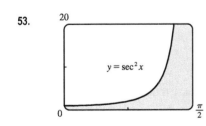

Area $= \int_0^{\pi/2} \sec^2 x \, dx = \lim\limits_{t \to (\pi/2)^-} \int_0^t \sec^2 x \, dx = \lim\limits_{t \to (\pi/2)^-} \left[\tan x\right]_0^t$

$= \lim\limits_{t \to (\pi/2)^-} (\tan t - 0) = \infty$

Infinite area

55. (a)

t	$\int_1^t g(x) \, dx$
2	0.447453
5	0.577101
10	0.621306
100	0.668479
1000	0.672957
10,000	0.673407

$g(x) = \frac{\sin^2 x}{x^2}.$

It appears that the integral is convergent.

(b) $-1 \leq \sin x \leq 1 \Rightarrow 0 \leq \sin^2 x \leq 1 \Rightarrow 0 \leq \dfrac{\sin^2 x}{x^2} \leq \dfrac{1}{x^2}$. Since $\int_1^\infty \dfrac{1}{x^2}\,dx$ is convergent

[Theorem 2 with $p = 2 > 1$], $\int_1^\infty \dfrac{\sin^2 x}{x^2}\,dx$ is convergent by the Comparison Theorem.

(c)

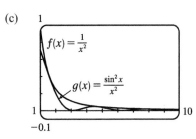

Since $\int_1^\infty f(x)\,dx$ is finite and the area under $g(x)$ is less than the area under $f(x)$ on any interval $[1, t]$, $\int_1^\infty g(x)\,dx$ must be finite; that is, the integral is convergent.

57. For $x > 0$, $\dfrac{x}{x^3+1} < \dfrac{x}{x^3} = \dfrac{1}{x^2}$. $\int_1^\infty \dfrac{1}{x^2}\,dx$ is convergent by Theorem 2 with $p = 2 > 1$, so $\int_1^\infty \dfrac{x}{x^3+1}\,dx$ is convergent

by the Comparison Theorem. $\int_0^1 \dfrac{x}{x^3+1}\,dx$ is a constant, so $\int_0^\infty \dfrac{x}{x^3+1}\,dx = \int_0^1 \dfrac{x}{x^3+1}\,dx + \int_1^\infty \dfrac{x}{x^3+1}\,dx$ is also

convergent.

59. For $x \geq 1$, $\dfrac{1}{x - \ln x} \geq \dfrac{1}{x}$. $\int_2^\infty \dfrac{1}{x}\,dx$ is divergent by Equation 2 with $p = 1 \leq 1$, so $\int_2^\infty \dfrac{1}{x - \ln x}\,dx$ is divergent by the

Comparison Theorem.

61. For $x > 1$, $f(x) = \dfrac{x+1}{\sqrt{x^4 - x}} > \dfrac{x+1}{\sqrt{x^4}} > \dfrac{x}{x^2} = \dfrac{1}{x}$, so $\int_2^\infty f(x)\,dx$ diverges by comparison with $\int_2^\infty \dfrac{1}{x}\,dx$, which diverges

by Theorem 2 with $p = 1 \leq 1$. Thus, $\int_1^\infty f(x)\,dx = \int_1^2 f(x)\,dx + \int_2^\infty f(x)\,dx$ also diverges.

63. For $0 < x \leq 1$, $\dfrac{\sec^2 x}{x\sqrt{x}} > \dfrac{1}{x^{3/2}}$. Now

$I = \int_0^1 x^{-3/2}\,dx = \lim\limits_{t \to 0^+} \int_t^1 x^{-3/2}\,dx = \lim\limits_{t \to 0^+} \left[-2x^{-1/2} \right]_t^1 = \lim\limits_{t \to 0^+} \left(-2 + \dfrac{2}{\sqrt{t}} \right) = \infty$, so I is divergent, and by

comparison, $\int_0^1 \dfrac{\sec^2 x}{x\sqrt{x}}$ is divergent.

65. $I = \int_0^\infty \dfrac{1}{x^2}\,dx = \int_0^1 \dfrac{1}{x^2}\,dx + \int_1^\infty \dfrac{1}{x^2}\,dx = I_1 + I_2$. Now,

$I_1 = \lim\limits_{t \to 0^+} \int_t^1 \dfrac{1}{x^2}\,dx = \lim\limits_{t \to 0^+} \left[-x^{-1} \right]_t^1 = \lim\limits_{t \to 0^+} \left[-1 + \dfrac{1}{t} \right] = \infty$. Since I_1 is divergent, I is divergent, and there is no need

to evaluate I_2.

67. $\int_0^\infty \dfrac{1}{\sqrt{x}\,(1+x)}\,dx = \int_0^1 \dfrac{dx}{\sqrt{x}\,(1+x)} + \int_1^\infty \dfrac{dx}{\sqrt{x}\,(1+x)} = \lim\limits_{t \to 0^+} \int_t^1 \dfrac{dx}{\sqrt{x}\,(1+x)} + \lim\limits_{t \to \infty} \int_1^t \dfrac{dx}{\sqrt{x}\,(1+x)}$. Now

$\displaystyle\int \dfrac{dx}{\sqrt{x}\,(1+x)} = \int \dfrac{2u\,du}{u(1+u^2)} \quad \left[\begin{array}{l} u = \sqrt{x},\, x = u^2, \\ dx = 2u\,du \end{array} \right] = 2\int \dfrac{du}{1+u^2} = 2\tan^{-1} u + C = 2\tan^{-1} \sqrt{x} + C$, so

$\displaystyle\int_0^\infty \dfrac{dx}{\sqrt{x}\,(1+x)} = \lim\limits_{t \to 0^+} \left[2\tan^{-1}\sqrt{x} \right]_t^1 + \lim\limits_{t \to \infty} \left[2\tan^{-1}\sqrt{x} \right]_1^t$

$= \lim\limits_{t \to 0^+} \left[2\left(\tfrac{\pi}{4}\right) - 2\tan^{-1}\sqrt{t} \right] + \lim\limits_{t \to \infty} \left[2\tan^{-1}\sqrt{t} - 2\left(\tfrac{\pi}{4}\right) \right] = \tfrac{\pi}{2} - 0 + 2\left(\tfrac{\pi}{2}\right) - \tfrac{\pi}{2} = \pi$.

69. If $p = 1$, then $\int_0^1 \frac{1}{x^p}\,dx = \lim_{t \to 0^+} \int_t^1 \frac{dx}{x} = \lim_{t \to 0^+} [\ln x]_t^1 = \infty.$ Divergent

If $p \neq 1$, then $\int_0^1 \frac{dx}{x^p} = \lim_{t \to 0^+} \int_t^1 \frac{dx}{x^p}$ [note that the integral is not improper if $p < 0$]

$$= \lim_{t \to 0^+} \left[\frac{x^{-p+1}}{-p+1}\right]_t^1 = \lim_{t \to 0^+} \frac{1}{1-p}\left[1 - \frac{1}{t^{p-1}}\right]$$

If $p > 1$, then $p - 1 > 0$, so $\frac{1}{t^{p-1}} \to \infty$ as $t \to 0^+$, and the integral diverges.

If $p < 1$, then $p - 1 < 0$, so $\frac{1}{t^{p-1}} \to 0$ as $t \to 0^+$ and $\int_0^1 \frac{dx}{x^p} = \frac{1}{1-p}\left[\lim_{t \to 0^+}(1 - t^{1-p})\right] = \frac{1}{1-p}.$

Thus, the integral converges if and only if $p < 1$, and in that case its value is $\frac{1}{1-p}$.

71. First suppose $p = -1$. Then

$$\int_0^1 x^p \ln x\,dx = \int_0^1 \frac{\ln x}{x}\,dx = \lim_{t \to 0^+} \int_t^1 \frac{\ln x}{x}\,dx = \lim_{t \to 0^+} \left[\tfrac{1}{2}(\ln x)^2\right]_t^1 = -\tfrac{1}{2}\lim_{t \to 0^+}(\ln t)^2 = -\infty,$$ so the

integral diverges. Now suppose $p \neq -1$. Then integration by parts gives

$$\int x^p \ln x\,dx = \frac{x^{p+1}}{p+1}\ln x - \int \frac{x^p}{p+1}\,dx = \frac{x^{p+1}}{p+1}\ln x - \frac{x^{p+1}}{(p+1)^2} + C.$$ If $p < -1$, then $p + 1 < 0$, so

$$\int_0^1 x^p \ln x\,dx = \lim_{t \to 0^+}\left[\frac{x^{p+1}}{p+1}\ln x - \frac{x^{p+1}}{(p+1)^2}\right]_t^1 = \frac{-1}{(p+1)^2} - \left(\frac{1}{p+1}\right)\lim_{t \to 0^+}\left[t^{p+1}\left(\ln t - \frac{1}{p+1}\right)\right] = \infty.$$

If $p > -1$, then $p + 1 > 0$ and

$$\int_0^1 x^p \ln x\,dx = \frac{-1}{(p+1)^2} - \left(\frac{1}{p+1}\right)\lim_{t \to 0^+}\frac{\ln t - 1/(p+1)}{t^{-(p+1)}} \stackrel{\text{H}}{=} \frac{-1}{(p+1)^2} - \left(\frac{1}{p+1}\right)\lim_{t \to 0^+}\frac{1/t}{-(p+1)t^{-(p+2)}}$$

$$= \frac{-1}{(p+1)^2} + \frac{1}{(p+1)^2}\lim_{t \to 0^+} t^{p+1} = \frac{-1}{(p+1)^2}$$

Thus, the integral converges to $-\frac{1}{(p+1)^2}$ if $p > -1$ and diverges otherwise.

73. $I = \int_{-\infty}^\infty x\,dx = \int_{-\infty}^0 x\,dx + \int_0^\infty x\,dx$ and $\int_0^\infty x\,dx = \lim_{t \to \infty}\int_0^t x\,dx = \lim_{t \to \infty}\left[\tfrac{1}{2}x^2\right]_0^t = \lim_{t \to \infty}\left[\tfrac{1}{2}t^2 - 0\right] = \infty,$

so I is divergent. The Cauchy principal value of I is given by

$\lim_{t \to \infty}\int_{-t}^t x\,dx = \lim_{t \to \infty}\left[\tfrac{1}{2}x^2\right]_{-t}^t = \lim_{t \to \infty}\left[\tfrac{1}{2}t^2 - \tfrac{1}{2}(-t)^2\right] = \lim_{t \to \infty}[0] = 0.$ Hence, I is divergent, but its Cauchy principal value is 0.

75. Volume $= \int_1^\infty \pi\left(\frac{1}{x}\right)^2 dx = \pi\lim_{t \to \infty}\int_1^t \frac{dx}{x^2} = \pi\lim_{t \to \infty}\left[-\frac{1}{x}\right]_1^t = \pi\lim_{t \to \infty}\left(1 - \frac{1}{t}\right) = \pi < \infty.$

77. Work $= \int_R^\infty F\,dr = \lim_{t \to \infty}\int_R^t \frac{GmM}{r^2}\,dr = \lim_{t \to \infty}GmM\left(\frac{1}{R} - \frac{1}{t}\right) = \frac{GmM}{R}$. The initial kinetic energy provides the work,

so $\tfrac{1}{2}mv_0^2 = \frac{GmM}{R} \;\Rightarrow\; v_0 = \sqrt{\frac{2GM}{R}}$.

79. We would expect a small percentage of bulbs to burn out in the first few hundred hours, most of the bulbs to burn out after close to 700 hours, and a few overachievers to burn on and on.

(a)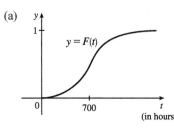

(b) $r(t) = F'(t)$ is the rate at which the fraction $F(t)$ of burnt-out bulbs increases as t increases. This could be interpreted as a fractional burnout rate.

(c) $\int_0^\infty r(t)\, dt = \lim_{x \to \infty} F(x) = 1$, since all of the bulbs will eventually burn out.

81. $\gamma = \int_0^\infty \dfrac{cN(1 - e^{-kt})}{k} e^{-\lambda t}\, dt = \dfrac{cN}{k} \lim_{x \to \infty} \int_0^x \left[e^{-\lambda t} - e^{(-k-\lambda)t} \right] dt$

$= \dfrac{cN}{k} \lim_{x \to \infty} \left[\dfrac{1}{-\lambda} e^{-\lambda t} - \dfrac{1}{-k-\lambda} e^{(-k-\lambda)t} \right]_0^x = \dfrac{cN}{k} \lim_{x \to \infty} \left[\dfrac{1}{-\lambda e^{\lambda x}} + \dfrac{1}{(k+\lambda) e^{(k+\lambda)x}} - \left(\dfrac{1}{-\lambda} + \dfrac{1}{k+\lambda} \right) \right]$

$= \dfrac{cN}{k} \left(\dfrac{1}{\lambda} - \dfrac{1}{k+\lambda} \right) = \dfrac{cN}{k} \left(\dfrac{k+\lambda-\lambda}{\lambda(k+\lambda)} \right) = \dfrac{cN}{\lambda(k+\lambda)}$

83. $I = \int_a^\infty \dfrac{1}{x^2+1}\, dx = \lim_{t \to \infty} \int_a^t \dfrac{1}{x^2+1}\, dx = \lim_{t \to \infty} \left[\tan^{-1} x \right]_a^t = \lim_{t \to \infty} (\tan^{-1} t - \tan^{-1} a) = \tfrac{\pi}{2} - \tan^{-1} a.$

$I < 0.001 \;\Rightarrow\; \tfrac{\pi}{2} - \tan^{-1} a < 0.001 \;\Rightarrow\; \tan^{-1} a > \tfrac{\pi}{2} - 0.001 \;\Rightarrow\; a > \tan\!\left(\tfrac{\pi}{2} - 0.001\right) \approx 1000.$

85. (a) $F(s) = \int_0^\infty f(t) e^{-st}\, dt = \int_0^\infty e^{-st}\, dt = \lim_{n \to \infty} \left[-\dfrac{e^{-st}}{s} \right]_0^n = \lim_{n \to \infty} \left(\dfrac{e^{-sn}}{-s} + \dfrac{1}{s} \right)$. This converges to $\dfrac{1}{s}$ only if $s > 0$.

Therefore $F(s) = \dfrac{1}{s}$ with domain $\{s \mid s > 0\}$.

(b) $F(s) = \int_0^\infty f(t) e^{-st}\, dt = \int_0^\infty e^t e^{-st}\, dt = \lim_{n \to \infty} \int_0^n e^{t(1-s)}\, dt = \lim_{n \to \infty} \left[\dfrac{1}{1-s} e^{t(1-s)} \right]_0^n$

$= \lim_{n \to \infty} \left(\dfrac{e^{(1-s)n}}{1-s} - \dfrac{1}{1-s} \right)$

This converges only if $1 - s < 0 \;\Rightarrow\; s > 1$, in which case $F(s) = \dfrac{1}{s-1}$ with domain $\{s \mid s > 1\}$.

(c) $F(s) = \int_0^\infty f(t) e^{-st}\, dt = \lim_{n \to \infty} \int_0^n t e^{-st}\, dt$. Use integration by parts: let $u = t$, $dv = e^{-st}\, dt \;\Rightarrow\; du = dt$,

$v = -\dfrac{e^{-st}}{s}$. Then $F(s) = \lim_{n \to \infty} \left[-\dfrac{t}{s} e^{-st} - \dfrac{1}{s^2} e^{-st} \right]_0^n = \lim_{n \to \infty} \left(\dfrac{-n}{s e^{sn}} - \dfrac{1}{s^2 e^{sn}} + 0 + \dfrac{1}{s^2} \right) = \dfrac{1}{s^2}$ only if $s > 0$.

Therefore, $F(s) = \dfrac{1}{s^2}$ and the domain of F is $\{s \mid s > 0\}$.

87. $G(s) = \int_0^\infty f'(t) e^{-st}\, dt$. Integrate by parts with $u = e^{-st}$, $dv = f'(t)\, dt \;\Rightarrow\; du = -s e^{-st}$, $v = f(t)$:

$G(s) = \lim_{n \to \infty} \left[f(t) e^{-st} \right]_0^n + s \int_0^\infty f(t) e^{-st}\, dt = \lim_{n \to \infty} f(n) e^{-sn} - f(0) + s F(s)$

But $0 \le f(t) \le M e^{at} \;\Rightarrow\; 0 \le f(t) e^{-st} \le M e^{at} e^{-st}$ and $\lim_{t \to \infty} M e^{t(a-s)} = 0$ for $s > a$. So by the Squeeze Theorem,

$\lim_{t \to \infty} f(t) e^{-st} = 0$ for $s > a \;\Rightarrow\; G(s) = 0 - f(0) + sF(s) = sF(s) - f(0)$ for $s > a$.

89. We use integration by parts: let $u = x$, $dv = xe^{-x^2}\, dx$ $\Rightarrow$ $du = dx$, $v = -\frac{1}{2}e^{-x^2}$. So

$$\int_0^\infty x^2 e^{-x^2}\, dx = \lim_{t\to\infty}\left[-\frac{1}{2}xe^{-x^2}\right]_0^t + \frac{1}{2}\int_0^\infty e^{-x^2}\, dx = \lim_{t\to\infty}\left[-\frac{t}{2e^{t^2}}\right] + \frac{1}{2}\int_0^\infty e^{-x^2}\, dx = \frac{1}{2}\int_0^\infty e^{-x^2}\, dx$$

(The limit is 0 by l'Hospital's Rule.)

91. For the first part of the integral, let $x = 2\tan\theta$ $\Rightarrow$ $dx = 2\sec^2\theta\, d\theta$.

$$\int \frac{1}{\sqrt{x^2 + 4}}\, dx = \int \frac{2\sec^2\theta}{2\sec\theta}\, d\theta = \int \sec\theta\, d\theta = \ln|\sec\theta + \tan\theta|.$$

From the figure, $\tan\theta = \frac{x}{2}$, and $\sec\theta = \frac{\sqrt{x^2+4}}{2}$. So

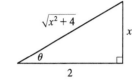

$$I = \int_0^\infty \left(\frac{1}{\sqrt{x^2+4}} - \frac{C}{x+2}\right) dx = \lim_{t\to\infty}\left[\ln\left|\frac{\sqrt{x^2+4}}{2} + \frac{x}{2}\right| - C\ln|x+2|\right]_0^t$$

$$= \lim_{t\to\infty}\left[\ln\frac{\sqrt{t^2+4}+t}{2} - C\ln(t+2) - (\ln 1 - C\ln 2)\right]$$

$$= \lim_{t\to\infty}\left[\ln\left(\frac{\sqrt{t^2+4}+t}{2(t+2)^C}\right) + \ln 2^C\right] = \ln\left(\lim_{t\to\infty}\frac{t + \sqrt{t^2+4}}{(t+2)^C}\right) + \ln 2^{C-1}$$

Now $L = \lim_{t\to\infty}\frac{t + \sqrt{t^2+4}}{(t+2)^C} \stackrel{H}{=} \lim_{t\to\infty}\frac{1 + t/\sqrt{t^2+4}}{C(t+2)^{C-1}} = \frac{2}{C\lim_{t\to\infty}(t+2)^{C-1}}.$

If $C < 1$, $L = \infty$ and I diverges.

If $C = 1$, $L = 2$ and I converges to $\ln 2 + \ln 2^0 = \ln 2$.

If $C > 1$, $L = 0$ and I diverges to $-\infty$.

93. No, $I = \int_0^\infty f(x)\, dx$ must be *divergent*. Since $\lim_{x\to\infty} f(x) = 1$, there must exist an N such that if $x \geq N$, then $f(x) \geq \frac{1}{2}$. Thus, $I = I_1 + I_2 = \int_0^N f(x)\, dx + \int_N^\infty f(x)\, dx$, where I_1 is an ordinary definite integral that has a finite value, and I_2 is improper and diverges by comparison with the divergent integral $\int_N^\infty \frac{1}{2}\, dx$.

7 Review

TRUE-FALSE QUIZ

1. True.   See Example 5 in Section 7.1.

3. False.   Substituting $x = 5\sin\theta$ into $\sqrt{25 + x^2}$ gives $\sqrt{25 + 25\sin^2\theta}$. This expression cannot be further simplified using a trigonometric identity. A more useful substitution would be $x = 5\tan\theta$.

5. False.   Since the numerator has a higher degree than the denominator, $\frac{x(x^2+4)}{x^2-4} = x + \frac{8x}{x^2-4} = x + \frac{A}{x+2} + \frac{B}{x-2}$.

7. False. $\frac{x^2+4}{x^2(x-4)}$ can be put in the form $\frac{A}{x} + \frac{B}{x^2} + \frac{C}{x-4}$.

9. False. This is an improper integral, since the denominator vanishes at $x = 1$.

$$\int_0^4 \frac{x}{x^2-1}\,dx = \int_0^1 \frac{x}{x^2-1}\,dx + \int_1^4 \frac{x}{x^2-1}\,dx \text{ and}$$

$$\int_0^1 \frac{x}{x^2-1}\,dx = \lim_{t\to 1^-}\int_0^t \frac{x}{x^2-1}\,dx = \lim_{t\to 1^-}\left[\tfrac{1}{2}\ln|x^2-1|\right]_0^t = \lim_{t\to 1^-}\tfrac{1}{2}\ln|t^2-1| = \infty$$

So the integral diverges.

11. True. $\int_{-\infty}^{\infty} f(x)\,dx = \int_{-\infty}^0 f(x)\,dx + \int_0^{\infty} f(x)\,dx = I_1 + I_2$. If $\int_{-\infty}^{\infty} f(x)\,dx$ is convergent, it follows that both I_1 and I_2 must be convergent.

13. (a) True. See the end of Section 7.5.

(b) False. Examples include the functions $f(x) = e^{x^2}$, $g(x) = \sin(x^2)$, and $h(x) = \dfrac{\sin x}{x}$.

15. False. If $f(x) = 1/x$, then f is continuous and decreasing on $[1, \infty)$ with $\lim_{x\to\infty} f(x) = 0$, but $\int_1^{\infty} f(x)\,dx$ is divergent.

17. False. Take $f(x) = 1$ for all x and $g(x) = -1$ for all x. Then $\int_a^{\infty} f(x)\,dx = \infty$ [divergent] and $\int_a^{\infty} g(x)\,dx = -\infty$ [divergent], but $\int_a^{\infty}[f(x) + g(x)]\,dx = 0$ [convergent].

EXERCISES

1. $\displaystyle\int_1^2 \frac{(x+1)^2}{x}\,dx = \int_1^2 \frac{x^2 + 2x + 1}{x}\,dx = \int_1^2 \left(x + 2 + \frac{1}{x}\right)dx = \left[\tfrac{1}{2}x^2 + 2x + \ln|x|\right]_1^2$
$= (2 + 4 + \ln 2) - (\tfrac{1}{2} + 2 + 0) = \tfrac{7}{2} + \ln 2$

3. $\displaystyle\int \frac{e^{\sin x}}{\sec x}\,dx = \int \cos x\, e^{\sin x}\,dx = \int e^u\,du \quad \begin{bmatrix} u = \sin x, \\ du = \cos x\,dx \end{bmatrix}$
$= e^u + C = e^{\sin x} + C$

5. $\displaystyle\int \frac{dt}{2t^2 + 3t + 1} = \int \frac{1}{(2t+1)(t+1)}\,dt = \int\left(\frac{2}{2t+1} - \frac{1}{t+1}\right)dt$ [partial fractions] $= \ln|2t+1| - \ln|t+1| + C$

7. $\int_0^{\pi/2}\sin^3\theta\,\cos^2\theta\,d\theta = \int_0^{\pi/2}(1 - \cos^2\theta)\cos^2\theta\,\sin\theta\,d\theta = \int_1^0 (1 - u^2)u^2(-du) \quad \begin{bmatrix} u = \cos\theta, \\ du = -\sin\theta\,d\theta \end{bmatrix}$
$= \int_0^1 (u^2 - u^4)\,du = \left[\tfrac{1}{3}u^3 - \tfrac{1}{5}u^5\right]_0^1 = \left(\tfrac{1}{3} - \tfrac{1}{5}\right) - 0 = \tfrac{2}{15}$

9. Let $u = \ln t$, $du = dt/t$. Then $\displaystyle\int \frac{\sin(\ln t)}{t}\,dt = \int \sin u\,du = -\cos u + C = -\cos(\ln t) + C$.

11. First let $u = (\ln x)^2$, $dv = x\,dx$ $\Rightarrow$ $du = \dfrac{2\ln x}{x}\,dx$, $v = \tfrac{1}{2}x^2$. Then $I = \displaystyle\int x\,(\ln x)^2\,dx = \tfrac{1}{2}x^2(\ln x)^2 - \int x\ln x\,dx$.

Next, let $U = \ln x$, $dV = x\,dx$ $\Rightarrow$ $dU = \dfrac{1}{x}\,dx$, $V = \tfrac{1}{2}x^2$, so $\displaystyle\int x\ln x\,dx = \tfrac{1}{2}x^2\ln x - \tfrac{1}{2}\int x\,dx = \tfrac{1}{2}x^2\ln x - \tfrac{1}{4}x^2$.

Substituting in the previous formula gives $I = \tfrac{1}{2}x^2(\ln x)^2 - \left(\tfrac{1}{2}x^2\ln x - \tfrac{1}{4}x^2\right) + C = \tfrac{1}{4}x^2[2(\ln x)^2 - 2\ln x + 1] + C$.

13. Let $x = \sec\theta$. Then

$$\int_1^2 \frac{\sqrt{x^2-1}}{x}\,dx = \int_0^{\pi/3} \frac{\tan\theta}{\sec\theta}\sec\theta\tan\theta\,d\theta = \int_0^{\pi/3} \tan^2\theta\,d\theta = \int_0^{\pi/3}(\sec^2\theta - 1)\,d\theta = \big[\tan\theta - \theta\big]_0^{\pi/3} = \sqrt{3} - \tfrac{\pi}{3}.$$

15. Let $w = \sqrt[3]{x}$. Then $w^3 = x$ and $3w^2\,dw = dx$, so $\int e^{\sqrt[3]{x}}\,dx = \int e^w \cdot 3w^2\,dw = 3I$. To evaluate I, let $u = w^2$, $dv = e^w\,dw \Rightarrow du = 2w\,dw$, $v = e^w$, so $I = \int w^2 e^w\,dw = w^2 e^w - \int 2we^w\,dw$. Now let $U = w$, $dV = e^w\,dw \Rightarrow dU = dw$, $V = e^w$. Thus, $I = w^2 e^w - 2\big[we^w - \int e^w\,dw\big] = w^2 e^w - 2we^w + 2e^w + C_1$, and hence

$$3I = 3e^w(w^2 - 2w + 2) + C = 3e^{\sqrt[3]{x}}(x^{2/3} - 2x^{1/3} + 2) + C.$$

17. Integrate by parts with $u = \tan^{-1} x$, $dv = x^2\,dx$, so that $du = \dfrac{1}{1+x^2}\,dx$, $v = \dfrac{1}{3}x^3$. Then

$$\int x^2 \tan^{-1} x\,dx = \tfrac{1}{3}x^3 \tan^{-1} x - \tfrac{1}{3}\int \frac{x^3}{1+x^2}\,dx = \tfrac{1}{3}x^3 \tan^{-1} x - \tfrac{1}{3}\cdot\tfrac{1}{2}\int \frac{y-1}{y}\,dy \quad \begin{bmatrix} y = 1 + x^2, \\ dy = 2x\,dx \end{bmatrix}$$

$$= \tfrac{1}{3}x^3 \tan^{-1} x - \tfrac{1}{6}\int\left(1 - \tfrac{1}{y}\right)dy = \tfrac{1}{3}x^3 \tan^{-1} x - \tfrac{1}{6}\big(y - \ln|y|\big) + C_1$$

$$= \tfrac{1}{3}x^3 \tan^{-1} x - \tfrac{1}{6}\big[1 + x^2 - \ln(1 + x^2)\big] + C_1$$

$$= \tfrac{1}{6}\big[2x^3 \tan^{-1} x - x^2 + \ln(1 + x^2)\big] + C, \text{ where } C = C_1 - \tfrac{1}{6}$$

19. $\dfrac{x-1}{x^2+2x} = \dfrac{x-1}{x(x+2)} = \dfrac{A}{x} + \dfrac{B}{x+2} \Rightarrow x - 1 = A(x+2) + Bx$. Set $x = -2$ to get $-3 = -2B$, so $B = \tfrac{3}{2}$. Set $x = 0$ to get $-1 = 2A$, so $A = -\tfrac{1}{2}$. Thus, $\displaystyle\int \frac{x-1}{x^2+2x}\,dx = \int\left(\frac{-\tfrac{1}{2}}{x} + \frac{\tfrac{3}{2}}{x+2}\right)dx = -\tfrac{1}{2}\ln|x| + \tfrac{3}{2}\ln|x+2| + C$.

21. $\int x\cosh x\,dx = x\sinh x - \int \sinh x\,dx \quad \begin{bmatrix} u = x, & dv = \cosh x\,dx \\ du = dx, & v = \sinh x \end{bmatrix}$

$$= x\sinh x - \cosh x + C$$

23. $\displaystyle\int \frac{dx}{\sqrt{x^2-4x}} = \int \frac{dx}{\sqrt{(x^2-4x+4)-4}} = \int \frac{dx}{\sqrt{(x-2)^2 - 2^2}}$

$$= \int \frac{2\sec\theta\tan\theta\,d\theta}{2\tan\theta} \quad \begin{bmatrix} x - 2 = 2\sec\theta, \\ dx = 2\sec\theta\tan\theta\,d\theta \end{bmatrix}$$

$$= \int \sec\theta\,d\theta = \ln|\sec\theta + \tan\theta| + C_1$$

$$= \ln\left|\frac{x-2}{2} + \frac{\sqrt{x^2-4x}}{2}\right| + C_1$$

$$= \ln\left|x - 2 + \sqrt{x^2-4x}\right| + C, \text{ where } C = C_1 - \ln 2$$

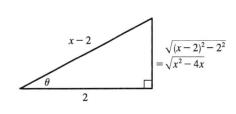

25. $\displaystyle\int \frac{x+1}{9x^2+6x+5}\,dx = \int \frac{x+1}{(9x^2+6x+1)+4}\,dx = \int \frac{x+1}{(3x+1)^2+4}\,dx \quad \begin{bmatrix} u = 3x + 1, \\ du = 3\,dx \end{bmatrix}$

$$= \int \frac{[\tfrac{1}{3}(u-1)] + 1}{u^2+4}\left(\tfrac{1}{3}du\right) = \tfrac{1}{3}\cdot\tfrac{1}{3}\int \frac{(u-1)+3}{u^2+4}\,du$$

$$= \tfrac{1}{9}\int \frac{u}{u^2+4}\,du + \tfrac{1}{9}\int \frac{2}{u^2+2^2}\,du = \tfrac{1}{9}\cdot\tfrac{1}{2}\ln(u^2+4) + \tfrac{2}{9}\cdot\tfrac{1}{2}\tan^{-1}\left(\tfrac{1}{2}u\right) + C$$

$$= \tfrac{1}{18}\ln(9x^2+6x+5) + \tfrac{1}{9}\tan^{-1}\big[\tfrac{1}{2}(3x+1)\big] + C$$

27. $\sqrt{x^2 - 2x + 2} = \sqrt{x^2 - 2x + 1 + 1} = \sqrt{(x-1)^2 + 1}$. Since this is a sum of squares,

we try the substitution $x - 1 = \tan\theta$, where $-\pi/2 < \theta < \pi/2$. Then $dx = \sec^2\theta\, d\theta$ and

$\sqrt{(x-1)^2 + 1} = \sqrt{\tan^2\theta + 1} = \sqrt{\sec^2\theta} = |\sec\theta| = \sec\theta$. Also, $x = 0 \Rightarrow \theta = -\pi/4$ and $x = 2 \Rightarrow \theta = \pi/4$.

Thus,

$$\int_0^2 \sqrt{x^2 - 2x + 2}\, dx = \int_{-\pi/4}^{\pi/4} \sec\theta\, (\sec^2\theta\, d\theta) = \int_{-\pi/4}^{\pi/4} \sec^3\theta\, d\theta$$

$$= \frac{1}{2}\Big[\sec\theta\tan\theta + \ln|\sec\theta + \tan\theta|\Big]_{-\pi/4}^{\pi/4} \quad \text{[by Example 8 in Section 7.2]}$$

$$= \frac{1}{2}[(\sqrt{2} + \ln(\sqrt{2}+1)) - (-\sqrt{2} + \ln(\sqrt{2}-1))] = \frac{1}{2}\left[2\sqrt{2} + \ln\left(\frac{\sqrt{2}+1}{\sqrt{2}-1}\right)\right]$$

$$= \frac{1}{2}\left[2\sqrt{2} + \ln\left(\frac{\sqrt{2}+1}{\sqrt{2}-1}\cdot\frac{\sqrt{2}+1}{\sqrt{2}+1}\right)\right] = \frac{1}{2}\left[2\sqrt{2} + \ln(\sqrt{2}+1)^2\right]$$

$$= \frac{1}{2}[2\sqrt{2} + 2\ln(\sqrt{2}+1)] = \sqrt{2} + \ln(\sqrt{2}+1)$$

29. Let $x = \tan\theta$, so that $dx = \sec^2\theta\, d\theta$. Then

$$\int \frac{dx}{x\sqrt{x^2+1}} = \int \frac{\sec^2\theta\, d\theta}{\tan\theta\,\sec\theta} = \int \frac{\sec\theta}{\tan\theta}\, d\theta$$

$$= \int \csc\theta\, d\theta = \ln|\csc\theta - \cot\theta| + C$$

$$= \ln\left|\frac{\sqrt{x^2+1}}{x} - \frac{1}{x}\right| + C = \ln\left|\frac{\sqrt{x^2+1}-1}{x}\right| + C$$

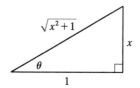

31. Let $u = \sqrt{1 + x^2}$, so that $du = \dfrac{x}{\sqrt{1+x^2}}\, dx$. Thus,

$$\int \frac{x\sin(\sqrt{1+x^2})}{\sqrt{1+x^2}}\, dx = \int \sin u\, du = -\cos u + C = -\cos\left(\sqrt{1+x^2}\right) + C.$$

33. $\dfrac{3x^3 - x^2 + 6x - 4}{(x^2+1)(x^2+2)} = \dfrac{Ax+B}{x^2+1} + \dfrac{Cx+D}{x^2+2} \Rightarrow 3x^3 - x^2 + 6x - 4 = (Ax+B)(x^2+2) + (Cx+D)(x^2+1)$.

Equating the coefficients gives $A + C = 3$, $B + D = -1$, $2A + C = 6$, and $2B + D = -4$ $\Rightarrow$

$A = 3$, $C = 0$, $B = -3$, and $D = 2$. Now

$$\int \frac{3x^3 - x^2 + 6x - 4}{(x^2+1)(x^2+2)}\, dx = 3\int \frac{x-1}{x^2+1}\, dx + 2\int \frac{dx}{x^2+2} = \frac{3}{2}\ln(x^2+1) - 3\tan^{-1}x + \sqrt{2}\tan^{-1}\left(\frac{x}{\sqrt{2}}\right) + C.$$

35. $\int_0^{\pi/2} \cos^3 x \sin 2x\, dx = \int_0^{\pi/2} \cos^3 x\,(2\sin x\cos x)\, dx = \int_0^{\pi/2} 2\cos^4 x \sin x\, dx = \left[-\frac{2}{5}\cos^5 x\right]_0^{\pi/2} = \frac{2}{5}$

37. The integrand is an odd function, so $\displaystyle\int_{-3}^{3} \frac{x}{1+|x|}\, dx = 0 \quad \text{[by 4.5.6]}.$

39. Let $u = \sqrt{e^x - 1}$. Then $u^2 = e^x - 1$ and $2u\,du = e^x\,dx$. Also, $e^x + 8 = u^2 + 9$. Thus,

$$\int_0^{\ln 10} \frac{e^x \sqrt{e^x - 1}}{e^x + 8}\,dx = \int_0^3 \frac{u \cdot 2u\,du}{u^2 + 9} = 2\int_0^3 \frac{u^2}{u^2 + 9}\,du = 2\int_0^3 \left(1 - \frac{9}{u^2 + 9}\right)du$$

$$= 2\left[u - \frac{9}{3}\tan^{-1}\left(\frac{u}{3}\right)\right]_0^3 = 2[(3 - 3\tan^{-1} 1) - 0] = 2\left(3 - 3 \cdot \frac{\pi}{4}\right) = 6 - \frac{3\pi}{2}$$

41. Let $x = 2\sin\theta \;\Rightarrow\; (4 - x^2)^{3/2} = (2\cos\theta)^3$, $dx = 2\cos\theta\,d\theta$, so

$$\int \frac{x^2}{(4 - x^2)^{3/2}}\,dx = \int \frac{4\sin^2\theta}{8\cos^3\theta} 2\cos\theta\,d\theta = \int \tan^2\theta\,d\theta = \int (\sec^2\theta - 1)\,d\theta$$

$$= \tan\theta - \theta + C = \frac{x}{\sqrt{4 - x^2}} - \sin^{-1}\left(\frac{x}{2}\right) + C$$

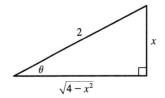

43. $\displaystyle\int \frac{1}{\sqrt{x + x^{3/2}}}\,dx = \int \frac{dx}{\sqrt{x(1 + \sqrt{x})}} = \int \frac{dx}{\sqrt{x}\sqrt{1 + \sqrt{x}}} \quad \begin{bmatrix} u = 1 + \sqrt{x}, \\ du = \dfrac{dx}{2\sqrt{x}} \end{bmatrix} = \int \frac{2\,du}{\sqrt{u}} = \int 2u^{-1/2}\,du$

$$= 4\sqrt{u} + C = 4\sqrt{1 + \sqrt{x}} + C$$

45. $\int (\cos x + \sin x)^2 \cos 2x\,dx = \int (\cos^2 x + 2\sin x\cos x + \sin^2 x)\cos 2x\,dx = \int (1 + \sin 2x)\cos 2x\,dx$

$$= \int \cos 2x\,dx + \tfrac{1}{2}\int \sin 4x\,dx = \tfrac{1}{2}\sin 2x - \tfrac{1}{8}\cos 4x + C$$

Or: $\int (\cos x + \sin x)^2 \cos 2x\,dx = \int (\cos x + \sin x)^2 (\cos^2 x - \sin^2 x)\,dx$

$$= \int (\cos x + \sin x)^3 (\cos x - \sin x)\,dx = \tfrac{1}{4}(\cos x + \sin x)^4 + C_1$$

47. We'll integrate $I = \displaystyle\int \frac{xe^{2x}}{(1 + 2x)^2}\,dx$ by parts with $u = xe^{2x}$ and $dv = \dfrac{dx}{(1 + 2x)^2}$. Then $du = (x \cdot 2e^{2x} + e^{2x} \cdot 1)\,dx$

and $v = -\dfrac{1}{2} \cdot \dfrac{1}{1 + 2x}$, so

$$I = -\frac{1}{2} \cdot \frac{xe^{2x}}{1 + 2x} - \int \left[-\frac{1}{2} \cdot \frac{e^{2x}(2x + 1)}{1 + 2x}\right]dx = -\frac{xe^{2x}}{4x + 2} + \frac{1}{2} \cdot \frac{1}{2}e^{2x} + C = e^{2x}\left(\frac{1}{4} - \frac{x}{4x + 2}\right) + C$$

Thus, $\displaystyle\int_0^{1/2} \frac{xe^{2x}}{(1 + 2x)^2}\,dx = \left[e^{2x}\left(\frac{1}{4} - \frac{x}{4x + 2}\right)\right]_0^{1/2} = e\left(\frac{1}{4} - \frac{1}{8}\right) - 1\left(\frac{1}{4} - 0\right) = \frac{1}{8}e - \frac{1}{4}$.

49. Let $u = \sqrt{e^x - 4}$, so that $e^x = u^2 + 4$ and $2u\,du = e^x\,dx = (u^2 + 4)\,dx \;\Leftrightarrow\; \dfrac{2u}{u^2 + 4}\,du = dx$. Thus,

$$\int \frac{1}{\sqrt{e^x - 4}}\,dx = \int \frac{1}{u} \cdot \frac{2u}{u^2 + 4}\,du = 2\int \frac{1}{u^2 + 4}\,du = 2\left(\frac{1}{2}\tan^{-1}\left(\frac{u}{2}\right)\right) + C = \tan^{-1}\left(\frac{\sqrt{e^x - 4}}{2}\right) + C.$$

51. $\displaystyle\int_1^\infty \frac{1}{(2x + 1)^3}\,dx = \lim_{t \to \infty}\int_1^t \frac{1}{(2x + 1)^3}\,dx = \lim_{t \to \infty}\int_1^t \tfrac{1}{2}(2x + 1)^{-3} 2\,dx = \lim_{t \to \infty}\left[-\frac{1}{4(2x + 1)^2}\right]_1^t$

$$= -\frac{1}{4}\lim_{t \to \infty}\left[\frac{1}{(2t + 1)^2} - \frac{1}{9}\right] = -\frac{1}{4}\left(0 - \frac{1}{9}\right) = \frac{1}{36}$$

53. $\displaystyle\int \frac{dx}{x \ln x} \quad \begin{bmatrix} u = \ln x, \\ du = dx/x \end{bmatrix} = \int \frac{du}{u} = \ln|u| + C = \ln|\ln x| + C$, so

$\displaystyle\int_2^\infty \frac{dx}{x \ln x} = \lim_{t\to\infty} \int_2^t \frac{dx}{x \ln x} = \lim_{t\to\infty}\Big[\ln|\ln x|\Big]_2^t = \lim_{t\to\infty}[\ln(\ln t) - \ln(\ln 2)] = \infty$, so the integral is divergent.

55. $\displaystyle\int_0^4 \frac{\ln x}{\sqrt{x}}\,dx = \lim_{t\to 0^+}\int_t^4 \frac{\ln x}{\sqrt{x}}\,dx \stackrel{\star}{=} \lim_{t\to 0^+}\Big[2\sqrt{x}\,\ln x - 4\sqrt{x}\Big]_t^4$

$\quad = \lim_{t\to 0^+}\big[(2\cdot 2\ln 4 - 4\cdot 2) - (2\sqrt{t}\,\ln t - 4\sqrt{t})\big] \stackrel{\star\star}{=} (4\ln 4 - 8) - (0 - 0) = 4\ln 4 - 8$

$(\star)$ Let $u = \ln x$, $dv = \dfrac{1}{\sqrt{x}}\,dx \;\Rightarrow\; du = \dfrac{1}{x}\,dx$, $v = 2\sqrt{x}$. Then

$$\int \frac{\ln x}{\sqrt{x}}\,dx = 2\sqrt{x}\,\ln x - 2\int \frac{dx}{\sqrt{x}} = 2\sqrt{x}\,\ln x - 4\sqrt{x} + C$$

$(\star\star)$ $\displaystyle\lim_{t\to 0^+}\left(2\sqrt{t}\,\ln t\right) = \lim_{t\to 0^+}\frac{2\ln t}{t^{-1/2}} \stackrel{H}{=} \lim_{t\to 0^+}\frac{2/t}{-\tfrac{1}{2}t^{-3/2}} = \lim_{t\to 0^+}(-4\sqrt{t}) = 0$

57. $\displaystyle\int_0^1 \frac{x-1}{\sqrt{x}}\,dx = \lim_{t\to 0^+}\int_t^1\left(\frac{x}{\sqrt{x}} - \frac{1}{\sqrt{x}}\right)dx = \lim_{t\to 0^+}\int_t^1 (x^{1/2} - x^{-1/2})\,dx = \lim_{t\to 0^+}\Big[\tfrac{2}{3}x^{3/2} - 2x^{1/2}\Big]_t^1$

$\quad = \displaystyle\lim_{t\to 0^+}\Big[\big(\tfrac{2}{3}-2\big) - \big(\tfrac{2}{3}t^{3/2} - 2t^{1/2}\big)\Big] = -\tfrac{4}{3} - 0 = -\tfrac{4}{3}$

59. Let $u = 2x+1$. Then

$\displaystyle\int_{-\infty}^\infty \frac{dx}{4x^2+4x+5} = \int_{-\infty}^\infty \frac{\tfrac{1}{2}\,du}{u^2+4} = \tfrac{1}{2}\int_{-\infty}^0 \frac{du}{u^2+4} + \tfrac{1}{2}\int_0^\infty \frac{du}{u^2+4}$

$\quad = \tfrac{1}{2}\displaystyle\lim_{t\to -\infty}\Big[\tfrac{1}{2}\tan^{-1}\big(\tfrac{1}{2}u\big)\Big]_t^0 + \tfrac{1}{2}\lim_{t\to\infty}\Big[\tfrac{1}{2}\tan^{-1}\big(\tfrac{1}{2}u\big)\Big]_0^t = \tfrac{1}{4}\big[0-(-\tfrac{\pi}{2})\big] + \tfrac{1}{4}\big[\tfrac{\pi}{2}-0\big] = \tfrac{\pi}{4}$.

61. We first make the substitution $t = x+1$, so $\ln(x^2+2x+2) = \ln[(x+1)^2 + 1] = \ln(t^2+1)$. Then we use parts with $u = \ln(t^2+1)$, $dv = dt$:

$\displaystyle\int \ln(t^2+1)\,dt = t\ln(t^2+1) - \int \frac{t(2t)\,dt}{t^2+1} = t\ln(t^2+1) - 2\int\frac{t^2\,dt}{t^2+1} = t\ln(t^2+1) - 2\int\left(1 - \frac{1}{t^2+1}\right)dt$

$\quad = t\ln(t^2+1) - 2t + 2\arctan t + C$

$\quad = (x+1)\ln(x^2+2x+2) - 2x + 2\arctan(x+1) + K$, where $K = C - 2$

[Alternatively, we could have integrated by parts immediately with $u = \ln(x^2+2x+2)$.] Notice from the graph that $f = 0$ where F has a horizontal tangent. Also, F is always increasing, and $f \ge 0$.

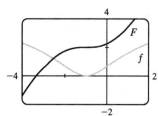

63. From the graph, it seems as though $\int_0^{2\pi} \cos^2 x\,\sin^3 x\,dx$ is equal to 0.

To evaluate the integral, we write the integral as

$I = \int_0^{2\pi}\cos^2 x\,(1-\cos^2 x)\sin x\,dx$ and let $u = \cos x \;\Rightarrow\;$

$du = -\sin x\,dx$. Thus, $I = \int_1^1 u^2(1-u^2)(-du) = 0$.

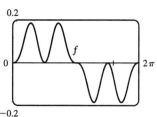

65. $\int \sqrt{4x^2 - 4x - 3}\, dx = \int \sqrt{(2x-1)^2 - 4}\, dx \quad \begin{bmatrix} u = 2x - 1, \\ du = 2\, dx \end{bmatrix} = \int \sqrt{u^2 - 2^2} \left(\tfrac{1}{2}\, du\right)$

$$\stackrel{39}{=} \tfrac{1}{2}\left(\tfrac{u}{2}\sqrt{u^2 - 2^2} - \tfrac{2^2}{2}\ln\left|u + \sqrt{u^2 - 2^2}\right|\right) + C = \tfrac{1}{4}u\sqrt{u^2 - 4} - \ln\left|u + \sqrt{u^2 - 4}\right| + C$$

$$= \tfrac{1}{4}(2x - 1)\sqrt{4x^2 - 4x - 3} - \ln\left|2x - 1 + \sqrt{4x^2 - 4x - 3}\right| + C$$

67. Let $u = \sin x$, so that $du = \cos x\, dx$. Then

$$\int \cos x \sqrt{4 + \sin^2 x}\, dx = \int \sqrt{2^2 + u^2}\, du \stackrel{21}{=} \tfrac{u}{2}\sqrt{2^2 + u^2} + \tfrac{2^2}{2}\ln\left(u + \sqrt{2^2 + u^2}\right) + C$$

$$= \tfrac{1}{2}\sin x\sqrt{4 + \sin^2 x} + 2\ln\left(\sin x + \sqrt{4 + \sin^2 x}\right) + C$$

69. (a) $\dfrac{d}{du}\left[-\dfrac{1}{u}\sqrt{a^2 - u^2} - \sin^{-1}\left(\dfrac{u}{a}\right) + C\right] = \dfrac{1}{u^2}\sqrt{a^2 - u^2} + \dfrac{1}{\sqrt{a^2 - u^2}} - \dfrac{1}{\sqrt{1 - u^2/a^2}} \cdot \dfrac{1}{a}$

$$= (a^2 - u^2)^{-1/2}\left[\dfrac{1}{u^2}(a^2 - u^2) + 1 - 1\right] = \dfrac{\sqrt{a^2 - u^2}}{u^2}$$

(b) Let $u = a\sin\theta \;\Rightarrow\; du = a\cos\theta\, d\theta$, $a^2 - u^2 = a^2(1 - \sin^2\theta) = a^2\cos^2\theta$.

$$\int \dfrac{\sqrt{a^2 - u^2}}{u^2}\, du = \int \dfrac{a^2\cos^2\theta}{a^2\sin^2\theta}\, d\theta = \int \dfrac{1 - \sin^2\theta}{\sin^2\theta}\, d\theta = \int (\csc^2\theta - 1)\, d\theta = -\cot\theta - \theta + C$$

$$= -\dfrac{\sqrt{a^2 - u^2}}{u} - \sin^{-1}\left(\dfrac{u}{a}\right) + C$$

71. For $n \geq 0$, $\int_0^\infty x^n\, dx = \lim_{t \to \infty}\left[x^{n+1}/(n+1)\right]_0^t = \infty$. For $n < 0$, $\int_0^\infty x^n\, dx = \int_0^1 x^n\, dx + \int_1^\infty x^n\, dx$. Both integrals are improper. By (7.8.2), the second integral diverges if $-1 \leq n < 0$. By Exercise 7.8.69, the first integral diverges if $n \leq -1$. Thus, $\int_0^\infty x^n\, dx$ is divergent for all values of n.

73. $f(x) = \dfrac{1}{\ln x}$, $\Delta x = \dfrac{b - a}{n} = \dfrac{4 - 2}{10} = \dfrac{1}{5}$

(a) $T_{10} = \dfrac{1}{5 \cdot 2}\{f(2) + 2[f(2.2) + f(2.4) + \cdots + f(3.8)] + f(4)\} \approx 1.925444$

(b) $M_{10} = \tfrac{1}{5}[f(2.1) + f(2.3) + f(2.5) + \cdots + f(3.9)] \approx 1.920915$

(c) $S_{10} = \dfrac{1}{5 \cdot 3}[f(2) + 4f(2.2) + 2f(2.4) + \cdots + 2f(3.6) + 4f(3.8) + f(4)] \approx 1.922470$

75. $f(x) = \dfrac{1}{\ln x} \;\Rightarrow\; f'(x) = -\dfrac{1}{x(\ln x)^2} \;\Rightarrow\; f''(x) = \dfrac{2 + \ln x}{x^2(\ln x)^3} = \dfrac{2}{x^2(\ln x)^3} + \dfrac{1}{x^2(\ln x)^2}$. Note that each term of

$f''(x)$ decreases on $[2, 4]$, so we'll take $K = f''(2) \approx 2.022$. $|E_T| \leq \dfrac{K(b-a)^3}{12n^2} \approx \dfrac{2.022(4-2)^3}{12(10)^2} = 0.01348$ and

$|E_M| \leq \dfrac{K(b-a)^3}{24n^2} = 0.00674$. $|E_T| \leq 0.00001 \;\Leftrightarrow\; \dfrac{2.022(8)}{12n^2} \leq \dfrac{1}{10^5} \;\Leftrightarrow\; n^2 \geq \dfrac{10^5(2.022)(8)}{12} \;\Rightarrow\; n \geq 367.2$.

Take $n = 368$ for T_n. $|E_M| \leq 0.00001 \;\Leftrightarrow\; n^2 \geq \dfrac{10^5(2.022)(8)}{24} \;\Rightarrow\; n \geq 259.6$. Take $n = 260$ for M_n.

77. $\Delta t = \left(\frac{10}{60} - 0\right)/10 = \frac{1}{60}$.

Distance traveled $= \int_0^{10} v\,dt \approx S_{10}$

$= \frac{1}{60 \cdot 3}[40 + 4(42) + 2(45) + 4(49) + 2(52) + 4(54) + 2(56) + 4(57) + 2(57) + 4(55) + 56]$

$= \frac{1}{180}(1544) = 8.5\overline{7}$ mi

79. (a) $f(x) = \sin(\sin x)$. A CAS gives

$f^{(4)}(x) = \sin(\sin x)[\cos^4 x + 7\cos^2 x - 3]$
$\quad + \cos(\sin x)[6\cos^2 x \sin x + \sin x]$

From the graph, we see that $\left|f^{(4)}(x)\right| < 3.8$ for $x \in [0, \pi]$.

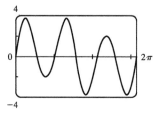

(b) We use Simpson's Rule with $f(x) = \sin(\sin x)$ and $\Delta x = \frac{\pi}{10}$:

$\int_0^\pi f(x)\,dx \approx \frac{\pi}{10 \cdot 3}\left[f(0) + 4f\left(\frac{\pi}{10}\right) + 2f\left(\frac{2\pi}{10}\right) + \cdots + 4f\left(\frac{9\pi}{10}\right) + f(\pi)\right] \approx 1.786721$

From part (a), we know that $\left|f^{(4)}(x)\right| < 3.8$ on $[0, \pi]$, so we use Theorem 7.7.4 with $K = 3.8$, and estimate the error

as $|E_S| \le \dfrac{3.8(\pi - 0)^5}{180(10)^4} \approx 0.000646$.

(c) If we want the error to be less than 0.00001, we must have $|E_S| \le \dfrac{3.8\pi^5}{180n^4} \le 0.00001$,

so $n^4 \ge \dfrac{3.8\pi^5}{180(0.00001)} \approx 646{,}041.6 \;\Rightarrow\; n \ge 28.35$. Since n must be even for Simpson's Rule, we must have $n \ge 30$ to ensure the desired accuracy.

81. (a) $\dfrac{2 + \sin x}{\sqrt{x}} \ge \dfrac{1}{\sqrt{x}}$ for x in $[1, \infty)$. $\int_1^\infty \dfrac{1}{\sqrt{x}}\,dx$ is divergent by (7.8.2) with $p = \dfrac{1}{2} \le 1$. Therefore, $\int_1^\infty \dfrac{2 + \sin x}{\sqrt{x}}\,dx$ is divergent by the Comparison Theorem.

(b) $\dfrac{1}{\sqrt{1 + x^4}} < \dfrac{1}{\sqrt{x^4}} = \dfrac{1}{x^2}$ for x in $[1, \infty)$. $\int_1^\infty \dfrac{1}{x^2}\,dx$ is convergent by (7.8.2) with $p = 2 > 1$. Therefore,

$\int_1^\infty \dfrac{1}{\sqrt{1 + x^4}}\,dx$ is convergent by the Comparison Theorem.

83. For x in $\left[0, \frac{\pi}{2}\right]$, $0 \le \cos^2 x \le \cos x$. For x in $\left[\frac{\pi}{2}, \pi\right]$, $\cos x \le 0 \le \cos^2 x$. Thus,

area $= \int_0^{\pi/2}(\cos x - \cos^2 x)\,dx + \int_{\pi/2}^\pi(\cos^2 x - \cos x)\,dx$

$= \left[\sin x - \tfrac{1}{2}x - \tfrac{1}{4}\sin 2x\right]_0^{\pi/2} + \left[\tfrac{1}{2}x + \tfrac{1}{4}\sin 2x - \sin x\right]_{\pi/2}^\pi = \left[\left(1 - \tfrac{\pi}{4}\right) - 0\right] + \left[\tfrac{\pi}{2} - \left(\tfrac{\pi}{4} - 1\right)\right] = 2$

85. Using the formula for disks, the volume is

$V = \int_0^{\pi/2} \pi\,[f(x)]^2\,dx = \pi\int_0^{\pi/2}(\cos^2 x)^2\,dx = \pi\int_0^{\pi/2}\left[\tfrac{1}{2}(1 + \cos 2x)\right]^2 dx$

$= \tfrac{\pi}{4}\int_0^{\pi/2}(1 + \cos^2 2x + 2\cos 2x)\,dx = \tfrac{\pi}{4}\int_0^{\pi/2}\left[1 + \tfrac{1}{2}(1 + \cos 4x) + 2\cos 2x\right] dx$

$= \tfrac{\pi}{4}\left[\tfrac{3}{2}x + \tfrac{1}{2}\left(\tfrac{1}{4}\sin 4x\right) + 2\left(\tfrac{1}{2}\sin 2x\right)\right]_0^{\pi/2} = \tfrac{\pi}{4}\left[\left(\tfrac{3\pi}{4} + \tfrac{1}{8}\cdot 0 + 0\right) - 0\right] = \tfrac{3}{16}\pi^2$

87. By the Fundamental Theorem of Calculus,

$\int_0^\infty f'(x)\, dx = \lim_{t \to \infty} \int_0^t f'(x)\, dx = \lim_{t \to \infty} [f(t) - f(0)] = \lim_{t \to \infty} f(t) - f(0) = 0 - f(0) = -f(0).$

89. Let $u = 1/x \;\Rightarrow\; x = 1/u \;\Rightarrow\; dx = -(1/u^2)\, du.$

$\int_0^\infty \dfrac{\ln x}{1+x^2}\, dx = \int_\infty^0 \dfrac{\ln(1/u)}{1+1/u^2}\left(-\dfrac{du}{u^2}\right) = \int_\infty^0 \dfrac{-\ln u}{u^2+1}(-du) = \int_\infty^0 \dfrac{\ln u}{1+u^2}\, du = -\int_0^\infty \dfrac{\ln u}{1+u^2}\, du$

Therefore, $\int_0^\infty \dfrac{\ln x}{1+x^2}\, dx = -\int_0^\infty \dfrac{\ln x}{1+x^2}\, dx = 0.$

PROBLEMS PLUS

1.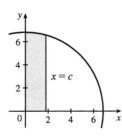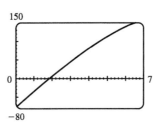

By symmetry, the problem can be reduced to finding the line $x = c$ such that the shaded area is one-third of the area of the quarter-circle. An equation of the semicircle is $y = \sqrt{49 - x^2}$, so we require that $\int_0^c \sqrt{49 - x^2}\, dx = \tfrac{1}{3} \cdot \tfrac{1}{4}\pi(7)^2$ $\Leftrightarrow$ $\left[\tfrac{1}{2}x\sqrt{49 - x^2} + \tfrac{49}{2}\sin^{-1}(x/7)\right]_0^c = \tfrac{49}{12}\pi$ [by Formula 30] $\Leftrightarrow$ $\tfrac{1}{2}c\sqrt{49 - c^2} + \tfrac{49}{2}\sin^{-1}(c/7) = \tfrac{49}{12}\pi$.
This equation would be difficult to solve exactly, so we plot the left-hand side as a function of c, and find that the equation holds for $c \approx 1.85$. So the cuts should be made at distances of about 1.85 inches from the center of the pizza.

3. The given integral represents the difference of the shaded areas, which appears to be 0. It can be calculated by integrating with respect to either x or y, so we find x in terms of y for each curve: $y = \sqrt[3]{1 - x^7}$ $\Rightarrow$ $x = \sqrt[7]{1 - y^3}$ and $y = \sqrt[7]{1 - x^3}$ $\Rightarrow$ $x = \sqrt[3]{1 - y^7}$, so
$\int_0^1 \left(\sqrt[3]{1 - y^7} - \sqrt[7]{1 - y^3}\right) dy = \int_0^1 \left(\sqrt[7]{1 - x^3} - \sqrt[3]{1 - x^7}\right) dx$. But this equation is of the form $z = -z$. So $\int_0^1 \left(\sqrt[3]{1 - x^7} - \sqrt[7]{1 - x^3}\right) dx = 0$.

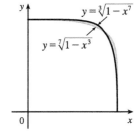

5. $I = \int_{-r}^{r} \dfrac{f(x)}{1 + a^x}\, dx = \int_{-r}^{0} \dfrac{f(x)}{1 + a^x}\, dx + \int_{0}^{r} \dfrac{f(x)}{1 + a^x}\, dx = I_1 + I_2$

Using the substitution $u = -x$, $du = -dx$ to evaluate I_1 gives

$$\int_{-r}^{0} \dfrac{f(x)}{1 + a^x}\, dx = \int_{r}^{0} \dfrac{f(-u)}{1 + a^{-u}}(-du) = \int_{0}^{r} f(-u)\left[\dfrac{1}{1 + a^{-u}}\right] du$$

$$= \int_{0}^{r} f(u) \left[\dfrac{1}{1 + a^{-u}}\right] du \quad \text{[since } f(x) \text{ is even]}$$

$$= \int_{0}^{r} f(u) \left[1 - \dfrac{1}{1 + a^u}\right] du \quad \text{[using the provided hint]}$$

$$= \int_{0}^{r} f(u)\, du - \int_{0}^{r} \dfrac{f(u)}{1 + a^u}\, du$$

Thus, $I = I_1 + I_2 = \left(\int_0^r f(u)\, du - \int_0^r \dfrac{f(u)}{1+a^u}\, du\right) + \int_0^r \dfrac{f(x)}{1+a^x}\, dx = \int_0^r f(u)\, du$.

7. The area A of the remaining part of the circle is given by

$$A = 4I = 4\int_0^a \left(\sqrt{a^2 - x^2} - \frac{b}{a}\sqrt{a^2 - x^2}\right) dx = 4\left(1 - \frac{b}{a}\right)\int_0^a \sqrt{a^2 - x^2}\, dx$$

$$\stackrel{30}{=} \frac{4}{a}(a - b)\left[\frac{x}{2}\sqrt{a^2 - x^2} + \frac{a^2}{2}\sin^{-1}\frac{x}{a}\right]_0^a$$

$$= \frac{4}{a}(a - b)\left[\left(0 + \frac{a^2}{2}\frac{\pi}{2}\right) - 0\right] = \frac{4}{a}(a - b)\left(\frac{a^2\pi}{4}\right) = \pi a(a - b),$$

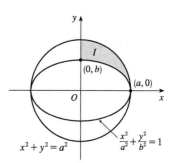

which is the area of an ellipse with semiaxes a and $a - b$.

Alternate solution: Subtracting the area of the ellipse from the area of the circle gives us $\pi a^2 - \pi ab = \pi a(a - b)$, as calculated above. (The formula for the area of an ellipse was derived in Example 2 in Section 7.3.)

9. Recall that $\cos A \cos B = \frac{1}{2}[\cos(A + B) + \cos(A - B)]$. So

$$f(x) = \int_0^\pi \cos t \cos(x - t)\, dt = \frac{1}{2}\int_0^\pi [\cos(t + x - t) + \cos(t - x + t)]\, dt = \frac{1}{2}\int_0^\pi [\cos x + \cos(2t - x)]\, dt$$

$$= \frac{1}{2}\left[t\cos x + \frac{1}{2}\sin(2t - x)\right]_0^\pi = \frac{\pi}{2}\cos x + \frac{1}{4}\sin(2\pi - x) - \frac{1}{4}\sin(-x)$$

$$= \frac{\pi}{2}\cos x + \frac{1}{4}\sin(-x) - \frac{1}{4}\sin(-x) = \frac{\pi}{2}\cos x$$

The minimum of $\cos x$ on this domain is -1, so the minimum value of $f(x)$ is $f(\pi) = -\frac{\pi}{2}$.

11. In accordance with the hint, we let $I_k = \int_0^1 (1 - x^2)^k\, dx$, and we find an expression for I_{k+1} in terms of I_k. We integrate I_{k+1} by parts with $u = (1 - x^2)^{k+1}\ \Rightarrow\ du = (k + 1)(1 - x^2)^k(-2x),\ dv = dx\ \Rightarrow\ v = x$, and then split the remaining integral into identifiable quantities:

$$I_{k+1} = x(1 - x^2)^{k+1}\big|_0^1 + 2(k + 1)\int_0^1 x^2(1 - x^2)^k\, dx = (2k + 2)\int_0^1 (1 - x^2)^k[1 - (1 - x^2)]\, dx$$

$$= (2k + 2)(I_k - I_{k+1})$$

So $I_{k+1}[1 + (2k + 2)] = (2k + 2)I_k\ \Rightarrow\ I_{k+1} = \dfrac{2k + 2}{2k + 3}I_k$. Now to complete the proof, we use induction:

$I_0 = 1 = \dfrac{2^0(0!)^2}{1!}$, so the formula holds for $n = 0$. Now suppose it holds for $n = k$. Then

$$I_{k+1} = \frac{2k + 2}{2k + 3}I_k = \frac{2k + 2}{2k + 3}\left[\frac{2^{2k}(k!)^2}{(2k + 1)!}\right] = \frac{2(k + 1)2^{2k}(k!)^2}{(2k + 3)(2k + 1)!} = \frac{2(k + 1)}{2k + 2}\cdot\frac{2(k + 1)2^{2k}(k!)^2}{(2k + 3)(2k + 1)!}$$

$$= \frac{[2(k + 1)]^2\, 2^{2k}(k!)^2}{(2k + 3)(2k + 2)(2k + 1)!} = \frac{2^{2(k+1)}[(k + 1)!]^2}{[2(k + 1) + 1]!}$$

So by induction, the formula holds for all integers $n \geq 0$.

13. $0 < a < b$. Now

$$\int_0^1 [bx + a(1 - x)]^t\, dx = \int_a^b \frac{u^t}{(b - a)}\, du\quad [u = bx + a(1 - x)] = \left[\frac{u^{t+1}}{(t + 1)(b - a)}\right]_a^b = \frac{b^{t+1} - a^{t+1}}{(t + 1)(b - a)}.$$

Now let $y = \lim\limits_{t \to 0}\left[\dfrac{b^{t+1} - a^{t+1}}{(t + 1)(b - a)}\right]^{1/t}$. Then $\ln y = \lim\limits_{t \to 0}\left[\dfrac{1}{t}\ln\dfrac{b^{t+1} - a^{t+1}}{(t + 1)(b - a)}\right]$. This limit is of the form $0/0$, so we can apply l'Hospital's Rule to get

$$\ln y = \lim_{t \to 0} \left[\frac{b^{t+1} \ln b - a^{t+1} \ln a}{b^{t+1} - a^{t+1}} - \frac{1}{t+1} \right] = \frac{b \ln b - a \ln a}{b - a} - 1 = \frac{b \ln b}{b-a} - \frac{a \ln a}{b-a} - \ln e = \ln \frac{b^{b/(b-a)}}{e a^{a/(b-a)}}.$$

Therefore, $y = e^{-1} \left(\dfrac{b^b}{a^a} \right)^{1/(b-a)}$.

15. Write $I = \displaystyle\int \frac{x^8}{(1+x^6)^2}\, dx = \int x^3 \cdot \frac{x^5}{(1+x^6)^2}\, dx$. Integrate by parts with $u = x^3$, $dv = \dfrac{x^5}{(1+x^6)^2}\, dx$. Then

$du = 3x^2\, dx$, $v = -\dfrac{1}{6(1+x^6)}$ $\Rightarrow$ $I = -\dfrac{x^3}{6(1+x^6)} + \dfrac{1}{2}\displaystyle\int \frac{x^2}{1+x^6}\, dx$. Substitute $t = x^3$ in this latter

integral. $\displaystyle\int \frac{x^2}{1+x^6}\, dx = \frac{1}{3}\int \frac{dt}{1+t^2} = \frac{1}{3}\tan^{-1} t + C = \frac{1}{3}\tan^{-1}(x^3) + C$. Therefore

$I = -\dfrac{x^3}{6(1+x^6)} + \dfrac{1}{6}\tan^{-1}(x^3) + C$. Returning to the improper integral,

$$\int_{-1}^{\infty} \left(\frac{x^4}{1+x^6} \right)^2 dx = \lim_{t \to \infty} \int_{-1}^{t} \frac{x^8}{(1+x^6)^2}\, dx = \lim_{t \to \infty} \left[-\frac{x^3}{6(1+x^6)} + \frac{1}{6}\tan^{-1}(x^3) \right]_{-1}^{t}$$

$$= \lim_{t \to \infty} \left(-\frac{t^3}{6(1+t^6)} + \frac{1}{6}\tan^{-1}(t^3) + \frac{-1}{6(1+1)} - \frac{1}{6}\tan^{-1}(-1) \right)$$

$$= 0 + \frac{1}{6}\left(\frac{\pi}{2}\right) - \frac{1}{12} - \frac{1}{6}\left(-\frac{\pi}{4}\right) = \frac{\pi}{12} - \frac{1}{12} + \frac{\pi}{24} = \frac{\pi}{8} - \frac{1}{12}$$

17. An equation of the circle with center $(0, c)$ and radius 1 is $x^2 + (y-c)^2 = 1^2$, so an equation of the lower semicircle is $y = c - \sqrt{1-x^2}$. At the points of tangency, the slopes of the line and semicircle must be equal. For $x \geq 0$, we must have

$y' = 2$ $\Rightarrow$ $\dfrac{x}{\sqrt{1-x^2}} = 2$ $\Rightarrow$ $x = 2\sqrt{1-x^2}$ $\Rightarrow$ $x^2 = 4(1-x^2)$ $\Rightarrow$

$5x^2 = 4$ $\Rightarrow$ $x^2 = \tfrac{4}{5}$ $\Rightarrow$ $x = \tfrac{2}{5}\sqrt{5}$ and so $y = 2\left(\tfrac{2}{5}\sqrt{5}\right) = \tfrac{4}{5}\sqrt{5}$.

The slope of the perpendicular line segment is $-\tfrac{1}{2}$, so an equation of the line segment is $y - \tfrac{4}{5}\sqrt{5} = -\tfrac{1}{2}\left(x - \tfrac{2}{5}\sqrt{5}\right)$ $\Leftrightarrow$
$y = -\tfrac{1}{2}x + \tfrac{1}{5}\sqrt{5} + \tfrac{4}{5}\sqrt{5}$ $\Leftrightarrow$ $y = -\tfrac{1}{2}x + \sqrt{5}$, so $c = \sqrt{5}$ and an equation of the lower semicircle is $y = \sqrt{5} - \sqrt{1-x^2}$.
Thus, the shaded area is

$$2\int_{0}^{(2/5)\sqrt{5}} \left[\left(\sqrt{5} - \sqrt{1-x^2}\right) - 2x \right] dx \stackrel{30}{=} 2\left[\sqrt{5}\,x - \frac{x}{2}\sqrt{1-x^2} - \frac{1}{2}\sin^{-1} x - x^2 \right]_{0}^{(2/5)\sqrt{5}}$$

$$= 2\left[2 - \frac{\sqrt{5}}{5} \cdot \frac{1}{\sqrt{5}} - \frac{1}{2}\sin^{-1}\left(\frac{2}{\sqrt{5}}\right) - \frac{4}{5} \right] - 2(0)$$

$$= 2\left[1 - \frac{1}{2}\sin^{-1}\left(\frac{2}{\sqrt{5}}\right) \right] = 2 - \sin^{-1}\left(\frac{2}{\sqrt{5}}\right)$$

8 ☐ FURTHER APPLICATIONS OF INTEGRATION

8.1 Arc Length

1. $y = 3 - 2x \Rightarrow L = \int_{-1}^{3} \sqrt{1 + (dy/dx)^2}\, dx = \int_{-1}^{3} \sqrt{1 + (-2)^2}\, dx = \sqrt{5}\, [\, x\,]_{-1}^{3} = \sqrt{5}\,[3 - (-1)] = 4\sqrt{5}.$

The arc length can be calculated using the distance formula, since the curve is a line segment, so

$L = [\text{distance from } (-1, 5) \text{ to } (3, -3)] = \sqrt{[3 - (-1)]^2 + (-3 - 5)^2} = \sqrt{80} = 4\sqrt{5}.$

3. $y = x^3 \Rightarrow dy/dx = 3x^2 \Rightarrow 1 + (dy/dx)^2 = 1 + (3x^2)^2.$ So $L = \int_0^2 \sqrt{1 + 9x^4}\, dx.$

5. $y = x - \ln x \Rightarrow dy/dx = 1 - 1/x \Rightarrow 1 + (dy/dx)^2 = 1 + (1 - 1/x)^2.$ So $L = \int_1^4 \sqrt{1 + (1 - 1/x)^2}\, dx.$

7. $x = \sin y \Rightarrow dx/dy = \cos y \Rightarrow 1 + (dx/dy)^2 = 1 + \cos^2 y.$ So $L = \int_0^{\pi/2} \sqrt{1 + \cos^2 y}\, dy.$

9. $y = \tfrac{2}{3} x^{3/2} \Rightarrow dy/dx = x^{1/2} \Rightarrow 1 + (dy/dx)^2 = 1 + x.$ So

$L = \int_0^2 \sqrt{1+x}\, dx = \int_0^2 (1+x)^{1/2}\, dx = \left[\tfrac{2}{3}(1+x)^{3/2}\right]_0^2 = \tfrac{2}{3}(3^{3/2} - 1^{3/2}) = \tfrac{2}{3}(3\sqrt{3} - 1) = 2\sqrt{3} - \tfrac{2}{3}.$

11. $y = \tfrac{2}{3}(1 + x^2)^{3/2} \Rightarrow dy/dx = 2x(1 + x^2)^{1/2} \Rightarrow 1 + (dy/dx)^2 = 1 + 4x^2(1 + x^2).$ So

$L = \int_0^1 \sqrt{1 + 4x^2(1 + x^2)}\, dx = \int_0^1 \sqrt{4x^4 + 4x^2 + 1}\, dx = \int_0^1 \sqrt{(2x^2 + 1)^2}\, dx = \int_0^1 |2x^2 + 1|\, dx$

$= \int_0^1 (2x^2 + 1)\, dx = \left[\tfrac{2}{3} x^3 + x\right]_0^1 = \left(\tfrac{2}{3} + 1\right) - 0 = \tfrac{5}{3}$

13. $y = \dfrac{x^3}{3} + \dfrac{1}{4x} \Rightarrow \dfrac{dy}{dx} = x^2 - \dfrac{1}{4x^2} \Rightarrow$

$1 + \left(\dfrac{dy}{dx}\right)^2 = 1 + \left(x^4 - \dfrac{1}{2} + \dfrac{1}{16x^4}\right) = x^4 + \dfrac{1}{2} + \dfrac{1}{16x^4} = \left(x^2 + \dfrac{1}{4x^2}\right)^2.$ So

$L = \int_1^2 \sqrt{1 + \left(\dfrac{dy}{dx}\right)^2}\, dx = \int_1^2 \left| x^2 + \dfrac{1}{4x^2} \right|\, dx = \int_1^2 \left(x^2 + \dfrac{1}{4x^2} \right) dx$

$= \left[\dfrac{1}{3} x^3 - \dfrac{1}{4x}\right]_1^2 = \left(\dfrac{8}{3} - \dfrac{1}{8}\right) - \left(\dfrac{1}{3} - \dfrac{1}{4}\right) = \dfrac{7}{3} + \dfrac{1}{8} = \dfrac{59}{24}$

15. $y = \tfrac{1}{2} \ln(\sin 2x) \Rightarrow \dfrac{dy}{dx} = \dfrac{\cos 2x}{\sin 2x} = \cot 2x \Rightarrow 1 + \left(\dfrac{dy}{dx}\right)^2 = 1 + \cot^2 2x = \csc^2 2x.$ So

$L = \int_{\pi/8}^{\pi/6} \sqrt{\csc^2 2x}\, dx = \int_{\pi/8}^{\pi/6} |\csc 2x|\, dx = \int_{\pi/8}^{\pi/6} \csc 2x\, dx = \dfrac{1}{2}\int_{\pi/4}^{\pi/3} \csc u\, du \quad \begin{bmatrix} u = 2x \\ du = 2\, dx \end{bmatrix}$

$= \dfrac{1}{2}\Big[\ln|\csc u - \cot u|\Big]_{\pi/4}^{\pi/3} = \dfrac{1}{2}\left[\ln\left(\dfrac{2}{\sqrt{3}} - \dfrac{1}{\sqrt{3}}\right) - \ln(\sqrt{2} - 1)\right] = \dfrac{1}{2}\left[\ln \dfrac{1}{\sqrt{3}} - \ln(\sqrt{2} - 1)\right]$

17. $y = \ln(\sec x) \;\Rightarrow\; \dfrac{dy}{dx} = \dfrac{\sec x \tan x}{\sec x} = \tan x \;\Rightarrow\; 1 + \left(\dfrac{dy}{dx}\right)^2 = 1 + \tan^2 x = \sec^2 x$. So

$L = \int_0^{\pi/4} \sqrt{\sec^2 x}\, dx = \int_0^{\pi/4} |\sec x|\, dx = \int_0^{\pi/4} \sec x\, dx = \Big[\ln|\sec x + \tan x|\Big]_0^{\pi/4}$
$= \ln(\sqrt{2} + 1) - \ln(1 + 0) = \ln(\sqrt{2} + 1)$

19. $x = \tfrac{1}{3}\sqrt{y}\,(y - 3) = \tfrac{1}{3}y^{3/2} - y^{1/2} \;\Rightarrow\; dx/dy = \tfrac{1}{2}y^{1/2} - \tfrac{1}{2}y^{-1/2} \;\Rightarrow$

$1 + (dx/dy)^2 = 1 + \tfrac{1}{4}y - \tfrac{1}{2} + \tfrac{1}{4}y^{-1} = \tfrac{1}{4}y + \tfrac{1}{2} + \tfrac{1}{4}y^{-1} = \left(\tfrac{1}{2}y^{1/2} + \tfrac{1}{2}y^{-1/2}\right)^2$. So

$L = \int_1^9 \left(\tfrac{1}{2}y^{1/2} + \tfrac{1}{2}y^{-1/2}\right) dy = \tfrac{1}{2}\Big[\tfrac{2}{3}y^{3/2} + 2y^{1/2}\Big]_1^9 = \tfrac{1}{2}\big[(\tfrac{2}{3}\cdot 27 + 2\cdot 3) - (\tfrac{2}{3}\cdot 1 + 2\cdot 1)\big]$
$= \tfrac{1}{2}(24 - \tfrac{8}{3}) = \tfrac{1}{2}(\tfrac{64}{3}) = \tfrac{32}{3}$.

21. $y = \dfrac{1}{4}x^2 - \dfrac{1}{2}\ln x \;\Rightarrow\; \dfrac{dy}{dx} = \dfrac{1}{2}x - \dfrac{1}{2x} \;\Rightarrow$

$1 + \left(\dfrac{dy}{dx}\right)^2 = 1 + \left(\dfrac{1}{4}x^2 - \dfrac{1}{2} + \dfrac{1}{4x^2}\right) = \dfrac{1}{4}x^2 + \dfrac{1}{2} + \dfrac{1}{4x^2} = \left(\dfrac{1}{2}x + \dfrac{1}{2x}\right)^2$. So

$L = \int_1^2 \sqrt{1 + \left(\dfrac{dy}{dx}\right)^2}\, dx = \int_1^2 \left|\dfrac{1}{2}x + \dfrac{1}{2x}\right| dx = \int_1^2 \left(\dfrac{1}{2}x + \dfrac{1}{2x}\right) dx$
$= \left[\dfrac{1}{4}x^2 + \dfrac{1}{2}\ln|x|\right]_1^2 = \left(1 + \dfrac{1}{2}\ln 2\right) - \left(\dfrac{1}{4} + 0\right) = \dfrac{3}{4} + \dfrac{1}{2}\ln 2$

23. $y = \ln(1 - x^2) \;\Rightarrow\; \dfrac{dy}{dx} = \dfrac{1}{1 - x^2}\cdot(-2x) \;\Rightarrow$

$1 + \left(\dfrac{dy}{dx}\right)^2 = 1 + \dfrac{4x^2}{(1-x^2)^2} = \dfrac{1 - 2x^2 + x^4 + 4x^2}{(1-x^2)^2} = \dfrac{1 + 2x^2 + x^4}{(1-x^2)^2} = \dfrac{(1+x^2)^2}{(1-x^2)^2} \;\Rightarrow$

$\sqrt{1 + \left(\dfrac{dy}{dx}\right)^2} = \sqrt{\left(\dfrac{1+x^2}{1-x^2}\right)^2} = \dfrac{1+x^2}{1-x^2} = -1 + \dfrac{2}{1-x^2}$ [by division] $= -1 + \dfrac{1}{1+x} + \dfrac{1}{1-x}$ [partial fractions].

So $L = \int_0^{1/2} \left(-1 + \dfrac{1}{1+x} + \dfrac{1}{1-x}\right) dx = \Big[-x + \ln|1+x| - \ln|1-x|\Big]_0^{1/2} = \left(-\tfrac{1}{2} + \ln\tfrac{3}{2} - \ln\tfrac{1}{2}\right) - 0 = \ln 3 - \tfrac{1}{2}$.

25. $y = \tfrac{1}{2}x^2 \;\Rightarrow\; dy/dx = x \;\Rightarrow\; 1 + (dy/dx)^2 = 1 + x^2$. So

$L = \int_{-1}^{1} \sqrt{1 + x^2}\, dx = 2\int_0^1 \sqrt{1+x^2}\, dx$ [by symmetry] $\overset{21}{=} 2\Big[\tfrac{x}{2}\sqrt{1+x^2} + \tfrac{1}{2}\ln(x + \sqrt{1+x^2})\Big]_0^1$ $\begin{bmatrix}\text{or substitute}\\ x = \tan\theta\end{bmatrix}$
$= 2\big[(\tfrac{1}{2}\sqrt{2} + \tfrac{1}{2}\ln(1+\sqrt{2})) - (0 + \tfrac{1}{2}\ln 1)\big] = \sqrt{2} + \ln(1 + \sqrt{2})$

27.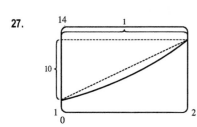

From the figure, the length of the curve is slightly larger than the hypotenuse of the triangle formed by the points $(1, 2)$, $(1, 12)$, and $(2, 12)$. This length is about $\sqrt{10^2 + 1^2} \approx 10$, so we might estimate the length to be 10.

$y = x^2 + x^3 \;\Rightarrow\; y' = 2x + 3x^2 \;\Rightarrow\; 1 + (y')^2 = 1 + (2x + 3x^2)^2$.

So $L = \int_1^2 \sqrt{1 + (2x + 3x^2)^2}\, dx \approx 10.0556$.

29. 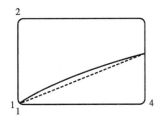 From the figure, the length of the curve is slightly larger than the line segment joining the points $(1, 1)$ and $(4, \sqrt[3]{4})$. This length is

$$\sqrt{(4-1)^2 + (\sqrt[3]{4}-1)^2} \approx 3.057,$$ so we might estimate the length of the curve to be 3.06. $y = \sqrt[3]{x} = x^{1/3} \Rightarrow y' = \tfrac{1}{3}x^{-2/3} \Rightarrow$

$1 + (y')^2 = 1 + \tfrac{1}{9}x^{-4/3}$. So $L = \int_1^4 \sqrt{1 + \tfrac{1}{9}x^{-4/3}}\, dx \approx 3.0609$.

31. 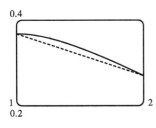 From the figure, the length of the curve is slightly larger than the line segment joining the points $(1, e^{-1})$ and $(2, 2e^{-2})$. This length is

$$\sqrt{(2-1)^2 + (2e^{-2} - e^{-1})^2} \approx 1.0047,$$ so we might estimate the length of the curve to be 1.005. $y = xe^{-x} \Rightarrow y' = -xe^{-x} + e^{-x} \Rightarrow$

$1 + (y')^2 = 1 + (e^{-x} - xe^{-x})^2$. So

$L = \int_1^2 \sqrt{1 + (e^{-x} - xe^{-x})^2}\, dx \approx 1.0054$.

33. $y = x\sin x \Rightarrow dy/dx = x\cos x + (\sin x)(1) \Rightarrow 1 + (dy/dx)^2 = 1 + (x\cos x + \sin x)^2$. Let $f(x) = \sqrt{1 + (dy/dx)^2} = \sqrt{1 + (x\cos x + \sin x)^2}$. Then $L = \int_0^{2\pi} f(x)\,dx$. Since $n = 10$, $\Delta x = \frac{2\pi - 0}{10} = \frac{\pi}{5}$. Now

$$L \approx S_{10} = \tfrac{\pi/5}{3}\bigl[f(0) + 4f\bigl(\tfrac{\pi}{5}\bigr) + 2f\bigl(\tfrac{2\pi}{5}\bigr) + 4f\bigl(\tfrac{3\pi}{5}\bigr) + 2f\bigl(\tfrac{4\pi}{5}\bigr) + 4f\bigl(\tfrac{5\pi}{5}\bigr) + 2f\bigl(\tfrac{6\pi}{5}\bigr)$$
$$+ 4f\bigl(\tfrac{7\pi}{5}\bigr) + 2f\bigl(\tfrac{8\pi}{5}\bigr) + 4f\bigl(\tfrac{9\pi}{5}\bigr) + f(2\pi)\bigr]$$
$$\approx 15.498085$$

The value of the integral produced by a calculator is 15.374568 (to six decimal places).

35. (a) Let $f(x) = y = x\sqrt[3]{4-x}$ with $0 \le x \le 4$.

(b) The polygon with one side is just the line segment joining the points $(0, f(0)) = (0, 0)$ and $(4, f(4)) = (4, 0)$, and its length $L_1 = 4$.

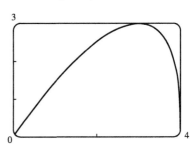

The polygon with two sides joins the points $(0,0)$, $(2, f(2)) = (2, 2\sqrt[3]{2})$ and $(4, 0)$. Its length

$$L_2 = \sqrt{(2-0)^2 + (2\sqrt[3]{2} - 0)^2} + \sqrt{(4-2)^2 + (0 - 2\sqrt[3]{2})^2} = 2\sqrt{4 + 2^{8/3}} \approx 6.43$$

[continued]

Similarly, the inscribed polygon with four sides joins the points $(0,0)$, $(1, \sqrt[3]{3})$, $(2, 2\sqrt[3]{2})$, $(3, 3)$, and $(4, 0)$, so its length

$$L_4 = \sqrt{1 + (\sqrt[3]{3})^2} + \sqrt{1 + (2\sqrt[3]{2} - \sqrt[3]{3})^2} + \sqrt{1 + (3 - 2\sqrt[3]{2})^2} + \sqrt{1 + 9} \approx 7.50$$

(c) Using the arc length formula with $\dfrac{dy}{dx} = x\left[\tfrac{1}{3}(4-x)^{-2/3}(-1)\right] + \sqrt[3]{4-x} = \dfrac{12 - 4x}{3(4-x)^{2/3}}$, the length of the curve is

$$L = \int_0^4 \sqrt{1 + \left(\frac{dy}{dx}\right)^2}\, dx = \int_0^4 \sqrt{1 + \left[\frac{12 - 4x}{3(4-x)^{2/3}}\right]^2}\, dx.$$

(d) According to a calculator, the length of the curve is $L \approx 7.7988$. The actual value is larger than any of the approximations in part (b). This is always true, since any approximating straight line between two points on the curve is shorter than the length of the curve between the two points.

37. $y = e^x \;\Rightarrow\; dy/dx = e^x \;\Rightarrow\; 1 + (dy/dx)^2 \;\Rightarrow\; 1 + e^{2x} \;\Rightarrow$

$$L = \int_0^2 \sqrt{1 + e^{2x}}\, dx = \int_1^{e^2} \sqrt{1 + u^2}\left(\frac{1}{u}\,du\right) \quad \begin{bmatrix} u = e^x, \\ du = e^x\, dx \end{bmatrix}$$

$$\overset{23}{=} \left[\sqrt{1 + u^2} - \ln\left|\frac{1 + \sqrt{1 + u^2}}{u}\right|\right]_1^{e^2} = \left(\sqrt{1 + e^4} - \ln\frac{1 + \sqrt{1 + e^4}}{e^2}\right) - \left(\sqrt{2} - \ln\frac{1 + \sqrt{2}}{1}\right)$$

$$= \sqrt{1 + e^4} - \ln(1 + \sqrt{1 + e^4}) + 2 - \sqrt{2} + \ln(1 + \sqrt{2})$$

An equivalent answer from a CAS is

$$-\sqrt{2} + \operatorname{arctanh}(\sqrt{2}/2) + \sqrt{e^4 + 1} - \operatorname{arctanh}(1/\sqrt{e^4 + 1}).$$

39. The astroid $x^{2/3} + y^{2/3} = 1$ has an equal length of arc in each quadrant. Thus, we can find the length of the curve in the first quadrant and then multiply by 4. The top half of the astroid has equation $y = (1 - x^{2/3})^{3/2}$. Then

$$dy/dx = -x^{-1/3}(1 - x^{2/3})^{1/2} \;\Rightarrow\; 1 + (dy/dx)^2 = 1 + \left[-x^{-1/3}(1 - x^{2/3})^{1/2}\right]^2 = 1 + x^{-2/3}(1 - x^{2/3}) = x^{-2/3}.$$

So the portion of the astroid in quadrant 1 has length $L = \int_0^1 \sqrt{x^{-2/3}}\, dx = \int_0^1 x^{-1/3}\, dx = \left[\tfrac{3}{2}x^{2/3}\right]_0^1 = \tfrac{3}{2} - 0 = \tfrac{3}{2}$. Thus, the astroid has length $4\left(\tfrac{3}{2}\right) = 6$.

41. $y = 2x^{3/2} \;\Rightarrow\; y' = 3x^{1/2} \;\Rightarrow\; 1 + (y')^2 = 1 + 9x$. The arc length function with starting point $P_0(1, 2)$ is

$$s(x) = \int_1^x \sqrt{1 + 9t}\, dt = \left[\tfrac{2}{27}(1 + 9t)^{3/2}\right]_1^x = \tfrac{2}{27}\left[(1 + 9x)^{3/2} - 10\sqrt{10}\right].$$

43. $y = \sin^{-1} x + \sqrt{1 - x^2} \;\Rightarrow\; y' = \dfrac{1}{\sqrt{1 - x^2}} - \dfrac{x}{\sqrt{1 - x^2}} = \dfrac{1 - x}{\sqrt{1 - x^2}} \;\Rightarrow$

$$1 + (y')^2 = 1 + \frac{(1 - x)^2}{1 - x^2} = \frac{1 - x^2 + 1 - 2x + x^2}{1 - x^2} = \frac{2 - 2x}{1 - x^2} = \frac{2(1 - x)}{(1 + x)(1 - x)} = \frac{2}{1 + x} \;\Rightarrow$$

$\sqrt{1 + (y')^2} = \sqrt{\frac{2}{1+x}}$. Thus, the arc length function with starting point $(0, 1)$ is given by

$s(x) = \int_0^x \sqrt{1 + [f'(t)]^2}\, dt = \int_0^x \sqrt{\frac{2}{1+t}}\, dt = \sqrt{2}\,[2\sqrt{1+t}]_0^x = 2\sqrt{2}\,(\sqrt{1+x} - 1)$.

45. The prey hits the ground when $y = 0 \Leftrightarrow 180 - \frac{1}{45}x^2 = 0 \Leftrightarrow x^2 = 45 \cdot 180 \Rightarrow x = \sqrt{8100} = 90$,

since x must be positive. $y' = -\frac{2}{45}x \Rightarrow 1 + (y')^2 = 1 + \frac{4}{45^2}x^2$, so the distance traveled by the prey is

$L = \int_0^{90} \sqrt{1 + \frac{4}{45^2}x^2}\, dx = \int_0^4 \sqrt{1+u^2}\,\left(\frac{45}{2}\,du\right) \quad \begin{bmatrix} u = \frac{2}{45}x, \\ du = \frac{2}{45}\,dx \end{bmatrix}$

$\overset{21}{=} \frac{45}{2}\left[\frac{1}{2}u\sqrt{1+u^2} + \frac{1}{2}\ln(u + \sqrt{1+u^2})\right]_0^4 = \frac{45}{2}\left[2\sqrt{17} + \frac{1}{2}\ln(4 + \sqrt{17})\right] = 45\sqrt{17} + \frac{45}{4}\ln(4 + \sqrt{17}) \approx 209.1$ m

47. The sine wave has amplitude 1 and period 14, since it goes through two periods in a distance of 28 in., so its equation is

$y = 1\sin\left(\frac{2\pi}{14}x\right) = \sin\left(\frac{\pi}{7}x\right)$. The width w of the flat metal sheet needed to make the panel is the arc length of the sine curve

from $x = 0$ to $x = 28$. We set up the integral to evaluate w using the arc length formula with $\frac{dy}{dx} = \frac{\pi}{7}\cos\left(\frac{\pi}{7}x\right)$:

$L = \int_0^{28} \sqrt{1 + \left[\frac{\pi}{7}\cos\left(\frac{\pi}{7}x\right)\right]^2}\, dx = 2\int_0^{14} \sqrt{1 + \left[\frac{\pi}{7}\cos\left(\frac{\pi}{7}x\right)\right]^2}\, dx$. This integral would be very difficult to evaluate exactly,

so we use a CAS, and find that $L \approx 29.36$ inches.

49. $y = c + a\cosh\left(\frac{x}{a}\right) \Rightarrow \frac{dy}{dx} = \sinh\left(\frac{x}{a}\right) \Rightarrow 1 + \left(\frac{dy}{dx}\right)^2 = 1 + \sinh^2\left(\frac{x}{a}\right) = \cosh^2\left(\frac{x}{a}\right)$. So

$L = \int_{-25}^{25} \sqrt{\cosh^2\left(\frac{x}{a}\right)}\, dx = 2\int_0^{25} \cosh\left(\frac{x}{a}\right) dx = 2\left[a\sinh\left(\frac{x}{a}\right)\right]_0^{25} = 2a\sinh\left(\frac{25}{a}\right)$. Now, $L = 51 \Rightarrow$

$51 = 2a\sinh\left(\frac{25}{a}\right)$. From the figure, we see that $y = 51$ intersects

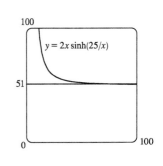

$y = 2x\sinh\left(\frac{25}{x}\right)$ at $x \approx 72.3843$ for $x > 0$. So $a \approx 72.3843$. At $x = 0$,

$y = c + a$, so $c + a = 20 \Rightarrow c = 20 - a$. Thus, the wire should be attached

at a distance of $y = c + a\cosh\left(\frac{25}{a}\right) = 20 - a + a\cosh\left(\frac{25}{a}\right) \approx 24.36$ ft

above the ground.

51. $f(x) = \frac{1}{4}e^x + e^{-x} \Rightarrow f'(x) = \frac{1}{4}e^x - e^{-x} \Rightarrow$

$1 + [f'(x)]^2 = 1 + \left(\frac{1}{4}e^x - e^{-x}\right)^2 = 1 + \frac{1}{16}e^{2x} - \frac{1}{2} + e^{-2x} = \frac{1}{16}e^{2x} + \frac{1}{2} + e^{-2x} = \left(\frac{1}{4}e^x + e^{-x}\right)^2 = [f(x)]^2$. The arc

length of the curve $y = f(x)$ on the interval $[a, b]$ is $L = \int_a^b \sqrt{1 + [f'(x)]^2}\, dx = \int_a^b \sqrt{[f(x)]^2}\, dx = \int_a^b f(x)\, dx$, which is

the area under the curve $y = f(x)$ on the interval $[a, b]$.

53. $y = \int_1^x \sqrt{t^3 - 1}\, dt \Rightarrow dy/dx = \sqrt{x^3 - 1}$ [by FTC1] $\Rightarrow 1 + (dy/dx)^2 = 1 + \left(\sqrt{x^3 - 1}\right)^2 = x^3 \Rightarrow$

$L = \int_1^4 \sqrt{x^3}\, dx = \int_1^4 x^{3/2}\, dx = \frac{2}{5}\left[x^{5/2}\right]_1^4 = \frac{2}{5}(32 - 1) = \frac{62}{5} = 12.4$

8.2 Area of a Surface of Revolution

1. (a) $y = \sqrt[3]{x} = x^{1/3} \;\Rightarrow\; dy/dx = \frac{1}{3}x^{-2/3} \;\Rightarrow\; ds = \sqrt{1 + (dy/dx)^2}\, dx = \sqrt{1 + \frac{1}{9}x^{-4/3}}\, dx.$

    By (7), $S = \int 2\pi y\, ds = \int_1^8 2\pi \sqrt[3]{x}\sqrt{1 + \frac{1}{9}x^{-4/3}}\, dx.$

    (b) $y = \sqrt[3]{x} \;\Rightarrow\; x = y^3 \;\Rightarrow\; dx/dy = 3y^2 \;\Rightarrow\; ds = \sqrt{1 + (dx/dy)^2}\, dy = \sqrt{1 + 9y^4}\, dy.$
    When $x = 1$, $y = 1$ and when $x = 8$, $y = 2$. Thus, $S = \int 2\pi y\, ds = \int_1^2 2\pi y\sqrt{1 + 9y^4}\, dy.$

3. (a) $x = \ln(2y+1) \;\Rightarrow\; e^x = 2y+1 \;\Rightarrow\; y = \frac{1}{2}e^x - \frac{1}{2} \;\Rightarrow\; dy/dx = \frac{1}{2}e^x \;\Rightarrow\;$
    $ds = \sqrt{1 + (dy/dx)^2}\, dx = \sqrt{1 + \frac{1}{4}e^{2x}}\, dx.$ When $y = 0$, $x = 0$ and when $y = 1$, $x = \ln 3$. Thus, by (7),
    $S = \int 2\pi y\, ds = \int_0^{\ln 3} 2\pi\left(\frac{1}{2}e^x - \frac{1}{2}\right)\sqrt{1 + \frac{1}{4}e^{2x}}\, dx = \int_0^{\ln 3} \pi(e^x - 1)\sqrt{1 + \frac{1}{4}e^{2x}}\, dx.$

    (b) $x = \ln(2y+1) \;\Rightarrow\; dx/dy = \dfrac{2}{2y+1} \;\Rightarrow\; ds = \sqrt{1 + (dx/dy)^2}\, dy = \sqrt{1 + 4/(2y+1)^2}\, dy.$ By (7),
    $S = \int 2\pi y\, ds = \int_0^1 2\pi y\sqrt{1 + 4/(2y+1)^2}\, dy.$

5. (a) $xy = 4 \;\Rightarrow\; y = \dfrac{4}{x} = 4x^{-1} \;\Rightarrow\; dy/dx = -4x^{-2} \;\Rightarrow\; ds = \sqrt{1 + (dy/dx)^2}\, dx = \sqrt{1 + 16/x^4}\, dx.$

    By (8), $S = \int 2\pi x\, ds = \int_1^8 2\pi x\sqrt{1 + 16/x^4}\, dx.$

    (b) $xy = 4 \;\Rightarrow\; x = \dfrac{4}{y} = 4y^{-1} \;\Rightarrow\; dx/dy = -4y^{-2} \;\Rightarrow\; ds = \sqrt{1 + (dx/dy)^2}\, dy = \sqrt{1 + 16/y^4}\, dy.$
    When $x = 1$, $y = 4$ and when $x = 8$, $y = \frac{1}{2}$. By (8), $S = \int 2\pi x\, ds = \int_{1/2}^4 \dfrac{8\pi}{y}\sqrt{1 + 16/y^4}\, dy.$

7. (a) $y = 1 + \sin x \;\Rightarrow\; dy/dx = \cos x \;\Rightarrow\; ds = \sqrt{1 + (dy/dx)^2}\, dx = \sqrt{1 + \cos^2 x}\, dx.$
    By (8), $S = \int 2\pi x\, ds = \int_0^{\pi/2} 2\pi x\sqrt{1 + \cos^2 x}\, dx.$

    (b) $y = 1 + \sin x \;\Rightarrow\; x = \sin^{-1}(y-1) \;\Rightarrow\; \dfrac{dx}{dy} = \dfrac{1}{\sqrt{1 - (y-1)^2}} \;\Rightarrow\;$
    $ds = \sqrt{1 + \left(\dfrac{dx}{dy}\right)^2}\, dy = \sqrt{1 + \dfrac{1}{1 - (y-1)^2}}\, dy.$ When $x = 0$, $y = 1$ and when $x = \pi/2$, $y = 2$.

    By (8), $S = \int 2\pi x\, ds = \int_1^2 2\pi \sin^{-1}(y-1)\sqrt{1 + \dfrac{1}{1 - (y-1)^2}}\, dy$ or $\int_1^2 2\pi \sin^{-1}(y-1)\sqrt{1 + \dfrac{1}{2y - y^2}}\, dy.$

9. $y = x^3 \;\Rightarrow\; y' = 3x^2.$ So

    $$S = \int_0^2 2\pi y\sqrt{1 + (y')^2}\, dx = 2\pi \int_0^2 x^3\sqrt{1 + 9x^4}\, dx$$

    $$= \dfrac{2\pi}{36}\int_1^{145} \sqrt{u}\, du \qquad [u = 1 + 9x^4,\; du = 36x^3\, dx]$$

    $$= \dfrac{\pi}{18}\left[\dfrac{2}{3}u^{3/2}\right]_1^{145} = \dfrac{\pi}{27}\left(145\sqrt{145} - 1\right)$$

11. $y^2 = x + 1 \;\Rightarrow\; y = \sqrt{x+1}\;$ (for $0 \le x \le 3$ and $1 \le y \le 2$) $\;\Rightarrow\; y' = 1/(2\sqrt{x+1})$. So

$$S = \int_0^3 2\pi y \sqrt{1 + (y')^2}\, dx = 2\pi \int_0^3 \sqrt{x+1}\sqrt{1 + \frac{1}{4(x+1)}}\, dx = 2\pi \int_0^3 \sqrt{x + 1 + \tfrac{1}{4}}\, dx$$

$$= 2\pi \int_0^3 \sqrt{x + \tfrac{5}{4}}\, dx = 2\pi \int_{5/4}^{17/4} \sqrt{u}\, du \qquad \begin{bmatrix} u = x + \tfrac{5}{4}, \\ du = dx \end{bmatrix}$$

$$= 2\pi \left[\tfrac{2}{3} u^{3/2}\right]_{5/4}^{17/4} = 2\pi \cdot \tfrac{2}{3}\left(\frac{17^{3/2}}{8} - \frac{5^{3/2}}{8}\right) = \frac{\pi}{6}(17\sqrt{17} - 5\sqrt{5}).$$

13. $y = \cos\!\left(\tfrac{1}{2}x\right) \;\Rightarrow\; y' = -\tfrac{1}{2}\sin\!\left(\tfrac{1}{2}x\right)$. So

$$S = \int_0^\pi 2\pi y \sqrt{1 + (y')^2}\, dx = 2\pi \int_0^\pi \cos\!\left(\tfrac{1}{2}x\right) \sqrt{1 + \tfrac{1}{4}\sin^2\!\left(\tfrac{1}{2}x\right)}\, dx$$

$$= 2\pi \int_0^1 \sqrt{1 + \tfrac{1}{4}u^2}\,(2\,du) \qquad \begin{bmatrix} u = \sin\!\left(\tfrac{1}{2}x\right), \\ du = \tfrac{1}{2}\cos\!\left(\tfrac{1}{2}x\right) dx \end{bmatrix}$$

$$= 2\pi \int_0^1 \sqrt{4 + u^2}\, du \overset{21}{=} 2\pi \left[\tfrac{u}{2}\sqrt{4+u^2} + 2\ln\!\left(u + \sqrt{4+u^2}\right)\right]_0^1$$

$$= 2\pi\left[\left(\tfrac{1}{2}\sqrt{5} + 2\ln(1 + \sqrt{5})\right) - (0 + 2\ln 2)\right] = \pi\sqrt{5} + 4\pi \ln\!\left(\frac{1+\sqrt{5}}{2}\right)$$

15. $x = \tfrac{1}{3}(y^2 + 2)^{3/2} \;\Rightarrow\; dx/dy = \tfrac{1}{2}(y^2+2)^{1/2}(2y) = y\sqrt{y^2+2} \;\Rightarrow\; 1 + (dx/dy)^2 = 1 + y^2(y^2+2) = (y^2+1)^2$.

So $S = 2\pi \int_1^2 y(y^2+1)\, dy = 2\pi\left[\tfrac{1}{4}y^4 + \tfrac{1}{2}y^2\right]_1^2 = 2\pi\left(4 + 2 - \tfrac{1}{4} - \tfrac{1}{2}\right) = \frac{21\pi}{2}$.

17. $y = \tfrac{1}{3}x^{3/2} \;\Rightarrow\; y' = \tfrac{1}{2}x^{1/2} \;\Rightarrow\; 1 + (y')^2 = 1 + \tfrac{1}{4}x$. So

$$S = \int_0^{12} 2\pi x\sqrt{1+(y')^2}\, dx = 2\pi \int_0^{12} x\sqrt{1 + \tfrac{1}{4}x}\, dx = 2\pi \int_0^{12} x\cdot\tfrac{1}{2}\sqrt{4+x}\, dx$$

$$= \pi \int_4^{16} (u-4)\sqrt{u}\, du \qquad \begin{bmatrix} u = x + 4, \\ du = dx \end{bmatrix}$$

$$= \pi \int_4^{16}\left(u^{3/2} - 4u^{1/2}\right) du = \pi\left[\tfrac{2}{5}u^{5/2} - \tfrac{8}{3}u^{3/2}\right]_4^{16} = \pi\left[\left(\tfrac{2}{5}\cdot 1024 - \tfrac{8}{3}\cdot 64\right) - \left(\tfrac{2}{5}\cdot 32 - \tfrac{8}{3}\cdot 8\right)\right]$$

$$= \pi\left(\tfrac{2}{5}\cdot 992 - \tfrac{8}{3}\cdot 56\right) = \pi\left(\tfrac{5952 - 2240}{15}\right) = \tfrac{3712\pi}{15}$$

19. $x = \sqrt{a^2 - y^2} \;\Rightarrow\; dx/dy = \tfrac{1}{2}(a^2-y^2)^{-1/2}(-2y) = -y/\sqrt{a^2-y^2} \;\Rightarrow\;$

$$1 + \left(\frac{dx}{dy}\right)^2 = 1 + \frac{y^2}{a^2-y^2} = \frac{a^2-y^2}{a^2-y^2} + \frac{y^2}{a^2-y^2} = \frac{a^2}{a^2-y^2}. \text{ So}$$

$$S = \int_0^{a/2} 2\pi \sqrt{a^2-y^2}\,\frac{a}{\sqrt{a^2-y^2}}\, dy = 2\pi \int_0^{a/2} a\, dy = 2\pi a\left[y\right]_0^{a/2} = 2\pi a\left(\tfrac{a}{2} - 0\right) = \pi a^2.$$

Note that this is $\tfrac{1}{4}$ the surface area of a sphere of radius a, and the length of the interval $y = 0$ to $y = a/2$ is $\tfrac{1}{4}$ the length of the interval $y = -a$ to $y = a$.

21. $y = e^{-x^2} \;\Rightarrow\; dy/dx = -2xe^{-x^2} \;\Rightarrow\; ds = \sqrt{1 + (dy/dx)^2}\, dx = \sqrt{1 + 4x^2 e^{-2x^2}}\, dx.$ By (7),

$$S = \int 2\pi y\, ds = \int_{-1}^1 2\pi e^{-x^2}\sqrt{1 + 4x^2 e^{-2x^2}}\, dx \approx 11.0753.$$

23. $x = y + y^3 \Rightarrow dx/dy = 1 + 3y^2 \Rightarrow ds = \sqrt{1 + (dx/dy)^2}\, dy = \sqrt{1 + (1 + 3y^2)^2}\, dy$.

By (8), $S = \int 2\pi x\, ds = \int_0^1 2\pi(y + y^3)\sqrt{1 + (1 + 3y^2)^2}\, dy \approx 13.5134$.

25. $\ln y = x - y^2 \Rightarrow x = \ln y + y^2 \Rightarrow \dfrac{dx}{dy} = \dfrac{1}{y} + 2y \Rightarrow ds = \sqrt{1 + (dx/dy)^2}\, dy = \sqrt{1 + (y^{-1} + 2y)^2}\, dy$.

By (7), $S = \int 2\pi y\, ds = \int_1^4 2\pi y\sqrt{1 + (y^{-1} + 2y)^2}\, dy \approx 286.9239$.

27. $y = 1/x \Rightarrow ds = \sqrt{1 + (dy/dx)^2}\, dx = \sqrt{1 + (-1/x^2)^2}\, dx = \sqrt{1 + 1/x^4}\, dx \Rightarrow$

$S = \displaystyle\int_1^2 2\pi \cdot \dfrac{1}{x}\sqrt{1 + \dfrac{1}{x^4}}\, dx = 2\pi \int_1^2 \dfrac{\sqrt{x^4 + 1}}{x^3}\, dx = 2\pi \int_1^4 \dfrac{\sqrt{u^2 + 1}}{u^2}\left(\tfrac{1}{2}\, du\right) \quad [u = x^2,\, du = 2x\, dx]$

$= \pi \displaystyle\int_1^4 \dfrac{\sqrt{1 + u^2}}{u^2}\, du \stackrel{24}{=} \pi \left[-\dfrac{\sqrt{1 + u^2}}{u} + \ln\left(u + \sqrt{1 + u^2}\right)\right]_1^4$

$= \pi\left[-\dfrac{\sqrt{17}}{4} + \ln(4 + \sqrt{17}) + \dfrac{\sqrt{2}}{1} - \ln(1 + \sqrt{2})\right] = \dfrac{\pi}{4}\left[4\ln(\sqrt{17} + 4) - 4\ln(\sqrt{2} + 1) - \sqrt{17} + 4\sqrt{2}\right]$

29. $y = x^3$ and $0 \le y \le 1 \Rightarrow y' = 3x^2$ and $0 \le x \le 1$.

$S = \int_0^1 2\pi x\sqrt{1 + (3x^2)^2}\, dx = 2\pi \int_0^3 \sqrt{1 + u^2}\, \tfrac{1}{6}\, du \quad \begin{bmatrix} u = 3x^2, \\ du = 6x\, dx \end{bmatrix}$

$= \dfrac{\pi}{3}\int_0^3 \sqrt{1 + u^2}\, du \stackrel{21}{=}$ [or use CAS] $\dfrac{\pi}{3}\left[\tfrac{1}{2}u\sqrt{1 + u^2} + \tfrac{1}{2}\ln\left(u + \sqrt{1 + u^2}\right)\right]_0^3$

$= \dfrac{\pi}{3}\left[\tfrac{3}{2}\sqrt{10} + \tfrac{1}{2}\ln(3 + \sqrt{10})\right] = \dfrac{\pi}{6}\left[3\sqrt{10} + \ln(3 + \sqrt{10})\right]$

31. $y = \tfrac{1}{5}x^5 \Rightarrow dy/dx = x^4 \Rightarrow 1 + (dy/dx)^2 = 1 + x^8 \Rightarrow S = \int_0^5 2\pi\left(\tfrac{1}{5}x^5\right)\sqrt{1 + x^8}\, dx$.

Let $f(x) = \tfrac{2}{5}\pi x^5\sqrt{1 + x^8}$. Since $n = 10$, $\Delta x = \dfrac{5 - 0}{10} = \dfrac{1}{2}$. Then

$S \approx S_{10} = \dfrac{1/2}{3}[f(0) + 4f(0.5) + 2f(1) + 4f(1.5) + 2f(2) + 4f(2.5) + 2f(3)$
$\qquad\qquad\qquad + 4f(3.5) + 2f(4) + 4f(4.5) + f(5)]$

$\approx 1{,}230{,}507$

The value of the integral produced by a calculator is approximately $1{,}227{,}192$.

33. $S = 2\pi \displaystyle\int_1^\infty y\sqrt{1 + \left(\dfrac{dy}{dx}\right)^2}\, dx = 2\pi \int_1^\infty \dfrac{1}{x}\sqrt{1 + \dfrac{1}{x^4}}\, dx = 2\pi \int_1^\infty \dfrac{\sqrt{x^4 + 1}}{x^3}\, dx$. Rather than trying to evaluate this

integral, note that $\sqrt{x^4 + 1} > \sqrt{x^4} = x^2$ for $x > 0$. Thus, if the area is finite,

$S = 2\pi \displaystyle\int_1^\infty \dfrac{\sqrt{x^4 + 1}}{x^3}\, dx > 2\pi \int_1^\infty \dfrac{x^2}{x^3}\, dx = 2\pi \int_1^\infty \dfrac{1}{x}\, dx$. But we know that this integral diverges, so the area S

is infinite.

35. As seen in the exercise figure, the loop of the curve $3ay^2 = x(a - x)^2$ extends from $x = 0$ to $x = a$. The top half of the loop

is given by $y = \sqrt{\dfrac{1}{3a}x(a - x)^2} = \dfrac{1}{\sqrt{3a}}\sqrt{x}\,|a - x| = \dfrac{1}{\sqrt{3a}}\sqrt{x}(a - x)$, since $x \le a$. Now,

$$\frac{dy}{dx} = \frac{1}{\sqrt{3a}}\left[\sqrt{x}(-1) + \frac{1}{2\sqrt{x}}(a-x)\right] = \frac{1}{\sqrt{3a}}\left[-\frac{2x}{2\sqrt{x}} + \frac{a-x}{2\sqrt{x}}\right] = \frac{1}{\sqrt{3a}}\left[\frac{a-3x}{2\sqrt{x}}\right] \Rightarrow$$

$$1 + \left(\frac{dy}{dx}\right)^2 = 1 + \frac{(a-3x)^2}{12ax} = \frac{12ax}{12ax} + \frac{a^2 - 6ax + 9x^2}{12ax} = \frac{a^2 + 6ax + 9x^2}{12ax} = \frac{(a+3x)^2}{12ax}.$$

(a) $S = \int 2\pi y\, ds = 2\pi \int_0^a \frac{\sqrt{x}(a-x)}{\sqrt{3a}}\sqrt{\frac{(a+3x)^2}{12ax}}\, dx = 2\pi \int_0^a \frac{(a-x)(a+3x)}{6a}\, dx$

$$= \frac{\pi}{3a}\int_0^a (a^2 + 2ax - 3x^2)\, dx = \frac{\pi}{3a}\left[a^2 x + ax^2 - x^3\right]_0^a = \frac{\pi}{3a}(a^3 + a^3 - a^3)$$

$$= \frac{\pi}{3a}\cdot a^3 = \frac{\pi a^2}{3}$$

Note that the top half of the loop has been rotated about the x-axis, producing the full surface.

(b) We must rotate the full loop about the y-axis, so we get double the area obtained by rotating the top half of the loop:

$$S = 2\int_{x=0}^a 2\pi x\, ds = 4\pi \int_0^a x\frac{a+3x}{\sqrt{12ax}}\, dx = \frac{4\pi}{2\sqrt{3a}}\int_0^a x^{1/2}(a+3x)\, dx = \frac{2\pi}{\sqrt{3a}}\int_0^a (ax^{1/2} + 3x^{3/2})\, dx$$

$$= \frac{2\pi}{\sqrt{3a}}\left[\frac{2}{3}ax^{3/2} + \frac{6}{5}x^{5/2}\right]_0^a = \frac{2\pi\sqrt{3}}{3\sqrt{a}}\left(\frac{2}{3}a^{5/2} + \frac{6}{5}a^{5/2}\right) = \frac{2\pi\sqrt{3}}{3}\left(\frac{2}{3} + \frac{6}{5}\right)a^2 = \frac{2\pi\sqrt{3}}{3}\left(\frac{28}{15}\right)a^2$$

$$= \frac{56\pi\sqrt{3}a^2}{45}$$

37. (a) $\dfrac{x^2}{a^2} + \dfrac{y^2}{b^2} = 1 \Rightarrow \dfrac{y(dy/dx)}{b^2} = -\dfrac{x}{a^2} \Rightarrow \dfrac{dy}{dx} = -\dfrac{b^2 x}{a^2 y} \Rightarrow$

$$1 + \left(\frac{dy}{dx}\right)^2 = 1 + \frac{b^4 x^2}{a^4 y^2} = \frac{b^4 x^2 + a^4 y^2}{a^4 y^2} = \frac{b^4 x^2 + a^4 b^2(1 - x^2/a^2)}{a^4 b^2(1 - x^2/a^2)} = \frac{a^4 b^2 + b^4 x^2 - a^2 b^2 x^2}{a^4 b^2 - a^2 b^2 x^2}$$

$$= \frac{a^4 + b^2 x^2 - a^2 x^2}{a^4 - a^2 x^2} = \frac{a^4 - (a^2 - b^2)x^2}{a^2(a^2 - x^2)}$$

The ellipsoid's surface area is twice the area generated by rotating the first-quadrant portion of the ellipse about the x-axis. Thus,

$$S = 2\int_0^a 2\pi y\sqrt{1 + \left(\frac{dy}{dx}\right)^2}\, dx = 4\pi \int_0^a \frac{b}{a}\sqrt{a^2 - x^2}\,\frac{\sqrt{a^4 - (a^2 - b^2)x^2}}{a\sqrt{a^2 - x^2}}\, dx = \frac{4\pi b}{a^2}\int_0^a \sqrt{a^4 - (a^2 - b^2)x^2}\, dx$$

$$= \frac{4\pi b}{a^2}\int_0^{a\sqrt{a^2 - b^2}}\sqrt{a^4 - u^2}\,\frac{du}{\sqrt{a^2 - b^2}}\quad [u = \sqrt{a^2 - b^2}\,x]\stackrel{30}{=} \frac{4\pi b}{a^2\sqrt{a^2 - b^2}}\left[\frac{u}{2}\sqrt{a^4 - u^2} + \frac{a^4}{2}\sin^{-1}\left(\frac{u}{a^2}\right)\right]_0^{a\sqrt{a^2 - b^2}}$$

$$= \frac{4\pi b}{a^2\sqrt{a^2 - b^2}}\left[\frac{a\sqrt{a^2 - b^2}}{2}\sqrt{a^4 - a^2(a^2 - b^2)} + \frac{a^4}{2}\sin^{-1}\frac{\sqrt{a^2 - b^2}}{a}\right] = 2\pi\left[b^2 + \frac{a^2 b\sin^{-1}\frac{\sqrt{a^2 - b^2}}{a}}{\sqrt{a^2 - b^2}}\right]$$

(b) $\dfrac{x^2}{a^2} + \dfrac{y^2}{b^2} = 1 \Rightarrow \dfrac{x(dx/dy)}{a^2} = -\dfrac{y}{b^2} \Rightarrow \dfrac{dx}{dy} = -\dfrac{a^2 y}{b^2 x} \Rightarrow$

$$1 + \left(\frac{dx}{dy}\right)^2 = 1 + \frac{a^4 y^2}{b^4 x^2} = \frac{b^4 x^2 + a^4 y^2}{b^4 x^2} = \frac{b^4 a^2(1 - y^2/b^2) + a^4 y^2}{b^4 a^2(1 - y^2/b^2)} = \frac{a^2 b^4 - a^2 b^2 y^2 + a^4 y^2}{a^2 b^4 - a^2 b^2 y^2}$$

$$= \frac{b^4 - b^2 y^2 + a^2 y^2}{b^4 - b^2 y^2} = \frac{b^4 - (b^2 - a^2)y^2}{b^2(b^2 - y^2)} \qquad \text{[continued]}$$

426 ☐ **CHAPTER 8** FURTHER APPLICATIONS OF INTEGRATION

The oblate spheroid's surface area is twice the area generated by rotating the first-quadrant portion of the ellipse about the y-axis. Thus,

$$S = 2\int_0^b 2\pi x \sqrt{1+\left(\frac{dx}{dy}\right)^2}\,dy = 4\pi \int_0^b \frac{a}{b}\sqrt{b^2-y^2}\,\frac{\sqrt{b^4-(b^2-a^2)y^2}}{b\sqrt{b^2-y^2}}\,dy$$

$$= \frac{4\pi a}{b^2}\int_0^b \sqrt{b^4-(b^2-a^2)y^2}\,dy = \frac{4\pi a}{b^2}\int_0^b \sqrt{b^4+(a^2-b^2)y^2}\,dy \qquad [\text{since } a > b]$$

$$= \frac{4\pi a}{b^2}\int_0^{b\sqrt{a^2-b^2}}\sqrt{b^4+u^2}\,\frac{du}{\sqrt{a^2-b^2}} \qquad [u = \sqrt{a^2-b^2}\,y]$$

$$\stackrel{21}{=} \frac{4\pi a}{b^2\sqrt{a^2-b^2}}\left[\frac{u}{2}\sqrt{b^4+u^2}+\frac{b^4}{2}\ln\!\left(u+\sqrt{b^4+u^2}\right)\right]_0^{b\sqrt{a^2-b^2}}$$

$$= \frac{4\pi a}{b^2\sqrt{a^2-b^2}}\left\{\left[\frac{b\sqrt{a^2-b^2}}{2}(ab)+\frac{b^4}{2}\ln\!\left(b\sqrt{a^2-b^2}+ab\right)\right]-\left[0+\frac{b^4}{2}\ln(b^2)\right]\right\}$$

$$= \frac{4\pi a}{b^2\sqrt{a^2-b^2}}\left[\frac{ab^2\sqrt{a^2-b^2}}{2}+\frac{b^4}{2}\ln\frac{b\sqrt{a^2-b^2}+ab}{b^2}\right] = 2\pi a^2 + \frac{2\pi ab^2}{\sqrt{a^2-b^2}}\ln\frac{\sqrt{a^2-b^2}+a}{b}$$

39. (a) The analogue of $f(x_i^*)$ in the derivation of (4) is now $c - f(x_i^*)$, so

$$S = \lim_{n\to\infty}\sum_{i=1}^n 2\pi[c-f(x_i^*)]\sqrt{1+[f'(x_i^*)]^2}\,\Delta x = \int_a^b 2\pi[c-f(x)]\sqrt{1+[f'(x)]^2}\,dx.$$

(b) $y = x^{1/2} \Rightarrow y' = \tfrac{1}{2}x^{-1/2} \Rightarrow 1+(y')^2 = 1+1/4x$, so by part (a), $S = \int_0^4 2\pi\!\left(4-\sqrt{x}\right)\sqrt{1+1/(4x)}\,dx$.

Using a CAS, we get $S = 2\pi\ln(\sqrt{17}+4) + \tfrac{\pi}{6}(31\sqrt{17}+1) \approx 80.6095$.

41. For the upper semicircle, $y = \sqrt{r^2-x^2}$, $dy/dx = -x/\sqrt{r^2-x^2}$. The surface area generated is

$$S_1 = \int_{-r}^r 2\pi(r-y)\,ds = \int_{-r}^r 2\pi\!\left(r-\sqrt{r^2-x^2}\right)\sqrt{1+\frac{x^2}{r^2-x^2}}\,dx$$

$$= 4\pi\int_0^r \left(r-\sqrt{r^2-x^2}\right)\frac{r}{\sqrt{r^2-x^2}}\,dx = 4\pi\int_0^r\left(\frac{r^2}{\sqrt{r^2-x^2}}-r\right)dx$$

For the lower semicircle, $y = -\sqrt{r^2-x^2}$ and $\dfrac{dy}{dx} = \dfrac{x}{\sqrt{r^2-x^2}}$, so $S_2 = 4\pi\int_0^r\left(\dfrac{r^2}{\sqrt{r^2-x^2}}+r\right)dx$.

Thus, the total area is $S = S_1 + S_2 = 8\pi\int_0^r\left(\dfrac{r^2}{\sqrt{r^2-x^2}}\right)dx = 8\pi\left[r^2\sin^{-1}\!\left(\dfrac{x}{r}\right)\right]_0^r = 8\pi r^2\!\left(\dfrac{\pi}{2}\right) = 4\pi^2 r^2$.

43. Rotate $y = R$ with $0 \le x \le h$ about the x-axis to generate a zone of a cylinder. $y = R \Rightarrow y' = 0 \Rightarrow$
$ds = \sqrt{1+0^2}\,dx = dx$. The surface area is $S = \int_0^h 2\pi y\,ds = 2\pi\int_0^h R\,dx = 2\pi R\,[x]_0^h = 2\pi Rh$.

45. $y = e^{x/2}+e^{-x/2} \Rightarrow y' = \tfrac{1}{2}e^{x/2}-\tfrac{1}{2}e^{-x/2} \Rightarrow$

$1+(y')^2 = 1+\left(\tfrac{1}{2}e^{x/2}-\tfrac{1}{2}e^{-x/2}\right)^2 = 1+\tfrac{1}{4}e^x-\tfrac{1}{2}+\tfrac{1}{4}e^{-x} = \tfrac{1}{4}e^x+\tfrac{1}{2}+\tfrac{1}{4}e^{-x} = \left(\tfrac{1}{2}e^{x/2}+\tfrac{1}{2}e^{-x/2}\right)^2.$

If we rotate the curve about the x-axis on the interval $a \le x \le b$, the resulting surface area is

$S = \int_a^b 2\pi y\sqrt{1+(y')^2}\,dx = 2\pi \int_a^b (e^{x/2} + e^{-x/2})\left(\frac{1}{2}e^{x/2} + \frac{1}{2}e^{-x/2}\right)dx = \pi \int_a^b (e^{x/2} + e^{-x/2})^2\,dx$, which is the same as the volume obtained by rotating the curve y about the x-axis on the interval $a \leq x \leq b$, namely, $V = \pi \int_a^b y^2\,dx$.

8.3 Applications to Physics and Engineering

1. The weight density of water is $\delta = 62.5$ lb/ft^3.

(a) $P = \delta d \approx (62.5 \text{ lb/ft}^3)(3 \text{ ft}) = 187.5 \text{ lb/ft}^2$

(b) $F = PA \approx (187.5 \text{ lb/ft}^2)(5 \text{ ft})(2 \text{ ft}) = 1875 \text{ lb}$.   ($A$ is the area of the bottom of the tank.)

(c) By reasoning as in Example 1, the area of the ith strip is $2\,(\Delta x)$ and the pressure is $\delta d = \delta x_i$. Thus,

$F = \int_0^3 \delta x \cdot 2\, dx \approx (62.5)(2)\int_0^3 x\,dx = 125\left[\frac{1}{2}x^2\right]_0^3 = 125\left(\frac{9}{2}\right) = 562.5$ lb.

In Exercises 3–9, n is the number of subintervals of length Δx and x_i^* is a sample point in the ith subinterval $[x_{i-1}, x_i]$.

3. Set up a vertical x-axis as shown, with $x = 0$ at the water's surface and x increasing in the downward direction. Then the area of the ith rectangular strip is $2\,\Delta x$ and the pressure on the strip is δx_i^* (where $\delta \approx 62.5$ lb/ft^3). Thus, the hydrostatic force on the strip is

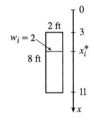

$\delta x_i^* \cdot 2\,\Delta x$ and the total hydrostatic force $\approx \sum_{i=1}^{n} \delta x_i^* \cdot 2\,\Delta x$. The total force is

$F = \lim_{n\to\infty}\sum_{i=1}^{n}\delta x_i^* \cdot 2\,\Delta x = \int_3^{11}\delta x \cdot 2\,dx = 2\delta \int_3^{11} x\,dx = 2\delta\left[\frac{1}{2}x^2\right]_3^{11} = \delta(121 - 9)$

$= 112\delta \approx 7000$ lb

5. Set up a coordinate system as shown. Then the area of the ith rectangular strip is $2\sqrt{8^2 - (y_i^*)^2}\,\Delta y$. The pressure on the strip is $\delta d_i = \rho g(12 - y_i^*)$, so the hydrostatic force on the strip is $\rho g(12 - y_i^*)\,2\sqrt{64 - (y_i^*)^2}\,\Delta y$ and the total hydrostatic force on the plate $\approx \sum_{i=1}^{n}\rho g(12 - y_i^*)\,2\sqrt{64 - (y_i^*)^2}\,\Delta y$.

The total force $F = \lim_{n\to\infty}\sum_{i=1}^{n}\rho g(12 - y_i^*)\,2\sqrt{64 - (y_i^*)^2}\,\Delta y = \int_{-8}^{8}\rho g(12 - y)\,2\sqrt{64 - y^2}\,dy$

$= 2\rho g \cdot 12 \int_{-8}^{8}\sqrt{64 - y^2}\,dy - 2\rho g\int_{-8}^{8} y\sqrt{64 - y^2}\,dy$.

The second integral is 0 because the integrand is an odd function. The first integral is the area of a semicircular disk with radius 8. Thus, $F = 24\rho g\left(\frac{1}{2}\pi(8)^2\right) = 768\pi\rho g \approx 768\pi(1000)(9.8) \approx 2.36 \times 10^7$ N.

7. Set up a vertical x-axis as shown. By similar triangles, $w_i/4 = x_i^*/6$, so $w_i = \frac{2}{3}x_i^*$, and the area of the ith rectangular strip is $\frac{2}{3}x_i^*\,\Delta x$. The pressure on the ith strip is $\rho g x_i^*$, so the hydrostatic force on the strip is $\rho g x_i^* \cdot \frac{2}{3}x_i^*\,\Delta x$, and the hydrostatic force on the plate is $\approx \sum_{i=1}^{n}\rho g x_i^* \cdot \frac{2}{3}x_i^*\,\Delta x$. The total force is

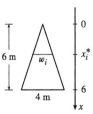

$F = \lim_{n\to\infty}\sum_{i=1}^{n}\rho g x_i^* \cdot \frac{2}{3}x_i^*\,\Delta x = \frac{2}{3}\rho g\int_0^6 x^2\,dx = \frac{2}{3}\rho g\left[\frac{1}{3}x^3\right]_0^6 = \frac{2}{9}\rho g(216 - 0)$

$= 48\rho g = 48(1000)(9.8) = 470{,}400$ N

9. Set up a vertical x-axis as shown. Then the area of the ith rectangular strip is $w_i \, \Delta x = \left(4 + 2 \cdot \frac{2}{3} x_i^*\right) \Delta x$. The pressure on the strip is $\delta(x_i^* - 1)$, so the hydrostatic force on the strip is $\delta(x_i^* - 1)\left(4 + \frac{4}{3} x_i^*\right) \Delta x$ and the hydrostatic force on the plate $\approx \sum_{i=1}^{n} \delta(x_i^* - 1)\left(4 + \frac{4}{3} x_i^*\right) \Delta x$. The total force is

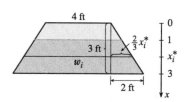

$$F = \lim_{n \to \infty} \sum_{i=1}^{n} \delta(x_i^* - 1)\left(4 + \tfrac{4}{3} x_i^*\right) \Delta x = \int_1^3 \delta(x-1)\left(4 + \tfrac{4}{3} x\right) dx$$

$$= \delta \int_1^3 \left(\tfrac{4}{3} x^2 + \tfrac{8}{3} x - 4\right) dx = \delta\left[\tfrac{4}{9} x^3 + \tfrac{4}{3} x^2 - 4x\right]_1^3$$

$$= \delta\left[(12 + 12 - 12) - \left(\tfrac{4}{9} + \tfrac{4}{3} - 4\right)\right] = \delta\left(\tfrac{128}{9}\right)$$

$$\approx 889 \text{ lb} \quad [\delta \approx 62.5]$$

11. Set up a vertical x-axis as shown. Then the area of the ith rectangular strip is $\frac{a}{h}(2h - x_i^*) \Delta x$. $\left[\text{By similar triangles, } \dfrac{w_i}{2h - x_i^*} = \dfrac{2a}{2h}, \text{ so } w_i = \dfrac{a}{h}(2h - x_i^*).\right]$

The pressure on the strip is δx_i^*, so the hydrostatic force on the plate

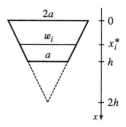

$\approx \sum_{i=1}^{n} \delta x_i^* \dfrac{a}{h}(2h - x_i^*) \Delta x$. The total force is

$$F = \lim_{n \to \infty} \sum_{i=1}^{n} \delta x_i^* \tfrac{a}{h}(2h - x_i^*) \Delta x = \delta \tfrac{a}{h} \int_0^h x(2h - x) \, dx = \tfrac{a\delta}{h} \int_0^h \left(2hx - x^2\right) dx$$

$$= \tfrac{a\delta}{h}\left[hx^2 - \tfrac{1}{3} x^3\right]_0^h = \tfrac{a\delta}{h}\left(h^3 - \tfrac{1}{3} h^3\right) = \tfrac{a\delta}{h}\left(\tfrac{2h^3}{3}\right) = \tfrac{2}{3}\delta a h^2$$

13. The solution is similar to the solution for Example 2. The pressure on a strip is approximately $\delta d_i = 47(4 - y_i^*)$ and the area of the ith rectangular strip is $2\sqrt{16 - (y_i^*)^2} \, \Delta y$, so the total force is

$$F = \lim_{n \to \infty} \sum_{i=1}^{n} 47(4 - y_i^*) \, 2\sqrt{16 - (y_i^*)^2} \, \Delta y = 94 \int_{-4}^{4} (4 - y)\sqrt{16 - y^2} \, dy$$

$$= 94 \cdot 4 \int_{-4}^{4} \sqrt{16 - y^2} \, dy - 94 \int_{-4}^{4} y\sqrt{16 - y^2} \, dy$$

$$= 376 \cdot \tfrac{1}{2} \pi (4)^2 - 0 \quad \left[\begin{array}{l}\text{The first integral is the area of a semicircular disk with radius 4} \\ \text{and the second integral is 0 because the integrand is an odd function.}\end{array}\right]$$

$$= 3008\pi \approx 9450 \text{ lb}$$

15. (a) The top of the cube has depth $d = 1 \text{ m} - 20 \text{ cm} = 80 \text{ cm} = 0.8 \text{ m}$.

$$F = \rho g d A \approx (1000)(9.8)(0.8)(0.2)^2 = 313.6 \approx 314 \text{ N}$$

(b) The area of a strip is $0.2 \, \Delta x$ and the pressure on it is $\rho g x_i^*$.

$$F = \int_{0.8}^{1} \rho g x (0.2) \, dx = 0.2 \rho g \left[\tfrac{1}{2} x^2\right]_{0.8}^{1} = (0.2 \rho g)(0.18) = 0.036 \rho g = 0.036(1000)(9.8) = 352.8 \approx 353 \text{ N}$$

17. (a) *Shallow end:* The area of a strip is $20 \, \Delta x$ and the pressure on it is δx_i.

$$F = \int_0^3 \delta x \, 20 \, dx = 20\delta\left[\tfrac{1}{2} x^2\right]_0^3 = 20\delta \cdot \tfrac{9}{2} = 90\delta$$

$$= 90(62.5) = 5625 \text{ lb} \approx 5.63 \times 10^3 \text{ lb}$$

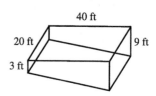

Deep end: $F = \int_0^9 \delta x 20\, dx = 20\delta\left[\frac{1}{2}x^2\right]_0^9 = 20\delta \cdot \frac{81}{2} = 810\delta = 810(62.5) = 50{,}625$ lb $\approx 5.06 \times 10^4$ lb.

Sides: For the first 3 ft, the length of a side is constant at 40 ft. For $3 < x \leq 9$, we can use similar triangles to find the length a: $\dfrac{a}{40} = \dfrac{9-x}{6} \Rightarrow a = 40 \cdot \dfrac{9-x}{6}$.

$F = \int_0^3 \delta x 40\, dx + \int_3^9 \delta x(40)\dfrac{9-x}{6}\, dx = 40\delta\left[\frac{1}{2}x^2\right]_0^3 + \frac{20}{3}\delta \int_3^9 (9x - x^2)\, dx = 180\delta + \frac{20}{3}\delta\left[\frac{9}{2}x^2 - \frac{1}{3}x^3\right]_3^9$

$= 180\delta + \frac{20}{3}\delta\left[\left(\frac{729}{2} - 243\right) - \left(\frac{81}{2} - 9\right)\right] = 180\delta + 600\delta = 780\delta = 780(62.5) = 48{,}750$ lb $\approx 4.88 \times 10^4$ lb

(b) For any right triangle with hypotenuse on the bottom,

$\sin\theta = \dfrac{\Delta x}{\text{hypotenuse}} \Rightarrow$

hypotenuse $= \Delta x \csc\theta = \Delta x \dfrac{\sqrt{40^2 + 6^2}}{6} = \dfrac{\sqrt{409}}{3}\Delta x$.

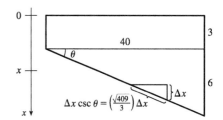

$F = \int_3^9 \delta x 20 \dfrac{\sqrt{409}}{3}\, dx = \frac{1}{3}(20\sqrt{409})\delta\left[\frac{1}{2}x^2\right]_3^9$

$= \frac{1}{3} \cdot 10\sqrt{409}\,\delta(81 - 9) \approx 303{,}356$ lb $\approx 3.03 \times 10^5$ lb

19. From Exercise 18, we have $F = \int_a^b \rho g x w(x)\, dx = \int_{7.0}^{9.4} 64xw(x)\, dx$. From the table, we see that $\Delta x = 0.4$, so using Simpson's Rule to estimate F, we get

$F \approx 64\,\frac{0.4}{3}[7.0w(7.0) + 4(7.4)w(7.4) + 2(7.8)w(7.8) + 4(8.2)w(8.2) + 2(8.6)w(8.6) + 4(9.0)w(9.0) + 9.4w(9.4)]$

$= \frac{25.6}{3}[7(1.2) + 29.6(1.8) + 15.6(2.9) + 32.8(3.8) + 17.2(3.6) + 36(4.2) + 9.4(4.4)]$

$= \frac{25.6}{3}(486.04) \approx 4148$ lb

21. The moment M of the system about the origin is $M = \sum_{i=1}^{2} m_i x_i = m_1 x_1 + m_2 x_2 = 6 \cdot 10 + 9 \cdot 30 = 330$.

The mass m of the system is $m = \sum_{i=1}^{2} m_i = m_1 + m_2 = 6 + 9 = 15$.

The center of mass of the system is $\overline{x} = M/m = \frac{330}{15} = 22$.

23. The mass is $m = \sum_{i=1}^{3} m_i = 5 + 8 + 7 = 20$. The moment about the x-axis is $M_x = \sum_{i=1}^{3} m_i y_i = 5(1) + 8(4) + 7(-2) = 23$.

The moment about the y-axis is $M_y = \sum_{i=1}^{3} m_i x_i = 5(3) + 8(0) + 7(-5) = -20$. The center of mass is

$(\overline{x}, \overline{y}) = \left(\dfrac{M_y}{m}, \dfrac{M_x}{m}\right) = \left(-\dfrac{20}{20}, \dfrac{23}{20}\right) = (-1, 1.15)$.

25. The region in the figure is "left-heavy" and "bottom-heavy," so we know that $\overline{x} < 1$ and $\overline{y} < 2$, and we might guess that $\overline{x} = 0.7$ and $\overline{y} = 1.3$. The line $2x + y = 4$ can be expressed as $y = 4 - 2x$, so $A = \int_0^2 (4 - 2x)\, dx = \left[4x - x^2\right]_0^2 = 8 - 4 = 4$.

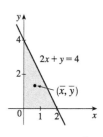

[continued]

$\bar{x} = \dfrac{1}{A}\displaystyle\int_0^2 x(4-2x)\,dx = \dfrac{1}{4}\int_0^2 (4x-2x^2)\,dx = \dfrac{1}{4}\left[2x^2 - \dfrac{2}{3}x^3\right]_0^2 = \dfrac{1}{4}\left(8-\dfrac{16}{3}\right) = \dfrac{2}{3}.$

$\bar{y} = \dfrac{1}{A}\displaystyle\int_0^2 \dfrac{1}{2}(4-2x)^2\,dx = \dfrac{1}{4}\int_0^2 \dfrac{1}{2}\cdot 4(2-x)^2\,dx = \dfrac{1}{2}\left[-\dfrac{1}{3}(2-x)^3\right]_0^2 = -\dfrac{1}{6}(0-8) = \dfrac{4}{3}.$

Thus, the centroid is $(\bar{x}, \bar{y}) = \left(\dfrac{2}{3}, \dfrac{4}{3}\right).$

27. The region in the figure is "right-heavy" and "bottom-heavy," so we know that $\bar{x} > 1$ and $\bar{y} < 1$, and we might guess that $\bar{x} = 1.5$ and $\bar{y} = 0.5$.

$A = \displaystyle\int_0^2 \dfrac{1}{2}x^2\,dx = \left[\dfrac{1}{6}x^3\right]_0^2 = \dfrac{8}{6} - 0 = \dfrac{4}{3}.$

$\bar{x} = \dfrac{1}{A}\displaystyle\int_0^2 x\cdot\dfrac{1}{2}x^2\,dx = \dfrac{3}{4}\int_0^2 \dfrac{1}{2}x^3\,dx = \dfrac{3}{8}\int_0^2 x^3\,dx = \dfrac{3}{8}\left[\dfrac{1}{4}x^4\right]_0^2 = \dfrac{3}{32}(16-0) = \dfrac{3}{2}.$

$\bar{y} = \dfrac{1}{A}\displaystyle\int_0^2 \dfrac{1}{2}\left(\dfrac{1}{2}x^2\right)^2 dx = \dfrac{3}{4}\int_0^2 \dfrac{1}{8}x^4\,dx = \dfrac{3}{32}\left[\dfrac{1}{5}x^5\right]_0^2 = \dfrac{3}{160}(32-0) = \dfrac{3}{5}.$

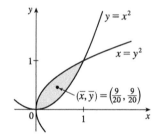

Thus, the centroid is $(\bar{x}, \bar{y}) = \left(\dfrac{3}{2}, \dfrac{3}{5}\right).$

29. $A = \displaystyle\int_0^1 (x^{1/2} - x^2)\,dx = \left[\dfrac{2}{3}x^{3/2} - \dfrac{1}{3}x^3\right]_0^1 = \left(\dfrac{2}{3} - \dfrac{1}{3}\right) - 0 = \dfrac{1}{3}.$

$\bar{x} = \dfrac{1}{A}\displaystyle\int_0^1 x(x^{1/2} - x^2)\,dx = 3\int_0^1 (x^{3/2} - x^3)\,dx = 3\left[\dfrac{2}{5}x^{5/2} - \dfrac{1}{4}x^4\right]_0^1$

$= 3\left(\dfrac{2}{5} - \dfrac{1}{4}\right) = 3\left(\dfrac{3}{20}\right) = \dfrac{9}{20}.$

$\bar{y} = \dfrac{1}{A}\displaystyle\int_0^1 \dfrac{1}{2}\left[(x^{1/2})^2 - (x^2)^2\right] dx = 3\left(\dfrac{1}{2}\right)\int_0^1 (x - x^4)\,dx$

$= \dfrac{3}{2}\left[\dfrac{1}{2}x^2 - \dfrac{1}{5}x^5\right]_0^1 = \dfrac{3}{2}\left(\dfrac{1}{2} - \dfrac{1}{5}\right) = \dfrac{3}{2}\left(\dfrac{3}{10}\right) = \dfrac{9}{20}.$

Thus, the centroid is $(\bar{x}, \bar{y}) = \left(\dfrac{9}{20}, \dfrac{9}{20}\right).$

31. $A = \displaystyle\int_0^{\pi/3}(\sin 2x - \sin x)\,dx = \left[-\dfrac{1}{2}\cos 2x + \cos x\right]_0^{\pi/3} = \left(\dfrac{1}{4} + \dfrac{1}{2}\right) - \left(-\dfrac{1}{2} + 1\right) = \dfrac{1}{4}.$

$\bar{x} = \dfrac{1}{A}\displaystyle\int_0^{\pi/3} x(\sin 2x - \sin x)\,dx = 4\left[x\left(-\dfrac{1}{2}\cos 2x + \cos x\right) - \left(-\dfrac{1}{4}\sin 2x + \sin x\right)\right]_0^{\pi/3} \quad \begin{bmatrix}\text{by parts with } u = x \text{ and}\\ dv = (\sin 2x - \sin x)\,dx\end{bmatrix}$

$= 4\left[\dfrac{\pi}{3}\left(\dfrac{1}{4} + \dfrac{1}{2}\right) - \left(-\dfrac{\sqrt{3}}{8} + \dfrac{\sqrt{3}}{2}\right)\right] = \pi - \dfrac{3\sqrt{3}}{2}.$

$\bar{y} = \dfrac{1}{A}\displaystyle\int_0^{\pi/3} \dfrac{1}{2}(\sin^2 2x - \sin^2 x)\,dx$

$= 4\displaystyle\int_0^{\pi/3} \dfrac{1}{2}(4\sin^2 x\cos^2 x - \sin^2 x)\,dx \quad \begin{bmatrix}\text{double-angle}\\ \text{identity}\end{bmatrix}$

$= 2\displaystyle\int_0^{\pi/3}[4\sin^2 x(1 - \sin^2 x) - \sin^2 x]\,dx$

$= 6\displaystyle\int_0^{\pi/3}\sin^2 x\,dx - 8\displaystyle\int_0^{\pi/3}\sin^4 x\,dx$

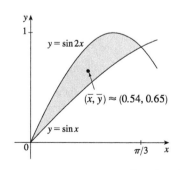

[continued]

Now, $\int_0^{\pi/3} \sin^2 x\, dx \stackrel{63}{=} \left[\frac{1}{2}x - \frac{1}{4}\sin 2x\right]_0^{\pi/3} = \frac{\pi}{6} - \frac{\sqrt{3}}{8}$ and

$$\int_0^{\pi/3} \sin^4 x\, dx = \frac{1}{4}\left[\frac{3}{2}x - \sin 2x + \frac{1}{8}\sin 4x\right]_0^{\pi/3} \quad \text{[by Example 7.2.4]}$$

$$= \frac{1}{4}\left[\frac{\pi}{2} - \frac{\sqrt{3}}{2} - \frac{\sqrt{3}}{16}\right] = \frac{\pi}{8} - \frac{9\sqrt{3}}{64}$$

So $\bar{y} = 6\left(\frac{\pi}{6} - \frac{\sqrt{3}}{8}\right) - 8\left(\frac{\pi}{8} - \frac{9\sqrt{3}}{64}\right) = \frac{3\sqrt{3}}{8}$. Thus, the centroid is $(\bar{x}, \bar{y}) = \left(\pi - \frac{3\sqrt{3}}{2}, \frac{3\sqrt{3}}{8}\right) \approx (0.54, 0.65)$.

33. The curves intersect when $2 - y = y^2 \iff 0 = y^2 + y - 2 \iff$
$0 = (y+2)(y-1) \iff y = -2$ or $y = 1$.

$A = \int_{-2}^{1} (2 - y - y^2)\, dy = \left[2y - \frac{1}{2}y^2 - \frac{1}{3}y^3\right]_{-2}^{1} = \frac{7}{6} - \left(-\frac{10}{3}\right) = \frac{9}{2}$.

$\bar{x} = \frac{1}{A}\int_{-2}^{1} \frac{1}{2}[(2-y)^2 - (y^2)^2]\, dy = \frac{2}{9}\cdot\frac{1}{2}\int_{-2}^{1}(4 - 4y + y^2 - y^4)\, dy$

$= \frac{1}{9}\left[4y - 2y^2 + \frac{1}{3}y^3 - \frac{1}{5}y^5\right]_{-2}^{1} = \frac{1}{9}\left[\frac{32}{15} - \left(-\frac{184}{15}\right)\right] = \frac{8}{5}$.

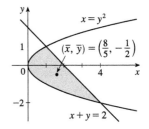

$\bar{y} = \frac{1}{A}\int_{-2}^{1} y(2 - y - y^2)\, dy = \frac{2}{9}\int_{-2}^{1}(2y - y^2 - y^3)\, dy$

$= \frac{2}{9}\left[y^2 - \frac{1}{3}y^3 - \frac{1}{4}y^4\right]_{-2}^{1} = \frac{2}{9}\left(\frac{5}{12} - \frac{8}{3}\right) = -\frac{1}{2}$.

Thus, the centroid is $(\bar{x}, \bar{y}) = \left(\frac{8}{5}, -\frac{1}{2}\right)$.

35. The quarter-circle has equation $y = \sqrt{4^2 - x^2}$ for $0 \le x \le 4$ and the line has equation $y = -2$.

$A = \frac{1}{4}\pi(4)^2 + 2(4) = 4\pi + 8 = 4(\pi + 2)$, so $m = \rho A = 6 \cdot 4(\pi + 2) = 24(\pi + 2)$.

$M_x = \rho\int_0^4 \frac{1}{2}\left[\left(\sqrt{16-x^2}\right)^2 - (-2)^2\right] dx = \frac{1}{2}\rho\int_0^4 (16 - x^2 - 4)\, dx = \frac{1}{2}(6)\left[12x - \frac{1}{3}x^3\right]_0^4 = 3\left(48 - \frac{64}{3}\right) = 80$.

$M_y = \rho\int_0^4 x\left[\sqrt{16-x^2} - (-2)\right] dx = \rho\int_0^4 x\sqrt{16-x^2}\, dx + \rho\int_0^4 2x\, dx = 6\left[-\frac{1}{3}(16-x^2)^{3/2}\right]_0^4 + 6\left[x^2\right]_0^4$

$= 6\left(0 + \frac{64}{3}\right) + 6(16) = 224$.

$\bar{x} = \frac{M_y}{m} = \frac{224}{24(\pi+2)} = \frac{28}{3(\pi+2)}$ and $\bar{y} = \frac{M_x}{m} = \frac{80}{24(\pi+2)} = \frac{10}{3(\pi+2)}$.

Thus, the center of mass is $\left(\frac{28}{3(\pi+2)}, \frac{10}{3(\pi+2)}\right) \approx (1.82, 0.65)$.

37. $A = \int_{-1}^{1}[(x^3 - x) - (x^2 - 1)]\, dx = \int_{-1}^{1}(1 - x^2)\, dx$ $\begin{bmatrix}\text{odd-degree terms}\\ \text{drop out}\end{bmatrix}$

$= 2\int_0^1 (1 - x^2)\, dx = 2\left[x - \frac{1}{3}x^3\right]_0^1 = 2\left(\frac{2}{3}\right) = \frac{4}{3}$.

$\bar{x} = \frac{1}{A}\int_{-1}^{1} x(x^3 - x - x^2 + 1)\, dx = \frac{3}{4}\int_{-1}^{1}(x^4 - x^2 - x^3 + x)\, dx$

$= \frac{3}{4}\int_{-1}^{1}(x^4 - x^2)\, dx = \frac{3}{4}\cdot 2\int_0^1 (x^4 - x^2)\, dx$

$= \frac{3}{2}\left[\frac{1}{5}x^5 - \frac{1}{3}x^3\right]_0^1 = \frac{3}{2}\left(-\frac{2}{15}\right) = -\frac{1}{5}$.

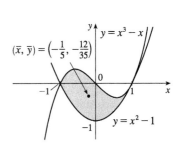

[continued]

$\bar{y} = \frac{1}{A}\int_{-1}^{1}\frac{1}{2}[(x^3-x)^2-(x^2-1)^2]\,dx = \frac{3}{4}\cdot\frac{1}{2}\int_{-1}^{1}(x^6-2x^4+x^2-x^4+2x^2-1)\,dx$

$= \frac{3}{8}\cdot 2\int_{0}^{1}(x^6-3x^4+3x^2-1)\,dx = \frac{3}{4}\left[\frac{1}{7}x^7-\frac{3}{5}x^5+x^3-x\right]_0^1 = \frac{3}{4}\left(-\frac{16}{35}\right) = -\frac{12}{35}$.

Thus, the centroid is $(\bar{x},\bar{y}) = \left(-\frac{1}{5},-\frac{12}{35}\right)$.

39. Choose x- and y-axes so that the base (one side of the triangle) lies along the x-axis with the other vertex along the positive y-axis as shown. From geometry, we know the medians intersect at a point $\frac{2}{3}$ of the way from each vertex (along the median) to the opposite side. The median from B goes to the midpoint $\left(\frac{1}{2}(a+c),0\right)$ of side AC, so the point of intersection of the medians is $\left(\frac{2}{3}\cdot\frac{1}{2}(a+c),\frac{1}{3}b\right) = \left(\frac{1}{3}(a+c),\frac{1}{3}b\right)$.

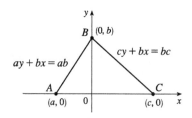

This can also be verified by finding the equations of two medians, and solving them simultaneously to find their point of intersection. Now let us compute the location of the centroid of the triangle. The area is $A = \frac{1}{2}(c-a)b$.

$\bar{x} = \frac{1}{A}\left[\int_a^0 x\cdot\frac{b}{a}(a-x)\,dx + \int_0^c x\cdot\frac{b}{c}(c-x)\,dx\right] = \frac{1}{A}\left[\frac{b}{a}\int_a^0 (ax-x^2)\,dx + \frac{b}{c}\int_0^c (cx-x^2)\,dx\right]$

$= \frac{b}{Aa}\left[\frac{1}{2}ax^2-\frac{1}{3}x^3\right]_a^0 + \frac{b}{Ac}\left[\frac{1}{2}cx^2-\frac{1}{3}x^3\right]_0^c = \frac{b}{Aa}\left[-\frac{1}{2}a^3+\frac{1}{3}a^3\right] + \frac{b}{Ac}\left[\frac{1}{2}c^3-\frac{1}{3}c^3\right]$

$= \frac{2}{a(c-a)}\cdot\frac{-a^3}{6} + \frac{2}{c(c-a)}\cdot\frac{c^3}{6} = \frac{1}{3(c-a)}(c^2-a^2) = \frac{a+c}{3}$

and $\bar{y} = \frac{1}{A}\left[\int_a^0 \frac{1}{2}\left(\frac{b}{a}(a-x)\right)^2 dx + \int_0^c \frac{1}{2}\left(\frac{b}{c}(c-x)\right)^2 dx\right]$

$= \frac{1}{A}\left[\frac{b^2}{2a^2}\int_a^0 (a^2-2ax+x^2)\,dx + \frac{b^2}{2c^2}\int_0^c (c^2-2cx+x^2)\,dx\right]$

$= \frac{1}{A}\left[\frac{b^2}{2a^2}[a^2x-ax^2+\frac{1}{3}x^3]_a^0 + \frac{b^2}{2c^2}[c^2x-cx^2+\frac{1}{3}x^3]_0^c\right]$

$= \frac{1}{A}\left[\frac{b^2}{2a^2}(-a^3+a^3-\frac{1}{3}a^3) + \frac{b^2}{2c^2}(c^3-c^3+\frac{1}{3}c^3)\right] = \frac{1}{A}\left[\frac{b^2}{6}(-a+c)\right] = \frac{2}{(c-a)b}\cdot\frac{(c-a)b^2}{6} = \frac{b}{3}$

Thus, the centroid is $(\bar{x},\bar{y}) = \left(\frac{a+c}{3},\frac{b}{3}\right)$, as claimed.

Remarks: Actually the computation of $\bar{y}$ is all that is needed. By considering each side of the triangle in turn to be the base, we see that the centroid is $\frac{1}{3}$ of the way from each side to the opposite vertex and must therefore be the intersection of the medians.

The computation of $\bar{y}$ in this problem (and many others) can be simplified by using horizontal rather than vertical approximating rectangles. If the length of a thin rectangle at coordinate y is $\ell(y)$, then its area is $\ell(y)\,\Delta y$, its mass is $\rho\ell(y)\,\Delta y$, and its moment about the x-axis is $\Delta M_x = \rho y \ell(y)\,\Delta y$. Thus,

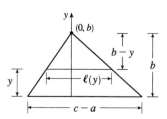

$$M_x = \int \rho y \ell(y)\,dy \quad \text{and} \quad \overline{y} = \frac{\int \rho y \ell(y)\,dy}{\rho A} = \frac{1}{A}\int y\ell(y)\,dy$$

In this problem, $\ell(y) = \dfrac{c-a}{b}(b-y)$ by similar triangles, so

$$\overline{y} = \frac{1}{A}\int_0^b \frac{c-a}{b} y(b-y)\,dy = \frac{2}{b^2}\int_0^b (by - y^2)\,dy = \frac{2}{b^2}\left[\tfrac{1}{2}by^2 - \tfrac{1}{3}y^3\right]_0^b = \frac{2}{b^2}\cdot\frac{b^3}{6} = \frac{b}{3}$$

Notice that only one integral is needed when this method is used.

41. Divide the lamina into two triangles and one rectangle with respective masses of 2, 2 and 4, so that the total mass is 8. Using the result of Exercise 39, the triangles have centroids $\left(-1, \tfrac{2}{3}\right)$ and $\left(1, \tfrac{2}{3}\right)$. The centroid of the rectangle (its center) is $\left(0, -\tfrac{1}{2}\right)$. So, using Formulas 6 and 7, we have $\overline{y} = \dfrac{M_x}{m} = \dfrac{1}{m}\sum_{i=1}^{3} m_i y_i = \tfrac{1}{8}\left[2\left(\tfrac{2}{3}\right) + 2\left(\tfrac{2}{3}\right) + 4\left(-\tfrac{1}{2}\right)\right] = \tfrac{1}{8}\left(\tfrac{2}{3}\right) = \tfrac{1}{12}$, and $\overline{x} = 0$, since the lamina is symmetric about the line $x = 0$. Thus, the centroid is $(\overline{x}, \overline{y}) = \left(0, \tfrac{1}{12}\right)$.

43. $\int_a^b (cx + d) f(x)\,dx = \int_a^b cx\,f(x)\,dx + \int_a^b d f(x)\,dx = c\int_a^b x f(x)\,dx + d\int_a^b f(x)\,dx = c\overline{x}A + d\int_a^b f(x)\,dx$ [by (8)]
$= c\overline{x}\int_a^b f(x)\,dx + d\int_a^b f(x)\,dx = (c\overline{x} + d)\int_a^b f(x)\,dx$

45. A cone of height h and radius r can be generated by rotating a right triangle about one of its legs as shown. By Exercise 39, $\overline{x} = \tfrac{1}{3}r$, so by the Theorem of Pappus, the volume of the cone is

$$V = Ad = \left(\tfrac{1}{2}\cdot \text{base}\cdot\text{height}\right)\cdot(2\pi\overline{x}) = \tfrac{1}{2}rh\cdot 2\pi\left(\tfrac{1}{3}r\right) = \tfrac{1}{3}\pi r^2 h.$$

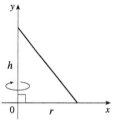

47. The curve C is the quarter-circle $y = \sqrt{16 - x^2}$, $0 \leq x \leq 4$. Its length L is $\tfrac{1}{4}(2\pi\cdot 4) = 2\pi$.

Now $y' = \tfrac{1}{2}(16 - x^2)^{-1/2}(-2x) = \dfrac{-x}{\sqrt{16 - x^2}} \Rightarrow 1 + (y')^2 = 1 + \dfrac{x^2}{16 - x^2} = \dfrac{16}{16 - x^2} \Rightarrow$

$ds = \sqrt{1 + (y')^2}\,dx = \dfrac{4}{\sqrt{16 - x^2}}\,dx$, so

$$\overline{x} = \frac{1}{L}\int x\,ds = \frac{1}{2\pi}\int_0^4 4x(16 - x^2)^{-1/2}\,dx = \frac{4}{2\pi}\left[-(16 - x^2)^{1/2}\right]_0^4 = \frac{2}{\pi}(0 + 4) = \frac{8}{\pi}\text{ and}$$

$$\overline{y} = \frac{1}{L}\int y\,ds = \frac{1}{2\pi}\int_0^4 \sqrt{16 - x^2}\cdot\frac{4}{\sqrt{16 - x^2}}\,dx = \frac{4}{2\pi}\int_0^4 dx = \frac{2}{\pi}\left[x\right]_0^4 = \frac{2}{\pi}(4 - 0) = \frac{8}{\pi}.\text{ Thus, the centroid}$$

is $\left(\dfrac{8}{\pi}, \dfrac{8}{\pi}\right)$. Note that the centroid does not lie on the curve, but does lie on the line $y = x$, as expected, due to the symmetry of the curve.

49. The circle has arc length (circumference) $L = 2\pi r$. As in Example 7, the distance traveled by the centroid during a rotation is $d = 2\pi R$. Therefore, by the Second Theorem of Pappus, the surface area is

$$S = Ld = (2\pi r)(2\pi R) = 4\pi^2 rR$$

51. Suppose the region lies between two curves $y = f(x)$ and $y = g(x)$ where $f(x) \geq g(x)$, as illustrated in Figure 13. Choose points x_i with $a = x_0 < x_1 < \cdots < x_n = b$ and choose x_i^* to be the midpoint of the ith subinterval; that is, $x_i^* = \bar{x}_i = \frac{1}{2}(x_{i-1} + x_i)$. Then the centroid of the ith approximating rectangle R_i is its center $C_i = \left(\bar{x}_i, \frac{1}{2}[f(\bar{x}_i) + g(\bar{x}_i)]\right)$. Its area is $[f(\bar{x}_i) - g(\bar{x}_i)] \Delta x$, so its mass is

$\rho[f(\bar{x}_i) - g(\bar{x}_i)] \Delta x$. Thus, $M_y(R_i) = \rho[f(\bar{x}_i) - g(\bar{x}_i)] \Delta x \cdot \bar{x}_i = \rho \bar{x}_i [f(\bar{x}_i) - g(\bar{x}_i)] \Delta x$ and

$M_x(R_i) = \rho[f(\bar{x}_i) - g(\bar{x}_i)] \Delta x \cdot \frac{1}{2}[f(\bar{x}_i) + g(\bar{x}_i)] = \rho \cdot \frac{1}{2}[f(\bar{x}_i)^2 - g(\bar{x}_i)^2] \Delta x$. Summing over i and taking the limit as $n \to \infty$, we get $M_y = \lim_{n \to \infty} \sum_i \rho \bar{x}_i [f(\bar{x}_i) - g(\bar{x}_i)] \Delta x = \rho \int_a^b x[f(x) - g(x)] \, dx$ and

$M_x = \lim_{n \to \infty} \sum_i \rho \cdot \frac{1}{2}[f(\bar{x}_i)^2 - g(\bar{x}_i)^2] \Delta x = \rho \int_a^b \frac{1}{2}[f(x)^2 - g(x)^2] \, dx$.

Thus, $\bar{x} = \dfrac{M_y}{m} = \dfrac{M_y}{\rho A} = \dfrac{1}{A} \int_a^b x[f(x) - g(x)] \, dx$ and $\bar{y} = \dfrac{M_x}{m} = \dfrac{M_x}{\rho A} = \dfrac{1}{A} \int_a^b \frac{1}{2}[f(x)^2 - g(x)^2] \, dx$.

8.4 Applications to Economics and Biology

1. By the Net Change Theorem, $C(4000) - C(0) = \int_0^{4000} C'(x) \, dx \Rightarrow$

$C(4000) = 18{,}000 + \int_0^{4000} (0.82 - 0.00003x + 0.000000003x^2) \, dx$

$= 18{,}000 + \left[0.82x - 0.000015x^2 + 0.000000001x^3\right]_0^{4000} = 18{,}000 + 3104 = \$21{,}104$

3. By the Net Change Theorem, $C(50) - C(0) = \int_0^{50} (0.6 + 0.008x) \, dx \Rightarrow$

$C(50) = 100 + \left[0.6x + 0.004x^2\right]_0^{50} = 100 + (40 - 0) = 140$, or $\$140{,}000$.

Similarly, $C(100) - C(50) = \left[0.6x + 0.004x^2\right]_{50}^{100} = 100 - 40 = 60$, or $\$60{,}000$.

5. Consumer surplus $= \int_0^{45} [p(x) - p(45)] \, dx$

$= \int_0^{45} \left(870 e^{-0.03x} - 870 e^{-0.03(45)}\right) dx$

$= 870 \left[-\dfrac{1}{0.03} e^{-0.03x} - e^{-1.35} x\right]_0^{45}$

$= 870 \left(-\dfrac{1}{0.03} e^{-1.35} - 45 e^{-1.35} + \dfrac{1}{0.03}\right) \approx \$11{,}332.78$

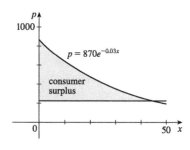

7. Since the demand increases by 30 for each dollar the price is lowered, the demand function, $p(x)$, is linear with slope $-\frac{1}{30}$. Also, $p(210) = 18$, so an equation for the demand is $p - 18 = -\frac{1}{30}(x - 210)$ or $p = -\frac{1}{30}x + 25$. A selling price of $\$15$ implies that $15 = -\frac{1}{30}x + 25 \Rightarrow \frac{1}{30}x = 10 \Rightarrow x = 300$.

Consumer surplus $= \int_0^{300} \left(-\frac{1}{30}x + 25 - 15\right) dx = \int_0^{300} \left(-\frac{1}{30}x + 10\right) dx = \left[-\frac{1}{60}x^2 + 10x\right]_0^{300} = \$1500$

9. $p_S(x) = 3 + 0.01x^2$. $P = p_S(10) = 3 + 1 = 4$.

Producer surplus $= \int_0^{10} [P - p_S(x)] \, dx = \int_0^{10} [4 - 3 - 0.01x^2] \, dx$

$= \left[x - \dfrac{0.01}{3}x^3\right]_0^{10} \approx 10 - 3.33 = \$6.67$

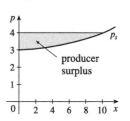

11. $p = \sqrt{30 + 0.01xe^{0.001x}} = 30$ when $x \approx 3278.5$ (using a graphing calculator or other computing device). Then the producer surplus is approximately $\int_0^{3278.5} \left[30 - \sqrt{30 + 0.01xe^{0.001x}}\right] dx \approx \$55{,}735$.

13. (a) Demand function $p(x) = $ supply function $p_S(x) \iff 228.4 - 18x = 27x + 57.4 \iff 171 = 45x \iff x = \frac{19}{5}$ [3.8 thousand]. $p(3.8) = 228.4 - 18(3.8) = 160$. The market for the stereos is in equilibrium when the quantity is 3800 and the price is \$160.

(b) Consumer surplus $= \int_0^{3.8}[p(x) - 160]\, dx = \int_0^{3.8}(228.4 - 18x - 160)\, dx = \int_0^{3.8}(68.4 - 18x)\, dx$

$$= \left[68.4x - 9x^2\right]_0^{3.8} = 68.4(3.8) - 9(3.8)^2 = 129.96$$

Producer surplus $= \int_0^{3.8}[160 - p_S(x)]\, dx = \int_0^{3.8}[160 - (27x + 57.4)]\, dx = \int_0^{3.8}(102.6 - 27x)\, dx$

$$= \left[102.6x - 13.5x^2\right]_0^{3.8} = 102.6(3.8) - 13.5(3.8)^2 = 194.94$$

Thus, the maximum total surplus for the stereos is $129.96 + 194.94 = 324.9$, or \$324,900.

15. $f(8) - f(4) = \int_4^8 f'(t)\, dt = \int_4^8 \sqrt{t}\, dt = \left[\frac{2}{3}t^{3/2}\right]_4^8 = \frac{2}{3}(16\sqrt{2} - 8) \approx \$9.75$ million

17. Future value $= \int_0^T f(t)\, e^{r(T-t)}\, dt = \int_0^6 8000 e^{0.04t}\, e^{0.062(6-t)}\, dt = 8000 \int_0^6 e^{0.04t}\, e^{0.372 - 0.062t}\, dt$

$$= 8000 \int_0^6 e^{0.372 - 0.022t}\, dt = 8000 e^{0.372} \int_0^6 e^{-0.022t}\, dt = 8000 e^{0.372} \left[\frac{e^{-0.022t}}{-0.022}\right]_0^6$$

$$= \frac{8000 e^{0.372}}{-0.022}(e^{-0.132} - 1) \approx \$65{,}230.48$$

19. $N = \int_a^b Ax^{-k}\, dx = A\left[\frac{x^{-k+1}}{-k+1}\right]_a^b = \frac{A}{1-k}(b^{1-k} - a^{1-k})$.

Similarly, $\int_a^b Ax^{1-k}\, dx = A\left[\frac{x^{2-k}}{2-k}\right]_a^b = \frac{A}{2-k}(b^{2-k} - a^{2-k})$.

Thus, $\overline{x} = \frac{1}{N}\int_a^b Ax^{1-k}\, dx = \frac{[A/(2-k)](b^{2-k} - a^{2-k})}{[A/(1-k)](b^{1-k} - a^{1-k})} = \frac{(1-k)(b^{2-k} - a^{2-k})}{(2-k)(b^{1-k} - a^{1-k})}$.

21. $F = \frac{\pi P R^4}{8\eta l} = \frac{\pi(4000)(0.008)^4}{8(0.027)(2)} \approx 1.19 \times 10^{-4}$ cm^3/s

23. From (3), $F = \dfrac{A}{\int_0^T c(t)\, dt} = \dfrac{6}{20 I}$, where

$$I = \int_0^{10} t e^{-0.6t}\, dt = \left[\frac{1}{(-0.6)^2}(-0.6t - 1)e^{-0.6t}\right]_0^{10} \quad \begin{bmatrix}\text{integrating} \\ \text{by parts}\end{bmatrix} = \frac{1}{0.36}(-7e^{-6} + 1)$$

Thus, $F = \dfrac{6(0.36)}{20(1 - 7e^{-6})} = \dfrac{0.108}{1 - 7e^{-6}} \approx 0.1099$ L/s or 6.594 L/min.

25. As in Example 2, we will estimate the cardiac output using Simpson's Rule with $\Delta t = (16 - 0)/8 = 2$.

$$\int_0^{16} c(t)\, dt \approx \tfrac{2}{3}[c(0) + 4c(2) + 2c(4) + 4c(6) + 2c(8) + 4c(10) + 2c(12) + 4c(14) + c(16)]$$
$$\approx \tfrac{2}{3}[0 + 4(6.1) + 2(7.4) + 4(6.7) + 2(5.4) + 4(4.1) + 2(3.0) + 4(2.1) + 1.5]$$
$$= \tfrac{2}{3}(109.1) = 72.7\overline{3}\ \text{mg}\cdot\text{s/L}$$

Therefore, $F \approx \dfrac{A}{72.7\overline{3}} = \dfrac{7}{72.7\overline{3}} \approx 0.0962$ L/s or 5.77 L/min.

8.5 Probability

1. (a) $\int_{30,000}^{40,000} f(x)\, dx$ is the probability that a randomly chosen tire will have a lifetime between 30,000 and 40,000 miles.

(b) $\int_{25,000}^{\infty} f(x)\, dx$ is the probability that a randomly chosen tire will have a lifetime of at least 25,000 miles.

3. (a) In general, we must satisfy the two conditions that are mentioned before Example 1—namely, **(1)** $f(x) \geq 0$ for all x, and **(2)** $\int_{-\infty}^{\infty} f(x)\, dx = 1$. For $0 \leq x \leq 1$, $f(x) = 30x^2(1-x)^2 \geq 0$ and $f(x) = 0$ for all other values of x, so $f(x) \geq 0$ for all x. Also,

$$\int_{-\infty}^{\infty} f(x)\, dx = \int_0^1 30x^2(1-x)^2\, dx = \int_0^1 30x^2(1 - 2x + x^2)\, dx = \int_0^1 (30x^2 - 60x^3 + 30x^4)\, dx$$
$$= \left[10x^3 - 15x^4 + 6x^5\right]_0^1 = 10 - 15 + 6 = 1$$

Therefore, f is a probability density function.

(b) $P(X \leq \tfrac{1}{3}) = \int_{-\infty}^{1/3} f(x)\, dx = \int_0^{1/3} 30x^2(1-x)^2\, dx = \left[10x^3 - 15x^4 + 6x^5\right]_0^{1/3} = \tfrac{10}{27} - \tfrac{15}{81} + \tfrac{6}{243} = \tfrac{17}{81}$

5. (a) In general, we must satisfy the two conditions that are mentioned before Example 1—namely, **(1)** $f(x) \geq 0$ for all x, and **(2)** $\int_{-\infty}^{\infty} f(x)\, dx = 1$. If $c \geq 0$, then $f(x) \geq 0$, so condition (1) is satisfied. For condition (2), we see that

$$\int_{-\infty}^{\infty} f(x)\, dx = \int_{-\infty}^{\infty} \frac{c}{1 + x^2}\, dx \quad \text{and}$$

$$\int_0^{\infty} \frac{c}{1+x^2}\, dx = \lim_{t \to \infty} \int_0^t \frac{c}{1+x^2}\, dx = c\lim_{t \to \infty} \left[\tan^{-1} x\right]_0^t = c \lim_{t \to \infty} \tan^{-1} t = c\left(\frac{\pi}{2}\right)$$

Similarly, $\displaystyle\int_{-\infty}^0 \frac{c}{1+x^2}\, dx = c\left(\frac{\pi}{2}\right)$, so $\displaystyle\int_{-\infty}^{\infty} \frac{c}{1+x^2}\, dx = 2c\left(\frac{\pi}{2}\right) = c\pi$.

Since $c\pi$ must equal 1, we must have $c = 1/\pi$ so that f is a probability density function.

(b) $P(-1 < X < 1) = \displaystyle\int_{-1}^1 \frac{1/\pi}{1+x^2}\, dx = \frac{2}{\pi}\int_0^1 \frac{1}{1+x^2}\, dx = \frac{2}{\pi}\left[\tan^{-1} x\right]_0^1 = \frac{2}{\pi}\left(\frac{\pi}{4} - 0\right) = \frac{1}{2}$

7. (a) In general, we must satisfy the two conditions that are mentioned before Example 1—namely, **(1)** $f(x) \geq 0$ for all x, and **(2)** $\int_{-\infty}^{\infty} f(x)\, dx = 1$. Since $f(x) = 0$ or $f(x) = 0.1$, condition (1) is satisfied. For condition (2), we see that

$$\int_{-\infty}^{\infty} f(x)\, dx = \int_0^{10} 0.1\, dx = \left[\tfrac{1}{10}x\right]_0^{10} = 1. \text{ Thus, } f(x) \text{ is a probability density function for the spinner's values.}$$

(b) Since all the numbers between 0 and 10 are equally likely to be selected, we expect the mean to be halfway between the endpoints of the interval; that is, $x = 5$.

$$\mu = \int_{-\infty}^{\infty} x f(x)\, dx = \int_0^{10} x(0.1)\, dx = \left[\tfrac{1}{20} x^2\right]_0^{10} = \tfrac{100}{20} = 5, \text{ as expected.}$$

9. We need to find m so that $\int_m^{\infty} f(t)\, dt = \tfrac{1}{2} \;\Rightarrow\; \lim_{x\to\infty} \int_m^x \tfrac{1}{5} e^{-t/5}\, dt = \tfrac{1}{2} \;\Rightarrow\; \lim_{x\to\infty} \left[\tfrac{1}{5}(-5)e^{-t/5}\right]_m^x = \tfrac{1}{2} \;\Rightarrow\;$
$(-1)(0 - e^{-m/5}) = \tfrac{1}{2} \;\Rightarrow\; e^{-m/5} = \tfrac{1}{2} \;\Rightarrow\; -m/5 = \ln \tfrac{1}{2} \;\Rightarrow\; m = -5 \ln \tfrac{1}{2} = 5\ln 2 \approx 3.47$ min.

11. (a) An exponential density function with $\mu = 1.6$ is $f(t) = \begin{cases} 0 & \text{if } t < 0 \\ \tfrac{1}{1.6} e^{-t/1.6} & \text{if } t \geq 0 \end{cases}$.

The probability that a customer waits less than a second is

$$P(X < 1) = \int_0^1 f(t)\, dt = \int_0^1 \tfrac{1}{1.6} e^{-t/1.6}\, dt = \left[-e^{-t/1.6}\right]_0^1 = -e^{-1/1.6} + 1 \approx 0.465.$$

(b) The probability that a customer waits more than 3 seconds is

$$P(X > 3) = \int_3^{\infty} f(t)\, dt = \lim_{s\to\infty} \int_3^s f(t)\, dt = \lim_{s\to\infty} \left[-e^{-t/1.6}\right]_3^s = \lim_{s\to\infty} \left(-e^{-s/1.6} + e^{-3/1.6}\right) = e^{-3/1.6} \approx 0.153.$$

Or: Calculate $1 - \int_0^3 f(t)\, dt$.

(c) We want to find b such that $P(X > b) = 0.05$. From part (b), $P(X > b) = e^{-b/1.6}$. Solving $e^{-b/1.6} = 0.05$ gives us
$-\tfrac{b}{1.6} = \ln 0.05 \;\Rightarrow\; b = -1.6 \ln 0.05 \approx 4.79$ seconds.

Or: Solve $\int_0^b f(t)\, dt = 0.95$ for b.

13. (a) $f(t) = \begin{cases} \tfrac{1}{1600} t & \text{if } 0 \leq t \leq 40 \\ \tfrac{1}{20} - \tfrac{1}{1600} t & \text{if } 40 < t \leq 80 \\ 0 & \text{otherwise} \end{cases}$

$$P(30 \leq T \leq 60) = \int_{30}^{60} f(t)\, dt = \int_{30}^{40} \tfrac{t}{1600}\, dt + \int_{40}^{60} \left(\tfrac{1}{20} - \tfrac{t}{1600}\right) dt = \left[\tfrac{t^2}{3200}\right]_{30}^{40} + \left[\tfrac{t}{20} - \tfrac{t^2}{3200}\right]_{40}^{60}$$

$$= \left(\tfrac{1600}{3200} - \tfrac{900}{3200}\right) + \left(\tfrac{60}{20} - \tfrac{3600}{3200}\right) - \left(\tfrac{40}{20} - \tfrac{1600}{3200}\right) = -\tfrac{1300}{3200} + 1 = \tfrac{19}{32}$$

The probability that the amount of REM sleep is between 30 and 60 minutes is $\tfrac{19}{32} \approx 59.4\%$.

(b) $\mu = \int_{-\infty}^{\infty} t f(t)\, dt = \int_0^{40} t\left(\tfrac{t}{1600}\right) dt + \int_{40}^{80} t\left(\tfrac{1}{20} - \tfrac{t}{1600}\right) dt = \left[\tfrac{t^3}{4800}\right]_0^{40} + \left[\tfrac{t^2}{40} - \tfrac{t^3}{4800}\right]_{40}^{80}$

$$= \tfrac{64{,}000}{4800} + \left(\tfrac{6400}{40} - \tfrac{512{,}000}{4800}\right) - \left(\tfrac{1600}{40} - \tfrac{64{,}000}{4800}\right) = -\tfrac{384{,}000}{4800} + 120 = 40$$

The mean amount of REM sleep is 40 minutes.

15. $P(X \geq 10) = \int_{10}^{\infty} \tfrac{1}{4.2\sqrt{2\pi}} \exp\!\left(-\tfrac{(x-9.4)^2}{2 \cdot 4.2^2}\right) dx$. To avoid the improper integral we approximate it by the integral from 10 to 100. Thus, $P(X \geq 10) \approx \int_{10}^{100} \tfrac{1}{4.2\sqrt{2\pi}} \exp\!\left(-\tfrac{(x-9.4)^2}{2 \cdot 4.2^2}\right) dx \approx 0.443$ (using a calculator or computer to estimate

17. (a) $P(0 \le X \le 100) = \int_0^{100} \dfrac{1}{8\sqrt{2\pi}} \exp\left(-\dfrac{(x-112)^2}{2\cdot 8^2}\right) dx \approx 0.0668$ (using a calculator or computer to estimate the integral), so there is about a 6.68% chance that a randomly chosen vehicle is traveling at a legal speed.

(b) $P(X \ge 125) = \int_{125}^{\infty} \dfrac{1}{8\sqrt{2\pi}} \exp\left(-\dfrac{(x-112)^2}{2\cdot 8^2}\right) dx = \int_{125}^{\infty} f(x)\, dx$. In this case, we could use a calculator or computer to estimate either $\int_{125}^{300} f(x)\, dx$ or $1 - \int_0^{125} f(x)\, dx$. Both are approximately 0.0521, so about 5.21% of the motorists are targeted.

19. $P(\mu - 2\sigma \le X \le \mu + 2\sigma) = \int_{\mu - 2\sigma}^{\mu + 2\sigma} \dfrac{1}{\sigma\sqrt{2\pi}} \exp\left(-\dfrac{(x-\mu)^2}{2\sigma^2}\right) dx$. Substituting $t = \dfrac{x-\mu}{\sigma}$ and $dt = \dfrac{1}{\sigma} dx$ gives us

$\int_{-2}^{2} \dfrac{1}{\sigma\sqrt{2\pi}} e^{-t^2/2}(\sigma\, dt) = \dfrac{1}{\sqrt{2\pi}} \int_{-2}^{2} e^{-t^2/2}\, dt \approx 0.9545$.

21. (a) First $p(r) = \dfrac{4}{a_0^3} r^2 e^{-2r/a_0} \ge 0$ for $r \ge 0$. Next,

$$\int_{-\infty}^{\infty} p(r)\, dr = \int_0^{\infty} \dfrac{4}{a_0^3} r^2 e^{-2r/a_0}\, dr = \dfrac{4}{a_0^3} \lim_{t \to \infty} \int_0^t r^2 e^{-2r/a_0}\, dr$$

By using parts, tables, or a CAS, we find that $\int x^2 e^{bx}\, dx = (e^{bx}/b^3)(b^2 x^2 - 2bx + 2)$. ($\star$)

Next, we use ($\star$) (with $b = -2/a_0$) and l'Hospital's Rule to get $\dfrac{4}{a_0^3}\left[\dfrac{a_0^3}{-8}(-2)\right] = 1$. This satisfies the second condition for a function to be a probability density function.

(b) Using l'Hospital's Rule, $\dfrac{4}{a_0^3} \lim_{r \to \infty} \dfrac{r^2}{e^{2r/a_0}} = \dfrac{4}{a_0^3} \lim_{r \to \infty} \dfrac{2r}{(2/a_0)e^{2r/a_0}} = \dfrac{2}{a_0^2} \lim_{r \to \infty} \dfrac{2}{(2/a_0)e^{2r/a_0}} = 0$.

To find the maximum of p, we differentiate:

$$p'(r) = \dfrac{4}{a_0^3}\left[r^2 e^{-2r/a_0}\left(-\dfrac{2}{a_0}\right) + e^{-2r/a_0}(2r)\right] = \dfrac{4}{a_0^3} e^{-2r/a_0}(2r)\left(-\dfrac{r}{a_0} + 1\right)$$

$p'(r) = 0 \Leftrightarrow r = 0$ or $1 = \dfrac{r}{a_0} \Leftrightarrow r = a_0$ $[a_0 \approx 5.59 \times 10^{-11}$ m$]$.

$p'(r)$ changes from positive to negative at $r = a_0$, so $p(r)$ has its maximum value at $r = a_0$.

(c) It is fairly difficult to find a viewing rectangle, but knowing the maximum value from part (b) helps.

$p(a_0) = \dfrac{4}{a_0^3} a_0^2 e^{-2a_0/a_0} = \dfrac{4}{a_0} e^{-2} \approx 9{,}684{,}098{,}979$

With a maximum of nearly 10 billion and a total area under the curve of 1, we know that the "hump" in the graph must be extremely narrow.

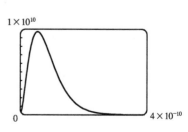

(d) $P(r) = \int_0^r \dfrac{4}{a_0^3} s^2 e^{-2s/a_0}\, ds \;\Rightarrow\; P(4a_0) = \int_0^{4a_0} \dfrac{4}{a_0^3} s^2 e^{-2s/a_0}\, ds$. Using $(\star)$ from part (a) [with $b = -2/a_0$],

$$P(4a_0) = \dfrac{4}{a_0^3}\left[\dfrac{e^{-2s/a_0}}{-8/a_0^3}\left(\dfrac{4}{a_0^2}s^2 + \dfrac{4}{a_0}s + 2\right)\right]_0^{4a_0} = \dfrac{4}{a_0^3}\left(\dfrac{a_0^3}{-8}\right)[e^{-8}(64 + 16 + 2) - 1(2)] = -\tfrac{1}{2}(82e^{-8} - 2)$$

$$= 1 - 41e^{-8} \approx 0.986$$

(e) $\mu = \displaystyle\int_{-\infty}^\infty rp(r)\,dr = \dfrac{4}{a_0^3}\lim_{t\to\infty}\int_0^t r^3 e^{-2r/a_0}\, dr$. Integrating by parts three times or using a CAS, we find that

$$\int x^3 e^{bx}\, dx = \dfrac{e^{bx}}{b^4}(b^3 x^3 - 3b^2 x^2 + 6bx - 6).\; \text{So with } b = -\dfrac{2}{a_0}, \text{ we use l'Hospital's Rule, and get}$$

$$\mu = \dfrac{4}{a_0^3}\left[-\dfrac{a_0^4}{16}(-6)\right] = \tfrac{3}{2}a_0.$$

8 Review

TRUE-FALSE QUIZ

1. True. The graph of $y = f(x) + c$ is obtained by vertically translating $y = f(x)$ by c units. The arc length over the interval $a \le x \le b$ will be unchanged by this transformation.

3. False. Suppose $f(x) = x$ and $g(x) = 1$ so that $f(x) \le g(x)$ in the interval $[0, 1]$. $f(x)$ is a straight line so its arc length over $[0, 1]$ is the distance between its endpoints and is given by $\sqrt{(1-0)^2 + [f(1) - f(0)]^2} = \sqrt{2}$. Similarly, the arc length of $g(x)$ over $[0, 1]$ is 1, which is less than $\sqrt{2}$.

5. True. The smallest possible length of arc between the points $(0, 0)$ and $(3, 4)$ is the length of a straight line segment connecting the two points. This length is $\sqrt{(3-0)^2 + (4-0)^2} = \sqrt{25} = 5$.

7. True. The hydrostatic pressure depends only on the depth of the fluid, d, and the fluid's weight density, δ, as given by $P = \delta d$. See margin note next to Equation 8.3.1.

EXERCISES

1. $y = 4(x-1)^{3/2} \;\Rightarrow\; \dfrac{dy}{dx} = 6(x-1)^{1/2} \;\Rightarrow\; 1 + \left(\dfrac{dy}{dx}\right)^2 = 1 + 36(x-1) = 36x - 35.$ Thus,

$$L = \int_1^4 \sqrt{36x - 35}\, dx = \int_1^{109} \sqrt{u}\left(\dfrac{1}{36}\, du\right) \quad \left[\begin{array}{l} u = 36x - 35, \\ du = 36\, dx \end{array}\right]$$

$$= \tfrac{1}{36}\left[\tfrac{2}{3}u^{3/2}\right]_1^{109} = \tfrac{1}{54}(109\sqrt{109} - 1)$$

3. $12x = 4y^3 + 3y^{-1}$ $\Rightarrow$ $x = \frac{1}{3}y^3 + \frac{1}{4}y^{-1}$ $\Rightarrow$ $\dfrac{dx}{dy} = y^2 - \frac{1}{4}y^{-2}$ $\Rightarrow$

$$1 + \left(\frac{dx}{dy}\right)^2 = 1 + y^4 - \tfrac{1}{2} + \tfrac{1}{16}y^{-4} = y^4 + \tfrac{1}{2} + \tfrac{1}{16}y^{-4} = (y^2 + \tfrac{1}{4}y^{-2})^2. \text{ Thus,}$$

$$L = \int_1^3 \sqrt{(y^2 + \tfrac{1}{4}y^{-2})^2}\, dy = \int_1^3 \left|y^2 + \tfrac{1}{4}y^{-2}\right| dy = \int_1^3 (y^2 + \tfrac{1}{4}y^{-2})\, dy = \left[\tfrac{1}{3}y^3 - \tfrac{1}{4}y^{-1}\right]_1^3$$

$$= (9 - \tfrac{1}{12}) - (\tfrac{1}{3} - \tfrac{1}{4}) = \tfrac{106}{12} = \tfrac{53}{6}$$

5. (a) $y = \dfrac{2}{x+1}$ $\Rightarrow$ $y' = \dfrac{-2}{(x+1)^2}$ $\Rightarrow$ $1 + (y')^2 = 1 + \dfrac{4}{(x+1)^4}$.

For $0 \leq x \leq 3$, $L = \int_0^3 \sqrt{1 + (y')^2}\, dx = \int_0^3 \sqrt{1 + 4/(x+1)^4}\, dx \approx 3.5121$.

(b) The area of the surface obtained by rotating C about the x-axis is

$$S = \int_0^3 2\pi y\, ds = 2\pi \int_0^3 \frac{2}{x+1} \sqrt{1 + 4/(x+1)^4}\, dx \approx 22.1391.$$

(c) The area of the surface obtained by rotating C about the y-axis is

$$S = \int_0^3 2\pi x\, ds = 2\pi \int_0^3 x\sqrt{1 + 4/(x+1)^4}\, dx \approx 29.8522.$$

7. $y = \sin x$ $\Rightarrow$ $y' = \cos x$ $\Rightarrow$ $1 + (y')^2 = 1 + \cos^2 x$. Let $f(x) = \sqrt{1 + \cos^2 x}$. Then

$L = \int_0^\pi f(x)\, dx \approx S_{10}$

$= \dfrac{(\pi - 0)/10}{3}\left[f(0) + 4f\left(\tfrac{\pi}{10}\right) + 2f\left(\tfrac{2\pi}{10}\right) + 4f\left(\tfrac{3\pi}{10}\right) + 2f\left(\tfrac{4\pi}{10}\right)\right.$

$\left. + 4f\left(\tfrac{5\pi}{10}\right) + 2f\left(\tfrac{6\pi}{10}\right) + 4f\left(\tfrac{7\pi}{10}\right) + 2f\left(\tfrac{8\pi}{10}\right) + 4f\left(\tfrac{9\pi}{10}\right) + f(\pi)\right]$

≈ 3.8202

9. $y = \int_1^x \sqrt{\sqrt{t} - 1}\, dt$ $\Rightarrow$ $dy/dx = \sqrt{\sqrt{x} - 1}$ $\Rightarrow$ $1 + (dy/dx)^2 = 1 + \left(\sqrt{x} - 1\right) = \sqrt{x}$.

Thus, $L = \int_1^{16} \sqrt{\sqrt{x}}\, dx = \int_1^{16} x^{1/4}\, dx = \tfrac{4}{5}\left[x^{5/4}\right]_1^{16} = \tfrac{4}{5}(32 - 1) = \tfrac{124}{5}$.

11. By reasoning as in Example 8.3.1, $\dfrac{a}{2-x} = \dfrac{1}{2}$ $\Rightarrow$ $2a = 2 - x$ and $w = 2(1.5 + a) = 3 + 2a = 3 + 2 - x = 5 - x$.

Thus, $F = \int_0^2 \delta x(5-x)\, dx = \delta\left[\tfrac{5}{2}x^2 - \tfrac{1}{3}x^3\right]_0^2 = \delta(10 - \tfrac{8}{3}) = \tfrac{22}{3}\delta \approx 458$ lb $[\delta \approx 62.5$ lb/ft$^3]$.

13. The area of the triangular region is $A = \tfrac{1}{2}(2)(4) = 4$. An equation of the line is $y = \tfrac{1}{2}x$ or $x = 2y$.

$$\bar{x} = \frac{1}{A}\int_0^2 \tfrac{1}{2}[f(y)]^2\, dy = \frac{1}{4}\int_0^2 \tfrac{1}{2}(2y)^2\, dy = \frac{1}{8}\int_0^2 4y^2\, dy = \frac{1}{8}\left[\tfrac{4}{3}y^3\right]_0^2 = \frac{1}{6}(8) = \frac{4}{3}$$

$$\bar{y} = \frac{1}{A}\int_0^2 y f(y)\, dy = \frac{1}{4}\int_0^2 y(2y)\, dy = \frac{1}{2}\int_0^2 y^2\, dy = \frac{1}{2}\left[\tfrac{1}{3}y^3\right]_0^2 = \frac{1}{6}(8) = \frac{4}{3}$$

The centroid of the region is $\left(\tfrac{4}{3}, \tfrac{4}{3}\right)$.

15. $A = \int_0^4 \left(\sqrt{x} - \frac{1}{2}x\right) dx = \left[\frac{2}{3}x^{3/2} - \frac{1}{4}x^2\right]_0^4 = \frac{16}{3} - 4 = \frac{4}{3}$

$\bar{x} = \frac{1}{A}\int_0^4 x\left(\sqrt{x} - \frac{1}{2}x\right) dx = \frac{3}{4}\int_0^4 \left(x^{3/2} - \frac{1}{2}x^2\right) dx$

$= \frac{3}{4}\left[\frac{2}{5}x^{5/2} - \frac{1}{6}x^3\right]_0^4 = \frac{3}{4}\left(\frac{64}{5} - \frac{64}{6}\right) = \frac{3}{4}\left(\frac{64}{30}\right) = \frac{8}{5}$

$\bar{y} = \frac{1}{A}\int_0^4 \frac{1}{2}\left[\left(\sqrt{x}\right)^2 - \left(\frac{1}{2}x\right)^2\right] dx = \frac{3}{4}\int_0^4 \frac{1}{2}\left(x - \frac{1}{4}x^2\right) dx = \frac{3}{8}\left[\frac{1}{2}x^2 - \frac{1}{12}x^3\right]_0^4 = \frac{3}{8}\left(8 - \frac{16}{3}\right) = \frac{3}{8}\left(\frac{8}{3}\right) = 1$

Thus, the centroid is $(\bar{x}, \bar{y}) = \left(\frac{8}{5}, 1\right)$.

17. The centroid of this circle, $(1, 0)$, travels a distance $2\pi(1)$ when the lamina is rotated about the y-axis. The area of the circle is $\pi(1)^2$. So by the Theorem of Pappus, $V = A(2\pi\bar{x}) = \pi(1)^2 2\pi(1) = 2\pi^2$.

19. $x = 100 \Rightarrow P = 2000 - 0.1(100) - 0.01(100)^2 = 1890$

Consumer surplus $= \int_0^{100}[p(x) - P] dx = \int_0^{100}\left(2000 - 0.1x - 0.01x^2 - 1890\right) dx$

$= \left[110x - 0.05x^2 - \frac{0.01}{3}x^3\right]_0^{100} = 11{,}000 - 500 - \frac{10{,}000}{3} \approx \$7166.67$

21. $f(x) = \begin{cases} \frac{\pi}{20}\sin\left(\frac{\pi}{10}x\right) & \text{if } 0 \leq x \leq 10 \\ 0 & \text{if } x < 0 \text{ or } x > 10 \end{cases}$

(a) $f(x) \geq 0$ for all real numbers x and

$\int_{-\infty}^{\infty} f(x) dx = \int_0^{10} \frac{\pi}{20}\sin\left(\frac{\pi}{10}x\right) dx = \frac{\pi}{20} \cdot \frac{10}{\pi}\left[-\cos\left(\frac{\pi}{10}x\right)\right]_0^{10} = \frac{1}{2}(-\cos\pi + \cos 0) = \frac{1}{2}(1 + 1) = 1$

Therefore, f is a probability density function.

(b) $P(X < 4) = \int_{-\infty}^4 f(x) dx = \int_0^4 \frac{\pi}{20}\sin\left(\frac{\pi}{10}x\right) dx = \frac{1}{2}\left[-\cos\left(\frac{\pi}{10}x\right)\right]_0^4 = \frac{1}{2}\left(-\cos\frac{2\pi}{5} + \cos 0\right)$

$\approx \frac{1}{2}(-0.309017 + 1) \approx 0.3455$

(c) $\mu = \int_{-\infty}^{\infty} xf(x) dx = \int_0^{10} \frac{\pi}{20}x\sin\left(\frac{\pi}{10}x\right) dx$

$= \int_0^{\pi} \frac{\pi}{20} \cdot \frac{10}{\pi}u(\sin u)\left(\frac{10}{\pi}\right) du \quad [u = \frac{\pi}{10}x, du = \frac{\pi}{10} dx]$

$= \frac{5}{\pi}\int_0^{\pi} u\sin u\, du \stackrel{82}{=} \frac{5}{\pi}[\sin u - u\cos u]_0^{\pi} = \frac{5}{\pi}[0 - \pi(-1)] = 5$

This answer is expected because the graph of f is symmetric about the line $x = 5$.

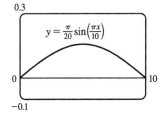

23. (a) The probability density function is $f(t) = \begin{cases} 0 & \text{if } t < 0 \\ \frac{1}{8}e^{-t/8} & \text{if } t \geq 0 \end{cases}$

$P(0 \leq X \leq 3) = \int_0^3 \frac{1}{8}e^{-t/8} dt = \left[-e^{-t/8}\right]_0^3 = -e^{-3/8} + 1 \approx 0.3127$

(b) $P(X > 10) = \int_{10}^{\infty} \frac{1}{8} e^{-t/8} \, dt = \lim_{x \to \infty} \left[-e^{-t/8} \right]_{10}^{x} = \lim_{x \to \infty} (-e^{-x/8} + e^{-10/8}) = 0 + e^{-5/4} \approx 0.2865$

(c) We need to find m such that $P(X \geq m) = \frac{1}{2}$ $\Rightarrow$ $\int_{m}^{\infty} \frac{1}{8} e^{-t/8} \, dt = \frac{1}{2}$ $\Rightarrow$ $\lim_{x \to \infty} \left[-e^{-t/8} \right]_{m}^{x} = \frac{1}{2}$ $\Rightarrow$

$\lim_{x \to \infty} (-e^{-x/8} + e^{-m/8}) = \frac{1}{2}$ $\Rightarrow$ $e^{-m/8} = \frac{1}{2}$ $\Rightarrow$ $-m/8 = \ln \frac{1}{2}$ $\Rightarrow$ $m = -8 \ln \frac{1}{2} = 8 \ln 2 \approx 5.55$ minutes.

PROBLEMS PLUS

1. $x^2 + y^2 \leq 4y \Leftrightarrow x^2 + (y-2)^2 \leq 4$, so S is part of a circle, as shown in the diagram. The area of S is

$$\int_0^1 \sqrt{4y - y^2}\, dy \stackrel{113}{=} \left[\tfrac{y-2}{2}\sqrt{4y-y^2} + 2\cos^{-1}\left(\tfrac{2-y}{2}\right)\right]_0^1 \quad [a=2]$$

$$= -\tfrac{1}{2}\sqrt{3} + 2\cos^{-1}\left(\tfrac{1}{2}\right) - 2\cos^{-1} 1$$

$$= -\tfrac{\sqrt{3}}{2} + 2\left(\tfrac{\pi}{3}\right) - 2(0) = \tfrac{2\pi}{3} - \tfrac{\sqrt{3}}{2}$$

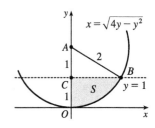

Another method (without calculus): Note that $\theta = \angle CAB = \tfrac{\pi}{3}$, so the area is

$$\text{(area of sector } OAB) - \text{(area of } \triangle ABC) = \tfrac{1}{2}(2^2)\tfrac{\pi}{3} - \tfrac{1}{2}(1)\sqrt{3} = \tfrac{2\pi}{3} - \tfrac{\sqrt{3}}{2}$$

3. (a) The two spherical zones, whose surface areas we will call S_1 and S_2, are generated by rotation about the y-axis of circular arcs, as indicated in the figure. The arcs are the upper and lower portions of the circle $x^2 + y^2 = r^2$ that are obtained when the circle is cut with the line $y = d$. The portion of the upper arc in the first quadrant is sufficient to generate the upper spherical zone. That portion of the arc can be described by the relation $x = \sqrt{r^2 - y^2}$ for $d \leq y \leq r$. Thus, $dx/dy = -y/\sqrt{r^2 - y^2}$ and

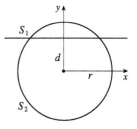

$$ds = \sqrt{1 + \left(\frac{dx}{dy}\right)^2}\, dy = \sqrt{1 + \frac{y^2}{r^2 - y^2}}\, dy = \sqrt{\frac{r^2}{r^2 - y^2}}\, dy = \frac{r\, dy}{\sqrt{r^2 - y^2}}$$

From Formula 8.2.8 we have

$$S_1 = \int_d^r 2\pi x \sqrt{1 + \left(\frac{dx}{dy}\right)^2}\, dy = \int_d^r 2\pi \sqrt{r^2 - y^2}\, \frac{r\, dy}{\sqrt{r^2 - y^2}} = \int_d^r 2\pi r\, dy = 2\pi r(r - d)$$

Similarly, we can compute $S_2 = \int_{-r}^d 2\pi x \sqrt{1 + (dx/dy)^2}\, dy = \int_{-r}^d 2\pi r\, dy = 2\pi r(r + d)$. Note that $S_1 + S_2 = 4\pi r^2$, the surface area of the entire sphere.

(b) $r = 3960$ mi and $d = r(\sin 75°) \approx 3825$ mi, so the surface area of the Arctic Ocean is about $2\pi r(r-d) \approx 2\pi(3960)(135) \approx 3.36 \times 10^6$ mi^2.

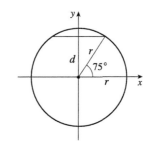

(c) The area on the sphere lies between planes $y = y_1$ and $y = y_2$, where $y_2 - y_1 = h$. Thus, we compute the surface area on the sphere to be $S = \int_{y_1}^{y_2} 2\pi x \sqrt{1 + \left(\dfrac{dx}{dy}\right)^2}\, dy = \int_{y_1}^{y_2} 2\pi r\, dy = 2\pi r(y_2 - y_1) = 2\pi rh$.

This equals the lateral area of a cylinder of radius r and height h, since such a cylinder is obtained by rotating the line $x = r$ about the y-axis, so the surface area of the cylinder between the planes $y = y_1$ and $y = y_2$ is

$$A = \int_{y_1}^{y_2} 2\pi x \sqrt{1 + \left(\dfrac{dx}{dy}\right)^2}\, dy = \int_{y_1}^{y_2} 2\pi r \sqrt{1 + 0^2}\, dy$$

$$= 2\pi r y \Big|_{y=y_1}^{y_2} = 2\pi r(y_2 - y_1) = 2\pi rh$$

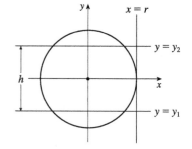

(d) $h = 2r \sin 23.45° \approx 3152$ mi, so the surface area of the Torrid Zone is $2\pi rh \approx 2\pi(3960)(3152) \approx 7.84 \times 10^7$ mi^2.

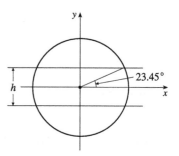

5. (a) Choose a vertical x-axis pointing downward with its origin at the surface. In order to calculate the pressure at depth z, consider n subintervals of the interval $[0, z]$ by points x_i and choose a point $x_i^* \in [x_{i-1}, x_i]$ for each i. The thin layer of water lying between depth x_{i-1} and depth x_i has a density of approximately $\rho(x_i^*)$, so the weight of a piece of that layer with unit cross-sectional area is $\rho(x_i^*)g\,\Delta x$. The total weight of a column of water extending from the surface to depth z (with unit cross-sectional area) would be approximately $\sum_{i=1}^{n} \rho(x_i^*)g\,\Delta x$. The estimate becomes exact if we take the limit as $n \to \infty$; weight (or force) per unit area at depth z is $W = \lim_{n\to\infty} \sum_{i=1}^{n} \rho(x_i^*)g\,\Delta x$. In other words, $P(z) = \int_0^z \rho(x)g\,dx$.

More generally, if we make no assumptions about the location of the origin, then $P(z) = P_0 + \int_0^z \rho(x)g\,dx$, where P_0 is the pressure at $x = 0$. Differentiating, we get $dP/dz = \rho(z)g$.

(b)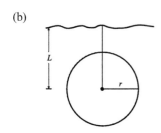

$F = \int_{-r}^{r} P(L + x) \cdot 2\sqrt{r^2 - x^2}\, dx$

$= \int_{-r}^{r} \left(P_0 + \int_0^{L+x} \rho_0 e^{z/H} g\, dz\right) \cdot 2\sqrt{r^2 - x^2}\, dx$

$= P_0 \int_{-r}^{r} 2\sqrt{r^2 - x^2}\, dx + \rho_0 g H \int_{-r}^{r} \left(e^{(L+x)/H} - 1\right) \cdot 2\sqrt{r^2 - x^2}\, dx$

$= (P_0 - \rho_0 g H) \int_{-r}^{r} 2\sqrt{r^2 - x^2}\, dx + \rho_0 g H \int_{-r}^{r} e^{(L+x)/H} \cdot 2\sqrt{r^2 - x^2}\, dx$

$= (P_0 - \rho_0 g H)(\pi r^2) + \rho_0 g H e^{L/H} \int_{-r}^{r} e^{x/H} \cdot 2\sqrt{r^2 - x^2}\, dx$

7. To find the height of the pyramid, we use similar triangles. The first figure shows a cross-section of the pyramid passing through the top and through two opposite corners of the square base. Now $|BD| = b$, since it is a radius of the sphere, which has diameter $2b$ since it is tangent to the opposite sides of the square base. Also, $|AD| = b$ since $\triangle ADB$ is isosceles. So the height is $|AB| = \sqrt{b^2 + b^2} = \sqrt{2}\,b$.

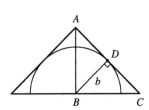

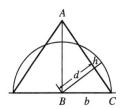

We first observe that the shared volume is equal to half the volume of the sphere, minus the sum of the four equal volumes (caps of the sphere) cut off by the triangular faces of the pyramid. See Exercise 5.2.61 for a derivation of the formula for the volume of a cap of a sphere. To use the formula, we need to find the perpendicular distance h of each triangular face from the surface of the sphere. We first find the distance d from the center of the sphere to one of the triangular faces. The third figure shows a cross-section of the pyramid through the top and through the midpoints of opposite sides of the square base. From similar triangles we find that

$$\frac{d}{b} = \frac{|AB|}{|AC|} = \frac{\sqrt{2}\,b}{\sqrt{b^2 + \left(\sqrt{2}\,b\right)^2}} \;\Rightarrow\; d = \frac{\sqrt{2}\,b^2}{\sqrt{3b^2}} = \frac{\sqrt{6}}{3}b$$

So $h = b - d = b - \frac{\sqrt{6}}{3}b = \frac{3-\sqrt{6}}{3}b$. So, using the formula $V = \pi h^2(r - h/3)$ from Exercise 5.2.61 with $r = b$, we find that the volume of each of the caps is $\pi\left(\frac{3-\sqrt{6}}{3}b\right)^2\left(b - \frac{3-\sqrt{6}}{3\cdot 3}b\right) = \frac{15-6\sqrt{6}}{9}\cdot\frac{6+\sqrt{6}}{9}\pi b^3 = \left(\frac{2}{3} - \frac{7}{27}\sqrt{6}\right)\pi b^3$. So, using our first observation, the shared volume is $V = \frac{1}{2}\left(\frac{4}{3}\pi b^3\right) - 4\left(\frac{2}{3} - \frac{7}{27}\sqrt{6}\right)\pi b^3 = \left(\frac{28}{27}\sqrt{6} - 2\right)\pi b^3$.

9. We can assume that the cut is made along a vertical line $x = b > 0$, that the disk's boundary is the circle $x^2 + y^2 = 1$, and that the center of mass of the smaller piece (to the right of $x = b$) is $\left(\frac{1}{2}, 0\right)$. We wish to find b to two decimal places. We have $\dfrac{1}{2} = \bar{x} = \dfrac{\int_b^1 x \cdot 2\sqrt{1-x^2}\,dx}{\int_b^1 2\sqrt{1-x^2}\,dx}$. Evaluating the

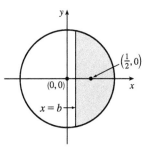

numerator gives us $-\int_b^1 (1-x^2)^{1/2}(-2x)\,dx = -\frac{2}{3}\left[(1-x^2)^{3/2}\right]_b^1 = -\frac{2}{3}\left[0 - (1-b^2)^{3/2}\right] = \frac{2}{3}(1-b^2)^{3/2}$.

Using Formula 30 in the table of integrals, we find that the denominator is

$\left[x\sqrt{1-x^2} + \sin^{-1}x\right]_b^1 = \left(0 + \frac{\pi}{2}\right) - \left(b\sqrt{1-b^2} + \sin^{-1}b\right)$. Thus, we have $\dfrac{1}{2} = \bar{x} = \dfrac{\frac{2}{3}(1-b^2)^{3/2}}{\frac{\pi}{2} - b\sqrt{1-b^2} - \sin^{-1}b}$, or,

equivalently, $\frac{2}{3}(1-b^2)^{3/2} = \frac{\pi}{4} - \frac{1}{2}b\sqrt{1-b^2} - \frac{1}{2}\sin^{-1}b$. Solving this equation numerically with a calculator or CAS, we obtain $b \approx 0.138173$, or $b = 0.14$ m to two decimal places.

11. If $h = L$, then $P = \dfrac{\text{area under } y = L\sin\theta}{\text{area of rectangle}} = \dfrac{\int_0^\pi L\sin\theta\, d\theta}{\pi L} = \dfrac{[-\cos\theta]_0^\pi}{\pi} = \dfrac{-(-1)+1}{\pi} = \dfrac{2}{\pi}$.

If $h = \tfrac{1}{2}L$, we replace L with $\tfrac{1}{2}L$ in the above calculation to get $P = \tfrac{1}{2}\left(\tfrac{2}{\pi}\right) = \tfrac{1}{\pi}$.

13. Solve for y: $x^2 + (x+y+1)^2 = 1 \;\Rightarrow\; (x+y+1)^2 = 1 - x^2 \;\Rightarrow\; x+y+1 = \pm\sqrt{1-x^2} \;\Rightarrow\;$
$y = -x - 1 \pm \sqrt{1-x^2}$.

$A = \displaystyle\int_{-1}^{1}\left[\left(-x-1+\sqrt{1-x^2}\right) - \left(-x-1-\sqrt{1-x^2}\right)\right]dx$

$= \displaystyle\int_{-1}^{1} 2\sqrt{1-x^2}\, dx = 2\left(\tfrac{\pi}{2}\right) \quad \begin{bmatrix}\text{area of}\\\text{semicircle}\end{bmatrix} = \pi$

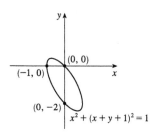

$\bar{x} = \dfrac{1}{A}\displaystyle\int_{-1}^{1} x\cdot 2\sqrt{1-x^2}\, dx = 0 \quad$ [odd integrand]

$\bar{y} = \dfrac{1}{A}\displaystyle\int_{-1}^{1}\dfrac{1}{2}\left[\left(-x-1+\sqrt{1-x^2}\right)^2 - \left(-x-1-\sqrt{1-x^2}\right)^2\right]dx = \dfrac{1}{\pi}\displaystyle\int_{-1}^{1}\dfrac{1}{2}\left(-4x\sqrt{1-x^2} - 4\sqrt{1-x^2}\right)dx$

$= -\dfrac{2}{\pi}\displaystyle\int_{-1}^{1}\left(x\sqrt{1-x^2} + \sqrt{1-x^2}\right)dx = -\dfrac{2}{\pi}\displaystyle\int_{-1}^{1} x\sqrt{1-x^2}\, dx - \dfrac{2}{\pi}\displaystyle\int_{-1}^{1}\sqrt{1-x^2}\, dx$

$= -\dfrac{2}{\pi}(0) \quad \text{[odd integrand]} \quad -\dfrac{2}{\pi}\left(\dfrac{\pi}{2}\right) \quad \begin{bmatrix}\text{area of}\\\text{semicircle}\end{bmatrix} = -1$

Thus, as expected, the centroid is $(\bar{x}, \bar{y}) = (0, -1)$. We might expect this result since the centroid of an ellipse is located at its center.

9 □ DIFFERENTIAL EQUATIONS

9.1 Modeling with Differential Equations

1. $\dfrac{dr}{dt} = \dfrac{k}{r}$, where k is a proportionality constant.

3. $\dfrac{dv}{dt} = k(M - v)$, where k is a proportionality constant.

5. The number of individuals who have *not* heard about the product is $N - y$. Thus, $\dfrac{dy}{dt} = k(N - y)$, where k is a proportionality constant.

7. $y = \tfrac{2}{3}e^x + e^{-2x} \;\Rightarrow\; y' = \tfrac{2}{3}e^x - 2e^{-2x}$. To determine whether y is a solution of the differential equation, we will substitute the expressions for y and y' in the left-hand side of the equation and see if the left-hand side is equal to the right-hand side.

   LHS $= y' + 2y = \tfrac{2}{3}e^x - 2e^{-2x} + 2(\tfrac{2}{3}e^x + e^{-2x}) = \tfrac{2}{3}e^x - 2e^{-2x} + \tfrac{4}{3}e^x + 2e^{-2x} = \tfrac{6}{3}e^x = 2e^x =$ RHS

   so $y = \tfrac{2}{3}e^x + e^{-2x}$ **is** a solution of the differential equation.

9. $y = \sqrt{x} = x^{1/2} \;\Rightarrow\; y' = \tfrac{1}{2}x^{-1/2}$.

   LHS $= xy' - y = x\left(\tfrac{1}{2}x^{-1/2}\right) - x^{1/2} = \tfrac{1}{2}x^{1/2} - x^{1/2} = -\tfrac{1}{2}x^{1/2} \neq 0$, so $y = \sqrt{x}$ **is not** a solution of the differential equation.

11. $y = x^3 \;\Rightarrow\; y' = 3x^2 \;\Rightarrow\; y'' = 6x$.

    LHS $= x^2 y'' - 6y = x^2 \cdot 6x - 6 \cdot x^3 = 6x^3 - 6x^3 = 0 =$ RHS, so $y = x^3$ **is** a solution of the differential equation.

13. $y = -t\cos t - t \;\Rightarrow\; dy/dt = -t(-\sin t) + \cos t(-1) - 1 = t\sin t - \cos t - 1$.

    LHS $= t\dfrac{dy}{dt} = t(t\sin t - \cos t - 1) = t^2 \sin t - t\cos t - t = t^2 \sin t + (-t\cos t - t) = t^2 \sin t + y =$ RHS,

    so y **is** a solution of the differential equation. Also, $y(\pi) = -\pi\cos\pi - \pi = -\pi(-1) - \pi = \pi - \pi = 0$, so the initial condition, $y(\pi) = 0$, is satisfied.

15. (a) $y = e^{rx} \;\Rightarrow\; y' = re^{rx} \;\Rightarrow\; y'' = r^2 e^{rx}$. Substituting these expressions into the differential equation $2y'' + y' - y = 0$, we get $2r^2 e^{rx} + re^{rx} - e^{rx} = 0 \;\Rightarrow\; (2r^2 + r - 1)e^{rx} = 0 \;\Rightarrow\;$
    $(2r - 1)(r + 1) = 0$ [since e^{rx} is never zero] $\;\Rightarrow\; r = \tfrac{1}{2}$ or -1.

    (b) Let $r_1 = \tfrac{1}{2}$ and $r_2 = -1$, so we need to show that every member of the family of functions $y = ae^{x/2} + be^{-x}$ is a solution of the differential equation $2y'' + y' - y = 0$.

    [continued]

$y = ae^{x/2} + be^{-x} \Rightarrow y' = \frac{1}{2}ae^{x/2} - be^{-x} \Rightarrow y'' = \frac{1}{4}ae^{x/2} + be^{-x}$.

$\text{LHS} = 2y'' + y' - y = 2\left(\frac{1}{4}ae^{x/2} + be^{-x}\right) + \left(\frac{1}{2}ae^{x/2} - be^{-x}\right) - (ae^{x/2} + be^{-x})$

$= \frac{1}{2}ae^{x/2} + 2be^{-x} + \frac{1}{2}ae^{x/2} - be^{-x} - ae^{x/2} - be^{-x}$

$= \left(\frac{1}{2}a + \frac{1}{2}a - a\right)e^{x/2} + (2b - b - b)e^{-x}$

$= 0 = \text{RHS}$

17. (a) $y = \sin x \Rightarrow y' = \cos x \Rightarrow y'' = -\sin x$.

$\text{LHS} = y'' + y = -\sin x + \sin x = 0 \neq \sin x$, so $y = \sin x$ **is not** a solution of the differential equation.

(b) $y = \cos x \Rightarrow y' = -\sin x \Rightarrow y'' = -\cos x$.

$\text{LHS} = y'' + y = -\cos x + \cos x = 0 \neq \sin x$, so $y = \cos x$ **is not** a solution of the differential equation.

(c) $y = \frac{1}{2}x\sin x \Rightarrow y' = \frac{1}{2}(x\cos x + \sin x) \Rightarrow y'' = \frac{1}{2}(-x\sin x + \cos x + \cos x)$.

$\text{LHS} = y'' + y = \frac{1}{2}(-x\sin x + 2\cos x) + \frac{1}{2}x\sin x = \cos x \neq \sin x$, so $y = \frac{1}{2}x\sin x$ **is not** a solution of the differential equation.

(d) $y = -\frac{1}{2}x\cos x \Rightarrow y' = -\frac{1}{2}(-x\sin x + \cos x) \Rightarrow y'' = -\frac{1}{2}(-x\cos x - \sin x - \sin x)$.

$\text{LHS} = y'' + y = -\frac{1}{2}(-x\cos x - 2\sin x) + \left(-\frac{1}{2}x\cos x\right) = \sin x = \text{RHS}$, so $y = -\frac{1}{2}x\cos x$ **is** a solution of the differential equation.

19. (a) Since the derivative $y' = -y^2$ is always negative (or 0, if $y = 0$), the function y must be decreasing (or equal to 0) on any interval on which it is defined.

(b) $y = \dfrac{1}{x+C} \Rightarrow y' = -\dfrac{1}{(x+C)^2}$. $\text{LHS} = y' = -\dfrac{1}{(x+C)^2} = -\left(\dfrac{1}{x+C}\right)^2 = -y^2 = \text{RHS}$

(c) $y = 0$ is a solution of $y' = -y^2$ that is not a member of the family in part (b).

(d) If $y(x) = \dfrac{1}{x+C}$, then $y(0) = \dfrac{1}{0+C} = \dfrac{1}{C}$. Since $y(0) = 0.5$, $\dfrac{1}{C} = \dfrac{1}{2} \Rightarrow C = 2$, so $y = \dfrac{1}{x+2}$.

21. (a) $\dfrac{dP}{dt} = 1.2P\left(1 - \dfrac{P}{4200}\right)$. Now $\dfrac{dP}{dt} > 0 \Rightarrow 1 - \dfrac{P}{4200} > 0$ [assuming that $P > 0$] $\Rightarrow \dfrac{P}{4200} < 1 \Rightarrow P < 4200 \Rightarrow$ the population is increasing for $0 < P < 4200$.

(b) $\dfrac{dP}{dt} < 0 \Rightarrow P > 4200$

(c) $\dfrac{dP}{dt} = 0 \Rightarrow P = 4200$ or $P = 0$

23. (a) This function is increasing *and* also decreasing. But $dy/dt = e^t(y-1)^2 \geq 0$ for all t, implying that the graph of the solution of the differential equation cannot be decreasing on any interval.

(b) When $y = 1$, $dy/dt = 0$, but the graph does not have a horizontal tangent line.

25. (a) $y' = 1 + x^2 + y^2 \geq 1$ and $y' \to \infty$ as $x \to \infty$. The only curve satisfying these conditions is labeled III.

(b) $y' = xe^{-x^2-y^2} > 0$ if $x > 0$ and $y' < 0$ if $x < 0$. The only curve with negative tangent slopes when $x < 0$ and positive tangent slopes when $x > 0$ is labeled I.

(c) $y' = \dfrac{1}{1 + e^{x^2+y^2}} > 0$ and $y' \to 0$ as $x \to \infty$. The only curve satisfying these conditions is labeled IV.

(d) $y' = \sin(xy)\cos(xy) = 0$ if $y = 0$, which is the solution graph labeled II.

27. (a) P increases most rapidly at the beginning, since there are usually many simple, easily-learned sub-skills associated with learning a skill. As t increases, we would expect dP/dt to remain positive, but decrease. This is because as time progresses, the only points left to learn are the more difficult ones.

(b) $\dfrac{dP}{dt} = k(M - P)$ is always positive, so the level of performance P is increasing. As P gets close to M, dP/dt gets close to 0; that is, the performance levels off, as explained in part (a).

(c)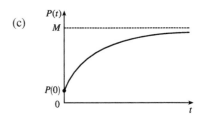

29. If $c(t) = c_s\left(1 - e^{-\alpha t^{1-b}}\right) = c_s - c_s e^{-\alpha t^{1-b}}$ for $t > 0$, where $k > 0$, $c_s > 0$, $0 < b < 1$, and $\alpha = k/(1-b)$, then

$$\dfrac{dc}{dt} = c_s\left[0 - e^{-\alpha t^{1-b}} \cdot \dfrac{d}{dt}\left(-\alpha t^{1-b}\right)\right] = -c_s e^{-\alpha t^{1-b}} \cdot (-\alpha)(1-b)t^{-b} = \dfrac{\alpha(1-b)}{t^b} c_s e^{-\alpha t^{1-b}} = \dfrac{k}{t^b}(c_s - c).$$

The equation for c indicates that as t increases, c approaches c_s. The differential equation indicates that as t increases, the rate of increase of c decreases steadily and approaches 0 as c approaches c_s.

9.2 Direction Fields and Euler's Method

1. (a)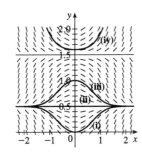

(b) It appears that the constant functions $y = 0.5$ and $y = 1.5$ are equilibrium solutions. Note that these two values of y satisfy the given differential equation $y' = x\cos\pi y$.

3. $y' = 2 - y$. The slopes at each point are independent of x, so the slopes are the same along each line parallel to the x-axis. Thus, III is the direction field for this equation. Note that for $y = 2$, $y' = 0$.

5. $y' = x + y - 1 = 0$ on the line $y = -x + 1$. Direction field IV satisfies this condition. Notice also that on the line $y = -x$ we have $y' = -1$, which is true in IV.

7.

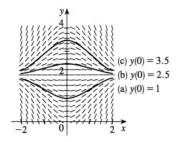

9.

x	y	$y' = \frac{1}{2}y$
0	0	0
0	1	0.5
0	2	1
0	−3	−1.5
0	−2	−1

Note that for $y = 0$, $y' = 0$. The three solution curves sketched go through $(0, 0)$, $(0, 1)$, and $(0, -1)$.

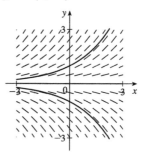

11.

x	y	$y' = y - 2x$
−2	−2	2
−2	2	6
2	2	−2
2	−2	−6

Note that $y' = 0$ for any point on the line $y = 2x$. The slopes are positive to the left of the line and negative to the right of the line. The solution curve in the graph passes through $(1, 0)$.

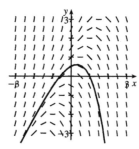

13.

x	y	$y' = y + xy$
0	±2	±2
1	±2	±4
−3	±2	∓4

Note that $y' = y(x + 1) = 0$ for any point on $y = 0$ or on $x = -1$. The slopes are positive when the factors y and $x + 1$ have the same sign and negative when they have opposite signs. The solution curve in the graph passes through $(0, 1)$.

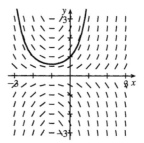

15. $y' = x^2 y - \frac{1}{2}y^2$ and $y(0) = 1$.

In Maple, use the following commands to obtain a similar figure.
```
with(DETools):
ODE:=diff(y(x),x)=x^2*y(x)-(1/2)*y(x)^2;
ivs:=[y(0)=1];
DEplot({ODE},y(x),x=-3..2,y=0..4,ivs,linecolor=black);
```

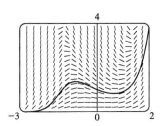

17.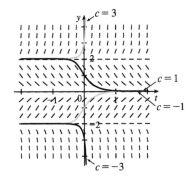

The direction field is for the differential equation $y' = y^3 - 4y$.

$L = \lim\limits_{t \to \infty} y(t)$ exists for $-2 \leq c \leq 2$;

$L = \pm 2$ for $c = \pm 2$ and $L = 0$ for $-2 < c < 2$.

For other values of c, L does not exist.

19. (a) $y' = F(x,y) = y$ and $y(0) = 1 \Rightarrow x_0 = 0, y_0 = 1$.

  (i) $h = 0.4$ and $y_1 = y_0 + hF(x_0, y_0) \Rightarrow y_1 = 1 + 0.4 \cdot 1 = 1.4$. $x_1 = x_0 + h = 0 + 0.4 = 0.4$,
  so $y_1 = y(0.4) = 1.4$.

  (ii) $h = 0.2 \Rightarrow x_1 = 0.2$ and $x_2 = 0.4$, so we need to find y_2.
  $y_1 = y_0 + hF(x_0, y_0) = 1 + 0.2y_0 = 1 + 0.2 \cdot 1 = 1.2$,
  $y_2 = y_1 + hF(x_1, y_1) = 1.2 + 0.2y_1 = 1.2 + 0.2 \cdot 1.2 = 1.44$.

  (iii) $h = 0.1 \Rightarrow x_4 = 0.4$, so we need to find y_4. $y_1 = y_0 + hF(x_0, y_0) = 1 + 0.1y_0 = 1 + 0.1 \cdot 1 = 1.1$,
  $y_2 = y_1 + hF(x_1, y_1) = 1.1 + 0.1y_1 = 1.1 + 0.1 \cdot 1.1 = 1.21$,
  $y_3 = y_2 + hF(x_2, y_2) = 1.21 + 0.1y_2 = 1.21 + 0.1 \cdot 1.21 = 1.331$,
  $y_4 = y_3 + hF(x_3, y_3) = 1.331 + 0.1y_3 = 1.331 + 0.1 \cdot 1.331 = 1.4641$.

(b)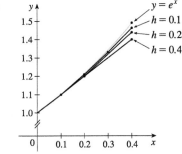

We see that the estimates are underestimates since they are all below the graph of $y = e^x$.

(c) (i) For $h = 0.4$: (exact value) − (approximate value) $= e^{0.4} - 1.4 \approx 0.0918$

  (ii) For $h = 0.2$: (exact value) − (approximate value) $= e^{0.4} - 1.44 \approx 0.0518$

  (iii) For $h = 0.1$: (exact value) − (approximate value) $= e^{0.4} - 1.4641 \approx 0.0277$

Each time the step size is halved, the error estimate also appears to be halved (approximately).

21. $h = 0.5$, $x_0 = 1$, $y_0 = 0$, and $F(x,y) = y - 2x$.

Note that $x_1 = x_0 + h = 1 + 0.5 = 1.5$, $x_2 = 2$, and $x_3 = 2.5$.

$y_1 = y_0 + hF(x_0, y_0) = 0 + 0.5F(1, 0) = 0.5[0 - 2(1)] = -1$.

$y_2 = y_1 + hF(x_1, y_1) = -1 + 0.5F(1.5, -1) = -1 + 0.5[-1 - 2(1.5)] = -3$.

$y_3 = y_2 + hF(x_2, y_2) = -3 + 0.5F(2, -3) = -3 + 0.5[-3 - 2(2)] = -6.5$.

$y_4 = y_3 + hF(x_3, y_3) = -6.5 + 0.5F(2.5, -6.5) = -6.5 + 0.5[-6.5 - 2(2.5)] = -12.25$.

23. $h = 0.1$, $x_0 = 0$, $y_0 = 1$, and $F(x,y) = y + xy$.

Note that $x_1 = x_0 + h = 0 + 0.1 = 0.1$, $x_2 = 0.2$, $x_3 = 0.3$, and $x_4 = 0.4$.

$y_1 = y_0 + hF(x_0, y_0) = 1 + 0.1F(0, 1) = 1 + 0.1[1 + (0)(1)] = 1.1$.

$y_2 = y_1 + hF(x_1, y_1) = 1.1 + 0.1F(0.1, 1.1) = 1.1 + 0.1[1.1 + (0.1)(1.1)] = 1.221$.

$y_3 = y_2 + hF(x_2, y_2) = 1.221 + 0.1F(0.2, 1.221) = 1.221 + 0.1[1.221 + (0.2)(1.221)] = 1.36752$.

$y_4 = y_3 + hF(x_3, y_3) = 1.36752 + 0.1F(0.3, 1.36752) = 1.36752 + 0.1[1.36752 + (0.3)(1.36752)]$
$= 1.5452976$.

$y_5 = y_4 + hF(x_4, y_4) = 1.5452976 + 0.1F(0.4, 1.5452976)$
$= 1.5452976 + 0.1[1.5452976 + (0.4)(1.5452976)] = 1.761639264$.

Thus, $y(0.5) \approx 1.7616$.

25. (a) $dy/dx + 3x^2 y = 6x^2 \Rightarrow y' = 6x^2 - 3x^2 y$. Store this expression in Y_1 and use the following simple program to evaluate $y(1)$ for each part, using $H = h = 1$ and $N = 1$ for part (i), $H = 0.1$ and $N = 10$ for part (ii), and so forth.

$h \to H$: $0 \to X$: $3 \to Y$:

For(I, 1, N): $Y + H \times Y_1 \to Y$: $X + H \to X$:

End(loop):

Display Y. [To see all iterations, include this statement in the loop.]

(i) $H = 1, N = 1 \Rightarrow y(1) = 3$

(ii) $H = 0.1, N = 10 \Rightarrow y(1) \approx 2.3928$

(iii) $H = 0.01, N = 100 \Rightarrow y(1) \approx 2.3701$

(iv) $H = 0.001, N = 1000 \Rightarrow y(1) \approx 2.3681$

(b) $y = 2 + e^{-x^3} \Rightarrow y' = -3x^2 e^{-x^3}$

LHS $= y' + 3x^2 y = -3x^2 e^{-x^3} + 3x^2\left(2 + e^{-x^3}\right) = -3x^2 e^{-x^3} + 6x^2 + 3x^2 e^{-x^3} = 6x^2 =$ RHS

$y(0) = 2 + e^{-0} = 2 + 1 = 3$

(c) The exact value of $y(1)$ is $2 + e^{-1^3} = 2 + e^{-1}$.

(i) For $h = 1$: (exact value) − (approximate value) $= 2 + e^{-1} - 3 \approx -0.6321$

(ii) For $h = 0.1$: (exact value) − (approximate value) $= 2 + e^{-1} - 2.3928 \approx -0.0249$

(iii) For $h = 0.01$: (exact value) − (approximate value) $= 2 + e^{-1} - 2.3701 \approx -0.0022$

(iv) For $h = 0.001$: (exact value) − (approximate value) $= 2 + e^{-1} - 2.3681 \approx -0.0002$

In (ii)–(iv), it seems that when the step size is divided by 10, the error estimate is also divided by 10 (approximately).

27. (a) $R\dfrac{dQ}{dt} + \dfrac{1}{C}Q = E(t)$ becomes $5Q' + \dfrac{1}{0.05}Q = 60$ or $Q' + 4Q = 12$.

(b) From the graph, it appears that the limiting value of the charge Q is about 3.

(c) If $Q' = 0$, then $4Q = 12 \;\Rightarrow\; Q = 3$ is an equilibrium solution.

(d)

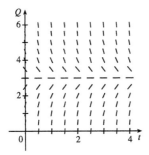
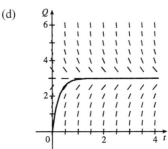

(e) $Q' + 4Q = 12 \;\Rightarrow\; Q' = 12 - 4Q$. Now $Q(0) = 0$, so $t_0 = 0$ and $Q_0 = 0$.

$$Q_1 = Q_0 + hF(t_0, Q_0) = 0 + 0.1(12 - 4 \cdot 0) = 1.2$$
$$Q_2 = Q_1 + hF(t_1, Q_1) = 1.2 + 0.1(12 - 4 \cdot 1.2) = 1.92$$
$$Q_3 = Q_2 + hF(t_2, Q_2) = 1.92 + 0.1(12 - 4 \cdot 1.92) = 2.352$$
$$Q_4 = Q_3 + hF(t_3, Q_3) = 2.352 + 0.1(12 - 4 \cdot 2.352) = 2.6112$$
$$Q_5 = Q_4 + hF(t_4, Q_4) = 2.6112 + 0.1(12 - 4 \cdot 2.6112) = 2.76672$$

Thus, $Q_5 = Q(0.5) \approx 2.77$ C.

9.3 Separable Equations

1. $\dfrac{dy}{dx} = 3x^2 y^2 \;\Rightarrow\; \dfrac{dy}{y^2} = 3x^2\,dx \;\; [y \ne 0] \;\Rightarrow\; \int y^{-2}\,dy = \int 3x^2\,dx \;\Rightarrow\; -y^{-1} = x^3 + C \;\Rightarrow\;$

$\dfrac{-1}{y} = x^3 + C \;\Rightarrow\; y = \dfrac{-1}{x^3 + C}. \;\; y = 0$ is also a solution.

3. $\dfrac{dy}{dx} = x\sqrt{y} \;\Rightarrow\; \dfrac{dy}{\sqrt{y}} = x\,dx \;\; [y \ne 0] \;\Rightarrow\; \int y^{-1/2}\,dy = \int x\,dx \;\Rightarrow\; 2y^{1/2} = \tfrac{1}{2}x^2 + K \;\Rightarrow\;$

$\sqrt{y} = \tfrac{1}{4}x^2 + \tfrac{1}{2}K \;\Rightarrow\; y = \left(\tfrac{1}{4}x^2 + C\right)^2$, where $C = \tfrac{1}{2}K$. $y = 0$ is also a solution.

5. $xy\,y' = x^2 + 1 \;\Rightarrow\; xy\dfrac{dy}{dx} = x^2 + 1 \;\Rightarrow\; y\,dy = \dfrac{x^2 + 1}{x}\,dx \;\; [x \ne 0] \;\Rightarrow\; \int y\,dy = \int \left(x + \dfrac{1}{x}\right) dx \;\Rightarrow\;$

$\tfrac{1}{2}y^2 = \tfrac{1}{2}x^2 + \ln|x| + K \;\Rightarrow\; y^2 = x^2 + 2\ln|x| + 2K \;\Rightarrow\; y = \pm\sqrt{x^2 + 2\ln|x| + C}$, where $C = 2K$.

454 ☐ **CHAPTER 9** DIFFERENTIAL EQUATIONS

7. $(e^y - 1) y' = 2 + \cos x \Rightarrow (e^y - 1) \dfrac{dy}{dx} = 2 + \cos x \Rightarrow (e^y - 1) \, dy = (2 + \cos x) \, dx \Rightarrow$

$\int (e^y - 1) \, dy = \int (2 + \cos x) \, dx \Rightarrow e^y - y = 2x + \sin x + C$. We cannot solve explicitly for y.

9. $\dfrac{dp}{dt} = t^2 p - p + t^2 - 1 = p(t^2 - 1) + 1(t^2 - 1) = (p+1)(t^2 - 1) \Rightarrow \dfrac{1}{p+1} \, dp = (t^2 - 1) \, dt \Rightarrow$

$\int \dfrac{1}{p+1} \, dp = \int (t^2 - 1) \, dt \Rightarrow \ln|p+1| = \tfrac{1}{3}t^3 - t + C \Rightarrow |p+1| = e^{t^3/3 - t + C} \Rightarrow p + 1 = \pm e^C e^{t^3/3 - t} \Rightarrow$

$p = Ke^{t^3/3 - t} - 1$, where $K = \pm e^C$. Since $p = -1$ is also a solution, K can equal 0, and hence, K can be any real number.

11. $\dfrac{d\theta}{dt} = \dfrac{t \sec \theta}{\theta e^{t^2}} \Rightarrow \theta \cos \theta \, d\theta = t e^{-t^2} \, dt \Rightarrow \int \theta \cos \theta \, d\theta = \int t e^{-t^2} \, dt \Rightarrow$

$\theta \sin \theta + \cos \theta = -\tfrac{1}{2} e^{-t^2} + C$ [by parts]. We cannot solve explicitly for θ.

13. $\dfrac{dy}{dx} = xe^y \Rightarrow e^{-y} \, dy = x \, dx \Rightarrow \int e^{-y} \, dy = \int x \, dx \Rightarrow -e^{-y} = \tfrac{1}{2}x^2 + C$.

$y(0) = 0 \Rightarrow -e^{-0} = \tfrac{1}{2}(0)^2 + C \Rightarrow C = -1$, so $-e^{-y} = \tfrac{1}{2}x^2 - 1 \Rightarrow e^{-y} = -\tfrac{1}{2}x^2 + 1 \Rightarrow$

$-y = \ln\left(1 - \tfrac{1}{2}x^2\right) \Rightarrow y = -\ln\left(1 - \tfrac{1}{2}x^2\right)$.

15. $\dfrac{dA}{dr} = Ab^2 \cos br \Rightarrow \dfrac{dA}{A} = b^2 \cos br \, dr \Rightarrow \int \dfrac{1}{A} \, dA = \int b^2 \cos br \, dr \Rightarrow \ln|A| = b \sin br + C$.

$A(0) = b^3 \Rightarrow \ln|b^3| = b \sin 0 + C \Rightarrow C = \ln|b^3|$, so $\ln|A| = b \sin br + \ln|b^3| \Rightarrow$

$|A| = e^{b \sin br + \ln|b^3|} = e^{b \sin br} e^{\ln|b^3|} = |b^3| e^{b \sin br} \Rightarrow A = \pm b^3 e^{b \sin br}$. Since $A(0) = b^3$, the solution is

$A = b^3 e^{b \sin br}$.

17. $\dfrac{du}{dt} = \dfrac{2t + \sec^2 t}{2u}$, $u(0) = -5$. $\int 2u \, du = \int (2t + \sec^2 t) \, dt \Rightarrow u^2 = t^2 + \tan t + C$,

where $[u(0)]^2 = 0^2 + \tan 0 + C \Rightarrow C = (-5)^2 = 25$. Therefore, $u^2 = t^2 + \tan t + 25$, so $u = \pm\sqrt{t^2 + \tan t + 25}$.

Since $u(0) = -5 < 0$, we must have $u = -\sqrt{t^2 + \tan t + 25}$.

19. $x \ln x = y\left(1 + \sqrt{3 + y^2}\right) y'$, $y(1) = 1$. $\int x \ln x \, dx = \int \left(y + y\sqrt{3 + y^2}\right) dy \Rightarrow \tfrac{1}{2}x^2 \ln x - \int \tfrac{1}{2}x \, dx$

[use parts with $u = \ln x$, $dv = x \, dx$] $= \tfrac{1}{2}y^2 + \tfrac{1}{3}(3 + y^2)^{3/2} \Rightarrow \tfrac{1}{2}x^2 \ln x - \tfrac{1}{4}x^2 + C = \tfrac{1}{2}y^2 + \tfrac{1}{3}(3 + y^2)^{3/2}$.

Now $y(1) = 1 \Rightarrow 0 - \tfrac{1}{4} + C = \tfrac{1}{2} + \tfrac{1}{3}(4)^{3/2} \Rightarrow C = \tfrac{1}{2} + \tfrac{8}{3} + \tfrac{1}{4} = \tfrac{41}{12}$, so

$\tfrac{1}{2}x^2 \ln x - \tfrac{1}{4}x^2 + \tfrac{41}{12} = \tfrac{1}{2}y^2 + \tfrac{1}{3}(3 + y^2)^{3/2}$. We do not solve explicitly for y.

21. $\dfrac{dy}{dx} = \dfrac{x}{y} \Rightarrow y \, dy = x \, dx \Rightarrow \int y \, dy = \int x \, dx \Rightarrow \tfrac{1}{2}y^2 = \tfrac{1}{2}x^2 + C$. $y(0) = 2 \Rightarrow \tfrac{1}{2}(2)^2 = \tfrac{1}{2}(0)^2 + C \Rightarrow$

$C = 2$, so $\tfrac{1}{2}y^2 = \tfrac{1}{2}x^2 + 2 \Rightarrow y^2 = x^2 + 4 \Rightarrow y = \sqrt{x^2 + 4}$ since $y(0) = 2 > 0$.

23. $u = x + y \;\Rightarrow\; \dfrac{d}{dx}(u) = \dfrac{d}{dx}(x+y) \;\Rightarrow\; \dfrac{du}{dx} = 1 + \dfrac{dy}{dx},$ but $\dfrac{dy}{dx} = x + y = u,$ so $\dfrac{du}{dx} = 1 + u \;\Rightarrow\;$

$\dfrac{du}{1+u} = dx \;\; [u \neq -1] \;\Rightarrow\; \displaystyle\int \dfrac{du}{1+u} = \int dx \;\Rightarrow\; \ln|1+u| = x + C \;\Rightarrow\; |1+u| = e^{x+C} \;\Rightarrow\;$

$1 + u = \pm e^C e^x \;\Rightarrow\; u = \pm e^C e^x - 1 \;\Rightarrow\; x + y = \pm e^C e^x - 1 \;\Rightarrow\; y = Ke^x - x - 1,$ where $K = \pm e^C \neq 0.$
If $u = -1,$ then $-1 = x + y \;\Rightarrow\; y = -x - 1,$ which is just $y = Ke^x - x - 1$ with $K = 0.$ Thus, the general solution is $y = Ke^x - x - 1,$ where $K \in \mathbb{R}.$

25. (a) $y' = 2x\sqrt{1-y^2} \;\Rightarrow\; \dfrac{dy}{dx} = 2x\sqrt{1-y^2} \;\Rightarrow\; \dfrac{dy}{\sqrt{1-y^2}} = 2x\,dx \;\Rightarrow\; \displaystyle\int \dfrac{dy}{\sqrt{1-y^2}} = \int 2x\,dx \;\Rightarrow\;$

$\sin^{-1} y = x^2 + C$ for $-\dfrac{\pi}{2} \leq x^2 + C \leq \dfrac{\pi}{2}.$

(b) $y(0) = 0 \;\Rightarrow\; \sin^{-1} 0 = 0^2 + C \;\Rightarrow\; C = 0,$

so $\sin^{-1} y = x^2$ and $y = \sin(x^2)$ for $-\sqrt{\pi/2} \leq x \leq \sqrt{\pi/2}.$

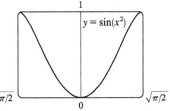

(c) For $\sqrt{1-y^2}$ to be a real number, we must have $-1 \leq y \leq 1;$ that is, $-1 \leq y(0) \leq 1.$ Thus, the initial-value problem $y' = 2x\sqrt{1-y^2}, \; y(0) = 2$ does *not* have a solution.

27. $\dfrac{dy}{dx} = \dfrac{\sin x}{\sin y}, \; y(0) = \dfrac{\pi}{2}.$ So $\int \sin y \, dy = \int \sin x \, dx \;\Leftrightarrow\; -\cos y = -\cos x + C \;\Leftrightarrow\; \cos y = \cos x - C.$ From the initial condition, we need $\cos \dfrac{\pi}{2} = \cos 0 - C \;\Rightarrow\; 0 = 1 - C \;\Rightarrow\; C = 1,$ so the solution is $\cos y = \cos x - 1.$ Note that we cannot take $\cos^{-1}$ of both sides, since that would unnecessarily restrict the solution to the case where $-1 \leq \cos x - 1 \;\Leftrightarrow\; 0 \leq \cos x,$ as $\cos^{-1}$ is defined only on $[-1, 1].$ Instead we plot the graph using Maple's `plots[implicitplot]` or Mathematica's `Plot[Evaluate[···]]`.

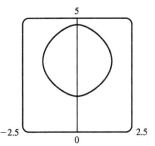

29. (a), (c)

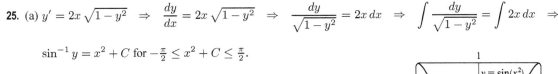

(b) $y' = y^2 \;\Rightarrow\; \dfrac{dy}{dx} = y^2 \;\Rightarrow\; \displaystyle\int y^{-2}\,dy = \int dx \;\Rightarrow\;$

$-y^{-1} = x + C \;\Rightarrow\; \dfrac{1}{y} = -x - C \;\Rightarrow\;$

$y = \dfrac{1}{K-x},$ where $K = -C.$ $y = 0$ is also a solution.

31. The curves $x^2 + 2y^2 = k^2$ form a family of ellipses with major axis on the x-axis. Differentiating gives

$\dfrac{d}{dx}(x^2 + 2y^2) = \dfrac{d}{dx}(k^2) \;\Rightarrow\; 2x + 4yy' = 0 \;\Rightarrow\; 4yy' = -2x \;\Rightarrow\; y' = \dfrac{-x}{2y}.$ Thus, the slope of the tangent line at any point (x, y) on one of the ellipses is $y' = \dfrac{-x}{2y}$, so the orthogonal trajectories

must satisfy $y' = \dfrac{2y}{x} \Leftrightarrow \dfrac{dy}{dx} = \dfrac{2y}{x} \Leftrightarrow \dfrac{dy}{y} = 2 = \dfrac{dx}{x} \Leftrightarrow$

$\displaystyle\int \dfrac{dy}{y} = 2\int \dfrac{dx}{x} \;\;\Leftrightarrow\;\; \ln|y| = 2\ln|x| + C_1 \;\;\Leftrightarrow\;\; \ln|y| = \ln|x|^2 + C_1 \;\;\Leftrightarrow\;\;$

$|y| = e^{\ln x^2 + C_1} \;\;\Leftrightarrow\;\; y = \pm x^2 \cdot e^{C_1} = Cx^2$. This is a family of parabolas.

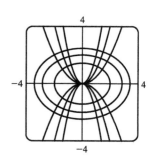

33. The curves $y = k/x$ form a family of hyperbolas with asymptotes $x = 0$ and $y = 0$. Differentiating gives

$\dfrac{d}{dx}(y) = \dfrac{d}{dx}\left(\dfrac{k}{x}\right) \;\Rightarrow\; y' = -\dfrac{k}{x^2} \;\Rightarrow\; y' = -\dfrac{xy}{x^2}$ [since $y = k/x \;\Rightarrow\; xy = k$] $\;\Rightarrow\; y' = -\dfrac{y}{x}$. Thus, the slope

of the tangent line at any point (x, y) on one of the hyperbolas is $y' = -y/x$,

so the orthogonal trajectories must satisfy $y' = x/y \;\Leftrightarrow\; \dfrac{dy}{dx} = \dfrac{x}{y} \;\Leftrightarrow\;$

$y\,dy = x\,dx \;\Leftrightarrow\; \int y\,dy = \int x\,dx \;\Leftrightarrow\; \tfrac{1}{2}y^2 = \tfrac{1}{2}x^2 + C_1 \;\Leftrightarrow\;$

$y^2 = x^2 + C_2 \;\Leftrightarrow\; x^2 - y^2 = C$. This is a family of hyperbolas with

asymptotes $y = \pm x$.

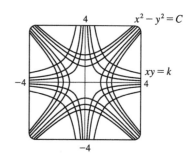

35. $y(x) = 2 + \displaystyle\int_2^x [t - ty(t)]\,dt \;\Rightarrow\; y'(x) = x - xy(x)$ [by FTC 1] $\;\Rightarrow\; \dfrac{dy}{dx} = x(1 - y) \;\Rightarrow\;$

$\displaystyle\int \dfrac{dy}{1 - y} = \int x\,dx \;\Rightarrow\; -\ln|1 - y| = \tfrac{1}{2}x^2 + C$. Letting $x = 2$ in the original integral equation

gives us $y(2) = 2 + 0 = 2$. Thus, $-\ln|1 - 2| = \tfrac{1}{2}(2)^2 + C \;\Rightarrow\; 0 = 2 + C \;\Rightarrow\; C = -2$.

Thus, $-\ln|1 - y| = \tfrac{1}{2}x^2 - 2 \;\Rightarrow\; \ln|1 - y| = 2 - \tfrac{1}{2}x^2 \;\Rightarrow\; |1 - y| = e^{2 - x^2/2} \;\Rightarrow\;$

$1 - y = \pm e^{2 - x^2/2} \;\Rightarrow\; y = 1 + e^{2 - x^2/2}$ [$y(2) = 2$].

37. $y(x) = 4 + \displaystyle\int_0^x 2t\sqrt{y(t)}\,dt \;\Rightarrow\; y'(x) = 2x\sqrt{y(x)} \;\Rightarrow\; \dfrac{dy}{dx} = 2x\sqrt{y} \;\Rightarrow\; \int \dfrac{dy}{\sqrt{y}} = \int 2x\,dx \;\Rightarrow\;$

$2\sqrt{y} = x^2 + C$. Letting $x = 0$ in the original integral equation gives us $y(0) = 4 + 0 = 4$.

Thus, $2\sqrt{4} = 0^2 + C \;\Rightarrow\; C = 4$. $2\sqrt{y} = x^2 + 4 \;\Rightarrow\; \sqrt{y} = \tfrac{1}{2}x^2 + 2 \;\Rightarrow\; y = \left(\tfrac{1}{2}x^2 + 2\right)^2$.

39. From Exercise 9.2.27, $\dfrac{dQ}{dt} = 12 - 4Q \;\Leftrightarrow\; \displaystyle\int \dfrac{dQ}{12 - 4Q} = \int dt \;\Leftrightarrow\; -\tfrac{1}{4}\ln|12 - 4Q| = t + C \;\Leftrightarrow\;$

$\ln|12 - 4Q| = -4t - 4C \;\Leftrightarrow\; |12 - 4Q| = e^{-4t - 4C} \;\Leftrightarrow\; 12 - 4Q = Ke^{-4t}$ [$K = \pm e^{-4C}$] $\;\Leftrightarrow\;$

$4Q = 12 - Ke^{-4t} \;\Leftrightarrow\; Q = 3 - Ae^{-4t}$ [$A = K/4$]. $Q(0) = 0 \;\Leftrightarrow\; 0 = 3 - A \;\Leftrightarrow\; A = 3 \;\Leftrightarrow\;$

$Q(t) = 3 - 3e^{-4t}$. As $t \to \infty$, $Q(t) \to 3 - 0 = 3$ (the limiting value).

41. $\dfrac{dP}{dt} = k(M - P) \Leftrightarrow \displaystyle\int \dfrac{dP}{P - M} = \int (-k)\, dt \Leftrightarrow \ln|P - M| = -kt + C \Leftrightarrow |P - M| = e^{-kt+C} \Leftrightarrow$

$P - M = Ae^{-kt}$ $[A = \pm e^C]$ $\Leftrightarrow$ $P = M + Ae^{-kt}$. If we assume that performance is at level 0 when $t = 0$, then

$P(0) = 0 \Leftrightarrow 0 = M + A \Leftrightarrow A = -M \Leftrightarrow P(t) = M - Me^{-kt}$. $\displaystyle\lim_{t \to \infty} P(t) = M - M \cdot 0 = M$.

43. (a) If $a = b$, then $\dfrac{dx}{dt} = k(a - x)(b - x)^{1/2}$ becomes $\dfrac{dx}{dt} = k(a - x)^{3/2} \Rightarrow (a - x)^{-3/2}\, dx = k\, dt \Rightarrow$

$\int (a - x)^{-3/2}\, dx = \int k\, dt \Rightarrow 2(a - x)^{-1/2} = kt + C$ [by substitution] $\Rightarrow \dfrac{2}{kt + C} = \sqrt{a - x} \Rightarrow$

$\left(\dfrac{2}{kt + C}\right)^2 = a - x \Rightarrow x(t) = a - \dfrac{4}{(kt + C)^2}$. The initial concentration of HBr is 0, so $x(0) = 0 \Rightarrow$

$0 = a - \dfrac{4}{C^2} \Rightarrow \dfrac{4}{C^2} = a \Rightarrow C^2 = \dfrac{4}{a} \Rightarrow C = 2/\sqrt{a}$ [C is positive since $kt + C = 2(a - x)^{-1/2} > 0$].

Thus, $x(t) = a - \dfrac{4}{(kt + 2/\sqrt{a})^2}$.

(b) $\dfrac{dx}{dt} = k(a - x)(b - x)^{1/2} \Rightarrow \dfrac{dx}{(a - x)\sqrt{b - x}} = k\, dt \Rightarrow \displaystyle\int \dfrac{dx}{(a - x)\sqrt{b - x}} = \int k\, dt$ $(\star)$.

From the hint, $u = \sqrt{b - x} \Rightarrow u^2 = b - x \Rightarrow 2u\, du = -dx$, so

$\displaystyle\int \dfrac{dx}{(a - x)\sqrt{b - x}} = \int \dfrac{-2u\, du}{[a - (b - u^2)]u} = -2\int \dfrac{du}{a - b + u^2} = -2\int \dfrac{du}{(\sqrt{a - b})^2 + u^2}$

$\overset{17}{=} -2\left(\dfrac{1}{\sqrt{a - b}} \tan^{-1} \dfrac{u}{\sqrt{a - b}}\right)$

So $(\star)$ becomes $\dfrac{-2}{\sqrt{a - b}} \tan^{-1} \dfrac{\sqrt{b - x}}{\sqrt{a - b}} = kt + C$. Now $x(0) = 0 \Rightarrow C = \dfrac{-2}{\sqrt{a - b}} \tan^{-1} \dfrac{\sqrt{b}}{\sqrt{a - b}}$ and we have

$\dfrac{-2}{\sqrt{a - b}} \tan^{-1} \dfrac{\sqrt{b - x}}{\sqrt{a - b}} = kt - \dfrac{2}{\sqrt{a - b}} \tan^{-1} \dfrac{\sqrt{b}}{\sqrt{a - b}} \Rightarrow \dfrac{2}{\sqrt{a - b}}\left(\tan^{-1}\sqrt{\dfrac{b}{a - b}} - \tan^{-1}\sqrt{\dfrac{b - x}{a - b}}\right) = kt \Rightarrow$

$t(x) = \dfrac{2}{k\sqrt{a - b}}\left(\tan^{-1}\sqrt{\dfrac{b}{a - b}} - \tan^{-1}\sqrt{\dfrac{b - x}{a - b}}\right)$.

45. (a) $\dfrac{dC}{dt} = r - kC \Rightarrow \dfrac{dC}{dt} = -(kC - r) \Rightarrow \displaystyle\int \dfrac{dC}{kC - r} = \int -dt \Rightarrow (1/k)\ln|kC - r| = -t + M_1 \Rightarrow$

$\ln|kC - r| = -kt + M_2 \Rightarrow |kC - r| = e^{-kt + M_2} \Rightarrow kC - r = M_3 e^{-kt} \Rightarrow kC = M_3 e^{-kt} + r \Rightarrow$

$C(t) = M_4 e^{-kt} + r/k$. $C(0) = C_0 \Rightarrow C_0 = M_4 + r/k \Rightarrow M_4 = C_0 - r/k \Rightarrow$

$C(t) = (C_0 - r/k)e^{-kt} + r/k$.

(b) If $C_0 < r/k$, then $C_0 - r/k < 0$ and the formula for $C(t)$ shows that $C(t)$ increases and $\displaystyle\lim_{t \to \infty} C(t) = r/k$.

As t increases, the formula for $C(t)$ shows how the role of C_0 steadily diminishes as that of r/k increases.

47. (a) Let $y(t)$ be the amount of salt (in kg) after t minutes. Then $y(0) = 15$. The amount of liquid in the tank is 1000 L at all times, so the concentration at time t (in minutes) is $y(t)/1000$ kg/L and $\dfrac{dy}{dt} = -\left[\dfrac{y(t)}{1000}\dfrac{\text{kg}}{\text{L}}\right]\left(10\,\dfrac{\text{L}}{\text{min}}\right) = -\dfrac{y(t)}{100}\dfrac{\text{kg}}{\text{min}}$.

$\displaystyle\int \dfrac{dy}{y} = -\dfrac{1}{100}\int dt \;\Rightarrow\; \ln y = -\dfrac{t}{100} + C$, and $y(0) = 15 \;\Rightarrow\; \ln 15 = C$, so $\ln y = \ln 15 - \dfrac{t}{100}$.

It follows that $\ln\left(\dfrac{y}{15}\right) = -\dfrac{t}{100}$ and $\dfrac{y}{15} = e^{-t/100}$, so $y = 15e^{-t/100}$ kg.

(b) After 20 minutes, $y = 15e^{-20/100} = 15e^{-0.2} \approx 12.3$ kg.

49. Let $y(t)$ be the amount of alcohol in the vat after t minutes. Then $y(0) = 0.04(500) = 20$ gal. The amount of beer in the vat is 500 gallons at all times, so the percentage at time t (in minutes) is $y(t)/500 \times 100$, and the change in the amount of alcohol with respect to time t is $\dfrac{dy}{dt} = \text{rate in} - \text{rate out} = 0.06\left(5\,\dfrac{\text{gal}}{\text{min}}\right) - \dfrac{y(t)}{500}\left(5\,\dfrac{\text{gal}}{\text{min}}\right) = 0.3 - \dfrac{y}{100} = \dfrac{30 - y}{100}\dfrac{\text{gal}}{\text{min}}$.

Hence, $\displaystyle\int \dfrac{dy}{30 - y} = \int \dfrac{dt}{100}$ and $-\ln|30 - y| = \tfrac{1}{100}t + C$. Because $y(0) = 20$, we have $-\ln 10 = C$, so

$-\ln|30 - y| = \tfrac{1}{100}t - \ln 10 \;\Rightarrow\; \ln|30 - y| = -t/100 + \ln 10 \;\Rightarrow\; \ln|30 - y| = \ln e^{-t/100} + \ln 10 \;\Rightarrow\;$

$\ln|30 - y| = \ln(10e^{-t/100}) \;\Rightarrow\; |30 - y| = 10e^{-t/100}$. Since y is continuous, $y(0) = 20$, and the right-hand side is never zero, we deduce that $30 - y$ is always positive. Thus, $30 - y = 10e^{-t/100} \;\Rightarrow\; y = 30 - 10e^{-t/100}$. The percentage of alcohol is $p(t) = y(t)/500 \times 100 = y(t)/5 = 6 - 2e^{-t/100}$. The percentage of alcohol after one hour is $p(60) = 6 - 2e^{-60/100} \approx 4.9$.

51. Assume that the raindrop begins at rest, so that $v(0) = 0$. $dm/dt = km$ and $(mv)' = gm \;\Rightarrow\; mv' + vm' = gm \;\Rightarrow\;$

$mv' + v(km) = gm \;\Rightarrow\; v' + vk = g \;\Rightarrow\; \dfrac{dv}{dt} = g - kv \;\Rightarrow\; \displaystyle\int \dfrac{dv}{g - kv} = \int dt \;\Rightarrow\;$

$-(1/k)\ln|g - kv| = t + C \;\Rightarrow\; \ln|g - kv| = -kt - kC \;\Rightarrow\; g - kv = Ae^{-kt}.\; v(0) = 0 \;\Rightarrow\; A = g$.

So $kv = g - ge^{-kt} \;\Rightarrow\; v = (g/k)(1 - e^{-kt})$. Since $k > 0$, as $t \to \infty$, $e^{-kt} \to 0$ and therefore, $\displaystyle\lim_{t\to\infty} v(t) = g/k$.

53. (a) $\dfrac{1}{L_1}\dfrac{dL_1}{dt} = k\dfrac{1}{L_2}\dfrac{dL_2}{dt} \;\Rightarrow\; \dfrac{d}{dt}(\ln L_1) = \dfrac{d}{dt}(k \ln L_2) \;\Rightarrow\; \displaystyle\int \dfrac{d}{dt}(\ln L_1)\,dt = \int \dfrac{d}{dt}(\ln L_2^k)\,dt \;\Rightarrow\;$

$\ln L_1 = \ln L_2^k + C \;\Rightarrow\; L_1 = e^{\ln L_2^k + C} = e^{\ln L_2^k}e^C \;\Rightarrow\; L_1 = KL_2^k$, where $K = e^C$.

(b) From part (a) with $L_1 = B$, $L_2 = V$, and $k = 0.0794$, we have $B = KV^{0.0794}$.

55. (a) The rate of growth of the area is jointly proportional to $\sqrt{A(t)}$ and $M - A(t)$; that is, the rate is proportional to the product of those two quantities. So for some constant k, $dA/dt = k\sqrt{A}\,(M - A)$. We are interested in the maximum of the function dA/dt (when the tissue grows the fastest), so we differentiate, using the Chain Rule and then substituting for

dA/dt from the differential equation:

$$\frac{d}{dt}\left(\frac{dA}{dt}\right) = k\left[\sqrt{A}\,(-1)\frac{dA}{dt} + (M-A)\cdot \tfrac{1}{2}A^{-1/2}\frac{dA}{dt}\right] = \tfrac{1}{2}kA^{-1/2}\frac{dA}{dt}[-2A + (M-A)]$$

$$= \tfrac{1}{2}kA^{-1/2}\left[k\sqrt{A}(M-A)\right][M-3A] = \tfrac{1}{2}k^2(M-A)(M-3A)$$

This is 0 when $M - A = 0$ [this situation never actually occurs, since the graph of $A(t)$ is asymptotic to the line $y = M$, as in the logistic model] and when $M - 3A = 0 \ \Leftrightarrow\ A(t) = M/3$. This represents a maximum by the First Derivative Test, since $\dfrac{d}{dt}\left(\dfrac{dA}{dt}\right)$ goes from positive to negative when $A(t) = M/3$.

(b) From the CAS, we get $A(t) = M\left(\dfrac{Ce^{\sqrt{M}kt} - 1}{Ce^{\sqrt{M}kt} + 1}\right)^2$. To get C in terms of the initial area A_0 and the maximum area M,

we substitute $t = 0$ and $A = A_0 = A(0)$: $A_0 = M\left(\dfrac{C-1}{C+1}\right)^2 \ \Leftrightarrow\ (C+1)\sqrt{A_0} = (C-1)\sqrt{M} \ \Leftrightarrow$

$C\sqrt{A_0} + \sqrt{A_0} = C\sqrt{M} - \sqrt{M} \ \Leftrightarrow\ \sqrt{M} + \sqrt{A_0} = C\sqrt{M} - C\sqrt{A_0} \ \Leftrightarrow$

$\sqrt{M} + \sqrt{A_0} = C\left(\sqrt{M} - \sqrt{A_0}\right) \ \Leftrightarrow\ C = \dfrac{\sqrt{M} + \sqrt{A_0}}{\sqrt{M} - \sqrt{A_0}}$. [Notice that if $A_0 = 0$, then $C = 1$.]

57. (a) According to the hint we use the Chain Rule: $m\dfrac{dv}{dt} = m\dfrac{dv}{dx}\cdot\dfrac{dx}{dt} = mv\dfrac{dv}{dx} = -\dfrac{mgR^2}{(x+R)^2} \ \Rightarrow$

$\displaystyle\int v\,dv = \int \dfrac{-gR^2\,dx}{(x+R)^2} \ \Rightarrow\ \dfrac{v^2}{2} = \dfrac{gR^2}{x+R} + C$. When $x = 0$, $v = v_0$, so $\dfrac{v_0^2}{2} = \dfrac{gR^2}{0+R} + C \ \Rightarrow$

$C = \tfrac{1}{2}v_0^2 - gR \ \Rightarrow\ \tfrac{1}{2}v^2 - \tfrac{1}{2}v_0^2 = \dfrac{gR^2}{x+R} - gR$. Now at the top of its flight, the rocket's velocity will be 0, and its

height will be $x = h$. Solving for v_0: $-\tfrac{1}{2}v_0^2 = \dfrac{gR^2}{h+R} - gR \ \Rightarrow\ \dfrac{v_0^2}{2} = g\left[-\dfrac{R^2}{R+h} + \dfrac{R(R+h)}{R+h}\right] = \dfrac{gRh}{R+h} \ \Rightarrow$

$v_0 = \sqrt{\dfrac{2gRh}{R+h}}$.

(b) $v_e = \displaystyle\lim_{h\to\infty} v_0 = \lim_{h\to\infty}\sqrt{\dfrac{2gRh}{R+h}} = \lim_{h\to\infty}\sqrt{\dfrac{2gR}{(R/h)+1}} = \sqrt{2gR}$

(c) $v_e = \sqrt{2\cdot 32\text{ ft/s}^2 \cdot 3960\text{ mi}\cdot 5280\text{ ft/mi}} \approx 36{,}581\text{ ft/s} \approx 6.93\text{ mi/s}$

9.4 Models for Population Growth

1. (a) Comparing the given equation, $\dfrac{dP}{dt} = 0.04P\left(1 - \dfrac{P}{1200}\right)$, to Equation 4, $\dfrac{dP}{dt} = kP\left(1 - \dfrac{P}{M}\right)$, we see that the carrying capacity is $M = 1200$ and the value of k is 0.04.

(b) By Equation 7, the solution of the equation is $P(t) = \dfrac{M}{1 + Ae^{-kt}}$, where $A = \dfrac{M - P_0}{P_0}$. Since $P(0) = P_0 = 60$, we have

$A = \dfrac{1200 - 60}{60} = 19$, and hence, $P(t) = \dfrac{1200}{1 + 19e^{-0.04t}}$.

(c) The population after 10 weeks is $P(10) = \dfrac{1200}{1 + 19e^{-0.04(10)}} \approx 87$.

3. (a) $dP/dt = 0.05P - 0.0005P^2 = 0.05P(1 - 0.01P) = 0.05P(1 - P/100)$. Comparing to Equation 4, $dP/dt = kP(1 - P/M)$, we see that the carrying capacity is $M = 100$ and the value of k is 0.05.

(b) The slopes close to 0 occur where P is near 0 or 100. The largest slopes appear to be on the line $P = 50$. The solutions are increasing for $0 < P_0 < 100$ and decreasing for $P_0 > 100$.

(c)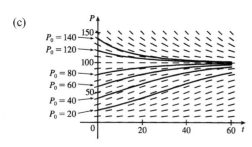

All of the solutions approach $P = 100$ as t increases. As in part (b), the solutions differ since for $0 < P_0 < 100$ they are increasing, and for $P_0 > 100$ they are decreasing. Also, some have an IP and some don't. It appears that the solutions which have $P_0 = 20$ and $P_0 = 40$ have inflection points at $P = 50$.

(d) The equilibrium solutions are $P = 0$ (trivial solution) and $P = 100$. The increasing solutions move away from $P = 0$ and all nonzero solutions approach $P = 100$ as $t \to \infty$.

5. (a) $\dfrac{dy}{dt} = ky\left(1 - \dfrac{y}{M}\right)$ ⇒ $y(t) = \dfrac{M}{1 + Ae^{-kt}}$ with $A = \dfrac{M - y(0)}{y(0)}$. With $M = 8 \times 10^7$, $k = 0.71$, and $y(0) = 2 \times 10^7$, we get the model $y(t) = \dfrac{8 \times 10^7}{1 + 3e^{-0.71t}}$, so $y(1) = \dfrac{8 \times 10^7}{1 + 3e^{-0.71}} \approx 3.23 \times 10^7$ kg.

(b) $y(t) = 4 \times 10^7$ ⇒ $\dfrac{8 \times 10^7}{1 + 3e^{-0.71t}} = 4 \times 10^7$ ⇒ $2 = 1 + 3e^{-0.71t}$ ⇒ $e^{-0.71t} = \tfrac{1}{3}$ ⇒

$-0.71t = \ln \tfrac{1}{3}$ ⇒ $t = \dfrac{\ln 3}{0.71} \approx 1.55$ years

7. Using Equation 7, $A = \dfrac{M - P_0}{P_0} = \dfrac{10{,}000 - 1000}{1000} = 9$, so $P(t) = \dfrac{10{,}000}{1 + 9e^{-kt}}$. $P(1) = 2500$ ⇒

$2500 = \dfrac{10{,}000}{1 + 9e^{-k(1)}}$ ⇒ $1 + 9e^{-k} = 4$ ⇒ $9e^{-k} = 3$ ⇒ $e^{-k} = \tfrac{1}{3}$ ⇒ $-k = \ln \tfrac{1}{3}$ ⇒ $k = \ln 3$. After

another three years, $t = 4$, and $P(4) = \dfrac{10{,}000}{1 + 9e^{-(\ln 3)4}} = \dfrac{10{,}000}{1 + 9(e^{\ln 3})^{-4}} = \dfrac{10{,}000}{1 + 9(3)^{-4}} = \dfrac{10{,}000}{1 + \tfrac{1}{9}} = \dfrac{10{,}000}{\tfrac{10}{9}} = 9000$.

9. (a) We will assume that the difference in birth and death rates is 20 million/year. Let $t = 0$ correspond to the year 2000. Thus,

$k \approx \dfrac{1}{P}\dfrac{dP}{dt} = \dfrac{1}{6.1 \text{ billion}}\left(\dfrac{20 \text{ million}}{\text{year}}\right) = \dfrac{1}{305}$, and $\dfrac{dP}{dt} = kP\left(1 - \dfrac{P}{M}\right) = \dfrac{1}{305}P\left(1 - \dfrac{P}{20}\right)$ with P in billions.

(b) $A = \dfrac{M - P_0}{P_0} = \dfrac{20 - 6.1}{6.1} = \dfrac{139}{61} \approx 2.2787$. $P(t) = \dfrac{M}{1 + Ae^{-kt}} = \dfrac{20}{1 + \frac{139}{61}e^{-t/305}}$, so

$P(10) = \dfrac{20}{1 + \frac{139}{61}e^{-10/305}} \approx 6.24$ billion, which underestimates the actual 2010 population of 6.9 billion.

(c) The years 2100 and 2500 correspond to $t = 100$ and $t = 500$, respectively. $P(100) = \dfrac{20}{1 + \frac{139}{61}e^{-100/305}} \approx 7.57$ billion

and $P(500) = \dfrac{20}{1 + \frac{139}{61}e^{-500/305}} \approx 13.87$ billion.

11. (a) Our assumption is that $\dfrac{dy}{dt} = ky(1 - y)$, where y is the fraction of the population that has heard the rumor.

(b) Using the logistic equation (4), $\dfrac{dP}{dt} = kP\left(1 - \dfrac{P}{M}\right)$, we substitute $y = \dfrac{P}{M}$, $P = My$, and $\dfrac{dP}{dt} = M\dfrac{dy}{dt}$,

to obtain $M\dfrac{dy}{dt} = k(My)(1 - y) \Leftrightarrow \dfrac{dy}{dt} = ky(1 - y)$, our equation in part (a).

Now the solution to (4) is $P(t) = \dfrac{M}{1 + Ae^{-kt}}$, where $A = \dfrac{M - P_0}{P_0}$.

We use the same substitution to obtain $My = \dfrac{M}{1 + \dfrac{M - My_0}{My_0}e^{-kt}} \Rightarrow y = \dfrac{y_0}{y_0 + (1 - y_0)e^{-kt}}$.

Alternatively, we could use the same steps as outlined in the solution of Equation 4.

(c) Let t be the number of hours since 8 AM. Then $y_0 = y(0) = \dfrac{80}{1000} = 0.08$ and $y(4) = \dfrac{1}{2}$, so

$\dfrac{1}{2} = y(4) = \dfrac{0.08}{0.08 + 0.92e^{-4k}}$. Thus, $0.08 + 0.92e^{-4k} = 0.16$, $e^{-4k} = \dfrac{0.08}{0.92} = \dfrac{2}{23}$, and $e^{-k} = \left(\dfrac{2}{23}\right)^{1/4}$,

so $y = \dfrac{0.08}{0.08 + 0.92(2/23)^{t/4}} = \dfrac{2}{2 + 23(2/23)^{t/4}}$. Solving this equation for t, we get

$2y + 23y\left(\dfrac{2}{23}\right)^{t/4} = 2 \Rightarrow \left(\dfrac{2}{23}\right)^{t/4} = \dfrac{2 - 2y}{23y} \Rightarrow \left(\dfrac{2}{23}\right)^{t/4} = \dfrac{2}{23} \cdot \dfrac{1 - y}{y} \Rightarrow \left(\dfrac{2}{23}\right)^{t/4 - 1} = \dfrac{1 - y}{y}$.

It follows that $\dfrac{t}{4} - 1 = \dfrac{\ln[(1 - y)/y]}{\ln \frac{2}{23}}$, so $t = 4\left[1 + \dfrac{\ln((1 - y)/y)}{\ln \frac{2}{23}}\right]$.

When $y = 0.9$, $\dfrac{1 - y}{y} = \dfrac{1}{9}$, so $t = 4\left(1 - \dfrac{\ln 9}{\ln \frac{2}{23}}\right) \approx 7.6$ h or 7 h 36 min. Thus, 90% of the population will have heard

the rumor by 3:36 PM.

13. (a) $\dfrac{dP}{dt} = kP\left(1 - \dfrac{P}{M}\right) \Rightarrow \dfrac{d^2P}{dt^2} = k\left[P\left(-\dfrac{1}{M}\dfrac{dP}{dt}\right) + \left(1 - \dfrac{P}{M}\right)\dfrac{dP}{dt}\right] = k\dfrac{dP}{dt}\left(-\dfrac{P}{M} + 1 - \dfrac{P}{M}\right)$

$= k\left[kP\left(1 - \dfrac{P}{M}\right)\right]\left(1 - \dfrac{2P}{M}\right) = k^2 P\left(1 - \dfrac{P}{M}\right)\left(1 - \dfrac{2P}{M}\right)$

(b) P grows fastest when P' has a maximum, that is, when $P'' = 0$. From part (a), $P'' = 0 \Leftrightarrow P = 0, P = M$,

or $P = M/2$. Since $0 < P < M$, we see that $P'' = 0 \Leftrightarrow P = M/2$.

15. (a)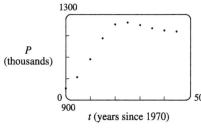

(b) After subtracting 900 from each value of P, we get the logistic model $f(t) = \dfrac{345.5899}{1 + 7.9977e^{-0.2482t}}$.

(c) $P(t) = 900 + \dfrac{345.5899}{1 + 7.9977e^{-0.2482t}}$

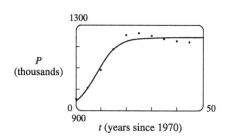

The model fits the data reasonably well, even though the data is decreasing for the last five values and a logistic function is never decreasing.

(d) As $t \to \infty$, $P(t) \to 900 + 345.6 = 1245.6$, so the population approaches 1.246 million.

17. (a) $\dfrac{dP}{dt} = kP - m = k\left(P - \dfrac{m}{k}\right)$. Let $y = P - \dfrac{m}{k}$, so $\dfrac{dy}{dt} = \dfrac{dP}{dt}$ and the differential equation becomes $\dfrac{dy}{dt} = ky$.

The solution is $y = y_0 e^{kt} \Rightarrow P - \dfrac{m}{k} = \left(P_0 - \dfrac{m}{k}\right)e^{kt} \Rightarrow P(t) = \dfrac{m}{k} + \left(P_0 - \dfrac{m}{k}\right)e^{kt}$.

(b) Since $k > 0$, there will be an exponential expansion $\Leftrightarrow P_0 - \dfrac{m}{k} > 0 \Leftrightarrow m < kP_0$.

(c) The population will be constant if $P_0 - \dfrac{m}{k} = 0 \Leftrightarrow m = kP_0$. It will decline if $P_0 - \dfrac{m}{k} < 0 \Leftrightarrow m > kP_0$.

(d) $P_0 = 8{,}000{,}000$, $k = \alpha - \beta = 0.016$, $m = 210{,}000 \Rightarrow m > kP_0\ (= 128{,}000)$, so by part (c), the population was declining.

19. (a) The term -15 represents a harvesting of fish at a constant rate—in this case, 15 fish/week. This is the rate at which fish are caught.

(b)

(c) From the graph in part (b), it appears that $P(t) = 250$ and $P(t) = 750$ are the equilibrium solutions. We confirm this analytically by solving the equation $dP/dt = 0$ as follows: $0.08P(1 - P/1000) - 15 = 0 \Rightarrow$
$0.08P - 0.00008P^2 - 15 = 0 \Rightarrow$
$-0.00008(P^2 - 1000P + 187{,}500) = 0 \Rightarrow$
$(P - 250)(P - 750) = 0 \Rightarrow P = 250$ or 750.

(d)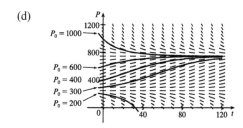

For $0 < P_0 < 250$, $P(t)$ decreases to 0. For $P_0 = 250$, $P(t)$ remains constant. For $250 < P_0 < 750$, $P(t)$ increases and approaches 750. For $P_0 = 750$, $P(t)$ remains constant. For $P_0 > 750$, $P(t)$ decreases and approaches 750.

(e) $\dfrac{dP}{dt} = 0.08P\left(1 - \dfrac{P}{1000}\right) - 15 \Leftrightarrow -\dfrac{100{,}000}{8} \cdot \dfrac{dP}{dt} = (0.08P - 0.00008P^2 - 15) \cdot \left(-\dfrac{100{,}000}{8}\right) \Leftrightarrow$

$-12{,}500 \dfrac{dP}{dt} = P^2 - 1000P + 187{,}500 \Leftrightarrow \dfrac{dP}{(P-250)(P-750)} = -\dfrac{1}{12{,}500} dt \Leftrightarrow$

$\displaystyle\int\left(\dfrac{-1/500}{P-250} + \dfrac{1/500}{P-750}\right) dP = -\dfrac{1}{12{,}500} dt \Leftrightarrow \int\left(\dfrac{1}{P-250} - \dfrac{1}{P-750}\right) dP = \tfrac{1}{25} dt \Leftrightarrow$

$\ln|P-250| - \ln|P-750| = \tfrac{1}{25}t + C \Leftrightarrow \ln\left|\dfrac{P-250}{P-750}\right| = \tfrac{1}{25}t + C \Leftrightarrow \left|\dfrac{P-250}{P-750}\right| = e^{t/25+C} = ke^{t/25} \Leftrightarrow$

$\dfrac{P-250}{P-750} = ke^{t/25} \Leftrightarrow P - 250 = Pke^{t/25} - 750ke^{t/25} \Leftrightarrow P - Pke^{t/25} = 250 - 750ke^{t/25} \Leftrightarrow$

$P(t) = \dfrac{250 - 750ke^{t/25}}{1 - ke^{t/25}}$. If $t = 0$ and $P = 200$, then $200 = \dfrac{250 - 750k}{1 - k} \Leftrightarrow 200 - 200k = 250 - 750k \Leftrightarrow$

$550k = 50 \Leftrightarrow k = \tfrac{1}{11}$. Similarly, if $t = 0$ and $P = 300$, then

$k = -\tfrac{1}{9}$. Simplifying P with these two values of k gives us

$P(t) = \dfrac{250(3e^{t/25} - 11)}{e^{t/25} - 11}$ and $P(t) = \dfrac{750(e^{t/25} + 3)}{e^{t/25} + 9}$.

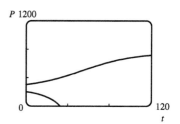

21. (a) $\dfrac{dP}{dt} = (kP)\left(1 - \dfrac{P}{M}\right)\left(1 - \dfrac{m}{P}\right)$. If $m < P < M$, then $dP/dt = (+)(+)(+) = + \Rightarrow$ P is increasing.

If $0 < P < m$, then $dP/dt = (+)(+)(-) = - \Rightarrow$ P is decreasing.

(b)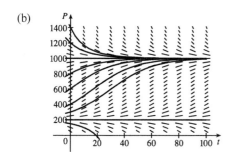

$k = 0.08$, $M = 1000$, and $m = 200 \Rightarrow$

$\dfrac{dP}{dt} = 0.08P\left(1 - \dfrac{P}{1000}\right)\left(1 - \dfrac{200}{P}\right)$

For $0 < P_0 < 200$, the population dies out. For $P_0 = 200$, the population is steady. For $200 < P_0 < 1000$, the population increases and approaches 1000. For $P_0 > 1000$, the population decreases and approaches 1000. The equilibrium solutions are $P(t) = 200$ and $P(t) = 1000$.

(c) $\dfrac{dP}{dt} = kP\left(1 - \dfrac{P}{M}\right)\left(1 - \dfrac{m}{P}\right) = kP\left(\dfrac{M-P}{M}\right)\left(\dfrac{P-m}{P}\right) = \dfrac{k}{M}(M-P)(P-m) \Leftrightarrow$

$\displaystyle\int\dfrac{dP}{(M-P)(P-m)} = \int\dfrac{k}{M} dt$. By partial fractions, $\dfrac{1}{(M-P)(P-m)} = \dfrac{A}{M-P} + \dfrac{B}{P-m}$, so

$A(P - m) + B(M - P) = 1$.

If $P = m$, $B = \dfrac{1}{M-m}$; if $P = M$, $A = \dfrac{1}{M-m}$, so $\dfrac{1}{M-m}\displaystyle\int\left(\dfrac{1}{M-P} + \dfrac{1}{P-m}\right) dP = \int \dfrac{k}{M} dt \Rightarrow$

$\dfrac{1}{M-m}(-\ln|M-P| + \ln|P-m|) = \dfrac{k}{M}t + C \Rightarrow \dfrac{1}{M-m}\ln\left|\dfrac{P-m}{M-P}\right| = \dfrac{k}{M}t + C \Rightarrow$

$$\ln\left|\frac{P-m}{M-P}\right| = (M-m)\frac{k}{M}t + C_1 \;\Leftrightarrow\; \frac{P-m}{M-P} = De^{(M-m)(k/M)t} \quad [D = \pm e^{C_1}].$$

Let $t = 0$: $\dfrac{P_0 - m}{M - P_0} = D$. So $\dfrac{P - m}{M - P} = \dfrac{P_0 - m}{M - P_0} e^{(M-m)(k/M)t}$.

Solving for P, we get $P(t) = \dfrac{m(M - P_0) + M(P_0 - m)e^{(M-m)(k/M)t}}{M - P_0 + (P_0 - m)e^{(M-m)(k/M)t}}$.

(d) If $P_0 < m$, then $P_0 - m < 0$. Let $N(t)$ be the numerator of the expression for $P(t)$ in part (c). Then

$N(0) = P_0(M - m) > 0$, and $P_0 - m < 0 \;\Leftrightarrow\; \lim_{t \to \infty} M(P_0 - m)e^{(M-m)(k/M)t} = -\infty \;\Rightarrow\; \lim_{t \to \infty} N(t) = -\infty$.

Since N is continuous, there is a number t such that $N(t) = 0$ and thus $P(t) = 0$. So the species will become extinct.

23. (a) $dP/dt = kP\cos(rt - \phi) \;\Rightarrow\; (dP)/P = k\cos(rt - \phi)\,dt \;\Rightarrow\; \int (dP)/P = k\int \cos(rt - \phi)\,dt \;\Rightarrow\;$

$\ln P = (k/r)\sin(rt - \phi) + C$. (Since this is a growth model, $P > 0$ and we can write $\ln P$ instead of $\ln|P|$.) Since

$P(0) = P_0$, we obtain $\ln P_0 = (k/r)\sin(-\phi) + C = -(k/r)\sin\phi + C \;\Rightarrow\; C = \ln P_0 + (k/r)\sin\phi$. Thus,

$\ln P = (k/r)\sin(rt - \phi) + \ln P_0 + (k/r)\sin\phi$, which we can rewrite as $\ln(P/P_0) = (k/r)[\sin(rt - \phi) + \sin\phi]$ or,

after exponentiation, $P(t) = P_0 e^{(k/r)[\sin(rt - \phi) + \sin\phi]}$.

(b) As k increases, the amplitude increases, but the minimum value stays the same.

As r increases, the amplitude and the period decrease.

A change in ϕ produces slight adjustments in the phase shift and amplitude.

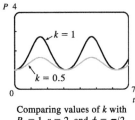
Comparing values of k with $P_0 = 1$, $r = 2$, and $\phi = \pi/2$

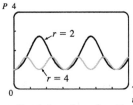
Comparing values of r with $P_0 = 1$, $k = 1$, and $\phi = \pi/2$

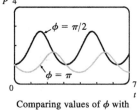
Comparing values of ϕ with $P_0 = 1$, $k = 1$, and $r = 2$

$P(t)$ oscillates between $P_0 e^{(k/r)(1 + \sin\phi)}$ and $P_0 e^{(k/r)(-1 + \sin\phi)}$ (the extreme values are attained when $rt - \phi$ is an odd multiple of $\frac{\pi}{2}$), so $\lim_{t \to \infty} P(t)$ does not exist.

25. By Equation 7, $P(t) = \dfrac{K}{1 + Ae^{-kt}}$. By comparison, if $c = (\ln A)/k$ and $u = \tfrac{1}{2}k(t - c)$, then

$$1 + \tanh u = 1 + \frac{e^u - e^{-u}}{e^u + e^{-u}} = \frac{e^u + e^{-u}}{e^u + e^{-u}} + \frac{e^u - e^{-u}}{e^u + e^{-u}} = \frac{2e^u}{e^u + e^{-u}} \cdot \frac{e^{-u}}{e^{-u}} = \frac{2}{1 + e^{-2u}}$$

and $e^{-2u} = e^{-k(t-c)} = e^{kc}e^{-kt} = e^{\ln A}e^{-kt} = Ae^{-kt}$, so

$$\tfrac{1}{2}K\left[1 + \tanh\left(\tfrac{1}{2}k(t-c)\right)\right] = \frac{K}{2}[1 + \tanh u] = \frac{K}{2} \cdot \frac{2}{1 + e^{-2u}} = \frac{K}{1 + e^{-2u}} = \frac{K}{1 + Ae^{-kt}} = P(t).$$

9.5 Linear Equations

1. $y' + x\sqrt{y} = x^2$ is not linear since it cannot be put into the standard form (1), $y' + P(x)y = Q(x)$.

3. $ue^t = t + \sqrt{t}\,\dfrac{du}{dt} \;\Leftrightarrow\; \sqrt{t}\,u' - e^t u = -t \;\Leftrightarrow\; u' - \dfrac{e^t}{\sqrt{t}}\,u = -\sqrt{t}$ is linear since it can be put into the standard form, $u' + P(t)u = Q(t)$.

5. Comparing the given equation, $y' + y = 1$, with the general form, $y' + P(x)y = Q(x)$, we see that $P(x) = 1$ and the integrating factor is $I(x) = e^{\int P(x)\,dx} = e^{\int 1\,dx} = e^x$. Multiplying the differential equation by $I(x)$ gives

$e^x y' + e^x y = e^x \;\Rightarrow\; (e^x y)' = e^x \;\Rightarrow\; e^x y = \int e^x\,dx \;\Rightarrow\; e^x y = e^x + C \;\Rightarrow\; \dfrac{e^x y}{e^x} = \dfrac{e^x}{e^x} + \dfrac{C}{e^x} \;\Rightarrow\;$

$y = 1 + Ce^{-x}$.

7. $y' = x - y \;\Rightarrow\; y' + y = x$ $(\star)$. $I(x) = e^{\int P(x)\,dx} = e^{\int 1\,dx} = e^x$. Multiplying the differential equation $(\star)$ by $I(x)$ gives

$e^x y' + e^x y = xe^x \;\Rightarrow\; (e^x y)' = xe^x \;\Rightarrow\; e^x y = \int xe^x\,dx \;\Rightarrow\; e^x y = xe^x - e^x + C$ [by parts] $\Rightarrow$

$y = x - 1 + Ce^{-x}$ [divide by e^x].

9. Since $P(x)$ is the derivative of the coefficient of y' [$P(x) = 1$ and the coefficient is x], we can write the differential equation $xy' + y = \sqrt{x}$ in the easily integrable form $(xy)' = \sqrt{x} \;\Rightarrow\; xy = \tfrac{2}{3}x^{3/2} + C \;\Rightarrow\; y = \tfrac{2}{3}\sqrt{x} + C/x$.

11. $xy' - 2y = x^2 \;\Rightarrow\; y' - \dfrac{2}{x}\,y = x \;\Rightarrow\; P(x) = -\dfrac{2}{x}$.

$I(x) = e^{\int P(x)\,dx} = e^{\int -2/x\,dx} = e^{-2\ln x}$ $[x > 0] = x^{-2} = \dfrac{1}{x^2}$. Multiplying the differential equation by $I(x)$ gives

$\dfrac{1}{x^2}y' - \dfrac{2}{x^3}y = \dfrac{1}{x} \;\Rightarrow\; \left(\dfrac{1}{x^2}y\right)' = \dfrac{1}{x} \;\Rightarrow\; \dfrac{1}{x^2}y = \int \dfrac{1}{x}\,dx \;\Rightarrow\; \dfrac{1}{x^2}y = \ln x + C \;\Rightarrow\; y = x^2(\ln x + C)$.

13. $t^2\,\dfrac{dy}{dt} + 3ty = \sqrt{1+t^2} \;\Rightarrow\; y' + \dfrac{3}{t}y = \dfrac{\sqrt{1+t^2}}{t^2} \;\Rightarrow\; P(t) = \dfrac{3}{t}$.

$I(t) = e^{\int P(t)\,dt} = e^{\int 3/t\,dt} = e^{3\ln t}$ $[t > 0] = t^3$. Multiplying by t^3 gives $t^3 y' + 3t^2 y = t\sqrt{1+t^2} \;\Rightarrow\;$

$(t^3 y)' = t\sqrt{1+t^2} \;\Rightarrow\; t^3 y = \int t\sqrt{1+t^2}\,dt \;\Rightarrow\; t^3 y = \tfrac{1}{3}(1+t^2)^{3/2} + C \;\Rightarrow\; y = \tfrac{1}{3}t^{-3}(1+t^2)^{3/2} + Ct^{-3}$.

15. $y' + y\cos x = x \;\Rightarrow\; P(x) = \cos x$. $I(x) = e^{\int P(x)\,dx} = e^{\int \cos x\,dx} = e^{\sin x}$. Multiplying the differential equation by $I(x)$ gives $e^{\sin x} y' + e^{\sin x}\cos x \cdot y = xe^{\sin x} \;\Rightarrow\; (e^{\sin x} y)' = xe^{\sin x} \;\Rightarrow\; e^{\sin x} y = \int xe^{\sin x}\,dx + C \;\Rightarrow\;$

$y = e^{-\sin x}\int xe^{\sin x}\,dx + Ce^{-\sin x}$. Note: $f(x) = xe^{\sin x}$ has an antiderivative F that is *not* an elementary function [see Section 7.5].

17. $xy' + y = 3x^2 \;\Rightarrow\; (xy)' = 3x^2 \;\Rightarrow\; xy = \int 3x^2\,dx \;\Rightarrow\; xy = x^3 + C \;\Rightarrow\; y = x^2 + \dfrac{C}{x}$. Since $y(1) = 4$,

$4 = 1^2 + \dfrac{C}{1} \;\Rightarrow\; C = 3$, so $y = x^2 + \dfrac{3}{x}$.

19. $x^2 y' + 2xy = \ln x \Rightarrow (x^2 y)' = \ln x \Rightarrow x^2 y = \int \ln x \, dx \Rightarrow x^2 y = x \ln x - x + C$ [by parts]. Since $y(1) = 2$, $1^2(2) = 1 \ln 1 - 1 + C \Rightarrow 2 = -1 + C \Rightarrow C = 3$, so $x^2 y = x \ln x - x + 3$, or $y = \dfrac{1}{x} \ln x - \dfrac{1}{x} + \dfrac{3}{x^2}$.

21. $t \dfrac{du}{dt} = t^2 + 3u \Rightarrow u' - \dfrac{3}{t} u = t$ $(\star)$. $I(t) = e^{\int -3/t \, dt} = e^{-3 \ln|t|} = (e^{\ln|t|})^{-3} = t^{-3}$ $[t > 0] = \dfrac{1}{t^3}$. Multiplying $(\star)$ by $I(t)$ gives $\dfrac{1}{t^3} u' - \dfrac{3}{t^4} u = \dfrac{1}{t^2} \Rightarrow \left(\dfrac{1}{t^3} u\right)' = \dfrac{1}{t^2} \Rightarrow \dfrac{1}{t^3} u = \int \dfrac{1}{t^2} dt \Rightarrow \dfrac{1}{t^3} u = -\dfrac{1}{t} + C$. Since $u(2) = 4$, $\dfrac{1}{2^3}(4) = -\dfrac{1}{2} + C \Rightarrow C = 1$, so $\dfrac{1}{t^3} u = -\dfrac{1}{t} + 1$, or $u = -t^2 + t^3$.

23. $xy' = y + x^2 \sin x \Rightarrow y' - \dfrac{1}{x} y = x \sin x$. $I(x) = e^{\int (-1/x) \, dx} = e^{-\ln x} = e^{\ln x^{-1}} = \dfrac{1}{x}$.

Multiplying by $\dfrac{1}{x}$ gives $\dfrac{1}{x} y' - \dfrac{1}{x^2} y = \sin x \Rightarrow \left(\dfrac{1}{x} y\right)' = \sin x \Rightarrow \dfrac{1}{x} y = -\cos x + C \Rightarrow y = -x \cos x + Cx$. $y(\pi) = 0 \Rightarrow -\pi \cdot (-1) + C\pi = 0 \Rightarrow C = -1$, so $y = -x \cos x - x$.

25. $xy' + 2y = e^x \Rightarrow y' + \dfrac{2}{x} y = \dfrac{e^x}{x}$.

$I(x) = e^{\int (2/x) \, dx} = e^{2 \ln|x|} = \left(e^{\ln|x|}\right)^2 = |x|^2 = x^2$.

Multiplying by $I(x)$ gives $x^2 y' + 2xy = xe^x \Rightarrow (x^2 y)' = xe^x \Rightarrow$ $x^2 y = \int xe^x \, dx = (x-1)e^x + C$ [by parts] $\Rightarrow$ $y = [(x-1)e^x + C]/x^2$. The graphs for $C = -5, -3, -1, 1, 3, 5$, and 7 are shown. $C = 1$ is a transitional value. For $C < 1$, there is an inflection point and for $C > 1$, there is a local minimum. As $|C|$ gets larger, the "branches" get further from the origin.

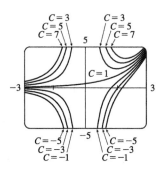

27. Setting $u = y^{1-n}$, $\dfrac{du}{dx} = (1-n) y^{-n} \dfrac{dy}{dx}$ or $\dfrac{dy}{dx} = \dfrac{y^n}{1-n} \dfrac{du}{dx} = \dfrac{u^{n/(1-n)}}{1-n} \dfrac{du}{dx}$. Then the Bernoulli differential equation becomes $\dfrac{u^{n/(1-n)}}{1-n} \dfrac{du}{dx} + P(x) u^{1/(1-n)} = Q(x) u^{n/(1-n)}$ or $\dfrac{du}{dx} + (1-n) P(x) u = Q(x)(1-n)$.

29. Here $y' + \dfrac{2}{x} y = \dfrac{y^3}{x^2}$, so $n = 3$, $P(x) = \dfrac{2}{x}$ and $Q(x) = \dfrac{1}{x^2}$. Setting $u = y^{-2}$, u satisfies $u' - \dfrac{4u}{x} = -\dfrac{2}{x^2}$.

Then $I(x) = e^{\int (-4/x) \, dx} = x^{-4}$ and $u = x^4 \left(\int -\dfrac{2}{x^6} \, dx + C\right) = x^4 \left(\dfrac{2}{5x^5} + C\right) = Cx^4 + \dfrac{2}{5x}$.

Thus, $y = \pm \left(Cx^4 + \dfrac{2}{5x}\right)^{-1/2}$.

31. (a) $2\dfrac{dI}{dt} + 10I = 40$ or $\dfrac{dI}{dt} + 5I = 20$. Then the integrating factor is $e^{\int 5 \, dt} = e^{5t}$. Multiplying the differential equation by the integrating factor gives $e^{5t} \dfrac{dI}{dt} + 5Ie^{5t} = 20e^{5t} \Rightarrow (e^{5t} I)' = 20e^{5t} \Rightarrow$

$I(t) = e^{-5t} \left[\int 20 e^{5t} \, dt + C\right] = 4 + Ce^{-5t}$. But $0 = I(0) = 4 + C$, so $I(t) = 4 - 4e^{-5t}$.

(b) $I(0.1) = 4 - 4e^{-0.5} \approx 1.57$ A

33. $5\dfrac{dQ}{dt} + 20Q = 60$ with $Q(0) = 0$ C. Then the integrating factor is $e^{\int 4\,dt} = e^{4t}$, and multiplying the differential equation by the integrating factor gives $e^{4t}\dfrac{dQ}{dt} + 4e^{4t}Q = 12e^{4t} \;\Rightarrow\; (e^{4t}Q)' = 12e^{4t} \;\Rightarrow\;$
$Q(t) = e^{-4t}\left[\int 12e^{4t}\,dt + C\right] = 3 + Ce^{-4t}$. But $0 = Q(0) = 3 + C$ so $Q(t) = 3(1 - e^{-4t})$ is the charge at time t and $I = dQ/dt = 12e^{-4t}$ is the current at time t.

35. $\dfrac{dP}{dt} = k[M - P(t)] \;\Rightarrow\; \dfrac{dP}{dt} + kP = kM$ $(\star)$, so $I(t) = e^{\int k\,dt} = e^{kt}$.

Multiplying $(\star)$ by $I(t)$ gives $e^{kt}\dfrac{dP}{dt} + kPe^{kt} = kMe^{kt} \;\Rightarrow\;$
$(e^{kt}P)' = kMe^{kt} \;\Rightarrow\; P(t) = e^{-kt}\left(\int kMe^{kt}\,dt + C\right) = M + Ce^{-kt},\; k > 0$.
Furthermore, it is reasonable to assume that $0 \le P(0) \le M$, so $-M \le C \le 0$.

37. $y(0) = 0$ kg. Salt is added at a rate of $\left(0.4\,\dfrac{\text{kg}}{\text{L}}\right)\left(5\,\dfrac{\text{L}}{\text{min}}\right) = 2\,\dfrac{\text{kg}}{\text{min}}$. Since solution is drained from the tank at a rate of 3 L/min, but salt solution is added at a rate of 5 L/min, the tank, which starts out with 100 L of water, contains $(100 + 2t)$ L of liquid after t min. Thus, the salt concentration at time t is $\dfrac{y(t)}{100 + 2t}\,\dfrac{\text{kg}}{\text{L}}$. Salt therefore leaves the tank at a rate of
$\left(\dfrac{y(t)}{100 + 2t}\,\dfrac{\text{kg}}{\text{L}}\right)\left(3\,\dfrac{\text{L}}{\text{min}}\right) = \dfrac{3y}{100 + 2t}\,\dfrac{\text{kg}}{\text{min}}$. Combining the rates at which salt enters and leaves the tank, we get
$\dfrac{dy}{dt} = 2 - \dfrac{3y}{100 + 2t}$. Rewriting this equation as $\dfrac{dy}{dt} + \left(\dfrac{3}{100 + 2t}\right)y = 2$, we see that it is linear.
$$I(t) = \exp\left(\int \dfrac{3\,dt}{100 + 2t}\right) = \exp\left(\tfrac{3}{2}\ln(100 + 2t)\right) = (100 + 2t)^{3/2}$$
Multiplying the differential equation by $I(t)$ gives $(100 + 2t)^{3/2}\dfrac{dy}{dt} + 3(100 + 2t)^{1/2}y = 2(100 + 2t)^{3/2} \;\Rightarrow\;$
$[(100 + 2t)^{3/2}y]' = 2(100 + 2t)^{3/2} \;\Rightarrow\; (100 + 2t)^{3/2}y = \tfrac{2}{5}(100 + 2t)^{5/2} + C \;\Rightarrow\;$
$y = \tfrac{2}{5}(100 + 2t) + C(100 + 2t)^{-3/2}$. Now $0 = y(0) = \tfrac{2}{5}(100) + C \cdot 100^{-3/2} = 40 + \tfrac{1}{1000}C \;\rightarrow\; C = -40{,}000$, so
$y = \left[\tfrac{2}{5}(100 + 2t) - 40{,}000(100 + 2t)^{-3/2}\right]$ kg. From this solution (no pun intended), we calculate the salt concentration
at time t to be $C(t) = \dfrac{y(t)}{100 + 2t} = \left[\dfrac{-40{,}000}{(100 + 2t)^{5/2}} + \tfrac{2}{5}\right]\dfrac{\text{kg}}{\text{L}}$. In particular, $C(20) = \dfrac{-40{,}000}{140^{5/2}} + \tfrac{2}{5} \approx 0.2275\,\dfrac{\text{kg}}{\text{L}}$
and $y(20) = \tfrac{2}{5}(140) - 40{,}000(140)^{-3/2} \approx 31.85$ kg.

39. (a) $m\dfrac{dv}{dt} = mg - cv \;\Rightarrow\; \dfrac{dv}{dt} + \dfrac{c}{m}v = g$ and $I(t) = e^{\int (c/m)\,dt} = e^{(c/m)t}$, and multiplying the last differential equation by $I(t)$ gives $e^{(c/m)t}\dfrac{dv}{dt} + \dfrac{vce^{(c/m)t}}{m} = ge^{(c/m)t} \;\Rightarrow\; \left[e^{(c/m)t}v\right]' = ge^{(c/m)t}$. Hence,
$v(t) = e^{-(c/m)t}\left[\int ge^{(c/m)t}\,dt + K\right] = mg/c + Ke^{-(c/m)t}$. But the object is dropped from rest, so $v(0) = 0$ and
$K = -mg/c$. Thus, the velocity at time t is $v(t) = (mg/c)[1 - e^{-(c/m)t}]$.

(b) $\lim_{t \to \infty} v(t) = mg/c$

(c) $s(t) = \int v(t)\,dt = (mg/c)[t + (m/c)e^{-(c/m)t}] + c_1$ where $c_1 = s(0) - m^2g/c^2$.

$s(0)$ is the initial position, so $s(0) = 0$ and $s(t) = (mg/c)[t + (m/c)e^{-(c/m)t}] - m^2g/c^2$.

41. (a) $z = \dfrac{1}{P} \;\Rightarrow\; P = \dfrac{1}{z} \;\Rightarrow\; P' = -\dfrac{z'}{z^2}$. Substituting into $P' = kP(1 - P/M)$ gives us $-\dfrac{z'}{z^2} = k\dfrac{1}{z}\left(1 - \dfrac{1}{zM}\right) \;\Rightarrow\;$

$z' = -kz\left(1 - \dfrac{1}{zM}\right) \;\Rightarrow\; z' = -kz + \dfrac{k}{M} \;\Rightarrow\; z' + kz = \dfrac{k}{M}$ $(\star)$.

(b) The integrating factor is $e^{\int k\,dt} = e^{kt}$. Multiplying $(\star)$ by e^{kt} gives $e^{kt}z' + ke^{kt}z = \dfrac{ke^{kt}}{M} \;\Rightarrow\; (e^{kt}z)' = \dfrac{k}{M}e^{kt} \;\Rightarrow\;$

$e^{kt}z = \int \dfrac{k}{M}e^{kt}\,dt \;\Rightarrow\; e^{kt}z = \dfrac{1}{M}e^{kt} + C \;\Rightarrow\; z = \dfrac{1}{M} + Ce^{-kt}$. Since $P = \dfrac{1}{z}$, we have

$P = \dfrac{1}{\dfrac{1}{M} + Ce^{-kt}} \;\Rightarrow\; P = \dfrac{M}{1 + MCe^{-kt}}$, which agrees with Equation 9.4.7, $P = \dfrac{M}{1 + Ae^{-kt}}$, when $MC = A$.

9.6 Predator-Prey Systems

1. (a) $dx/dt = -0.05x + 0.0001xy$. If $y = 0$, we have $dx/dt = -0.05x$, which indicates that in the absence of y, x declines at a rate proportional to itself. So x represents the predator population and y represents the prey population. The growth of the prey population, $0.1y$ (from $dy/dt = 0.1y - 0.005xy$), is restricted only by encounters with predators (the term $-0.005xy$). The predator population increases only through the term $0.0001xy$; that is, by encounters with the prey and not through additional food sources.

   (b) $dy/dt = -0.015y + 0.00008xy$. If $x = 0$, we have $dy/dt = -0.015y$, which indicates that in the absence of x, y would decline at a rate proportional to itself. So y represents the predator population and x represents the prey population. The growth of the prey population, $0.2x$ (from $dx/dt = 0.2x - 0.0002x^2 - 0.006xy = 0.2x(1 - 0.001x) - 0.006xy$), is restricted by a carrying capacity of 1000 [from the term $1 - 0.001x = 1 - x/1000$] and by encounters with predators (the term $-0.006xy$). The predator population increases only through the term $0.00008xy$; that is, by encounters with the prey and not through additional food sources.

3. (a) $dx/dt = 0.5x - 0.004x^2 - 0.001xy = 0.5x(1 - x/125) - 0.001xy$.
   $dy/dt = 0.4y - 0.001y^2 - 0.002xy = 0.4y(1 - y/400) - 0.002xy$.

   The system shows that x and y have carrying capacities of 125 and 400. An increase in x reduces the growth rate of y due to the negative term $-0.002xy$. An increase in y reduces the growth rate of x due to the negative term $-0.001xy$. Hence the system describes a competition model.

(b) $dx/dt = 0 \Rightarrow x(0.5 - 0.004x - 0.001y) = 0 \Rightarrow x(500 - 4x - y) = 0$ **(1)** and $dy/dt = 0 \Rightarrow$

$y(0.4 - 0.001y - 0.002x) = 0 \Rightarrow y(400 - y - 2x) = 0$ **(2)**.

From **(1)** and **(2)**, we get four equilibrium solutions.

(i) $x = 0$ and $y = 0$: If the populations are zero, there is no change.

(ii) $x = 0$ and $400 - y - 2x = 0 \Rightarrow x = 0$ and $y = 400$: In the absence of an x-population, the y-population stabilizes at 400.

(iii) $500 - 4x - y = 0$ and $y = 0 \Rightarrow x = 125$ and $y = 0$: In the absence of y-population, the x-population stabilizes at 125.

(iv) $500 - 4x - y = 0$ and $400 - y - 2x = 0 \Rightarrow y = 500 - 4x$ and $y = 400 - 2x \Rightarrow 500 - 4x = 400 - 2x \Rightarrow 100 = 2x \Rightarrow x = 50$ and $y = 300$: A y-population of 300 is just enough to support a constant x-population of 50.

5. (a) At $t = 0$, there are about 300 rabbits and 100 foxes. At $t = t_1$, the number of foxes reaches a minimum of about 20 while the number of rabbits is about 1000. At $t = t_2$, the number of rabbits reaches a maximum of about 2400, while the number of foxes rebounds to 100. At $t = t_3$, the number of rabbits decreases to about 1000 and the number of foxes reaches a maximum of about 315. As t increases, the number of foxes decreases greatly to 100, and the number of rabbits decreases to 300 (the initial populations), and the cycle starts again.

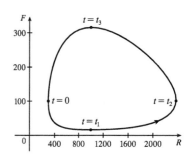

(b)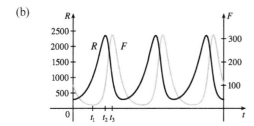

7.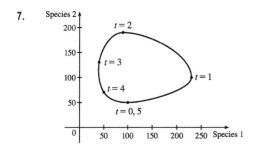

9. (a) $\dfrac{dW}{dR} = \dfrac{-0.02W + 0.00002RW}{0.08R - 0.001RW} \Leftrightarrow (0.08 - 0.001W)R\,dW = (-0.02 + 0.00002R)W\,dR \Leftrightarrow$

$\dfrac{0.08 - 0.001W}{W}\,dW = \dfrac{-0.02 + 0.00002R}{R}\,dR \Leftrightarrow \displaystyle\int \left(\dfrac{0.08}{W} - 0.001\right) dW = \int \left(-\dfrac{0.02}{R} + 0.00002\right) dR \Leftrightarrow$

$0.08 \ln|W| - 0.001W = -0.02 \ln|R| + 0.00002R + K \Leftrightarrow$

$0.08 \ln W + 0.02 \ln R = 0.001W + 0.00002R + K \Leftrightarrow \ln(W^{0.08}R^{0.02}) = 0.00002R + 0.001W + K \Leftrightarrow$

$W^{0.08}R^{0.02} = e^{0.00002R + 0.001W + K} \Leftrightarrow R^{0.02}W^{0.08} = Ce^{0.00002R}e^{0.001W} \Leftrightarrow \dfrac{R^{0.02}W^{0.08}}{e^{0.00002R}e^{0.001W}} = C.$

In general, if $\dfrac{dy}{dx} = \dfrac{-ry + bxy}{kx - axy}$, then $C = \dfrac{x^r y^k}{e^{bx}e^{ay}}$.

(b)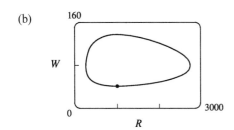

11. (a) Letting $W = 0$ gives us $dR/dt = 0.08R(1 - 0.0002R)$. $dR/dt = 0 \Leftrightarrow R = 0$ or 5000. Since $dR/dt > 0$ for $0 < R < 5000$, we would expect the rabbit population to *increase* to 5000 for these values of R. Since $dR/dt < 0$ for $R > 5000$, we would expect the rabbit population to *decrease* to 5000 for these values of R. Hence, in the absence of wolves, we would expect the rabbit population to stabilize at 5000.

(b) R and W are constant $\Rightarrow R' = 0$ and $W' = 0 \Rightarrow$

$\begin{cases} 0 = 0.08R(1 - 0.0002R) - 0.001RW \\ 0 = -0.02W + 0.00002RW \end{cases} \Rightarrow \begin{cases} 0 = R[0.08(1 - 0.0002R) - 0.001W] \\ 0 = W(-0.02 + 0.00002R) \end{cases}$

The second equation is true if $W = 0$ or $R = \dfrac{0.02}{0.00002} = 1000$. If $W = 0$ in the first equation, then either $R = 0$ or $R = \dfrac{1}{0.0002} = 5000$ [as in part (a)]. If $R = 1000$, then $0 = 1000[0.08(1 - 0.0002 \cdot 1000) - 0.001W] \Leftrightarrow 0 = 80(1 - 0.2) - W \Leftrightarrow W = 64$.

Case (i): $W = 0, R = 0$: both populations are zero

Case (ii): $W = 0, R = 5000$: see part (a)

Case (iii): $R = 1000, W = 64$: the predator/prey interaction balances and the populations are stable.

(c) The populations of wolves and rabbits fluctuate around 64 and 1000, respectively, and eventually stabilize at those values.

(d)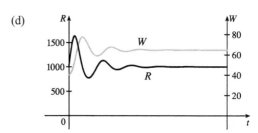

9 Review

TRUE-FALSE QUIZ

1. True. Since $y^4 \geq 0$, $y' = -1 - y^4 < 0$ and the solutions are decreasing functions.

3. False. $y = 3e^{2x} - 1 \;\Rightarrow\; y' = 6e^{2x}$. LHS $= y' - 2y = 6e^{2x} - 2(3e^{2x} - 1) = 6e^{2x} - 6e^{2x} + 2 = 2 \neq 1$, so $y = 3e^{2x} - 1$ is not a solution to the initial-value problem.

5. True. $y' = 3y - 2x + 6xy - 1 = 6xy - 2x + 3y - 1 = 2x(3y - 1) + 1(3y - 1) = (2x + 1)(3y - 1)$, so y' can be written in the form $g(x)f(y)$, and hence, is separable.

7. False. $y' + xy = e^y$ cannot be put in the form $y' + P(x)y = Q(x)$, so it is not linear.

9. True. By comparing $\dfrac{dy}{dt} = 2y\left(1 - \dfrac{y}{5}\right)$ with the logistic differential equation (9.4.4), we see that the carrying capacity is 5; that is, $\lim\limits_{t\to\infty} y = 5$.

EXERCISES

1. (a) $y' = y(y-2)(y-4)$

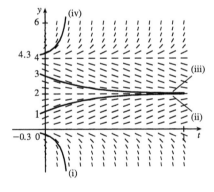

(b) $\lim\limits_{t\to\infty} y(t)$ appears to be finite for $0 \leq c \leq 4$. In fact $\lim\limits_{t\to\infty} y(t) = 4$ for $c = 4$, $\lim\limits_{t\to\infty} y(t) = 2$ for $0 < c < 4$, and $\lim\limits_{t\to\infty} y(t) = 0$ for $c = 0$. The equilibrium solutions are $y(t) = 0$, $y(t) = 2$, and $y(t) = 4$.

3. (a) $y' = x^2 - y^2$

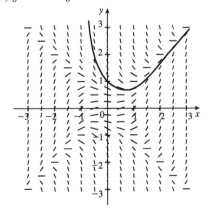

We estimate that when $x = 0.3$, $y = 0.8$, so $y(0.3) \approx 0.8$.

(b) $h = 0.1$, $x_0 = 0$, $y_0 = 1$ and $F(x,y) = x^2 - y^2$. So $y_n = y_{n-1} + 0.1(x_{n-1}^2 - y_{n-1}^2)$. Thus,

$y_1 = 1 + 0.1(0^2 - 1^2) = 0.9$, $y_2 = 0.9 + 0.1(0.1^2 - 0.9^2) = 0.82$, $y_3 = 0.82 + 0.1(0.2^2 - 0.82^2) = 0.75676$.

This is close to our graphical estimate of $y(0.3) \approx 0.8$.

(c) The centers of the horizontal line segments of the direction field are located on the lines $y = x$ and $y = -x$. When a solution curve crosses one of these lines, it has a local maximum or minimum.

5. $y' = xe^{-\sin x} - y\cos x \;\Rightarrow\; y' + (\cos x)y = xe^{-\sin x}$ $(\star)$. This is a linear equation and the integrating factor is
$I(x) = e^{\int \cos x\, dx} = e^{\sin x}$. Multiplying $(\star)$ by $e^{\sin x}$ gives $e^{\sin x} y' + e^{\sin x}(\cos x)y = x \;\Rightarrow\; (e^{\sin x} y)' = x \;\Rightarrow\;$
$e^{\sin x} y = \frac{1}{2}x^2 + C \;\Rightarrow\; y = \left(\frac{1}{2}x^2 + C\right)e^{-\sin x}$.

7. $2ye^{y^2} y' = 2x + 3\sqrt{x} \;\Rightarrow\; 2ye^{y^2} \dfrac{dy}{dx} = 2x + 3\sqrt{x} \;\Rightarrow\; 2ye^{y^2}\, dy = \left(2x + 3\sqrt{x}\right)dx \;\Rightarrow\;$
$\int 2ye^{y^2}\, dy = \int \left(2x + 3\sqrt{x}\right)dx \;\Rightarrow\; e^{y^2} = x^2 + 2x^{3/2} + C \;\Rightarrow\; y^2 = \ln(x^2 + 2x^{3/2} + C) \;\Rightarrow\;$
$y = \pm\sqrt{\ln(x^2 + 2x^{3/2} + C)}$

9. $\dfrac{dr}{dt} + 2tr = r \;\Rightarrow\; \dfrac{dr}{dt} = r - 2tr = r(1 - 2t) \;\Rightarrow\; \displaystyle\int \dfrac{dr}{r} = \int (1 - 2t)\, dt \;\Rightarrow\; \ln|r| = t - t^2 + C \;\Rightarrow\;$
$|r| = e^{t - t^2 + C} = ke^{t - t^2}$. Since $r(0) = 5$, $5 = ke^0 = k$. Thus, $r(t) = 5e^{t - t^2}$.

11. $xy' - y = x\ln x \;\Rightarrow\; y' - \dfrac{1}{x}y = \ln x$. $I(x) = e^{\int (-1/x)\, dx} = e^{-\ln|x|} = \left(e^{\ln|x|}\right)^{-1} = |x|^{-1} = 1/x$ since the condition $y(1) = 2$ implies that we want a solution with $x > 0$. Multiplying the last differential equation by $I(x)$ gives
$\dfrac{1}{x}y' - \dfrac{1}{x^2}y = \dfrac{1}{x}\ln x \;\Rightarrow\; \left(\dfrac{1}{x}y\right)' = \dfrac{1}{x}\ln x \;\Rightarrow\; \dfrac{1}{x}y = \int \dfrac{\ln x}{x}\, dx \;\Rightarrow\; \dfrac{1}{x}y = \dfrac{1}{2}(\ln x)^2 + C \;\Rightarrow\;$
$y = \dfrac{1}{2}x(\ln x)^2 + Cx$. Now $y(1) = 2 \;\Rightarrow\; 2 = 0 + C \;\Rightarrow\; C = 2$, so $y = \dfrac{1}{2}x(\ln x)^2 + 2x$.

13. $\dfrac{d}{dx}(y) = \dfrac{d}{dx}(ke^x) \;\Rightarrow\; y' = ke^x = y$, so the orthogonal trajectories must have $y' = -\dfrac{1}{y} \;\Rightarrow\; \dfrac{dy}{dx} = -\dfrac{1}{y} \;\Rightarrow\;$
$y\, dy = -dx \;\Rightarrow\; \int y\, dy = -\int dx \;\Rightarrow\; \dfrac{1}{2}y^2 = -x + C \;\Rightarrow\; x = C - \dfrac{1}{2}y^2$, which are parabolas with a horizontal axis.

15. (a) Using (4) and (7) in Section 9.4, we see that for $\dfrac{dP}{dt} = 0.1P\left(1 - \dfrac{P}{2000}\right)$ with $P(0) = 100$, we have $k = 0.1$,
$M = 2000$, $P_0 = 100$, and $A = \dfrac{2000 - 100}{100} = 19$. Thus, the solution of the initial-value problem is
$P(t) = \dfrac{2000}{1 + 19e^{-0.1t}}$ and $P(20) = \dfrac{2000}{1 + 19e^{-2}} \approx 560$.

(b) $P = 1200 \;\Leftrightarrow\; 1200 = \dfrac{2000}{1 + 19e^{-0.1t}} \;\Leftrightarrow\; 1 + 19e^{-0.1t} = \dfrac{2000}{1200} \;\Leftrightarrow\; 19e^{-0.1t} = \dfrac{5}{3} - 1 \;\Leftrightarrow\;$
$e^{-0.1t} = \left(\dfrac{2}{3}\right)/19 \;\Leftrightarrow\; -0.1t = \ln\dfrac{2}{57} \;\Leftrightarrow\; t = -10\ln\dfrac{2}{57} \approx 33.5$.

17. (a) $\dfrac{dL}{dt} \propto L_\infty - L \;\Rightarrow\; \dfrac{dL}{dt} = k(L_\infty - L) \;\Rightarrow\; \displaystyle\int \dfrac{dL}{L_\infty - L} = \int k\,dt \;\Rightarrow\; -\ln|L_\infty - L| = kt + C \;\Rightarrow\;$

$\ln|L_\infty - L| = -kt - C \;\Rightarrow\; |L_\infty - L| = e^{-kt-C} \;\Rightarrow\; L_\infty - L = Ae^{-kt} \;\Rightarrow\; L = L_\infty - Ae^{-kt}.$

At $t = 0$, $L = L(0) = L_\infty - A \;\Rightarrow\; A = L_\infty - L(0) \;\Rightarrow\; L(t) = L_\infty - [L_\infty - L(0)]e^{-kt}.$

(b) $L_\infty = 53$ cm, $L(0) = 10$ cm, and $k = 0.2 \;\Rightarrow\; L(t) = 53 - (53 - 10)e^{-0.2t} = 53 - 43e^{-0.2t}.$

19. Let P represent the population and I the number of infected people. The rate of spread dI/dt is jointly proportional to I and to $P - I$, so for some constant k, $\dfrac{dI}{dt} = kI(P - I) \;\Rightarrow\; I(t) = \dfrac{I_0 P}{I_0 + (P - I_0)e^{-kPt}}$ [from the discussion of logistic growth in Section 9.4].

Now, measuring t in days, we substitute $t = 7$, $P = 5000$, $I_0 = 160$ and $I(7) = 1200$ to find k:

$1200 = \dfrac{160 \cdot 5000}{160 + (5000 - 160)e^{-5000 \cdot 7 \cdot k}} \;\Leftrightarrow\; 3 = \dfrac{2000}{160 + 4840e^{-35{,}000k}} \;\Leftrightarrow\; 480 + 14{,}520 e^{-35{,}000k} = 2000 \;\Leftrightarrow\;$

$e^{-35{,}000k} = \dfrac{2000 - 480}{14{,}520} \;\Leftrightarrow\; -35{,}000k = \ln \dfrac{38}{363} \;\Leftrightarrow\; k = \dfrac{-1}{35{,}000} \ln \dfrac{38}{363} \approx 0.00006448.$ Next, let

$I = 5000 \times 80\% = 4000$, and solve for t: $4000 = \dfrac{160 \cdot 5000}{160 + (5000 - 160)e^{-k \cdot 5000 \cdot t}} \;\Leftrightarrow\; 1 = \dfrac{200}{160 + 4840e^{-5000kt}} \;\Leftrightarrow\;$

$160 + 4840 e^{-5000kt} = 200 \;\Leftrightarrow\; e^{-5000kt} = \dfrac{200 - 160}{4840} \;\Leftrightarrow\; -5000kt = \ln \dfrac{1}{121} \;\Leftrightarrow\;$

$t = \dfrac{-1}{5000k} \ln \dfrac{1}{121} = \dfrac{1}{\frac{1}{7} \ln \frac{38}{363}} \cdot \ln \dfrac{1}{121} = 7 \cdot \dfrac{\ln 121}{\ln \frac{363}{38}} \approx 14.875.$ So it takes about 15 days for 80% of the population to be infected.

21. $\dfrac{dh}{dt} = -\dfrac{R}{V}\left(\dfrac{h}{k+h}\right) \;\Rightarrow\; \displaystyle\int \dfrac{k+h}{h}\,dh = \int \left(-\dfrac{R}{V}\right)dt \;\Rightarrow\; \int \left(1 + \dfrac{k}{h}\right)dh = -\dfrac{R}{V}\int 1\,dt \;\Rightarrow\;$

$h + k\ln h = -\dfrac{R}{V}t + C$. This equation gives a relationship between h and t, but it is not possible to isolate h and express it in terms of t.

23. (a) $dx/dt = 0.4x(1 - 0.000005x) - 0.002xy$, $dy/dt = -0.2y + 0.000008xy$. If $y = 0$, then

$dx/dt = 0.4x(1 - 0.000005x)$, so $dx/dt = 0 \;\Leftrightarrow\; x = 0$ or $x = 200{,}000$, which shows that the insect population increases logistically with a carrying capacity of 200,000. Since $dx/dt > 0$ for $0 < x < 200{,}000$ and $dx/dt < 0$ for $x > 200{,}000$, we expect the insect population to stabilize at 200,000.

(b) x and y are constant $\;\Rightarrow\; x' = 0$ and $y' = 0 \;\Rightarrow\;$

$\begin{cases} 0 = 0.4x(1 - 0.000005x) - 0.002xy \\ 0 = -0.2y + 0.000008xy \end{cases} \;\Rightarrow\; \begin{cases} 0 = 0.4x[(1 - 0.000005x) - 0.005y] \\ 0 = y(-0.2 + 0.000008x) \end{cases}$

The second equation is true if $y = 0$ or $x = \dfrac{0.2}{0.000008} = 25{,}000$. If $y = 0$ in the first equation, then either $x = 0$

or $x = \frac{1}{0.000005} = 200{,}000$. If $x = 25{,}000$, then $0 = 0.4(25{,}000)[(1 - 0.000005 \cdot 25{,}000) - 0.005y]$ $\Rightarrow$
$0 = 10{,}000[(1 - 0.125) - 0.005y]$ $\Rightarrow$ $0 = 8750 - 50y$ $\Rightarrow$ $y = 175$.

Case (i): $y = 0, x = 0$: Zero populations

Case (ii): $y = 0, x = 200{,}000$: In the absence of birds, the insect population is always 200,000.

Case (iii): $x = 25{,}000, y = 175$: The predator/prey interaction balances and the populations are stable.

(c) The populations of the birds and insects fluctuate around 175 and 25,000, respectively, and eventually stabilize at those values.

(d)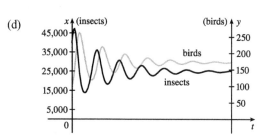

PROBLEMS PLUS

1. We use the Fundamental Theorem of Calculus to differentiate the given equation:

$$[f(x)]^2 = 100 + \int_0^x \left\{[f(t)]^2 + [f'(t)]^2\right\} dt \quad \Rightarrow \quad 2f(x)f'(x) = [f(x)]^2 + [f'(x)]^2 \quad \Rightarrow$$

$$[f(x)]^2 + [f'(x)]^2 - 2f(x)f'(x) = 0 \quad \Rightarrow \quad [f(x) - f'(x)]^2 = 0 \quad \Leftrightarrow \quad f(x) = f'(x).$$ We can solve this as a separable equation, or else use Theorem 9.4.2 with $k=1$, which says that the solutions are $f(x) = Ce^x$. Now $[f(0)]^2 = 100$, so $f(0) = C = \pm 10$, and hence $f(x) = \pm 10e^x$ are the only functions satisfying the given equation.

3. $f'(x) = \lim_{h \to 0} \dfrac{f(x+h) - f(x)}{h} = \lim_{h \to 0} \dfrac{f(x)[f(h) - 1]}{h}$ [since $f(x+h) = f(x)f(h)$]

$= f(x) \lim_{h \to 0} \dfrac{f(h) - 1}{h} = f(x) \lim_{h \to 0} \dfrac{f(h) - f(0)}{h - 0} = f(x)f'(0) = f(x) \cdot 1 = f(x)$

Therefore, $f'(x) = f(x)$ for all x and from Theorem 9.4.2 we get $f(x) = Ae^x$.

Now $f(0) = 1 \quad \Rightarrow \quad A = 1 \quad \Rightarrow \quad f(x) = e^x$.

5. "The area under the graph of f from 0 to x is proportional to the $(n+1)$st power of $f(x)$" translates to

$\int_0^x f(t)\,dt = k[f(x)]^{n+1}$ for some constant k. By FTC1, $\dfrac{d}{dx}\int_0^x f(t)\,dt = \dfrac{d}{dx}\left\{k[f(x)]^{n+1}\right\} \quad \Rightarrow$

$f(x) = k(n+1)[f(x)]^n f'(x) \quad \Rightarrow \quad 1 = k(n+1)[f(x)]^{n-1} f'(x) \quad \Rightarrow \quad 1 = k(n+1)y^{n-1}\dfrac{dy}{dx} \quad \Rightarrow$

$k(n+1)y^{n-1}\,dy = dx \quad \Rightarrow \quad \int k(n+1)y^{n-1}\,dy = \int dx \quad \Rightarrow \quad k(n+1)\dfrac{1}{n}y^n = x + C.$

Now $f(0) = 0 \quad \Rightarrow \quad 0 = 0 + C \quad \Rightarrow \quad C = 0$ and then $f(1) = 1 \quad \Rightarrow \quad k(n+1)\dfrac{1}{n} = 1 \quad \Rightarrow \quad k = \dfrac{n}{n+1}$,

so $y^n = x$ and $y = f(x) = x^{1/n}$.

7. Let $y(t)$ denote the temperature of the peach pie t minutes after 5:00 PM and R the temperature of the room. Newton's Law of Cooling gives us $dy/dt = k(y - R)$. Solving for y we get $\dfrac{dy}{y - R} = k\,dt \quad \Rightarrow \quad \ln|y - R| = kt + C \quad \Rightarrow$

$|y - R| = e^{kt + C} \quad \Rightarrow \quad y - R = \pm e^{kt} \cdot e^C \quad \Rightarrow \quad y = Me^{kt} + R$, where M is a nonzero constant. We are given temperatures at three times.

$$y(0) = 100 \quad \Rightarrow \quad 100 = M + R \quad \Rightarrow \quad R = 100 - M$$
$$y(10) = 80 \quad \Rightarrow \quad 80 = Me^{10k} + R \quad \quad (1)$$
$$y(20) = 65 \quad \Rightarrow \quad 65 = Me^{20k} + R \quad \quad (2)$$

Substituting $100 - M$ for R in **(1)** and **(2)** gives us

$$-20 = Me^{10k} - M \quad \textbf{(3)} \quad \text{and} \quad -35 = Me^{20k} - M \quad \textbf{(4)}$$

[continued]

Dividing **(3)** by **(4)** gives us $\dfrac{-20}{-35} = \dfrac{M(e^{10k}-1)}{M(e^{20k}-1)} \;\Rightarrow\; \dfrac{4}{7} = \dfrac{e^{10k}-1}{e^{20k}-1} \;\Rightarrow\; 4e^{20k}-4 = 7e^{10k}-7 \;\Rightarrow\;$
$4e^{20k} - 7e^{10k} + 3 = 0$. This is a quadratic equation in e^{10k}. $(4e^{10k}-3)(e^{10k}-1) = 0 \;\Rightarrow\; e^{10k} = \tfrac{3}{4}$ or $1 \;\Rightarrow\;$
$10k = \ln\tfrac{3}{4}$ or $\ln 1 \;\Rightarrow\; k = \tfrac{1}{10}\ln\tfrac{3}{4}$ since k is a nonzero constant of proportionality. Substituting $\tfrac{3}{4}$ for e^{10k} in **(3)** gives us
$-20 = M \cdot \tfrac{3}{4} - M \;\Rightarrow\; -20 = -\tfrac{1}{4}M \;\Rightarrow\; M = 80$. Now $R = 100 - M$ so $R = 20°C$.

9. (a) While running from $(L, 0)$ to (x, y), the dog travels a distance

$s = \int_x^L \sqrt{1+(dy/dx)^2}\,dx = -\int_L^x \sqrt{1+(dy/dx)^2}\,dx$, so

$\dfrac{ds}{dx} = -\sqrt{1+(dy/dx)^2}$. The dog and rabbit run at the same speed, so the

rabbit's position when the dog has traveled a distance s is $(0, s)$. Since the

dog runs straight for the rabbit, $\dfrac{dy}{dx} = \dfrac{s-y}{0-x}$ (see the figure).

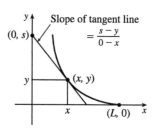

Thus, $s = y - x\dfrac{dy}{dx} \;\Rightarrow\; \dfrac{ds}{dx} = \dfrac{dy}{dx} - \left(x\dfrac{d^2y}{dx^2} + 1\cdot\dfrac{dy}{dx}\right) = -x\dfrac{d^2y}{dx^2}$. Equating the two expressions for $\dfrac{ds}{dx}$

gives us $x\dfrac{d^2y}{dx^2} = \sqrt{1+\left(\dfrac{dy}{dx}\right)^2}$, as claimed.

(b) Letting $z = \dfrac{dy}{dx}$, we obtain the differential equation $x\dfrac{dz}{dx} = \sqrt{1+z^2}$, or $\dfrac{dz}{\sqrt{1+z^2}} = \dfrac{dx}{x}$. Integrating:

$\ln x = \displaystyle\int \dfrac{dz}{\sqrt{1+z^2}} \stackrel{25}{=} \ln\left(z + \sqrt{1+z^2}\right) + C$. When $x = L$, $z = dy/dx = 0$, so $\ln L = \ln 1 + C$. Therefore,

$C = \ln L$, so $\ln x = \ln\left(\sqrt{1+z^2}+z\right) + \ln L = \ln\left[L\left(\sqrt{1+z^2}+z\right)\right] \;\Rightarrow\; x = L\left(\sqrt{1+z^2}+z\right) \;\Rightarrow\;$

$\sqrt{1+z^2} = \dfrac{x}{L} - z \;\Rightarrow\; 1 + z^2 = \left(\dfrac{x}{L}\right)^2 - \dfrac{2xz}{L} + z^2 \;\Rightarrow\; \left(\dfrac{x}{L}\right)^2 - 2z\left(\dfrac{x}{L}\right) - 1 = 0 \;\Rightarrow\;$

$z = \dfrac{(x/L)^2 - 1}{2(x/L)} = \dfrac{x^2 - L^2}{2Lx} = \dfrac{x}{2L} - \dfrac{L}{2}\dfrac{1}{x}$ [for $x > 0$]. Since $z = \dfrac{dy}{dx}$, $y = \dfrac{x^2}{4L} - \dfrac{L}{2}\ln x + C_1$.

Since $y = 0$ when $x = L$, $0 = \dfrac{L}{4} - \dfrac{L}{2}\ln L + C_1 \;\Rightarrow\; C_1 = \dfrac{L}{2}\ln L - \dfrac{L}{4}$. Thus,

$y = \dfrac{x^2}{4L} - \dfrac{L}{2}\ln x + \dfrac{L}{2}\ln L - \dfrac{L}{4} = \dfrac{x^2 - L^2}{4L} - \dfrac{L}{2}\ln\left(\dfrac{x}{L}\right)$.

(c) As $x \to 0^+$, $y \to \infty$, so the dog never catches the rabbit.

11. (a) We are given that $V = \tfrac{1}{3}\pi r^2 h$, $dV/dt = 60{,}000\pi$ ft^3/h, and $r = 1.5h = \tfrac{3}{2}h$. So $V = \tfrac{1}{3}\pi\left(\tfrac{3}{2}h\right)^2 h = \tfrac{3}{4}\pi h^3 \;\Rightarrow\;$

$\dfrac{dV}{dt} = \tfrac{3}{4}\pi \cdot 3h^2 \dfrac{dh}{dt} = \tfrac{9}{4}\pi h^2 \dfrac{dh}{dt}$. Therefore, $\dfrac{dh}{dt} = \dfrac{4(dV/dt)}{9\pi h^2} = \dfrac{240{,}000\pi}{9\pi h^2} = \dfrac{80{,}000}{3h^2}$ ($\star$) $\;\Rightarrow\;$

$\int 3h^2\,dh = \int 80{,}000\,dt \;\Rightarrow\; h^3 = 80{,}000t + C$. When $t = 0$, $h = 60$. Thus, $C = 60^3 = 216{,}000$, so

$h^3 = 80{,}000t + 216{,}000$. Let $h = 100$. Then $100^3 = 1{,}000{,}000 = 80{,}000t + 216{,}000 \;\Rightarrow\;$

$80{,}000t = 784{,}000 \;\Rightarrow\; t = 9.8$, so the time required is 9.8 hours.

(b) The floor area of the silo is $F = \pi \cdot 200^2 = 40{,}000\pi$ ft^2, and the area of the base of the pile is

$A = \pi r^2 = \pi\left(\tfrac{3}{2}h\right)^2 = \tfrac{9\pi}{4}h^2$. So the area of the floor which is not covered when $h = 60$ is

$F - A = 40{,}000\pi - 8100\pi = 31{,}900\pi \approx 100{,}217$ ft^2. Now $A = \tfrac{9\pi}{4}h^2 \;\Rightarrow\; dA/dt = \tfrac{9\pi}{4} \cdot 2h\,(dh/dt)$,

and from $(\star)$ in part (a) we know that when $h = 60$, $dh/dt = \tfrac{80{,}000}{3(60)^2} = \tfrac{200}{27}$ ft/h. Therefore,

$dA/dt = \tfrac{9\pi}{4}(2)(60)\left(\tfrac{200}{27}\right) = 2000\pi \approx 6283$ ft^2/h.

(c) At $h = 90$ ft, $dV/dt = 60{,}000\pi - 20{,}000\pi = 40{,}000\pi$ ft^3/h. From $(\star)$ in part (a),

$\dfrac{dh}{dt} = \dfrac{4(dV/dt)}{9\pi h^2} = \dfrac{4(40{,}000\pi)}{9\pi h^2} = \dfrac{160{,}000}{9h^2} \;\Rightarrow\; \int 9h^2\,dh = \int 160{,}000\,dt \;\Rightarrow\; 3h^3 = 160{,}000t + C$. When $t = 0$,

$h = 90$; therefore, $C = 3 \cdot 729{,}000 = 2{,}187{,}000$. So $3h^3 = 160{,}000t + 2{,}187{,}000$. At the top, $h = 100 \;\Rightarrow\;$

$3(100)^3 = 160{,}000t + 2{,}187{,}000 \;\Rightarrow\; t = \tfrac{813{,}000}{160{,}000} \approx 5.1$. The pile reaches the top after about 5.1 h.

13. Let $P(a, b)$ be any point on the curve. If m is the slope of the tangent line at P, then $m = y'(a)$, and an equation of the

normal line at P is $y - b = -\dfrac{1}{m}(x - a)$, or equivalently, $y = -\dfrac{1}{m}x + b + \dfrac{a}{m}$. The y-intercept is always 6, so

$b + \dfrac{a}{m} = 6 \;\Rightarrow\; \dfrac{a}{m} = 6 - b \;\Rightarrow\; m = \dfrac{a}{6 - b}$. We will solve the equivalent differential equation $\dfrac{dy}{dx} = \dfrac{x}{6 - y} \;\Rightarrow\;$

$(6 - y)\,dy = x\,dx \;\Rightarrow\; \int (6 - y)\,dy = \int x\,dx \;\Rightarrow\; 6y - \tfrac{1}{2}y^2 = \tfrac{1}{2}x^2 + C \;\Rightarrow\; 12y - y^2 = x^2 + K$.

Since $(3, 2)$ is on the curve, $12(2) - 2^2 = 3^2 + K \;\Rightarrow\; K = 11$. So the curve is given by $12y - y^2 = x^2 + 11 \;\Rightarrow\;$

$x^2 + y^2 - 12y + 36 = -11 + 36 \;\Rightarrow\; x^2 + (y - 6)^2 = 25$, a circle with center $(0, 6)$ and radius 5.

15. From the figure, slope $OA = \dfrac{y}{x}$. If triangle OAB is isosceles, then slope

AB must be $-\dfrac{y}{x}$, the negative of slope OA. This slope is also equal to $f'(x)$,

so we have $\dfrac{dy}{dx} = -\dfrac{y}{x} \;\Rightarrow\; \int \dfrac{dy}{y} = -\int \dfrac{dx}{x} \;\Rightarrow\;$

$\ln|y| = -\ln|x| + C \;\Rightarrow\; |y| = e^{-\ln|x| + C} \;\Rightarrow\;$

$|y| = (e^{\ln|x|})^{-1}e^C \;\Rightarrow\; |y| = \dfrac{1}{|x|}e^C \;\Rightarrow\; y = \dfrac{K}{x},\; K \neq 0.$

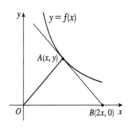

10 ☐ PARAMETRIC EQUATIONS AND POLAR COORDINATES

10.1 Curves Defined by Parametric Equations

1. $x = t^2 + t$, $y = 3^{t+1}$, $t = -2, -1, 0, 1, 2$

t	-2	-1	0	1	2
x	2	0	0	2	6
y	$\frac{1}{3}$	1	3	9	27

Therefore, the coordinates are $\left(2, \frac{1}{3}\right)$, $(0, 1)$, $(0, 3)$, $(2, 9)$, and $(6, 27)$.

3. $x = 1 - t^2$, $y = 2t - t^2$, $-1 \leq t \leq 2$

t	-1	0	1	2
x	0	1	0	-3
y	-3	0	1	0

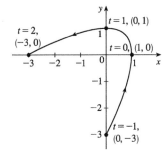

5. $x = 2^t - t$, $y = 2^{-t} + t$, $-3 \leq t \leq 3$

t	-3	-2	-1	0	1	2	3
x	3.125	2.25	1.5	1	1	2	5
y	5	2	1	1	1.5	2.25	3.125

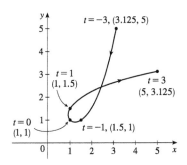

7. $x = 2t - 1$, $y = \frac{1}{2}t + 1$

(a)

t	-4	-2	0	2	4
x	-9	-5	-1	3	7
y	-1	0	1	2	3

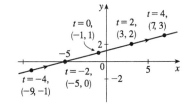

(b) $x = 2t - 1 \Rightarrow 2t = x + 1 \Rightarrow t = \frac{1}{2}x + \frac{1}{2}$, so

$y = \frac{1}{2}t + 1 = \frac{1}{2}\left(\frac{1}{2}x + \frac{1}{2}\right) + 1 = \frac{1}{4}x + \frac{1}{4} + 1 \Rightarrow y = \frac{1}{4}x + \frac{5}{4}$

480 ☐ **CHAPTER 10** PARAMETRIC EQUATIONS AND POLAR COORDINATES

9. $x = t^2 - 3$, $y = t + 2$, $-3 \leq t \leq 3$

(a)

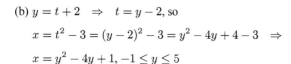

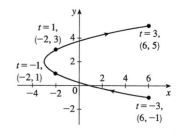

(b) $y = t + 2 \Rightarrow t = y - 2$, so

$x = t^2 - 3 = (y - 2)^2 - 3 = y^2 - 4y + 4 - 3 \Rightarrow$

$x = y^2 - 4y + 1$, $-1 \leq y \leq 5$

11. $x = \sqrt{t}$, $y = 1 - t$

(a)

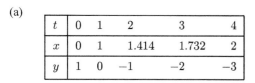

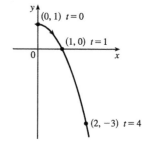

(b) $x = \sqrt{t} \Rightarrow t = x^2 \Rightarrow y = 1 - t = 1 - x^2$. Since $t \geq 0$, $x \geq 0$.

So the curve is the right half of the parabola $y = 1 - x^2$.

13. (a) $x = 3\cos t$, $y = 3\sin t$, $0 \leq t \leq \pi$

$x^2 + y^2 = 9\cos^2 t + 9\sin^2 t = 9(\cos^2 t + \sin^2 t) = 9$, which is the equation

of a circle with radius 3. For $0 \leq t \leq \pi/2$, we have $3 \geq x \geq 0$ and

$0 \leq y \leq 3$. For $\pi/2 < t \leq \pi$, we have $0 > x \geq -3$ and $3 > y \geq 0$. Thus,

the curve is the top half of the circle $x^2 + y^2 = 9$ traced counterclockwise.

(b)

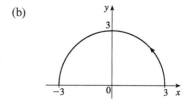

15. (a) $x = \cos\theta$, $y = \sec^2\theta$, $0 \leq \theta < \pi/2$.

$y = \sec^2\theta = \dfrac{1}{\cos^2\theta} = \dfrac{1}{x^2}$. For $0 \leq \theta < \pi/2$, we have $1 \geq x > 0$

and $1 \leq y$.

(b)

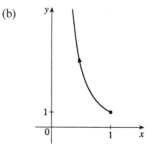

17. (a) $y = e^t = 1/e^{-t} = 1/x$ for $x > 0$ since $x = e^{-t}$. Thus, the curve is the

portion of the hyperbola $y = 1/x$ with $x > 0$.

(b)

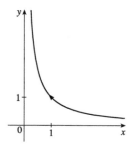

19. (a) $x = \ln t$, $y = \sqrt{t}$, $t \geq 1$.

$x = \ln t \Rightarrow t = e^x \Rightarrow y = \sqrt{t} = e^{x/2}$, $x \geq 0$.

(b)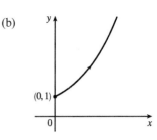

21. (a) $x = \sin^2 t$, $y = \cos^2 t$. $x + y = \sin^2 t + \cos^2 t = 1$. For all t, we have $0 \leq x \leq 1$ and $0 \leq y \leq 1$. Thus, the curve is the portion of the line $x + y = 1$ or $y = -x + 1$ in the first quadrant.

(b)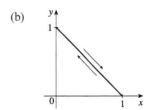

23. The parametric equations $x = 5\cos t$ and $y = -5\sin t$ both have period 2π. When $t = 0$, we have $x = 5$ and $y = 0$. When $t = \pi/2$, we have $x = 0$ and $y = -5$. This is one-fourth of a circle. Thus, the object completes one revolution in $4 \cdot \frac{\pi}{2} = 2\pi$ seconds following a clockwise path.

25. $x = 5 + 2\cos \pi t$, $y = 3 + 2\sin \pi t \Rightarrow \cos \pi t = \dfrac{x-5}{2}$, $\sin \pi t = \dfrac{y-3}{2}$. $\cos^2(\pi t) + \sin^2(\pi t) = 1 \Rightarrow$

$\left(\dfrac{x-5}{2}\right)^2 + \left(\dfrac{y-3}{2}\right)^2 = 1$. The motion of the particle takes place on a circle centered at $(5, 3)$ with a radius 2. As t goes from 1 to 2, the particle starts at the point $(3, 3)$ and moves counterclockwise along the circle $\left(\dfrac{x-5}{2}\right)^2 + \left(\dfrac{y-3}{2}\right)^2 = 1$ to $(7, 3)$ [one-half of a circle].

27. $x = 5\sin t$, $y = 2\cos t \Rightarrow \sin t = \dfrac{x}{5}$, $\cos t = \dfrac{y}{2}$. $\sin^2 t + \cos^2 t = 1 \Rightarrow \left(\dfrac{x}{5}\right)^2 + \left(\dfrac{y}{2}\right)^2 = 1$. The motion of the particle takes place on an ellipse centered at $(0, 0)$. As t goes from $-\pi$ to 5π, the particle starts at the point $(0, -2)$ and moves clockwise around the ellipse 3 times.

29. We must have $1 \leq x \leq 4$ and $2 \leq y \leq 3$. So the graph of the curve must be contained in the rectangle $[1, 4]$ by $[2, 3]$.

31. When $t = -1$, $(x, y) = (1, 1)$. As t increases to 0, x and y both decrease to 0. As t increases from 0 to 1, x increases from 0 to 1 and y decreases from 0 to -1. As t increases beyond 1, x continues to increase and y continues to decrease. For $t < -1$, x and y are both positive and decreasing. We could achieve greater accuracy by estimating x- and y-values for selected values of t from the given graphs and plotting the corresponding points.

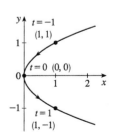

33. When $t = -1$, $(x, y) = (0, 1)$. As t increases to 0, x increases from 0 to 1 and y decreases from 1 to 0. As t increases from 0 to 1, the curve is retraced in the opposite direction with x decreasing from 1 to 0 and y increasing from 0 to 1. We could achieve greater accuracy by estimating x- and y-values for selected values of t from the given graphs and plotting the corresponding points.

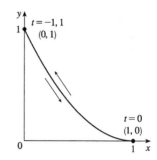

35. Use $y = t$ and $x = t - 2\sin \pi t$ with a t-interval of $[-\pi, \pi]$.

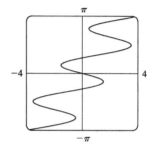

37. (a) $x = x_1 + (x_2 - x_1)t$, $y = y_1 + (y_2 - y_1)t$, $0 \le t \le 1$. Clearly the curve passes through $P_1(x_1, y_1)$ when $t = 0$ and through $P_2(x_2, y_2)$ when $t = 1$. For $0 < t < 1$, x is strictly between x_1 and x_2 and y is strictly between y_1 and y_2. For every value of t, x and y satisfy the relation $y - y_1 = \dfrac{y_2 - y_1}{x_2 - x_1}(x - x_1)$, which is the equation of the line through $P_1(x_1, y_1)$ and $P_2(x_2, y_2)$.

Finally, any point (x, y) on that line satisfies $\dfrac{y - y_1}{y_2 - y_1} = \dfrac{x - x_1}{x_2 - x_1}$; if we call that common value t, then the given parametric equations yield the point (x, y); and any (x, y) on the line between $P_1(x_1, y_1)$ and $P_2(x_2, y_2)$ yields a value of t in $[0, 1]$. So the given parametric equations exactly specify the line segment from $P_1(x_1, y_1)$ to $P_2(x_2, y_2)$.

(b) $x = -2 + [3 - (-2)]t = -2 + 5t$ and $y = 7 + (-1 - 7)t = 7 - 8t$ for $0 \le t \le 1$.

39. The result in Example 4 indicates the parametric equations have the form $x = h + r\sin bt$ and $y = k + r\cos bt$ where (h, k) is the center of the circle with radius r and $b = 2\pi/\text{period}$. (The use of positive sine in the x-equation and positive cosine in the y-equation results in a clockwise motion.) With $h = 0$, $k = 0$ and $b = 2\pi/4\pi = 1/2$, we have $x = 5\sin(\tfrac{1}{2}t)$, $y = 5\cos(\tfrac{1}{2}t)$.

41. The circle $x^2 + (y - 1)^2 = 4$ has center $(0, 1)$ and radius 2, so by Example 4 it can be represented by $x = 2\cos t$, $y = 1 + 2\sin t$, $0 \le t \le 2\pi$. This representation gives us the circle with a counterclockwise orientation starting at $(2, 1)$.

(a) To get a clockwise orientation, we could change the equations to $x = 2\cos t$, $y = 1 - 2\sin t$, $0 \le t \le 2\pi$.

(b) To get three times around in the counterclockwise direction, we use the original equations $x = 2\cos t$, $y = 1 + 2\sin t$ with the domain expanded to $0 \le t \le 6\pi$.

(c) To start at $(0, 3)$ using the original equations, we must have $x_1 = 0$; that is, $2 \cos t = 0$. Hence, $t = \frac{\pi}{2}$. So we use

$x = 2 \cos t, y = 1 + 2 \sin t, \frac{\pi}{2} \leq t \leq \frac{3\pi}{2}$.

Alternatively, if we want t to start at 0, we could change the equations of the curve. For example, we could use

$x = -2 \sin t, y = 1 + 2 \cos t, 0 \leq t \leq \pi$.

43. *Big circle:* It's centered at $(2, 2)$ with a radius of 2, so by Example 4, parametric equations are

$$x = 2 + 2 \cos t, \qquad y = 2 + 2 \sin t, \qquad 0 \leq t \leq 2\pi$$

Small circles: They are centered at $(1, 3)$ and $(3, 3)$ with a radius of 0.1. By Example 4, parametric equations are

and
(left) $x = 1 + 0.1 \cos t, \qquad y = 3 + 0.1 \sin t, \qquad 0 \leq t \leq 2\pi$
(right) $x = 3 + 0.1 \cos t, \qquad y = 3 + 0.1 \sin t, \qquad 0 \leq t \leq 2\pi$

Semicircle: It's the lower half of a circle centered at $(2, 2)$ with radius 1. By Example 4, parametric equations are

$$x = 2 + 1 \cos t, \qquad y = 2 + 1 \sin t, \qquad \pi \leq t \leq 2\pi$$

To get all four graphs on the same screen with a typical graphing calculator, we need to change the last t-interval to $[0, 2\pi]$ in order to match the others. We can do this by changing t to $0.5t$. This change gives us the upper half. There are several ways to get the lower half—one is to change the "+" to a "−" in the y-assignment, giving us

$$x = 2 + 1 \cos(0.5t), \qquad y = 2 - 1 \sin(0.5t), \qquad 0 \leq t \leq 2\pi$$

45. (a) (i) $x = t^2, y = t \;\Rightarrow\; y^2 = t^2 = x$ (ii) $x = t, y = \sqrt{t} \;\Rightarrow\; y^2 = t = x$

(iii) $x = \cos^2 t, y = \cos t \;\Rightarrow\; y^2 = \cos^2 t = x$ (iv) $x = 3^{2t}, y = 3^t \;\Rightarrow\; y^2 = (3^t)^2 = 3^{2t} = x$.

Thus, the points on all four of the given parametric curves satisfy the Cartesian equation $y^2 = x$.

(b) The graph of $y^2 = x$ is a right-opening parabola with vertex at the origin. For curve (i), $x \geq 0$ and y is unbounded so the graph contains the entire parabola. For (ii), $y = \sqrt{t}$ requires that $t \geq 0$, so that both $x \geq 0$ and $y \geq 0$, which captures the upper half of the parabola, including the origin. For (iii), $-1 \leq \cos t \leq 1$ so the graph is the portion of the parabola contained in the intervals $0 \leq x \leq 1$ and $-1 \leq y \leq 1$. For (iv), $x > 0$ and $y > 0$, which captures the upper half of the parabola excluding the origin.

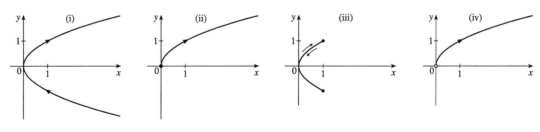

47. (a) $x = t^3 \Rightarrow t = x^{1/3}$, so $y = t^2 = x^{2/3}$.

We get the entire curve $y = x^{2/3}$ traversed in a left to right direction.

(b) $x = t^6 \Rightarrow t = x^{1/6}$, so $y = t^4 = x^{4/6} = x^{2/3}$.

Since $x = t^6 \geq 0$, we only get the right half of the curve $y = x^{2/3}$.

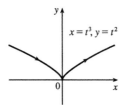

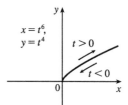

(c) $x = e^{-3t} = (e^{-t})^3$ [so $e^{-t} = x^{1/3}$],

$y = e^{-2t} = (e^{-t})^2 = (x^{1/3})^2 = x^{2/3}$.

If $t < 0$, then x and y are both larger than 1. If $t > 0$, then x and y are between 0 and 1. Since $x > 0$ and $y > 0$, the curve never quite reaches the origin.

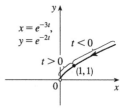

49. The first two diagrams depict the case $\pi < \theta < \frac{3\pi}{2}$, $d < r$. As in Example 7, C has coordinates $(r\theta, r)$. Now Q (in the second diagram) has coordinates $(r\theta, r + d\cos(\theta - \pi)) = (r\theta, r - d\cos\theta)$, so a typical point P of the trochoid has coordinates $(r\theta + d\sin(\theta - \pi), r - d\cos\theta)$. That is, P has coordinates (x, y), where $x = r\theta - d\sin\theta$ and $y = r - d\cos\theta$. When $d = r$, these equations agree with those of the cycloid.

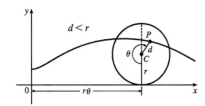 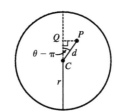 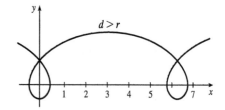

51. It is apparent that $x = |OQ|$ and $y = |QP| = |ST|$. From the diagram, $x = |OQ| = a\cos\theta$ and $y = |ST| = b\sin\theta$. Thus, the parametric equations are $x = a\cos\theta$ and $y = b\sin\theta$. To eliminate θ we rearrange: $\sin\theta = y/b \Rightarrow \sin^2\theta = (y/b)^2$ and $\cos\theta = x/a \Rightarrow \cos^2\theta = (x/a)^2$. Adding the two equations: $\sin^2\theta + \cos^2\theta = 1 = x^2/a^2 + y^2/b^2$. Thus, we have an ellipse.

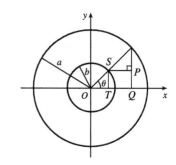

53. $C = (2a\cot\theta, 2a)$, so the x-coordinate of P is $x = 2a\cot\theta$. Let $B = (0, 2a)$. Then $\angle OAB$ is a right angle and $\angle OBA = \theta$, so $|OA| = 2a\sin\theta$ and $A = ((2a\sin\theta)\cos\theta, (2a\sin\theta)\sin\theta)$. Thus, the y-coordinate of P is $y = 2a\sin^2\theta$.

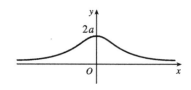

55. (a) *Red particle*: $x = t + 5$, $y = t^2 + 4t + 6$

Blue particle: $x = 2t + 1$, $y = 2t + 6$

Substituting $x = 1$ and $y = 6$ into the parametric equations for the red particle gives $1 = t + 5$ and $6 = t^2 + 4t + 6$, which are both satisfied when $t = -4$. Making the same substitution for the blue particle gives $1 = 2t + 1$ and $6 = 2t + 6$, which are both satisfied when $t = 0$. Repeating the process for $x = 6$ and $y = 11$, the red particle's equations become $6 = t + 5$ and $11 = t^2 + 4t + 6$, which are both satisfied when $t = 1$. Similarly, the blue particle's equations become $6 = 2t + 1$ and $11 = 2t + 6$, which are both satisfied when $t = 2.5$. Thus, $(1, 6)$ and $(6, 11)$ are both intersection points, but they are not collision points, since the particles reach each of these points at different times.

(b) *Blue particle*: $x = 2t + 1$ $\Rightarrow$ $t = \frac{1}{2}(x - 1)$.

Substituting into the equation for y gives $y = 2t + 6 = 2\left[\frac{1}{2}(x - 1)\right] + 6 = x + 5$.

Green particle: $x = 2t + 4$ $\Rightarrow$ $t = \frac{1}{2}(x - 4)$.

Substituting into the equation for y gives $y = 2t + 9 = 2\left[\frac{1}{2}(x - 4)\right] + 9 = x + 5$.

Thus, the green and blue particles both move along the line $y = x + 5$.

Now, the red and green particles will collide if there is a time t when both particles are at the same point. Equating the x parametric equations, we find $t + 5 = 2t + 4$, which is satisfied when $t = 1$, and gives $x = 1 + 5 = 6$. Substituting $t = 1$ into the red and green particles' y equations gives $y = (1)^2 + 4(1) + 6 = 11$ and $y = 2(1) + 9 = 11$, respectively. Thus, the red and green particles collide at the point $(6, 11)$ when $t = 1$.

57. (a) $x = 1 - t^2$, $y = t - t^3$. The curve intersects itself if there are two distinct times $t = a$ and $t = b$ (with $a < b$) such that $x(a) = x(b)$ and $y(a) = y(b)$. The equation $x(a) = x(b)$ gives $1 - a^2 = 1 - b^2$ so that $a^2 = b^2$. Since $a \neq b$ by assumption, we must have $a = -b$. Substituting into the equation for y gives $y(-b) = y(b)$ $\Rightarrow$

$-b - (-b)^3 = b - b^3$ $\Rightarrow$ $2b^3 - 2b = 0$ $\Rightarrow$ $2b(b - 1)(b + 1) = 0$ $\Rightarrow$ $b = -1, 0, 1$. Since $a < b$, the only valid solution is $b = 1$, which corresponds to $a = -1$ and results in the coordinates $x = 0$ and $y = 0$. Thus, the curve intersects itself at $(0, 0)$ when $t = -1$ and $t = 1$.

(b) $x = 2t - t^3$, $y = t - t^2$. Similar to part (a), we try to find the times $t = a$ and $t = b$ with $a < b$ such that $x(a) = x(b)$ and $y(a) = y(b)$. The equation $y(a) = y(b)$ gives $a - a^2 = b - b^2$ $\Rightarrow$ $0 = a^2 - a + (b - b^2)$. Using the quadratic formula to solve for a, we get

$a = \dfrac{1 \pm \sqrt{1 - 4(b - b^2)}}{2} = \dfrac{1 \pm \sqrt{4b^2 - 4b + 1}}{2} = \dfrac{1 \pm \sqrt{(2b - 1)^2}}{2} = \dfrac{1 \pm (2b - 1)}{2}$ $\Rightarrow$ $a = b$ or $a = 1 - b$. Since

$a < b$ by assumption, we reject the first solution and substitute $a = 1 - b$ into $x(a) = x(b)$ $\Rightarrow$ $x(1 - b) = x(b)$ $\Rightarrow$ $2(1 - b) - (1 - b)^3 = 2b - b^3$. Expanding and simplifying gives $2b^3 - 3b^2 - b + 1 = 0$. By graphing the equation, we see that $b = \frac{1}{2}$ is a zero, so $2b - 1$ is a factor, and by long division $b^2 - b - 1$ is another factor. Hence, the solutions are $b = \frac{1}{2}$ and $b = \frac{1}{2} \pm \frac{1}{2}\sqrt{5}$ (found using the quadratic formula). Since $a = 1 - b$ and we require $a < b$, the only valid solution is $b = \frac{1}{2} + \frac{1}{2}\sqrt{5}$, which corresponds to $a = \frac{1}{2} - \frac{1}{2}\sqrt{5}$ and results in the coordinates

$x = 2\left(\frac{1}{2} - \frac{1}{2}\sqrt{5}\right) - \left(\frac{1}{2} - \frac{1}{2}\sqrt{5}\right)^3 = -1$ and $y = \frac{1}{2} - \frac{1}{2}\sqrt{5} - \left(\frac{1}{2} - \frac{1}{2}\sqrt{5}\right)^2 = -1$. Thus, the curve intersects itself at $(-1, -1)$ when $t = \frac{1}{2} - \frac{1}{2}\sqrt{5}$ and $t = \frac{1}{2} + \frac{1}{2}\sqrt{5}$.

59. $x = t^2, y = t^3 - ct$. We use a graphing device to produce the graphs for various values of c with $-\pi \leq t \leq \pi$. Note that all the members of the family are symmetric about the x-axis. For $c < 0$, the graph does not cross itself, but for $c = 0$ it has a cusp at $(0, 0)$ and for $c > 0$ the graph crosses itself at $x = c$, so the loop grows larger as c increases.

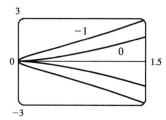
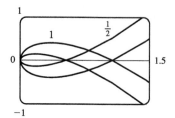

61. $x = t + a\cos t, y = t + a\sin t, a > 0$. From the first figure, we see that curves roughly follow the line $y = x$, and they start having loops when a is between 1.4 and 1.6. The loops increase in size as a increases.

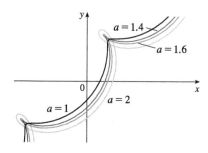

While not required, the following is a solution to determine the *exact* values for which the curve has a loop, that is, we seek the values of a for which there exist parameter values t and u such that $t < u$ and $(t + a\cos t, t + a\sin t) = (u + a\cos u, u + a\sin u)$.

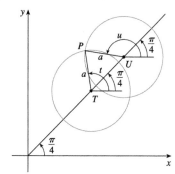

In the diagram at the left, T denotes the point (t, t), U the point (u, u), and P the point $(t + a\cos t, t + a\sin t) = (u + a\cos u, u + a\sin u)$. Since $\overline{PT} = \overline{PU} = a$, the triangle PTU is isosceles. Therefore its base angles, $\alpha = \angle PTU$ and $\beta = \angle PUT$ are equal. Since $\alpha = t - \frac{\pi}{4}$ and $\beta = 2\pi - \frac{3\pi}{4} - u = \frac{5\pi}{4} - u$, the relation $\alpha = \beta$ implies that $u + t = \frac{3\pi}{2}$ **(1)**.

Since $\overline{TU} = \text{distance}((t, t), (u, u)) = \sqrt{2(u-t)^2} = \sqrt{2}(u - t)$, we see that $\cos\alpha = \frac{\frac{1}{2}\overline{TU}}{\overline{PT}} = \frac{(u-t)/\sqrt{2}}{a}$, so $u - t = \sqrt{2}\,a\cos\alpha$, that is,

$u - t = \sqrt{2}\,a\cos\left(t - \frac{\pi}{4}\right)$ **(2)**. Now $\cos\left(t - \frac{\pi}{4}\right) = \sin\left[\frac{\pi}{2} - \left(t - \frac{\pi}{4}\right)\right] = \sin\left(\frac{3\pi}{4} - t\right)$, so we can rewrite **(2)** as $u - t = \sqrt{2}\,a\sin\left(\frac{3\pi}{4} - t\right)$ **(2′)**. Subtracting **(2′)** from **(1)** and dividing by 2, we obtain $t = \frac{3\pi}{4} - \frac{\sqrt{2}}{2}a\sin\left(\frac{3\pi}{4} - t\right)$, or $\frac{3\pi}{4} - t = \frac{a}{\sqrt{2}}\sin\left(\frac{3\pi}{4} - t\right)$ **(3)**.

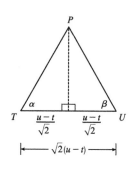

[continued]

Since $a > 0$ and $t < u$, it follows from (2′) that $\sin\left(\frac{3\pi}{4} - t\right) > 0$. Thus from (3) we see that $t < \frac{3\pi}{4}$. [We have implicitly assumed that $0 < t < \pi$ by the way we drew our diagram, but we lost no generality by doing so since replacing t by $t + 2\pi$ merely increases x and y by 2π. The curve's basic shape repeats every time we change t by 2π.] Solving for a in (3), we get $a = \dfrac{\sqrt{2}\left(\frac{3\pi}{4} - t\right)}{\sin\left(\frac{3\pi}{4} - t\right)}$. Write $z = \frac{3\pi}{4} - t$. Then $a = \dfrac{\sqrt{2}\,z}{\sin z}$, where $z > 0$. Now $\sin z < z$ for $z > 0$, so $a > \sqrt{2}$. $\left[\text{As } z \to 0^+,\ \text{that is, as } t \to \left(\tfrac{3\pi}{4}\right)^-,\ a \to \sqrt{2}\right].$

63. Note that all the Lissajous figures are symmetric about the x-axis. The parameters a and b simply stretch the graph in the x- and y-directions respectively. For $a = b = n = 1$ the graph is simply a circle with radius 1. For $n = 2$ the graph crosses itself at the origin and there are loops above and below the x-axis. In general, the figures have $n - 1$ points of intersection, all of which are on the y-axis, and a total of n closed loops.

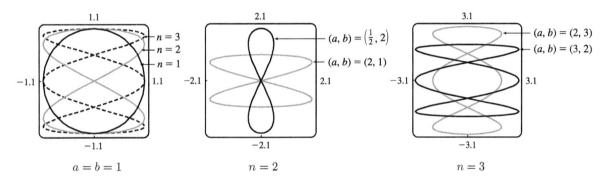

10.2 Calculus with Parametric Curves

1. $x = 2t^3 + 3t,\ y = 4t - 5t^2\ \Rightarrow\ \dfrac{dx}{dt} = 6t^2 + 3,\ \dfrac{dy}{dt} = 4 - 10t,\ \text{and}\ \dfrac{dy}{dx} = \dfrac{dy/dt}{dx/dt} = \dfrac{4 - 10t}{6t^2 + 3}$.

3. $x = te^t,\ y = t + \sin t\ \Rightarrow\ \dfrac{dx}{dt} = te^t + e^t = e^t(t+1),\ \dfrac{dy}{dt} = 1 + \cos t,\ \text{and}\ \dfrac{dy}{dx} = \dfrac{dy/dt}{dx/dt} = \dfrac{1 + \cos t}{e^t(t+1)}$.

5. $x = t^2 + 2t,\ y = 2^t - 2t;\ (15, 2).\ \dfrac{dy}{dt} = 2^t \ln 2 - 2,\ \dfrac{dx}{dt} = 2t + 2,\ \text{and}\ \dfrac{dy}{dx} = \dfrac{dy/dt}{dx/dt} = \dfrac{2^t \ln 2 - 2}{2t + 2}$.

At $(15, 2)$, $x = t^2 + 2t = 15\ \Rightarrow\ t^2 + 2t - 15 = 0\ \Rightarrow\ (t+5)(t-3) = 0\ \Rightarrow\ t = -5$ or $t = 3$. Only $t = 3$ gives $y = 2$. With $t = 3$, $\dfrac{dy}{dx} = \dfrac{2^3 \ln 2 - 2}{2(3) + 2} = \dfrac{4 \ln 2 - 1}{4} = \ln 2 - \dfrac{1}{4} \approx 0.44$.

7. $x = t^3 + 1,\ y = t^4 + t;\ t = -1.\ \dfrac{dy}{dt} = 4t^3 + 1,\ \dfrac{dx}{dt} = 3t^2,\ \text{and}\ \dfrac{dy}{dx} = \dfrac{dy/dt}{dx/dt} = \dfrac{4t^3 + 1}{3t^2}$. When $t = -1$, $(x, y) = (0, 0)$ and $dy/dx = -3/3 = -1$, so an equation of the tangent to the curve at the point corresponding to $t = -1$ is $y - 0 = -1(x - 0)$, or $y = -x$.

9. $x = \sin 2t + \cos t$, $y = \cos 2t - \sin t$; $t = \pi$. $\dfrac{dy}{dt} = -2\sin 2t - \cos t$, $\dfrac{dx}{dt} = 2\cos 2t - \sin t$, and

$\dfrac{dy}{dx} = \dfrac{dy/dt}{dx/dt} = \dfrac{-2\sin 2t - \cos t}{2\cos 2t - \sin t}$. When $t = \pi$, $(x,y) = (-1, 1)$, and $\dfrac{dy}{dx} = \dfrac{1}{2}$, so an equation of the tangent to the curve at the point corresponding to $t = \pi$ is $y - 1 = \dfrac{1}{2}[x - (-1)]$, or $y = \dfrac{1}{2}x + \dfrac{3}{2}$.

11. (a) $x = \sin t$, $y = \cos^2 t$; $\left(\dfrac{1}{2}, \dfrac{3}{4}\right)$. $\dfrac{dy}{dt} = 2\cos t(-\sin t)$, $\dfrac{dx}{dt} = \cos t$, and $\dfrac{dy}{dx} = \dfrac{dy/dt}{dx/dt} = \dfrac{-2\sin t\cos t}{\cos t} = -2\sin t$.

At $\left(\dfrac{1}{2}, \dfrac{3}{4}\right)$, $x = \sin t = \dfrac{1}{2}$ $\Rightarrow$ $t = \dfrac{\pi}{6}$, so $dy/dx = -2\sin\dfrac{\pi}{6} = -2\left(\dfrac{1}{2}\right) = -1$, and an equation of the tangent is $y - \dfrac{3}{4} = -1\left(x - \dfrac{1}{2}\right)$, or $y = -x + \dfrac{5}{4}$.

(b) $x = \sin t$ $\Rightarrow$ $x^2 = \sin^2 t = 1 - \cos^2 t = 1 - y$, so $y = 1 - x^2$, and $y' = -2x$. At $\left(\dfrac{1}{2}, \dfrac{3}{4}\right)$, $y' = -2 \cdot \dfrac{1}{2} = -1$, so an equation of the tangent is $y - \dfrac{3}{4} = -1\left(x - \dfrac{1}{2}\right)$, or $y = -x + \dfrac{5}{4}$.

13. $x = t^2 - t$, $y = t^2 + t + 1$; $(0, 3)$. $\dfrac{dy}{dx} = \dfrac{dy/dt}{dx/dt} = \dfrac{2t+1}{2t-1}$. To find the value of t corresponding to the point $(0, 3)$, solve $x = 0$ $\Rightarrow$ $t^2 - t = 0$ $\Rightarrow$ $t(t-1) = 0$ $\Rightarrow$ $t = 0$ or $t = 1$. Only $t = 1$ gives $y = 3$. With $t = 1$, $dy/dx = 3$, and an equation of the tangent is $y - 3 = 3(x - 0)$, or $y = 3x + 3$.

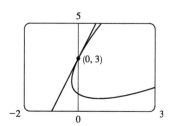

15. $x = t^2 + 1$, $y = t^2 + t$ $\Rightarrow$ $\dfrac{dy}{dx} = \dfrac{dy/dt}{dx/dt} = \dfrac{2t+1}{2t} = 1 + \dfrac{1}{2t}$ $\Rightarrow$ $\dfrac{d^2y}{dx^2} = \dfrac{\dfrac{d}{dt}\left(\dfrac{dy}{dx}\right)}{dx/dt} = \dfrac{-1/(2t^2)}{2t} = -\dfrac{1}{4t^3}$.

The curve is CU when $\dfrac{d^2y}{dx^2} > 0$, that is, when $t < 0$.

17. $x = e^t$, $y = te^{-t}$ $\Rightarrow$ $\dfrac{dy}{dx} = \dfrac{dy/dt}{dx/dt} = \dfrac{-te^{-t} + e^{-t}}{e^t} = \dfrac{e^{-t}(1-t)}{e^t} = e^{-2t}(1-t)$ $\Rightarrow$

$\dfrac{d^2y}{dx^2} = \dfrac{\dfrac{d}{dt}\left(\dfrac{dy}{dx}\right)}{dx/dt} = \dfrac{e^{-2t}(-1) + (1-t)(-2e^{-2t})}{e^t} = \dfrac{e^{-2t}(-1 - 2 + 2t)}{e^t} = e^{-3t}(2t - 3)$. The curve is CU when $\dfrac{d^2y}{dx^2} > 0$, that is, when $t > \dfrac{3}{2}$.

19. $x = t - \ln t$, $y = t + \ln t$ [note that $t > 0$] $\Rightarrow$ $\dfrac{dy}{dx} = \dfrac{dy/dt}{dx/dt} = \dfrac{1 + 1/t}{1 - 1/t} = \dfrac{t+1}{t-1}$ $\Rightarrow$

$\dfrac{d^2y}{dx^2} = \dfrac{\dfrac{d}{dt}\left(\dfrac{dy}{dx}\right)}{dx/dt} = \dfrac{\dfrac{(t-1)(1) - (t+1)(1)}{(t-1)^2}}{(t-1)/t} = \dfrac{-2t}{(t-1)^3}$. The curve is CU when $\dfrac{d^2y}{dx^2} > 0$, that is, when $0 < t < 1$.

21. $x = t^3 - 3t$, $y = t^2 - 3$. $\frac{dy}{dt} = 2t$, so $\frac{dy}{dt} = 0 \Leftrightarrow t = 0 \Leftrightarrow$
$(x, y) = (0, -3)$. $\frac{dx}{dt} = 3t^2 - 3 = 3(t+1)(t-1)$, so $\frac{dx}{dt} = 0 \Leftrightarrow$
$t = -1$ or $1 \Leftrightarrow (x, y) = (2, -2)$ or $(-2, -2)$. The curve has a horizontal
tangent at $(0, -3)$ and vertical tangents at $(2, -2)$ and $(-2, -2)$.

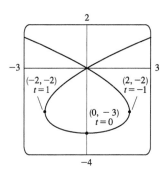

23. $x = \cos\theta$, $y = \cos 3\theta$. The whole curve is traced out for $0 \leq \theta \leq \pi$.

$\frac{dy}{d\theta} = -3\sin 3\theta$, so $\frac{dy}{d\theta} = 0 \Leftrightarrow \sin 3\theta = 0 \Leftrightarrow 3\theta = 0, \pi, 2\pi$, or $3\pi \Leftrightarrow$

$\theta = 0, \frac{\pi}{3}, \frac{2\pi}{3}$, or $\pi \Leftrightarrow (x, y) = (1, 1), (\frac{1}{2}, -1), (-\frac{1}{2}, 1)$, or $(-1, -1)$.

$\frac{dx}{d\theta} = -\sin\theta$, so $\frac{dx}{d\theta} = 0 \Leftrightarrow \sin\theta = 0 \Leftrightarrow \theta = 0$ or $\pi \Leftrightarrow$

$(x, y) = (1, 1)$ or $(-1, -1)$. Both $\frac{dy}{d\theta}$ and $\frac{dx}{d\theta}$ equal 0 when $\theta = 0$ and π.

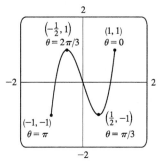

To find the slope when $\theta = 0$, we find $\lim_{\theta \to 0} \frac{dy}{dx} = \lim_{\theta \to 0} \frac{-3\sin 3\theta}{-\sin\theta} \stackrel{H}{=} \lim_{\theta \to 0} \frac{-9\cos 3\theta}{-\cos\theta} = 9$, which is the same slope when $\theta = \pi$.

Thus, the curve has horizontal tangents at $(\frac{1}{2}, -1)$ and $(-\frac{1}{2}, 1)$, and there are no vertical tangents.

25. From the graph, it appears that the rightmost point on the curve $x = t - t^6$, $y = e^t$
is about $(0.6, 2)$. To find the exact coordinates, we find the value of t for which the
graph has a vertical tangent, that is, $0 = dx/dt = 1 - 6t^5 \Leftrightarrow t = 1/\sqrt[5]{6}$.
Hence, the rightmost point is
$\left(1/\sqrt[5]{6} - 1/\left(6\sqrt[5]{6}\right), e^{1/\sqrt[5]{6}}\right) = \left(5 \cdot 6^{-6/5}, e^{6^{-1/5}}\right) \approx (0.58, 2.01)$.

27. We graph the curve $x = t^4 - 2t^3 - 2t^2$, $y = t^3 - t$ in the viewing rectangle $[-2, 1.1]$ by $[-0.5, 0.5]$. This rectangle
corresponds approximately to $t \in [-1, 0.8]$.

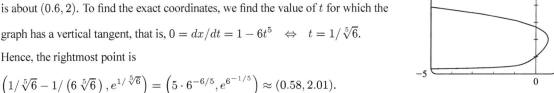

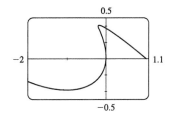
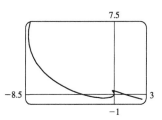

We estimate that the curve has horizontal tangents at about $(-1, -0.4)$ and $(-0.17, 0.39)$ and vertical tangents at
about $(0, 0)$ and $(-0.19, 0.37)$. We calculate $\frac{dy}{dx} = \frac{dy/dt}{dx/dt} = \frac{3t^2 - 1}{4t^3 - 6t^2 - 4t}$. The horizontal tangents occur when

$dy/dt = 3t^2 - 1 = 0 \Leftrightarrow t = \pm\frac{1}{\sqrt{3}}$, so both horizontal tangents are shown in our graph. The vertical tangents occur when

$dx/dt = 2t(2t^2 - 3t - 2) = 0 \iff 2t(2t+1)(t-2) = 0 \iff t = 0, -\frac{1}{2}$ or 2. It seems that we have missed one vertical tangent, and indeed if we plot the curve on the t-interval $[-1.2, 2.2]$ we see that there is another vertical tangent at $(-8, 6)$.

29. $x = \cos t$, $y = \sin t \cos t$. $dx/dt = -\sin t$,

$dy/dt = -\sin^2 t + \cos^2 t = \cos 2t$. $(x, y) = (0, 0) \iff \cos t = 0 \iff t$ is an odd multiple of $\frac{\pi}{2}$. When $t = \frac{\pi}{2}$, $dx/dt = -1$ and $dy/dt = -1$, so $dy/dx = 1$. When $t = \frac{3\pi}{2}$, $dx/dt = 1$ and $dy/dt = -1$. So $dy/dx = -1$. Thus, $y = x$ and $y = -x$ are both tangent to the curve at $(0, 0)$.

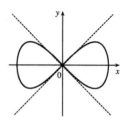

31. $x = r\theta - d\sin\theta$, $y = r - d\cos\theta$.

(a) $\dfrac{dx}{d\theta} = r - d\cos\theta$, $\dfrac{dy}{d\theta} = d\sin\theta$, so $\dfrac{dy}{dx} = \dfrac{d\sin\theta}{r - d\cos\theta}$.

(b) If $0 < d < r$, then $|d\cos\theta| \le d < r$, so $r - d\cos\theta \ge r - d > 0$. This shows that $dx/d\theta$ never vanishes, so the trochoid can have no vertical tangent if $d < r$.

33. $x = 3t^2 + 1$, $y = t^3 - 1 \implies \dfrac{dy}{dx} = \dfrac{dy/dt}{dx/dt} = \dfrac{3t^2}{6t} = \dfrac{t}{2}$. The tangent line has slope $\dfrac{1}{2}$ when $\dfrac{t}{2} = \dfrac{1}{2} \iff t = 1$, so the point is $(4, 0)$.

35. The curve $x = t^3 + 1$, $y = 2t - t^2 = t(2-t)$ intersects the x-axis when $y = 0$, that is, when $t = 0$ and $t = 2$. The corresponding values of x are 1 and 9. The shaded area is given by

$\displaystyle\int_{x=1}^{x=9}(y_T - y_B)\,dx = \int_{t=0}^{t=2}[y(t) - 0]\,x'(t)\,dt = \int_0^2 (2t - t^2)(3t^2)\,dt$

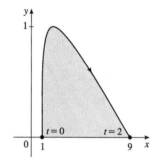

$= 3\int_0^2 (2t^3 - t^4)\,dt = 3\left[\tfrac{1}{2}t^4 - \tfrac{1}{5}t^5\right]_0^2 = 3\left(8 - \tfrac{32}{5}\right) = \tfrac{24}{5}$

37. The curve $x = \sin^2 t$, $y = \cos t$ intersects the y-axis when $x = 0$, that is, when $t = 0$ and $t = \pi$. (Any integer multiple of π will result in $x = 0$, though we choose two values of t over which the curve is traced out once.) The corresponding values of y are 1 and -1, so the area enclosed by the curve and the y-axis is given by

$\displaystyle\int_{y=-1}^{y=1} x\,dy = \int_{t=\pi}^{t=0} x(t)\,y'(t)\,dt = \int_\pi^0 \sin^2 t\,(-\sin t)\,dt = \int_0^\pi \sin^3 t\,dt \stackrel{67}{=} \left[-\tfrac{1}{3}(2 + \sin^2 t)\cos t\right]_0^\pi$

$= \tfrac{2}{3} - \left(-\tfrac{2}{3}\right) = \tfrac{4}{3}$

39. $x = a\cos\theta$, $y = b\sin\theta$, $0 \le \theta \le 2\pi$. By symmetry of the ellipse about the x- and y-axes,

$A = 4\displaystyle\int_{x=0}^{x=a} y\,dx = 4\int_{\theta=\pi/2}^{\theta=0} b\sin\theta\,(-a\sin\theta)\,d\theta = 4ab\int_0^{\pi/2}\sin^2\theta\,d\theta = 4ab\int_0^{\pi/2}\tfrac{1}{2}(1 - \cos 2\theta)\,d\theta$

$= 2ab\left[\theta - \tfrac{1}{2}\sin 2\theta\right]_0^{\pi/2} = 2ab\left(\tfrac{\pi}{2}\right) = \pi ab$

41. $x = r\theta - d\sin\theta$, $y = r - d\cos\theta$.

$A = \int_0^{2\pi r} y\, dx = \int_0^{2\pi}(r - d\cos\theta)(r - d\cos\theta)\, d\theta = \int_0^{2\pi}(r^2 - 2dr\cos\theta + d^2\cos^2\theta)\, d\theta$

$= \left[r^2\theta - 2dr\sin\theta + \tfrac{1}{2}d^2\left(\theta + \tfrac{1}{2}\sin 2\theta\right)\right]_0^{2\pi} = 2\pi r^2 + \pi d^2$

43. $x = 3t^2 - t^3$, $y = t^2 - 2t$. $dx/dt = 6t - 3t^2$ and $dy/dt = 2t - 2$, so

$(dx/dt)^2 + (dy/dt)^2 = (6t - 3t^2)^2 + (2t - 2)^2 = 36t^2 - 36t^3 + 9t^4 + 4t^2 - 8t + 4 = 9t^4 - 36t^3 + 40t^2 - 8t + 4$. The

endpoints of the curve both have $y = 3$, so the value of t at these points must satisfy $t^2 - 2t = 3 \Rightarrow t^2 - 2t - 3 = 0 \Rightarrow$

$(t + 1)(t - 3) = 0 \Rightarrow t = -1$ or $t = 3$. Thus,

$$L = \int_a^b \sqrt{(dx/dt)^2 + (dy/dt)^2}\, dt = \int_{-1}^3 \sqrt{9t^4 - 36t^3 + 40t^2 - 8t + 4}\, dt \approx 15.2092$$

45. $x = t - 2\sin t$, $y = 1 - 2\cos t$, $0 \le t \le 4\pi$. $dx/dt = 1 - 2\cos t$ and $dy/dt = 2\sin t$, so

$(dx/dt)^2 + (dy/dt)^2 = (1 - 2\cos t)^2 + (2\sin t)^2 = 1 - 4\cos t + 4\cos^2 t + 4\sin^2 t = 5 - 4\cos t$. Thus,

$L = \int_a^b \sqrt{(dx/dt)^2 + (dy/dt)^2}\, dt = \int_0^{4\pi} \sqrt{5 - 4\cos t}\, dt \approx 26.7298$.

47. $x = \tfrac{2}{3}t^3$, $y = t^2 - 2$, $0 \le t \le 3$. $dx/dt = 2t^2$ and $dy/dt = 2t$, so $(dx/dt)^2 + (dy/dt)^2 = 4t^4 + 4t^2 = 4t^2(t^2 + 1)$.

Thus,

$$L = \int_a^b \sqrt{(dx/dt)^2 + (dy/dt)^2}\, dt = \int_0^3 \sqrt{4t^2(t^2 + 1)}\, dt = \int_0^3 2t\sqrt{t^2 + 1}\, dt$$

$$= \int_1^{10} \sqrt{u}\, du \quad [u = t^2 + 1, du = 2t\, dt] \quad = \left[\tfrac{2}{3}u^{3/2}\right]_1^{10} = \tfrac{2}{3}(10^{3/2} - 1) = \tfrac{2}{3}\left(10\sqrt{10} - 1\right)$$

49. $x = t\sin t$, $y = t\cos t$, $0 \le t \le 1$. $\dfrac{dx}{dt} = t\cos t + \sin t$ and $\dfrac{dy}{dt} = -t\sin t + \cos t$, so

$$\left(\frac{dx}{dt}\right)^2 + \left(\frac{dy}{dt}\right)^2 = t^2\cos^2 t + 2t\sin t\cos t + \sin^2 t + t^2\sin^2 t - 2t\sin t\cos t + \cos^2 t$$

$$= t^2(\cos^2 t + \sin^2 t) + \sin^2 t + \cos^2 t = t^2 + 1.$$

Thus, $L = \int_0^1 \sqrt{t^2 + 1}\, dt \stackrel{21}{=} \left[\tfrac{1}{2}t\sqrt{t^2 + 1} + \tfrac{1}{2}\ln\left(t + \sqrt{t^2 + 1}\right)\right]_0^1 = \tfrac{1}{2}\sqrt{2} + \tfrac{1}{2}\ln\left(1 + \sqrt{2}\right)$.

51.

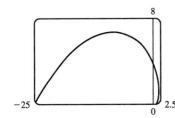

$x = e^t\cos t$, $y = e^t\sin t$, $0 \le t \le \pi$.

$\left(\dfrac{dx}{dt}\right)^2 + \left(\dfrac{dy}{dt}\right)^2 = [e^t(\cos t - \sin t)]^2 + [e^t(\sin t + \cos t)]^2$

$= (e^t)^2(\cos^2 t - 2\cos t\sin t + \sin^2 t)$

$\quad + (e^t)^2(\sin^2 t + 2\sin t\cos t + \cos^2 t)$

$= e^{2t}(2\cos^2 t + 2\sin^2 t) = 2e^{2t}$

Thus, $L = \int_0^\pi \sqrt{2e^{2t}}\, dt = \int_0^\pi \sqrt{2}\, e^t\, dt = \sqrt{2}\left[e^t\right]_0^\pi = \sqrt{2}\,(e^\pi - 1)$.

53. 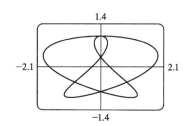 The figure shows the curve $x = \sin t + \sin 1.5t$, $y = \cos t$ for $0 \le t \le 4\pi$.
$dx/dt = \cos t + 1.5 \cos 1.5t$ and $dy/dt = -\sin t$, so
$(dx/dt)^2 + (dy/dt)^2 = \cos^2 t + 3 \cos t \cos 1.5t + 2.25 \cos^2 1.5t + \sin^2 t$.
Thus, $L = \int_0^{4\pi} \sqrt{1 + 3 \cos t \cos 1.5t + 2.25 \cos^2 1.5t} \, dt \approx 16.7102$.

55. $x = \sin^2 t$, $y = \cos^2 t$, $0 \le t \le 3\pi$.
$(dx/dt)^2 + (dy/dt)^2 = (2 \sin t \cos t)^2 + (-2 \cos t \sin t)^2 = 8 \sin^2 t \cos^2 t = 2 \sin^2 2t \Rightarrow$

Distance $= \int_0^{3\pi} \sqrt{2} |\sin 2t| \, dt = 6\sqrt{2} \int_0^{\pi/2} \sin 2t \, dt$ [by symmetry] $= -3\sqrt{2} \left[\cos 2t\right]_0^{\pi/2} = -3\sqrt{2}(-1 - 1) = 6\sqrt{2}$.

The full curve is traversed as t goes from 0 to $\frac{\pi}{2}$, because the curve is the segment of $x + y = 1$ that lies in the first quadrant (since $x, y \ge 0$), and this segment is completely traversed as t goes from 0 to $\frac{\pi}{2}$. Thus, $L = \int_0^{\pi/2} \sin 2t \, dt = \sqrt{2}$, as above.

57. $x = 2t - 3$, $y = 2t^2 - 3t + 6$. $dx/dt = 2$ and $dy/dt = 4t - 3$, so $v(t) = s'(t) = \sqrt{2^2 + (4t-3)^2}$. Thus, the speed of the particle at $t = 5$ is $v(5) = \sqrt{4 + (4 \cdot 5 - 3)^2} = \sqrt{293} \approx 17.12$ m/s.

59. $x = e^t$, $y = te^t$. $dx/dt = e^t$ and $dy/dt = te^t + e^t$, so $v(t) = s'(t) = \sqrt{(e^t)^2 + (te^t + e^t)^2}$. At (e, e), $x = e^t = e \Rightarrow t = 1$. Thus, the speed of the particle at (e, e) is $v(1) = \sqrt{e^2 + (e + e)^2} = \sqrt{5e^2} = \sqrt{5}\, e \approx 6.08$ m/s.

61. $x = (v_0 \cos \alpha)t$, $y = (v_0 \sin \alpha)t - \frac{1}{2}gt^2$. $dx/dt = v_0 \cos \alpha$ and $dy/dt = v_0 \sin \alpha - gt$, so
speed $= v(t) = s'(t) = \sqrt{(dx/dt)^2 + (dy/dt)^2} = \sqrt{v_0^2 \cos^2 \alpha + (v_0 \sin \alpha - gt)^2}$.

(a) The projectile hits the ground when $y = 0 \Rightarrow (v_0 \sin \alpha)t - \frac{1}{2}gt^2 = 0 \Rightarrow t(v_0 \sin \alpha - \frac{1}{2}gt) = 0 \Rightarrow t = 0$
or $t = \dfrac{2v_0 \sin \alpha}{g}$. The second solution gives the time at which the projectile hits the ground, and at this time it will have a speed of

$$v\left(\frac{2v_0 \sin \alpha}{g}\right) = \sqrt{v_0^2 \cos^2 \alpha + \left[v_0 \sin \alpha - g\left(\frac{2v_0 \sin \alpha}{g}\right)\right]^2} = \sqrt{v_0^2 \cos^2 \alpha + (-v_0 \sin \alpha)^2}$$
$$= \sqrt{v_0^2 \cos^2 \alpha + v_0^2 \sin^2 \alpha} = \sqrt{v_0^2(\cos^2 \alpha + \sin^2 \alpha)} = \sqrt{v_0^2} = v_0 \text{ m/s.}$$

Thus, the projectile hits the ground with the same speed at which it was fired.

(b) The projectile is at its highest point (maximum height) when $dy/dt = 0$. Thus, the speed of the projectile at this time is
$v(t) = \sqrt{(dx/dt)^2 + (dy/dt)^2} = \sqrt{v_0^2 \cos^2 \alpha + (0)^2} = v_0 \cos \alpha$ m/s.

63. (a) $x = 11 \cos t - 4 \cos(11t/2)$, $y = 11 \sin t - 4 \sin(11t/2)$.

Notice that $0 \le t \le 2\pi$ does not give the complete curve because $x(0) \ne x(2\pi)$. In fact, we must take $t \in [0, 4\pi]$ in order to obtain the complete curve, since the first term in each of the parametric equations has period 2π and the second has period $\frac{2\pi}{11/2} = \frac{4\pi}{11}$, and the least common integer multiple of these two numbers is 4π.

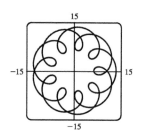

(b) We use the CAS to find the derivatives dx/dt and dy/dt, and then use Theorem 5 to find the arc length. Recent versions of Maple express the integral $\int_0^{4\pi}\sqrt{(dx/dt)^2+(dy/dt)^2}\,dt$ as $88E(2\sqrt{2}\,i)$, where $E(x)$ is the elliptic integral $\int_0^1 \frac{\sqrt{1-x^2t^2}}{\sqrt{1-t^2}}\,dt$ and i is the imaginary number $\sqrt{-1}$.

Some earlier versions of Maple (as well as Mathematica) cannot do the integral exactly, so we use the command `evalf(Int(sqrt(diff(x,t)^2+diff(y,t)^2),t=0..4*Pi));` to estimate the length, and find that the arc length is approximately 294.03.

65. $x = a\cos^3\theta$, $y = a\sin^3\theta$. By symmetry,

$A = 4\int_0^a y\,dx = 4\int_{\pi/2}^0 a\sin^3\theta(-3a\cos^2\theta\sin\theta)\,d\theta = 12a^2\int_0^{\pi/2}\sin^4\theta\cos^2\theta\,d\theta$. Now

$\int \sin^4\theta\cos^2\theta\,d\theta = \int \sin^2\theta(\frac{1}{4}\sin^2 2\theta)\,d\theta = \frac{1}{8}\int(1-\cos 2\theta)\sin^2 2\theta\,d\theta$

$\quad = \frac{1}{8}\int\left[\frac{1}{2}(1-\cos 4\theta)-\sin^2 2\theta\cos 2\theta\right]d\theta = \frac{1}{16}\theta - \frac{1}{64}\sin 4\theta - \frac{1}{48}\sin^3 2\theta + C$

so $\int_0^{\pi/2}\sin^4\theta\cos^2\theta\,d\theta = \left[\frac{1}{16}\theta - \frac{1}{64}\sin 4\theta - \frac{1}{48}\sin^3 2\theta\right]_0^{\pi/2} = \frac{\pi}{32}$. Thus, $A = 12a^2\left(\frac{\pi}{32}\right) = \frac{3}{8}\pi a^2$.

67. $x = t\sin t$, $y = t\cos t$, $0 \le t \le \pi/2$. $dx/dt = t\cos t + \sin t$ and $dy/dt = -t\sin t + \cos t$, so

$(dx/dt)^2 + (dy/dt)^2 = t^2\cos^2 t + 2t\sin t\cos t + \sin^2 t + t^2\sin^2 t - 2t\sin t\cos t + \cos^2 t$
$\quad = t^2(\cos^2 t + \sin^2 t) + \sin^2 t + \cos^2 t = t^2 + 1$

$S = \int 2\pi y\,ds = \int_0^{\pi/2} 2\pi t\cos t\sqrt{t^2+1}\,dt \approx 4.7394$.

69. $x = t + e^t$, $y = e^{-t}$, $0 \le t \le 1$.

$dx/dt = 1 + e^t$ and $dy/dt = -e^{-t}$, so $(dx/dt)^2 + (dy/dt)^2 = (1+e^t)^2 + (-e^{-t})^2 = 1 + 2e^t + e^{2t} + e^{-2t}$.

$S = \int 2\pi y\,ds = \int_0^1 2\pi e^{-t}\sqrt{1 + 2e^t + e^{2t} + e^{-2t}}\,dt \approx 10.6705$.

71. $x = t^3$, $y = t^2$, $0 \le t \le 1$. $\left(\frac{dx}{dt}\right)^2 + \left(\frac{dy}{dt}\right)^2 = (3t^2)^2 + (2t)^2 = 9t^4 + 4t^2$.

$S = \int_0^1 2\pi y\sqrt{\left(\frac{dx}{dt}\right)^2 + \left(\frac{dy}{dt}\right)^2}\,dt = \int_0^1 2\pi t^2\sqrt{9t^4 + 4t^2}\,dt = 2\pi\int_0^1 t^2\sqrt{t^2(9t^2+4)}\,dt$

$\quad = 2\pi\int_4^{13}\left(\frac{u-4}{9}\right)\sqrt{u}\left(\frac{1}{18}du\right) \quad \begin{bmatrix} u = 9t^2 + 4,\ t^2 = (u-4)/9, \\ du = 18t\,dt,\text{ so } t\,dt = \frac{1}{18}du \end{bmatrix} = \frac{2\pi}{9\cdot 18}\int_4^{13}(u^{3/2} - 4u^{1/2})\,du$

$\quad = \frac{\pi}{81}\left[\frac{2}{5}u^{5/2} - \frac{8}{3}u^{3/2}\right]_4^{13} = \frac{\pi}{81}\cdot\frac{2}{15}\left[3u^{5/2} - 20u^{3/2}\right]_4^{13}$

$\quad = \frac{2\pi}{1215}\left[(3\cdot 13^2\sqrt{13} - 20\cdot 13\sqrt{13}) - (3\cdot 32 - 20\cdot 8)\right] = \frac{2\pi}{1215}(247\sqrt{13} + 64)$

73. $x = a\cos^3\theta$, $y = a\sin^3\theta$, $0 \le \theta \le \frac{\pi}{2}$. $\left(\frac{dx}{d\theta}\right)^2 + \left(\frac{dy}{d\theta}\right)^2 = (-3a\cos^2\theta\sin\theta)^2 + (3a\sin^2\theta\cos\theta)^2 = 9a^2\sin^2\theta\cos^2\theta$.

$S = \int_0^{\pi/2} 2\pi\cdot a\sin^3\theta\cdot 3a\sin\theta\cos\theta\,d\theta = 6\pi a^2\int_0^{\pi/2}\sin^4\theta\cos\theta\,d\theta = \frac{6}{5}\pi a^2\left[\sin^5\theta\right]_0^{\pi/2} = \frac{6}{5}\pi a^2$

75. $x = 3t^2$, $y = 2t^3$, $0 \le t \le 5$ $\Rightarrow$ $\left(\frac{dx}{dt}\right)^2 + \left(\frac{dy}{dt}\right)^2 = (6t)^2 + (6t^2)^2 = 36t^2(1+t^2)$ $\Rightarrow$

$S = \int_0^5 2\pi x \sqrt{(dx/dt)^2 + (dy/dt)^2}\, dt = \int_0^5 2\pi(3t^2) 6t\sqrt{1+t^2}\, dt = 18\pi \int_0^5 t^2 \sqrt{1+t^2}\, 2t\, dt$

$= 18\pi \int_1^{26} (u-1)\sqrt{u}\, du$ $\begin{bmatrix} u = 1+t^2, \\ du = 2t\, dt \end{bmatrix}$ $= 18\pi \int_1^{26} (u^{3/2} - u^{1/2})\, du = 18\pi \left[\frac{2}{5} u^{5/2} - \frac{2}{3} u^{3/2}\right]_1^{26}$

$= 18\pi \left[\left(\frac{2}{5} \cdot 676\sqrt{26} - \frac{2}{3} \cdot 26\sqrt{26}\right) - \left(\frac{2}{5} - \frac{2}{3}\right)\right] = \frac{24}{5}\pi (949\sqrt{26} + 1)$

77. If f' is continuous and $f'(t) \ne 0$ for $a \le t \le b$, then either $f'(t) > 0$ for all t in $[a,b]$ or $f'(t) < 0$ for all t in $[a,b]$. Thus, f is monotonic (in fact, strictly increasing or strictly decreasing) on $[a,b]$. It follows that f has an inverse. Set $F = g \circ f^{-1}$, that is, define F by $F(x) = g(f^{-1}(x))$. Then $x = f(t)$ $\Rightarrow$ $f^{-1}(x) = t$, so $y = g(t) = g(f^{-1}(x)) = F(x)$.

79. $\phi = \tan^{-1}\left(\frac{dy}{dx}\right)$ $\Rightarrow$ $\frac{d\phi}{dt} = \frac{d}{dt}\tan^{-1}\left(\frac{dy}{dx}\right) = \frac{1}{1+(dy/dx)^2}\left[\frac{d}{dt}\left(\frac{dy}{dx}\right)\right]$. But $\frac{dy}{dx} = \frac{dy/dt}{dx/dt} = \frac{\dot{y}}{\dot{x}}$ $\Rightarrow$

$\frac{d}{dt}\left(\frac{dy}{dx}\right) = \frac{d}{dt}\left(\frac{\dot{y}}{\dot{x}}\right) = \frac{\ddot{y}\dot{x} - \ddot{x}\dot{y}}{\dot{x}^2}$ $\Rightarrow$ $\frac{d\phi}{dt} = \frac{1}{1+(\dot{y}/\dot{x})^2}\left(\frac{\ddot{y}\dot{x} - \ddot{x}\dot{y}}{\dot{x}^2}\right) = \frac{\dot{x}\ddot{y} - \ddot{x}\dot{y}}{\dot{x}^2 + \dot{y}^2}$. Using the Chain Rule, and the fact

that $s = \int_0^t \sqrt{\left(\frac{dx}{dt}\right)^2 + \left(\frac{dy}{dt}\right)^2}\, dt$ $\Rightarrow$ $\frac{ds}{dt} = \sqrt{\left(\frac{dx}{dt}\right)^2 + \left(\frac{dy}{dt}\right)^2} = (\dot{x}^2 + \dot{y}^2)^{1/2}$, we have that

$\frac{d\phi}{ds} = \frac{d\phi/dt}{ds/dt} = \left(\frac{\dot{x}\ddot{y} - \ddot{x}\dot{y}}{\dot{x}^2 + \dot{y}^2}\right)\frac{1}{(\dot{x}^2+\dot{y}^2)^{1/2}} = \frac{\dot{x}\ddot{y} - \ddot{x}\dot{y}}{(\dot{x}^2+\dot{y}^2)^{3/2}}$. So $\kappa = \left|\frac{d\phi}{ds}\right| = \left|\frac{\dot{x}\ddot{y} - \ddot{x}\dot{y}}{(\dot{x}^2+\dot{y}^2)^{3/2}}\right| = \frac{|\dot{x}\ddot{y} - \ddot{x}\dot{y}|}{(\dot{x}^2+\dot{y}^2)^{3/2}}$.

81. $x = \theta - \sin\theta$ $\Rightarrow$ $\dot{x} = 1 - \cos\theta$ $\Rightarrow$ $\ddot{x} = \sin\theta$, and $y = 1 - \cos\theta$ $\Rightarrow$ $\dot{y} = \sin\theta$ $\Rightarrow$ $\ddot{y} = \cos\theta$. Therefore,

$\kappa = \frac{|\cos\theta - \cos^2\theta - \sin^2\theta|}{[(1-\cos\theta)^2 + \sin^2\theta]^{3/2}} = \frac{|\cos\theta - (\cos^2\theta + \sin^2\theta)|}{(1 - 2\cos\theta + \cos^2\theta + \sin^2\theta)^{3/2}} = \frac{|\cos\theta - 1|}{(2 - 2\cos\theta)^{3/2}}$. The top of the arch is

characterized by a horizontal tangent, and from Example 2(b) in Section 10.2, the tangent is horizontal when $\theta = (2n-1)\pi$,

so take $n = 1$ and substitute $\theta = \pi$ into the expression for κ: $\kappa = \frac{|\cos\pi - 1|}{(2 - 2\cos\pi)^{3/2}} = \frac{|-1-1|}{[2-2(-1)]^{3/2}} = \frac{1}{4}$.

83. (a) Every straight line has parametrizations of the form $x = a + vt$, $y = b + wt$, where a, b are arbitrary and $v, w \ne 0$.

For example, a straight line passing through distinct points (a, b) and (c, d) can be described as the parametrized curve $x = a + (c-a)t$, $y = b + (d-b)t$. Starting with $x = a + vt$, $y = b + wt$, we compute $\dot{x} = v$, $\dot{y} = w$, $\ddot{x} = \ddot{y} = 0$,

and $\kappa = \frac{|v \cdot 0 - w \cdot 0|}{(v^2 + w^2)^{3/2}} = 0$.

(b) Parametric equations for a circle of radius r are $x = r\cos\theta$ and $y = r\sin\theta$. We can take the center to be the origin.
So $\dot{x} = -r\sin\theta$ $\Rightarrow$ $\ddot{x} = -r\cos\theta$ and $\dot{y} = r\cos\theta$ $\Rightarrow$ $\ddot{y} = -r\sin\theta$. Therefore,

$\kappa = \frac{|r^2\sin^2\theta + r^2\cos^2\theta|}{(r^2\sin^2\theta + r^2\cos^2\theta)^{3/2}} = \frac{r^2}{r^3} = \frac{1}{r}$. And so for any θ (and thus any point), $\kappa = \frac{1}{r}$.

85. The coordinates of T are $(r\cos\theta, r\sin\theta)$. Since TP was unwound from arc TA, TP has length $r\theta$. Also $\angle PTQ = \angle PTR - \angle QTR = \frac{1}{2}\pi - \theta$, so P has coordinates $x = r\cos\theta + r\theta\cos(\frac{1}{2}\pi - \theta) = r(\cos\theta + \theta\sin\theta)$, $y = r\sin\theta - r\theta\sin(\frac{1}{2}\pi - \theta) = r(\sin\theta - \theta\cos\theta)$.

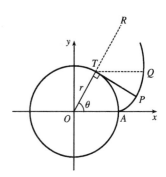

10.3 Polar Coordinates

1. (a) $\left(1, \frac{\pi}{4}\right)$

By adding 2π to $\frac{\pi}{4}$, we obtain the point $\left(1, \frac{9\pi}{4}\right)$, which satisfies the $r > 0$ requirement. The direction opposite $\frac{\pi}{4}$ is $\frac{5\pi}{4}$, so $\left(-1, \frac{5\pi}{4}\right)$ is a point that satisfies the $r < 0$ requirement.

(b) $\left(-2, \frac{3\pi}{2}\right)$

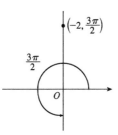

$r > 0$: $\left(-(-2), \frac{3\pi}{2} - \pi\right) = \left(2, \frac{\pi}{2}\right)$

$r < 0$: $\left(-2, \frac{3\pi}{2} + 2\pi\right) = \left(-2, \frac{7\pi}{2}\right)$

(c) $\left(3, -\frac{\pi}{3}\right)$

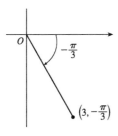

$r > 0$: $\left(3, -\frac{\pi}{3} + 2\pi\right) = \left(3, \frac{5\pi}{3}\right)$

$r < 0$: $\left(-3, -\frac{\pi}{3} + \pi\right) = \left(-3, \frac{2\pi}{3}\right)$

3. (a)

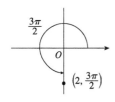

$x = 2\cos\frac{3\pi}{2} = 2(0) = 0$ and $y = 2\sin\frac{3\pi}{2} = 2(-1) = -2$ give us the Cartesian coordinates $(0, -2)$.

(b)

$x = \sqrt{2}\cos\frac{\pi}{4} = \sqrt{2}\left(\frac{1}{\sqrt{2}}\right) = 1$ and $y = \sqrt{2}\sin\frac{\pi}{4} = \sqrt{2}\left(\frac{1}{\sqrt{2}}\right) = 1$ give us the Cartesian coordinates $(1, 1)$.

(c) $x = -1\cos\left(-\frac{\pi}{6}\right) = -1\left(\frac{\sqrt{3}}{2}\right) = -\frac{\sqrt{3}}{2}$ and

$y = -1\sin\left(-\frac{\pi}{6}\right) = -1\left(-\frac{1}{2}\right) = \frac{1}{2}$ give us the Cartesian

coordinates $\left(-\frac{\sqrt{3}}{2}, \frac{1}{2}\right)$.

5. (a) $x = -4$ and $y = 4$ $\Rightarrow$ $r = \sqrt{(-4)^2 + 4^2} = 4\sqrt{2}$ and $\tan\theta = \frac{4}{-4} = -1$ $[\theta = -\frac{\pi}{4} + n\pi]$. Since $(-4, 4)$ is in the second quadrant, the polar coordinates are (i) $\left(4\sqrt{2}, \frac{3\pi}{4}\right)$ and (ii) $\left(-4\sqrt{2}, \frac{7\pi}{4}\right)$.

(b) $x = 3$ and $y = 3\sqrt{3}$ $\Rightarrow$ $r = \sqrt{3^2 + \left(3\sqrt{3}\right)^2} = \sqrt{9 + 27} = 6$ and $\tan\theta = \frac{3\sqrt{3}}{3} = \sqrt{3}$ $[\theta = \frac{\pi}{3} + n\pi]$.

Since $(3, 3\sqrt{3})$ is in the first quadrant, the polar coordinates are (i) $\left(6, \frac{\pi}{3}\right)$ and (ii) $\left(-6, \frac{4\pi}{3}\right)$.

7. $1 < r \le 3$. The curves $r = 1$ and $r = 3$ represent circles centered at O with radius 1 and 3, respectively. So $1 < r \le 3$ represents the region outside the radius 1 circle and on or inside the radius 3 circle. Note that θ can take on any value.

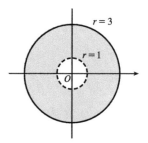

9. $0 \le r \le 1$, $-\pi/2 \le \theta \le \pi/2$. This is the region on or inside the circle $r = 1$ in the first and fourth quadrants.

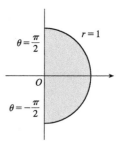

11. $2 \le r < 4$, $3\pi/4 \le \theta \le 7\pi/4$

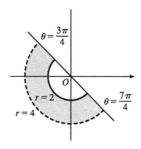

13. Converting the polar coordinates $\left(4, \frac{4\pi}{3}\right)$ and $\left(6, \frac{5\pi}{3}\right)$ to Cartesian coordinates gives us $\left(4\cos\frac{4\pi}{3}, 4\sin\frac{4\pi}{3}\right) = \left(-2, -2\sqrt{3}\right)$ and $\left(6\cos\frac{5\pi}{3}, 6\sin\frac{5\pi}{3}\right) = \left(3, -3\sqrt{3}\right)$. Now use the distance formula

$$d = \sqrt{(x_2 - x_1)^2 + (y_2 - y_1)^2} = \sqrt{[3 - (-2)]^2 + \left[-3\sqrt{3} - \left(-2\sqrt{3}\right)\right]^2}$$

$$= \sqrt{5^2 + \left(-\sqrt{3}\right)^2} = \sqrt{25 + 3} = \sqrt{28} = 2\sqrt{7}$$

15. $r^2 = 5 \Leftrightarrow x^2 + y^2 = 5$, a circle of radius $\sqrt{5}$ centered at the origin.

17. $r = 5\cos\theta \Rightarrow r^2 = 5r\cos\theta \Leftrightarrow x^2 + y^2 = 5x \Leftrightarrow x^2 - 5x + \frac{25}{4} + y^2 = \frac{25}{4} \Leftrightarrow \left(x - \frac{5}{2}\right)^2 + y^2 = \frac{25}{4}$,
a circle of radius $\frac{5}{2}$ centered at $\left(\frac{5}{2}, 0\right)$. The first two equations are actually equivalent since $r^2 = 5r\cos\theta \Rightarrow r(r - 5\cos\theta) = 0 \Rightarrow r = 0$ or $r = 5\cos\theta$. But $r = 5\cos\theta$ gives the point $r = 0$ (the pole) when $\theta = 0$. Thus, the equation $r = 5\cos\theta$ is equivalent to the compound condition ($r = 0$ or $r = 5\cos\theta$).

19. $r^2 \cos 2\theta = 1 \Leftrightarrow r^2(\cos^2\theta - \sin^2\theta) = 1 \Leftrightarrow (r\cos\theta)^2 - (r\sin\theta)^2 = 1 \Leftrightarrow x^2 - y^2 = 1$, a hyperbola centered at the origin with foci on the x-axis.

21. $x^2 + y^2 = 7 \Rightarrow (r\cos\theta)^2 + (r\sin\theta)^2 = 7 \Rightarrow r^2(\cos^2\theta + \sin^2\theta) = 7 \Rightarrow r^2 = 7 \Rightarrow r = \sqrt{7}$.
Note that $r = -\sqrt{7}$ produces the same curve as $r = \sqrt{7}$.

23. $y = \sqrt{3}\,x \Rightarrow \frac{y}{x} = \sqrt{3} \;[x \neq 0] \Rightarrow \tan\theta = \sqrt{3} \Rightarrow \theta = \frac{\pi}{3}$ or $\frac{4\pi}{3}$ [either includes the pole]

25. $x^2 + y^2 = 4y \Rightarrow r^2 = 4r\sin\theta \Rightarrow r^2 - 4r\sin\theta = 0 \Rightarrow r(r - 4\sin\theta) = 0 \Rightarrow r = 0$ or $r = 4\sin\theta$. $r = 0$ is included in $r = 4\sin\theta$ when $\theta = 0$, so the curve is represented by the single equation $r = 4\sin\theta$.

27. (a) The description leads immediately to the polar equation $\theta = \frac{\pi}{6}$, and the Cartesian equation $y = \tan\left(\frac{\pi}{6}\right)x = \frac{1}{\sqrt{3}}x$ is slightly more difficult to derive.

(b) The easier description here is the Cartesian equation $x = 3$.

29. For $\theta = 0, \pi$, and 2π, r has its minimum value of about 0.5. For $\theta = \frac{\pi}{2}$ and $\frac{3\pi}{2}$, r attains its maximum value of 2. We see that the graph has a similar shape for $0 \leq \theta \leq \pi$ and $\pi \leq \theta \leq 2\pi$.

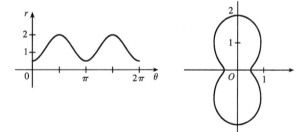

31. r has a maximum value of approximately 1 slightly before $\theta = \frac{\pi}{4}$ and slightly after $\theta = \frac{7\pi}{4}$. r has a minimum value of -2 when $\theta = \pi$. The graph touches the pole ($r = 0$) when $\theta = 0, \frac{\pi}{2}, \frac{3\pi}{2}$, and 2π. Since r is positive in the θ-intervals $\left(0, \frac{\pi}{2}\right)$ and $\left(\frac{3\pi}{2}, 2\pi\right)$, and negative in the interval $\left(\frac{\pi}{2}, \frac{3\pi}{2}\right)$, the graph lies entirely in the first and fourth quadrants.

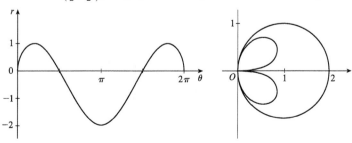

33. $r = -2\sin\theta$

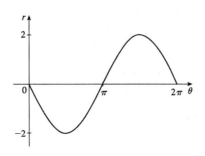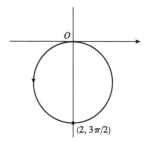

35. $r = 2(1 + \cos\theta)$

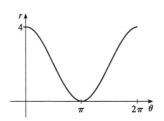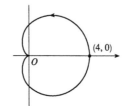

37. $r = \theta$, $\theta \geq 0$

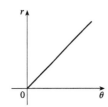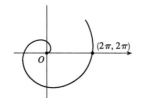

39. $r = 3\cos 3\theta$

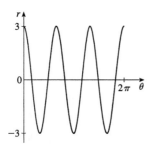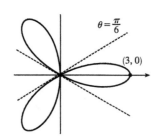

41. $r = 2\cos 4\theta$

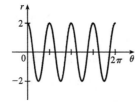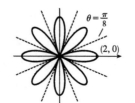

43. $r = 1 + 3\cos\theta$

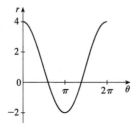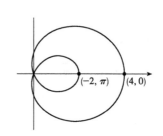

45. $r^2 = 9\sin 2\theta$

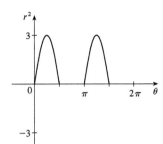
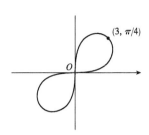

47. $r = 2 + \sin 3\theta$

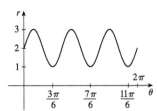
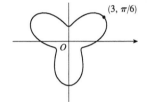

49. $r = \sin(\theta/2)$

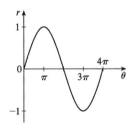
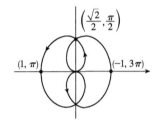

51. $x = r\cos\theta = (4 + 2\sec\theta)\cos\theta = 4\cos\theta + 2$. Now, $r \to \infty$ $\Rightarrow$

$(4 + 2\sec\theta) \to \infty$ $\Rightarrow$ $\theta \to \left(\frac{\pi}{2}\right)^-$ or $\theta \to \left(\frac{3\pi}{2}\right)^+$ [since we need only

consider $0 \le \theta < 2\pi$], so $\lim\limits_{r \to \infty} x = \lim\limits_{\theta \to \pi/2^-}(4\cos\theta + 2) = 2$. Also,

$r \to -\infty$ $\Rightarrow$ $(4 + 2\sec\theta) \to -\infty$ $\Rightarrow$ $\theta \to \left(\frac{\pi}{2}\right)^+$ or $\theta \to \left(\frac{3\pi}{2}\right)^-$, so

$\lim\limits_{r \to -\infty} x = \lim\limits_{\theta \to \pi/2^+}(4\cos\theta + 2) = 2$. Therefore, $\lim\limits_{r \to \pm\infty} x = 2$ $\Rightarrow$ $x = 2$ is a vertical asymptote.

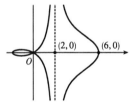

53. To show that $x = 1$ is an asymptote, we must prove $\lim\limits_{r \to \pm\infty} x = 1$.

$x = (r)\cos\theta = (\sin\theta \tan\theta)\cos\theta = \sin^2\theta$. Now, $r \to \infty$ $\Rightarrow$ $\sin\theta \tan\theta \to \infty$ $\Rightarrow$

$\theta \to \left(\frac{\pi}{2}\right)^-$, so $\lim\limits_{r \to \infty} x = \lim\limits_{\theta \to \pi/2^-}\sin^2\theta = 1$. Also, $r \to -\infty$ $\Rightarrow$ $\sin\theta \tan\theta \to -\infty$ $\Rightarrow$

$\theta \to \left(\frac{\pi}{2}\right)^+$, so $\lim\limits_{r \to -\infty} x = \lim\limits_{\theta \to \pi/2^+}\sin^2\theta = 1$. Therefore, $\lim\limits_{r \to \pm\infty} x = 1$ $\Rightarrow$ $x = 1$ is

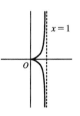

a vertical asymptote. Also notice that $x = \sin^2\theta \ge 0$ for all θ, and $x = \sin^2\theta \le 1$ for all θ. And $x \ne 1$, since the curve is not defined at odd multiples of $\frac{\pi}{2}$. Therefore, the curve lies entirely within the vertical strip $0 \le x < 1$.

55. (a) We see that the curve $r = 1 + c\sin\theta$ crosses itself at the origin, where $r = 0$ (in fact the inner loop corresponds to negative r-values,) so we solve the equation of the limaçon for $r = 0 \Leftrightarrow c\sin\theta = -1 \Leftrightarrow \sin\theta = -1/c$. Now if $|c| < 1$, then this equation has no solution and hence there is no inner loop. But if $c < -1$, then on the interval $(0, 2\pi)$ the equation has the two solutions $\theta = \sin^{-1}(-1/c)$ and $\theta = \pi - \sin^{-1}(-1/c)$, and if $c > 1$, the solutions are $\theta = \pi + \sin^{-1}(1/c)$ and $\theta = 2\pi - \sin^{-1}(1/c)$. In each case, $r < 0$ for θ between the two solutions, indicating a loop.

(b) For $0 < c < 1$, the dimple (if it exists) is characterized by the fact that y has a local maximum at $\theta = \frac{3\pi}{2}$. So we determine for what c-values $\dfrac{d^2y}{d\theta^2}$ is negative at $\theta = \frac{3\pi}{2}$, since by the Second Derivative Test this indicates a maximum:

$$y = r\sin\theta = \sin\theta + c\sin^2\theta \Rightarrow \frac{dy}{d\theta} = \cos\theta + 2c\sin\theta\cos\theta = \cos\theta + c\sin 2\theta \Rightarrow \frac{d^2y}{d\theta^2} = -\sin\theta + 2c\cos 2\theta.$$

At $\theta = \frac{3\pi}{2}$, this is equal to $-(-1) + 2c(-1) = 1 - 2c$, which is negative only for $c > \frac{1}{2}$. A similar argument shows that for $-1 < c < 0$, y only has a local minimum at $\theta = \frac{\pi}{2}$ (indicating a dimple) for $c < -\frac{1}{2}$.

57. $r = a\sin\theta + b\cos\theta \Rightarrow r^2 = ar\sin\theta + br\cos\theta \Rightarrow x^2 + y^2 = ay + bx \Rightarrow$
$x^2 - bx + \left(\frac{1}{2}b\right)^2 + y^2 - ay + \left(\frac{1}{2}a\right)^2 = \left(\frac{1}{2}b\right)^2 + \left(\frac{1}{2}a\right)^2 \Rightarrow \left(x - \frac{1}{2}b\right)^2 + \left(y - \frac{1}{2}a\right)^2 = \frac{1}{4}(a^2 + b^2)$, and this is a circle with center $\left(\frac{1}{2}b, \frac{1}{2}a\right)$ and radius $\frac{1}{2}\sqrt{a^2 + b^2}$.

59. $r = 1 + 2\sin(\theta/2)$. The parameter interval is $[0, 4\pi]$.

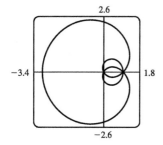

61. $r = e^{\sin\theta} - 2\cos(4\theta)$.

The parameter interval is $[0, 2\pi]$.

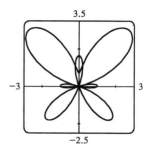

63. $r = 1 + \cos^{999}\theta$. The parameter interval is $[0, 2\pi]$.

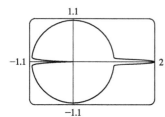

65. It appears that the graph of $r = 1 + \sin\left(\theta - \frac{\pi}{6}\right)$ is the same shape as the graph of $r = 1 + \sin\theta$, but rotated counterclockwise about the origin by $\frac{\pi}{6}$. Similarly, the graph of $r = 1 + \sin\left(\theta - \frac{\pi}{3}\right)$ is rotated by $\frac{\pi}{3}$. In general, the graph of $r = f(\theta - \alpha)$ is the same shape as that of $r = f(\theta)$, but rotated counterclockwise through α about the origin. That is, for any point (r_0, θ_0) on the curve $r = f(\theta)$, the point $(r_0, \theta_0 + \alpha)$ is on the curve $r = f(\theta - \alpha)$, since $r_0 = f(\theta_0) = f((\theta_0 + \alpha) - \alpha)$.

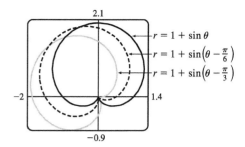

67. Consider curves with polar equation $r = 1 + c\cos\theta$, where c is a real number. If $c = 0$, we get a circle of radius 1 centered at the pole. For $0 < c \le 0.5$, the curve gets slightly larger, moves right, and flattens out a bit on the left side. For $0.5 < c < 1$, the left side has a dimple shape. For $c = 1$, the dimple becomes a cusp. For $c > 1$, there is an internal loop. For $c \ge 0$, the rightmost point on the curve is $(1 + c, 0)$. For $c < 0$, the curves are reflections through the vertical axis of the curves with $c > 0$.

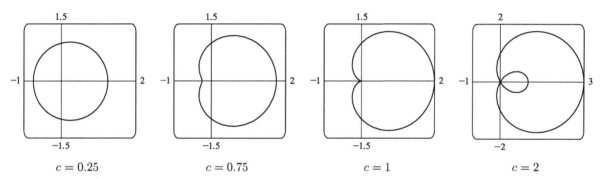

10.4 Calculus in Polar Coordinates

1. $r = \sqrt{2\theta}$, $0 \le \theta \le \pi/2$.

$$A = \int_0^{\pi/2} \frac{1}{2} r^2 \, d\theta = \int_0^{\pi/2} \frac{1}{2} \left(\sqrt{2\theta}\right)^2 d\theta = \int_0^{\pi/2} \theta \, d\theta = \left[\frac{1}{2}\theta^2\right]_0^{\pi/2} = \frac{1}{2}\left(\frac{\pi}{2}\right)^2 = \frac{\pi^2}{8}$$

3. $r = \sin\theta + \cos\theta$, $0 \le \theta \le \pi$.

$$A = \int_0^\pi \frac{1}{2} r^2 \, d\theta = \int_0^\pi \frac{1}{2}(\sin\theta + \cos\theta)^2 \, d\theta = \int_0^\pi \frac{1}{2}(\sin^2\theta + 2\sin\theta\cos\theta + \cos^2\theta) \, d\theta = \int_0^\pi \frac{1}{2}(1 + \sin 2\theta) \, d\theta$$

$$= \frac{1}{2}\left[\theta - \frac{1}{2}\cos 2\theta\right]_0^\pi = \frac{1}{2}\left[\left(\pi - \frac{1}{2}\right) - \left(0 - \frac{1}{2}\right)\right] = \frac{\pi}{2}$$

5. $r^2 = \sin 2\theta$, $0 \le \theta \le \pi/2$.

$$A = \int_0^{\pi/2} \frac{1}{2} r^2 \, d\theta = \int_0^{\pi/2} \frac{1}{2}\sin 2\theta \, d\theta = \left[-\frac{1}{4}\cos 2\theta\right]_0^{\pi/2} = -\frac{1}{4}(\cos\pi - \cos 0) = -\frac{1}{4}(-1 - 1) = \frac{1}{2}$$

7. $r = 4 + 3\sin\theta$, $-\frac{\pi}{2} \le \theta \le \frac{\pi}{2}$.

$$A = \int_{-\pi/2}^{\pi/2} \tfrac{1}{2}(4+3\sin\theta)^2\, d\theta = \tfrac{1}{2}\int_{-\pi/2}^{\pi/2}(16+24\sin\theta+9\sin^2\theta)\,d\theta$$

$$= \tfrac{1}{2}\int_{-\pi/2}^{\pi/2}(16+9\sin^2\theta)\,d\theta \quad \text{[by Theorem 4.5.6]}$$

$$= \tfrac{1}{2}\cdot 2\int_{0}^{\pi/2}\left[16+9\cdot\tfrac{1}{2}(1-\cos 2\theta)\right]d\theta \quad \text{[by Theorem 4.5.6]}$$

$$= \int_{0}^{\pi/2}\left(\tfrac{41}{2}-\tfrac{9}{2}\cos 2\theta\right)d\theta = \left[\tfrac{41}{2}\theta - \tfrac{9}{4}\sin 2\theta\right]_0^{\pi/2} = \left(\tfrac{41\pi}{4}-0\right)-(0-0) = \tfrac{41\pi}{4}$$

9. The area is bounded by $r = 4\cos\theta$ for $\theta = 0$ to $\theta = \pi$.

$$A = \int_0^\pi \tfrac{1}{2}r^2\, d\theta = \int_0^\pi \tfrac{1}{2}(4\cos\theta)^2 d\theta = \int_0^\pi 8\cos^2\theta\, d\theta$$

$$= 8\int_0^\pi \tfrac{1}{2}(1+\cos 2\theta)\,d\theta = 4\left[\theta + \tfrac{1}{2}\sin 2\theta\right]_0^\pi = 4\pi$$

Also, note that this is a circle with radius 2, so its area is $\pi(2)^2 = 4\pi$.

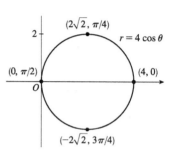

11. $A = \int_0^{2\pi}\tfrac{1}{2}r^2\, d\theta = \int_0^{2\pi}\tfrac{1}{2}(3-2\sin\theta)^2\,d\theta$

$$= \tfrac{1}{2}\int_0^{2\pi}(9-12\sin\theta+4\sin^2\theta)\,d\theta$$

$$= \tfrac{1}{2}\int_0^{2\pi}\left[9-12\sin\theta+4\cdot\tfrac{1}{2}(1-\cos 2\theta)\right]d\theta$$

$$= \tfrac{1}{2}\int_0^{2\pi}(11-12\sin\theta-2\cos 2\theta)\,d\theta = \tfrac{1}{2}\left[11\theta+12\cos\theta-\sin 2\theta\right]_0^{2\pi}$$

$$= \tfrac{1}{2}[(22\pi+12)-12] = 11\pi$$

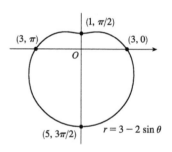

13. $A = \int_0^{2\pi}\tfrac{1}{2}r^2\,d\theta = \int_0^{2\pi}\tfrac{1}{2}(2+\sin 4\theta)^2\,d\theta = \tfrac{1}{2}\int_0^{2\pi}(4+4\sin 4\theta+\sin^2 4\theta)\,d\theta$

$$= \tfrac{1}{2}\int_0^{2\pi}\left[4+4\sin 4\theta + \tfrac{1}{2}(1-\cos 8\theta)\right]d\theta$$

$$= \tfrac{1}{2}\int_0^{2\pi}\left(\tfrac{9}{2}+4\sin 4\theta - \tfrac{1}{2}\cos 8\theta\right)d\theta = \tfrac{1}{2}\left[\tfrac{9}{2}\theta - \cos 4\theta - \tfrac{1}{16}\sin 8\theta\right]_0^{2\pi}$$

$$= \tfrac{1}{2}[(9\pi - 1) - (-1)] = \tfrac{9}{2}\pi$$

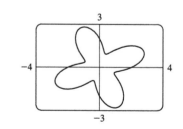

15. $A = \int_0^{2\pi}\tfrac{1}{2}r^2\,d\theta = \int_0^{2\pi}\tfrac{1}{2}\left(\sqrt{1+\cos^2 5\theta}\right)^2 d\theta$

$$= \tfrac{1}{2}\int_0^{2\pi}(1+\cos^2 5\theta)\,d\theta = \tfrac{1}{2}\int_0^{2\pi}\left[1+\tfrac{1}{2}(1+\cos 10\theta)\right]d\theta$$

$$= \tfrac{1}{2}\left[\tfrac{3}{2}\theta + \tfrac{1}{20}\sin 10\theta\right]_0^{2\pi} = \tfrac{1}{2}(3\pi) = \tfrac{3}{2}\pi$$

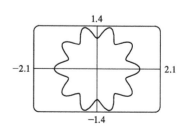

17. The curve passes through the pole when $r = 0 \Rightarrow 4\cos 3\theta = 0 \Rightarrow \cos 3\theta = 0 \Rightarrow 3\theta = \frac{\pi}{2} + \pi n \Rightarrow$

$\theta = \frac{\pi}{6} + \frac{\pi}{3}n$. The part of the shaded loop above the polar axis is traced out for $\theta = 0$ to $\theta = \pi/6$, so we'll use $-\pi/6$ and $\pi/6$ as our limits of integration.

$A = \int_{-\pi/6}^{\pi/6} \frac{1}{2}(4\cos 3\theta)^2 \, d\theta = 2\int_{0}^{\pi/6} \frac{1}{2}(16\cos^2 3\theta) \, d\theta$

$= 16\int_{0}^{\pi/6} \frac{1}{2}(1 + \cos 6\theta) \, d\theta = 8\left[\theta + \frac{1}{6}\sin 6\theta\right]_{0}^{\pi/6} = 8\left(\frac{\pi}{6}\right) = \frac{4}{3}\pi$

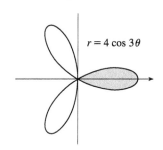

19. $r = 0 \Rightarrow \sin 4\theta = 0 \Rightarrow 4\theta = \pi n \Rightarrow \theta = \frac{\pi}{4}n$.

$A = \int_{0}^{\pi/4} \frac{1}{2}(\sin 4\theta)^2 \, d\theta = \frac{1}{2}\int_{0}^{\pi/4} \sin^2 4\theta \, d\theta = \frac{1}{2}\int_{0}^{\pi/4} \frac{1}{2}(1 - \cos 8\theta) \, d\theta$

$= \frac{1}{4}\left[\theta - \frac{1}{8}\sin 8\theta\right]_{0}^{\pi/4} = \frac{1}{4}\left(\frac{\pi}{4}\right) = \frac{1}{16}\pi$

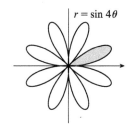

21.

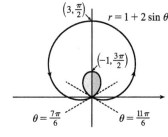

This is a limaçon, with inner loop traced out between $\theta = \frac{7\pi}{6}$ and $\frac{11\pi}{6}$ [found by solving $r = 0$].

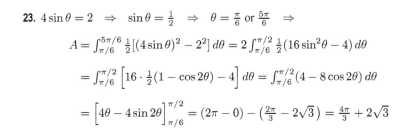

$A = 2\int_{7\pi/6}^{3\pi/2} \frac{1}{2}(1 + 2\sin\theta)^2 \, d\theta = \int_{7\pi/6}^{3\pi/2} (1 + 4\sin\theta + 4\sin^2\theta) \, d\theta = \int_{7\pi/6}^{3\pi/2} \left[1 + 4\sin\theta + 4\cdot\frac{1}{2}(1 - \cos 2\theta)\right] d\theta$

$= \left[\theta - 4\cos\theta + 2\theta - \sin 2\theta\right]_{7\pi/6}^{3\pi/2} = \left(\frac{9\pi}{2}\right) - \left(\frac{7\pi}{2} + 2\sqrt{3} - \frac{\sqrt{3}}{2}\right) = \pi - \frac{3\sqrt{3}}{2}$

23. $4\sin\theta = 2 \Rightarrow \sin\theta = \frac{1}{2} \Rightarrow \theta = \frac{\pi}{6}$ or $\frac{5\pi}{6} \Rightarrow$

$A = \int_{\pi/6}^{5\pi/6} \frac{1}{2}[(4\sin\theta)^2 - 2^2] \, d\theta = 2\int_{\pi/6}^{\pi/2} \frac{1}{2}(16\sin^2\theta - 4) \, d\theta$

$= \int_{\pi/6}^{\pi/2}\left[16\cdot\frac{1}{2}(1 - \cos 2\theta) - 4\right] d\theta = \int_{\pi/6}^{\pi/2}(4 - 8\cos 2\theta) \, d\theta$

$= \left[4\theta - 4\sin 2\theta\right]_{\pi/6}^{\pi/2} = (2\pi - 0) - \left(\frac{2\pi}{3} - 2\sqrt{3}\right) = \frac{4\pi}{3} + 2\sqrt{3}$

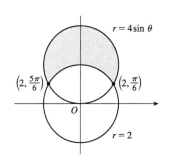

25. To find the area inside the leminiscate $r^2 = 8\cos 2\theta$ and outside the circle $r = 2$, we first note that the two curves intersect when $r^2 = 8\cos 2\theta$ and $r = 2$, that is, when $\cos 2\theta = \frac{1}{2}$. For $-\pi < \theta \leq \pi$, $\cos 2\theta = \frac{1}{2} \Leftrightarrow 2\theta = \pm\pi/3$ or $\pm 5\pi/3 \Leftrightarrow \theta = \pm\pi/6$ or $\pm 5\pi/6$. The figure shows that the desired area is 4 times the area between the curves from 0 to $\pi/6$. Thus,

$$A = 4\int_0^{\pi/6}\left[\tfrac{1}{2}(8\cos 2\theta) - \tfrac{1}{2}(2)^2\right]d\theta = 8\int_0^{\pi/6}(2\cos 2\theta - 1)\,d\theta$$

$$= 8\left[\sin 2\theta - \theta\right]_0^{\pi/6} = 8(\sqrt{3}/2 - \pi/6) = 4\sqrt{3} - 4\pi/3$$

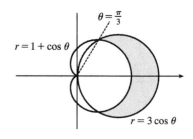

27. $3\cos\theta = 1 + \cos\theta \Leftrightarrow \cos\theta = \tfrac{1}{2} \Rightarrow \theta = \tfrac{\pi}{3}$ or $-\tfrac{\pi}{3}$.

$$A = 2\int_0^{\pi/3}\tfrac{1}{2}[(3\cos\theta)^2 - (1+\cos\theta)^2]\,d\theta$$

$$= \int_0^{\pi/3}(8\cos^2\theta - 2\cos\theta - 1)\,d\theta = \int_0^{\pi/3}[4(1+\cos 2\theta) - 2\cos\theta - 1]\,d\theta$$

$$= \int_0^{\pi/3}(3 + 4\cos 2\theta - 2\cos\theta)\,d\theta = \left[3\theta + 2\sin 2\theta - 2\sin\theta\right]_0^{\pi/3}$$

$$= \pi + \sqrt{3} - \sqrt{3} = \pi$$

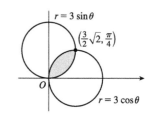

29. $3\sin\theta = 3\cos\theta \Rightarrow \dfrac{3\sin\theta}{3\cos\theta} = 1 \Rightarrow \tan\theta = 1 \Rightarrow \theta = \tfrac{\pi}{4} \Rightarrow$

$$A = 2\int_0^{\pi/4}\tfrac{1}{2}(3\sin\theta)^2\,d\theta = \int_0^{\pi/4}9\sin^2\theta\,d\theta = \int_0^{\pi/4}9\cdot\tfrac{1}{2}(1-\cos 2\theta)\,d\theta$$

$$= \int_0^{\pi/4}\left(\tfrac{9}{2} - \tfrac{9}{2}\cos 2\theta\right)d\theta = \left[\tfrac{9}{2}\theta - \tfrac{9}{4}\sin 2\theta\right]_0^{\pi/4} = \left(\tfrac{9\pi}{8} - \tfrac{9}{4}\right) - (0-0)$$

$$= \tfrac{9\pi}{8} - \tfrac{9}{4}$$

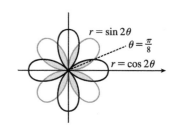

31. $\sin 2\theta = \cos 2\theta \Rightarrow \dfrac{\sin 2\theta}{\cos 2\theta} = 1 \Rightarrow \tan 2\theta = 1 \Rightarrow 2\theta = \tfrac{\pi}{4} \Rightarrow$
$\theta = \tfrac{\pi}{8} \Rightarrow$

$$A = 8\cdot 2\int_0^{\pi/8}\tfrac{1}{2}\sin^2 2\theta\,d\theta = 8\int_0^{\pi/8}\tfrac{1}{2}(1-\cos 4\theta)\,d\theta$$

$$= 4\left[\theta - \tfrac{1}{4}\sin 4\theta\right]_0^{\pi/8} = 4\left(\tfrac{\pi}{8} - \tfrac{1}{4}\cdot 1\right) = \tfrac{\pi}{2} - 1$$

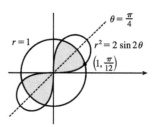

33. From the figure, we see that the shaded region is 4 times the shaded region from $\theta = 0$ to $\theta = \pi/4$. $r^2 = 2\sin 2\theta$ and $r = 1 \Rightarrow$
$2\sin 2\theta = 1^2 \Rightarrow \sin 2\theta = \tfrac{1}{2} \Rightarrow 2\theta = \tfrac{\pi}{6} \Rightarrow \theta = \tfrac{\pi}{12}$.

$$A = 4\int_0^{\pi/12}\tfrac{1}{2}(2\sin 2\theta)\,d\theta + 4\int_{\pi/12}^{\pi/4}\tfrac{1}{2}(1)^2\,d\theta$$

$$= \int_0^{\pi/12}4\sin 2\theta\,d\theta + \int_{\pi/12}^{\pi/4}2\,d\theta = \left[-2\cos 2\theta\right]_0^{\pi/12} + \left[2\theta\right]_{\pi/12}^{\pi/4}$$

$$= (-\sqrt{3} + 2) + \left(\tfrac{\pi}{2} - \tfrac{\pi}{6}\right) = -\sqrt{3} + 2 + \tfrac{\pi}{3}$$

35. The darker shaded region (from $\theta = 0$ to $\theta = 2\pi/3$) represents $\frac{1}{2}$ of the desired area plus $\frac{1}{2}$ of the area of the inner loop. From this area, we'll subtract $\frac{1}{2}$ of the area of the inner loop (the lighter shaded region from $\theta = 2\pi/3$ to $\theta = \pi$), and then double that difference to obtain the desired area.

$$A = 2\left[\int_0^{2\pi/3} \tfrac{1}{2}\left(\tfrac{1}{2} + \cos\theta\right)^2 d\theta - \int_{2\pi/3}^{\pi} \tfrac{1}{2}\left(\tfrac{1}{2} + \cos\theta\right)^2 d\theta\right]$$

$$= \int_0^{2\pi/3}\left(\tfrac{1}{4} + \cos\theta + \cos^2\theta\right) d\theta - \int_{2\pi/3}^{\pi}\left(\tfrac{1}{4} + \cos\theta + \cos^2\theta\right) d\theta$$

$$= \int_0^{2\pi/3}\left[\tfrac{1}{4} + \cos\theta + \tfrac{1}{2}(1 + \cos 2\theta)\right] d\theta$$
$$\quad - \int_{2\pi/3}^{\pi}\left[\tfrac{1}{4} + \cos\theta + \tfrac{1}{2}(1 + \cos 2\theta)\right] d\theta$$

$$= \left[\tfrac{\theta}{4} + \sin\theta + \tfrac{\theta}{2} + \tfrac{\sin 2\theta}{4}\right]_0^{2\pi/3} - \left[\tfrac{\theta}{4} + \sin\theta + \tfrac{\theta}{2} + \tfrac{\sin 2\theta}{4}\right]_{2\pi/3}^{\pi}$$

$$= \left(\tfrac{\pi}{6} + \tfrac{\sqrt{3}}{2} + \tfrac{\pi}{3} - \tfrac{\sqrt{3}}{8}\right) - \left(\tfrac{\pi}{4} + \tfrac{\pi}{2}\right) + \left(\tfrac{\pi}{6} + \tfrac{\sqrt{3}}{2} + \tfrac{\pi}{3} - \tfrac{\sqrt{3}}{8}\right)$$

$$= \tfrac{\pi}{4} + \tfrac{3}{4}\sqrt{3} = \tfrac{1}{4}\left(\pi + 3\sqrt{3}\right)$$

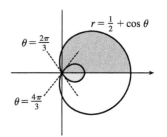

37. The pole is a point of intersection. $\sin\theta = 1 - \sin\theta \;\Rightarrow\; 2\sin\theta = 1 \;\Rightarrow\;$ $\sin\theta = \tfrac{1}{2} \;\Rightarrow\; \theta = \tfrac{\pi}{6}$ or $\tfrac{5\pi}{6}$. So the other points of intersection are $\left(\tfrac{1}{2}, \tfrac{\pi}{6}\right)$ and $\left(\tfrac{1}{2}, \tfrac{5\pi}{6}\right)$.

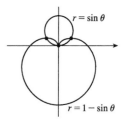

39. $2\sin 2\theta = 1 \;\Rightarrow\; \sin 2\theta = \tfrac{1}{2} \;\Rightarrow\; 2\theta = \tfrac{\pi}{6}, \tfrac{5\pi}{6}, \tfrac{13\pi}{6}$, or $\tfrac{17\pi}{6}$.

By symmetry, the eight points of intersection are given by

$(1, \theta)$, where $\theta = \tfrac{\pi}{12}, \tfrac{5\pi}{12}, \tfrac{13\pi}{12}$, and $\tfrac{17\pi}{12}$, and

$(-1, \theta)$, where $\theta = \tfrac{7\pi}{12}, \tfrac{11\pi}{12}, \tfrac{19\pi}{12}$, and $\tfrac{23\pi}{12}$.

[There are many ways to describe these points.]

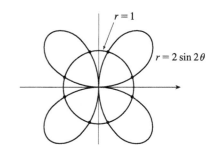

41. $r = 1$ and $r^2 = 2\cos 2\theta \;\Rightarrow\; 1^2 = 2\cos 2\theta \;\Rightarrow\; \cos 2\theta = \tfrac{1}{2} \;\Rightarrow\;$ $2\theta = \tfrac{\pi}{3}, \tfrac{5\pi}{3}, \tfrac{7\pi}{3}$, or $\tfrac{11\pi}{3} \;\Rightarrow\; \theta = \tfrac{\pi}{6}, \tfrac{5\pi}{6}, \tfrac{7\pi}{6}$, or $\tfrac{11\pi}{6}$. Thus, the four points of intersection are $\left(1, \tfrac{\pi}{6}\right), \left(1, \tfrac{5\pi}{6}\right), \left(1, \tfrac{7\pi}{6}\right)$, and $\left(1, \tfrac{11\pi}{6}\right)$.

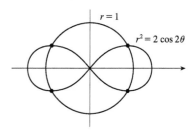

43. The shaded region lies outside the rose $r = \sin 2\theta$ and inside the limaçon $r = 3 + 2\cos\theta$, so its area is given by

$$A = \int_0^{2\pi} \tfrac{1}{2}\left[(3+2\cos\theta)^2 - (\sin 2\theta)^2\right] d\theta = \tfrac{1}{2}\int_0^{2\pi} (9 + 12\cos\theta + 4\cos^2\theta - \sin^2 2\theta)\, d\theta$$

$$= \tfrac{1}{2}\int_0^{2\pi}\left[9 + 12\cos\theta + 4 \cdot \tfrac{1}{2}(1+\cos 2\theta) - \tfrac{1}{2}(1-\cos 4\theta)\right] d\theta$$

$$= \tfrac{1}{2}\int_0^{2\pi}\left[\tfrac{21}{2} + 12\cos\theta + 2\cos 2\theta + \tfrac{1}{2}\cos 4\theta\right] d\theta$$

$$= \tfrac{1}{2}\left[\tfrac{21}{2}\theta + 12\sin\theta + \sin 2\theta + \tfrac{1}{8}\sin 4\theta\right]_0^{2\pi} = \tfrac{1}{2}(21\pi) = \tfrac{21\pi}{2}$$

45. $1 + \cos\theta = 3\cos\theta \Rightarrow 1 = 2\cos\theta \Rightarrow \cos\theta = \tfrac{1}{2} \Rightarrow \theta = \tfrac{\pi}{3}$. The area swept out by $r = 1 + \cos\theta$, $\pi/3 \le \theta \le \pi$, contains the shaded region plus the portion of the circle $r = 3\cos\theta$, $\pi/3 \le \theta \le \pi/2$. Thus, the area of the shaded region is given by

$$A = \int_{\pi/3}^{\pi} \tfrac{1}{2}(1+\cos\theta)^2\, d\theta - \int_{\pi/3}^{\pi/2} \tfrac{1}{2}(3\cos\theta)^2\, d\theta = \tfrac{1}{2}\int_{\pi/3}^{\pi}(1 + 2\cos\theta + \cos^2\theta)\, d\theta - \tfrac{9}{2}\int_{\pi/3}^{\pi/2} \cos^2\theta\, d\theta$$

$$= \tfrac{1}{2}\int_{\pi/3}^{\pi}\left[1 + 2\cos\theta + \tfrac{1}{2}(1+\cos 2\theta)\right] d\theta - \tfrac{9}{2}\int_{\pi/3}^{\pi/2} \tfrac{1}{2}(1+\cos 2\theta)\, d\theta$$

$$= \tfrac{1}{2}\int_{\pi/3}^{\pi}\left(\tfrac{3}{2} + 2\cos\theta + \tfrac{1}{2}\cos 2\theta\right) d\theta - \tfrac{9}{4}\int_{\pi/3}^{\pi/2}(1+\cos 2\theta)\, d\theta$$

$$= \tfrac{1}{2}\left[\tfrac{3}{2}\theta + 2\sin\theta + \tfrac{1}{4}\sin 2\theta\right]_{\pi/3}^{\pi} - \tfrac{9}{4}\left[\theta + \tfrac{1}{2}\sin 2\theta\right]_{\pi/3}^{\pi/2}$$

$$= \tfrac{1}{2}\left[\tfrac{3\pi}{2} - \left(\tfrac{\pi}{2} + \sqrt{3} + \tfrac{\sqrt{3}}{8}\right)\right] - \tfrac{9}{4}\left[\tfrac{\pi}{2} - \left(\tfrac{\pi}{3} + \tfrac{\sqrt{3}}{4}\right)\right] = \tfrac{1}{2}\left(\pi - \tfrac{9\sqrt{3}}{8}\right) - \tfrac{9}{4}\left(\tfrac{\pi}{6} - \tfrac{\sqrt{3}}{4}\right) = \tfrac{\pi}{2} - \tfrac{3\pi}{8} = \tfrac{\pi}{8}$$

47.

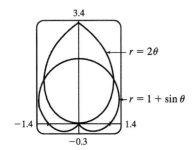
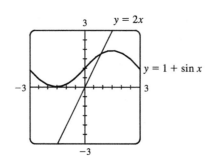

From the first graph, we see that the pole is one point of intersection. By zooming in or using the cursor, we find the θ-values of the intersection points to be $\alpha \approx 0.88786 \approx 0.89$ and $\pi - \alpha \approx 2.25$. (The first of these values may be more easily estimated by plotting $y = 1 + \sin x$ and $y = 2x$ in rectangular coordinates; see the second graph.) By symmetry, the total area contained is twice the area contained in the first quadrant, that is,

$$A = 2\int_0^{\alpha} \tfrac{1}{2}(2\theta)^2\, d\theta + 2\int_{\alpha}^{\pi/2} \tfrac{1}{2}(1+\sin\theta)^2\, d\theta = \int_0^{\alpha} 4\theta^2\, d\theta + \int_{\alpha}^{\pi/2}\left[1 + 2\sin\theta + \tfrac{1}{2}(1 - \cos 2\theta)\right] d\theta$$

$$= \left[\tfrac{4}{3}\theta^3\right]_0^{\alpha} + \left[\theta - 2\cos\theta + \left(\tfrac{1}{2}\theta - \tfrac{1}{4}\sin 2\theta\right)\right]_{\alpha}^{\pi/2} = \tfrac{4}{3}\alpha^3 + \left[\left(\tfrac{\pi}{2} + \tfrac{\pi}{4}\right) - \left(\alpha - 2\cos\alpha + \tfrac{1}{2}\alpha - \tfrac{1}{4}\sin 2\alpha\right)\right] \approx 3.4645$$

49. $L = \int_a^b \sqrt{r^2 + (dr/d\theta)^2}\, d\theta = \int_0^\pi \sqrt{(2\cos\theta)^2 + (-2\sin\theta)^2}\, d\theta$

$= \int_0^\pi \sqrt{4(\cos^2\theta + \sin^2\theta)}\, d\theta = \int_0^\pi \sqrt{4}\, d\theta = [2\theta]_0^\pi = 2\pi$

As a check, note that the curve is a circle of radius 1, so its circumference is $2\pi(1) = 2\pi$.

51. $L = \int_a^b \sqrt{r^2 + (dr/d\theta)^2}\, d\theta = \int_0^{2\pi} \sqrt{(\theta^2)^2 + (2\theta)^2}\, d\theta = \int_0^{2\pi} \sqrt{\theta^4 + 4\theta^2}\, d\theta$

$= \int_0^{2\pi} \sqrt{\theta^2(\theta^2 + 4)}\, d\theta = \int_0^{2\pi} \theta\sqrt{\theta^2 + 4}\, d\theta$

Now let $u = \theta^2 + 4$, so that $du = 2\theta\, d\theta$ $\left[\theta\, d\theta = \tfrac{1}{2}\, du\right]$ and

$\int_0^{2\pi} \theta\sqrt{\theta^2 + 4}\, d\theta = \int_4^{4\pi^2 + 4} \tfrac{1}{2}\sqrt{u}\, du = \tfrac{1}{2} \cdot \tfrac{2}{3}\left[u^{3/2}\right]_4^{4(\pi^2+1)} = \tfrac{1}{3}[4^{3/2}(\pi^2+1)^{3/2} - 4^{3/2}] = \tfrac{8}{3}[(\pi^2+1)^{3/2} - 1]$

53. The blue section of the curve $r = 3 + 3\sin\theta$ is traced from $\theta = -\pi/2$ to $\theta = \pi$.

$r^2 + (dr/d\theta)^2 = (3 + 3\sin\theta)^2 + (3\cos\theta)^2 = 9 + 18\sin\theta + 9\sin^2\theta + 9\cos^2\theta = 18 + 18\sin\theta$

$L = \int_{-\pi/2}^\pi \sqrt{18 + 18\sin\theta}\, d\theta = \sqrt{18} \int_{-\pi/2}^\pi \sqrt{1 + \sin\theta}\, d\theta = \sqrt{18} \int_{-\pi/2}^\pi \sqrt{\dfrac{(1+\sin\theta)(1-\sin\theta)}{1-\sin\theta}}\, d\theta$

$= \sqrt{18} \int_{-\pi/2}^\pi \sqrt{\dfrac{1-\sin^2\theta}{1-\sin\theta}}\, d\theta = \sqrt{18} \int_{-\pi/2}^\pi \sqrt{\dfrac{\cos^2\theta}{1-\sin\theta}}\, d\theta = \sqrt{18} \int_{-\pi/2}^\pi \dfrac{|\cos\theta|}{\sqrt{1-\sin\theta}}\, d\theta$

$= \sqrt{18} \left(\int_{-\pi/2}^{\pi/2} \dfrac{\cos\theta}{\sqrt{1-\sin\theta}}\, d\theta + \int_{\pi/2}^\pi \dfrac{-\cos\theta}{\sqrt{1-\sin\theta}}\, d\theta \right)$

$\stackrel{s}{=} \sqrt{18}\left(\int_{-1}^1 \dfrac{1}{\sqrt{1-u}}\, du - \int_1^0 \dfrac{1}{\sqrt{1-u}}\, du \right) = \sqrt{18}\left(\left[-2(1-u)^{1/2}\right]_{-1}^1 - \left[-2(1-u)^{1/2}\right]_1^0 \right)$

$= \sqrt{18}\,(2\sqrt{2} + 2) = 12 + 6\sqrt{2}$

55. The curve $r = \cos^4(\theta/4)$ is completely traced with $0 \le \theta \le 4\pi$.

$r^2 + (dr/d\theta)^2 = [\cos^4(\theta/4)]^2 + \left[4\cos^3(\theta/4) \cdot (-\sin(\theta/4)) \cdot \tfrac{1}{4}\right]^2$

$= \cos^8(\theta/4) + \cos^6(\theta/4)\sin^2(\theta/4)$

$= \cos^6(\theta/4)[\cos^2(\theta/4) + \sin^2(\theta/4)] = \cos^6(\theta/4)$

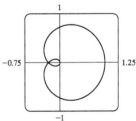

$L = \int_0^{4\pi} \sqrt{\cos^6(\theta/4)}\, d\theta = \int_0^{4\pi} |\cos^3(\theta/4)|\, d\theta$

$= 2\int_0^{2\pi} \cos^3(\theta/4)\, d\theta$ [since $\cos^3(\theta/4) \ge 0$ for $0 \le \theta \le 2\pi$] $= 8\int_0^{\pi/2} \cos^3 u\, du$ $\left[u = \tfrac{1}{4}\theta\right]$

$= 8\int_0^{\pi/2}(1 - \sin^2 u)\cos u\, du = 8\int_0^1 (1 - x^2)\, dx$ $\begin{bmatrix} x = \sin u,\\ dx = \cos u\, du \end{bmatrix}$

$= 8\left[x - \tfrac{1}{3}x^3\right]_0^1 = 8\left(1 - \tfrac{1}{3}\right) = \tfrac{16}{3}$

57. The graph is symmetric about the polar axis and touches the pole when $r = \cos(\theta/5) = 0$, that is, when $\theta/5 = \pi/2$ or $\theta = 5\pi/2$. $|r|$ is a maximum when $\theta = 0$, so the top half of the red section of the curve is traced starting from the polar axis $\theta = 0$ to $\theta = \pi$. Half of the blue section of the curve is then traced from $\theta = \pi$ to $\theta = 5\pi/2$ (the pole), and, by symmetry, the entire blue section is traced from $\theta = \pi$ to $\theta = 5\pi/2 + (5\pi/2 - \pi) = 4\pi$.

$$r^2 + \left(\frac{dr}{d\theta}\right)^2 = \left(\cos\frac{\theta}{5}\right)^2 + \left(-\frac{1}{5}\sin\frac{\theta}{5}\right)^2 = \cos^2\left(\frac{\theta}{5}\right) + \frac{1}{25}\sin^2\left(\frac{\theta}{5}\right)$$

Thus, $L = \int_\pi^{4\pi} \sqrt{r^2 + (dr/d\theta)^2}\, d\theta = \int_\pi^{4\pi} \sqrt{\cos^2\left(\frac{\theta}{5}\right) + \frac{1}{25}\sin^2\left(\frac{\theta}{5}\right)}\, d\theta$.

59. One loop of the curve $r = \cos 2\theta$ is traced with $-\pi/4 \leq \theta \leq \pi/4$.

$$r^2 + \left(\frac{dr}{d\theta}\right)^2 = \cos^2 2\theta + (-2\sin 2\theta)^2 = \cos^2 2\theta + 4\sin^2 2\theta = 1 + 3\sin^2 2\theta \Rightarrow$$

$$L = \int_{-\pi/4}^{\pi/4} \sqrt{1 + 3\sin^2 2\theta}\, d\theta \approx 2.4221.$$

61. The curve $r = \sin(6\sin\theta)$ is completely traced with $0 \leq \theta \leq \pi$. $r = \sin(6\sin\theta) \Rightarrow$

$\frac{dr}{d\theta} = \cos(6\sin\theta) \cdot 6\cos\theta$, so $r^2 + \left(\frac{dr}{d\theta}\right)^2 = \sin^2(6\sin\theta) + 36\cos^2\theta \cos^2(6\sin\theta) \Rightarrow$

$$L = \int_0^\pi \sqrt{\sin^2(6\sin\theta) + 36\cos^2\theta\, \cos^2(6\sin\theta)}\, d\theta \approx 8.0091.$$

63. $r = 2\cos\theta \Rightarrow x = r\cos\theta = 2\cos^2\theta$, $y = r\sin\theta = 2\sin\theta\cos\theta = \sin 2\theta \Rightarrow$

$$\frac{dy}{dx} = \frac{dy/d\theta}{dx/d\theta} = \frac{2\cos 2\theta}{2 \cdot 2\cos\theta\,(-\sin\theta)} = \frac{\cos 2\theta}{-\sin 2\theta} = -\cot 2\theta$$

When $\theta = \frac{\pi}{3}$, $\frac{dy}{dx} = -\cot\left(2 \cdot \frac{\pi}{3}\right) = \cot\frac{\pi}{3} = \frac{1}{\sqrt{3}}$. [*Another method:* Use Equation 3.]

65. $r = 1/\theta \Rightarrow x = r\cos\theta = (\cos\theta)/\theta$, $y = r\sin\theta = (\sin\theta)/\theta \Rightarrow$

$$\frac{dy}{dx} = \frac{dy/d\theta}{dx/d\theta} = \frac{\sin\theta(-1/\theta^2) + (1/\theta)\cos\theta}{\cos\theta(-1/\theta^2) - (1/\theta)\sin\theta} \cdot \frac{\theta^2}{\theta^2} = \frac{-\sin\theta + \theta\cos\theta}{-\cos\theta - \theta\sin\theta}$$

When $\theta = \pi$, $\frac{dy}{dx} = \frac{-0 + \pi(-1)}{-(-1) - \pi(0)} = \frac{-\pi}{1} = -\pi$.

67. $r = \cos 2\theta \Rightarrow x = r\cos\theta = \cos 2\theta \cos\theta$, $y = r\sin\theta = \cos 2\theta \sin\theta \Rightarrow$

$$\frac{dy}{dx} = \frac{dy/d\theta}{dx/d\theta} = \frac{\cos 2\theta \cos\theta + \sin\theta\,(-2\sin 2\theta)}{\cos 2\theta\,(-\sin\theta) + \cos\theta\,(-2\sin 2\theta)}$$

When $\theta = \frac{\pi}{4}$, $\frac{dy}{dx} = \frac{0(\sqrt{2}/2) + (\sqrt{2}/2)(-2)}{0(-\sqrt{2}/2) + (\sqrt{2}/2)(-2)} = \frac{-\sqrt{2}}{-\sqrt{2}} = 1$.

69. $r = \sin\theta \;\Rightarrow\; x = r\cos\theta = \sin\theta\cos\theta,\; y = r\sin\theta = \sin^2\theta \;\Rightarrow\; \frac{dy}{d\theta} = 2\sin\theta\cos\theta = \sin 2\theta = 0 \;\Rightarrow\;$
$2\theta = 0$ or $\pi \;\Rightarrow\; \theta = 0$ or $\frac{\pi}{2} \;\Rightarrow\;$ horizontal tangent at $(0,0)$, and $\left(1, \frac{\pi}{2}\right)$.
$\frac{dx}{d\theta} = -\sin^2\theta + \cos^2\theta = \cos 2\theta = 0 \;\Rightarrow\; 2\theta = \frac{\pi}{2}$ or $\frac{3\pi}{2} \;\Rightarrow\; \theta = \frac{\pi}{4}$ or $\frac{3\pi}{4} \;\Rightarrow\;$ vertical tangent at $\left(\frac{1}{\sqrt{2}}, \frac{\pi}{4}\right)$
and $\left(\frac{1}{\sqrt{2}}, \frac{3\pi}{4}\right)$.

71. $r = 1 + \cos\theta \;\Rightarrow\; x = r\cos\theta = \cos\theta(1+\cos\theta),\; y = r\sin\theta = \sin\theta(1+\cos\theta) \;\Rightarrow\;$
$\frac{dy}{d\theta} = (1+\cos\theta)\cos\theta - \sin^2\theta = 2\cos^2\theta + \cos\theta - 1 = (2\cos\theta-1)(\cos\theta+1) = 0 \;\Rightarrow\; \cos\theta = \frac{1}{2}$ or $-1 \;\Rightarrow\;$
$\theta = \frac{\pi}{3}, \pi,$ or $\frac{5\pi}{3} \;\Rightarrow\;$ horizontal tangent at $\left(\frac{3}{2}, \frac{\pi}{3}\right), (0, \pi),$ and $\left(\frac{3}{2}, \frac{5\pi}{3}\right)$.
$\frac{dx}{d\theta} = -(1+\cos\theta)\sin\theta - \cos\theta\sin\theta = -\sin\theta(1+2\cos\theta) = 0 \;\Rightarrow\; \sin\theta = 0$ or $\cos\theta = -\frac{1}{2} \;\Rightarrow\;$
$\theta = 0, \pi, \frac{2\pi}{3},$ or $\frac{4\pi}{3} \;\Rightarrow\;$ vertical tangent at $(2,0), \left(\frac{1}{2}, \frac{2\pi}{3}\right),$ and $\left(\frac{1}{2}, \frac{4\pi}{3}\right)$.

Note that the tangent is horizontal, not vertical when $\theta = \pi$, since $\displaystyle\lim_{\theta\to\pi}\frac{dy/d\theta}{dx/d\theta} = 0$.

73. $\tan\psi = \tan(\phi - \theta) = \dfrac{\tan\phi - \tan\theta}{1 + \tan\phi\tan\theta} = \dfrac{\dfrac{dy}{dx} - \tan\theta}{1 + \dfrac{dy}{dx}\tan\theta} = \dfrac{\dfrac{dy/d\theta}{dx/d\theta} - \tan\theta}{1 + \dfrac{dy/d\theta}{dx/d\theta}\tan\theta}$

$= \dfrac{\dfrac{dy}{d\theta} - \dfrac{dx}{d\theta}\tan\theta}{\dfrac{dx}{d\theta} + \dfrac{dy}{d\theta}\tan\theta} = \dfrac{\left(\dfrac{dr}{d\theta}\sin\theta + r\cos\theta\right) - \tan\theta\left(\dfrac{dr}{d\theta}\cos\theta - r\sin\theta\right)}{\left(\dfrac{dr}{d\theta}\cos\theta - r\sin\theta\right) + \tan\theta\left(\dfrac{dr}{d\theta}\sin\theta + r\cos\theta\right)} = \dfrac{r\cos\theta + r\cdot\dfrac{\sin^2\theta}{\cos\theta}}{\dfrac{dr}{d\theta}\cos\theta + \dfrac{dr}{d\theta}\cdot\dfrac{\sin^2\theta}{\cos\theta}}$

$= \dfrac{r\cos^2\theta + r\sin^2\theta}{\dfrac{dr}{d\theta}\cos^2\theta + \dfrac{dr}{d\theta}\sin^2\theta} = \dfrac{r}{dr/d\theta}$

75. (a) From (10.2.9),
$$S = \int_a^b 2\pi y\sqrt{(dx/d\theta)^2 + (dy/d\theta)^2}\,d\theta$$
$$= \int_a^b 2\pi y\sqrt{r^2 + (dr/d\theta)^2}\,d\theta \quad \text{[from the derivation of Equation 10.4.6]}$$
$$= \int_a^b 2\pi r\sin\theta\sqrt{r^2 + (dr/d\theta)^2}\,d\theta$$

(b) The curve $r^2 = \cos 2\theta$ goes through the pole when $\cos 2\theta = 0 \;\Rightarrow\;$
$2\theta = \frac{\pi}{2} \;\Rightarrow\; \theta = \frac{\pi}{4}$. We'll rotate the curve from $\theta = 0$ to $\theta = \frac{\pi}{4}$ and double
this value to obtain the total surface area generated.
$r^2 = \cos 2\theta \;\Rightarrow\; 2r\dfrac{dr}{d\theta} = -2\sin 2\theta \;\Rightarrow\; \left(\dfrac{dr}{d\theta}\right)^2 = \dfrac{\sin^2 2\theta}{r^2} = \dfrac{\sin^2 2\theta}{\cos 2\theta}.$

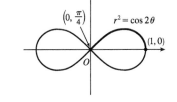

$S = 2\displaystyle\int_0^{\pi/4} 2\pi\sqrt{\cos 2\theta}\sin\theta\sqrt{\cos 2\theta + (\sin^2 2\theta)/\cos 2\theta}\,d\theta = 4\pi\int_0^{\pi/4}\sqrt{\cos 2\theta}\sin\theta\sqrt{\dfrac{\cos^2 2\theta + \sin^2 2\theta}{\cos 2\theta}}\,d\theta$

$= 4\pi\displaystyle\int_0^{\pi/4}\sqrt{\cos 2\theta}\sin\theta\dfrac{1}{\sqrt{\cos 2\theta}}\,d\theta = 4\pi\int_0^{\pi/4}\sin\theta\,d\theta = 4\pi\bigl[-\cos\theta\bigr]_0^{\pi/4} = -4\pi\left(\dfrac{\sqrt{2}}{2} - 1\right) = 2\pi\left(2 - \sqrt{2}\right)$

10.5 Conic Sections

1. $x^2 = 8y$ and $x^2 = 4py$ $\Rightarrow$ $4p = 8$ $\Leftrightarrow$ $p = 2$. The vertex is $(0, 0)$, the focus is $(0, 2)$, and the directrix is $y = -2$.

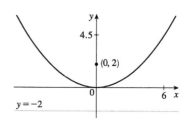

3. $5x + 3y^2 = 0$ $\Leftrightarrow$ $y^2 = -\frac{5}{3}x$, so $4p = -\frac{5}{3}$ $\Leftrightarrow$ $p = -\frac{5}{12}$.

   The vertex is $(0, 0)$, the focus is $\left(-\frac{5}{12}, 0\right)$, and the directrix is $x = \frac{5}{12}$.

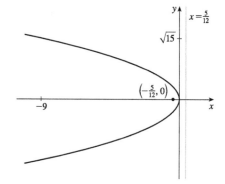

5. $(y + 1)^2 = 16(x - 3)$, so $4p = 16$ $\Leftrightarrow$ $p = 4$. The vertex is $(3, -1)$, the focus is $(3 + 4, -1) = (7, -1)$, and the directrix is $x = 3 - 4 = -1$.

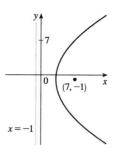

7. $y^2 + 6y + 2x + 1 = 0$ $\Leftrightarrow$ $y^2 + 6y = -2x - 1$ $\Leftrightarrow$
   $y^2 + 6y + 9 = -2x + 8$ $\Leftrightarrow$ $(y + 3)^2 = -2(x - 4)$, so $4p = -2$ $\Leftrightarrow$
   $p = -\frac{1}{2}$. The vertex is $(4, -3)$, the focus is $\left(\frac{7}{2}, -3\right)$, and the directrix is $x = \frac{9}{2}$.

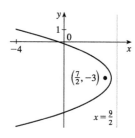

9. The equation has the form $y^2 = 4px$, where $p < 0$. Since the parabola passes through $(-1, 1)$, we have $1^2 = 4p(-1)$, so $4p = -1$ and an equation is $y^2 = -x$ or $x = -y^2$. $4p = -1$, so $p = -\frac{1}{4}$ and the focus is $\left(-\frac{1}{4}, 0\right)$ while the directrix is $x = \frac{1}{4}$.

11. $\dfrac{x^2}{16}+\dfrac{y^2}{25}=1 \Rightarrow a=\sqrt{25}=5, b=\sqrt{16}=4,$

$c=\sqrt{a^2-b^2}=\sqrt{25-16}=\sqrt{9}=3.$ The ellipse is centered at $(0,0)$ with vertices $(0,\pm 5)$. The foci are $(0,\pm 3)$.

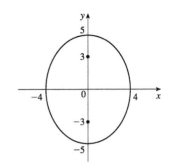

13. $x^2+3y^2=9 \Leftrightarrow \dfrac{x^2}{9}+\dfrac{y^2}{3}=1 \Rightarrow a=\sqrt{9}=3, b=\sqrt{3},$

$c=\sqrt{a^2-b^2}=\sqrt{9-3}=\sqrt{6}.$ The ellipse is centered at $(0,0)$ with vertices $(\pm 3,0)$. The foci are $(\pm\sqrt{6},0)$.

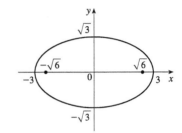

15. $4x^2+25y^2-50y=75 \Leftrightarrow 4x^2+25(y^2-2y+1)=75+25 \Leftrightarrow$

$4x^2+25(y-1)^2=100 \Leftrightarrow \dfrac{x^2}{25}+\dfrac{(y-1)^2}{4}=1 \Rightarrow a=\sqrt{25}=5,$

$b=\sqrt{4}=2, c=\sqrt{25-4}=\sqrt{21}.$ The ellipse is centered at $(0,1)$ with vertices $(\pm 5,1)$. The foci are $(\pm\sqrt{21},1)$.

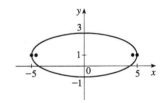

17. The center is $(0,0)$, $a=3$, and $b=2$, so an equation is $\dfrac{x^2}{4}+\dfrac{y^2}{9}=1$. $c=\sqrt{a^2-b^2}=\sqrt{5}$, so the foci are $(0,\pm\sqrt{5})$.

19. $\dfrac{y^2}{25}-\dfrac{x^2}{9}=1 \Rightarrow a=5, b=3, c=\sqrt{a^2+b^2}=\sqrt{25+9}=\sqrt{34} \Rightarrow$

center $(0,0)$, vertices $(0,\pm 5)$, foci $(0,\pm\sqrt{34})$, asymptotes $y=\pm\tfrac{5}{3}x$.

Note: It is helpful to draw a $2a$-by-$2b$ rectangle whose center is the center of the hyperbola. The asymptotes are the extended diagonals of the rectangle.

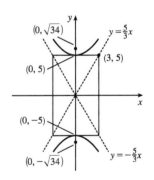

21. $x^2-y^2=100 \Leftrightarrow \dfrac{x^2}{100}-\dfrac{y^2}{100}=1 \Rightarrow a=b=10,$

$c=\sqrt{100+100}=10\sqrt{2} \Rightarrow$ center $(0,0)$, vertices $(\pm 10,0)$, foci $(\pm 10\sqrt{2},0)$, asymptotes $y=\pm\tfrac{10}{10}x=\pm x$

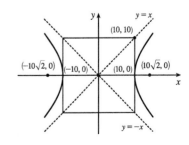

23. $x^2 - y^2 + 2y = 2 \Leftrightarrow x^2 - (y^2 - 2y + 1) = 2 - 1 \Leftrightarrow$

$\dfrac{x^2}{1} - \dfrac{(y-1)^2}{1} = 1 \Rightarrow a = b = 1, c = \sqrt{1+1} = \sqrt{2} \Rightarrow$

center $(0, 1)$, vertices $(\pm 1, 1)$, foci $(\pm\sqrt{2}, 1)$,

asymptotes $y - 1 = \pm\frac{1}{1}x = \pm x$.

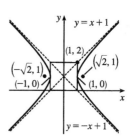

25. The hyperbola has vertices $(\pm 3, 0)$, which lie on the x-axis, so the equation has the form $\dfrac{x^2}{a^2} - \dfrac{y^2}{b^2} = 1$ with $a = 3$. The point

$(5, 4)$ is on the hyperbola, so $\dfrac{5^2}{3^2} - \dfrac{4^2}{b^2} = 1 \Rightarrow \dfrac{25}{9} - \dfrac{16}{b^2} = 1 \Rightarrow \dfrac{16}{9} = \dfrac{16}{b^2} \Rightarrow b^2 = 9 \Rightarrow b = 3$. Thus, an

equation is $\dfrac{x^2}{9} - \dfrac{y^2}{9} = 1$. Now, $c^2 = a^2 + b^2 = 9 + 9 = 18 \Rightarrow c = \sqrt{18} = 3\sqrt{2}$, so the foci are $(\pm 3\sqrt{2}, 0)$, while the

asymptotes are $y = \pm\dfrac{a}{b}x = \pm\dfrac{3}{3}x = \pm x$.

27. $4x^2 = y^2 + 4 \Leftrightarrow 4x^2 - y^2 = 4 \Leftrightarrow \dfrac{x^2}{1} - \dfrac{y^2}{4} = 1$. This is an equation of a *hyperbola* with vertices $(\pm 1, 0)$.

The foci are at $(\pm\sqrt{1+4}, 0) = (\pm\sqrt{5}, 0)$.

29. $x^2 = 4y - 2y^2 \Leftrightarrow x^2 + 2y^2 - 4y = 0 \Leftrightarrow x^2 + 2(y^2 - 2y + 1) = 2 \Leftrightarrow x^2 + 2(y-1)^2 = 2 \Leftrightarrow$

$\dfrac{x^2}{2} + \dfrac{(y-1)^2}{1} = 1$. This is an equation of an *ellipse* with vertices at $(\pm\sqrt{2}, 1)$. The foci are at $(\pm\sqrt{2-1}, 1) = (\pm 1, 1)$.

31. $3x^2 - 6x - 2y = 1 \Leftrightarrow 3x^2 - 6x = 2y + 1 \Leftrightarrow 3(x^2 - 2x + 1) = 2y + 1 + 3 \Leftrightarrow 3(x-1)^2 = 2y + 4 \Leftrightarrow$

$(x-1)^2 = \frac{2}{3}(y+2)$. This is an equation of a *parabola* with $4p = \frac{2}{3}$, so $p = \frac{1}{6}$. The vertex is $(1, -2)$ and the focus is

$(1, -2 + \frac{1}{6}) = (1, -\frac{11}{6})$.

33. The parabola with vertex $(0, 0)$ and focus $(1, 0)$ opens to the right and has $p = 1$, so its equation is $y^2 = 4px$, or $y^2 = 4x$.

35. The distance from the focus $(-4, 0)$ to the directrix $x = 2$ is $2 - (-4) = 6$, so the distance from the focus to the vertex is

$\frac{1}{2}(6) = 3$ and the vertex is $(-1, 0)$. Since the focus is to the left of the vertex, $p = -3$. An equation is $y^2 = 4p(x+1) \Rightarrow$

$y^2 = -12(x+1)$.

37. The parabola with vertex $(3, -1)$ having a horizontal axis has equation $[y - (-1)]^2 = 4p(x - 3)$. Since it passes through

$(-15, 2)$, $(2 + 1)^2 = 4p(-15 - 3) \Rightarrow 9 = 4p(-18) \Rightarrow 4p = -\frac{1}{2}$. An equation is $(y+1)^2 = -\frac{1}{2}(x - 3)$.

39. The ellipse with foci $(\pm 2, 0)$ and vertices $(\pm 5, 0)$ has center $(0, 0)$ and a horizontal major axis, with $a = 5$ and $c = 2$,

so $b^2 = a^2 - c^2 = 25 - 4 = 21$. An equation is $\dfrac{x^2}{25} + \dfrac{y^2}{21} = 1$.

41. Since the vertices are $(0, 0)$ and $(0, 8)$, the ellipse has center $(0, 4)$ with a vertical axis and $a = 4$. The foci at $(0, 2)$ and $(0, 6)$ are 2 units from the center, so $c = 2$ and $b = \sqrt{a^2 - c^2} = \sqrt{4^2 - 2^2} = \sqrt{12}$. An equation is $\dfrac{(x - 0)^2}{b^2} + \dfrac{(y - 4)^2}{a^2} = 1 \Rightarrow \dfrac{x^2}{12} + \dfrac{(y - 4)^2}{16} = 1$.

43. An equation of an ellipse with center $(-1, 4)$ and vertex $(-1, 0)$ is $\dfrac{(x + 1)^2}{b^2} + \dfrac{(y - 4)^2}{4^2} = 1$. The focus $(-1, 6)$ is 2 units from the center, so $c = 2$. Thus, $b^2 + 2^2 = 4^2 \Rightarrow b^2 = 12$, and the equation is $\dfrac{(x + 1)^2}{12} + \dfrac{(y - 4)^2}{16} = 1$.

45. An equation of a hyperbola with vertices $(\pm 3, 0)$ is $\dfrac{x^2}{3^2} - \dfrac{y^2}{b^2} = 1$. Foci $(\pm 5, 0) \Rightarrow c = 5$ and $3^2 + b^2 = 5^2 \Rightarrow b^2 = 25 - 9 = 16$, so the equation is $\dfrac{x^2}{9} - \dfrac{y^2}{16} = 1$.

47. The center of a hyperbola with vertices $(-3, -4)$ and $(-3, 6)$ is $(-3, 1)$, so $a = 5$ and an equation is $\dfrac{(y - 1)^2}{5^2} - \dfrac{(x + 3)^2}{b^2} = 1$. Foci $(-3, -7)$ and $(-3, 9) \Rightarrow c = 8$, so $5^2 + b^2 = 8^2 \Rightarrow b^2 = 64 - 25 = 39$ and the equation is $\dfrac{(y - 1)^2}{25} - \dfrac{(x + 3)^2}{39} = 1$.

49. The center of a hyperbola with vertices $(\pm 3, 0)$ is $(0, 0)$, so $a = 3$ and an equation is $\dfrac{x^2}{3^2} - \dfrac{y^2}{b^2} = 1$.

Asymptotes $y = \pm 2x \Rightarrow \dfrac{b}{a} = 2 \Rightarrow b = 2(3) = 6$ and the equation is $\dfrac{x^2}{9} - \dfrac{y^2}{36} = 1$.

51. In Figure 8, we see that the point on the ellipse closest to a focus is the closer vertex (which is a distance $a - c$ from it) while the farthest point is the other vertex (at a distance of $a + c$). So for this lunar orbit, $(a - c) + (a + c) = 2a = (1728 + 110) + (1728 + 314)$, or $a = 1940$; and $(a + c) - (a - c) = 2c = 314 - 110$, or $c = 102$. Thus, $b^2 = a^2 - c^2 = 3{,}753{,}196$, and the equation is $\dfrac{x^2}{3{,}763{,}600} + \dfrac{y^2}{3{,}753{,}196} = 1$.

53. (a) Set up the coordinate system so that A is $(-200, 0)$ and B is $(200, 0)$.

$|PA| - |PB| = (1200)(980) = 1{,}176{,}000 \text{ ft} = \dfrac{2450}{11} \text{ mi} = 2a \Rightarrow a = \dfrac{1225}{11}$, and $c = 200$ so

$b^2 = c^2 - a^2 = \dfrac{3{,}339{,}375}{121} \Rightarrow \dfrac{121x^2}{1{,}500{,}625} - \dfrac{121y^2}{3{,}339{,}375} = 1$.

(b) Due north of $B \Rightarrow x = 200 \Rightarrow \dfrac{(121)(200)^2}{1{,}500{,}625} - \dfrac{121y^2}{3{,}339{,}375} = 1 \Rightarrow y = \dfrac{133{,}575}{539} \approx 248 \text{ mi}$

55. The function whose graph is the upper branch of this hyperbola is concave upward. The function is

$y = f(x) = a\sqrt{1 + \dfrac{x^2}{b^2}} = \dfrac{a}{b}\sqrt{b^2 + x^2}$, so $y' = \dfrac{a}{b}x(b^2 + x^2)^{-1/2}$ and

$y'' = \dfrac{a}{b}\left[(b^2 + x^2)^{-1/2} - x^2(b^2 + x^2)^{-3/2}\right] = ab(b^2 + x^2)^{-3/2} > 0$ for all x, and so f is concave upward.

57. (a) If $k > 16$, then $k - 16 > 0$, and $\dfrac{x^2}{k} + \dfrac{y^2}{k-16} = 1$ is an *ellipse* since it is the sum of two squares on the left side.

(b) If $0 < k < 16$, then $k - 16 < 0$, and $\dfrac{x^2}{k} + \dfrac{y^2}{k-16} = 1$ is a *hyperbola* since it is the difference of two squares on the left side.

(c) If $k < 0$, then $k - 16 < 0$, and there is *no curve* since the left side is the sum of two negative terms, which cannot equal 1.

(d) In case (a), $a^2 = k$, $b^2 = k - 16$, and $c^2 = a^2 - b^2 = 16$, so the foci are at $(\pm 4, 0)$. In case (b), $k - 16 < 0$, so $a^2 = k$, $b^2 = 16 - k$, and $c^2 = a^2 + b^2 = 16$, and so again the foci are at $(\pm 4, 0)$.

59. $x^2 = 4py \;\Rightarrow\; 2x = 4py' \;\Rightarrow\; y' = \dfrac{x}{2p}$, so the tangent line at (x_0, y_0) is $y - \dfrac{x_0^2}{4p} = \dfrac{x_0}{2p}(x - x_0)$. This line passes through the point $(a, -p)$ on the directrix, so $-p - \dfrac{x_0^2}{4p} = \dfrac{x_0}{2p}(a - x_0) \;\Rightarrow\; -4p^2 - x_0^2 = 2ax_0 - 2x_0^2 \;\Leftrightarrow\;$

$x_0^2 - 2ax_0 - 4p^2 = 0 \;\Leftrightarrow\; x_0^2 - 2ax_0 + a^2 = a^2 + 4p^2 \;\Leftrightarrow\;$

$(x_0 - a)^2 = a^2 + 4p^2 \;\Leftrightarrow\; x_0 = a \pm \sqrt{a^2 + 4p^2}$. The slopes of the tangent lines at $x = a \pm \sqrt{a^2 + 4p^2}$ are $\dfrac{a \pm \sqrt{a^2 + 4p^2}}{2p}$, so the product of the two slopes is

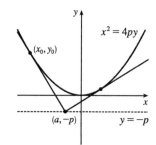

$\dfrac{a + \sqrt{a^2 + 4p^2}}{2p} \cdot \dfrac{a - \sqrt{a^2 + 4p^2}}{2p} = \dfrac{a^2 - (a^2 + 4p^2)}{4p^2} = \dfrac{-4p^2}{4p^2} = -1$,

showing that the tangent lines are perpendicular.

61. $9x^2 + 4y^2 = 36 \;\Leftrightarrow\; \dfrac{x^2}{4} + \dfrac{y^2}{9} = 1$. We use the parametrization $x = 2\cos t$, $y = 3\sin t$, $0 \le t \le 2\pi$. The circumference is given by

$$L = \int_0^{2\pi} \sqrt{(dx/dt)^2 + (dy/dt)^2}\, dt = \int_0^{2\pi} \sqrt{(-2\sin t)^2 + (3\cos t)^2}\, dt$$
$$= \int_0^{2\pi} \sqrt{4\sin^2 t + 9\cos^2 t}\, dt = \int_0^{2\pi} \sqrt{4 + 5\cos^2 t}\, dt$$

Now use Simpson's Rule with $n = 8$, $\Delta t = \dfrac{2\pi - 0}{8} = \dfrac{\pi}{4}$, and $f(t) = \sqrt{4 + 5\cos^2 t}$ to get

$$L \approx S_8 = \tfrac{\pi/4}{3}\left[f(0) + 4f\!\left(\tfrac{\pi}{4}\right) + 2f\!\left(\tfrac{\pi}{2}\right) + 4f\!\left(\tfrac{3\pi}{4}\right) + 2f(\pi) + 4f\!\left(\tfrac{5\pi}{4}\right) + 2f\!\left(\tfrac{3\pi}{2}\right) + 4f\!\left(\tfrac{7\pi}{4}\right) + f(2\pi)\right] \approx 15.9.$$

63. $\dfrac{x^2}{a^2} - \dfrac{y^2}{b^2} = 1 \;\Rightarrow\; \dfrac{y^2}{b^2} = \dfrac{x^2 - a^2}{a^2} \;\Rightarrow\; y = \pm\dfrac{b}{a}\sqrt{x^2 - a^2}.$

$A = 2\displaystyle\int_a^c \dfrac{b}{a}\sqrt{x^2 - a^2}\, dx \stackrel{39}{=} \dfrac{2b}{a}\left[\dfrac{x}{2}\sqrt{x^2 - a^2} - \dfrac{a^2}{2}\ln\left|x + \sqrt{x^2 - a^2}\right|\right]_a^c$

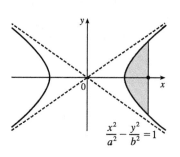

$= \dfrac{b}{a}\left[c\sqrt{c^2 - a^2} - a^2\ln\left|c + \sqrt{c^2 - a^2}\right| + a^2\ln|a|\right]$

Since $a^2 + b^2 = c^2$, $c^2 - a^2 = b^2$, and $\sqrt{c^2 - a^2} = b$.

$= \dfrac{b}{a}\left[cb - a^2\ln(c + b) + a^2\ln a\right] = \dfrac{b}{a}\left[cb + a^2(\ln a - \ln(b + c))\right]$

$= b^2 c/a + ab\ln[a/(b + c)]$, where $c^2 = a^2 + b^2$.

65. $9x^2 + 4y^2 = 36$ $\Leftrightarrow$ $\dfrac{x^2}{4} + \dfrac{y^2}{9} = 1$ $\Rightarrow$ $a = 3, b = 2$. By symmetry, $\bar{x} = 0$. By Example 2 in Section 7.3, the area of the top half of the ellipse is $\frac{1}{2}(\pi ab) = 3\pi$. Solve $9x^2 + 4y^2 = 36$ for y to get an equation for the top half of the ellipse:

$9x^2 + 4y^2 = 36$ $\Leftrightarrow$ $4y^2 = 36 - 9x^2$ $\Leftrightarrow$ $y^2 = \frac{9}{4}(4 - x^2)$ $\Rightarrow$ $y = \frac{3}{2}\sqrt{4 - x^2}$. Now

$\bar{y} = \dfrac{1}{A}\displaystyle\int_a^b \dfrac{1}{2}[f(x)]^2\,dx = \dfrac{1}{3\pi}\int_{-2}^{2}\dfrac{1}{2}\left(\dfrac{3}{2}\sqrt{4-x^2}\right)^2 dx = \dfrac{3}{8\pi}\int_{-2}^{2}(4-x^2)\,dx$

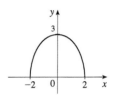

$= \dfrac{3}{8\pi} \cdot 2\displaystyle\int_0^2(4-x^2)\,dx = \dfrac{3}{4\pi}\left[4x - \dfrac{1}{3}x^3\right]_0^2 = \dfrac{3}{4\pi}\left(\dfrac{16}{3}\right) = \dfrac{4}{\pi}$

so the centroid is $(0, 4/\pi)$.

67. Differentiating implicitly, $\dfrac{x^2}{a^2} + \dfrac{y^2}{b^2} = 1$ $\Rightarrow$ $\dfrac{2x}{a^2} + \dfrac{2yy'}{b^2} = 0$ $\Rightarrow$ $y' = -\dfrac{b^2 x}{a^2 y}$ $[y \neq 0]$. Thus, the slope of the tangent line at P is $-\dfrac{b^2 x_1}{a^2 y_1}$. The slope of $F_1 P$ is $\dfrac{y_1}{x_1 + c}$ and of $F_2 P$ is $\dfrac{y_1}{x_1 - c}$. By the formula in Problem 17 in Problems Plus following Chapter 2, we have

$\tan\alpha = \dfrac{\dfrac{y_1}{x_1 + c} + \dfrac{b^2 x_1}{a^2 y_1}}{1 - \dfrac{b^2 x_1 y_1}{a^2 y_1 (x_1 + c)}} = \dfrac{a^2 y_1^2 + b^2 x_1(x_1 + c)}{a^2 y_1(x_1 + c) - b^2 x_1 y_1} = \dfrac{a^2 b^2 + b^2 c x_1}{c^2 x_1 y_1 + a^2 c y_1}$ $\left[\begin{array}{l}\text{using } b^2 x_1^2 + a^2 y_1^2 = a^2 b^2, \\ \text{and } a^2 - b^2 = c^2\end{array}\right]$

$= \dfrac{b^2(cx_1 + a^2)}{cy_1(cx_1 + a^2)} = \dfrac{b^2}{cy_1}$

and $\tan\beta = \dfrac{-\dfrac{b^2 x_1}{a^2 y_1} - \dfrac{y_1}{x_1 - c}}{1 - \dfrac{b^2 x_1 y_1}{a^2 y_1 (x_1 - c)}} = \dfrac{-a^2 y_1^2 - b^2 x_1(x_1 - c)}{a^2 y_1 (x_1 - c) - b^2 x_1 y_1} = \dfrac{-a^2 b^2 + b^2 c x_1}{c^2 x_1 y_1 - a^2 c y_1} = \dfrac{b^2(cx_1 - a^2)}{cy_1(cx_1 - a^2)} = \dfrac{b^2}{cy_1}$

Thus, $\alpha = \beta$.

69. Let C be the center of a circle (gray) with radius r that is tangent to both black circles (see the figure). We wish to show that $AC + BC$ is constant for all values of r, that is, for any circle drawn tangent to both black circles. The smaller black circle has radius 3, so $AC = 3 + r$, and the larger black circle has radius 5, so $BC = 5 - r$. Hence, $AC + BC = 3 + r + 5 - r = 8$, which is a constant. Since the sum of the distances from C to $(-1, 0)$ and $(1, 0)$ is constant, the centers of all the circles lie on an ellipse with foci $(\pm 1, 0)$ $[c = 1]$. The sum of the distances from the foci to any point on the ellipse is $2a$, so $2a = 8$ $\Rightarrow$ $a = 4$. Now, $c^2 = a^2 - b^2$ $\Rightarrow$ $1^2 = 4^2 - b^2$ $\Rightarrow$ $b^2 = 15$. Thus, the ellipse has equation $\dfrac{x^2}{16} + \dfrac{y^2}{15} = 1$.

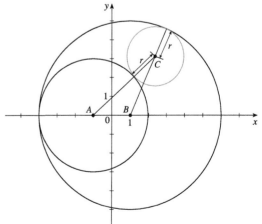

10.6 Conic Sections in Polar Coordinates

1. The directrix $x = 2$ is to the right of the focus at the origin, so we use the form with "$+e\cos\theta$" in the denominator. (See Theorem 6 and Figure 2.) $e = 1$ for a parabola, so an equation is $r = \dfrac{ed}{1 + e\cos\theta} = \dfrac{1 \cdot 2}{1 + 1\cos\theta} = \dfrac{2}{1 + \cos\theta}$.

3. The directrix $y = -4$ is below the focus at the origin, so we use the form with "$-e\sin\theta$" in the denominator. An equation of the hyperbola is $r = \dfrac{ed}{1 - e\sin\theta} = \dfrac{2 \cdot 4}{1 - 2\sin\theta} = \dfrac{8}{1 - 2\sin\theta}$.

5. The vertex $(2, \pi)$ is to the left of the focus at the origin, so we use the form with "$-e\cos\theta$" in the denominator. An equation of the ellipse is $r = \dfrac{ed}{1 - e\cos\theta}$. Using eccentricity $e = \dfrac{2}{3}$ with $\theta = \pi$ and $r = 2$, we get $2 = \dfrac{\frac{2}{3}d}{1 - \frac{2}{3}(-1)}$ $\Rightarrow$ $2 = \dfrac{2d}{5}$ $\Rightarrow$ $d = 5$, so we have $r = \dfrac{\frac{2}{3}(5)}{1 - \frac{2}{3}\cos\theta} = \dfrac{10}{3 - 2\cos\theta}$.

7. The vertex $\left(3, \frac{\pi}{2}\right)$ is 3 units above the focus at the origin, so the directrix is 6 units above the focus ($d = 6$), and we use the form "$+e\sin\theta$" in the denominator. $e = 1$ for a parabola, so an equation is $r = \dfrac{ed}{1 + e\sin\theta} = \dfrac{1(6)}{1 + 1\sin\theta} = \dfrac{6}{1 + \sin\theta}$.

9. $r = \dfrac{3}{1 - \sin\theta}$, where $e = 1$, so the conic is a parabola. If $\sin\theta$ appears in the denominator, use 0 and π for θ. If $\cos\theta$ appears in the denominator, use $\frac{\pi}{2}$ and $\frac{3\pi}{2}$ for θ. Thus, when $\theta = 0$ or π, $r = 3$. Hence, the equation is matched with graph VI.

11. $r = \dfrac{12}{8 - 7\cos\theta} \cdot \dfrac{1/8}{1/8} = \dfrac{3/2}{1 - \frac{7}{8}\cos\theta}$, where $e = \dfrac{7}{8} < 1$, so the conic is an ellipse. When $\theta = \frac{\pi}{2}$ or $\frac{3\pi}{2}$, $r = \dfrac{12}{8}$. Hence, the equation is matched with graph II.

13. $r = \dfrac{5}{2 + 3\sin\theta} \cdot \dfrac{1/2}{1/2} = \dfrac{5/2}{1 + \frac{3}{2}\sin\theta}$, where $e = \dfrac{3}{2} > 1$, so the conic is a hyperbola. When $\theta = 0$ or π, $r = \dfrac{5}{2}$. Hence, the equation is matched with graph IV.

15. $r = \dfrac{4}{5 - 4\sin\theta} \cdot \dfrac{1/5}{1/5} = \dfrac{4/5}{1 - \frac{4}{5}\sin\theta}$, where $e = \dfrac{4}{5}$ and $ed = \dfrac{4}{5}$ $\Rightarrow$ $d = 1$.

   (a) Eccentricity $= e = \dfrac{4}{5}$

   (b) Since $e = \dfrac{4}{5} < 1$, the conic is an ellipse.

   (c) Since "$-e\sin\theta$" appears in the denominator, the directrix is below the focus at the origin, $d = |Fl| = 1$, so an equation of the directrix is $y = -1$.

   (d) The vertices are $\left(4, \frac{\pi}{2}\right)$ and $\left(\frac{4}{9}, \frac{3\pi}{2}\right)$.

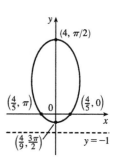

17. $r = \dfrac{2}{3+3\sin\theta} \cdot \dfrac{1/3}{1/3} = \dfrac{2/3}{1+1\sin\theta}$, where $e=1$ and $ed = \tfrac{2}{3}$ $\Rightarrow$ $d = \tfrac{2}{3}$.

(a) Eccentricity $= e = 1$

(b) Since $e=1$, the conic is a parabola.

(c) Since "$+e\sin\theta$" appears in the denominator, the directrix is above the focus at the origin. $d = |Fl| = \tfrac{2}{3}$, so an equation of the directrix is $y = \tfrac{2}{3}$.

(d) The vertex is at $\left(\tfrac{1}{3}, \tfrac{\pi}{2}\right)$, midway between the focus and directrix.

19. $r = \dfrac{9}{6+2\cos\theta} \cdot \dfrac{1/6}{1/6} = \dfrac{3/2}{1+\tfrac{1}{3}\cos\theta}$, where $e = \tfrac{1}{3}$ and $ed = \tfrac{3}{2}$ $\Rightarrow$ $d = \tfrac{9}{2}$.

(a) Eccentricity $= e = \tfrac{1}{3}$

(b) Since $e = \tfrac{1}{3} < 1$, the conic is an ellipse.

(c) Since "$+e\cos\theta$" appears in the denominator, the directrix is to the right of the focus at the origin. $d = |Fl| = \tfrac{9}{2}$, so an equation of the directrix is $x = \tfrac{9}{2}$.

(d) The vertices are $\left(\tfrac{9}{8}, 0\right)$ and $\left(\tfrac{9}{4}, \pi\right)$, so the center is midway between them, that is, $\left(\tfrac{9}{16}, \pi\right)$.

21. $r = \dfrac{3}{4-8\cos\theta} \cdot \dfrac{1/4}{1/4} = \dfrac{3/4}{1-2\cos\theta}$, where $e = 2$ and $ed = \tfrac{3}{4}$ $\Rightarrow$ $d = \tfrac{3}{8}$.

(a) Eccentricity $= e = 2$

(b) Since $e = 2 > 1$, the conic is a hyperbola.

(c) Since "$-e\cos\theta$" appears in the denominator, the directrix is to the left of the focus at the origin. $d = |Fl| = \tfrac{3}{8}$, so an equation of the directrix is $x = -\tfrac{3}{8}$.

(d) The vertices are $\left(-\tfrac{3}{4}, 0\right)$ and $\left(\tfrac{1}{4}, \pi\right)$, so the center is midway between them, that is, $\left(\tfrac{1}{2}, \pi\right)$.

23. (a) $r = \dfrac{1}{1-2\sin\theta}$, where $e = 2$ and $ed = 1$ $\Rightarrow$ $d = \tfrac{1}{2}$. The eccentricity $e = 2 > 1$, so the conic is a hyperbola. Since "$-e\sin\theta$" appears in the denominator, the directrix is below the focus at the origin. $d = |Fl| = \tfrac{1}{2}$, so an equation of the directrix is $y = -\tfrac{1}{2}$. The vertices are $\left(-1, \tfrac{\pi}{2}\right)$ and $\left(\tfrac{1}{3}, \tfrac{3\pi}{2}\right)$, so the center is midway between them, that is, $\left(\tfrac{2}{3}, \tfrac{3\pi}{2}\right)$.

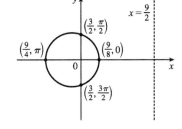

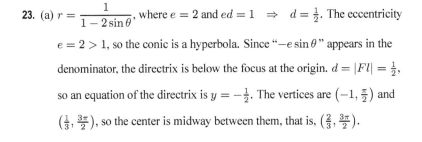

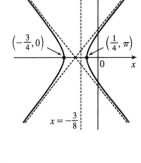

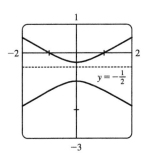

(b) By the discussion that precedes Example 4, the equation

is $r = \dfrac{1}{1 - 2\sin\left(\theta - \frac{3\pi}{4}\right)}$.

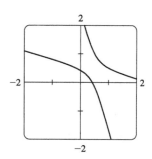

25. $r = \dfrac{e}{1 - e\cos\theta}$. For $e < 1$ the curve is an ellipse. It is nearly circular when e is close to 0. As e increases, the graph is stretched out to the right, and grows larger (that is, its right-hand focus moves to the right while its left-hand focus remains at the origin.) At $e = 1$, the curve becomes a parabola with focus at the origin.

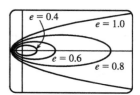

27. $|PF| = e|Pl| \Rightarrow r = e[d - r\cos(\pi - \theta)] = e(d + r\cos\theta) \Rightarrow$
$r(1 - e\cos\theta) = ed \Rightarrow r = \dfrac{ed}{1 - e\cos\theta}$

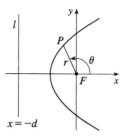

29. $|PF| = e|Pl| \Rightarrow r = e[d - r\sin(\theta - \pi)] = e(d + r\sin\theta) \Rightarrow$
$r(1 - e\sin\theta) = ed \Rightarrow r = \dfrac{ed}{1 - e\sin\theta}$

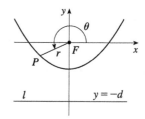

31. We are given $e = 0.093$ and $a = 2.28 \times 10^8$. By (7), we have

$$r = \dfrac{a(1 - e^2)}{1 + e\cos\theta} = \dfrac{2.28 \times 10^8 [1 - (0.093)^2]}{1 + 0.093\cos\theta} \approx \dfrac{2.26 \times 10^8}{1 + 0.093\cos\theta}$$

33. Here $2a$ = length of major axis = 36.18 AU $\Rightarrow$ $a = 18.09$ AU and $e = 0.97$. By (7), the equation of the orbit is

$r = \dfrac{18.09[1 - (0.97)^2]}{1 + 0.97\cos\theta} \approx \dfrac{1.07}{1 + 0.97\cos\theta}$. By (8), the maximum distance from the comet to the sun is

$18.09(1 + 0.97) \approx 35.64$ AU or about 3.314 billion miles.

35. The minimum distance is at perihelion, where $4.6 \times 10^7 = r = a(1-e) = a(1-0.206) = a(0.794) \Rightarrow$
$a = 4.6 \times 10^7/0.794$. So the maximum distance, which is at aphelion, is
$r = a(1+e) = (4.6 \times 10^7/0.794)(1.206) \approx 7.0 \times 10^7$ km.

37. From Exercise 35, we have $e = 0.206$ and $a(1-e) = 4.6 \times 10^7$ km. Thus, $a = 4.6 \times 10^7/0.794$. From (7), we can write the equation of Mercury's orbit as $r = a\dfrac{1-e^2}{1+e\cos\theta}$. So since

$$\frac{dr}{d\theta} = \frac{a(1-e^2)e\sin\theta}{(1+e\cos\theta)^2} \Rightarrow$$

$$r^2 + \left(\frac{dr}{d\theta}\right)^2 = \frac{a^2(1-e^2)^2}{(1+e\cos\theta)^2} + \frac{a^2(1-e^2)^2 e^2 \sin^2\theta}{(1+e\cos\theta)^4} = \frac{a^2(1-e^2)^2}{(1+e\cos\theta)^4}(1+2e\cos\theta+e^2)$$

the length of the orbit is

$$L = \int_0^{2\pi} \sqrt{r^2 + (dr/d\theta)^2}\, d\theta = a(1-e^2)\int_0^{2\pi} \frac{\sqrt{1+e^2+2e\cos\theta}}{(1+e\cos\theta)^2}\, d\theta \approx 3.6 \times 10^8 \text{ km}$$

This seems reasonable, since Mercury's orbit is nearly circular, and the circumference of a circle of radius a is $2\pi a \approx 3.6 \times 10^8$ km.

10 Review

TRUE-FALSE QUIZ

1. False. Consider the curve defined by $x = f(t) = (t-1)^3$ and $y = g(t) = (t-1)^2$. Then $g'(t) = 2(t-1)$, so $g'(1) = 0$, but its graph has a *vertical* tangent when $t = 1$. *Note:* The statement is true if $f'(1) \neq 0$ when $g'(1) = 0$.

3. False. For example, if $f(t) = \cos t$ and $g(t) = \sin t$ for $0 \leq t \leq 4\pi$, then the curve is a circle of radius 1, hence its length is 2π, but $\int_0^{4\pi}\sqrt{[f'(t)]^2+[g'(t)]^2}\,dt = \int_0^{4\pi}\sqrt{(-\sin t)^2+(\cos t)^2}\,dt = \int_0^{4\pi} 1\,dt = 4\pi$, since as t increases from 0 to 4π, the circle is traversed twice.

5. False. If $(r, \theta) = (1, \pi)$, then $(x, y) = (-1, 0)$, so $\tan^{-1}(y/x) = \tan^{-1} 0 = 0 \neq \theta$. The statement is true for points in quadrants I and IV.

7. True. The polar equation $r = 2$, the Cartesian equation $x^2 + y^2 = 4$, and the parametric equations $x = 2\sin 3t$, $y = 2\cos 3t$ $[0 \leq t \leq 2\pi]$ all describe the circle of radius 2 centered at the origin.

9. True. $y^2 = 2y + 3x \Leftrightarrow (y-1)^2 = 3x + 1 = 3\left(x + \tfrac{1}{3}\right) = 4\left(\tfrac{3}{4}\right)\left(x + \tfrac{1}{3}\right)$, which is the equation of a parabola with vertex $\left(-\tfrac{1}{3}, 1\right)$ and focus $\left(-\tfrac{1}{3} + \tfrac{3}{4}, 1\right)$, opening to the right.

11. True. Consider a hyperbola with focus at the origin, oriented so that its polar equation is $r = \dfrac{ed}{1+e\cos\theta}$, where $e > 1$.

The directrix is $x = d$, but along the hyperbola we have $x = r\cos\theta = \dfrac{ed\cos\theta}{1+e\cos\theta} = d\left(\dfrac{e\cos\theta}{1+e\cos\theta}\right) \neq d$.

EXERCISES

1. $x = t^2 + 4t$, $y = 2 - t$, $-4 \le t \le 1$.

   Since $y = 2 - t$ and $-4 \le t \le 1$, we have $1 \le y \le 6$. $t = 2 - y$, so
   $x = (2-y)^2 + 4(2-y) = 4 - 4y + y^2 + 8 - 4y = y^2 - 8y + 12$ $\Leftrightarrow$
   $x + 4 = y^2 - 8y + 16 = (y-4)^2$. This is part of a parabola with vertex
   $(-4, 4)$, opening to the right.

   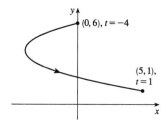

3. $x = \ln t$, $t > 0$ $\Rightarrow$ $t = e^x$.
   $y = t^2 = (e^x)^2 = e^{2x}$.

   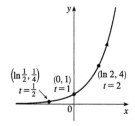

5. $y = \sec\theta = \dfrac{1}{\cos\theta} = \dfrac{1}{x}$. Since $0 \le \theta \le \pi/2$, $0 < x \le 1$ and $y \ge 1$.

   This is part of the hyperbola $y = 1/x$.

   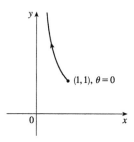

7. Three different sets of parametric equations for the curve $y = \sqrt{x}$ are

   (i) $x = t$, $y = \sqrt{t}$

   (ii) $x = t^4$, $y = t^2$

   (iii) $x = \tan^2 t$, $y = \tan t$, $0 \le t < \pi/2$

   There are many other sets of equations that also give this curve.

9. (a) 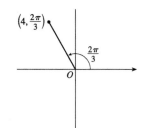 The Cartesian coordinates are $x = 4\cos\frac{2\pi}{3} = 4(-\frac{1}{2}) = -2$ and
   $y = 4\sin\frac{2\pi}{3} = 4\left(\frac{\sqrt{3}}{2}\right) = 2\sqrt{3}$, that is, the point $(-2, 2\sqrt{3})$.

   (b) Given $x = -3$ and $y = 3$, we have $r = \sqrt{(-3)^2 + 3^2} = \sqrt{18} = 3\sqrt{2}$. Also, $\tan\theta = \dfrac{y}{x}$ $\Rightarrow$ $\tan\theta = \dfrac{3}{-3}$, and since
   $(-3, 3)$ is in the second quadrant, $\theta = \frac{3\pi}{4}$. Thus, one set of polar coordinates for $(-3, 3)$ is $\left(3\sqrt{2}, \frac{3\pi}{4}\right)$, and two others are
   $\left(3\sqrt{2}, \frac{11\pi}{4}\right)$ and $\left(-3\sqrt{2}, \frac{7\pi}{4}\right)$.

11. $r = 1 + \sin\theta$. This cardioid is symmetric about the $\theta = \pi/2$ axis.

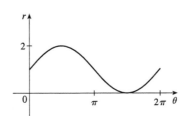
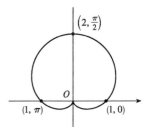

13. $r = \cos 3\theta$. This is a three-leaved rose. The curve is traced twice.

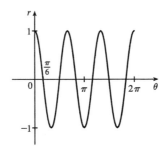
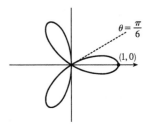

15. $r = 1 + \cos 2\theta$. The curve is symmetric about the pole and both the horizontal and vertical axes.

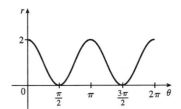
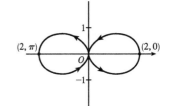

17. $r = \dfrac{3}{1 + 2\sin\theta}$ $\Rightarrow$ $e = 2 > 1$, so the conic is a hyperbola. $de = 3$ $\Rightarrow$ $d = \tfrac{3}{2}$ and the form "$+2\sin\theta$" imply that the directrix is above the focus at the origin and has equation $y = \tfrac{3}{2}$. The vertices are $\left(1, \tfrac{\pi}{2}\right)$ and $\left(-3, \tfrac{3\pi}{2}\right)$.

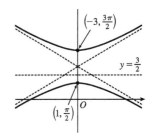

19. $x + y = 2$ $\Leftrightarrow$ $r\cos\theta + r\sin\theta = 2$ $\Leftrightarrow$ $r(\cos\theta + \sin\theta) = 2$ $\Leftrightarrow$ $r = \dfrac{2}{\cos\theta + \sin\theta}$

21. $r = (\sin\theta)/\theta$. As $\theta \to \pm\infty$, $r \to 0$. As $\theta \to 0$, $r \to 1$. In the first figure, there are an infinite number of x-intercepts at $x = \pi n$, n a nonzero integer. These correspond to pole points in the second figure.

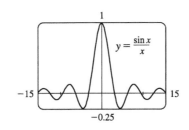
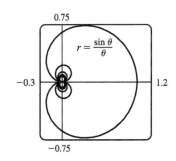

23. $x = \ln t$, $y = 1 + t^2$; $t = 1$. $\dfrac{dy}{dt} = 2t$ and $\dfrac{dx}{dt} = \dfrac{1}{t}$, so $\dfrac{dy}{dx} = \dfrac{dy/dt}{dx/dt} = \dfrac{2t}{1/t} = 2t^2$.

When $t = 1$, $(x, y) = (0, 2)$ and $dy/dx = 2$.

25. $r = e^{-\theta}$ $\Rightarrow$ $y = r\sin\theta = e^{-\theta}\sin\theta$ and $x = r\cos\theta = e^{-\theta}\cos\theta$ $\Rightarrow$

$$\dfrac{dy}{dx} = \dfrac{dy/d\theta}{dx/d\theta} = \dfrac{\frac{dr}{d\theta}\sin\theta + r\cos\theta}{\frac{dr}{d\theta}\cos\theta - r\sin\theta} = \dfrac{-e^{-\theta}\sin\theta + e^{-\theta}\cos\theta}{-e^{-\theta}\cos\theta - e^{-\theta}\sin\theta} \cdot \dfrac{-e^{\theta}}{-e^{\theta}} = \dfrac{\sin\theta - \cos\theta}{\cos\theta + \sin\theta}.$$

When $\theta = \pi$, $\dfrac{dy}{dx} = \dfrac{0 - (-1)}{-1 + 0} = \dfrac{1}{-1} = -1$.

27. $x = t + \sin t$, $y = t - \cos t$ $\Rightarrow$ $\dfrac{dy}{dx} = \dfrac{dy/dt}{dx/dt} = \dfrac{1 + \sin t}{1 + \cos t}$ $\Rightarrow$

$$\dfrac{d^2y}{dx^2} = \dfrac{\frac{d}{dt}\left(\frac{dy}{dx}\right)}{dx/dt} = \dfrac{\frac{(1+\cos t)\cos t - (1+\sin t)(-\sin t)}{(1+\cos t)^2}}{1 + \cos t} = \dfrac{\cos t + \cos^2 t + \sin t + \sin^2 t}{(1+\cos t)^3} = \dfrac{1 + \cos t + \sin t}{(1+\cos t)^3}$$

29. We graph the curve $x = t^3 - 3t$, $y = t^2 + t + 1$ for $-2.2 \le t \le 1.2$.
By zooming in or using a cursor, we find that the lowest point is about
$(1.4, 0.75)$. To find the exact values, we find the t-value at which
$dy/dt = 2t + 1 = 0$ $\Leftrightarrow$ $t = -\tfrac{1}{2}$ $\Leftrightarrow$ $(x, y) = \left(\tfrac{11}{8}, \tfrac{3}{4}\right)$.

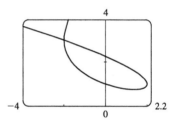

31. $x = 2a\cos t - a\cos 2t$ $\Rightarrow$ $\dfrac{dx}{dt} = -2a\sin t + 2a\sin 2t = 2a\sin t(2\cos t - 1) = 0$ $\Leftrightarrow$

$\sin t = 0$ or $\cos t = \tfrac{1}{2}$ $\Rightarrow$ $t = 0, \tfrac{\pi}{3}, \pi$, or $\tfrac{5\pi}{3}$.

$y = 2a\sin t - a\sin 2t$ $\Rightarrow$ $\dfrac{dy}{dt} = 2a\cos t - 2a\cos 2t = 2a(1 + \cos t - 2\cos^2 t) = 2a(1 - \cos t)(1 + 2\cos t) = 0$ $\Rightarrow$

$t = 0, \tfrac{2\pi}{3}$, or $\tfrac{4\pi}{3}$.

Thus the graph has vertical tangents where $t = \tfrac{\pi}{3}, \pi$ and $\tfrac{5\pi}{3}$, and horizontal tangents where $t = \tfrac{2\pi}{3}$ and $\tfrac{4\pi}{3}$. To determine

what the slope is where $t = 0$, we use l'Hospital's Rule to evaluate $\displaystyle\lim_{t\to 0} \dfrac{dy/dt}{dx/dt} = 0$, so there is a horizontal tangent there.

t	x	y
0	a	0
$\tfrac{\pi}{3}$	$\tfrac{3}{2}a$	$\tfrac{\sqrt{3}}{2}a$
$\tfrac{2\pi}{3}$	$-\tfrac{1}{2}a$	$\tfrac{3\sqrt{3}}{2}a$
π	$-3a$	0
$\tfrac{4\pi}{3}$	$-\tfrac{1}{2}a$	$-\tfrac{3\sqrt{3}}{2}a$
$\tfrac{5\pi}{3}$	$\tfrac{3}{2}a$	$-\tfrac{\sqrt{3}}{2}a$

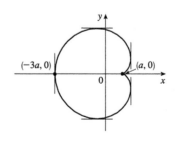

33. The curve $r^2 = 9\cos 5\theta$ has 10 "petals." For instance, for $-\frac{\pi}{10} \le \theta \le \frac{\pi}{10}$, there are two petals, one with $r > 0$ and one with $r < 0$.

$$A = 10 \int_{-\pi/10}^{\pi/10} \tfrac{1}{2} r^2 \, d\theta = 5 \int_{-\pi/10}^{\pi/10} 9 \cos 5\theta \, d\theta = 5 \cdot 9 \cdot 2 \int_{0}^{\pi/10} \cos 5\theta \, d\theta = 18 \big[\sin 5\theta\big]_0^{\pi/10} = 18$$

35. The curves intersect when $4\cos\theta = 2 \ \Rightarrow\ \cos\theta = \tfrac{1}{2} \ \Rightarrow\ \theta = \pm\tfrac{\pi}{3}$ for $-\pi \le \theta \le \pi$. The points of intersection are $\left(2, \tfrac{\pi}{3}\right)$ and $\left(2, -\tfrac{\pi}{3}\right)$.

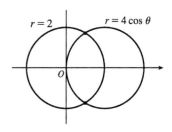

37. The curves intersect where $2\sin\theta = \sin\theta + \cos\theta \ \Rightarrow\ \sin\theta = \cos\theta \ \Rightarrow\ \theta = \tfrac{\pi}{4}$, and also at the origin (at which $\theta = \tfrac{3\pi}{4}$ on the second curve).

$$A = \int_0^{\pi/4} \tfrac{1}{2}(2\sin\theta)^2 \, d\theta + \int_{\pi/4}^{3\pi/4} \tfrac{1}{2}(\sin\theta + \cos\theta)^2 \, d\theta$$
$$= \int_0^{\pi/4} (1 - \cos 2\theta) \, d\theta + \tfrac{1}{2}\int_{\pi/4}^{3\pi/4} (1 + \sin 2\theta) \, d\theta$$
$$= \left[\theta - \tfrac{1}{2}\sin 2\theta\right]_0^{\pi/4} + \left[\tfrac{1}{2}\theta - \tfrac{1}{4}\cos 2\theta\right]_{\pi/4}^{3\pi/4} = \tfrac{1}{2}(\pi - 1)$$

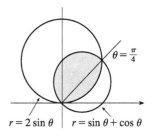

39. $x = 3t^2$, $y = 2t^3$.

$$L = \int_0^2 \sqrt{(dx/dt)^2 + (dy/dt)^2} \, dt = \int_0^2 \sqrt{(6t)^2 + (6t^2)^2} \, dt = \int_0^2 \sqrt{36t^2 + 36t^4} \, dt = \int_0^2 \sqrt{36t^2}\sqrt{1+t^2} \, dt$$
$$= \int_0^2 6|t|\sqrt{1+t^2} \, dt = 6\int_0^2 t\sqrt{1+t^2} \, dt = 6\int_1^5 u^{1/2}\left(\tfrac{1}{2}du\right) \qquad [u = 1+t^2, \, du = 2t\,dt]$$
$$= 6 \cdot \tfrac{1}{2} \cdot \tfrac{2}{3}\left[u^{3/2}\right]_1^5 = 2(5^{3/2} - 1) = 2(5\sqrt{5} - 1)$$

41. $L = \int_\pi^{2\pi} \sqrt{r^2 + (dr/d\theta)^2} \, d\theta = \int_\pi^{2\pi} \sqrt{(1/\theta)^2 + (-1/\theta^2)^2} \, d\theta = \int_\pi^{2\pi} \dfrac{\sqrt{\theta^2 + 1}}{\theta^2} \, d\theta$

$$\overset{24}{=} \left[-\dfrac{\sqrt{\theta^2+1}}{\theta} + \ln\left(\theta + \sqrt{\theta^2+1}\right)\right]_\pi^{2\pi} = \dfrac{\sqrt{\pi^2+1}}{\pi} - \dfrac{\sqrt{4\pi^2+1}}{2\pi} + \ln\left(\dfrac{2\pi + \sqrt{4\pi^2+1}}{\pi + \sqrt{\pi^2+1}}\right)$$
$$= \dfrac{2\sqrt{\pi^2+1} - \sqrt{4\pi^2+1}}{2\pi} + \ln\left(\dfrac{2\pi + \sqrt{4\pi^2+1}}{\pi + \sqrt{\pi^2+1}}\right)$$

43. (a) $x = \tfrac{1}{2}(t^2 + 3)$, $y = 5 - \tfrac{1}{3}t^3$. $dx/dt = t$ and $dy/dt = -t^2$, so the speed at time t is the function

$v(t) = s'(t) = \sqrt{t^2 + (-t^2)^2}$. At the point $(6, -4)$, $y = 5 - \tfrac{1}{3}t^3 = -4 \ \Rightarrow\ 9 = \tfrac{1}{3}t^3 \ \Rightarrow\ t^3 = 27 \ \Rightarrow\ t = 3$.

Thus, the speed of the particle at the point $(6, -4)$ is $v(3) = \sqrt{3^2 + 3^4} = \sqrt{90} \approx 9.49 \text{ m/s}$.

(b) To find the average speed of the particle for $0 \le t \le 8$, we find the total distance L traveled in this time, and divide it by the length of the interval. By Theorem 10.2.5,

$$L = \int_0^8 \sqrt{\left(\frac{dx}{dt}\right)^2 + \left(\frac{dy}{dt}\right)^2}\, dt = \int_0^8 \sqrt{t^2 + t^4}\, dt = \int_0^8 t\sqrt{1+t^2}\, dt$$

$$= \tfrac{1}{2}\int_1^{65} \sqrt{u}\, du \quad [u = 1 + t^2, du = 2t\, dt]$$

$$= \tfrac{1}{2}\left[\tfrac{2}{3} u^{3/2}\right]_1^{65} = \tfrac{1}{3}(65^{3/2} - 1)$$

Thus, the average speed is $\dfrac{L}{8} = \dfrac{1}{24}(65\sqrt{65} - 1) \approx 21.79$ m/s.

45. $x = 4\sqrt{t},\ y = \dfrac{t^3}{3} + \dfrac{1}{2t^2},\ 1 \le t \le 4 \ \Rightarrow$

$S = \int_1^4 2\pi y \sqrt{(dx/dt)^2 + (dy/dt)^2}\, dt = \int_1^4 2\pi\left(\tfrac{1}{3}t^3 + \tfrac{1}{2}t^{-2}\right)\sqrt{\left(2/\sqrt{t}\right)^2 + (t^2 - t^{-3})^2}\, dt$

$= 2\pi \int_1^4 \left(\tfrac{1}{3}t^3 + \tfrac{1}{2}t^{-2}\right)\sqrt{(t^2 + t^{-3})^2}\, dt = 2\pi \int_1^4 \left(\tfrac{1}{3}t^5 + \tfrac{5}{6} + \tfrac{1}{2}t^{-5}\right)dt = 2\pi\left[\tfrac{1}{18}t^6 + \tfrac{5}{6}t - \tfrac{1}{8}t^{-4}\right]_1^4 = \dfrac{471{,}295}{1024}\pi$

47. $x = \dfrac{t^2 - c}{t^2 + 1},\ y = \dfrac{t(t^2 - c)}{t^2 + 1}$. For all c except -1, the curve is asymptotic to the line $x = 1$. For $c < -1$, the curve bulges to the right near $y = 0$. As c increases, the bulge becomes smaller, until at $c = -1$ the curve is the straight line $x = 1$. As c continues to increase, the curve bulges to the left, until at $c = 0$ there is a cusp at the origin. For $c > 0$, there is a loop to the left of the origin, whose size and roundness increase as c increases. Note that the x-intercept of the curve is always $-c$.

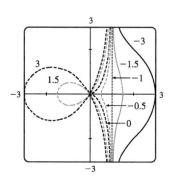

49. $\dfrac{x^2}{9} + \dfrac{y^2}{8} = 1$ is an ellipse with center $(0, 0)$.

$a = 3,\ b = 2\sqrt{2},\ c = 1\ \Rightarrow$ foci $(\pm 1, 0)$, vertices $(\pm 3, 0)$.

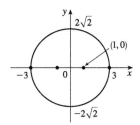

51. $6y^2 + x - 36y + 55 = 0\ \Leftrightarrow$

$6(y^2 - 6y + 9) = -(x + 1)\ \Leftrightarrow$

$(y - 3)^2 = -\tfrac{1}{6}(x + 1)$, a parabola with vertex $(-1, 3)$, opening to the left, $p = -\tfrac{1}{24}\ \Rightarrow$ focus $\left(-\tfrac{25}{24}, 3\right)$ and directrix $x = -\tfrac{23}{24}$.

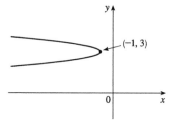

53. The ellipse with foci $(\pm 4, 0)$ and vertices $(\pm 5, 0)$ has center $(0, 0)$ and a horizontal major axis, with $a = 5$ and $c = 4$, so $b^2 = a^2 - c^2 = 5^2 - 4^2 = 9$. An equation is $\dfrac{x^2}{25} + \dfrac{y^2}{9} = 1$.

55. The center of a hyperbola with foci $(0, \pm 4)$ is $(0,0)$, so $c = 4$ and an equation is $\dfrac{y^2}{a^2} - \dfrac{x^2}{b^2} = 1$.

The asymptote $y = 3x$ has slope 3, so $\dfrac{a}{b} = \dfrac{3}{1} \;\Rightarrow\; a = 3b$ and $a^2 + b^2 = c^2 \;\Rightarrow\; (3b)^2 + b^2 = 4^2 \;\Rightarrow\;$

$10b^2 = 16 \;\Rightarrow\; b^2 = \tfrac{8}{5}$ and so $a^2 = 16 - \tfrac{8}{5} = \tfrac{72}{5}$. Thus, an equation is $\dfrac{y^2}{72/5} - \dfrac{x^2}{8/5} = 1$, or $\dfrac{5y^2}{72} - \dfrac{5x^2}{8} = 1$.

57. $x^2 + y = 100 \;\Leftrightarrow\; x^2 = -(y - 100)$ has its vertex at $(0, 100)$, so one of the vertices of the ellipse is $(0, 100)$. Another form of the equation of a parabola is $x^2 = 4p(y - 100)$ so $4p(y - 100) = -(y - 100) \;\Rightarrow\; 4p = -1 \;\Rightarrow\; p = -\tfrac{1}{4}$.
Therefore the shared focus is found at $\left(0, \tfrac{399}{4}\right)$ so $2c = \tfrac{399}{4} - 0 \;\Rightarrow\; c = \tfrac{399}{8}$ and the center of the ellipse is $\left(0, \tfrac{399}{8}\right)$. So
$a = 100 - \tfrac{399}{8} = \tfrac{401}{8}$ and $b^2 = a^2 - c^2 = \dfrac{401^2 - 399^2}{8^2} = 25$. So the equation of the ellipse is $\dfrac{x^2}{b^2} + \dfrac{\left(y - \tfrac{399}{8}\right)^2}{a^2} = 1 \;\Rightarrow\;$
$\dfrac{x^2}{25} + \dfrac{\left(y - \tfrac{399}{8}\right)^2}{\left(\tfrac{401}{8}\right)^2} = 1$, or $\dfrac{x^2}{25} + \dfrac{(8y - 399)^2}{160{,}801} = 1$.

59. Directrix $x = 4 \;\Rightarrow\; d = 4$, so $e = \tfrac{1}{3} \;\Rightarrow\; r = \dfrac{ed}{1 + e \cos\theta} = \dfrac{4}{3 + \cos\theta}$.

61. See the end of the proof of Theorem 10.6.1. If $e > 1$, then $1 - e^2 < 0$ and Equations 10.6.4 become $a^2 = \dfrac{e^2 d^2}{(e^2 - 1)^2}$ and
$b^2 = \dfrac{e^2 d^2}{e^2 - 1}$, so $\dfrac{b^2}{a^2} = e^2 - 1$. The asymptotes $y = \pm\dfrac{b}{a}x$ have slopes $\pm\dfrac{b}{a} = \pm\sqrt{e^2 - 1}$, so the angles they make with the polar axis are $\pm\tan^{-1}\left[\sqrt{e^2 - 1}\right] = \cos^{-1}(\pm 1/e)$.

☐ PROBLEMS PLUS

1. See the figure. The circle with center $(-1, 0)$ and radius $\sqrt{2}$ has equation $(x+1)^2 + y^2 = 2$ and describes the circular arc from $(0, -1)$ to $(0, 1)$. Converting the equation to polar coordinates gives us

$(r\cos\theta + 1)^2 + (r\sin\theta)^2 = 2 \Rightarrow$
$r^2\cos^2\theta + 2r\cos\theta + 1 + r^2\sin^2\theta = 2 \Rightarrow$
$r^2(\cos^2\theta + \sin^2\theta) + 2r\cos\theta = 1 \Rightarrow r^2 + 2r\cos\theta = 1$. Using the quadratic formula to solve for r gives us

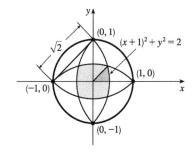

$$r = \frac{-2\cos\theta \pm \sqrt{4\cos^2\theta + 4}}{2} = -\cos\theta + \sqrt{\cos^2\theta + 1} \text{ for } r > 0.$$

The darkest shaded region is $\frac{1}{8}$ of the entire shaded region A, so $\frac{1}{8}A = \int_0^{\pi/4} \frac{1}{2}r^2\,d\theta = \frac{1}{2}\int_0^{\pi/4} (1 - 2r\cos\theta)\,d\theta \Rightarrow$

$$\tfrac{1}{4}A = \int_0^{\pi/4}\left[1 - 2\cos\theta\left(-\cos\theta + \sqrt{\cos^2\theta + 1}\right)\right]d\theta = \int_0^{\pi/4}\left(1 + 2\cos^2\theta - 2\cos\theta\sqrt{\cos^2\theta + 1}\right)d\theta$$

$$= \int_0^{\pi/4}\left[1 + 2\cdot\tfrac{1}{2}(1 + \cos 2\theta) - 2\cos\theta\sqrt{(1 - \sin^2\theta) + 1}\right]d\theta$$

$$= \int_0^{\pi/4}(2 + \cos 2\theta)\,d\theta - 2\int_0^{\pi/4}\cos\theta\sqrt{2 - \sin^2\theta}\,d\theta$$

$$= \left[2\theta + \tfrac{1}{2}\sin 2\theta\right]_0^{\pi/4} - 2\int_0^{1/\sqrt{2}}\sqrt{2 - u^2}\,du \quad \begin{bmatrix} u = \sin\theta, \\ du = \cos\theta\,d\theta \end{bmatrix}$$

$$= \left(\tfrac{\pi}{2} + \tfrac{1}{2}\right) - (0 + 0) - 2\left[\tfrac{u}{2}\sqrt{2 - u^2} + \sin^{-1}\tfrac{u}{\sqrt{2}}\right]_0^{1/\sqrt{2}} \quad \begin{bmatrix} \text{Formula 30,} \\ a = \sqrt{2} \end{bmatrix}$$

$$= \tfrac{\pi}{2} + \tfrac{1}{2} - 2\left(\tfrac{1}{2\sqrt{2}}\cdot\tfrac{\sqrt{3}}{\sqrt{2}} + \tfrac{\pi}{6}\right) = \tfrac{\pi}{2} + \tfrac{1}{2} - \tfrac{1}{2}\sqrt{3} - \tfrac{\pi}{3} = \tfrac{\pi}{6} + \tfrac{1}{2} - \tfrac{1}{2}\sqrt{3}.$$

Thus, $A = 4\left(\tfrac{\pi}{6} + \tfrac{1}{2} - \tfrac{1}{2}\sqrt{3}\right) = \tfrac{2\pi}{3} + 2 - 2\sqrt{3}.$

3. In terms of x and y, we have $x = r\cos\theta = (1 + c\sin\theta)\cos\theta = \cos\theta + c\sin\theta\cos\theta = \cos\theta + \tfrac{1}{2}c\sin 2\theta$ and $y = r\sin\theta = (1 + c\sin\theta)\sin\theta = \sin\theta + c\sin^2\theta$. Now $-1 \leq \sin\theta \leq 1 \Rightarrow -1 \leq \sin\theta + c\sin^2\theta \leq 1 + c \leq 2$, so $-1 \leq y \leq 2$. Furthermore, $y = 2$ when $c = 1$ and $\theta = \tfrac{\pi}{2}$, while $y = -1$ for $c = 0$ and $\theta = \tfrac{3\pi}{2}$. Therefore, we need a viewing rectangle with $-1 \leq y \leq 2$.

To find the x-values, look at the equation $x = \cos\theta + \tfrac{1}{2}c\sin 2\theta$ and use the fact that $\sin 2\theta \geq 0$ for $0 \leq \theta \leq \tfrac{\pi}{2}$ and $\sin 2\theta \leq 0$ for $-\tfrac{\pi}{2} \leq \theta \leq 0$. [Because $r = 1 + c\sin\theta$ is symmetric about the y-axis, we only need to consider $-\tfrac{\pi}{2} \leq \theta \leq \tfrac{\pi}{2}$.] So for $-\tfrac{\pi}{2} \leq \theta \leq 0$, x has a maximum value when $c = 0$ and then $x = \cos\theta$ has a maximum value of 1 at $\theta = 0$. Thus, the maximum value of x must occur on $\left[0, \tfrac{\pi}{2}\right]$ with $c = 1$. Then $x = \cos\theta + \tfrac{1}{2}\sin 2\theta \Rightarrow$

$\frac{dx}{d\theta} = -\sin\theta + \cos 2\theta = -\sin\theta + 1 - 2\sin^2\theta \implies \frac{dx}{d\theta} = -(2\sin\theta - 1)(\sin\theta + 1) = 0$ when $\sin\theta = -1$ or $\frac{1}{2}$ [but $\sin\theta \neq -1$ for $0 \leq \theta \leq \frac{\pi}{2}$]. If $\sin\theta = \frac{1}{2}$, then $\theta = \frac{\pi}{6}$ and $x = \cos\frac{\pi}{6} + \frac{1}{2}\sin\frac{\pi}{3} = \frac{3}{4}\sqrt{3}$. Thus, the maximum value of x is $\frac{3}{4}\sqrt{3}$, and, by symmetry, the minimum value is $-\frac{3}{4}\sqrt{3}$. Therefore, the smallest viewing rectangle that contains every member of the family of polar curves $r = 1 + c\sin\theta$, where $0 \leq c \leq 1$, is $\left[-\frac{3}{4}\sqrt{3}, \frac{3}{4}\sqrt{3}\right] \times [-1, 2]$.

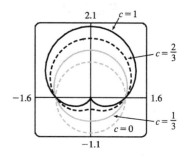

5. Without loss of generality, assume the hyperbola has equation $\frac{x^2}{a^2} - \frac{y^2}{b^2} = 1$. Use implicit differentiation to get $\frac{2x}{a^2} - \frac{2yy'}{b^2} = 0$, so $y' = \frac{b^2 x}{a^2 y}$. The tangent line at the point (c, d) on the hyperbola has equation $y - d = \frac{b^2 c}{a^2 d}(x - c)$.

The tangent line intersects the asymptote $y = \frac{b}{a}x$ when $\frac{b}{a}x - d = \frac{b^2 c}{a^2 d}(x - c) \implies abdx - a^2 d^2 = b^2 cx - b^2 c^2 \implies$

$abdx - b^2 cx = a^2 d^2 - b^2 c^2 \implies x = \frac{a^2 d^2 - b^2 c^2}{b(ad - bc)} = \frac{ad + bc}{b}$ and the y-value is $\frac{b}{a} \cdot \frac{ad + bc}{b} = \frac{ad + bc}{a}$.

Similarly, the tangent line intersects $y = -\frac{b}{a}x$ at $\left(\frac{bc - ad}{b}, \frac{ad - bc}{a}\right)$. The midpoint of these intersection points is

$\left(\frac{1}{2}\left(\frac{ad + bc}{b} + \frac{bc - ad}{b}\right), \frac{1}{2}\left(\frac{ad + bc}{a} + \frac{ad - bc}{a}\right)\right) = \left(\frac{1}{2}\frac{2bc}{b}, \frac{1}{2}\frac{2ad}{a}\right) = (c, d)$, the point of tangency.

Note: If $y = 0$, then at $(\pm a, 0)$, the tangent line is $x = \pm a$, and the points of intersection are clearly equidistant from the point of tangency.

11 □ SEQUENCES, SERIES, AND POWER SERIES

11.1 Sequences

1. (a) A sequence is an ordered list of numbers. It can also be defined as a function whose domain is the set of positive integers.

   (b) The terms a_n approach 8 as n becomes large. In fact, we can make a_n as close to 8 as we like by taking n sufficiently large.

   (c) The terms a_n become large as n becomes large. In fact, we can make a_n as large as we like by taking n sufficiently large.

3. $a_n = n^3 - 1$, so the sequence is $\{1^3 - 1, 2^3 - 1, 3^3 - 1, 4^3 - 1, 5^3 - 1, \ldots\} = \{0, 7, 26, 63, 124, \ldots\}$.

5. $\{2^n + n\}_{n=2}^{\infty}$, so the sequence is $\{2^2 + 2, 2^3 + 3, 2^4 + 4, 2^5 + 5, 2^6 + 6, \ldots\} = \{6, 11, 20, 37, 70, \ldots\}$.

7. $a_n = \dfrac{(-1)^{n-1}}{n^2}$, so the sequence is

   $$\left\{\frac{(-1)^{1-1}}{1^2}, \frac{(-1)^{2-1}}{2^2}, \frac{(-1)^{3-1}}{3^2}, \frac{(-1)^{4-1}}{4^2}, \frac{(-1)^{5-1}}{5^2}, \ldots\right\} = \left\{1, -\frac{1}{4}, \frac{1}{9}, -\frac{1}{16}, \frac{1}{25}, \ldots\right\}.$$

9. $a_n = \cos n\pi$, so the sequence is $\{\cos \pi, \cos 2\pi, \cos 3\pi, \cos 4\pi, \cos 5\pi, \ldots\} = \{-1, 1, -1, 1, -1, \ldots\}$.

11. $a_n = \dfrac{(-2)^n}{(n+1)!}$, so the sequence is

    $$\left\{\frac{(-2)^1}{2!}, \frac{(-2)^2}{3!}, \frac{(-2)^3}{4!}, \frac{(-2)^4}{5!}, \frac{(-2)^5}{6!}, \ldots\right\} = \left\{-\frac{2}{2}, \frac{4}{6}, -\frac{8}{24}, \frac{16}{120}, -\frac{32}{720}, \ldots\right\} = \left\{-1, \frac{2}{3}, -\frac{1}{3}, \frac{2}{15}, -\frac{2}{45}, \ldots\right\}.$$

13. $a_1 = 1$, $a_{n+1} = 2a_n + 1$. $a_2 = 2a_1 + 1 = 2 \cdot 1 + 1 = 3$. $a_3 = 2a_2 + 1 = 2 \cdot 3 + 1 = 7$. $a_4 = 2a_3 + 1 = 2 \cdot 7 + 1 = 15$. $a_5 = 2a_4 + 1 = 2 \cdot 15 + 1 = 31$. The sequence is $\{1, 3, 7, 15, 31, \ldots\}$.

15. $a_1 = 2$, $a_{n+1} = \dfrac{a_n}{1 + a_n}$. $a_2 = \dfrac{a_1}{1 + a_1} = \dfrac{2}{1 + 2} = \dfrac{2}{3}$. $a_3 = \dfrac{a_2}{1 + a_2} = \dfrac{2/3}{1 + 2/3} = \dfrac{2}{5}$. $a_4 = \dfrac{a_3}{1 + a_3} = \dfrac{2/5}{1 + 2/5} = \dfrac{2}{7}$. $a_5 = \dfrac{a_4}{1 + a_4} = \dfrac{2/7}{1 + 2/7} = \dfrac{2}{9}$. The sequence is $\{2, \frac{2}{3}, \frac{2}{5}, \frac{2}{7}, \frac{2}{9}, \ldots\}$.

17. $\{\frac{1}{2}, \frac{1}{4}, \frac{1}{6}, \frac{1}{8}, \frac{1}{10}, \ldots\}$. The denominator is two times the number of the term, n, so $a_n = \dfrac{1}{2n}$.

19. $\{-3, 2, -\frac{4}{3}, \frac{8}{9}, -\frac{16}{27}, \ldots\}$. The first term is -3 and each term is $-\frac{2}{3}$ times the preceding one, so $a_n = -3\left(-\frac{2}{3}\right)^{n-1}$.

21. $\{\frac{1}{2}, -\frac{4}{3}, \frac{9}{4}, -\frac{16}{5}, \frac{25}{6}, \ldots\}$. The numerator of the nth term is n^2 and its denominator is $n + 1$. Including the alternating signs, we get $a_n = (-1)^{n+1} \dfrac{n^2}{n+1}$.

23.

n	$a_n = \dfrac{3n}{1+6n}$
1	0.4286
2	0.4615
3	0.4737
4	0.4800
5	0.4839
6	0.4865
7	0.4884
8	0.4898
9	0.4909
10	0.4918

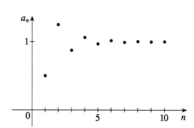

It appears that $\lim_{n\to\infty} a_n = 0.5$.

$$\lim_{n\to\infty} \frac{3n}{1+6n} = \lim_{n\to\infty} \frac{(3n)/n}{(1+6n)/n} = \lim_{n\to\infty} \frac{3}{1/n+6} = \frac{3}{6} = \frac{1}{2}$$

25.

n	$a_n = 1 + \left(-\frac{1}{2}\right)^n$
1	0.5000
2	1.2500
3	0.8750
4	1.0625
5	0.9688
6	1.0156
7	0.9922
8	1.0039
9	0.9980
10	1.0010

It appears that $\lim_{n\to\infty} a_n = 1$.

$\lim_{n\to\infty}\left(1+\left(-\frac{1}{2}\right)^n\right) = \lim_{n\to\infty} 1 + \lim_{n\to\infty}\left(-\frac{1}{2}\right)^n = 1 + 0 = 1$ since $\lim_{n\to\infty}\left(-\frac{1}{2}\right)^n = 0$ by (9).

27. $a_n = \dfrac{5}{n+2} = \dfrac{5/n}{(n+2)/n} = \dfrac{5/n}{1+2/n}$, so $a_n \to \dfrac{0}{1+0} = 0$ as $n \to \infty$. Converges

29. $a_n = \dfrac{4n^2-3n}{2n^2+1} = \dfrac{(4n^2-3n)/n^2}{(2n^2+1)/n^2} = \dfrac{4-3/n}{2+1/n^2}$, so $a_n \to \dfrac{4-0}{2+0} = 2$ as $n \to \infty$. Converges

31. $a_n = \dfrac{n^4}{n^3-2n} = \dfrac{n^4/n^3}{(n^3-2n)/n^3} = \dfrac{n}{1-2/n^2}$, so $a_n \to \infty$ as $n \to \infty$ since $\lim_{n\to\infty} n = \infty$ and $\lim_{n\to\infty}\left(1-\dfrac{2}{n^2}\right) = 1-0 = 1$. Diverges

33. $a_n = 3^n 7^{-n} = \dfrac{3^n}{7^n} = \left(\dfrac{3}{7}\right)^n$, so $\lim_{n\to\infty} a_n = 0$ by (9) with $r = \dfrac{3}{7}$. Converges

35. Because the natural exponential function is continuous at 0, Theorem 7 enables us to write

$$\lim_{n\to\infty} a_n = \lim_{n\to\infty} e^{-1/\sqrt{n}} = e^{\lim_{n\to\infty}(-1/\sqrt{n})} = e^0 = 1. \text{ Converges}$$

37. $a_n = \sqrt{\dfrac{1+4n^2}{1+n^2}} = \sqrt{\dfrac{(1+4n^2)/n^2}{(1+n^2)/n^2}} = \sqrt{\dfrac{(1/n^2)+4}{(1/n^2)+1}} \to \sqrt{4} = 2$ as $n \to \infty$ since $\lim\limits_{n\to\infty}(1/n^2) = 0$. Converges

39. $a_n = \dfrac{n^2}{\sqrt{n^3+4n}} = \dfrac{n^2/\sqrt{n^3}}{\sqrt{n^3+4n}/\sqrt{n^3}} = \dfrac{\sqrt{n}}{\sqrt{1+4/n^2}}$, so $a_n \to \infty$ as $n \to \infty$ since $\lim\limits_{n\to\infty}\sqrt{n} = \infty$ and

$\lim\limits_{n\to\infty}\sqrt{1+4/n^2} = 1$. Diverges

41. $\lim\limits_{n\to\infty}|a_n| = \lim\limits_{n\to\infty}\left|\dfrac{(-1)^n}{2\sqrt{n}}\right| = \dfrac{1}{2}\lim\limits_{n\to\infty}\dfrac{1}{n^{1/2}} = \dfrac{1}{2}(0) = 0$, so $\lim\limits_{n\to\infty}a_n = 0$ by (6). Converges

43. $a_n = \dfrac{(2n-1)!}{(2n+1)!} = \dfrac{(2n-1)!}{(2n+1)(2n)(2n-1)!} = \dfrac{1}{(2n+1)(2n)} \to 0$ as $n \to \infty$. Converges

45. $a_n = \sin n$. This sequence diverges since the terms don't approach any particular real number as $n \to \infty$. The terms take on values between -1 and 1. Diverges

47. $a_n = n^2 e^{-n} = \dfrac{n^2}{e^n}$. Since $\lim\limits_{x\to\infty}\dfrac{x^2}{e^x} \overset{H}{=} \lim\limits_{x\to\infty}\dfrac{2x}{e^x} \overset{H}{=} \lim\limits_{x\to\infty}\dfrac{2}{e^x} = 0$, it follows from Theorem 4 that $\lim\limits_{n\to\infty}a_n = 0$. Converges

49. $0 \le \dfrac{\cos^2 n}{2^n} \le \dfrac{1}{2^n}$ [since $0 \le \cos^2 n \le 1$], so since $\lim\limits_{n\to\infty}\dfrac{1}{2^n} = 0$, $\left\{\dfrac{\cos^2 n}{2^n}\right\}$ converges to 0 by the Squeeze Theorem.

51. $a_n = n\sin(1/n) = \dfrac{\sin(1/n)}{1/n}$. Since $\lim\limits_{x\to\infty}\dfrac{\sin(1/x)}{1/x} = \lim\limits_{t\to 0^+}\dfrac{\sin t}{t}$ [where $t = 1/x$] $= 1$, it follows from Theorem 4

that $\{a_n\}$ converges to 1.

53. $y = \left(1+\dfrac{2}{x}\right)^x \;\Rightarrow\; \ln y = x\ln\left(1+\dfrac{2}{x}\right)$, so

$\lim\limits_{x\to\infty}\ln y = \lim\limits_{x\to\infty}\dfrac{\ln(1+2/x)}{1/x} \overset{H}{=} \lim\limits_{x\to\infty}\dfrac{\left(\dfrac{1}{1+2/x}\right)\left(-\dfrac{2}{x^2}\right)}{-1/x^2} = \lim\limits_{x\to\infty}\dfrac{2}{1+2/x} = 2 \;\Rightarrow\;$

$\lim\limits_{x\to\infty}\left(1+\dfrac{2}{x}\right)^x = \lim\limits_{x\to\infty} e^{\ln y} = e^2$, so by Theorem 4, $\lim\limits_{n\to\infty}\left(1+\dfrac{2}{n}\right)^n = e^2$. Converges

55. $a_n = \ln(2n^2+1) - \ln(n^2+1) = \ln\left(\dfrac{2n^2+1}{n^2+1}\right) = \ln\left(\dfrac{2+1/n^2}{1+1/n^2}\right) \to \ln 2$ as $n \to \infty$. Converges

57. $a_n = \arctan(\ln n)$. Let $f(x) = \arctan(\ln x)$. Then $\lim\limits_{x\to\infty} f(x) = \dfrac{\pi}{2}$ since $\ln x \to \infty$ as $x \to \infty$ and $\arctan$ is continuous.

Thus, $\lim\limits_{n\to\infty} a_n = \lim\limits_{n\to\infty} f(n) = \dfrac{\pi}{2}$. Converges

59. $\{0,1,0,0,1,0,0,0,1,\ldots\}$ diverges since the sequence takes on only two values, 0 and 1, and never stays arbitrarily close to either value (or any other value) for n sufficiently large.

61. $a_n = \dfrac{n!}{2^n} = \dfrac{1}{2}\cdot\dfrac{2}{2}\cdot\dfrac{3}{2}\cdot\ldots\cdot\dfrac{(n-1)}{2}\cdot\dfrac{n}{2} \ge \dfrac{1}{2}\cdot\dfrac{n}{2}$ [for $n > 1$] $= \dfrac{n}{4} \to \infty$ as $n \to \infty$, so $\{a_n\}$ diverges.

63.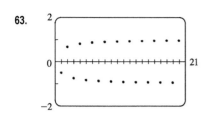

From the graph, it appears that the sequence $\{a_n\} = \left\{(-1)^n \dfrac{n}{n+1}\right\}$ is divergent, since it oscillates between 1 and -1 (approximately). To prove this, suppose that $\{a_n\}$ converges to L. If $b_n = \dfrac{n}{n+1}$, then $\{b_n\}$ converges to 1, and $\lim\limits_{n \to \infty} \dfrac{a_n}{b_n} = \dfrac{L}{1} = L$. But $\dfrac{a_n}{b_n} = (-1)^n$, so $\lim\limits_{n \to \infty} \dfrac{a_n}{b_n}$ does not exist. This contradiction shows that $\{a_n\}$ diverges.

65.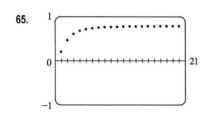

From the graph, it appears that the sequence converges to a number between 0.7 and 0.8.

$$a_n = \arctan\left(\dfrac{n^2}{n^2+4}\right) = \arctan\left(\dfrac{n^2/n^2}{(n^2+4)/n^2}\right) = \arctan\left(\dfrac{1}{1+4/n^2}\right) \to$$

$\arctan 1 = \dfrac{\pi}{4}$ $[\approx 0.785]$ as $n \to \infty$.

67.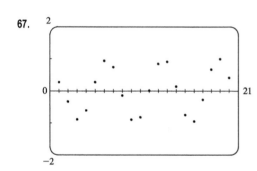

From the graph, it appears that the sequence $\{a_n\} = \left\{\dfrac{n^2 \cos n}{1+n^2}\right\}$ is divergent, since it oscillates between 1 and -1 (approximately). To prove this, suppose that $\{a_n\}$ converges to L. If $b_n = \dfrac{n^2}{1+n^2}$, then $\{b_n\}$ converges to 1, and $\lim\limits_{n \to \infty} \dfrac{a_n}{b_n} = \dfrac{L}{1} = L$. But $\dfrac{a_n}{b_n} = \cos n$, so $\lim\limits_{n \to \infty} \dfrac{a_n}{b_n}$ does not exist. This contradiction shows that $\{a_n\}$ diverges.

69.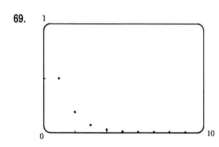

From the graph, it appears that the sequence approaches 0.

$$0 < a_n = \dfrac{1 \cdot 3 \cdot 5 \cdots (2n-1)}{(2n)^n} = \dfrac{1}{2n} \cdot \dfrac{3}{2n} \cdot \dfrac{5}{2n} \cdots \dfrac{2n-1}{2n}$$

$$\leq \dfrac{1}{2n} \cdot (1) \cdot (1) \cdots (1) = \dfrac{1}{2n} \to 0 \text{ as } n \to \infty$$

So by the Squeeze Theorem, $\left\{\dfrac{1 \cdot 3 \cdot 5 \cdots (2n-1)}{(2n)^n}\right\}$ converges to 0.

71. (a) $a_n = 1000(1.06)^n \Rightarrow a_1 = 1060$, $a_2 = 1123.60$, $a_3 = 1191.02$, $a_4 = 1262.48$, and $a_5 = 1338.23$.

(b) $\lim\limits_{n \to \infty} a_n = 1000 \lim\limits_{n \to \infty} (1.06)^n$, so the sequence diverges by (9) with $r = 1.06 > 1$.

73. (a) We are given that the initial population is 5000, so $P_0 = 5000$. The number of catfish increases by 8% per month and is decreased by 300 per month, so $P_1 = P_0 + 8\% P_0 - 300 = 1.08 P_0 - 300$, $P_2 = 1.08 P_1 - 300$, and so on. Thus, $P_n = 1.08 P_{n-1} - 300$.

(b) Using the recursive formula with $P_0 = 5000$, we get $P_1 = 5100$, $P_2 = 5208$, $P_3 = 5325$ (rounding any portion of a catfish), $P_4 = 5451$, $P_5 = 5587$, and $P_6 = 5734$, which is the number of catfish in the pond after six months.

75. If $|r| \geq 1$, then $\{r^n\}$ diverges by (9), so $\{nr^n\}$ diverges also, since $|nr^n| = n\,|r^n| \geq |r^n|$. If $|r| < 1$ then

$$\lim_{x \to \infty} xr^x = \lim_{x \to \infty} \frac{x}{r^{-x}} \stackrel{\mathrm{H}}{=} \lim_{x \to \infty} \frac{1}{(-\ln r)\,r^{-x}} = \lim_{x \to \infty} \frac{r^x}{-\ln r} = 0, \text{ so } \lim_{n \to \infty} nr^n = 0, \text{ and hence } \{nr^n\} \text{ converges}$$

whenever $|r| < 1$.

77. Since $\{a_n\}$ is a decreasing sequence, $a_n > a_{n+1}$ for all $n \geq 1$. Because all of its terms lie between 5 and 8, $\{a_n\}$ is a bounded sequence. By the Monotonic Sequence Theorem, $\{a_n\}$ is convergent; that is, $\{a_n\}$ has a limit L. L must be less than 8 since $\{a_n\}$ is decreasing, so $5 \leq L < 8$.

79. $a_n = \dfrac{1}{2n+3}$ is decreasing since $a_{n+1} = \dfrac{1}{2(n+1)+3} = \dfrac{1}{2n+5} < \dfrac{1}{2n+3} = a_n$ for each $n \geq 1$. The sequence is bounded since $0 < a_n \leq \frac{1}{5}$ for all $n \geq 1$. Note that $a_1 = \frac{1}{5}$.

81. The terms of $a_n = n(-1)^n$ alternate in sign, so the sequence is not monotonic. The first five terms are $-1, 2, -3, 4,$ and -5. Since $\lim\limits_{n \to \infty} |a_n| = \lim\limits_{n \to \infty} n = \infty$, the sequence is not bounded.

83. $a_n = 3 - 2ne^{-n}$. Let $f(x) = 3 - 2xe^{-x}$. Then $f'(x) = 0 - 2[x(-e^{-x}) + e^{-x}] = 2e^{-x}(x-1)$, which is positive for $x > 1$, so f is increasing on $(1, \infty)$. It follows that the sequence $\{a_n\} = \{f(n)\}$ is increasing. The sequence is bounded below by $a_1 = 3 - 2e^{-1} \approx 2.26$ and above by 3, so the sequence is bounded.

85. For $\left\{\sqrt{2}, \sqrt{2\sqrt{2}}, \sqrt{2\sqrt{2\sqrt{2}}}, \ldots\right\}$, $a_1 = 2^{1/2}$, $a_2 = 2^{3/4}$, $a_3 = 2^{7/8}, \ldots$, so $a_n = 2^{(2^n - 1)/2^n} = 2^{1-(1/2^n)}$.

$\lim\limits_{n \to \infty} a_n = \lim\limits_{n \to \infty} 2^{1-(1/2^n)} = 2^1 = 2$.

Alternate solution: Let $L = \lim\limits_{n \to \infty} a_n$. (We could show the limit exists by showing that $\{a_n\}$ is bounded and increasing.) Then L must satisfy $L = \sqrt{2 \cdot L} \;\Rightarrow\; L^2 = 2L \;\Rightarrow\; L(L-2) = 0$. $L \neq 0$ since the sequence increases, so $L = 2$.

87. $a_1 = 1$, $a_{n+1} = 3 - \dfrac{1}{a_n}$. We show by induction that $\{a_n\}$ is increasing and bounded above by 3. Let P_n be the proposition that $a_{n+1} > a_n$ and $0 < a_n < 3$. Clearly P_1 is true. Assume that P_n is true. Then $a_{n+1} > a_n \;\Rightarrow\; \dfrac{1}{a_{n+1}} < \dfrac{1}{a_n} \;\Rightarrow\;$

$-\dfrac{1}{a_{n+1}} > -\dfrac{1}{a_n}$. Now $a_{n+2} = 3 - \dfrac{1}{a_{n+1}} > 3 - \dfrac{1}{a_n} = a_{n+1} \;\Leftrightarrow\; P_{n+1}$. This proves that $\{a_n\}$ is increasing and bounded above by 3, so $1 = a_1 < a_n < 3$, that is, $\{a_n\}$ is bounded, and hence convergent by the Monotonic Sequence Theorem.

If $L = \lim\limits_{n \to \infty} a_n$, then $\lim\limits_{n \to \infty} a_{n+1} = L$ also, so L must satisfy $L = 3 - 1/L \;\Rightarrow\; L^2 - 3L + 1 = 0 \;\Rightarrow\; L = \dfrac{3 \pm \sqrt{5}}{2}$.

But $L > 1$, so $L = \dfrac{3+\sqrt{5}}{2}$.

89. (a) Let a_n be the number of rabbit pairs in the nth month. Clearly $a_1 = 1 = a_2$. In the nth month, each pair that is 2 or more months old (that is, a_{n-2} pairs) will produce a new pair to add to the a_{n-1} pairs already present. Thus, $a_n = a_{n-1} + a_{n-2}$, so that $\{a_n\} = \{f_n\}$, the Fibonacci sequence.

(b) $a_n = \dfrac{f_{n+1}}{f_n} \Rightarrow a_{n-1} = \dfrac{f_n}{f_{n-1}} = \dfrac{f_{n-1} + f_{n-2}}{f_{n-1}} = 1 + \dfrac{f_{n-2}}{f_{n-1}} = 1 + \dfrac{1}{f_{n-1}/f_{n-2}} = 1 + \dfrac{1}{a_{n-2}}$. If $L = \lim\limits_{n \to \infty} a_n$,

then $L = \lim\limits_{n \to \infty} a_{n-1}$ and $L = \lim\limits_{n \to \infty} a_{n-2}$, so L must satisfy $L = 1 + \dfrac{1}{L} \Rightarrow L^2 - L - 1 = 0 \Rightarrow L = \dfrac{1 + \sqrt{5}}{2}$

[since L must be positive].

91. (a) 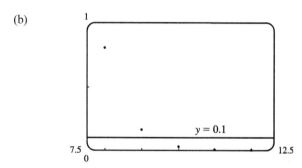 From the graph, it appears that the sequence $\left\{ \dfrac{n^5}{n!} \right\}$

converges to 0, that is, $\lim\limits_{n \to \infty} \dfrac{n^5}{n!} = 0$.

(b)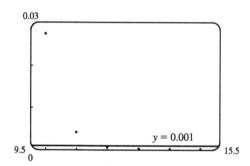

From the first graph, it seems that the smallest possible value of N corresponding to $\varepsilon = 0.1$ is 9, since $n^5/n! < 0.1$ whenever $n \geq 10$, but $9^5/9! > 0.1$. From the second graph, it seems that for $\varepsilon = 0.001$, the smallest possible value for N is 11 since $n^5/n! < 0.001$ whenever $n \geq 12$.

93. **Theorem 6:** If $\lim\limits_{n \to \infty} |a_n| = 0$ then $\lim\limits_{n \to \infty} -|a_n| = 0$, and since $-|a_n| \leq a_n \leq |a_n|$, we have that $\lim\limits_{n \to \infty} a_n = 0$ by the Squeeze Theorem.

95. **To Prove:** If $\lim\limits_{n \to \infty} a_n = 0$ and $\{b_n\}$ is bounded, then $\lim\limits_{n \to \infty} (a_n b_n) = 0$.

Proof: Since $\{b_n\}$ is bounded, there is a positive number M such that $|b_n| \leq M$ and hence, $|a_n|\,|b_n| \leq |a_n|\,M$ for all $n \geq 1$. Let $\varepsilon > 0$ be given. Since $\lim\limits_{n \to \infty} a_n = 0$, there is an integer N such that $|a_n - 0| < \dfrac{\varepsilon}{M}$ if $n > N$. Then

$|a_n b_n - 0| = |a_n b_n| = |a_n|\,|b_n| \leq |a_n|\,M = |a_n - 0|\,M < \dfrac{\varepsilon}{M} \cdot M = \varepsilon$ for all $n > N$. Since ε was arbitrary,

$\lim\limits_{n \to \infty} (a_n b_n) = 0$.

97. (a) First we show that $a > a_1 > b_1 > b$.

$a_1 - b_1 = \dfrac{a+b}{2} - \sqrt{ab} = \tfrac{1}{2}\left(a - 2\sqrt{ab} + b\right) = \tfrac{1}{2}\left(\sqrt{a} - \sqrt{b}\right)^2 > 0$ [since $a > b$] $\Rightarrow a_1 > b_1$. Also

$a - a_1 = a - \tfrac{1}{2}(a+b) = \tfrac{1}{2}(a-b) > 0$ and $b - b_1 = b - \sqrt{ab} = \sqrt{b}\left(\sqrt{b} - \sqrt{a}\right) < 0$, so $a > a_1 > b_1 > b$. In the same way we can show that $a_1 > a_2 > b_2 > b_1$ and so the given assertion is true for $n = 1$. Suppose it is true for $n = k$, that is, $a_k > a_{k+1} > b_{k+1} > b_k$. Then

$$a_{k+2} - b_{k+2} = \tfrac{1}{2}(a_{k+1} + b_{k+1}) - \sqrt{a_{k+1}b_{k+1}} = \tfrac{1}{2}\left(a_{k+1} - 2\sqrt{a_{k+1}b_{k+1}} + b_{k+1}\right) = \tfrac{1}{2}\left(\sqrt{a_{k+1}} - \sqrt{b_{k+1}}\right)^2 > 0,$$

$$a_{k+1} - a_{k+2} = a_{k+1} - \tfrac{1}{2}(a_{k+1} + b_{k+1}) = \tfrac{1}{2}(a_{k+1} - b_{k+1}) > 0, \text{ and}$$

$$b_{k+1} - b_{k+2} = b_{k+1} - \sqrt{a_{k+1}b_{k+1}} = \sqrt{b_{k+1}}\left(\sqrt{b_{k+1}} - \sqrt{a_{k+1}}\right) < 0 \quad \Rightarrow \quad a_{k+1} > a_{k+2} > b_{k+2} > b_{k+1},$$

so the assertion is true for $n = k + 1$. Thus, it is true for all n by mathematical induction.

(b) From part (a) we have $a > a_n > a_{n+1} > b_{n+1} > b_n > b$, which shows that both sequences, $\{a_n\}$ and $\{b_n\}$, are monotonic and bounded. So they are both convergent by the Monotonic Sequence Theorem.

(c) Let $\lim\limits_{n\to\infty} a_n = \alpha$ and $\lim\limits_{n\to\infty} b_n = \beta$. Then $\lim\limits_{n\to\infty} a_{n+1} = \lim\limits_{n\to\infty} \dfrac{a_n + b_n}{2} \;\Rightarrow\; \alpha = \dfrac{\alpha + \beta}{2} \;\Rightarrow\;$
$2\alpha = \alpha + \beta \;\Rightarrow\; \alpha = \beta$.

99. (a) Suppose $\{p_n\}$ converges to p. Then $p_{n+1} = \dfrac{bp_n}{a + p_n} \;\Rightarrow\; \lim\limits_{n\to\infty} p_{n+1} = \dfrac{b \lim\limits_{n\to\infty} p_n}{a + \lim\limits_{n\to\infty} p_n} \;\Rightarrow\; p = \dfrac{bp}{a + p} \;\Rightarrow\;$

$p^2 + ap = bp \;\Rightarrow\; p(p + a - b) = 0 \;\Rightarrow\; p = 0$ or $p = b - a$.

(b) $p_{n+1} = \dfrac{bp_n}{a + p_n} = \dfrac{\left(\dfrac{b}{a}\right)p_n}{1 + \dfrac{p_n}{a}} < \left(\dfrac{b}{a}\right)p_n$ since $1 + \dfrac{p_n}{a} > 1$.

(c) By part (b), $p_1 < \left(\dfrac{b}{a}\right)p_0$, $p_2 < \left(\dfrac{b}{a}\right)p_1 < \left(\dfrac{b}{a}\right)^2 p_0$, $p_3 < \left(\dfrac{b}{a}\right)p_2 < \left(\dfrac{b}{a}\right)^3 p_0$, etc. In general, $p_n < \left(\dfrac{b}{a}\right)^n p_0$,

so $\lim\limits_{n\to\infty} p_n \le \lim\limits_{n\to\infty} \left(\dfrac{b}{a}\right)^n \cdot p_0 = 0$ since $b < a$. $\left[\text{By (9)}, \lim\limits_{n\to\infty} r^n = 0 \text{ if } -1 < r < 1. \text{ Here } r = \dfrac{b}{a} \in (0, 1).\right]$

(d) Let $a < b$. We first show, by induction, that if $p_0 < b - a$, then $p_n < b - a$ and $p_{n+1} > p_n$.

For $n = 0$, we have $p_1 - p_0 = \dfrac{bp_0}{a + p_0} - p_0 = \dfrac{p_0(b - a - p_0)}{a + p_0} > 0$ since $p_0 < b - a$. So $p_1 > p_0$.

Now we suppose the assertion is true for $n = k$, that is, $p_k < b - a$ and $p_{k+1} > p_k$. Then

$b - a - p_{k+1} = b - a - \dfrac{bp_k}{a + p_k} = \dfrac{a(b-a) + bp_k - ap_k - bp_k}{a + p_k} = \dfrac{a(b - a - p_k)}{a + p_k} > 0$ because $p_k < b - a$. So

$p_{k+1} < b - a$. And $p_{k+2} - p_{k+1} = \dfrac{bp_{k+1}}{a + p_{k+1}} - p_{k+1} = \dfrac{p_{k+1}(b - a - p_{k+1})}{a + p_{k+1}} > 0$ since $p_{k+1} < b - a$. Therefore,

$p_{k+2} > p_{k+1}$. Thus, the assertion is true for $n = k + 1$. It is therefore true for all n by mathematical induction.

A similar proof by induction shows that if $p_0 > b - a$, then $p_n > b - a$ and $\{p_n\}$ is decreasing.

In either case the sequence $\{p_n\}$ is bounded and monotonic, so it is convergent by the Monotonic Sequence Theorem. It then follows from part (a) that $\lim\limits_{n\to\infty} p_n = b - a$.

11.2 Series

1. (a) A sequence is an ordered list of numbers whereas a series is the *sum* of a list of numbers.

(b) A series is convergent if the sequence of partial sums is a convergent sequence. A series is divergent if it is not convergent.

3. $\sum_{n=1}^{\infty} a_n = \lim_{n \to \infty} s_n = \lim_{n \to \infty} [2 - 3(0.8)^n] = \lim_{n \to \infty} 2 - 3 \lim_{n \to \infty} (0.8)^n = 2 - 3(0) = 2$

5. For $\sum_{n=1}^{\infty} \frac{1}{n^3}$, $a_n = \frac{1}{n^3}$. $s_1 = a_1 = \frac{1}{1^3} = 1$, $s_2 = s_1 + a_2 = 1 + \frac{1}{2^3} = 1.125$,

$s_3 = s_2 + a_3 \approx 1.1620$, $s_4 = s_3 + a_4 \approx 1.1777$, $s_5 = s_4 + a_5 \approx 1.1857$, $s_6 = s_5 + a_6 \approx 1.1903$,

$s_7 = s_6 + a_7 \approx 1.1932$, $s_8 = s_7 + a_8 \approx 1.1952$. It appears that the series is convergent.

7. For $\sum_{n=1}^{\infty} \sin n$, $a_n = \sin n$. $s_1 = a_1 = \sin 1 \approx 0.8415$, $s_2 = s_1 + a_2 \approx 1.7508$,

$s_3 = s_2 + a_3 \approx 1.8919$, $s_4 = s_3 + a_4 \approx 1.1351$, $s_5 = s_4 + a_5 \approx 0.1762$, $s_6 = s_5 + a_6 \approx -0.1033$,

$s_7 = s_6 + a_7 \approx 0.5537$, and $s_8 = s_7 + a_8 \approx 1.5431$. It appears that the series is divergent.

9. For $\sum_{n=1}^{\infty} \frac{1}{n^4 + n^2}$, $a_n = \frac{1}{n^4 + n^2}$. $s_1 = a_1 = \frac{1}{1^4 + 1^2} = \frac{1}{2} = 0.5$, $s_2 = s_1 + a_2 = \frac{1}{2} + \frac{1}{16 + 4} = 0.55$,

$s_3 = s_2 + a_3 \approx 0.5611$, $s_4 = s_3 + a_4 \approx 0.5648$, $s_5 = s_4 + a_5 \approx 0.5663$, $s_6 = s_5 + a_6 \approx 0.5671$,

$s_7 = s_6 + a_7 \approx 0.5675$, and $s_8 = s_7 + a_8 \approx 0.5677$. It appears that the series is convergent.

11.

n	s_n
1	-2
2	-1.33333
3	-1.55556
4	-1.48148
5	-1.50617
6	-1.49794
7	-1.50069
8	-1.49977
9	-1.50008
10	-1.49997

From the graph and the table, it seems that the series converges to -1.5. In fact, it is a geometric series with $a = -2$ and $r = -\frac{1}{3}$, so its sum is $\sum_{n=1}^{\infty} \frac{6}{(-3)^n} = \frac{-2}{1 - \left(-\frac{1}{3}\right)} = -1.5$.

Note that the point corresponding to $n = 1$ is part of both $\{a_n\}$ and $\{s_n\}$.

13.

n	s_n
1	0.44721
2	1.15432
3	1.98637
4	2.88080
5	3.80927
6	4.75796
7	5.71948
8	6.68962
9	7.66581
10	8.64639

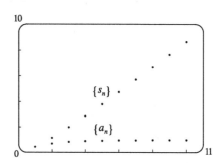

The series $\sum_{n=1}^{\infty} \frac{n}{\sqrt{n^2 + 4}}$ diverges, since its terms do not approach 0.

15. (a) $\lim\limits_{n\to\infty} a_n = \lim\limits_{n\to\infty} \dfrac{2n}{3n+1} = \dfrac{2}{3}$, so the *sequence* $\{a_n\}$ is convergent by (11.1.1).

(b) Since $\lim\limits_{n\to\infty} a_n = \dfrac{2}{3} \neq 0$, the *series* $\sum\limits_{n=1}^{\infty} a_n$ is divergent by the Test for Divergence.

17. For the series $\sum\limits_{n=1}^{\infty} \left(\dfrac{1}{n+2} - \dfrac{1}{n}\right)$,

$$s_n = \sum_{i=1}^{n} \left(\dfrac{1}{i+2} - \dfrac{1}{i}\right)$$

$$= \left(\dfrac{1}{3} - 1\right) + \left(\dfrac{1}{4} - \dfrac{1}{2}\right) + \left(\dfrac{1}{5} - \dfrac{1}{3}\right) + \left(\dfrac{1}{6} - \dfrac{1}{4}\right) + \cdots + \left(\dfrac{1}{n} - \dfrac{1}{n-2}\right) + \left(\dfrac{1}{n+1} - \dfrac{1}{n-1}\right) + \left(\dfrac{1}{n+2} - \dfrac{1}{n}\right)$$

$$= -1 - \dfrac{1}{2} + \dfrac{1}{n+1} + \dfrac{1}{n+2} \quad \text{[telescoping series]}$$

Thus, $\sum\limits_{n=1}^{\infty} \left(\dfrac{1}{n+2} - \dfrac{1}{n}\right) = \lim\limits_{n\to\infty} s_n = \lim\limits_{n\to\infty} \left(-1 - \dfrac{1}{2} + \dfrac{1}{n+1} + \dfrac{1}{n+2}\right) = -1 - \dfrac{1}{2} = -\dfrac{3}{2}$. Converges

19. For the series $\sum\limits_{n=1}^{\infty} \dfrac{3}{n(n+3)}$, $s_n = \sum\limits_{i=1}^{n} \dfrac{3}{i(i+3)} = \sum\limits_{i=1}^{n} \left(\dfrac{1}{i} - \dfrac{1}{i+3}\right)$ [using partial fractions]. The latter sum is

$$\left(1 - \tfrac{1}{4}\right) + \left(\tfrac{1}{2} - \tfrac{1}{5}\right) + \left(\tfrac{1}{3} - \tfrac{1}{6}\right) + \left(\tfrac{1}{4} - \tfrac{1}{7}\right) + \cdots + \left(\tfrac{1}{n-3} - \tfrac{1}{n}\right) + \left(\tfrac{1}{n-2} - \tfrac{1}{n+1}\right) + \left(\tfrac{1}{n-1} - \tfrac{1}{n+2}\right) + \left(\tfrac{1}{n} - \tfrac{1}{n+3}\right)$$

$$= 1 + \tfrac{1}{2} + \tfrac{1}{3} - \tfrac{1}{n+1} - \tfrac{1}{n+2} - \tfrac{1}{n+3} \quad \text{[telescoping series]}$$

Thus, $\sum\limits_{n=1}^{\infty} \dfrac{3}{n(n+3)} = \lim\limits_{n\to\infty} s_n = \lim\limits_{n\to\infty}\left(1 + \tfrac{1}{2} + \tfrac{1}{3} - \tfrac{1}{n+1} - \tfrac{1}{n+2} - \tfrac{1}{n+3}\right) = 1 + \tfrac{1}{2} + \tfrac{1}{3} = \tfrac{11}{6}$. Converges

21. For the series $\sum\limits_{n=1}^{\infty} \left(e^{1/n} - e^{1/(n+1)}\right)$,

$$s_n = \sum_{i=1}^{n} \left(e^{1/i} - e^{1/(i+1)}\right) = (e^1 - e^{1/2}) + (e^{1/2} - e^{1/3}) + \cdots + \left(e^{1/n} - e^{1/(n+1)}\right) = e - e^{1/(n+1)}$$

[telescoping series]

Thus, $\sum\limits_{n=1}^{\infty} \left(e^{1/n} - e^{1/(n+1)}\right) = \lim\limits_{n\to\infty} s_n = \lim\limits_{n\to\infty} \left(e - e^{1/(n+1)}\right) = e - e^0 = e - 1$. Converges

23. $3 - 4 + \dfrac{16}{3} - \dfrac{64}{9} + \cdots$ is a geometric series with ratio $r = -\dfrac{4}{3}$. Since $|r| = \dfrac{4}{3} > 1$, the series diverges.

25. $10 - 2 + 0.4 - 0.08 + \cdots$ is a geometric series with ratio $-\dfrac{2}{10} = -\dfrac{1}{5}$. Since $|r| = \dfrac{1}{5} < 1$, the series converges to $\dfrac{a}{1-r} = \dfrac{10}{1-(-1/5)} = \dfrac{10}{6/5} = \dfrac{50}{6} = \dfrac{25}{3}$.

27. $\sum\limits_{n=1}^{\infty} 12(0.73)^{n-1}$ is a geometric series with first term $a = 12$ and ratio $r = 0.73$. Since $|r| = 0.73 < 1$, the series converges to $\dfrac{a}{1-r} = \dfrac{12}{1-0.73} = \dfrac{12}{0.27} = \dfrac{12(100)}{27} = \dfrac{400}{9}$.

29. $\sum\limits_{n=1}^{\infty} \dfrac{(-3)^{n-1}}{4^n} = \dfrac{1}{4} \sum\limits_{n=1}^{\infty} \left(-\dfrac{3}{4}\right)^{n-1}$. The latter series is geometric with $a = 1$ and ratio $r = -\dfrac{3}{4}$. Since $|r| = \dfrac{3}{4} < 1$, it converges to $\dfrac{1}{1-(-3/4)} = \dfrac{4}{7}$. Thus, the given series converges to $\left(\dfrac{1}{4}\right)\left(\dfrac{4}{7}\right) = \dfrac{1}{7}$.

31. $\sum_{n=1}^{\infty} \dfrac{e^{2n}}{6^{n-1}} = \sum_{n=1}^{\infty} \dfrac{(e^2)^n}{6^n 6^{-1}} = 6 \sum_{n=1}^{\infty} \left(\dfrac{e^2}{6}\right)^n$ is a geometric series with ratio $r = \dfrac{e^2}{6}$. Since $|r| = \dfrac{e^2}{6} [\approx 1.23] > 1$, the series diverges.

33. $\dfrac{1}{3} + \dfrac{1}{6} + \dfrac{1}{9} + \dfrac{1}{12} + \dfrac{1}{15} + \cdots = \sum_{n=1}^{\infty} \dfrac{1}{3n} = \dfrac{1}{3} \sum_{n=1}^{\infty} \dfrac{1}{n}$. This is a constant multiple of the divergent harmonic series, so it diverges.

35. $\dfrac{2}{5} + \dfrac{4}{25} + \dfrac{8}{125} + \dfrac{16}{625} + \dfrac{32}{3125} + \cdots = \sum_{n=1}^{\infty} \left(\dfrac{2}{5}\right)^n$. This series is geometric with $a = \dfrac{2}{5}$ and ratio $r = \dfrac{2}{5}$. Since $|r| = \dfrac{2}{5} < 1$, it converges to $\dfrac{2/5}{1 - 2/5} = \dfrac{2}{3}$.

37. $\sum_{n=1}^{\infty} \dfrac{2+n}{1-2n}$ diverges by the Test for Divergence since $\lim\limits_{n \to \infty} a_n = \lim\limits_{n \to \infty} \dfrac{2+n}{1-2n} = \lim\limits_{n \to \infty} \dfrac{2/n + 1}{1/n - 2} = -\dfrac{1}{2} \neq 0$.

39. $\sum_{n=1}^{\infty} 3^{n+1} 4^{-n} = \sum_{n=1}^{\infty} \dfrac{3^n \cdot 3^1}{4^n} = 3 \sum_{n=1}^{\infty} \left(\dfrac{3}{4}\right)^n$. The latter series is geometric with $a = \dfrac{3}{4}$ and ratio $r = \dfrac{3}{4}$. Since $|r| = \dfrac{3}{4} < 1$, it converges to $\dfrac{3/4}{1 - 3/4} = 3$. Thus, the given series converges to $3(3) = 9$.

41. $\sum_{n=1}^{\infty} \dfrac{1}{4 + e^{-n}}$ diverges by the Test for Divergence since $\lim\limits_{n \to \infty} \dfrac{1}{4 + e^{-n}} = \dfrac{1}{4 + 0} = \dfrac{1}{4} \neq 0$.

43. $\sum_{k=1}^{\infty} (\sin 100)^k$ is a geometric series with first term $a = \sin 100 \ [\approx -0.506]$ and ratio $r = \sin 100$. Since $|r| < 1$, the series converges to $\dfrac{\sin 100}{1 - \sin 100} \approx -0.336$.

45. $\sum_{n=1}^{\infty} \ln\left(\dfrac{n^2 + 1}{2n^2 + 1}\right)$ diverges by the Test for Divergence since

$\lim\limits_{n \to \infty} a_n = \lim\limits_{n \to \infty} \ln\left(\dfrac{n^2 + 1}{2n^2 + 1}\right) = \ln\left(\lim\limits_{n \to \infty} \dfrac{n^2 + 1}{2n^2 + 1}\right) = \ln \dfrac{1}{2} \neq 0$.

47. $\sum_{n=1}^{\infty} \arctan n$ diverges by the Test for Divergence since $\lim\limits_{n \to \infty} a_n = \lim\limits_{n \to \infty} \arctan n = \dfrac{\pi}{2} \neq 0$.

49. $\sum_{n=1}^{\infty} \dfrac{1}{e^n} = \sum_{n=1}^{\infty} \left(\dfrac{1}{e}\right)^n$ is a geometric series with first term $a = \dfrac{1}{e}$ and ratio $r = \dfrac{1}{e}$. Since $|r| = \dfrac{1}{e} < 1$, the series converges

to $\dfrac{1/e}{1 - 1/e} = \dfrac{1/e}{1 - 1/e} \cdot \dfrac{e}{e} = \dfrac{1}{e - 1}$. By Example 8, $\sum_{n=1}^{\infty} \dfrac{1}{n(n+1)} = 1$. Thus, by Theorem 8(ii),

$\sum_{n=1}^{\infty} \left(\dfrac{1}{e^n} + \dfrac{1}{n(n+1)}\right) = \sum_{n=1}^{\infty} \dfrac{1}{e^n} + \sum_{n=1}^{\infty} \dfrac{1}{n(n+1)} = \dfrac{1}{e-1} + 1 = \dfrac{1}{e-1} + \dfrac{e-1}{e-1} = \dfrac{e}{e-1}$.

51. (a) Many people would guess that $x < 1$, but note that x consists of an infinite number of 9s.

(b) $x = 0.99999\ldots = \dfrac{9}{10} + \dfrac{9}{100} + \dfrac{9}{1000} + \dfrac{9}{10{,}000} + \cdots = \sum_{n=1}^{\infty} \dfrac{9}{10^n}$, which is a geometric series with $a_1 = 0.9$ and

$r = 0.1$. Its sum is $\dfrac{0.9}{1 - 0.1} = \dfrac{0.9}{0.9} = 1$, that is, $x = 1$.

(c) The number 1 has two decimal representations, $1.00000\ldots$ and $0.99999\ldots$.

(d) Except for 0, all rational numbers that have a terminating decimal representation can be written in more than one way. For example, 0.5 can be written as $0.49999\ldots$ as well as $0.50000\ldots$.

53. $0.\overline{8} = \dfrac{8}{10} + \dfrac{8}{10^2} + \cdots$ is a geometric series with $a = \dfrac{8}{10}$ and $r = \dfrac{1}{10}$. It converges to $\dfrac{a}{1-r} = \dfrac{8/10}{1 - 1/10} = \dfrac{8}{9}$.

55. $2.\overline{516} = 2 + \dfrac{516}{10^3} + \dfrac{516}{10^6} + \cdots$. Now $\dfrac{516}{10^3} + \dfrac{516}{10^6} + \cdots$ is a geometric series with $a = \dfrac{516}{10^3}$ and $r = \dfrac{1}{10^3}$. It converges to

$\dfrac{a}{1-r} = \dfrac{516/10^3}{1 - 1/10^3} = \dfrac{516/10^3}{999/10^3} = \dfrac{516}{999}$. Thus, $2.\overline{516} = 2 + \dfrac{516}{999} = \dfrac{2514}{999} = \dfrac{838}{333}$.

57. $1.234\overline{567} = 1.234 + \dfrac{567}{10^6} + \dfrac{567}{10^9} + \cdots$. Now $\dfrac{567}{10^6} + \dfrac{567}{10^9} + \cdots$ is a geometric series with $a = \dfrac{567}{10^6}$ and

$r = \dfrac{1}{10^3}$. It converges to $\dfrac{a}{1-r} = \dfrac{567/10^6}{1 - 1/10^3} = \dfrac{567/10^6}{999/10^3} = \dfrac{567}{999{,}000} = \dfrac{21}{37{,}000}$. Thus,

$1.234\overline{567} = 1.234 + \dfrac{21}{37{,}000} = \dfrac{1234}{1000} + \dfrac{21}{37{,}000} = \dfrac{45{,}658}{37{,}000} + \dfrac{21}{37{,}000} = \dfrac{45{,}679}{37{,}000}$.

59. $\sum_{n=1}^{\infty} (-5)^n x^n = \sum_{n=1}^{\infty} (-5x)^n$ is a geometric series with $r = -5x$, so the series converges $\Leftrightarrow |r| < 1 \Leftrightarrow$

$|-5x| < 1 \Leftrightarrow |x| < \tfrac{1}{5}$, that is, $-\tfrac{1}{5} < x < \tfrac{1}{5}$. In that case, the sum of the series is $\dfrac{a}{1-r} = \dfrac{-5x}{1-(-5x)} = \dfrac{-5x}{1+5x}$.

61. $\sum_{n=0}^{\infty} \dfrac{(x-2)^n}{3^n} = \sum_{n=0}^{\infty} \left(\dfrac{x-2}{3}\right)^n$ is a geometric series with $r = \dfrac{x-2}{3}$, so the series converges $\Leftrightarrow |r| < 1 \Leftrightarrow$

$\left|\dfrac{x-2}{3}\right| < 1 \Leftrightarrow -1 < \dfrac{x-2}{3} < 1 \Leftrightarrow -3 < x - 2 < 3 \Leftrightarrow -1 < x < 5$. In that case, the sum of the series is

$\dfrac{a}{1-r} = \dfrac{1}{1 - \dfrac{x-2}{3}} = \dfrac{1}{\dfrac{3-(x-2)}{3}} = \dfrac{3}{5-x}$.

63. $\sum_{n=0}^{\infty} \dfrac{2^n}{x^n} = \sum_{n=0}^{\infty} \left(\dfrac{2}{x}\right)^n$ is a geometric series with $r = \dfrac{2}{x}$, so the series converges $\Leftrightarrow |r| < 1 \Leftrightarrow \left|\dfrac{2}{x}\right| < 1 \Leftrightarrow$

$2 < |x| \Leftrightarrow x > 2$ or $x < -2$. In that case, the sum of the series is $\dfrac{a}{1-r} = \dfrac{1}{1 - 2/x} = \dfrac{x}{x-2}$.

65. $\sum_{n=0}^{\infty} e^{nx} = \sum_{n=0}^{\infty} (e^x)^n$ is a geometric series with $r = e^x$, so the series converges $\Leftrightarrow |r| < 1 \Leftrightarrow |e^x| < 1 \Leftrightarrow$

$-1 < e^x < 1 \Leftrightarrow 0 < e^x < 1 \Leftrightarrow x < 0$. In that case, the sum of the series is $\dfrac{a}{1-r} = \dfrac{1}{1-e^x}$.

67. After defining f, We use `convert(f,parfrac);` in Maple or `Apart` in Mathematica to find that the general term is $\dfrac{3n^2+3n+1}{(n^2+n)^3} = \dfrac{1}{n^3} - \dfrac{1}{(n+1)^3}$. So the nth partial sum is

$$s_n = \sum_{k=1}^{n}\left(\frac{1}{k^3}-\frac{1}{(k+1)^3}\right) = \left(1-\frac{1}{2^3}\right)+\left(\frac{1}{2^3}-\frac{1}{3^3}\right)+\cdots+\left(\frac{1}{n^3}-\frac{1}{(n+1)^3}\right) = 1-\frac{1}{(n+1)^3}$$

The series converges to $\lim_{n\to\infty} s_n = 1$. This can be confirmed by directly computing the sum using `sum(f,n=1..infinity);` (in Maple) or `Sum[f,{n,1,Infinity}]` (in Mathematica).

69. For $n=1$, $a_1 = 0$ since $s_1 = 0$. For $n > 1$,

$$a_n = s_n - s_{n-1} = \frac{n-1}{n+1} - \frac{(n-1)-1}{(n-1)+1} = \frac{(n-1)n-(n+1)(n-2)}{(n+1)n} = \frac{2}{n(n+1)}$$

Also, $\displaystyle\sum_{n=1}^{\infty} a_n = \lim_{n\to\infty} s_n = \lim_{n\to\infty} \frac{1-1/n}{1+1/n} = 1$.

71. (a) The quantity of the drug in the body after the first tablet is 100 mg. After 8 hours, the body eliminates 75% of the drug, which means 25% remains. Thus, after the second tablet, there is 100 mg plus 25% of the first 100-mg tablet, that is, $100 + 0.25(100) = 125$ mg. After the third tablet, the quantity is $100 + 0.25(125)$ or, equivalently, $100 + 100(0.25) + 100(0.25)^2$. Either expression gives 131.25 mg.

(b) From part (a), we see that $Q_{n+1} = 100 + 0.25\, Q_n$.

(c) $Q_n = 100 + 100(0.25)^1 + 100(0.25)^2 + \cdots + 100(0.25)^{n-1}$

$\quad = \displaystyle\sum_{i=1}^{n} 100(0.25)^{i-1}$ [geometric with $a=100$ and $r=0.25$]

The quantity of the antibiotic that remains in the body in the long run is $\displaystyle\lim_{n\to\infty} Q_n = \frac{100}{1-0.25} = \frac{100}{0.75} = 133.\overline{3}$ mg.

73. (a) The quantity of the drug in the body after the first tablet is 150 mg. "Eliminates 95%" is the same as retains 5%, so after the second tablet, there is 150 mg plus 5% of the first 150-mg tablet, that is, $[150 + 150(0.05)]$ mg. After the third tablet, the quantity is $[150 + 150(0.05) + 150(0.05)^2] = 157.875$ mg. After n tablets, the quantity (in mg) is $150 + 150(0.05) + \cdots + 150(0.05)^{n-1}$. We can use Formula 3 to write this as $\dfrac{150(1-0.05^n)}{1-0.05} = \dfrac{3000}{19}(1-0.05^n)$.

(b) The number of milligrams remaining in the body in the long run is $\displaystyle\lim_{n\to\infty}\left[\tfrac{3000}{19}(1-0.05^n)\right] = \tfrac{3000}{19}(1-0) \approx 157.895$, only 0.02 mg more than the amount after 3 tablets.

75. (a) The first step in the chain occurs when the local government spends D dollars. The people who receive it spend a fraction c of those D dollars, that is, Dc dollars. Those who receive the Dc dollars spend a fraction c of it, that is, Dc^2 dollars. Continuing in this way, we see that the total spending after n transactions is

$$S_n = D + Dc + Dc^2 + \cdots + Dc^{n-1} = \frac{D(1-c^n)}{1-c} \text{ by (3)}.$$

(b) $\lim\limits_{n\to\infty} S_n = \lim\limits_{n\to\infty} \dfrac{D(1-c^n)}{1-c} = \dfrac{D}{1-c}\lim\limits_{n\to\infty}(1-c^n) = \dfrac{D}{1-c}$ $\left[\text{since } 0 < c < 1 \;\Rightarrow\; \lim\limits_{n\to\infty} c^n = 0\right]$

$= \dfrac{D}{s}$ [since $c + s = 1$] $= kD$ [since $k = 1/s$]

If $c = 0.8$, then $s = 1 - c = 0.2$ and the multiplier is $k = 1/s = 5$.

77. $\sum\limits_{n=2}^{\infty}(1+c)^{-n}$ is a geometric series with $a = (1+c)^{-2}$ and $r = (1+c)^{-1}$, so the series converges when

$|(1+c)^{-1}| < 1 \;\Leftrightarrow\; |1+c| > 1 \;\Leftrightarrow\; 1+c > 1 \text{ or } 1+c < -1 \;\Leftrightarrow\; c > 0 \text{ or } c < -2$. We calculate the sum of the

series and set it equal to 2: $\dfrac{(1+c)^{-2}}{1-(1+c)^{-1}} = 2 \;\Leftrightarrow\; \left(\dfrac{1}{1+c}\right)^2 = 2 - 2\left(\dfrac{1}{1+c}\right) \;\Leftrightarrow\; 1 = 2(1+c)^2 - 2(1+c) \;\Leftrightarrow\;$

$2c^2 + 2c - 1 = 0 \;\Leftrightarrow\; c = \dfrac{-2 \pm \sqrt{12}}{4} = \dfrac{\pm\sqrt{3}-1}{2}$. However, the negative root is inadmissible because $-2 < \dfrac{-\sqrt{3}-1}{2} < 0$.

So $c = \dfrac{\sqrt{3}-1}{2}$.

79. Assume the harmonic series converges with sum S. Following the outlined method of proof, we have

$S = \left(1 + \dfrac{1}{2}\right) + \left(\dfrac{1}{3} + \dfrac{1}{4}\right) + \left(\dfrac{1}{5} + \dfrac{1}{6}\right) + \cdots > \left(\dfrac{1}{2} + \dfrac{1}{2}\right) + \left(\dfrac{1}{4} + \dfrac{1}{4}\right) + \left(\dfrac{1}{6} + \dfrac{1}{6}\right) + \cdots = 1 + \dfrac{1}{2} + \dfrac{1}{3} + \cdots = S$.

This indicates that $S > S \;\Leftrightarrow\; 0 > 0$, which is a contradiction. Thus, the assumption was incorrect and the harmonic series diverges.

81. From the hint, we'll show that $e^x > 1 + x$ for $x > 0$ by proving that $f(x) = e^x - (1+x) > 0$ for $x > 0$. Since

$f(0) = e^0 - 1 = 0$ and $f'(x) = e^x - 1 > 0$ when $x > 0$, f is increasing on $(0, \infty)$ $\;\Rightarrow\;$ when $x > 0$, $f(x) > 0$ $\;\Rightarrow\;$

$e^x - (1+x) > 0$ for $x > 0$ $\;\Rightarrow\;$ $e^x > 1 + x$ for $x > 0$. Now,

$e^{1+(1/2)+(1/3)+\cdots+(1/n)} = e^1 e^{1/2} e^{1/3} \cdots e^{1/n} > (1+1)\left(1+\dfrac{1}{2}\right)\left(1+\dfrac{1}{3}\right)\cdots\left(1+\dfrac{1}{n}\right)$ $\quad [\text{since } e^x > 1 + x]$

$= \left(\dfrac{2}{1}\right)\left(\dfrac{3}{2}\right)\left(\dfrac{4}{3}\right)\cdots\left(\dfrac{n+1}{n}\right)$ $\quad \begin{bmatrix}\text{each denominator cancels with the}\\\text{numerator of the preceding fraction}\end{bmatrix}$

$= n + 1$

Assuming the harmonic series converges with sum S, we take the limit as $n \to \infty$ to get

$\lim\limits_{n\to\infty} e^{1+(1/2)+(1/3)+\cdots+(1/n)} > \lim\limits_{n\to\infty}(n+1) \;\Rightarrow\;$

$e^{\lim\limits_{n\to\infty}[1+(1/2)+(1/3)+\cdots+(1/n)]} > \lim\limits_{n\to\infty}(n+1)$ [since e^x is continuous] $\;\Rightarrow\;$ $e^S > \lim\limits_{n\to\infty}(n+1)$. The limit does not exist since

$n + 1 \to \infty$ as $n \to \infty$. e^S is larger than the value of the limit, so S is not finite valued, which is a contradiction. Thus, the

assumption was incorrect and the harmonic series diverges.

83. Let d_n be the diameter of C_n. We draw lines from the centers of the C_i to

the center of D (or C), and using the Pythagorean Theorem, we can write

$1^2 + \left(1 - \tfrac{1}{2}d_1\right)^2 = \left(1 + \tfrac{1}{2}d_1\right)^2 \;\Leftrightarrow\;$

$1 = \left(1 + \tfrac{1}{2}d_1\right)^2 - \left(1 - \tfrac{1}{2}d_1\right)^2 = 2d_1$ [difference of squares] $\;\Rightarrow\; d_1 = \tfrac{1}{2}$.

Similarly,

$1 = \left(1 + \tfrac{1}{2}d_2\right)^2 - \left(1 - d_1 - \tfrac{1}{2}d_2\right)^2 = 2d_2 + 2d_1 - d_1^2 - d_1 d_2$

$= (2 - d_1)(d_1 + d_2) \;\Leftrightarrow\;$

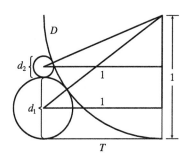

$d_2 = \dfrac{1}{2-d_1} - d_1 = \dfrac{(1-d_1)^2}{2-d_1}$, $1 = \left(1 + \tfrac{1}{2}d_3\right)^2 - \left(1 - d_1 - d_2 - \tfrac{1}{2}d_3\right)^2 \Leftrightarrow d_3 = \dfrac{[1-(d_1+d_2)]^2}{2-(d_1+d_2)}$, and in general,

$d_{n+1} = \dfrac{\left(1 - \sum_{i=1}^{n} d_i\right)^2}{2 - \sum_{i=1}^{n} d_i}$. If we actually calculate d_2 and d_3 from the formulas above, we find that they are $\dfrac{1}{6} = \dfrac{1}{2 \cdot 3}$ and $\dfrac{1}{12} = \dfrac{1}{3 \cdot 4}$ respectively, so we suspect that in general, $d_n = \dfrac{1}{n(n+1)}$. To prove this, we use induction: Assume that for all $k \leq n$, $d_k = \dfrac{1}{k(k+1)} = \dfrac{1}{k} - \dfrac{1}{k+1}$. Then $\sum_{i=1}^{n} d_i = 1 - \dfrac{1}{n+1} = \dfrac{n}{n+1}$ [telescoping sum]. Substituting this into our formula for d_{n+1}, we get $d_{n+1} = \dfrac{\left[1 - \dfrac{n}{n+1}\right]^2}{2 - \left(\dfrac{n}{n+1}\right)} = \dfrac{\dfrac{1}{(n+1)^2}}{\dfrac{n+2}{n+1}} = \dfrac{1}{(n+1)(n+2)}$, and the induction is complete.

Now, we observe that the partial sums $\sum_{i=1}^{n} d_i$ of the diameters of the circles approach 1 as $n \to \infty$; that is,

$\sum_{n=1}^{\infty} a_n = \sum_{n=1}^{\infty} \dfrac{1}{n(n+1)} = 1$, which is what we wanted to prove.

85. The series $1 - 1 + 1 - 1 + 1 - 1 + \cdots$ diverges (geometric series with $r = -1$), so we cannot say that $0 = 1 - 1 + 1 - 1 + 1 - 1 + \cdots$.

87. (a) $\sum_{n=1}^{\infty} ca_n = \lim_{n \to \infty} \sum_{i=1}^{n} ca_i = \lim_{n \to \infty} c \sum_{i=1}^{n} a_i = c \lim_{n \to \infty} \sum_{i=1}^{n} a_i = c \sum_{n=1}^{\infty} a_n$, which exists by hypothesis.

(b) $\sum_{n=1}^{\infty} (a_n - b_n) = \lim_{n \to \infty} \sum_{i=1}^{n} (a_i - b_i) = \lim_{n \to \infty} \left(\sum_{i=1}^{n} a_i - \sum_{i=1}^{n} b_i\right) = \lim_{n \to \infty} \sum_{i=1}^{n} a_i - \lim_{n \to \infty} \sum_{i=1}^{n} b_i$

$= \sum_{n=1}^{\infty} a_n - \sum_{n=1}^{\infty} b_n$, which exists by hypothesis.

89. Suppose on the contrary that $\sum(a_n + b_n)$ converges. Then $\sum(a_n + b_n)$ and $\sum a_n$ are convergent series. So by Theorem 8(iii), $\sum[(a_n + b_n) - a_n]$ would also be convergent. But $\sum[(a_n + b_n) - a_n] = \sum b_n$, a contradiction, since $\sum b_n$ is given to be divergent.

91. The partial sums $\{s_n\}$ form an increasing sequence, since $s_n - s_{n-1} = a_n > 0$ for all n. Also, the sequence $\{s_n\}$ is bounded since $s_n \leq 1000$ for all n. So by the Monotonic Sequence Theorem, the sequence of partial sums converges, that is, the series $\sum a_n$ is convergent.

93. (a) At the first step, only the interval $\left(\tfrac{1}{3}, \tfrac{2}{3}\right)$ (length $\tfrac{1}{3}$) is removed. At the second step, we remove the intervals $\left(\tfrac{1}{9}, \tfrac{2}{9}\right)$ and $\left(\tfrac{7}{9}, \tfrac{8}{9}\right)$, which have a total length of $2 \cdot \left(\tfrac{1}{3}\right)^2$. At the third step, we remove 2^2 intervals, each of length $\left(\tfrac{1}{3}\right)^3$. In general, at the nth step we remove 2^{n-1} intervals, each of length $\left(\tfrac{1}{3}\right)^n$, for a length of $2^{n-1} \cdot \left(\tfrac{1}{3}\right)^n = \tfrac{1}{3}\left(\tfrac{2}{3}\right)^{n-1}$. Thus, the total length of all removed intervals is $\sum_{n=1}^{\infty} \tfrac{1}{3}\left(\tfrac{2}{3}\right)^{n-1} = \dfrac{1/3}{1 - 2/3} = 1$ [geometric series with $a = \tfrac{1}{3}$ and $r = \tfrac{2}{3}$]. Notice that at the nth step, the leftmost interval that is removed is $\left(\left(\tfrac{1}{3}\right)^n, \left(\tfrac{2}{3}\right)^n\right)$, so we never remove 0, and 0 is in the Cantor set. Also, the rightmost interval removed is $\left(1 - \left(\tfrac{2}{3}\right)^n, 1 - \left(\tfrac{1}{3}\right)^n\right)$, so 1 is never removed. Some other numbers in the Cantor set are $\tfrac{1}{3}, \tfrac{2}{3}, \tfrac{1}{9}, \tfrac{2}{9}, \tfrac{7}{9}$, and $\tfrac{8}{9}$.

(b) The area removed at the first step is $\frac{1}{9}$; at the second step, $8 \cdot \left(\frac{1}{9}\right)^2$; at the third step, $(8)^2 \cdot \left(\frac{1}{9}\right)^3$. In general, the area removed at the nth step is $(8)^{n-1}\left(\frac{1}{9}\right)^n = \frac{1}{9}\left(\frac{8}{9}\right)^{n-1}$, so the total area of all removed squares is

$$\sum_{n=1}^{\infty} \frac{1}{9}\left(\frac{8}{9}\right)^{n-1} = \frac{1/9}{1-8/9} = 1.$$

95. (a) For $\sum_{n=1}^{\infty} \frac{n}{(n+1)!}$, $s_1 = \frac{1}{1 \cdot 2} = \frac{1}{2}$, $s_2 = \frac{1}{2} + \frac{2}{1 \cdot 2 \cdot 3} = \frac{5}{6}$, $s_3 = \frac{5}{6} + \frac{3}{1 \cdot 2 \cdot 3 \cdot 4} = \frac{23}{24}$,

$s_4 = \frac{23}{24} + \frac{4}{1 \cdot 2 \cdot 3 \cdot 4 \cdot 5} = \frac{119}{120}$. The denominators are $(n+1)!$, so a guess would be $s_n = \frac{(n+1)!-1}{(n+1)!}$.

(b) For $n=1$, $s_1 = \frac{1}{2} = \frac{2!-1}{2!}$, so the formula holds for $n=1$. Assume $s_k = \frac{(k+1)!-1}{(k+1)!}$. Then

$$s_{k+1} = \frac{(k+1)!-1}{(k+1)!} + \frac{k+1}{(k+2)!} = \frac{(k+1)!-1}{(k+1)!} + \frac{k+1}{(k+1)!(k+2)} = \frac{(k+2)!-(k+2)+k+1}{(k+2)!}$$

$$= \frac{(k+2)!-1}{(k+2)!}$$

Thus, the formula is true for $n = k+1$. So by induction, the guess is correct.

(c) $\lim_{n \to \infty} s_n = \lim_{n \to \infty} \frac{(n+1)!-1}{(n+1)!} = \lim_{n \to \infty}\left[1 - \frac{1}{(n+1)!}\right] = 1$ and so $\sum_{n=1}^{\infty} \frac{n}{(n+1)!} = 1$.

11.3 The Integral Test and Estimates of Sums

1. The picture shows that $a_2 = \frac{1}{2^{1.5}} < \int_1^2 \frac{1}{x^{1.5}}\, dx$,

$a_3 = \frac{1}{3^{1.5}} < \int_2^3 \frac{1}{x^{1.5}}\, dx$, and so on, so $\sum_{n=2}^{\infty} \frac{1}{n^{1.5}} < \int_1^{\infty} \frac{1}{x^{1.5}}\, dx$.

The integral converges by (7.8.2) with $p = 1.5 > 1$, so the series converges.

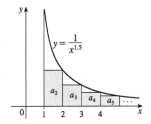

3. The function $f(x) = x^{-3}$ is continuous, positive, and decreasing on $[1, \infty)$, so the Integral Test applies.

$$\int_1^{\infty} x^{-3}\, dx = \lim_{t \to \infty} \int_1^t x^{-3}\, dx = \lim_{t \to \infty} \left[\frac{x^{-2}}{-2}\right]_1^t = \lim_{t \to \infty}\left(-\frac{1}{2t^2} + \frac{1}{2}\right) = \frac{1}{2}.$$

Since this improper integral is convergent, the series $\sum_{n=1}^{\infty} n^{-3}$ is also convergent by the Integral Test.

5. The function $f(x) = \frac{2}{5x-1}$ is continuous, positive, and decreasing on $[1, \infty)$, so the Integral Test applies.

$$\int_1^{\infty} \frac{2}{5x-1}\, dx = \lim_{t \to \infty} \int_1^t \frac{2}{5x-1}\, dx = \lim_{t \to \infty}\left[\frac{2}{5}\ln(5x-1)\right]_1^t = \lim_{t \to \infty}\left[\frac{2}{5}\ln(5t-1) - \frac{2}{5}\ln 4\right] = \infty.$$

Since this improper integral is divergent, the series $\sum_{n=1}^{\infty} \frac{2}{5n-1}$ is also divergent by the Integral Test.

7. The function $f(x) = \dfrac{x^2}{x^3+1}$ is continuous, positive, and decreasing ($\star$) on $[2, \infty)$, so the Integral Test applies.

$$\int_2^\infty \frac{x^2}{x^3+1}\, dx = \lim_{t\to\infty} \int_2^t \frac{x^2}{x^3+1}\, dx = \lim_{t\to\infty} \left[\frac{1}{3}\ln|x^3+1|\right]_2^t = \lim_{t\to\infty}\left[\frac{1}{3}\ln(t^3+1) - \frac{1}{3}\ln 9\right] = \infty.$$

Since the improper integral is divergent, the series $\displaystyle\sum_{n=2}^\infty \frac{n^2}{n^3+1}$ is also divergent by the Integral Test.

($\star$): $f'(x) = \dfrac{(x^3+1)(2x) - x^2(3x^2)}{(x^3+1)^2} = \dfrac{2x - x^4}{(x^3+1)^2} = -\dfrac{x(x^3-2)}{(x^3+1)^2} < 0$ for $x \geq 2$.

9. The function $f(x) = \dfrac{1}{x(\ln x)^3}$ is continuous, positive, and decreasing on $[2, \infty)$, so the Integral Test applies.

$$\int_2^\infty \frac{1}{x(\ln x)^3}\, dx = \lim_{t\to\infty}\int_2^t \frac{1}{x(\ln x)^3}\, dx = \lim_{t\to\infty}\int_{\ln 2}^{\ln t} \frac{du}{u^3}\quad \left[u = \ln x,\ du = \frac{dx}{x}\right] = \lim_{t\to\infty}\left[-\frac{1}{2}u^{-2}\right]_{\ln 2}^{\ln t}$$

$$= \lim_{t\to\infty}\left[-\frac{1}{2(\ln t)^2} + \frac{1}{2(\ln 2)^2}\right] = 0 + \frac{1}{2(\ln 2)^2} = \frac{1}{2(\ln 2)^2}$$

Since the improper integral is convergent, the series $\displaystyle\sum_{n=2}^\infty \frac{1}{n(\ln n)^3}$ is also convergent by the Integral Test.

11. $\displaystyle\sum_{n=1}^\infty \frac{1}{n^{\sqrt{2}}}$ is a p-series with $p = \sqrt{2} > 1$, so it converges by (1).

13. $1 + \dfrac{1}{8} + \dfrac{1}{27} + \dfrac{1}{64} + \dfrac{1}{125} + \cdots = \displaystyle\sum_{n=1}^\infty \frac{1}{n^3}$. This is a p-series with $p = 3 > 1$, so it converges by (1).

15. $\dfrac{1}{3} + \dfrac{1}{7} + \dfrac{1}{11} + \dfrac{1}{15} + \dfrac{1}{19} + \cdots = \displaystyle\sum_{n=1}^\infty \frac{1}{4n-1}$. The function $f(x) = \dfrac{1}{4x-1}$ is continuous, positive, and decreasing on $[1, \infty)$, so the Integral Test applies.

$$\int_1^\infty \frac{1}{4x-1}\, dx = \lim_{t\to\infty}\int_1^t \frac{1}{4x-1}\, dx = \lim_{t\to\infty}\left[\tfrac{1}{4}\ln(4x-1)\right]_1^t = \lim_{t\to\infty}\left[\tfrac{1}{4}\ln(4t-1) - \tfrac{1}{4}\ln 3\right] = \infty,\ \text{so the series}$$

$\displaystyle\sum_{n=1}^\infty \frac{1}{4n-1}$ diverges.

17. $\displaystyle\sum_{n=1}^\infty \frac{\sqrt{n}+4}{n^2} = \sum_{n=1}^\infty\left(\frac{\sqrt{n}}{n^2} + \frac{4}{n^2}\right) = \sum_{n=1}^\infty \frac{1}{n^{3/2}} + \sum_{n=1}^\infty \frac{4}{n^2}$. $\displaystyle\sum_{n=1}^\infty \frac{1}{n^{3/2}}$ is a convergent p-series with $p = \tfrac{3}{2} > 1$.

$\displaystyle\sum_{n=1}^\infty \frac{4}{n^2} = 4\sum_{n=1}^\infty \frac{1}{n^2}$ is a constant multiple of a convergent p-series with $p = 2 > 1$, so it converges. The sum of two convergent series is convergent, so the original series is convergent.

19. The function $f(x) = \dfrac{1}{x^2+4}$ is continuous, positive, and decreasing on $[1, \infty)$, so we can apply the Integral Test.

$$\int_1^\infty \frac{1}{x^2+4}\, dx = \lim_{t\to\infty}\int_1^t \frac{1}{x^2+4}\, dx = \lim_{t\to\infty}\left[\frac{1}{2}\tan^{-1}\frac{x}{2}\right]_1^t = \frac{1}{2}\lim_{t\to\infty}\left[\tan^{-1}\left(\frac{t}{2}\right) - \tan^{-1}\left(\frac{1}{2}\right)\right]$$

$$= \frac{1}{2}\left[\frac{\pi}{2} - \tan^{-1}\left(\frac{1}{2}\right)\right]$$

Therefore, the series $\displaystyle\sum_{n=1}^\infty \frac{1}{n^2+4}$ converges.

21. The function $f(x) = \dfrac{x^3}{x^4 + 4}$ is continuous and positive on $[2, \infty)$, and is also decreasing since

$$f'(x) = \dfrac{(x^4 + 4)(3x^2) - x^3(4x^3)}{(x^4 + 4)^2} = \dfrac{12x^2 - x^6}{(x^4 + 4)^2} = \dfrac{x^2(12 - x^4)}{(x^4 + 4)^2} < 0 \text{ for } x > \sqrt[4]{12} \approx 1.86, \text{ so we can use the}$$

Integral Test on $[2, \infty)$.

$$\int_2^\infty \dfrac{x^3}{x^4 + 4}\, dx = \lim_{t \to \infty} \int_2^t \dfrac{x^3}{x^4 + 4}\, dx = \lim_{t \to \infty} \left[\tfrac{1}{4}\ln(x^4 + 4)\right]_2^t = \lim_{t \to \infty} \left[\tfrac{1}{4}\ln(t^4 + 4) - \tfrac{1}{4}\ln 20\right] = \infty, \text{ so the series}$$

$\displaystyle\sum_{n=2}^\infty \dfrac{n^3}{n^4 + 4}$ diverges, and it follows that $\displaystyle\sum_{n=1}^\infty \dfrac{n^3}{n^4 + 4}$ diverges as well.

23. $f(x) = \dfrac{1}{x \ln x}$ is continuous and positive on $[2, \infty)$, and also decreasing since $f'(x) = -\dfrac{1 + \ln x}{x^2(\ln x)^2} < 0$ for $x > 2$, so we can

use the Integral Test. $\displaystyle\int_2^\infty \dfrac{1}{x \ln x}\, dx = \lim_{t \to \infty} \left[\ln(\ln x)\right]_2^t = \lim_{t \to \infty} \left[\ln(\ln t) - \ln(\ln 2)\right] = \infty$, so the series $\displaystyle\sum_{n=2}^\infty \dfrac{1}{n \ln n}$ diverges.

25. The function $f(x) = xe^{-x} = \dfrac{x}{e^x}$ is continuous and positive on $[1, \infty)$, and also decreasing since

$$f'(x) = \dfrac{e^x \cdot 1 - xe^x}{(e^x)^2} = \dfrac{e^x(1 - x)}{(e^x)^2} = \dfrac{1 - x}{e^x} < 0 \text{ for } x > 1 \text{ [and } f(1) > f(2)\text{], so we can use the Integral Test on } [1, \infty).$$

$$\int_1^\infty xe^{-x}\, dx = \lim_{t \to \infty} \int_1^t xe^{-x}\, dx = \lim_{t \to \infty} \left(\left[-xe^{-x}\right]_1^t + \int_1^t e^{-x}\, dx\right) \quad \left[\begin{array}{l}\text{by parts with}\\ u = x,\ dv = e^{-x}\, dx\end{array}\right]$$

$$= \lim_{t \to \infty} \left(-te^{-t} + e^{-1} + \left[-e^{-x}\right]_1^t\right) = \lim_{t \to \infty} \left(-\dfrac{t}{e^t} + \dfrac{1}{e} - \dfrac{1}{e^t} + \dfrac{1}{e}\right)$$

$$\stackrel{\text{H}}{=} \lim_{t \to \infty} \left(-\dfrac{1}{e^t} + \dfrac{1}{e} - 0 + \dfrac{1}{e}\right) = \dfrac{2}{e},$$

so the series $\displaystyle\sum_{k=1}^\infty ke^{-k}$ converges.

27. The function $f(x) = \dfrac{1}{x^2 + x^3} = \dfrac{1}{x^2} - \dfrac{1}{x} + \dfrac{1}{x + 1}$ [by partial fractions] is continuous, positive and decreasing on $[1, \infty)$,

so the Integral Test applies.

$$\int_1^\infty f(x)\, dx = \lim_{t \to \infty} \int_1^t \left(\dfrac{1}{x^2} - \dfrac{1}{x} + \dfrac{1}{x + 1}\right) dx = \lim_{t \to \infty} \left[-\dfrac{1}{x} - \ln x + \ln(x + 1)\right]_1^t$$

$$= \lim_{t \to \infty} \left[-\dfrac{1}{t} + \ln \dfrac{t + 1}{t} + 1 - \ln 2\right] = 0 + 0 + 1 - \ln 2$$

The integral converges, so the series $\displaystyle\sum_{n=1}^\infty \dfrac{1}{n^2 + n^3}$ converges.

29. The function $f(x) = \dfrac{\cos \pi x}{\sqrt{x}}$ is neither positive nor decreasing on $[1, \infty)$, so the hypotheses of the Integral Test are not

satisfied for the series $\displaystyle\sum_{n=1}^\infty \dfrac{\cos \pi n}{\sqrt{n}}$.

31. We have already shown (in Exercise 23) that when $p = 1$ the series $\displaystyle\sum_{n=2}^\infty \dfrac{1}{n(\ln n)^p}$ diverges, so assume that $p \neq 1$.

$f(x) = \dfrac{1}{x(\ln x)^p}$ is continuous and positive on $[2, \infty)$, and $f'(x) = -\dfrac{p + \ln x}{x^2(\ln x)^{p+1}} < 0$ if $x > e^{-p}$, so that f is eventually

546 ☐ **CHAPTER 11** SEQUENCES, SERIES, AND POWER SERIES

decreasing and we can use the Integral Test.

$$\int_2^\infty \frac{1}{x(\ln x)^p} \, dx = \lim_{t \to \infty} \left[\frac{(\ln x)^{1-p}}{1-p}\right]_2^t \text{ [for } p \neq 1\text{]} = \lim_{t \to \infty} \left[\frac{(\ln t)^{1-p}}{1-p} - \frac{(\ln 2)^{1-p}}{1-p}\right]$$

This limit exists whenever $1 - p < 0 \Leftrightarrow p > 1$, so the series converges for $p > 1$.

33. Clearly the series cannot converge if $p \geq -\frac{1}{2}$, because then $\lim_{n \to \infty} n(1+n^2)^p \neq 0$. So assume $p < -\frac{1}{2}$. Then

$f(x) = x(1+x^2)^p$ is continuous, positive, and eventually decreasing on $[1, \infty)$, and we can use the Integral Test.

$$\int_1^\infty x(1+x^2)^p \, dx = \lim_{t \to \infty} \left[\frac{1}{2} \cdot \frac{(1+x^2)^{p+1}}{p+1}\right]_1^t = \frac{1}{2(p+1)} \lim_{t \to \infty} [(1+t^2)^{p+1} - 2^{p+1}].$$

This limit exists and is finite $\Leftrightarrow p+1 < 0 \Leftrightarrow p < -1$, so the series $\sum_{n=1}^\infty n(1+n^2)^p$ converges whenever $p < -1$.

35. Since this is a p-series with $p = x$, $\zeta(x)$ is defined when $x > 1$. Unless specified otherwise, the domain of a function f is the set of real numbers x such that the expression for $f(x)$ makes sense and defines a real number. So, in the case of a series, it's the set of real numbers x such that the series is convergent.

37. (a) $\sum_{n=1}^\infty \left(\frac{3}{n}\right)^4 = \sum_{n=1}^\infty \frac{81}{n^4} = 81 \sum_{n=1}^\infty \frac{1}{n^4} = 81\left(\frac{\pi^4}{90}\right) = \frac{9\pi^4}{10}$

(b) $\sum_{k=5}^\infty \frac{1}{(k-2)^4} = \frac{1}{3^4} + \frac{1}{4^4} + \frac{1}{5^4} + \cdots = \sum_{k=3}^\infty \frac{1}{k^4} = \frac{\pi^4}{90} - \left(\frac{1}{1^4} + \frac{1}{2^4}\right)$ [subtract a_1 and a_2] $= \frac{\pi^4}{90} - \frac{17}{16}$

39. (a) $f(x) = \frac{1}{x^2}$ is positive and continuous and $f'(x) = -\frac{2}{x^3}$ is negative for $x > 0$, and so the Integral Test applies.

$\sum_{n=1}^\infty \frac{1}{n^2} \approx s_{10} = \frac{1}{1^2} + \frac{1}{2^2} + \frac{1}{3^2} + \cdots + \frac{1}{10^2} \approx 1.549768$.

$R_{10} \leq \int_{10}^\infty \frac{1}{x^2} \, dx = \lim_{t \to \infty}\left[\frac{-1}{x}\right]_{10}^t = \lim_{t \to \infty}\left(-\frac{1}{t} + \frac{1}{10}\right) = \frac{1}{10}$, so the error is at most 0.1.

(b) $s_{10} + \int_{11}^\infty \frac{1}{x^2} \, dx \leq s \leq s_{10} + \int_{10}^\infty \frac{1}{x^2} \, dx \Rightarrow s_{10} + \frac{1}{11} \leq s \leq s_{10} + \frac{1}{10} \Rightarrow$

$1.549768 + 0.090909 = 1.640677 \leq s \leq 1.549768 + 0.1 = 1.649768$, so we get $s \approx 1.64522$ (the average of 1.640677 and 1.649768) with error ≤ 0.005 (the maximum of $1.649768 - 1.64522$ and $1.64522 - 1.640677$, rounded up).

(c) The estimate in part (b) is $s \approx 1.64522$ with error ≤ 0.005. The exact value given in Exercise 36 is $\pi^2/6 \approx 1.644934$. The difference is less than 0.0003.

(d) $R_n \leq \int_n^\infty \frac{1}{x^2} \, dx = \frac{1}{n}$. So $R_n < 0.001$ if $\frac{1}{n} < \frac{1}{1000} \Leftrightarrow n > 1000$.

41. $\sum_{n=1}^\infty (2n+1)^{-6}$. $f(x) = 1/(2x+1)^6$ is continuous, positive, and decreasing on $[1, \infty)$, so the Integral Test applies.

Using (2), $R_n \leq \int_n^\infty (2x+1)^{-6} \, dx = \lim_{t \to \infty}\left[\frac{-1}{10(2x+1)^5}\right]_n^t = \frac{1}{10(2n+1)^5}$. To be correct to five decimal places, we want $\dfrac{1}{10(2n+1)^5} \leq \dfrac{5}{10^6}$ $\Leftrightarrow$ $(2n+1)^5 \geq 20{,}000$ $\Leftrightarrow$ $n \geq \tfrac{1}{2}(\sqrt[5]{20{,}000} - 1) \approx 3.12$, so use $n = 4$.

$s_4 = \displaystyle\sum_{n=1}^{4} \dfrac{1}{(2n+1)^6} = \dfrac{1}{3^6} + \dfrac{1}{5^6} + \dfrac{1}{7^6} + \dfrac{1}{9^6} \approx 0.001\,446 \approx 0.00145$.

43. $\displaystyle\sum_{n=1}^{\infty} n^{-1.001} = \sum_{n=1}^{\infty} \dfrac{1}{n^{1.001}}$ is a convergent p-series with $p = 1.001 > 1$. Using (2), we get

$R_n \leq \displaystyle\int_{n}^{\infty} x^{-1.001}\,dx = \lim_{t \to \infty}\left[\dfrac{x^{-0.001}}{-0.001}\right]_n^t = -1000 \lim_{t \to \infty}\left[\dfrac{1}{x^{0.001}}\right]_n^t = -1000\left(-\dfrac{1}{n^{0.001}}\right) = \dfrac{1000}{n^{0.001}}$. We want

$R_n < 0.000\,000\,005$ $\Leftrightarrow$ $\dfrac{1000}{n^{0.001}} < 5 \times 10^{-9}$ $\Leftrightarrow$ $n^{0.001} > \dfrac{1000}{5 \times 10^{-9}}$ $\Leftrightarrow$

$n > (2 \times 10^{11})^{1000} = 2^{1000} \times 10^{11{,}000} \approx 1.07 \times 10^{301} \times 10^{11{,}000} = 1.07 \times 10^{11{,}301}$.

45. (a) From the figure, $a_2 + a_3 + \cdots + a_n \leq \int_1^n f(x)\,dx$, so with

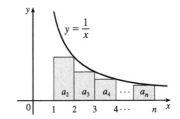

$f(x) = \dfrac{1}{x}$, $\dfrac{1}{2} + \dfrac{1}{3} + \dfrac{1}{4} + \cdots + \dfrac{1}{n} \leq \displaystyle\int_1^n \dfrac{1}{x}\,dx = \ln n$.

Thus, $s_n = 1 + \dfrac{1}{2} + \dfrac{1}{3} + \dfrac{1}{4} + \cdots + \dfrac{1}{n} \leq 1 + \ln n$.

(b) By part (a), $s_{10^6} \leq 1 + \ln 10^6 \approx 14.82 < 15$ and

$s_{10^9} \leq 1 + \ln 10^9 \approx 21.72 < 22$.

47. $b^{\ln n} = \left(e^{\ln b}\right)^{\ln n} = \left(e^{\ln n}\right)^{\ln b} = n^{\ln b} = \dfrac{1}{n^{-\ln b}}$. $\displaystyle\sum_{n=1}^{\infty} b^{\ln n}$ is a p-series, which converges for all b such that $-\ln b > 1$ $\Leftrightarrow$

$\ln b < -1$ $\Leftrightarrow$ $b < e^{-1}$ $\Leftrightarrow$ $b < 1/e$ [with $b > 0$].

11.4 The Comparison Tests

1. (a) We cannot say anything about $\sum a_n$. If $a_n > b_n$ for all n and $\sum b_n$ is convergent, then $\sum a_n$ could be convergent or divergent. (See the discusssion preceding the box titled "The Limit Comparison Test.")

(b) If $a_n < b_n$ for all n, then $\sum a_n$ is convergent. [This is part (i) of the Direct Comparison Test.]

3. (a) $\dfrac{n}{n^3 + 5} < \dfrac{n}{n^3} = \dfrac{1}{n^2}$ for all $n \geq 2$. $\displaystyle\sum_{n=2}^{\infty} \dfrac{1}{n^2}$ converges because it is a p-series with $p = 2 > 1$, so $\displaystyle\sum_{n=2}^{\infty} \dfrac{n}{n^3 + 5}$ converges by part (i) of the Direct Comparison Test.

(b) Use the Limit Comparison Test with $a_n = \dfrac{n}{n^3 - 5}$ and $b_n = \dfrac{1}{n^2}$:

$\displaystyle\lim_{n \to \infty} \dfrac{a_n}{b_n} = \lim_{n \to \infty} \dfrac{n}{n^3 - 5} \cdot \dfrac{n^2}{1} = \lim_{n \to \infty} \dfrac{n^3}{n^3(1 - 5/n^3)} = \lim_{n \to \infty} \dfrac{1}{1 - 5/n^3} = \dfrac{1}{1 - 0} = 1 > 0$

Since $\displaystyle\sum_{n=2}^{\infty} \dfrac{1}{n^2}$ is a convergent (partial) p-series $[p = 2 > 1]$, the series $\displaystyle\sum_{n=2}^{\infty} \dfrac{n}{n^3 - 5}$ also converges.

5. An inequality can be used to show that a series converges if its general term can be shown to be less than or equal to the general term of a known convergent series. The only inequality that satisfies this condition is given in part (c) since $\sum_{n=1}^{\infty} \frac{1}{n^2}$ is a convergent p-series $[p = 2 > 1]$.

7. $\frac{1}{n^3 + 8} < \frac{1}{n^3}$ for all $n \geq 1$, so $\sum_{n=1}^{\infty} \frac{1}{n^3 + 8}$ converges by direct comparison with $\sum_{n=1}^{\infty} \frac{1}{n^3}$, which converges because it is a p-series with $p = 3 > 1$.

9. $\frac{n+1}{n\sqrt{n}} > \frac{n}{n\sqrt{n}} = \frac{1}{\sqrt{n}}$ for all $n \geq 1$, so $\sum_{n=1}^{\infty} \frac{n+1}{n\sqrt{n}}$ diverges by direct comparison with $\sum_{n=1}^{\infty} \frac{1}{\sqrt{n}}$, which diverges because it is a p-series with $p = \frac{1}{2} \leq 1$.

11. $\frac{9^n}{3 + 10^n} < \frac{9^n}{10^n} = \left(\frac{9}{10}\right)^n$ for all $n \geq 1$. $\sum_{n=1}^{\infty} \left(\frac{9}{10}\right)^n$ is a convergent geometric series $(|r| = \frac{9}{10} < 1)$, so $\sum_{n=1}^{\infty} \frac{9^n}{3 + 10^n}$ converges by the Direct Comparison Test.

13. For $n \geq 2$, $\ln n < n$, so $\frac{1}{\ln n} > \frac{1}{n}$. Thus, $\sum_{n=2}^{\infty} \frac{1}{\ln n}$ diverges by direct comparison with $\sum_{n=1}^{\infty} \frac{1}{n}$, which diverges because it is a p-series with $p = 1 \leq 1$ (the harmonic series).

15. $\frac{\sqrt[3]{k}}{\sqrt{k^3 + 4k + 3}} < \frac{\sqrt[3]{k}}{\sqrt{k^3}} = \frac{k^{1/3}}{k^{3/2}} = \frac{1}{k^{7/6}}$ for all $k \geq 1$, so $\sum_{k=1}^{\infty} \frac{\sqrt[3]{k}}{\sqrt{k^3 + 4k + 3}}$ converges by direct comparison with $\sum_{k=1}^{\infty} \frac{1}{k^{7/6}}$, which converges because it is a p-series with $p = \frac{7}{6} > 1$.

17. $\frac{1 + \cos n}{e^n} < \frac{2}{e^n}$ for all $n \geq 1$. $\sum_{n=1}^{\infty} \frac{2}{e^n}$ is a convergent geometric series $(|r| = \frac{1}{e} < 1)$, so $\sum_{n=1}^{\infty} \frac{1 + \cos n}{e^n}$ converges by the Direct Comparison Test.

19. $\frac{4^{n+1}}{3^n - 2} > \frac{4 \cdot 4^n}{3^n} = 4\left(\frac{4}{3}\right)^n$ for all $n \geq 1$. $\sum_{n=1}^{\infty} 4\left(\frac{4}{3}\right)^n = 4\sum_{n=1}^{\infty} \left(\frac{4}{3}\right)^n$ is a divergent geometric series $(|r| = \frac{4}{3} > 1)$, so $\sum_{n=1}^{\infty} \frac{4^{n+1}}{3^n - 2}$ diverges by the Direct Comparison Test.

21. Use the Limit Comparison Test with $a_n = \frac{1}{\sqrt{n^2 + 1}}$ and $b_n = \frac{1}{n}$:

$\lim_{n \to \infty} \frac{a_n}{b_n} = \lim_{n \to \infty} \frac{n}{\sqrt{n^2 + 1}} = \lim_{n \to \infty} \frac{1}{\sqrt{1 + (1/n^2)}} = 1 > 0$. Since the harmonic series $\sum_{n=1}^{\infty} \frac{1}{n}$ diverges, so does $\sum_{n=1}^{\infty} \frac{1}{\sqrt{n^2 + 1}}$.

23. Use the Limit Comparison Test with $a_n = \frac{n+1}{n^3 + n}$ and $b_n = \frac{1}{n^2}$:

$\lim_{n \to \infty} \frac{a_n}{b_n} = \lim_{n \to \infty} \frac{(n+1)n^2}{n(n^2 + 1)} = \lim_{n \to \infty} \frac{n^2 + n}{n^2 + 1} = \lim_{n \to \infty} \frac{1 + 1/n}{1 + 1/n^2} = 1 > 0$. Since $\sum_{n=1}^{\infty} \frac{1}{n^2}$ is a convergent p-series $[p = 2 > 1]$, the series $\sum_{n=1}^{\infty} \frac{n+1}{n^3 + n}$ also converges.

25. Use the Limit Comparison Test with $a_n = \dfrac{\sqrt{1+n}}{2+n}$ and $b_n = \dfrac{1}{\sqrt{n}}$:

$$\lim_{n\to\infty} \frac{a_n}{b_n} = \lim_{n\to\infty} \frac{\sqrt{1+n}\sqrt{n}}{2+n} = \lim_{n\to\infty} \frac{\sqrt{n^2+n}/\sqrt{n^2}}{(2+n)/n} = \lim_{n\to\infty} \frac{\sqrt{1+1/n}}{2/n+1} = 1 > 0.$$ Since $\sum_{n=1}^{\infty} \dfrac{1}{\sqrt{n}}$ is a divergent p-series $[p = \tfrac{1}{2} \le 1]$, the series $\sum_{n=1}^{\infty} \dfrac{\sqrt{1+n}}{2+n}$ also diverges.

27. Use the Limit Comparison Test with $a_n = \dfrac{5+2n}{(1+n^2)^2}$ and $b_n = \dfrac{1}{n^3}$:

$$\lim_{n\to\infty} \frac{a_n}{b_n} = \lim_{n\to\infty} \frac{n^3(5+2n)}{(1+n^2)^2} = \lim_{n\to\infty} \frac{5n^3+2n^4}{(1+n^2)^2} \cdot \frac{1/n^4}{1/(n^2)^2} = \lim_{n\to\infty} \frac{\frac{5}{n}+2}{\left(\frac{1}{n^2}+1\right)^2} = 2 > 0.$$ Since $\sum_{n=1}^{\infty} \dfrac{1}{n^3}$ is a convergent p-series $[p = 3 > 1]$, the series $\sum_{n=1}^{\infty} \dfrac{5+2n}{(1+n^2)^2}$ also converges.

29. $\dfrac{e^n+1}{ne^n+1} \ge \dfrac{e^n+1}{ne^n+n} = \dfrac{e^n+1}{n(e^n+1)} = \dfrac{1}{n}$ for $n \ge 1$, so the series $\sum_{n=1}^{\infty} \dfrac{e^n+1}{ne^n+1}$ diverges by direct comparison with the divergent harmonic series $\sum_{n=1}^{\infty} \dfrac{1}{n}$. *Or:* Use the Limit Comparison Test with $a_n = \dfrac{e^n+1}{ne^n+1}$ and $b_n = \dfrac{1}{n}$.

31. $\dfrac{2+\sin n}{n^2} \le \dfrac{2+1}{n^2} = \dfrac{3}{n^2}$, for all $n \ge 1$, so $\sum_{n=1}^{\infty} \dfrac{2+\sin n}{n^2}$ converges by direct comparison with $3\sum_{n=1}^{\infty} \dfrac{1}{n^2}$, which converges because it is a constant multiple of a p-series with $p = 2 > 1$.

33. Use the Limit Comparison Test with $a_n = \left(1+\dfrac{1}{n}\right)^2 e^{-n}$ and $b_n = e^{-n}$: $\lim_{n\to\infty} \dfrac{a_n}{b_n} = \lim_{n\to\infty}\left(1+\dfrac{1}{n}\right)^2 = 1 > 0$. Since $\sum_{n=1}^{\infty} e^{-n} = \sum_{n=1}^{\infty} \dfrac{1}{e^n}$ is a convergent geometric series $\left[|r| = \tfrac{1}{e} < 1\right]$, the series $\sum_{n=1}^{\infty}\left(1+\dfrac{1}{n}\right)^2 e^{-n}$ also converges.

35. Clearly $n! = n(n-1)(n-2)\cdots(3)(2) \ge 2\cdot 2\cdot 2 \cdots\cdot 2\cdot 2 = 2^{n-1}$, so $\dfrac{1}{n!} \le \dfrac{1}{2^{n-1}}$. $\sum_{n=1}^{\infty} \dfrac{1}{2^{n-1}}$ is a convergent geometric series $\left[|r| = \tfrac{1}{2} < 1\right]$, so $\sum_{n=1}^{\infty} \dfrac{1}{n!}$ converges by the Direct Comparison Test.

37. Use the Limit Comparison Test with $a_n = \sin\left(\dfrac{1}{n}\right)$ and $b_n = \dfrac{1}{n}$. Then $\sum a_n$ and $\sum b_n$ are series with positive terms and $\lim_{n\to\infty} \dfrac{a_n}{b_n} = \lim_{n\to\infty} \dfrac{\sin(1/n)}{1/n} = \lim_{\theta\to 0} \dfrac{\sin\theta}{\theta} = 1 > 0$. Since $\sum_{n=1}^{\infty} b_n$ is the divergent harmonic series, $\sum_{n=1}^{\infty} \sin(1/n)$ also diverges. [Note that we could also use l'Hospital's Rule to evaluate the limit:

$$\lim_{x\to\infty} \frac{\sin(1/x)}{1/x} \stackrel{H}{=} \lim_{x\to\infty} \frac{\cos(1/x)\cdot(-1/x^2)}{-1/x^2} = \lim_{x\to\infty} \cos\frac{1}{x} = \cos 0 = 1.]$$

39. Use the Limit Comparison Test with $a_n = \frac{1}{n}\tan\left(\frac{1}{n}\right)$ and $b_n = \frac{1}{n^2}$:

$$\lim_{n\to\infty}\frac{a_n}{b_n} = \lim_{n\to\infty}\frac{\frac{1}{n}\tan\left(\frac{1}{n}\right)}{\frac{1}{n^2}} = \lim_{n\to\infty} n\tan\left(\frac{1}{n}\right) = \lim_{x\to 0}\frac{\tan x}{x}\ \left[\text{where } x = \frac{1}{n}\right] \stackrel{\text{H}}{=} \lim_{x\to 0}\frac{\sec^2 x}{1} = \frac{1}{1} = 1 > 0$$

Since $\sum_{n=1}^{\infty}\frac{1}{n^2}$ is a convergent p-series [$p = 2 > 1$], the series $\sum_{n=1}^{\infty}\frac{1}{n}\tan\left(\frac{1}{n}\right)$ also converges.

41. $\sum_{n=1}^{10}\frac{1}{5+n^5} = \frac{1}{5+1^5} + \frac{1}{5+2^5} + \frac{1}{5+3^5} + \cdots + \frac{1}{5+10^5} \approx 0.19926$. Now $\frac{1}{5+n^5} < \frac{1}{n^5}$, so the error is

$$R_{10} \leq T_{10} \leq \int_{10}^{\infty}\frac{1}{x^5}\,dx = \lim_{t\to\infty}\int_{10}^{t} x^{-5}\,dx = \lim_{t\to\infty}\left[\frac{-1}{4x^4}\right]_{10}^{t} = \lim_{t\to\infty}\left(\frac{-1}{4t^4} + \frac{1}{40{,}000}\right) = \frac{1}{40{,}000} = 0.000\,025.$$

43. $\sum_{n=1}^{10} 5^{-n}\cos^2 n = \frac{\cos^2 1}{5} + \frac{\cos^2 2}{5^2} + \frac{\cos^2 3}{5^3} + \cdots + \frac{\cos^2 10}{5^{10}} \approx 0.07393$. Now $\frac{\cos^2 n}{5^n} \leq \frac{1}{5^n}$, so the error is

$$R_{10} \leq T_{10} \leq \int_{10}^{\infty}\frac{1}{5^x}\,dx = \lim_{t\to\infty}\int_{10}^{t} 5^{-x}\,dx = \lim_{t\to\infty}\left[-\frac{5^{-x}}{\ln 5}\right]_{10}^{t} = \lim_{t\to\infty}\left(-\frac{5^{-t}}{\ln 5} + \frac{5^{-10}}{\ln 5}\right) = \frac{1}{5^{10}\ln 5} < 6.4\times 10^{-8}.$$

45. Since $\frac{d_n}{10^n} \leq \frac{9}{10^n}$ for each n, and since $\sum_{n=1}^{\infty}\frac{9}{10^n}$ is a convergent geometric series ($|r| = \frac{1}{10} < 1$), $0.d_1 d_2 d_3\ldots = \sum_{n=1}^{\infty}\frac{d_n}{10^n}$ will always converge by the Comparison Test.

47. Using Formula 6.3.7 or combining Property 6.3*.4 with Law 3 of Theorem 6.2*.3, we have

$(\ln n)^{\ln \ln n} = e^{(\ln \ln n)(\ln \ln n)}$ [$n > 1$] $= e^{(\ln \ln n)^2} < e^{(\sqrt{\ln n})^2} = e^{\ln n} = n$. So $\frac{1}{(\ln n)^{\ln \ln n}} > \frac{1}{n}$ for all $n \geq 2$. Thus,

the series $\sum_{n=2}^{\infty}\frac{1}{(\ln n)^{\ln \ln n}}$ diverges by comparison with the partial harmonic series $\sum_{n=2}^{\infty}\frac{1}{n}$, which diverges.

49. (a) Since $\lim_{n\to\infty}\frac{a_n}{b_n} = \infty$, there is an integer N such that $\frac{a_n}{b_n} > 1$ whenever $n > N$. (Take $M = 1$ in Definition 11.1.3.)

Then $a_n > b_n$ whenever $n > N$ and since $\sum b_n$ is divergent, $\sum a_n$ is also divergent by the Comparison Test.

(b) (i) If $a_n = \frac{1}{\ln n}$ and $b_n = \frac{1}{n}$ for $n \geq 2$, then $\lim_{n\to\infty}\frac{a_n}{b_n} = \lim_{n\to\infty}\frac{n}{\ln n} = \lim_{x\to\infty}\frac{x}{\ln x} \stackrel{\text{H}}{=} \lim_{x\to\infty}\frac{1}{1/x} = \lim_{x\to\infty} x = \infty$,

so by part (a), $\sum_{n=2}^{\infty}\frac{1}{\ln n}$ is divergent.

(ii) If $a_n = \frac{\ln n}{n}$ and $b_n = \frac{1}{n}$, then $\sum_{n=1}^{\infty} b_n$ is the divergent harmonic series and $\lim_{n\to\infty}\frac{a_n}{b_n} = \lim_{n\to\infty} \ln n = \lim_{x\to\infty}\ln x = \infty$,

so $\sum_{n=1}^{\infty} a_n$ diverges by part (a).

51. $\lim_{n\to\infty} na_n = \lim_{n\to\infty}\frac{a_n}{1/n}$, so we apply the Limit Comparison Test with $b_n = \frac{1}{n}$. Since $\lim_{n\to\infty} na_n > 0$ we know that either both series converge or both series diverge, and we also know that $\sum_{n=1}^{\infty}\frac{1}{n}$ diverges [p-series with $p = 1$]. Therefore, $\sum a_n$ must be divergent.

53. Yes. Since $\sum a_n$ is a convergent series with positive terms, $\lim\limits_{n\to\infty} a_n = 0$ by Theorem 11.2.6, and $\sum b_n = \sum \sin(a_n)$ is a series with positive terms (for large enough n). We have $\lim\limits_{n\to\infty} \dfrac{b_n}{a_n} = \lim\limits_{n\to\infty} \dfrac{\sin(a_n)}{a_n} = 1 > 0$ by Theorem 2.4.5. Thus, $\sum b_n$ is also convergent by the Limit Comparison Test.

55. (a) False. The series $\sum a_n = \sum \dfrac{1}{n}$ and $\sum b_n = \sum \dfrac{1}{n}$ are both divergent, but $\sum a_n b_n = \sum \dfrac{1}{n} \cdot \dfrac{1}{n} = \sum \dfrac{1}{n^2}$ is convergent.

(b) False. The series $\sum a_n = \sum \dfrac{1}{n^2}$ converges and $\sum b_n = \sum \dfrac{1}{n}$ diverges, but $\sum a_n b_n = \sum \dfrac{1}{n^2} \cdot \dfrac{1}{n} = \sum \dfrac{1}{n^3}$ converges.

(c) True. Since $\sum a_n$ converges, for suffficiently large n, $a_n < 1 \Rightarrow a_n b_n < b_n$. Since $\sum b_n$ converges, it follows by the Direct Comparison Test that $\sum a_n b_n$ converges.

11.5 Alternating Series and Absolute Convergence

1. (a) An alternating series is a series whose terms are alternately positive and negative.

(b) An alternating series $\sum\limits_{n=1}^{\infty} a_n = \sum\limits_{n=1}^{\infty} (-1)^{n-1} b_n$, where $b_n = |a_n|$, converges if $0 < b_{n+1} \le b_n$ for all n and $\lim\limits_{n\to\infty} b_n = 0$.

(This is the Alternating Series Test.)

(c) The error involved in using the partial sum s_n as an approximation to the total sum s is the remainder $R_n = s - s_n$ and the size of the error is smaller than b_{n+1}; that is, $|R_n| \le b_{n+1}$. (This is the Alternating Series Estimation Theorem.)

3. $-\dfrac{2}{5} + \dfrac{4}{6} - \dfrac{6}{7} + \dfrac{8}{8} - \dfrac{10}{9} + \cdots = \sum\limits_{n=1}^{\infty} (-1)^n \dfrac{2n}{n+4}$. Now $\lim\limits_{n\to\infty} b_n = \lim\limits_{n\to\infty} \dfrac{2n}{n+4} = \lim\limits_{n\to\infty} \dfrac{2}{1+4/n} = \dfrac{2}{1} \ne 0$. Since $\lim\limits_{n\to\infty} a_n \ne 0$ (in fact the limit does not exist), the series diverges by the Test for Divergence.

5. $\sum\limits_{n=1}^{\infty} a_n = \sum\limits_{n=1}^{\infty} \dfrac{(-1)^{n-1}}{3+5n} = \sum\limits_{n=1}^{\infty} (-1)^{n-1} b_n$. Now $b_n = \dfrac{1}{3+5n} > 0$, $\{b_n\}$ is decreasing, and $\lim\limits_{n\to\infty} b_n = 0$, so the series converges by the Alternating Series Test.

7. $\sum\limits_{n=1}^{\infty} a_n = \sum\limits_{n=1}^{\infty} (-1)^n \dfrac{3n-1}{2n+1} = \sum\limits_{n=1}^{\infty} (-1)^n b_n$. Now $\lim\limits_{n\to\infty} b_n = \lim\limits_{n\to\infty} \dfrac{3-1/n}{2+1/n} = \dfrac{3}{2} \ne 0$. Since $\lim\limits_{n\to\infty} a_n \ne 0$ (in fact the limit does not exist), the series diverges by the Test for Divergence.

9. $\sum\limits_{n=1}^{\infty} a_n = \sum\limits_{n=1}^{\infty} (-1)^n e^{-n} = \sum\limits_{n=1}^{\infty} (-1)^n b_n$. Now $b_n = \dfrac{1}{e^n} > 0$, $\{b_n\}$ is decreasing, and $\lim\limits_{n\to\infty} b_n = 0$, so the series converges by the Alternating Series Test.

11. $b_n = \dfrac{n^2}{n^3+4} > 0$ for $n \ge 1$. $\{b_n\}$ is decreasing for $n \ge 2$ since

$$\left(\dfrac{x^2}{x^3+4}\right)' = \dfrac{(x^3+4)(2x) - x^2(3x^2)}{(x^3+4)^2} = \dfrac{x(2x^3+8-3x^3)}{(x^3+4)^2} = \dfrac{x(8-x^3)}{(x^3+4)^2} < 0 \text{ for } x > 2.$$ Also,

$\lim\limits_{n\to\infty} b_n = \lim\limits_{n\to\infty} \dfrac{1/n}{1+4/n^3} = 0$. Thus, the series $\sum\limits_{n=1}^{\infty} (-1)^{n+1} \dfrac{n^2}{n^3+4}$ converges by the Alternating Series Test.

13. $\lim\limits_{n\to\infty} b_n = \lim\limits_{n\to\infty} e^{2/n} = e^0 = 1$, so $\lim\limits_{n\to\infty}(-1)^{n-1}e^{2/n}$ does not exist. Thus, the series $\sum\limits_{n=1}^{\infty}(-1)^{n-1}e^{2/n}$ diverges by the Test for Divergence.

15. $a_n = \dfrac{\sin\left(n+\frac{1}{2}\right)\pi}{1+\sqrt{n}} = \dfrac{(-1)^n}{1+\sqrt{n}}$. Now $b_n = \dfrac{1}{1+\sqrt{n}} > 0$ for $n \geq 0$, $\{b_n\}$ is decreasing, and $\lim\limits_{n\to\infty} b_n = 0$, so the series $\sum\limits_{n=0}^{\infty} \dfrac{\sin\left(n+\frac{1}{2}\right)\pi}{1+\sqrt{n}}$ converges by the Alternating Series Test.

17. $\sum\limits_{n=1}^{\infty}(-1)^n \sin\dfrac{\pi}{n}$. $b_n = \sin\dfrac{\pi}{n} > 0$ for $n \geq 2$ and $\sin\dfrac{\pi}{n} \geq \sin\dfrac{\pi}{n+1}$, and $\lim\limits_{n\to\infty}\sin\dfrac{\pi}{n} = \sin 0 = 0$, so the series converges by the Alternating Series Test.

19. $b_n = \dfrac{n^2}{5^n} > 0$ for $n \geq 1$. $\{b_n\}$ is decreasing for $n \geq 2$ since

$$\left(\dfrac{x^2}{5^x}\right)' = \dfrac{5^x \cdot 2x - x^2 \, 5^x \ln 5}{(5^x)^2} = \dfrac{x\, 5^x(2 - x\ln 5)}{(5^x)^2} = \dfrac{x(2 - x\ln 5)}{5^x} < 0 \text{ for } x > \dfrac{2}{\ln 5} \approx 1.2.$$ Also,

$\lim\limits_{n\to\infty} b_n = \lim\limits_{n\to\infty} \dfrac{n^2}{5^n} \stackrel{\text{H}}{=} \lim\limits_{n\to\infty} \dfrac{2n}{5^n \ln 5} \stackrel{\text{H}}{=} \lim\limits_{n\to\infty} \dfrac{2}{5^n(\ln 5)^2} = 0$. Thus, the series $\sum\limits_{n=1}^{\infty}(-1)^n \dfrac{n^2}{5^n}$ converges by the Alternating Series Test.

21. (a) A series $\sum a_n$ is called absolutely convergent if the series of absolute values $\sum |a_n|$ is convergent. If a series is absolutely convergent, then it is convergent.

(b) A series $\sum a_n$ is called conditionally convergent if it is convergent but not absolutely convergent; that is, if $\sum a_n$ converges, but $\sum |a_n|$ diverges.

(c) Suppose the series of positive terms $\sum_{n=1}^{\infty} b_n$ converges. Then $\sum |(-1)^n b_n| = \sum |b_n| = \sum b_n$ also converges, so $\sum_{n=1}^{\infty}(-1)^n b_n$ is absolutely convergent (and therefore convergent).

23. $b_n = \dfrac{1}{\sqrt[3]{n^2}} > 0$ for $n \geq 1$. $\{b_n\}$ is decreasing for $n \geq 1$, and $\lim\limits_{n\to\infty}\dfrac{1}{\sqrt[3]{n^2}} = 0$, so the series $\sum\limits_{n=1}^{\infty}\dfrac{(-1)^{n-1}}{\sqrt[3]{n^2}}$ converges by the Alternating Series Test. Also, observe that $\sum\limits_{n=1}^{\infty}\left|\dfrac{(-1)^{n-1}}{\sqrt[3]{n^2}}\right| = \sum\limits_{n=1}^{\infty}\dfrac{1}{n^{2/3}}$ is divergent since it is a p-series with $p = \frac{2}{3} \leq 1$. Thus, the series $\sum\limits_{n=1}^{\infty}\dfrac{(-1)^{n-1}}{\sqrt[3]{n^2}}$ is conditionally convergent.

25. $b_n = \dfrac{1}{5n+1} > 0$ for $n \geq 1$, $\{b_n\}$ is decreasing for $n \geq 1$, and $\lim\limits_{n\to\infty} b_n = 0$, so $\sum\limits_{n=1}^{\infty}\dfrac{(-1)^n}{5n+1}$ converges by the Alternating Series Test. To determine absolute convergence, choose $a_n = \dfrac{1}{n}$ to get

$\lim\limits_{n\to\infty}\dfrac{a_n}{b_n} = \lim\limits_{n\to\infty}\dfrac{1/n}{1/(5n+1)} = \lim\limits_{n\to\infty}\dfrac{5n+1}{n} = 5 > 0$, so $\sum\limits_{n=1}^{\infty}\dfrac{1}{5n+1}$ diverges by the Limit Comparison Test with the harmonic series. Thus, the series $\sum\limits_{n=1}^{\infty}\dfrac{(-1)^n}{5n+1}$ is conditionally convergent.

27. $\sum_{n=1}^{\infty} |a_n| = \sum_{n=1}^{\infty} \left|\frac{(-1)^n}{n^2+1}\right| = \sum_{n=1}^{\infty} \frac{1}{n^2+1}$. Since $\frac{1}{n^2+1} < \frac{1}{n^2}$ and $\sum_{n=1}^{\infty} \frac{1}{n^2}$ is a convergent p-series [$p = 2 > 1$], the series $\sum_{n=1}^{\infty} \frac{1}{n^2+1}$ is convergent by the Direct Comparison Test. Thus, $\sum_{n=1}^{\infty} \frac{(-1)^n}{n^2+1}$ is absolutely convergent.

29. $0 < \left|\frac{1+2\sin n}{n^3}\right| < \frac{3}{n^3}$ for $n \geq 1$ and $3\sum_{n=1}^{\infty} \frac{1}{n^3}$ is a constant times a convergent p-series [$p = 3 > 1$], so $\sum_{n=1}^{\infty} \left|\frac{1+2\sin n}{n^3}\right|$ converges by direct comparison and the series $\sum_{n=1}^{\infty} \frac{1+2\sin n}{n^3}$ is absolutely convergent.

31. $\sum_{n=2}^{\infty} \frac{(-1)^n}{\ln n}$ converges by the Alternating Series Test since $\lim_{n \to \infty} \frac{1}{\ln n} = 0$ and $\left\{\frac{1}{\ln n}\right\}$ is decreasing. Now $\ln n < n$, so $\frac{1}{\ln n} > \frac{1}{n}$, and since $\sum_{n=2}^{\infty} \frac{1}{n}$ is the divergent (partial) harmonic series, $\sum_{n=2}^{\infty} \frac{1}{\ln n}$ diverges by the Direct Comparison Test. Thus, $\sum_{n=2}^{\infty} \frac{(-1)^n}{\ln n}$ is conditionally convergent.

33. $a_n = \frac{\cos n\pi}{3n+2} = (-1)^n \frac{1}{3n+2} = (-1)^n b_n$. $\{b_n\}$ is decreasing for $n \geq 1$, and $\lim_{n \to \infty} b_n = 0$, so $\sum_{n=1}^{\infty} \frac{\cos n\pi}{3n+2}$ converges by the Alternating Series Test. To determine absolute convergence, use the Limit Comparison Test with $a_n = \frac{1}{n}$:

$\lim_{n \to \infty} \frac{a_n}{b_n} = \lim_{n \to \infty} \frac{1/n}{1/(3n+2)} = \lim_{n \to \infty} \frac{3n+2}{n} = \lim_{n \to \infty} \frac{3+2/n}{1} = 3 > 0$. Since the harmonic series diverges, so does $\sum_{n=1}^{\infty} \frac{1}{3n+2}$. Thus, the series $\sum_{n=1}^{\infty} \frac{\cos n\pi}{3n+2}$ is conditionally convergent.

35.

The graph gives us an estimate for the sum of the series $\sum_{n=1}^{\infty} \frac{(-0.8)^n}{n!}$ of -0.55.

$b_8 = \frac{(0.8)^n}{8!} \approx 0.000\,004$, so

$\sum_{n=1}^{\infty} \frac{(-0.8)^n}{n!} \approx s_7 = \sum_{n=1}^{7} \frac{(-0.8)^n}{n!}$

$\approx -0.8 + 0.32 - 0.08\overline{53} + 0.017\overline{06} - 0.002\,731 + 0.000\,364 - 0.000\,042 \approx -0.5507$

Adding b_8 to s_7 does not change the fourth decimal place of s_7, so the sum of the series, correct to four decimal places, is -0.5507.

37. The series $\sum_{n=1}^{\infty} \frac{(-1)^{n+1}}{n^6}$ satisfies (i) of the Alternating Series Test because $\frac{1}{(n+1)^6} < \frac{1}{n^6}$ and (ii) $\lim_{n \to \infty} \frac{1}{n^6} = 0$, so the series is convergent. Now $b_5 = \frac{1}{5^6} = 0.000064 > 0.00005$ and $b_6 = \frac{1}{6^6} \approx 0.00002 < 0.00005$, so by the Alternating Series

39. The series $\sum_{n=1}^{\infty} \frac{(-1)^{n-1}}{n^2 2^n}$ satisfies (i) of the Alternating Series Test because $\frac{1}{(n+1)^2 2^{n+1}} < \frac{1}{n^2 2^n}$ and (ii) $\lim_{n\to\infty} \frac{1}{n^2 2^n} = 0$, so the series is convergent. Now $b_5 = \frac{1}{5^2 2^5} = 0.00125 > 0.0005$ and $b_6 = \frac{1}{6^2 2^6} \approx 0.0004 < 0.0005$, so by the Alternating Series Estimation Theorem, $n = 5$. (That is, since the 6th term is less than the desired error, we need to add the first 5 terms to get the sum to the desired accuracy.)

41. $b_4 = \frac{1}{8!} = \frac{1}{40{,}320} \approx 0.000\,025$, so

$$\sum_{n=1}^{\infty} \frac{(-1)^n}{(2n)!} \approx s_3 = \sum_{n=1}^{3} \frac{(-1)^n}{(2n)!} = -\frac{1}{2} + \frac{1}{24} - \frac{1}{720} \approx -0.459\,722$$

Adding b_4 to s_3 does not change the fourth decimal place of s_3, so by the Alternating Series Estimation Theorem, the sum of the series, correct to four decimal places, is -0.4597.

43. $\sum_{n=1}^{\infty} (-1)^n n e^{-2n} \approx s_5 = -\frac{1}{e^2} + \frac{2}{e^4} - \frac{3}{e^6} + \frac{4}{e^8} - \frac{5}{e^{10}} \approx -0.105\,025$. Adding $b_6 = 6/e^{12} \approx 0.000\,037$ to s_5 does not change the fourth decimal place of s_5, so by the Alternating Series Estimation Theorem, the sum of the series, correct to four decimal places, is -0.1050.

45. $\sum_{n=1}^{\infty} \frac{(-1)^{n-1}}{n} = 1 - \frac{1}{2} + \frac{1}{3} - \frac{1}{4} + \cdots + \frac{1}{49} - \frac{1}{50} + \frac{1}{51} - \frac{1}{52} + \cdots$. The 50th partial sum of this series is an underestimate, since $\sum_{n=1}^{\infty} \frac{(-1)^{n-1}}{n} = s_{50} + \left(\frac{1}{51} - \frac{1}{52}\right) + \left(\frac{1}{53} - \frac{1}{54}\right) + \cdots$, and the terms in parentheses are all positive. The result can be seen geometrically in Figure 1.

47. Clearly $b_n = \frac{1}{n+p}$ is decreasing and eventually positive and $\lim_{n\to\infty} b_n = 0$ for any p. So the series $\sum_{n=1}^{\infty} \frac{(-1)^n}{n+p}$ converges (by the Alternating Series Test) for any p for which every b_n is defined, that is, $n + p \neq 0$ for $n \geq 1$, or p is not a negative integer.

49. $\sum b_{2n} = \sum 1/(2n)^2$ clearly converges (by direct comparison with the p-series for $p = 2$). So suppose that $\sum (-1)^{n-1} b_n$ converges. Then by Theorem 11.2.8(ii), so does $\sum \left[(-1)^{n-1} b_n + b_n\right] = 2\left(1 + \frac{1}{3} + \frac{1}{5} + \cdots\right) = 2\sum \frac{1}{2n-1}$. But this diverges by direct comparison with the harmonic series, a contradiction. Therefore, $\sum (-1)^{n-1} b_n$ must diverge. The Alternating Series Test does not apply since $\{b_n\}$ is not decreasing.

51. (a) Since $\sum a_n$ is absolutely convergent, and since $|a_n^+| \leq |a_n|$ and $|a_n^-| \leq |a_n|$ (because a_n^+ and a_n^- each equal either a_n or 0), we conclude by the Direct Comparison Test that both $\sum a_n^+$ and $\sum a_n^-$ must be absolutely convergent.

Or: Use Theorem 11.2.8.

(b) We will show by contradiction that both $\sum a_n^+$ and $\sum a_n^-$ must diverge. For suppose that $\sum a_n^+$ converged. Then so would $\sum \left(a_n^+ - \frac{1}{2}a_n\right)$ by Theorem 11.2.8. But $\sum \left(a_n^+ - \frac{1}{2}a_n\right) = \sum \left[\frac{1}{2}\left(a_n + |a_n|\right)\right] - \frac{1}{2}a_n] = \frac{1}{2}\sum |a_n|$, which diverges because $\sum a_n$ is only conditionally convergent. Hence, $\sum a_n^+$ can't converge. Similarly, neither can $\sum a_n^-$.

53. Suppose that $\sum a_n$ is conditionally convergent.

(a) $\sum n^2 a_n$ is divergent: Suppose $\sum n^2 a_n$ converges. Then $\lim\limits_{n\to\infty} n^2 a_n = 0$ by Theorem 6 in Section 11.2, so there is an integer $N > 0$ such that $n > N \;\Rightarrow\; n^2 |a_n| < 1$. For $n > N$, we have $|a_n| < \dfrac{1}{n^2}$, so $\sum\limits_{n>N} |a_n|$ converges by comparison with the convergent p-series $\sum\limits_{n>N} \dfrac{1}{n^2}$. In other words, $\sum a_n$ converges absolutely, contradicting the assumption that $\sum a_n$ is conditionally convergent. This contradiction shows that $\sum n^2 a_n$ diverges.

Remark: The same argument shows that $\sum n^p a_n$ diverges for any $p > 1$.

(b) $\sum\limits_{n=2}^{\infty} \dfrac{(-1)^n}{n \ln n}$ is conditionally convergent. It converges by the Alternating Series Test, but does not converge absolutely

$\left[\text{by the Integral Test, since the function } f(x) = \dfrac{1}{x \ln x} \text{ is continuous, positive, and decreasing on } [2, \infty) \text{ and}\right.$

$\left. \displaystyle\int_2^\infty \dfrac{dx}{x \ln x} = \lim_{t\to\infty} \int_2^t \dfrac{dx}{x \ln x} = \lim_{t\to\infty} \Big[\ln(\ln x)\Big]_2^t = \infty \right]$. Setting $a_n = \dfrac{(-1)^n}{n \ln n}$ for $n \geq 2$, we find that

$\displaystyle\sum_{n=2}^{\infty} n a_n = \sum_{n=2}^{\infty} \dfrac{(-1)^n}{\ln n}$ converges by the Alternating Series Test.

It is easy to find conditionally convergent series $\sum a_n$ such that $\sum n a_n$ diverges. Two examples are $\displaystyle\sum_{n=1}^{\infty} \dfrac{(-1)^{n-1}}{n}$ and $\displaystyle\sum_{n=1}^{\infty} \dfrac{(-1)^{n-1}}{\sqrt{n}}$, both of which converge by the Alternating Series Test and fail to converge absolutely because $\sum |a_n|$ is a p-series with $p \leq 1$. In both cases, $\sum n a_n$ diverges by the Test for Divergence.

11.6 The Ratio and Root Tests

1. (a) Since $\lim\limits_{n\to\infty} \left|\dfrac{a_{n+1}}{a_n}\right| = 8 > 1$, part (b) of the Ratio Test tells us that the series $\sum a_n$ is divergent.

(b) Since $\lim\limits_{n\to\infty} \left|\dfrac{a_{n+1}}{a_n}\right| = 0.8 < 1$, part (a) of the Ratio Test tells us that the series $\sum a_n$ is absolutely convergent (and therefore convergent).

(c) Since $\lim\limits_{n\to\infty} \left|\dfrac{a_{n+1}}{a_n}\right| = 1$, the Ratio Test fails and the series $\sum a_n$ might converge or it might diverge.

3. $\lim\limits_{n\to\infty} \left|\dfrac{a_{n+1}}{a_n}\right| = \lim\limits_{n\to\infty} \left|\dfrac{n+1}{5^{n+1}} \cdot \dfrac{5^n}{n}\right| = \lim\limits_{n\to\infty} \left|\dfrac{1}{5} \cdot \dfrac{n+1}{n}\right| = \dfrac{1}{5} \lim\limits_{n\to\infty} \dfrac{1+1/n}{1} = \dfrac{1}{5}(1) = \dfrac{1}{5} < 1$, so the series $\displaystyle\sum_{n=1}^{\infty} \dfrac{n}{5^n}$ is absolutely convergent by the Ratio Test.

5. $\lim_{n\to\infty} \left|\dfrac{a_{n+1}}{a_n}\right| = \lim_{n\to\infty} \left|\dfrac{(-1)^n 3^{n+1}}{2^{n+1}(n+1)^3} \cdot \dfrac{2^n n^3}{(-1)^{n-1} 3^n}\right| = \lim_{n\to\infty} \left|\left(-\dfrac{3}{2}\right) \dfrac{n^3}{(n+1)^3}\right| = \dfrac{3}{2} \lim_{n\to\infty} \dfrac{1}{(1+1/n)^3} = \dfrac{3}{2}(1) = \dfrac{3}{2} > 1$,

so the series $\sum_{n=1}^{\infty} (-1)^{n-1} \dfrac{3^n}{2^n n^3}$ is divergent by the Ratio Test.

7. $\lim_{k\to\infty} \left|\dfrac{a_{k+1}}{a_k}\right| = \lim_{k\to\infty} \left|\dfrac{1}{(k+1)!} \cdot \dfrac{k!}{1}\right| = \lim_{k\to\infty} \dfrac{1}{k+1} = 0 < 1$, so the series $\sum_{k=1}^{\infty} \dfrac{1}{k!}$ is absolutely convergent by the Ratio Test.

Since the terms of this series are positive, absolute convergence is the same as convergence.

9. $\lim_{n\to\infty} \left|\dfrac{a_{n+1}}{a_n}\right| = \lim_{n\to\infty} \left[\dfrac{10^{n+1}}{(n+2)\, 4^{2n+3}} \cdot \dfrac{(n+1)\, 4^{2n+1}}{10^n}\right] = \lim_{n\to\infty} \left(\dfrac{10}{4^2} \cdot \dfrac{n+1}{n+2}\right) = \dfrac{5}{8} < 1$, so the series $\sum_{n=1}^{\infty} \dfrac{10^n}{(n+1) 4^{2n+1}}$

is absolutely convergent by the Ratio Test. Since the terms of this series are positive, absolute convergence is the same as convergence.

11. $\lim_{n\to\infty} \left|\dfrac{a_{n+1}}{a_n}\right| = \lim_{n\to\infty} \left|\dfrac{(n+1)\pi^{n+1}}{(-3)^n} \cdot \dfrac{(-3)^{n-1}}{n\pi^n}\right| = \lim_{n\to\infty} \left|\dfrac{\pi}{-3} \cdot \dfrac{n+1}{n}\right| = \dfrac{\pi}{3} \lim_{n\to\infty} \dfrac{1+1/n}{1} = \dfrac{\pi}{3}(1) = \dfrac{\pi}{3} > 1$, so the

series $\sum_{n=1}^{\infty} \dfrac{n\pi^n}{(-3)^{n-1}}$ diverges by the Ratio Test. *Or:* Since $\lim_{n\to\infty} |a_n| = \infty$, the series diverges by the Test for Divergence.

13. $\lim_{n\to\infty} \left|\dfrac{a_{n+1}}{a_n}\right| = \lim_{n\to\infty} \left|\dfrac{\cos[(n+1)\pi/3]}{(n+1)!} \cdot \dfrac{n!}{\cos(n\pi/3)}\right| = \lim_{n\to\infty} \left|\dfrac{\cos[(n+1)\pi/3]}{(n+1)\cos(n\pi/3)}\right| = \lim_{n\to\infty} \dfrac{c}{n+1} = 0 < 1$ (where

$0 < c \le 2$ for all positive integers n), so the series $\sum_{n=1}^{\infty} \dfrac{\cos(n\pi/3)}{n!}$ is absolutely convergent by the Ratio Test.

15. $\lim_{n\to\infty} \left|\dfrac{a_{n+1}}{a_n}\right| = \lim_{n\to\infty} \left|\dfrac{(n+1)^{100} 100^{n+1}}{(n+1)!} \cdot \dfrac{n!}{n^{100} 100^n}\right| = \lim_{n\to\infty} \dfrac{100}{n+1} \left(\dfrac{n+1}{n}\right)^{100} = \lim_{n\to\infty} \dfrac{100}{n+1} \left(1 + \dfrac{1}{n}\right)^{100}$

$= 0 \cdot 1 = 0 < 1$

so the series $\sum_{n=1}^{\infty} \dfrac{n^{100} 100^n}{n!}$ is absolutely convergent by the Ratio Test.

17. $\lim_{n\to\infty} \left|\dfrac{a_{n+1}}{a_n}\right| = \lim_{n\to\infty} \left|\dfrac{(-1)^n (n+1)!}{1\cdot 3\cdot 5\cdots(2n-1)(2n+1)} \cdot \dfrac{1\cdot 3\cdot 5\cdots(2n-1)}{(-1)^{n-1} n!}\right| = \lim_{n\to\infty} \dfrac{n+1}{2n+1}$

$= \lim_{n\to\infty} \dfrac{1+1/n}{2+1/n} = \dfrac{1}{2} < 1$,

so the series $1 - \dfrac{2!}{1\cdot 3} + \dfrac{3!}{1\cdot 3\cdot 5} - \dfrac{4!}{1\cdot 3\cdot 5\cdot 7} + \cdots + (-1)^{n-1}\dfrac{n!}{1\cdot 3\cdot 5\cdots(2n-1)} + \cdots$ is absolutely convergent by the Ratio Test.

19. $\lim_{n\to\infty} \left|\dfrac{a_{n+1}}{a_n}\right| = \lim_{n\to\infty} \left|\dfrac{2\cdot 4\cdot 6\cdots(2n)(2n+2)}{(n+1)!} \cdot \dfrac{n!}{2\cdot 4\cdot 6\cdots(2n)}\right| = \lim_{n\to\infty} \dfrac{2n+2}{n+1} = \lim_{n\to\infty} \dfrac{2(n+1)}{n+1} = 2 > 1$, so

the series $\sum_{n=1}^{\infty} \dfrac{2\cdot 4\cdot 6\cdots(2n)}{n!}$ diverges by the Ratio Test.

21. $\lim_{n\to\infty} \sqrt[n]{|a_n|} = \lim_{n\to\infty} \dfrac{n^2+1}{2n^2+1} = \lim_{n\to\infty} \dfrac{1+1/n^2}{2+1/n^2} = \dfrac{1}{2} < 1$, so the series $\sum_{n=1}^{\infty} \left(\dfrac{n^2+1}{2n^2+1}\right)^n$ is absolutely convergent by the Root Test.

23. $\lim\limits_{n\to\infty} \sqrt[n]{|a_n|} = \lim\limits_{n\to\infty} \sqrt[n]{\left|\dfrac{(-1)^{n-1}}{(\ln n)^n}\right|} = \lim\limits_{n\to\infty} \dfrac{1}{\ln n} = 0 < 1$, so the series $\sum\limits_{n=2}^{\infty} \dfrac{(-1)^{n-1}}{(\ln n)^n}$ is absolutely convergent by the Root Test.

25. $\lim\limits_{n\to\infty} \sqrt[n]{|a_n|} = \lim\limits_{n\to\infty} \sqrt[n]{\left(1+\dfrac{1}{n}\right)^{n^2}} = \lim\limits_{n\to\infty} \left(1+\dfrac{1}{n}\right)^n = e > 1$ [by Equation 6.4.9 (or 6.4*.9)], so the series

$\sum\limits_{n=1}^{\infty} \left(1+\dfrac{1}{n}\right)^{n^2}$ diverges by the Root Test.

27. $\sum\limits_{n=2}^{\infty} a_n = \sum\limits_{n=2}^{\infty} \dfrac{(-1)^n \ln n}{n} = \sum\limits_{n=2}^{\infty} (-1)^n b_n$. Now $b_n = \dfrac{\ln n}{n} > 0$ for $n \ge 2$, and $\{b_n\}$ is decreasing for $n \ge 3$ since

$\left(\dfrac{\ln x}{x}\right)' = \dfrac{x \cdot \frac{1}{x} - \ln x \cdot 1}{x^2} = \dfrac{1 - \ln x}{x^2} < 0$ when $\ln x > 1$ or $x > e \approx 2.7$. Also, $\lim\limits_{n\to\infty} b_n = \lim\limits_{n\to\infty} \dfrac{\ln n}{n} \stackrel{H}{=} \lim\limits_{n\to\infty} \dfrac{1/n}{1} = 0$,

so the series $\sum\limits_{n=2}^{\infty} \dfrac{(-1)^n \ln n}{n}$ converges by the Alternating Series Test. To determine absolute convergence, note that

$\left|\dfrac{(-1)^n \ln n}{n}\right| = \dfrac{\ln n}{n} > \dfrac{1}{n}$ for $n \ge 3$, so $\sum\limits_{n=2}^{\infty} \left|\dfrac{(-1)^n \ln n}{n}\right|$ is divergent by direct comparison with $\sum\limits_{n=2}^{\infty} \dfrac{1}{n}$, which is divergent.

Hence, $\sum\limits_{n=2}^{\infty} \dfrac{(-1)^n \ln n}{n}$ is conditionally convergent.

29. $\lim\limits_{n\to\infty} \left|\dfrac{a_{n+1}}{a_n}\right| = \lim\limits_{n\to\infty} \left|\dfrac{(-9)^{n+1}}{(n+1)10^{n+2}} \cdot \dfrac{n 10^{n+1}}{(-9)^n}\right| = \lim\limits_{n\to\infty} \left|\dfrac{(-9)n}{10(n+1)}\right| = \dfrac{9}{10} \lim\limits_{n\to\infty} \dfrac{1}{1+1/n} = \dfrac{9}{10}(1) = \dfrac{9}{10} < 1$, so the

series $\sum\limits_{n=1}^{\infty} \dfrac{(-9)^n}{n 10^{n+1}}$ is absolutely convergent by the Ratio Test.

31. $\lim\limits_{n\to\infty} \sqrt[n]{|a_n|} = \lim\limits_{n\to\infty} \sqrt[n]{\left|\left(\dfrac{n}{\ln n}\right)^n\right|} = \lim\limits_{n\to\infty} \dfrac{n}{\ln n} = \lim\limits_{x\to\infty} \dfrac{x}{\ln x} \stackrel{H}{=} \lim\limits_{x\to\infty} \dfrac{1}{1/x} = \lim\limits_{x\to\infty} x = \infty$, so the series $\sum\limits_{n=2}^{\infty} \left(\dfrac{n}{\ln n}\right)^n$

diverges by the Root Test.

33. $\left|\dfrac{(-1)^n \arctan n}{n^2}\right| < \dfrac{\pi/2}{n^2}$, so since $\sum\limits_{n=1}^{\infty} \dfrac{\pi/2}{n^2} = \dfrac{\pi}{2} \sum\limits_{n=1}^{\infty} \dfrac{1}{n^2}$ converges ($p = 2 > 1$), the given series $\sum\limits_{n=1}^{\infty} \dfrac{(-1)^n \arctan n}{n^2}$

converges absolutely by the Direct Comparison Test.

35. By the recursive definition, $\lim\limits_{n\to\infty} \left|\dfrac{a_{n+1}}{a_n}\right| = \lim\limits_{n\to\infty} \left|\dfrac{5n+1}{4n+3}\right| = \dfrac{5}{4} > 1$, so the series diverges by the Ratio Test.

37. The series $\sum\limits_{n=1}^{\infty} \dfrac{b_n^n \cos n\pi}{n} = \sum\limits_{n=1}^{\infty} (-1)^n \dfrac{b_n^n}{n}$, where $b_n > 0$ for $n \ge 1$ and $\lim\limits_{n\to\infty} b_n = \dfrac{1}{2}$.

$\lim\limits_{n\to\infty} \left|\dfrac{a_{n+1}}{a_n}\right| = \lim\limits_{n\to\infty} \left|\dfrac{(-1)^{n+1} b_n^{n+1}}{n+1} \cdot \dfrac{n}{(-1)^n b_n^n}\right| = \lim\limits_{n\to\infty} b_n \dfrac{n}{n+1} = \dfrac{1}{2}(1) = \dfrac{1}{2} < 1$, so the series $\sum\limits_{n=1}^{\infty} \dfrac{b_n^n \cos n\pi}{n}$ is

absolutely convergent by the Ratio Test.

39. (a) $\lim\limits_{n\to\infty} \left| \dfrac{1/(n+1)^3}{1/n^3} \right| = \lim\limits_{n\to\infty} \dfrac{n^3}{(n+1)^3} = \lim\limits_{n\to\infty} \dfrac{1}{(1+1/n)^3} = 1$. Inconclusive for $\sum\limits_{n=1}^{\infty} \dfrac{1}{n^3}$.

(b) $\lim\limits_{n\to\infty} \left| \dfrac{(n+1)}{2^{n+1}} \cdot \dfrac{2^n}{n} \right| = \lim\limits_{n\to\infty} \dfrac{n+1}{2n} = \lim\limits_{n\to\infty} \left(\dfrac{1}{2} + \dfrac{1}{2n} \right) = \dfrac{1}{2}$. Conclusive (convergent) for $\sum\limits_{n=1}^{\infty} \dfrac{n}{2^n}$.

(c) $\lim\limits_{n\to\infty} \left| \dfrac{(-3)^n}{\sqrt{n+1}} \cdot \dfrac{\sqrt{n}}{(-3)^{n-1}} \right| = 3 \lim\limits_{n\to\infty} \sqrt{\dfrac{n}{n+1}} = 3 \lim\limits_{n\to\infty} \sqrt{\dfrac{1}{1+1/n}} = 3$. Conclusive (divergent) for $\sum\limits_{n=1}^{\infty} \dfrac{(-3)^{n-1}}{\sqrt{n}}$.

(d) $\lim\limits_{n\to\infty} \left| \dfrac{\sqrt{n+1}}{1+(n+1)^2} \cdot \dfrac{1+n^2}{\sqrt{n}} \right| = \lim\limits_{n\to\infty} \left[\sqrt{1 + \dfrac{1}{n}} \cdot \dfrac{1/n^2 + 1}{1/n^2 + (1+1/n)^2} \right] = 1$. Inconclusive for $\sum\limits_{n=1}^{\infty} \dfrac{\sqrt{n}}{1+n^2}$.

41. (a) $\lim\limits_{n\to\infty} \left| \dfrac{a_{n+1}}{a_n} \right| = \lim\limits_{n\to\infty} \left| \dfrac{x^{n+1}}{(n+1)!} \cdot \dfrac{n!}{x^n} \right| = \lim\limits_{n\to\infty} \left| \dfrac{x}{n+1} \right| = |x| \lim\limits_{n\to\infty} \dfrac{1}{n+1} = |x| \cdot 0 = 0 < 1$, so by the Ratio Test the series $\sum\limits_{n=0}^{\infty} \dfrac{x^n}{n!}$ converges for all x.

(b) Since the series of part (a) always converges, we must have $\lim\limits_{n\to\infty} \dfrac{x^n}{n!} = 0$ by Theorem 11.2.6.

43. (a) $s_5 = \sum\limits_{n=1}^{5} \dfrac{1}{n2^n} = \dfrac{1}{2} + \dfrac{1}{8} + \dfrac{1}{24} + \dfrac{1}{64} + \dfrac{1}{160} = \dfrac{661}{960} \approx 0.68854$. Now the ratios

$r_n = \dfrac{a_{n+1}}{a_n} = \dfrac{n2^n}{(n+1)2^{n+1}} = \dfrac{n}{2(n+1)}$ form an increasing sequence, since

$r_{n+1} - r_n = \dfrac{n+1}{2(n+2)} - \dfrac{n}{2(n+1)} = \dfrac{(n+1)^2 - n(n+2)}{2(n+1)(n+2)} = \dfrac{1}{2(n+1)(n+2)} > 0$. So by Exercise 42(b), the error in using s_5 is $R_5 \leq \dfrac{a_6}{1 - \lim\limits_{n\to\infty} r_n} = \dfrac{1/(6 \cdot 2^6)}{1 - 1/2} = \dfrac{1}{192} \approx 0.00521$.

(b) The error in using s_n as an approximation to the sum is $R_n = \dfrac{a_{n+1}}{1 - \frac{1}{2}} = \dfrac{2}{(n+1)2^{n+1}}$. We want $R_n < 0.00005 \Leftrightarrow \dfrac{1}{(n+1)2^n} < 0.00005 \Leftrightarrow (n+1)2^n > 20{,}000$. To find such an n we can use trial and error or a graph. We calculate $(11+1)2^{11} = 24{,}576$, so $s_{11} = \sum\limits_{n=1}^{11} \dfrac{1}{n2^n} \approx 0.693109$ is within 0.00005 of the actual sum.

45. (i) Following the hint, we get that $|a_n| < r^n$ for $n \geq N$, and so since the geometric series $\sum_{n=1}^{\infty} r^n$ converges $[0 < r < 1]$, the series $\sum_{n=N}^{\infty} |a_n|$ converges as well by the Direct Comparison Test, and hence so does $\sum_{n=1}^{\infty} |a_n|$, so $\sum_{n=1}^{\infty} a_n$ is absolutely convergent.

(ii) If $\lim\limits_{n\to\infty} \sqrt[n]{|a_n|} = L > 1$, then there is an integer N such that $\sqrt[n]{|a_n|} > 1$ for all $n \geq N$, so $|a_n| > 1$ for $n \geq N$. Thus, $\lim\limits_{n\to\infty} a_n \neq 0$, so $\sum_{n=1}^{\infty} a_n$ diverges by the Test for Divergence.

(iii) Consider $\sum\limits_{n=1}^{\infty} \dfrac{1}{n}$ [diverges] and $\sum\limits_{n=1}^{\infty} \dfrac{1}{n^2}$ [converges]. For each sum, $\lim\limits_{n\to\infty} \sqrt[n]{|a_n|} = 1$, so the Root Test is inconclusive.

11.7 Strategy for Testing Series

1. (a) $\sum_{n=1}^{\infty} \frac{1}{5^n} = \sum_{n=1}^{\infty} \left(\frac{1}{5}\right)^n$ is a geometric series with ratio $r = \frac{1}{5}$. Since $|r| = \frac{1}{5} < 1$, the series converges.

(b) $\frac{1}{5^n + n} < \frac{1}{5^n}$ for $n \geq 1$, so $\sum_{n=1}^{\infty} \frac{1}{5^n + n}$ converges by direct comparison with $\sum_{n=1}^{\infty} \frac{1}{5^n}$, which converges because it is a geometric series with $|r| = \frac{1}{5} < 1$.

3. (a) $\lim_{n \to \infty} \left|\frac{a_{n+1}}{a_n}\right| = \lim_{n \to \infty} \left|\frac{n+1}{3^{n+1}} \cdot \frac{3^n}{n}\right| = \lim_{n \to \infty} \frac{n+1}{3n} = \lim_{n \to \infty} \left(\frac{n}{3n} + \frac{1}{3n}\right) = \lim_{n \to \infty} \left(\frac{1}{3} + \frac{1}{3n}\right) = \frac{1}{3} < 1$,

so the series $\sum_{n=1}^{\infty} \frac{n}{3^n}$ is absolutely convergent (and therefore convergent) by the Ratio Test.

(b) $\lim_{n \to \infty} \frac{3^n}{n} \stackrel{H}{=} \lim_{n \to \infty} \frac{3^n \ln 3}{1} = \infty$, so the series $\sum_{n=1}^{\infty} \frac{3^n}{n}$ diverges by the Test for Divergence.

5. (a) Use the Limit Comparison Test with $a_n = \frac{n}{n^2 + 1}$ and $b_n = \frac{1}{n}$.

$\lim_{n \to \infty} \frac{a_n}{b_n} = \lim_{n \to \infty} \frac{n^2}{n^2 + 1} = \lim_{n \to \infty} \frac{1}{1 + 1/n^2} = 1 > 0$. Since $\sum_{n=1}^{\infty} \frac{1}{n}$ is the divergent harmonic series, the series

$\sum_{n=1}^{\infty} \frac{n^2}{n^2 + 1}$ also diverges.

(b) $\lim_{n \to \infty} \sqrt[n]{|a_n|} = \lim_{n \to \infty} \sqrt[n]{\left(\frac{n}{n^2 + 1}\right)^n} = \lim_{n \to \infty} \frac{n}{n^2 + 1} = \lim_{n \to \infty} \frac{1/n}{1 + 1/n^2} = \frac{0}{1} = 0 < 1$, so the series $\sum_{n=1}^{\infty} \left(\frac{n}{n^2 + 1}\right)^n$

converges by the Root Test.

7. (a) Since $n! > n^2$ for $n \geq 4$, we have $\frac{1}{n + n!} < \frac{1}{n + n^2} < \frac{1}{n^2}$ for $n \geq 4$. Thus, $\sum_{n=1}^{\infty} \frac{1}{n + n!}$ converges by direct comparison

with $\sum_{n=1}^{\infty} \frac{1}{n^2}$, which converges because it is a p-series with $p = 2 > 1$.

(b) $\frac{1}{n} + \frac{1}{n!} > \frac{1}{n}$ for $n \geq 1$, so $\sum_{n=1}^{\infty} \left(\frac{1}{n} + \frac{1}{n!}\right)$ diverges by direct comparison with $\sum_{n=1}^{\infty} \frac{1}{n}$, which diverges because it is a p-series with $p = 1 \leq 1$.

9. Use the Limit Comparison Test with $a_n = \frac{n^2 - 1}{n^3 + 1}$ and $b_n = \frac{1}{n}$:

$\lim_{n \to \infty} \frac{a_n}{b_n} = \lim_{n \to \infty} \frac{(n^2 - 1)n}{n^3 + 1} = \lim_{n \to \infty} \frac{n^3 - n}{n^3 + 1} = \lim_{n \to \infty} \frac{1 - 1/n^2}{1 + 1/n^3} = 1 > 0$. Since $\sum_{n=1}^{\infty} \frac{1}{n}$ is the divergent harmonic series, the

series $\sum_{n=1}^{\infty} \frac{n^2 - 1}{n^3 + 1}$ also diverges.

11. $\sum_{n=1}^{\infty} (-1)^n \dfrac{n^2-1}{n^3+1} = \sum_{n=1}^{\infty} (-1)^n b_n$. Now $b_n = \dfrac{n^2-1}{n^3+1} > 0$ for $n \geq 2$, $\{b_n\}$ is decreasing for $n \geq 2$, and $\lim_{n\to\infty} b_n = 0$, so the series $\sum_{n=1}^{\infty} (-1)^n \dfrac{n^2-1}{n^3+1}$ converges by the Alternating Series Test. By Exercise 9, $\sum_{n=1}^{\infty} \dfrac{n^2-1}{n^3+1}$ diverges, so the series $\sum_{n=1}^{\infty} (-1)^n \dfrac{n^2-1}{n^3+1}$ is conditionally convergent.

13. $\lim_{x\to\infty} \dfrac{e^x}{x^2} \stackrel{H}{=} \lim_{x\to\infty} \dfrac{e^x}{2x} \stackrel{H}{=} \lim_{x\to\infty} \dfrac{e^x}{2} = \infty$, so $\lim_{n\to\infty} \dfrac{e^n}{n^2} = \infty$. Thus, the series $\sum_{n=1}^{\infty} \dfrac{e^n}{n^2}$ diverges by the Test for Divergence.

15. Let $f(x) = \dfrac{1}{x\sqrt{\ln x}}$. Then f is positive, continuous, and decreasing on $[2, \infty)$, so we can apply the Integral Test.

Since $\displaystyle\int \dfrac{1}{x\sqrt{\ln x}}\, dx$ $\begin{bmatrix} u = \ln x, \\ du = dx/x \end{bmatrix}$ $= \displaystyle\int u^{-1/2}\, du = 2u^{1/2} + C = 2\sqrt{\ln x} + C$, we find

$\displaystyle\int_2^\infty \dfrac{dx}{x\sqrt{\ln x}} = \lim_{t\to\infty} \int_2^t \dfrac{dx}{x\sqrt{\ln x}} = \lim_{t\to\infty} \left[2\sqrt{\ln x}\right]_2^t = \lim_{t\to\infty} \left(2\sqrt{\ln t} - 2\sqrt{\ln 2}\right) = \infty$. Since the integral diverges, the given series $\displaystyle\sum_{n=2}^{\infty} \dfrac{1}{n\sqrt{\ln n}}$ diverges.

17. $\lim_{n\to\infty} \left|\dfrac{a_{n+1}}{a_n}\right| = \lim_{n\to\infty} \left|\dfrac{\pi^{2n+2}}{(2n+2)!} \cdot \dfrac{(2n)!}{\pi^{2n}}\right| = \lim_{n\to\infty} \dfrac{\pi^2}{(2n+2)(2n+1)} = 0 < 1$, so the series $\sum_{n=0}^{\infty} (-1)^n \dfrac{\pi^{2n}}{(2n)!}$ is absolutely convergent (and therefore convergent) by the Ratio Test.

19. $\sum_{n=1}^{\infty} \left(\dfrac{1}{n^3} + \dfrac{1}{3^n}\right) = \sum_{n=1}^{\infty} \dfrac{1}{n^3} + \sum_{n=1}^{\infty} \left(\dfrac{1}{3}\right)^n$. The first series converges since it is a p-series with $p = 3 > 1$ and the second series converges since it is geometric with $|r| = \tfrac{1}{3} < 1$. The sum of two convergent series is convergent.

21. $\lim_{n\to\infty} \left|\dfrac{a_{n+1}}{a_n}\right| = \lim_{n\to\infty} \left|\dfrac{3^{n+1}(n+1)^2}{(n+1)!} \cdot \dfrac{n!}{3^n n^2}\right| = \lim_{n\to\infty} \dfrac{3(n+1)^2}{(n+1)n^2} = 3\lim_{n\to\infty} \dfrac{n+1}{n^2} = 0 < 1$, so the series $\sum_{n=1}^{\infty} \dfrac{3^n n^2}{n!}$ converges by the Ratio Test.

23. $a_k = \dfrac{2^{k-1} 3^{k+1}}{k^k} = \dfrac{2^k 2^{-1} 3^k 3^1}{k^k} = \dfrac{3}{2}\left(\dfrac{2\cdot 3}{k}\right)^k$. By the Root Test, $\lim_{k\to\infty} \sqrt[k]{\left(\dfrac{6}{k}\right)^k} = \lim_{k\to\infty} \dfrac{6}{k} = 0 < 1$, so the series $\sum_{k=1}^{\infty} \left(\dfrac{6}{k}\right)^k$ converges. It follows from Theorem 8(i) in Section 11.2 that the given series, $\sum_{k=1}^{\infty} \dfrac{2^{k-1} 3^{k+1}}{k^k} = \sum_{k=1}^{\infty} \dfrac{3}{2}\left(\dfrac{6}{k}\right)^k$, also converges.

25. $\lim_{n\to\infty} \left|\dfrac{a_{n+1}}{a_n}\right| = \lim_{n\to\infty} \left|\dfrac{1\cdot 3\cdot 5\cdots (2n-1)(2n+1)}{2\cdot 5\cdot 8\cdots (3n-1)(3n+2)} \cdot \dfrac{2\cdot 5\cdot 8\cdots (3n-1)}{1\cdot 3\cdot 5\cdots (2n-1)}\right| = \lim_{n\to\infty} \dfrac{2n+1}{3n+2}$

$= \lim_{n\to\infty} \dfrac{2 + 1/n}{3 + 2/n} = \dfrac{2}{3} < 1$,

so the series $\sum_{n=1}^{\infty} \dfrac{1\cdot 3\cdot 5\cdots (2n-1)}{2\cdot 5\cdot 8\cdots (3n-1)}$ converges by the Ratio Test.

27. Let $f(x) = \dfrac{\ln x}{\sqrt{x}}$. Then $f'(x) = \dfrac{2 - \ln x}{2x^{3/2}} < 0$ when $\ln x > 2$ or $x > e^2$, so $\dfrac{\ln n}{\sqrt{n}}$ is decreasing for $n > e^2$.

By l'Hospital's Rule, $\lim\limits_{n\to\infty} \dfrac{\ln n}{\sqrt{n}} = \lim\limits_{n\to\infty} \dfrac{1/n}{1/(2\sqrt{n})} = \lim\limits_{n\to\infty} \dfrac{2}{\sqrt{n}} = 0$, so the series $\sum\limits_{n=1}^{\infty} (-1)^n \dfrac{\ln n}{\sqrt{n}}$ converges by the Alternating Series Test.

29. $\lim\limits_{n\to\infty} |a_n| = \lim\limits_{n\to\infty} |(-1)^n \cos(1/n^2)| = \lim\limits_{n\to\infty} |\cos(1/n^2)| = \cos 0 = 1$, so the series $\sum\limits_{n=1}^{\infty} (-1)^n \cos(1/n^2)$ diverges by the Test for Divergence.

31. Using the Limit Comparison Test with $a_n = \tan\left(\dfrac{1}{n}\right)$ and $b_n = \dfrac{1}{n}$, we have

$\lim\limits_{n\to\infty} \dfrac{a_n}{b_n} = \lim\limits_{n\to\infty} \dfrac{\tan(1/n)}{1/n} = \lim\limits_{x\to\infty} \dfrac{\tan(1/x)}{1/x} \stackrel{H}{=} \lim\limits_{x\to\infty} \dfrac{\sec^2(1/x) \cdot (-1/x^2)}{-1/x^2} = \lim\limits_{x\to\infty} \sec^2(1/x) = 1^2 = 1 > 0$. Since

$\sum\limits_{n=1}^{\infty} b_n$ is the divergent harmonic series, $\sum\limits_{n=1}^{\infty} a_n$ is also divergent.

33. $\dfrac{4 - \cos n}{\sqrt{n}} \geq \dfrac{4 - 1}{\sqrt{n}} = \dfrac{3}{n^{1/2}}$ for $n \geq 1$, so $\sum\limits_{n=1}^{\infty} \dfrac{4 - \cos n}{\sqrt{n}}$ diverges by direct comparison with $\sum\limits_{n=1}^{\infty} \dfrac{3}{n^{1/2}}$, which diverges because it is a constant multiple of a p-series with $p = \tfrac{1}{2} \leq 1$.

35. Use the Ratio Test. $\lim\limits_{n\to\infty} \left|\dfrac{a_{n+1}}{a_n}\right| = \lim\limits_{n\to\infty} \left|\dfrac{(n+1)!}{e^{(n+1)^2}} \cdot \dfrac{e^{n^2}}{n!}\right| = \lim\limits_{n\to\infty} \dfrac{(n+1)n! \cdot e^{n^2}}{e^{n^2 + 2n + 1} n!} = \lim\limits_{n\to\infty} \dfrac{n+1}{e^{2n+1}} = 0 < 1$, so $\sum\limits_{n=1}^{\infty} \dfrac{n!}{e^{n^2}}$ converges.

37. $\displaystyle\int_2^{\infty} \dfrac{\ln x}{x^2} dx = \lim\limits_{t\to\infty} \left[-\dfrac{\ln x}{x} - \dfrac{1}{x}\right]_1^t$ [using integration by parts] $\stackrel{H}{=} 1$. So $\sum\limits_{n=1}^{\infty} \dfrac{\ln n}{n^2}$ converges by the Integral Test, and since

$\dfrac{k \ln k}{(k+1)^3} < \dfrac{k \ln k}{k^3} = \dfrac{\ln k}{k^2}$, the given series $\sum\limits_{k=1}^{\infty} \dfrac{k \ln k}{(k+1)^3}$ converges by the Direct Comparison Test.

39. $\sum\limits_{n=1}^{\infty} a_n = \sum\limits_{n=1}^{\infty} (-1)^n \dfrac{1}{\cosh n} = \sum\limits_{n=1}^{\infty} (-1)^n b_n$. Now $b_n = \dfrac{1}{\cosh n} > 0$, $\{b_n\}$ is decreasing, and $\lim\limits_{n\to\infty} b_n = 0$, so the series converges by the Alternating Series Test.

Or: Write $\dfrac{1}{\cosh n} = \dfrac{2}{e^n + e^{-n}} < \dfrac{2}{e^n}$ and $\sum\limits_{n=1}^{\infty} \dfrac{1}{e^n}$ is a convergent geometric series, so $\sum\limits_{n=1}^{\infty} \dfrac{1}{\cosh n}$ is convergent by the

Direct Comparison Test. So $\sum\limits_{n=1}^{\infty} (-1)^n \dfrac{1}{\cosh n}$ is absolutely convergent and therefore convergent.

41. $\lim\limits_{k\to\infty} a_k = \lim\limits_{k\to\infty} \dfrac{5^k}{3^k + 4^k} =$ [divide by 4^k] $\lim\limits_{k\to\infty} \dfrac{(5/4)^k}{(3/4)^k + 1} = \infty$ since $\lim\limits_{k\to\infty} \left(\dfrac{3}{4}\right)^k = 0$ and $\lim\limits_{k\to\infty} \left(\dfrac{5}{4}\right)^k = \infty$.

Thus, $\sum\limits_{k=1}^{\infty} \dfrac{5^k}{3^k + 4^k}$ diverges by the Test for Divergence.

43. $\lim\limits_{n \to \infty} \sqrt[n]{|a_n|} = \lim\limits_{n \to \infty} \left(\dfrac{n}{n+1}\right)^{n^2/n} = \lim\limits_{n \to \infty} \dfrac{1}{[(n+1)/n]^n} = \dfrac{1}{\lim\limits_{n \to \infty}(1+1/n)^n} = \dfrac{1}{e} < 1$, so the series $\sum\limits_{n=1}^{\infty}\left(\dfrac{n}{n+1}\right)^{n^2}$

converges by the Root Test.

45. $a_n = \dfrac{1}{n^{1+1/n}} = \dfrac{1}{n \cdot n^{1/n}}$, so let $b_n = \dfrac{1}{n}$ and use the Limit Comparison Test. $\lim\limits_{n \to \infty} \dfrac{a_n}{b_n} = \lim\limits_{n \to \infty} \dfrac{1}{n^{1/n}} = 1 > 0$

[see Exercise 6.8.63], so the series $\sum\limits_{n=1}^{\infty} \dfrac{1}{n^{1+1/n}}$ diverges by comparison with the divergent harmonic series.

47. $\lim\limits_{n \to \infty} \sqrt[n]{|a_n|} = \lim\limits_{n \to \infty}(2^{1/n} - 1) = 1 - 1 = 0 < 1$, so the series $\sum\limits_{n=1}^{\infty}\left(\sqrt[n]{2} - 1\right)^n$ converges by the Root Test.

11.8 Power Series

1. A power series is a series of the form $\sum_{n=0}^{\infty} c_n x^n = c_0 + c_1 x + c_2 x^2 + c_3 x^3 + \cdots$, where x is a variable and the c_n's are constants called the coefficients of the series.

More generally, a series of the form $\sum_{n=0}^{\infty} c_n (x-a)^n = c_0 + c_1(x-a) + c_2(x-a)^2 + \cdots$ is called a power series in $(x-a)$ or a power series centered at a or a power series about a, where a is a constant.

3. If $a_n = \dfrac{x^n}{n}$, then $\lim\limits_{n \to \infty}\left|\dfrac{a_{n+1}}{a_n}\right| = \lim\limits_{n \to \infty}\left|\dfrac{x^{n+1}}{n+1} \cdot \dfrac{n}{x^n}\right| = \lim\limits_{n \to \infty}\left|\dfrac{nx}{n+1}\right| = \lim\limits_{n \to \infty}\left(\dfrac{1}{1+1/n}|x|\right) = |x|$. By the Ratio Test,

the series $\sum\limits_{n=1}^{\infty} \dfrac{x^n}{n}$ converges when $|x| < 1$, so the radius of convergence is $R = 1$. Now we'll check the endpoints, that is,

$x = \pm 1$. When $x = 1$, the series $\sum\limits_{n=1}^{\infty} \dfrac{1}{n}$ diverges since it is the harmonic series. When $x = -1$, the series $\sum\limits_{n=1}^{\infty} \dfrac{(-1)^n}{n}$

converges by the Alternating Series Test. Thus, the interval of convergence is $[-1, 1)$.

5. If $a_n = \sqrt{n}\, x^n$, then $\lim\limits_{n \to \infty}\left|\dfrac{a_{n+1}}{a_n}\right| = \lim\limits_{n \to \infty}\left|\dfrac{\sqrt{n+1}\, x^{n+1}}{\sqrt{n}\, x^n}\right| = \lim\limits_{n \to \infty}\sqrt{\dfrac{n+1}{n}}\,|x| = \lim\limits_{n \to \infty}\sqrt{1+\dfrac{1}{n}}\,|x| = |x|$. By the Ratio

Test, the series $\sum\limits_{n=1}^{\infty} \sqrt{n}\, x^n$ converges when $|x| < 1$, so $R = 1$. When $x = \pm 1$, both series $\sum\limits_{n=1}^{\infty} \sqrt{n}\,(\pm 1)^n$ diverge by the Test

for Divergence since $\lim\limits_{n \to \infty}\left|\sqrt{n}\,(\pm 1)^n\right| = \infty$. Thus, the interval of convergence is $(-1, 1)$.

7. If $a_n = \dfrac{n}{5^n} x^n$, then $\lim\limits_{n \to \infty}\left|\dfrac{a_{n+1}}{a_n}\right| = \lim\limits_{n \to \infty}\left|\dfrac{(n+1)x^{n+1}}{5^{n+1}} \cdot \dfrac{5^n}{nx^n}\right| = \lim\limits_{n \to \infty}\left|\dfrac{n+1}{5n}x\right| = \lim\limits_{n \to \infty}\left(\dfrac{1}{5}+\dfrac{1}{5n}\right)|x| = \dfrac{|x|}{5}$. By the

Ratio Test, the series $\sum\limits_{n=1}^{\infty} \dfrac{n}{5^n} x^n$ converges when $\dfrac{|x|}{5} < 1 \Leftrightarrow |x| < 5$, so $R = 5$. When $x = \pm 5$, both series

$\sum\limits_{n=1}^{\infty} \dfrac{n(\pm 5)^n}{5^n} = \sum\limits_{n=1}^{\infty}(\pm 1)^n n$ diverge by the Test for Divergence since $\lim\limits_{n \to \infty}|(\pm 1)^n n| = \infty$. Thus, the interval of convergence

is $(-5, 5)$.

SECTION 11.8   POWER SERIES   □   563

9. If $a_n = \dfrac{x^n}{n\,3^n}$, then $\lim\limits_{n\to\infty}\left|\dfrac{a_{n+1}}{a_n}\right| = \lim\limits_{n\to\infty}\left|\dfrac{x^{n+1}}{(n+1)\,3^{n+1}}\cdot\dfrac{n\,3^n}{x^n}\right| = \lim\limits_{n\to\infty}\left|\dfrac{n}{3(n+1)}\,x\right| = \lim\limits_{n\to\infty}\left(\dfrac{1}{3+3/n}\,|x|\right) = \dfrac{|x|}{3}$. By

the Ratio Test, the series $\sum\limits_{n=1}^{\infty}\dfrac{x^n}{n\,3^n}$ converges when $\dfrac{|x|}{3}<1$ ⇔ $|x|<3$, so $R=3$. When $x=3$, the series $\sum\limits_{n=1}^{\infty}\dfrac{1}{n}$ diverges

since it is the harmonic series. When $x=-3$, the series $\sum\limits_{n=1}^{\infty}\dfrac{(-1)^n}{n}$ converges by the Alternating Series Test. Thus, the

interval of convergence is $[-3,3)$.

11. If $a_n = \dfrac{x^n}{2n-1}$, then $\lim\limits_{n\to\infty}\left|\dfrac{a_{n+1}}{a_n}\right| = \lim\limits_{n\to\infty}\left|\dfrac{x^{n+1}}{2n+1}\cdot\dfrac{2n-1}{x^n}\right| = \lim\limits_{n\to\infty}\left(\dfrac{2n-1}{2n+1}\,|x|\right) = \lim\limits_{n\to\infty}\left(\dfrac{2-1/n}{2+1/n}\,|x|\right) = |x|$. By

the Ratio Test, the series $\sum\limits_{n=1}^{\infty}\dfrac{x^n}{2n-1}$ converges when $|x|<1$, so $R=1$. When $x=1$, the series $\sum\limits_{n=1}^{\infty}\dfrac{1}{2n-1}$ diverges by

direct comparison with $\sum\limits_{n=1}^{\infty}\dfrac{1}{2n}$ since $\dfrac{1}{2n-1}>\dfrac{1}{2n}$ and $\dfrac{1}{2}\sum\limits_{n=1}^{\infty}\dfrac{1}{n}$ diverges since it is a constant multiple of the harmonic

series. When $x=-1$, the series $\sum\limits_{n=1}^{\infty}\dfrac{(-1)^n}{2n-1}$ converges by the Alternating Series Test. Thus, the interval of convergence

is $[-1,1)$.

13. If $a_n = \dfrac{x^n}{n!}$, then $\lim\limits_{n\to\infty}\left|\dfrac{a_{n+1}}{a_n}\right| = \lim\limits_{n\to\infty}\left|\dfrac{x^{n+1}}{(n+1)!}\cdot\dfrac{n!}{x^n}\right| = \lim\limits_{n\to\infty}\left|\dfrac{x}{n+1}\right| = |x|\lim\limits_{n\to\infty}\dfrac{1}{n+1} = |x|\cdot 0 = 0 < 1$ for *all* real x.

So, by the Ratio Test, $R=\infty$ and $I=(-\infty,\infty)$.

15. If $a_n = \dfrac{x^n}{n^4\,4^n}$, then

$\lim\limits_{n\to\infty}\left|\dfrac{a_{n+1}}{a_n}\right| = \lim\limits_{n\to\infty}\left|\dfrac{x^{n+1}}{(n+1)^4\,4^{n+1}}\cdot\dfrac{n^4\,4^n}{x^n}\right| = \lim\limits_{n\to\infty}\left|\dfrac{n^4}{(n+1)^4}\cdot\dfrac{x}{4}\right| = \lim\limits_{n\to\infty}\left(\dfrac{n}{n+1}\right)^4\dfrac{|x|}{4} = 1^4\cdot\dfrac{|x|}{4} = \dfrac{|x|}{4}$. By the

Ratio Test, the series $\sum\limits_{n=1}^{\infty}\dfrac{x^n}{n^4\,4^n}$ converges when $\dfrac{|x|}{4}<1$ ⇔ $|x|<4$, so $R=4$. When $x=4$, the series $\sum\limits_{n=1}^{\infty}\dfrac{1}{n^4}$

converges since it is a p-series ($p=4>1$). When $x=-4$, the series $\sum\limits_{n=1}^{\infty}\dfrac{(-1)^n}{n^4}$ converges by the Alternating Series Test.

Thus, the interval of convergence is $[-4,4]$.

17. If $a_n = \dfrac{(-1)^n\,4^n}{\sqrt{n}}\,x^n$, then $\lim\limits_{n\to\infty}\left|\dfrac{a_{n+1}}{a_n}\right| = \lim\limits_{n\to\infty}\left|\dfrac{(-1)^{n+1}\,4^{n+1}\,x^{n+1}}{\sqrt{n+1}}\cdot\dfrac{\sqrt{n}}{(-1)^n\,4^n\,x^n}\right| = \lim\limits_{n\to\infty}\sqrt{\dfrac{n}{n+1}}\cdot 4\,|x| = 4\,|x|$.

By the Ratio Test, the series $\sum\limits_{n=1}^{\infty}\dfrac{(-1)^n\,4^n}{\sqrt{n}}\,x^n$ converges when $4\,|x|<1$ ⇔ $|x|<\tfrac{1}{4}$, so $R=\tfrac{1}{4}$. When $x=\tfrac{1}{4}$, the series

$\sum\limits_{n=1}^{\infty}\dfrac{(-1)^n}{\sqrt{n}}$ converges by the Alternating Series Test. When $x=-\tfrac{1}{4}$, the series $\sum\limits_{n=1}^{\infty}\dfrac{1}{\sqrt{n}}$ diverges since it is a p-series

$(p=\tfrac{1}{2}\le 1)$. Thus, the interval of convergence is $\left(-\tfrac{1}{4},\tfrac{1}{4}\right]$.

19. If $a_n = \dfrac{n}{2^n(n^2+1)} x^n$, then

$$\lim_{n\to\infty} \left|\dfrac{a_{n+1}}{a_n}\right| = \lim_{n\to\infty} \left|\dfrac{(n+1)x^{n+1}}{2^{n+1}(n^2+2n+2)} \cdot \dfrac{2^n(n^2+1)}{nx^n}\right| = \lim_{n\to\infty} \dfrac{n^3+n^2+n+1}{n^3+2n^2+2n} \cdot \dfrac{|x|}{2}$$

$$= \lim_{n\to\infty} \dfrac{1+1/n+1/n^2+1/n^3}{1+2/n+2/n^2} \cdot \dfrac{|x|}{2} = \dfrac{|x|}{2}$$

By the Ratio Test, the series $\sum_{n=1}^{\infty} \dfrac{n}{2^n(n^2+1)} x^n$ converges when $\dfrac{|x|}{2} < 1 \Leftrightarrow |x| < 2$, so $R = 2$. When $x = 2$, the series $\sum_{n=1}^{\infty} \dfrac{n}{n^2+1}$ diverges by the Limit Comparison Test with $b_n = \dfrac{1}{n}$. When $x = -2$, the series $\sum_{n=1}^{\infty} \dfrac{(-1)^n n}{n^2+1}$ converges by the Alternating Series Test. Thus, the interval of convergence is $[-2, 2)$.

21. If $a_n = \dfrac{(x-2)^n}{n^2+1}$, then $\lim_{n\to\infty} \left|\dfrac{a_{n+1}}{a_n}\right| = \lim_{n\to\infty} \left|\dfrac{(x-2)^{n+1}}{(n+1)^2+1} \cdot \dfrac{n^2+1}{(x-2)^n}\right| = |x-2| \lim_{n\to\infty} \dfrac{n^2+1}{(n+1)^2+1} = |x-2|$. By the

Ratio Test, the series $\sum_{n=0}^{\infty} \dfrac{(x-2)^n}{n^2+1}$ converges when $|x-2| < 1$ $[R=1]$ $\Leftrightarrow$ $-1 < x-2 < 1$ $\Leftrightarrow$ $1 < x < 3$. When $x=1$, the series $\sum_{n=0}^{\infty} (-1)^n \dfrac{1}{n^2+1}$ converges by the Alternating Series Test; when $x=3$, the series $\sum_{n=0}^{\infty} \dfrac{1}{n^2+1}$ converges by direct comparison with the p-series $\sum_{n=1}^{\infty} \dfrac{1}{n^2}$ $[p=2>1]$. Thus, the interval of convergence is $I = [1, 3]$.

23. If $a_n = \dfrac{(x+2)^n}{2^n \ln n}$, then $\lim_{n\to\infty} \left|\dfrac{(x+2)^{n+1}}{2^{n+1} \ln(n+1)} \cdot \dfrac{2^n \ln n}{(x+2)^n}\right| = \lim_{n\to\infty} \dfrac{\ln n}{\ln(n+1)} \cdot \dfrac{|x+2|}{2} = \dfrac{|x+2|}{2}$ since

$\lim_{n\to\infty} \dfrac{\ln n}{\ln(n+1)} = \lim_{x\to\infty} \dfrac{\ln x}{\ln(x+1)} \stackrel{H}{=} \lim_{x\to\infty} \dfrac{1/x}{1/(x+1)} = \lim_{x\to\infty} \dfrac{x+1}{x} = \lim_{x\to\infty} \left(1 + \dfrac{1}{x}\right) = 1$. By the Ratio Test, the series

$\sum_{n=2}^{\infty} \dfrac{(x+2)^n}{2^n \ln n}$ converges when $\dfrac{|x+2|}{2} < 1 \Leftrightarrow |x+2| < 2$ $[R=2]$ $\Leftrightarrow$ $-2 < x+2 < 2$ $\Leftrightarrow$ $-4 < x < 0$.

When $x = -4$, the series $\sum_{n=2}^{\infty} \dfrac{(-1)^n}{\ln n}$ converges by the Alternating Series Test. When $x = 0$, the series $\sum_{n=2}^{\infty} \dfrac{1}{\ln n}$ diverges by the Limit Comparison Test with $b_n = \dfrac{1}{n}$ (or by direct comparison with the harmonic series). Thus, the interval of convergence is $[-4, 0)$.

25. If $a_n = \dfrac{(x-2)^n}{n^n}$, then $\lim_{n\to\infty} \sqrt[n]{|a_n|} = \lim_{n\to\infty} \dfrac{|x-2|}{n} = 0$, so the series converges for all x (by the Root Test). $R = \infty$ and $I = (-\infty, \infty)$.

27. If $a_n = \dfrac{\ln n}{n} x^n$, then

$$\lim_{n\to\infty} \left|\dfrac{a_{n+1}}{a_n}\right| = \lim_{n\to\infty} \left|\dfrac{\ln(n+1) x^{n+1}}{n+1} \cdot \dfrac{n}{(\ln n) x^n}\right| = \lim_{n\to\infty} \left|\dfrac{n}{n+1} \cdot \dfrac{\ln(n+1)}{\ln n} x\right|$$
$$= 1 \cdot 1 \cdot |x| = |x|$$

since $\lim_{n\to\infty} \dfrac{\ln(n+1)}{\ln n} \stackrel{H}{=} \lim_{n\to\infty} \dfrac{1/(n+1)}{1/n} = \lim_{n\to\infty} \dfrac{n}{n+1} = \lim_{n\to\infty} \dfrac{1}{1+1/n} = 1$. By the Ratio Test, the series $\sum_{n=4}^{\infty} \dfrac{\ln n}{n} x^n$

converges when $|x| < 1$, so $R = 1$. When $x = 1$, the series $\sum_{n=4}^{\infty} \frac{\ln n}{n}$ diverges by direct comparison with the (partial) harmonic series $\sum_{n=4}^{\infty} \frac{1}{n}$. When $x = -1$, the series $\sum_{n=4}^{\infty} \frac{(-1)^n \ln n}{n}$ converges by the Alternating Series Test. Thus, the interval of convergence is $[-1, 1)$.

29. $a_n = \frac{n}{b^n}(x-a)^n$, where $b > 0$.

$$\lim_{n \to \infty} \left| \frac{a_{n+1}}{a_n} \right| = \lim_{n \to \infty} \frac{(n+1)|x-a|^{n+1}}{b^{n+1}} \cdot \frac{b^n}{n|x-a|^n} = \lim_{n \to \infty} \left(1 + \frac{1}{n}\right) \frac{|x-a|}{b} = \frac{|x-a|}{b}.$$

By the Ratio Test, the series converges when $\frac{|x-a|}{b} < 1 \Leftrightarrow |x-a| < b$ [so $R = b$] $\Leftrightarrow -b < x - a < b \Leftrightarrow a - b < x < a + b$. When $|x - a| = b$, $\lim_{n \to \infty} |a_n| = \lim_{n \to \infty} n = \infty$, so the series diverges. Thus, $I = (a-b, a+b)$.

31. If $a_n = n!(2x-1)^n$, then $\lim_{n \to \infty} \left| \frac{a_{n+1}}{a_n} \right| = \lim_{n \to \infty} \left| \frac{(n+1)!(2x-1)^{n+1}}{n!(2x-1)^n} \right| = \lim_{n \to \infty} (n+1)|2x-1| \to \infty$ as $n \to \infty$

for all $x \neq \frac{1}{2}$. Since the series diverges for all $x \neq \frac{1}{2}$, $R = 0$ and $I = \{\frac{1}{2}\}$.

33. If $a_n = \frac{(5x-4)^n}{n^3}$, then

$$\lim_{n \to \infty} \left| \frac{a_{n+1}}{a_n} \right| = \lim_{n \to \infty} \left| \frac{(5x-4)^{n+1}}{(n+1)^3} \cdot \frac{n^3}{(5x-4)^n} \right| = \lim_{n \to \infty} |5x-4| \left(\frac{n}{n+1} \right)^3 = \lim_{n \to \infty} |5x-4| \left(\frac{1}{1+1/n} \right)^3$$

$$= |5x-4| \cdot 1 = |5x-4|$$

By the Ratio Test, $\sum_{n=1}^{\infty} \frac{(5x-4)^n}{n^3}$ converges when $|5x-4| < 1 \Leftrightarrow |x - \frac{4}{5}| < \frac{1}{5} \Leftrightarrow -\frac{1}{5} < x - \frac{4}{5} < \frac{1}{5} \Leftrightarrow$

$\frac{3}{5} < x < 1$, so $R = \frac{1}{5}$. When $x = 1$, the series $\sum_{n=1}^{\infty} \frac{1}{n^3}$ is a convergent p-series ($p = 3 > 1$). When $x = \frac{3}{5}$, the series

$\sum_{n=1}^{\infty} \frac{(-1)^n}{n^3}$ converges by the Alternating Series Test. Thus, the interval of convergence is $I = [\frac{3}{5}, 1]$.

35. If $a_n = \frac{x^n}{1 \cdot 3 \cdot 5 \cdot \cdots \cdot (2n-1)}$, then

$$\lim_{n \to \infty} \left| \frac{a_{n+1}}{a_n} \right| = \lim_{n \to \infty} \left| \frac{x^{n+1}}{1 \cdot 3 \cdot 5 \cdot \cdots \cdot (2n-1)(2n+1)} \cdot \frac{1 \cdot 3 \cdot 5 \cdot \cdots \cdot (2n-1)}{x^n} \right| = \lim_{n \to \infty} \frac{|x|}{2n+1} = 0 < 1.$$ Thus, by

the Ratio Test, the series $\sum_{n=1}^{\infty} \frac{x^n}{1 \cdot 3 \cdot 5 \cdot \cdots \cdot (2n-1)}$ converges for *all* real x and we have $R = \infty$ and $I = (-\infty, \infty)$.

37. (a) We are given that the power series $\sum_{n=0}^{\infty} c_n x^n$ is convergent for $x = 4$. So by Theorem 4, it must converge for at least $-4 < x \leq 4$. In particular, it converges when $x = -2$; that is, $\sum_{n=0}^{\infty} c_n(-2)^n$ is convergent.

(b) It does not follow that $\sum_{n=0}^{\infty} c_n(-4)^n$ is necessarily convergent. [See the comments after Theorem 4 about convergence at the endpoint of an interval. An example is $c_n = (-1)^n/(n4^n)$.]

39. If $a_n = \dfrac{(n!)^k}{(kn)!}x^n$, then

$$\lim_{n\to\infty}\left|\dfrac{a_{n+1}}{a_n}\right| = \lim_{n\to\infty}\dfrac{[(n+1)!]^k\,(kn)!}{(n!)^k\,[k(n+1)]!}|x| = \lim_{n\to\infty}\dfrac{(n+1)^k}{(kn+k)(kn+k-1)\cdots(kn+2)(kn+1)}|x|$$

$$= \lim_{n\to\infty}\left[\dfrac{(n+1)}{(kn+1)}\dfrac{(n+1)}{(kn+2)}\cdots\dfrac{(n+1)}{(kn+k)}\right]|x|$$

$$= \lim_{n\to\infty}\left[\dfrac{n+1}{kn+1}\right]\lim_{n\to\infty}\left[\dfrac{n+1}{kn+2}\right]\cdots\lim_{n\to\infty}\left[\dfrac{n+1}{kn+k}\right]|x|$$

$$= \left(\dfrac{1}{k}\right)^k |x| < 1 \;\;\Leftrightarrow\;\; |x| < k^k \text{ for convergence, and the radius of convergence is } R = k^k.$$

41. No. If a power series is centered at a, its interval of convergence is symmetric about a. If a power series has an infinite radius of convergence, then its interval of convergence must be $(-\infty, \infty)$, not $[0, \infty)$.

43. We use the Root Test on the series $\sum c_n x^n$. We need $\lim_{n\to\infty}\sqrt[n]{|c_n x^n|} = |x|\lim_{n\to\infty}\sqrt[n]{|c_n|} = c|x| < 1$ for convergence, or $|x| < 1/c$, so $R = 1/c$.

45. For $2 < x < 3$, $\sum c_n x^n$ diverges and $\sum d_n x^n$ converges. By Exercise 11.2.89, $\sum(c_n + d_n)x^n$ diverges. Since both series converge for $|x| < 2$, the radius of convergence of $\sum(c_n + d_n)x^n$ is 2.

11.9 Representations of Functions as Power Series

1. If $f(x) = \sum\limits_{n=0}^{\infty} c_n x^n$ has radius of convergence 10, then $f'(x) = \sum\limits_{n=1}^{\infty} nc_n x^{n-1}$ also has radius of convergence 10 by Theorem 2.

3. Our goal is to write the function in the form $\dfrac{1}{1-r}$, and then use Equation 1 to represent the function as a sum of a power series. $f(x) = \dfrac{1}{1+x} = \dfrac{1}{1-(-x)} = \sum\limits_{n=0}^{\infty}(-x)^n = \sum\limits_{n=0}^{\infty}(-1)^n x^n$ with $|-x| < 1 \;\Leftrightarrow\; |x| < 1$, so $R = 1$ and $I = (-1, 1)$.

5. $f(x) = \dfrac{1}{1-x^2} = \sum\limits_{n=0}^{\infty}(x^2)^n = \sum\limits_{n=0}^{\infty} x^{2n}$. The series converges when $|x^2| < 1 \;\Leftrightarrow\; |x| < 1$, so $R = 1$ and $I = (-1, 1)$.

7. $f(x) = \dfrac{2}{3-x} = \dfrac{2}{3}\left(\dfrac{1}{1-x/3}\right) = \dfrac{2}{3}\sum\limits_{n=0}^{\infty}\left(\dfrac{x}{3}\right)^n$ or, equivalently, $2\sum\limits_{n=0}^{\infty}\dfrac{1}{3^{n+1}}x^n$. The series converges when $\left|\dfrac{x}{3}\right| < 1$, that is, when $|x| < 3$, so $R = 3$ and $I = (-3, 3)$.

9. $f(x) = \dfrac{x^2}{x^4 + 16} = \dfrac{x^2}{16}\left(\dfrac{1}{1+x^4/16}\right) = \dfrac{x^2}{16}\left(\dfrac{1}{1-[-(x/2)^4]}\right) = \dfrac{x^2}{16}\sum\limits_{n=0}^{\infty}\left[-\left(\dfrac{x}{2}\right)^4\right]^n$ or, equivalently, $\sum\limits_{n=0}^{\infty}\dfrac{(-1)^n x^{4n+2}}{2^{4n+4}}$.

The series converges when $\left|-\left(\dfrac{x}{2}\right)^4\right| < 1 \;\Rightarrow\; \left|\dfrac{x}{2}\right| < 1 \;\Rightarrow\; |x| < 2$, so $R = 2$ and $I = (-2, 2)$.

11. $f(x) = \dfrac{x-1}{x+2} = \dfrac{x+2-3}{x+2} = 1 - \dfrac{3}{x+2} = 1 - \dfrac{3/2}{x/2+1} = 1 - \dfrac{3}{2} \cdot \dfrac{1}{1-(-x/2)}$

$= 1 - \dfrac{3}{2} \sum\limits_{n=0}^{\infty} \left(-\dfrac{x}{2}\right)^n = 1 - \dfrac{3}{2} - \dfrac{3}{2} \sum\limits_{n=1}^{\infty} \left(-\dfrac{x}{2}\right)^n = -\dfrac{1}{2} - \sum\limits_{n=1}^{\infty} \dfrac{(-1)^n 3x^n}{2^{n+1}}.$

The geometric series $\sum\limits_{n=0}^{\infty} \left(-\dfrac{x}{2}\right)^n$ converges when $\left|-\dfrac{x}{2}\right| < 1 \Leftrightarrow |x| < 2$, so $R = 2$ and $I = (-2, 2)$.

Alternatively, you could write $f(x) = 1 - 3\left(\dfrac{1}{x+2}\right)$ and use the series for $\dfrac{1}{x+2}$ found in Example 2.

13. $f(x) = \dfrac{2x-4}{x^2-4x+3} = \dfrac{2x-4}{(x-1)(x-3)} = \dfrac{A}{x-1} + \dfrac{B}{x-3} \Rightarrow 2x-4 = A(x-3) + B(x-1)$. Let $x = 1$ to get $-2 = -2A \Leftrightarrow A = 1$ and $x = 3$ to get $2 = 2B \Leftrightarrow B = 1$. Thus,

$\dfrac{2x-4}{x^2-4x+3} = \dfrac{1}{x-1} + \dfrac{1}{x-3} = \dfrac{-1}{1-x} + \dfrac{1}{-3}\left[\dfrac{1}{1-(x/3)}\right] = -\sum\limits_{n=0}^{\infty} x^n - \dfrac{1}{3}\sum\limits_{n=0}^{\infty}\left(\dfrac{x}{3}\right)^n = \sum\limits_{n=0}^{\infty}\left(-1 - \dfrac{1}{3^{n+1}}\right)x^n.$

We represented f as the sum of two geometric series; the first converges for $x \in (-1, 1)$ and the second converges for $x \in (-3, 3)$. Thus, the sum converges for $x \in (-1, 1) = I$.

15. (a) $f(x) = \dfrac{1}{(1+x)^2} = \dfrac{d}{dx}\left(\dfrac{-1}{1+x}\right) = -\dfrac{d}{dx}\left[\sum\limits_{n=0}^{\infty}(-1)^n x^n\right]$ [from Exercise 3]

$= \sum\limits_{n=1}^{\infty}(-1)^{n+1} n x^{n-1}$ [from Theorem 2(i)] $= \sum\limits_{n=0}^{\infty}(-1)^n (n+1) x^n$ with $R = 1$.

In the last step, note that we *decreased* the initial value of the summation variable n by 1, and then *increased* each occurrence of n in the term by 1 [also note that $(-1)^{n+2} = (-1)^n$].

(b) $f(x) = \dfrac{1}{(1+x)^3} = -\dfrac{1}{2}\dfrac{d}{dx}\left[\dfrac{1}{(1+x)^2}\right] = -\dfrac{1}{2}\dfrac{d}{dx}\left[\sum\limits_{n=0}^{\infty}(-1)^n(n+1)x^n\right]$ [from part (a)]

$= -\dfrac{1}{2}\sum\limits_{n=1}^{\infty}(-1)^n(n+1)nx^{n-1} = \dfrac{1}{2}\sum\limits_{n=0}^{\infty}(-1)^n(n+2)(n+1)x^n$ with $R = 1$.

(c) $f(x) = \dfrac{x^2}{(1+x)^3} = x^2 \cdot \dfrac{1}{(1+x)^3} = x^2 \cdot \dfrac{1}{2}\sum\limits_{n=0}^{\infty}(-1)^n(n+2)(n+1)x^n$ [from part (b)]

$= \dfrac{1}{2}\sum\limits_{n=0}^{\infty}(-1)^n(n+2)(n+1)x^{n+2}$

To write the power series with x^n rather than x^{n+2}, we will *decrease* each occurrence of n in the term by 2 and *increase* the initial value of the summation variable by 2. This gives us $\dfrac{1}{2}\sum\limits_{n=2}^{\infty}(-1)^n(n)(n-1)x^n$ with $R = 1$.

17. We know that $\dfrac{1}{1+4x} = \dfrac{1}{1-(-4x)} = \sum\limits_{n=0}^{\infty}(-4x)^n$. Differentiating, we get

$\dfrac{-4}{(1+4x)^2} = \sum\limits_{n=1}^{\infty}(-4)^n n x^{n-1} = \sum\limits_{n=0}^{\infty}(-4)^{n+1}(n+1)x^n$, so

$f(x) = \dfrac{x}{(1+4x)^2} = \dfrac{-x}{4} \cdot \dfrac{-4}{(1+4x)^2} = \dfrac{-x}{4}\sum\limits_{n=0}^{\infty}(-4)^{n+1}(n+1)x^n = \sum\limits_{n=0}^{\infty}(-1)^n 4^n(n+1)x^{n+1}$

for $|-4x| < 1 \Leftrightarrow |x| < \tfrac{1}{4}$, so $R = \tfrac{1}{4}$.

19. By Example 4, $\dfrac{1}{(1-x)^2} = \sum_{n=0}^{\infty}(n+1)x^n$. Thus,

$$f(x) = \dfrac{1+x}{(1-x)^2} = \dfrac{1}{(1-x)^2} + \dfrac{x}{(1-x)^2} = \sum_{n=0}^{\infty}(n+1)x^n + \sum_{n=0}^{\infty}(n+1)x^{n+1}$$

$$= \sum_{n=0}^{\infty}(n+1)x^n + \sum_{n=1}^{\infty}nx^n \quad \text{[make the starting values equal]}$$

$$= 1 + \sum_{n=1}^{\infty}[(n+1)+n]x^n = 1 + \sum_{n=1}^{\infty}(2n+1)x^n = \sum_{n=0}^{\infty}(2n+1)x^n \text{ with } R = 1.$$

21. $f(x) = \ln(5-x) = -\displaystyle\int \dfrac{dx}{5-x} = -\dfrac{1}{5}\int \dfrac{dx}{1-x/5} = -\dfrac{1}{5}\int \left[\sum_{n=0}^{\infty}\left(\dfrac{x}{5}\right)^n\right] dx$

$= C - \dfrac{1}{5}\displaystyle\sum_{n=0}^{\infty}\dfrac{x^{n+1}}{5^n(n+1)} = C - \displaystyle\sum_{n=1}^{\infty}\dfrac{x^n}{n5^n}$

Putting $x = 0$, we get $C = \ln 5$. The series converges for $|x/5| < 1 \Leftrightarrow |x| < 5$, so $R = 5$.

23. $f(x) = \dfrac{x^2}{x^2+1} = x^2\left(\dfrac{1}{1-(-x^2)}\right) = x^2\displaystyle\sum_{n=0}^{\infty}(-x^2)^n = \displaystyle\sum_{n=0}^{\infty}(-1)^n x^{2n+2}$. This series converges when $\left|-x^2\right| < 1 \Leftrightarrow$

$x^2 < 1 \Leftrightarrow |x| < 1$, so $R = 1$. The partial sums are $s_1 = x^2$,

$s_2 = s_1 - x^4$, $s_3 = s_2 + x^6$, $s_4 = s_3 - x^8$, $s_5 = s_4 + x^{10}, \ldots$.

Note that s_1 corresponds to the first term of the infinite sum,
regardless of the value of the summation variable and the value of the
exponent. As n increases, $s_n(x)$ approximates f better on the
interval of convergence, which is $(-1, 1)$.

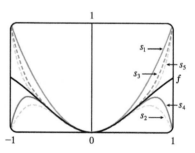

25. $f(x) = \ln\left(\dfrac{1+x}{1-x}\right) = \ln(1+x) - \ln(1-x) = \displaystyle\int \dfrac{dx}{1+x} + \displaystyle\int \dfrac{dx}{1-x} = \displaystyle\int \dfrac{dx}{1-(-x)} + \displaystyle\int \dfrac{dx}{1-x}$

$= \displaystyle\int \left[\sum_{n=0}^{\infty}(-1)^n x^n + \sum_{n=0}^{\infty}x^n\right] dx = \displaystyle\int [(1-x+x^2-x^3+x^4-\cdots)+(1+x+x^2+x^3+x^4+\cdots)]\, dx$

$= \displaystyle\int (2+2x^2+2x^4+\cdots)\, dx = \displaystyle\int \sum_{n=0}^{\infty}2x^{2n}\, dx = C + \sum_{n=0}^{\infty}\dfrac{2x^{2n+1}}{2n+1}$

But $f(0) = \ln\dfrac{1}{1} = 0$, so $C = 0$ and we have $f(x) = \displaystyle\sum_{n=0}^{\infty}\dfrac{2x^{2n+1}}{2n+1}$ with $R = 1$. If $x = \pm 1$, then $f(x) = \pm 2\displaystyle\sum_{n=0}^{\infty}\dfrac{1}{2n+1}$,

which both diverge by the Limit Comparison Test with $b_n = \dfrac{1}{n}$.

The partial sums are $s_1 = \dfrac{2x}{1}$, $s_2 = s_1 + \dfrac{2x^3}{3}$, $s_3 = s_2 + \dfrac{2x^5}{5}, \ldots$.

As n increases, $s_n(x)$ approximates f better on the interval of
convergence, which is $(-1, 1)$.

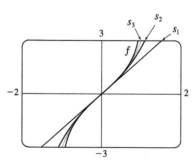

27. $\dfrac{t}{1-t^8} = t \cdot \dfrac{1}{1-t^8} = t \sum_{n=0}^{\infty} (t^8)^n = \sum_{n=0}^{\infty} t^{8n+1} \Rightarrow \int \dfrac{t}{1-t^8}\, dt = C + \sum_{n=0}^{\infty} \dfrac{t^{8n+2}}{8n+2}$. The series for $\dfrac{1}{1-t^8}$ converges when $|t^8| < 1 \Leftrightarrow |t| < 1$, so $R = 1$ for that series and also the series for $t/(1-t^8)$. By Theorem 2, the series for

$\displaystyle\int \dfrac{t}{1-t^8}\, dt$ also has $R = 1$.

29. From Example 5, $\ln(1+x) = \sum_{n=1}^{\infty} (-1)^{n-1} \dfrac{x^n}{n}$ for $|x| < 1$, so $x^2 \ln(1+x) = \sum_{n=1}^{\infty} (-1)^{n-1} \dfrac{x^{n+2}}{n}$ and

$\displaystyle\int x^2 \ln(1+x)\, dx = C + \sum_{n=1}^{\infty} (-1)^{n-1} \dfrac{x^{n+3}}{n(n+3)}$. $R = 1$ for the series for $\ln(1+x)$, so $R = 1$ for the series representing

$x^2 \ln(1+x)$ as well. By Theorem 2, the series for $\displaystyle\int x^2 \ln(1+x)\, dx$ also has $R = 1$.

31. $\dfrac{x}{1+x^3} = x\left[\dfrac{1}{1-(-x^3)}\right] = x \sum_{n=0}^{\infty} (-x^3)^n = \sum_{n=0}^{\infty} (-1)^n x^{3n+1} \Rightarrow$

$\displaystyle\int \dfrac{x}{1+x^3}\, dx = \int \sum_{n=0}^{\infty} (-1)^n x^{3n+1}\, dx = C + \sum_{n=0}^{\infty} (-1)^n \dfrac{x^{3n+2}}{3n+2}$. Thus,

$I = \displaystyle\int_0^{0.3} \dfrac{x}{1+x^3}\, dx = \left[\dfrac{x^2}{2} - \dfrac{x^5}{5} + \dfrac{x^8}{8} - \dfrac{x^{11}}{11} + \cdots\right]_0^{0.3} = \dfrac{(0.3)^2}{2} - \dfrac{(0.3)^5}{5} + \dfrac{(0.3)^8}{8} - \dfrac{(0.3)^{11}}{11} + \cdots$.

The series is alternating, so if we use the first three terms, the error is at most $(0.3)^{11}/11 \approx 1.6 \times 10^{-7}$.

So $I \approx (0.3)^2/2 - (0.3)^5/5 + (0.3)^8/8 \approx 0.044\,522$ to six decimal places.

33. We substitute x^2 for x in Example 5, and find that

$\displaystyle\int x \ln(1+x^2)\, dx = \int x \sum_{n=1}^{\infty} (-1)^{n-1} \dfrac{(x^2)^n}{n}\, dx = \int \sum_{n=1}^{\infty} (-1)^{n-1} \dfrac{x^{2n+1}}{n}\, dx = C + \sum_{n=1}^{\infty} (-1)^{n-1} \dfrac{x^{2n+2}}{n(2n+2)}$

Thus,

$I \approx \displaystyle\int_0^{0.2} x \ln(1+x^2)\, dx = \left[\dfrac{x^4}{1(4)} - \dfrac{x^6}{2(6)} + \dfrac{x^8}{3(8)} - \dfrac{x^{10}}{4(10)} + \cdots\right]_0^{0.2} = \dfrac{(0.2)^4}{4} - \dfrac{(0.2)^6}{12} + \dfrac{(0.2)^8}{24} - \dfrac{(0.2)^{10}}{40} + \cdots$

The series is alternating, so if we use two terms, the error is at most $(0.2)^8/24 \approx 1.1 \times 10^{-7}$. So

$I \approx \dfrac{(0.2)^4}{4} - \dfrac{(0.2)^6}{12} \approx 0.000\,395$ to six decimal places.

35. By Example 6, $\arctan x = x - \dfrac{x^3}{3} + \dfrac{x^5}{5} - \dfrac{x^7}{7} + \cdots$, so $\arctan 0.2 = 0.2 - \dfrac{(0.2)^3}{3} + \dfrac{(0.2)^5}{5} - \dfrac{(0.2)^7}{7} + \cdots$.

The series is alternating, so if we use three terms, the error is at most $\dfrac{(0.2)^7}{7} \approx 0.000\,002$.

Thus, to five decimal places, $\arctan 0.2 \approx 0.2 - \dfrac{(0.2)^3}{3} + \dfrac{(0.2)^5}{5} \approx 0.197\,40$.

37. (a) $f(x) = \sum_{n=0}^{\infty} \dfrac{x^n}{n!} \Rightarrow f'(x) = \sum_{n=1}^{\infty} \dfrac{nx^{n-1}}{n!} = \sum_{n=1}^{\infty} \dfrac{x^{n-1}}{(n-1)!} = \sum_{n=0}^{\infty} \dfrac{x^n}{n!} = f(x)$

(b) By Theorem 9.4.2, the only solution to the differential equation $df(x)/dx = f(x)$ is $f(x) = Ke^x$, but $f(0) = 1$, so $K = 1$ and $f(x) = e^x$.

Or: We could solve the equation $df(x)/dx = f(x)$ as a separable differential equation.

39. (a) $J_0(x) = \sum_{n=0}^{\infty} \frac{(-1)^n x^{2n}}{2^{2n}(n!)^2}$, $J_0'(x) = \sum_{n=1}^{\infty} \frac{(-1)^n 2nx^{2n-1}}{2^{2n}(n!)^2}$, and $J_0''(x) = \sum_{n=1}^{\infty} \frac{(-1)^n 2n(2n-1)x^{2n-2}}{2^{2n}(n!)^2}$, so

$$x^2 J_0''(x) + x J_0'(x) + x^2 J_0(x) = \sum_{n=1}^{\infty} \frac{(-1)^n 2n(2n-1)x^{2n}}{2^{2n}(n!)^2} + \sum_{n=1}^{\infty} \frac{(-1)^n 2nx^{2n}}{2^{2n}(n!)^2} + \sum_{n=0}^{\infty} \frac{(-1)^n x^{2n+2}}{2^{2n}(n!)^2}$$

$$= \sum_{n=1}^{\infty} \frac{(-1)^n 2n(2n-1)x^{2n}}{2^{2n}(n!)^2} + \sum_{n=1}^{\infty} \frac{(-1)^n 2nx^{2n}}{2^{2n}(n!)^2} + \sum_{n=1}^{\infty} \frac{(-1)^{n-1} x^{2n}}{2^{2n-2}[(n-1)!]^2}$$

$$= \sum_{n=1}^{\infty} \frac{(-1)^n 2n(2n-1)x^{2n}}{2^{2n}(n!)^2} + \sum_{n=1}^{\infty} \frac{(-1)^n 2nx^{2n}}{2^{2n}(n!)^2} + \sum_{n=1}^{\infty} \frac{(-1)^n(-1)^{-1} 2^2 n^2 x^{2n}}{2^{2n}(n!)^2}$$

$$= \sum_{n=1}^{\infty} (-1)^n \left[\frac{2n(2n-1) + 2n - 2^2 n^2}{2^{2n}(n!)^2} \right] x^{2n}$$

$$= \sum_{n=1}^{\infty} (-1)^n \left[\frac{4n^2 - 2n + 2n - 4n^2}{2^{2n}(n!)^2} \right] x^{2n} = 0$$

(b) $\int_0^1 J_0(x)\, dx = \int_0^1 \left[\sum_{n=0}^{\infty} \frac{(-1)^n x^{2n}}{2^{2n}(n!)^2} \right] dx = \int_0^1 \left(1 - \frac{x^2}{4} + \frac{x^4}{64} - \frac{x^6}{2304} + \cdots \right) dx$

$$= \left[x - \frac{x^3}{3 \cdot 4} + \frac{x^5}{5 \cdot 64} - \frac{x^7}{7 \cdot 2304} + \cdots \right]_0^1 = 1 - \frac{1}{12} + \frac{1}{320} - \frac{1}{16{,}128} + \cdots$$

Since $\frac{1}{16{,}128} \approx 0.000062$, it follows from The Alternating Series Estimation Theorem that, correct to three decimal places, $\int_0^1 J_0(x)\, dx \approx 1 - \frac{1}{12} + \frac{1}{320} \approx 0.920$.

41. (a) $A(x) = 1 + \sum_{n=1}^{\infty} a_n$, where $a_n = \frac{x^{3n}}{2 \cdot 3 \cdot 5 \cdot 6 \cdots (3n-1)(3n)}$, so $\lim_{n \to \infty} \left| \frac{a_{n+1}}{a_n} \right| = |x|^3 \lim_{n \to \infty} \frac{1}{(3n+2)(3n+3)} = 0$

for all x, so the domain is $\mathbb{R}$.

(b), (c)

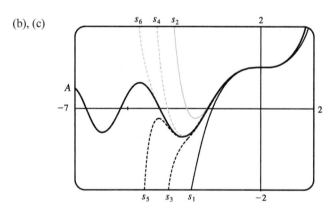

$s_0 = 1$ has been omitted from the graph. The partial sums seem to approximate $A(x)$ well near the origin, but as $|x|$ increases, we need to take a large number of terms to get a good approximation.

To plot A, we must first define $A(x)$ for the CAS. Note that for $n \geq 1$, the denominator of a_n is

$2 \cdot 3 \cdot 5 \cdot 6 \cdots (3n-1) \cdot 3n = \frac{(3n)!}{1 \cdot 4 \cdot 7 \cdots (3n-2)} = \frac{(3n)!}{\prod_{k=1}^n (3k-2)}$, so $a_n = \frac{\prod_{k=1}^n (3k-2)}{(3n)!} x^{3n}$ and thus

$A(x) = 1 + \sum_{n=1}^{\infty} \frac{\prod_{k=1}^n (3k-2)}{(3n)!} x^{3n}$. Both Maple and Mathematica are able to plot A if we define it this way.

[continued]

Maple and Mathematica have two initially known Airy functions, called `AI·SERIES(z,m)` and `BI·SERIES(z,m)` from `AiryAi` and `AiryBi` in Maple and Mathematica (just `Ai` and `Bi` in older versions of Maple). However, it is very difficult to solve for A in terms of the CAS's Airy functions, although in fact $A(x) = \dfrac{\sqrt{3}\,\text{AiryAi}(x) + \text{AiryBi}(x)}{\sqrt{3}\,\text{AiryAi}(0) + \text{AiryBi}(0)}$.

43. $f(x) = 1 + 2x + x^2 + 2x^3 + x^4 + \cdots$, where $c_{2n} = 1$ and $c_{2n+1} = 2$ for all $n \geq 0$. So

$$s_{2n-1} = 1 + 2x + x^2 + 2x^3 + x^4 + 2x^5 + \cdots + x^{2n-2} + 2x^{2n-1}$$
$$= 1(1+2x) + x^2(1+2x) + x^4(1+2x) + \cdots + x^{2n-2}(1+2x) = (1+2x)(1 + x^2 + x^4 + \cdots + x^{2n-2})$$
$$= (1+2x)\frac{1-x^{2n}}{1-x^2} \quad [\text{by (11.2.3) with } r = x^2] \;\to\; \frac{1+2x}{1-x^2} \text{ as } n \to \infty \text{ by (11.2.4), when } |x| < 1.$$

Also $s_{2n} = s_{2n-1} + x^{2n} \to \dfrac{1+2x}{1-x^2}$ since $x^{2n} \to 0$ for $|x| < 1$. Therefore, $s_n \to \dfrac{1+2x}{1-x^2}$ since s_{2n} and s_{2n-1} both approach $\dfrac{1+2x}{1-x^2}$ as $n \to \infty$. Thus, the interval of convergence is $(-1, 1)$ and $f(x) = \dfrac{1+2x}{1-x^2}$.

45. If $a_n = \dfrac{x^n}{n^2}$, then by the Ratio Test, $\lim\limits_{n\to\infty}\left|\dfrac{a_{n+1}}{a_n}\right| = \lim\limits_{n\to\infty}\left|\dfrac{x^{n+1}}{(n+1)^2} \cdot \dfrac{n^2}{x^n}\right| = |x|\lim\limits_{n\to\infty}\left(\dfrac{n}{n+1}\right)^2 = |x| < 1$ for convergence, so $R = 1$. When $x = \pm 1$, $\sum\limits_{n=1}^{\infty}\left|\dfrac{x^n}{n^2}\right| = \sum\limits_{n=1}^{\infty}\dfrac{1}{n^2}$ which is a convergent p-series ($p = 2 > 1$), so the interval of convergence for f is $[-1, 1]$. By Theorem 2, the radii of convergence of f' and f'' are both 1, so we need only check the endpoints. $f(x) = \sum\limits_{n=1}^{\infty}\dfrac{x^n}{n^2} \;\Rightarrow\; f'(x) = \sum\limits_{n=1}^{\infty}\dfrac{nx^{n-1}}{n^2} = \sum\limits_{n=0}^{\infty}\dfrac{x^n}{n+1}$, and this series diverges for $x = 1$ (harmonic series) and converges for $x = -1$ (Alternating Series Test), so the interval of convergence is $[-1, 1)$. $f''(x) = \sum\limits_{n=1}^{\infty}\dfrac{nx^{n-1}}{n+1}$ diverges at both 1 and -1 (Test for Divergence) since $\lim\limits_{n\to\infty}\dfrac{n}{n+1} = 1 \neq 0$, so its interval of convergence is $(-1, 1)$.

47. $f(x) = \dfrac{1}{1-x} = \sum\limits_{n=0}^{\infty}x^n \;\Rightarrow\; f'(x) = \dfrac{1}{(1-x)^2} = \sum\limits_{n=1}^{\infty}nx^{n-1} \;\Rightarrow\; f''(x) = \dfrac{2}{(1-x)^3} = \sum\limits_{n=2}^{\infty}n(n-1)x^{n-2}$

The power series representation of $h(x)$ is

$$h(x) = xf'(x) + x^2 f''(x) = x\sum_{n=1}^{\infty}nx^{n-1} + x^2\sum_{n=2}^{\infty}n(n-1)x^{n-2} = \sum_{n=1}^{\infty}nx^n + \sum_{n=2}^{\infty}n(n-1)x^n$$
$$= x + \sum_{n=2}^{\infty}nx^n + \sum_{n=2}^{\infty}n(n-1)x^n = x + \sum_{n=2}^{\infty}[nx^n + n(n-1)x^n]$$
$$= x + \sum_{n=2}^{\infty}(1 + n - 1)nx^n = x + \sum_{n=2}^{\infty}n^2 x^n = \sum_{n=1}^{\infty}n^2 x^n$$

$h(x)$ has the same radius of convergence as the power series representation of $f(x)$, that is, $R = 1$.

Now, $h\left(\dfrac{1}{2}\right) = \sum\limits_{n=1}^{\infty}\dfrac{n^2}{2^n}$ and using the function representation, we have

$h\left(\dfrac{1}{2}\right) = \dfrac{1}{2}f'\left(\dfrac{1}{2}\right) + \left(\dfrac{1}{2}\right)^2 f''\left(\dfrac{1}{2}\right) = \dfrac{1}{2} \cdot \dfrac{1}{(1-1/2)^2} + \dfrac{1}{4} \cdot \dfrac{2}{(1-1/2)^3} = 6$. Thus, $\sum\limits_{n=1}^{\infty}\dfrac{n^2}{2^n} = 6$.

49. By Example 6, $\tan^{-1} x = \sum_{n=0}^{\infty} (-1)^n \dfrac{x^{2n+1}}{2n+1}$ for $|x| < 1$. In particular, for $x = \dfrac{1}{\sqrt{3}}$, we

have $\dfrac{\pi}{6} = \tan^{-1}\left(\dfrac{1}{\sqrt{3}}\right) = \sum_{n=0}^{\infty} (-1)^n \dfrac{(1/\sqrt{3})^{2n+1}}{2n+1} = \sum_{n=0}^{\infty} (-1)^n \left(\dfrac{1}{3}\right)^n \dfrac{1}{\sqrt{3}} \dfrac{1}{2n+1}$, so

$\pi = \dfrac{6}{\sqrt{3}} \sum_{n=0}^{\infty} \dfrac{(-1)^n}{(2n+1)3^n} = 2\sqrt{3} \sum_{n=0}^{\infty} \dfrac{(-1)^n}{(2n+1)3^n}$.

51. Using the Ratio Test, the series $\sum_{n=0}^{\infty} c_n x^n$ converges when $\lim\limits_{n\to\infty} \left|\dfrac{a_{n+1}}{a_n}\right| < 1 \;\Rightarrow\;$

$\lim\limits_{n\to\infty} \left|\dfrac{c_{n+1} x^{n+1}}{c_n x^n}\right| = |x| \lim\limits_{n\to\infty} \left|\dfrac{c_{n+1}}{c_n}\right| < 1 \;\Rightarrow\; |x| < \dfrac{1}{\lim\limits_{n\to\infty}\left|\dfrac{c_{n+1}}{c_n}\right|} = R \;\Rightarrow\; \lim\limits_{n\to\infty} \left|\dfrac{c_{n+1}}{c_n}\right| = \dfrac{1}{R}$.

Now, using the Ratio Test for the series $\sum_{n=1}^{\infty} n c_n x^{n-1}$, we find

$\lim\limits_{n\to\infty} \left|\dfrac{a_{n+1}}{a_n}\right| = \lim\limits_{n\to\infty} \left|\dfrac{(n+1)c_{n+1} x^n}{n c_n x^{n-1}}\right| = \lim\limits_{n\to\infty} \left|\left(1+\dfrac{1}{n}\right)\dfrac{c_{n+1}}{c_n} x\right| = |x|\lim\limits_{n\to\infty}\left|\dfrac{c_{n+1}}{c_n}\right| = \dfrac{|x|}{R}$. Hence, the series

$\sum_{n=1}^{\infty} n c_n x^{n-1}$ converges when $\dfrac{|x|}{R} < 1$ or $|x| < R$. Finally, testing the series $\sum_{n=0}^{\infty} c_n \dfrac{x^{n+1}}{n+1}$ using the Ratio Test, we have

$\lim\limits_{n\to\infty} \left|\dfrac{a_{n+1}}{a_n}\right| = \lim\limits_{n\to\infty} \left|\dfrac{c_{n+1} x^{n+2}}{n+2} \cdot \dfrac{n+1}{c_n x^{n+1}}\right| = \lim\limits_{n\to\infty} \left|\dfrac{(n+1)c_{n+1} x}{(n+2) c_n}\right| = |x|\lim\limits_{n\to\infty} \left|\dfrac{(1+1/n)c_{n+1}}{(1+2/n)c_n}\right|$

$= |x|\lim\limits_{n\to\infty} \left|\dfrac{c_{n+1}}{c_n}\right| = \dfrac{|x|}{R}$

so $\sum_{n=0}^{\infty} c_n \dfrac{x^{n+1}}{n+1}$ also converges when $|x| < R$. Thus, both $\sum_{n=1}^{\infty} n c_n x^{n-1}$ and $\sum_{n=0}^{\infty} c_n \dfrac{x^{n+1}}{n+1}$ have radius of convergence R

when $\sum_{n=0}^{\infty} c_n x^n$ has radius of convergence R.

11.10 Taylor and Maclaurin Series

1. Using Theorem 5 with $\sum_{n=0}^{\infty} b_n (x-5)^n$, $b_n = \dfrac{f^{(n)}(a)}{n!}$, so $b_8 = \dfrac{f^{(8)}(5)}{8!}$.

3. Since $f^{(n)}(0) = (n+1)!$, Equation 7 gives the Maclaurin series

$\sum_{n=0}^{\infty} \dfrac{f^{(n)}(0)}{n!} x^n = \sum_{n=0}^{\infty} \dfrac{(n+1)!}{n!} x^n = \sum_{n=0}^{\infty} (n+1)x^n$. Applying the Ratio Test with $a_n = (n+1)x^n$ gives us

$\lim\limits_{n\to\infty} \left|\dfrac{a_{n+1}}{a_n}\right| = \lim\limits_{n\to\infty} \left|\dfrac{(n+2)x^{n+1}}{(n+1)x^n}\right| = |x| \lim\limits_{n\to\infty} \dfrac{n+2}{n+1} = |x| \cdot 1 = |x|$. For convergence, we must have $|x| < 1$, so the

radius of convergence $R = 1$.

5.

n	$f^{(n)}(x)$	$f^{(n)}(0)$
0	xe^x	0
1	$(x+1)e^x$	1
2	$(x+2)e^x$	2
3	$(x+3)e^x$	3
4	$(x+4)e^x$	4

Using Equation 6 with $n = 0$ to 4 and $a = 0$, we get

$$\sum_{n=0}^{4} \frac{f^{(n)}(0)}{n!}(x-0)^n = \frac{0}{0!}x^0 + \frac{1}{1!}x^1 + \frac{2}{2!}x^2 + \frac{3}{3!}x^3 + \frac{4}{4!}x^4$$

$$= x + x^2 + \tfrac{1}{2}x^3 + \tfrac{1}{6}x^4$$

7.

n	$f^{(n)}(x)$	$f^{(n)}(8)$
0	$\sqrt[3]{x}$	2
1	$\dfrac{1}{3x^{2/3}}$	$\dfrac{1}{12}$
2	$-\dfrac{2}{9x^{5/3}}$	$-\dfrac{2}{288}$
3	$\dfrac{10}{27x^{8/3}}$	$\dfrac{10}{6912}$

$$\sum_{n=0}^{3} \frac{f^{(n)}(8)}{n!}(x-8)^n = \frac{2}{0!}(x-8)^0 + \frac{\frac{1}{12}}{1!}(x-8)^1$$

$$- \frac{\frac{2}{288}}{2!}(x-8)^2 + \frac{\frac{10}{6912}}{3!}(x-8)^3$$

$$= 2 + \tfrac{1}{12}(x-8) - \tfrac{1}{288}(x-8)^2 + \tfrac{5}{20{,}736}(x-8)^3$$

9.

n	$f^{(n)}(x)$	$f^{(n)}(\pi/6)$
0	$\sin x$	$1/2$
1	$\cos x$	$\sqrt{3}/2$
2	$-\sin x$	$-1/2$
3	$-\cos x$	$-\sqrt{3}/2$

$$\sum_{n=0}^{3} \frac{f^{(n)}(\pi/6)}{n!}\left(x - \frac{\pi}{6}\right)^n = \frac{1/2}{0!}\left(x - \frac{\pi}{6}\right)^0 + \frac{\sqrt{3}/2}{1!}\left(x - \frac{\pi}{6}\right)^1 - \frac{1/2}{2!}\left(x - \frac{\pi}{6}\right)^2 - \frac{\sqrt{3}/2}{3!}\left(x - \frac{\pi}{6}\right)^3$$

$$= \frac{1}{2} + \frac{\sqrt{3}}{2}\left(x - \frac{\pi}{6}\right) - \frac{1}{4}\left(x - \frac{\pi}{6}\right)^2 - \frac{\sqrt{3}}{12}\left(x - \frac{\pi}{6}\right)^3$$

11.

n	$f^{(n)}(x)$	$f^{(n)}(0)$
0	$(1-x)^{-2}$	1
1	$2(1-x)^{-3}$	2
2	$6(1-x)^{-4}$	6
3	$24(1-x)^{-5}$	24
4	$120(1-x)^{-6}$	120
$\vdots$	$\vdots$	$\vdots$

$$(1-x)^{-2} = f(0) + f'(0)x + \frac{f''(0)}{2!}x^2 + \frac{f'''(0)}{3!}x^3 + \frac{f^{(4)}(0)}{4!}x^4 + \cdots$$

$$= 1 + 2x + \tfrac{6}{2}x^2 + \tfrac{24}{6}x^3 + \tfrac{120}{24}x^4 + \cdots$$

$$= 1 + 2x + 3x^2 + 4x^3 + 5x^4 + \cdots = \sum_{n=0}^{\infty}(n+1)x^n$$

$$\lim_{n \to \infty}\left|\frac{a_{n+1}}{a_n}\right| = \lim_{n \to \infty}\left|\frac{(n+2)x^{n+1}}{(n+1)x^n}\right| = |x|\lim_{n \to \infty}\frac{n+2}{n+1} = |x|(1) = |x| < 1$$

for convergence, so $R = 1$.

13.

n	$f^{(n)}(x)$	$f^{(n)}(0)$
0	$\cos x$	1
1	$-\sin x$	0
2	$-\cos x$	-1
3	$\sin x$	0
4	$\cos x$	1
$\vdots$	$\vdots$	$\vdots$

$$\cos x = f(0) + f'(0)x + \frac{f''(0)}{2!}x^2 + \frac{f'''(0)}{3!}x^3 + \frac{f^{(4)}(0)}{4!}x^4 + \cdots$$

$$= 1 - \frac{1}{2!}x^2 + \frac{1}{4!}x^4 - \cdots$$

$$= \sum_{n=0}^{\infty} (-1)^n \frac{x^{2n}}{(2n)!} \qquad \text{[Agrees with (16).]}$$

$$\lim_{n\to\infty}\left|\frac{a_{n+1}}{a_n}\right| = \lim_{n\to\infty}\left|\frac{x^{2n+2}}{(2n+2)!}\cdot\frac{(2n)!}{x^{2n}}\right| = \lim_{n\to\infty}\frac{x^2}{(2n+2)(2n+1)} = 0 < 1$$

for all x, so $R = \infty$.

15.

n	$f^{(n)}(x)$	$f^{(n)}(0)$
0	$2x^4 - 3x^2 + 3$	3
1	$8x^3 - 6x$	0
2	$24x^2 - 6$	-6
3	$48x$	0
4	48	48
5	0	0
6	0	0
$\vdots$	$\vdots$	$\vdots$

$f^{(n)}(x) = 0$ for $n \geq 5$, so f has a finite Maclaurin series.

$f(x) = 2x^4 - 3x^2 + 3$

$$= f(0) + \frac{f'(0)}{1!}x + \frac{f''(0)}{2!}x^2 + \frac{f'''(0)}{3!}x^3 + \frac{f^{(4)}(0)}{4!}x^4$$

$$= 3 + 0x + \frac{-6}{2}x^2 + 0x^3 + \frac{48}{24}x^4 = 3 - 3x^2 + 2x^4$$

A finite series converges for all x, so $R = \infty$.

17.

n	$f^{(n)}(x)$	$f^{(n)}(0)$
0	2^x	1
1	$2^x(\ln 2)$	$\ln 2$
2	$2^x(\ln 2)^2$	$(\ln 2)^2$
3	$2^x(\ln 2)^3$	$(\ln 2)^3$
4	$2^x(\ln 2)^4$	$(\ln 2)^4$
$\vdots$	$\vdots$	$\vdots$

$$2^x = \sum_{n=0}^{\infty}\frac{f^{(n)}(0)}{n!}x^n = \sum_{n=0}^{\infty}\frac{(\ln 2)^n}{n!}x^n.$$

$$\lim_{n\to\infty}\left|\frac{a_{n+1}}{a_n}\right| = \lim_{n\to\infty}\left|\frac{(\ln 2)^{n+1}x^{n+1}}{(n+1)!}\cdot\frac{n!}{(\ln 2)^n x^n}\right|$$

$$= \lim_{n\to\infty}\frac{(\ln 2)|x|}{n+1} = 0 < 1 \quad \text{for all } x, \text{ so } R = \infty.$$

19.

n	$f^{(n)}(x)$	$f^{(n)}(0)$
0	$\sinh x$	0
1	$\cosh x$	1
2	$\sinh x$	0
3	$\cosh x$	1
4	$\sinh x$	0
$\vdots$	$\vdots$	$\vdots$

$$f^{(n)}(0) = \begin{cases} 0 & \text{if } n \text{ is even} \\ 1 & \text{if } n \text{ is odd} \end{cases} \quad \text{so } \sinh x = \sum_{n=0}^{\infty}\frac{x^{2n+1}}{(2n+1)!}.$$

Use the Ratio Test to find R. If $a_n = \dfrac{x^{2n+1}}{(2n+1)!}$, then

$$\lim_{n\to\infty}\left|\frac{a_{n+1}}{a_n}\right| = \lim_{n\to\infty}\left|\frac{x^{2n+3}}{(2n+3)!}\cdot\frac{(2n+1)!}{x^{2n+1}}\right| = x^2 \cdot \lim_{n\to\infty}\frac{1}{(2n+3)(2n+2)}$$

$$= 0 < 1 \quad \text{for all } x, \text{ so } R = \infty.$$

21.

n	$f^{(n)}(x)$	$f^{(n)}(2)$
0	$x^5 + 2x^3 + x$	50
1	$5x^4 + 6x^2 + 1$	105
2	$20x^3 + 12x$	184
3	$60x^2 + 12$	252
4	$120x$	240
5	120	120
6	0	0
7	0	0
$\vdots$	$\vdots$	$\vdots$

$f^{(n)}(x) = 0$ for $n \geq 6$, so f has a finite expansion about $a = 2$.

$$f(x) = x^5 + 2x^3 + x = \sum_{n=0}^{5} \frac{f^{(n)}(2)}{n!}(x-2)^n$$

$$= \frac{50}{0!}(x-2)^0 + \frac{105}{1!}(x-2)^1 + \frac{184}{2!}(x-2)^2 + \frac{252}{3!}(x-2)^3$$

$$+ \frac{240}{4!}(x-2)^4 + \frac{120}{5!}(x-2)^5$$

$$= 50 + 105(x-2) + 92(x-2)^2 + 42(x-2)^3$$

$$+ 10(x-2)^4 + (x-2)^5$$

A finite series converges for all x, so $R = \infty$.

23.

n	$f^{(n)}(x)$	$f^{(n)}(2)$
0	$\ln x$	$\ln 2$
1	$1/x$	$1/2$
2	$-1/x^2$	$-1/2^2$
3	$2/x^3$	$2/2^3$
4	$-6/x^4$	$-6/2^4$
5	$24/x^5$	$24/2^5$
$\vdots$	$\vdots$	$\vdots$

$$f(x) = \ln x = \sum_{n=0}^{\infty} \frac{f^{(n)}(2)}{n!}(x-2)^n$$

$$= \frac{\ln 2}{0!}(x-2)^0 + \frac{1}{1!\,2^1}(x-2)^1 + \frac{-1}{2!\,2^2}(x-2)^2 + \frac{2}{3!\,2^3}(x-2)^3$$

$$+ \frac{-6}{4!\,2^4}(x-2)^4 + \frac{24}{5!\,2^5}(x-2)^5 + \cdots$$

$$= \ln 2 + \sum_{n=1}^{\infty} (-1)^{n+1} \frac{(n-1)!}{n!\,2^n}(x-2)^n$$

$$= \ln 2 + \sum_{n=1}^{\infty} (-1)^{n+1} \frac{1}{n\,2^n}(x-2)^n$$

$$\lim_{n \to \infty}\left|\frac{a_{n+1}}{a_n}\right| = \lim_{n \to \infty}\left|\frac{(-1)^{n+2}(x-2)^{n+1}}{(n+1)\,2^{n+1}} \cdot \frac{n\,2^n}{(-1)^{n+1}(x-2)^n}\right| = \lim_{n \to \infty}\left|\frac{(-1)(x-2)n}{(n+1)2}\right| = \lim_{n \to \infty}\left(\frac{n}{n+1}\right)\frac{|x-2|}{2}$$

$$= \frac{|x-2|}{2} < 1 \quad \text{for convergence, so } |x-2| < 2 \text{ and } R = 2.$$

25.

n	$f^{(n)}(x)$	$f^{(n)}(3)$
0	e^{2x}	e^6
1	$2e^{2x}$	$2e^6$
2	$2^2 e^{2x}$	$4e^6$
3	$2^3 e^{2x}$	$8e^6$
4	$2^4 e^{2x}$	$16e^6$
$\vdots$	$\vdots$	$\vdots$

$$f(x) = e^{2x} = \sum_{n=0}^{\infty} \frac{f^{(n)}(3)}{n!}(x-3)^n$$

$$= \frac{e^6}{0!}(x-3)^0 + \frac{2e^6}{1!}(x-3)^1 + \frac{4e^6}{2!}(x-3)^2$$

$$+ \frac{8e^6}{3!}(x-3)^3 + \frac{16e^6}{4!}(x-3)^4 + \cdots$$

$$= \sum_{n=0}^{\infty} \frac{2^n e^6}{n!}(x-3)^n$$

$$\lim_{n \to \infty}\left|\frac{a_{n+1}}{a_n}\right| = \lim_{n \to \infty}\left|\frac{2^{n+1} e^6 (x-3)^{n+1}}{(n+1)!} \cdot \frac{n!}{2^n e^6 (x-3)^n}\right| = \lim_{n \to \infty} \frac{2|x-3|}{n+1} = 0 < 1 \quad \text{for all } x, \text{ so } R = \infty.$$

27.

n	$f^{(n)}(x)$	$f^{(n)}(\pi)$
0	$\sin x$	0
1	$\cos x$	-1
2	$-\sin x$	0
3	$-\cos x$	1
4	$\sin x$	0
5	$\cos x$	-1
6	$-\sin x$	0
7	$-\cos x$	1
$\vdots$	$\vdots$	$\vdots$

$$f(x) = \sin x = \sum_{n=0}^{\infty} \frac{f^{(n)}(\pi)}{n!}(x-\pi)^n$$

$$= \frac{-1}{1!}(x-\pi)^1 + \frac{1}{3!}(x-\pi)^3 + \frac{-1}{5!}(x-\pi)^5 + \frac{1}{7!}(x-\pi)^7 + \cdots$$

$$= \sum_{n=0}^{\infty} \frac{(-1)^{n+1}}{(2n+1)!}(x-\pi)^{2n+1}$$

$$\lim_{n \to \infty} \left| \frac{a_{n+1}}{a_n} \right| = \lim_{n \to \infty} \left| \frac{(-1)^{n+2}(x-\pi)^{2n+3}}{(2n+3)!} \cdot \frac{(2n+1)!}{(-1)^{n+1}(x-\pi)^{2n+1}} \right|$$

$$= \lim_{n \to \infty} \frac{(x-\pi)^2}{(2n+3)(2n+2)} = 0 < 1 \quad \text{for all } x, \text{ so } R = \infty.$$

29.

n	$f^{(n)}(x)$	$f^{(n)}(\pi)$
0	$\sin 2x$	0
1	$2\cos 2x$	2
2	$-4\sin 2x$	0
3	$-8\cos 2x$	-8
4	$16\sin 2x$	0
5	$32\cos 2x$	32
$\vdots$	$\vdots$	$\vdots$

$$f(x) = \sin 2x = f(\pi) + \frac{f'(\pi)}{1!}(x-\pi) + \frac{f''(\pi)}{2!}(x-\pi)^2$$

$$+ \frac{f'''(\pi)}{3!}(x-\pi)^3 + \frac{f^{(4)}(\pi)}{4!}(x-\pi)^4 + \cdots$$

$$= 0 + \frac{2}{1!}(x-\pi) + 0 - \frac{8}{3!}(x-\pi)^3 + 0 + \frac{32}{5!}(x-\pi)^5 - \cdots$$

$$= \frac{2}{1!}(x-\pi) - \frac{8}{3!}(x-\pi)^3 + \frac{32}{5!}(x-\pi)^5 - \cdots$$

$$= \sum_{n=0}^{\infty} \frac{(-1)^n \, 2^{2n+1}}{(2n+1)!}(x-\pi)^{2n+1}$$

$$\lim_{n \to \infty} \left| \frac{a_{n+1}}{a_n} \right| = \lim_{n \to \infty} \left| \frac{(-1)^{n+1} \, 2^{2n+3}(x-\pi)^{2n+3}}{(2n+3)!} \cdot \frac{(2n+1)!}{(-1)^n \, 2^{2n+1}(x-\pi)^{2n+1}} \right|$$

$$= (x-\pi)^2 \lim_{n \to \infty} \frac{2^2}{(2n+3)(2n+2)} = 0 < 1 \text{ for all } x, \text{ so } R = \infty.$$

31. If $f(x) = \cos x$, then $f^{(n+1)}(x) = \pm \sin x$ or $\pm \cos x$. In each case, $\left| f^{(n+1)}(x) \right| \leq 1$, so by Formula 9 with $a = 0$ and $M = 1$, $|R_n(x)| \leq \frac{1}{(n+1)!}|x|^{n+1}$. Thus, $|R_n(x)| \to 0$ as $n \to \infty$ by Equation 10. So $\lim_{n \to \infty} R_n(x) = 0$ and, by Theorem 8, the series in Exercise 13 represents $\cos x$ for all x.

33. If $f(x) = \sinh x$, then for all n, $f^{(n+1)}(x) = \cosh x$ or $\sinh x$. Since $|\sinh x| < |\cosh x| = \cosh x$ for all x, we have $\left| f^{(n+1)}(x) \right| \leq \cosh x$ for all n. If d is any positive number and $|x| \leq d$, then $\left| f^{(n+1)}(x) \right| \leq \cosh x \leq \cosh d$, so by Formula 9 with $a = 0$ and $M = \cosh d$, we have $|R_n(x)| \leq \frac{\cosh d}{(n+1)!}|x|^{n+1}$. It follows that $|R_n(x)| \to 0$ as $n \to \infty$ for $|x| \leq d$ (by Equation 10). But d was an arbitrary positive number. So by Theorem 8, the series represents $\sinh x$ for all x.

35. $\sqrt[4]{1-x} = [1+(-x)]^{1/4} = \sum_{n=0}^{\infty} \binom{1/4}{n}(-x)^n$

$= 1 + \frac{1}{4}(-x) + \frac{\frac{1}{4}\left(-\frac{3}{4}\right)}{2!}(-x)^2 + \frac{\frac{1}{4}\left(-\frac{3}{4}\right)\left(-\frac{7}{4}\right)}{3!}(-x)^3 + \cdots$

$= 1 - \frac{1}{4}x + \sum_{n=2}^{\infty} \frac{(-1)^{n-1}(-1)^n \cdot [3 \cdot 7 \cdot \cdots \cdot (4n-5)]}{4^n \cdot n!} x^n$

$= 1 - \frac{1}{4}x - \sum_{n=2}^{\infty} \frac{3 \cdot 7 \cdot \cdots \cdot (4n-5)}{4^n \cdot n!} x^n$

and $|-x| < 1 \Leftrightarrow |x| < 1$, so $R = 1$.

37. $\frac{1}{(2+x)^3} = \frac{1}{[2(1+x/2)]^3} = \frac{1}{8}\left(1+\frac{x}{2}\right)^{-3} = \frac{1}{8}\sum_{n=0}^{\infty}\binom{-3}{n}\left(\frac{x}{2}\right)^n$. The binomial coefficient is

$\binom{-3}{n} = \frac{(-3)(-4)(-5)\cdot\cdots\cdot(-3-n+1)}{n!} = \frac{(-3)(-4)(-5)\cdot\cdots\cdot[-(n+2)]}{n!}$

$= \frac{(-1)^n \cdot 2 \cdot 3 \cdot 4 \cdot 5 \cdot \cdots \cdot (n+1)(n+2)}{2 \cdot n!} = \frac{(-1)^n (n+1)(n+2)}{2}$

Thus, $\frac{1}{(2+x)^3} = \frac{1}{8}\sum_{n=0}^{\infty}\frac{(-1)^n(n+1)(n+2)}{2}\frac{x^n}{2^n} = \sum_{n=0}^{\infty}\frac{(-1)^n(n+1)(n+2)x^n}{2^{n+4}}$ for $\left|\frac{x}{2}\right| < 1 \Leftrightarrow |x| < 2$, so $R = 2$.

39. $\arctan x = \sum_{n=0}^{\infty}(-1)^n\frac{x^{2n+1}}{2n+1}$, so $f(x) = \arctan(x^2) = \sum_{n=0}^{\infty}(-1)^n\frac{(x^2)^{2n+1}}{2n+1} = \sum_{n=0}^{\infty}(-1)^n\frac{1}{2n+1}x^{4n+2}$, $R = 1$.

41. $\cos x = \sum_{n=0}^{\infty}(-1)^n\frac{x^{2n}}{(2n)!} \Rightarrow \cos 2x = \sum_{n=0}^{\infty}(-1)^n\frac{(2x)^{2n}}{(2n)!} = \sum_{n=0}^{\infty}(-1)^n\frac{2^{2n}x^{2n}}{(2n)!}$, so

$f(x) = x\cos 2x = \sum_{n=0}^{\infty}(-1)^n\frac{2^{2n}}{(2n)!}x^{2n+1}$, $R = \infty$.

43. $\cos x = \sum_{n=0}^{\infty}(-1)^n\frac{x^{2n}}{(2n)!} \Rightarrow \cos(\tfrac{1}{2}x^2) = \sum_{n=0}^{\infty}(-1)^n\frac{\left(\tfrac{1}{2}x^2\right)^{2n}}{(2n)!} = \sum_{n=0}^{\infty}(-1)^n\frac{x^{4n}}{2^{2n}(2n)!}$, so

$f(x) = x\cos(\tfrac{1}{2}x^2) = \sum_{n=0}^{\infty}(-1)^n\frac{1}{2^{2n}(2n)!}x^{4n+1}$, $R = \infty$.

45. We must write the binomial in the form $(1+\text{expression})$, so we'll factor out a 4.

$\frac{x}{\sqrt{4+x^2}} = \frac{x}{\sqrt{4(1+x^2/4)}} = \frac{x}{2\sqrt{1+x^2/4}} = \frac{x}{2}\left(1+\frac{x^2}{4}\right)^{-1/2} = \frac{x}{2}\sum_{n=0}^{\infty}\binom{-\tfrac{1}{2}}{n}\left(\frac{x^2}{4}\right)^n$

$= \frac{x}{2}\left[1 + \left(-\tfrac{1}{2}\right)\frac{x^2}{4} + \frac{\left(-\tfrac{1}{2}\right)\left(-\tfrac{3}{2}\right)}{2!}\left(\frac{x^2}{4}\right)^2 + \frac{\left(-\tfrac{1}{2}\right)\left(-\tfrac{3}{2}\right)\left(-\tfrac{5}{2}\right)}{3!}\left(\frac{x^2}{4}\right)^3 + \cdots\right]$

$= \frac{x}{2} + \frac{x}{2}\sum_{n=1}^{\infty}(-1)^n\frac{1\cdot 3\cdot 5\cdot\cdots\cdot(2n-1)}{2^n\cdot 4^n\cdot n!}x^{2n}$

$= \frac{x}{2} + \sum_{n=1}^{\infty}(-1)^n\frac{1\cdot 3\cdot 5\cdot\cdots\cdot(2n-1)}{n!\,2^{3n+1}}x^{2n+1}$ and $\frac{x^2}{4} < 1 \Leftrightarrow \frac{|x|}{2} < 1 \Leftrightarrow |x| < 2$, so $R = 2$.

47. $\sin^2 x = \frac{1}{2}(1-\cos 2x) = \frac{1}{2}\left[1 - \sum_{n=0}^{\infty}\frac{(-1)^n (2x)^{2n}}{(2n)!}\right] = \frac{1}{2}\left[1 - 1 - \sum_{n=1}^{\infty}\frac{(-1)^n (2x)^{2n}}{(2n)!}\right] = \sum_{n=1}^{\infty}\frac{(-1)^{n+1} 2^{2n-1} x^{2n}}{(2n)!}$,
$R = \infty$

49. (a) The Maclaurin series for e^x is $e^x = 1 + x + \frac{1}{2!}x^2 + \frac{1}{3!}x^3 + \cdots$, so

$e^{-x} = 1 - x + \frac{1}{2!}(-x)^2 + \frac{1}{3!}(-x)^3 + \cdots = 1 - x + \frac{1}{2!}x^2 - \frac{1}{3!}x^3 + \cdots$, and

$\sinh x = \frac{1}{2}(e^x - e^{-x}) = \frac{1}{2}\left[\left(1 + x + \frac{1}{2!}x^2 + \frac{1}{3!}x^3 + \cdots\right) - \left(1 - x + \frac{1}{2!}x^2 - \frac{1}{3!}x^3 + \cdots\right)\right]$

$= \frac{1}{2}\left[(1-1) + (x+x) + \left(\frac{1}{2!}x^2 - \frac{1}{2!}x^2\right) + \left(\frac{1}{3!}x^3 + \frac{1}{3!}x^3\right) + \cdots\right]$

$= \frac{1}{2}\left[2x + \frac{2}{3!}x^3 + \frac{2}{5!}x^5 + \cdots\right] = x + \frac{1}{3!}x^3 + \frac{1}{5!}x^5 + \cdots = \sum_{n=0}^{\infty}\frac{x^{2n+1}}{(2n+1)!}$

(b) $\cosh x = \frac{1}{2}(e^x + e^{-x}) = \frac{1}{2}\left[\left(1 + x + \frac{1}{2!}x^2 + \frac{1}{3!}x^3 + \cdots\right) + \left(1 - x + \frac{1}{2!}x^2 - \frac{1}{3!}x^3 + \cdots\right)\right]$

$= \frac{1}{2}\left[(1+1) + (x-x) + \left(\frac{1}{2!}x^2 + \frac{1}{2!}x^2\right) + \left(\frac{1}{3!}x^3 - \frac{1}{3!}x^3\right) + \cdots\right]$

$= \frac{1}{2}\left[2 + \frac{2}{2!}x^2 + \frac{2}{4!}x^4 + \cdots\right] = 1 + \frac{1}{2!}x^2 + \frac{1}{4!}x^4 + \cdots = \sum_{n=0}^{\infty}\frac{x^{2n}}{(2n)!}$

51. $\cos x \overset{(16)}{=} \sum_{n=0}^{\infty}(-1)^n \frac{x^{2n}}{(2n)!} \Rightarrow$

$f(x) = \cos(x^2) = \sum_{n=0}^{\infty}\frac{(-1)^n (x^2)^{2n}}{(2n)!} = \sum_{n=0}^{\infty}\frac{(-1)^n x^{4n}}{(2n)!}$

$= 1 - \frac{1}{2}x^4 + \frac{1}{24}x^8 - \frac{1}{720}x^{12} + \cdots$

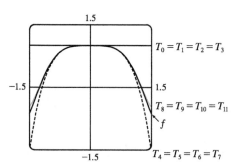

The series for $\cos x$ converges for all x, so the same is true of the series for $f(x)$, that is, $R = \infty$. Notice that, as n increases, $T_n(x)$ becomes a better approximation to $f(x)$.

53. $e^x \overset{(11)}{=} \sum_{n=0}^{\infty}\frac{x^n}{n!}$, so $e^{-x} = \sum_{n=0}^{\infty}\frac{(-x)^n}{n!} = \sum_{n=0}^{\infty}(-1)^n \frac{x^n}{n!}$, so

$f(x) = xe^{-x} = \sum_{n=0}^{\infty}(-1)^n \frac{1}{n!}x^{n+1}$

$= x - x^2 + \frac{1}{2}x^3 - \frac{1}{6}x^4 + \frac{1}{24}x^5 - \frac{1}{120}x^6 + \cdots$

$= \sum_{n=1}^{\infty}(-1)^{n-1}\frac{x^n}{(n-1)!}$

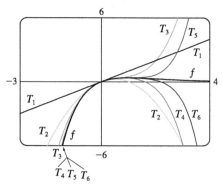

The series for e^x converges for all x, so the same is true of the series for $f(x)$; that is, $R = \infty$. From the graphs of f and the first few Taylor polynomials, we see that $T_n(x)$ provides a closer fit to $f(x)$ near 0 as n increases.

55. $5° = 5° \left(\dfrac{\pi}{180°}\right) = \dfrac{\pi}{36}$ radians and $\cos x = \sum\limits_{n=0}^{\infty} (-1)^n \dfrac{x^{2n}}{(2n)!} = 1 - \dfrac{x^2}{2!} + \dfrac{x^4}{4!} - \dfrac{x^6}{6!} + \cdots$, so

$\cos \dfrac{\pi}{36} = 1 - \dfrac{(\pi/36)^2}{2!} + \dfrac{(\pi/36)^4}{4!} - \dfrac{(\pi/36)^6}{6!} + \cdots$. Now $1 - \dfrac{(\pi/36)^2}{2!} \approx 0.99619$ and adding $\dfrac{(\pi/36)^4}{4!} \approx 2.4 \times 10^{-6}$

does not affect the fifth decimal place, so $\cos 5° \approx 0.99619$ by the Alternating Series Estimation Theorem.

57. (a) $1/\sqrt{1-x^2} = [1 + (-x^2)]^{-1/2} = 1 + \left(-\dfrac{1}{2}\right)(-x^2) + \dfrac{\left(-\frac{1}{2}\right)\left(-\frac{3}{2}\right)}{2!}(-x^2)^2 + \dfrac{\left(-\frac{1}{2}\right)\left(-\frac{3}{2}\right)\left(-\frac{5}{2}\right)}{3!}(-x^2)^3 + \cdots$

$= 1 + \sum\limits_{n=1}^{\infty} \dfrac{1 \cdot 3 \cdot 5 \cdots (2n-1)}{2^n \cdot n!} x^{2n}$

(b) $\sin^{-1} x = \displaystyle\int \dfrac{1}{\sqrt{1-x^2}}\, dx = C + x + \sum\limits_{n=1}^{\infty} \dfrac{1 \cdot 3 \cdot 5 \cdots (2n-1)}{(2n+1)2^n \cdot n!} x^{2n+1}$

$= x + \sum\limits_{n=1}^{\infty} \dfrac{1 \cdot 3 \cdot 5 \cdots (2n-1)}{(2n+1)2^n \cdot n!} x^{2n+1}$ since $0 = \sin^{-1} 0 = C$.

59. $\sqrt{1+x^3} = (1+x^3)^{1/2} = \sum\limits_{n=0}^{\infty} \binom{\frac{1}{2}}{n}(x^3)^n = \sum\limits_{n=0}^{\infty} \binom{\frac{1}{2}}{n} x^{3n} \;\Rightarrow\; \displaystyle\int \sqrt{1+x^3}\, dx = C + \sum\limits_{n=0}^{\infty} \binom{\frac{1}{2}}{n} \dfrac{x^{3n+1}}{3n+1}$,

with $R = 1$.

61. $\cos x \stackrel{(16)}{=} \sum\limits_{n=0}^{\infty} (-1)^n \dfrac{x^{2n}}{(2n)!} \;\Rightarrow\; \cos x - 1 = \sum\limits_{n=1}^{\infty} (-1)^n \dfrac{x^{2n}}{(2n)!} \;\Rightarrow\; \dfrac{\cos x - 1}{x} = \sum\limits_{n=1}^{\infty} (-1)^n \dfrac{x^{2n-1}}{(2n)!} \;\Rightarrow\;$

$\displaystyle\int \dfrac{\cos x - 1}{x}\, dx = C + \sum\limits_{n=1}^{\infty} (-1)^n \dfrac{x^{2n}}{2n \cdot (2n)!}$, with $R = \infty$.

63. $\arctan x = \sum\limits_{n=0}^{\infty} (-1)^n \dfrac{x^{2n+1}}{2n+1}$ for $|x| < 1$, so $x^3 \arctan x = \sum\limits_{n=0}^{\infty} (-1)^n \dfrac{x^{2n+4}}{2n+1}$ for $|x| < 1$ and

$\displaystyle\int x^3 \arctan x\, dx = C + \sum\limits_{n=0}^{\infty} (-1)^n \dfrac{x^{2n+5}}{(2n+1)(2n+5)}$. Since $\dfrac{1}{2} < 1$, we have

$\displaystyle\int_0^{1/2} x^3 \arctan x\, dx = \sum\limits_{n=0}^{\infty} (-1)^n \dfrac{(1/2)^{2n+5}}{(2n+1)(2n+5)} = \dfrac{(1/2)^5}{1 \cdot 5} - \dfrac{(1/2)^7}{3 \cdot 7} + \dfrac{(1/2)^9}{5 \cdot 9} - \dfrac{(1/2)^{11}}{7 \cdot 11} + \cdots$. Now

$\dfrac{(1/2)^5}{1 \cdot 5} - \dfrac{(1/2)^7}{3 \cdot 7} + \dfrac{(1/2)^9}{5 \cdot 9} \approx 0.0059$ and subtracting $\dfrac{(1/2)^{11}}{7 \cdot 11} \approx 6.3 \times 10^{-6}$ does not affect the fourth decimal place,

so $\int_0^{1/2} x^3 \arctan x\, dx \approx 0.0059$ by the Alternating Series Estimation Theorem.

65. $\sqrt{1+x^4} = (1+x^4)^{1/2} = \sum\limits_{n=0}^{\infty} \binom{\frac{1}{2}}{n}(x^4)^n$, so $\displaystyle\int \sqrt{1+x^4}\, dx = C + \sum\limits_{n=0}^{\infty} \binom{\frac{1}{2}}{n} \dfrac{x^{4n+1}}{4n+1}$ and hence, since $0.4 < 1$,

we have

$I = \displaystyle\int_0^{0.4} \sqrt{1+x^4}\, dx = \sum\limits_{n=0}^{\infty} \binom{\frac{1}{2}}{n} \dfrac{(0.4)^{4n+1}}{4n+1}$

$= (1)\dfrac{(0.4)^1}{0!} + \dfrac{\frac{1}{2}}{1!}\dfrac{(0.4)^5}{5} + \dfrac{\frac{1}{2}\left(-\frac{1}{2}\right)}{2!}\dfrac{(0.4)^9}{9} + \dfrac{\frac{1}{2}\left(-\frac{1}{2}\right)\left(-\frac{3}{2}\right)}{3!}\dfrac{(0.4)^{13}}{13} + \dfrac{\frac{1}{2}\left(-\frac{1}{2}\right)\left(-\frac{3}{2}\right)\left(-\frac{5}{2}\right)}{4!}\dfrac{(0.4)^{17}}{17} + \cdots$

$= 0.4 + \dfrac{(0.4)^5}{10} - \dfrac{(0.4)^9}{72} + \dfrac{(0.4)^{13}}{208} - \dfrac{5(0.4)^{17}}{2176} + \cdots$

[continued]

Now $\dfrac{(0.4)^9}{72} \approx 3.6 \times 10^{-6} < 5 \times 10^{-6}$, so by the Alternating Series Estimation Theorem, $I \approx 0.4 + \dfrac{(0.4)^5}{10} \approx 0.40102$ (correct to five decimal places).

67. $\lim\limits_{x \to 0} \dfrac{x - \ln(1+x)}{x^2} = \lim\limits_{x \to 0} \dfrac{x - (x - \frac{1}{2}x^2 + \frac{1}{3}x^3 - \frac{1}{4}x^4 + \frac{1}{5}x^5 - \cdots)}{x^2} = \lim\limits_{x \to 0} \dfrac{\frac{1}{2}x^2 - \frac{1}{3}x^3 + \frac{1}{4}x^4 - \frac{1}{5}x^5 + \cdots}{x^2}$

$= \lim\limits_{x \to 0} \left(\frac{1}{2} - \frac{1}{3}x + \frac{1}{4}x^2 - \frac{1}{5}x^3 + \cdots \right) = \frac{1}{2}$

since power series are continuous functions.

69. $\lim\limits_{x \to 0} \dfrac{\sin x - x + \frac{1}{6}x^3}{x^5} = \lim\limits_{x \to 0} \dfrac{\left(x - \frac{1}{3!}x^3 + \frac{1}{5!}x^5 - \frac{1}{7!}x^7 + \cdots\right) - x + \frac{1}{6}x^3}{x^5}$

$= \lim\limits_{x \to 0} \dfrac{\frac{1}{5!}x^5 - \frac{1}{7!}x^7 + \cdots}{x^5} = \lim\limits_{x \to 0} \left(\dfrac{1}{5!} - \dfrac{x^2}{7!} + \dfrac{x^4}{9!} - \cdots \right) = \dfrac{1}{5!} = \dfrac{1}{120}$

since power series are continuous functions.

71. $\lim\limits_{x \to 0} \dfrac{x^3 - 3x + 3\tan^{-1}x}{x^5} = \lim\limits_{x \to 0} \dfrac{x^3 - 3x + 3\left(x - \frac{1}{3}x^3 + \frac{1}{5}x^5 - \frac{1}{7}x^7 + \cdots\right)}{x^5}$

$= \lim\limits_{x \to 0} \dfrac{x^3 - 3x + 3x - x^3 + \frac{3}{5}x^5 - \frac{3}{7}x^7 + \cdots}{x^5} = \lim\limits_{x \to 0} \dfrac{\frac{3}{5}x^5 - \frac{3}{7}x^7 + \cdots}{x^5}$

$= \lim\limits_{x \to 0} \left(\frac{3}{5} - \frac{3}{7}x^2 + \cdots \right) = \frac{3}{5}$ since power series are continuous functions.

73. From Equation 11, we have $e^{-x^2} = 1 - \dfrac{x^2}{1!} + \dfrac{x^4}{2!} - \dfrac{x^6}{3!} + \cdots$ and we know that $\cos x = 1 - \dfrac{x^2}{2!} + \dfrac{x^4}{4!} - \cdots$ from Equation 16. Therefore, $e^{-x^2} \cos x = \left(1 - x^2 + \frac{1}{2}x^4 - \cdots\right)\left(1 - \frac{1}{2}x^2 + \frac{1}{24}x^4 - \cdots\right)$. Writing only the terms with degree ≤ 4, we get $e^{-x^2} \cos x = 1 - \frac{1}{2}x^2 + \frac{1}{24}x^4 - x^2 + \frac{1}{2}x^4 + \frac{1}{2}x^4 + \cdots = 1 - \frac{3}{2}x^2 + \frac{25}{24}x^4 + \cdots$.

75. $\dfrac{x}{\sin x} \stackrel{(15)}{=} \dfrac{x}{x - \frac{1}{6}x^3 + \frac{1}{120}x^5 - \cdots}$.

$\quad\quad\quad\quad\quad\quad\quad\quad\quad\quad\quad\quad\quad\quad 1 + \frac{1}{6}x^2 + \frac{7}{360}x^4 + \cdots$

$x - \frac{1}{6}x^3 + \frac{1}{120}x^5 - \cdots \overline{\big) \; x}$

$\quad\quad\quad\quad\quad\quad\quad\quad\quad\quad\;\; x - \frac{1}{6}x^3 + \frac{1}{120}x^5 - \cdots$

$\quad\quad\quad\quad\quad\quad\quad\quad\quad\quad\quad\quad\quad\;\; \frac{1}{6}x^3 - \frac{1}{120}x^5 + \cdots$

$\quad\quad\quad\quad\quad\quad\quad\quad\quad\quad\quad\quad\quad\;\; \frac{1}{6}x^3 - \frac{1}{36}x^5 + \cdots$

$\quad\quad\quad\quad\quad\quad\quad\quad\quad\quad\quad\quad\quad\quad\quad\quad\;\; \frac{7}{360}x^5 + \cdots$

$\quad\quad\quad\quad\quad\quad\quad\quad\quad\quad\quad\quad\quad\quad\quad\quad\;\; \frac{7}{360}x^5 + \cdots$

$\quad\quad\quad\quad\quad\quad\quad\quad\quad\quad\quad\quad\quad\quad\quad\quad\quad\quad\quad \cdots$

From the long division above, $\dfrac{x}{\sin x} = 1 + \frac{1}{6}x^2 + \frac{7}{360}x^4 + \cdots$.

77. $y = (\arctan x)^2 = \left(x - \frac{1}{3}x^3 + \frac{1}{5}x^5 - \frac{1}{7}x^7 + \cdots\right)\left(x - \frac{1}{3}x^3 + \frac{1}{5}x^5 - \frac{1}{7}x^7 + \cdots\right)$. Writing only the terms with degree ≤ 6, we get $(\arctan x)^2 = x^2 - \frac{1}{3}x^4 + \frac{1}{5}x^6 - \frac{1}{3}x^4 + \frac{1}{9}x^6 + \frac{1}{5}x^6 + \cdots = x^2 - \frac{2}{3}x^4 + \frac{23}{45}x^6 + \cdots$.

79. $\sum\limits_{n=0}^{\infty} (-1)^n \dfrac{x^{4n}}{n!} = \sum\limits_{n=0}^{\infty} \dfrac{(-x^4)^n}{n!} = e^{-x^4}$, by (11).

81. $\sum_{n=0}^{\infty} (-1)^n \dfrac{x^{2n+1}}{2^{2n+1}(2n+1)} = \sum_{n=0}^{\infty} (-1)^n \dfrac{(x/2)^{2n+1}}{2n+1} = \tan^{-1}\left(\dfrac{x}{2}\right)$ [from Table 1]

83. $\sum_{n=0}^{\infty} \dfrac{(-1)^n}{n!}$ is the Maclaurin series for e^x evaluated at $x = -1$. Thus, $\sum_{n=0}^{\infty} \dfrac{(-1)^n}{n!} = e^{-1}$ by (11).

85. $\sum_{n=1}^{\infty} (-1)^{n-1} \dfrac{3^n}{n 5^n} = \sum_{n=1}^{\infty} (-1)^{n-1} \dfrac{(3/5)^n}{n} = \ln\left(1 + \dfrac{3}{5}\right)$ [from Table 1] $= \ln \dfrac{8}{5}$

87. $\sum_{n=0}^{\infty} \dfrac{(-1)^n \pi^{2n+1}}{4^{2n+1}(2n+1)!} = \sum_{n=0}^{\infty} \dfrac{(-1)^n \left(\frac{\pi}{4}\right)^{2n+1}}{(2n+1)!} = \sin \dfrac{\pi}{4} = \dfrac{1}{\sqrt{2}}$, by (15).

89. $3 + \dfrac{9}{2!} + \dfrac{27}{3!} + \dfrac{81}{4!} + \cdots = \dfrac{3^1}{1!} + \dfrac{3^2}{2!} + \dfrac{3^3}{3!} + \dfrac{3^4}{4!} + \cdots = \sum_{n=1}^{\infty} \dfrac{3^n}{n!} = \sum_{n=0}^{\infty} \dfrac{3^n}{n!} - 1 = e^3 - 1$, by (11).

91. If p is an nth-degree polynomial, then $p^{(i)}(x) = 0$ for $i > n$, so its Taylor series at a is $p(x) = \sum_{i=0}^{n} \dfrac{p^{(i)}(a)}{i!}(x-a)^i$.

Put $x - a = 1$, so that $x = a + 1$. Then $p(a+1) = \sum_{i=0}^{n} \dfrac{p^{(i)}(a)}{i!}$.

This is true for any a, so replace a by x: $p(x+1) = \sum_{i=0}^{n} \dfrac{p^{(i)}(x)}{i!}$

93. Using Equation 15, we have

$$x \sin(x^2) = x \cdot \sum_{n=0}^{\infty} (-1)^n \dfrac{(x^2)^{2n+1}}{(2n+1)!} = x \cdot \sum_{n=0}^{\infty} (-1)^n \dfrac{x^{4n+2}}{(2n+1)!} = \sum_{n=0}^{\infty} (-1)^n \dfrac{x^{4n+3}}{(2n+1)!}$$

The x^{203} term is obtained when $n = 50$ and is $(-1)^{50} \dfrac{x^{203}}{101!} = \dfrac{1}{101!} x^{203}$. So the coefficient of x^{203} is $\dfrac{1}{101!}$, which, by

Equation 7, must equal $\dfrac{f^{(203)}(0)}{203!}$. Thus, $\dfrac{f^{(203)}(0)}{203!} = \dfrac{1}{101!} \;\Rightarrow\; f^{(203)}(0) = \dfrac{203!}{101!}$.

95. Assume that $|f'''(x)| \leq M$, so $f'''(x) \leq M$ for $a \leq x \leq a + d$. Now $\int_a^x f'''(t)\, dt \leq \int_a^x M\, dt \;\Rightarrow\;$

$f''(x) - f''(a) \leq M(x - a) \;\Rightarrow\; f''(x) \leq f''(a) + M(x - a)$. Thus, $\int_a^x f''(t)\, dt \leq \int_a^x [f''(a) + M(t-a)]\, dt \;\Rightarrow\;$

$f'(x) - f'(a) \leq f''(a)(x-a) + \tfrac{1}{2} M(x-a)^2 \;\Rightarrow\; f'(x) \leq f'(a) + f''(a)(x-a) + \tfrac{1}{2} M(x-a)^2 \;\Rightarrow\;$

$\int_a^x f'(t)\, dt \leq \int_a^x \left[f'(a) + f''(a)(t-a) + \tfrac{1}{2} M(t-a)^2 \right] dt \;\Rightarrow\;$

$f(x) - f(a) \leq f'(a)(x-a) + \tfrac{1}{2} f''(a)(x-a)^2 + \tfrac{1}{6} M(x-a)^3$. So

$f(x) - f(a) - f'(a)(x-a) - \tfrac{1}{2} f''(a)(x-a)^2 \leq \tfrac{1}{6} M(x-a)^3$. But

$R_2(x) = f(x) - T_2(x) = f(x) - f(a) - f'(a)(x-a) - \tfrac{1}{2} f''(a)(x-a)^2$, so $R_2(x) \leq \tfrac{1}{6} M(x-a)^3$.

A similar argument using $f'''(x) \geq -M$ shows that $R_2(x) \geq -\tfrac{1}{6} M(x-a)^3$. So $|R_2(x_2)| \leq \tfrac{1}{6} M |x-a|^3$.

Although we have assumed that $x > a$, a similar calculation shows that this inequality is also true if $x < a$.

97. (a) $g(x) = \sum_{n=0}^{\infty} \binom{k}{n} x^n \;\Rightarrow\; g'(x) = \sum_{n=1}^{\infty} \binom{k}{n} n x^{n-1}$, so

$$(1+x)g'(x) = (1+x) \sum_{n=1}^{\infty} \binom{k}{n} n x^{n-1} = \sum_{n=1}^{\infty} \binom{k}{n} n x^{n-1} + \sum_{n=1}^{\infty} \binom{k}{n} n x^n$$

$$= \sum_{n=0}^{\infty} \binom{k}{n+1} (n+1) x^n + \sum_{n=0}^{\infty} \binom{k}{n} n x^n \quad \left[\begin{array}{l}\text{Replace } n \text{ with } n+1 \\ \text{in the first series}\end{array}\right]$$

$$= \sum_{n=0}^{\infty} (n+1) \frac{k(k-1)(k-2)\cdots(k-n+1)(k-n)}{(n+1)!} x^n + \sum_{n=0}^{\infty} \left[(n) \frac{k(k-1)(k-2)\cdots(k-n+1)}{n!}\right] x^n$$

$$= \sum_{n=0}^{\infty} \frac{(n+1) k(k-1)(k-2)\cdots(k-n+1)}{(n+1)!} [(k-n)+n] x^n$$

$$= k \sum_{n=0}^{\infty} \frac{k(k-1)(k-2)\cdots(k-n+1)}{n!} x^n = k \sum_{n=0}^{\infty} \binom{k}{n} x^n = k g(x)$$

Thus, $g'(x) = \dfrac{k g(x)}{1+x}$.

(b) $h(x) = (1+x)^{-k} g(x) \;\Rightarrow\;$

$$h'(x) = -k(1+x)^{-k-1} g(x) + (1+x)^{-k} g'(x) \quad \text{[Product Rule]}$$

$$= -k(1+x)^{-k-1} g(x) + (1+x)^{-k} \frac{k g(x)}{1+x} \quad \text{[from part (a)]}$$

$$= -k(1+x)^{-k-1} g(x) + k(1+x)^{-k-1} g(x) = 0$$

(c) From part (b) we see that $h(x)$ must be constant for $x \in (-1, 1)$, so $h(x) = h(0) = 1$ for $x \in (-1, 1)$.

Thus, $h(x) = 1 = (1+x)^{-k} g(x) \;\Leftrightarrow\; g(x) = (1+x)^k$ for $x \in (-1, 1)$.

11.11 Applications of Taylor Polynomials

1. (a)

n	$f^{(n)}(x)$	$f^{(n)}(0)$	$T_n(x)$
0	$\sin x$	0	0
1	$\cos x$	1	x
2	$-\sin x$	0	x
3	$-\cos x$	-1	$x - \tfrac{1}{6}x^3$
4	$\sin x$	0	$x - \tfrac{1}{6}x^3$
5	$\cos x$	1	$x - \tfrac{1}{6}x^3 + \tfrac{1}{120}x^5$

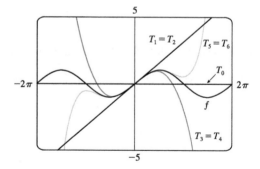

Note: $T_n(x) = \sum_{k=0}^{n} \dfrac{f^{(k)}(0)}{k!} x^k$

(b)

x	f	$T_0(x)$	$T_1(x) = T_2(x)$	$T_3(x) = T_4(x)$	$T_5(x)$
$\frac{\pi}{4}$	0.7071	0	0.7854	0.7047	0.7071
$\frac{\pi}{2}$	1	0	1.5708	0.9248	1.0045
π	0	0	3.1416	-2.0261	0.5240

(c) As n increases, $T_n(x)$ is a good approximation to $f(x)$ on a larger and larger interval.

3.

n	$f^{(n)}(x)$	$f^{(n)}(1)$
0	e^x	e
1	e^x	e
2	e^x	e
3	e^x	e

$T_3(x) = \sum_{n=0}^{3} \dfrac{f^{(n)}(1)}{n!}(x-1)^n$

$ = \dfrac{e}{0!}(x-1)^0 + \dfrac{e}{1!}(x-1)^1 + \dfrac{e}{2!}(x-1)^2 + \dfrac{e}{3!}(x-1)^3$

$ = e + e(x-1) + \tfrac{1}{2}e(x-1)^2 + \tfrac{1}{6}e(x-1)^3$

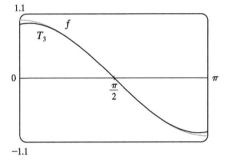

5.

n	$f^{(n)}(x)$	$f^{(n)}(\pi/2)$
0	$\cos x$	0
1	$-\sin x$	-1
2	$-\cos x$	0
3	$\sin x$	1

$T_3(x) = \sum_{n=0}^{3} \dfrac{f^{(n)}(\pi/2)}{n!}\left(x-\tfrac{\pi}{2}\right)^n$

$ = -\left(x-\tfrac{\pi}{2}\right) + \tfrac{1}{6}\left(x-\tfrac{\pi}{2}\right)^3$

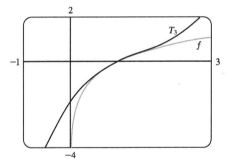

7.

n	$f^{(n)}(x)$	$f^{(n)}(1)$
0	$\ln x$	0
1	$1/x$	1
2	$-1/x^2$	-1
3	$2/x^3$	2

$T_3(x) = \sum_{n=0}^{3} \dfrac{f^{(n)}(1)}{n!}(x-1)^n$

$ = 0 + \dfrac{1}{1!}(x-1) + \dfrac{-1}{2!}(x-1)^2 + \dfrac{2}{3!}(x-1)^3$

$ = (x-1) - \tfrac{1}{2}(x-1)^2 + \tfrac{1}{3}(x-1)^3$

9.

n	$f^{(n)}(x)$	$f^{(n)}(0)$
0	xe^{-2x}	0
1	$(1-2x)e^{-2x}$	1
2	$4(x-1)e^{-2x}$	-4
3	$4(3-2x)e^{-2x}$	12

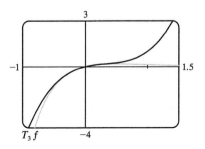

$T_3(x) = \sum_{n=0}^{3} \dfrac{f^{(n)}(0)}{n!}x^n = \tfrac{0}{1}\cdot 1 + \tfrac{1}{1}x^1 + \tfrac{-4}{2}x^2 + \tfrac{12}{6}x^3 = x - 2x^2 + 2x^3$

11. You may be able to simply find the Taylor polynomials for $f(x) = \cot x$ using your CAS. We will list the values of $f^{(n)}(\pi/4)$ for $n = 0$ to $n = 5$.

n	0	1	2	3	4	5
$f^{(n)}(\pi/4)$	1	-2	4	-16	80	-512

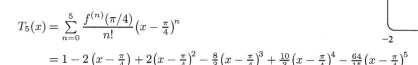

$$T_5(x) = \sum_{n=0}^{5} \frac{f^{(n)}(\pi/4)}{n!} \left(x - \tfrac{\pi}{4}\right)^n$$

$$= 1 - 2\left(x - \tfrac{\pi}{4}\right) + 2\left(x - \tfrac{\pi}{4}\right)^2 - \tfrac{8}{3}\left(x - \tfrac{\pi}{4}\right)^3 + \tfrac{10}{3}\left(x - \tfrac{\pi}{4}\right)^4 - \tfrac{64}{15}\left(x - \tfrac{\pi}{4}\right)^5$$

For $n = 2$ to $n = 5$, $T_n(x)$ is the polynomial consisting of all the terms up to and including the $\left(x - \tfrac{\pi}{4}\right)^n$ term.

13. (a)

n	$f^{(n)}(x)$	$f^{(n)}(1)$
0	$1/x$	1
1	$-1/x^2$	-1
2	$2/x^3$	2
3	$-6/x^4$	

$f(x) = 1/x \approx T_2(x)$

$$= \frac{1}{0!}(x-1)^0 - \frac{1}{1!}(x-1)^1 + \frac{2}{2!}(x-1)^2$$

$$= 1 - (x-1) + (x-1)^2$$

(b) $|R_2(x)| \leq \dfrac{M}{3!}|x-1|^3$, where $|f'''(x)| \leq M$. Now $0.7 \leq x \leq 1.3 \;\Rightarrow\; |x-1| \leq 0.3 \;\Rightarrow\; |x-1|^3 \leq 0.027$.

Since $|f'''(x)|$ is decreasing on $[0.7, 1.3]$, we can take $M = |f'''(0.7)| = 6/(0.7)^4$, so

$$|R_2(x)| \leq \frac{6/(0.7)^4}{6}(0.027) = 0.112\,453\,1.$$

(c)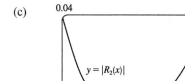

From the graph of $|R_2(x)| = \left|\dfrac{1}{x} - T_2(x)\right|$, it seems that the error is less than $0.038\,571$ on $[0.7, 1.3]$.

15.

n	$f^{(n)}(x)$	$f^{(n)}(1)$
0	$x^{2/3}$	1
1	$\tfrac{2}{3}x^{-1/3}$	$\tfrac{2}{3}$
2	$-\tfrac{2}{9}x^{-4/3}$	$-\tfrac{2}{9}$
3	$\tfrac{8}{27}x^{-7/3}$	$\tfrac{8}{27}$
4	$-\tfrac{56}{81}x^{-10/3}$	

(a) $f(x) = x^{2/3} \approx T_3(x) = 1 + \tfrac{2}{3}(x-1) - \dfrac{2/9}{2!}(x-1)^2 + \dfrac{8/27}{3!}(x-1)^3$

$= 1 + \tfrac{2}{3}(x-1) - \tfrac{1}{9}(x-1)^2 + \tfrac{4}{81}(x-1)^3$

(b) $|R_3(x)| \leq \dfrac{M}{4!}|x-1|^4$, where $\left|f^{(4)}(x)\right| \leq M$. Now $0.8 \leq x \leq 1.2 \;\Rightarrow\;$

$|x-1| \leq 0.2 \;\Rightarrow\; |x-1|^4 \leq 0.0016$. Since $\left|f^{(4)}(x)\right|$ is decreasing

on $[0.8, 1.2]$, we can take $M = \left|f^{(4)}(0.8)\right| = \tfrac{56}{81}(0.8)^{-10/3}$, so

$$|R_3(x)| \leq \frac{\tfrac{56}{81}(0.8)^{-10/3}}{24}(0.0016) \approx 0.000\,096\,97.$$

(c)

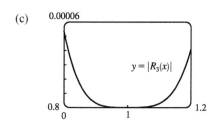

From the graph of $|R_3(x)| = \left|x^{2/3} - T_3(x)\right|$, it seems that the error is less than 0.0000533 on $[0.8, 1.2]$.

17.

n	$f^{(n)}(x)$	$f^{(n)}(0)$
0	$\sec x$	1
1	$\sec x \tan x$	0
2	$\sec x (2\sec^2 x - 1)$	1
3	$\sec x \tan x (6\sec^2 x - 1)$	

(a) $f(x) = \sec x \approx T_2(x) = 1 + \tfrac{1}{2}x^2$

(b) $|R_2(x)| \leq \dfrac{M}{3!}|x|^3$, where $\left|f^{(3)}(x)\right| \leq M$. Now $-0.2 \leq x \leq 0.2 \;\Rightarrow\; |x| \leq 0.2 \;\Rightarrow\; |x|^3 \leq (0.2)^3$.

$f^{(3)}(x)$ is an odd function and it is increasing on $[0, 0.2]$ since $\sec x$ and $\tan x$ are increasing on $[0, 0.2]$, so $\left|f^{(3)}(x)\right| \leq f^{(3)}(0.2) \approx 1.085\,158\,892$. Thus, $|R_2(x)| \leq \dfrac{f^{(3)}(0.2)}{3!}(0.2)^3 \approx 0.001\,447$.

(c)

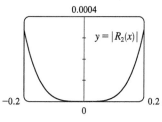

From the graph of $|R_2(x)| = |\sec x - T_2(x)|$, it seems that the error is less than $0.000\,339$ on $[-0.2, 0.2]$.

19.

n	$f^{(n)}(x)$	$f^{(n)}(0)$
0	e^{x^2}	1
1	$e^{x^2}(2x)$	0
2	$e^{x^2}(2 + 4x^2)$	2
3	$e^{x^2}(12x + 8x^3)$	0
4	$e^{x^2}(12 + 48x^2 + 16x^4)$	

(a) $f(x) = e^{x^2} \approx T_3(x) = 1 + \dfrac{2}{2!}x^2 = 1 + x^2$

(b) $|R_3(x)| \leq \dfrac{M}{4!}|x|^4$, where $\left|f^{(4)}(x)\right| \leq M$. Now $0 \leq x \leq 0.1 \;\Rightarrow\; x^4 \leq (0.1)^4$, and letting $x = 0.1$ gives

$|R_3(x)| \leq \dfrac{e^{0.01}(12 + 0.48 + 0.0016)}{24}(0.1)^4 \approx 0.000\,053$.

(c)

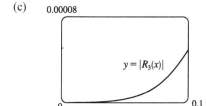

From the graph of $|R_3(x)| = \left|e^{x^2} - T_3(x)\right|$, it appears that the error is less than $0.000\,051$ on $[0, 0.1]$.

21.

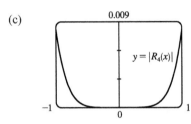

n	$f^{(n)}(x)$	$f^{(n)}(0)$
0	$x \sin x$	0
1	$\sin x + x \cos x$	0
2	$2\cos x - x \sin x$	2
3	$-3 \sin x - x \cos x$	0
4	$-4\cos x + x \sin x$	-4
5	$5 \sin x + x \cos x$	

(a) $f(x) = x \sin x \approx T_4(x) = \dfrac{2}{2!}(x - 0)^2 + \dfrac{-4}{4!}(x - 0)^4 = x^2 - \dfrac{1}{6}x^4$

(b) $|R_4(x)| \leq \dfrac{M}{5!}|x|^5$, where $\left|f^{(5)}(x)\right| \leq M$. Now $-1 \leq x \leq 1 \Rightarrow$ $|x| \leq 1$, and a graph of $f^{(5)}(x)$ shows that $\left|f^{(5)}(x)\right| \leq 5$ for $-1 \leq x \leq 1$.

Thus, we can take $M = 5$ and get $|R_4(x)| \leq \dfrac{5}{5!} \cdot 1^5 = \dfrac{1}{24} = 0.041\overline{6}$.

(c)

From the graph of $|R_4(x)| = |x \sin x - T_4(x)|$, it seems that the error is less than 0.0082 on $[-1, 1]$.

23. From Exercise 5, $\cos x = -\left(x - \dfrac{\pi}{2}\right) + \dfrac{1}{6}\left(x - \dfrac{\pi}{2}\right)^3 + R_3(x)$, where $|R_3(x)| \leq \dfrac{M}{4!}\left|x - \dfrac{\pi}{2}\right|^4$ with

$\left|f^{(4)}(x)\right| = |\cos x| \leq M = 1$. Now $x = 80° = (90° - 10°) = \left(\dfrac{\pi}{2} - \dfrac{\pi}{18}\right) = \dfrac{4\pi}{9}$ radians, so the error is

$\left|R_3\left(\dfrac{4\pi}{9}\right)\right| \leq \dfrac{1}{24}\left(\dfrac{\pi}{18}\right)^4 \approx 0.000\,039$, which means our estimate would *not* be accurate to five decimal places. However,

$T_3 = T_4$, so we can use $\left|R_4\left(\dfrac{4\pi}{9}\right)\right| \leq \dfrac{1}{120}\left(\dfrac{\pi}{18}\right)^5 \approx 0.000\,001$. Therefore, to five decimal places,

$\cos 80° \approx -\left(-\dfrac{\pi}{18}\right) + \dfrac{1}{6}\left(-\dfrac{\pi}{18}\right)^3 \approx 0.17365$.

25. All derivatives of e^x are e^x, so $|R_n(x)| \leq \dfrac{e^x}{(n+1)!}|x|^{n+1}$, where $0 < x < 0.1$. Letting $x = 0.1$,

$R_n(0.1) \leq \dfrac{e^{0.1}}{(n+1)!}(0.1)^{n+1} < 0.00001$, and by trial and error we find that $n = 3$ satisfies this inequality since

$R_3(0.1) < 0.0000046$. Thus, by adding the four terms of the Maclaurin series for e^x corresponding to $n = 0, 1, 2,$ and 3,

we can estimate $e^{0.1}$ to within 0.00001. (In fact, this sum is $1.1051\overline{6}$ and $e^{0.1} \approx 1.10517$.)

27. $\sin x = x - \dfrac{1}{3!}x^3 + \dfrac{1}{5!}x^5 - \cdots$. By the Alternating Series

Estimation Theorem, the error in the approximation

$\sin x = x - \dfrac{1}{3!}x^3$ is less than $\left|\dfrac{1}{5!}x^5\right| < 0.01 \Leftrightarrow$

$|x^5| < 120(0.01) \Leftrightarrow |x| < (1.2)^{1/5} \approx 1.037$. The curves

$y = x - \dfrac{1}{6}x^3$ and $y = \sin x - 0.01$ intersect at $x \approx 1.043$, so

the graph confirms our estimate. Since both the sine function

and the given approximation are odd functions, we need to check the estimate only for $x > 0$. Thus, the desired range of values for x is $-1.037 < x < 1.037$.

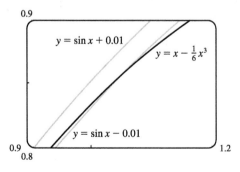

29. $\arctan x = x - \dfrac{x^3}{3} + \dfrac{x^5}{5} - \dfrac{x^7}{7} + \cdots$. By the Alternating Series Estimation Theorem, the error is less than $\left|-\dfrac{1}{7}x^7\right| < 0.05 \Leftrightarrow$ $\left|x^7\right| < 0.35 \Leftrightarrow |x| < (0.35)^{1/7} \approx 0.8607$. The curves $y = x - \dfrac{1}{3}x^3 + \dfrac{1}{5}x^5$ and $y = \arctan x + 0.05$ intersect at $x \approx 0.9245$, so the graph confirms our estimate. Since both the arctangent function and the given approximation are odd functions, we need to check the estimate only for $x > 0$. Thus, the desired range of values for x is $-0.86 < x < 0.86$.

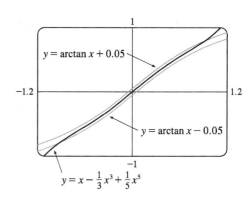

31. Let $s(t)$ be the position function of the car, and for convenience set $s(0) = 0$. The velocity of the car is $v(t) = s'(t)$ and the acceleration is $a(t) = s''(t)$, so the second degree Taylor polynomial is $T_2(t) = s(0) + v(0)t + \dfrac{a(0)}{2}t^2 = 20t + t^2$. We estimate the distance traveled during the next second to be $s(1) \approx T_2(1) = 20 + 1 = 21$ m. The function $T_2(t)$ would not be accurate over a full minute, since the car could not possibly maintain an acceleration of 2 m/s^2 for that long (if it did, its final speed would be 140 m/s $\approx$ 313 mi/h!).

33. $E = \dfrac{q}{D^2} - \dfrac{q}{(D+d)^2} = \dfrac{q}{D^2} - \dfrac{q}{D^2(1+d/D)^2} = \dfrac{q}{D^2}\left[1 - \left(1 + \dfrac{d}{D}\right)^{-2}\right]$.

We use the Binomial Series to expand $(1 + d/D)^{-2}$:

$$E = \dfrac{q}{D^2}\left[1 - \left(1 - 2\left(\dfrac{d}{D}\right) + \dfrac{2\cdot 3}{2!}\left(\dfrac{d}{D}\right)^2 - \dfrac{2\cdot 3\cdot 4}{3!}\left(\dfrac{d}{D}\right)^3 + \cdots\right)\right] = \dfrac{q}{D^2}\left[2\left(\dfrac{d}{D}\right) - 3\left(\dfrac{d}{D}\right)^2 + 4\left(\dfrac{d}{D}\right)^3 - \cdots\right]$$

$$\approx \dfrac{q}{D^2}\cdot 2\left(\dfrac{d}{D}\right) = 2qd\cdot\dfrac{1}{D^3}$$

when D is much larger than d; that is, when P is far away from the dipole.

35. (a) If the water is deep, then $2\pi d/L$ is large, and we know that $\tanh x \to 1$ as $x \to \infty$. So we can approximate $\tanh(2\pi d/L) \approx 1$, and so $v^2 \approx gL/(2\pi) \Leftrightarrow v \approx \sqrt{gL/(2\pi)}$.

(b) From the table, the first term in the Maclaurin series of $\tanh x$ is x, so if the water is shallow, we can approximate $\tanh\dfrac{2\pi d}{L} \approx \dfrac{2\pi d}{L}$, and so $v^2 \approx \dfrac{gL}{2\pi}\cdot\dfrac{2\pi d}{L} \Leftrightarrow v \approx \sqrt{gd}$.

n	$f^{(n)}(x)$	$f^{(n)}(0)$
0	$\tanh x$	0
1	$\text{sech}^2 x$	1
2	$-2\,\text{sech}^2 x\,\tanh x$	0
3	$2\,\text{sech}^2 x\,(3\tanh^2 x - 1)$	-2

(c) Since $\tanh x$ is an odd function, its Maclaurin series is alternating, so the error in the approximation $\tanh\dfrac{2\pi d}{L} \approx \dfrac{2\pi d}{L}$ is less than the first neglected term, which is $\dfrac{|f'''(0)|}{3!}\left(\dfrac{2\pi d}{L}\right)^3 = \dfrac{1}{3}\left(\dfrac{2\pi d}{L}\right)^3$.

[continued]

If $L > 10d$, then $\dfrac{1}{3}\left(\dfrac{2\pi d}{L}\right)^3 < \dfrac{1}{3}\left(2\pi \cdot \dfrac{1}{10}\right)^3 = \dfrac{\pi^3}{375}$, so the error in the approximation $v^2 = gd$ is less than $\dfrac{gL}{2\pi} \cdot \dfrac{\pi^3}{375} \approx 0.0132gL$.

37. (a) L is the length of the arc subtended by the angle θ, so $L = R\theta \Rightarrow$
$\theta = L/R$. Now $\sec\theta = (R+C)/R \Rightarrow R\sec\theta = R + C \Rightarrow$
$C = R\sec\theta - R = R\sec(L/R) - R$.

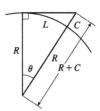

(b) First we'll find a Taylor polynomial $T_4(x)$ for $f(x) = \sec x$ at $x = 0$.

n	$f^{(n)}(x)$	$f^{(n)}(0)$
0	$\sec x$	1
1	$\sec x \tan x$	0
2	$\sec x(2\tan^2 x + 1)$	1
3	$\sec x \tan x(6\tan^2 x + 5)$	0
4	$\sec x(24\tan^4 x + 28\tan^2 x + 5)$	5

Thus, $f(x) = \sec x \approx T_4(x) = 1 + \dfrac{1}{2!}(x-0)^2 + \dfrac{5}{4!}(x-0)^4 = 1 + \dfrac{1}{2}x^2 + \dfrac{5}{24}x^4$. By part (a),

$$C \approx R\left[1 + \dfrac{1}{2}\left(\dfrac{L}{R}\right)^2 + \dfrac{5}{24}\left(\dfrac{L}{R}\right)^4\right] - R = R + \dfrac{1}{2}R\cdot\dfrac{L^2}{R^2} + \dfrac{5}{24}R\cdot\dfrac{L^4}{R^4} - R = \dfrac{L^2}{2R} + \dfrac{5L^4}{24R^3}.$$

(c) Taking $L = 100$ km and $R = 6370$ km, the formula in part (a) says that

$C = R\sec(L/R) - R = 6370\sec(100/6370) - 6370 \approx 0.785\,009\,965\,44$ km.

The formula in part (b) says that $C \approx \dfrac{L^2}{2R} + \dfrac{5L^4}{24R^3} = \dfrac{100^2}{2\cdot 6370} + \dfrac{5\cdot 100^4}{24\cdot 6370^3} \approx 0.785\,009\,957\,36$ km.

The difference between these two results is only $0.000\,000\,008\,08$ km, or $0.000\,008\,08$ m!

39. Using $f(x) = T_n(x) + R_n(x)$ with $n = 1$ and $x = r$, we have $f(r) = T_1(r) + R_1(r)$, where T_1 is the first-degree Taylor polynomial of f at a. Because $a = x_n$, $f(r) = f(x_n) + f'(x_n)(r - x_n) + R_1(r)$. But r is a zero of f, so $f(r) = 0$ and we have $0 = f(x_n) + f'(x_n)(r - x_n) + R_1(r)$. Taking the first two terms to the left side gives us

$f'(x_n)(x_n - r) - f(x_n) = R_1(r)$. Dividing by $f'(x_n)$, we get $x_n - r - \dfrac{f(x_n)}{f'(x_n)} = \dfrac{R_1(r)}{f'(x_n)}$. By the formula for Newton's method, the left side of the preceding equation is $x_{n+1} - r$, so $|x_{n+1} - r| = \left|\dfrac{R_1(r)}{f'(x_n)}\right|$. Taylor's Inequality gives us

$|R_1(r)| \le \dfrac{|f''(r)|}{2!}|r - x_n|^2$. Combining this inequality with the facts $|f''(x)| \le M$ and $|f'(x)| \ge K$ gives us

$|x_{n+1} - r| \le \dfrac{M}{2K}|x_n - r|^2$.

11 Review

TRUE-FALSE QUIZ

1. **False.** See the WARNING after Theorem 11.2.6.

3. **True.** If $\lim_{n\to\infty} a_n = L$, then as $n \to \infty$, $2n+1 \to \infty$, so $a_{2n+1} \to L$.

5. **False.** For example, take $c_n = (-1)^n/(n6^n)$.

7. **False**, since $\lim_{n\to\infty}\left|\dfrac{a_{n+1}}{a_n}\right| = \lim_{n\to\infty}\left|\dfrac{1}{(n+1)^3}\cdot\dfrac{n^3}{1}\right| = \lim_{n\to\infty}\left|\dfrac{n^3}{(n+1)^3}\cdot\dfrac{1/n^3}{1/n^3}\right| = \lim_{n\to\infty}\dfrac{1}{(1+1/n)^3} = 1$.

9. **False.** See the paragraph preceding "The Limit Comparison Test" in Section 11.4.

11. **True.** See (9) in Section 11.1.

13. **True.** By Theorem 11.10.5 the coefficient of x^3 is $\dfrac{f'''(0)}{3!} = \dfrac{1}{3} \Rightarrow f'''(0) = 2$.

    Or: Use Theorem 11.9.2 to differentiate f three times.

15. **False.** For example, let $a_n = b_n = (-1)^n$. Then $\{a_n\}$ and $\{b_n\}$ are divergent, but $a_n b_n = 1$, so $\{a_n b_n\}$ is convergent.

17. **True** by Theorem 11.5.3. $[\sum (-1)^n a_n$ is absolutely convergent and hence convergent.]

19. **True.** $0.99999\ldots = 0.9 + 0.9(0.1)^1 + 0.9(0.1)^2 + 0.9(0.1)^3 + \cdots = \sum_{n=1}^{\infty}(0.9)(0.1)^{n-1} = \dfrac{0.9}{1-0.1} = 1$ by the formula for the sum of a geometric series $[S = a_1/(1-r)]$ with ratio r satisfying $|r| < 1$.

21. **True.** A finite number of terms doesn't affect convergence or divergence of a series.

EXERCISES

1. $\left\{\dfrac{2+n^3}{1+2n^3}\right\}$ converges since $\lim_{n\to\infty}\dfrac{2+n^3}{1+2n^3} = \lim_{n\to\infty}\dfrac{2/n^3+1}{1/n^3+2} = \dfrac{1}{2}$.

3. $\lim_{n\to\infty} a_n = \lim_{n\to\infty}\dfrac{n^3}{1+n^2} = \lim_{n\to\infty}\dfrac{n}{1/n^2+1} = \infty$, so the sequence diverges.

5. $|a_n| = \left|\dfrac{n\sin n}{n^2+1}\right| \leq \dfrac{n}{n^2+1} < \dfrac{1}{n}$, so $|a_n| \to 0$ as $n \to \infty$. Thus, $\lim_{n\to\infty} a_n = 0$. The sequence $\{a_n\}$ is convergent.

7. $\left\{\left(1+\dfrac{3}{n}\right)^{4n}\right\}$ is convergent. Let $y = \left(1+\dfrac{3}{x}\right)^{4x}$. Then

$\lim_{x\to\infty}\ln y = \lim_{x\to\infty} 4x\ln(1+3/x) = \lim_{x\to\infty}\dfrac{\ln(1+3/x)}{1/(4x)} \overset{\text{H}}{=} \lim_{x\to\infty}\dfrac{\dfrac{1}{1+3/x}\left(-\dfrac{3}{x^2}\right)}{-1/(4x^2)} = \lim_{x\to\infty}\dfrac{12}{1+3/x} = 12$, so

$\lim_{x\to\infty} y = \lim_{n\to\infty}\left(1+\dfrac{3}{n}\right)^{4n} = e^{12}$.

9. We use induction, hypothesizing that $a_{n-1} < a_n < 2$. Note first that $1 < a_2 = \frac{1}{3}(1+4) = \frac{5}{3} < 2$, so the hypothesis holds for $n = 2$. Now assume that $a_{k-1} < a_k < 2$. Then $a_k = \frac{1}{3}(a_{k-1} + 4) < \frac{1}{3}(a_k + 4) < \frac{1}{3}(2+4) = 2$. So $a_k < a_{k+1} < 2$, and the induction is complete. To find the limit of the sequence, we note that $L = \lim_{n \to \infty} a_n = \lim_{n \to \infty} a_{n+1} \Rightarrow L = \frac{1}{3}(L+4) \Rightarrow L = 2$.

11. $\dfrac{n}{n^3 + 1} < \dfrac{n}{n^3} = \dfrac{1}{n^2}$, so $\sum\limits_{n=1}^{\infty} \dfrac{n}{n^3 + 1}$ converges by the Direct Comparison Test with the convergent p-series $\sum\limits_{n=1}^{\infty} \dfrac{1}{n^2}$ $[p = 2 > 1]$.

13. $\lim\limits_{n \to \infty} \left| \dfrac{a_{n+1}}{a_n} \right| = \lim\limits_{n \to \infty} \left[\dfrac{(n+1)^3}{5^{n+1}} \cdot \dfrac{5^n}{n^3} \right] = \lim\limits_{n \to \infty} \left(1 + \dfrac{1}{n}\right)^3 \cdot \dfrac{1}{5} = \dfrac{1}{5} < 1$, so $\sum\limits_{n=1}^{\infty} \dfrac{n^3}{5^n}$ converges by the Ratio Test.

15. Let $f(x) = \dfrac{1}{x\sqrt{\ln x}}$. Then f is continuous, positive, and decreasing on $[2, \infty)$, so the Integral Test applies.

$$\int_2^{\infty} f(x)\,dx = \lim_{t \to \infty} \int_2^t \dfrac{1}{x\sqrt{\ln x}}\,dx \left[u = \ln x, du = \dfrac{1}{x}dx \right] = \lim_{t \to \infty} \int_{\ln 2}^{\ln t} u^{-1/2}\,du = \lim_{t \to \infty} \left[2\sqrt{u} \right]_{\ln 2}^{\ln t}$$
$$= \lim_{t \to \infty} \left(2\sqrt{\ln t} - 2\sqrt{\ln 2} \right) = \infty,$$

so the series $\sum\limits_{n=2}^{\infty} \dfrac{1}{n\sqrt{\ln n}}$ diverges.

17. $|a_n| = \left| \dfrac{\cos 3n}{1 + (1.2)^n} \right| \leq \dfrac{1}{1 + (1.2)^n} < \dfrac{1}{(1.2)^n} = \left(\dfrac{5}{6}\right)^n$, so $\sum\limits_{n=1}^{\infty} |a_n|$ converges by direct comparison with the convergent geometric series $\sum\limits_{n=1}^{\infty} \left(\dfrac{5}{6}\right)^n$ $\left[r = \dfrac{5}{6} < 1 \right]$. It follows that $\sum\limits_{n=1}^{\infty} a_n$ converges (by Theorem 11.5.3).

19. $\lim\limits_{n \to \infty} \left| \dfrac{a_{n+1}}{a_n} \right| = \lim\limits_{n \to \infty} \dfrac{1 \cdot 3 \cdot 5 \cdots (2n-1)(2n+1)}{5^{n+1}(n+1)!} \cdot \dfrac{5^n\, n!}{1 \cdot 3 \cdot 5 \cdots (2n-1)} = \lim\limits_{n \to \infty} \dfrac{2n+1}{5(n+1)} = \dfrac{2}{5} < 1$, so the series $\sum\limits_{n=1}^{\infty} \dfrac{1 \cdot 3 \cdot 5 \cdots (2n-1)}{5^n\, n!}$ converges by the Ratio Test.

21. $b_n = \dfrac{\sqrt{n}}{n+1} > 0$, $\{b_n\}$ is decreasing, and $\lim\limits_{n \to \infty} b_n = 0$, so the series $\sum\limits_{n=1}^{\infty} (-1)^{n-1} \dfrac{\sqrt{n}}{n+1}$ converges by the Alternating Series Test.

23. Consider the series of absolute values: $\sum\limits_{n=1}^{\infty} n^{-1/3}$ is a p-series with $p = \frac{1}{3} \leq 1$ and is therefore divergent. But if we apply the Alternating Series Test, we see that $b_n = \dfrac{1}{\sqrt[3]{n}} > 0$, $\{b_n\}$ is decreasing, and $\lim\limits_{n \to \infty} b_n = 0$, so the series $\sum\limits_{n=1}^{\infty} (-1)^{n-1} n^{-1/3}$ converges. Thus, $\sum\limits_{n=1}^{\infty} (-1)^{n-1} n^{-1/3}$ is conditionally convergent.

25. $\left| \dfrac{a_{n+1}}{a_n} \right| = \left| \dfrac{(-1)^{n+1}(n+2)3^{n+1}}{2^{2n+3}} \cdot \dfrac{2^{2n+1}}{(-1)^n(n+1)3^n} \right| = \dfrac{n+2}{n+1} \cdot \dfrac{3}{4} = \dfrac{1 + (2/n)}{1 + (1/n)} \cdot \dfrac{3}{4} \to \dfrac{3}{4} < 1$ as $n \to \infty$, so by the Ratio Test, $\sum\limits_{n=1}^{\infty} \dfrac{(-1)^n(n+1)3^n}{2^{2n+1}}$ is absolutely convergent.

27. $\sum_{n=1}^{\infty} \frac{(-3)^{n-1}}{2^{3n}} = \sum_{n=1}^{\infty} \frac{(-3)^{n-1}}{(2^3)^n} = \sum_{n=1}^{\infty} \frac{(-3)^{n-1}}{8^n} = \frac{1}{8}\sum_{n=1}^{\infty} \frac{(-3)^{n-1}}{8^{n-1}} = \frac{1}{8}\sum_{n=1}^{\infty} \left(-\frac{3}{8}\right)^{n-1} = \frac{1}{8}\left(\frac{1}{1-(-3/8)}\right)$

$= \frac{1}{8} \cdot \frac{8}{11} = \frac{1}{11}$

29. $\sum_{n=1}^{\infty}[\tan^{-1}(n+1) - \tan^{-1} n] = \lim_{n\to\infty} s_n$

$= \lim_{n\to\infty}[(\tan^{-1} 2 - \tan^{-1} 1) + (\tan^{-1} 3 - \tan^{-1} 2) + \cdots + (\tan^{-1}(n+1) - \tan^{-1} n)]$

$= \lim_{n\to\infty}[\tan^{-1}(n+1) - \tan^{-1} 1] = \frac{\pi}{2} - \frac{\pi}{4} = \frac{\pi}{4}$

31. $1 - e + \frac{e^2}{2!} - \frac{e^3}{3!} + \frac{e^4}{4!} - \cdots = \sum_{n=0}^{\infty}(-1)^n \frac{e^n}{n!} = \sum_{n=0}^{\infty} \frac{(-e)^n}{n!} = e^{-e}$ since $e^x = \sum_{n=0}^{\infty} \frac{x^n}{n!}$ for all x.

33. $\cosh x = \frac{1}{2}(e^x + e^{-x}) = \frac{1}{2}\left(\sum_{n=0}^{\infty} \frac{x^n}{n!} + \sum_{n=0}^{\infty} \frac{(-x)^n}{n!}\right)$

$= \frac{1}{2}\left[\left(1 + x + \frac{x^2}{2!} + \frac{x^3}{3!} + \frac{x^4}{4!} + \cdots\right) + \left(1 - x + \frac{x^2}{2!} - \frac{x^3}{3!} + \frac{x^4}{4!} - \cdots\right)\right]$

$= \frac{1}{2}\left(2 + 2\cdot\frac{x^2}{2!} + 2\cdot\frac{x^4}{4!} + \cdots\right) = 1 + \frac{1}{2}x^2 + \sum_{n=2}^{\infty} \frac{x^{2n}}{(2n)!} \geq 1 + \frac{1}{2}x^2$ for all x

35. $\sum_{n=1}^{\infty} \frac{(-1)^{n+1}}{n^5} = 1 - \frac{1}{32} + \frac{1}{243} - \frac{1}{1024} + \frac{1}{3125} - \frac{1}{7776} + \frac{1}{16{,}807} - \frac{1}{32{,}768} + \cdots$.

Since $b_8 = \frac{1}{8^5} = \frac{1}{32{,}768} < 0.000031$, $\sum_{n=1}^{\infty} \frac{(-1)^{n+1}}{n^5} \approx \sum_{n=1}^{7} \frac{(-1)^{n+1}}{n^5} \approx 0.9721$.

37. $\sum_{n=1}^{\infty} \frac{1}{2+5^n} \approx \sum_{n=1}^{8} \frac{1}{2+5^n} \approx 0.18976224$. To estimate the error, note that $\frac{1}{2+5^n} < \frac{1}{5^n}$, so the remainder term is

$R_8 = \sum_{n=9}^{\infty} \frac{1}{2+5^n} < \sum_{n=9}^{\infty} \frac{1}{5^n} = \frac{1/5^9}{1-1/5} = 6.4\times 10^{-7}$ [geometric series with $a = \frac{1}{5^9}$ and $r = \frac{1}{5}$].

39. Use the Limit Comparison Test. $\lim_{n\to\infty}\left|\frac{\left(\frac{n+1}{n}\right)a_n}{a_n}\right| = \lim_{n\to\infty} \frac{n+1}{n} = \lim_{n\to\infty}\left(1 + \frac{1}{n}\right) = 1 > 0$.

Since $\sum |a_n|$ is convergent, so is $\sum \left|\left(\frac{n+1}{n}\right)a_n\right|$, by the Limit Comparison Test.

41. $\lim_{n\to\infty}\left|\frac{a_{n+1}}{a_n}\right| = \lim_{n\to\infty}\left[\frac{|x+2|^{n+1}}{(n+1)\,4^{n+1}} \cdot \frac{n\,4^n}{|x+2|^n}\right] = \lim_{n\to\infty}\left[\frac{n}{n+1} \cdot \frac{|x+2|}{4}\right] = \frac{|x+2|}{4} < 1 \Leftrightarrow |x+2| < 4$, so $R = 4$.

$|x+2| < 4 \Leftrightarrow -4 < x+2 < 4 \Leftrightarrow -6 < x < 2$. If $x = -6$, then the series $\sum_{n=1}^{\infty} \frac{(x+2)^n}{n\,4^n}$ becomes

$\sum_{n=1}^{\infty} \frac{(-4)^n}{n\,4^n} = \sum_{n=1}^{\infty} \frac{(-1)^n}{n}$, the alternating harmonic series, which converges by the Alternating Series Test. When $x = 2$, the

series becomes the harmonic series $\sum_{n=1}^{\infty} \frac{1}{n}$, which diverges. Thus, $I = [-6, 2)$.

43. $\lim\limits_{n\to\infty}\left|\dfrac{a_{n+1}}{a_n}\right| = \lim\limits_{n\to\infty}\left|\dfrac{2^{n+1}(x-3)^{n+1}}{\sqrt{n+4}} \cdot \dfrac{\sqrt{n+3}}{2^n(x-3)^n}\right| = 2\,|x-3|\lim\limits_{n\to\infty}\sqrt{\dfrac{n+3}{n+4}} = 2\,|x-3| < 1 \;\Leftrightarrow\; |x-3| < \tfrac{1}{2}$,

so $R = \tfrac{1}{2}$. $|x-3| < \tfrac{1}{2} \;\Leftrightarrow\; -\tfrac{1}{2} < x-3 < \tfrac{1}{2} \;\Leftrightarrow\; \tfrac{5}{2} < x < \tfrac{7}{2}$. For $x = \tfrac{7}{2}$, the series $\sum\limits_{n=1}^{\infty}\dfrac{2^n(x-3)^n}{\sqrt{n+3}}$ becomes

$\sum\limits_{n=0}^{\infty}\dfrac{1}{\sqrt{n+3}} = \sum\limits_{n=3}^{\infty}\dfrac{1}{n^{1/2}}$, which diverges $\left[p = \tfrac{1}{2} \le 1\right]$, but for $x = \tfrac{5}{2}$, we get $\sum\limits_{n=0}^{\infty}\dfrac{(-1)^n}{\sqrt{n+3}}$, which is a convergent alternating series, so $I = \left[\tfrac{5}{2}, \tfrac{7}{2}\right)$.

45.

n	$f^{(n)}(x)$	$f^{(n)}\!\left(\tfrac{\pi}{6}\right)$
0	$\sin x$	$\tfrac{1}{2}$
1	$\cos x$	$\tfrac{\sqrt{3}}{2}$
2	$-\sin x$	$-\tfrac{1}{2}$
3	$-\cos x$	$-\tfrac{\sqrt{3}}{2}$
4	$\sin x$	$\tfrac{1}{2}$
$\vdots$	$\vdots$	$\vdots$

$\sin x = f\!\left(\tfrac{\pi}{6}\right) + f'\!\left(\tfrac{\pi}{6}\right)\!\left(x - \tfrac{\pi}{6}\right) + \dfrac{f''\!\left(\tfrac{\pi}{6}\right)}{2!}\!\left(x - \tfrac{\pi}{6}\right)^2 + \dfrac{f^{(3)}\!\left(\tfrac{\pi}{6}\right)}{3!}\!\left(x - \tfrac{\pi}{6}\right)^3 + \dfrac{f^{(4)}\!\left(\tfrac{\pi}{6}\right)}{4!}\!\left(x - \tfrac{\pi}{6}\right)^4 + \cdots$

$= \dfrac{1}{2}\!\left[1 - \dfrac{1}{2!}\!\left(x - \tfrac{\pi}{6}\right)^2 + \dfrac{1}{4!}\!\left(x - \tfrac{\pi}{6}\right)^4 - \cdots\right] + \dfrac{\sqrt{3}}{2}\!\left[\left(x - \tfrac{\pi}{6}\right) - \dfrac{1}{3!}\!\left(x - \tfrac{\pi}{6}\right)^3 + \cdots\right]$

$= \dfrac{1}{2}\sum\limits_{n=0}^{\infty}(-1)^n\dfrac{1}{(2n)!}\!\left(x - \tfrac{\pi}{6}\right)^{2n} + \dfrac{\sqrt{3}}{2}\sum\limits_{n=0}^{\infty}(-1)^n\dfrac{1}{(2n+1)!}\!\left(x - \tfrac{\pi}{6}\right)^{2n+1}$

47. $\dfrac{1}{1+x} = \dfrac{1}{1-(-x)} = \sum\limits_{n=0}^{\infty}(-x)^n = \sum\limits_{n=0}^{\infty}(-1)^n x^n$ for $|x| < 1 \;\Rightarrow\; \dfrac{x^2}{1+x} = \sum\limits_{n=0}^{\infty}(-1)^n x^{n+2}$ with $R = 1$.

49. $\displaystyle\int\dfrac{1}{4-x}\,dx = -\ln(4-x) + C$ and

$\displaystyle\int\dfrac{1}{4-x}\,dx = \dfrac{1}{4}\int\dfrac{1}{1-x/4}\,dx = \dfrac{1}{4}\int\sum\limits_{n=0}^{\infty}\!\left(\dfrac{x}{4}\right)^n dx = \dfrac{1}{4}\int\sum\limits_{n=0}^{\infty}\dfrac{x^n}{4^n}\,dx = \dfrac{1}{4}\sum\limits_{n=0}^{\infty}\dfrac{x^{n+1}}{4^n(n+1)} + C$. So

$\ln(4-x) = -\dfrac{1}{4}\sum\limits_{n=0}^{\infty}\dfrac{x^{n+1}}{4^n(n+1)} + C = -\sum\limits_{n=0}^{\infty}\dfrac{x^{n+1}}{4^{n+1}(n+1)} + C = -\sum\limits_{n=1}^{\infty}\dfrac{x^n}{n4^n} + C$. Putting $x = 0$, we get $C = \ln 4$.

Thus, $f(x) = \ln(4-x) = \ln 4 - \sum\limits_{n=1}^{\infty}\dfrac{x^n}{n4^n}$. The series converges for $|x/4| < 1 \;\Leftrightarrow\; |x| < 4$, so $R = 4$.

Another solution:

$\ln(4-x) = \ln[4(1-x/4)] = \ln 4 + \ln(1-x/4) = \ln 4 + \ln[1 + (-x/4)]$

$= \ln 4 + \sum\limits_{n=1}^{\infty}(-1)^{n+1}\dfrac{(-x/4)^n}{n}$ [from Table 1 in Section 11.10]

$= \ln 4 + \sum\limits_{n=1}^{\infty}(-1)^{2n+1}\dfrac{x^n}{n4^n} = \ln 4 - \sum\limits_{n=1}^{\infty}\dfrac{x^n}{n4^n}$.

51. $\sin x = \sum_{n=0}^{\infty} \frac{(-1)^n x^{2n+1}}{(2n+1)!} \Rightarrow \sin(x^4) = \sum_{n=0}^{\infty} \frac{(-1)^n (x^4)^{2n+1}}{(2n+1)!} = \sum_{n=0}^{\infty} \frac{(-1)^n x^{8n+4}}{(2n+1)!}$ for all x, so the radius of convergence is ∞.

53. $f(x) = \frac{1}{\sqrt[4]{16-x}} = \frac{1}{\sqrt[4]{16(1-x/16)}} = \frac{1}{\sqrt[4]{16}\left(1-\frac{1}{16}x\right)^{1/4}} = \frac{1}{2}\left(1-\frac{1}{16}x\right)^{-1/4}$

$= \frac{1}{2}\left[1 + \left(-\frac{1}{4}\right)\left(-\frac{x}{16}\right) + \frac{\left(-\frac{1}{4}\right)\left(-\frac{5}{4}\right)}{2!}\left(-\frac{x}{16}\right)^2 + \frac{\left(-\frac{1}{4}\right)\left(-\frac{5}{4}\right)\left(-\frac{9}{4}\right)}{3!}\left(-\frac{x}{16}\right)^3 + \cdots\right]$

$= \frac{1}{2} + \sum_{n=1}^{\infty} \frac{1 \cdot 5 \cdot 9 \cdot \cdots \cdot (4n-3)}{2 \cdot 4^n \cdot n! \cdot 16^n} x^n = \frac{1}{2} + \sum_{n=1}^{\infty} \frac{1 \cdot 5 \cdot 9 \cdot \cdots \cdot (4n-3)}{2^{6n+1}\, n!} x^n$

for $\left|-\frac{x}{16}\right| < 1 \Leftrightarrow |x| < 16$, so $R = 16$.

55. $e^x = \sum_{n=0}^{\infty} \frac{x^n}{n!}$, so $\frac{e^x}{x} = \frac{1}{x}\sum_{n=0}^{\infty} \frac{x^n}{n!} = \sum_{n=0}^{\infty} \frac{x^{n-1}}{n!} = x^{-1} + \sum_{n=1}^{\infty} \frac{x^{n-1}}{n!} = \frac{1}{x} + \sum_{n=1}^{\infty} \frac{x^{n-1}}{n!}$ and

$\int \frac{e^x}{x}\, dx = C + \ln|x| + \sum_{n=1}^{\infty} \frac{x^n}{n \cdot n!}.$

57. (a)

n	$f^{(n)}(x)$	$f^{(n)}(1)$
0	$x^{1/2}$	1
1	$\frac{1}{2}x^{-1/2}$	$\frac{1}{2}$
2	$-\frac{1}{4}x^{-3/2}$	$-\frac{1}{4}$
3	$\frac{3}{8}x^{-5/2}$	$\frac{3}{8}$
4	$-\frac{15}{16}x^{-7/2}$	$-\frac{15}{16}$
$\vdots$	$\vdots$	$\vdots$

$\sqrt{x} \approx T_3(x) = 1 + \frac{1/2}{1!}(x-1) - \frac{1/4}{2!}(x-1)^2 + \frac{3/8}{3!}(x-1)^3$

$= 1 + \frac{1}{2}(x-1) - \frac{1}{8}(x-1)^2 + \frac{1}{16}(x-1)^3$

(b)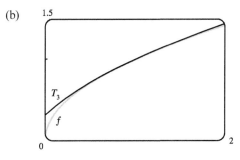

(c) $|R_3(x)| \leq \frac{M}{4!}|x-1|^4$, where $\left|f^{(4)}(x)\right| \leq M$ with $f^{(4)}(x) = -\frac{15}{16}x^{-7/2}$. Now $0.9 \leq x \leq 1.1 \Rightarrow$

$-0.1 \leq x - 1 \leq 0.1 \Rightarrow (x-1)^4 \leq (0.1)^4$, and letting $x = 0.9$ gives $M = \frac{15}{16(0.9)^{7/2}}$, so

$|R_3(x)| \leq \frac{15}{16(0.9)^{7/2}\, 4!}(0.1)^4 \approx 0.000\,005\,648 \approx 0.000\,006 = 6 \times 10^{-6}.$

(d)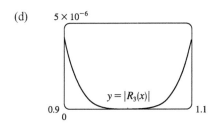

From the graph of $|R_3(x)| = |\sqrt{x} - T_3(x)|$, it appears that the error is less than 5×10^{-6} on $[0.9, 1.1]$.

59. $\sin x = \sum_{n=0}^{\infty} (-1)^n \dfrac{x^{2n+1}}{(2n+1)!} = x - \dfrac{x^3}{3!} + \dfrac{x^5}{5!} - \dfrac{x^7}{7!} + \cdots$, so $\sin x - x = -\dfrac{x^3}{3!} + \dfrac{x^5}{5!} - \dfrac{x^7}{7!} + \cdots$ and

$\dfrac{\sin x - x}{x^3} = -\dfrac{1}{3!} + \dfrac{x^2}{5!} - \dfrac{x^4}{7!} + \cdots$. Thus, $\lim_{x \to 0} \dfrac{\sin x - x}{x^3} = \lim_{x \to 0} \left(-\dfrac{1}{6} + \dfrac{x^2}{120} - \dfrac{x^4}{5040} + \cdots \right) = -\dfrac{1}{6}$.

61. $f(x) = \sum_{n=0}^{\infty} c_n x^n \Rightarrow f(-x) = \sum_{n=0}^{\infty} c_n (-x)^n = \sum_{n=0}^{\infty} (-1)^n c_n x^n$

(a) If f is an odd function, then $f(-x) = -f(x) \Rightarrow \sum_{n=0}^{\infty} (-1)^n c_n x^n = \sum_{n=0}^{\infty} -c_n x^n$. The coefficients of any power series are uniquely determined (by Theorem 11.10.5), so $(-1)^n c_n = -c_n$.

If n is even, then $(-1)^n = 1$, so $c_n = -c_n \Rightarrow 2c_n = 0 \Rightarrow c_n = 0$. Thus, all even coefficients are 0, that is, $c_0 = c_2 = c_4 = \cdots = 0$.

(b) If f is even, then $f(-x) = f(x) \Rightarrow \sum_{n=0}^{\infty} (-1)^n c_n x^n = \sum_{n=0}^{\infty} c_n x^n \Rightarrow (-1)^n c_n = c_n$.

If n is odd, then $(-1)^n = -1$, so $-c_n = c_n \Rightarrow 2c_n = 0 \Rightarrow c_n = 0$. Thus, all odd coefficients are 0, that is, $c_1 = c_3 = c_5 = \cdots = 0$.

PROBLEMS PLUS

1. (a) From Formula 15a in Appendix D, with $x = y = \theta$, we get $\tan 2\theta = \dfrac{2 \tan \theta}{1 - \tan^2 \theta}$, so $\cot 2\theta = \dfrac{1 - \tan^2 \theta}{2 \tan \theta}$ $\Rightarrow$

$2 \cot 2\theta = \dfrac{1 - \tan^2 \theta}{\tan \theta} = \cot \theta - \tan \theta$. Replacing θ by $\tfrac{1}{2}x$, we get $2 \cot x = \cot \tfrac{1}{2}x - \tan \tfrac{1}{2}x$, or

$\tan \tfrac{1}{2}x = \cot \tfrac{1}{2}x - 2 \cot x$.

(b) From part (a) with $\dfrac{x}{2^{n-1}}$ in place of x, $\tan \dfrac{x}{2^n} = \cot \dfrac{x}{2^n} - 2 \cot \dfrac{x}{2^{n-1}}$, so the nth partial sum of $\displaystyle\sum_{n=1}^{\infty} \dfrac{1}{2^n} \tan \dfrac{x}{2^n}$ is

$$s_n = \dfrac{\tan(x/2)}{2} + \dfrac{\tan(x/4)}{4} + \dfrac{\tan(x/8)}{8} + \cdots + \dfrac{\tan(x/2^n)}{2^n}$$

$$= \left[\dfrac{\cot(x/2)}{2} - \cot x\right] + \left[\dfrac{\cot(x/4)}{4} - \dfrac{\cot(x/2)}{2}\right] + \left[\dfrac{\cot(x/8)}{8} - \dfrac{\cot(x/4)}{4}\right] + \cdots$$

$$+ \left[\dfrac{\cot(x/2^n)}{2^n} - \dfrac{\cot(x/2^{n-1})}{2^{n-1}}\right] = -\cot x + \dfrac{\cot(x/2^n)}{2^n} \quad \text{[telescoping sum]}$$

Now $\dfrac{\cot(x/2^n)}{2^n} = \dfrac{\cos(x/2^n)}{2^n \sin(x/2^n)} = \dfrac{\cos(x/2^n)}{x} \cdot \dfrac{x/2^n}{\sin(x/2^n)} \to \dfrac{1}{x} \cdot 1 = \dfrac{1}{x}$ as $n \to \infty$ since $x/2^n \to 0$

for $x \neq 0$. Therefore, if $x \neq 0$ and $x \neq k\pi$ where k is any integer, then

$$\sum_{n=1}^{\infty} \dfrac{1}{2^n} \tan \dfrac{x}{2^n} = \lim_{n \to \infty} s_n = \lim_{n \to \infty} \left(-\cot x + \dfrac{1}{2^n} \cot \dfrac{x}{2^n}\right) = -\cot x + \dfrac{1}{x}$$

If $x = 0$, then all terms in the series are 0, so the sum is 0.

3. (a) At each stage, each side is replaced by four shorter sides, each of length $\tfrac{1}{3}$ of the side length at the preceding stage. Writing s_0 and ℓ_0 for the number of sides and the length of the side of the initial triangle, we generate the table at right. In general, we have $s_n = 3 \cdot 4^n$ and $\ell_n = \left(\tfrac{1}{3}\right)^n$, so the length of the perimeter at the nth stage of construction is $p_n = s_n \ell_n = 3 \cdot 4^n \cdot \left(\tfrac{1}{3}\right)^n = 3 \cdot \left(\tfrac{4}{3}\right)^n$.

$s_0 = 3$	$\ell_0 = 1$
$s_1 = 3 \cdot 4$	$\ell_1 = 1/3$
$s_2 = 3 \cdot 4^2$	$\ell_2 = 1/3^2$
$s_3 = 3 \cdot 4^3$	$\ell_3 = 1/3^3$
$\vdots$	$\vdots$

(b) $p_n = \dfrac{4^n}{3^{n-1}} = 4\left(\dfrac{4}{3}\right)^{n-1}$. Since $\tfrac{4}{3} > 1$, $p_n \to \infty$ as $n \to \infty$.

(c) The area of each of the small triangles added at a given stage is one-ninth of the area of the triangle added at the preceding stage. Let a be the area of the original triangle. Then the area a_n of each of the small triangles added at stage n is

$a_n = a \cdot \dfrac{1}{9^n} = \dfrac{a}{9^n}$. Since a small triangle is added to each side at every stage, it follows that the total area A_n added to the

figure at the nth stage is $A_n = s_{n-1} \cdot a_n = 3 \cdot 4^{n-1} \cdot \dfrac{a}{9^n} = a \cdot \dfrac{4^{n-1}}{3^{2n-1}}$. Then the total area enclosed by the snowflake

curve is $A = a + A_1 + A_2 + A_3 + \cdots = a + a \cdot \dfrac{1}{3} + a \cdot \dfrac{4}{3^3} + a \cdot \dfrac{4^2}{3^5} + a \cdot \dfrac{4^3}{3^7} + \cdots$. After the first term, this is a geometric series with common ratio $\frac{4}{9}$, so $A = a + \frac{a/3}{1 - \frac{4}{9}} = a + \frac{a}{3} \cdot \frac{9}{5} = \frac{8a}{5}$. But the area of the original equilateral triangle with side 1 is $a = \frac{1}{2} \cdot 1 \cdot \sin \frac{\pi}{3} = \frac{\sqrt{3}}{4}$. So the area enclosed by the snowflake curve is $\frac{8}{5} \cdot \frac{\sqrt{3}}{4} = \frac{2\sqrt{3}}{5}$.

5. (a) Let $a = \arctan x$ and $b = \arctan y$. Then, from Formula 15b in Appendix D,

$$\tan(a - b) = \frac{\tan a - \tan b}{1 + \tan a \tan b} = \frac{\tan(\arctan x) - \tan(\arctan y)}{1 + \tan(\arctan x) \tan(\arctan y)} = \frac{x - y}{1 + xy}$$

Now $\arctan x - \arctan y = a - b = \arctan(\tan(a - b)) = \arctan \frac{x - y}{1 + xy}$ since $-\frac{\pi}{2} < a - b < \frac{\pi}{2}$.

(b) From part (a) we have

$$\arctan \tfrac{120}{119} - \arctan \tfrac{1}{239} = \arctan \frac{\tfrac{120}{119} - \tfrac{1}{239}}{1 + \tfrac{120}{119} \cdot \tfrac{1}{239}} = \arctan \frac{\tfrac{28{,}561}{28{,}441}}{\tfrac{28{,}561}{28{,}441}} = \arctan 1 = \tfrac{\pi}{4}$$

(c) Replacing y by $-y$ in the formula of part (a), we get $\arctan x + \arctan y = \arctan \frac{x + y}{1 - xy}$. So

$$4 \arctan \tfrac{1}{5} = 2\left(\arctan \tfrac{1}{5} + \arctan \tfrac{1}{5}\right) = 2 \arctan \frac{\tfrac{1}{5} + \tfrac{1}{5}}{1 - \tfrac{1}{5} \cdot \tfrac{1}{5}} = 2 \arctan \tfrac{5}{12} = \arctan \tfrac{5}{12} + \arctan \tfrac{5}{12}$$

$$= \arctan \frac{\tfrac{5}{12} + \tfrac{5}{12}}{1 - \tfrac{5}{12} \cdot \tfrac{5}{12}} = \arctan \tfrac{120}{119}$$

Thus, from part (b), we have $4 \arctan \tfrac{1}{5} - \arctan \tfrac{1}{239} = \arctan \tfrac{120}{119} - \arctan \tfrac{1}{239} = \tfrac{\pi}{4}$.

(d) From Example 11.9.6 we have $\arctan x = x - \frac{x^3}{3} + \frac{x^5}{5} - \frac{x^7}{7} + \frac{x^9}{9} - \frac{x^{11}}{11} + \cdots$, so

$$\arctan \tfrac{1}{5} = \tfrac{1}{5} - \frac{1}{3 \cdot 5^3} + \frac{1}{5 \cdot 5^5} - \frac{1}{7 \cdot 5^7} + \frac{1}{9 \cdot 5^9} - \frac{1}{11 \cdot 5^{11}} + \cdots$$

This is an alternating series and the size of the terms decreases to 0, so by the Alternating Series Estimation Theorem, the sum lies between s_5 and s_6, that is, $0.197395560 < \arctan \tfrac{1}{5} < 0.197395562$.

(e) From the series in part (d) we get $\arctan \tfrac{1}{239} = \tfrac{1}{239} - \frac{1}{3 \cdot 239^3} + \frac{1}{5 \cdot 239^5} - \cdots$. The third term is less than 2.6×10^{-13}, so by the Alternating Series Estimation Theorem, we have, to nine decimal places, $\arctan \tfrac{1}{239} \approx s_2 \approx 0.004184076$. Thus, $0.004184075 < \arctan \tfrac{1}{239} < 0.004184077$.

(f) From part (c) we have $\pi = 16 \arctan \tfrac{1}{5} - 4 \arctan \tfrac{1}{239}$, so from parts (d) and (e) we have

$16(0.197395560) - 4(0.004184077) < \pi < 16(0.197395562) - 4(0.004184075)$ $\Rightarrow$
$3.141592652 < \pi < 3.141592692$. So, to 7 decimal places, $\pi \approx 3.1415927$.

7. We want $\arctan\left(\frac{2}{n^2}\right)$ to equal $\arctan \frac{x - y}{1 + xy}$. Note that $1 + xy = n^2 \Leftrightarrow xy = n^2 - 1 = (n+1)(n-1)$, so if we let $x = n + 1$ and $y = n - 1$, then $x - y = 2$ and $xy \neq -1$. Thus, from Problem 5(a),

$\arctan\left(\dfrac{2}{n^2}\right) = \arctan\dfrac{x-y}{1+xy} = \arctan x - \arctan y = \arctan(n+1) - \arctan(n-1)$. Therefore,

$$\sum_{n=1}^{k} \arctan\left(\dfrac{2}{n^2}\right) = \sum_{n=1}^{k} [\arctan(n+1) - \arctan(n-1)]$$

$$= \sum_{n=1}^{k} [\arctan(n+1) - \arctan n + \arctan n - \arctan(n-1)]$$

$$= \sum_{n=1}^{k} [\arctan(n+1) - \arctan n] + \sum_{n=1}^{k} [\arctan n - \arctan(n-1)]$$

$$= [\arctan(k+1) - \arctan 1] + [\arctan k - \arctan 0] \quad \text{[since both sums are telescoping]}$$

$$= \arctan(k+1) - \tfrac{\pi}{4} + \arctan k - 0$$

Now $\sum_{n=1}^{k} \arctan\left(\dfrac{2}{n^2}\right) = \lim_{k\to\infty} \sum_{n=1}^{k} \arctan\left(\dfrac{2}{n^2}\right) = \lim_{k\to\infty} \left[\arctan(k+1) - \dfrac{\pi}{4} + \arctan k\right] = \dfrac{\pi}{2} - \dfrac{\pi}{4} + \dfrac{\pi}{2} = \dfrac{3\pi}{4}$.

Note: For all $n \geq 1$, $0 \leq \arctan(n-1) < \arctan(n+1) < \tfrac{\pi}{2}$, so $-\tfrac{\pi}{2} < \arctan(n+1) - \arctan(n-1) < \tfrac{\pi}{2}$, and the identity in Problem 5(a) holds.

9. We start with the geometric series $\sum_{n=0}^{\infty} x^n = \dfrac{1}{1-x}$, $|x| < 1$, and differentiate:

$$\sum_{n=1}^{\infty} nx^{n-1} = \dfrac{d}{dx}\left(\sum_{n=0}^{\infty} x^n\right) = \dfrac{d}{dx}\left(\dfrac{1}{1-x}\right) = \dfrac{1}{(1-x)^2} \text{ for } |x| < 1 \implies \sum_{n=1}^{\infty} nx^n = x\sum_{n=1}^{\infty} nx^{n-1} = \dfrac{x}{(1-x)^2}$$

for $|x| < 1$. Differentiate again:

$$\sum_{n=1}^{\infty} n^2 x^{n-1} = \dfrac{d}{dx}\dfrac{x}{(1-x)^2} = \dfrac{(1-x)^2 - x\cdot 2(1-x)(-1)}{(1-x)^4} = \dfrac{x+1}{(1-x)^3} \implies \sum_{n=1}^{\infty} n^2 x^n = \dfrac{x^2+x}{(1-x)^3} \implies$$

$$\sum_{n=1}^{\infty} n^3 x^{n-1} = \dfrac{d}{dx}\dfrac{x^2+x}{(1-x)^3} = \dfrac{(1-x)^3(2x+1) - (x^2+x)3(1-x)^2(-1)}{(1-x)^6} = \dfrac{x^2+4x+1}{(1-x)^4} \implies$$

$$\sum_{n=1}^{\infty} n^3 x^n = \dfrac{x^3+4x^2+x}{(1-x)^4}, \ |x| < 1.$$ The radius of convergence is 1 because that is the radius of convergence for the geometric series we started with. If $x = \pm 1$, the series is $\sum n^3 (\pm 1)^n$, which diverges by the Test For Divergence, so the interval of convergence is $(-1, 1)$.

11. $\ln\left(1 - \dfrac{1}{n^2}\right) = \ln\left(\dfrac{n^2-1}{n^2}\right) = \ln\dfrac{(n+1)(n-1)}{n^2} = \ln[(n+1)(n-1)] - \ln n^2$

$$= \ln(n+1) + \ln(n-1) - 2\ln n = \ln(n-1) - \ln n - \ln n + \ln(n+1)$$

$$= \ln\dfrac{n-1}{n} - [\ln n - \ln(n+1)] = \ln\dfrac{n-1}{n} - \ln\dfrac{n}{n+1}.$$

Let $s_k = \sum_{n=2}^{k} \ln\left(1 - \dfrac{1}{n^2}\right) = \sum_{n=2}^{k}\left(\ln\dfrac{n-1}{n} - \ln\dfrac{n}{n+1}\right)$ for $k \geq 2$. Then

$$s_k = \left(\ln\dfrac{1}{2} - \ln\dfrac{2}{3}\right) + \left(\ln\dfrac{2}{3} - \ln\dfrac{3}{4}\right) + \cdots + \left(\ln\dfrac{k-1}{k} - \ln\dfrac{k}{k+1}\right) = \ln\dfrac{1}{2} - \ln\dfrac{k}{k+1}, \text{ so}$$

$$\sum_{n=2}^{\infty} \ln\left(1 - \dfrac{1}{n^2}\right) = \lim_{k\to\infty} s_k = \lim_{k\to\infty}\left(\ln\dfrac{1}{2} - \ln\dfrac{k}{k+1}\right) = \ln\dfrac{1}{2} - \ln 1 = \ln 1 - \ln 2 - \ln 1 = -\ln 2 \ (\text{or } \ln\tfrac{1}{2}).$$

13. If L is the length of a side of the equilateral triangle, then the area is $A = \frac{1}{2}L \cdot \frac{\sqrt{3}}{2}L = \frac{\sqrt{3}}{4}L^2$ and so $L^2 = \frac{4}{\sqrt{3}}A$.

Let r be the radius of one of the circles. When there are n rows of circles, the figure shows that

$$L = \sqrt{3}r + r + (n-2)(2r) + r + \sqrt{3}r = r(2n - 2 + 2\sqrt{3}), \text{ so } r = \frac{L}{2(n + \sqrt{3} - 1)}.$$

The number of circles is $1 + 2 + \cdots + n = \dfrac{n(n+1)}{2}$, and so the total area of the circles is

$$A_n = \frac{n(n+1)}{2}\pi r^2 = \frac{n(n+1)}{2}\pi \frac{L^2}{4(n+\sqrt{3}-1)^2}$$

$$= \frac{n(n+1)}{2}\pi \frac{4A/\sqrt{3}}{4(n+\sqrt{3}-1)^2} = \frac{n(n+1)}{(n+\sqrt{3}-1)^2}\frac{\pi A}{2\sqrt{3}} \Rightarrow$$

$$\frac{A_n}{A} = \frac{n(n+1)}{(n+\sqrt{3}-1)^2}\frac{\pi}{2\sqrt{3}}$$

$$= \frac{1 + 1/n}{[1 + (\sqrt{3}-1)/n]^2}\frac{\pi}{2\sqrt{3}} \to \frac{\pi}{2\sqrt{3}} \text{ as } n \to \infty$$

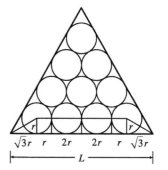

15. (a) The x-intercepts of the curve occur where $\sin x = 0 \Leftrightarrow x = n\pi$, n an integer. So using the formula for disks (and either a CAS or $\sin^2 x = \frac{1}{2}(1 - \cos 2x)$ and Formula 99 to evaluate the integral), the volume of the nth bead is

$$V_n = \pi \int_{(n-1)\pi}^{n\pi} (e^{-x/10}\sin x)^2\, dx = \pi \int_{(n-1)\pi}^{n\pi} e^{-x/5}\sin^2 x\, dx$$

$$= \frac{250\pi}{101}\left(e^{-(n-1)\pi/5} - e^{-n\pi/5}\right)$$

(b) The total volume is

$$\pi \int_0^\infty e^{-x/5}\sin^2 x\, dx = \sum_{n=1}^\infty V_n = \frac{250\pi}{101}\sum_{n=1}^\infty \left[e^{-(n-1)\pi/5} - e^{-n\pi/5}\right] = \frac{250\pi}{101} \quad \text{[telescoping sum]}.$$

Another method: If the volume in part (a) has been written as $V_n = \frac{250\pi}{101}e^{-n\pi/5}(e^{\pi/5} - 1)$, then we recognize $\sum_{n=1}^\infty V_n$ as a geometric series with $a = \frac{250\pi}{101}(1 - e^{-\pi/5})$ and $r = e^{-\pi/5}$.

17. By Table 1 in Section 11.10, $\tan^{-1}x = \sum_{n=0}^\infty (-1)^n \dfrac{x^{2n+1}}{2n+1}$ for $|x| < 1$. In particular, for $x = \dfrac{1}{\sqrt{3}}$, we

have $\dfrac{\pi}{6} = \tan^{-1}\left(\dfrac{1}{\sqrt{3}}\right) = \sum_{n=0}^\infty (-1)^n \dfrac{(1/\sqrt{3})^{2n+1}}{2n+1} = \sum_{n=0}^\infty (-1)^n \left(\dfrac{1}{3}\right)^n \dfrac{1}{\sqrt{3}}\dfrac{1}{2n+1}$, so

$$\pi = \frac{6}{\sqrt{3}}\sum_{n=0}^\infty \frac{(-1)^n}{(2n+1)3^n} = 2\sqrt{3}\sum_{n=0}^\infty \frac{(-1)^n}{(2n+1)3^n} = 2\sqrt{3}\left(1 + \sum_{n=1}^\infty \frac{(-1)^n}{(2n+1)3^n}\right) \Rightarrow \sum_{n=1}^\infty \frac{(-1)^n}{(2n+1)3^n} = \frac{\pi}{2\sqrt{3}} - 1.$$

19. Let $f(x)$ denote the left-hand side of the equation $1 + \dfrac{x}{2!} + \dfrac{x^2}{4!} + \dfrac{x^3}{6!} + \dfrac{x^4}{8!} + \cdots = 0$. If $x \geq 0$, then $f(x) \geq 1$ and there are no solutions of the equation. Note that $f(-x^2) = 1 - \dfrac{x^2}{2!} + \dfrac{x^4}{4!} - \dfrac{x^6}{6!} + \dfrac{x^8}{8!} - \cdots = \cos x$. The solutions of $\cos x = 0$ for $x < 0$ are given by $x = \dfrac{\pi}{2} - \pi k$, where k is a positive integer. Thus, the solutions of $f(x) = 0$ are $x = -\left(\dfrac{\pi}{2} - \pi k\right)^2$, where k is a positive integer.

21. Call the series S. We group the terms according to the number of digits in their denominators:

$$S = \underbrace{\left(\tfrac{1}{1} + \tfrac{1}{2} + \cdots + \tfrac{1}{8} + \tfrac{1}{9}\right)}_{g_1} + \underbrace{\left(\tfrac{1}{11} + \cdots + \tfrac{1}{99}\right)}_{g_2} + \underbrace{\left(\tfrac{1}{111} + \cdots + \tfrac{1}{999}\right)}_{g_3} + \cdots$$

Now in the group g_n, since we have 9 choices for each of the n digits in the denominator, there are 9^n terms. Furthermore, each term in g_n is less than $\dfrac{1}{10^{n-1}}$ [except for the first term in g_1]. So $g_n < 9^n \cdot \dfrac{1}{10^{n-1}} = 9\left(\dfrac{9}{10}\right)^{n-1}$.

Now $\displaystyle\sum_{n=1}^{\infty} 9\left(\dfrac{9}{10}\right)^{n-1}$ is a geometric series with $a = 9$ and $r = \dfrac{9}{10} < 1$. Therefore, by the Comparison Test,

$$S = \sum_{n=1}^{\infty} g_n < \sum_{n=1}^{\infty} 9\left(\tfrac{9}{10}\right)^{n-1} = \dfrac{9}{1 - 9/10} = 90.$$

23. $u = 1 + \dfrac{x^3}{3!} + \dfrac{x^6}{6!} + \dfrac{x^9}{9!} + \cdots,\ v = x + \dfrac{x^4}{4!} + \dfrac{x^7}{7!} + \dfrac{x^{10}}{10!} + \cdots,\ w = \dfrac{x^2}{2!} + \dfrac{x^5}{5!} + \dfrac{x^8}{8!} + \cdots.$

Use the Ratio Test to show that the series for u, v, and w have positive radii of convergence (∞ in each case), so Theorem 11.9.2 applies, and hence, we may differentiate each of these series:

$$\dfrac{du}{dx} = \dfrac{3x^2}{3!} + \dfrac{6x^5}{6!} + \dfrac{9x^8}{9!} + \cdots = \dfrac{x^2}{2!} + \dfrac{x^5}{5!} + \dfrac{x^8}{8!} + \cdots = w$$

Similarly, $\dfrac{dv}{dx} = 1 + \dfrac{x^3}{3!} + \dfrac{x^6}{6!} + \dfrac{x^9}{9!} + \cdots = u$, and $\dfrac{dw}{dx} = x + \dfrac{x^4}{4!} + \dfrac{x^7}{7!} + \dfrac{x^{10}}{10!} + \cdots = v$.

So $u' = w$, $v' = u$, and $w' = v$. Now differentiate the left-hand side of the desired equation:

$$\dfrac{d}{dx}(u^3 + v^3 + w^3 - 3uvw) = 3u^2 u' + 3v^2 v' + 3w^2 w' - 3(u'vw + uv'w + uvw')$$
$$= 3u^2 w + 3v^2 u + 3w^2 v - 3(vw^2 + u^2 w + uv^2) = 0 \;\Rightarrow$$

$u^3 + v^3 + w^3 - 3uvw = C$. To find the value of the constant C, we put $x = 0$ in the last equation and get $1^3 + 0^3 + 0^3 - 3(1 \cdot 0 \cdot 0) = C \;\Rightarrow\; C = 1$, so $u^3 + v^3 + w^3 - 3uvw = 1$.

APPENDIXES

A Numbers, Inequalities, and Absolute Values

1. $|5 - 23| = |-18| = 18$

3. $|-\pi| = \pi$ because $\pi > 0$.

5. $|\sqrt{5} - 5| = -(\sqrt{5} - 5) = 5 - \sqrt{5}$ because $\sqrt{5} - 5 < 0$.

7. If $x < 2$, $x - 2 < 0$, so $|x - 2| = -(x - 2) = 2 - x$.

9. $|x + 1| = \begin{cases} x + 1 & \text{if } x + 1 \geq 0 \\ -(x + 1) & \text{if } x + 1 < 0 \end{cases} = \begin{cases} x + 1 & \text{if } x \geq -1 \\ -x - 1 & \text{if } x < -1 \end{cases}$

11. $|x^2 + 1| = x^2 + 1$ [since $x^2 + 1 \geq 0$ for all x].

13. $2x + 7 > 3 \Leftrightarrow 2x > -4 \Leftrightarrow x > -2$, so $x \in (-2, \infty)$.

15. $1 - x \leq 2 \Leftrightarrow -x \leq 1 \Leftrightarrow x \geq -1$, so $x \in [-1, \infty)$.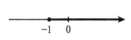

17. $2x + 1 < 5x - 8 \Leftrightarrow 9 < 3x \Leftrightarrow 3 < x$, so $x \in (3, \infty)$.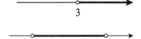

19. $-1 < 2x - 5 < 7 \Leftrightarrow 4 < 2x < 12 \Leftrightarrow 2 < x < 6$, so $x \in (2, 6)$.

21. $0 \leq 1 - x < 1 \Leftrightarrow -1 \leq -x < 0 \Leftrightarrow 1 \geq x > 0$, so $x \in (0, 1]$.

23. $4x < 2x + 1 \leq 3x + 2$. So $4x < 2x + 1 \Leftrightarrow 2x < 1 \Leftrightarrow x < \frac{1}{2}$, and $2x + 1 \leq 3x + 2 \Leftrightarrow -1 \leq x$. Thus, $x \in [-1, \frac{1}{2})$.

25. $(x - 1)(x - 2) > 0$.

    Case 1: (both factors are positive, so their product is positive) $x - 1 > 0 \Leftrightarrow x > 1$, and $x - 2 > 0 \Leftrightarrow x > 2$, so $x \in (2, \infty)$.

    Case 2: (both factors are negative, so their product is positive) $x - 1 < 0 \Leftrightarrow x < 1$, and $x - 2 < 0 \Leftrightarrow x < 2$, so $x \in (-\infty, 1)$.

    Thus, the solution set is $(-\infty, 1) \cup (2, \infty)$.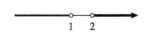

27. $2x^2 + x \leq 1 \Leftrightarrow 2x^2 + x - 1 \leq 0 \Leftrightarrow (2x - 1)(x + 1) \leq 0$.

    Case 1: $2x - 1 \geq 0 \Leftrightarrow x \geq \frac{1}{2}$, and $x + 1 \leq 0 \Leftrightarrow x \leq -1$, which is an impossible combination.

    Case 2: $2x - 1 \leq 0 \Leftrightarrow x \leq \frac{1}{2}$, and $x + 1 \geq 0 \Leftrightarrow x \geq -1$, so $x \in [-1, \frac{1}{2}]$.

    Thus, the solution set is $[-1, \frac{1}{2}]$.

29. $x^2 + x + 1 > 0 \Leftrightarrow x^2 + x + \frac{1}{4} + \frac{3}{4} > 0 \Leftrightarrow \left(x + \frac{1}{2}\right)^2 + \frac{3}{4} > 0$. But since $\left(x + \frac{1}{2}\right)^2 \geq 0$ for every real x, the original inequality will be true for all real x as well.

Thus, the solution set is $(-\infty, \infty)$.

31. $x^2 < 3 \Leftrightarrow x^2 - 3 < 0 \Leftrightarrow (x - \sqrt{3})(x + \sqrt{3}) < 0$.

Case 1: $x > \sqrt{3}$ and $x < -\sqrt{3}$, which is impossible.

Case 2: $x < \sqrt{3}$ and $x > -\sqrt{3}$.

Thus, the solution set is $(-\sqrt{3}, \sqrt{3})$.

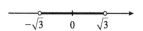

Another method: $x^2 < 3 \Leftrightarrow |x| < \sqrt{3} \Leftrightarrow -\sqrt{3} < x < \sqrt{3}$.

33. $x^3 - x^2 \leq 0 \Leftrightarrow x^2(x - 1) \leq 0$. Since $x^2 \geq 0$ for all x, the inequality is satisfied when $x - 1 \leq 0 \Leftrightarrow x \leq 1$.

Thus, the solution set is $(-\infty, 1]$.

35. $x^3 > x \Leftrightarrow x^3 - x > 0 \Leftrightarrow x(x^2 - 1) > 0 \Leftrightarrow x(x-1)(x+1) > 0$. Construct a chart:

Interval	x	$x-1$	$x+1$	$x(x-1)(x+1)$
$x < -1$	−	−	−	−
$-1 < x < 0$	−	−	+	+
$0 < x < 1$	+	−	+	−
$x > 1$	+	+	+	+

Since $x^3 > x$ when the last column is positive, the solution set is $(-1, 0) \cup (1, \infty)$.

37. $1/x < 4$. This is clearly true for $x < 0$. So suppose $x > 0$. then $1/x < 4 \Leftrightarrow 1 < 4x \Leftrightarrow \frac{1}{4} < x$. Thus, the solution set is $(-\infty, 0) \cup \left(\frac{1}{4}, \infty\right)$.

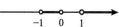

39. $C = \frac{5}{9}(F - 32) \Rightarrow F = \frac{9}{5}C + 32$. So $50 \leq F \leq 95 \Rightarrow 50 \leq \frac{9}{5}C + 32 \leq 95 \Rightarrow 18 \leq \frac{9}{5}C \leq 63 \Rightarrow 10 \leq C \leq 35$. So the interval is $[10, 35]$.

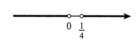

41. (a) Let T represent the temperature in degrees Celsius and h the height in km. $T = 20$ when $h = 0$ and T decreases by $10°C$ for every km ($1°C$ for each 100-m rise). Thus, $T = 20 - 10h$ when $0 \leq h \leq 12$.

(b) From part (a), $T = 20 - 10h \Rightarrow 10h = 20 - T \Rightarrow h = 2 - T/10$. So $0 \leq h \leq 5 \Rightarrow 0 \leq 2 - T/10 \leq 5 \Rightarrow -2 \leq -T/10 \leq 3 \Rightarrow -20 \leq -T \leq 30 \Rightarrow 20 \geq T \geq -30 \Rightarrow -30 \leq T \leq 20$. Thus, the range of temperatures (in $°C$) to be expected is $[-30, 20]$.

43. $|2x| = 3 \Leftrightarrow$ either $2x = 3$ or $2x = -3 \Leftrightarrow x = \frac{3}{2}$ or $x = -\frac{3}{2}$.

45. $|x + 3| = |2x + 1| \Leftrightarrow$ either $x + 3 = 2x + 1$ or $x + 3 = -(2x + 1)$. In the first case, $x = 2$, and in the second case, $x + 3 = -2x - 1 \Leftrightarrow 3x = -4 \Leftrightarrow x = -\frac{4}{3}$. So the solutions are $-\frac{4}{3}$ and 2.

47. By Property 5 of absolute values, $|x| < 3 \Leftrightarrow -3 < x < 3$, so $x \in (-3, 3)$.

49. $|x - 4| < 1 \Leftrightarrow -1 < x - 4 < 1 \Leftrightarrow 3 < x < 5$, so $x \in (3, 5)$.

51. $|x + 5| \geq 2 \Leftrightarrow x + 5 \geq 2$ or $x + 5 \leq -2 \Leftrightarrow x \geq -3$ or $x \leq -7$, so $x \in (-\infty, -7] \cup [-3, \infty)$.

53. $|2x - 3| \leq 0.4 \Leftrightarrow -0.4 \leq 2x - 3 \leq 0.4 \Leftrightarrow 2.6 \leq 2x \leq 3.4 \Leftrightarrow 1.3 \leq x \leq 1.7$, so $x \in [1.3, 1.7]$.

55. $1 \leq |x| \leq 4$. So either $1 \leq x \leq 4$ or $1 \leq -x \leq 4 \Leftrightarrow -1 \geq x \geq -4$. Thus, $x \in [-4, -1] \cup [1, 4]$.

57. $a(bx - c) \geq bc \Leftrightarrow bx - c \geq \dfrac{bc}{a} \Leftrightarrow bx \geq \dfrac{bc}{a} + c = \dfrac{bc + ac}{a} \Leftrightarrow x \geq \dfrac{bc + ac}{ab}$

59. $ax + b < c \Leftrightarrow ax < c - b \Leftrightarrow x > \dfrac{c - b}{a}$ [since $a < 0$]

61. $|(x + y) - 5| = |(x - 2) + (y - 3)| \leq |x - 2| + |y - 3| < 0.01 + 0.04 = 0.05$

63. If $a < b$ then $a + a < a + b$ and $a + b < b + b$. So $2a < a + b < 2b$. Dividing by 2, we get $a < \frac{1}{2}(a + b) < b$.

65. $|ab| = \sqrt{(ab)^2} = \sqrt{a^2b^2} = \sqrt{a^2}\sqrt{b^2} = |a||b|$

67. If $0 < a < b$, then $a \cdot a < a \cdot b$ and $a \cdot b < b \cdot b$ [using Rule 3 of Inequalities]. So $a^2 < ab < b^2$ and hence $a^2 < b^2$.

69. Observe that the sum, difference and product of two integers is always an integer. Let the rational numbers be represented by $r = m/n$ and $s = p/q$ (where m, n, p and q are integers with $n \neq 0, q \neq 0$). Now $r + s = \dfrac{m}{n} + \dfrac{p}{q} = \dfrac{mq + pn}{nq}$, but $mq + pn$ and nq are both integers, so $\dfrac{mq + pn}{nq} = r + s$ is a rational number by definition. Similarly, $r - s = \dfrac{m}{n} - \dfrac{p}{q} = \dfrac{mq - pn}{nq}$ is a rational number. Finally, $r \cdot s = \dfrac{m}{n} \cdot \dfrac{p}{q} = \dfrac{mp}{nq}$ but mp and nq are both integers, so $\dfrac{mp}{nq} = r \cdot s$ is a rational number by definition.

B Coordinate Geometry and Lines

1. Use the distance formula with $P_1(x_1, y_1) = (1, 1)$ and $P_2(x_2, y_2) = (4, 5)$ to get
$$|P_1P_2| = \sqrt{(4 - 1)^2 + (5 - 1)^2} = \sqrt{3^2 + 4^2} = \sqrt{25} = 5$$

3. The distance from $(6, -2)$ to $(-1, 3)$ is $\sqrt{(-1 - 6)^2 + [3 - (-2)]^2} = \sqrt{(-7)^2 + 5^2} = \sqrt{74}$.

5. The distance from $(2, 5)$ to $(4, -7)$ is $\sqrt{(4 - 2)^2 + (-7 - 5)^2} = \sqrt{2^2 + (-12)^2} = \sqrt{148} = 2\sqrt{37}$.

7. The slope m of the line through $P(1, 5)$ and $Q(4, 11)$ is $m = \dfrac{11 - 5}{4 - 1} = \dfrac{6}{3} = 2$.

9. The slope m of the line through $P(-3, 3)$ and $Q(-1, -6)$ is $m = \dfrac{-6 - 3}{-1 - (-3)} = -\dfrac{9}{2}$.

11. Using $A(0, 2)$, $B(-3, -1)$, and $C(-4, 3)$, we have $|AC| = \sqrt{(-4 - 0)^2 + (3 - 2)^2} = \sqrt{(-4)^2 + 1^2} = \sqrt{17}$ and $|BC| = \sqrt{[-4 - (-3)]^2 + [3 - (-1)]^2} = \sqrt{(-1)^2 + 4^2} = \sqrt{17}$, so the triangle has two sides of equal length, and is isosceles.

13. Using $A(-2,9)$, $B(4,6)$, $C(1,0)$, and $D(-5,3)$, we have

$|AB| = \sqrt{[4-(-2)]^2 + (6-9)^2} = \sqrt{6^2 + (-3)^2} = \sqrt{45} = \sqrt{9}\sqrt{5} = 3\sqrt{5}$,

$|BC| = \sqrt{(1-4)^2 + (0-6)^2} = \sqrt{(-3)^2 + (-6)^2} = \sqrt{45} = \sqrt{9}\sqrt{5} = 3\sqrt{5}$,

$|CD| = \sqrt{(-5-1)^2 + (3-0)^2} = \sqrt{(-6)^2 + 3^2} = \sqrt{45} = \sqrt{9}\sqrt{5} = 3\sqrt{5}$, and

$|DA| = \sqrt{[-2-(-5)]^2 + (9-3)^2} = \sqrt{3^2 + 6^2} = \sqrt{45} = \sqrt{9}\sqrt{5} = 3\sqrt{5}$. So all sides are of equal length and we have a rhombus. Moreover, $m_{AB} = \dfrac{6-9}{4-(-2)} = -\dfrac{1}{2}$, $m_{BC} = \dfrac{0-6}{1-4} = 2$, $m_{CD} = \dfrac{3-0}{-5-1} = -\dfrac{1}{2}$, and

$m_{DA} = \dfrac{9-3}{-2-(-5)} = 2$, so the sides are perpendicular. Thus, A, B, C, and D are vertices of a square.

15. For the vertices $A(1,1)$, $B(7,4)$, $C(5,10)$, and $D(-1,7)$, the slope of the line segment AB is $\dfrac{4-1}{7-1} = \dfrac{1}{2}$, the slope of CD is $\dfrac{7-10}{-1-5} = \dfrac{1}{2}$, the slope of BC is $\dfrac{10-4}{5-7} = -3$, and the slope of DA is $\dfrac{1-7}{1-(-1)} = -3$. So AB is parallel to CD and BC is parallel to DA. Hence $ABCD$ is a parallelogram.

17. The graph of the equation $x=3$ is a vertical line with x-intercept 3. The line does not have a slope.

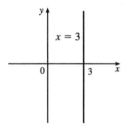

19. $xy = 0 \Leftrightarrow x = 0$ or $y = 0$. The graph consists of the coordinate axes.

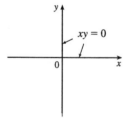

21. By the point-slope form of the equation of a line, an equation of the line through $(2,-3)$ with slope 6 is

$y - (-3) = 6(x-2)$ or $y = 6x - 15$.

23. $y - 7 = \frac{2}{3}(x-1)$ or $y = \frac{2}{3}x + \frac{19}{3}$

25. The slope of the line through $(2,1)$ and $(1,6)$ is $m = \dfrac{6-1}{1-2} = -5$, so an equation of the line is

$y - 1 = -5(x-2)$ or $y = -5x + 11$.

27. By the slope-intercept form of the equation of a line, an equation of the line is $y = 3x - 2$.

29. Since the line passes through $(1,0)$ and $(0,-3)$, its slope is $m = \dfrac{-3-0}{0-1} = 3$, so an equation is $y = 3x - 3$.

Another method: From Exercise 61, $\dfrac{x}{1} + \dfrac{y}{-3} = 1 \Rightarrow -3x + y = -3 \Rightarrow y = 3x - 3$.

31. The line is parallel to the x-axis, so it is horizontal and must have the form $y = k$. Since it goes through the point $(x, y) = (4, 5)$, the equation is $y = 5$.

33. Putting the line $x + 2y = 6$ into its slope-intercept form gives us $y = -\frac{1}{2}x + 3$, so we see that this line has slope $-\frac{1}{2}$. Thus, we want the line of slope $-\frac{1}{2}$ that passes through the point $(1, -6)$: $y - (-6) = -\frac{1}{2}(x - 1) \Leftrightarrow y = -\frac{1}{2}x - \frac{11}{2}$.

35. $2x + 5y + 8 = 0 \Leftrightarrow y = -\frac{2}{5}x - \frac{8}{5}$. Since this line has slope $-\frac{2}{5}$, a line perpendicular to it would have slope $\frac{5}{2}$, so the required line is $y - (-2) = \frac{5}{2}[x - (-1)] \Leftrightarrow y = \frac{5}{2}x + \frac{1}{2}$.

37. $x + 3y = 0 \Leftrightarrow y = -\frac{1}{3}x$, so the slope is $-\frac{1}{3}$ and the y-intercept is 0.

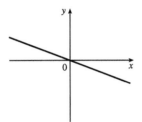

39. $y = -2$ is a horizontal line with slope 0 and y-intercept -2.

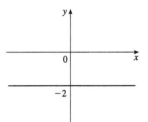

41. $3x - 4y = 12 \Leftrightarrow y = \frac{3}{4}x - 3$, so the slope is $\frac{3}{4}$ and the y-intercept is -3.

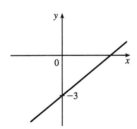

43. $\{(x, y) \mid x < 0\}$

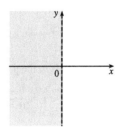

45. $\{(x, y) \mid xy < 0\} = \{(x, y) \mid x < 0 \text{ and } y > 0\}$
$\cup \{(x, y) \mid x > 0 \text{ and } y < 0\}$

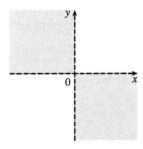

47. $\{(x, y) \mid |x| \leq 2\} = \{(x, y) \mid -2 \leq x \leq 2\}$

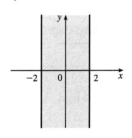

49. $\{(x,y) \mid 0 \le y \le 4, x \le 2\}$

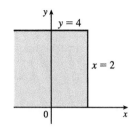

51. $\{(x,y) \mid 1+x \le y \le 1-2x\}$

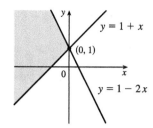

53. Let $P(0, y)$ be a point on the y-axis. The distance from P to $(5, -5)$ is $\sqrt{(5-0)^2 + (-5-y)^2} = \sqrt{5^2 + (y+5)^2}$. The distance from P to $(1, 1)$ is $\sqrt{(1-0)^2 + (1-y)^2} = \sqrt{1^2 + (y-1)^2}$. We want these distances to be equal: $\sqrt{5^2 + (y+5)^2} = \sqrt{1^2 + (y-1)^2} \Leftrightarrow 5^2 + (y+5)^2 = 1^2 + (y-1)^2 \Leftrightarrow 25 + (y^2 + 10y + 25) = 1 + (y^2 - 2y + 1) \Leftrightarrow 12y = -48 \Leftrightarrow y = -4$. So the desired point is $(0, -4)$.

55. (a) Using the midpoint formula from Exercise 54 with $(1, 3)$ and $(7, 15)$, we get $\left(\frac{1+7}{2}, \frac{3+15}{2}\right) = (4, 9)$.

(b) Using the midpoint formula from Exercise 54 with $(-1, 6)$ and $(8, -12)$, we get $\left(\frac{-1+8}{2}, \frac{6+(-12)}{2}\right) = \left(\frac{7}{2}, -3\right)$.

57. $2x - y = 4 \Leftrightarrow y = 2x - 4 \Rightarrow m_1 = 2$ and $6x - 2y = 10 \Leftrightarrow 2y = 6x - 10 \Leftrightarrow y = 3x - 5 \Rightarrow m_2 = 3$. Since $m_1 \ne m_2$, the two lines are not parallel. To find the point of intersection: $2x - 4 = 3x - 5 \Leftrightarrow x = 1 \Rightarrow y = -2$. Thus, the point of intersection is $(1, -2)$.

59. With $A(1, 4)$ and $B(7, -2)$, the slope of segment AB is $\frac{-2-4}{7-1} = -1$, so its perpendicular bisector has slope 1. The midpoint of AB is $\left(\frac{1+7}{2}, \frac{4+(-2)}{2}\right) = (4, 1)$, so an equation of the perpendicular bisector is $y - 1 = 1(x - 4)$ or $y = x - 3$.

61. (a) Since the x-intercept is a, the point $(a, 0)$ is on the line, and similarly since the y-intercept is b, $(0, b)$ is on the line. Hence, the slope of the line is $m = \dfrac{b-0}{0-a} = -\dfrac{b}{a}$. Substituting into $y = mx + b$ gives $y = -\dfrac{b}{a}x + b \Leftrightarrow \dfrac{b}{a}x + y = b \Leftrightarrow \dfrac{x}{a} + \dfrac{y}{b} = 1$.

(b) Letting $a = 6$ and $b = -8$ gives $\dfrac{x}{6} + \dfrac{y}{-8} = 1 \Leftrightarrow -8x + 6y = -48$ [multiply by -48] $\Leftrightarrow 6y = 8x - 48 \Leftrightarrow 3y = 4x - 24 \Leftrightarrow y = \frac{4}{3}x - 8$.

C Graphs of Second-Degree Equations

1. An equation of the circle with center $(3, -1)$ and radius 5 is $(x - 3)^2 + (y + 1)^2 = 5^2 = 25$.

3. The equation has the form $x^2 + y^2 = r^2$. Since $(4, 7)$ lies on the circle, we have $4^2 + 7^2 = r^2 \Rightarrow r^2 = 65$. So the required equation is $x^2 + y^2 = 65$.

5. $x^2 + y^2 - 4x + 10y + 13 = 0 \Leftrightarrow x^2 - 4x + y^2 + 10y = -13 \Leftrightarrow (x^2 - 4x + 4) + (y^2 + 10y + 25) = -13 + 4 + 25 = 16 \Leftrightarrow (x - 2)^2 + (y + 5)^2 = 4^2$. Thus, we have a circle with center $(2, -5)$ and radius 4.

7. $x^2 + y^2 + x = 0$ $\Leftrightarrow$ $\left(x^2 + x + \frac{1}{4}\right) + y^2 = \frac{1}{4}$ $\Leftrightarrow$ $\left(x + \frac{1}{2}\right)^2 + y^2 = \left(\frac{1}{2}\right)^2$. Thus, we have a circle with center $\left(-\frac{1}{2}, 0\right)$ and radius $\frac{1}{2}$.

9. $2x^2 + 2y^2 - x + y = 1$ $\Leftrightarrow$ $2\left(x^2 - \frac{1}{2}x + \frac{1}{16}\right) + 2\left(y^2 + \frac{1}{2}y + \frac{1}{16}\right) = 1 + \frac{1}{8} + \frac{1}{8}$ $\Leftrightarrow$
$2\left(x - \frac{1}{4}\right)^2 + 2\left(y + \frac{1}{4}\right)^2 = \frac{5}{4}$ $\Leftrightarrow$ $\left(x - \frac{1}{4}\right)^2 + \left(y + \frac{1}{4}\right)^2 = \frac{5}{8}$. Thus, we have a circle with center $\left(\frac{1}{4}, -\frac{1}{4}\right)$ and radius $\frac{\sqrt{5}}{2\sqrt{2}} = \frac{\sqrt{10}}{4}$.

11. $y = -x^2$. Parabola

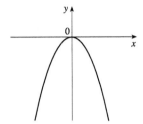

13. $x^2 + 4y^2 = 16$ $\Leftrightarrow$ $\frac{x^2}{16} + \frac{y^2}{4} = 1$. Ellipse

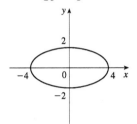

15. $16x^2 - 25y^2 = 400$ $\Leftrightarrow$ $\frac{x^2}{25} - \frac{y^2}{16} = 1$. Hyperbola

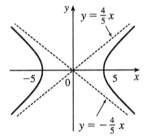

17. $4x^2 + y^2 = 1$ $\Leftrightarrow$ $\frac{x^2}{1/4} + y^2 = 1$. Ellipse

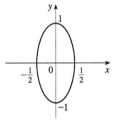

19. $x = y^2 - 1$. Parabola with vertex at $(-1, 0)$

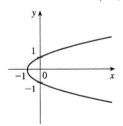

21. $9y^2 - x^2 = 9$ $\Leftrightarrow$ $y^2 - \frac{x^2}{9} = 1$. Hyperbola

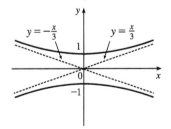

23. $xy = 4$. Hyperbola

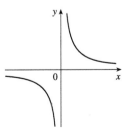

25. $9(x - 1)^2 + 4(y - 2)^2 = 36$ $\Leftrightarrow$
$\frac{(x-1)^2}{4} + \frac{(y-2)^2}{9} = 1$. Ellipse centered at $(1, 2)$

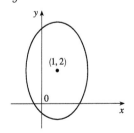

27. $y = x^2 - 6x + 13 = (x^2 - 6x + 9) + 4 = (x - 3)^2 + 4$.
Parabola with vertex at $(3, 4)$.

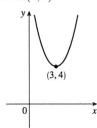

29. $x = 4 - y^2 = -y^2 + 4$. Parabola with vertex at $(4, 0)$

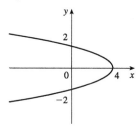

31. $x^2 + 4y^2 - 6x + 5 = 0 \Leftrightarrow$
$(x^2 - 6x + 9) + 4y^2 = -5 + 9 = 4 \Leftrightarrow$
$\dfrac{(x - 3)^2}{4} + y^2 = 1$. Ellipse centered at $(3, 0)$.

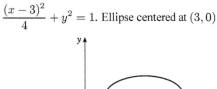

33. $y = 3x$ and $y = x^2$ intersect where $3x = x^2 \Leftrightarrow$
$0 = x^2 - 3x = x(x - 3)$, that is, at $(0, 0)$ and $(3, 9)$.

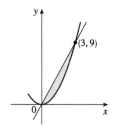

35. The parabola must have an equation of the form $y = a(x - 1)^2 - 1$. Substituting $x = 3$ and $y = 3$ into the equation gives $3 = a(3 - 1)^2 - 1$, so $a = 1$, and the equation is $y = (x - 1)^2 - 1 = x^2 - 2x$. Note that using the other point $(-1, 3)$ would have given the same value for a, and hence the same equation.

37. $\{(x, y) \mid x^2 + y^2 \leq 1\}$

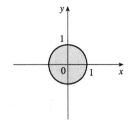

39. $\{(x, y) \mid y \geq x^2 - 1\}$

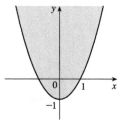

D Trigonometry

1. $210° = 210°\left(\dfrac{\pi}{180°}\right) = \dfrac{7\pi}{6}$ rad

3. $9° = 9°\left(\dfrac{\pi}{180°}\right) = \dfrac{\pi}{20}$ rad

5. $900° = 900°\left(\dfrac{\pi}{180°}\right) = 5\pi$ rad

7. 4π rad $= 4\pi\left(\dfrac{180°}{\pi}\right) = 720°$

9. $\dfrac{5\pi}{12}$ rad $= \dfrac{5\pi}{12}\left(\dfrac{180°}{\pi}\right) = 75°$

11. $-\dfrac{3\pi}{8}$ rad $= -\dfrac{3\pi}{8}\left(\dfrac{180°}{\pi}\right) = -67.5°$

13. Using Formula 3, $a = r\theta = 36 \cdot \dfrac{\pi}{12} = 3\pi$ cm.

15. Using Formula 3, $\theta = a/r = \frac{1}{1.5} = \frac{2}{3}$ rad $= \frac{2}{3}\left(\frac{180°}{\pi}\right) = \left(\frac{120}{\pi}\right)° \approx 38.2°$.

17.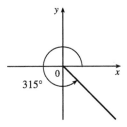

19.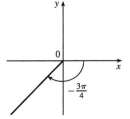

21.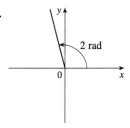

23.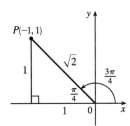

From the diagram we see that a point on the terminal side is $P(-1, 1)$. Therefore, taking $x = -1, y = 1, r = \sqrt{2}$ in the definitions of the trigonometric ratios, we have $\sin\frac{3\pi}{4} = \frac{1}{\sqrt{2}}$, $\cos\frac{3\pi}{4} = -\frac{1}{\sqrt{2}}$, $\tan\frac{3\pi}{4} = -1$, $\csc\frac{3\pi}{4} = \sqrt{2}$, $\sec\frac{3\pi}{4} = -\sqrt{2}$, and $\cot\frac{3\pi}{4} = -1$.

25.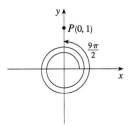

From the diagram we see that a point on the terminal side is $P(0, 1)$. Therefore taking $x = 0, y = 1, r = 1$ in the definitions of the trigonometric ratios, we have $\sin\frac{9\pi}{2} = 1$, $\cos\frac{9\pi}{2} = 0$, $\tan\frac{9\pi}{2} = y/x$ is undefined since $x = 0$, $\csc\frac{9\pi}{2} = 1$, $\sec\frac{9\pi}{2} = r/x$ is undefined since $x = 0$, and $\cot\frac{9\pi}{2} = 0$.

27.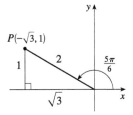

Using Figure 9 we see that a point on the terminal side is $P(-\sqrt{3}, 1)$. Therefore taking $x = -\sqrt{3}, y = 1, r = 2$ in the definitions of the trigonometric ratios, we have $\sin\frac{5\pi}{6} = \frac{1}{2}$, $\cos\frac{5\pi}{6} = -\frac{\sqrt{3}}{2}$, $\tan\frac{5\pi}{6} = -\frac{1}{\sqrt{3}}$, $\csc\frac{5\pi}{6} = 2$, $\sec\frac{5\pi}{6} = -\frac{2}{\sqrt{3}}$, and $\cot\frac{5\pi}{6} = -\sqrt{3}$.

29. $\sin\theta = y/r = \frac{3}{5} \Rightarrow y = 3, r = 5$, and $x = \sqrt{r^2 - y^2} = 4$ (since $0 < \theta < \frac{\pi}{2}$). Therefore taking $x = 4, y = 3, r = 5$ in the definitions of the trigonometric ratios, we have $\cos\theta = \frac{4}{5}$, $\tan\theta = \frac{3}{4}$, $\csc\theta = \frac{5}{3}$, $\sec\theta = \frac{5}{4}$, and $\cot\theta = \frac{4}{3}$.

31. $\frac{\pi}{2} < \phi < \pi \Rightarrow \phi$ is in the second quadrant, where x is negative and y is positive. Therefore $\sec\phi = r/x = -1.5 = -\frac{3}{2} \Rightarrow r = 3, x = -2$, and $y = \sqrt{r^2 - x^2} = \sqrt{5}$. Taking $x = -2, y = \sqrt{5}$, and $r = 3$ in the definitions of the trigonometric ratios, we have $\sin\phi = \frac{\sqrt{5}}{3}$, $\cos\phi = -\frac{2}{3}$, $\tan\phi = -\frac{\sqrt{5}}{2}$, $\csc\phi = \frac{3}{\sqrt{5}}$, and $\cot\theta = -\frac{2}{\sqrt{5}}$.

33. $\pi < \beta < 2\pi$ means that β is in the third or fourth quadrant where y is negative. Also since $\cot\beta = x/y = 3$ which is positive, x must also be negative. Therefore $\cot\beta = x/y = \frac{3}{1} \Rightarrow x = -3, y = -1$, and $r = \sqrt{x^2 + y^2} = \sqrt{10}$. Taking $x = -3, y = -1$ and $r = \sqrt{10}$ in the definitions of the trigonometric ratios, we have $\sin\beta = -\frac{1}{\sqrt{10}}$, $\cos\beta = -\frac{3}{\sqrt{10}}$, $\tan\beta = \frac{1}{3}$, $\csc\beta = -\sqrt{10}$, and $\sec\beta = -\frac{\sqrt{10}}{3}$.

35. $\sin 35° = \dfrac{x}{10} \;\Rightarrow\; x = 10\sin 35° \approx 5.73576$ cm

37. $\tan\dfrac{2\pi}{5} = \dfrac{x}{8} \;\Rightarrow\; x = 8\tan\dfrac{2\pi}{5} \approx 24.62147$ cm

39.

(a) From the diagram we see that $\sin\theta = \dfrac{y}{r} = \dfrac{a}{c}$, and $\sin(-\theta) = \dfrac{-a}{c} = -\dfrac{a}{c} = -\sin\theta$.

(b) Again from the diagram we see that $\cos\theta = \dfrac{x}{r} = \dfrac{b}{c} = \cos(-\theta)$.

41. (a) Using (13a) and (14a), we have

$$\tfrac{1}{2}[\sin(x+y)+\sin(x-y)] = \tfrac{1}{2}[\sin x\cos y+\cos x\sin y+\sin x\cos y-\cos x\sin y] = \tfrac{1}{2}(2\sin x\cos y) = \sin x\cos y.$$

(b) This time, using (13b) and (14b), we have

$$\tfrac{1}{2}[\cos(x+y)+\cos(x-y)] = \tfrac{1}{2}[\cos x\cos y-\sin x\sin y+\cos x\cos y+\sin x\sin y] = \tfrac{1}{2}(2\cos x\cos y) = \cos x\cos y.$$

(c) Again using (13b) and (14b), we have

$$\tfrac{1}{2}[\cos(x-y)-\cos(x+y)] = \tfrac{1}{2}[\cos x\cos y+\sin x\sin y-\cos x\cos y+\sin x\sin y]$$
$$= \tfrac{1}{2}(2\sin x\sin y) = \sin x\sin y$$

43. Using (13a), we have $\sin\left(\dfrac{\pi}{2}+x\right) = \sin\dfrac{\pi}{2}\cos x + \cos\dfrac{\pi}{2}\sin x = 1\cdot\cos x + 0\cdot\sin x = \cos x$.

45. Using (7), we have $\sin\theta\cot\theta = \sin\theta\cdot\dfrac{\cos\theta}{\sin\theta} = \cos\theta$.

47. $\sec y - \cos y = \dfrac{1}{\cos y} - \cos y$ [by (7)] $= \dfrac{1-\cos^2 y}{\cos y} = \dfrac{\sin^2 y}{\cos y}$ [by (8)] $= \dfrac{\sin y}{\cos y}\sin y = \tan y\sin y$ [by (7)]

49. $\cot^2\theta + \sec^2\theta = \dfrac{\cos^2\theta}{\sin^2\theta} + \dfrac{1}{\cos^2\theta}$ [by (7)] $= \dfrac{\cos^2\theta\cos^2\theta+\sin^2\theta}{\sin^2\theta\cos^2\theta}$

$= \dfrac{(1-\sin^2\theta)(1-\sin^2\theta)+\sin^2\theta}{\sin^2\theta\cos^2\theta}$ [by (8)] $= \dfrac{1-\sin^2\theta+\sin^4\theta}{\sin^2\theta\cos^2\theta}$

$= \dfrac{\cos^2\theta+\sin^4\theta}{\sin^2\theta\cos^2\theta}$ [by (7)] $= \dfrac{1}{\sin^2\theta}+\dfrac{\sin^2\theta}{\cos^2\theta} = \csc^2\theta+\tan^2\theta$ [by (7)]

51. Using (15a), we have $\tan 2\theta = \tan(\theta+\theta) = \dfrac{\tan\theta+\tan\theta}{1-\tan\theta\tan\theta} = \dfrac{2\tan\theta}{1-\tan^2\theta}$.

53. Using (16a) and (17a),

$$\sin x\sin 2x + \cos x\cos 2x = \sin x\,(2\sin x\cos x) + \cos x\,(2\cos^2 x - 1) = 2\sin^2 x\cos x + 2\cos^3 x - \cos x$$
$$= 2(1-\cos^2 x)\cos x + 2\cos^3 x - \cos x \quad\text{[by (8)]}$$
$$= 2\cos x - 2\cos^3 x + 2\cos^3 x - \cos x = \cos x$$

Or: $\sin x\sin 2x + \cos x\cos 2x = \cos(2x-x)$ [by 14(b)] $= \cos x$

55. $\dfrac{\sin\phi}{1-\cos\phi} = \dfrac{\sin\phi}{1-\cos\phi} \cdot \dfrac{1+\cos\phi}{1+\cos\phi} = \dfrac{\sin\phi(1+\cos\phi)}{1-\cos^2\phi} = \dfrac{\sin\phi(1+\cos\phi)}{\sin^2\phi}$ [by (8)]

$= \dfrac{1+\cos\phi}{\sin\phi} = \dfrac{1}{\sin\phi} + \dfrac{\cos\phi}{\sin\phi} = \csc\phi + \cot\phi$ [by (7)]

57. Using (13a),

$\sin 3\theta + \sin\theta = \sin(2\theta+\theta) + \sin\theta = \sin 2\theta\,\cos\theta + \cos 2\theta\,\sin\theta + \sin\theta$

$= \sin 2\theta\,\cos\theta + (2\cos^2\theta - 1)\sin\theta + \sin\theta$ [by (17a)]

$= \sin 2\theta\,\cos\theta + 2\cos^2\theta\,\sin\theta - \sin\theta + \sin\theta = \sin 2\theta\,\cos\theta + \sin 2\theta\,\cos\theta$ [by (16a)]

$= 2\sin 2\theta\,\cos\theta$

59. Since $\sin x = \tfrac{1}{3}$ we can label the opposite side as having length 1, the hypotenuse as having length 3, and use the Pythagorean Theorem to get that the adjacent side has length $\sqrt{8}$. Then, from the diagram, $\cos x = \tfrac{\sqrt{8}}{3}$. Similarly we have that $\sin y = \tfrac{3}{5}$. Now use (13a):

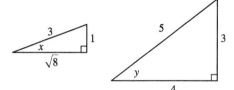

$\sin(x+y) = \sin x\,\cos y + \cos x\,\sin y = \tfrac{1}{3}\cdot\tfrac{4}{5} + \tfrac{\sqrt{8}}{3}\cdot\tfrac{3}{5} = \tfrac{4}{15} + \tfrac{3\sqrt{8}}{15} = \tfrac{4+6\sqrt{2}}{15}$.

61. Using (14b) and the values for $\cos x$ and $\sin y$ obtained in Exercise 59, we have

$\cos(x-y) = \cos x\,\cos y + \sin x\,\sin y = \tfrac{\sqrt{8}}{3}\cdot\tfrac{4}{5} + \tfrac{1}{3}\cdot\tfrac{3}{5} = \tfrac{8\sqrt{2}+3}{15}$

63. Using (16a) and the values for $\sin y$ and $\cos y$ obtained in Exercise 59, we have $\sin 2y = 2\sin y\,\cos y = 2\cdot\tfrac{3}{5}\cdot\tfrac{4}{5} = \tfrac{24}{25}$.

65. $2\cos x - 1 = 0 \;\Leftrightarrow\; \cos x = \tfrac{1}{2} \;\Rightarrow\; x = \tfrac{\pi}{3}, \tfrac{5\pi}{3}$ for $x \in [0, 2\pi]$.

67. $2\sin^2 x = 1 \;\Leftrightarrow\; \sin^2 x = \tfrac{1}{2} \;\Leftrightarrow\; \sin x = \pm\tfrac{1}{\sqrt{2}} \;\Rightarrow\; x = \tfrac{\pi}{4}, \tfrac{3\pi}{4}, \tfrac{5\pi}{4}, \tfrac{7\pi}{4}$.

69. Using (16a), we have $\sin 2x = \cos x \;\Leftrightarrow\; 2\sin x\,\cos x - \cos x = 0 \;\Leftrightarrow\; \cos x(2\sin x - 1) = 0 \;\Leftrightarrow\; \cos x = 0$ or $2\sin x - 1 = 0 \;\Rightarrow\; x = \tfrac{\pi}{2}, \tfrac{3\pi}{2}$ or $\sin x = \tfrac{1}{2} \;\Rightarrow\; x = \tfrac{\pi}{6}$ or $\tfrac{5\pi}{6}$. Therefore, the solutions are $x = \tfrac{\pi}{6}, \tfrac{\pi}{2}, \tfrac{5\pi}{6}, \tfrac{3\pi}{2}$.

71. $\sin x = \tan x \;\Leftrightarrow\; \sin x - \tan x = 0 \;\Leftrightarrow\; \sin x - \dfrac{\sin x}{\cos x} = 0 \;\Leftrightarrow\; \sin x\left(1 - \dfrac{1}{\cos x}\right) = 0 \;\Leftrightarrow\; \sin x = 0$ or $1 - \dfrac{1}{\cos x} = 0 \;\Rightarrow\; x = 0, \pi, 2\pi$ or $1 = \dfrac{1}{\cos x} \;\Rightarrow\; \cos x = 1 \;\Rightarrow\; x = 0, 2\pi$. Therefore the solutions are $x = 0, \pi, 2\pi$.

73. We know that $\sin x = \tfrac{1}{2}$ when $x = \tfrac{\pi}{6}$ or $\tfrac{5\pi}{6}$, and from Figure 18(a), we see that $\sin x \leq \tfrac{1}{2} \;\Rightarrow\; 0 \leq x \leq \tfrac{\pi}{6}$ or $\tfrac{5\pi}{6} \leq x \leq 2\pi$ for $x \in [0, 2\pi]$.

75. $\tan x = -1$ when $x = \tfrac{3\pi}{4}, \tfrac{7\pi}{4}$, and $\tan x = 1$ when $x = \tfrac{\pi}{4}$ or $\tfrac{5\pi}{4}$. From Figure 19(a) we see that $-1 < \tan x < 1 \;\Rightarrow\; 0 \leq x < \tfrac{\pi}{4}, \tfrac{3\pi}{4} < x < \tfrac{5\pi}{4}$, and $\tfrac{7\pi}{4} < x \leq 2\pi$.

77. $\angle C = 180° - 50° - 68° = 62°$.

Then $a = \dfrac{230 \sin 50°}{\sin 62°} \approx 200$ and $b = \dfrac{230 \sin 68°}{\sin 62°} \approx 242$.

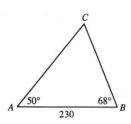

79. $|AB|^2 = |AC|^2 + |BC|^2 - 2|AC||BC|\cos \angle C = (820)^2 + (910)^2 - 2(820)(910)\cos 103° \approx 1{,}836{,}217 \Rightarrow$
$|AB| \approx 1355$ m

81. Using the formula from (6), the area of the triangle is $\frac{1}{2}(10)(3) \sin 107° \approx 14.34457$ cm^2.

83. $y = \cos\left(x - \frac{\pi}{3}\right)$. We start with the graph of $y = \cos x$ and shift it $\frac{\pi}{3}$ units to the right.

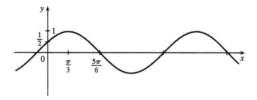

85. $y = \frac{1}{3}\tan\left(x - \frac{\pi}{2}\right)$. We start with the graph of $y = \tan x$, shift it $\frac{\pi}{2}$ units to the right and compress it to $\frac{1}{3}$ of its original vertical size.

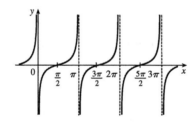

87. $y = |\sin x|$. We start with the graph of $y = \sin x$ and reflect the parts below the x-axis about the x-axis.

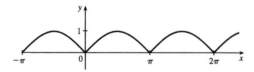

89. Using the Law of Cosines, we have $c^2 = 1^2 + 1^2 - 2(1)(1)\cos(\alpha - \beta) = 2[1 - \cos(\alpha - \beta)]$. Now, using the distance formula, $c^2 = |AB|^2 = (\cos\alpha - \cos\beta)^2 + (\sin\alpha - \sin\beta)^2$. Equating these two expressions for c^2, we get
$2[1 - \cos(\alpha - \beta)] = \cos^2\alpha + \sin^2\alpha + \cos^2\beta + \sin^2\beta - 2\cos\alpha\cos\beta - 2\sin\alpha\sin\beta \Rightarrow$
$1 - \cos(\alpha - \beta) = 1 - \cos\alpha\cos\beta - \sin\alpha\sin\beta \Rightarrow \cos(\alpha - \beta) = \cos\alpha\cos\beta + \sin\alpha\sin\beta$.

91. In Exercise 90 we used the subtraction formula for cosine to prove the addition formula for cosine. Using that formula with $x = \frac{\pi}{2} - \alpha$, $y = \beta$, we get $\cos\left[\left(\frac{\pi}{2} - \alpha\right) + \beta\right] = \cos\left(\frac{\pi}{2} - \alpha\right)\cos\beta - \sin\left(\frac{\pi}{2} - \alpha\right)\sin\beta \Rightarrow$
$\cos\left[\frac{\pi}{2} - (\alpha - \beta)\right] = \cos\left(\frac{\pi}{2} - \alpha\right)\cos\beta - \sin\left(\frac{\pi}{2} - \alpha\right)\sin\beta$. Now we use the identities given in the problem,
$\cos\left(\frac{\pi}{2} - \theta\right) = \sin\theta$ and $\sin\left(\frac{\pi}{2} - \theta\right) = \cos\theta$, to get $\sin(\alpha - \beta) = \sin\alpha\cos\beta - \cos\alpha\sin\beta$.

E Sigma Notation

1. $\sum_{i=1}^{5} \sqrt{i} = \sqrt{1} + \sqrt{2} + \sqrt{3} + \sqrt{4} + \sqrt{5}$

3. $\sum_{i=4}^{6} 3^i = 3^4 + 3^5 + 3^6$

5. $\sum_{k=0}^{4} \frac{2k-1}{2k+1} = -1 + \frac{1}{3} + \frac{3}{5} + \frac{5}{7} + \frac{7}{9}$

7. $\sum_{i=1}^{n} i^{10} = 1^{10} + 2^{10} + 3^{10} + \cdots + n^{10}$

9. $\sum_{j=0}^{n-1} (-1)^j = 1 - 1 + 1 - 1 + \cdots + (-1)^{n-1}$

11. $1 + 2 + 3 + 4 + \cdots + 10 = \sum_{i=1}^{10} i$

13. $\frac{1}{2} + \frac{2}{3} + \frac{3}{4} + \frac{4}{5} + \cdots + \frac{19}{20} = \sum_{i=1}^{19} \frac{i}{i+1}$

15. $2 + 4 + 6 + 8 + \cdots + 2n = \sum_{i=1}^{n} 2i$

17. $1 + 2 + 4 + 8 + 16 + 32 = \sum_{i=0}^{5} 2^i$

19. $x + x^2 + x^3 + \cdots + x^n = \sum_{i=1}^{n} x^i$

21. $\sum_{i=4}^{8} (3i - 2) = [3(4) - 2] + [3(5) - 2] + [3(6) - 2] + [3(7) - 2] + [3(8) - 2] = 10 + 13 + 16 + 19 + 22 = 80$

23. $\sum_{j=1}^{6} 3^{j+1} = 3^2 + 3^3 + 3^4 + 3^5 + 3^6 + 3^7 = 9 + 27 + 81 + 243 + 729 + 2187 = 3276$

(For a more general method, see Exercise 47.)

25. $\sum_{n=1}^{20} (-1)^n = -1 + 1 - 1 + 1 - 1 + 1 - 1 + 1 - 1 + 1 - 1 + 1 - 1 + 1 - 1 + 1 - 1 + 1 - 1 + 1 = 0$

27. $\sum_{i=0}^{4} (2^i + i^2) = (1 + 0) + (2 + 1) + (4 + 4) + (8 + 9) + (16 + 16) = 61$

29. $\sum_{i=1}^{n} 2i = 2 \sum_{i=1}^{n} i = 2 \cdot \frac{n(n+1)}{2}$ [by Theorem 3(c)] $= n(n+1)$

31. $\sum_{i=1}^{n} (i^2 + 3i + 4) = \sum_{i=1}^{n} i^2 + 3 \sum_{i=1}^{n} i + \sum_{i=1}^{n} 4 = \frac{n(n+1)(2n+1)}{6} + \frac{3n(n+1)}{2} + 4n$

$= \frac{1}{6}[(2n^3 + 3n^2 + n) + (9n^2 + 9n) + 24n] = \frac{1}{6}(2n^3 + 12n^2 + 34n) = \frac{1}{3}n(n^2 + 6n + 17)$

33. $\sum_{i=1}^{n} (i+1)(i+2) = \sum_{i=1}^{n} (i^2 + 3i + 2) = \sum_{i=1}^{n} i^2 + 3 \sum_{i=1}^{n} i + \sum_{i=1}^{n} 2 = \frac{n(n+1)(2n+1)}{6} + \frac{3n(n+1)}{2} + 2n$

$= \frac{n(n+1)}{6}[(2n+1) + 9] + 2n = \frac{n(n+1)}{3}(n+5) + 2n$

$= \frac{n}{3}[(n+1)(n+5) + 6] = \frac{n}{3}(n^2 + 6n + 11)$

35. $\sum_{i=1}^{n} (i^3 - i - 2) = \sum_{i=1}^{n} i^3 - \sum_{i=1}^{n} i - \sum_{i=1}^{n} 2 = \left[\frac{n(n+1)}{2}\right]^2 - \frac{n(n+1)}{2} - 2n$

$= \frac{1}{4}n(n+1)[n(n+1) - 2] - 2n = \frac{1}{4}n(n+1)(n+2)(n-1) - 2n$

$= \frac{1}{4}n[(n+1)(n-1)(n+2) - 8] = \frac{1}{4}n[(n^2-1)(n+2) - 8] = \frac{1}{4}n(n^3 + 2n^2 - n - 10)$

37. By Theorem 2(a) and Example 3, $\sum_{i=1}^{n} c = c \sum_{i=1}^{n} 1 = cn$.

39. $\sum_{i=1}^{n}[(i+1)^4 - i^4] = (2^4 - 1^4) + (3^4 - 2^4) + (4^4 - 3^4) + \cdots + [(n+1)^4 - n^4]$

$= (n+1)^4 - 1^4 = n^4 + 4n^3 + 6n^2 + 4n$

On the other hand,

$$\sum_{i=1}^{n}[(i+1)^4 - i^4] = \sum_{i=1}^{n}(4i^3 + 6i^2 + 4i + 1) = 4\sum_{i=1}^{n} i^3 + 6\sum_{i=1}^{n} i^2 + 4\sum_{i=1}^{n} i + \sum_{i=1}^{n} 1$$

$= 4S + n(n+1)(2n+1) + 2n(n+1) + n \quad \left[\text{where } S = \sum_{i=1}^{n} i^3\right]$

$= 4S + 2n^3 + 3n^2 + n + 2n^2 + 2n + n = 4S + 2n^3 + 5n^2 + 4n$

Thus, $n^4 + 4n^3 + 6n^2 + 4n = 4S + 2n^3 + 5n^2 + 4n$, from which it follows that

$4S = n^4 + 2n^3 + n^2 = n^2(n^2 + 2n + 1) = n^2(n+1)^2$ and $S = \left[\dfrac{n(n+1)}{2}\right]^2$.

41. (a) $\sum_{i=1}^{n}[i^4 - (i-1)^4] = (1^4 - 0^4) + (2^4 - 1^4) + (3^4 - 2^4) + \cdots + [n^4 - (n-1)^4] = n^4 - 0 = n^4$

(b) $\sum_{i=1}^{100}(5^i - 5^{i-1}) = (5^1 - 5^0) + (5^2 - 5^1) + (5^3 - 5^2) + \cdots + (5^{100} - 5^{99}) = 5^{100} - 5^0 = 5^{100} - 1$

(c) $\sum_{i=3}^{99}\left(\dfrac{1}{i} - \dfrac{1}{i+1}\right) = \left(\dfrac{1}{3} - \dfrac{1}{4}\right) + \left(\dfrac{1}{4} - \dfrac{1}{5}\right) + \left(\dfrac{1}{5} - \dfrac{1}{6}\right) + \cdots + \left(\dfrac{1}{99} - \dfrac{1}{100}\right) = \dfrac{1}{3} - \dfrac{1}{100} = \dfrac{97}{300}$

(d) $\sum_{i=1}^{n}(a_i - a_{i-1}) = (a_1 - a_0) + (a_2 - a_1) + (a_3 - a_2) + \cdots + (a_n - a_{n-1}) = a_n - a_0$

43. $\lim_{n\to\infty}\sum_{i=1}^{n}\dfrac{1}{n}\left(\dfrac{i}{n}\right)^2 = \lim_{n\to\infty}\dfrac{1}{n^3}\sum_{i=1}^{n} i^2 = \lim_{n\to\infty}\dfrac{1}{n^3}\dfrac{n(n+1)(2n+1)}{6} = \lim_{n\to\infty}\dfrac{1}{6}\left(1+\dfrac{1}{n}\right)\left(2+\dfrac{1}{n}\right) = \tfrac{1}{6}(1)(2) = \tfrac{1}{3}$

45. $\lim_{n\to\infty}\sum_{i=1}^{n}\dfrac{2}{n}\left[\left(\dfrac{2i}{n}\right)^3 + 5\left(\dfrac{2i}{n}\right)\right] = \lim_{n\to\infty}\sum_{i=1}^{n}\left[\dfrac{16}{n^4}i^3 + \dfrac{20}{n^2}i\right] = \lim_{n\to\infty}\left[\dfrac{16}{n^4}\sum_{i=1}^{n} i^3 + \dfrac{20}{n^2}\sum_{i=1}^{n} i\right]$

$= \lim_{n\to\infty}\left[\dfrac{16}{n^4}\dfrac{n^2(n+1)^2}{4} + \dfrac{20}{n^2}\dfrac{n(n+1)}{2}\right] = \lim_{n\to\infty}\left[\dfrac{4(n+1)^2}{n^2} + \dfrac{10n(n+1)}{n^2}\right]$

$= \lim_{n\to\infty}\left[4\left(1+\dfrac{1}{n}\right)^2 + 10\left(1+\dfrac{1}{n}\right)\right] = 4\cdot 1 + 10\cdot 1 = 14$

47. Let $S = \sum_{i=1}^{n} ar^{i-1} = a + ar + ar^2 + \cdots + ar^{n-1}$. Multiplying both sides by r gives us

$rS = ar + ar^2 + \cdots + ar^{n-1} + ar^n$. Subtracting the first equation from the second, we find

$(r-1)S = ar^n - a = a(r^n - 1)$, so $S = \dfrac{a(r^n - 1)}{r - 1}$ [since $r \neq 1$].

49. $\sum_{i=1}^{n}(2i + 2^i) = 2\sum_{i=1}^{n} i + \sum_{i=1}^{n} 2\cdot 2^{i-1} = 2\dfrac{n(n+1)}{2} + \dfrac{2(2^n - 1)}{2 - 1} = 2^{n+1} + n^2 + n - 2$.

For the first sum we have used Theorems 2(a) and 3(c), and for the second, Exercise 47 with $a = r = 2$.